D1553292

Methods in Enzymology

Volume 292

ABC TRANSPORTERS: BIOCHEMICAL, CELLULAR, AND MOLECULAR ASPECTS

METHODS IN ENZYMOLOGY

Methods in Enzymology

Volume 292

ABC Transporters: Biochemical, Cellular, and Molecular Aspects

EDITED BY

Suresh V. Ambudkar

NATIONAL CANCER INSTITUTE
NATIONAL INSTITUTES OF HEALTH
BETHESDA, MARYLAND

Michael M. Gottesman

NATIONAL CANCER INSTITUTE
NATIONAL INSTITUTES OF HEALTH
BETHESDA, MARYLAND

ACADEMIC PRESS
San Diego London Boston New York Sydney Tokyo Toronto

This book is printed on acid-free paper. 

Academic Press
525 B Street, Suite 1900, San Diego, California 92101-4495, USA
http://www.academicpress.com

Academic Press Limited
24-28 Oval Road, London NW1 7DX, UK
http://www.hbuk.co.uk/ap/

International Standard Book Number: 0-12-182193-5

PRINTED IN THE UNITED STATES OF AMERICA
98 99 00 01 02 03 MM 9 8 7 6 5 4 3 2 1

Table of Contents

Contributors to Volume 292 xi

Preface xvii

Volumes in Series xix

Section I. Prokaryotic ABC Transporters

1. Overview of Bacterial ABC Transporters — Hiroshi Nikaido and Jason A. Hall — 3
2. Preparation and Reconstitution of Membrane-Associated Maltose Transporter Complex of *Escherichia coli* — Jason A. Hall, Amy L. Davidson, and Hiroshi Nikaido — 20
3. Maltose Transport in *Escherichia coli:* Mutations That Uncouple ATP Hydrolysis from Transport — Cynthia H. Panagiotidis and Howard A. Shuman — 30
4. Binding Protein-Dependent ABC Transport System for Glycerol 3-Phosphate of *Escherichia coli* — Winfried Boos — 40
5. Structure–Function Analysis of Hemolysin B — Fang Zhang, Jonathan A. Sheps, and Victor Ling — 51
6. *Erwinia* Metalloprotease Permease: Aspects of Secretion Pathway and Secretion Functions — Philippe Delepelaire — 67
7. Arsenical Pumps in Prokaryotes and Eukaryotes — Masayuki Kuroda, Hiranmoy Bhattacharjee, and Barry P. Rosen — 82

Section II. Eukaryotic ABC Transporters

8. Evolutionary Relationships among ABC Transporters — James M. Croop — 101
9. Cloning of Novel ABC Transporter Genes — Rando Allikmets and Michael Dean — 116
10. *Saccharomyces cerevisiae* ABC Proteins and Their Relevance to Human Health and Disease — Daniel Taglicht and Susan Michaelis — 130

A. Nonmammalian ABC Transport Systems

11. *Arabidopsis MDR* Genes: Molecular Cloning and Protein Chemical Aspects — Robert Dudler and Michael Sidler — 162

12. Functional Analysis of *pfmdr1* Gene of *Plasmodium falciparum* — SARAH VOLKMAN AND DYANN WIRTH — 174

13. Amplification of ABC Transporter Gene *pgpA* and of Other Heavy Metal Resistance Genes in *Leishmania tarentolae* and Their Study by Gene Transfection and Gene Disruption — MARC OUELLETTE, ANASS HAIMEUR, KATHERINE GRONDIN, DANIELLE LÉGARÉ, AND BARBARA PAPADOPOULOU — 182

14. Functional Assay for Analysis of Yeast *ste6* Mutants — GABY L. NIJBROEK AND SUSAN MICHAELIS — 193

15. ABC Transporters Involved in Transport of Eye Pigment Precursors in *Drosophila melanogaster* — GARY D. EWART AND ANTHONY J. HOWELLS — 213

B. Mammalian P-Glycoproteins

16. Isolation of Altered-Function Mutants and Genetic Suppressor Elements of Multidrug Transporter P-Glycoprotein by Expression Selection from Retroviral Libraries — IGOR B. RONINSON, DONALD ZUHN, ADAM RUTH, AND DAVID DE GRAAF — 225

17. Selection and Maintenance of Multidrug-Resistant Cells — MICHAEL M. GOTTESMAN, CAROL CARDARELLI, SARAH GOLDENBERG, THOMAS LICHT, AND IRA PASTAN — 248

18. Monoclonal Antibodies to P-Glycoprotein: Preparation and Applications to Basic and Clinical Research — MIKIHIKO NAITO AND TAKASHI TSURUO — 258

19. Topology of P-Glycoproteins — WILLIAM R. SKACH — 265

20. Use of Cell-Free Systems to Determine P-Glycoprotein Transmembrane Topology — JIAN-TING ZHANG — 279

21. Photoaffinity Labels for Characterizing Drug Interaction Sites of P-Glycoprotein — AHMAD R. SAFA — 289

22. Identification of Drug Interaction Sites in P-Glycoprotein — LEE M. GREENBERGER — 307

23. Photoaffinity Labeling of Human P-Glycoprotein: Effect of Modulator Interaction and ATP Hydrolysis on Substrate Binding — SAIBAL DEY, MURALIDHARA RAMACHANDRA, IRA PASTAN, MICHAEL M. GOTTESMAN, AND SURESH V. AMBUDKAR — 318

24. Identification of Phosphorylation Sites in Human MDR1 P-Glycoprotein — TIMOTHY C. CHAMBERS — 328

25. Identification of *in Vivo* Phosphorylation Sites for Basic-Directed Kinases in Murine *mdr1b* P-Glycoprotein by Combination of Mass Spectrometry and Site-Directed Mutagenesis — J. S. GLAVY, M. WOLFSON, E. NIEVES, E.-K. HAN, C.-P. H. YANG, S. B. HORWITZ, AND G. A. ORR — 342

26. P-Glycoprotein and Swelling-Activated Chloride Channels — TAMARA D. BOND, CHRISTOPHER F. HIGGINS, AND MIGUEL A. VALVERDE — 359

27. Functional Expression of *mdr* and *mdr*-like cDNAs in *Escherichia coli* — EITAN BIBI, ROTEM EDGAR, AND ODED BÉJÀ — 370

28. Yeast Secretory Vesicle System for Expression and Functional Characterization of P-Glycoproteins — STEPHAN RUETZ — 382

29. High-Level Expression of Mouse Mdr3 P-Glycoprotein in Yeast *Pichia pastoris* and Characterization of ATPase Activity — LUCILLE BEAUDET, INA L. URBATSCH, AND PHILIPPE GROS — 397

30. Mutational Analysis of P-Glycoprotein in Yeast *Saccharomyces cerevisiae* — LUCILLE BEAUDET AND PHILIPPE GROS — 414

31. Baculovirus-Mediated Expression of Human Multidrug Resistance cDNA in Insect Cells and Functional Analysis of Recombinant P-Glycoprotein — URSULA A. GERMANN — 427

32. Recombinant Vaccinia Virus Vectors for Functional Expression of P-Glycoprotein in Mammalian Cells — MURALIDHARA RAMACHANDRA, MICHAEL M. GOTTESMAN, AND IRA PASTAN — 441

33. Functional Expression of Human P-Glycoprotein from Plasmids Using Vaccinia Virus–Bacteriophage T7 RNA Polymerase System — CHRISTINE A. HRYCYNA, MURALIDHARA RAMACHANDRA, IRA PASTAN, AND MICHAEL M. GOTTESMAN — 456

34. pHaMDR-DHFR Bicistronic Expression System for Mutation Analysis of P-Glycoprotein — SHUDONG ZHANG, YOSHIKAZU SUGIMOTO, TZIPORA SHOSHANI, IRA PASTAN, AND MICHAEL M. GOTTESMAN — 474

35. Mutational Analysis of Human P-Glycoprotein — TIP W. LOO AND DAVID M. CLARKE — 480

36. Purification and Reconstitution of Human P-Glycoprotein — SURESH V. AMBUDKAR, ISABELLE H. LELONG, JIAPING ZHANG, AND CAROL CARDARELLI — 492

37. Drug-Stimulatable ATPase Activity in Crude Membranes of Human *MDR1*-Transfected Mammalian Cells — SURESH V. AMBUDKAR — 504

38. ATPase Activity of Chinese Hamster P-Glycoprotein — ALAN E. SENIOR, MARWAN K. AL-SHAWI, AND INA L. URBATSCH — 514

39. Construction of *MDR1* Vectors for Gene Therapy — YOSHIKAZU SUGIMOTO, MICHAEL M. GOTTESMAN, IRA PASTAN, AND TAKASHI TSURUO — 523

40. Construction of *MDR1* Adeno-Associated Virus Vectors for Gene Therapy — MARION BAUDARD — 538

41. Retroviral Transfer of Multidrug Transporter to Murine Hematopoietic Stem Cells — THOMAS LICHT, MICHAEL M. GOTTESMAN, AND IRA PASTAN — 546

42. Retroviral Transfer of Human *MDR1* Gene into Human T Lymphocytes — CAROLINE G. L. LEE, IRA PASTAN, AND MICHAEL M. GOTTESMAN — 557

43. Construction and Analysis of Multidrug Resistance Transgenic Mice — GREGORY L. EVANS — 572

C. Multidrug Resistance Associated Protein

44. Cloning, Transfer, and Characterization of Multidrug Resistance Protein — CAROLINE E. GRANT, GABU BHARDWAJ, SUSAN P. C. COLE, AND ROGER G. DEELEY — 594

45. Transport Function and Substrate Specificity of Multidrug Resistance Protein — DIETRICH KEPPLER, GABRIELE JEDLITSCHKY, AND INKA LEIER — 607

D. Cystic Fibrosis Transmembrane Conductance Regulator

46. Heterologous Expression Systems for Study of Cystic Fibrosis Transmembrane Conductance Regulator — XIU-BAO CHANG, NORBERT KARTNER, FABIAN S. SEIBERT, ANDREI A. ALEKSANDROV, ANDREW W. KLOSER, GRETCHEN L. KISER, AND JOHN R. RIORDAN — 616

47. Characterization of Polyclonal and Monoclonal Antibodies to Cystic Fibrosis Transmembrane Conductance Regulator — NORBERT KARTNER AND JOHN R. RIORDAN — 629

48. Identification of Cystic Fibrosis Transmembrane Conductance Regulator in Renal Endosomes — ISABELLE T. CRAWFORD AND PETER C. MALONEY — 652

49. Assays of Dynamics, Mechanisms, and Regulation of ATP Transport and Release: Implications for Study of ABC Transporter Function — ERIK M. SCHWIEBERT, MARIE E. EGAN, AND WILLIAM B. GUGGINO — 664

50. Overexpression, Purification, and Function of First Nucleotide-Binding Fold of Cystic Fibrosis Transmembrane Conductance Regulator — YOUNG HEE KO AND PETER L. PEDERSEN — 675

51. Recombinant Synthesis of Cystic Fibrosis Transmembrane Conductance Regulator and Functional Nucleotide-Binding Domains — SCOTT A. KING AND ERIC J. SORSCHER — 686

52. Cationic Lipid Formulations for Intracellular Gene Delivery of Cystic Fibrosis Transmembrane Conductance Regulator to Airway Epithelia — SENG HING CHENG, JOHN MARSHALL, RONALD K. SCHEULE, AND ALAN E. SMITH — 697

53. Adeno-Associated Virus Vectors for Gene Therapy of Cystic Fibrosis — TERENCE R. FLOTTE AND BARRIE J. CARTER — 717

E. Sulfonylurea Receptor

54. Sulfonylurea Receptors and ATP-Sensitive Potassium Ion Channels — LYDIA AGUILAR-BRYAN, JOHN P. CLEMENT IV, AND DANIEL A. NELSON — 732

F. Intracellular ABC Transporters

55. Peptide Transport Assay for TAP Function — YE WANG, DAVID S. GUTTOH, AND MATTHEW J. ANDROLEWICZ — 745

56. Peroxisomal ABC Transporters — NOAM SHANI AND DAVID VALLE — 753

57. Mitochondrial ABC Transporters — JONATHAN LEIGHTON — 776

AUTHOR INDEX 789

SUBJECT INDEX 833

Contributors to Volume 292

Article numbers are in parentheses following the names of contributors.
Affiliations listed are current.

LYDIA AGUILAR-BRYAN (54), *Department of Medicine, Baylor College of Medicine, Houston, Texas 77030*

ANDREI A. ALEKSANDROV (46), *Mayo Clinic, Scottsdale, Scottsdale, Arizona 85259*

RANDO ALLIKMETS (9), *Intramural Research Support Program, SAIC–Frederick, NCI–Frederick Cancer Research and Development Center, Frederick, Maryland 21702*

MARWAN K. AL-SHAWI (38), *Department of Molecular Physiology and Biological Physics, University of Virginia, Charlottesville, Virginia 22908*

SURESH V. AMBUDKAR (23, 36, 37), *Laboratory of Cell Biology, Division of Basic Sciences, National Cancer Institute, National Institutes of Health, Bethesda, Maryland 20892*

MATTHEW J. ANDROLEWICZ (55), *Immunology Program, H. Lee Moffitt Cancer Center and Research Institute, and Department of Biochemistry and Molecular Biology, University of South Florida College of Medicine, Tampa, Florida 33612*

MARION BAUDARD (40), *Department of Hematology, Hôpital Hôtel-Dieu, 75004 Paris, France*

LUCILLE BEAUDET (29, 30), *Department of Biochemistry, McGill University, Montréal, Québec H3G 1Y6, Canada*

ODED BÉJÀ (27), *Department of Biological Chemistry, Weizmann Institute of Science, Rehovot 76100, Israel*

GABU BHARDWAJ (44), *Department of Gastroenterology, Hepatology, and Nutrition, University of Florida College of Medicine, Gainesville, Florida 32610*

HIRANMOY BHATTACHARJEE (7), *Department of Biochemistry and Molecular Biology, Wayne State University School of Medicine, Detroit, Michigan 48201*

EITAN BIBI (27), *Department of Biological Chemistry, Weizmann Institute of Science, Rehovot 76100, Israel*

TAMARA D. BOND (26), *Department of Anesthesiology, Vanderbilt University, Nashville, Tennessee 37232*

WINFRIED BOOS (4), *Faculty of Biology, University of Konstanz, D-78457 Konstanz, Germany*

CAROL CARDARELLI (17, 36), *Laboratory of Molecular Biology, Division of Basic Sciences, National Cancer Institute, National Institutes of Health, Bethesda, Maryland 20892*

BARRIE J. CARTER (53), *Targeted Genetics Corporation, Seattle, Washington 98101*

TIMOTHY C. CHAMBERS (24), *Department of Biochemistry and Molecular Biology, University of Arkansas for Medical Sciences, Little Rock, Arkansas 72205*

XIU-BAO CHANG (46), *Mayo Clinic, Scottsdale, Scottsdale, Arizona 85259*

SENG HING CHENG (52), *Genzyme Corporation, Framingham, Massachusetts 01701*

DAVID M. CLARKE (35), *Departments of Medicine and Biochemistry, Medical Research Council Group in Membrane Biology, University of Toronto, Toronto, Ontario M5S 1A8, Canada*

JOHN P. CLEMENT IV (54), *Department of Cell Biology, Baylor College of Medicine, Houston, Texas 77030*

SUSAN P. C. COLE (44), *Cancer Research Laboratories, Queen's University, Kingston, Ontario K7L 3N6, Canada*

ISABELLE T. CRAWFORD (48), *Science Applications International Corporation, Joppa, Maryland 21085*

JAMES M. CROOP (8), *Section of Pediatric Hematology/Oncology, James Whitcomb Riley Hospital for Children, Indianapolis, Indiana 46202*

AMY L. DAVIDSON (2), *Department of Microbiology and Immunology, Baylor College of Medicine, Houston, Texas 77030*

MICHAEL DEAN (9), *Laboratory of Genomic Diversity, NCI–Frederick Cancer Research and Development Center, Frederick, Maryland 21702*

ROGER G. DEELEY (44), *Cancer Research Laboratories, Queen's University, Kingston, Ontario K7L 3N6, Canada*

DAVID DE GRAAF (16), *MIT/Whitehead Institute Center for Genome Research, Functional Genomics, Cambridge, Massachusetts 02139*

PHILIPPE DELEPELAIRE (6), *Unité de Physiologie Cellulaire, Département des Biotechnologies, Institut Pasteur, 75724 Paris cédex 15, France*

SAIBAL DEY (23), *Laboratory of Cell Biology, Division of Basic Sciences, National Cancer Institute, National Institutes of Health, Bethesda, Maryland 20892*

ROBERT DUDLER (11), *Institute of Plant Biology, University of Zurich, CH-8008 Zurich, Switzerland*

ROTEM EDGAR (27), *Department of Biological Chemistry, Weizmann Institute of Science, Rehovot 76100, Israel*

MARIE E. EGAN (49), *Department of Pediatrics, Johns Hopkins University School of Medicine, Baltimore, Maryland 21205*

GREGORY L. EVANS (43), *Clinical Gene Therapy Branch, National Center for Human Genome Research, Bethesda, Maryland 20892*

GARY D. EWART (15), *John Curtin School of Medical Research, Australian National University, Canberra ACT 2601, Australia*

TERENCE R. FLOTTE (53), *Gene Therapy Center, University of Florida, Gainesville, Florida 33610*

URSULA A. GERMANN (31), *Vertex Pharmaceuticals Incorporated, Cambridge, Massachusetts 02139*

J. S. GLAVY (25), *Albert Einstein College of Medicine, Bronx, New York 10461*

SARAH GOLDENBERG (17), *Laboratory of Cell Biology, Division of Basic Sciences, National Cancer Institute, National Institutes of Health, Bethesda, Maryland 20892*

MICHAEL M. GOTTESMAN (17, 23, 32–34, 39, 41, 42), *Laboratory of Cell Biology, Division of Basic Sciences, National Cancer Institute, National Institutes of Health, Bethesda, Maryland 20892*

CAROLINE E. GRANT (44), *Cancer Research Laboratories, Queen's University, Kingston, Ontario K7L 3N6, Canada*

LEE M. GREENBERGER (22), *Oncology and Immunology Section, Wyeth-Ayerst Research, Pearl River, New York 10965*

KATHERINE GRONDIN (13), *Department of Microbiology, Université Laval, Québec G1V 4G2, Canada*

PHILIPPE GROS (29, 30), *Department of Biochemistry, McGill University, Montréal, Québec H3G 1Y6, Canada*

WILLIAM B. GUGGINO (49), *Department of Physiology, Johns Hopkins University School of Medicine, Baltimore, Maryland 21205*

DAVID S. GUTTOH (55), *Immunology Program, H. Lee Moffitt Cancer Center and Research Institute, and Department of Biochemistry and Molecular Biology, University of South Florida College of Medicine, Tampa, Florida 33612*

ANASS HAIMEUR (13), *Department of Microbiology, Université Laval, Québec G1V 4G2, Canada*

JASON A. HALL (1, 2), *Department of Physiology, Johns Hopkins University School of Medicine, Baltimore, Maryland 21205*

E. K. HAN (25), *Albert Einstein College of Medicine, Bronx, New York 10461*

CHRISTOPHER F. HIGGINS (26), *MRC Clinical Sciences Centre, Imperial College School of Medicine, Hammersmith Hospital, London W12 ONN, United Kingdom*

S. B. HORWITZ (25), *Albert Einstein College of Medicine, Bronx, New York 10461*

ANTHONY J. HOWELLS (15), *Division of Biochemistry and Molecular Biology, Australian National University, Canberra ACT 2601, Australia*

CHRISTINE A. HRYCYNA (33), *Laboratory of Cell Biology, Division of Basic Sciences, National Cancer Institute, National Institutes of Health, Bethesda, Maryland 20892*

GABRIELE JEDLITSCHKY (45), *Division of Tumor Biochemistry, Deutsches Krebsforschungszentrum, D-69120 Heidelberg, Germany*

NORBERT KARTNER (46, 47), *Department of Pharmacology, Faculty of Medicine, University of Toronto, Toronto, Ontario M5S 1A8, Canada*

DIETRICH KEPPLER (45), *Division of Tumor Biochemistry, Deutsches Krebsforschungszentrum, D-69120 Heidelberg, Germany*

SCOTT A. KING (51), *Gregory Fleming James Cystic Fibrosis Research Center and Department of Physiology and Biophysics, University of Alabama at Birmingham, Birmingham, Alabama 35294*

GRETCHEN L. KISER (46), *Mayo Clinic, Scottsdale, Scottsdale, Arizona 85259*

ANDREW W. KLOSER (46), *Mayo Clinic, Scottsdale, Scottsdale, Arizona 85259*

YOUNG HEE KO (50), *Department of Biological Chemistry, Johns Hopkins University School of Medicine, Baltimore, Maryland 21205*

MASAYUKI KURODA (7), *Department of Biochemistry and Molecular Biology, Wayne State University School of Medicine, Detroit, Michigan 48201*

CAROLINE G. L. LEE (42), *Laboratory of Cell Biology, Division of Basic Sciences, National Cancer Institute, National Institutes of Health, Bethesda, Maryland 20892*

DANIELLE LÉGARÉ (13), *Department of Microbiology, Université Laval, Québec G1V 4G2, Canada*

INKA LEIER (45), *Division of Tumor Biochemistry, Deutsches Krebsforschungszentrum, D-69120 Heidelberg, Germany*

JONATHAN LEIGHTON (57), *Biozentrum, University of Basel, CH-4056 Basel, Switzerland*

ISABELLE H. LELONG (36), *CNRS, I.R. C.A.D., Hôpital Cival BP-426, F-67091, Strasburg, France*

THOMAS LICHT (17, 41), *Laboratory of Molecular Biology, Division of Basic Sciences, National Cancer Institute, National Institutes of Health, Bethesda, Maryland 20892*

VICTOR LING (5), *British Columbia Cancer Research Center, Vancouver, British Columbia V5Z 1L3, Canada*

TIP W. LOO (35), *Departments of Medicine and Biochemistry, Medical Research Council Group in Membrane Biology, University of Toronto, Toronto, Ontario M5S 1A8, Canada*

PETER C. MALONEY (48), *Department of Physiology, Johns Hopkins University School of Medicine, Baltimore, Maryland 21205*

JOHN MARSHALL (52), *Genzyme Corporation, Framingham, Massachusetts 01701*

SUSAN MICHAELIS (10, 14), *Department of Cell Biology and Anatomy, Johns Hopkins University School of Medicine, Baltimore, Maryland 21205*

MIKIHIKO NAITO (18), *Institute of Molecular and Cellular Biosciences, The University of Tokyo, Yayoi, Bunkyo-ku, Tokyo 113, Japan*

DANIEL A. NELSON (54), *Department of Biology, University of North Carolina at Charlotte, Charlotte, North Carolina 28223*

E. NIEVES (25), *Albert Einstein College of Medicine, Bronx, New York 10461*

GABY L. NIJBROEK (14), *Department of Cell Biology and Anatomy, Johns Hopkins University School of Medicine, Baltimore, Maryland 21205*

HIROSHI NIKAIDO (1, 2), *Department of Molecular and Cell Biology, University of California, Berkeley, California 94720*

G. A. ORR (25), *Albert Einstein College of Medicine, Bronx, New York 10461*

MARC OUELLETTE (13), *Department of Microbiology, Université Laval, Québec G1V 4G2, Canada*

CYNTHIA H. PANAGIOTIDIS (3), *Department of Microbiology, College of Physicians and Surgeons, Columbia University, New York, New York 10032*

BARBARA PAPADOPOULOU (13), *Department of Microbiology, Université Laval, Québec G1V 4G2, Canada*

IRA PASTAN (17, 23, 32–34, 39, 41, 42), *Laboratory of Molecular Biology, Division of Basic Sciences, National Cancer Institute, National Institutes of Health, Bethesda, Maryland 20892*

PETER L. PEDERSEN (50), *Department of Biological Chemistry, Johns Hopkins University School of Medicine, Baltimore, Maryland 21205*

MURALIDHARA RAMACHANDRA (23, 32, 33), *Canji Inc., San Diego, California 92121*

JOHN R. RIORDAN (46, 47), *Mayo Clinic, Scottsdale, Scottsdale, Arizona 85259*

IGOR B. RONINSON (16), *Department of Molecular Genetics, University of Illinois at Chicago, Chicago, Illinois 60607*

BARRY P. ROSEN (7), *Department of Biochemistry and Molecular Biology, Wayne State University School of Medicine, Detroit, Michigan 48201*

STEPHAN RUETZ (28), *Novartis Pharma Inc., Oncology Research, CH-4002 Basel, Switzerland*

ADAM RUTH (16), *Department of Molecular Genetics, University of Illinois at Chicago, Chicago, Illinois 60607*

AHMAD R. SAFA (21), *Department of Experimental Oncology and Hollings Cancer Center, Medical University of South Carolina, Charleston, South Carolina 29425*

RONALD K. SCHEULE (52), *Genzyme Corporation, Framingham, Massachusetts 01701*

ERIK M. SCHWIEBERT (49), *Department of Physiology, Johns Hopkins University School of Medicine, Baltimore, Maryland 21205*

FABIAN S. SEIBERT (46), *Department of Biochemistry, Faculty of Medicine, University of Toronto, Toronto, Ontario M5S 1A8, Canada*

ALAN E. SENIOR (38), *Department of Biochemistry and Biophysics, University of Rochester Medical Center, Rochester, New York 14642*

NOAM SHANI (56), *Kennedy Krieger Institute, Department of Pediatrics, Johns Hopkins University School of Medicine, Baltimore, Maryland 21205*

JONATHAN A. SHEPS (5), *British Columbia Cancer Research Center, Vancouver, British Columbia V5Z 1L3, Canada*

TZIPORA SHOSHANI (34), *Laboratory of Cell Biology, Division of Basic Sciences, National Cancer Institute, National Institutes of Health, Bethesda, Maryland 20892*

HOWARD A. SHUMAN (3), *Department of Microbiology, College of Physicians and Surgeons, Columbia University, New York, New York 10032*

MICHAEL SIDLER (11), *Institute of Plant Biology, University of Zurich, CH-8008 Zurich, Switzerland*

WILLIAM R. SKACH (19), *Department of Molecular and Cellular Engineering, University of Pennsylvania, Philadelphia, Pennsylvania 19104*

ALAN E. SMITH (52), *Genzyme Corporation, Framingham, Massachusetts 01701*

ERIC J. SORSCHER (51), *Gregory Fleming James Cystic Fibrosis Research Center and Departments of Physiology and Biophysics, and Medicine, University of Alabama at Birmingham, Birmingham, Alabama 35294*

YOSHIKAZU SUGIMOTO (34, 39), *Cancer Chemotherapy Center, Japanese Foundation for Cancer Research, Kami-ikebukuro, Toshima-ku, Tokyo 170, Japan*

DANIEL TAGLICHT (10), *Sigma Israel Chemicals Limited, Jerusalem 91064, Israel*

TAKASHI TSURUO (18, 39), *Institute of Molecular and Cellular Biosciences, The University of Tokyo, Yayoi, Bunkyo-ku, Tokyo 113, Japan*

INA L. URBATSCH (29, 38), *Department of Biochemistry, McGill University, Montréal, Québec H3G 1Y6, Canada*

DAVID VALLE (56), *Department of Pediatrics, Howard Hughes Medical Institute, Johns Hopkins University School of Medicine, Baltimore, Maryland 21205*

MIGUEL A. VALVERDE (26), *Department of Physiology, King's College, London, Strand, London WC2R 2LS, United Kingdom*

SARAH VOLKMAN (12), *Department of Immunology and Infectious Diseases, Harvard School of Public Health, Boston, Massachusetts 02115*

YE WANG (55), *Immunology Program, H. Lee Moffitt Cancer Center and Research Institute, and Department of Biochemistry and Molecular Biology, University of South Florida College of Medicine, Tampa, Florida 33612*

DYANN WIRTH (12), *Department of Immunology and Infectious Diseases, Harvard School of Public Health, Boston, Massachusetts 02115*

M. WOLFSON (25), *Albert Einstein College of Medicine, Bronx, New York 10461*

C.-P. H. YANG (25), *Albert Einstein College of Medicine, Bronx, New York 10461*

FANG ZHANG (5), *British Columbia Cancer Research Center, Vancouver, British Columbia V5Z 1L3, Canada*

JAIPING ZHANG (36), *Division of Nephrology, Department of Medicine, Johns Hopkins University School of Medicine, Baltimore, Maryland 21205*

JIAN-TING ZHANG (20), *Department of Physiology and Biophysics, University of Texas Medical Branch, Galveston, Texas 77555*

SHUDONG ZHANG (34), *Division of Oncology, The Hospital for Sick Children, Toronto, Ontario M5G 1X8, Canada*

DONALD ZUHN (16), *Department of Molecular Genetics, University of Illinois at Chicago, Chicago, Illinois 60607*

Preface

Since the cloning of mammalian multidrug resistance (MDR) genes in 1986, there has been tremendous progress in the identification and characterization of the members of the superfamily of ATP-binding cassette (ABC) transporters also known as traffic ATPases. More than eighty ABC transporters have been identified. However, the function of many of these transporters is unknown, and as genome mapping of a number of organisms proceeds, open reading frames encoding new family members continue to be identified. The ABC superfamily members transport a wide variety of substrates, which include ions, sugars, amino acids, glycans, peptides, proteins, phospholipids, toxins, antibiotics, and hydrophobic cytotoxic natural product drugs. The minimal functional transport unit appears to be composed of two membrane-associated domains, each containing six putative transmembrane helices and two nucleotide-binding domains or ATP binding/utilization sites. Among these family members, multiple structural configurations have been found, including a single polypeptide chain, homo- or heterodimers, and multisubunit systems, all of which include both the required membrane and ATP domains. Several of the ABC transporters are linked to diseases in humans, including cancer, cystic fibrosis, Stargardt disease, age-related macular degeneration (AMD), adrenoleukodystrophy, Zellweger's syndrome, rheumatoid arthritis, insulin-dependent diabetes, persistent hypoglycemia of infancy, Dubin-Johnson syndrome, and progressive familial intrahepatic cholestasis.

With the increased interest in ABC transporters, there was a real need for a guide that presents step-by-step protocols and methods that can be used for studying various aspects—including their physiological role—of these proteins. The articles in this *Methods in Enzymology* volume provide detailed descriptions of such methods. In addition, some articles include overviews from experts in the field which provide detailed summaries of various aspects of these transporters that should be useful not only to the novice but also to researchers interested in entering this field.

We would like to express our sincere thanks to all the contributing authors for their efforts and cooperation during the production of this book. We also thank Shirley Light of Academic Press for providing advice and encouragement in the preparation of this volume.

Suresh V. Ambudkar
Michael M. Gottesman

METHODS IN ENZYMOLOGY

VOLUME I. Preparation and Assay of Enzymes
Edited by SIDNEY P. COLOWICK AND NATHAN O. KAPLAN

VOLUME II. Preparation and Assay of Enzymes
Edited by SIDNEY P. COLOWICK AND NATHAN O. KAPLAN

VOLUME III. Preparation and Assay of Substrates
Edited by SIDNEY P. COLOWICK AND NATHAN O. KAPLAN

VOLUME IV. Special Techniques for the Enzymologist
Edited by SIDNEY P. COLOWICK AND NATHAN O. KAPLAN

VOLUME V. Preparation and Assay of Enzymes
Edited by SIDNEY P. COLOWICK AND NATHAN O. KAPLAN

VOLUME VI. Preparation and Assay of Enzymes (*Continued*)
Preparation and Assay of Substrates
Special Techniques
Edited by SIDNEY P. COLOWICK AND NATHAN O. KAPLAN

VOLUME VII. Cumulative Subject Index
Edited by SIDNEY P. COLOWICK AND NATHAN O. KAPLAN

VOLUME VIII. Complex Carbohydrates
Edited by ELIZABETH F. NEUFELD AND VICTOR GINSBURG

VOLUME IX. Carbohydrate Metabolism
Edited by WILLIS A. WOOD

VOLUME X. Oxidation and Phosphorylation
Edited by RONALD W. ESTABROOK AND MAYNARD E. PULLMAN

VOLUME XI. Enzyme Structure
Edited by C. H. W. HIRS

VOLUME XII. Nucleic Acids (Parts A and B)
Edited by LAWRENCE GROSSMAN AND KIVIE MOLDAVE

VOLUME XIII. Citric Acid Cycle
Edited by J. M. LOWENSTEIN

VOLUME XIV. Lipids
Edited by J. M. LOWENSTEIN

VOLUME XV. Steroids and Terpenoids
Edited by RAYMOND B. CLAYTON

VOLUME XVI. Fast Reactions
Edited by KENNETH KUSTIN

Volume XVII. Metabolism of Amino Acids and Amines (Parts A and B)
Edited by Herbert Tabor and Celia White Tabor

Volume XVIII. Vitamins and Coenzymes (Parts A, B, and C)
Edited by Donald B. McCormick and Lemuel D. Wright

Volume XIX. Proteolytic Enzymes
Edited by Gertrude E. Perlmann and Laszlo Lorand

Volume XX. Nucleic Acids and Protein Synthesis (Part C)
Edited by Kivie Moldave and Lawrence Grossman

Volume XXI. Nucleic Acids (Part D)
Edited by Lawrence Grossman and Kivie Moldave

Volume XXII. Enzyme Purification and Related Techniques
Edited by William B. Jakoby

Volume XXIII. Photosynthesis (Part A)
Edited by Anthony San Pietro

Volume XXIV. Photosynthesis and Nitrogen Fixation (Part B)
Edited by Anthony San Pietro

Volume XXV. Enzyme Structure (Part B)
Edited by C. H. W. Hirs and Serge N. Timasheff

Volume XXVI. Enzyme Structure (Part C)
Edited by C. H. W. Hirs and Serge N. Timasheff

Volume XXVII. Enzyme Structure (Part D)
Edited by C. H. W. Hirs and Serge N. Timasheff

Volume XXVIII. Complex Carbohydrates (Part B)
Edited by Victor Ginsburg

Volume XXIX. Nucleic Acids and Protein Synthesis (Part E)
Edited by Lawrence Grossman and Kivie Moldave

Volume XXX. Nucleic Acids and Protein Synthesis (Part F)
Edited by Kivie Moldave and Lawrence Grossman

Volume XXXI. Biomembranes (Part A)
Edited by Sidney Fleischer and Lester Packer

Volume XXXII. Biomembranes (Part B)
Edited by Sidney Fleischer and Lester Packer

Volume XXXIII. Cumulative Subject Index Volumes I–XXX
Edited by Martha G. Dennis and Edward A. Dennis

Volume XXXIV. Affinity Techniques (Enzyme Purification: Part B)
Edited by William B. Jakoby and Meir Wilchek

Volume XXXV. Lipids (Part B)
Edited by John M. Lowenstein

VOLUME XXXVI. Hormone Action (Part A: Steroid Hormones)
Edited by BERT W. O'MALLEY AND JOEL G. HARDMAN

VOLUME XXXVII. Hormone Action (Part B: Peptide Hormones)
Edited by BERT W. O'MALLEY AND JOEL G. HARDMAN

VOLUME XXXVIII. Hormone Action (Part C: Cyclic Nucleotides)
Edited by JOEL G. HARDMAN AND BERT W. O'MALLEY

VOLUME XXXIX. Hormone Action (Part D: Isolated Cells, Tissues, and Organ Systems)
Edited by JOEL G. HARDMAN AND BERT W. O'MALLEY

VOLUME XL. Hormone Action (Part E: Nuclear Structure and Function)
Edited by BERT W. O'MALLEY AND JOEL G. HARDMAN

VOLUME XLI. Carbohydrate Metabolism (Part B)
Edited by W. A. WOOD

VOLUME XLII. Carbohydrate Metabolism (Part C)
Edited by W. A. WOOD

VOLUME XLIII. Antibiotics
Edited by JOHN H. HASH

VOLUME XLIV. Immobilized Enzymes
Edited by KLAUS MOSBACH

VOLUME XLV. Proteolytic Enzymes (Part B)
Edited by LASZLO LORAND

VOLUME XLVI. Affinity Labeling
Edited by WILLIAM B. JAKOBY AND MEIR WILCHEK

VOLUME XLVII. Enzyme Structure (Part E)
Edited by C. H. W. HIRS AND SERGE N. TIMASHEFF

VOLUME XLVIII. Enzyme Structure (Part F)
Edited by C. H. W. HIRS AND SERGE N. TIMASHEFF

VOLUME XLIX. Enzyme Structure (Part G)
Edited by C. H. W. HIRS AND SERGE N. TIMASHEFF

VOLUME L. Complex Carbohydrates (Part C)
Edited by VICTOR GINSBURG

VOLUME LI. Purine and Pyrimidine Nucleotide Metabolism
Edited by PATRICIA A. HOFFEE AND MARY ELLEN JONES

VOLUME LII. Biomembranes (Part C: Biological Oxidations)
Edited by SIDNEY FLEISCHER AND LESTER PACKER

VOLUME LIII. Biomembranes (Part D: Biological Oxidations)
Edited by SIDNEY FLEISCHER AND LESTER PACKER

VOLUME LIV. Biomembranes (Part E: Biological Oxidations)
Edited by SIDNEY FLEISCHER AND LESTER PACKER

VOLUME LV. Biomembranes (Part F: Bioenergetics)
Edited by SIDNEY FLEISCHER AND LESTER PACKER

VOLUME LVI. Biomembranes (Part G: Bioenergetics)
Edited by SIDNEY FLEISCHER AND LESTER PACKER

VOLUME LVII. Bioluminescence and Chemiluminescence
Edited by MARLENE A. DELUCA

VOLUME LVIII. Cell Culture
Edited by WILLIAM B. JAKOBY AND IRA PASTAN

VOLUME LIX. Nucleic Acids and Protein Synthesis (Part G)
Edited by KIVIE MOLDAVE AND LAWRENCE GROSSMAN

VOLUME LX. Nucleic Acids and Protein Synthesis (Part H)
Edited by KIVIE MOLDAVE AND LAWRENCE GROSSMAN

VOLUME 61. Enzyme Structure (Part H)
Edited by C. H. W. HIRS AND SERGE N. TIMASHEFF

VOLUME 62. Vitamins and Coenzymes (Part D)
Edited by DONALD B. MCCORMICK AND LEMUEL D. WRIGHT

VOLUME 63. Enzyme Kinetics and Mechanism (Part A: Initial Rate and Inhibitor Methods)
Edited by DANIEL L. PURICH

VOLUME 64. Enzyme Kinetics and Mechanism (Part B: Isotopic Probes and Complex Enzyme Systems)
Edited by DANIEL L. PURICH

VOLUME 65. Nucleic Acids (Part I)
Edited by LAWRENCE GROSSMAN AND KIVIE MOLDAVE

VOLUME 66. Vitamins and Coenzymes (Part E)
Edited by DONALD B. MCCORMICK AND LEMUEL D. WRIGHT

VOLUME 67. Vitamins and Coenzymes (Part F)
Edited by DONALD B. MCCORMICK AND LEMUEL D. WRIGHT

VOLUME 68. Recombinant DNA
Edited by RAY WU

VOLUME 69. Photosynthesis and Nitrogen Fixation (Part C)
Edited by ANTHONY SAN PIETRO

VOLUME 70. Immunochemical Techniques (Part A)
Edited by HELEN VAN VUNAKIS AND JOHN J. LANGONE

VOLUME 71. Lipids (Part C)
Edited by JOHN M. LOWENSTEIN

VOLUME 72. Lipids (Part D)
Edited by JOHN M. LOWENSTEIN

VOLUME 73. Immunochemical Techniques (Part B)
Edited by JOHN J. LANGONE AND HELEN VAN VUNAKIS

VOLUME 74. Immunochemical Techniques (Part C)
Edited by JOHN J. LANGONE AND HELEN VAN VUNAKIS

VOLUME 75. Cumulative Subject Index Volumes XXXI, XXXII, XXXIV–LX
Edited by EDWARD A. DENNIS AND MARTHA G. DENNIS

VOLUME 76. Hemoglobins
Edited by ERALDO ANTONINI, LUIGI ROSSI-BERNARDI, AND EMILIA CHIANCONE

VOLUME 77. Detoxication and Drug Metabolism
Edited by WILLIAM B. JAKOBY

VOLUME 78. Interferons (Part A)
Edited by SIDNEY PESTKA

VOLUME 79. Interferons (Part B)
Edited by SIDNEY PESTKA

VOLUME 80. Proteolytic Enzymes (Part C)
Edited by LASZLO LORAND

VOLUME 81. Biomembranes (Part H: Visual Pigments and Purple Membranes, I)
Edited by LESTER PACKER

VOLUME 82. Structural and Contractile Proteins (Part A: Extracellular Matrix)
Edited by LEON W. CUNNINGHAM AND DIXIE W. FREDERIKSEN

VOLUME 83. Complex Carbohydrates (Part D)
Edited by VICTOR GINSBURG

VOLUME 84. Immunochemical Techniques (Part D: Selected Immunoassays)
Edited by JOHN J. LANGONE AND HELEN VAN VUNAKIS

VOLUME 85. Structural and Contractile Proteins (Part B: The Contractile Apparatus and the Cytoskeleton)
Edited by DIXIE W. FREDERIKSEN AND LEON W. CUNNINGHAM

VOLUME 86. Prostaglandins and Arachidonate Metabolites
Edited by WILLIAM E. M. LANDS AND WILLIAM L. SMITH

VOLUME 87. Enzyme Kinetics and Mechanism (Part C: Intermediates, Stereochemistry, and Rate Studies)
Edited by DANIEL L. PURICH

VOLUME 88. Biomembranes (Part I: Visual Pigments and Purple Membranes, II)
Edited by LESTER PACKER

VOLUME 89. Carbohydrate Metabolism (Part D)
Edited by WILLIS A. WOOD

VOLUME 90. Carbohydrate Metabolism (Part E)
Edited by WILLIS A. WOOD

VOLUME 91. Enzyme Structure (Part I)
Edited by C. H. W. HIRS AND SERGE N. TIMASHEFF

VOLUME 92. Immunochemical Techniques (Part E: Monoclonal Antibodies and General Immunoassay Methods)
Edited by JOHN J. LANGONE AND HELEN VAN VUNAKIS

VOLUME 93. Immunochemical Techniques (Part F: Conventional Antibodies, Fc Receptors, and Cytotoxicity)
Edited by JOHN J. LANGONE AND HELEN VAN VUNAKIS

VOLUME 94. Polyamines
Edited by HERBERT TABOR AND CELIA WHITE TABOR

VOLUME 95. Cumulative Subject Index Volumes 61–74, 76–80
Edited by EDWARD A. DENNIS AND MARTHA G. DENNIS

VOLUME 96. Biomembranes [Part J: Membrane Biogenesis: Assembly and Targeting (General Methods; Eukaryotes)]
Edited by SIDNEY FLEISCHER AND BECCA FLEISCHER

VOLUME 97. Biomembranes [Part K: Membrane Biogenesis: Assembly and Targeting (Prokaryotes, Mitochondria, and Chloroplasts)]
Edited by SIDNEY FLEISCHER AND BECCA FLEISCHER

VOLUME 98. Biomembranes (Part L: Membrane Biogenesis: Processing and Recycling)
Edited by SIDNEY FLEISCHER AND BECCA FLEISCHER

VOLUME 99. Hormone Action (Part F: Protein Kinases)
Edited by JACKIE D. CORBIN AND JOEL G. HARDMAN

VOLUME 100. Recombinant DNA (Part B)
Edited by RAY WU, LAWRENCE GROSSMAN, AND KIVIE MOLDAVE

VOLUME 101. Recombinant DNA (Part C)
Edited by RAY WU, LAWRENCE GROSSMAN, AND KIVIE MOLDAVE

VOLUME 102. Hormone Action (Part G: Calmodulin and Calcium-Binding Proteins)
Edited by ANTHONY R. MEANS AND BERT W. O'MALLEY

VOLUME 103. Hormone Action (Part H: Neuroendocrine Peptides)
Edited by P. MICHAEL CONN

VOLUME 104. Enzyme Purification and Related Techniques (Part C)
Edited by WILLIAM B. JAKOBY

VOLUME 105. Oxygen Radicals in Biological Systems
Edited by LESTER PACKER

VOLUME 106. Posttranslational Modifications (Part A)
Edited by FINN WOLD AND KIVIE MOLDAVE

VOLUME 107. Posttranslational Modifications (Part B)
Edited by FINN WOLD AND KIVIE MOLDAVE

VOLUME 108. Immunochemical Techniques (Part G: Separation and Characterization of Lymphoid Cells)
Edited by GIOVANNI DI SABATO, JOHN J. LANGONE, AND HELEN VAN VUNAKIS

VOLUME 109. Hormone Action (Part I: Peptide Hormones)
Edited by LUTZ BIRNBAUMER AND BERT W. O'MALLEY

VOLUME 110. Steroids and Isoprenoids (Part A)
Edited by JOHN H. LAW AND HANS C. RILLING

VOLUME 111. Steroids and Isoprenoids (Part B)
Edited by JOHN H. LAW AND HANS C. RILLING

VOLUME 112. Drug and Enzyme Targeting (Part A)
Edited by KENNETH J. WIDDER AND RALPH GREEN

VOLUME 113. Glutamate, Glutamine, Glutathione, and Related Compounds
Edited by ALTON MEISTER

VOLUME 114. Diffraction Methods for Biological Macromolecules (Part A)
Edited by HAROLD W. WYCKOFF, C. H. W. HIRS, AND SERGE N. TIMASHEFF

VOLUME 115. Diffraction Methods for Biological Macromolecules (Part B)
Edited by HAROLD W. WYCKOFF, C. H. W. HIRS, AND SERGE N. TIMASHEFF

VOLUME 116. Immunochemical Techniques (Part H: Effectors and Mediators of Lymphoid Cell Functions)
Edited by GIOVANNI DI SABATO, JOHN J. LANGONE, AND HELEN VAN VUNAKIS

VOLUME 117. Enzyme Structure (Part J)
Edited by C. H. W. HIRS AND SERGE N. TIMASHEFF

VOLUME 118. Plant Molecular Biology
Edited by ARTHUR WEISSBACH AND HERBERT WEISSBACH

VOLUME 119. Interferons (Part C)
Edited by SIDNEY PESTKA

VOLUME 120. Cumulative Subject Index Volumes 81–94, 96–101

VOLUME 121. Immunochemical Techniques (Part I: Hybridoma Technology and Monoclonal Antibodies)
Edited by JOHN J. LANGONE AND HELEN VAN VUNAKIS

VOLUME 122. Vitamins and Coenzymes (Part G)
Edited by FRANK CHYTIL AND DONALD B. MCCORMICK

VOLUME 123. Vitamins and Coenzymes (Part H)
Edited by FRANK CHYTIL AND DONALD B. MCCORMICK

VOLUME 124. Hormone Action (Part J: Neuroendocrine Peptides)
Edited by P. MICHAEL CONN

VOLUME 125. Biomembranes (Part M: Transport in Bacteria, Mitochondria, and Chloroplasts: General Approaches and Transport Systems)
Edited by SIDNEY FLEISCHER AND BECCA FLEISCHER

VOLUME 126. Biomembranes (Part N: Transport in Bacteria, Mitochondria, and Chloroplasts: Protonmotive Force)
Edited by SIDNEY FLEISCHER AND BECCA FLEISCHER

VOLUME 127. Biomembranes (Part O: Protons and Water: Structure and Translocation)
Edited by LESTER PACKER

VOLUME 128. Plasma Lipoproteins (Part A: Preparation, Structure, and Molecular Biology)
Edited by JERE P. SEGREST AND JOHN J. ALBERS

VOLUME 129. Plasma Lipoproteins (Part B: Characterization, Cell Biology, and Metabolism)
Edited by JOHN J. ALBERS AND JERE P. SEGREST

VOLUME 130. Enzyme Structure (Part K)
Edited by C. H. W. HIRS AND SERGE N. TIMASHEFF

VOLUME 131. Enzyme Structure (Part L)
Edited by C. H. W. HIRS AND SERGE N. TIMASHEFF

VOLUME 132. Immunochemical Techniques (Part J: Phagocytosis and Cell-Mediated Cytotoxicity)
Edited by GIOVANNI DI SABATO AND JOHANNES EVERSE

VOLUME 133. Bioluminescence and Chemiluminescence (Part B)
Edited by MARLENE DELUCA AND WILLIAM D. MCELROY

VOLUME 134. Structural and Contractile Proteins (Part C: The Contractile Apparatus and the Cytoskeleton)
Edited by RICHARD B. VALLEE

VOLUME 135. Immobilized Enzymes and Cells (Part B)
Edited by KLAUS MOSBACH

VOLUME 136. Immobilized Enzymes and Cells (Part C)
Edited by KLAUS MOSBACH

VOLUME 137. Immobilized Enzymes and Cells (Part D)
Edited by KLAUS MOSBACH

VOLUME 138. Complex Carbohydrates (Part E)
Edited by VICTOR GINSBURG

VOLUME 139. Cellular Regulators (Part A: Calcium- and Calmodulin-Binding Proteins)
Edited by ANTHONY R. MEANS AND P. MICHAEL CONN

VOLUME 140. Cumulative Subject Index Volumes 102–119, 121–134

VOLUME 141. Cellular Regulators (Part B: Calcium and Lipids)
Edited by P. MICHAEL CONN AND ANTHONY R. MEANS

VOLUME 142. Metabolism of Aromatic Amino Acids and Amines
Edited by SEYMOUR KAUFMAN

VOLUME 143. Sulfur and Sulfur Amino Acids
Edited by WILLIAM B. JAKOBY AND OWEN GRIFFITH

VOLUME 144. Structural and Contractile Proteins (Part D: Extracellular Matrix)
Edited by LEON W. CUNNINGHAM

VOLUME 145. Structural and Contractile Proteins (Part E: Extracellular Matrix)
Edited by LEON W. CUNNINGHAM

VOLUME 146. Peptide Growth Factors (Part A)
Edited by DAVID BARNES AND DAVID A. SIRBASKU

VOLUME 147. Peptide Growth Factors (Part B)
Edited by DAVID BARNES AND DAVID A. SIRBASKU

VOLUME 148. Plant Cell Membranes
Edited by LESTER PACKER AND ROLAND DOUCE

VOLUME 149. Drug and Enzyme Targeting (Part B)
Edited by RALPH GREEN AND KENNETH J. WIDDER

VOLUME 150. Immunochemical Techniques (Part K: *In Vitro* Models of B and T Cell Functions and Lymphoid Cell Receptors)
Edited by GIOVANNI DI SABATO

VOLUME 151. Molecular Genetics of Mammalian Cells
Edited by MICHAEL M. GOTTESMAN

VOLUME 152. Guide to Molecular Cloning Techniques
Edited by SHELBY L. BERGER AND ALAN R. KIMMEL

VOLUME 153. Recombinant DNA (Part D)
Edited by RAY WU AND LAWRENCE GROSSMAN

VOLUME 154. Recombinant DNA (Part E)
Edited by RAY WU AND LAWRENCE GROSSMAN

VOLUME 155. Recombinant DNA (Part F)
Edited by RAY WU

VOLUME 156. Biomembranes (Part P: ATP-Driven Pumps and Related Transport: The Na,K-Pump)
Edited by SIDNEY FLEISCHER AND BECCA FLEISCHER

VOLUME 157. Biomembranes (Part Q: ATP-Driven Pumps and Related Transport: Calcium, Proton, and Potassium Pumps)
Edited by SIDNEY FLEISCHER AND BECCA FLEISCHER

VOLUME 158. Metalloproteins (Part A)
Edited by JAMES F. RIORDAN AND BERT L. VALLEE

VOLUME 159. Initiation and Termination of Cyclic Nucleotide Action
Edited by JACKIE D. CORBIN AND ROGER A. JOHNSON

VOLUME 160. Biomass (Part A: Cellulose and Hemicellulose)
Edited by WILLIS A. WOOD AND SCOTT T. KELLOGG

VOLUME 161. Biomass (Part B: Lignin, Pectin, and Chitin)
Edited by WILLIS A. WOOD AND SCOTT T. KELLOGG

VOLUME 162. Immunochemical Techniques (Part L: Chemotaxis and Inflammation)
Edited by GIOVANNI DI SABATO

VOLUME 163. Immunochemical Techniques (Part M: Chemotaxis and Inflammation)
Edited by GIOVANNI DI SABATO

VOLUME 164. Ribosomes
Edited by HARRY F. NOLLER, JR., AND KIVIE MOLDAVE

VOLUME 165. Microbial Toxins: Tools for Enzymology
Edited by SIDNEY HARSHMAN

VOLUME 166. Branched-Chain Amino Acids
Edited by ROBERT HARRIS AND JOHN R. SOKATCH

VOLUME 167. Cyanobacteria
Edited by LESTER PACKER AND ALEXANDER N. GLAZER

VOLUME 168. Hormone Action (Part K: Neuroendocrine Peptides)
Edited by P. MICHAEL CONN

VOLUME 169. Platelets: Receptors, Adhesion, Secretion (Part A)
Edited by JACEK HAWIGER

VOLUME 170. Nucleosomes
Edited by PAUL M. WASSARMAN AND ROGER D. KORNBERG

VOLUME 171. Biomembranes (Part R: Transport Theory: Cells and Model Membranes)
Edited by SIDNEY FLEISCHER AND BECCA FLEISCHER

VOLUME 172. Biomembranes (Part S: Transport: Membrane Isolation and Characterization)
Edited by SIDNEY FLEISCHER AND BECCA FLEISCHER

VOLUME 173. Biomembranes [Part T: Cellular and Subcellular Transport: Eukaryotic (Nonepithelial) Cells]
Edited by SIDNEY FLEISCHER AND BECCA FLEISCHER

VOLUME 174. Biomembranes [Part U: Cellular and Subcellular Transport: Eukaryotic (Nonepithelial) Cells]
Edited by SIDNEY FLEISCHER AND BECCA FLEISCHER

VOLUME 175. Cumulative Subject Index Volumes 135–139, 141–167

VOLUME 176. Nuclear Magnetic Resonance (Part A: Spectral Techniques and Dynamics)
Edited by NORMAN J. OPPENHEIMER AND THOMAS L. JAMES

VOLUME 177. Nuclear Magnetic Resonance (Part B: Structure and Mechanism)
Edited by NORMAN J. OPPENHEIMER AND THOMAS L. JAMES

VOLUME 178. Antibodies, Antigens, and Molecular Mimicry
Edited by JOHN J. LANGONE

VOLUME 179. Complex Carbohydrates (Part F)
Edited by VICTOR GINSBURG

VOLUME 180. RNA Processing (Part A: General Methods)
Edited by JAMES E. DAHLBERG AND JOHN N. ABELSON

VOLUME 181. RNA Processing (Part B: Specific Methods)
Edited by JAMES E. DAHLBERG AND JOHN N. ABELSON

VOLUME 182. Guide to Protein Purification
Edited by MURRAY P. DEUTSCHER

VOLUME 183. Molecular Evolution: Computer Analysis of Protein and Nucleic Acid Sequences
Edited by RUSSELL F. DOOLITTLE

VOLUME 184. Avidin–Biotin Technology
Edited by MEIR WILCHEK AND EDWARD A. BAYER

VOLUME 185. Gene Expression Technology
Edited by DAVID V. GOEDDEL

VOLUME 186. Oxygen Radicals in Biological Systems (Part B: Oxygen Radicals and Antioxidants)
Edited by LESTER PACKER AND ALEXANDER N. GLAZER

VOLUME 187. Arachidonate Related Lipid Mediators
Edited by ROBERT C. MURPHY AND FRANK A. FITZPATRICK

VOLUME 188. Hydrocarbons and Methylotrophy
Edited by MARY E. LIDSTROM

VOLUME 189. Retinoids (Part A: Molecular and Metabolic Aspects)
Edited by LESTER PACKER

VOLUME 190. Retinoids (Part B: Cell Differentiation and Clinical Applications)
Edited by LESTER PACKER

VOLUME 191. Biomembranes (Part V: Cellular and Subcellular Transport: Epithelial Cells)
Edited by SIDNEY FLEISCHER AND BECCA FLEISCHER

VOLUME 192. Biomembranes (Part W: Cellular and Subcellular Transport: Epithelial Cells)
Edited by SIDNEY FLEISCHER AND BECCA FLEISCHER

VOLUME 193. Mass Spectrometry
Edited by JAMES A. MCCLOSKEY

VOLUME 194. Guide to Yeast Genetics and Molecular Biology
Edited by CHRISTINE GUTHRIE AND GERALD R. FINK

VOLUME 195. Adenylyl Cyclase, G Proteins, and Guanylyl Cyclase
Edited by ROGER A. JOHNSON AND JACKIE D. CORBIN

VOLUME 196. Molecular Motors and the Cytoskeleton
Edited by RICHARD B. VALLEE

VOLUME 197. Phospholipases
Edited by EDWARD A. DENNIS

VOLUME 198. Peptide Growth Factors (Part C)
Edited by DAVID BARNES, J. P. MATHER, AND GORDON H. SATO

VOLUME 199. Cumulative Subject Index Volumes 168–174, 176–194

VOLUME 200. Protein Phosphorylation (Part A: Protein Kinases: Assays, Purification, Antibodies, Functional Analysis, Cloning, and Expression)
Edited by TONY HUNTER AND BARTHOLOMEW M. SEFTON

VOLUME 201. Protein Phosphorylation (Part B: Analysis of Protein Phosphorylation, Protein Kinase Inhibitors, and Protein Phosphatases)
Edited by TONY HUNTER AND BARTHOLOMEW M. SEFTON

VOLUME 202. Molecular Design and Modeling: Concepts and Applications (Part A: Proteins, Peptides, and Enzymes)
Edited by JOHN J. LANGONE

VOLUME 203. Molecular Design and Modeling: Concepts and Applications (Part B: Antibodies and Antigens, Nucleic Acids, Polysaccharides, and Drugs)
Edited by JOHN J. LANGONE

VOLUME 204. Bacterial Genetic Systems
Edited by JEFFREY H. MILLER

VOLUME 205. Metallobiochemistry (Part B: Metallothionein and Related Molecules)
Edited by JAMES F. RIORDAN AND BERT L. VALLEE

VOLUME 206. Cytochrome P450
Edited by MICHAEL R. WATERMAN AND ERIC F. JOHNSON

VOLUME 207. Ion Channels
Edited by BERNARDO RUDY AND LINDA E. IVERSON

VOLUME 208. Protein–DNA Interactions
Edited by ROBERT T. SAUER

VOLUME 209. Phospholipid Biosynthesis
Edited by EDWARD A. DENNIS AND DENNIS E. VANCE

VOLUME 210. Numerical Computer Methods
Edited by LUDWIG BRAND AND MICHAEL L. JOHNSON

VOLUME 211. DNA Structures (Part A: Synthesis and Physical Analysis of DNA)
Edited by DAVID M. J. LILLEY AND JAMES E. DAHLBERG

VOLUME 212. DNA Structures (Part B: Chemical and Electrophoretic Analysis of DNA)
Edited by DAVID M. J. LILLEY AND JAMES E. DAHLBERG

Volume 213. Carotenoids (Part A: Chemistry, Separation, Quantitation, and Antioxidation)
Edited by Lester Packer

Volume 214. Carotenoids (Part B: Metabolism, Genetics, and Biosynthesis)
Edited by Lester Packer

Volume 215. Platelets: Receptors, Adhesion, Secretion (Part B)
Edited by Jacek J. Hawiger

Volume 216. Recombinant DNA (Part G)
Edited by Ray Wu

Volume 217. Recombinant DNA (Part H)
Edited by Ray Wu

Volume 218. Recombinant DNA (Part I)
Edited by Ray Wu

Volume 219. Reconstitution of Intracellular Transport
Edited by James E. Rothman

Volume 220. Membrane Fusion Techniques (Part A)
Edited by Nejat Düzgüneş

Volume 221. Membrane Fusion Techniques (Part B)
Edited by Nejat Düzgüneş

Volume 222. Proteolytic Enzymes in Coagulation, Fibrinolysis, and Complement Activation (Part A: Mammalian Blood Coagulation Factors and Inhibitors)
Edited by Laszlo Lorand and Kenneth G. Mann

Volume 223. Proteolytic Enzymes in Coagulation, Fibrinolysis, and Complement Activation (Part B: Complement Activation, Fibrinolysis, and Nonmammalian Blood Coagulation Factors)
Edited by Laszlo Lorand and Kenneth G. Mann

Volume 224. Molecular Evolution: Producing the Biochemical Data
Edited by Elizabeth Anne Zimmer, Thomas J. White, Rebecca L. Cann, and Allan C. Wilson

Volume 225. Guide to Techniques in Mouse Development
Edited by Paul M. Wassarman and Melvin L. DePamphilis

Volume 226. Metallobiochemistry (Part C: Spectroscopic and Physical Methods for Probing Metal Ion Environments in Metalloenzymes and Metalloproteins)
Edited by James F. Riordan and Bert L. Vallee

Volume 227. Metallobiochemistry (Part D: Physical and Spectroscopic Methods for Probing Metal Ion Environments in Metalloproteins)
Edited by James F. Riordan and Bert L. Vallee

Volume 228. Aqueous Two-Phase Systems
Edited by Harry Walter and Göte Johansson

Volume 229. Cumulative Subject Index Volumes 195–198, 200–227

VOLUME 230. Guide to Techniques in Glycobiology
Edited by WILLIAM J. LENNARZ AND GERALD W. HART

VOLUME 231. Hemoglobins (Part B: Biochemical and Analytical Methods)
Edited by JOHANNES EVERSE, KIM D. VANDEGRIFF, AND ROBERT M. WINSLOW

VOLUME 232. Hemoglobins (Part C: Biophysical Methods)
Edited by JOHANNES EVERSE, KIM D. VANDEGRIFF, AND ROBERT M. WINSLOW

VOLUME 233. Oxygen Radicals in Biological Systems (Part C)
Edited by LESTER PACKER

VOLUME 234. Oxygen Radicals in Biological Systems (Part D)
Edited by LESTER PACKER

VOLUME 235. Bacterial Pathogenesis (Part A: Identification and Regulation of Virulence Factors)
Edited by VIRGINIA L. CLARK AND PATRIK M. BAVOIL

VOLUME 236. Bacterial Pathogenesis (Part B: Integration of Pathogenic Bacteria with Host Cells)
Edited by VIRGINIA L. CLARK AND PATRIK M. BAVOIL

VOLUME 237. Heterotrimeric G Proteins
Edited by RAVI IYENGAR

VOLUME 238. Heterotrimeric G-Protein Effectors
Edited by RAVI IYENGAR

VOLUME 239. Nuclear Magnetic Resonance (Part C)
Edited by THOMAS L. JAMES AND NORMAN J. OPPENHEIMER

VOLUME 240. Numerical Computer Methods (Part B)
Edited by MICHAEL L. JOHNSON AND LUDWIG BRAND

VOLUME 241. Retroviral Proteases
Edited by LAWRENCE C. KUO AND JULES A. SHAFER

VOLUME 242. Neoglycoconjugates (Part A)
Edited by Y. C. LEE AND REIKO T. LEE

VOLUME 243. Inorganic Microbial Sulfur Metabolism
Edited by HARRY D. PECK, JR., AND JEAN LEGALL

VOLUME 244. Proteolytic Enzymes: Serine and Cysteine Peptidases
Edited by ALAN J. BARRETT

VOLUME 245. Extracellular Matrix Components
Edited by E. RUOSLAHTI AND E. ENGVALL

VOLUME 246. Biochemical Spectroscopy
Edited by KENNETH SAUER

VOLUME 247. Neoglycoconjugates (Part B: Biomedical Applications)
Edited by Y. C. LEE AND REIKO T. LEE

VOLUME 248. Proteolytic Enzymes: Aspartic and Metallo Peptidases
Edited by ALAN J. BARRETT

VOLUME 249. Enzyme Kinetics and Mechanism (Part D: Developments in Enzyme Dynamics)
Edited by DANIEL L. PURICH

VOLUME 250. Lipid Modifications of Proteins
Edited by PATRICK J. CASEY AND JANICE E. BUSS

VOLUME 251. Biothiols (Part A: Monothiols and Dithiols, Protein Thiols, and Thiyl Radicals)
Edited by LESTER PACKER

VOLUME 252. Biothiols (Part B: Glutathione and Thioredoxin; Thiols in Signal Transduction and Gene Regulation)
Edited by LESTER PACKER

VOLUME 253. Adhesion of Microbial Pathogens
Edited by RON J. DOYLE AND ITZHAK OFEK

VOLUME 254. Oncogene Techniques
Edited by PETER K. VOGT AND INDER M. VERMA

VOLUME 255. Small GTPases and Their Regulators (Part A: Ras Family)
Edited by W. E. BALCH, CHANNING J. DER, AND ALAN HALL

VOLUME 256. Small GTPases and Their Regulators (Part B: Rho Family)
Edited by W. E. BALCH, CHANNING J. DER, AND ALAN HALL

VOLUME 257. Small GTPases and Their Regulators (Part C: Proteins Involved in Transport)
Edited by W. E. BALCH, CHANNING J. DER, AND ALAN HALL

VOLUME 258. Redox-Active Amino Acids in Biology
Edited by JUDITH P. KLINMAN

VOLUME 259. Energetics of Biological Macromolecules
Edited by MICHAEL L. JOHNSON AND GARY K. ACKERS

VOLUME 260. Mitochondrial Biogenesis and Genetics (Part A)
Edited by GIUSEPPE M. ATTARDI AND ANNE CHOMYN

VOLUME 261. Nuclear Magnetic Resonance and Nucleic Acids
Edited by THOMAS L. JAMES

VOLUME 262. DNA Replication
Edited by JUDITH L. CAMPBELL

VOLUME 263. Plasma Lipoproteins (Part C: Quantitation)
Edited by WILLIAM A. BRADLEY, SANDRA H. GIANTURCO, AND JERE P. SEGREST

VOLUME 264. Mitochondrial Biogenesis and Genetics (Part B)
Edited by GIUSEPPE M. ATTARDI AND ANNE CHOMYN

VOLUME 265. Cumulative Subject Index Volumes 228, 230–262

VOLUME 266. Computer Methods for Macromolecular Sequence Analysis
Edited by RUSSELL F. DOOLITTLE

VOLUME 267. Combinatorial Chemistry
Edited by JOHN N. ABELSON

VOLUME 268. Nitric Oxide (Part A: Sources and Detection of NO; NO Synthase)
Edited by LESTER PACKER

VOLUME 269. Nitric Oxide (Part B: Physiological and Pathological Processes)
Edited by LESTER PACKER

VOLUME 270. High Resolution Separation and Analysis of Biological Macromolecules (Part A: Fundamentals)
Edited by BARRY L. KARGER AND WILLIAM S. HANCOCK

VOLUME 271. High Resolution Separation and Analysis of Biological Macromolecules (Part B: Applications)
Edited by BARRY L. KARGER AND WILLIAM S. HANCOCK

VOLUME 272. Cytochrome P450 (Part B)
Edited by ERIC F. JOHNSON AND MICHAEL R. WATERMAN

VOLUME 273. RNA Polymerase and Associated Factors (Part A)
Edited by SANKAR ADHYA

VOLUME 274. RNA Polymerase and Associated Factors (Part B)
Edited by SANKAR ADHYA

VOLUME 275. Viral Polymerases and Related Proteins
Edited by LAWRENCE C. KUO, DAVID B. OLSEN, AND STEVEN S. CARROLL

VOLUME 276. Macromolecular Crystallography (Part A)
Edited by CHARLES W. CARTER, JR., AND ROBERT M. SWEET

VOLUME 277. Macromolecular Crystallography (Part B)
Edited by CHARLES W. CARTER, JR., AND ROBERT M. SWEET

VOLUME 278. Fluorescence Spectroscopy
Edited by LUDWIG BRAND AND MICHAEL L. JOHNSON

VOLUME 279. Vitamins and Coenzymes, Part I
Edited by DONALD B. MCCORMICK, JOHN W. SUTTIE, AND CONRAD WAGNER

VOLUME 280. Vitamins and Coenzymes, Part J
Edited by DONALD B. MCCORMICK, JOHN W. SUTTIE, AND CONRAD WAGNER

VOLUME 281. Vitamins and Coenzymes, Part K
Edited by DONALD B. MCCORMICK, JOHN W. SUTTIE, AND CONRAD WAGNER

VOLUME 282. Vitamins and Coenzymes, Part L
Edited by DONALD B. MCCORMICK, JOHN W. SUTTIE, AND CONRAD WAGNER

VOLUME 283. Cell Cycle Control
Edited by WILLIAM G. DUNPHY

VOLUME 284. Lipases (Part A: Biotechnology)
Edited by BYRON RUBIN AND EDWARD A. DENNIS

VOLUME 285. Cumulative Subject Index Volumes 263, 264, 266–289

Volume 286. Lipases (Part B: Enzyme Characterization and Utilization)
Edited by Byron Rubin and Edward A. Dennis

Volume 287. Chemokines
Edited by Richard Horuk

Volume 288. Chemokine Receptors
Edited by Richard Horuk

Volume 289. Solid Phase Peptide Synthesis
Edited by Gregg B. Fields

Volume 290. Molecular Chaperones
Edited by George H. Lorimer and Thomas Baldwin

Volume 291. Caged Compounds
Edited by Gerard Marriott

Volume 292. ABC Transporters: Biochemical, Cellular, and Molecular Aspects
Edited by Suresh V. Ambudkar and Michael M. Gottsman

Volume 293. Ion Channels (Part B)
Edited by P. Michael Conn

Volume 294. Ion Channels (Part C) (in preparation)
Edited by P. Michael Conn

Volume 295. Energetics of Biological Macromolecules (Part B)
Edited by Gary K. Ackers and Michael L. Johnson

Volume 296. Neurotransmitter Transporters (in preparation)
Edited by Susan G. Amara

Volume 297. Photosynthesis: Molecular Biology of Energy Capture (in preparation)
Edited by Lee McIntosh

Volume 298. Molecular Motors and the Cytoskeleton (Part B) (in preparation)
Edited by Richard B. Vallee

Volume 299. Oxidants and Antioxidants (Part A) (in preparation)
Edited by Lester Packer

Volume 300. Oxidants and Antioxidants (Part B) (in preparation)
Edited by Lester Packer

Volume 301. Nitric Oxide: Biological and Antioxidant Activities (Part C: Biological and Antioxidant Activities) (in preparation)
Edited by Lester Packer

Volume 302. Green Fluorescent Protein (in preparation)
Edited by P. Michael Conn

Volume 303. cDNA Preparation and Display (in preparation)
Edited by Sherman M. Weissman

Volume 304. Chromatin (in preparation)
Edited by Paul M. Wasserman and Alan P. Wolffe

Section I

Prokaryotic ABC Transporters

[1] Overview of Bacterial ABC Transporters

By HIROSHI NIKAIDO and JASON A. HALL

Introduction

ABC transporters were so named because one of the domains or component proteins of the systems contains a homologous adenosine triphosphate (ATP)-binding sequence. Note, however, that the short ATP-binding motifs (so-called Walker motifs[1]) also occur on transporters other than the ATP-binding cassette (ABC) transporters, such as F-type ATPases (for example, F_1F_0-ATPase), P-type ATPases (for example, the Kdp potassium transport system of *Escherichia coli*), and the Ars arsenite/arsenate export system of *E. coli.* The ATP-binding domains or components of ABC transporters, however, share homology in a larger region of about 200 amino acids, extending far beyond the Walker motifs or the ATP-binding cassettes in the narrow sense, and the ABC family is defined on the basis of this and other evolutionary relationships.

Several different types of ABC transporters are found in bacteria, and they carry out a wide range of transport functions[2] (see Table I). (1) Many of them transport small molecules into the cell in association with the binding protein (BP) that occurs in the periplasm of gram-negative bacteria or anchored via a lipid moiety to the cytoplasmic membrane in gram-positive bacteria. The solutes that are transported by such systems include sugars, amino acids, peptides, phosphate esters, inorganic phosphate, sulfate, phosphonates, metal cations, iron–chelator complexes, vitamins, and polyamines.[3] (2) Some of them export proteins directly into the media[4-6]; these systems occur together usually with two "helper" proteins, an outer membrane channel[4] and a "linker" protein that belongs to the membrane fusion protein (MFP) family[4,7] and connects the transporter to the outer membrane. (3) Finally, some of them export drug molecules or polysaccharides across the cytoplasmic membrane. This last group is often called

[1] J. E. Walker, M. Saraste, and N. J. Gay, *Biochim. Biophys. Acta* **768,** 164 (1984).
[2] C. F. Higgins, *Annu. Rev. Cell Biol.* **8,** 67 (1992).
[3] W. Boos and J. M. Lucht, *in* "*Escherichia coli* and *Salmonella*: Cellular and Molecular Biology" (F. C. Neidhardt, ed.), 2nd Ed., p. 1175. ASM Press, Washington, DC, 1996.
[4] C. Wandersman, *Trends Genet.* **8,** 317 (1992).
[5] M. J. Fath and R. Kolter, *Microbiol. Rev.* **57,** 995 (1993).
[6] C. Wandersman, *in* "*Escherichia coli* and *Salmonella*: Cellular and Molecular Biology" (F. C. Neidhardt, ed.), 2nd Ed., p. 955. ASM Press, Washington, DC, 1996.
[7] T. Dinh, I. T. Paulsen, and M. H. Saier, Jr., *J. Bacteriol.* **176,** 3825 (1994).

0076-6879/98 $25.00

TABLE I
EXAMPLES OF BACTERIAL ABC TRANSPORTERS

Substrate	Bacteria	ATPase (α)	Channel (β)	BP	MFP	OM[a] channel
ABC Transporters for import[b]						
Maltose	*Escherichia coli*	MalK	MalF, G	MalE		
Histidine	*Salmonella typhimurium*	HisP	HisM, Q	HisJ		
Oligopeptides	*S. typhimurium*	OppD, F	OppB, C	OppA		
Galactose	*E. coli*	MglA	MglC	MglB		
Ribose	*E. coli*	RbsA	RbsC	RbsB		
Arabinose	*E. coli*	AraG	AraH	AraF		
Phosphate	*E. coli*	PstB	PstA, C	PstS		
Ribose	*E. coli*	RbsA	RbsC	RbsB		
Leucine/isoleucine/valine	*E. coli*	LivF, G	LivH, M	LivJ		
Leucine/isoleucine/valine-I	*Pseudomonas aeruginosa*	BraF, G	BraD, E	BraC		
Vitamin B_{12}	*E. coli*	BtuD	BtuC	BtuE		
Fe-enterobactin	*E. coli*	FecC	FepD, G	FepB		
Fe-hydroxamate	*E. coli*	FhuC	FhuB	FhuD		
Fe-dicitrate	*E. coli*	FecE	FecC, D	FecB		
Glycerol 3-phosphate	*E. coli*	UgpC	UgpA, E	UgpE		
ABC Transporters for Protein Export[c]						
Hemolysin	*E. coli*	HlyB (C)[d]	HlyB (N)[d]	—	HlyD	TolC
Cyclolysin	*Bordetella pertussi*	CyaB (C)	CyaB (N)	—	CyaD	CyaE
Proteases A, B, C	*Erwinia chrysanthemi*	PrtD (C)	PrtD (N)	—	PrtE	PrtF
Colicin V	*E. coli*	CvaB (C)	CvaB (N)	—	CvaA	TolC
ABC-2 Transporters[e]						
Daunomycin	*Streptomyces peucetius*	DrrA	DrrB	—		
Capsular polysaccharide	*E. coli*	KpsT	KpsM	—		
Capsular polysaccharide	*Haemophilus influenzae*	BexA	BexB	—		
Capsular polysaccharide	*Neisseria meningitidis*	CtrD	CtrC	—		
Lipooligosaccharide	*Rhizobium meliloti*	NodI	NodJ	—		

[a] Outer membrane.
[b] For reference see W. Boos and J. M. Lucht, *in* "*Escherichia coli* and *Salmonella*: Cellular and Molecular Biology" (F. C. Neidhardt, ed.), 2nd Ed., p. 1175. ASM Press, Washington, DC, 1996.
[c] For reference see M. J. Fath and R. Kolter, *Microbiol. Rev.* **57,** 995 (1993).
[d] C and N denote the C- and N-terminal portions of a protein, respectively.
[e] For reference see J. Reizer, A. Reizer, and M. H. Saier, Jr., *Prot. Sci.* **1,** 1326 (1992).

ABC-2 transporters, because the presumed channel-forming components are homologous to each other, but not to the components of other members of the ABC family.[8] (In this article, the word *channel* is used in a way that is nearly interchangeable with the word *transporter*, because in terms of structure a clear distinction between the two does not seem to exist.[9]) There are reviews that cover all types of ABC transporters,[2] BP-dependent import

[8] J. Reizer, A. Reizer, and M. H. Saier, Jr., *Prot. Sci.* **1,** 1326 (1992).
[9] H. Nikaido and M. H. Saier, Jr., *Science* **258,** 936 (1992).

transporters,[3] those involved in export processes,[4–6] ABC-2 type systems,[8] or binding proteins.[10] ABC transporters are also called traffic ATPases.[11]

Among many ABC transporters found in bacteria, two systems have been studied extensively, and reference is frequently made to these systems in this article. These are the maltose importer of *E. coli*, composed of the maltose BP (MalE) and the $MalK_2FG$ transporter, and the histidine importer of *Salmonella typhimurium*, composed of the histidine BP (HisJ) and the $HisP_2MQ$ transporter.

Domain Organization

Maltose transporter of *E. coli* has been purified to near homogeneity, and the complex was shown to contain two copies (α_2; for nomenclature see Table I) of the ATP-binding protein (MalK), and one copy each of the two intrinsic membrane proteins that are homologous to each other ($\beta\beta'$), that is, MalF and MalG.[12] Chemical cross-linking and immunoprecipitation of the histidine transporter from *S. typhimurium* led to a similar conclusion that the transporter has the $\alpha_2\beta\beta'$ composition, containing two copies of ATP-binding protein HisP and one copy each of transmembrane components HisM and HisQ.[13] It is reasonable to expect that similar construction, containing two ATP-binding domains and two membrane-traversing domains, exists with all other ABC transporters, although hard evidence is not available in most cases.

Some ABC transporters have two homologous, but nonidentical ATP-binding subunits, with the composition $\alpha\alpha'\beta\beta'$. An example is the OppDFBC oligopeptide transporter of *S. typhimurium* (Table I). In some transporters, only one gene is known for the membrane channel component, and we would expect $\alpha_2\beta_2$ (or possibly $\alpha\alpha'\beta_2$) construction in such cases. Examples include the BtuDC vitamin B_{12} transporter and MglAC galactose transporter of *E. coli*, as well as DrrAB daunorubicin exporter of *Streptomyces peucetius* and KpsTM capsular polysaccharide exporter of *E. coli* (Table I).

In the transporters mentioned so far, each domain occurs as a separate protein. However, in many transporters the domains are fused to form multidomain protein components. For example, the two ATP-binding domains of the ribose system are fused to produce a single protein, RbsA,

[10] R. Tam and M. H. Saier, Jr., *Microbiol. Rev.* **57,** 320 (1993).

[11] G. F.-L. Ames, C. S. Mimura, S. R. Holbrook, and V. Shyamala, *Adv. Enzymol.* **65,** 1 (1992).

[12] A. L. Davidson and H. Nikaido, *J. Biol. Chem.* **266,** 8946 (1991).

[13] R. E. Kerppola, V. K. Shyamala, P. Klebba, and G. F.-L. Ames, *J. Biol. Chem.* **266,** 9857 (1991).

so the composition of the transporter can be denoted as α-$\alpha'\beta_2$, following the nomenclature of Tam and Saier.[10] The FhuB membrane channel protein for the transport of ferric hydroxamate appears to contain a duplication of the usual channel domain, so in this case the composition is $\alpha_2\beta$-β'. Indeed the Fhu system was functional when the *fhuB* gene was cut in two nearly equal halves, and the N-terminal and C-terminal halves were expressed as two separate genes.[14] In most of the protein/peptide-exporting ABC transporters, the membrane-spanning domain and the ATP-binding domain are fused, so their composition is $(\alpha$-$\beta)_2$ (see the hemolysin exporter HlyB, protease exporter PrtD, etc., in Table I). In eukaryotic ABC transporters, all of the domains are frequently fused to produce one giant protein, as in the P-glycoproteins of animal and human cells.

Membrane Channel Proteins

Domains or component proteins that are thought to form the membrane-spanning channels are typically predicted to contain six transmembrane α helices. Because each transporter complex contains two such domains, this produces 12 transmembrane α helices, an architecture also found in transport proteins belonging to other classes, such as proton symporters.[15] Indeed OppB and OppC, studied by construction of β-lactamase gene fusion, were each confirmed to contain six transmembrane helices,[16] and the same conclusion was obtained with MalG, studied by using alkaline phosphatase gene fusion.[17] However, variations of this theme have also been observed. MalF, which is significantly larger than the usual membrane-spanning components, was shown to contain eight, rather than six, membrane-spanning helices, on the basis of alkaline phosphatase and β-galactosidase gene fusion experiments.[18] Alignment of the MalF sequence with the sequences of other ABC transporters showed that the two membrane-spanning helices closest to the N terminus did not have counterparts in other proteins; indeed the first of these helices could be deleted without eliminating the transport activity.[19] HisQ and HisM, with 228 and 235 amino acid residues, respectively, are among the smallest examples of membrane channel proteins of the ABC family. Alkaline phosphatase fusion and

[14] W. Köster and V. Braun, *Mol. Gen. Genet.* **233,** 379 (1990).

[15] P. J. F. Henderson, *Curr. Opin. Cell Biol.* **5,** 708 (1993).

[16] S. R. Pearce, M. L. Mimmack, M. P. Gallagher, U. Gileadi, S. C. Hyde, and C. F. Higgins, *Mol. Microbiol.* **6,** 47 (1992).

[17] E. Dassa and S. Muir, *Mol. Microbiol.* **7,** 29 (1993).

[18] S. Froshauer, G. N. Green, D. Boyd, K. McGovern and J. Beckwith, *J. Mol. Biol.* **200,** 501 (1988).

[19] M. Ehrmann and J. Beckwith, *J. Biol. Chem.* **266,** 16530 (1991).

proteolysis experiments indeed showed that each of these proteins crosses the membrane only five times, and their N terminus is found in the periplasm, rather than the cytoplasm as in most other homologous proteins.[20]

Dassa and Hofnung[21] noted that there is a conserved sequence in the cytoplasmic loop between the second and third transmembrane helices counting from the C terminus. Although the conservation of the sequence is not absolute, the sequence EAA-X_3-G-X_9-I-X-LP or its variant occurs in all of the BP-dependent importers of the ABC family,[3,22] but not in other classes of ABC transporters. This homologous stretch is functionally essential, because genetic alterations here produce nonfunctional transporters.[23] Overall, some homology can be detected among the sequences of the channel components of the BP-dependent importers, but little homology can be detected between these and the exporter members of the ABC family.[3,24] Whenever a system contains two channel components ($\beta\beta'$ construction), these two components tend to be more closely related to each other than to components of other systems, but there are exceptions. The systems that show particularly close homology include (1) channels of TonB-dependent transporters of vitamin B_{12} and iron–siderophore complexes (BtuC, FhuB, FecCD, and FepDG), (2) channel components of histidine and arginine transporters (HisMQ and ArtMQ), and (3) transporters of monosaccharides (MglC, RbsC, XylH and AraH).[3,22] Channel proteins of exporters of ABC-2 family share homology among them, as described earlier.[8]

Adenosine Triphosphate-Binding Proteins

The ATP-binding proteins or domains of ABC transporters share extensive sequence homology in a stretch (about 200 amino acids or more) that spans the region between the two ATP-binding Walker motifes (Fig. 1). This extensive homology was first noted by comparison of the sequences of MalK and HisP,[25] and forms the basis for the grouping of ABC transporters. The homology of these ATP-binding subunits extends into the protein exporters of the family, but those belonging to the ABC-2 family are somewhat distant in their relation to the rest of the family.[8] The two short ATP-binding motifs[1] were subsequently discovered by Higgins and co-

[20] R. E. Kerppola and G. F.-L. Ames, *J. Biol. Chem.* **267,** 2329 (1992).

[21] E. Dassa and M. Hofnung, *EMBO J.* **4,** 2287 (1985).

[22] W. Saurin, W. Köster, and E. Dassa, *Mol. Microbiol.* **12,** 993 (1994).

[23] E. Dassa, *Mol. Microbiol.* **7,** 39 (1993).

[24] W. Saurin and E. Dassa, *Prot. Sci.* **3,** 325 (1994).

[25] E. Gilson, C. F. Higgins, M. Hofnung, G. F.-L. Ames, and H. Nikaido, *J. Biol. Chem.* **257,** 9915 (1982).

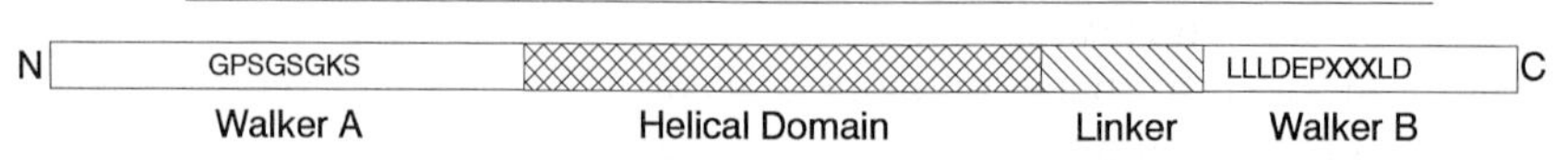

FIG. 1. Primary sequence of a typical ATP-binding subunit. The two nucleotide-binding motifs (Walker A and Walker B) surround the 90-residue helical domain and the 9-residue linker domain that is rich in glycine and glutamine residues. The region that is homologous among ATP-binding subunits of various ABC transporters is denoted by a continuous line at the top and the words "Homologous Region." This scheme applies to smaller members, with about 250 amino acid residues. When the subunit is much larger, it usually contains an added, nonhomologous extension at the C terminus, as in MalK (371 residues).

workers[26] by comparison of the transporter sequences with other nucleotide-binding proteins including the α and β subunits of F_1-ATPase, a finding that had a major impact on the field.

The ATP-binding subunits have hydrophilic sequences throughout, and do not contain signal sequences. The simplest assumption, therefore, is that they occur as peripheral membrane proteins, associated with the cytoplasmic surface of the membrane through their interaction with the membrane-spanning, channel components. Indeed, when the channel component MalF is absent, the ATP-binding component MalK is found free in the cytoplasm.[27,28] In contrast, HisP remains associated with the cytoplasmic membrane even in strains lacking the integral membrane components of the system,[13] and evidence has been presented that suggests the extension of a part of HisP through the thickness of the membrane and its exposure on the peripheral face.[29] Although it is somewhat difficult to accept that various ATP-binding subunits assume very different conformations in spite of the extensive homology among them, it may be that the HisP constitutes an exceptional case, perhaps owing to the unusually small sizes of the channel-forming components of this system (see earlier discussion). Earlier, the genetic suppression of a HisJ BP mutation by mutations in HisP[30] was taken to indicate the direct interaction between these two proteins and thus the exposure of the HisP protein on the external, periplasmic face of the membrane. However, more recently it became clear that these *hisP* mutations were those that made the histidine transport independent of

[26] C. F. Higgins, I. D. Hiles, G. P. C. Salmond, D. R. Gill, J. A. Downie, I. J. Evans, I. B. Holland, L. Gray, S. D. Buckel, A. W. Bell, and A. M. Hermodson, *Nature* **323,** 448 (1986).
[27] H. A. Shuman and T. J. Silhavy, *J. Biol. Chem.* **256,** 560 (1981).
[28] C. H. Panagiotidis, M. Reyes, A. Sievertsen, W. Boos, and H. A. Shuman, *J. Biol. Chem.* **268,** 23685 (1993).
[29] V. Baichwal, D. Liu, and G. F.-L. Ames, *Proc. Natl. Acad. Sci. USA* **90,** 820 (1993).
[30] G. F.-L. Ames and E. N. Spudich, *Proc. Natl. Acad. Sci. USA* **73,** 1877 (1976).

the presence of HisJ BP[31] (see later discussion), and that the suppression phenomenon in this case is not a proof of the direct interaction between HisJ and HisP. It has been claimed that a part of MalK is also exposed on the external surface of the cytoplasmic membrane.[32] However, the experimental approach involved the digestion of membrane vesicles with an extraordinarily high concentration (1 mg/ml) of proteinase K for several hours, and the conclusion is not persuasive.

The binding of ATP and analogs to these subunits has been shown in numerous systems (for example, see Refs. 28 and 33). In addition, ATP-hydrolyzing activity has been detected, for example, with the MalK protein that has been produced as inclusion bodies, denatured with urea, and subsequently renatured.[34] Although such results confirm the assumption that these subunits hydrolyze ATP in living cells, their physiological significance is questionable because the intact maltose transporter complex does not hydrolyze ATP unless stimulated by the maltose-binding protein and maltose.[35] In the membrane vesicles, even the photoaffinity labeling of MalK did not occur unless MalK was associated with MalF.[28] Finally, the specific ATPase activity of such renatured MalK protein was at least an order of magnitude lower than that of the fully stimulated intact complex. These observations all suggest that ATP hydrolysis by isolated MalK is likely to be a consequence of the improper folding of this subunit.

In one interesting case, the ATP-binding subunit of the maltose system, MalK, could be exchanged genetically with that of the glycerol 3-phosphate transporter, UgpC, with the conservation of the transport functions.[36] This shows clearly that the substrate specificity of the system comes from other components. Construction of chimera containing the first Walker motif and the helical domain (see below) of HisP and the C-terminal domain of MalK produced a subunit that functioned in the transport of maltose.[37]

None of the ATP-binding subunits appears to have been crystallized. Two laboratories, however, attempted to predict the three-dimensional structures of the HisP subunits by using adenylate kinase as a model.[38,39]

[31] D. M. Speiser and G. F.-L. Ames, *J. Bacteriol.* **173,** 1444 (1991).
[32] E. Schneider, S. Hunke, and S. Tebbe, *J. Bacteriol.* **177,** 5364 (1995).
[33] A. C. Hobson, R. Weatherwax, and G. F.-L. Ames, *Proc. Natl. Acad. Sci. USA* **81,** 7333 (1984).
[34] C. Walter, K. Hörner zu Bentrup, and E. Schneider, *J. Biol. Chem.* **267,** 8863 (1992).
[35] A. L. Davidson, H. A. Shuman, and H. Nikaido, *Proc. Natl. Acad. Sci. USA* **89,** 2360 (1992).
[36] D. Hoekstra and J. Tommassen, *J. Bacteriol.* **175,** 6546 (1993).
[37] E. Schneider and C. Walter, *Mol. Microbiol.* **5,** 1375 (1991).
[38] S. C. Hyde, P. Emsley, M. J. Hartshorn, M. M. Mimmack, U. Gileadi, S. R. Pearce, M. P. Gallagher, D. R. Gill, R. E. Hubbard, and C. Higgins, *Nature* **345,** 362 (1990).
[39] C. S. Mimura, S. R. Holbrook, and G. F.-L. Ames, *Proc. Natl. Acad. Sci. USA* **88,** 84 (1991).

The N-terminal part containing the Walker motif A and the C-terminal part containing the Walker motif B (Fig. 1) are predicted to fold together in a way similar to the model protein and to produce the ATP-binding pocket, but the large, middle part of the protein containing about 90 amino acid residues was excluded entirely from this pocket structure in at least one model.[39] This portion, often called the *helical domain*, and a linker region that connects the C-terminal end of the helical domain to the ATPase portion of the molecule (Fig. 1) are hypothesized to be important in coupling ATP hydrolysis to conformational changes in the channel portion of the transporter.[11] Interestingly, several mutations in HisP that uncouple ATP hydrolysis from solute translocation are indeed located in this linker region as well as at the beginning of the second ATP-binding region.[40]

Some of the ATP-binding subunits apparently perform additional functions. The accumulation of maltose by *E. coli* is inhibited in the presence of glucose in the medium, because the dephosphorylated form of enzyme IIA^{Glc} inhibits the function of MalK. Analysis of glucose-resistant mutant forms of MalK showed that the mutation occurred in the COOH-terminal segment of MalK,[41] an additional, unique segment that does not exist in ATP-binding subunits of other systems.

Binding Proteins

Practically all of the bacterial importers of small molecules occur together with the ligand-binding proteins. It used to be thought that BPs existed only in the periplasmic space of gram-negative bacteria. However, it is now clear that many gram-positive bacteria produce BPs, which are located at the external face of the cytoplasmic membrane, anchored to the membrane through their lipid extensions at their N termini (see Ref. 42).

Binding proteins have been classified into eight evolutionary clusters by comparison of their sequences.[10] Many BPs have been crystallized with and without ligands, and their structures solved most prominently by the laboratories of F. Quiocho[43] and S. Mowbray. All BPs appear to consist of two domains that are connected by a hinge. The ligands bind to a cleft between these two domains, and induce a large conformational change that consists of the closing of the cleft.

The maltose-binding protein (MBP) has been studied in great detail.

[40] V. Shyamala, V. Baichwal, E. Beall, and G. F.-L. Ames, *J. Biol. Chem.* **266,** 18714 (1991).
[41] D. A. Dean, J. Reizer, H. Nikaido, and M. H. Saier, Jr., *J. Biol. Chem.* **265,** 21005 (1990).
[42] E. Gilson, G. Alloing, T. Schmidt, J. P. Claverys, R. Dudier, and M. Hofnung, *EMBO J.* **7,** 3971 (1986).
[43] F. Quiocho, *Phil Trans. Roy. Soc.* (*London*), **326,** 341 (1990).

An interesting feature of MBP is its affinity not only to maltose but also to linear and cyclic maltodextrins. The crystal structure of the maltose–MBP complex shows a classical closed structure,[44] whereas the binding of β-cyclodextrin produces a largely open structure.[45] Interestingly, ^{3}H NMR studies of ligand binding using maltose and maltodextrins labeled with ^{3}H at C-1 showed that maltose binds to MBP in a way that involves a tight association of the reducing glucose moiety ("end-on mode"), whereas maltodextrins bind also in a way that does not involve the tight association of the reducing glucose residue ("middle mode").[46] The end-on binding is correlated with the red shift of the fluorescence emission spectrum of MBP, and the middle binding is correlated with its blue shift.[47] Reduced or oxidized maltodextrins appear to bind to MBP exclusively via the middle mode.[48] Because these ligands are not transported into the cell in spite of their tight binding to MBP,[49] it seems likely that the middle binding corresponds to a nonproductive binding mode. It is not yet clear what the middle mode binding means in terms of crystallographic structure.

Ligand Transport and the Role of Binding Proteins

It has been always assumed that a main function of BPs is to confer high affinity to the transport process. This hypothesis is supported by comparing the transport K_m values of the BP-dependent ABC transporters, which are usually around 1 μM or less, with the K_m values of proton symporters, which are commonly in the 0.1- to 1-mM range. In some cases, the BP clearly determines the final specificity of the transport process. Thus the same transporter, $HisP_2MQ$, catalyzes the accumulation of either histidine or lysine/arginine/ornithine, depending on whether it interacts with the HisJ BP or ArgT (LAO) BP, respectively.[50] Similarly, the LivFGHM transporter of *E. coli* transports either leucine exclusively or leucine/isoleucine/valine, depending on whether it operates in association with LivK BP or LivJ BP, respectively.[51] These results raised the question of whether the BP entirely determines the specificity of ABC transporters (see later discussion).

[44] J. C. Spurlino, G. Y. Lu, and F. A. Quiocho, *J. Biol. Chem.* **266,** 5202 (1991).

[45] A. J. Sharff, L. E. Rodseth, and F. A. Quiocho, *Biochemistry* **32,** 10553 (1993).

[46] K. Gehring, P. G. Williams, J. G. Pelton, and D. E. Wemmer, *Biochemistry* **30,** 5524 (1991).

[47] K. Gehring, K. Bao, and H. Nikaido, *FEBS Lett.* **300,** 33 (1992).

[48] H. Nikaido, I. D. Pokrovskaya, L. Reyes, A. K. Ganesan, and J. A. Hall, *in* "Phosphate in Microorganisms: Cellular and Molecular Biology" (A. Torriani-Gorini, E. Yagil, and S. Silver, eds.), p. 91. ASM Press, Washington, DC, 1994.

[49] T. Ferenci, M. Muir, K.-S. Lee, and D. Maris, *Biochim. Biophys. Acta* **860,** 44 (1986).

[50] S. G. Kustu and G. F.-L. Ames, *J. Bacteriol.* **116,** 107 (1973).

[51] R. Landick and D. L. Oxender, *J. Biol. Chem.* **260,** 8257 (1985).

It has been demonstrated in various systems that mutations that inhibit the expression of the functional BP abolish the transport process. Thus BP is not only needed to increase the affinity and the specificity of the system, but also is essential for transport. Major progress was brought about by the isolation, by H. A. Shuman and co-workers,[52] of *E. coli* mutants that no longer require the presence of MBP for the transport of maltose. These mutants are altered in the integral membrane proteins MalF or MalG, and transport maltose specifically. The affinity for maltose, however, is much lower, and the transport K_m for maltose is about 1 mM, in contrast to the transport K_m, about 1 μM, of the wild-type transport system that includes MBP. All MBP-independent mutants contain two mutations either in *malF* or *malG*. One of them occurs in the "distal" region, predicted to be buried deep in the membrane interior, and the other occurs in the "proximal" region, close to the periplasmic surface of the protein.[53]

Interestingly, expression of the wild-type MBP in these *malF* or *malG* mutants inhibited their growth on maltose minimal media.[52] This is apparently due to the excessively tight binding of the MBP to the mutant transporter complex.[54] By taking advantage of this inhibition, suppressor mutants of MBP can be isolated that no longer inhibit the growth of these *malF* and *malG* mutants on maltose even when the (mutated) MBPs are expressed.[55] Location of altered amino acids on MBP shows that the N-terminal lobe of MBP interacts with the mutated MalG protein, and that the C-terminal lobe with the mutated MalF protein.[56]

Maltose transport by MBP-independent mutants showed for the first time that the channel itself does show specificity. The specificity may not be very high, as one can imagine from the examples of the histidine and leucine transporters. But the presence of specificity in the channels of BP-dependent systems is consistent with the fact that the ABC exporters that do not contain BP do show specificity in the selection of ligands (Table I) (see also Refs. 5 and 6). Among the protein exporter systems, the protease secretion system of *Erwinia chrysanthemi*, Prt (Table I), is highly homologous to the hemoprotein secretion system of *Serratia marcescens*, Has, yet the latter system can only secrete its proper substrate. Use of hybrids between these systems showed that the specificity is determined by the

[52] N. A. Treptow and H. A. Shuman, *J. Bacteriol.* **163,** 654 (1985).

[53] K.-M. Y. Covitz, C. H. Panagiotidis, L.-I. Hor, M. Reyes, N. A. Treptow, and H. A. Shuman, *EMBO J.* **13,** 1752 (1994).

[54] D. A. Dean, L. I. Hor, H. A. Shuman, and H. Nikaido, *Mol. Microbiol.* **6,** 2033 (1992).

[55] N. A. Treptow and H. A. Shuman, *J. Mol. Biol.* **202,** 809 (1988).

[56] L.-I. Hor and H. A. Shuman, *J. Mol. Biol.* **233,** 659 (1993).

ABC transporters, possibly their channel domains, and the helper proteins (see Table I) play no role in the specificity.[57]

Recently, random mutagenesis of helices 6, 7, and 8 of MalF was carried out by Ehrle *et al.*[58] Many mutations, clustered on one face of helix 6, changed the specificity of transport, again showing that the channel protein contributes to substrate specificity. Many of them transported maltoheptaose but not maltose. One of them grew on maltose, but its growth on maltose was inhibited by the simultaneous presence of maltoheptaose, suggesting a tight binding of the latter to the channel.

Binding protein-independent mutants were also isolated in the histidine transport system of *S. typhimurium.*[31] Interestingly, in this system such mutations were always found within the *hisP* gene, coding for the ATP-binding subunit.

Binding proteins undergo a large conformational change upon the binding of ligands, as described earlier. It has often been assumed that only the liganded BPs in their closed forms interacted with the membrane-associated transporter complexes. However, kinetic analysis of the maltose transport process in strains expressing MBP to different levels was compatible only with the model in which both unliganded and liganded MBP interacted with the transporter.[59,60]

Biochemical Studies of Transport

Because a majority of the biochemical studies of transport processes were carried out using BP-dependent import systems, most of the following discussion is devoted to such systems.

Accumulation in Intact Cells

Following accumulation of ligands (usually labeled with radioactivity) by intact cells is the first step in the study of transport processes. It is desirable, or even essential, to use mutants that cannot metabolize the accumulated substrates, so that the assay of the total radioactivity inside the cell is not complicated by incorporation into macromolecules or release as CO_2 and other small products. The methodology using membrane filters to determine the accumulated radioactivity is well established. The extent of accumulation is very high. For example, when incubated with 1 μM

[57] R. Binet and C. Wandersman, *EMBO J.* **14,** 2298 (1995).
[58] R. Ehrle, C. Pick, R. Ulrich, E. Hofmann, and M. Ehrmann, *J. Bacteriol.* **178,** 2255 (1996).
[59] E. Bohl, H. A. Shuman, and W. Boos, *J. Theor. Biol.* **172,** 83 (1995).
[60] G. Merino, W. Boos, H. A. Shuman, and E. Bohl, *J. Theor. Biol.* **177,** 171 (1995).

maltose, *E. coli* cells that were unable to metabolize maltose accumulated maltose to an estimated intracellular concentration of 12 mM, a concentration ratio in excess of 10^5.[61] This is far in excess of what a proton symporter can achieve, but it is what is expected for systems coupled with ATP hydrolysis, where a -7.5 kcal mol^{-1} free-energy change can theoretically produce a concentration gradient of 2.5×10^5. Also unlike proton symporters, the maltose transporter did not catalyze an efflux of intracellular maltose, although there appears to be a separate system that catalyzes this process.[62] The transport rates measured in this way are also close to the expected rate in growing cells. For example, *E. coli* growing aerobically on maltose as the sole carbon source has a doubling time of about 90 min, that is, the mass of 1-mg cells is incremented by $(2^{(1/90)} - 1) = 0.0077$ mg/min. Since about 45% of sugars become incorporated into cellular macromolecules under aerobic conditions,[63] maltose must be transported and consumed by growing cells at the rate of 45 nmol min^{-1} mg^{-1}. Assaying accumulation under standard conditions gives values very close to this rate for maltose,[61] if we assume that 1-mg dry weight contains 3.5×10^9 cells.[64]

Addition of Binding Protein to Cells from External Medium

Binding protein-dependent import systems require the presence of periplasmic BP. The next step toward the reconstitution of the ABC transporter systems is an attempt to let BP enter the periplasm of BP-deficient mutants from the external medium; this would produce positive evidence that BP is required and that the isolated BP is functional. A reproducible result in this direction was first obtained by Brass *et al.*,[65] who permeabilized the outer membrane by treatment with high concentrations of Ca^{2+}. This system proved to be useful for other purposes, for example, in introducing fluorescent-labeled BP into the periplasm and measuring its rate of lateral diffusion.[66]

[61] S. Szmelcman, M. Schwartz, T. J. Silhavy, and W. Boos, *Eur. J. Biochem.* **65,** 13 (1976).

[62] T. Ferenci, W. Boos, M. Schwartz, and S. Szmelcman, *Eur. J. Biochem.* **75,** 187 (1977).

[63] O. M. Neissel, M. J. Teixeira de Mattos, and D. W. Tempest, *in* "*Escherichia coli* and *Salmonella*: Cellular and Molecular Biology" (F. C. Neidhardt, ed.) 2nd Ed., p. 1683. ASM Press, Washington, DC, 1996.

[64] F. C. Neidhardt and H. E. Umbarger, *in* "*Escherichia coli* and *Salmonella*: Cellular and Molecular Biology" (F. C. Neidhardt, ed.), 2nd Ed., p. 13. ASM Press, Washington, DC, 1996.

[65] J. M. Brass, U. Ehmann, and B. Bukau, *J. Bacteriol.* **155,** 97 (1983).

[66] J. M. Brass, C. F. Higgins, M. Foley, P. A. Rugman, J. Birmingham, and P. B. Garland, *J. Bacteriol.* **165,** 787 (1986).

Addition of Binding Protein to Right-Side-Out Membrane Vesicles

This next step was achieved by Hunt and Hong[67] in 1981 using the glutamine transport system of *E. coli*. They added purified glutamine BP at very high concentrations (close to 100 μM) to right-side-out vesicles containing NAD^+ inside, and showed that energization with pyruvate resulted in the BP-dependent uptake of glutamine. This pioneering effort nevertheless had less impact on the field than expected, and only one laboratory reported success in similar experiments[68] in the next 8 years. Possibly this was because the nature of the immediate energy source was not clarified by these experiments, as the metabolism of pyruvate in the presence of NAD^+ gave rise to many possible candidates.[69] It is also possible that addition of BP at high concentrations sometimes resulted in the introduction of contaminants that inhibited transport. Dean *et al.*[70] overcame this problem with the maltose transport system of *E. coli* by utilizing a mutant producing a "tethered" mutant MBP that remained associated with the external surface of the cytoplasmic membrane. In this system, which mimics the situation in gram-positive ABC importers, there was no need to add purified BP from the outside.

With intact cells, it was difficult to determine the energy source for the transport, because there was a rapid and extensive interconversion among the metabolites, for example, leading to the conclusion that the energy source was acetyl phosphate[71] or reduced lipoic acid.[72] A more reliable answer came with the use of membrane vesicles that were prepared from a mutant defective in F_1F_0-ATPase and that were devoid of most of the cytosolic enzymes.[73] Experiments with the addition of soluble, wild-type MBP showed conclusively that hydrolysis of ATP accompanied the accumulation of maltose. Reconstitution with the right-side-out vesicles was also achieved with the histidine transport system of *S. typhimurium*.[74] However, in this case the vesicles from an ATPase-deficient strain could not be energized by ATP, and thus the results were not conclusive.

[67] A. G. Hunt and J.-S. Hong, *J. Biol. Chem.* **256,** 11988 (1981).

[68] B. Rotman and R. Guzman, *in* "Microbiology–1984" (L. Leive and D. Schlessinger, eds.), p. 57. American Society for Microbiology, Washington, DC, 1985.

[69] A. G. Hunt and J.-S. Hong, *Biochemistry* **22,** 844 (1983).

[70] D. A. Dean, J. D. Fikes, K. Gehring, P. J. Bassford, Jr., and H. Nikaido, *J. Bacteriol.* **171,** 503 (1989).

[71] J.-S. Hong, A. G. Hunt, P. S. Masters, and M. A. Lieberman, *Proc. Natl. Acad. Sci. USA* **76,** 1213 (1979).

[72] G. Richarme, *J. Bacteriol.* **162,** 286 (1985).

[73] D. Dean, A. L. Davidson, and H. Nikaido, *Proc. Natl. Acad. Sci. USA* **86,** 9134 (1989).

[74] E. Prossnitz, A. Gee, and G. F.-L. Ames, *J. Biol. Chem.* **264,** 5006 (1989).

Addition of Adenosine Triphosphate to Inside-Out Vesicles

Ames *et al.*[75] prepared inverted membrane vesicles by French press disruption of cells in the presence of labeled histidine and 0.2 m*M* HisJ BP. After extensive washing, the vesicles pumped out histidine on the addition of ATP, but not of nonhydrolyzable ATP analogs or acetyl phosphate. Furthermore, addition of the proton conductor FCCP had no effect on the efflux process. These results provided strong evidence that the transport is indeed energized by ATP hydrolysis.

Reconstitution into Proteoliposomes with Solubilized or Purified Transporters

The ultimate step was the reconstitution into proteoliposomes. This has been achieved by utilizing crude detergent extracts of the cytoplasmic membrane with both the histidine and the maltose systems.[76,77] Furthermore, Davidson and Nikaido[78] achieved reconstitution subsequently with a maltose transporter complex that was purified to near homogeneity. These proteoliposomes were fully functional in the accumulation of ligands, hydrolyzing intravesicular ATP at the same time. In particular, the proteoliposomes reconstituted from the purified transporter did not contain anything else except ATP, and this result showed conclusively that ABC transporters can accumulate their substrates only by using ATP, without any additional factors, including the proton-motive force. Similar reconstitution into proteoliposomes has been achieved since then, for the Mgl galactose transport system of *S. typhimurium*[79] and for the branched-chain amino acid transport system of *Pseudomonas aeruginosa.*[80] In the former case, a purified transporter complex was used.

The transport activity observed was 11 and 0.5 nmol of ligands transported per minute per milligram of protein with the purified maltose transporter[78] and the purified MglAC transporter,[79] respectively. The very low activity of the latter system is likely to be due to the fact that denaturation/renaturation of inclusion bodies was used to purify the transporter. Presumably a large fraction of the transporter was not refolded in the correct conformation. When crude extracts of cytoplasmic membranes containing overproduced transporters were used for reconstitution, rates of 3 and 0.6 nmol ligands transported per minute per milligram of protein have been

[75] G. F.-L. Ames, K. Nikaido, J. Groarke, and J. Petithory, *J. Biol. Chem.* **264,** 3998 (1989).

[76] L. Bishop, R. Agbayani, Jr., S. V. Ambudkar, P. C. Maloney, and G. F.-L. Ames, *Proc. Natl. Acad. Sci. USA* **86,** 6953 (1989).

[77] A. L. Davidson and H. Nikaido, *J. Biol. Chem.* **265,** 4254 (1990).

[78] A. L. Davidson and H. Nikaido, *J. Biol. Chem.* **266,** 8946 (1991).

[79] G. Richarme and M. Kohiyama, *FEBS Lett.* **304,** 167 (1992).

[80] T. Hoshino, K. Kose-Terai, and K. Sato, *J. Biol. Chem.* **267,** 21313 (1992).

reported for the branched amino acid system[80] and histidine system,[76] respectively.

It was thought that these systems might be ideal for the determination of stoichiometry between ATP hydrolysis and the solute transport. However, the stoichiometric ratios varied widely from experiment to experiment, from 1.4 to 17 for maltose[77] and about 5 for histidine.[76] In the more direct approach, Mimmack *et al.*[81] determined the ratio in intact cells that are lacking the F_1F_0-ATPase and are poisoned with iodoacetamide so that ATP cannot be generated by glycolysis either. In this experiment a decrease of 2 ATP molecules was observed per molecule of maltose transported. Although this fits with the presence of two ATP-binding subunits per transporter, it is questionable whether ATP assay in intact cells can produce data of sufficient precision for the determination of this stoichiometric ratio. In contrast, Muir *et al.*[82] measured the anaerobic growth yields of *E. coli* growing on various carbon sources, and showed that only one molecule of ATP is used for the uptake of one molecule of maltose or maltodextrin. Anaerobic growth yields are highly precise, and we believe that this 1:1 stoichiometric ratio is the most reliable estimate at present. It leaves open the question, however, of why there are two copies of the ATP-binding units in a transporter. In this connection, it is of interest that the two MalK subunits of the maltose transporter complex were recently shown to exhibit cooperative interaction in ATP hydrolysis.[83] Although the molecular basis of such an interaction is not clear at present, the cooperativity is reminiscent of that of F_1-ATPase, where "rotational catalysis" has been suggested in the hydrolysis and synthesis of ATP.[84]

ATP Hydrolysis Assays with Reconstituted Proteolipsomes

The transporter is likely to be inserted in a more or less random orientation into the proteoliposome membrane. Thus a significant fraction of the transporters must exist with their ATP-binding subunits exposed on the outer surface. When the reconstitution is carried out in the presence of BP and the ligand, and ATP is added to the proteoliposomes, such transporters hydrolyze ATP. This ATPase activity requires the presence, in the intravesicular space, of both MBP and maltose, when wild-type maltose transporter is used for reconstitution.[35] However, MBP alone, without maltose, stimulates ATP hydrolysis to a small extent; this finding is noteworthy in view

[81] M. L. Mimmack, M. P. Gallagher, S. C. Hyde, S. R. Pearce, I. R. Booth, and C. F. Higgins, *Proc. Natl. Acad. Sci. USA* **86,** 8257 (1989).

[82] M. Muir, L. Williams and T. Ferenci, *J. Bacteriol.* **163,** 1237 (1985).

[83] A. L. Davidson, S. S. Laghaeian, and D. E. Mannering, *J. Biol. Chem.* **271,** 4858 (1996).

[84] J. P. Abrahams, A. G. W. Leslie, R. Lutter, and J. E. Walker, *Nature* **370,** 621 (1994).

of the possible presence of closed conformers among ligand-free MBP molecules.

Interesting results were obtained when the ATP hydrolysis assay was performed by using proteoliposomes containing MBP-independent mutants of *malF* and *malG*.[35] Unlike the wild-type transporters, these mutant transporters hydrolyzed ATP in the total absence of both maltose and MBP. Because maltose transport obviously does not occur under these conditions, this result suggests that in the mutant transporter, ligand translocation and ATP hydrolysis are not tightly coupled. Addition of liganded MBP stimulated ATP hydrolysis with some mutants, but caused inhibition in a few mutants. A model of the mechanism of transport suggested by these data is described in the following section.

Mechanism of Transport through ABC Transporters

In the proteoliposomes containing the wild-type transporter, ATP hydrolysis does not occur unless liganded MBPs are present on the other side of the membranes. The simplest interpretation is that liganded MBPs bind to the periplasmic surface of the transporter, and this association creates a wave of conformational changes that travels the thickness of the membrane through the integral membrane proteins MalF and MalG, so that the ATP hydrolysis can now be catalyzed by the ATP-binding subunit MalK, located on the other, cytoplasmic, side of the membrane (see Fig. 2, upper part). In other words, the liganded MBP acts as a signaling molecule, just as it does in binding to the chemotactic receptor, Tar. This signaling mechanism would allow bacterial cells to avoid the unnecessary hydrolysis of ATP, when there are no ligand molecules to be transported. There are at least two possible ways in which ATP hydrolysis becomes activated. In our original model, we assumed that ATP is bound to conformations **I** and **II** of the complex, and that the signaling produced the ATP hydrolysis, accompanied by the translocation of maltose and transition to conformation **III** (Fig. 2, upper part). Boos and Lucht[3] adopt our model almost entirely, except to assume that the binding of ATP occurs at conformation **III**, and that ATP hydrolysis is used to reset the complex into the original starting conformation, **I**. This model is more elegant because **I** and **II** would now become activated conformations, and the solute translocation can now be understood as the spontaneous stabilization of the complex into its lowest energy form. Activation of ATP hydrolysis (the "signaling") then becomes an indirect result of the initiation of cyclic conformational changes. At present, no data are available that favor one model over the other.

In contrast, in the MBP-independent mutants, the ATP hydrolysis occurs without the signaling from the liganded MBP, presumably because the conformation of the mutated MalF-G channel is already at its activated

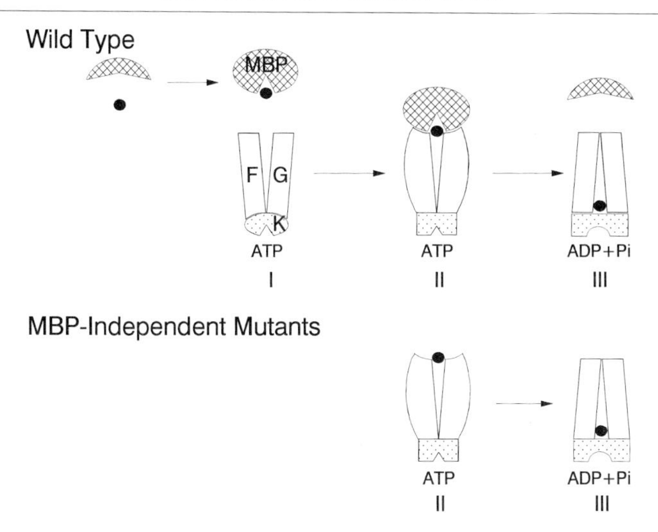

FIG. 2. Proposed functional cycle of the maltose transporter. The wild-type transporter (composed of MalFG transmembrane component and MalK ATPase) in conformation **I** binds liganded MBP at low affinity. The binding changes the conformation of MalF and G, and subsequently of K (conformation **II**), so that ATP hydrolysis is initiated and solute translocation occurs (conformation **III**). In contrast, with the MBP-independent mutant transporters, conformation **I** is unstable and is spontaneously converted into conformation **II**. Thus the ATP hydrolysis and accompanying cyclic conformational changes occur without MBP, and maltose is transported in the absence of MBP. Reprinted from Davidson, A. L., Shuman, H. A., and Nikaido, H. *Proc. Natl. Acad. Sci. USA* **89,** 2360 (1992). Although there are two MalK units in a single transporter complex, only one is drawn as the stoichiometry of transport vs. ATP hydrolysis appears to be 1 : 1 (see text).

state, that is, conformation **II**. This will result in the futile hydrolysis of some ATP molecules and, more importantly, also in the uptake of maltose molecules even in the absence of MBP, because cyclic conformational changes of the transporter, induced by ATP hydrolysis, are all that is needed for solute translocation (Fig. 2, lower part). According to this model, transport absolutely requires the presence of BP in the wild type, not because BP is involved in translocation itself, but because BP is needed as a signal to start the cycle, via ATP hydrolysis.

In this view, it would also be possible to hydrolyze ATP without the signal by mutation of MalK. It is unclear why MBP-independent mutations of *malK* have not been encountered. However, in the histidine system, all BP-independent mutations did indeed occur in the ATP-binding subunit, HisP.[31]

Why would the production of MBP inhibit growth of the MBP-independent mutants? When transport by right-side-out membrane vesicles, from such mutants, was measured in the presence of various concentrations of added MBP, MBP at low concentrations was found to stimulate maltose transport, whereas at high concentrations pronounced inhibition was observed.[54] In quantitative terms, the half-maximal stimulation of transport occurred at about a 25–50 μM concentration of liganded MBP with the wild-type transporter, whereas with the mutant transporter it occurred at 3 μM. Thus the mutant transporter appeared to have a much higher affinity for the maltose-MBP complex. In our model, much of the mutant transporter population already has an activated conformation **II**, which is likely to have a much higher affinity for the liganded MBP than conformation **I** (presumed to be unstable in the mutant). This would certainly explain the observed differences in affinity. The MBP concentrations in induced cells are thought to be around 1 mM, 300-fold higher than the apparent K_m of the mutant transporter. It is a common occurrence that substrates present in such overwhelming excess inhibit the enzyme reaction ("substrate inhibition"), although mechanisms may vary from one case to another.

We believe that this model can be profitably used for the study of other BP-dependent transporters. For exporters, the ligands are in the cytoplasm, and obviously the transmembrane signaling mechanism is not necessary. The cyclic conformational change, concomitant with ATP hydrolysis, will probably be initiated by the binding of the ligand to the channel, but little experimental evidence is available at present.

Note Added in Proof

This chapter was completed and submitted for publication in May, 1996. The authors regret that more recent publications could not be cited because of these circumstances.

Acknowledgment

The experimental studies in the authors' laboratory have been supported in part by a grant from the National Institutes of Health (AI-09644).

[2] Preparation and Reconstitution of Membrane-Associated Maltose Transporter Complex of *Escherichia coli*

By JASON A. HALL, AMY L. DAVIDSON, and HIROSHI NIKAIDO

Introduction

Maltose and its higher homologs, the maltodextrins, are actively transported across the cytoplasmic membrane of *Escherichia coli* via a high-

0076-6879/98 $25.00

affinity, periplasmic binding protein-dependent transport system.[1,2] The components of this transport system are the soluble maltose-binding protein (MBP) and an inner membrane-associated transporter complex. The transporter complex is a member of the large family of ABC (ATP-binding cassette) transporters and is composed of two transmembrane proteins (MalF and MalG) together with two copies of the ATP-binding subunit (MalK).[3,4] MBP, the periplasmic component of the transport system, recognizes and binds maltose and maltodextrins tightly, and thus confers high affinity to the transport system. On binding either maltose or maltodextrin, MBP undergoes a conformational change that is necessary for its productive interaction with the inner membrane-associated transporter complex such that substrate transport into the cytoplasm can occur. In wild-type cells, MBP is absolutely required for substrate transport across the membrane. There are, however, mutant strains of *E. coli* that do not require MBP for maltose transport.[5] These "MBP-independent" mutants all have mutations that map to either *malF* or *malG*. Such MBP-independent transporter complexes are also capable of interacting productively with MBP at micromolar concentrations such that substrate transport can occur, whereas at MBP concentrations found in the periplasm of induced cells (~1 m*M*) transport is inhibited.[6,7]

In this chapter, an *in vitro* method for reconstitution of transporter complexes into proteoliposome vesicles is described. This method[8] is based on the Ambudkar–Maloney modification[9,10] of the octylglucoside dilution procedure of Racker *et al.*[11] Such a method is convenient for studying the interaction of MBP with either wild-type or MBP-independent transporter complexes. Using this method it can be shown that MBP-independent transporter complexes hydrolyze ATP in the absence of both MBP and maltose whereas the wild-type transporter requires the presence of both

[1] R. Hengge and W. Boos, *Biochim. Biophys. Acta* **737,** 443 (1983).
[2] M. Schwartz, *in* "*Escherichia coli* and *Salmonella typhimurium*: Cellular and Molecular Biology" (F. C. Neidhardt, ed.), Vol. 2, p. 1482. American Society for Microbiology, Washington, DC, 1987.
[3] C. F. Higgins, *Annu. Rev. Cell Biol.* **8,** 67 (1992).
[4] A. L. Davidson and H. Nikaido, *J. Biol. Chem.* **266,** 8946 (1991).
[5] N. A. Treptow and H. A. Shuman, *J. Bacteriol.* **163,** 654 (1985).
[6] D. A. Dean, L. I. Hor, H. A. Shuman, and H. Nikaido, *Mol. Microbiol.* **6,** 2033 (1992).
[7] J. A. Hall, J. L. Chen, and H. Nikaido, unpublished data (1995).
[8] A. L. Davidson and H. Nikaido, *J. Biol. Chem.* **265,** 4254 (1990).
[9] S. V. Ambudkar and P. C. Maloney, *Methods Enzymol.* **125,** 558 (1986).
[10] S. V. Ambudkar and P. C. Maloney, *J. Biol. Chem.* **261,** 10079 (1986).
[11] E. Racker, B. Violand, S. O'Neal, M. Alfonzo, and J. Telford, *Arch. Biochem. Biophys.* **198,** 470 (1979).

for ATP hydrolysis to occur.[12] This method also allows for kinetic studies of ATP hydrolysis[13] and substrate uptake.

Preparation of Inner Membrane Transport Complex

Growth of Bacteria

The inner membrane transport complex of the maltose transport system is prepared from strain HN741 (K-12 *argH his rpsL1 malT*^{c} Δ*malB13* Δ*uncBC ilv*::Tn*10*/F′ *lacI*^{q} Tn*5*) containing pMR11 (*malK*$^+$ *cat*$^+$), and either pFG23 (*malF*$^+$ *malG*$^+$ *bla*$^+$) or pLH33 (*malF*$^+$ *malG511 bla*$^+$).[8,12,14] This strain has a deletion encompassing the *malB* region of the chromosome such that those genes necessary for maltose/maltodextrin transport (*malE, malF, malG, malK*) are absent. Copies of wild-type *malF*, wild-type *malG* or *malG511*, and wild-type *malK* are present on plasmids in which *mal* gene expression is under control of the *trc* promoter. This is done in order to increase the level of expression of the maltose transporter complex. The *malG511* allele encodes a mutant MalG protein, which confers on the transporter complex MBP-independent transport capabilities. Strain HN741 also contains a deletion in the genes for F_0F_1-ATPase (Δ*uncBC*) such that the basal rate of ATP hydrolysis in reconstituted proteoliposomes would be minimized, and a F′ plasmid with a gene coding for a high-affinity lactose repressor (LacI^q) in order to decrease the uninduced level of expression of the maltose transporter genes.

Cells from an overnight culture are diluted (1 : 40) and grown at 37° in Terrific Broth (12 g Bacto-tryptone, 24 g Bacto-yeast extract, 4 ml glycerol, 2.31 g KH_2PO_4, and 12.54 g K_2HPO_4 per 1.0 liter) containing kanamycin (50 μg/ml), chloramphenicol (34 μg/ml), and ampicillin (100 μg/ml), with aeration by shaking. When the OD_{600} of the culture reaches 0.20 to 0.25, expression of the *mal* genes is induced by the addition of isopropyl-β-D-thiogalactoside (IPTG) to the medium to a final concentration of 0.1 m*M*. Cells are harvested 3–4 hr after addition of IPTG (OD_{600} between 1.00 and 1.25), and washed once with 50 m*M* Tris-HCl/1 m*M* EDTA, pH 8.0. Cells are then frozen and stored at −70° until needed. Cell yield is approximately 2.0 g (wet weight) per liter of culture media.

[12] A. L. Davidson, H. A. Shuman, and H. Nikaido, *Proc. Natl. Acad. Sci. USA* **89,** 2360 (1992).
[13] A. L. Davidson, S. S. Laghaeian, and D. E. Mannering, *J. Biol. Chem.* **271,** 4858 (1996).
[14] M. Reyes and H. A. Shuman, *J. Bacteriol.* **170,** 4598 (1988).

Preparation of Membrane Vesicles

Membrane vesicles are prepared as follows. All procedures are carried out at 4° unless stated otherwise. Cells (2–4 g, wet weight) are thawed and resuspended in 30 ml 50 m*M* Tris-HCl, pH 8.0, 1 m*M* EDTA, 1 m*M* phenylmethylsulfonyl fluoride (PMSF), 50 μg/ml DNase, 50 μg/ml RNase A and passed through a precooled French pressure cell at 12,000 psi twice. Unbroken cells and cellular debris are removed by centrifugation at 5000*g* for 15 min. The supernatant is transferred to a clean tube, and recentrifuged as described earlier. Membrane vesicles are then pelleted from the supernatant by centrifugation at 160,000*g* for 1 hr, resuspended in 24 ml of 50 m*M* Tris-HCl, pH 8.0, and centrifuged again. Total protein concentration of membrane vesicles is determined by using the BCA protein assay (Pierce Chemical Co., Rockford, IL). Vesicles are then either solubilized immediately or stored at −70° until needed.

Solubilization of Transporter Complex

Two methods have been used for the solubilization of membrane vesicles containing the transporter complex. Method 1 has been utilized for the immediate use of solubilized proteins for reconstitution and assaying of both wild-type transporter and MalG511 transporter activities, and makes use of the nonionic detergent octyl-β-D-glucopyranoside (OG). Method 2 has been developed as an initial step in the purification of wild-type MalF, MalG, and MalK as the intact transporter complex, and uses the detergent dodecyl-β-D-maltoside (DM) for vesicle solubilization.

Method 1. Membrane vesicles are resuspended in 20 m*M* KPO_4, pH 6.2, 20% (w/v) glycerol, 5 m*M* $MgCl_2$, 1 m*M* dithiothreitol (DTT), 1.1% OG using a bath sonicator, and are kept on ice for 30 min. The volume of the solubilization mixture is set so that the weight ratio of OG to membrane protein is between 3 and 4 to 1. The MalG511 transporter complex tends to be less stable than the wild-type complex. For this reason, the mutant complex is solubilized as described earlier except that bath-sonicated *E. coli* total lipids (acetone-precipitated, ether-washed, Avanti Polar Lipids, Inc., Alabaster, AL) are added to the solubilization mixture to a final concentration of about 8 mg/ml. As was true for the LacY protein,[15] addition of lipids during solubilization enhanced the recovery of transport activity by the MalG511 mutant transporter following reconstitution. The OG-containing solubilization mixture is then centrifuged at 160,000*g* for 1 hr at 4°, and the supernatant is divided into portions that are stored at −70°

[15] M. J. Newman and T. H. Wilson, *J. Biol. Chem.* **255,** 10583 (1980).

until use. Protein concentration in the OG-solubilized supernatant is determined using the method of Brown *et al.*[16]

Method 2. Membrane vesicles (30 mg total protein, 3 ml) are mixed with 1.5 volumes of 10 *M* urea, and the suspension is kept on ice for 10 min. This mixture is then centrifuged at 4° for 30 min at 100,000*g*, washed once with 20 m*M* Tris–SO_4, pH 8.0, and resuspended in 20 m*M* Tris–SO_4, pH 8.0, 5 m*M* $MgCl_2$, 1 m*M* DTT, 20% glycerol to a concentration of approximately 2.5 mg (protein)/ml. Then DM is added to a concentration of 1.0% (w/v). The mixture is kept for 20 min on ice, centrifuged at 4° for 30 min at 100,000*g*, and the resulting supernatant, containing the wild-type transporter complex, is used for subsequent purification.

Purification

The wild-type transporter complex is purified using high-performance liquid chromatography (HPLC).[4] Columns are connected to a Perkin-Elmer (Norwalk, CT) Series 3B HPLC pump, and all chromatographic procedures are carried out at room temperature. The solubilized membrane vesicles (i.e., the supernatant fraction obtained by method 2) is loaded onto a DEAE-5 PW column (7.5 × 75 mm, TosoHaas, Montgomeryville, PA) equilibrated with 20 m*M* Tris–SO_4, pH 8.0, 1 m*M* DTT, 20% glycerol, 0.01% DM (buffer A). The column is washed with 10 ml of buffer A and then eluted with a 30-ml gradient (from 0 to 250 m*M* Na_2SO_4 in buffer A) at a flow rate of 0.5 ml/min. Absorbance of the eluate is measured continuously at 280 nm. Fractions (0.5 ml) are collected, and maltose transport activity is determined by reconstituting proteins into proteoliposome vesicles as described later. Fractions exhibiting transport activity are pooled and concentrated by centrifugation in a Centricon-30 (molecular weight cutoff 30,000, Amicon, Danvers, MA). A portion of this concentrated pooled fraction (100 μl, 3 mg/ml) is then loaded onto a Superose 12 gel filtration column (HR 10/30, Pharmacia LKB Biotechnology Inc., Piscataway, NJ), which is equilibrated with buffer B (50 m*M* KPO_4, pH 7.0, 50 m*M* Na_2SO_4, 1 m*M* DTT, 20% glycerol, 0.01% DM). The column is eluted at a flow rate of 0.2 ml/min and column fractions of 0.2 ml are assayed for maltose transport activity.

Preparation of Maltose-Binding Protein

Wild-type MBP is prepared from *E. coli* strain HS2019 (K-12 F^- *araD139* Δ*lacU169 rpsL thi* Δ*malE444*) containing pPD1 (*malE*$^+$ *bla*) by a cold

[16] R. E. Brown, K. L. Jarvis, and K. J. Hyland, *Anal. Biochem.* **180,** 136 (1989).

osmotic shock treatment followed by affinity chromatography using cross-linked amylose.[17–19] Cells from an overnight culture are diluted (1 : 40) and grown at 37°, with aeration by shaking, to late log phase/early stationary phase in LB medium (10 g Bacto-tryptone, 5 g yeast extract, and 5 g NaCl per liter) containing 0.4% maltose and 100 mg/liter ampicillin. Cells are then harvested and resuspended in 10 m*M* Tris-HCl, pH 7.5, 2 m*M* EDTA, 20% sucrose (one-tenth the volume of the culture media), and stirred for 10 min at 4°. Cells are again collected by centrifugation, resuspended in the aforementioned buffer, and rapidly dispersed into a 20-fold volume of cold distilled water (4°) with rapid stirring. After 5 min, $MgCl_2$, NaN_3, and DNase are added to final concentrations of 5 m*M*, 3 m*M*, and 1000 U/liter, respectively. The periplasmic shock fluid is cleared of cells via centrifugation and passage through Whatman (Clifton, NJ) 1 filter paper, followed by filtration through a Millipore 0.22-μm filter. The shock fluid is then loaded onto a column containing cross-linked amylose, prepared as described by Ferenci and Klotz,[20] equilibrated in 10 m*M* Tris-HCl, pH 7.5, 3 m*M* NaN_3 at a flow rate of 25–75 ml/hr. MBP is eluted from the column with 10 m*M* maltose. MBP is then precipitated by the addition of $(NH_4)_2SO_4$ to 90% saturation. The resultant precipitate is dissolved in a small volume of 10 m*M* KPO_4, pH 7.0, and bound maltose is removed either by extensive dialysis[21,22] or by a denaturation–renaturation procedure, as follows. Precipitated MBP, originally from 4 liter of culture, is dissolved in 65 ml of 10 m*M* KPO_4, pH 7.0, and dialyzed against 6 liter of the same buffer overnight at 4°. To the dialyzed MBP, guanidine hydrochloride (Boehringer Mannheim Corp., Germany) is added to 6 *M*. This solution is then dialyzed against 2 liter of 6 *M* guanidine hydrochloride overnight at 4° in order to remove maltose. The MBP is then renatured by dialysis against 10 liter of 10 m*M* KPO_4, pH 7.0, at 4°, for 5 days with daily changes of the dialysis buffer. Maltose-free MBP, prepared by either one of these methods, is then concentrated by ultrafiltration using an Amicon membrane with a molecular weight cutoff of 10,000, aliquoted, and then quickly frozen for storage at −70°. Protein concentrations are calculated using the UV extinction coefficient of MBP of 1.7 ($\varepsilon^{0.1\%}_{1\,cm}$) at 280 nm, reported by Gehring *et al.*[22]

[17] H. A. Shuman, *J. Biol. Chem.* **257,** 5455 (1982).

[18] P. Duplay, H. Bedouelle, A. Fowler, I. Zabin, W. Saurin, and M. Hofnung, *J. Biol. Chem.* **259,** 10606 (1984).

[19] H. C. Neu and L. A. Heppel, *J. Biol. Chem.* **140,** 3685 (1965).

[20] T. Ferenci and U. Klotz, *FEBS Lett.* **94,** 212 (1978).

[21] T. J. Silhavy, S. Szmelcman, W. Boos, and M. Schwartz, *Proc. Natl. Acad. Sci. USA* **72,** 2120 (1975).

[22] K. Gehring, P. G. Williams, J. G. Pelton, H. Morimoto, and D. E. Wemmer, *Biochemistry* **30,** 5524 (1991).

Fluorescence emission spectroscopy is used to show that MBP is free of maltose and is able to bind maltose with the expected affinity (K_d of about 1 μM).[21]

Reconstitution

Solubilized membrane fractions containing the transporter complex are reconstituted into phospholiplid vesicles using a modified form of the detergent dilution method. Total *E. coli* lipids (from the same source mentioned earlier) are dispersed in 20 mM KPO_4, pH 6.2, 2 mM 2-mercaptoethanol to a concentration of 50 mg/ml via bath sonication. It appears essential to carry out sonication until the suspension becomes translucent, an indication of the formation of liposomes of the smallest possible sizes. To a centrifuge tube (for a Beckman 60Ti rotor) chilled on ice, 90 μl of the lipid solution is then added. Subsequently, OG (10% aqueous solution) is added followed by the addition of OG-solubilized proteins (90 μg) to give a final OG concentration of 1.1%. For the preparation of proteoliposomes for measurement of ATP hydrolytic activity of the transporter complex, MBP and maltose are also added such that on dilution their concentrations will be 1.5–4 μM and 100–800 μM. (When MalG511 transporter is used, the concentration of MBP is kept within a concentration range that does not inhibit the activities of this mutant complex.[6,7]) In experiments using OG-solubilized membrane fractions, the weight ratio of *E. coli* phospholipids/MBP (if added)/membrane protein is typically kept at 50:8:1. Based on the K_d value of 1.0 μM of maltose for MBP, these ratios ensure that at least 90% of the MBP present will be in the liganded form. Proteoliposome vesicles for the measurement of the substrate uptake are made without adding either MBP or maltose, but are loaded with Na-ATP (usually 5 mM). Finally, precooled 20 mM KPO_4, pH 6.2, 1 mM DTT is added to the centrifuge tube to bring the contents to a final volume of 0.54 ml. The mixture is then kept on ice for 30 min, quickly diluted by the addition of 14 ml of precooled 20 mM KPO_4, pH 6.2, 1 mM DTT (containing 5 mM Na-ATP for sugar uptake assays), and centrifuged at 160,000g for 1 hr at 4°. The proteoliposome pellet is easily disturbed, so we turn off the brake at the end of centrifugation when the speed comes down to about 10,000 rpm, and remove the tubes immediately when the rotor stops. The supernatant is gently removed with a Pasteur pipette, and the proteoliposome pellet is used immediately. When the reconstitution is successful, the pellet is translucent. When the pellet is whitish and strongly turbid, it usually means that proteoliposomes were not produced efficiently.

Activity Assays

Assay of ATP Hydrolysis

Proteoliposomes are resuspended in 200 μl precooled 20 m*M* KPO_4, pH 6.2, 3 m*M* $MgCl_2$, 10 μM maltose (if added when preparing proteoliposomes) and kept on ice until use. Resuspension in this volume gives a protein concentration of 0.45 mg/ml assuming 100% incorporation of all protein during the reconstitution procedure. This was confirmed by the protein determination method of Brown *et al.*[16] To start an assay, 15 μl of 1.0 m*M* [γ-^{32}P]ATP (50 mCi/mmol) is added to 135 μl of the proteoliposome suspension, and the mixture is incubated at room temperature. At given time points 25-μl portions are removed from the mixture, and the reaction is stopped by mixing each portion with 175 μl of 1 *M* perchloric acid, 1 m*M* KPO_4 in a 13 $\times$ 100-mm tube. To measure the release of [γ-^{32}P]P_i due to ATP hydrolysis activity of the transporter complex, the procedure of Lill *et al.*[23] is used. Isopropyl acetate (0.5 ml) is added to the stopped reactions followed by the addition of 0.5 ml 20 m*M* ammonium molybdate. This mixture is then vigorously vortexed for 10 sec and left at room temperature to allow the separation of the phases. Then 200 μl of the organic (upper) phase, containing the phosphomolybdate complex, is removed from the microfuge tube, placed in a scintillation vial containing 10 ml Ecolume (ICN Chemicals, Costa Mesa, CA) scintillant, and counted in a liquid scintillation counter.

With the wild-type transporter complex, full activity (in the range of several hundred nanomoles of ATP hydrolyzed per minute per milligram of protein when crude OG extracts of the membrane are used for reconstitution[12]) is observed only when both maltose and MBP are added at the time of reconstitution. In contrast, with the MBP-independent transporter complexes, significant ATP hydrolysis occurs without the addition of maltose and MBP, although their addition often stimulates the hydrolytic activity.[12]

Assay of Substrate Uptake

The accumulation of maltose inside proteoliposome vesicles is measured using a modified form of the procedure of Davidson and Nikaido.[4] Proteoliposome pellets are resuspended in 200 μl precooled 20 m*M* KPO_4, pH 6.2, 3 m*M* $MgCl_2$ to a final concentration of 0.45 mg protein/ml, with or without

[23] R. Lill, K. Cunningham, L. A. Brundate, K. Ito, D. Oliver, and W. Wickner, *EMBO J.* **8,** 961 (1989).

MBP. MBP concentration of 10 μM was used earlier.[4] Because the half-maximal stimulation of maltose transport occurs at about 25 μM MBP in membrane vesicles,[6] it is expected that even 10 μM MBP will not give maximal activity. On the other hand, use of high concentrations of MBP may introduce potentially inhibitory contaminants into the system, and also raises the baseline in the assay through the adsorption of liganded MBP to membrane filters. For these reasons, routine assays can be carried out by using a lower concentration of MBP, typically 1 μM.

The assay is initiated by the addition of 15 μl 10 μM [^{14}C]maltose (150 mCi/mmol) to 135 μl of the suspension of proteoliposomes containing ATP in the intravesicular space (see earlier discussion) at room temperature. At specified times, 25-μl portions of the reaction mixture are diluted 1:10 into precooled 20 mM KPO_4, pH 6.2, 3 mM $MgCl_2$ and immediately filtered through a Millipore filter (0.22-μm GSTF). The filter is then quickly washed with 5 ml ice-cold 50 mM LiCl. Filters are air dried and then counted in a liquid scintillation counter using 10 ml Ecolume as scintillant.

With the wild-type transporter, initial rates of uptake corresponded to the specific activity of 3 and 11 nmol per minute per milligram of protein (in the presence of 10 μM MBP), when crude OG extract and the purified preparation were used for reconstitution, respectively.[4,8] The uptake was approximately linear with time, up to 30 sec, sometimes 1 min.

Comments

We can routinely show transport and ATPase activities of the transporters by these assays and, in the same experiment, can compare the activity of different strains or the activity of one transporter under different conditions. It is sometimes difficult, however, to maintain the quantitative reproducibility of the data over a long time span. Because the assay involves many ingredients, frequently it is difficult to pinpoint the cause of variability. When poor activity is observed, it may become necessary to examine systematically different batches of various ingredients, especially the phospholipids.

With the MalG511 transporter complex, we have sometimes encountered significant reduction in transport activity, without concomitant decreases in the ATPase activity.[7] In certain cases, this could be due to the consumption of the trapped ATP by the constitutive ATPase activity of the mutant transporter. However, the mutant transporter appears to be less stable during extraction, storage, and purification, exhibiting a strong tendency to become dissociated. As mentioned earlier, the addition of excess *E. coli* phospholipids during solubilization enhanced the recovery of transport activity. Partial dissociation (or weakening of subunit interactions) could contribute to uncoupling between ATPase activity and solute translocation activity, because the former can be catalyzed by MalK alone under certain conditions[24] whereas the latter requires the participation of all subunits.

We have also experienced decreases in the MBP-stimulated ATPase activity of the wild-type transporter complex under experimental conditions in which the high endogenous ATPase activity of another MBP-independent mutant complex was unaffected.[25] Here again, the former activity, like solute translocation activity, may be more sensitive to suboptimal reconstitution conditions than the latter activity, that is, the uncoupled ATP hydrolysis by a mutant transporter.

It is possible to vary the ratio of protein to lipid in the reconstitution experiments to generate greater or fewer numbers of transport complexes per lipid vesicle. However, with higher concentrations of transporters, it is necessary to ensure that intravesicular maltose does not become depleted during the ATPase assay. Decreasing the levels of MBP much below the recommended levels will also adversely affect the activity observed. Because of the inherent complexities of the MBP-stimulated ATPase assays, it is essential to make sure that the activity is not limited by the paucity of unsuspected constituents under any of the chosen reconstitution conditions.

In the procedure described, proteoliposomes are used without washing, after their harvest by centrifugation. Although this is adequate for most routine assays, obviously it is more desirable to use washed proteoliposomes[8] for critical experiments, especially in those experiments where the carryover of the constituents of the OG dilution mixture becomes an issue.

Acknowledgments

The studies in our laboratories have been supported by grants from the National Institutes of Health (AI-09644 to H. Nikaido and GM-49261 to A. L. Davidson).

[24] C. Walter, K. Hörner zu Bentrup, and E. Schneider, *J. Biol. Chem.* **267,** 8863 (1992).
[25] A. L. Davidson, unpublished results (1995).

[3] Maltose Transport in *Escherichia coli:* Mutations That Uncouple ATP Hydrolysis from Transport

By CYNTHIA H. PANAGIOTIDIS and HOWARD A. SHUMAN

Introduction

Role of Binding Protein-Dependent ABC Transporters

Bacteria use adenosine triphosphate-binding cassette (ABC) transporters both for exporting macromolecules[1,2] and accumulating nutrients and ions.[3] Some of the earliest reports of specific transporters from bacteria include the description of the maltose "permease."[4] The maltose transport system and similar periplasmic (or extrinsic, in the case of gram-positive organisms) binding protein-dependent systems, described both in this volume[5] and in Boos and Lucht[3] are responsible for accumulating a wide variety of substances in many different prokaryotes. These systems provide a high affinity with apparent $K_{0.5}$ values that are usually at or below micromolar concentrations.[3,6] In addition, the transporters are often quite specific for the limited number of ligands of the binding protein (BP) component.[3] Importantly, the systems can pump the concentration of the substrate inside the cytosol to as much as 10^5-fold higher than the external medium.[3,7] These properties must confer a significant growth advantage given the wide distribution of these transport systems among all prokaryotes. These advantages must be so significant that they outweigh the complexity of these systems. Each is composed of: a water-soluble or extrinsic ligand BP, one or two integral membrane proteins usually with a total of 12 membrane-spanning segments, and a peripheral membrane component that bears the ABC signature.[8,9] Although very little direct information is available about

[1] F. Zang, J. A. Sheps, and V. Ling, *Methods Enzymol.* **292,** [5], 1998 (this volume).

[2] R. Binet, S. Letoffe, J. Ghigo, P. Delepelaire, and C. Wandersman, *Gene* **192,** 7 (1997).

[3] W. Boos and J. M. Lucht, *in* "*Escherichia coli* and *Salmonella*" (F. C. Neidhardt, ed.) 2nd Ed., p. 1175. ASM Press, Washington, DC, 1996.

[4] H. Wiesmeyer and M. Cohn, *Biochim. Biophys. Acta* **39,** 440 (1960).

[5] W. Boos, *Methods Enzymol.* **292,** [4], 1998 (this volume).

[6] T. J. Silhavy, E. Brickman, P. J. Bassford, M. J. Casadaban, H. A. Shuman, V. Schwartz, L. Guarente, M. Schwartz, and J. R. Beckwith, *Mol. Gen. Genet.* **174,** 249 (1979).

[7] S. Szmelcman and M. Hofnung, *J. Bacteriol.* **124,** 112 (1975).

[8] C. Higgins, I. Hiles, G. Salmond, D. Gill, J. Downie, and I. Evans, *Nature* (*London*) **323,** 448 (1986).

[9] A. C. Doige and G. F. L. Ames, *Annu. Rev. Microbiol.* **47,** 291 (1993).

0076-6879/98 $25.00

the mechanism of these transporters, it is possible to speculate about the roles of the different component subunits.

Role of Binding Proteins

The strict correlation between the presence of a water-soluble or extrinsic ligand binding component and a signature ABC sequence in these transporters indicates that some fundamental aspect of the transport mechanism must rely on the BP. As we discuss in detail later, there is evidence that the binding protein has four potential functions: (1) binding the correct ligands to be transported, (2) communicating its state of occupancy to the membrane components, (3) triggering the transport/ATPase cycle, and (4) delivering its ligand to the membrane components.

Role of Other Subunits

Analysis of the genes that encode this transporter was done during the 1970s in conjunction with the study of the *mal* regulon and the receptor for bacteriophage λ.[6,10–12] These studies established that at least four gene products were directly involved in the transport of maltose and longer linear $\alpha(1\rightarrow4)$-malto-oligosaccharides. The identification of the *malEFG* and *malK* gene products indicated that a multisubunit complex was likely to be the functional unit in the cytoplasmic membrane together with a periplasmic maltose-binding protein (MBP), which is the product of the *malE* gene. The *malF* and *malG* genes encode very hydrophobic polypeptides[13,14] that contain eight and six transmembrane segments, respectively.[15,16] In addition, each of these proteins contains a large periplasmic domain of unknown function. The *malK* gene[17] encodes a product that does not contain any obvious hydrophobic regions and has been shown to possess ATPase activity.[18] Indeed the family resemblance with the members of the prokaryotic and eukaryotic branches of the ABC transporter family lies within the MalK protein sequence.[8] This subunit anchors to the MalF and/or MalG subunits[19,20] and is required for their stable insertion in the

[10] M. Hofnung, *Genetics* **76,** 169 (1974).
[11] O. Kellermann and S. Szmelcman, *Eur. J. Biochem.* **47,** 139 (1974).
[12] L. L. Randall-Hazelbauer and M. Schwartz, *J. Bacteriol.* **116,** 1436 (1973).
[13] S. Froshauer and J. Beckwith, *J. Biol. Chem.* **259,** 10896 (1984).
[14] E. Dassa and M. Hofnung, *EMBO J.* **4,** 2287 (1985).
[15] D. Boyd, C. Manoil, and J. Beckwith, *Proc. Natl. Acad. Sci. USA* **84,** 8525 (1987).
[16] E. Dassa and S. Muir, *Mol. Microbiol.* **7,** 29 (1993).
[17] E. Gilson, H. Nikaido, and M. Hofnung, *Nucl. Acids Res.* **10,** 7449 (1982).
[18] S. Morbach, S. Tebbe, and E. Schneider, *J. Biol. Chem.* **268,** 18617 (1993).
[19] P. Bavoil, M. Hofnung, and H. Nikaido, *J. Biol. Chem.* **255,** 8366 (1980).
[20] H. A. Shuman and T. J. Silhavy, *J. Biol. Chem.* **256,** 560 (1981).

membrane.[21] Purification of the $MalFGK_2$ complex has been described,[22,23] and the stoichiometry and enzymatic properties of the complex have been investigated.[24,25] The MalF and MalG proteins are likely to constitute the path through which the substrate molecules traverse the membrane. Because maltose transport is functionally unidirectional, this path may either be a "pore"-type structure that is gated or a binding site that is alternately exposed to either side of the membrane. In addition, the MalF and MalG proteins are likely to communicate with both the MBP on the periplasmic side of the membrane and the MalK ATPase on the cytoplasmic side of the membrane so that the rate of adenosine triphosphate (ATP) hydrolysis is maximal when liganded MBP is present.[25] The histidine transporter of *Salmonella typhimurium* is thought to operate in a different way with the periplasmic binding component, HisJ, communicating directly with the HisP ATPase subunit.[26]

The MalK subunit is the actual site of ATP hydrolysis[18] and is unlikely to participate directly in substrate recognition or movement. This is most clearly demonstrated by the fact that the MalK subunit and the corresponding UgpC subunit of the glycerol–phosphate transporter are interchangeable for both maltose transport and glycerol–phosphate transport without any change in the substrate specificity of either system.[27] MalK also plays two other important roles. First, in the absence of MalK, the MalF and MalG proteins are not stably inserted and assembled in the membrane.[21,28] Second, MalK is able to communicate whether or not the maltose is being transported to the transcriptional activator of the *mal* regulon, MalT. If the transporter is inactive, MalK can decrease MalT activity, but is unable to do so if the transporter is actively transporting.[29]

Transport Activity

Genetic analysis has been useful in studying the interactions among the subunits and the periplasmic MBP.[30–33] The MBP plays an essential role

[21] C. H. Panagiotidis, M. Reyes, A. Sievertsen, W. Boos, and H. A. Shuman, *J. Biol. Chem.* **268,** 23685 (1993).
[22] A. L. Davidson and H. Nikaido, *J. Biol. Chem.* **265,** 4254 (1990).
[23] A. L. Davidson and H. Nikaido, *J. Biol. Chem.* **266,** 8946 (1991).
[24] H. Nikaido, *Methods Enzymol.* **292,** [1, 2], 1998 (this volume).
[25] A. L. Davidson, H. A. Shuman, and H. Nikaido, *Proc. Natl. Acad. Sci. USA* **89,** 2360 (1992).
[26] V. Baichwal, D. Liu, and G. F. Ames, *Proc. Natl. Acad. Sci. USA* **90,** 620 (1993).
[27] D. Hekstra and J. Tommassen, *J. Bacteriol.* **175,** 6546 (1993).
[28] J. Lippincott and B. Traxler, *J. Bacteriol.* **179,** 1337 (1997).
[29] C. Panagiotidis, W. Boos, and H. A. Shuman, unpublished results (1997).
[30] N. A. Treptow and H. A. Shuman, *J. Bacteriol.* **163,** 654 (1985).
[31] N. A. Treptow and H. A. Shuman, *J. Mol. Biol.* **202,** 809 (1988).

in the transport of both the disaccharide, maltose, and the longer polymers of glucose. This was established by studying[30,33] the properties of a strain in which the *malFG* and *malK* genes were constitutively expressed in the complete absence of the MBP. In this strain, the MalFGK$_2$ complex was unable to carry out detectable transport even at very high (10–100 m*M*) external concentrations of sugar.[34] It was possible, however, to isolate mutants that acquired the ability to transport maltose without the MBP.[30,33] These results were used to argue that periplasmic binding proteins, in addition to being the primary receptors for substrates, act as allosteric effectors that trigger substrate transport across the inner membrane through the cognate ABC transporter.[25] In the case of the maltose transporter, it was shown that the ATPase activity of the purified wild-type MalFGK$_2$ complex is stimulated greater than 10-fold by the presence of liganded MBP.[25] Unliganded MBP also interacts with the MalFGK$_2$ complex,[35] and may either trigger very low levels or no ATP hydrolysis. Kinetic models for the interaction of the liganded and unliganded forms of MBP indicate that the K_D's of the interaction of the MalFGK$_2$ complex with the two forms of MBP are similar.[36] Therefore, transport and ATP hydrolysis appear to be triggered by a liganded MBP docking onto the MalFGK$_2$ complex or by a sugar molecule binding to the unliganded MBP that is already docked. In either case, interaction of the MBP with the complex is an essential step for ATP hydrolysis and transport. A simple interpretation is that the MalFGK$_2$ complex can exist in either a closed or open conformation and the MBP acts as an allosteric effector that favors formation of the open conformation.[36] ATP binding and hydrolysis might be required for returning the complex to the closed conformation.

Isolation of Mutants with Uncoupled ATPase Activity

Selection for Restored Transport Activity in malE Strains

To select for mutants that can transport in the absence of MBP, an *Escherichia coli* was constructed that contained a deletion of the MBP structural gene, *malE,* which did not interfere with the expression of the other subunits.[34] This Δ*malE* strain is completely unable to transport malt-

[32] L. I. Hor and H. A. Shuman, *J. Mol. Biol.* **233,** 659 (1993).

[33] K. M. Covitz, C. H. Panagiotidis, L. I. Hor, M. Reyes, N. A. Treptow, and H. A. Shuman, *EMBO J.* **13,** 1752 (1994).

[34] H. A. Shuman, *J. Biol. Chem.* **257,** 5455 (1982).

[35] E. Bohl, H. A. Shuman, and W. Boos, *J. Theor. Biol.* **172,** 83 (1995).

[36] G. Merino, W. Boos, H. A. Shuman, and E. Bohl, *J. Theor. Biol.* **177,** 171 (1995).

ose or longer maltodextrins and as a consequence does not grow on either sugar as the sole source of carbon and energy.[34] Because transport is rate limiting for growth in most instances, it is possible to select for restoration of transport activity by selecting Mal^+ revertants of the Δ*malE* strain on minimal salts maltose agar plates. Originally, this was done in a strain that contained all of the *mal* genes on the chromosome.[30] Subsequently, it was possible to do the same with plasmid-encoded *mal* genes to facilitate the sequence analysis of the mutations.[33] In both cases, mutagenesis was required to isolate Mal^+ revertants. Standard genetic analysis using complementation and recombinational mapping of the Mal^+ revertants indicated that the mutations were in the *malF* and *malG* genes and that two mutations were present in the strains.[30] When the sequence analysis of the mutations was completed it supported the genetic data.[33] Indeed, all of the mutants (except one), have two mutations[33]; in most cases these are both in the *malF* gene in two distinct regions, and in some cases both mutations are in the *malG* gene. In one case one of the mutations is in *malF* and the other in *malG*. These are illustrated in Fig. 1.

Most of the mutations alter residues in the membrane-spanning segments (MSS) and cluster in a few regions. One region is near the periplasmic side of MSS 5 of MalF and includes residues 334–338. The other region is more spread out and contains residues in MSS 6, 7, and 8 of MalF. These positions form a "path" leading from the periplasmic side of MalF to the cytosolic side of MSS 8. There are fewer *malG* mutations and these alter residue L135 in MSS 3 of MalG and residue I154 in the periplasmic loop between MSS3 and 4. Mixing and matching different mutations indicated that some combinations resulted in much more effective transporters than others[33] (see later discussion). To enlarge the diversity of the collection, individual mutations of plasmids were used to select for "secondary" mutations that would result in the MBP-independent phenotype. This resulted in both reisolation of previously identified alleles and some that were previously unknown.[33] The fact that some alleles were isolated repeatedly indicated that the system was unlikely to yield many more new mutations.

Characterization of Transport Properties of Maltose-Binding Protein-Independent Mutants

Maltose transport in the MBP-independent mutants was found to exhibit a lower affinity than wild type, with $K_{0.5}$ values in the 1–2 mM range.[30,33] Maximal rates of transport in the different mutants were found to vary from as high as wild type to about 5–10% of wild type.[30] The relative rates of transport are indicated in Fig. 2. In all cases, the specificity for maltooligosaccharides was retained, but 4-nitrophenylmaltoside, which is not a

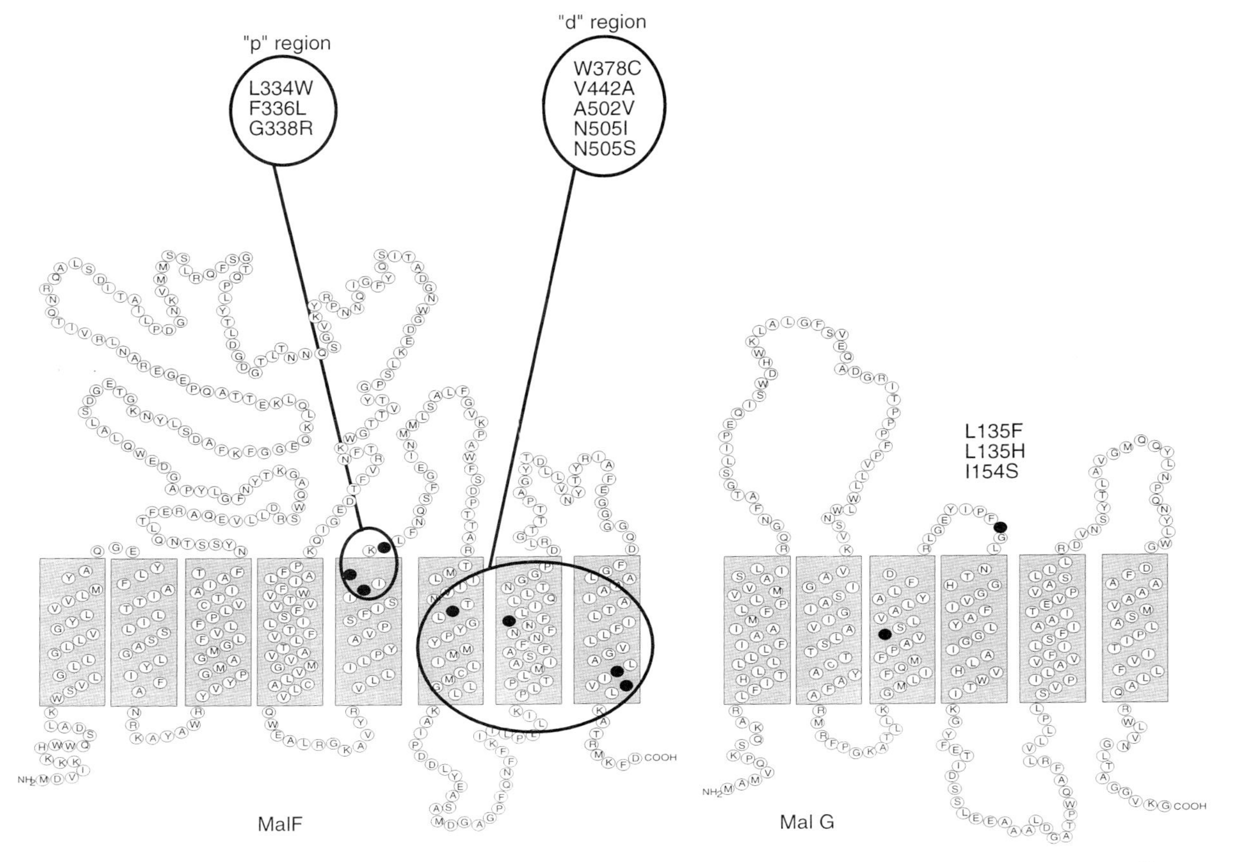

FIG. 1. Topological model of the MalF and MalG proteins. This model is based on the published sequences of MalF[13] and MalG[14] and results obtained with *malF-phoA* [15,45,46] and *malG-phoA* [16] hybrids. The positions of amino acid residues altered in the MBP-independent mutants are indicated by filled circles. (From K. M. Covitz, C. H. Panagiotidis, L. I. Hor, M. Reyes, N. A. Treptow, and H. A. Shuman, *EMBO J.* **13,** 1752 (1994), with permission, Oxford University Press.)

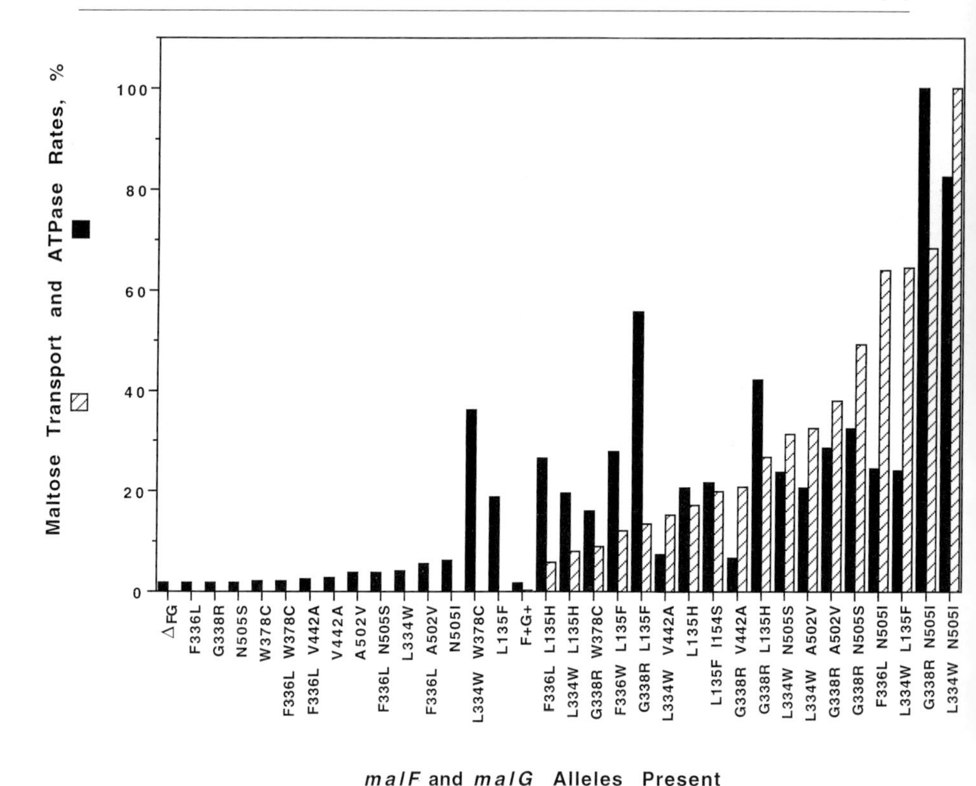

FIG. 2. Comparison of maltose transport and ATPase activities. The values for transport and ATPase are displayed as percentages, using the highest value for transport (*malF526* allele with the two amino acid changes, L334W and N505I) and ATPase activity (*malF500* with the two amino acid changes, G338R and N505I) as 100%. The mutants are ordered in increasing transport activity from left to right. The values for transport are shown as hatched bars and values for ATP hydrolysis are indicated by solid bars. (From K. M. Covitz, C. H. Panagiotidis, L. I. Hor, M. Reyes, N. A. Treptow, and H. A. Shuman, *EMBO J.* **13,** 1752 (1994), with permission, Oxford University Press.)

substrate of the wild-type system was found to be both a competitive inhibitor of maltose transport and transported by all of the mutants.[37] These results suggest that there is an additional substrate recognition site within the $MalFGK_2$ complex. In the different mutants this same site may function with different efficiencies and may account for the fact that the mutants exhibit similar $K_{0.5}$ values and variable V_{max} values.

To measure the ATPase activity of the $MalFGK_2$ complex it is necessary to overproduce the complex from plasmids in cells that are defective for the F_1F_0 H^+-translocating ATPase and lack the chromosomal *malFGK*

[37] M. Reyes, N. A. Treptow, and H. A. Shuman, *J. Bacteriol.* **165,** 918 (1986).

TABLE I
TECHNIQUES FOR MEASUREMENT OF ATP HYDROLYSIS BY $MalFGK_2$ COMPLEX IN EVERTED MEMBRANE VESICLES

$MalFGK_2$ complex	Initial rate of ATP hydrolysis[a]	
	Malachite green assay[b]	^{32}P release from $[\gamma\text{-}^{32}P]ATP$[c]
tMBP with $MalF^+G^+K^+$	44	36
MalF500[d] G^+K^+	347	287
MalF526[e] G^+K^+	312	237

[a] Initial rates of ATP hydrolysis are given as nmol/min/mg protein.
[b] Time courses of orthophosphate release were measured using the malachite green assay described in the text.
[c] Release of ^{32}P from $[\gamma\text{-}^{32}P]ATP$ was measured in triplicate using the method described by Lill *et al.*[40]
[d] The *malF500* allele has two amino acid changes: G338R and N505I.
[e] The *malF526* allele has two amino acid changes: L334W and N505I.

genes.[25,21] These cells are then converted to everted or inside-out membrane vesicles (IOV) by passage of the cells in a French pressure cell (Spectronic Instruments, Rochester, NY).[21] In the case of the wild-type $MalFGK_2$ complex, MBP is required for ATPase activity and can be supplied by including purified MBP in the cell suspension during IOV formation so that some is trapped in the lumen of the IOV.[25] Alternatively, we have used "tethered" MBP (tMBP) in which the uncleaved signal peptide anchors tMBP to the periplasmic or internal side of the IOV following passage in the French pressure cell.[21] The tMBP is supplied by the *malE*24-1 mutation that alters the signal sequence so that it is not cleaved by signal peptidase.[38] The presence of tMBP ensures that all IOV have equivalent amounts of MBP per vesicle. For the MBP-independent mutants, tMBP is omitted.

Adenosine triphosphate hydrolysis can be measured by a variety of techniques including the release of ^{32}P from $[\gamma\text{-}^{32}P]ATP$ or the formation of a dye–phosphomolybdate complex from released P_i[39] (see Table I for a comparison). The latter method is conveniently based on the formation of a colored product with the dye malachite green. When these methods[40,41] were used to study the ATPase activity of the MBP-independent mutant complexes, researchers found that in contrast to the wild-type complex, they were able to hydrolyze ATP in the absence of MBP.[25] In some cases

[38] J. Fikes, V. Bankaitis, and J. Ryan, *J. Bacteriol.* **169,** 2345 (1987).
[39] K. Itaya and M. Ui, *Clin. Chim. Acta* **14,** 361 (1966).
[40] R. Lill, K. Cunningham, L. A. Brundage, K. Ito, D. Oliver, and W. Wickner, *EMBO J.* **8,** 961 (1989).
[41] P. A. Lanzetta, L. J. Alvarez, P. S. Reinach, and O. A. Candia, *Anal. Biochem.* **100,** 95 (1979).

the rates of hydrolysis were close to those found for the maximal observed rates of the purified wild-type complex, ca. 1 μmol/min/mg protein.[25] These results are also shown in Fig. 2.

The individual mutations that make up the MBP-independent mutants resulted in very low values for both transport in cells and ATPase activity in IOV. Although the two measurements are made on different materials, the relative activities showed a good, if not perfect, correlation. The only exception is the *malG510* mutation, which resulted in measurable activities in the absence of a second mutation.

Malachite Green Dye-Binding Assay for Adenosine Triphosphate Hydrolysis

Principle. This colorimetric method[41,42] for detection of nanomole amounts of inorganic phosphate, developed by Itaya and Ui,[39] is based on the change in the absorption spectrum of the basic dye malachite green when it reacts to form a complex with phosphomolybdate. Free dye in acid solution is orange, whereas the malachite green–phosphomolybdate complex is green in color. Modifications of the assay, first by Hess and Derr,[42] and later by Lanzetta *et al.*,[41] improved the sensitivity and color stability of the method. More recently, Henkel *et al.*[43] have described a microassay using malachite green to detect nanomole amounts of inorganic phosphate in individual wells of a 96-well microplate.

Reagents

- Vesicle dilution buffer: 40 m*M* Tris-HCl, pH 7.1, 4 m*M* $MgCl_2$, 5% glycerol (w/v)
- ATP: 40 m*M* ATP (<1 ppm vanadium), pH 7.0, with NaOH
- Malachite green (MG) solution: 0.045% (w/v) malachite green (oxalate salt) in 1 *N* HCl
- Ammonium molybdate (AM) solution: 4.2% (w/v) ammonium molybdate in 4 *N* HCl
- Nonidet P-40
- Citric acid: 34% (w/v) citric acid

All chemicals may be purchased from Sigma Chemical Co. (Saint Louis, MO).

Procedure. On the day of assay a color reagent is prepared fresh by mixing 0.01 ml Nonidet P-40 with 10 ml MG solution and then adding 3.3 ml AM.

[42] H. H. Hess and J. E. Derr, *Anal. Biochem.* **63,** 607 (1975).
[43] R. D. Henkel, J. L. VandeBerg, and R. A. Walsh, *Anal. Biochem.* **169,** 312 (1988).

Freshly prepared IOV membrane vesicles[21] are diluted to a protein concentration of about 0.04 mg/ml with vesicle dilution buffer just prior to assay. The vesicles have been loaded with 20 m*M* maltose (Pfanstiehl Laboratories, Waukegan, IL) or sucrose during their formation in the French pressure cell by including the appropriate sugar in the cell resuspension buffer.[21] If an inhibitor such as sodium orthovanadate is to be added, the addition is done at this point and the vesicle solution is preincubated (along with the untreated vesicles) for 10–20 min. Aliquots (90 μl) of the diluted vesicles are pipetted into individual assay tubes and ATP hydrolysis is initiated by the addition of 10 μl of 40 m*M* ATP. Hydrolysis of ATP is terminated at the desired time by the addition of 800 μl of the color reagent. After 1 min 800 μl of citric acid is added to stabilize the color. After 10 min the absorbance is read at 660 nm and the phosphate content is calculated using a molar absorptivity value of 80,000 for the malachite green–phosphomolybdate complex.[41]

Conclusion

ABC transporters are widespread throughout nature; they are used to accomplish many different tasks in all different types of environments.[3] Nevertheless, there must be some mechanistic similarities that derive from their common structural organization and sequence homology.[8] A fundamental aspect is how ATP binding and hydrolysis are used to drive the movement of other substances, either through channels or other means. We expect that the mutational changes we have described in the bacterial system will have counterparts in other ABC transporters. Bacterial genetics has provided a powerful tool to examine the molecular basis of energy coupling (reviewed by Harold and Maloney[44]). The amino acid changes in MalF and MalG that alter coupling of ATP hydrolysis to transport would never have been made by site-directed mutagenesis. To understand how these amino acid alterations affect energy coupling will, of course, require additional structural information about the transporter. It will be particularly intriguing to discover how interactions among the transmembrane segments are able to transmit information from the periplasmic side of the transporter to the MalK subunit and thus regulate its activity.

[44] F. M. Harold and P. C. Maloney, *in* "*Escherichia coli* and *Salmonella*" (F. C. Neidhardt, ed.) 2nd Ed., p. 283. ASM Press, Washington, DC, 1996.
[45] D. Boyd and J. Beckwith, *Proc. Natl. Acad. Sci. USA* **86,** 9446 (1989).
[46] D. Boyd, B. Traxler, and J. Beckwith, *J. Bacteriol.* **175,** 553 (1993).

[4] Binding Protein-Dependent ABC Transport System for Glycerol 3-Phosphate of *Escherichia coli*

By WINFRIED BOOS

Introduction

sn-Glycerol 3-phosphate (G3P) is an essential intermediate in the biosynthesis of phospholipids. When *Escherichia coli* grows on a carbon source other than G3P it synthesizes it by the NADH-dependent reduction of glyceraldehyde 3-phosphate. On the other hand, the presence of G3P in the medium results in its uptake and direct utilization as a lipid precursor.[1] G3P is also a carbon source. It enters glycolysis by its oxidation to dihydroxyacetone phosphate.[2–4] Using G3P labeled with tritium at the C-2 position, one can easily follow the portion of G3P that is incorporated exclusively into lipids and their derivatives since oxidation of G3P to dihydroxyacetone phosphate results in the loss of the tritium at the C-2 position. The enzymes geared for the utilization of G3P, glycerol, or glycerol phosphate phosphodiesters are encoded by the *glp* genes.[4] They are negatively regulated by the GlpR repressor[5] and consist of a typical catabolite repressible regulon, the inducer being G3P.[1]

The major uptake system for G3P is the *glp* regulon-dependent and *glpT*-encoded GlpT transport system.[1,6] It is a secondary tranport system in which the inward movement of G3P is coupled to the outward movement of inorganic phosphate (P_i).[7,8] When growing on minimal media containing P_i at concentrations >1 m*M* the GlpT transport system is the only G3P transport system and *glpT* mutants cannot grow on G3P as the sole source of carbon.

The second transport system for G3P has been discovered in phenotypic revertants of *glpT* mutants that are again able to grow on G3P. They constitutively synthesized a novel transport system for G3P, the Ugp (for

[1] S. Hayashi, J. P. Koch, and E. C. C. Lin, *J. Biol. Chem.* **239,** 3098 (1964).
[2] D. Austin and T. J. Larson, *J. Bacteriol.* **173,** 101 (1991).
[3] S. T. Cole, K. Eiglmeier, S. Ahmed, N. Honore, L. Elmes, W. F. Anderson, and J. H. Weiner, *J. Bacteriol.* **170,** 2448 (1988).
[4] E. C. C. Lin, *Ann. Rev. Microbiol.* **30,** 535 (1976).
[5] T. J. Larson, S. Ye, D. L. Weissenborn, H. J. Hoffmann, and H. Schweizer, *J. Biol. Chem.* **262,** 15869 (1987).
[6] T. J. Larson, G. Schumacher, and W. Boos, *J. Bacteriol.* **152,** 1008 (1982).
[7] C. M. Elvin, C. M. Hardy, and H. Rosenberg, *J. Bacteriol.* **161,** 1054 (1985).
[8] S. V. Ambudkar, T. H. Larson, and P. C. Maloney, *J. Biol. Chem.* **261,** 9083 (1986).

METHODS IN ENZYMOLOGY, VOL. 292
0076-6879/98 $25.00

uptake **g**lycerol **p**hosphate) system, which seemed to be absent in wild-type strains under normal growth conditions.[9] Simultaneously with the appearance of the transport system the suppressor strains synthesized several periplasmic proteins and one outer membrane protein that were not present in the parent strain under identical growth conditions. It was then discovered that the suppressor mutations were in fact mutations leading to the constitutivity of the *pho* (phosphate) regulon, and, consequently, the Ugp system was recognized as part of the Pho system.[10–12] Ironically, even though the suppressor strains were now able to grow on G3P as a sole source of carbon, this was not due to the Ugp transport system but due to the derepression of alkaline phosphatase (a known member of the *pho* regulon) hydrolyzing G3P in the periplasm to glycerol followed by its uptake and subsequent internal phosphorylation to G3P. What has become known as the Ugp paradox is the observation that the Ugp system, even though capable of transporting G3P very well, is unable to supply enough carbon for growth.[13,14]

Studies with the Ugp system have revealed it to be a typical binding protein (BP)-dependent adenosine triphosphate-binding cassette (ABC) transport system.[15–17] Even though no reconstitution of Ugp activity in the *in vitro* system of liposomes has been attempted as yet, the system offers some interesting aspects with regard to its activity and its feedback regulation, for instance, the exchangeability of the UgpC subunit with the corresponding MalK subunit of the maltose system, as well as its role in gene regulation.

Substrates of Ugp System, Transport-Associated Phosphoglycerol Diester Phosphodiesterase, and Transport Assay

G3P is not the only substrate that is transported by the Ugp system. Additional substrates were discovered when it was found that Luria Bertani broth (LB) contained compounds that were inhibitory to the uptake of

[9] M. Argast, D. Ludtke, T. J. Silhavy, and W. Boos, *J. Bacteriol.* **136,** 1070 (1978).
[10] J. Tommassen and B. Lugtenberg, *J. Bacteriol.* **143,** 151 (1980).
[11] M. Argast and W. Boos, *J. Bacteriol.* **143,** 142 (1980).
[12] H. Schweizer and W. Boos, *J. Bacteriol.* **163,** 392 (1985).
[13] H. Schweizer, M. Argast, and W. Boos, *J. Bacteriol.* **150,** 1154 (1982).
[14] P. Brzoska, M. Rimmele, K. Brzostek, and W. Boos, *in* "Molecular Biology of Phosphate in Microorganisms" (A. M. Torriani, S. Silver, and E. Yagil, eds.), p. 30. American Society of Microbiology, Washington, DC, 1994.
[15] M. Argast and W. Boos, *J. Biol. Chem.* **254,** 10931 (1979).
[16] H. Schweizer and W. Boos, *Mol. Gen. Genet.* **192,** 177 (1983).
[17] P. Overduin, W. Boos, and J. Tommassen, *Mol. Microbiol.* **2,** 767 (1988).

G3P by the Ugp system. The inhibition was not relieved after treatment of LB with alkaline phosphatase (thus it was not G3P that caused inhibition), but it was relieved after treatment with phosphodiesterase. The inhibitory compounds were subsequently identified as phosphoglycerol phosphodiesters, the diacylated products of phospholipids. Thus, glycerophosphorylethanolamine (GPE), glycerophosphorylglycerol (GPG), and glycerophosphorylcholine (GPC) are substrates of the Ugp system with K_m values of uptake in the micromolar range. When these phosphodiesters were used as substrates labeled in the G3P portion, accumulation of radioactivity could readily be observed. This was not the case when phosphodiesters were used that were labeled in the alcohol portion. This is due to an effective phosphoglycerol phosphodiesterase associated with the transport system that cleaves the phosphodiesters followed by the release of the alcohol portion (ethanol, glycerol, but not choline) into the medium. The enzyme responsible for the cleavage of the phosphoglycerol phosphodiesters is localized at the inner phase of the cytoplasmic membrane, most likely associated with the Ugp transport system. It only hydrolyzes the phosphodiesters entering the Ugp system but it does not hydrolyze cytoplasmic phosphoglycerol phosphodiesters. The enzyme is not connected to the mechanism of Ugp-mediated transport *per se* and *ugpQ* mutants lacking the enzyme are not defective in transport.[18,19] In contrast, the accumulation of GPE or GPG labeled in the alcohol portion can be observed in *ugpQ* mutants. *ugpQ* is the last gene in the *ugp* operon.[17]

To measure the Ugp transport system under optimized conditions[20,21] strains should be grown under aeration in Tris-based G+L medium[22] with 0.2% glucose as carbon source and 0.1 mM P_i as the only source of phosphate. The culture should be grown until the phosphate limitation prevents further growth (optical density at 600 nm of about 0.7–0.8). The cells are harvested and resuspended at room temperature in G+L medium without P_i to an optical density at 600 nm of 0.5. The assay is initiated by the addition of 30 μl of 10 μM [^{14}C]G3P (144 mCi/mmol) to 3 ml of culture. The 0.5-ml samples are taken at short time intervals (five samples within 120 sec) and filtered through Millipore filters with a 45-μm pore size. The filters are washed once with 10 ml G+L medium and counted in a liquid scintillation counter. Under these conditions, more than 20% of the total amounts of cpms are used up after 2 min. This assay is most sensitive for the activity of the Ugp system. It measures uptake far below the K_m value

[18] P. Brzoska and W. Boos, *J. Bacteriol.* **170,** 4125 (1988).

[19] J. Tommassen, K. Eiglmeier, S. T. Cole, P. Overduin, T. J. Larson, and W. Boos, *Mol. Gen. Genet.* **226,** 321 (1991).

[20] P. Brzoska, M. Rimmele, K. Brzostek, and W. Boos, *J. Bacteriol.* **176,** 15 (1994).

[21] K. B. Xavier, M. Kossmann, H. Santos, and W. Boos, *J. Bacteriol.* **177,** 699 (1995).

[22] A. Garen and C. Levinthal, *Biochim. Biophys. Acta* **38,** 470 (1960).

(1–2 μM). The assay is strongly inhibited in the presence of P_i, and transport is essentially absent when the strain has not been grown to the limitation of P_i (derepression of the *pho* regulon is necessary). Glycerol as carbon source should not be used unless the strain carries a *glpT* mutation. When glycerol is used as a carbon source in a *glpT* mutant the strain should nevertheless be depleted of P_i to ensure the derepression of the Ugp system. Even if *pst* mutants[23] are used that are constitutive for the *pho* system,[24] it is worthwhile to deplete the strain of P_i to ensure highly reproducible transport assays. This is due to the transinhibition of G3P uptake by internal P_i.[20,21]

The active (accumulating) nature of the Ugp transport system can only be seen in mutants that prevent the metabolism of G3P. *glpD* mutants lacking the respiratory chain-linked G3P dehydrogenase are suffucient to observe accumulation of G3P after Ugp-mediated uptake under aerobic conditions. The transport characteristics are not influenced by the presence or absence of the metabolizing enzymes.

ugp Genes and Their Regulation

The *ugp* genes are located at 77.3 min (kb 3596.0–3600.0 of the physical map)[25–27] on the *E. coli* linkage map immediately downstream of the *liv* operon. *ugpB ugpA ugpE ugpC* and *ugpQ* form an operon with *ugpB* as the promoter proximal gene with counterclockwise direction of transcription.[17] The first four genes are essential for transport and encode proteins necessary for a typical BP-dependent ABC transport system.

The analysis of suppressor mutations for *glpT* mutants had revealed that all suppressor mutations causing constitutive expression of the Ugp system were identical with mutants constitutive for the *pho* regulon. They were either located in *pst*,[23] the genes encoding the *pho*-controlled and BP-dependent ABC transport system for P_i (Pst system), or in *phoR* encoding a regulatory component of the *pho* system.[28,29] Therefore, the *ugp* genes are typical members of the *pho* regulon. The regulation of the *pho* system occurs by a classical two-component system with the membrane-bound

[23] D. C. Webb, H. Rosenberg, and G. B. Cox, *J. Biol. Chem.* **267,** 24661 (1992).
[24] B. L. Wanner, *J. Cell. Biochem.* **51,** 47 (1993).
[25] H. Schweizer, T. Grussenmeyer, and W. Boos, *J. Bacteriol.* **150,** 1164 (1982).
[26] H. Schweizer and W. Boos, *Mol. Gen. Genet.* **197,** 161 (1984).
[27] M. K. B. Berlyn, B. Low, K. E. Rudd, and M. Singer, *in* "*Escherichia coli* and *Salmonella typhimurium:* Cellular and Molecular Biology" (F. C. Neidhardt, ed.), 2nd Ed. American Society of Microbiology, Washington, DC, 1996.
[28] B. L. Wanner and B. Chang, *J. Bacteriol.* **169,** 5569 (1987).
[29] K. H. Makino, H. Shinagawa, M. Amemura, and A. Nakata, *J. Mol. Biol.* **192,** 549 (1986).

phoR gene product as the sensor kinase and the *phoB*-encoded protein as the cytoplasmic response regulator. PhoB in its phosphorylated form recognizes conserved sequences, the *pho* boxes, in the control region of all *pho*-regulated genes or operons and is necessary for the initiation of transcription.[24] The system responds to the limitation of external P_i, which is sensed by the *pst*-encoded P_i transport system. Mutations in any of the *pst* genes lead to a derepressed state of the system, high phosphorylating activity of PhoR, and, consequently, high phosphorylated levels of PhoB. Because the interruption of the *Pst* activity causes constitutivity, it seems likely that the phosphorylating state of PhoR is the default state of the system that is inhibited by the event of P_i transport through the Pst system. The mechanism of this signal transduction from the transport system to the sensor kinase PhoR is unclear but requires the presence of the *phoU* gene product whose function is also still unclear.[30,31] *phoU* is the last gene in the *pst* operon in analogy to *ugpQ* in the *ugp* operon. Experiments to demonstrate a signal transduction pathway from Ugp-mediated G3P transport to PhoR, the sensor kinase of the system, did not reveal any role of the Ugp system in *pho* regulation.[20]

Mutants lacking PhoR function are also partially constitutive. This is due to cross-phosphorylation of PhoB by another sensor kinase, CreC (previously called PhoM[32,33]), that is not under the control of P_i limitation. Because PhoR under conditions of excess Pi is also an effective phosphatase for PhoB phosphate, CreC-dependent PhoB phosphorylation is only significant in the absence of PhoR.[34,35]

Despite the presence of typical PhoB boxes in the regulatory region of the *ugp* operon this region also contains a cAMP/CAP-binding site indicating an alternate mode of *ugp* expression. Thus, under conditions of low catabolite repression and high internal cAMP concentration (for instance, during the onset of carbon starvation), Ugp is expressed to some extent even in the presence of P_i. This is corroborated by the finding of two alternate transcripts of the operon under different growth conditions.[36,37] The *ugp* genes have been cloned and *ugpC* has been subcloned under the

[30] M. Muda, N. N. Rao, and A. Torriani, *J. Bacteriol.* **174,** 8057 (1992).

[31] P. M. Steed and B. L. Wanner, *J. Bacteriol.* **175,** 6797 (1993).

[32] D. Ludtke, J. Bernstein, C. Hamilton, and A. Torriani, *J. Bacteriol.* **159,** 19 (1984).

[33] K. Makino, H. Shinagawa, and A. Nakata, *Mol. Gen. Genet.* **195,** 381 (1984).

[34] B. L. Wanner, *J. Mol. Biol.* **166,** 283 (1983).

[35] B. L. Wanner, *J. Bacteriol.* **174,** 2053 (1992).

[36] T. Z. Su, H. P. Schweizer, and D. L. Oxender, *Mol. Gen. Genet.* **230,** 28 (1991).

[37] M. Kasahara, K. Makino, M. Amemura, A. Nakata, and H. Shinagawa, *J. Bacteriol.* **173,** 549 (1991).

IPTG-inducible *tac* promoter.[38] Antibiotic resistance insertion mutants in all *ugp* genes are available as well as *lacZ* fusions to some genes.

Physiological Role of Ugp System as Phosphate Scavenger

The role of the Ugp system is best characterized by comparison to the *glp* regulon-dependent GlpT transport system. Uptake of G3P via the GlpT transporter leads to the simultaneous counterflow of P_i from the cell.[7,8] Because the metabolism of G3P produces an excess of internal P_i, through the stoichiometric exchange of G3P against P_i during its uptake, the GlpT system appears optimally geared for the use of G3P as a carbon source. Indeed, following the concentration of internal P_i by nuclear magnetic resonance (NMR) imaging, one can observe that the uptake of G3P via the GlpT system does not lead to a change in the internal P_i concentration, be it initially low or high, even though high internal P_i levels greatly stimulate the exchange reaction.[21] In contrast, uptake of G3P via the Ugp system in a strain that can metabolize G3P always leads to a dramatic increase in the internal P_i concentration up to about 20 m*M*. This, in turn, inhibits the Ugp-mediated uptake of G3P.[20] This increase in internal P_i concentration also takes place at the simultaneously operating GlpT system.[21] The feedback inhibition of Ugp by internal P_i is the explanation for the Ugp paradox, the phenomenon that G3P transported exclusively by the Ugp system can serve as the sole source of P_i but not as the sole source of carbon, even though the initial rate of transport via the Ugp system should be sufficient for the flow of carbon necessary for growth.[14] Thus, the Ugp system, in line with its high substrate affinity and broad range substrate specificity, is ideally geared for scavenging P_i containing compounds. The "disregard" of the system for the carbon portion of its substrates manifests itself by the phenomenon that during uptake of PGE and PGG the carbon portion of the substrate, ethanolamine or glycerol, is expelled.[18]

Composition of Transport System

The complete system is composed of four proteins, corresponding to the canonical composition of BP-dependent ABC transporters.[39] The first protein that was recognized and purified was the high-affinity BP for G3P and phosphoglycerol phosphodiesters, encoded by *ugpB*.[15] It is located in

[38] D. Hekstra and J. Tommassen, *J. Bacteriol.* **175,** 6546 (1993).

[39] W. Boos and J. Lucht, *in* "*Escherichia coli* and *Salmonella typhimurium:* Cellular and Molecular Biology" (F. C. Neidhardt, ed.), 2nd Ed. American Society of Microbiology, Washington, DC, 1996.

the periplasm and represents the primary recognition site of the system. Judging from the appearance of the protein from the set of periplasmic proteins of fully derepressed strains on sodium dodecyl sulfate–polyacrylamide gel electrophoresis (SDS–PAGE), and comparing it with the measured levels of the maltose-binding protein (MBP), one can estimate that it constitutes one-third to one-half of the amount of MBP, that is, it amounts to about 0.3 to 0.5 m*M* in the periplasm.

UgpC, the ATP-binding site-carrying subunit of the system, has been isolated from inclusion bodies of an overproducing strain and purified by SDS–PAGE in inactive form for the production of antibodies.[38] All other Ugp proteins have not been isolated in purified form but are known to exist by the expression of the relevant gene in minicells[26] or are only deduced from the DNA sequence.[17] The computer-derived analysis of the deduced sequences, mainly the hydropathy plots, allowed their allocation to the different functions. Thus, *ugpA* and *ugpE,* the analogs of the corresponding *malF* and *malG* genes of the maltose system, encode the tightly membrane-bound proteins of the system, forming the transport gate proper through the membrane. With this membrane complex interacts the BP on the periplasmic side and the *ugpC*-encoded and ATP-binding site-carrying UgpC subunit on the cytoplasmic side. The repeatedly mentioned comparison to the *E. coli* BP-dependent ABC transport system for maltose is not without reason. Unusual for unrelated systems, the sequence of the Ugp and the maltose transport proteins can be seen not only in restricted consensus sequences around the ATP-binding fold of the energy subunit, or the EAA–G sequence found in all membrane components[40] but throughout all the components (with the exception of the large periplasmic loop in MalF that is not found in any other membrane component of BP-dependent ABC transporters). The reason for this homology is entirely unclear. Yet there is the curious connection of maltose with P_i. Certain mutations in *envZ* (encoding the sensor kinase of the osmolarity-sensing two-component system that regulates the relative composition of the outer membrane porins OmpC and OmpF) are negatively dominant on the *pho* regulon as well as the maltose regulon[41,42] even though their respective regulation is entirely different.[43] Also, maltose metabolism is keyed to the function of phosphorylases. Thus, the level of internal P_i may be important for controlling maltose and maltodextrin metabolism as well as regulation. In analogy to

[40] W. Saurin, W. Köster, and E. Dassa, *Mol. Microbiol.* **12,** 993 (1994).
[41] B. L. Wanner, A. Sarthy, and J. Beckwith, *J. Bacteriol.* **140,** 229 (1979).
[42] C. Wandersman, F. Moreno, and M. Schwartz, *J. Bacteriol.* **143,** 1374 (1980).
[43] C. C. Case, B. Bukau, S. Granett, M. R. Villarejo, and W. Boos, *J. Bacteriol.* **166,** 706 (1986).

the *lamB*-encoded λ receptor or maltoporin of the outer membrane[44] that is part of the maltose regulon and which functions as a specific pore for maltose and maltodextrins,[45,46] there is an equivalent outer membrane porin, the PhoE pore, whose expression is under *pho* regulon control.[10,11] However, even though this pore favors negative charges in the *in vitro* assay of conductivity measurements through planar lipids, its essential role for the uptake of G3P or glycerol phosphodiesters, or even P_i, through the outer membrane could never be demonstrated.[47]

G3P/Glycerophosphoryl Diester-Binding Protein and Measurement of Its Binding Affinity

Binding protein was originally isolated and purified from the osmotic shock fluid of a strain that was constitutive for the *pho* system. The major purification step was preparative flat-bed electrofocusing followed by molecular sieve chromatography. The yield starting with 9 liters of culture was only 4 mg. Today plasmids harboring *ugpB* encoding the BP are available for a more convenient preparation with the expectation of a much higher yield. A convenient binding assay is available for the G3P-binding protein (as well as other periplasmic BPs) that is not commonly used and is therefore described in some detail.

The K_d value is determined by the use of the retention phenomenon.[15,48] A small piece of dialysis tubing, open at one end, is tightly fit with its open end onto a bluntly cut plastic pipetting tip (the size fitting a 1-ml automatic pipette). Then 250–300 μl of a protein solution containing about 0.5–1 mg BP/ml that had previously been dialyzed overnight against 100 m*M* Tris-HCl, pH 7.0, is added into the dialysis tubing through the pipetting tip. Tip and tubing are placed in a tightly fitting Styrofoam plate (about 1 cm thick). The plate with its attached tubing is placed on top of a 1- to 2-liter Erlenmeyer flask containing 100 m*M* Tris-HCl, pH 7.0, that is slowly stirred at 4°. The tip with the attached tubing is adjusted so that the tubing is immersed in the external fluid and that the level of internal and external fluid are at the same height. Next, 0.1 μCi (in 10–20 μl, about 160,000 cpm) ^{14}C-labeled G3P (144 mCi/mmol) (corresponding to about 2.8 μ*M* final concentration) is added from above into the dialysis tubing at time

[44] T. Schirmer, T. A. Keller, Y. F. Wang, and J. P. Rosenbusch, *Science* **267,** 512 (1995).

[45] M. Hofnung, *Science* **267,** 473 (1995).

[46] S. Szmelcman, M. Schwartz, T. J. Silhavy, and W. Boos, *Eur. J. Biochem.* **65,** 13 (1976).

[47] K. Bauer, K. Auer, P. van der Ley, R. Benz, and J. Tommassen, *J. Biol. Chem.* **263,** 13046 (1988).

[48] T. J. Silhavy, W. Boos, S. Szmelcman, and M. Schwartz, *Proc. Natl. Acad. Sci. USA* **72,** 2120 (1975).

zero and carefully mixed by the use of the automatic pipette. Subsequently 20-μl samples are withdrawn at 10- to 20-min time intervals and counted in a scintillation counter. Care should be taken that the level of internal fluid is always adjusted to the level of the external fluid either by adjusting the position of the tip or by adjusting the external fluid.

When the radioactivity is plotted half-logarithmically with time, a straight line is obtained giving the time at which half the substrate has left the dialysis bag. The same bag should be used in the repeat of the experiment done in the absence of BP. Time points should be taken every 5 min. The half-lifetime of G3P in the absence of BP is, of course, independent of the initial G3P concentration and is about 40 min but depends to some extent on the physical shape of the dialysis tubing. The prerequisite of the experiment is that the free substrate concentration in the dialysis bag be below the K_d of the BP. In the case given, the K_d of the G3P-binding protein is between 0.12 and 0.25 μM. Even though 2.8 μM is added in the dialysis bag, this is reduced far below the K_d because the BP is present at about 15 μM concentration. Even if too much substrate is present initially, this is not a serious problem since excess substrate will leave the dialysis bag initially fast before attaining the slow rate characteristic of the BP-dependent retention process that can be measured subsequently. For calculating the K_d from the slope of the exit process the following equation can be applied [Eq. (1)]:

$$T_p = T_s(1 + ([\mathrm{P}]/K_d) \tag{1}$$

where T_p stands for the time when half the substrate has left the bag in the presence of the binding protein, T_s stands for the time when half the substrate has left the bag in the absence of the binding protein, [P] is the binding protein in molar concentration (assuming one binding site), and K_d is also in the dimension of molarity. The method allows us to determine the K_d when the concentration of the BP is known or it allows us to determine the amount of BP in crude extracts when the K_d is known. In the case of the G3P-binding protein with a BP concentration of 0.67 mg/ml (corresponding to a 14.5 μM solution), the half-lifetimes of substrate for leaving the bag was 116 hr versus 36 min in the presence and absence of BP, respectively, yielding 0.12 μM for the K_d. The limit of binding affinity that can be measured by this technique using a higher concentration of binding protein (about 7 mg/ml) would be a K_d as high as 0.1 mM. Under these conditions, the half-lifetime of substrate in the bag would be about 2 hr. The advantage of this method is that the concentration of the substrate does not need to be determined (as long as it is below the K_d of the BP). The presence of equimolar "nondialyzable" unlabeled substrate bound tightly to the BP does not interfere, and, in contrast to the classical equilibrium dialysis, the measurement does not require reaching the time-consuming equilibrium of binding.

Other convenient procedures are available for measuring binding affinity. One that needs near-homogeneous protein is to follow the change in the intrinsic tryptophan fluorescence on addition of substrate. Because this method is widely used, it is not described here. Examples of this technique can be found elsewhere.[46,49,50]

UgpC: The ATP-Binding Site-Containing Subunit That Can Functionally Exchange the MalK Subunit of Maltose System and Is Involved in Regulation of *pho* Regulon

The high homology between UgpC and MalK prompted an attempt to test whether or not these subunits could be exchanged functionally. A *malK-lacZ* mutant cannot grow on maltose. A *pst* derivative of this mutant constitutively expressing the Ugp transport system also cannot grow on maltose. However, the additional introduction of a mutation in *ugpA* or *ugpE* allowed growth on maltose. Similarly, when UgpC was overproduced from an IPTG-inducible promoter, growth on maltose was even possible in the ugp^+ background. This demonstrates that UgpC can functionally replace the MalK subunit in its transport-energizing function. The concentration of UgpC is critical in this experiment. Judging from the wild-type level of UgpC in fully derepressed strains, the protein is apparently well bound by the UgpA and UgpE subunits, and no excess UgpC is available for binding to the MalF and MalG subunits of the maltose system. Only the absence of the cognate partners UgpA or UgpE or massive overproduction of UgpC allows the functional interaction with the heterologous partner. The reciprocal experiment, the complementation of a missing UgpC subunit by the overproduction of MalK, gave similar results. In this case, the ability to use G3P as the sole source of P_i was tested. Here, the particular property of MalK to be inhibited by the unphosphorylated enzyme IIA^{Glc} of the phosphotransferase system (due to the presence of a substrate of this transport system) was maintained in its Ugp-complementing function.[38]

MalK is known not only for its role in transport but also in regulation.[51,52] Strains lacking MalK are constitutive for the expression of *mal* genes and overexpression of MalK strongly represses *mal* gene expression. The regu-

[49] W. Boos, A. S. Gordon, R. E. Hall, and H. D. Price, *J. Biol. Chem.* **247,** 917 (1972).
[50] M. R. Rohrbach, V. Braun, and W. Köster, *J. Bacteriol.* **177,** 7186 (1995).
[51] B. Bukau, M. Ehrmann, and W. Boos, *J. Bacteriol.* **166,** 884 (1986).
[52] M. Reyes and H. A. Shuman, *J. Bacteriol.* **170,** 4598 (1988).

lating domain of MalK is located in the C terminus,[53] which is unusually long in comparison to ATP-binding subunits of other BP-dependent systems, but it is conserved in UgpC. However, the overproduction of UgpC has no down-regulating activity on the maltose system.

The control of the association of UgpC to its membrane-bound partners UgpA and UgpE might be the explanation for the mode of the feedback inhibition of G3P transport by internal P_i. Inhibition of Ugp-mediated G3P uptake by internal P_i can be realized in the following way: *glpT phoA* mutants are grown on G+L medium with 0.2% glucose and 0.1 m*M* P_i, and standard assays for the uptake of G3P are performed with and without incubation with 1 m*M* P_i 10 min prior to the transport assay. The fully induced Pst transport system for P_i leads to the accumulation of internal P_i, which, in turn, reduces G3P uptake about sixfold. The simultaneous overproduction of UgpC completely abolishes the feedback inhibition by P_i. This might be interpreted by P_i binding to UgpC and reducing its affinity for UgpA and UgpE. Thus, the overproduction of UgpC will overcome the inhibition by P_i by allowing interaction despite the lowered affinity. Similarly, maltose transport complemented in a *malK*-deficient strain by the overproduction of UgpC cannot be inhibited by P_i.[54]

The experiment measuring G3P transport in strains overproducing UgpC is somewhat compromised by the fact that in analogy to the role of MalK in the regulation of the *mal* system the overproduction of UgpC leads to a fivefold reduction in the expression of *ugp* itself as well as of *phoA*, another member of the *pho* regulon encoding alkaline phosphatase.[55] Thus, the accumulation of P_i by the Pst system may also be reduced, not allowing the same internal P_i concentrations to be achieved. Yet, even incubations with high external levels of P_i do not lead to an inhibition of G3P uptake.

Nevertheless, it has become clear that overproduction of UgpC reduces the expression of the *pho* regulon. Currently, it is not clear whether or not all members of the regulon are affected; only the *ugp* and the *phoA* genes have been tested so far. The phenomenon is very similar to the observation that overproduction of MalK reduces the expression of maltose transport genes. In the *mal* system it can also be observed that mutants lacking *malK* function are constitutive for the expression of the remaining transport genes. This is not the case for *ugpC* mutants. Also, the degree of repression by the ABC subunit in the *mal* system is much stronger than in the Ugp

[53] S. Kühnau, M. Reyes, A. Sievertsen, H. A. Shuman, and W. Boos, *J. Bacteriol.* **173,** 2180 (1991).

[54] W. Boos, unpublished observations (1992).

[55] The regulating effect of UgpC was first noticed by J. Tommassen and co-workers.

system. The mechanism of the regulatory function of MalK or UgpC is not clear. In the *mal* system, it is rather likely that MalK interacts with the positive activator of the system, MalT, reducing its activity. The *pho* system is also controlled by a positive activator, PhoB, which is a classical response regulator of two-component systems.

Conclusion and Prospects

The study of the Ugp system, particularly in comparison with the maltose system, has led to new insights into the function of BP-dependent ABC transport systems. In particular, the exchangeability of its ATP-hydrolyzing subunits may shed some light on the mechanism of energy input in this multicomponent transport system as well as on the role that this subunit plays in regulation. One also may attempt to study the transfer of substrate from the binding protein to the membrane components using the heterologous approach. Thus, can maltose be transported by the membrane-bound components of the Ugp system in the presence of high concentrations of maltose-binding protein and in the absence of the G3P-binding protein?

Acknowledgments

Studies in the author's laboratory were supported by grants from the Deutsche Forschungsgemeinschaft (SFB 156) and the Fonds der Chemischen Industrie.

[5] Structure–Function Analysis of Hemolysin B

By FANG ZHANG, JONATHAN A. SHEPS, and VICTOR LING

Introduction

The *Escherichia coli* hemolysin transport system is part of a family of related protein toxin secretion systems found in gram-negative bacteria.[1] This family of toxins, named RTX (for repeat toxin), includes *Pasteurella* leukotoxin, *Bordetella* cyclolysin, and *Erwinia* metalloprotease.[2] The export of the 107-kDa hemolysin A (HlyA) protein occurs in a single step across both inner and outer membranes of *E. coli.* This translocation involves

[1] M. J. Fath and R. Kolter, *Microbiol. Rev.* **57,** 995 (1993).

[2] G. Menestrina, C. Moser, S. Pellet, and R. Welch, *Toxicology* **87,** 249 (1994).

0076-6879/98 $25.00

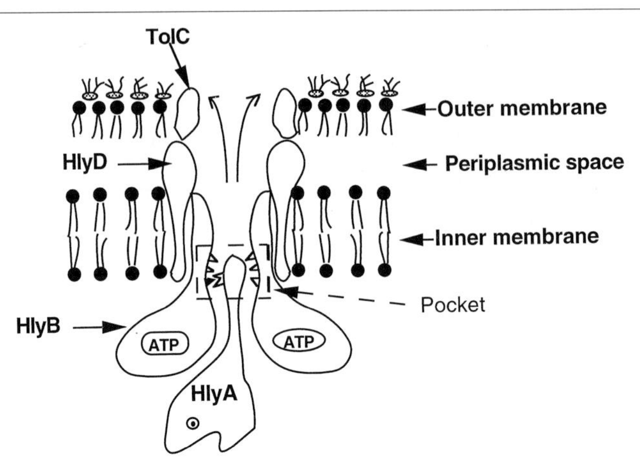

FIG. 1. A model of the hemolysin transporter system. The dashed box encloses the hypothetical binding pocket.

three proteins, hemolysin B (HlyB), hemolysin D (HlyD) and TolC (Fig. 1). HlyB is an inner membrane protein with eight transmembrane domains and an adenosine triphosphate (ATP)-binding domain.[3,4] HlyB is thought to recognize the signal sequence of HlyA and transduce energy from ATP hydrolysis for transport. TolC is an outer membrane porin protein.[5] HlyD is mainly exposed to the periplasmic space with one transmembrane domain anchored in the inner membrane. HlyD is thought to connect HlyB and TolC in forming the transport complex.[6]

HlyB and other RTX transporters are part of the ATP-binding cassette (ABC) transporter superfamily. Phylogenetic analysis of selected members of the ABC family based on amino acid sequences has revealed that HlyB is closely related to the multidrug resistance P-glycoprotein (P-gp) (Fig. 2).[7] It has been shown that the ATP-binding domain is able to bind and hydrolyze ATP *in vitro*, and this activity is required for the transport of HlyA *in vivo*.[8,9] Like P-gp and many ABC transporters, HlyB has an apparent

[3] I. Gentschev and W. Goebel, *Mol. Gen. Genet.* **232,** 40 (1992).

[4] R. C. Wang, S. J. Seror, M. Blight, J. M. Pratt, J. K. Broome-Smith, and I. B. Holland, *J. Mol. Biol.* **217,** 441 (1991).

[5] C. Wandersman and P. Delepelaire, *Proc. Natl. Acad. Sci. USA* **87,** 4776 (1990).

[6] R. Schulein, I. Gentschev, S. Schlor, R. Gross, and W. Goebel, *Mol. Gen. Genet.* **245,** 203 (1994).

[7] J. A. Sheps, F. Zhang, and V. Ling, *in* "Membrane Protein Transport" (S. R. Rothman, ed.), Vol. 3, p. 81. JAI Press, Greenwich, Connecticut, 1996.

[8] V. Koronakis, C. Hughes, and E. Koronakis, *Mol. Microbiol.* **8,** 1163 (1993).

[9] E. Koronakis, C. Hughes, I. Milisav, and V. Koronakis, *Mol. Microbiol.* **16,** 87 (1995).

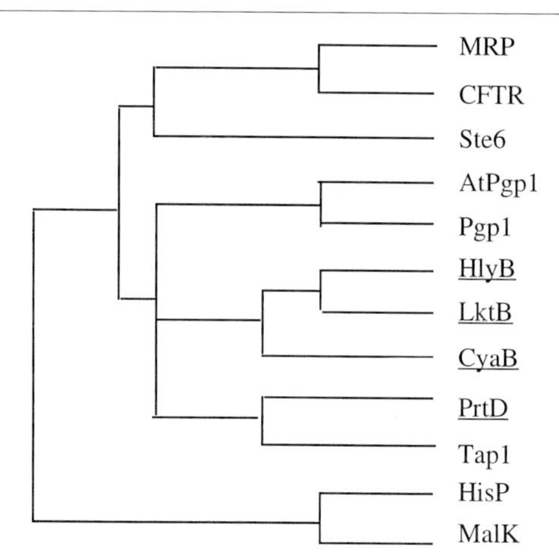

FIG. 2. Phylogenetic tree of ABC transporters. This tree is modified from a more extensive analysis in Sheps *et al.*[7] and is based on protein sequences from the C-terminal ATP-binding domains. The RTX transporters are underlined. LktB, *Pasteurella* leukotoxin transporter; CyaB, *Bordetella* cyclolysin transporter; PrtD, *Erwinia* metalloprotease transporter.

broad substrate specificity. It is capable of transporting different RTXs having transport signal sequences that are not conserved at the amino acid level.[7]

The transport signal sequence of HlyA is located within the C-terminal 60 amino acids. This region has been extensively studied by mutational analyses.[10–12] Few amino acids dispersed in the signal region have been identified as important for secretion.[11] Furthermore, when the C-terminal 58-amino-acid signal sequence of HlyA is replaced by the C-terminal 70-amino-acid sequence from the LktA protein of *Pasteurella hemolytica*, the chimeric protein possesses transport competence equivalent to that of wild-type HlyA.[12] The C-terminal sequence of LktA contains very little sequence similarity to that of HlyA. However, in a membrane mimetic environment both sequences form a similar secondary structure containing two helices.[13] Thus, it has been postulated that the distinguishing feature of the signal sequence recognized by the hemolysin transporter is its higher order structure.

In this chapter, we focus on genetic complementation analysis of the interaction of HlyB with the HlyA signal sequence. We chose a genetic approach because of the ease of genetic analysis in this bacterial system. The

[10] P. Stanley, V. Koronakis, and C. Hughes, *Mol. Microbiol.* **5,** 2391 (1991).
[11] B. Kenny, S. Taylor, and I. B. Holland, *Mol. Microbiol.* **6,** 1477 (1992).
[12] F. Zhang, D. I. Greig, and V. Ling, *Proc. Natl. Acad. Sci. USA* **90,** 4211 (1993).
[13] Y. Yin, F. Zhang, V. Ling, and C. H. Arrowsmith, *FEBS Lett.* **366,** 1 (1995).

rapid generation time and the ability to screen large numbers of colonies by means of a plate-based hemolytic assay provide an opportunity to perform complex screens for complementing mutations of HlyB that compensate for the transport deficiency of HlyA signal mutations. The power of this approach allows us to study the interaction of HlyB and the signal sequence of HlyA in greater detail than has generally been the case in studies of substrate recognition by other ABC transporters.

Overall Strategy

This genetic approach to identify interacting loci is by complementation, in which a phenotype caused by mutation of a specific region of one gene (signal sequence of HlyA) can be compensated for by mutations of the second gene (HlyB). The complementation analysis is facilitated by the fact that HlyA is hemolytic, and transport activity in different clones can be readily estimated by the size of the hemolytic zones surrounding colonies growing on blood agar. Two major steps are involved in this approach (Fig. 3). In the first step, the signal sequence of HlyA is mutated by site-directed mutagenesis. As a result, transport activity of these mutants is impaired, represented by a reduction in the size of hemolytic zones. In the second step, *hlyB* DNA is randomly mutated by chemical mutagens and then transformed into *E. coli* cells expressing the other components of the transport machinery and *one* of the HlyA mutants. Colonies that exhibit larger hemolytic zones are selected and the *hlyB* plasmids isolated from these colonies are further analyzed by sequencing. These mutations in HlyB compensate for the deficiency in transport caused by the HlyA mutant used for selection. This procedure biases the range of mutations to those sites on HlyB that either interact directly with the signal sequence of HlyA or dictate substrate specificity. Using this genetic approach, we have been able

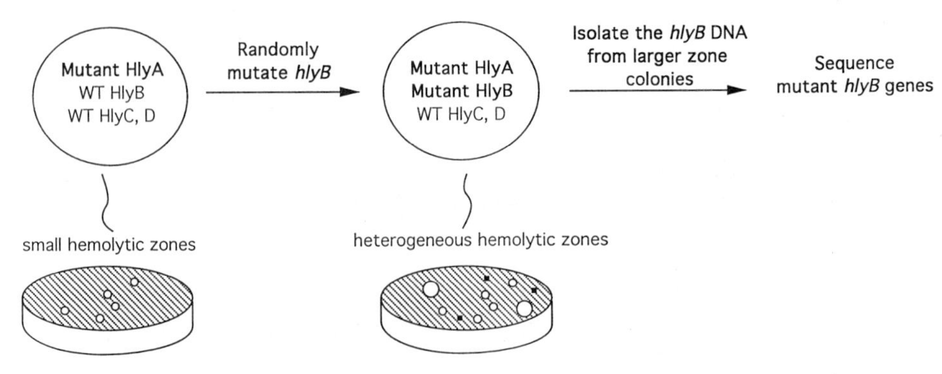

FIG. 3. Procedure for the isolation of HlyB mutants that complement HlyA mutations.

(1) to demonstrate that HlyB interacts with the C-terminal sequence of HlyA, (2) to identify regions of HlyB likely to be involved in this interaction, and (3) to show that single-point mutations in HlyB can change its substrate specificity.

Transport Assays

Transport activity of HlyA can be measured by three assays.[12,14] Two measure the hemolytic activity of secreted HlyA and the third measures secreted HlyA protein by sodium dodecyl sulfate–polyacrylamide gel electrophoresis (SDS–PAGE). Four genes are required for all three transport assays, the hemolysin gene *hlyA*, and three transporter genes (*hlyB, hlyD,* and *tolC*). In addition, the hemolytic assay requires the *hlyC* gene for the addition of a fatty acyl moiety to HlyA protein to render it hemolytic.[15] The *tolC* gene is located in the wild-type *E. coli* chromosome. HlyC and HlyD are expressed from the pACYC-derived plasmid pLGCD.[12] HlyA is expressed from the pSC101-derived plasmid pLG583sk, and HlyB from the pUC-derived plasmid pTG653.[12] These three plasmids contain different origins for DNA replication and different antibiotic (chloramphenicol, kanamycin, and ampicillin) resistance genes.

Hemolytic Zone or "Halo" Assay

Escherichia coli, which expresses the *hly* genes, secretes HlyA and produces a hemolytic zone or halo around colonies when grown on blood agar plates (Fig. 4). JM83 (λ^- *ara*Δ(*pro-lac*)*rps*L*thi*ϕ80*dlac*ZΔM15λ^-) cells harboring the desired plasmids containing *hlyA,B,C* and *D* are plated on 5% sheep blood Luria-Bertani (LB) agar plates containing 25 μg/ml chloramphenicol, 50 μg/ml ampicillin, and 50 μg/ml kanamycin, and grown at 37°. The thickness of the agar, the incubation time, and the quality of blood affect the apparent size of a zone. We routinely use 20 ml agar (measured) for each 100-mm petri dish to ensure a consistent thickness. We have found that using 10 ml plain LB agar on the bottom and 10 ml 5% blood LB agar on top increases the sensitivity for observing differences among hemolytic zones. Only defibrinated blood is suitable for the hemolytic assay. There is significant variation in the quality of the sheep blood available for this purpose. Testing of several suppliers may be necessary to determine the best source. Incubation at 37° for 8–9 hr reveals a suitable range of colony and zone sizes because too long an incubation period causes the large zones

[14] F. Zhang, J. A. Sheps, and V. Ling, *J. Biol. Chem.* **268,** 19889 (1993).

[15] C. Hughes, J.-P. Issartel, K. Hardie, P. Stanley, E. Koronakis, and V. Koronakis, *FEMS Microbiol. Immunol.* **5,** 37 (1992).

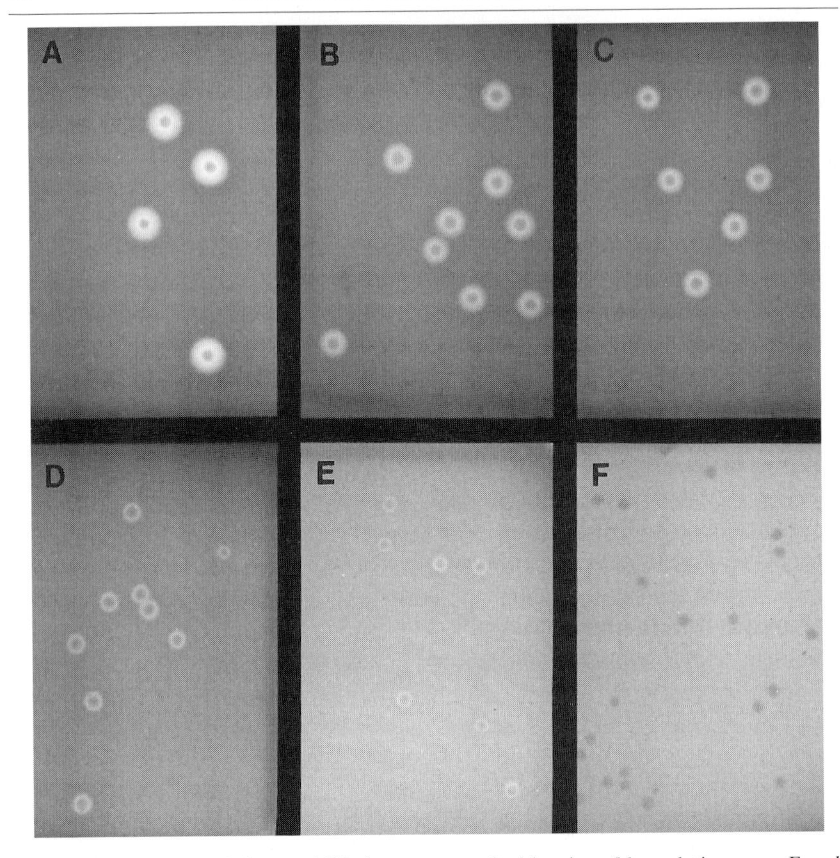

FIG. 4. Transport efficiency of HlyA mutants ranked by size of hemolytic zones. *E. coli* cells expressing HlyB, C, D, and various HlyA mutants were plated on blood agar. The measurements by liquid hemolytic assay of the same clones are given in parentheses: (A) rank 7, wild-type HlyA (100%), (B) rank 5 (36%), (C) rank 4 (19%), (D) rank 2 (1%), (E) rank 1 (<0.5%). The *E. coli* cells in (F) were expressing only *hlyA, C, D* but not *hlyB*. Therefore HlyA could not be secreted from these cells, and no hemolytic zone is observed. (Adapted with permission from Zhang, F., Sheps, J. A., and Ling, V. *J. Biol. Chem.* **268,** 19889 [1993].)

to be saturated, and when too short, the smaller zones are not detectable. A timer-controlled incubator is very useful.

The zone assay is very convenient for selecting complementing mutants. The sizes of the hemolytic zones on blood plates are examined and ranked by visual inspection. Comparisons are made between the halos of colonies of similar sizes. Rankings are defined relative to the other samples in a single experiment and are not regarded as absolute values. Colonies of cells that are missing any one of the five genes do not exhibit any hemolytic zone. We normally plate 300–400 colonies in a single 100-mm petri dish. This range allows colonies to grow to sufficient size and still produce non-overlapping hemolytic zones for screening.

Measurement of Secreted HlyA in Liquid Culture

JM83 cells harboring the desired plasmids are grown in liquid culture with the appropriate antibiotics and samples are taken at different points in the growth curve (A_{600}). Cells are removed by a brief centrifugation and 100 μl of supernatant, either undiluted or diluted with LB, is added to 100 μl hemolysis buffer (40 mM $CaCl_2$, 20 mM Tris-Cl, pH 7.4, 160 mM NaCl, and 4% sheep blood). We have found that a linear relationship between the amount of HlyA and hemolysis is only in reactions that yield less than 70% of complete blood lysis (complete blood lysis is defined as resulting from a 2% suspension of sheep blood in H_2O). Therefore, assaying of several dilutions of culture supernatants is necessary for testing wild-type HlyA and mutants of HlyA that are not severely defective in transport. We use 100 μl LB plus 100 μl hemolysis buffer as a blank. Samples are incubated for 30 min at 37°. After a brief centrifugation to remove unlysed blood cells, the lysate is diluted 20-fold with H_2O, and measured by absorbance at 420 nm. Hemolytic activity in the medium is plotted as a function of cell density (Fig. 5). For convenience, an optical density of $A_{600} = 0.7$ is chosen as an arbitrary point at which to compare the relative transport efficiencies of mutants. Hemolytic activities at this point are determined by interpolation. There can be significant day-to-day variations in the absolute hemolytic activity measured, and a wild-type control is always included for comparison. The hemolytic activity is calculated as a percentage of wild-type activity in each experiment.

Detection of Secreted HlyA Protein

JM83 cells harboring the desired plasmids are grown in liquid culture with the appropriate antibiotics and harvested at A_{600} of 1.0. Cells are removed by centrifugation. Supernatant (20 ml) is removed into a 30-ml Corex tube to which 3 ml 100% (w/v) trichloroacetic acid is added, and incubated on ice for >1 hr. Samples are centrifuged at 4°, 8000g for 15 min. The precipitated protein sample is dissolved in 50 μl 100 mM Tris-HCl, 500 mM NaOH and transferred to an Eppendorf tube. Then 50 μl H_2O is added to the Corex tube and transferred to the same Eppendorf tube. Next 50 μl of sample (equivalent to 10 ml culture supernatant) is loaded on an SDS–PAGE gel (3% stacking and 11% separating gel). Proteins are visualized by staining with Coomassie Brilliant Blue R-250. Wild-type HlyA migrates between the 97- and 116-kDa molecular mass markers (Fig. 6).

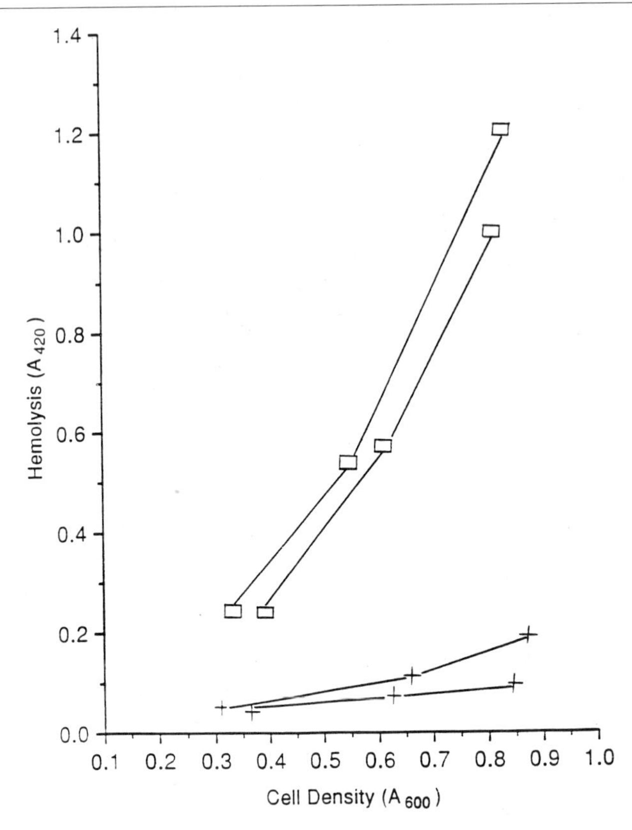

FIG. 5. Accumulation of secreted HlyA as a function of cell growth *E. coli* cells expressing *hlyB, C, D,* with the wild-type *hlyA* (□) or a mutant *hlyA* (+) are grown in LB medium. Cell growth is monitored by absorption at 600 nm. Samples are removed at different times for measurement of hemolytic activity in the medium. Hemolysis is determined by absorption at 420 nm. For each of the wild-type and mutant HlyA two different colonies are analyzed. (Adapted with permission from Zhang, F. Sheps, J. A., and Ling, V. *J. Biol. Chem.* **268,** 19889 [1993].)

For transport assays, certain controls need to be taken into consideration. To ensure that the hemolytic assays only reflect changes in transport activity, we check that mutations in HlyA do not affect their hemolytic activities. This is done by expressing wild-type *hlyA* or each HlyA mutant with *hlyC, D* without *hlyB* (therefore no secretion occurs), and measuring their hemolytic activity from cell lysates.[16] We also monitor the growth curve of cells harboring HlyB or HlyA variants to see whether they maintain similar levels of growth to the wild-type control.

[16] N. Mackman and I. B. Holland, *Mol. Gen. Genet.* **196,** 129 (1984).

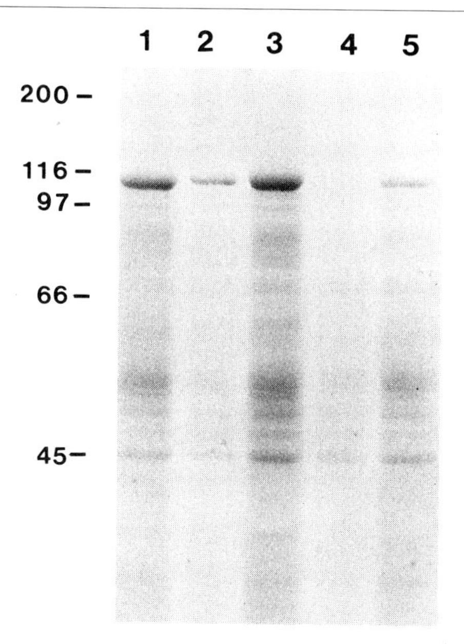

FIG. 6. An SDS–PAGE gel of secreted HlyA proteins. *Escherichia coli* cells expressing *hlyB, C, D* and wild-type *hlyA* or various *hlyA* mutants were grown in liquid culture and harvested at cell density A_{600} of 1. Culture supernatant samples were subjected to SDS–PAGE. The measurements by liquid hemolytic assay of the same clones are given in parenthesis. Lanes: 1, *hlyA* mutant (71%); 2, one-fifth the amount of lane 3; 3, wild-type *hlyA* (100%); 4, supernatant from cells expressing *hlyA, C, D,* but not *hlyB*; 5, *hlyA* mutant (19%).

Among these three transport assays, the zone assay is most sensitive for detecting changes at low levels of transport activity, but can only be quantified by visual ranking. The SDS–PAGE assay cannot detect secretions at less than 4% of the wild-type level. The liquid hemolytic assay is quantitative, but has more intrinsic variation and is more labor intensive. Nevertheless, in our hands, these three assays have generally been in agreement with each other. [Authors' addition: Recently we have found that an ELISA assay using an antiserum for HlyA is highly sensitive and quantitative for detecting the secretion of HlyA (Carla Morden, personal communications).]

Mutagenesis and Complementation

Site-directed mutagenesis is described in Refs. 12 and 14. Routine cloning procedures essentially follow Sambrook *et al.*[17]

[17] J. Sambrook, E. F. Fritsch, and T. Maniatis, "Molecular Cloning: A Laboratory Manual." Cold Spring Harbor Press, Cold Spring Harbor, New York, 1989.

Random Mutagenesis

The conditions used in the random mutagenesis procedure are designed to achieve a 5–10% rate of knockout mutations (defined as a colony with no hemolytic zone when the mutant HlyB is expressed with HlyA, C, and D). Plasmid DNA is incubated with 0.8 *M* hydroxylamine at 70° for 1 hr. A stock solution of hydroxylamine (2 *M* hydroxylamine, 0.1 *M* sodium pyrophosphate, 2 m*M* NaCl, pH 6.0) is prepared fresh each day. This is diluted 1.25-fold in 0.1× SSC and then 10 μl is added to 10 μl of DNA (2 μg/μl) in 0.1× SSC (1× SSC is 150 m*M* NaCl, 15 m*M* sodium citrate). After incubation the DNA is precipitated by the addition of 1.05 ml of stop buffer (70% ethanol, 0.1 *M* sodium acetate, 0.03× SSC) and dissolved in 50 μl TE (10 m*M* Tris-Cl, 1 m*M* Na_4 EDTA, pH 8.0). Transforming 1 μl of a 100-fold dilution of this mutated DNA stock into *E. coli* yields about 600 colonies. We have found that the transformation efficiency of this mutated DNA can be increased at least 10-fold after an additional NaI–glass beads cleaning up (Geneclean kit, Bio 101 Inc., Vista, CA).

Screening for Complementing Mutants

Randomly mutated plasmid DNA containing *hlyB* is transformed into the *E. coli* JM83 cells expressing *hlyC, D* and a mutant *hlyA*. It is advisable to have a control plate of cells transformed with wild-type *hlyB* as a reference while searching for mutants. Colonies that produce larger hemolytic zones are restreaked three times on blood agar plates (to ensure a consistent phenotype) before plasmid DNA is isolated. To exclude mutations in sequences outside of the *hlyB* gene, *hlyB* mutants are subcloned into new vectors that are distinguishable from the original vectors by a restriction site polymorphism.[12] The *hlyB* gene is sequenced with a set of 13 oligonucleotide primers which span the *hlyB* gene and flanking sequences.[18]

Results and Conclusions

Mutations in Signal Sequence of HlyA Resulting in Transport Deficiency

A complementation analysis involving the selection of compensatory HlyB mutations to single-point mutations in HlyA allows for the simplest interpretation of interactions between these two molecules. However, HlyB is able to tolerate many single amino acid changes in the HlyA signal. The most defective single-point mutation in HlyA that we have constructed

[18] P. Juranka, F. Zhang, J. Kulpa, J. Endicott, M. Blight, and I. B. Holland, *J. Biol. Chem.* **267,** 3764 (1992).

still maintains 30% of the wild-type transport efficiency.[14] However, other laboratories have reported individual point mutants causing defects of 50% and a combination of three point mutations reducing secretion to 0.6% of wild-type levels.[10,11,19] We have found that the zone assay is more sensitive for detecting changes in small hemolytic zones (ranking 1 to 3, or <10% wild-type transport, Fig. 4). Therefore, HlyA mutants with severe transport deficiencies are the ideal candidates for the screening of the complementing mutations in HlyB. We chose two nonoverlapping deletion mutants, HlyAcr-2 and HlyAαd (Fig. 7). In HlyAcr-2, the C-terminal 29 amino acids of the wild-type HlyA sequence are deleted and replaced by the 6-amino-acid sequence, DIDRCP. In addition, the phenylalanine in position -35 is substituted by leucine (this mutation alone has very little effect on transport efficiency). Transport activity of HlyAcr-2 is severely defective, less than 0.5% of wild-type. HlyAαd has an internal 29-amino-acid deletion at -58 to -30, which yields about 5% of wild-type transport activity.[14] Structural analysis of the HlyA signal peptide by NMR has indicated that two α helices are formed in a membrane mimetic environment. The first helix, located at -49 to -38, is missing in HlyAαd, and the second, at -27 to -17, is missing in HlyAcr-2 (Fig. 7).[13]

Compensatory Mutations in HlyB

Our initial random mutagenesis procedure was to grow an *E. coli* culture harboring the *hlyB* containing plasmid in the presence of *N*-methyl-*N'*-nitro-*N*-nitrosoguanidine (MNNG).[14] Using this method, six mutants have been selected that compensated for the mutation in HlyAcr-2. These mutants all contain single amino acid changes, which map to three positions on HlyB (Table I and Fig. 8). Multiple clones mutated at the same site were selected. Because the DNA was treated in a growing culture, it was difficult to distinguish whether the multiple isolates resulted from independent hits by the mutagen or were siblings from DNA replication. To circumvent this problem, we have adopted the *in vitro* mutagenesis procedure using hydroxylamine. The frequency of compensatory mutants is less than 10^{-4}. Among 3×10^5 colonies, 10 colonies have been found to exhibit larger zones. One clone lost the phenotype on subcloning and was not analyzed further. Sequencing the *hlyB* gene from the rest of the clones reveals that all contained single amino acid changes which map to a further three positions (Table I and Fig. 8). In this case, the selection of multiple independent clones yielding mutations at the same position suggests that mutagenesis is saturated.

[19] B. Kenny, C. Chervaux, and I. B. Holland, *Mol. Microbiol.* **11,** 99 (1994).

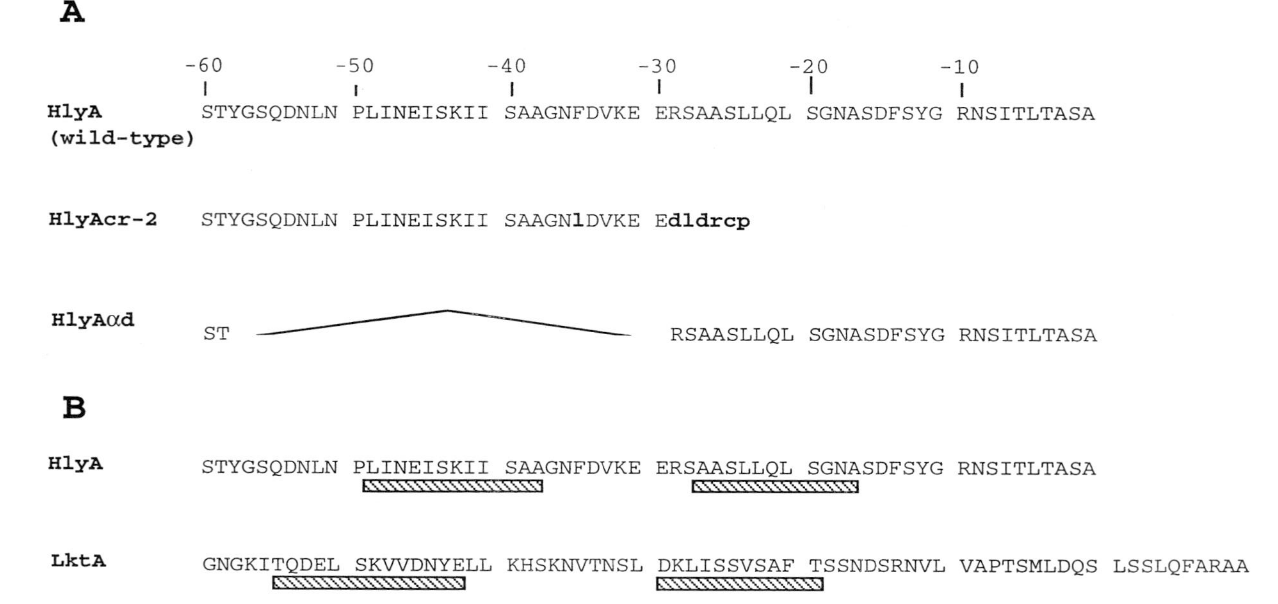

FIG. 7. (A) C-terminal sequences of HlyA and its mutants. The C-terminal 60 amino acids of HlyA are shown. The altered amino acids are printed in lowercase letters; ∧ indicates the region of amino acids that have been deleted. (B) The C-terminal signal sequences of HlyA and LktA. Hatched rectangles below the sequences indicate the positions of the putative α helices observed by nuclear magnetic resonance spectroscopy.[13]

TABLE I
HlyB Compensatory Mutations

HlyA mutant against which compensators are selected	Codon change	Amino acid change	Mutagen	No. of isolates	Increase in efficiency[a] (-fold)
HlyAcr-2	GCC→GTC	A146V	MNNG	3	4
	CTT→TTT	L158F	MNNG	1	12
	ACT→ATT	T251I	MNNG	2	2.5
	GAC→AAC	D259N	Hydroxylamine	4	18
	GCA→GTA	A269V	Hydroxylamine	1	10
	GAT→AAT	D433N	Hydroxylamine	4	4.6
HlyAαd	AGA→AAA	R212K	Hydroxylamine	1	3.7
	ACA→ATA	T220I	Hydroxylamine	1	1.5
	GCA→GTA	A269V	Hydroxylamine	1	3.5
	GCA→ACA	A269T	Hydroxylamine	1	3.2
	CTT→TTT	L427F	Hydroxylamine	1	2.7
	GGT→AGT	G445S	Hydroxylamine	1	2.4
	GTC→ATC	V599I	Hydroxylamine	1	2.5

[a] The increase-fold of efficiency in HlyB mutations as compared with wild-type HlyB is determined by the hemolytic assay of HlyA secreted into the growth medium.

With HlyAαd, nine colonies have been selected, which exhibit larger hemolytic zones. Seven of these have maintained the compensatory phenotype after subcloning. All are single amino acid changes, which map to six positions (Table I and Fig. 8). Two clones have mutations at alanine-269; one is changed to valine and the other to threonine. The change of alanine-269 to valine is also found in a HlyB compensator selected against HlyAcr-2 (Table I). One clone has a double mutation: a sense mutation (Gly-445→Ser) and one C-to-T transition in the upstream noncoding region. This isolate has been subcloned again to separate these two mutations, and the phenotype follows the sense mutation (Gly-445→Ser). Twelve out of 13 mutations are located in the transmembrane loops near the cytoplasmic face of the membrane (Fig. 8). This suggests that the site of interaction is located within the transmembrane region of HlyB. The compensatory mutations seem to cluster at a few sites and we speculate that these mutations are located in a substrate "binding pocket" in HlyB.

The ability to select allele-specific HlyB mutants that compensate for the transport deficiency caused by specific subsets of mutations in the HlyA signal sequence provides genetic evidence that HlyB interacts directly with the signal region of HlyA. The allele specificity of these HlyB mutants is tested by expressing them in combination with different HlyA signal mutants. A detailed listing of these experiments is found in Zhang *et al.*[14] and

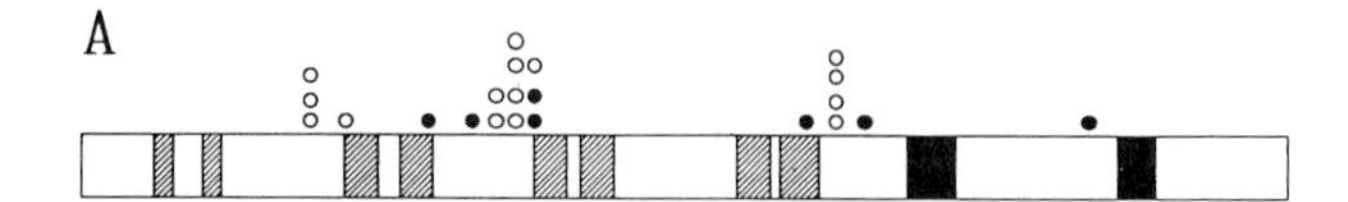

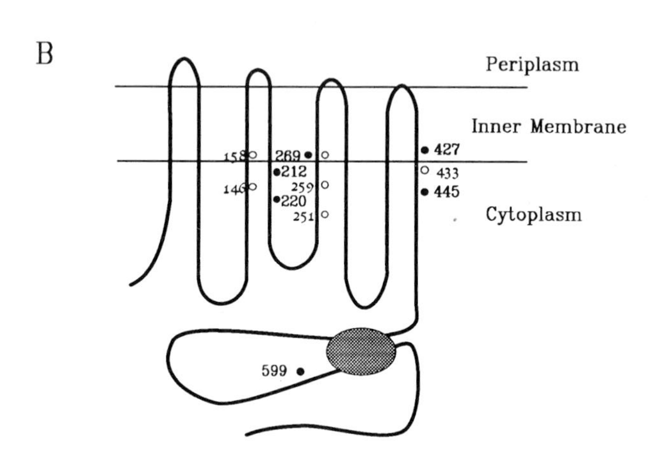

● Sites complementing N-terminal deletion in signal

○ Sites complementing C-terminal deletion in signal

FIG. 8. Location of compensatory mutations in HlyB. (A) Mutations in HlyB plotted on a linear representation of the protein. The *number of circles* at a location corresponds to the number of isolates of each genotype. Open and filled circles represent mutations selected for their ability to compensate for the transport defects of HlyAcr-2 and HlyAαd, respectively. Shaded boxes indicate transmembrane spans and black boxes represent the ATP-binding sites. (B) The topology of HlyB in the inner membrane of *E. coli* based on topological studies by Gentschev and Goebel.[3] The sites of amino acid changes in HlyB mutants are marked by open and filled circles as described in (A). The ATP-binding site is highlighted with a stippled circle. (Adapted with permission from Sheps J. A., Cheung, I., and Ling, V. *J. Biol. Chem.* **270,** 14829 [1995].)

Sheps *et al.*[20] In Table II we summarize some of the results. First, all the mutants have no detectable effect on transport of wild-type HlyA, and each mutant has a different efficiency profile in transporting the set of HlyA mutants.[14,20] This indicates that these mutations do not cause a nonspecific generalized induction of transport. The different transport efficiency pro-

[20] J. A. Sheps, I. Cheung, and V. Ling, *J. Biol. Chem.* **270,** 14829 (1995).

TABLE II
SUBSTRATE SPECIFICITY OF HLYB MUTANTS[a]

HlyA	Wt (%)	HlyB: HlyAcr-2 compensator: A146V	L158F	T251L	D259N	A269V	D433N
Wild-type	100	NC	NC	NC	NC	NC	NC
HlyAαd	6	↑	↑	↑	↑	↑	NC
HlyAcr-2	0.5	↑	↑	↑	↑	↑	↑
HylA/lkt70	100	NC	NC	NC	↓	NC	↓

HlyA	Wt (%)	HlyAαd compensator: R212K	T220I	A269V	A269T	L427F	G445S	V599I
Wild-type	100	NC	NC	NC	NC	NC	NC	NC
HlyAαd	6	↑	↑	↑	↑	↑	↑	↑
HlyAcr-2	0.5	↑	↑	↑	↑	↑	↑	NC
HlyA/ikt70	100	NC	NC	NC	NC	NC	NC	NC

[a] In each row the *E. coli* cells expressing *hlyC, D,* and an *hlyA* (as indicated in the leftmost column) and a *hlyB* mutant (as indicated at the top of each column) are plated on blood agar plates in triplicate. The size of hemolytic halos of colonies from each mutant is visually compared with that of the same *hlyA* expressed with wild-type *hlyB* in the same experiment. ↑, increase, ↓, decrease; NC, no change in the size of zones by comparison with wild-type HlyB. Transport efficiency of different HlyA mutants by wild-type HlyB is listed in the second column from the left, which is determined by the hemolytic assay of HlyA secreted in medium.

files of these mutants represent examples of single-point mutations in the HlyB transmembrane region causing a substrate specificity shift. The ease with which this shift is accomplished may reflect the mechanism by which the diverse substrate specificity of ABC transporters has evolved. Second, most HlyAcr-2 compensators can suppress the mutation of HlyAαd, and vice versa. This suggests that those amino acid changes in HlyB may cause some structural changes in the "binding pocket" that make transport easier to activate by particular mutant signal sequences. Third, two of the compensators have a decreased ability to transport HlyA/lkt70, and both of these have amino acids changed from Asp to Asn (from negatively charged to neutral).

This third observation leads us to suggest the following model. The "binding pocket" of HlyB contains multiple residues that can be "triggered" by close contact with the signal sequence to initiate transport (Fig. 1). As mentioned earlier HlyA/lkt70 can be transported by HlyB/D with efficiency equal to wild-type HlyA, and both signal sequences from similar secondary structures (Fig. 7). The initial recognition between HlyB and a signal sequence may be primarily dependent on the shape of the signal. HlyA and LktA signal sequences can fit into the "binding pocket" of HlyB because they both have an appropriate higher order structure containing two helices. The subsequent step may involve close contacts between specific amino acid residues in the binding pocket of HlyB and the signal sequence. These two signal sequences likely interact with different amino acids, since they have different primary sequences. The aspartic acid residues at positions 259 and 433 in HlyB may interact with the LktA signal but not the HlyA signal. Therefore, changes in these two positions decrease the transport efficiency of HlyA/lkt70 but have no effect on HlyA. This model provides a guide for further site-directed mutagenesis to locate precisely the amino acid interactions between HlyB and LktA signal sequence.

The hypothesis that HlyB substrate recognition involves a "pocket" containing multiple contact sites may be a model that is generally applicable to substrate recognition by ABC transporters. Different substrates may interact with different contact points to trigger the subsequent transport process, and this may explain the broad substrate specificity that characterizes many ABC transporters. The study of complementing mutations has provided an approach that can be used for future studies to understand the substrate specificity of ABC transporters.

Acknowledgments

We wish to thank Dr. Sarah Childs and Carla Morden for reading earlier versions of this chapter. This work was supported by the Medical Research Council of Canada. J. A. Sheps was supported by a graduate studentship from the Natural Sciences and Engineering Research Council of Canada.

[6] *Erwinia* Metalloprotease Permease: Aspects of Secretion Pathway and Secretion Functions

By PHILIPPE DELEPELAIRE

Introduction

Erwinia chrysanthemi is a phytopathogenic gram-negative enterobacterium that causes soft rot disease among various plants.[1] It secretes several hydrolytic enzymes among which pectinases and cellulases are the main constituents of its pathogenicity. Pectinases and cellulases are secreted by the general secretory pathway found in most gram-negative bacteria and allow proteins made as precursors with an N-terminal signal peptide to first cross the cytoplasmic membrane via the *sec* pathway and the outer membrane by a specific machinery.[2,3] By employing an adenosine triphosphate-binding cassette (ABC) transporter *E. chrysanthemi* also secretes several metalloproteases whose function in the pathogenic power has not been demonstrated.[4] This article reviews our current knowledge of the *E. chrysanthemi* metalloprotease secretion pathway and includes practical aspects (1) of different techniques used to study this pathway and (2) for the characterization of the secretion functions of this transporter with an emphasis on the ABC protein. All the data presented here are relevant to the metalloprotease secretion system from *E. chrysanthemi* reconstituted in *Escherichia coli.*

Protein secretion by the ABC pathway in gram-negative bacteria is widespread in many species,[3,5,6] allowing the secretion of several classes of proteins (toxin class exemplified by the *E. coli* α hemolysin, the protease class, the lipase class, and the iron-regulated class) and presents several distinctive features. The secretion apparatus is always made up of three proteins, the ABC protein itself being made of the ABC domain fused to a hydrophobic membrane domain and believed to function as a dimer and localized in the cytoplasmic membrane, another cytoplasmic membrane

[1] A. K. Chatterjee and M. P. Starr, *Ann. Rev. Microbiol.* **34,** 645 (1980).
[2] A. P. Pugsley, *Microbiol. Rev.* **57,** 50 (1993).
[3] C. Wandersman, *in* "*Escherichia coli* and *Salmonella*" (F. C. Neidhardt, ed.), 2nd Ed. ASM Press, Washington, DC, 1996.
[4] G. S. Dahler, F. Barras, and N. T. Keen, *J. Bacteriol.* **172,** 5803 (1990).
[5] M. J. Fath and R. Kolter, *Microbiol. Rev.* **57,** 995 (1993).
[6] C. Wandersman, *Trends Genet.* **8,** 317 (1992).

METHODS IN ENZYMOLOGY, VOL. 292
0076-6879/98 $25.00

protein belonging to the membrane fusion protein (MFP) family,[7] so-called because it is thought to establish specific contacts between the inner and the outer membrane, with one transmembrane segment and a large periplasmic domain and an outer membrane protein. These transporters are specific for their substrates or class of substrates and do not transport or transport very inefficiently substrates of another class.[3] All three proteins are required for secretion to occur. The secreted protein does not have a classical N-terminal hydrophobic signal peptide. Instead it presents in most cases an uncleaved C-terminal secretion signal, which is necessary and sufficient to allow secretion by the secretion apparatus.[8] This C-terminal secretion signal can be fused to the passenger protein and in some cases allows its secretion. In contrast to what is observed for the general secretion pathway, the periplasmic secretion intermediate has not been detected for the ABC secretion pathway, and whenever one of the secretion proteins is missing, the exoprotein is not secreted and does not accumulate in the periplasm.[3] The energetics of the secretion has been studied in only the case of the α hemolysin from *E. coli* and both adenosine triphosphate (ATP) and $\Delta\mu_{H}$ play a role in the secretion process.[9] The genetic organization in most cases associates the structural genes for the secretion functions with the structural genes for the secreted protein.

Metalloproteases are divided into several classes and those secreted by an ABC-dependent pathway by gram-negative bacteria belong to the serralysin subclass[10]; they are present in several species including *Pseudomonas aeruginosa* (alkaline protease),[11] *Serratia marcescens* (metalloprotease),[12] *Proteus mirabilis* (IgA_1, A_2, G protease),[13] and *E. chrysanthemi*. In the case of the *E. chrysanthemi* B374, four distinct metalloproteases are secreted by an ABC transporter, as has been shown by the functional reconstitution of the metalloprotease secretion pathway in *E. coli.*[14–16] The genetic organization of the gene cluster allowing the synthesis and secretion in *E. coli* of the four *E. chrysanthemi* metalloproteases is shown Fig. 1: *prtG*, which encodes the metalloprotease G, is the first gene of an operon

[7] T. Dinh, I. T. Paulsen, and M. H. Jr. Saier, *J. Bacteriol.* **176,** 3825 (1994).

[8] L. Gray, K. Baker, B. Kenny, N. Mackman, R. Haigh, and I. B. Holland, *J. Cell Sci. Suppl.* **11,** 45 (1989).

[9] V. Koronakis, C. Hughes, and E. Koronakis, *EMBO J.* **10,** 3263 (1991).

[10] W. Stöcker and W. Bode, *Curr. Opin. Struct. Biol.* **5,** 383 (1995).

[11] J. Guzzo, F. Duong, C. Wandersman, M. Murgier, and A. Lazdunski, *Mol. Microbiol.* **5,** 447 (1991).

[12] S. Létoffé, P. Delepelaire, and C. Wandersman, *J. Bacteriol.* **173,** 2160 (1991).

[13] C. Wassif, D. Cheek, and R. Belas, *J. Bacteriol.* **177,** 5790 (1995).

[14] C. Wandersman, P. Delepelaire, S. Létoffé, and M. Schwartz, *J. Bacteriol.* **169,** 5046 (1987).

[15] J. M. Ghigo and C. Wandersman, *Res. Microbiol.* **143,** 857 (1992).

[16] J. M. Chigo and C. Wandersman, *Mol. Gen. Genet.* **236,** 135 (1992).

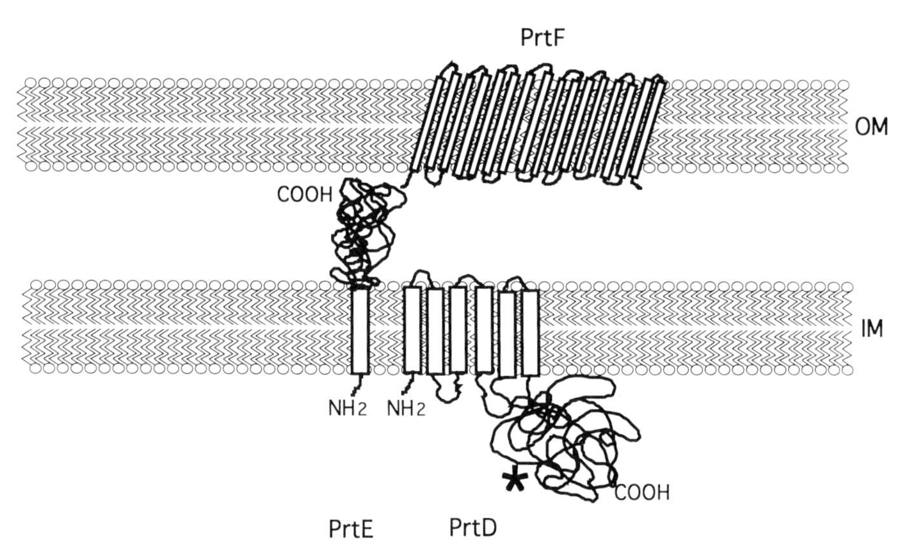

FIG. 1. *Upper*: Genetic organization of the cluster allowing the synthesis and secretion of metalloproteases A, B, C and G from *E. chrysanthemi* B374; the arrows correspond to distinct transcription units; the size of the region is ca. 12 kb (see text for details). *Lower*: Schematic representation of the topology of the three secretion functions from the *E. chrysanthemi* metalloprotease transporter; the number and the orientation of the transmembrane segments of PrtF are arbitrary, as is the lipid representation of the outer membrane. OM, outer membrane; IM, inner membrane; *, the ATP-binding site of PrtD.

that also contains *inh* (coding for a specific metalloprotease inhibitor[17] not required for secretion) and the three secretion function structural genes *prtD, prtE,* and *prtF,* which respectively encode the ABC component, the MFP component, and the outer membrane component[18]; three independent transcription units *prtB, prtC,* and *prtA,* respectively, encode each of the metalloproteases B, C, and A. Most of our studies on metalloprotease secretion have been carried out in an *E. coli* C600 recombinant strain expressing the secretion functions from a low copy number plasmid and one of the secreted metalloproteases under the control of *lac* promoter on a compatible high copy number plasmid.

[17] S. Létoffé, P. Delepelaire, and C. Wandersman, *Mol. Microbiol.* **3,** 79 (1989).
[18] S. Létoffé, P. Delepelaire, and C. Wandersman, *EMBO J.* **9,** 1375 (1990).

The PrtA, B, C, and G proteases (ca. 450 amino acids long) are highly homologous (50% identity, PrtB and C being even more similar) and homologous to the above-mentioned proteases from *P. aeruginosa, S. marcescens,* and *P. mirabilis,* which are secreted by similar pathways. They are secreted as zymogens in the external medium where activation occurs by the autocatalytic cleavage of a N-terminal propeptide, 8 to 17 amino acids long. Analysis of the crystal structure of two of them, the alkaline protease from *P. aeruginosa*[19] and the metalloprotease from *S. marcescens*,[20] has shown that they contain three structural domains that correspond to functional ones: The N-terminal part is the catalytic domain; it is followed by a glycine-rich repeat domain folded into a β-roll and to which are tightly bound calcium ions; and finally the C-terminal domain, which contains the secretion signal. The glycine-rich domain, which contain repeats of the GGXGXDXXX motif, is found in almost all the proteins secreted by such a pathway and it is not known when and where calcium binding occurs, whether in the cytoplasm or in the external medium; the folded structure of this domain is very tight, stabilized by the calcium ions that are not exchangeable in the mature protein. It might play a role in the presentation of the secretion signal to the secretion apparatus and in the activation of the zymogen.[19]

Detection of Protease Activity

The study of a secretion pathway requires (1) detection of the secreted protein in the various cellular compartments and (2) assessment of the specificity of the secretion and of the absence of cell lysis. Protease activity can be assayed by different ways either directly around the bacterial colonies, in the culture supernatant, or after electrophoresis of supernatant.

Skim Milk Agar Plates for Bacterial Colonies

Autoclave separately (20 min, 120°):

1 vol of 2% skim milk (Difco) in 10 m*M* Tris-HCl, pH 7.5

1 vol of 2% Tryptone, 3% agar in 10 m*M* Tris-HCl, pH 7.5

Once the solutions have cooled to 50–55°, mix them, add the appropriate antibiotics, and pour the plates. After bacterial growth, proteolytic colonies are surrounded by casein hydrolysis halos. This is a low ionic strength medium; the skim milk concentration can be raised from 2–4%, resulting in clearer halos around the colonies.

[19] U. Baumann, S. Wu, K. M. Flaherty, and D. B. McKay, *EMBO J.* **12,** 3357 (1993).

[20] U. Baumann, *J. Mol. Biol.* **242,** 244 (1994).

Cup–Plate Assay

Autoclave separately:

1 vol of 2% skim milk (Difco) in 100 m*M* Tris-HCl, pH 7.5

1 vol of 3% agar noble (Difco) in 100 m*M* Tris-HCl, pH 7.5

Once the solutions have cooled to 50–55°, mix them and pour the plates. Make wells with a Pasteur pipette (10–15 wells can easily be made in one plate). Culture supernatants or cellular fractions can be assayed for proteolytic activity or for the presence of inhibitory activity by putting them in the wells and incubating the plates at 37° from one to several hours, depending on the specific activity of the proteases. This is a very versatile and sensitive assay that can be used for semiquantitative determination of proteolytic activity or inhibitory activity by making serial dilutions of the solutions (from two by two, for example) and measuring the size of the cleared halos around the wells. Care should be taken always to include a control solution because some chemicals give clearing halos (ammonium sulfate for example).

Assay of Protease Activity after Gel Electrophoresis

Proteolytic activity of the *E. chrysanthemi* metalloproteases can also be assayed directly after sodium dodecyl sulfate–polyacril amide gel electrophoresis (SDS–PAGE) of proteins from culture supernatants precipitated with trichloracetic acid on gels containing copolymerized gelatin[21] (0.1% in the resolving gel of an usual Laemmli gel). After electrophoresis at 4°, the gel is incubated in 2.5% Triton X-100 for 1 hr at room temperature with gentle shaking, during which SDS diffuses out and proteases are renatured, then for 3–4 hr at 37° in 0.1 *M* glycin–NaOH, pH 8.3, 1 m*M* $MgCl_2$, 1 m*M* $CaCl_2$ during which the proteases hydrolyze the gelatin. The gel is then stained with 0.1% Amido black (in 10% acetic acid, 40% methanol, 50% water) and destained as usual. Proteolytic bands are white on a blue background. Alternatively they can be revealed after electrophoresis either on SDS–PAGE or on native gels by overlaying the gel on an agar gel cast in a 2-mm-thick gel mold of similar composition to the one used for cup–plate and incubating the two gels in contact at 37° for several hours. Proteins diffuse from the acrylamide gel to the agar gel and eventually renature and degrade the milk casein, giving rise to clearing zones.

Secretion Specificity

Although all of these techniques are useful for detecting proteolytic activities (or their inhibitors) in various cellular compartments and in the

[21] C. Heussen and E. B. Dowdle, *Anal. Biochem.* **102,** 196 (1980).

extracellular medium, they should be completed by the assessment of the specificity of the secretion and the absence of cell lysis, which can only be checked by measuring bona fide markers of the different cell compartments, like β-galactoside[22] for the cytoplasmic compartment and β-lactamase[23] for the periplasmic one, provided they are expressed in the strain under study. Alternatively, it can also be checked by the absence in the stained gel of a supernatant of bands other than those of the secreted proteins; this essentially applies when exoproteins are secreted at a high level. Also keep in mind that secreted proteases might degrade other proteins in the supernatant and thus lead to an underestimation of the cell lysis. Cells in exponential growth phase should be used to minimize perturbation by lysis.

Metalloproteases from *E. chrysanthemi* are easily detected by PAGE of the culture supernatants from different strains. They have been purified from the culture supernatant and used to obtain antibodies from rabbits.

Secretion Signal

All four metalloproteases contain a C-terminal secretion signal, that can function as an autonomous secretion signal, the shortest one being 30 amino acids long, and as a secretion signal for passenger proteins. The longer the passenger protein is, the less efficient the secretion[24]; in this case, the glycine-rich repeat domain enhances the secretion of some passenger proteins, possibly by allowing a better presentation of the secretion signal to the secretion machinery. Detailed studies on the C-terminal secretion signal of protease G[25] have emphasized the role of the last four amino acids of the signal (conserved in all proteases and consisting of an aspartate residue followed by three hydrophobic ones, the last one being a valine), which has to be exposed to the C terminus to be functional. The PrtG protease deleted of its last 48 residues and to which are fused its last 15 residues at its C terminus is efficiently secreted by the secretion functions. Structural studies on the purified C-terminal secretion signal of protease G have shown that these last residues do not adopt a preferential conformation and that two α helices induced in an apolar environment exist in front of this motif,[26] the functional significance of which has not been ascertained.

[22] J. H. Miller, "A Short Course in Bacterial Genetics." Cold Spring Harbor Laboratory Press, Cold Spring Harbor, New York, 1992.

[23] R. Labia and M. Barthelemy, *Ann. Inst. Pasteur Microbiol.* **130B,** 295 (1979).

[24] S. Létoffé and C. Wandersman, *J. Bacteriol.* **174,** 4920 (1992).

[25] J. M. Ghigo and C. Wandersman, *J. Biol. Chem.* **269,** 8979 (1994).

[26] N. Wolff, J. M. Ghigo, P. Delepelaire, C. Wandersman, and M. Delepierre, *Biochemistry* **33,** 6792 (1994).

Secretion Functions

The identification of the secretion functions first came from complementation studies that identified the structural genes for two of the proteases, *prtB* and *prtC*, on the one hand and functions required for their secretion on the other.[14] Sequencing studies then identified three components in the gene cluster involved in protease secretion.[18] This was followed by the identification of mutants in each of these components or subcloning, which established that each of these components is strictly required for secretion to occur.

Use of Antibodies against Secretion Functions

The determination of the sequence of the secretion functions and of the proteases allows the construction of chimeric proteins between the first amino acids from β-galactosidase, a part of each of the secretion functions, and the C-terminal end of one of the proteases, PrtB which are subsequently used to immunize rabbits.[27] Potential membrane-spanning segments are avoided in these constructions and care is taken that the fusion proteins have molecular weights different from those of the secretion proteins to avoid copurification on gel of proteins having the same molecular weight. The fusion protein aggregates inside the cell and this offers a convenient way to purify them.

An *E. coli* C600 strain carrying a recombinant plasmid expressing the fusion protein under the control of the *lac* promoter is grown at 37° up to an $OD_{600\ nm}$ of 1 in 100 ml of LB medium, harvested by centrifugation and washed once in 100 m*M* Tris-HCl, pH 8.0, 1 m*M* EDTA, 1 m*M* phenylmethylsulfonyl fluoride (PMSF), 5 m*M* ε-aminocaproic acid, and resuspended in 1 ml of the same buffer; they are sonicated (Branson sonifier, small probe, power 5–6, using several 10-sec strokes separated by cooling of the sample); the insoluble material is collected by centrifugation (30 min, Eppendorf centrifuge, 4°) and resuspended in 0.5 ml of the same buffer to which is added 0.5 ml of the same buffer containing 8 *M* urea. After incubation for 30 min at 4°, the insoluble material is removed by centrifugation (30 min, Eppendorf centrifuge, 4°) and the supernatant precipitated with 10% trichloroacetic acid (TCA) (1 h, 4°). The pellet is collected by centrifugation (10 min in an Eppendorf), washed with 80% acetone, and dissolved in gel sample buffer. Two cycles of preparative gel electrophoresis are carried out to purify the fusion protein, which is identified with antiprotease antibodies by Western blotting. The fusion protein is then used to immunize rabbits. The equivalent of 100 ml of a cell culture is sufficient

[27] P. Delepelaire and C. Wandersman, *Mol. Microbiol.* **5,** 2427 (1991).

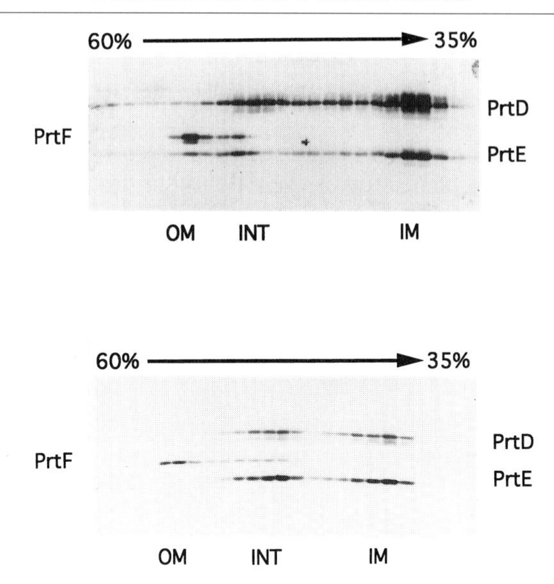

FIG. 2. Analysis of French press-derived membrane vesicles after centrifugation on a sucrose density gradient (35–60% sucrose, w/w) from a recombinant *E. coli* C600 strain expressing the three secretion functions PrtD, PrtE, and PrtF. The various fractions collected from the gradient were run on SDS–PAGE and tested for the presence of the PrtD, E, and F proteins by immunoblotting using the anti-PrtD, anti-PrtE, and anti-PrtF antibodies. Only the relevant part of the gel is shown. OM, Outer membrane vesicles fractions; INT, intermediate-density vesicles fractions; IM, inner membrane vesicles fractions. Two immunoblots are shown to illustrate the variation of the profiles observed from one experiment to another.

for one injection. With this method we obtain antibodies directed against each of the secretion functions of the metalloprotease transporter, PrtD, PrtE, and PrtF.[27]

In a recombinant *E. coli* C600 strain expressing the secretion functions, analysis of the membrane fraction after centrifugation at equilibrium on a sucrose density gradient, following the technique of Ishidate *et al.,*[28] shows that PrtD (60 kDa) and PrtE (50 kDa) are inner membrane proteins, whereas PrtF 56 kDa) is an outer membrane protein, and that all three proteins are also found in intermediate-density vesicles (Fig. 2). This is slightly different from what we have previously observed[27] and might come from the difference of the fractionation methods. The sequence data together with the analysis of protease accessibility of the secretion functions on whole cells or spheroplasts or inside-out inner membrane vesicles has

[28] K. Ishidate, E. S. Creeger, J. Zrike, S. Deb, B. Glauner, T. J. MacAlister, and L. I. Rothfield, *J. Biol. Chem.* **261,** 428 (1986).

allowed us to propose a topology for these three proteins. Prt D has six potential membrane-spanning segments in its N-terminal part and a large cytoplasmic domain containing the consensus Walker A and B box motifs involved in nucleotide binding; it is not accessible from the periplasmic side with trypsin or proteinase K. PrtE has one potential transmembrane segment at its N terminus and protease accessibility data show it has a short segment in the cytoplasm and most of the sequence in the periplasm. Sequence analysis of PrtE shows it has a high propensity to adopt a coiled structure in the central part of its periplasmic domain. PrtF is an outer membrane protein with either one or the two extremities exposed in the periplasm (see Fig. 1).

In Vivo Studies

The precise role of the secretion functions is not known at the present time; however, the study of another bacterial ABC secretion system, that of HasA from *S. marcescens*, has provided important information regarding this point. HasA is a heme-binding protein secreted by *S. marcescens* that allows it to grow on heme or hemoglobin as iron source[29]; it is secreted by an ABC pathway that has been reconstituted in *E. coli* comprising HasD, the ABC protein, HasE, the MFP component, and the resident *E. coli* TolC protein fulfills the role of the outer membrane component.[30] HasA presents a C-terminal secretion signal and has no glycine-rich repeat region. The HasA secretion system allows secretion of both HasA and metalloproteases from either *S. marcescens* or *E. chrysanthemi*; the Prt secretion system only allows the secretion of proteases; furthermore, HasA, coexpressed with the metalloprotease transporter inhibits protease secretion. Exchange of the components of the Prt and Has transporters has allowed the construction of functional hybrid transporters and shows that, *in vivo*, the ABC protein recognizes the substrate and that specific interactions exist between the MFP component and the outer membrane component.[31]

Characterization of PrtD, the ABC Protein of *E. chrysanthemi* Metalloprotease Transporter

Whereas there are a number of questions that can be addressed *in vivo* by combining molecular genetics and physiological studies, and since the availability of a successful reconstituted *in vitro* secretion system is hampered by the apparent coupling of the two membranes in this secretion

[29] S. Létoffé, J. M. Ghigo, and C. Wandersman, *Proc. Natl. Acad. Sci. USA* **91,** 9876 (1994).
[30] S. Létoffé, J. M. Ghigo, and C. Wandersman, *J. Bacteriol.* **176,** 5372 (1994).
[31] R. Binet and C. Wandersman, *EMBO J.* **14,** 2298 (1995).

system, the availability of the purified components might permit answers to some questions. In particular, the question of the ATPase activity of the ABC component could more easily be studied with either membrane vesicles or purified ABC protein. We have obtained membrane vesicles highly enriched in PrtD from an overproducing strain, partially purified it from these vesicles, and carried preliminary experiments to reconstitute it into phospholipid vesicles.

Overexpression of PrtD; Site-Directed Mutants of Conserved Lysine of Walker A Domain

A high-copy-number plasmid pPrtD/pBGS was constructed that leads to a final overproduction factor of PrtD of about 10 as compared to expression on a low-copy-number plasmid under the control of a *tet* promoter. Three variants of the conserved lysine of the Walker A box were constructed by site-directed mutagenesis, giving rise to three mutant proteins PrtD/K370R, PrtD/K370T, and PrtD/K370E. The details of the cloning and the reconstitution of the mutated gene have already been published for one of the mutants,[32] the only difference being the mutagenic oligonucleotide for the other two. In all cases the mutated protein was made and inserted in the membrane, although the expression level progressively decreased in the order lysine arginine, threonine, and glutamate. The PrtD/K370T and PrtD/K370E mutant proteins are synthesized at a level comparable to that of the wild-type PrtD on a low-copy-number plasmid. In terms of secretion efficiency, the PrtD/K370R variant promotes secretion at 5–10% of the wild-type level, whereas the PrtD/K370T variant is at 1% of the wild-type level; secretion is completely abolished in the PrtD/K370E mutant. To correlate the effect of the mutation on the secretion with a possible effect on the ATPase activity, it was necessary to be able to measure the PrtD-associated ATPase activity, either using vesicles or purified protein. Due to their lower expression level, membrane vesicles from a PrtD/K370T or PrtD/370E mutant strain were not isolated.

Cytoplasmic Membrane Vesicle Purification

Membrane vesicles have been purified from *E. coli* C600 strain harboring different recombinant plasmids grown at 30°. Temperature of growth affects the final concentration of PrtD in the membrane preparation of the overproducing strain, the relative PrtD amount being greater at 30° than at 37°. No attempt has been made to change other culture parameters, like recipient strain (we always use C600), temperature (below 30°), and medium (we

[32] P. Delepelaire, *J. Biol. Chem.* **269,** 27952 (1994).

always use LB medium). For membrane vesicle isolation, C600 (pPRTD/pBGS) is grown in 2-liter Fernbach flasks containing 500 ml of LB medium up to an OD_{600} of 1–1.5 at 30° with gentle agitation. All subsequent operations are carried out at 4°. Cells are harvested by centrifugation (5000 rpm for 15 min), washed once in 10 m*M* HEPES–NaOH, pH 7.5, and resuspended in the same buffer at 50 $OD_{600\ nm}$ units/ml. They are broken by one passage through a French press cell operated at 10,000 psi. DNase and RNase are added to final concentrations of 10 μg/ml each and incubated, with occasional agitation, with the lysate until viscosity is greatly reduced (10–15 min). EDTA (0.5 *M*, pH 8.0) is then added at a final concentration of 1 m*M*. The lysate (30 ml for each centrifuge tube) is loaded on a two-step sucrose cushion [4 ml of 42.5% sucrose (w/w) in 10 m*M* HEPES–NaOH, pH 7.5, 5 m*M* EDTA, 2 ml of 30% sucrose (w/w) in the same buffer] in 40-ml polyallomer tubes. After centrifugation (4 hr, 27,000 rpm, SW28 rotor, 6°) the upper layer is aspirated up to the 42.5% layer and the brownish cytoplasmic membrane interface collected, diluted five times with 100 m*M* Tris-HCl, pH 8.0, 1 m*M* EDTA, and centrifuged for 75 min at 20,000 rpm in a SS34 rotor at 4°. At this stage the membranes are resuspended and carefully dispersed in the minimal volume of 100 m*M* Tris-HCl, pH 8.0, 1 m*M* EDTA, and frozen in liquid nitrogen. After thawing, the membranes are resuspended in 0.1 *M* Na_2CO_3, 10 m*M* dithiothreitol (DTT), and incubated for 45 min at 4° (25 ml for 1500 OD_{600} units initial material). This treatment removes most of the extrinsic membrane proteins. The membrane vesicles are then centrifuged 75 min at 20,000 rpm in the SS34 rotor, washed once in 100 m*M* Tris-HCl, pH 8.0, 1 m*M* EDTA, and finally resuspended in 50 m*M* HEPES–NaOH, pH 7.5, 1 m*M* $MgCl_2$, 10 m*M* DTT. (DTT is omitted from the membranes used for ATPase activity measurements.) The protein concentration is measured by using the Bradford reagent from Bio-Rad (Richmond, CA) with bovine serum albumin (BSA) as a standard and at this stage the yield is around 1.5–2.0 mg protein/1000 OD_{600} units of starting material. The membrane vesicles are then frozen in liquid nitrogen in 2.5- to 3-mg protein aliquots (200–300 μl) and kept at −80° until use. The same protocol is applied for a control strain not expressing PrtD and for a strain expressing the mutated PrtD/K370R (Fig. 3).

PrtD Purification

Membrane vesicles (2.5–3 mg protein) are solubilized in 1 ml final volume of 50 m*M* HEPES–NaOH, pH 7.5, 1 m*M* $MgCl_2$, 10 m*M* DTT, 20% glycerol, 100 m*M* NaCl, 0.6% dodecyl-β-D-maltoside (Calbiochem, La Jolla, CA) at 4° for 1 hr. There is a compromise between efficiency of solubilization and subsequent binding of PrtD onto the phosphocellulose

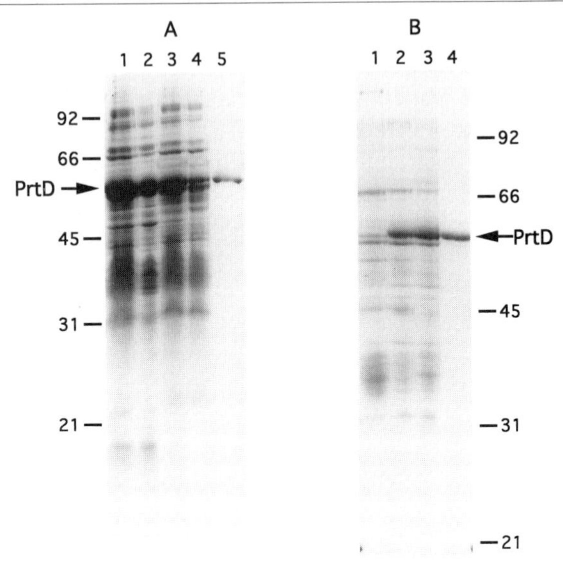

FIG. 3. (A) SDS–PAGE of the various fractions obtained during purification of PrtD. Lane 1, total membrane vesicles (80 μg); lane 2, insoluble material after solubilization; lane 3, soluble material after solubilization; lane 4, unadsorbed fraction after phosphocellulose column; lane 5, adsorbed fraction from the phosphocellulose. (B) SDS–PAGE of the inner membrane vesicles purified either from the control strain (lane 1), from the strain expressing the PrtD/K370R mutant (lane 2), or the strain expressing the wild-type PrtD protein (lane 3). In each case 20 μg of proteins were loaded per lane. Lane 4, 5 μg of purified PrtD protein. The scales on the right-hand (A) and left-hand (B) side indicate position of molecular mass markers in kilodaltons.

with respect to the contaminants; 0.6–0.8% detergent is found to be a convenient concentration, although the PrtD solubilization only reached 50% as judged from staining on gel (Fig. 3). After solubilization, the mixture is centrifuged for 1 hr at 4° at 40,000 rpm in a Ti 50 rotor, and the supernatant containing the solubilized proteins is brought to 1 m*M* EDTA (2 μl of a 0.5 *M* EDTA, pH 8.0, stock solution). This increases the adsorption of PrtD on the phosphocellulose column. The supernatant is loaded at 1 ml/hr on a 2-ml phosphocellulose column (phosphocellulose P11 from Whatman first equilibrated with 10 ml 0.5 *M* HEPES–NaOH, pH 7.5, and then with 10 ml 50 m*M* HEPES–NaOH, pH 7.5, 10 m*M* DTT, 20% glycerol, 0.02% dodecyl-β-D-maltoside, 100 m*M* NaCl). The column is then washed with 10 ml of the equilibration buffer then with 10 ml of the same buffer containing 0.3 *M* NaCl and 0.5 m*M* $MgCl_2$ and no DTT at 5 ml/hr. The bound proteins are then eluted with the same buffer containing 0.8 *M* NaCl, 2 m*M* $MgCl_2$ at 2 ml/hr and 0.3 ml fractions are collected. The PrtD-enriched fractions, which represent the bound fraction, are aliquoted into 20–30 μl, frozen into liquid nitrogen, and kept at −80° until use.

A representative gel of an usual purification is shown in Fig. 3. The yield is usually 200 μg of partially purified PrtD protein from 3 mg starting material. PrtD could be reconstituted into phospholipid vesicles according to the procedure of Davidson and Nikaido[33] for the MalFGK2 complex involving detergent exchange and removal in a second step.

Measuring ATPase Activity

ATPase activity is measured by using tritiated ATP (Amersham, [2,5′,8-^{3}H]ATP, ca. 40 Ci/mmol). ATP is separated from ADP by thin-layer chromatography on PEI-cellulose plates (Schleicher & Schuell, Keane, NH) followed by liquid scintillation counting.[34] For membrane vesicles, the reaction mixture is 50 m*M* HEPES–NaOH, pH 7.5, 2 m*M* $MgCl_2$, 20% glycerol, 50 m*M* NaCl, 10 m*M* NaN_3 to inhibit the residual F_1-ATPase activity. The protein concentration is 10 μg/ml, the reaction volume 20 μl. ATP [a mixture of cold and tritiated ATP (9 : 1) in relative concentration] is added at a final concentration of 10 μ*M* and the reaction is allowed to proceed for 1 hr at 25°, during which it is linear with time; a 3-μl aliquot is then withdrawn and mixed with an equal volume of 5 m*M* ATP, 5 m*M* ADP, 50 m*M* EDTA; 1.7–2 μl are spotted on a polyethyleneimine (PEI)-cellulose plate. The chromatography is developed with LiCl/formic acid and after chromatography, the ATP and ADP spots are revealed with a short-wavelength UV lamp, cut with scissors, and counted with 8 ml of scintillation fluid. In the case of the purified protein the protocol is similar, except that the reaction mixture is 50 m*M* HEPES–NaOH, pH 7.5, 2 m*M* $MgCl_2$, 20% glycerol, 50 m*M* NaCl, 0.015% dodecyl-β-D-maltoside, and 0.001% *E. coli* phospholipids (Avanti Polar Lipids Birmingham, AL). Protein concentration is 5–10 μg/ml and ATP concentration is also 10 μ*M*.

ATPase Activity of Various Fractions

The major problem of our previous study[32] was the absence of a reliable assay of PrtD activity in the starting membrane vesicles; hence the observed ATPase activity of the purified protein and its inhibition by the C-terminal secretion signal of one of the metalloproteases was subject to some caution as well as the meaning of this inhibition, which could have been due to detergent effect on the protein. The preparation of membrane vesicles highly enriched in PrtD as the starting material for the purification allowed

[33] A. L. Davidson and H. Nikaido, *J. Biol. Chem.* **266,** 8946 (1991).

[34] E. Richet and O. Raibaud, *EMBO J.* **8,** 981 (1989).

TABLE I
ATPase Activity of Various Preparations[a]

Preparation	Control vesicles	PrtD/K370R vesicles	PrtD vesicles	PrtD purified	PrtD reconstituted
No addition	840	1230	2250	1980	3130
100 μM VO_3	440	580	520	280	1030
10 mM VO_4	230	480	480	1910	1000[b]
5 μM CterG	710	1360	1460	580	2400

[a] The numbers refer to the specific activities of the various fractions and are expressed as picomoles of ATP hydrolyzed/mg protein/minute under our experimental conditions; the experimental error is about 100 pmol ATP hydrolyzed/mg protein/minute; CterG is the purified C-terminal fragment of PrtG.[26]

[b] Obtained with 20 mM VO_4.

us to carry such measurement on intact membrane vesicles and to compare the ATPase activity in vesicles, detergent extract and proteoliposomes. Table I presents the results of the ATPase activity from the different PrtD preparations and their respective characteristics.

The membrane vesicles exhibit different types of ATPase activities, which can be distinguished on the basis of their sensitivity to different inhibitors. The azide-sensitive component was not taken into account since 10 mM azide was present for measurements on membrane vesicles: It accounted for ca. 20% of the total activity of the control membranes; most of the remaining activity was inhibited by vanadate. Orthovanadate is inhibitory at 10 mM, and metavanadate is inhibitory at 100 μM with an old solution that most likely contains a high proportion of decavanadate, since the stock solution (1 mM) has a yellow-orange color.[35] Freshly prepared and boiled solutions of metavanadate or orthovanadate at 1 mM do not inhibit the ATPase activity. The C-terminal fragment of PrtG at 5 μM has little or no effect on control vesicles but partially inhibits the ATPase activity of PrtD-enriched vesicles and does not inhibit the ATPase activity of the PrtD/K370R mutant vesicles. The specific ATPase activity of PrtD membrane vesicles is greater than the one of control vesicles and than that of PrtD/K370R mutant vesicles. Because the level of background ATPase activity might vary somewhat from preparation to preparation, it is difficult to get a more quantitative measurement of the contribution of PrtD to the ATPase activity of the vesicles. If one simply subtracts the background ATPase activity from that of the PrtD and PrtD/K370R vesicles, the ATPase activity of the PrtD vesicles is three to four times higher than that

[35] B. R. Nechay, *Ann. Rev. Pharmacol. Toxicol.* **24,** 501 (1984).

of PrtD/K370R vesicles, which correlates with the lower secretion efficiency of the mutated protein.

The ATPase activity of the purified protein is completely inhibited by 100 μM metavanadate but not by 10 mM orthovanadate. It is also severely inhibited by 5 μM of the C-terminal peptide of the protease G; it is not inhibited by azide (10 mM), *N*-ethylmaleimide, or DCCD. Ca^{2+} does not support the ATPase activity of PrtD. The extent of inhibition by the C-terminal fragment of protease G might vary from preparation to preparation from 70% to nearly complete inhibition. The K_m of the purified protein for ATP is 12 μM and the V_{max} is around 0.2 μmol ATP hydrolyzed/mg protein/hr (there was an error of a factor 10 in our previous determination[32]). If one assumes that the ATPase activity of the PrtD membrane vesicles above the background is exclusively due to PrtD, one can calculate that the specific activity of the purified protein is three times lower than in the membrane environment, given the relative specific activities of the two fractions and the proportion of PrtD in the whole membranes (Fig. 3).

After reconstitution into phospholipid vesicles, the specific ATPase activity increases by 1.5-fold. Sidedness of reconstituted protein has not been determined and this value might thus be a low estimate if orientation is not 100% of the ATP-binding domain facing the outside of the vesicle. This ATPase activity is inhibited by both metavanadate at 100 μM and orthovanadate at 20 mM, although the extent of inhibition reaches a plateau. It is also partially inhibited by the C-terminal secretion signal of the protease G. It thus seems that the purified and solubilized protein has slightly different characteristics from the protein in its native environment (greater inhibition by the C-terminal fragment of the protease G and no longer inhibition by orthovanadate) and that reconstitution restores these characteristics at least partially. This opens the way to an improvement of the reconstitution conditions and to test the effect of the other secretion functions on the ATPase activity.

Acknowledgments

I wish to express my thanks to C. Wandersman for the opportunity to contribute this chapter and to C. Wandersman and J.-M. Ghigo for critical reading on the manuscript.

[7] Arsenical Pumps in Prokaryotes and Eukaryotes

By MASAYUKI KURODA, HIRANMOY BHATTACHARJEE, and BARRY P. ROSEN

Introduction

Both prokaryotes and eukaryotes have evolved transport systems that produce drug resistance by extrusion of the drug from the cell, reducing the intracellular concentration to subtoxic levels.[1] Knowledge of molecular mechanisms of these transport systems is important for the design of drugs to treat drug resistance. Arsenicals and antimonials were among the first chemotherapeutic agents used to treat infectious diseases, and resistance to those compounds was noted soon after their introduction into clinical use nearly a century ago.[2] In our laboratory we have investigated the mechanism of resistance to arsenicals and antimonials in prokaryotes and eukaryotes. Interestingly, although the overall scheme is remarkably similar between kingdoms, the specific mechanisms and the proteins that produce them appear to be the results of independent, convergent evolution (Fig. 1). This chapter provides a brief review of those systems and provide the details of the assays developed to measure their activities in cells, in membrane preparations, and with purified proteins.

Prokaryotic Arsenite Efflux Systems

Bacterial resistance to oxyanions of As(III) or Sb(III) is due to the action of extrusion systems for arsenite or antimonite encoded by arsenical resistance (*ars*) operons.[1] *We emphasize that the gene products of the ars operon are evolutionarily unrelated to the members of the ABC superfamily of transporters.* However, they have structural and functional similarities, including adenosine triphospate (ATP)-binding cassette (ABC) motifs and 12 membrane-spanning α helices. This system is unique in simultaneously having the ability to function as primary ATP-coupled pumps or secondary anion permeases, depending on the subunit composition.[3,4] The *ars* operons

[1] S. Dey and B. P. Rosen, *in* "Drug Transport in Antimicrobial and Anticancer Chemotherapy" (N. H. Georgopapadakou, ed.), pp. 103–132. Dekker, 1995.

[2] P. Ehrlich, *in* "Collected Papers of Paul Ehrlich" (F. Himmelweit, ed.), pp. 183–194. Pergammon Press, London, 1960.

[3] S. Dey and B. P. Rosen, *J. Bacteriol.* **177,** 385 (1995).

[4] Y. Chen, S. Dey, and B. P. Rosen, *J. Bacteriol.* **178,** 911 (1996).

METHODS IN ENZYMOLOGY, VOL. 292
0076-6879/98 $25.00

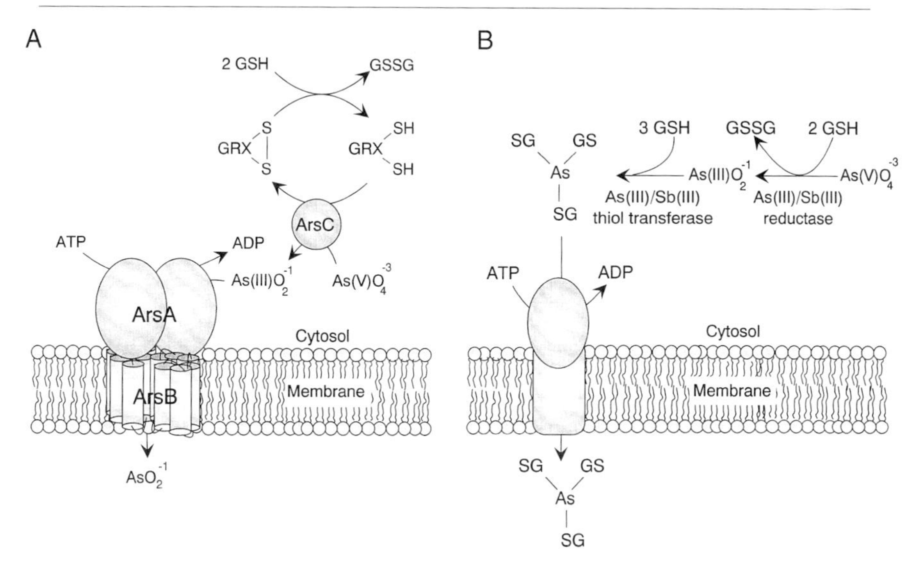

FIG. 1. Prokaryotic and eukaryotic As(III) pumps. Bacteria and eukaryotes have convergently evolved different resistance pumps for the ATP-coupled extrusion of arsenicals and antimonials from cells. (A) In bacteria ArsA and ArsB form a membrane-bound oxyanion-translocating ATPase for extrusion of arsenite and antimonite. The 63-kDa ArsA ATPase is the catalytic subunit. The 45.5-kDa ArsB subunit, an integral membrane protein located in the inner membrane of *E. coli,* has conductivity for arsenite or antimonite. ArsC is an arsenate reductase that expands the range of resistance by reducing arsenate (As(V)) to arsenite (As(III)). Electrons are transferred from glutaredoxin, which is re-reduced by glutathione. (B) Resistance to arsenite and antimonite in eukaryotic cells occurs by the ATP-coupled extrusion of a soft metal–thiol complex, where As(III) or Sb(III) are postulated to react with a physiologic thiol such as glutathione in mammalian cells or trypanothione in *Leishmania.* Although arsenate and antimonite are slowly reduced nonenzymatically by glutathione and other thiols, arsenate and the antileishmanial drug Pentostam, which contains Sb(V), may require a reductase to convert them to the trivalent form at physiologically significant rates. Sb(III) and As(III) also react spontaneously with thiols, but enzymatic conjugation may be required for reaction to produce resistance. The metalloid–thiol complex is proposed to be the substrate of a transport system that extrudes the conjugate, thus producing resistance.

of conjugative plasmids R773 and R46 have five genes, two of which, *arsA* and *arsB,* encode a pump for extrusion of arsenical oxyanions from *Escherichia coli.*[5–7] The pump is composed of the 63,169-Da hydrophilic ArsA protein, the catalytic subunit, and the hydrophobic 45,577-Da inner

[5] C. M. Chen, T. Misra, S. Silver, and B. P. Rosen, *J. Biol. Chem.* **261,** 15,030 (1986).

[6] M. J. D. San Francisco, C. L. Hope, J. B. Owolabi, L. S. Tisa, and B. P. Rosen, *Nucl. Acids Res.* **18,** 619 (1990).

[7] D. F. Bruhn, J. Li, S. Silver, F. Roberto, and B. P. Rosen, *FEMS Microbiol. Lett.* **139,** 149 (1996).

membrane ArsB protein, which serves as a membrane anchor for ArsA and forms the anion-conducting pathway. Other *ars* operons lack the *arsA* gene but still produce resistance to arsenicals through their extrusion.[8–11] In intact cells the ArsA–ArsB pump is an obligatory ATP-coupled pump and is independent of the electrochemical gradient of the cell.[12,13] In contrast, ArsB alone catalyzes oxyanion extrusion coupled to the electrochemical gradient and is independent of ATP.[3]

Assay of Arsenite Extrusion from Intact Cells.

Energy-dependent arsenite extrusion is assayed in cells preloaded with radioactive arsenite.[3,12,13] To dissociate the equilibrium between electrochemical and chemical energy in the cell, *E. coli* strain LE392Δ*uncIC*, containing a deletion of the *unc* operon that encodes the F_0F_1, is used. The cells are grown at 37° in TEA medium [50 m*M* triethanolamine hydrochloride, pH 6.9, 15 m*M* KCl, 10 m*M* $(NH_4)_2SO_4$, and 1 m*M* $MgSO_4$] supplemented with 0.5% (v/v) glycerol, 2.5 μg/ml thiamin, 0.5% (w/v) peptone, and 0.15% (w/v) succinate. At an OD_{600} of 0.8 the cultures are induced with 0.1 m*M* $NaAsO_2$ for 1 hr. The cells are washed three times with TEA medium lacking a carbon source, and cells are depleted of endogenous energy reserves by futile cycling with 2,4-dinitrophenol (DNP). The washed cells are incubated in TEA medium supplemented with 5 m*M* DNP for 2 hr, followed by washing three times with TEA medium to remove the DNP. The cells are then arsenite-loaded by incubation with radioactive $^{73}AsO_2^-$, which is prepared by reduction of $^{73}AsO_4^{3-}$ (Los Alamos National Laboratories).[14] A solution (40 μl) consisting of 0.1 m*M* $NaAsO_2$, 66 m*M* $Na_2S_2O_5$, 27 m*M* $Na_2S_2O_3$, and 82 m*M* H_2SO_4 is mixed with 40 μl of carrier-free $H_3^{73}AsO_4$ and incubated for 40 min. Efflux is initiated by dilution of the loaded cells into media without arsenite but containing either 20 m*M* glucose or 20 m*M* sodium succinate with or without metabolic inhibitors. To inhibit formation of an electrochemical proton gradient through respiration, the respiratory chain inhibitor KCN is added at 20 m*M*. To dissipate the protonmotive force, uncouplers such as 10 μ*M* carbonyl cyanide *m*-chlorophenylhydrazone (CCCP), can be added. Samples (0.1 ml) are

[8] G. Ji and S. Silver, *J. Bacteriol.* **174,** 3684 (1992).

[9] R. Rosenstein, P. Peschel, B. Wieland, and F. Götz, *J. Bacteriol.* **174,** 3676 (1992).

[10] H. J. Sofia, V. Burland, D. L. Daniels, G. Plunkett III, and F. R. Blattner. *Nucl. Acids Res.* **22,** 2576 (1994).

[11] A. Carlin, W. Shi, S. Dey, and B. P. Rosen, *J. Bacteriol.* **177,** 981 (1995).

[12] H. L. T. Mobley and B. P. Rosen, *Proc. Natl. Acad. Sci. U.S.A.* **79,** 6119 (1982).

[13] B. P. Rosen and M. G. Borbolla, *Biochem. Biophys. Res. Commun.* **124,** 760 (1984).

[14] P. F. Reay and C. J. Asher, *Anal. Biochem.* **78,** 557 (1977).

withdrawn at intervals, filtered on 0.45-μm pore diameter nitrocellulose filters (Whatman, Clifton, NJ), and washed with 5 ml of TEA medium. The filters are dried. In this and all subsequent assays radioactive arsenite is quantified by liquid scintillation counting. Figure 2 illustrates the results of an efflux assay.

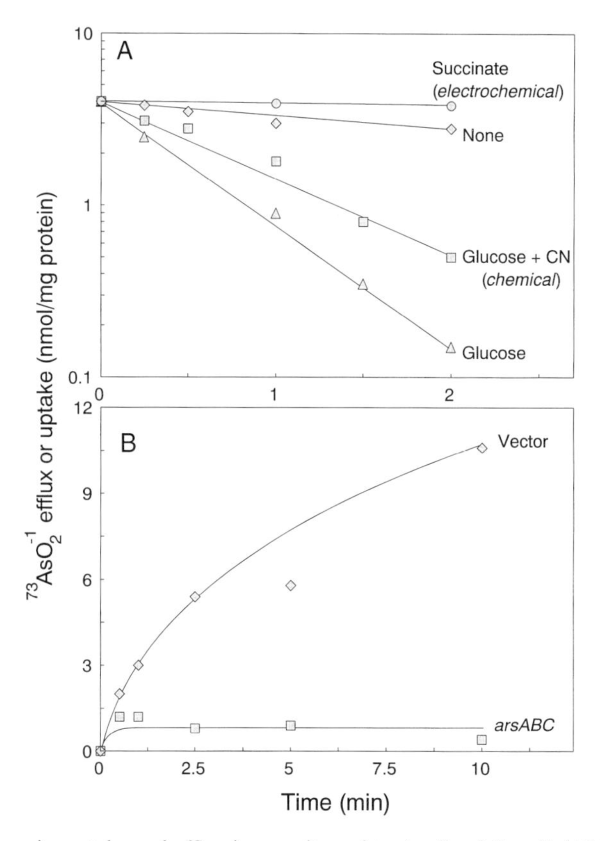

FIG. 2. Arsenite uptake and efflux in arsenite-resistant cells of *E. coli.* (A) Efflux: Cells of an *unc* strain expressing the *arsABC* genes are depleted of endogenous energy reserves and loaded with $Na^{73}AsO_2$. Efflux is initiated by 100-fold dilution of the cells into medium containing the following energy sources: ◆, none; ▲, 10 m*M* glucose, which produces both chemical and electrochemical energy; ■, 10 m*M* glucose and 10 m*M* NaCN, which produces only chemical energy in an *unc* strain; and ●, 10 m*M* sodium succinate, which produces only electrochemical energy in an *unc* strain. (B) Untake: Uptake of $^{73}AsO_2^-$ is assayed in energy-depleted cells with either the vector plasmid HB101 (◆) or plasmid pUM3 (*arsABC*) (■) with 10 m*M* glucose added as an energy source.

Assay of Arsenite Uptake in Intact Cells

Efflux from preloaded cells and lack of uptake into cells reflect two aspects of the same phenomenon.[1] Inability to accumulate is less specific in that it could be either due to blockage of uptake or to increased efflux, but it is technically less complicated to assay than efflux. First, cells do not have to be depleted of endogenous energy. Second, preloading is not required. The cells are grown and prepared in the same way as for efflux.[3] To initiate the transport assay, 50 μl of cells are diluted into 0.6 ml of TEA medium containing 20 mM glucose, 0.1 mM $NaAsO_2$, and 1.25 μCi of $H^{73}AsO_2^-$. Again, samples are withdrawn, filtered, and the radioactivity quantified. Figure 2 shows a comparison of arsenite uptake and efflux in arsenite-resistant cells.

Assay of ATP-Dependent Arsenite Uptake in Everted Membrane Vesicles

Expression of the *arsB* gene is limiting for synthesis of a functional pump. Fusion between *arsB* and a portion of *arsA* produced a hybrid *arsAB2* gene that was produced in amounts sufficient for assays of transport in membrane vesicles.[15] Arsenite transport can be examined in everted membrane vesicles containing the ArsA and ArsAB2 proteins.[16] To coexpress the *arsA* gene, the compatible plasmid pArsA from which *arsA* gene is constitutively expressed, is used. Cells of *E. coli* harboring both plasmids pJUN4, which contains the *arsAB2* gene, and pArsA are grown in LB medium at 37° and induced with 0.1 mM $NaAsO_2$ for 1 hr. The cells are harvested and washed once with the buffer of 10 mM Tris–HCl, pH 7.5, 0.25 M sucrose, and then suspended in same buffer containing 1 mM dithiothreitol (DTT). The cells are lysed by a single passage through a French pressure cell at 4000 psi. Immediately after lysis, diisopropyl fluorophosphate is added (2.5 μl/g of wet cells). DNase I (20 μg/ml) is added, and the suspension is incubated on ice for 40 min. Unbroken cells are removed by centrifugation at 10,000g for 30 min at 4°. Membrane vesicles are pelleted by centrifugation at 100,000g for 60 min at 4°, washed and suspended in the same buffer without DTT. Because of loss of transport activity on storage at −70°, membrane vesicles should be prepared fresh.

The assay mixture contains 10 mM Tris-HCl, pH 7.5, 0.1 M KCl, and 0.25 M sucrose. Membranes are added to 1 mg/ml, and $Na^{73}AsO_2$ to 0.1 mM. ATP (5 mM) is used as an energy source with an ATP regenerating system consisting of 20 units of pyruvate kinase and 10 mM phosphoenolpyruvate. The transport buffer is degassed to reduce oxidation of arsenite. The trans-

[15] D. Dou, J. B. Owolabi, S. Dey, and B. P. Rosen, *J. Biol. Chem.* **267,** 25,768 (1992).
[16] S. Dey, D. Dou, and B. P. Rosen, *J. Biol. Chem.* **269,** 25,442 (1994).

port reaction is initiated with either 5 m*M* $MgCl_2$ or $MgSO_4$. At intervals, 0.1-ml samples are withdrawn, filtered through nitrocellulose membrane filters (0.2-μm pore size), and washed with 5 ml of the same buffer. The filters are dried, and the radioactivity quantified. Figure 3 illustrates uptake of arsenite in everted vesicles from cells expressing the ArsA and ArsAB2 proteins.

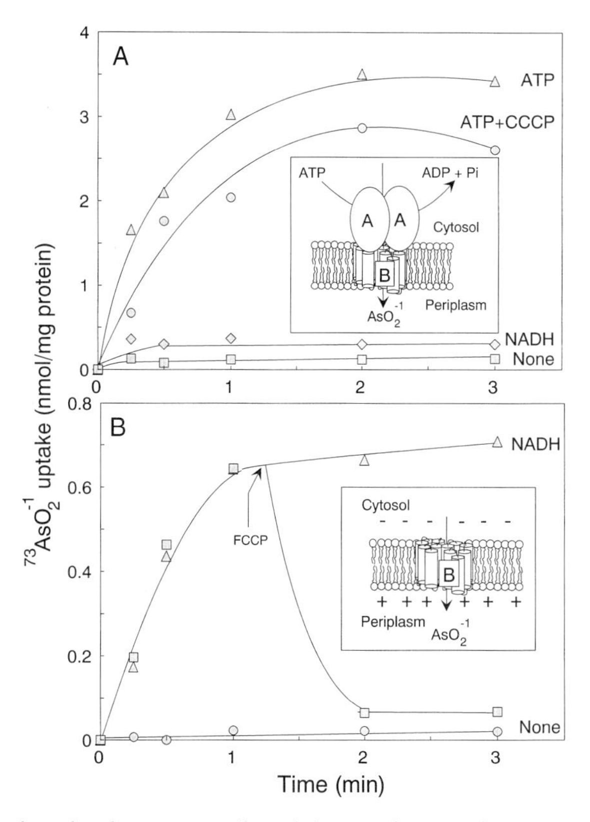

FIG. 3. Dual mode of energy coupling of the arsenite extrusion system. The arsenite permease can function as a primary ATP-driven pump (A, inset) or secondary porter coupled to the electrochemical proton gradient (B, inset), depending on subunit composition. (A) The ArsA–ArsB pump catalyzes ATP-driven $^{73}AsO_2^-$ uptake in everted membrane vesicles. Vesicles are prepared from cells of *E. coli* expressing both the *arsA* and *arsAB2* genes. Additions: ■, none; ▲, 5 m*M* MgATP; ●, 5 m*M* MgATP + 100 μ*M* carbonyl cyanide *m*-chlorophenylhydrazone (CCCP); ◆, 5 m*M* NADH. (B) ArsB catalyzes $^{73}AsO_2^-$ uptake coupled to the electrochemical proton gradient. Vesicles are prepared from cells expressing the *arsB* gene. Additions: ●, none; ▲, 5 m*M* NADH; ■, 5 m*M* NADH with 10 μ*M* carbonyl cyanide *p*-(trifluoromethoxy)phenylhydrazone (FCCP) added at the indicated time.

Assay of NADH-Dependent Arsenite Uptake in Everted Membrane Vesicles

As described earlier, in the absence of ArsA, the Ars system functions as a secondary carrier driven by the protonmotive force. Electrochemical energy-dependent uptake is difficult to observe in membrane vesicles because the amount of ArsB in the membranes is limiting. However, we have recently developed methods for expressing wild-type ArsB protein in sufficient amounts for assay of NADH-driven $^{73}AsO_2^-$ uptake in everted membrane vesicles.[17] The major differences between this assay and that for ATP-driven transport is that the buffer contains no depolarizing anions such as chloride, and, since respiration requires O_2, the buffer is not degassed. In the NADH-coupled assay the cells are lysed in a buffer consisting of 75 m*M* HEPES–KOH, pH 7.5, 0.15 *M* K_2SO_4, 2.5 m*M* $MgSO_4$, containing 0.25 *M* sucrose, and the membrane vesicles are assayed in the same buffer with 5 m*M* NADH in place of ATP. Otherwise the assay is the same as that for ATP-driven transport. Comparisons of the effects of inhibitors clearly demonstrate that ATP-driven transport by the ArsA–ArsB complex is independent of electrochemical energy (Fig. 3), while NADH-driven transport by ArsB alone requires electrochemical energy. The absolute differences in ATP-driven versus NADH-driven accumulation shown in Fig. 3 are due to several factors. First, pumps are thermodynamically more efficient mechanisms than secondary carriers and are capable of catalyzing accumulation ratios on the order of 10^6, compared with approximately 10^3 for secondary carriers.[1] Second, the ArsAB2 protein that was used for ATP-driven transport in Fig. 3A is produced in higher amounts than the wild-type ArsB protein in Fig. 3B.

Purification and Assay of ArsA ATPase

Under normal physiologic conditions ArsA is found as a membrane-bound complex with ArsB. When expressed at high levels, ArsA is mainly found as a soluble protein in the cytosol. This property has allowed large-scale purification of ArsA, facilitating biochemical characterization. The protein can be purified by conventional chromatographic methods, as described later, but more recently the sequence for a histidine hexalinker was cloned on to the 3′ end of the *arsA* gene, allowing purification in a single step with Ni^{2+} affinity chromatography.[18]

For purification of ArsA by conventional chromatographic methods, cells of *E. coli* strain SG20043 (*lon*$^-$) bearing plasmid pUM3, which constitu-

[17] M. Kuroda, S. Dey, O. I. Sanders, and B. P. Rosen, *J. Biol. Chem.* **272,** 326 (1997).

[18] T. Zhou and B. P. Rosen, *J. Biol. Chem.* **272,** 19,731 (1997).

tively expresses ArsA, are grown at 37° overnight with aeration in 1 liter of LB medium containing 100 μg/ml ampicillin and 1 m*M* $NaAsO_2$.[19] The cultures are diluted into 10 liters of prewarmed LB medium lacking arsenite and grown at 37°C for 3 hr with aeration. The cells are chilled in an ice–water bath and harvested by centrifugation at 4°, 7000*g* for 10 min. All subsequent steps are performed at 4° unless stated otherwise. Cells are suspended in 5 ml/g of wet cells in buffer A [25 m*M* Tris-HCl, pH 7.5, 20% (v/v) glycerol, 2 m*M* EDTA, 1 m*M* DTT] and lysed by a single passage through a French pressure cell at 20,000 psi. The serine protease inhibitor diisopropyl fluorophosphate (2.5 μl per gram of wet cells) is added to the lysate as quickly as possible with rapid mixing. Unbroken cells and membranes are removed by centrifugation at 150,000*g* for 1 hr. The cytosol is applied at a rate of 1 ml/min onto a 2.5-cm-diameter column packed with 150 ml of Q-Sepharose anion-exchanger preequilibrated with buffer A. The column is eluted first with 100 ml of buffer A and then with a 650-ml linear gradient of 0 to 0.3 *M* NaCl at a flow rate of 1 ml/min. Fractions of 8 ml are collected and analyzed by sodium dodecyl sulfate–polyacrylamide gel electrophoresis (SDS–PAGE).[20] The fractions containing the peak of ArsA are pooled and made 60% saturated with ammonium sulfate, kept on ice for a few minutes, and the precipitate collected by centrifugation at 10,000*g* for 15 min. The pellet is dissolved in 5 ml of buffer B (20 m*M* MOPS–KOH, pH 7.5, 10% (v/v) glycerol, 2 m*M* EDTA, 0.25 *M* NaCl, 1 m*M* DTT) and applied at a flow rate of 0.4 ml/min onto a 2.5-cm-diameter column packed with 350 ml of Sephacryl S-200, pre-equilibrated with buffer B. The protein is eluted in buffer B at a flow rate of 0.4 ml/min; 1.5-ml fractions are collected and analyzed by SDS–PAGE. The peak fractions are pooled and concentrated 10-fold in Centriprep-30 concentrator (Amicon, Beverly, MA). The enzyme preparation can be stored for several months at −70°. The concentration of ArsA in purified preparations is determined by the absorption at 280 nm using a molar extinction coefficient of 33,480.

ArsA ATPase activity is measured spectrophotometrically at 340 nm in an assay in which the regeneration of ATP is coupled to the oxidation of NADH via the pyruvate kinase and lactate dehydrogenase reactions.[21] The ATPase reaction mixture contains 50 m*M* MOPS–KOH buffer, pH 7.5, 0.25 m*M* EDTA, 5 m*M* ATP, 1.5 m*M* phosphoenolpyruvate, 0.2 m*M* NADH, 0.1 m*M* potassium antimonyl tartrate, 10 units of pyruvate kinase (EC 2.7.1.40), 10 units of lactate dehydrogenase (EC 1.1.1.27), and 10–20 μg of purified ArsA in a total volume of 1.0 ml. The assay mixture

[19] C. M. Hsu and B. P. Rosen, *J. Biol. Chem.* **264,** 17,349 (1989).
[20] U. K. Laemmli, *Nature,* **227,** 680 (1970).
[21] G. Vogel and R. Steinhart, *Biochemistry* **16,** 208 (1976).

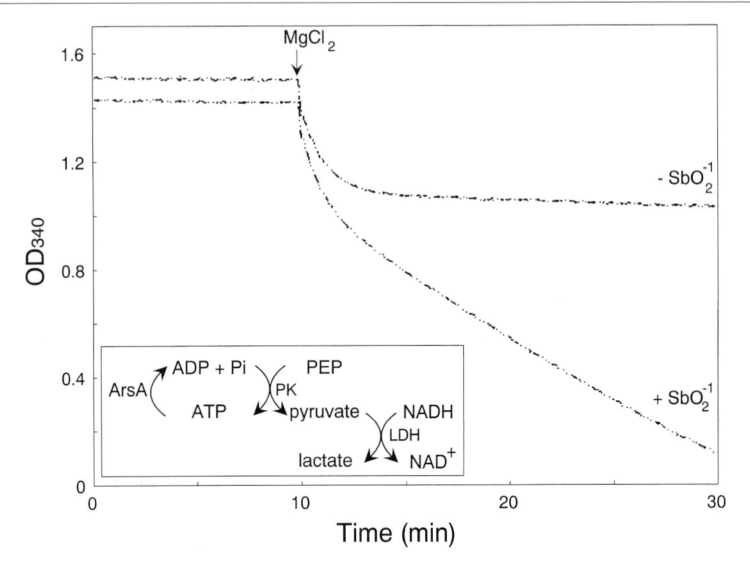

FIG. 4. Assay of ArsA ATPase activity. ArsA (13 μg/ml) is incubated with 5 mM ATP and with or without 0.1 mM potassium antimonyl tartrate at 37° for 10 min, following which the reaction is initiated by addition of 2.5 mM $MgCl_2$. Using a coupled assay with pyruvate kinase (PK) and lactate dehydrogenase (LDH), ATPase activity is estimated from the linear portion of the steady-state rate of NADH utilization, which is monitored at 340 nm.

is incubated at 37° for 10 min, and the reaction is initiated by addition of 2.5 mM $MgCl_2$. The reaction is linear for 5–15 min, and the linear steady-state rate is used to calculate the specific activity (Fig. 4).

Purified ArsA is allosterically activated 3- to 5-fold by arsenite or 15- to 20-fold by antimonite.[19] The apparent K_m for ATP is 100 μM. The concentration of oxyanion that produces 50% of V_{max} is 2 μM for antimonite and 600 μM for arsenite. The V_{max} of the antimonite-stimulated ATPase activity ranges from 0.6 to 1.0 μmol/min/mg of protein, while the arsenite-stimulated activity has a V_{max} between 0.2 and 0.3 μmol/min/mg. The optimal pH range for ATP hydrolysis is 7.5 to 7.8. ATPase activity requires Mg^{2+} at a molar ratio of 2 ATP : 1 Mg^{2+}. Inhibitors of other classes of ion-translocating ATPases have no effect on ArsA ATPase activity, including N,N'-dicyclohexylcarbodiimide, azide, vanadate, and nitrate. The active species of the ArsA ATPase is a homodimer with four nucleotide binding sites, one A1 and one A2 site per monomer. Genetic and biochemical studies have suggested that a catalytic site requires interaction of the A1 site in one subunit with the A2 site in the other subunit.[22] Recent studies

[22] P. Kaur and B. P. Rosen, *J. Bacteriol.* **175,** 351 (1993).

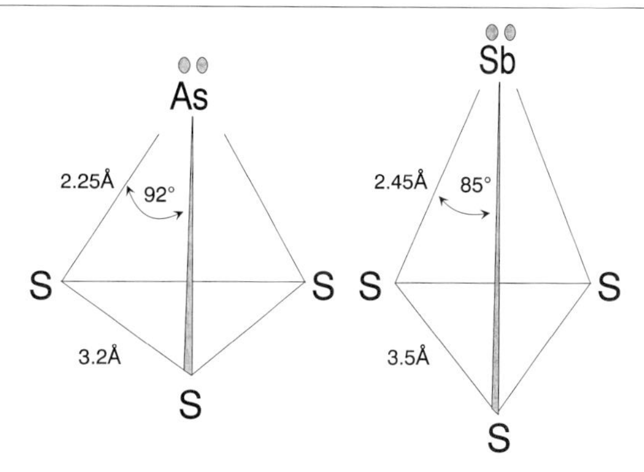

FIG. 5. Geometry of the allosteric site in ArsA with either As(III) or Sb(III). The arsenite/antimonite binding site is proposed to be a trigonal pyramidal structure that is three coordinate with the cysteine thiolates of ArsA residues 113, 172, and 422. Bond angles and distances are taken from crystallographic data of small molecules containing As–S or Sb–S bonds.

have shown that tryptophan-159 is conformationally coupled to one or both nucleotide-binding sites, providing a sensitive intrinsic probe for ATP binding.[23] By site-directed mutagenesis cysteines 113, 172, and 422 have been shown to be involved in the allosteric activation by arsenite or antimonite.[24] From distance measurements using a bifunctional thiol cross-linker these three cysteine residues must be within 3–6 Å of each other, suggesting that the allosteric site in ArsA is a metal–thiol cage for As(III) or Sb(III) (Fig. 5).[25]

Interaction between the Catalytic and the Membrane Subunits of the Pump

To function as a subunit of the arsenical resistant pump, ArsA, the catalytic subunit of the pump has to interact with ArsB, the membrane subunit of the pump. The properties of this interaction can be measured *in vitro* by binding purified ArsA to everted membrane vesicles containing ArsB.[26] For expression of *arsB* gene, plasmid pT7-5-16, in which *arsB* gene is under the control of T7 promoter, is used. Membrane vesicles prepared as described earlier are suspended in the buffer containing 10 m*M* Tris-HCl, pH 8.0, 50% glycerol, 5 m*M* EDTA, rapidly frozen in a dry ice–ethanol bath and stored at −70° until use.

[23] T. Zhou, S. Liu and B. P. Rosen, *Biochemistry* **34,** 13,622 (1995).
[24] H. Bhattacharjee, J. Li, M. Y. Ksenzenko and B. P. Rosen, *J. Biol. Chem.* **270,** 11,345 (1995).
[25] H. Bhattacharjee and B. P. Rosen, *J. Biol. Chem.* **271,** 24,465 (1996).
[26] S. Dey, D. Dou, L. S. Tisa, and B. P. Rosen, *Arch. Biochem. Biophys.* **311,** 418 (1994).

Purified ArsA is added to 1 mg of membrane protein in 0.2 ml of 25 m*M* Tris-HCl, pH 7.5, 0.2 *M* KCl. The membrane vesicle–ArsA mixture is incubated at room temperature for 20 min and diluted 10- to 20-fold with a buffer consisting of 10 m*M* Tris-HCl, pH 7.5, 20% (v/v) glycerol, and 0.2 *M* KCl. To separate the membrane-bound ArsA from unbound ArsA, the membrane vesicles are pelleted by centrifugation at 100,000*g* for 1 hr at 4° and washed once with the same buffer, which contains 0.1 mg/ml bovine serum albumin (BSA). To correct for nonspecific binding of ArsA, control assays with membrane vesicles lacking ArsB are performed. The binding affinity of ArsA is enhanced by arsenite or antimonite. ArsA is incubated with 1 m*M* sodium arsenite or 0.5 m*M* potassium antimonyl tartrate in the assay buffer at 30° for 30 min to enhance binding. To quantify the amount of ArsA bound to the membranes, immunoblotting is done with anti-ArsA serum. Two methods were used to blot proteins to the nitrocellulose membrane filter. First is electrophoretic transfer, and the second dot blotting. In the former, ArsA-bound membrane vesicles are solubilized in 0.1 ml of SDS–sample buffer overnight and treated at 90° for 5 min. The proteins are separated by SDS–PAGE with various amounts of purified ArsA standards. Then proteins are electrophoretically transferred to nitrocellulose membrane (0.2-μm pore size) overnight at 25 mV at 4°.

In the latter method, everted membranes with bound ArsA are suspended in 0.64 ml of 20 m*M* Tris-HCl, pH 7.5, 0.5 *M* NaCl, and 8 *M* urea and incubated at room temperature for 30 min. To separate the urea-extracted ArsA from membranes, the membranes are pelleted by centrifugation at 100,000*g* for 90 min, and supernatant solution is collected. For dot-blot analysis, methanol is added to the supernatant to a final concentration of 20% (v/v). Known amounts of purified ArsA in the same buffer are used as standards. A nitrocellulose membrane is soaked in the buffer of 25 m*M* Tris-HCl, pH 7.5, 0.5 *M* NaCl, and 20% (v/v) methanol. Samples are blotted to the nitrocellulose membrane using a Bio-Rad (Richmond, CA) microfiltration apparatus. The samples are filtered with gravity for 1 hr and then by vacuum. The nitrocellulose membrane is washed twice with the same buffer.

In both assays the amount of ArsA is quantified by immunoblotting. The nitrocellulose membrane is treated with 5% (w/v) nonfat dry milk in a buffer containing 21 m*M* KH_2PO_4, 11 m*M* Na_2HPO_4, 0.14 *M* NaCl, and 2.5 m*M* KCl, pH 7.4, to block unoccupied sites. The filter is incubated with rabbit anti-ArsA serum (1:2000) for 1 hr at 37°, then washed three times by incubation for 10 min each time with the same buffer containing 5% nonfat dry milk at 37°. The filter is treated with goat anti-rabbit immunoglobulin G (IgG) conjugated to horseradish peroxidase for 1 hr at 37°, washed three times with the same buffer, and ArsA visualized by enhanced chemiluminescence (ECL). The X-ray films are quantified densitometrically using a 2D densitometric analytical program (Stratagene). Figure 6 illustrates the results of a dot-blot assay of ArsA binding to ArsB-containing everted membrane vesicles.

Eukaryotic Arsenite Efflux Systems

Nearly a century ago Paul Ehrlich received the Nobel Prize in medicine for his synthesis of the first chemically synthesized chemotherapeutic drug, the organic arsenical Salvarsan. Ehrlich recognized that drug action required drug accumulation, and that drug resistance could ensue from reduction in accumulation.[2] The organic arsenicals he synthesized are related to drugs still used for the treatment of parasitic diseases. Today drug resistance is a major problem in cancer chemotherapy, and the importance of transport-mediated resistances is widely recognized.[1]

As useful nonmammalian eukaryotic models for analysis of mechanisms of resistance to chemotherapeutic agents, we have chosen two test organisms. One is a Chinese hamster cell line selected for arsenic resistance,

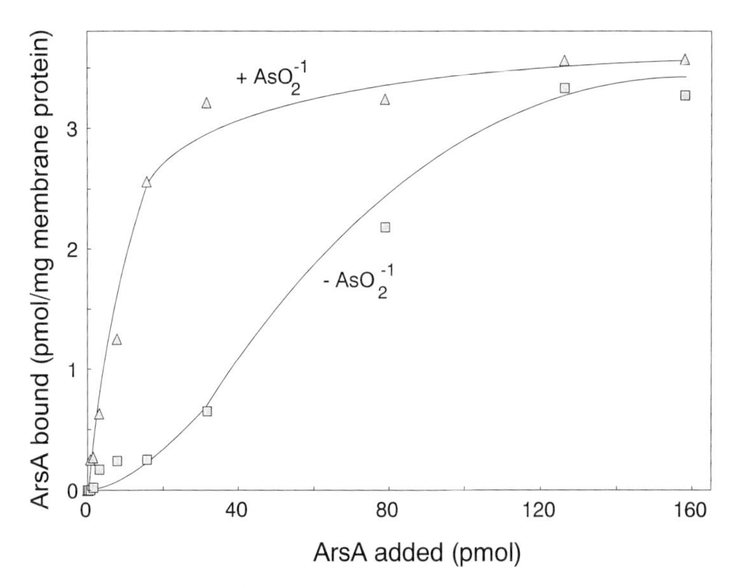

FIG. 6. *In vitro* interaction of the ArsA and ArsB subunits of the arsenite pump. Purified ArsA is preincubated for 30 min at 30° without additions (■) or with 1 m*M* $NaAsO_2$ arsenite (▲). The ATPase is then mixed with 1 mg of membrane protein prepared from cells expressing *arsB*. The membranes are washed, and the amount of bound ArsA determined by dot-blot analysis.

where we have shown that resistance involves active extrusion of As(III).[27] The second model is the trypanosomatid protozoan parasite *Leishmania,* the causative agent of kala azar and other forms of leishmaniasis. The treatment of choice is the Sb(V)-containing drug sodium stibogluconate (Pentostam). Unresponsiveness to Pentostam is becoming a common problem, occurring in 5% of patients. Resistance rates as high as 70% have been described in some endemic areas. In strains of *Leishmania tarentolae* selected *in vitro* for resistance to trivalent arsenicals and antimonials and cross-resistant to Pentostam, we have shown that high-level resistance is related to efflux.[28] We have recently identified an ATP-dependent As(III)–thiol pump in plasma membranes, suggesting that *in vivo* it is not free arsenite but rather the thiol adducts that are extruded.[29] At this point it is not known whether the arsenite pumps in either Chinese hamster cells or *Leishmania* are members of the ABC superfamily.

We have recently demonstrated that arsenite-resistant cells have increased levels of intracellular thiols.[30] By a combination of high-performance liquid chromatography and mass spectroscopy we showed that the glutathione derivative trypanothione was elevated 40-fold in resistant cells. The trypanothione adduct of arsenite was effectively transported by the As-thiol pump. We have proposed that this drug resistance results from conjugation to a thiol, either glutathione in mammalian cells or trypanothione in parasites, followed by extrusion of the xenobiotic by the As–thiol pump (Fig. 1).

Assays of Arsenite Uptake and Efflux in Chinese Hamster Cells

Arsenite-resistant sublines of Chinese hamster V79 cells are grown in monolayer cultures in F12 medium (Gibco, Grand Island, NY) supplemented with 5% (v/v) fetal calf serum, 100 units/ml of penicillin, and 100 μg/ml of streptomycin.[27] $NaAsO_2$ is added to the medium 24 hr after seeding and remains in the medium throughout the incubation. For uptake assays carrier-free $H^{73}AsO_2$ is added to medium containing 20 mM unlabeled $NaAsO_2$ to a specific activity of 400 cpm/pmol. Subconfluent (90% confluent) monolayer cultures are incubated in the $^{73}AsO_2^-$-containing medium for various periods of time and then washed three times with ice-

[27] Z. Wang, S. Dey, B. P. Rosen, and T. G. Rossman, *Toxicol. Appl. Pharmacol.* **137,** 112 (1996).

[28] S. Dey, B. Papadopoulou, G. Roy, K. Grondin, D. Dou, B. P. Rosen, and M. Ouellette, *Molec. Biol. Parasitol.* **67,** 49 (1994).

[29] S. Dey, M. Ouellette, J. Lightbody, B. Papadopoulou, and B. P. Rosen, *Proc. Natl. Acad. Sci. U.S.A.* **93,** 2192 (1996).

[30] R. Mukhopadhyay, S. Dey, N. Xu, D. Gage, J. Lightbody, M. Ouellette, and B. P. Rosen, *Proc. Natl. Acad. Sci. U.S.A.* **93,** 10,383 (1996).

cold Earle's balanced salt solution (EBSS). Cells are lysed with 0.1% (v/v) Triton X-100 in 0.05 *N* NaOH, and the radioactivity quantified.

For efflux assays cells are depleted of endogenous energy reserves to allow loading with $^{73}AsO_2^-$. Sodium azide (20 m*M*) is added to the $^{73}AsO_2^-$-containing medium, and the cells incubated in this medium for 2 hr. The cultures are then washed three times with ice-cold EBSS and incubated in arsenic-free medium for various periods of time. The cells are lysed, and the radioactivity quantified.

Assays of Arsenite Uptake and Efflux in Promastigotes of Leishmania

Cells of *L. tarentolae* TarII wild type and As20.3 or As50.1, which had been selected for resistance to 20 or 50 μM $NaAsO_2$, respectively, are grown in SDM-79 medium at 29° to mid-log phase, washed twice in a buffer C (50 m*M* triethanolamine hydrochloride, 15 m*M* KCl, 10 m*M* $(NH_4)_2SO_4$, and 1 m*M* $MgSO_4$, pH 6.9), and suspended at a density of 1×10^8 cells/ml in degassed buffer C.[28] To initiate uptake assays, 50 μl of cells are diluted into 0.45 ml of buffer C containing 10 m*M* glucose, 10 μM $NaAsO_2$, and 2.5 μCi of $H^{73}AsO_2$. Samples were withdrawn at intervals, filtered through nitrocellulose filters (0.45-μm pore, Whatman), washed with 5 ml of buffer C, and the radioactivity quantified.

For efflux studies, cells in degassed buffer C are depleted of energy reserves by incubation with DNP for 90 min at room temperature. Cells are washed three times, suspended in buffer C at 1×10^8 cells/ml, and loaded with arsenite on ice for 15 min in the presence of 0.2 m*M* $NaAsO_2$ with 1.25 μCi of $H^{73}AsO_2$. Efflux is initiated by diluting 50 μl of preloaded cells into 5 ml of buffer C containing 10 m*M* glucose as an energy source at room temperature. Samples (1 ml) are filtered, washed with 5 ml of buffer C, and the radioactivity quantified.

Assays of Arsenite Uptake in Plasma Membrane-Enriched Vesicles of Leishmania

Promastigotes of wild-type *L. tarentolae* TarII are grown in 2 liters of SDM-79 medium supplemented with 10% (v/v) fetal bovine serum (Gibco-BRL, Gaithersburg, MD) at 29° with gentle shaking to 5–6 $\times 10^7$ cells/ml, as determined by counting in a hemocytometer.[29] The cells are harvested at 4000*g* at 4°, washed twice with a buffer containing 75 m*M* Tris-HCl, pH 7.6, 0.14 *M* NaCl, 11 m*M* KCl, and once with a buffer D (10 m*M* HEPES–KOH, pH 7.4, 10 m*M* KCl, 1 m*M* magnesium acetate, 0.4 *M* mannitol, 1 m*M* phenylmethylsulfonyl fluoride and 0.15 mg/ml soybean trypsin inhibitor). After the final wash, cells are suspended in 2.5 ml of buffer D per milligram of wet cells and mixed with acid-washed glass beads

(150–212 mm in diameter, Sigma Chemicals, St. Louis, MO) at a ratio of 1:10 (wet weight of cells per weight of beads) and disrupted with a chilled morter for 10 min. Unbroken cells, cell debris, and glass beads are removed by centrifugation at 3000*g* for 10 min. To remove organelles and other intracellular membranes, the supernatant solution is centrifuged at 17,000*g* for 40 min at 6°. The membrane vesicles, highly enriched in plasma membranes, are isolated by centrifugation for 1 hr at 140,000*g*, washed, and suspended in 1 ml of buffer E (75 m*M* HEPES–KOH, pH 7.4, 0.15 *M* KCl, and 2 m*M* $MgCl_2$). The suspension is gently passed three times through a Wheaton homogenizer immersed in ice water. The plasma membrane vesicles are rapidly frozen in liquid nitrogen and stored at −70° until use.

Accumulation of $^{73}AsO_2^-$ and $^{73}As(GS)_3$ by membrane vesicles was assayed in 0.3 ml of buffer E lacking $MgCl_2$. Vesicles (0.5 mg/ml of membrane protein) and 5 m*M* $MgCl_2$ are added, and the mixture is incubated at 23° for 3 min. The $^{73}As(GS)_3$ (0.1 m*M* final concentration) is added, and the reaction initiated by addition of 10 m*M* ATP. At intervals 50-μl portions are filtered, washed with 5 ml of cold buffer E, and the radioactivity quantified (Fig. 7).

Conclusions

Whenever organisms are put under stress, they develop compensatory mechanisms. Interestingly, bacteria, lower eukaryotes and mammals have

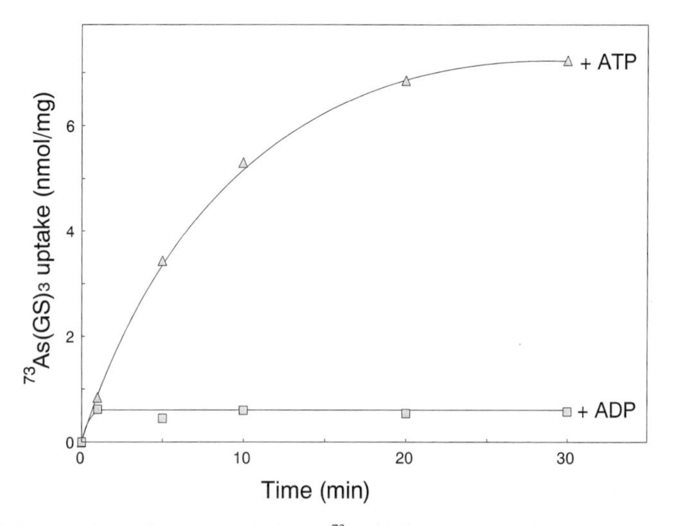

FIG. 7. Energy-dependent accumulation of $^{73}As(GS)_3$ in everted plasma membrane vesicles of *L. tarentolae.* Transport is measured in a buffer containing 5 m*M* $MgCl_2$ and 10 m*M* of either ADP (■) or ATP (▲).

all developed resistance mechanisms to arsenicals and antimonials. The stratagem is the same: each has evolved transport systems for the extrusion of the metalloids. However, the details of those mechanisms appears to differ, reflecting convergent evolution. Bacteria have evolved a secondary oxyanion carrier that can convert to a primary ATP-coupled pump in association with an ABC protein that is unrelated to members of the P-glycoprotein family. In contrast, *Leishmania* have evolved an ATP-coupled As–thiol pump. Thus the substrates of the prokaryotic and eukaryotic systems are different, but the results are the same: lowering of the intracellular metalloid concentration to produce resistance. Whether this pump also exists in mammalian cells and whether it is related to other GS-X pumps or to ABC transporters remains to be determined.

Section II

Eukaryotic ABC Transporters

A. Nonmammalian ABC Transport Systems
Articles 11 through 15

B. Mammalian P-Glycoproteins
Articles 16 through 43

C. Multidrug Resistance Associated Protein
Articles 44 and 45

D. Cystic Fibrosis Transmembrane Conductance Regulator
Articles 46 through 53

E. Sulfonylurea Receptor
Article 54

F. Intracellular ABC Transporters
Articles 55 through 57

[8] Evolutionary Relationships among ABC Transporters

By JAMES M. CROOP

Introduction

ATP-binding cassette (ABC) transporters[1] or traffic ATPases[2] comprise a superfamily of proteins that translocates a wide range of substrates across a variety of cellular membranes. The substrates transported include sugars, amino acids, peptides, proteins, metals, inorganic ions, toxins, and antibiotics. This family of proteins is defined by a functional transport unit that is believed to include two adenosine triphosphate (ATP)-binding domains and two hydrophobic domains, each containing five to eight membrane-spanning regions. The consensus sequence for the nucleotide-binding domain is comprised of two motifs separated by approximately 120 amino acids.[3] These motifs are termed motif A or the P-loop with a consensus sequence of [AG]-X(4)-G-K-[ST],[4] which is surrounded by a cluster of hydrophobic residues, and motif B with a consensus sequence of [RK]-X(3)-G-X(3)-L-hydrophobic(4)-D, which is often slightly variant (Fig. 1). Proposals have been made that the binding of the nucleotide is coordinated through P-loop binding of the phosphate groups, whereas the negatively charged aspartic acid in motif B interacts with the nucleotide associated Mg^{2+}.[5,6] Although this ATP-binding domain is found in a wide range of ATPases and partially defines a number of other superfamilies, the ABC transporters have an additional signature sequence between motifs A and B comprised of [LIVMFYC]-[SA]-[SAPGVYKQ]-G-[DENQMW]-[KRQAPCLW]-[KRNQSTAVM]-[KRACLVM]-[LIVMYPAN]-{PHY}-[LIVMFW]-[SAGCLIVP]-{FYWHP}-{KRHP}-[LIVMFYWSTA].[7,8] The complete functional transport unit is comprised of one to four polypeptides, which assemble in the membrane. Several configurations have been identified, including single proteins, homo/heterodimers, and multicomponent

[1] C. F. Higgins, *Ann. Rev. Cell Biol.* **8,** 67 (1992).
[2] G. F. Ames, C. S. Mimura, S. R. Holbrook, and V. Shyamala, *Adv. Enzym. Rel. Areas Mol. Biol.* **65,** 1 (1992).
[3] J. E. Walker, M. Saraste, M. J. Runswick, and N. J. Gay, *Embo J.* **1,** 945 (1982).
[4] Alternative amino acids at any one position []; amino acids not accepted at a position {}; X, amino acid; number of amino acids ().
[5] T. W. Traut, *Eur. J. Biochem.* **222,** 9 (1994).
[6] M. Saraste, P. R. Sibbald, and A. Wittinghofer, *Trends Biochem. Sci.* **15,** 430 (1990).
[7] A. Bairoch, *Nucl. Acids Res.* **20,** Suppl. 2013 (1992).
[8] A. E. Gorbalenya and E. V. Koonin, *J. Mol. Biol.* **213,** 583 (1990).

METHODS IN ENZYMOLOGY, VOL. 292
0076-6879/98 $25.00

```
       1                                                                     70
drrA   RAVDGLDLNVPAGLVYGILGPNGAGKSTTIRMLA.....TLLRPDGGTARVFGHD.VTSEPDTVRRRISVT
carA   DRLSVDSLHLGPGERLLVTGPNGAGKTTLLRVLS.....GELEPDSGSLLVSGRV.GHLRQEQTPWRPGMT
Hmdr1  V.LQGLSLEVKKGQTLALVGSSGCGKSTVVQLLE.....RFYDPLAGKVLLDGKEIKRLNVQWLRAHLGIV
hlyB   VILDNINLSIKQGEVIGIVGRSGSGKSTLTKLIQ.....RFYIPENGQVLIDGHDLALADPNWLRRQVGVV
Tap1   LVLQGLTFTLRPGEVTALVGPNGSGKSTVAALLQ.....NLYQPTGGQLLLDGKPLPQYEHRYLHRQVAAV
Hcftr  AILENISFSISPGQRVGLLGRTGSGKSTLLSAFL.....RLLN.TEGEIQIDGVSWDSITLQQWRKAFGVI
malK   VVSKDINLDIHDGEFVVFVGPSGCGKSTLLRMIA.....GLETITSGDLFI.GETRMNDIPPA.ERGVGMV
white  HLLKNVCGVAYPGELLAVMGSSGAGKTTLLNALAFRSPQGIQVSPSGMRLLNGQP...VDAKEMQARCAYV

       104                                                                170
drrA   ARERAAELIDGFGLG.....DARDRLLKTYSGGMRRRLDIAASIVVTPDLLFLDEPTTGLDPRSRNQVW
carA   AGDIDEHTEALLSLGLFSPDDLRQR.VQDLSYGQRRRIELARLVTEPVDLLLLDEPTNHL...SPALVE
Hmdr1  AKEANIHAFIESLPNKYST..KVGDKGTQLSGGQKQRIAIARALVRQPHILLLDEATSALDTESEKVVQ
hlyB   AKLAGAHDFISELREGYNT..IVGEQGAGLSGGQRQRIAIARALVNNPKILIFDEATSALDYESEHVIM
Tap1   AVKSGAHSFISGLPQGYDT..EVDEAGSQLSGGQRQAVALARALIRKPCVLILDDATSALDANSQLQVE
Hcftr  ADEVGLRSVIEQFPGKLDF..VLVDGGCVLSHGHKQLMCLARSVLSKAKILLLDEPSAHLDPVTYQIIR
malK   AKKEVMNQRVNQVAEVLQLAHLLERKPKALSGGQRQAVAIGRTLVAEPRVFLLDEPLSNLDAALRVQMR
white  VARVDQVIQELSLSKCQHTIIGVPGRVKGLSGGERKRLAFASEALTDPPLLICDEPTSGLDSFTAHSVV
```

FIG. 1. Nucleotide binding domain. Alignment of amino acid sequences surrounding the nucleotide binding domain of representative members of the ABC transporter superfamily. Motifs A and B are identified by solid underlines and the ABC transporter signature sequence is identified by a shaded underline. Identical residues among this group of polypeptides are shaded. Alignment was performed using the UWGCG Gap program [J. Devereux, P. Haeberli, and O. Smithies, *Nucl. Acids Res.* **12,** 387 (1984)]. Consecutive numbering of sequences is shown.

systems, which include all of the required domains (Fig. 2). The function of many of these systems is unknown and as genome mapping proceeds in a number of organisms, open reading frames encoding new family members continue to be identified.

In addition to the consensus sequences that define the ABC transporters, high sequence similarity occurs among all of the family members in the 200 amino acids surrounding the ATP binding site.[9] It is not unusual to find 30% identical residues and another 25% conserved residues between prokaryotic and eukaryotic family members. A comparison of amino acid similarity among the hydrophilic domains of a large group of ABC transporters reveals clusters that segregate family members by structural organization, functional activity, and speciation (Fig. 3). Table I provides the names and known functions of these proteins, demonstrating the wide range of substrates translocated by this superfamily.

[9] S. Childs and V. Ling, *in* "Important Advances in Oncology" (V. DeVita, S. Hellman, and S. Rosenberg, eds.), p. 21. Lippincott, Philadelphia, 1994.

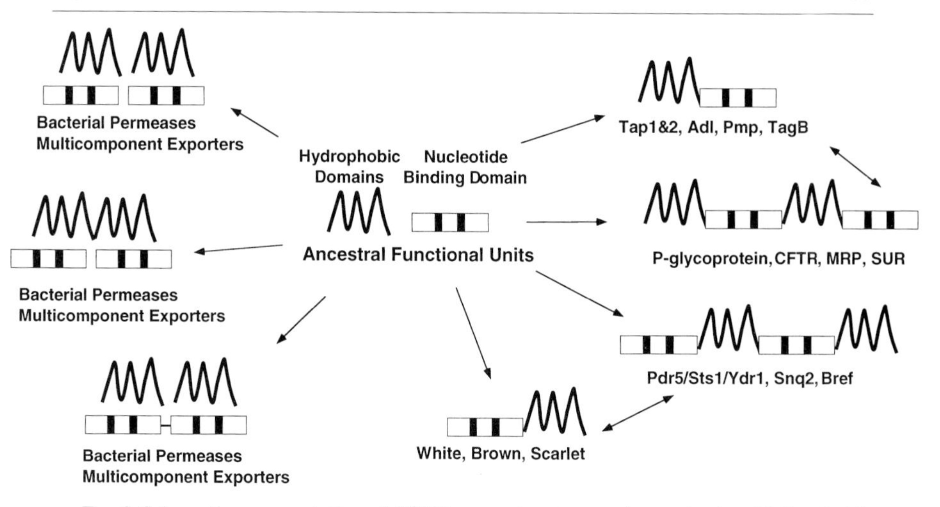

FIG. 2. Schematic representation of ABC transporter structural organization. Hydrophobic domains with transmembrane segments and hydrophilic domains with nucleotide binding site identified. See Table I for a description of proteins and accession numbers for references.

Single Polypeptides

ABC transporters with all of the components required to form a functional unit on a single polypeptide can be subdivided into three broad categories based on sequence similarity and structural arrangement. The P-glycoproteins, or multidrug-resistant proteins, are a highly conserved group of ABC transporters found as multigene families in a wide range of species. Their structural organization includes an N-terminal hydrophobic domain with six transmembrane segments followed by a hydrophilic domain with a nucleotide binding fold. This combination is repeated in the second half of the polypeptide. A second group with a similar structure but considerable sequence divergence is comprised predominantly of cystic fibrosis transmembrane regulators (CFTRs), multidrug-resistant associated proteins (MRPs), and the sulfonylurea receptor gene (SUR). Finally, the third group is organized with a hydrophilic domain at the N terminus followed by the hydrophobic domain, a structure that is repeated in the second half of the polypeptide producing a mirror image of the P-glycoprotein.

The mammalian P-glycoproteins have been extensively studied using genetic, biochemical, and physiologic analyses. The P-glycoproteins are encoded in a multigene family with two members, classes I and II, which convey multidrug resistance and a third family member, class III, which functions as a phospholipid flippase. The rodent P-glycoproteins conveying drug resistance are approximately 95% similar at the amino acid level.

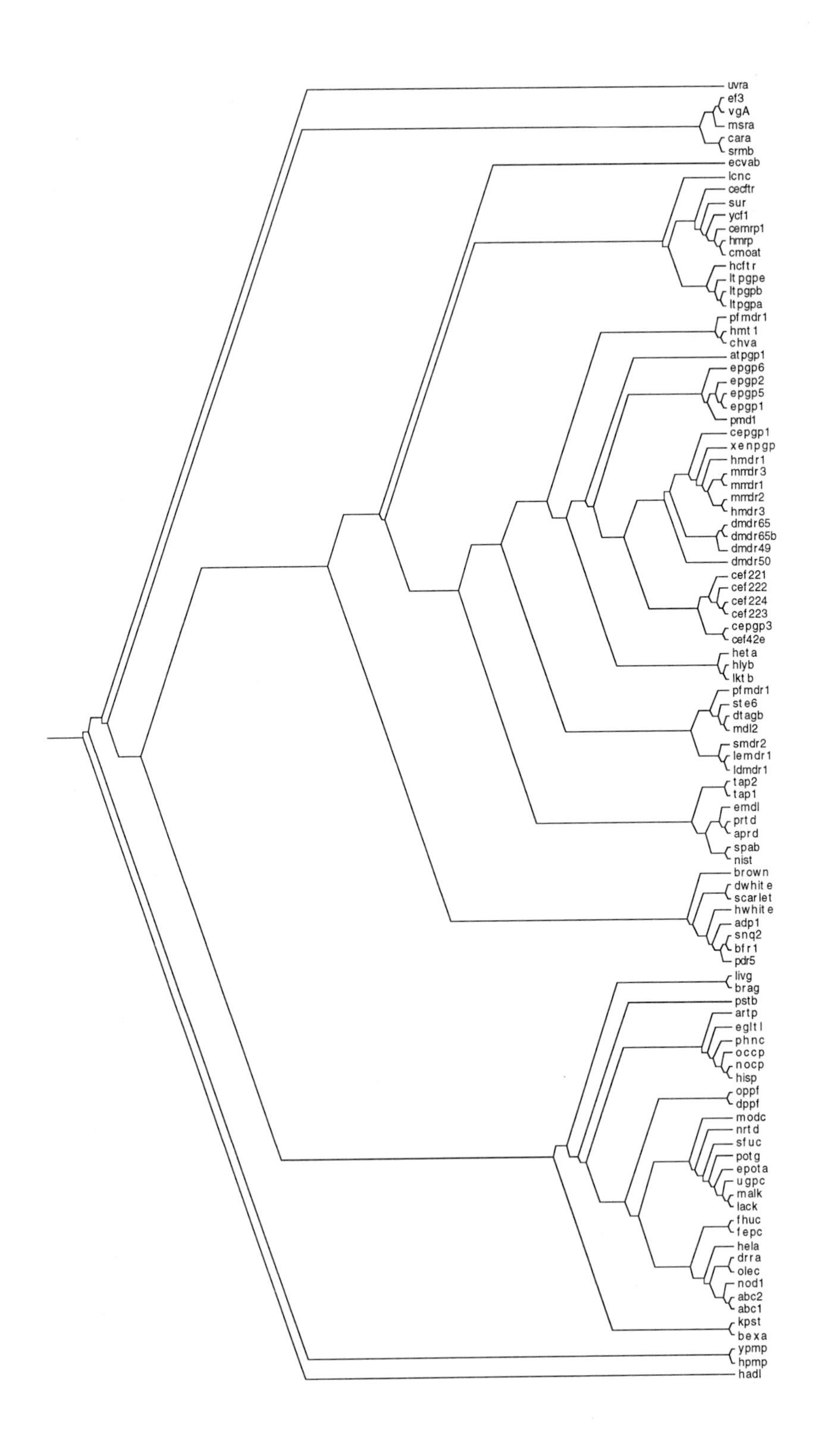

uvra
ef3
vgA
msra
cara
srmb
ecvab
lcnc
cecftr
sur
ycf1
cemrp1
hmrp
cmoat
hcftr
ltpgpe
ltpgpb
ltpgpa
pfmdr1
hmt1
chva
atpgp1
epgp6
epgp2
epgp5
epgp1
pmd1
cepgp1
xenpgp
hmdr1
mmdr3
mmdr1
mmdr2
hmdr3
dmdr65
dmdr65b
dmdr49
dmdr50
cef221
cef222
cef224
cef223
cepgp3
cef42e
heta
hlyb
lktb
pfmdr1
ste6
dtagb
mdl2
smdr2
lemdr1
ldmdr1
tap2
tap1
emdl
prtd
aprd
spab
nist
brown
dwhite
scarlet
hwhite
adp1
snq2
bfr1
pdr5
livg
brag
pstb
artp
egltl
phnc
occp
nocp
hisp
oppf
dppf
modc
nrtd
sfuc
potg
epota
ugpc
malk
lack
fhuc
fepc
hela
drra
olec
nod1
abc2
abc1
kpst
bexa
ypmp
hpmp
hadl

Humans do not have a class II P-glycoprotein, suggesting gene duplication in rodents after divergence from humans. The human class I MDR1 has approximately 92% amino acid similarity with rodent class I. Class III P-glycoprotein has approximately 85% similar amino acids with the members that convey drug resistance. The mammalian P-glycoproteins display approximately 60–65% amino acid similarity with most P-glycoproteins from other species.

There are four members of the *Drosophila* P-glycoprotein multigene family. Unlike the mammalian genes, which are clustered in tandem on a single chromosome, the *Drosophila* multigene family has a pair located in tandem on chromosome 3, Mdr65 and Mdr65b, and two members located far apart on chromosone 2, Mdr49 and Mdr50. Although Mdr65 and Mdr65b are most similar, there is only 65–72% amino acid similarity among the entire family. Flies with misexpression of Mdr49 are fertile and viable, but can be more sensitive to growth on colchicine. Preliminary observations indicate that both Mdr49 and Mdr65 can transport substrates from the multidrug resistance phenotype and are labeled with the same photoaffinity agents that label mammalian class I and II P-glycoproteins.[10]

The *Caenorhabditis elegans* P-glycoprotein multigene family may contain as many as nine family members. One homolog characterized in detail, Cepgp3, shows tissue-specific expression on the apical membranes of the excretory cell and the intestine.[11,12] Disruption of this homolog, but not Cepgp1, which is also expressed in the intestine, increases sensitivity of the worms to colchicine and chloroquine, suggesting a protective role against xenobiotics. Seven additional homologs have been identified on chromosome III, four on a single cosmid.

P-Glycoprotein homologs have been identified in several protozoa species. The P-glycoprotein identified in *Plasmodium falciparum*, pfmdr1, can be found amplified in chloroquine resistance isolates; however, its role in resistance remains controversial.[13] Located on the surface of the food vacu-

[10] I. Bosch and J. Croop, unpublished observations (1996).
[11] A. Broeks, H. W. Janssen, J. Calafat, and R. H. Plasterk, *Embo J.* **14,** 1858 (1995).
[12] C. R. Lincke, A. Broeks, I. The, R. H. Plasterk, and P. Borst, *Embo J.* **12,** 1615 (1993).
[13] P. Borst and M. Ouellette, *Ann. Rev. Micro.* **49,** 427 (1995).

FIG. 3. Cladogram of ABC transporters. Amino acid sequences surrounding the nucleotide binding site (approximately 230 amino acids) of representative ABC transporters aligned using the UWGCG Pileup program. Unrooted phylogeny are inferred using Protpars and the corresponding tree is plotted using Drawgram from PHYLIP, version 3.5. [J. Felsenstein, *Cladistics.* **5,** 164 (1989)]. Lowercase names are used for graphical presentation. See Table I for a description of proteins and accession numbers for references.

TABLE I
ABC TRANSPORTERS[a]

Protein	Source	Accession number
Single Polypeptides		
Hmdr1	Human Mdr1	P08183
Hmdr3	Human Mdr3	P21439
Mmdr1	Mouse Mdr1	P06795
Mmdr2	Mouse Mdr2	P21440
Mmdr3	Mouse Mdr3	P21447
Hmpgp1	Hamster Pgp1	B38696
Hmpgp2	Hamster Pgp2	B27126
Hmpgp3	Hamster Pgp3	A38696
Rpgp1	Rat Pgp1	P43245
Rpgp2	Rat Pgp2	Q08201
Dmdr49	*Drosophila* Mdr49	Q00449
Dmdr50	*Drosophila* Mdr50	A47377
Dmdr65	*Drosophila* Mdr65	Q00748
Epgp1	*E. histolytica* Pgp1	S30327
Epgp2	*E. histolytica* Pgp2	S30328
Epgp5	*E. histolytica* Pgp5	L23922
Epgp6	*E. histolytica* Pgp6	V01056
Cepgp1	*C. elegans* Pgp	X65054
Cepgp3	*C. elegans* Pgp	X65055
Cef221	*C. elegans* Pgp	Z67882
Cef222	*C. elegans* Pgp	Z67882
Cef223	*C. elegans* Pgp	Z67882
Cef224	*C. elegans* Pgp	Z67882
Cef42e	*C. elegans* Pgp	Z66562
Ldmdr1	*Leishmania* Mdr1	Q06034
Lemdr1	*Leishmania* Mdr1	Q06034
Xenpgp	*Xenopus* Pgp	U17608
pfmdr1	*P. falciparum*	P13568
Ste6	*S. cerevisiae*	P12866
pmd1	*S. pombe*	P36619
Smdr2	*S. mansoni*	L26287
Atpgp1	*A. thaliana*	A42150
Emd1	*E. coli* Mdl	P30751
Spgp	Sister of Pgp	U20587
Hcftr	Human CFTR	P13569
Hmrp	Human MRP	P33527
Cemrp1	*C. elegans* Mrp1	U41554
Cemrp2	*C. elegans* Mrp2	Z68113
cMOAT	Anion transport	L49379
ltpgpA	*Leishmania*	A34207
ltpgpB	*Leishmania*	L29484
ltpgpE	*Leishmania*	L29485

TABLE I (*Continued*)

Protein	Source	Accession number
Ycf1	*S. cerevisiae*	P39109
SUR	Sulfonylurea	Q09428
Pdr5C	*S. cerevisiae*	P33302
Bref	*S. pombe*	P41820
Snq2	*S. cerevisiae*	X66732
abc1	Mouse	P41233
abc2	Mouse	P41234
Half-Molecules		
Tap1	MHC peptide transport	Q03518
Tap2	MHC peptide transport	Q03519
hlyB	Hemolysin Toxin Export	P08716
cyaB	Cyclolysin Toxin Export	P18770
lktB	Leukotoxin Export	P16532
lcnC	Lactococcin A Export	Q00564
dtagB	*Dictyostelium*	U20432
md12	*S. cerevisiae*	P33311
chvA	β(1, 2) Glucan Export	P18768
ndvA	β(1, 2) Glucan Export	P18767
hetA	Heterocyst Development	P22638
hmt1	Cadmium resistance	Q02592
dwhite	*Drosophila* white	P10090
hwhite	Human white	P45844
scarlet	*Drosophila* scarlet	U39739
brown	*Drosophila* brown	P12428
Adp1	*Drosophila* ORF	P25371
Ald	Adrenoleukodystrophy	P33897
Hpmp	Peroxisomal protein	P28288
Ypmp	*S. cerevisiae*	P34230
Multicomponent Bacterial Permeases		
hisP	Histidine	P07109
fhuC	Fe(III)-Ferrichrome	P07821
fepC	Ferric enterobactin	P23878
sfuc	Fe(III)	P21410
epotA	Spermidine/putrescine	P23858
egltL	Glutamate/aspartate	P41076
cysA	Sulfate/thiosulfate	P14788
braG	Branch Chained AA	P21630
livG	Branch Chained AA	P30293
artP	Arginine	P30858
mglA	Methylgalactoside	U45323
araG	Arabinose	X06091
upgC	Glycerol 3-phosphate	S47669
sapF	Peptide	P36637
rbsA	Ribose	P04983

(*Continued*)

TABLE I (*Continued*)

Protein	Source	Accession number
pstB	Phosphate	P07655
potG	Putrescine	P31134
phnC	Alkyl phosphonate	P16677
oppF	Oligopeptide	P24137
dppF	Dipeptides	P37313
occP	Octopine	P35117
nocP	Nopaline/octopine	P35116
nrtD	Nitrate	P38046
modC	Molybdenum	Q08381
malK	Maltose/maltodextrin	P02914
lacK	Lactose	Q01937
btuD	Vitamin B_{12}	M14031
Multicomponent Exporters		
helA	Heme	P29959
arsA	Arsenicals	J02591
kpsT	Polysialic acid	P23888
bexA	Polysaccharide	P10640
nod1	Oligosaccharide	P26050
aprD	Alkaline protease	Q03024
prtD	Proteases	P23596
drrA	Anthracycline	M73758
ecvab	Colicin V	P22520
msrA	Erythromycin	P23212
srmB	Spiromycin	S25202
vgA	Virginiamycin	JC1204
oleC	Oleandomycin	L06249
spaB	Lantibiotic subtilin	P33116
nisT	Lantibiotic nisin	Q03203
P29	Mycoplasma	M37339
mbcF	Microcin B17	X07875
epiY	Epidermin	X63286
Other Proteins		
uvra	Excision nuclease	P13567
ef3	Elongation factor	P16521
rpoN	Sigma factor	P33527
fimA	Fimbral protein	M26130

[a] Representative proteins with species or function when known. See text for discussion. Accession numbers from Swiss Protein or GenBank database listed for retrieval of sequences and references.

ole, it is believed to transport substrates into this organelle. Although overexpression in mammalian cells lead to increased sensitivity to chloroquine,[14] evidence has suggested a role for pfmdr1 in resistance to a second antimalarial agent, mefloquine.[15,16] Four members of the P-glycoprotein multigene family have been identified in *Entomoeba histolytica.* At least one of these genes is associated with increased resistance of the amoeba to emetine. Similarly, a P-glycoprotein homolog has been identified in two species of *Leishmania* that convey resistance to vinblastine, lemdr1 and ldmdr1. A P-glycoprotein homolog, Smdr2, has also been identified in schistosomes.

STE6 encodes a P-glycoprotein homolog, which transports the **a** factor peptide, which is required for mating in *Saccharomyces cerevisiae.* It is of interest that **a** factor undergoes hydrophobic modification prior to export, suggesting that lipophilicity may be a general property of P-glycoprotein transport substrates. Indeed, both mouse mdr3 and pfmdr1 have been shown to be able to transport **a** factor when expressed in yeast.[17,18] A P-glycoprotein homolog, pmd1$^+$, has also been isolated from a leptomycin resistant strain of *Schizosaccharomyces pombe.*

Two general observations can be made in comparing the primary amino acid sequence of the P-glycoproteins. First, gene duplication with modification of substrate specificity seems to be a common phenomenon in a wide range of species. Homologs that are in tandem on the same chromosome display the highest similarity. Second, sequence similarity along the entire polypeptide is approximately 65% among most of the homologs across large evolutionary distances.[19] The most notable exceptions are Ste6 and pfmdr1, which are more divergent with only 50–55% similar residues. Indeed, the question of which polypeptides should be considered P-glycoproteins becomes problematic. It is not clear whether functional activities should be considered or whether sequence similarity and secondary structure alone is sufficient.

The second group of ABC transporters, which have an organization similar to the P-glycoproteins, includes CFTR, MRP, and SUR. The mammalian proteins display approximately 50% amino acids similarity with the

[14] H. H. G. van Es, S. Karcz, F. Chu, A. Cowman, S. Vidal, P. Gros, and E. Schurr, *Mol. Cell Biol.* **14,** 2419 (1994).

[15] C. Wilson, S. K. Volkman, S. Thaithong, R. K. Martin, D. E. Kyle, W. K. Milhous, and D. Wirth, *Mol. Biochem. Para.* **57,** 151 (1993).

[16] S. A. Peel, P. Bright, B. Yount, J. Handy, and R. S. Baric, *Am. J. Trop. Med. Hyg.* **51,** 648 (1994).

[17] M. Raymond, P. Gros, M. Whiteway, and D. Y. Thomas, *Science* **256,** 232 (1992).

[18] S. K. Volkman, A. F. Cowman, and D. F. Wirth, *Proc. Natl. Acad. Sci. U.S.A.* **92,** 8921 (1995).

[19] J. Croop, *Cytotechnology* **12,** 1 (1993).

mammalian P-glycoproteins and approximately 60% similarity with each other. CFTR is a chloride channel that is activated by cAMP-mediated phosphorylation of the R region, a highly charged domain with protein kinase consensus sequences located between the two halves of the polypeptide. Mutations that decrease CFTR functional activity are responsible for cystic fibrosis, one of the most common genetic diseases. MRP, responsible for one form of multidrug resistance, has been demonstrated to have a role in the export of certain glutathione conjugates.[20] Decreased expression of a rat liver specific MRP homolog, the canalicular multispecific organic transporter, cMOAT, has been associated with a defect in excretion of conjugated bilirubin and organic acids. Sulfonylureas, oral hypoglycemic agents that decrease serum glucose, stimulate the secretion of insulin through inhibition of SUR functional activity, the regulation of an ATP-dependent potassium channel. Consistent with this observation, familial persistant hyperinsulinemic hypoglycemia of infancy has been associated with mutations that decrease SUR functional activity.[21] Similarly, both P-glycoprotein and CFTR have been proposed to regulate chloride channels, suggesting a common functional property of these ABC transporters.[22] Five homologs with sequence similarity to this group of ABC transporters have been identified in *Leishmania.* Overexpression of 1tpgpA, which displays 50% amino acid similarity with both MRP and CFTR, conveys resistance to oxyanions such as arsenite, antimonite, and pentavalent antimony, which is the drug of choice for treatment leishmaniasis.[13] Similarly the *S. cerevisiae* homolog ycf1 has been shown to convey resistance to the heavy metal cadmium.

A third group of ABC transporters that encode all of domains required for a functional unit on a single polypeptide is related to the *Drosophila* white protein. The hydrophilic domain, which includes the nucleotide binding fold, is proximal to the hydrophobic domain, as is found in the white protein, and this configuration is then repeated on the second half of the molecule. Several transporters that export specific or multiple toxins from yeast have been identified. Overexpression of pdr5/sts1/ydr1, snq2, and bfr1 conveys resistance to multiple antiseptics and other toxic compounds. Each transporter has a distinct drug-resistant phenotype, although there is some overlap in specificity. Disruption of these genes results in increased sensitivity to the same compounds.

[20] G. J. R. Zaman, J. Lankelma, O. Vantellingen, J. Beijnen, H. Dekker, C. Paulusma, R. P. J. Oudeelferink, F. Baas, and P. Borst, *Proc. Natl. Acad. Sci. U.S.A.* **92,** 7690 (1995).

[21] P. M. Thomas, G. J. Cote, N. Wohllk, B. Haddad, P. M. Mathew, W. Rabl, L. Aguilar-Bryan, R. F. Gagel, and J. Bryan, *Science* **268,** 426 (1995).

[22] C. F. Higgins, *Cell* **82,** 693 (1995).

Half-Molecules, Homodimers, and Heterodimers

ABC transporters comprised of polypeptides, which include a hydrophobic domain with transmembrane segments and a hydrophilic domain with the consensus for the nucleotide binding domain, have similarly been identified. Although it is presumed that these transporters function as dimers or heterodimers to include all the necessary motifs of the ABC transporters, supporting experimental evidence has not yet been presented for most of the transport systems. Two main configurations have been identified, the hydrophobic domain follow by the hydrophilic domain with the nucleotide binding site and the reverse orientation with the nucleotide binding site at the amino terminus. In addition, several of the polypeptides have novel domains that convey specific functionality.

Polypeptides consisting of a series of transmembrane loops followed by a hydrophilic domain with a nucleotide-binding fold, similar to half of a P-glycoprotein, have been identified in organisms across large evolutionary distances. Transport systems that export virulence toxins (hylB, ltkB, cyaB), antibiotics (cvaB, spaB, lcnC, syrD, nisT), polysaccharides (ndvA, chvA, hetA), proteases (prtD), and heavy metals (hmt) have been identified in a wide range of prokaryotes, These polypeptides have hydrophobicity plots that are remarkably similar to either half of the P-glycoprotein, and approximately 50% of their amino acids are conserved with the P-glycoprotein along the entire polypeptide. The hydrophilic domains display 40–45% amino acid identity with the analogous region of the P-glycoprotein and several of the prokaryote sequences display considerable nucleic acid identity around the nucleotide binding consensus. The mammalian P-glycoproteins display higher levels of amino acid similarity with several of the prokaryotic transporters (hlyB, ltka, chvA, hetA) than with several family members considered P-glycoproteins such as Ste6 and pfmdr1.

The mammalian major histocompatibility (MHC) peptide transport system is comprised of two polypeptides, Tap1 and Tap2, which also have the structure of half of a P-glycoprotein. These proteins have been implicated in the transport of short peptides from the cytosol into the endoplasmic reticulum where they are inserted into the groove of the MHC class I protein for antigen presentation at the cell surface. Both polypeptides are required, suggesting that they function as a heterodimer. These two proteins share approximately 60% similar amino acid residues between themselves and 55% similar amino acid residues with the P-glycoproteins. Both are slightly more similar to the carboxy half of the P-glycoprotein. A second group of mammalian half P-glycoproteins is expressed as integral membrane proteins in peroxisomes. Adrenoleukodystrophy is associated with a partially deleted half P-glycoprotein, Ald, in some patients. The disorder is characterized by the accumulation of very long chain fatty acids, presumably due to a defect in translocation of a component of the oxidative metabolic pathway into the peroxisome by Adl. Two other ABC transporters with high amino acid similarity to Ald, the related AldR and the peroxisomal membrane protein, Pmp, with approximately 80 and 60% conserved residues, respectively, are also associated with peroxisomes althogh the tissue-specific distribution appears distinct from Ald. Zellweger syndrome, a disorder of peroxisome biogenesis, has been linked to Pmp but the precise relationship of Pmp, or the other ABC transporters, to the disorder remains uncertain.[23]

The *Drosophila* white, brown, and scarlet genes are ABC transporters that are mirror images to the half P-glycoproteins with the hydrophilic domain followed by the transmembrane domains. Genetic, biochemical, and molecular evidence indicates that these proteins function as heterodimers to transport pigment precursors to the eye. White and brown are responsible for guanine uptake and white and scarlet are responsible for tryptophan uptake. The transposition of the hydrophilic and hydrophobic domains in a number of ABC transporters suggests that the ancestral functional units that comprise the functional transport system originally had significant latitude as building blocks.

ABC transporters with specific motifs in addition to the components required for substrate translocation have been identified. Disruption of the prestalk specific *tagB* gene in *Dictyostelium* results in a block in development and differentiation. This ABC transporter family member is arranged like a half P-glycoprotein, but has an additional 1000 amino acids at the N terminus, including sequence similarity to a family of serine proteases. In addition, partial sequence of *tagC* suggests a polypeptide with high similarity to *tagB* and a similar phenotype when disrupted, suggesting possible functioning as a heterodimer. Adp1, an open reading frame in the yeast genome, has the hydrophilic and hydrophobic domains arranged as is found in the white protein. However, there are additional 300 amino acids at the N terminus, including two epidermal growth factor cysteine pattern signatures in an additional predicted extracellular domain at the amino terminus.

Multicomponment Bacterial Permeases

The bacterial permeases comprise a family of transport systems for the energy-dependent uptake of nutrients in gram-negative bacteria.[24] Permeases move nutrients from the periplasmic space, between the outer cell wall

[23] J. Gartner, H. Moser, and D. Valle, *Nature Genet.* **1,** 16 (1992).

[24] G. Ames, *Annu. Rev. Biochem.* **55,** 397 (1986).

and the relatively impermeable cytoplasmic membrane, into the cytoplasm. These systems are comprised of a soluble substrate-binding protein and a membrane-bound complex encoded in one or two operons. The substrate-binding protein is located in the periplasmic space where it undergoes a conformational change on ligand binding. Although there is little amino acid homology among the substrate-binding proteins, their tertiary structures are remarkably similar based on X-ray crystallography analysis with a binding site located in a cleft between two lobes.[25] The receptor–ligand complex is then able to interact with the membrane-bound components comprised of proteins encoding two hydrophobic domains and two hydrophilic domains. The hydrophilic components include the highly conserved ATP binding fold. Substrate translocation requires a series of confirmational changes and ATP hydrolysis with movement hypothesized to occur either through a pore or multiple substrate-binding sites.[26] Mutants with constitutive ATPase activity have gained the ability to transport substrate without the binding protein, suggesting that one function of binding proteins is to signal ATP hydrolysis.[27]

Although there is a high degree of similarity in the functional transport units of the bacterial permeases, the genetic organization is quite complex. The highest level of amino acid similarity, approximately 50%, is found among the hydrophilic domains, which include the ATP-binding fold. This sequence level of similarity extends across large evolutionary distances to all ABC transporters. Although these proteins are described as hydrophilic, they are tightly bound to the membrane. Indeed, HisP, the hydrophilic component of the histidine permease, is accessible to proteolysis and biotinylation on the external surface of the membrane.[28] Although it is not clear if partial or complete membrane penetration occurs, it is interesting to note that overexpression of the analogous region of CFTR is sufficient to produce an anion channel.[29] The ATP-binding domains required for transport can be encoded in a polypeptide present twice in the functional unit (hisP), in two similar polypeptides, both required for transport as in the oligopeptide transport system (oppD and oppF), or with the nucleotide-binding domain duplicated on a single protein that is found in several sugar transporters (mglA, rbsA, araG). The two hydrophobic domains in the transport unit are each comprised of five or six predicted transmembrane domains suggesting conservation of secondary structure along with some

[25] M. D. Adams and D. L. Oxender, *J. Biol. Chem.* **264,** 15739 (1989).

[26] C. A. Doige and G. F. Ames, *Ann. Rev. Micro.* **47,** 291 (1993).

[27] A. L. Davidson, H. A. Shuman, and H. Nikaido, *Proc. Natl. Acad. Sci. U.S.A.* **89,** 2360 (1992).

[28] V. Baichwal, D. Liu, and G. F. Ames, *Proc. Natl. Acad. Sci. U.S.A.* **90,** 620 (1993).

[29] N. Arispe, E. Rojas, J. Hartman, E. Sorscher, and H. B. Pollard, *Proc. Natl. Acad. Sci U.S.A.* **89,** 1539 (1992).

conservation of primary sequence.[30] Although these domains are usually encoded by two separate proteins, the iron hydroxamate transporter (fhuB) and a high-affinity *Mycoplasma* transport system[31] have both hydrophobic domains encoded on a single polypeptide. Many of the permeases are organized as a single operon. However, the maltose transporter is comprised of two transcription units.

Multicomponent Exporters

Multicomponent ABC transporters that export a variety of substrates have been identified in both eukaryotic and prokaryotic organisms. Although not characterized as well as many of the bacterial permeases, the multicomponent exporters also appear to have several different types of polypeptide arrangements to form a similar functional unit. Several of these transport systems export cytotoxic drugs, providing a mechanism for drug resistance in microorganisms that produce these antibiotics. Included in this group is the resistance loci for daunorubicin and doxorubicin in *Streptomyces* species producing these compounds. The drrAB locus is linked to the anthracycline biosynthetic genes encoding a single transcription unit, which is expressed during antibiotic synthesis. The drrA protein is hydrophilic with the consensus sequence for a nucleotide-binding domain, whereas drrB is hydrophobic with six potential transmembrane domains. Although the tertiary structure is not known at this time, it is tempting to envision a transport unit comprised of these subunits arranged similarly to the P-glycoprotein that translocates the same substrates. The oleandomycin resistance operon encodes a similar pair of polypeptides. The resistance determinants to the macrolide antibiotics spiramycin, carbomycin, and tylosin, also produced in *Streptomyces,* include polypeptides with two nucleotide-binding domains linked in tandem, srmB, carA, and tlrC, respectively. Other polypeptides that could complete the functional unit have not yet been identified, but it is presumed that antibiotics are transported out of the cell.

Antibiotic resistance determinants in bacteria that are ABC transporters have been similarly identified. Transport systems with a single hydrophilic domain that includes the nucleotide-binding site have been identified for resistance to the peptide antibiotics microcin B17 (mbcF) and epidermin (epiY). Transport systems that include a hydrophilic domain with two linked nucleotide-binding sites have been identified as inducible mechanisms of

[30] R. E. Kerppola and G. F. Ames, *J. Biol. Chem.* **267,** 2239 (1992).

[31] R. Dudler, C. Schmidhauser, R. W. Parish, R. E. Wettenhall, and T. Schmidt, *EMBO J.* **7,** 3963 (1988).

erythromycin (msrA) and virginiamycin (vgA) resistance. Overexpression of msrA has been demonstrated to decrease erythromycin accumulation. This duplicated structure is similarly found in arsA, which is responsible for arsenical resistance.

A group of transport systems that translocate substrates required for assembly of the bacterial capsule and are responsible for virulence has been identified. The *bexA* and *ctrD* genes encode hydrophilic polypeptides containing a nucleotide-binding domain and are required for polysaccharide incorporation into the capsules of *Haemophilus influenzae* type b and *Neisseria meningitides*, respectively. The Nod1 protein from *Rhizobium* is thought to function similarly. The *kpsT* gene has a similar structure and is responsible for polysialic acid incorporation into the capsule of *Escherichia coli*. Each interacts with a hydrophobic protein comprised of transmembrane domains as is found in other ABC transporters. These transport systems provide potential targets for the therapy of pathogenic microorganisms.

Other Related Proteins

A small group of diverse proteins, which include nucleotide-binding domains and the ABC transporter signature sequence but do not appear to transport substrates, has also been identified. A subunit of the ABC excision nuclease, uvrA, which functions as a DNA repair enzyme, includes two nucleotide-binding domains, as does elongation factor-3, which is involved in the energy transduction required for aminoacyl-tRNA binding to the ribosome. Whether these proteins represent convergent or divergent evolution from the other ABC transporters is not clear.

Future Considerations

The ABC transporters represent a fascinating superfamily of proteins that transports a wide range of substrates across a variety of biological membranes. A number of issues remain unanswered, however. It is not totally clear how similar the transport system organization is among the different family members. A more complete understanding of the three-dimensional topology, protein–protein interactions, and tertiary structure is required. An understanding of whether there are common determinants for substrate specificity and whether multiple functions are included within each system is still required. Identification of the structural elements that account for the direction of translocation and subcellular localization remains elusive. Finally, the fundamental question of how the substrates move across membranes is unknown. Further analysis of the ABC transporters should provide information on basic biological processes and their effect on physiological functions.

Acknowledgments

I would like to thank Irene Bosch for critically reviewing the manuscript and Tom Graf for help with the sequence analysis. JMC is supported in part by a grant from Sandoz Pharmaceuticals, the Dana-Farber Cancer Institute, and licensing agreements on patents related to the multidrug resistance gene.

[9] Cloning of Novel ABC Transporter Genes

By RANDO ALLIKMETS and MICHAEL DEAN

Introduction

The multidrug resistance/ATP-binding cassette (MDR/ABC) superfamily includes genes whose products represent membrane proteins involved in energy-dependent transport of a wide variety of substrates across membranes (for review, see Refs. 1 and 2). The ABC family (also called traffic ATPases) is one of the largest superfamilies of proteins known in both prokaryotic and eukaryotic organisms and can be clearly distinguished from other ATP-binding protein families such as kinases.[3] In eukaryotes, ABC genes typically encode four domains consisting of two ATP-binding segments, and two transmembrane (TM) segments; half-molecules, containing one ATP and one TM domain also exist.[4] The ATP-binding domains of ABC genes contain characteristic conserved residues that have allowed new ABC genes to be isolated by hybridization, degenerate PCR (polymerase chain reaction), and inspection of DNA sequence databases.[5–10]

[1] S. Childs and V. Ling, *in* "Important Advances in Oncology" (V. T. DeVita, S. Hellman, and S. A. Rosenberg, eds), pp. 21–36. Lippincott, Philadelphia, 1994.

[2] M. Dean and R. Allikmets, *Curr. Opin. Genet. Devel.* **5,** 79 (1995).

[3] C. F. Higgins, I. D. Hiles, G. P. C. Salmond, D. R. Gill, J. A. Downie, I. J. Evans, B. Holland, L. Gray, S. D. Buckel, A. W. Bell, and M. A. Hermodson, *Nature* **323,** 448 (1986).

[4] S. C. Hyde, P. Emsley, M. J. Hartshorn, M. M. Mimmack, U. Gileadi, S. R. Pearce, M. P. Gallagher, D. R. Gill, R. E. Hubbard and C. F. Higgins, *Nature* **346,** 362 (1990).

[5] M. F. Luciani, F. Denizot, S. Savary, M. G. Mattei and G. Chimini, *Genomics* **21,** 150 (1994).

[6] M. Dean, R. Allikmets, B. Gerrard, C. Stewart, A. Kistler, B. Shafer, S. Michaelis and J. Strathern, *Yeast* **10,** 377 (1994).

[7] R. Allikmets, B. Gerrard, D. Glavač, M. Ravnik-Glavač, N. A. Jenkins, D. J. Gilbert, N. G. Copeland, W. Modi and M. Dean, *Mammal. Genome* **6,** 114 (1995).

[8] R. Allikmets, B. Gerrard, D. Court and M. Dean, *Gene* **136,** 231 (1993).

0076-6879/98 $25.00

Previously characterized members of the superfamily include those transporting chemotherapeutic drugs (PGP1/3, MRP[11-14]), peptide transporters involved in antigen presentation (TAP1/2[15]), and those involved in a number of inherited human diseases such as cystic fibrosis transmembrane regulator (CFTR), adrenoleukodystrophy gene (ALD), sulfonyl urea receptor (SUR), peroxisomal membrane protein (PMP70)[16-19]. To date, only 12 ABC genes have been identified in humans, although more than 25 members of this highly conserved superfamily have been described in *Escherichia coli* alone.[20,21] The fact that prokaryotes contain a large number of ABC genes suggests that many mammalian members of the superfamily remain uncharacterized. This article focuses on analyses of different methods used to characterize new genes of the ABC superfamily structurally and, ultimately, functionally.

Function/Positional Cloning

Most ABC genes have been identified based on their function or through the cloning of disease genes.[1,2] Most bacterial ABC genes are involved in the import of essential nutrients, and the genes encoding these transporters were identified through genetic mapping of phenotypes.[21] For example, the *araG* operon is required for the import of arabinose. Mutation of the genes in this operon impair the ability of *E. coli* to utilize arabinose as a carbon source.

---

[9] J. Leighton and G. Schatz, *EMBO J.* **14,** 188 (1995).

[10] B. Gerrard, C. Stewart and M. Dean *Genomics* **17,** 83 (1993).

[11] J. H. Gerlach, N. Kartner, D. R. Bell and V. Ling, *Cancer Surv.* **5,** 25 (1986).

[12] P. Gros, J. Croop and D. Housman, *Cell* **47,** 317 (1986).

[13] J. Croop, P. Gros and D. Housman, *J. Clin. Invest.* **81,** 1303 (1988).

[14] S. P. C. Cole, G. Bhardwaj, J. H. Gerlach, J. E. Machie, C. E. Grant, K. C. Almquist, A. J. Stewart, E. U. Kurz, A. M. V. Duncan and R. G. Deeley, *Science* **258,** 1650 (1992).

[15] T. Spies, M. Bresnahan, S. Bahram, D. Arnold, G. Blanck, E. Mellins, D. Pious and R. DeMars, *Nature* **348,** 744 (1990).

[16] J. R. Riordan, J. M. Rommens, B.-S. Kerem, N. Alon, R. Rozmahel, Z. Grzelczak, J. Zielenski, S. Lok, N. Plavsic, J.-L. Chou, M. L. Drumm, M. C. Ianuzzi, F. S. Collins and L.-C. Tsui, *Science* **245,** 1066 (1989).

[17] J. Mosser, A.-M. Douar, C.-O. Sarde, P. Kioschis, R. Feil, H. Moser, A.-M. Poustka, J.-L. Mandel and P. Aubourg, *Nature* **361,** 726 (1993).

[18] P. M. Thomas, G. J. Cote, N. Wohllk, B. Haddad, P. M. Mathew, W. Rabl, L. Aguilar-Bryan, R. F. Gagel and J. Bryan, *Science* **268,** 426 (1995).

[19] N. Shimozawa, T. Tsukamoto, Y. Suzuki, T. Orii, Y. Shirayoshi, T. Mori and Y. Fujiki, *Science* **255,** 1132 (1992).

[20] G. Ames and H. Lecar, *FASEB J.* **6,** 2660 (1992).

[21] C. F. Higgins, *Annu. Rev. Cell. Biol.* **8,** 67 (1992).

Many eukaryotic ABC genes have also been identified based on phenotype. A number of rodent cell lines selected for resistance to chemotherapy drugs overexpress a 170-kDa protein (P-glycoprotein), often due to amplification of the corresponding gene. The amplified gene was cloned and shown to be an ABC protein also known as the multidrug resistance (MDR) gene.[11] A number of genes encoding for visible phenotypes in experimental organisms have been identified as ABC genes. For example, one of the first identified mutants is the white locus of *Drosophila.* White encodes an ABC half-molecule that pairs up with either the brown or scarlet gene product (also half-molecules) to form a transporter for the pterin molecules that are the precursors for the eye pigments.[22] Sterile 6 (STE6) in the yeast *Saccharomyces cerevisiae* encodes an ABC protein that is the transporter for the mating factor, a short modified peptide.[23] When characterizing a gene based on its phenotype one has to be careful with nomenclature. Several, mainly prokaryotic, proteins have been termed "multidrug resistant" because they confer resistance to multiple antibiotics. However, many of these genes are not members of the ABC superfamily.[24,25] At the same time a number of proteins exist in eukaryotes that transport metals and ions across membranes but do not qualify as the ABC genes because of their structure.[26] The sole criteria for the ABC genes are the highly conserved sequences in the ATP-binding domains, the Walker A and B motifs, and the signature (C) domain (Fig. 1).

Several human disease genes, once cloned, were identified as ABC genes. The cystic fibrosis gene, CFTR, encodes the only ABC gene to date that has been shown to be a channel instead of a pump. CFTR encodes for a cAMP-regulated chloride channel that serves to regulate fluid secretion from exocrine tissues.[27] The gene that causes ALD, a disorder characterized by abnormal metabolism of long-chain fatty acids, is encoded by a half-molecule that probably functions as a partner with PMP70.[28] The immune system depends on the processing of antigenic peptides and the transport of these foreign peptides into the endoplasmic reticulum where they are complexed with the Class I encoded human leukocyte antigen (HLA) proteins. The transport of these peptides is accomplished by a dimer

[22] G. D. Ewart, D. Cannell, B. C. Cox and A. J. Howells, *J. Biol. Chem.* **269,** 10,370 (1994).

[23] J. P. McGrath and A. Varshavsky, *Nature* **340,** 400 (1989).

[24] O. Lomovskaya and K. Lewis, *Proc. Natl. Acad. Sci. U.S.A.* **89,** 8938 (1992).

[25] V. Blanc, K. Salah-Bey, M. Folcher and C. J. Thompson, *Mol. Microbiol.* **17**(5), 989 (1995).

[26] K. Petrukhin, S. Lutsenko, I. Chernov, B. M. Ross, J. H. Kaplan and T. C. Gilliam, *Hum. Mol. Genet.* **3,** 1647 (1994).

[27] E. M. Schwiebert, M. E. Egan, T.-H. Hwang, S. B. Fulmer, S. S. Allen, G. R. Cutting and W. B. Guggino, *Cell* **81,** 1063 (1995).

[28] D. Valle and J. Gartner, *Nature* **361,** 682 (1993).

```
                                                                                     MDRSEQ
           L  S  G  G  Q  K  Q  R  I  A  I  A  R  A  L  V  R       N  P  K  I  L  L  L  D  E  A  T  S  A  L  D

HuPGY1C TTGAGTGGTGGGCAGAAGCAGAGGATCGCCATTGCACGTGCCCTGGTTCGC   AACCCCAAGATCCTCCTGCTGGATGAGGCCACGTCAGCCTTGGAC
      N CTCTCTGGTGGCCAGAAACAACGCATTGCGATAGCTCGTGCCCTTGTTAGA   CAGCCTCATATTTTGCTTTTGGATGAAGCCACGTCAGCTCTGGAT
STE6  C CTAAGTGGCGGGCAACAACAAAGAGTTGCTATAGCACGTGCATTCATCAGA   GATACTCCAATATTATTCTTAGACGAAGCTGTATCGGCTCTAGAT
      N CTGTCAGGTGGACAAGCGCAAAGGCTTTGCATAGCCAGAGCACTTCTGAGA   AAATCAAAAATTCTGATTTTAGATGAGTGTACTTCAGCCTTGGAT
DmBr    CTGTCGGGCGGAGAGCGAAAGCGACTCAGCTTGGCCGAGGAGCTGATCACC   GATCCCATATTCCTGTTCTGCGATGAACCCACCACAGGCTTGGAC
PFMDR N TTATCAGGTGGACAAAAACAAAGAATATCCATTGCAAGAGCAATTATGAGA   AATCCTAAAATTCTAATTCTTGATGAAGCTACATCTTCTTTAGAT
      C TTATCAGGTGGACAAAAACAGAGAATAGCTATAGCTAGAGCATTATTAAGA   GAACCTAAAATATTATTATTAGATGAAGCAACATCATCACTTGAT
HYLB    TTATCCGGAGGTCAACGTCAACGCATCGCAATTGCAAGGGCGCTGGTGAAC   AACCCTAAAATACTCATCTTTGATGAAGCAACCAGTGCTCTGGAT

                  GGICAGAAACAGCGIATIGC                                         GATGAAGCNACCTCAGCTCT
                      AC G  AA  C                                                 G    AA T  AT
                    Linker region
                                                                          CTACTTCGITGIAGACGIAA-5'
                                                                              B box

                 S   G   C   G   K   S   T
                 TCN GGN TGT GGN AAA TCN AC
                 AG        C       G AG

                 TCI GGI TGT GGI AAA TCI AC
                         C           G AG
                         A box
```

Fig. 1. Degenerate primers used for amplifying ABC gene sequences. Alignments are shown of the linker region, including the C domain, and the B boxes of several ABC transporters used to derive degenerate primers. Above the alignments are the one-letter codes for the amino acids and below are the primers that were designed from the sequence. Forward and reverse primers are shown for the B box. The upper part shows the alignments of the linker region and the B box for the human multidrug resistance (PGY1) gene, yeast sterile 6 (STE6), *Drosophila* brown (DmBr), *Plasmodium* transporter (PFMDR), and the *E. coli* hemolysin B (HlyB). C represents the C-terminal ATP-binding domain, and N, the N-terminal. Below, all possible codons for the most conserved residues in the A box shown.

TABLE I
STRATEGIES FOR CLONING ABC GENES

Method	Description
Function/positional cloning	Identify gene by functional complementation or through cloning of mutant/disease alleles
Hybridization	Low stringency hybridization to identify related sequences
Degenerate PCR	Design degenerate primers to conserved portion of the ATP-binding domains
Database screening	Identify putative ABC genes from genomic and/or cDNA sequences deposited in sequence databases
Antibody[a]	Use antibodies to obtain cross-reacting (related proteins) or to identify associated proteins
Yeast 2-hybrid[a]	Genetic screen for associated proteins

[a] These methods have been used to study other gene families and protein interactions, but have not yet been used to identify ABC genes/proteins.

of two ABC half-molecules, TAP1 and TAP2.[15] The TAP genes are also the target for inactivation by the herpes simplex virus, a mechanism for the virus to escape immune surveillance.[29,30]

From the wide range of vital and biologically interesting phenotypes encoded by ABC genes it is clear that identification of additional members of the gene family is important. Therefore a number of groups have sought to clone additional ABC genes based. Most of the strategies (see Table I) to clone these genes have relied on the conservation of the ATP-binding domains.

Methods

Hybridization

Many of the ABC genes can be grouped into subfamilies that are more closely related to each other at the nucleotide and amino acid level than they are to other genes in the family. Low stringency hybridization of single ABC genes back to genomic or cDNA libraries is used to identify additional subfamily members. Using the PGP/MDR genes as probes, several groups cloned related genes from both rodent and human genomes. There are two human genes (PGY1, PGY3) and three genes in both mice and rats (*Mdr1a,*

[29] A. Hill, P. Jugovic, I. York, G. Russ, J. Bennink, J. Yewdell, H. Ploegh and D. Johnson, *Nature* **375,** 411 (1995).

[30] K. Früh, K. Ahn, H. Djaballah, P. Sempé, P. M. van Endert, R. Tampé, P. A. Peterson and Y. Yang, *Nature* **375,** 415 (1995).

Mdr1b, Mdr2). In each case the genes are found adjacent to each other on the chromosome, suggesting that they underwent a recent duplication.

Using ABC gene clones we are able to identify additional gene sequences from the yeast and *E. coli* genomes by hybridization. In yeast the *Mdl1* gene is used to clone the closely related *Mdl2* sequence, which maps to a different chromosome.[6] The *E. coli mdl* gene (the only known prokaryotic ABC full molecule) weakly hybridizes to another sequence, the half-molecule abc.[8] However, despite the fact that the ATP-binding domains of ABC genes all have characteristic conserved residues, the conservation is not high enough to identify all of a species of ABC genes by hybridization.

To identify related genes standard relaxed hybridization conditions are used. Probes are hybridized in 30% (v/v) formamide containing solutions at 37°, and washed in 2X SSC at 50°. If these conditions produce a high background, the stringency is gradually raised by washing the filters at higher temperature and/or by using lower concentrations of SSC.

Degenerate Polymerase Chain Reaction

Degenerate PCR primers that correspond to conserved domains of proteins, are used extensively to expand gene families. By using primer mixtures that are degenerate at variable sites (typically third positions of codons) or by incorporating inosine at the variable sites, it is possible to amplify simultaneously multiple gene sequences. Using conserved regions of the Walker A and B boxes, as well as the C domain sequences, additional ABC genes are cloned from the mouse, yeast, and *E. coli* genomes.[6,8] Luciani *et al.*[5] utilized this approach to identify the ABC1 and ABC2 genes that represent a new subfamily of transporters of unknown function. By extending this approach they have identified additional murine ABC genes.

However, even within the A, B, and C boxes, there is considerable divergence between ABC gene subfamilies at the nucleotide level, and degenerate PCR may only detect a subset of the genes. In addition, all known higher eukaryotic ABC genes are disrupted by introns, making cloning from genomic DNA difficult. Degenerate PCR of cDNA can be used to clone expressed genes, but because some ABC genes display tissue-specific expression, this approach also misses some of the genes.

To clone ABC genes by degenerate PCR, we align sequences from the A, B and signature (C) domains. The A box primer is 5′ TCI GGI TGT/C GGI AAA/G C/GIAC; the C box primer GGI CAG/A A/CAA/G CAG/A C/AGI A/CTI GC; and the B box primer 5′ CAG/A AG/AC A/TGA/T IGT IGC T/CTC ATC, where I corresponds to inosine. Polymerase chain reactions are performed with 1–5 O.D./ml of each primer, with annealing temperatures ranging from 37 to 45°. Reactions yielding a correctly sized product are cloned into a pBS-KS$^+$ digested with *Eco*RV and treated with TTP and *Taq* polymerase to add a single T residue.[31] Multiple clones from each reaction are sequenced to identify all species. Following a database search to establish that a clone is unique, the insert can be used as a hybridization probe.

Database Searching

As the sequence of the expressed genes and the genomic sequence of more organisms becomes available, additional ABC genes are being identified. The complete sequence of the yeast genome is providing the first complete characterization of all of the ABC genes of an eukaryotic organism. The availability of DNA sequence from a large number of expressed genes, in the form of expressed sequence tags (ESTs), allows for the identification of new members of gene families.[32,33]

To search for new ABC genes in the EST database, individual ATP-binding domains of known ABC genes are used in a BLAST search of the translated database.[34,35] This program translates the DNA sequences (in this case ESTs) in all six frames, and compares each sequence to the test sequence. Sequences that have statistically significant matches are reported. BLAST is designed to find short stretches of high homology. The first numerical value is the BLAST score, a numerical representation of the size and identify of the match. The highest score for any of the matching segments is reported. The second reported value is the smallest sum probability and is the product of the probability of each of the matches. If a sequence shows several stretches of high homology to the query sequence, these are all reported, and all of the matches contribute to the probability. BLAST reports back the NCBI EST number, the GenBank accession number, a portion of the description, the reading frame analyzed, the two numerical scores for the match, and the number of segments that matched. For amino acid searches a value of greater than 80 for the BLAST score is considered significant. In some cases matches can be found that have scores less than 80, but probabilities that are very low (less than 10^{-5}). Although these may represent truly significant matches, they should be carefully examined. These matches can represent artifacts where several segments of a sequence display weak matches, and the sum of the probabili-

[31] D. Marchuk, M. Drumm, A. Saulino and F. S. Collins, *Nucl. Acids Res.* **19,** 1154 (1991).
[32] M. S. Boguski, T. M. Lowe and C. M. Tolstoshev, *Nature Genet* **4,** 332 (1993).
[33] G. G. Lennon, C. Auffray, M. Polymeropoulos and M. B. Soares, *Genomics* **33,** 151 (1996).
[34] S. F. Altschul, W. Gish, W. Miller, E. W. Myers and D. J. Lipman, *J. Mol. Biol.* **215,** 403 (1990).
[35] W. Gish and D. J. States, *Nature Genet.* **3,** 266 (1993).

ties is low. Newer releases of BLAST allow for filtering out of low-complexity elements of the sequence.

For database searches across gene families it is also possible to find valid matches that show very low scores and probabilities from the BLAST search. Although there is considerable homology between the ATP-binding domains of the ABC genes, the homology is spread into several domains and is lower between subfamilies. In our searches we examine every sequence reported back by BLAST, and several real hits are found in sequences with scores of less than 70 and probabilities greater than 0.90. In each case we visually examine the matches in the region of the A, B, or C boxes, and suspected ABC sequences are searched back against the database for homology to known genes. For example, a search with an ATP-binding domain of MDR might identify very weak homology to a EST sequence from a gene from the MRP subfamily. When the EST sequence is searched against the entire database, homology to MRP-related genes is apparent.

Once an EST sequence is identified that is a putative ABC gene, several steps are taken to verify it. A search of the DNA sequence is performed against the DNA database to see if it is identical to a known ABC gene or overlaps with another EST clone. The sequences from the same gene are assembled into a contig, along with the corresponding sequences from the other end of the clone. The longest cDNA clone of the contig is completely sequenced, and sequence from the 3′ UTR is used to map the gene in the genome.

Examples of ABC Gene Identification from Expressed Sequence Tags

Contig Assembly

By assembling the overlapping sequence of several ESTs, it is possible to obtain contigs of greater than 1000 bp (Fig. 2). In addition to the forward sequence, most EST clones are sequenced from both ends. By obtaining and assembling the 3′ end sequences it is possible to identify additional clones from the gene of interest and assemble larger contigs (Fig. 2). Since most cDNAs are primed from the poly(A) tail they start at the 3′ end of the gene, and extend for different lengths in the 5′ direction. In the case of several ABC gene contigs ESTs that had homology in the ATP-binding domain to a known ABC gene led us to identify a large number of clones that spanned a total of 2–3 kb.

Once all of the EST clones to a given gene are assembled, the final sequence is searched against the DNA database to make sure all overlapping clones are identified. The cDNA clones that extend the farthest are

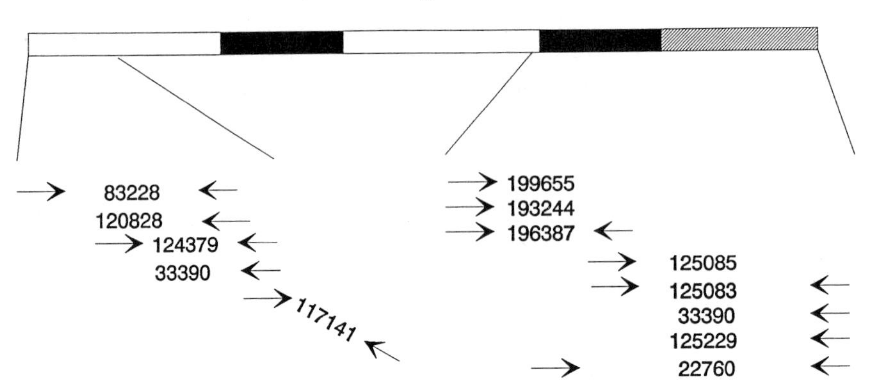

FIG. 2. EST contig of the human mMOAT homolog. A diagram of the human cMOAT gene, which encodes a multiorganic anion transporter, is shown. The filled boxes represent the ATP-binding domains, the open boxes transmembrane domains, and the hatched box the 3′ untranslated region. Below the diagram are two contigs of cDNA clones. The arrows represent sequences from the ends of the clone, and the numbers represent IMAGE consortium clone numbers. Clone 22760 is a putatively spliced clone. The homology with the rate cMOAT sequence ends at a consensus splice donor site.

then obtained (Table II) and the sequence of the clone is filled in. For each EST clone that is obtained, it is essential to sequence in from both ends to confirm the identity of the clone. Due to the large number of clones handled by the EST projects occasional errors in clone and sequence match up have occurred. In the case of clones sequenced in only the 5′ direction, when the 3′ end of the clone is sequenced and used to search the database, additional clones can be identified. Similarly the completion of the sequence of the clone can reveal other overlapping clones.

Xref Database

Many of the database searching procedures discussed here have been automated by the Xref project.[36] A query amino acid sequence in the user's Xref account is used to search both the databases of known genes and the mammalian EST sequences. The search is automatically performed monthly, as long as the user maintains the account, and any new hits are reported. The only limitation of Xref is that only the top 20 matches to a given gene are reported. Because the ABC gene family is so large this could lead to some genes being missed. However, by placing genes from

[36] D. E. Bassett, M. S. Boguski, F. Spencer, R. Reeves, M. Goebl and P. Hieter, *Trends Genet.* **11,** 372 (1995).

TABLE II
SOURCES OF EST CLONES[a]

Research Genetics	http://www.resgen.com 2130 Memorial Parkway Huntsville, AL 800-533-4363 205-536-9016 (FAX) info@www.resgen.com
Genome Systems	http://www.genome.systems.com 8620 Pennell Drive St. Louis, MO 314-692-0033 314-692-0044 (FAX) genome@mo.net
American Type Culture Collection	http://atcc.org 12301 Parklawn Drive Rockville, MD 20852 301-881-2600 301-816-4361 (FAX) request@atcc.org
UK HGMP Resource Centre	http://www/hgmp.mcr.ac.uk Hinxton Hall Hinxton, Cambridge CB10 1RQ, UK 44 1223 494500 44 1223 4944512 (FAX) support@hgmp.mcr.ac.uk
RLDB	Max Planck Institute Berlin Germany

[a] The largest depositor in GenBank is the IMAGE consortium (http://www.bio.llnl.gov/bbrp/image), and those clones are available from all of the sources given here. The ATCC also distributes clones described by other EST projects. Additional clones can also be obtained from The Institute for Genomic Research (http://www.tigr.org).

each of the ABC subfamilies in an Xref account we have been able to identify new ABC ESTs as they have become available.

Full-Length Sequence

Due to the 3′ bias of EST clones, and the fact that most ATP-binding domains are at the C terminus of ABC genes, the resulting sequence of the EST contig is almost always incomplete. The full-length sequence of the gene can be obtained by (1) screening an appropriate cDNA library,

TABLE III
HUMAN LIBRARIES WITH BACTERIAL CONTAMINATION

Manufacturer	Source	Catalog number
Clontech	CCRF-CEM	HL 1063g
Clontech	Adult heart	(42)
Clontech	Fetal brain[a]	6535

[a] Not confirmed. Four ESTs were found with high homology to a *Bacillus subtilis* ABC gene.

and isolating additional clones, (2) using 5′ RACE (rapid amplification of cDNA ends) to extend the sequence in the 5′ direction, and eventually obtain the full-length clone by PCR using primers from both ends of the sequence. Clontech[37] introduced a cDNA cloning strategy that allows the 5′ end of RNAs to be primed from the CAP sequence. A library made by this strategy should be enriched for full-length clones and represents an excellent tool for completing the structural characterization of a gene.

Pitfalls to Database Searching

By definition EST clones are incompletely characterized, and thus each sequence needs to be confirmed before a new gene can be reported. We have encountered a number of stumbling blocks in this process, and outline the steps needed to verify genes obtained from ESTs. The most severe problem with early versions of the EST database was the presence of sequences from a human cDNA library contaminated with yeast and bacterial sequences. This is particularly a problem with ABC genes, as they are abundant in prokaryotic genomes. The majority of the contaminating sequences were eventually identified as derived from a species of *Leuconostoc.*[38] However, sequences from the contaminated libraries remain in the database as human clones. A list of libraries used for EST sequences that contain bacterial contamination is shown in Table III. If an investigator comes across an EST derived from one of these libraries care must be taken in further characterization of the clone before drawing any conclusions.

The confirmation that an EST clones derives from the human genome can be obtained in several ways. The 3′ end of the sequence is often within the 3′ untranslated region, which is typically devoid of introns.[39] Primers

[37] *Clontechniques* **XI**(1), 2 (1996).
[38] M. Dean and R. Allikmets, *Am. J. Hum. Genet.* **57,** 1254 (1995).
[39] S. Poduslo, M. Dean, U. Kolch and S. O'Brien, *Am. J. Hum. Gen.* **49,** 106 (1991).

designed from the 3′ end sequence should amplify the correctly sized product from human DNA. This product should be specific to only one gene, and should be positive in a single monochromosomal hybrid. The EST clone should identify one or more RNA species when hybridized to a Northern blot, and should hybridize at high stringency to human genomic DNA. At least one of the tests discussed here should be performed on one clone from each EST contig.

A significant percentage of clones in a cDNA library are either chimeric or incompletely spliced. Incompletely spliced clones are usually revealed when the sequence is completed. In this case one typically observes an end to the open reading frame at the position of a putative splice site. An example of an incompletely spliced clone is shown in Fig. 3A. EST T87741 displays a high homology to the rat cMOAT gene at both the nucleotide and amino acid level (data not shown). At a position in which the homology ends there is a consensus splice donor site, consistent with this clone being incompletely spliced.

Chimeric clones in which the 5′ half derive from an ABC gene, and the 3′ half from another gene, can lead to the misassembly of the contig. In addition, if the 3′ end of a chimeric clone is used as a sequence tagged site (STS) the clone may be incorrectly mapped. Chimeric clones can be identified during contig assembly and/or sequencing. In the example in Fig. 3B, a chimeric clone was revealed on assembly of the full contig.

Alternate Strategies

A number of other strategies could potentially be employed to identify additional ABC genes. Antibodies that recognize epitopes in the ATP-binding domains could cross-react with other ABC proteins and be used to screen expression libraries. In addition, since ABC members that are half-molecules often function as heterodimers, the two half-molecule proteins could be immunoprecipitated with antibodies to one of the proteins. In a similar strategy the yeast two-hybrid system could be used to identify associated proteins. These associated proteins could either represent half-molecule proteins or associated/regulatory molecules. In some cases cellular transport functions have been recognized and shown to be ATP dependent. The corresponding transport proteins (ABC proteins) could be identified by expression of cDNA in *Xenopus* oocytes, and assayed for transport. For example, an ATP-dependent export pump for the excretion of glutathione conjugates in the liver has been functionally characterized. Although this

A

```
        701                                                750
RcMOAT  CATCCACTGA CACTAGAAGA TGTCTGGGAT ATCGATGAAG GGTTTAAAAC
T87741                                                    ....

        751                                                800
RcMOAT  AAGGTCAGTC ACCAGCAAGT TTGAGGCGGC CATGACAAAG GACCTGCAGA
T87741  C.A.A..T.A GTG....... ....A...CA .....AG.GA ..G.......

        801                                                850
RcMOAT  AAGCCAGGCA GGCTTTTCAG AGGCGGCTGC AGAAG   TC CCAGCGGAAA
T87741  .........G ...AC.C... ..A....A.G .....AGC.. .....A...C

        851                                                900
RcMOAT  CCTGAGGCCA CACTACACGG ACTGAACAAG AAGCAGAGTC AGAGCCAAGA
T87741  T...GA.... GG..G.CT.. CT........ ..T....... .A........

        901                                                950
RcMOAT  CGTTCTCGCC CTG GAAGAAG CGAAAAAGAA GTCTGAGAAG ACCACCAAAG
T87741  T.CC..T.T. .../GTAACTT TCCCTTGAGT GTCTGTGTGA GCGCGCTGCA
```

B

```
EST45597  CGTCGCCATT GCCCGCACC. .ATCCTCAAG GCTCCGGGCA TCATTCTGCT
  T19369  CTGAGCTCTC GGTGGCGCCT GAGAAGTGAA AGAGCAGCAG TTTCAAA...

EST45597  GGATGAGGCA ACGTCAGCGC TGGATACATC TAATGAGAGG GCCATCCAGG
 T193691  .......... .......... .......... .......... ..........
```

FIG. 3. Examples of incompletely spliced and chimeric ESTs. (A) A portion of the rat cMOAT (RcMOAT) gene (GenBank X96191) is shown aligned with an EST from the human homolog (GenBank T87741). The cMOAT gene is an ABC gene that encodes a liver-specific anion transporter.[41] Residues of the human EST that are identical to the rat sequence are shown as dots. The gap in the rat sequence at position 836 corresponds to a 3-bp insertion in the human sequence. The underlined residues correspond to the consensus splice donor site in the EST clone. (B) Alignment of the sequence of the human EST45597 EST contig with one of a chimeric clone that is part of the contig. The underlined sequences derive from another gene.

gene was identified by molecular techniques,[40,41] it could have been cloned by functional assays.

[40] R. Mayer, J. Kartenbeck, M. Buchler, G. Jedlitschky, I. Leier and D. Keppler, *J. Cell Biol.* **131,** 137 (1995).

[41] C. C. Paulusma, P. J. Bosma, G. J. R. Zaman, C. T. M. Bakker, M. Otter, G. L. Scheffer, R. J. Scheper, P. Borst and R. P. J. Oude Elferink, *Science* **271,** 1126 (1996).

[42] C. C. Liew, D. M. Hwang, Y. W. Fung, C. Laurenssen, E. Cukerman, S. Tsui and C. Y. Lee, *Proc. Natl. Acad. Sci. U.S.A.* **91,** 10645 (1994).

TABLE IV
CLASSIFICATION OF KNOWN HUMAN ABC GENES

Subfamily	Gene/EST	Location	Class	Function/phenotype
PGP/TAP	PGP1	7q21	FM	Lipid transport/multidrug resistance
	PGP3	7q21	FM	Phosphatidylcholine transport
	TAP1/2	6p21.3	HM	Peptide transport
	45597	2q35	HM	?
	122234	12q24-qter	HM	?
	ABC7/140535	Xq21-q22	HM	?
	157481	4q22-q23	HM	?
	328128	7q35-q36	HM	?
	422562	7p14	FM	?
	20237	1q42	HM	?
CFTR/MRP	CFTR	7q31	FM	Chloride channel
	MRP1	16p13.1	FM	Glutathione conjugate transporter
	SUR1	11p15.1	FM	Potassium channel regulator
	MRP3/90757	17q11-q12	FM	Multidrug resistance (?)
	155051	N/A	FM	?
	MRP2/cMOAT	10q12-q24	FM	Organic anion transporter
	MRP4/170205	13q31	FM	?
	182763	6p21	FM	?
	205858	N/A	FM	?
	MRP5/277145	3q25-q26	FM	?
	MRP6/349056	16p13.1	FM	?
ALDP	ALDP	Xq28	HM	Fatty acid transporter/adrenoleukodystrophy
	ALDR	12q11	HM	
	PMP70	1p21-p22	HM	Zellweger syndrome
	P7OR/PMP69	14q24	HM	?
ABC1	ABC1	9q22-31	FM	Involved in phagocytosis
	ABC2	9q34	FM	?
	ABC-C	16p13.3	FM	?
	ABCR	1p13-p21	FM	Retina degeneration
	90625	17p21-q24	FM	?
GCN20	123147	6p21.3	ATP	?
	133090	7q35-q36	ATP	?
	201864	3q25.1-q25.2	ATP	?
OAB	OAB	4q31	ATP	2′,5′-Oligoadenylate binding

Conclusions

Multigene families can be characterized by a variety of different methods. The most popular ones were analyzed briefly in this article, the emphasis being on evolution of the methods used. Certain methodologies, for example, hybridization-based, that were widely used not so long ago have given way to the other, more powerful and faster techniques, mainly based on sequence database searches. We have shown that the analysis of the EST database is probably the most effective strategy for an exhaustive characterization of a gene family (see Table IV). In this analysis we were aided by the fact that ABC genes have a large domain that is highly conserved, and there are a number of characterized genes from this superfamily to use in database searches. In addition, the ATP-binding domain is almost always in the 3′ end of the gene, and EST sequences are biased toward the 3′ end. As the EST database grows to better represent the full spectrum of mammalian cell types, this resource will becoming increasing more valuable for identifying biologically important genes.

At the same time the other techniques discussed, especially those based on functional analyses, are complementing the newer methods extremely well. In the case when the ultimate goal is a complete structure–function analyses of all genes, we must use the full arsenal of the available methods to accomplish it.

[10] *Saccharomyces cerevisiae* ABC Proteins and Their Relevance to Human Health and Disease

By DANIEL TAGLICHT and SUSAN MICHAELIS

Introduction

The ATP-binding cassette (ABC) superfamily, also called the *traffic ATPases,* comprises more than 200 proteins, the majority of which mediate the selective transport of substrates across biological membranes. ABC proteins are found in all organisms, including prokaryotes, archaea, and eukaryotes.[1–4] The prototypical ABC protein is large (~140 kDa) and contains four modules: two nucleotide-binding domains (NBDs) that bind and hydrolyze ATP and two membrane-spanning domains (MSDs), each containing multiple (~6) transmembrane segments. These four ABC modules can be encoded by a single gene or by separate genes.

As a family, ABC proteins can transport substances that differ markedly

[1] M. J. Fath and R. Kolter, *Microbiol. Rev.* **57,** 995 (1993).

[2] C. F. Higgins, *Annu. Rev. Cell Biol.* **8,** 67 (1992).

[3] M. Dean and R. Allikmets, *Curr. Opin. Genet. Dev.* **5,** 779 (1995).

[4] S. Michaelis and C. Berkower, *in* "Cold Spring Harbor Symposium on Quantitative Biology, Vol. LX: Protein Kinesis—The Dynamics of Protein Trafficking and Stability," p. 291. Cold Spring Harbor Press, Cold Spring Harbor, 1995.

METHODS IN ENZYMOLOGY, VOL. 292
0076-6879/98 $25.00

from one another in their chemical structure and size, ranging from small molecules such as ions, sugars, amino acids, peptides, and phospholipids, to complex hydrophobic drugs, lipopeptides, and large proteins (>100,000 Da). Individual ABC proteins vary in the degree of specificity they have for their substrates. Some are highly specific for a single compound, for example, bacterial amino acid and sugar permeases.[5,6] In contrast, others are promiscuous, like the mammalian multidrug resistance protein, which can transport a variety of chemically unrelated lipophilic compounds.[7,8] The direction of transport is unique for a particular ABC protein, and all are thought to be oriented with their NBDs cytosolically disposed; however, certain ABC proteins function as permeases to move solutes into the cytosol, while others operate as efflux pumps to transport substrates out of the cytoplasm to the external milieu or into intracellular organelles. Substrate-stimulated ATPase activity has been directly demonstrated for several ABC proteins[9–11]; nevertheless, the manner in which ATP hydrolysis is coupled to substrate translocation remains elusive, as does our understanding of substrate selectivity.

Note that not all ABC proteins are energy-driven transporters; instead, some function as channels or as channel regulators.[12,13] Furthermore, a subset of ABC proteins lacks MSDs altogether, but do contain NBDs. These members of the ABC superfamily (discussed further in a later section) are unlikely to be involved in transport.

In this study, we present a compilation of the complete inventory of ABC proteins in the budding yeast *Saccharomyces cerevisiae.* This study was made possible by completion of the entire genome sequence of yeast.[14] We examine the yeast ABC proteins in light of what is known about other eukaryotic members of the superfamily, particularly those involved in human health and disease. We also discuss how the powerful experimental

[5] A. L. Davidson, H. A. Shuman, and H. Nikaido, *Proc. Natl. Acad. Sci. U.S.A.* **89,** 2360 (1992).

[6] G. F.-L. Ames, *in* "Molecular Biology and Function of Carrier Proteins" (L. Reuss, J. M. Russell, and M. L. Jennings, eds.), Vol. 48, p. 77. Rockefeller University Press, New York, 1993.

[7] M. M. Gottesman and I. Pastan, *Annu. Rev. Biochem.* **62,** 385 (1993).

[8] M. M. Gottesman, I. Pastan, and S. V. Ambudkar, *Curr. Opin. Genet. Dev.* **6,** 610 (1996).

[9] S. V. Ambudkar, I. H. Lelong, J. Zhang, C. O. Cardarelli, M. M. Gottesman, and I. Pastan, *Proc. Natl. Acad. Sci. U.S.A.* **89,** 8472 (1992).

[10] C. A. Doige, X. Yu, and F. J. Sharom, *Biochim. Biophys. Acta* **1109,** 149 (1992).

[11] F. J. Sharom, X. Yu, and C. A. Doige, *J. Biol. Chem.* **268,** 24,197 (1993).

[12] C. F. Higgins, *Cell* **82,** 693 (1996).

[13] S. Demolombe and D. Escande, *TIPS* **17,** 273 (1996).

[14] A. Goffeau, B. G. Barrell, H. Bussey, R. W. Davis, B. Dujon, H. Feldmann, F. Galibert, J. D. Hoheisel, C. Jacq, M. Johnston, E. J. Louis, H. W. Mewes, Y. Murakami, P. Philippsen, H. Tettelin, and S. G. Oliver, *Science* **274,** 546 (1996).

strategies available in yeast can be used for dissecting the structure and function of native and heterologously expressed mammalian ABC proteins.

Connection Between Mammalian ABC Proteins Involved in Human Health and Disease and Yeast ABC Proteins

Clinical Importance of ABC Proteins

It is becoming increasingly clear that ABC proteins play a significant role in human health and illness. Several genetic diseases can be attributed to defects in ABC proteins. These diseases and their corresponding ABC proteins include the following examples: cystic fibrosis (CFTR,[15,16] an ion channel); the peroxisomal disorders adrenoleukodystrophy and Zellweger syndrome (ALDP[17] and PMP70,[18] respectively); hyperinsulinemic hypoglycemia of infancy, in which the release of insulin is abnormal (SUR,[19,20] the sulfonylurea receptor); and Dubin–Johnson syndrome, exemplified by the naturally occurring mutant Wistar rat strain (the TR–rat model) in which bile transport is compromised, resulting in congenital jaundice (cMOAT,[21,22] the canalicular multispecific organic ion transporter). The clinical importance of ABC proteins is not confined to disease genes per se, but extends to normal cellular physiology. ABC proteins function to transport peptides into the lumen of the ER for antigen presentation, a process that recruits the immune system to kill virus-infected and otherwise abnormal cells (TAP1 and TAP2[23], transporters associated with antigen presentation). In addition, by having the capacity to function as a drug efflux pump, two distinct ABC proteins can cause drug resistance in tumor cells when overexpressed due to gene amplification (MDR,[7,24] the multi-

[15] J. R. Riordan, J. M. Rommens, B. S. Kerem, N. Alon, R. Rozmahel, Z. Grzelczak, J. Zielenski, S. Lok, N. Plavsic, J. L. Chou, M. L. Drumm, M. C. Iannuzzi, F. S. Collins, and L. C. Tsui, *Science* **245,** 1066 (1989).

[16] F. Collins, *Science* **256,** 774 (1992).

[17] J. Mosser, A.-M. Douar, C.-O. Sarde, P. Kioschis, R. Feil, H. Moser, A.-M. Poustka, J.-L. Mandel, and P. Aubourg, *Nature* **361,** 726 (1993).

[18] J. Gartner, H. Mosser, and D. Valle, *Nature Genet.* **1,** 6 (1992).

[19] L. Aguilar-Bryan, C. G. Nichols, S. W. Wechsler, J. P. Clement, A. E. Boyd, G. Gonzalez, H. Herrera-Sosa, K. Nguy, J. Bryan, and D. A. Nelson, *Science* **268,** 423 (1995).

[20] P. Thomas, G. Cote, N. Wohllk, B. Haddad, P. Mathew, W. Rabl, L. Aguilar-Bryan, R. Gagel, and J. Bryan, *Science* **268,** 426 (1995).

[21] C. C. Paulusma, P. J. Bosma, G. J. R. Zaman, C. T. M. Bakker, M. Otter, G. L. Scheffer, R. J. Scheper, P. Borst, and O. Elferink, *Science* **271,** 1126 (1996).

[22] M. Buchler, J. Konig, M. Brom, J. Kartenbeck, H. Spring, T. Horie, and D. Keppler, *J. Biol. Chem.* **271,** 15,091 (1996).

[23] F. Momburg, J. J. Neefjes, and G. J. Hämmerling, *Curr. Opin. Immunol.* **6,** 32 (1994).

[24] J. A. Endicott and V. Ling, *Annu. Rev. Biochem.* **58,** 137 (1989).

drug resistance protein (also known as P-glycoprotein), and MRP,[25] the multidrug resistance-associated protein).

Because of the clinical significance of ABC proteins, it is critical to determine their role in normal cellular physiology and disease. This presents a challenge, in that the normal physiological transport substrate(s) for most ABC proteins are not yet known. For instance, the normal role of mammalian MDR1 could be for clearance of xenobiotic compounds, as suggested by the knockout mouse[26]; alternatively, it may transport an as-yet undetermined native hormone. An important emerging view that warrants further experimentation is the notion that certain ABC transporters, in particular MDR1 and CFTR, may play a regulatory role in modulating the efficiency of other transporters, in addition to functioning as transporters themselves.[12] It is clear that gaining an understanding of the physiological role of these ABC transporters presents a considerable scientific challenge that has important clinical implications. As discussed later, the study of ABC transporters in a tractable model organism such as yeast provides a promising avenue of research for studying this issue.

Yeast Biology and Human Disease: Utility of Yeast as Model Organism for Studying ABC Proteins

Saccharomyces cerevisiae is an ideal model organism for the functional dissection of disease-related genes such as those of the ABC superfamily, because of the ease of molecular manipulation of genes in yeast. Starting with a gene of interest, phenotypic analysis can readily be carried out by generating null and conditional alleles. Establishing the pattern of expression, cellular localization properties, and interaction partners of a gene product can also be easily achieved, as can optimization of expression to facilitate biochemical studies.[27] These procedures can be carried out using classical and recombinant techniques that are standard in most yeast laboratories and involve epitope tagging, site-directed or random mutagenesis, overexpression suppression, two-hybrid studies, and polymerase chain reaction (PCR)-mediated gene replacement.[27-29] If the yeast gene being ana-

[25] S. P. C. Cole, G. Bhardwaj, J. H. Gerlach, J. E. Mackie, C. E. Grant, K. C. Almquist, A. J. Stewart, E. U. Kurz, A. M. Duncan, and R. G. Deeley, *Science* **258,** 1650 (1992).

[26] A. H. Schinkel, J. J. M. Smit, O. van Telligen, J. H. Beijnen, E. Wagenaar, L. van Deemter, C. A. A. M. Mol, M. A. van der Valk, E. C. Robanus-Maandag, H. P. J. te Riele, A. J. M. Berns, and P. Borst, *Cell* **77,** 491 (1994).

[27] C. Guthrie, G. R. Fink, *Methods Enzymol.* **194,** (1991).

[28] A. Wach, A. Brachat, R. Pohlmann, and P. Philippsen, *Yeast* **10,** 1793 (1994).

[29] A. Wach, *Yeast* **12,** 259 (1996).

lyzed is a homolog of a mammalian disease gene, the information gained in yeast may guide mammalian studies. An excellent example of the value of this cross-species viewpoint is provided by a pair of yeast peroxisomal ABC proteins, PXA1 and PXA2, which were identified on the basis of their homology to the mammalian peroxisomal disease genes ALDP and PMP70.[30–32] In the initial stages of this research, it was knowledge about the human disease genes that fueled the identification of the yeast homologs. At present, the reverse is true; namely, the wealth of information gained in *S. cerevisiae* about the structure and function of PXA1 and PXA2 is guiding studies on mammalian ALDP and PMP70.

The completion of the entire genome sequence of *S. cerevisiae* in the spring of 1996 is a landmark achievement in modern biology, because it represents the first complete sequence of a eukaryotic genome.[14,33] Given the well-proven power of multiorganismal approaches in dissecting the basic functional mechanisms of a protein,[34,35] the *S. cerevisiae* genome sequence can be expected to provide an invaluable resource not only to yeast laboratories, but to mammalian researchers, too. For instance, a mammalian disease gene identified by positional cloning can now be used in a similarity search to find unambiguously the most closely related protein in yeast, as discussed in the concluding section of this article. Establishing such a cross-species link between a mammalian and yeast gene can provide an immediate experimental paradigm for further analysis, as noted above for PXA1 and PXA2.

Another potentially powerful application of a cross-species connection is testing functional complementation of a yeast mutant, using the corresponding mammalian gene. A compelling case in point is the cross-species complementation involving the mouse MDR3 gene (mMDR3) and its closest yeast match, STE6, which encodes the transporter for the mating pheromone **a**-factor.[36] The ability of mMDR3 to complement a *ste6* deletion mutant by restoring **a**-factor transport and mating provides a powerful assay that has facilitated the structure–function analysis of mammalian MDR genes. The heterologous expression of mMDR3 also confers drug resistance to yeast, an activity lacking for native STE6; thus, yeast drug resistance provides an additional assay for mammalian mMDR3 functional

[30] N. Shani, P. A. Watkins, and D. Valle, *Proc. Natl. Acad. Sci. U.S.A.* **92,** 6012 (1995).
[31] N. Shani, A. Sapag, and D. Valle, *J. Biol. Chem.* **271,** 8725 (1996).
[32] E. Swartzman, M. Viswanathan, and J. Thorner, *J. Cell Biol.* **132,** 549 (1996).
[33] P. Hieter, D. E. Jr. Bassett, and D. Valle, *Nature Genet.* **13,** 253 (1996).
[34] D. E. Jr. Bassett, M. Boguski, and P. Hieter, *Nature* **379,** 589 (1996).
[35] D. E. Jr. Bassett, M. Boguski, F. Spencer, R. Reeves, M. Goebl, and P. Hieter, *Trends Genet.* **11,** 372 (1996).
[36] M. Raymond, P. Gros, M. Whiteway, and D. Y. Thomas, *Science* **256,** 232 (1992).

studies.[37] Such phenotypic approaches can allow clinical researchers to use yeast as a "test tube" to dissect the biological, biochemical, and localization properties of mammalian disease genes.[34,35,38] Even in cases where heterologous complementation has not yet been demonstrated, a yeast–human homology connection provides a source of ideas for formulating hypotheses testable by other methods. In addition to its direct applicability for research on human disease genes, the complete inventory of yeast ABC genes provides direct information on what appears to be the minimal set of ABC proteins necessary for the viability of a simple eukaryotic cell.

Here we present a complete inventory of the ABC proteins in the yeast *S. cerevisiae* (shown later in Table I). We describe the existence of six distinct ABC subfamilies revealed by phylogenetic analysis (Fig. 1).[4] We also provide a comparison of the complete collection of yeast ABC proteins to a partial, but fast-growing list of human ABC proteins that have been deduced from the human expressed sequence tag (EST) database.[39] Through the functional studies of the yeast ABC genes, together with genetic and biochemical analysis of mammalian ABC genes heterologously expressed in yeast, we can anticipate obtaining a wealth of information that is directly applicable to understanding the role of ABC proteins in human health and disease.

Hallmarks of ABC Proteins Important for Sequence Analysis and Comparison

Nucleotide-Binding Domains

The NBDs of ABC proteins are their most characteristic feature and provide the main basis for sequence comparison among members of the superfamily.[2] All ABC proteins possess one or two NBDs, each approximately 200 residues in length with five conserved regions: (1, 2) the Walker A and B motifs, which are separated by ~90–120 residues: (3) a diagnostic "signature motif" (also called the C motif) with the consensus LSGGQ, which is just upstream of the Walker B region; (4) a less highly conserved "center region" approximately midway between the Walker A and B regions [the center region has been a focus of attention in the ABC field, because the most common disease allele of CFTR (ΔF508) lies within it[40]];

[37] M. Raymond, S. Ruetz, D. Y. Thomas, and P. Gros, *Mol. Cell. Biol.* **14,** 277 (1994).
[38] D. E. Jr. Bassett, M. S. Boguski, F. Spencer, R. Reeves, S. Kim, T. Weaver, and P. Hieter, **15,** 339 (1997).
[39] R. Allikmets, B. Gerrard, A. Hutchinson, and M. Dean, *Hum. Mol. Genet.* **5,** 1649 (1996).
[40] B. S. Kerem, J. M. Rommens, J. A. Buchanan, D. Markiewicz, T. K. Cox, A. Chakravarti, M. Buchwald, and L. C. Tsui, *Science* **245,** 1073 (1989).

and (5) a region extending downstream of the Walker B motif, which exhibits modest conservation. These regions of the NBD are described in more detail elsewhere[4] and are indicated in the multiple sequence alignment shown in Fig. 2. Whereas a host of other nucleotide-binding proteins, including kinases, also contain Walker A and B sites, the ABC proteins are distinguished by the spacing between these sites (90–120 amino acids) and by the additional presence of the signature and center motifs.[4]

Modular Architecture

A distinctive feature of the ABC superfamily is the modular architecture of its members, nearly all of which contain two nucleotide-binding domains (NBD1 and NBD2) and two membrane-spanning domains (MSD1 and MSD2).[2,41] For most eukaryotic ABC proteins all four of these core domains are present on a single polypeptide, thus constituting a *full-length transporter.* Typically the modules are in the "forward" order MSD1–NBD1–MSD2–NBD2, but can also be in the "reverse order" with the NBDs preceding the MSDs. Certain eukaryotic ABC proteins, such as the *Drosophila* pigment transporters *white, brown,* and *scarlet,* are expressed as *half-transporters* (each with a single NBD and MSD).[42] It is thought that half-transporters must coassemble into heterodimers or homodimers to form a functional transporter. Whereas prokaryotic ABC transporters are frequently further subdivided into *quarter molecules* composed of individual MSDs and NBDs, the quarter-molecular architecture has not been found in eukaryotic ABC proteins.

For full-length transporters, a relatively charged region known as the *linker* separates the two halves. The residues of the linker are not highly conserved among ABC proteins. An additional region, called the regulatory (R) domain is present in a subset of ABC proteins, most notably CFTR and related family members. The R domain is phosphorylated, and is thought to function in a regulatory capacity.[43]

Membrane-Spanning Domains

The MSDs of ABC proteins are predicted by hydropathy analysis to contain multiple membrane-spanning stretches, generally six, although four to eight predicted spans are not uncommon. Detailed topological studies

[41] S. C. Hyde, P. Emsley, M. J. Hartshorn, M. M. Mimmack, U. Gileadi, S. R. Pearce, M. P. Gallagher, D. R. Gill, R. E. Hubbard, and C. F. Higgins, *Nature* **346,** 362 (1990).

[42] G. D. Ewart, D. Cannell, G. B. Cox, and A. J. Howells, *J. Biol. Chem.* **269,** 10,370 (1994).

[43] S. H. Cheng, D. P. Rich, J. Marshall, R. J. Gregory, M. J. Welsh, and A. E. Smith, *Cell* **66,** 1027 (1991).

have been carried out for only a few of these, most notably P-glycoprotein.[44] Among yeast ABC proteins, topology has been examined directly only for the N-terminal MSD of STE6, in which the six experimentally determined spans match theoretical hydropathy predictions.[45] In contrast to the strong homology observed among the NBDs of ABC proteins, only a low amount of homology is apparent even among the MSDs of closely related ABC proteins, and between more distantly related members of the superfamily such homology is barely discernible. It is important to reiterate that while most ABC proteins possess MSDs, certain subfamilies do not (as described in a section that follows). Whereas the latter ABC proteins are unlikely to function as transporters, they are nevertheless members of the ABC superfamily on the basis of their canonical NBDs.

Inventory of Yeast ABC Proteins

Yeast ABC Proteins Classified into Distinct Subfamilies

We previously reported the results of a BLAST search of yeast ABC proteins, which was carried out at a time (early 1995) when the yeast genome sequence had not yet been completed. In that study we identified 17 ABC proteins in *S. cerevisiae*, plus several ABC proteins from other fungi (three from the *Schizosaccharomyces pombe* and two from *Candida albicans*).[4] Phylogenetic analysis of these 22 proteins was performed together with a set of well-characterized human and bacterial "reference proteins." Our analysis revealed that the available subset of yeast ABC proteins could be classified by sequence similarity into five groups, designated the PDR, ALDP, CFTR/MRP, MDR, and YEF3 subfamilies. These designations reflect the most prominent mammalian (ALDP, CFTR, MRP, MDR) or yeast (PDR5, YEF3) member(s) of each group. Proteins within the same subfamily presumably share not only a common evolutionary origin, but may also possess common functional and biochemical features. The phylogenetic analysis also revealed the unexpected finding that within some, but not all subfamilies, the N- and C-terminal NBDs comprise separate subclasses based on sequence similarity.[4] This point is discussed in more detail in the next section.

Search of Complete Genome of Saccharomyces cerevisiae for ABC genes

On release of the complete genome sequence of *S. cerevisiae*,[14] we carried out a BLAST search aimed at generating a comprehensive inventory

[44] J.-T. Zhang and V. Ling, *Biochim. Biophys. Acta* **1153,** 191 (1993).
[45] D. Geller, D. Taglicht, R. Edgar, A. Tam, O. Pines, S. Michaelis, and E. Bibi, *J. Biol. Chem.* **271,** 13,746 (1996).

of ABC proteins in yeast.[46] As our query, we used the NBD1 amino acid sequence of STE6, from which portions showing little homology to other ABC proteins were removed. We queried the *S. cerevisiae* genome sequence using the *Saccharomyces* Genome Database (SGD) server (http://genome-www.stanford.edu/Saccharomyces/). The resultant BLAST hits were screened manually for significant homology to the Walker A, Walker B, signature, and center motifs, and for the distance between the Walker A and B sites. This search revealed 31 separate open reading frames that included at least one ABC-type NBD. Two of these (YKR103w and YKR104w) are directly adjacent to one another on chromosome XI, separated only by a stop codon. We presume these ORFs actually represent portions of a single protein (Table I), artifactually separated by a mutation in the DNA that was used for the sequencing project or by a sequencing error, bringing the total number of ABC proteins in *S. cerevisiae* to 30. Note also that for one gene (YOL075c), the submitted ORF starts in the middle of an NBD; however, manual inspection of the DNA sequence indicated that if a termination codon immediately upstream of the putative initiator ATG is ignored, a longer ORF is found that includes two complete NBDs and MSDs. In our analysis we have used this extended ORF as YOL075c (Table I). To make certain that we did not miss any proteins, we repeated the BLAST search using both NBD1 and NBD2 of YNR070w, NBD1 of YOR1, and NBD1 of YNL014w, which represent NBDs from several distinct subfamilies. No additional proteins were revealed by these searches. All the protein sequences were subjected to a hydropathy analysis using the algorithm of Kyte and Doolittle[47] to predict the presence and location of transmembrane regions. Based on this analysis, of the 30 *S. cerevisiae* ABC proteins we have identified, 22 are putative transporters with multiple membrane spans, whereas the remaining 8 are likely to be soluble proteins.

The complete inventory of the 30 *S. cerevisiae* ABC proteins is presented in Table I. The table shows the ORF number, GenBank accession number, schematic structure predicted by hydropathy analysis, and closest relatives as determined by individual BLAST searches, for the entire collection of yeast ABC proteins. Where established, gene names and functions are also provided in Table I. Gene names in yeast are generally designated only when some form of phenotypic analysis has been carried out. Of the 30 yeast ABC genes, 7 had been previously identified by genetic screens, based on their phenotypic association with drug or heavy metal resistance (PDR5, SNQ2, YCF1, and YOR1), **a**-factor export (STE6), or involvement in the

[46] S. F. Altschul, W. Gish, W. Miller, E. W. Myers, and D. J. Lipman, *J. Mol. Biol.* **215,** 403 (1990).

[47] J. Kyte and R. F. Doolittle, *J. Mol. Biol.* **157,** 105 (1982).

regulation of translation (YEF3 and GCN20). Six others (ADP1, PXA1, PXA2, MDL1, MDL2, and ATM1) are known from PCR-based strategies or sequence analyses and have been studied to varying extents. These 13 genes have been assigned standard gene names (see Table I for references). In addition, PDR10, 11, and 15 have been assigned gene names in the literature, available in the Yeast Protein Database (YPD) (http://www.proteome.com/YPDhome.html), even though the phenotypes of the corresponding mutants have not been reported. The remaining 18 genes are known only by their ORF or locus name and await phenotypic analysis, a prospect that promises to extend significantly our understanding of the function of ABC proteins in eukaryotic cellular biology.

General Features of Yeast ABC Subfamilies

Six ABC Subfamilies in Yeast

To perform phylogenetic analysis, individual NBDs from all 30 yeast sequences were extracted and compared by multiple sequence analysis, generating the dendogram shown in Fig. 1. This analysis revealed that in addition to the five previously identified subfamilies (PDR5, ALDP, CFTR/MRP, MDR, YEF3),[4] there is a sixth subfamily (RLI), which contains a single member. Two additional distantly related ABC proteins that do not clearly belong to any of the six major subfamilies and have no known matches from other organisms are designated as "other" (Table I). For each subfamily, a multiple sequence alignment is shown in Fig. 2, highlighting the Walker A, center, signature, Walker B, and Walker B downstream regions. This alignment reveals multiple distinct consensus sequences that are diagnostic for a particular subfamily.

Within Several Subfamilies, N- and C-Terminal NBDs Form Two Separate Clusters

A striking feature of members of three subfamilies (CFTR, PDR5, and YEF3) involves the relationship between the N- and C-terminal NBDs, as we have noted previously.[4] Within each of these subfamilies, the N- and C-terminal NBDs form two distinct clusters, indicated as separate classes (N and C) in Fig. 1. In contrast, for members of the MDR subfamily, the N- and C-terminal NBDs are paired, which becomes quite evident when additional members of this subfamily from other fungi are examined (see Ref. 4). Interestingly, the C-terminal NBDs of CFTR subfamily members are even more similar to the NBDs of MDR proteins than they are to their own corresponding N-terminal NBDs (Fig. 1). Allikmets *et al.*[39] have noted

TABLE I

INVENTORY OF *Saccharomyces cerevisiae* ABC PROTEINS

Subfamily[a]	Gene name[b]	ORF number[c]	Accession number[d]	Structure[e]	Nearest matches (*P*-value)[f]	Other names/function/localization/essentiality
PDR	*PDR5*	YOR153w	Z75061		Other PDR proteins	*SSTS1, YDR5, LEM1*, pleiotropic drug resistance, plasma membrane localized[j,k,l,m,n]
	PDR15	YDR406w	U32274		Other PDR proteins	
	PDR10	YOR328w	Z75236		Other PDR proteins	
	SNQ2	YDR011w	Z48008		Other PDR proteins White, *D. melanogaster* ($1.7e^{-36}$)	Pleiotropic drug resistance[o,p]
		YNR070w	Z71685		Other PDR proteins White, *D. melanogaster* ($1.1e^{-39}$)	
	PDR12	YPL058c	U39205		Other PDR proteins White homologue, *Human* ($2.4e^{-38}$)	
	PDR11	YIL013c	Z47047		Other PDR proteins White homologue, *Human* ($3.4e^{-27}$)	
		YOR011w	Z74919		Other PDR proteins White homologue, *Human* ($1.4e^{-24}$)	
	ADP1	YCR011c	X59720		White, *D. melanogaster* ($6.5e^{-54}$) ABC8, *Mouse* ($5.2e^{-41}$) White homologue, *Human* ($1.6e^{-47}$) Scarlet, *D. melanogaster* ($2.4e^{-31}$)	Nonessential for viability[q]
		YOL075c[g]	Z74817		White, *Human* ($4.9e^{-48}$) White, *D. melanogaster* ($1.6e^{-40}$) ABC8, *Mouse* ($5.2e^{-41}$)	
RLI		SC6652X[hh]	Z50111		RNase L inhibitor, *Human* ($8.0e^{-301}$)	
ALDP	*PXA1*	YPL147w	Z73503		ALDR, *Human* ($7.2e^{-110}$) ALDP, *Human* ($4.8e^{-107}$) PMP70, *Human* ($1.2e^{-104}$)	*PAL1, SSH2*, Peroxisomal protein required for growth on oleic acid[r,s]
	PXA2	YKL188c	Z28188		ALDR, *Mouse* ($5.7e^{-95}$) ALDP, *Human* ($1.7e^{-85}$) PMP70, *Human* ($1.2e^{-73}$)	*YKL741*, Peroxisomal protein required for growth on oleic acid[t]

MRP/CFTR		YHL035c	U11583		MRP1, *Human* ($3.8e^{-214}$) cMOAT, *Rat* ($3.7e^{-208}$) SUR, *Rat* ($1.4e^{-156}$) CFTR, *Human* ($1.7e^{-100}$)	
		YLL048	Z73153		cMOAT, *Rat* ($1.0e^{-239}$) MRP1, *Human* ($9.6e^{-236}$) SUR, *Rat* ($1.5e^{-187}$) CFTR, *Human* ($1.0e^{-114}$)	
		YKR103w YKR104w[h]	Z28328 Z28329	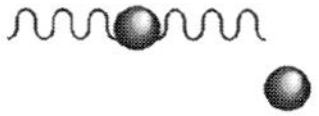	cMOAT, *Rat* ($1.6e^{-95}$) MRP, *Human* ($2.1e^{-86}$) SUR, *Rat* ($2.4e^{-73}$) CFTR, *Human* ($1.1e^{-50}$)	
	YCF1	YDR135c	Z48179		MRP1 *Human* (0.0) cMOAT, *Rat* (0.0) SUR, *Rat* ($1.5e^{-215}$) CFTR, *Human* ($3.6e^{-171}$)	Vacuolar protein required for cadmium resistance, mediates glutathione S-conjugate transport[u,v,w]
		YLL015w	Z73120		MRP1, *Human* ($3.3e^{-297}$) cMOAT, *Rat* ($4.0e^{-246}$) SUR, *Rat* ($5.1e^{-179}$) CFTR, *Human* ($2.7e^{-119}$)	
	YOR1	YGR281w	Z73066		MRP1, *Human* ($2.3e^{-217}$) cMOAT, *Rat* ($1.0e^{-196}$) SUR, *Rat* ($1.0e^{-150}$) CFTR, *Human* ($3.0e^{-120}$)	Oligomycin resistance[x]
MDR	*MDL1*	YLR188w	U17246		MDR3, *Human* ($5.0e^{-80}$) MDR2, *Mouse* ($3.0e^{-78}$) MDR1, *Human* ($7.8e^{-77}$)	Nonessential for viability[y]
	MDL2	YPL270w	Z73626		TAP1, *Gorilla* ($1.2e^{-86}$) TAP1, *Human* ($9.2e^{-85}$) MDR1, *Human* ($3.7e^{-82}$)	*SSH1*, Nonessential for viability[z]
	ATM1	YMR301c	Z49212	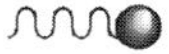	ABC-7, *Mouse* ($9.4e^{-211}$) HMT1, *S. pombe* ($2.7e^{-113}$) Pfmdr, *P. falciparum* ($4.1e^{-97}$) HlyB, *E. coli* ($1.3e^{-72}$)	Essential mitochondrial protein[aa]

(*Continued*)

TABLE I (*Continued*)

Subfamily[a]	Gene name[b]	ORF number[c]	Accession number[d]	Structure[e]	Nearest matches (*P*-value)[f]	Other names/function/localization/essentiality
	STE6	YKL209c	Z28209		HST6, *Candida albicans* (0.0) MDR1, *Mouse* (1.2e − 135) MDR1, *Human* (5.8e − 131 MDR2, *Hamster* (5.7e − 134)	Exporter of the **a**-factor mating pheromone, localized to the plasma membrane and Golgi[bb,cc,dd,ee]
YEF3	*YEF3*	YLR249w	U20865		YHES, *E. coli* ($4.6e^{-64}$) TlrC, *Streptomyces fradiae* ($6.1e^{-36}$) CarA, *Streptomyces sp.* ($6.0e^{-35}$)	Translation elongation factor[ff]
		YNL014w	Z71290		YJJK, *E. coli* ($2.6e^{-40}$) CarA, *Streptomyces sp.* ($4.1e^{-37}$) TlrC, *Streptomyces fradiae* ($9.8e^{-36}$)	
		YPL226w	Z73582		YJJK, *E. coli* ($2.0e^{-35}$) ertX, *Streptomyces erythraeus* ($1.7e^{-34}$)	
	GCN20	YFR009w	D50617		TlrC, *Streptomyces fradiae* ($1.1e^{-40}$) YJJK, *E. coli* ($2.6e^{-38}$)	Required for derepression of Gcn4p translation[gg]
		YER036c	U18796		CarA, *Streptomyces sp.* ($1.9e^{-50}$) YJJK, *E. coli* ($1.2e^{-49}$) SrmB, *Strep. ambo.* ($4.1e^{-48}$)	
Other[i]		YFL028c	D50617		None with *P*-value < 1.0 e-2	
		YDR061w	Z49209		ModF, *E. coli* ($2.5e^{-33}$) PhrA, *E. coli* ($1.4e^{-26}$)	

[a] Subfamilies are defined in Fig. 1 and described in the text.

[b] Generally, a yeast gene name (three letters followed by a number) is assigned only to proteins for which there is experimental data. Exceptions are *PDR10, PDR11, PDR12,* and *PDR15,* which were named based on their homology to *PDR5.* Gene names listed here were obtained either from the literature, or from one of two databases: SGD (http://genome-www.stanford.edu/Saccharomyces/) and YPD(http://www.proteome.com/YPDhome.html). When multiple names exist, the first one published is indicated on the left and others are listed in the last column.

[c] The systematic ORF number is designated by Martinsried Institute for Protein Sequences (MIPS), Max Planck Institute for Biochemistry (http://www.mips.biochem.mpg.de/yeast/). All ORF numbers begin with Y for yeast, the second letter is the chromosome number, the third letter (R or L) is for the right or left arm of the chromosome, and is followed by a three-digit number given sequentially from the centromere toward the telomere; the last letter is w or c, designating the coding strand.

[d] The accession number is from GenBank or EMBL.
[e] The structure of the protein is shown schematically. The spheres designate NBDs, the wavy lines designate MSDs (but are not meant to imply a specific number of transmembrane helices), and the curved line preceding the NBD of ADP1 represents the pair of EGF motifs residing between two predicted membrane spans. Straight lines designate non-membrane-spanning, and thus presumably soluble, domains.
[f] BLAST searches were performed using the XREF interface (http.//www.ncbi.nlm.nih.gov/XREFdb). Only several of the most significant matches to fungal or nonfungal proteins are listed, with preference to sequences of proteins with known function. The *P* value for each match is shown in parentheses. The lower the *P* value, the higher the degree of similarity. The number 0.0 indicates a number smaller than $1.0e^{-301}$.
[g] The YOL075c ORF is considered here to comprise a full-length ABC protein with two NBDs and two MSDs, and to include sequences adjacent to those initially reported for this ORF, as described in the text.
[h] The adjacent ORFs YKR103w and YKR104w are considered here to comprise a single ABC protein, as described in the text.
[i] The two ORFs in the "other" category do not fall into any of the subfamilies shown and are only distantly related to the other ABC proteins.
[j] Ref. 51.
[k] Ref. 50.
[l] Ref. 90.
[m] Ref. 91.
[n] Ref. 54.
[o] Ref. 53.
[p] Ref. 52.
[q] Ref. 56.
[r] Refs. 30 and 56.
[s] Ref. 32.
[t] Refs. 31 and 92.
[u] Ref. 64.
[v] Ref. 67.
[w] Ref. 93.
[x] Ref. 68.
[y] Ref. 81.
[z] Refs. 81 and 94.
[aa] Ref. 80.
[bb] Ref. 71.
[cc] Ref. 72.
[dd] Ref. 95.
[ee] Ref. 96.
[ff] Ref. 48.
[gg] Ref. 49.
[hh] An ORF number was not available at the time this paper was completed. The designation SC6652X is a locus number. The ORF number is YDR091c.

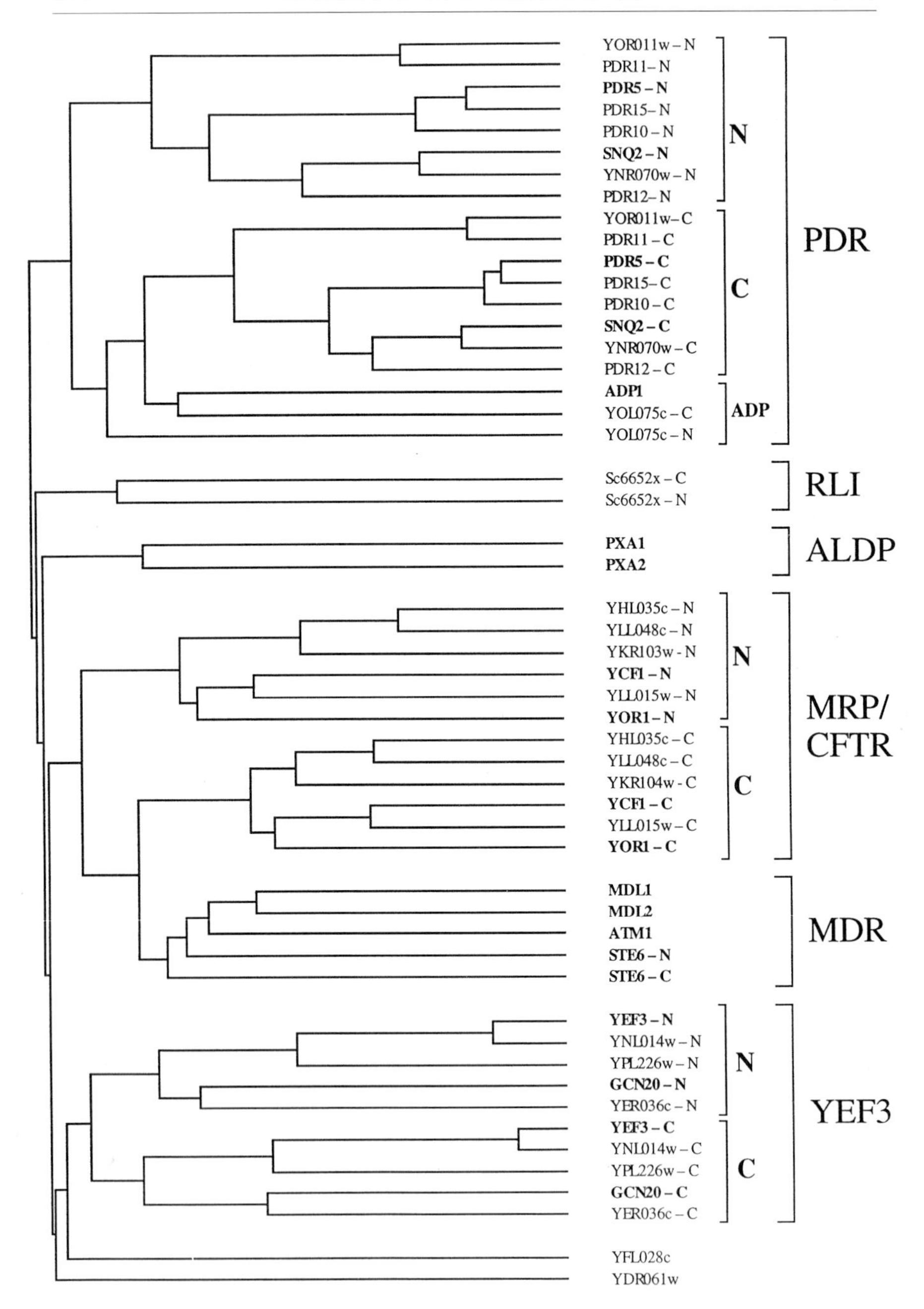
YOR011w – N
PDR11 – N
PDR5 – N
PDR15 – N
PDR10 – N
SNQ2 – N
YNR070w – N
PDR12 – N
YOR011w – C
PDR11 – C
PDR5 – C
PDR15 – C
PDR10 – C
SNQ2 – C
YNR070w – C
PDR12 – C
ADP1
YOL075c – C
YOL075c – N
N
C
ADP
PDR
Sc6652x – C
Sc6652x – N
RLI
PXA1
PXA2
ALDP
YHL035c – N
YLL048c – N
YKR103w - N
YCF1 – N
YLL015w – N
YOR1 – N
YHL035c – C
YLL048c – C
YKR104w - C
YCF1 – C
YLL015w – C
YOR1 – C
N
C
MRP/
CFTR
MDL1
MDL2
ATM1
STE6 – N
STE6 – C
MDR
YEF3 – N
YNL014w – N
YPL226w – N
GCN20 – N
YER036c – N
YEF3 – C
YNL014w – C
YPL226w – C
GCN20 – C
YER036c – C
N
C
YEF3
YFL028c
YDR061w

a similar relationship among additional human ABC proteins included in their study.

Full-length Transporters Predominating over Half-Transporters among Yeast ABC Proteins

It is likely that the yeast half-transporters form homo- or heterodimers. Direct evidence from immunoprecipitation experiments indicate that two of the half-transporters, PXA1 and PXA2, heterodimerize. MDL1 and MDL2 are half-transporters in the same subfamily; it will be of interest to determine whether they can pair together. Some half-transporters, such as ATM1 and ADP1, can be predicted to homodimerize, since they represent the sole half-transporters in their respective families.

Not All ABC Proteins Contain Membrane Spans

Members of four of the six subfamilies (PDR, ALDP, MRP/CFTR, and MDR) are clearly integral membrane proteins whose MSDs contain multiple membrane spans; all of these ABC proteins are likely to function as transporters (Table I). On the other hand, members of the YEF3 subfamily do not possess any predicted membrane spans and are probably not involved in transport. Experimental data for YEF3 and GCN20 indicate that they are soluble proteins involved in regulation of protein translation.[48,49] Nevertheless, the NBDs of this subfamily bear all the hallmarks of the ABC superfamily, and thus are considered a part of it. Similarly, the two ORFs that do not belong to any of the subfamilies (see "Other," Table I)

[48] G. P. Belfield and M. F. Tuite, *Mol. Microbiol.* **9,** 411 (1993).
[49] C. R. Vazquez de Aldana, M. J. Marton, and A. G. Hinnebusch, *EMBO J.* **14,** 3184 (1995).

FIG. 1. Phylogenetic relationship between the NBDs of *S. cerevisiae* ABC proteins. Relationships between the NBDs of the yeast ABC proteins listed in Table I are shown in the form of a dendogram. The most closely related NBDs branch from common or nearby points; the most divergent branch from distant points. The portion of each NBD that was used to generate this tree extends from 50 amino acids upstream of the Walker A site to 50 amino acids downstream of the Walker B site. For proteins that contain two NBDs, each was analyzed separately and is designated NBD-N or NBD-C, according to its position (N- or C-terminal) in the full-length protein. The dendogram was derived using the PILEUP program from the University of Wisconsin Genetics Computer Group (GCG) package.[88] The yeast ABC proteins are classified into six subfamilies[4,89] whose names derive from a human (RLI, ALDP, MRP/CFTR, MDR) or yeast protein (PDR, YEF3) that also falls within the subfamily. The N- and C-terminal NBDs form separate subclusters within several subfamilies, as indicated in the dendogram. In addition, an ADP1 subcluster is noted in the PDR subfamily. Two ORFs (YFL028c and YDR061w, bottom) do not belong to any of the six major subfamilies.

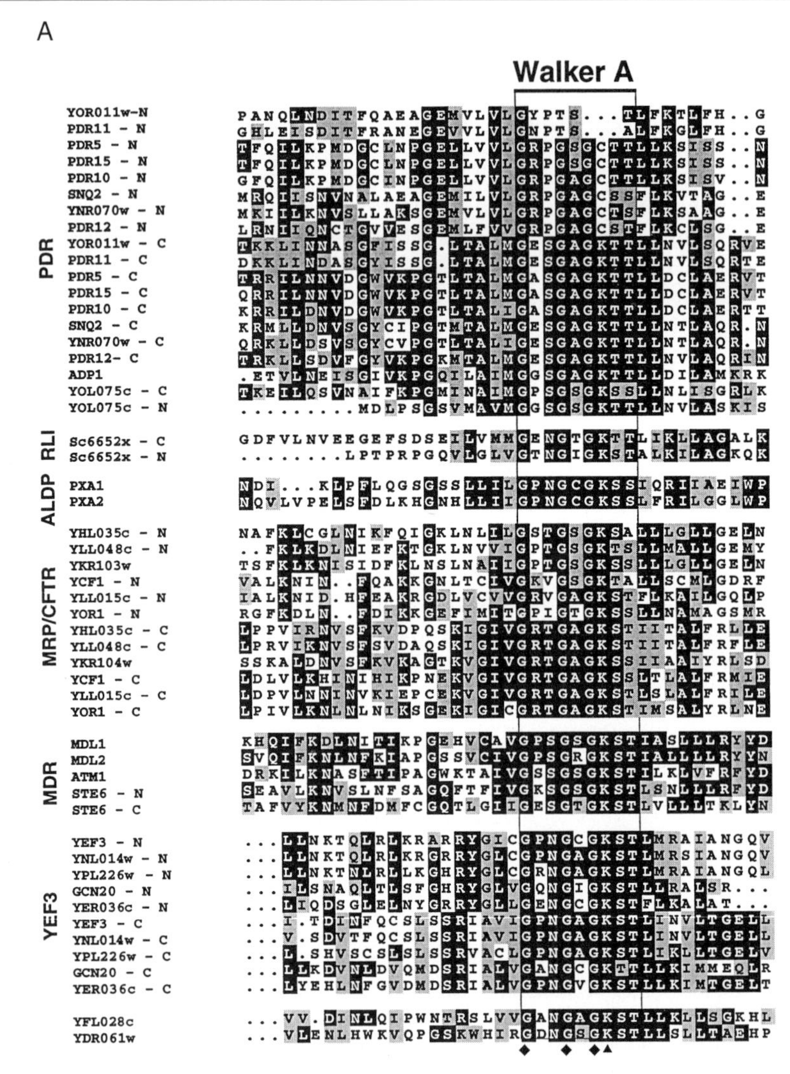

FIG. 2. Multiple sequence alignment of conserved regions of the NBDs of yeast ABC proteins. Distinctive patches of homology are evident between members of each subfamily. Fifty-two NBDs derived from the complete set of 30 yeast ABC proteins were aligned, using the PILEUP program from the University of Wisconsin Genetics Computer Group (GCG) package.[88] The alignment is shown for the subregions of the NBD that contain the highest number of conserved residues: (A) the Walker A region, (B) the center region, (C) the signature and Walker B regions, and (D) the Walker B downstream region. Alignments were shaded to emphasize similar residues, using Boxshade 3.0 (http://ulrec3.unil.ch/software/BOX_form.html), treating each subfamily as a single unit. Residues that are identical in more

B

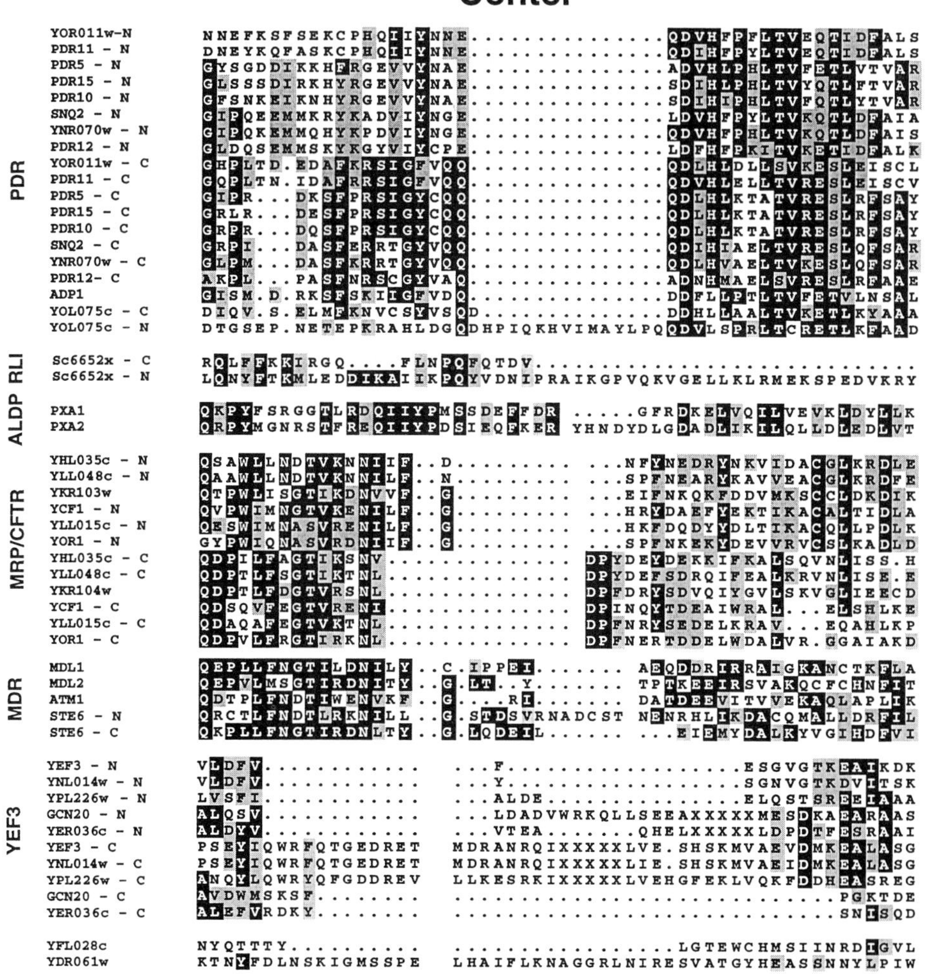

than 40% of the sequences of the subfamily are shaded with a black background, whereas residues that are conserved, but nonidentical, in more than 40% of the sequences of the subfamily are shaded gray. In the Walker A and Walker B regions, residues that are invariant are denoted by diamonds and residues that are nearly invariant among all NBDs are denoted by triangles (*bottom*). To examine an alignment of the uninterrupted NBDs of 17 *S. cerevisiae* ABC proteins, together with mammalian reference proteins in the same subfamily, the reader is referred to our previous study.[4]

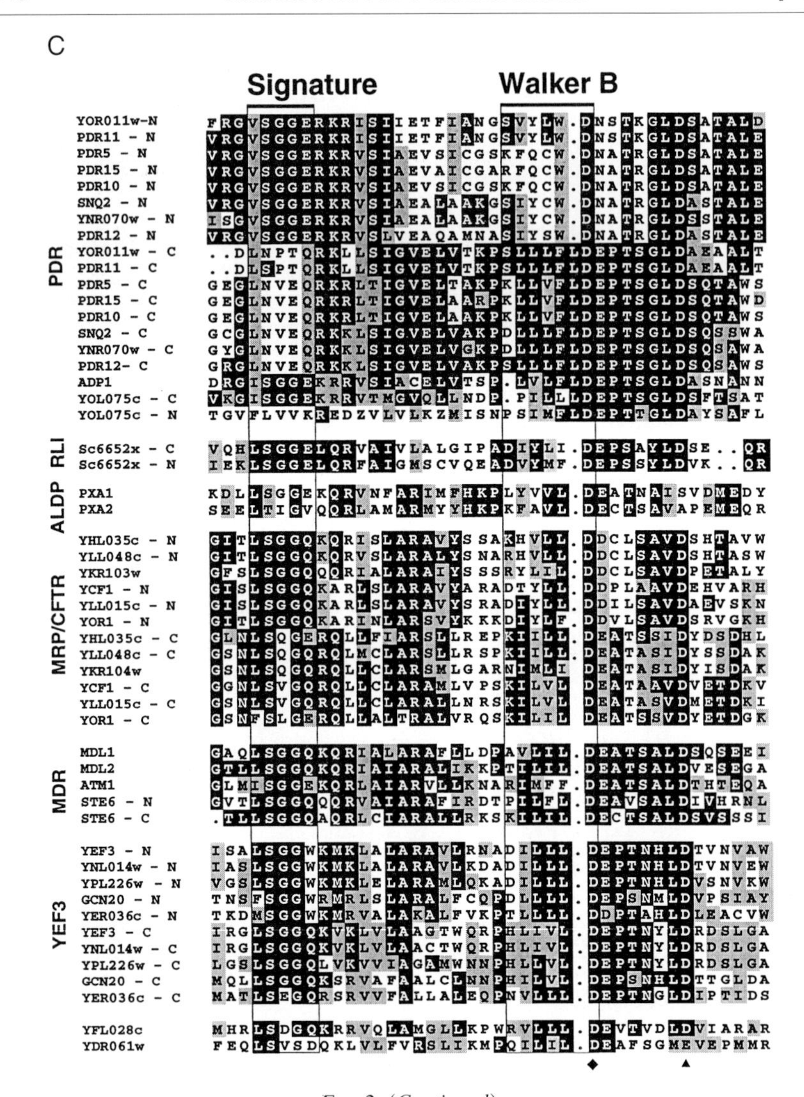

FIG. 2. (*Continued*)

both consist of only a single NBD and no predicted membrane spans. It is possible that one or both of these ORFs could interact with an integral membrane protein to form a transporter. However, because both of these proteins are most similar to the members of the YEF3 subfamily, it is probable that they function more like YEF3 than like a transporter. In addition, the single member (SC6652X) of the new subfamily, RLI1, is unlikely to be a membrane protein. Taken together, it appears that most

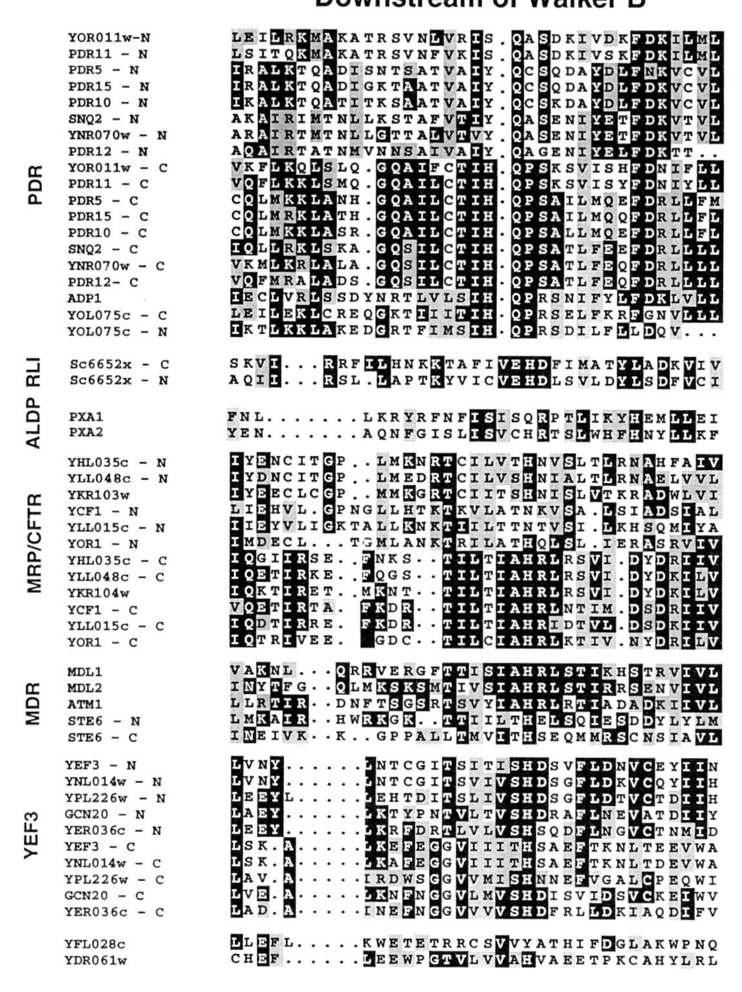

FIG. 2. (*Continued*)

yeast ABC proteins are membrane associated, although there are clearly a significant number of exceptions.

Features Revealed by Multiple Sequence Alignment

Among the yeast ABC proteins, only a few residues are invariant. These are indicated in Fig. 2. In the Walker A region are three glycine (◆) residues that are completely conserved and a lysine (▲) that is conserved in all the subfamilies except for the N-terminal NBDs of the PDR subfamily, where it is always a cysteine. The Walker B region contains a conserved aspartate (◆) and a nearby downstream aspartate (▲) that is conserved in all the subfamilies, except ALDP.

In general, particular positions are invariantly conserved within a subfamily, but these are occupied by distinctive amino acids in each different subfamily. A good example is the "signature region" in which particular clusters of residues are diagnostic for a subfamily, or a subdivision of a subfamily (Fig. 2C); the signature sequence is LSGGQ in the NBDs of several subfamilies (MDR, YEF3-C, and MRP/CFTR-N), LSGGW in YEF3-N, VSGGE in PDR-N, and LNVEQ in PDR-C. Thus, whereas none of the five residues of the signature is the same in all subfamilies, this entire 5-amino-acid stretch is nearly completely conserved within each subfamily or a division of it. A similar situation holds for the regions including and surrounding the Walker A, center, Walker B, and Walker B downstream regions (Figs. 2A–D).

Six ABC Subfamilies of Yeast

PDR Subfamily

PDR is the largest of the yeast subfamilies, containing nine members. The architecture of PDR subfamily members is distinctive, because their modules are arranged in "reverse order" (NBD1–MSD1–NBD2–MSD2) with the NBDs preceding the MSDS. This reverse order arrangement also occurs in the *Drosophila* pigment precursor transporters (*white*, *brown*, and *scarlet*), which are half-molecule transporters.[42] However, nearly all members of the yeast PDR subfamily are full length (with the exception of ADP1, discussed later). Nevertheless, in a BLAST search, *Drosophila white*, *brown*, and *scarlet*, together with the human homolog of *white* comprise the set of nonyeast proteins that are most closely related to yeast PDR subfamily members (Table I).

Members of the PDR5 subfamily also display a notable sequence feature, namely, the degeneracy of the Walker A motif of NBD1, in which the conserved lysine is replaced by cysteine (GX_4G**C**S/T instead of GX_4G**K**S/T) (Fig. 2). This substitution is unexpected because the lysine residue of Walker A is thought to interact directly with ATP. In ABC proteins that contain a lysine in this position, mutations that alter this conserved lysine disrupt function. Nonetheless, all PDR NBD1 sequences are distinguished by the natural absence of a lysine in this position. Indeed, within two of the newly sequenced ORFs (YOR011c and PDR11) there is a 3-amino-acid deletion that encompasses this critical Walker A lysine residue (Fig. 2A). Whether their distinctive Walker A motif influences particular biochemical properties of PDR proteins remains to be determined. However, this feature provides a convenient diagnostic attribute for members of this subfamily.

The phenotypically best characterized PDR gene products are PDR5[50,51] and SNQ2.[52] Both confer pleiotropic drug resistance when overexpressed and presumably function as drug transporters. A topic of intense investigation is the mechanism whereby yeast PDR proteins, like mammalian MDR proteins, can transport a structurally diverse array of substrates, yet exhibit specificity for a preferred hierarchy of substrates. Both PDR5 and SNQ2 have been partially purified and shown to have nucleotide triphosphatase (NTPase) activity.[53,54] However, the NTPase and inhibitor spectrum of these two proteins are distinct. Efforts are currently under way to demonstrate an *in vitro* transport activity for PDR5 to examine how these transporters utilize nucleotide hydrolysis for drug transport. The biochemical characterization of yeast PDR proteins promises to reveal insights relevant to the other multidrug transporters, like mammalian MDR and MRP, despite the fact that PDR, MDR, and MRP represent distinct ABC subfamilies.

It is notable that although PDR5 and SNQ2 are implicated in pleiotropic drug resistance, and the expression of both are under the control of a common transcriptional regulator, PDR1, their true physiological roles are not yet known. Deletion mutants of each are viable and have no detectable phenotype in the absence of drugs. A reasonable speculation is that these transporters may function to protect yeast from xenobiotic compounds that they may encounter. The finding that mouse MDR1 may constitute a component of the blood–brain barrier, functioning to eliminate xenobiotic compounds such as drugs, provides a compelling precedent for the role of PDR proteins in yeast.[26] It has also been speculated that PDR5 may be involved in the elimination of toxic compounds that could potentially accumulate within yeast cells during stationary phase.[55] It will be interesting to determine whether the new members of the yeast PDR subfamily can also confer drug resistance.

One PDR5 family member that is worthy of special mention is ADP1,[56] which together with the NBD1 and NBD2 of YOL075c (which shows

[50] P. H. Bissinger and K. Kuchler, *J. Biol. Chem.* **269,** 4180 (1994).

[51] E. Balzi, M. Wang, S. Leterme, L. Van Dyck, and A. Goffeau, *J. Biol. Chem.* **269,** 2206 (1994).

[52] J. Servos, E. Haase, and M. Brendel, *Mol. Gen. Genet.* **236,** 214 (1993).

[53] A. Decottignies, L. Lambert, P. Catty, H. Degand, E. A. Epping, W. S. Moye-Rowley, E. Balzi, and A. Goffeau, *J. Biol. Chem.* **270,** 18,150 (1995).

[54] A. Decottignies, M. Kolaczkowski, E. Balzi, and A. Goffeau, *J. Biol. Chem.* **269,** 12,797 (1994).

[55] R. Egner and K. Kuchler, *FEBS Lett.* **378,** 177 (1996).

[56] B. Purnelle, J. Skala, and A. Goffeau, *Yeast* **7,** 867 (1991).

homology to mouse ABC8),[39] forms a separate subdivision of the PDR5 subfamily (Fig. 1). In addition to its phylogenetic distance, ADP1 is structurally distinct from other members of the PDR subfamily in possessing only a single NBD, and a modified MSD1 that contains only two predicted membrane spans separated by a loop with two cysteine-rich repeats that correspond to epidermal growth factor motifs. In mammalian cells, EGF repeats are involved in cell adhesion or differentiation. Interestingly, ADP1 represents the closest match to a mammalian ABC protein (EST 157481) detected in the EST database, which appears to define a gene expressed solely in placenta.[39]

RLI Subfamily

RLI represents a new subfamily, not identified in our previous analysis, which contains only a single member (defined by the SC6625X locus, Table I). The structure of this ORF is noncanonical in terms of ABC proteins, in that there are two NBDs but no true MSDs, making it improbable that SC6625X functions as a transporter. SC6625X exhibits remarkable homology to a human protein of known biochemical function, the RNase L inhibitor (RLI).[57] The 2′,5′-oligoadenylate-activated RNase L is an interferon-stimulated RNase that has been postulated to play a central role in mediating mRNA turnover in normal and interferon-treated cells. RNase L is a dimer composed of a regulatory and catalytic subunit.[58] The regulatory subunit is inhibitory, hence the designation RNase L inhibitor (RLI). Because of the high homology score (P value of e^{-301}, Table I) between the yeast and human sequences, it is quite likely that they are functional homologs, although no studies have addressed this point to date. Nevertheless, it is of great interest that its human homolog plays a regulatory role, since there is a growing body of evidence that in addition to their role as transporters, certain ABC proteins (notably, human CFTR and MDR) can regulate the activity of other cellular proteins.[12] A role in regulation is not uncommon for nucleotide-binding proteins, for instance G proteins, in which nucleotide hydrolysis is postulated to represent a timing or proofreading mechanism that operates as a regulatory device.[59] It will be important to examine the phenotype of SC6625X mutants in yeast, to determine whether they are altered in their pattern of mRNA turnover due to loss of the putative RNase inhibitor activity in the mutant.

[57] C. Bisbal, C. Martinand, M. Silhol, B. Lebleu, and T. Salehzada, *J. Biol. Chem.* **270,** 13,308 (1995).
[58] T. Salehzada, M. Silhol, A. Steff, B. Lebleu, and C. Bisbal, *J. Biol. Chem.* **268,** 7733 (1993).
[59] H. Bourne, D. A. Sanders, and F. McCormick, *Nature* **349,** 117 (1991).

ALDP Subfamily

The two sole members of the ALDP subfamily, PXA1 and PXA2, are half-transporters with a forward orientation (MSD–NBD). They exhibit compelling similarities to their human homologs, ALDP and PMP70 (*P* values ranging from e^{-72} to e^{-103}, Table I), which are associated with adrenoleukodystrophy and possibly Zellweger syndrome, respectively. In human cell lines from adrenoleukodystrophy patients, there is a defect in the β-oxidation of very long-chain fatty acids (VLCFA), a process that occurs in the peroxisome.[60] The peroxisomal structure and function of these cells are otherwise normal. PXA1 and PXA2 are also localized to the peroxisomal membrane in yeast. Both single mutants (*pxa1* or *pxa2*) are unable to grow normally on medium containing the long-chain fatty acid oleate and exhibit a defect in oxidation of VLCFA.[30,61] Thus, from both structural and functional considerations, these yeast and human genes appear to be true homologs.

Coimmunoprecipitation studies indicate that PXA1 and PXA2 can directly interact with one another *in vivo* and that the metabolic stability of PXA1 is markedly reduced in the absence of PXA2.[61] These observations together with the mutant phenotypes (discussed earlier), suggest that the PXA1 and PXA2 half-transporters may form a single heterodimeric ABC transporter, which could also occur for human ALDP and PMP70. Two conserved MSD motifs (loop 1 and the EAA box) lie between the transmembrane spans of yeast PXA1, PXA2, and several human ABC proteins. Mutational analysis indicates that both of these motifs are critical for function.

The precise biochemical function of yeast PXA1, PXA2, and their mammalian counterparts remains an important unanswered question. One role proposed for these transporters is to mediate the transport of VLFCA into the peroxisome, although direct evidence for this model is lacking.[31] Nevertheless it is highly likely that yeast PXA1 and PXA2 function in the same way as human ALDP and PMP70, and will thus provide a powerful yeast model for examining the cell biological and biochemical basis of human peroxisomal disorders.

MRP/CFTR Subfamily

The MRP/CFTR subfamily is the second largest group of ABC proteins in *S. cerevisiae*, comprising six members, which are full-length molecules with the typical structural organization (MSD1–NBD1–MSD2–NBD2, Ta-

[60] J. Mosser, Y. Lutz, M. E. Stoeckel, C.-O. Sarde, C. Kretz, A.-M. Douar, J. Lopez, P. Aubourg, and J.-L. Mandel, *Hum. Molec. Genet.* **3,** 265 (1994).

[61] N. Shani and D. Valle, *Proc. Natl. Acad. Sci. U.S.A.* **93,** 11,901 (1996).

ble I). All members of this subfamily exhibit striking homology to four mammalian proteins of clinical significance: the cystic fibrosis chloride channel (CFTR[15]), the human multidrug resistance-associated protein (MRP[25]), the sulfonylurea receptor implicated in insulin secretion (SUR[19,20]), and the organic anion transporter involved in bile secretion (cMOAT[21,62]). The remarkably high degree of homology of the yeast MRP/CFTR subfamily members to these mammalian proteins (e^{-106} to e^{-301}, Table I) strongly suggests functional as well as structural homology. To date, YCF1 is the only member of this yeast subfamily that has been well studied, as described later. The exciting finding that mammalian MRP (see later discussion) can functionally complement a yeast *ycf1* mutant is likely to stimulate intensive research on other members of the yeast MRP/CFTR subfamily.[63] Aside from YCF1 and YOR1, the other members of this subfamily were discovered solely through the yeast genome project.

YCF1 was identified on the basis of conferring cadmium resistance to yeast when overexpressed.[64] The designation YCF1 reflects both its function as a yeast cadmium resistance factor and its structural resemblance to CFTR, including the presence of a domain with homology to the R domain that is characteristic of CFTR. Yeast YCF1 may be an excellent model for examining structure–function issues relating to human CFTR and related ABC proteins, since the *ycf1* mutation ΔF713, which corresponds to ΔF508 in NBD1 of CFTR, results in loss of YCF1 function for cadmium detoxification.[64]

Importantly, YCF1 also appears to be an ideal model for the human multidrug resistance-associated protein, MRP. A significant finding for the ABC field was the biochemical demonstration that human MRP is an ATP-dependent transporter for glutathione S-conjugates such as the leukotrienes.[65,66] The resistance of MRP-overproducing human tumor cells to lipophilic drugs is likely to be due to the conjugation of these drugs to glutathione and their subsequent MRP-mediated efflux. Motivated by the structural similarity between human MRP and yeast YCF1, Li *et al.*[67] demonstrated that yeast YCF1 also transports glutathione S-conjugates, and

[62] M. Buchler, J. Konig, M. Brom, J. Kartenbeck, H. Spring, T. Horie, and D. Keppler, *J. Biol. Chem.* **271,** 15,091 (1996).

[63] R. Tommasini, R. Evers, E. Vogt, C. Mornet, G. Zaman, A. Schinkel, P. Borst, and E. Martinoia, *Proc. Natl. Acad. Sci. U.S.A.* **93,** 6743 (1996).

[64] M. S. Szczypka, J. A. Wemmie, W. S. Moye-Rowley, and D. J. Thiele, *J. Biol. Chem.* **269,** 22,853 (1994).

[65] I. Leier, G. Jeditschky, U. Bucholz, S. P. C. Cole, R. G. Deeley, and D. Keppler, *J. Biol. Chem.* **269,** 27,807 (1994).

[66] M. Muller, C. Meijer, G. J. R. Zaman, P. Borst, R. J. Scheper, N. H. Mulder, E. G. E. De Vries, and P. L. M. Jansen, *Proc. Natl. Acad. Sci. U.S.A.* **91,** 13,033 (1994).

[67] Z. Li, M. Szczypka, Y. Lu, D. Thiele, and P. Rea, *J. Biol. Chem.* **271,** 6509 (1996).

possesses energy requirements, kinetics, substrate specificity, and inhibitor profiles similar to human MRP. Most excitingly, it has been shown that the expression of human MRP in a yeast *ycf1* mutant can complement its cadmium sensitivity and can restore glutathione S-conjugate transport activity.[63] This finding provides dramatic evidence that the sequence similarity between yeast YCF1 and human MRP reflects a striking functional resemblance.

YCF1 is localized to the vacuolar membrane in yeast.[67] YCF1 appears to provide a detoxification function by clearance from the cytosol of glutathione S-conjugates that might otherwise be toxic. It is not clear why yeast sequester these conjugates within the vacuole, rather than excreting them. Interestingly, although human MRP is predominantly at the plasma membrane in drug-resistant lung cancer cells, in normal lung cells it is also found in intracellular compartments. Thus MRP in mammalian cells may have a role as an intracellular pump, as does its counterpart, YCF1, in yeast.

There are five members of the yeast MRP/CFTR subfamily in addition to YCF1. One of these, YOR1, is required for conferring oligomycin resistance in yeast and could be an oligomycin transporter.[68,69] Interestingly, YOR1 transcription is partially regulated by the PDR1 and PDR3 transcription factors that also modulate the expression of PDR5 and other members of the PDR family. No functional studies have been carried out to date with the remaining four members of the MRP/CFTR subfamily. However, given their similarity to one another and to MRP, CFTR, cMOAT, and SUR1 these four newly discovered yeast ABC proteins promise to be the focus of intensive investigation. It will be interesting to see if any of them, like CFTR, exhibit channel function and/or regulate the function of other transporters, which are two distinctive features of CFTR.

MDR Subfamily

The modules of the four members of the yeast MDR subfamily are in the forward orientation. This subfamily is distinctive in containing both full-length (MSD1–NBD1–MSD2–NBD2) and half-transporters (MSD1–NBD1) (Table I). As discussed in a previous section, a notable feature of MDR subfamily members from fungi to mammals is that their N- and C-terminal NBDs are highly similar to one another and do not fall into the separate NBD-N and NBD-C classes seen for the PDR, MRP/CFTR, and YEF3 subfamilies. This phylogenetic pattern suggests a relatively recent duplication event, and might even hint at the possibility that the half-molecules in this subfamily could form functional homodimers.

[68] D. Katzmann, T. Hallstrom, M. Voet, W. Wysock, J. Golin, G. x. Volckaert, and W. Moye-Rowley, *Mol. Cell. Biol.* **15,** 6875 (1995).

[69] Z. Cui, D. Hirata, E. Tsuchiya, H. Osada, and T. Miyakawa, *J. Biol. Chem.* **271,** 14,712 (1996).

The MDR subfamily member STE6 is the best characterized of the yeast ABC proteins.[70–72] Its function is to mediate the transport of the mating pheromone **a**-factor. As such, it is one of the few ABC proteins, in addition to YCF1, for which a true physiological substrate is established. The **a**-factor mating pheromone is extremely hydrophobic and contains a C-terminal isoprenylated and carboxylmethylated cysteine.[70] The export of **a**-factor does not require components of the "classical secretory pathway," as shown by the relatively normal export of **a**-factor in yeast *sec* mutants.[72] Instead, the translocation of **a**-factor across the cellular membrane is carried out by the "nonclassical exporter," STE6. The *S. cerevisiae* **a**-factor pheromone and related pheromones from other fungi are the only prenylated molecules that are known to be exported, although the discovery of mammalian examples would not be entirely unexpected. The *S. pombe* M-factor transporter is highly homologous to STE6.[72a] A *C. albicans* protein HST6 is also strikingly homologous (e^{-300}, Table I).[73]

The export of **a**-factor is necessary for yeast mating. A straightforward petri plate assay for mating provides a robust assay for the study of STE6.[73a,74] This simple functional assay has facilitated structure–function analysis of STE6. These studies have demonstrated the importance of conserved residues in the Walker A and B motifs in the NBDs of STE6.[75] and have shown that STE6 half-molecules, which show no function when expressed individually, can promote efficient **a**-factor export when coexpressed. In addition, a half-molecule was shown to rescue the mating defect of a mutant full-length form of STE6. Coimmunoprecipitation experiments provide evidence, independent of the mating assay, that STE6 partial molecules can physically interact.[76] Mutational analysis of STE6 has also shown that in certain contexts nonsense mutations are efficiently suppressed in yeast, which may also occur for mutant forms of mammalian ABC proteins.[77] Studies on STE6 trafficking-defective mutants have also been informative. In this laboratory, we have generated mutant forms of STE6 that

[70] S. Michaelis, *Semin. Cell Biol.* **4,** 17 (1993).
[71] K. Kuchler, R. E. Sterne, and J. Thorner, *EMBO J.* **8,** 3973 (1989).
[72] J. P. McGrath and A. Varshavsky, *Nature* **340,** 400 (1989).
[72a] J. Davey, P. U. Christensen, and O. Nielsen, *Biochem. Soc. Trans.* **25,** 224S, (1997).
[73] M. Raymond, A. M. Alarco, D. Dignard, B. B. Magee, and D. Y. Thomas, *Yeast* **11,** S85 (1995).
[73a] G. L. Nijbroek and S. Michaelis, *Methods Enzymol.* **292,** [14] 1998 (this volume).
[74] C. Berkower and S. Michaelis, *in* "Molecular Biology and Function of Carrier Proteins" (J. Reuss, J. M. Russell, and M. L. Jennings, eds.), Vol. 48, p. 130. Rockefeller University Press, New York, 1993.
[75] C. Berkower and S. Michaelis, *EMBO J.* **10,** 3777 (1991).
[76] C. Berkower, D. Taglicht, and S. Michaelis, *J. Biol. Chem.* **271,** 22,983 (1996).
[77] K. Fearon, V. McClendon, B. Bonetti, and D. M. Bedwell, *J. Biol. Chem.* **269,** 17,802 (1994).

are specifically retained in the ER.[77a] Such mutants are expected to be useful tools in the identification of cellular components, which are likely to be important for the proper intracellular trafficking and ER quality control of ABC proteins and other membrane proteins in general.

The similarity between STE6 and MDR and the resemblance in terms of the hydrophobicity between their substrates, **a**-factor and lipophilic drugs, respectively, prompted Gros and co-workers[36] to test for functional complementation of a *ste6* mutant, by mouse MDR3, using the mating assay.[37] Strikingly, mMDR3 could partially complement a *ste6* mutant for mating. This finding is important for two reasons: First, it suggests that STE6 and MDR share mechanistic features that make them particularly suitable for the translocation of highly hydrophobic substrates. Second, the yeast mating assay provides a straightforward test system for analysis of mammalian MDR mutants. The expression of mMDR3 was also shown to confer drug resistance to yeast, providing an additional functional assay. Subsequent studies demonstrated that, like mouse MDR3, the *Plasmodium flaciparum* MDR and human *MRP1* genes also confer drug resistance to yeast and complement a *ste6* mutant for **a**-factor transport.[78,79] It is likely that other mammalian ABC proteins can also promote mating or drug resistance in yeast.

The function of the other members of the yeast MDR subfamily is not well understood. ATM1 is a mitochondrial inner membrane protein required for cell viability.[80] The mitochondrion carries out unidentified essential metabolic functions in yeast that are unrelated to respiration. These are apparently compromised in the *atm1* mutant. In yeast lacking ATM1 function, but kept alive by a suppressor mutation, there are pleiotropic defects that include the lack of cytochromes and a high degree of loss of mitochondrial DNA. Further studies are necessary to determine whether ATM1 participates in the transport of essential metabolic compounds into and out of the mitochondrion or, alternatively, if it transports toxic metabolites out of mitochondria. The two other *S. cerevisiae* MDR subfamily members, MDL1 and MDL2, are nonessential genes for which no function is known.[81] However, these are potentially quite interesting in

[77a] D. Loayza and S. Michaelis, submitted (1998); G. L. Nijbroek and S. Michaelis, manuscript in preparation (1998).

[78] S. K. Volkman, A. F. Cowman, and D. F. Wirth, *Proc. Natl. Acad. Sci. U.S.A.* **92,** 8,921 (1995).

[79] S. Ruetz, U. Delling, M. Brault, E. Schurr, and P. Gros, *Proc. Natl. Acad. Sci. U.S.A.* **93,** 9942 (1996).

[80] J. Leighton and G. Schatz, *EMBO J.* **14,** 188 (1995).

[81] M. Dean, R. Allikmets, B. Gerrard, C. Stewart, A. Kistler, B. Shafer, S. Michaelis, and J. Strathern, *Yeast* **10,** 377 (1994).

light of their strong homology (e^{-80} to e^{-86}, Table I) to the mammalian TAP1 and TAP2 peptide transporters (Table I).

YEF3 Subfamily

Although the vast majority of ABC proteins are associated with membrane transport events, a few are involved in functions that appear to be quite unrelated to transport. All members of the YEF3 subfamily lack MSDs, and thus can be categorized as nontransporter ABC proteins. Another feature of YEF3 subfamily members is that they lack homology to other ABC proteins in the center region and in some cases contain a large insertion (up to 50 amino acids, denoted by X's in Fig. 2) that considerably lengthens the distance between the Walker A and B motifs. YEF3 and GCN20 are the two members of this subfamily that have been well studied.[48,49] Both appear to be soluble proteins involved in some aspect of the regulation of protein translation.

YEF3, also called yeast elongation factor 3 (EF-3), stimulates the binding of aminoacyl-tRNA to the A site of the ribosome, by promoting the release of deacylated tRNA from the B site. This reaction requires NTP hydrolysis, presumably provided by the NBDs of YEF3. YEF3 may also enhance the fidelity of protein translation by stimulating the binding of cognate aminoacyl-tRNAs to the A site at the expense of noncognate aminoacyl tRNAs. GCN20 plays a key role in general amino acid control of yeast and may be involved in sensing the availability of amino acids in the cell, which in turn influences the activity of a key kinase. The sense that emerges from what we know about these YEF3 subfamily members is that they utilize ATP hydrolysis in proofreading and fidelity, rather than to mediate the movement of molecules across the membrane. Such a function has also been attributed to the bacterial DNA repair enzyme UvrA, which was one of the earliest described ABC proteins.[82]

Nothing is yet known about the function of the other three YEF3 subfamily members. Two other ORFs that do not belong to any of the subfamilies (and are classified as "Other" in Table I) both consist of only a single NBD. Because these other proteins are more similar to the YEF3 subfamily than to any other subfamily, they could also carry out related functions.

Expression of Heterologous ABC Proteins in Yeast

Yeast has proven to be a highly effective system for both the functional and biochemical dissection of ABC proteins.[27] Functional assays take advantage of the ease of carrying out phenotypic analysis in yeast, in contrast

[82] C. F. Higgins, I. D. Hiles, G. P. C. Salmond, D. R. Gill, J. A. Downie, I. J. Evans, I. B. Holland, L. Gray, S. D. Buckel, A. W. Bell, and M. A. Hermondson, *Nature* **323,** 448 (1986).

to the difficulties associated with the phenotypic analysis of tissue culture cells. To date, the only examples of functional complementation in yeast by mammalian ABC proteins are mouse MDR3 which complements a *ste6* mutant for mating, and human MRP which complements both the *ste6* and *ycf1* mutants, for mating and cadmium resistance, respectively. However, other examples are sure to follow. In particular, mammalian ALD and PMP70 represent promising candidates for the complementation of *pxa1* and *pxa2* mutants. The possibility of using yeast as well as human cells for the study of these and other proteins would greatly expand the presently available assays.

Saccharomyces cerevisiae has provided an ideal system for examining the biochemical transport properties of ABC proteins *in vitro*, in addition to its utility for phenotypic analysis. One effective system for this purpose is the *sec* vesicle system developed by Nakamoto and co-workers.[83] Briefly, newly synthesized plasma membrane proteins move from the Golgi to the cell surface by secretory vesicles. In a *sec6* mutant, these vesicles fail to fuse with the plasma membrane, accumulate intracellularly to high levels, and can be isolated with ease. Using such secretory vesicles to analyze heterologously expressed proteins, it was demonstrated that mouse MDR2 can function as a phosphatidyl choline flippase, whereas mouse MDR3 could transport drugs into these vesicles.[84,85] Crude vesicle preparations from yeast or fractionated organelles are also effective for transport studies, and can be used to study proteins that are not plasma membrane localized, for instance the vacuolar glutathione S-conjugate transporter, YCF1. Plasma membrane preparations from which transport proteins are solubilized have been used to measure the nucleotide hydrolysis activities of the plasma membrane proteins PDR5 and SNQ2.[53,54]

Comparison of Yeast ABC Proteins with Human ABC Proteins Deduced from Analysis of EST Database

The study reported in this article indicates that the genome of the simple eukaryote *S. cerevisiae* encodes 30 ABC proteins, of which 28 can be classified into six subfamilies whose members are phylogenetically and, presumably, functionally related. How does this compare to the more complex genome of humans? Obviously, since only a small part of the genome sequence is complete, the answer to this question is presently incomplete. Nevertheless, an ever-growing number of human cDNAs whose sequence is at least partially known is available through the characterization of human

[83] R. K. Nakamoto, R. Rao, and C. W. Slayman, *J. Biol. Chem.* **266,** 7940 (1991).
[84] S. Ruetz and P. Gros, *Cell* **77,** 1071 (1994).
[85] S. Ruetz and P. Gros, *J. Biol. Chem.* **269,** 12,277 (1994).

ESTs.[86] As an approach to determine the identity of all human ABC proteins, Allikmets and co-workers searched the human EST database.[39] In their study, a total of 141 clones containing sequences of potential ABC genes were identified. These sequences represent segments of 13 known and 21 new ABC genes, bringing the number of ABC members in humans, to date, to 34, which is close to the number found in yeast. In their study, Allikmets and co-workers observed at least one EST hit to each of the 13 known human genes, documenting that the EST base is becoming highly representative of all human genes. Thus, it may well be that only a small number of human ABC genes remain unidentified. Interestingly, phylogenetic analysis indicates that well-defined subfamilies of human ABC proteins exist that correspond generally to the subfamilies we have described for yeast, with seven subfamilies in humans, instead of six as in yeast. Two interesting differences are notable. Humans contain an additional subfamily, ABC1, whose two members (ABC1 and ABC2[87]) are not represented in yeast but show high matches to bacterial genes involved in the specialized process of nodulation. Second, the PDR family, which is the largest group in yeast, with 10 members, contains only a single member (ABC8) thus far in humans, and this protein is more similar to the ADP1 subcluster than to the main PDR subfamily. As described earlier, 2 of the 34 mammalian ABC genes (MDR3 and MRP1) have been shown to functionally complement a yeast defect, whereas others are certain to be demonstrated to do so in the near future.

Yeast as Model Organism for Studying Mechanisms of Disease

A basic tenet of the value of cross-species biology is that a similarity in sequence points to a similarity in mechanism, and that this is conserved across species.[33–35,38] As indicated by the yeast–human connections in this article, the ABC proteins provide a compelling example of this principle.

[86] M. S. Boguski, T. M. J. Lowe, and C. M. Tolstoshev, *Nature Genet.* **4,** 331 (1993).
[87] M. Luciani, F. Denizot, S. Savary, M. Mattei, and G. Chimini, *Genomics* **21,** 150 (1994).
[88] J. Devereaux, P. Haeberli, and O. Smithies, *Nucl. Acids Res.* **12,** 387 (1984).
[89] C. Berkower and S. Michaelis, *in* "Membrane Protein Transport" (S. Rothman, ed.), Vol. 3, pp. 231–277. JAI Press, Greenwich, Connecticut, 1996.
[90] A. Kralli, S. P. Bohen, and K. R. Yamamoto, *Proc. Natl. Acad. Sci. U.S.A.* **92,** 4701 (1995).
[91] D. Hirata, K. Yano, K. Miyahara, and T. Miyakawa, *Curr. Genet.* **26,** 285 (1994).
[92] P. Bossier, L. Fernandes, C. Vilela, and C. Rodrigues-Pousada, *Yeast* **10,** 681 (1994).
[93] J. Wemmie, M. Szczypka, D. Thiele, and W. Moye-Rowley, *J. Biol. Chem.* **269,** 32,592 (1994).
[94] K. Kuchler, H. M. Goransson, M. N. Viswanathan, and J. Thorner, *Cold Spring Harbor Symp. Quantitative Biol.* **57,** 579 (1992).
[95] C. Berkower, D. Loayza, and S. Michaelis, *Mol. Biol. Cell* **5,** 1185 (1994).
[96] R. Kölling and C. P. Hollenberg, *EMBO J.* **13,** 3261 (1994).

TABLE II
CLINICALLY SIGNIFICANT HUMAN GENES AND BEST MATCHES AMONG *S. cerevisiae* GENES

Subfamily	Human gene[a]	Disease or physiological significance[a]	*S. cerevisiae* best matches[b]	*P* value
ALDP	ALDP1	Adrenoleukodystrophy	*PXA1*	e^{-106}
			PXA2	e^{-92}
	PMP70	Zellweger syndrome	*PXA1*	e^{-83}
			PXA2	e^{-72}
CFTR/MRP	CFTR	Cystic fibrosis	*YCF1*	e^{-166}
			YHL035c	e^{-106}
	MRP1	Multidrug resistance-associated protein	*YCF1*	$e^{-0.0}$
			YHL035c	e^{-193}
	SUR1	Persistent familial insulinemia/sulfonylurea receptor	*YCF1*	e^{-182}
			YLL015c	e^{-173}
			YLL048c	e^{-169}
	cMOAT	Dubin–Johnson syndrome/canalicular multispecific organic anion transporter	*YCF1*	$e^{-0.0}$
			YLL015c	e^{-257}
MDR	MDR1	Multidrug resistance	*STE6*	e^{-127}
			MDL2	e^{-81}
	MDR3[c]	PC flippase in bile canalicular membrane	*STE6*	e^{-170}
			MDL1	e^{-79}
	TAP1	Antigen presentation	*MDL2*	e^{-81}
			MDL1	e^{-61}
	TAP2	Antigen presentation	*MDL1*	e^{-83}
			MDL2	e^{-73}

[a] See text for references.
[b] Only the two best matches are given.
[c] Human MDR3 is also referred to as MDR2.

The *S. cerevisiae* genome sequence provides an unparalleled resource for cross-species comparisons. It is now possible, starting with a human protein of interest, to determine the most closely related protein(s) in yeast and focus attention on those particular genes, knowing with certainty that these are indeed the best matches.

To illustrate this point, we have carried out a BLAST search of 10 human ABC proteins involved in human health and disease, to identify the two most closely related *S. cerevisiae* proteins. These results are reported in Table II. This table reinforces the notion that human CFTR, human MRP, and yeast YCF1 are highly related. It also focuses attention on several newly sequenced proteins as potentially important ones to study, for example, YHL035C, an ORF known solely from the sequencing project that also exhibits a high match to both CFTR and MRP. Indeed, the closest yeast matches for each disease gene listed here, as well as new disease genes that may be discovered, are likely to generate considerable interest during the coming years. Because of the availability of powerful genetic, molecular, biochemical, and cell biological methods, yeast provides an ideal organism for addressing questions regarding the substrate specificity, mechanism of action, and trafficking of ABC proteins. The intensive study of the 30 yeast ABC proteins discussed in this chapter is certain to provide fundamental insights into the ABC superfamily that will ultimately contribute to our understanding of the role of these proteins not only in a unicellular organism, but also in human health and disease.

Notes Added in Proof

1. Since the submission of this paper, an independent analysis of yeast ABC proteins similar to that presented here has been published by another group [A. Decottignies and A. Goffeau, *Nature Genetics* **15,** 137 (1997)].
2. The ORF YLL048 has now been given the gene name BAT1 or YBT1. The product of this gene is nonessential for viability and can mediate the transport of bile acids [D. F. Oritz, M. V. St. Pierre, A. Abdulmessih, and I. M. Arias, *J. Biol. Chem.* **272,** 15358 (1997)].
3. The gene referred to herein by its locus number SC6625X has been designated the ORF number YDR091C.

Acknowledgments

We thank R. Rao, A. Heinzer, and G. L. Nijbroek for critical comments on this article. SM is supported in these studies by grants (GM51508 and DK48977) from the National Institutes of Health.

[11] *Arabidopsis MDR* Genes: Molecular Cloning and Protein Chemical Aspects

By Robert Dudler and Michael Sidler

Introduction

The ATP-binding cassette (ABC) superfamily of proteins includes an ever-growing number of ATP-driven transporters in both prokaryotes and eukaryotes that are structurally characterized by a conserved ATP-binding

0076-6879/98 $25.00

domain. A great variety of specific substrates is transported by members of this family of transport proteins, including drugs, anorganic ions, amino acids, proteins, sugars, and polysaccharides. For example, the human P-glycoprotein (Pgp) is a drug efflux pump that can cause multidrug resistance in tumor cells,[1] while a Pgp homolog encoded by the yeast *STE6* gene mediates the export of the proteinaceous **a**-factor mating pheromone,[2,3] and the products of *mdr*-like genes located in the mammalian Class II region of the major histocompatibility complex are involved in the transport of processed antigens to the surface of antigen presenting cells.[4–7] Thus, some Pgp-like proteins provide a signal peptide-independent export mechanism for polypeptides. In fission yeast, a Pgp-like protein is associated with the sequestration of heavy metal–phytochelatin complexes to the vacuole and is thus involved in heavy metal tolerance.[8]

We became interested in possible Pgp homologs of plants because of phenomena described in the literature in which such homologs may play a role. These phenomena include instances of cross-resistance of weed plants to different herbicides with dissimilar sites and modes of action,[9,10] heavy metal detoxification in plants by phytochelatins,[11–13] and signal peptide-independent transmembrane transport of peptide hormones such as systemin, an oligopeptide proposed to be the systemic signal in tomato that moves from mechanically wounded leaves through the phloem and leads to gene activation at remote sites.[14,15]

[1] P. Gros, J. Croop, and D. Housman, *Cell* **47,** 371 (1986).

[2] K. Kuchler and J. Thorner, *Curr. Opin. Cell Biol.* **2,** 617 (1990).

[3] J. P. McGrath and A. Varshavsky, *Nature* **340,** 400 (1989).

[4] S. Bahram, D. Arnold, M. Bresnahan, J. L. Strominger, and T. Spies, *Proc. Natl. Acad. Sci. U.S.A.* **88,** 10,094 (1991).

[5] E. V. Deverson, I. R. Gow, W. J. Coadwell, J. J. Monaco, G. W. Butcher, and J. C. Howard, *Nature* **348,** 738 (1990).

[6] T. Spies, M. Bresnahan, S. Bahram, D. Arnold, G. Blanck, E. Mellins, D. Pious, and R. DeMars, *Nature* **348,** 744 (1990).

[7] J. Trowsdale, I. Hanson, I. Mockridge, S. Beck, A. Towsend, and A. Kelly, *Nature* **348,** 741 (1990).

[8] D. F. Ortiz, L. Kreppel, D. M. Speiser, G. Scheel, G. McDonald, and D. W. Ow, *EMBO J.* **11,** 3491 (1992).

[9] J. Heap and R. Knight, *Aust. J. Agric. Res.* **37,** 149 (1986).

[10] S. R. Moss and G. W. Cussans, *in* "Biological and Chemical Approaches to Combating Resistances to Xenobiotics" (M. Ford, D. Holloman, B. Khambay, and R. Sawicki, eds.), p. 200. Society for Chemical Industry, London, 1987.

[11] E. Grill, E.-L. Winnacker, and M. H. Zenk, *Science* **230,** 674 (1985).

[12] E. Grill, E.-L. Winnacker, and M. H. Zenk, *Proc. Natl. Acad. Sci. U.S.A.* **84,** 439 (1987).

[13] R. Vögeli-Lange and G. J. Wagner, *Plant Physiol.* **92,** 1086 (1990).

[14] G. Pearce, D. Strydom, S. Johnson, and C. A. Ryan, *Science* **253,** 895 (1991).

[15] B. McGurl, G. Pearce, M. Orozco-Cardenas, and C. A. Ryan, *Science* **255,** 1570 (1992).

We have attempted without success to clone genes encoding Pgp homologs from plants using either available heterologous probes for library screening experiments or approaches based on the polymerase chain reaction (PCR). Here, we describe the methods successfully used to isolate MDR-like genes from the plant *Arabidopsis thaliana* and the development of tools for the characterization of the gene products.

Isolation of *mdr*-like Genes Using Degenerate Oligonucleotides as Hybridization Probes

A widely employed method to clone genes encoding proteins that belong to a family whose members share conserved sequence motifs is to use degenerate primers encoding such motifs in conjunction with PCR for the amplification of a partial gene or cDNA sequence. For the PCR amplification to succeed, primer sites have to be separated by a suitable distance, usually between one and a few hundred base pairs. If the latter condition is not fulfilled, the PCR method is not applicable or unlikely to succeed, and direct screening of cDNA or gene libraries with radiolabeled oligonucleotides may be the method of choice.

Design of Oligonucleotide Probes

To isolate *mdr*-like genes from *A. thaliana,* advantage is taken of the high sequence conservation around the second nucleotide-binding fold of the protein products of *mdr*-like genes from organisms as diverse as *Escherichia coli* and humans (Fig. 1). Thus, three pools of degenerate 17-mer oligonucleotides encoding hexameric peptides that are conserved between the *E. coli* HlyB protein[16] and both homologous halves of the human MDR1 protein[1] are synthesized (Fig. 1). To reduce degeneracy, the wobble base in the codon of the sixth amino acid is not included. These degenerate primers are judged to be located too close to each other (or too far apart, if the respective sites in different homologous halves of the MDR1 protein are considered) to allow PCR amplification of a gene fragment of suitable size, and thus the oligonucleotides are radiolabeled to high specific activity by T4 polynucleotide kinase with [γ-^{32}P]ATP (6000 Ci/mmol; DuPont-New England Nuclear, Boston, MA) according to standard procedures[17] and directly used as hybridization probes to screen a genomic λEMBL3 library of *A. thaliana.*

[16] T. Felmlee, S. Pellett, and R. A. Welch, *J. Bacteriol.* **163,** 94 (1985).

[17] T. Maniatis, E. F. Fritsch, and J. Sambrook, "Molecular Cloning: A Laboratory Manual." Cold Spring Harbor Laboratory, Cold Spring Harbor, New York, 1989.

```
                                  NB-2
MDR1   529   QLSGGQKQRIAIARALVRNPKILLLDEATSALDTESE
MDR1  1173   QLSGGQKQRIAIARALVRQPHILLLDEATSALDTESE
HLYB   605   GLSGGQRQRIAIARALVNNPKILIFDEATSALDYESE
Consensus    -LSGGQ-QRIAIARALV--P-IL--DEATSALD-ESE
Pool             1       2                3
```

Pool 1: 384 17-mers	Pool 2: 864 17-mers	Pool 3: 384 17mers
SerGlyGlyGlnLysGln G G 5'TCNGGNGGNCAAAAACA T G G 5'AGCGGNGGNCAAAAACA	IleAlaIleAlaArgAla C C 5'ATAGCNATAGCNCGNGC T T C C G 5'ATAGCNATAGCNAGAGC T T	AspGluAlaThrSerAla T G 5'GACGAAGCNACNTCNGC T G T 5'GACGAAGCNACNAGCGC

FIG. 1. Oligonucleotides used as hybridization probes. (*Top*) Alignment of the conserved sequences around nucleotide-binding fold 2 (NB-2) of the homologous halves of the human MDR1 [P. Gros, J. Croop, and D. Housman, *Cell* **47,** 371 (1986)] and the *E. coli* HLYB [T. Felmlee, S. Pellett, and R. A. Welch, *J. Bacteriol.* **163,** 94 (1985)] proteins. (*Bottom*) Pools of degenerate oligonucleotides encoding the hexameric sequences underlined in the top figure.

Screening Procedure

An *A. thaliana* genomic λEMBL3 library is plated on 150-mm plates at a density of 2×10^4 phages per plate, and triplicate filters (Colony/Plaque Screen, DuPont-New England Nuclear; 137 mm in diameter) are lifted as described.[17] Each triplicate set of filters is separately hybridized to one of the three oligonucleotide pools. Because these pools consist of 384 (pools 1 and 3) and 864 (pool 2) different 17-mers with varying GC contents, the minimization of the influence of base composition on the oligonucleotide dissociation temperature (T_d) is essential. Thus, hybridization of the filters is performed in a buffer containing 3 *M* tetramethylammonium chloride, which results in a base composition-independent T_d of about 55° for 17-mer oligonucleotides.[18] The detailed hybridization and washing conditions are given below.

1. After lifting, place filters twice for 3 min on Whatman (Clifton, NJ) 3MM paper soaked with 0.5 *N* NaOH for denaturation and then twice for 3 min on 3MM soaked with 1 *M* Tris-HCl, pH 7.5, for neutralization. After air drying, wash filters in a large volume of washing solution (6 × SSC, 1% sodium *N*-lauroylsarcosinate) in a shaking water bath at 60° for 2 hr.

[18] W. I. Wood, J. Gitschier, L. A. Lasky, and R. M. Lawn, *Proc. Natl. Acad. Sci. U.S.A.* **82,** 1585 (1985).

2. Prehybridize filters in washing solution containing 100 μg/ml denatured salmon sperm DNA at 65° for 2 hr.
3. Equilibrate filters with hybridization buffer (3 *M* tetramethylammonium chloride, 10 m*M* sodium phosphate, pH 6.5, 1 m*M* EDTA, 2% sodium *N*-lauroylsarcosinate) at room temperature.
4. Seal each of the three replica sets of filters in a separate plastic bag; add hybridization buffer and 10 ng/ml ^{32}P-labeled oligonucleotides of one of the three pools (each pool contains equimolar amounts of the degenerate 17-mers).
5. Hybridize over night at 50°.
6. Wash filters 5 × 10 min at room temperature in washing solution on a shaker.
7. Wash filters 3 × 30 min in prewarmed washing solution in a shaking water bath at 37°.
8. Air dry and autoradiograph filters.

In a library screen of about 3×10^5 plaques, 18 positive candidate clones are identified, of which 8 hybridized with all three oligonucleotide pools. Rescreening of these clones with probe 3 (which gave the strongest hybridization signals of the three probes in most cases in the first round) under conditions identical to the ones used in the primary screen result in 9 positive clones that are then purified. Restriction analysis and DNA gel blot hybridization to probe 3 of these clones reveal that they can be divided into two classes, each one probably representing a different locus. Sequence analysis of one clone is each class confirms in both cases the presence of an *mdr*-like gene (*atpgp 1*[19] and *atpgp2*[20]). Both of the genes encode the three hexameric peptide sequences from which the degenerate oligonucleotide probes are derived.

Sequence Comparison of ATPGP1 and ATPGP2 with MDR-like Gene Products of Other Organisms

Both *A. thaliana* genes encode proteins of 1286 (ATPGP1) and 1229 (ATPGP2) amino acids that exhibit sequence similarity to each other and to MDR-like proteins from other organisms (Table I). Both proteins are internally duplicated, each homologous half containing six putative transmembrane domains and a conserved ATP-binding domain. The genes *atpgp 1* and *atpgp2* contain 9 and 10 introns, respectively, of which 4 share identical positions relative to the sequence.

[19] R. Dudler and C. Hertig, *J. Biol. Chem.* **267,** 5882 (1992).
[20] R. Dudler and M. Sidler, unpublished results (1996).

TABLE I
SEQUENCE CONSERVATION OF MDR-LIKE PROTEINS

	Percentage[a] of identical (conserved) amino acids					
Sequence[b]	ATPGP1	ATPGP2	*H.s.* MDR1	*C.e.* MDR1	*D.m.* MDR5	*S.c.* STE6
ATPGP2	44 (65)	100				
H.s. MDR1	42 (64)	42 (66)	100			
C.e. MDR1	36 (58)	37 (61)	46 (65)	100		
D.m. MDR5	37 (58)	39 (61)	43 (63)	38 (59)	100	
S.c. STE6	26 (50)	28 (52)	30 (51)	29 (52)	27 (50)	100
P.f. MDR1	30 (54)	31 (59)	33 (57)	30 (55)	25 (49)	29 (53)

[a] Pairwise sequence comparisons were performed with the GAP program of the Genetics Computer Group Inc. sequence analysis package.

[b] *H.s.* MDR1, human MDR1 [accession number P08183; C.-J. Chen, J. E. Chin, K. Ueda, D. P. Clark, I. Pastan, M. M. Gottesman, and I. B. Roninson, *Cell* **44,** 381 (1986)]; *C.e.* MDR1, *Caenorhabditis elegans* MDR1 [P34712; C. R. Lincke, I. The, M. Van Groenigen, and P. Borst, *J. Mol. Biol.* **228,** 701 (1992)]; *D.m.* MDR5, *Drosophila melanogaster* MDR65 [Q00748; C.-T. Wu, M. Budding, M. S. Griffin, and J. M. Croop, *Mol. Cell. Biol.* **11,** 3940 (1991)]; *S.c. STE6, Saccharomyces cerevisiae* STE6 [P12866; J. P. McGrath, and A. Varshavsky, *Nature* **340,** 400 (1989)]; *P.f.* MDR1, *Plasmodium falciparum* MDR1 [P13568; T. Triglia, S. J. Foote, D. J. Kemp, and A. F. Cowman, *Mol. Cell. Biol.* **11,** 5244 (1991)].

Generation of ATPGP1-Specific Antibody and Epitope-Tagging of ATPGP1

Two strategies are employed to obtain antibodies specifically recognizing the *atpgp 1* gene product. The first consisted in the expression in *E. coli* of part of ATPGP1 as a fusion protein that is used as an antigen to raise ATPGP1-specific antisera in rabbits. This approach has the potential disadvantage that the resulting antibodies may also cross-react with other members of the MDR-like protein family in *Arabidopsis.* This potential source of difficulties in the interpretation of results obtained by immunological methods is overcome by the second approach, in which the ATPGP1 protein is tagged with an epitope recognized by a commercially available monoclonal antibody. With this method, care must be taken to exclude that the tag causes alterations in the properties of the protein to be analyzed.

Preparation of ATPGP1 Fusion Protein

Construction of an Expression Plasmid. To obtain an antigen for the immunization of rabbits, part of the ATPGP1 protein is expressed as a protein in *E. coli.* For this purpose, *atpgp1* cDNA is needed because introns interrupted suitable regions (encoding hydrophilic protein domains) in the gene. Since screening of several cDNA libraries with the *atpgp1* gene as a probe remains unsuccessful, a partial cDNA sequence is obtained with a reverse transcriptase–polymerase chain reaction (RT–PCR) procedure.[21] Two PCR primers are designed for the amplification of an 890-bp cDNA fragment encoding transmembrane domains 5 and 6 and the first ATP-binding domain of ATPGP1. The 26-mer upstream primer 5′-aa*tctAGA*GCTTCACAAGCTTATTCA corresponds to the gene sequence at nucleotide position 2033-2054.[19] The nucleotides indicated in lowercase are added to give an *Xba*I site (underlined) to facilitate later cloning. The 23-mer downstream primer 5′-tt*gGATCC*AAAGCCTCTTGTACA is complementary to the gene sequence at position 3438–3457[19] and contains additional nucleotides (indicated in lowercase) at the 5′ end resulting in a *Bam*HI site (underlined) for cloning. The reverse transcription reaction is performed with 25 μg of total LiCl-precipitated RNA in a volume of 100 μl containing 10 μl 10 × PCR buffer (500 m*M* KCl, 200 m*M* Tris-HCl, pH 8.4, 25 m*M* $MgCl_2$, 1 mg/ml nuclease-free bovine serum albumin), 1 m*M* deoxynucleotide triphosphates, 100 units RNase inhibitor (Böehringer, Mannheim, Germany), 100 pmol random hexamer oligonucleotides, and 1000 units of Moloney murine leukemia virus reverse transcriptase (GIBCO/BRL, Gaithersburg, MD) for 10 min at room temperature and for an additional 60 min at 37°. After heat inactivation (3 min at 90°), a 10-μl aliquot is PCR amplified in a reaction volume of 100 μl containing 9 μl 10 × PCR buffer, 8 μl of 1.25 m*M* deoxynucleotide triphosphates, 100 pmol of each oligonucleotide primer, and 2.5 units of *Taq* polymerase (Perkin-Elmer, Foster City, CA) for 50 cycles (1 min at 94°, 30 sec at 51°, 1 min at 72°). Under these conditions, only a single band of the expected size of 890 bp is obtained, indicating that a cDNA has been amplified and not the corresponding region of the gene, which contains three introns and has a size of about 1.4 kb. The nucleotide sequence of the 890-bp PCR fragment is verified after subcloning. Cleaving the 890-bp PCR fragment at *Hin*dIII sites near both ends yields an 843 bp *Hin*dIII fragment that is cloned into the unique *Hin*dIII site of the bacterial expression vector p6xHis-DHFRS(−2).[22] Recombinant plasmids containing the insert in the correct orientation allow β-D-thiogalactoside (IPTG)-inducible expression in *E. coli* of a fusion protein consisting of an N-terminal histidine hexamer followed by a derivative of the mouse dihydrofolate reductase fused to part of the ATPGP1 protein that contained transmembrane domains 5 and 6 and the first ATP-binding domain (amino acids 262–544). The histidine

[21] E. S. Kawasaki and A. M. Wang, *in* "PCR Technology" (H. A. Erlich, ed.), p. 89. Stockton Press, New York, 1989.

[22] D. Stüber, H. Matile, and G. Garotta, *in* "Immunological Methods," Vol. IV (I. Levkovits and B. Pernis, eds.), p. 121. Academic Press, New York, 1990.

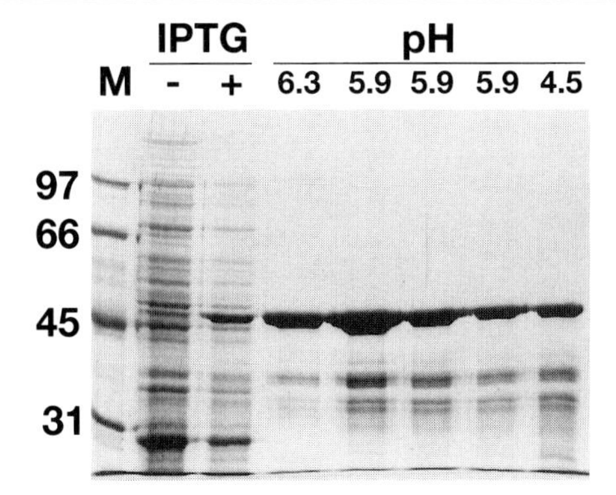

FIG. 2. Production and isolation of ATPGP1 fusion protein. Lysates of *E. coli* expressing the ATPGP1 fusion protein were chromatographed over a Ni-NTA column. Aliquots (10 μl) of fractions eluted at pH 6.3 (6 ml), pH 5.9 (3×3.5 ml), and pH 4.5 (6 ml) were separated by SDS–PAGE on a 10% polyacrylamide gel and proteins were stained with Coomassie blue. Also shown are aliquots of lysates with or without IPTG induction. The fusion protein is indicated by an arrow and the molecular weight ($\times 10^{-3}$) of marker proteins (M) is indicated on the left.

tail allows nickel chelate affinity purification of the recombinant protein.[22]

Production of Antigen. For the production and purification of fusion protein, the construct is introduced into the *lon* mutant *E. coli* strain SG13009, which also contains the repressor plasmid pDMI.[22,23] Bacteria are grown in LB medium containing 100 μg/ml ampicillin and 25 μg/ml kanamycin to an A_{600} of 0.8, on which IPTG is added to a final concentration of 2 mM and incubation is continued at 37° for another 4–5 hr. Cells are harvested by centrifugation and lysed in 6 M guanidine hydrochloride, 100 mM NaH_2PO_4, 10 mM Tris (pH adjusted to 8) by shaking for 1 hr at room temperature. Ni-NTA (Quiagen, Chatsworth, CA) affinity chromatography is performed exactly as described in the manufacturer's manual. Briefly, the lysate is cleared by centrifugation at 10,000*g* for 15 min at room temperature and the supernatant loaded on a Ni-NTA column (bed volume 1 ml) equilibrated with lysis buffer. After washing the column with 10 volumes of lysis buffer and 5 volumes of 8 M urea, 100 mM NaH_2PO_4, 10 mM Tris (pH adjusted to 8), bound proteins are eluted with the same buffer adjusted to pH 6.3, 5.9, and 4.5, respectively. Aliquots of eluted fractions are subjected to sodium dodecyl sulfate–polyacrylamide gel electrophoresis (SDS–PAGE) and Coomassie blue staining (Fig. 2). Gel slices containing the fusion protein are cut out, crushed, and directly used to immunize rabbits.

[23] S. Gottesman, E. Halpern, and P. Trisler, *J. Bacteriol.* **148,** 265 (1981).

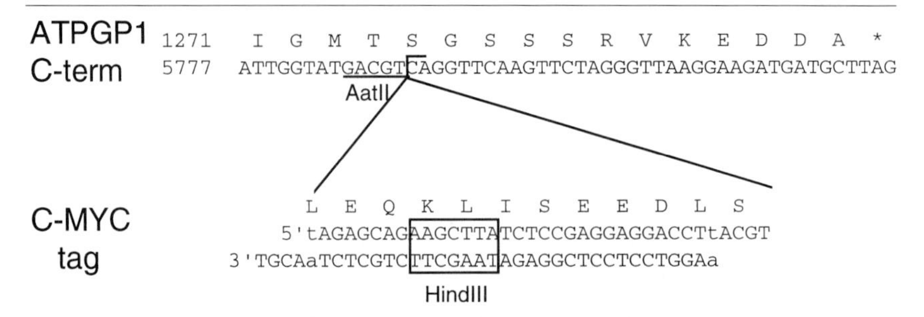

FIG. 3. Construction of c-MYC epitope-tagged ATPGP1. The double-stranded oligonucleotide encoding the c-MYC epitope sequence (given in the single-letter amino acid code above the nucleotide sequence) with *Aat*II compatible sticky ends was inserted into the unique *Aat*II site in the *atpgp1* gene. The *Hin*dIII site (boxed) was introduced for diagnosis, and nucleotides altered to destroy the *Aat*II site on insertion are indicated in lowercase letters.

Comment. It is worth mentioning that, while high level expression of the fusion protein was achieved in the case just described, a number of attempts to express other parts of ATPGP1 as fusion proteins were less successful. For example, it was not possible to obtain recombinant clones with different PCR products encoding the second ATP-binding domain (amino acids 1016–1254), possibly due to a toxic effect of the fusion product. In other cases, the fusion protein was only very weakly expressed or not detected at all after IPTG induction [e.g., fusion proteins containing ATPGP1 amino acids 340–815 (encoding the first ATP-binding domain and putative transmembrane domains 7 and 8) or amino acids 814–1286 (encoding putative transmembrane domains 9 to 12 and the second ATP-binding domain)].

Tagging of ATPGP1 with an Epitope of the Human c-MYC Protein

The human c-MYC-specific monoclonal mouse antibody Myc1-9E10 recognizes an epitope contained in the oligopeptide EQKLISEEDL.[24] The ATPGP1 protein is tagged with this epitope near its C terminus. Advantage is taken of a unique *Aat*II restriction site in the *atpgp1* gene that allows insertion of an epitope-encoding oligonucleotide between position 1274 and 1275 in the corresponding amino acid sequence (Fig. 3). The oligonucleotides are designed such that insertion destroys the unique *Aat*II site of the *atpgp1* gene and introduces a new *Hin*dIII site (Fig. 3). This arrangement allows both selection of recombinant molecules by digestion with *Aat*II after ligation but before transformation, and verification of oligonucleotide insertion by checking for the presence of the diagnostic *Hin*dIII site. Selec-

[24] G. I. Evan, G. K. Lewis, G. Ramsay, and J. M. Bishop, *Mol. Cell. Biol.* **5,** 3610 (1985).

tion is helpful since the *Aat*II ends of the vector are left phosphorylated (resulting in a high rate of self-ligation) and the oligonucleotides are left unphosphorylated in order to preclude insertion of multiple copies. The resulting recombinant clones are verified by sequence analysis of the engineered region and clones containing the tag-encoding oligonucleotide in the correct orientation are selected.

Western Blot Analysis of ATPGP1 Expression

Protein Extraction and Western Blotting

For the efficient extraction and solubilization of ATPGP1 protein from *Arabidopsis* tissue, it is necessary to prepare first a membrane fraction. All steps are carried out at 4°. About 10 g of leaves from *Arabidopsis* plants (at the transition to the reproductive stage) are homogenized with 3 volumes of buffer (50 m*M* HEPES–KOH, pH 7.5, 5 m*M* EDTA, 0.1% (w/v) bovine serum albumun, 1 m*M* phenylmethanesulfonyl fluoride, 2 m*M* dithiothreitol (DTT), 1% polyvinylpyrrolidone, 0.1 mg/ml butylated hydroxytoluene, 0.5 *M* sucrose) in a Waring blender for 1 min and afterward in a Sorvall Omni-Mixer (Sorvall, Newtown, CT) for 1 min. The homogenate is centrifuged at 7000*g* for 15 min to pellet cell walls and debris. Membranes (and chloroplasts) are pelleted from the supernatant by ultracentrifugation in a Beckman SW28 (Beckman, Palo Alto, CA) rotor at 100,000*g* for 30 min. The microsomal pellet is resuspended in 1–2 ml 5 m*M* Tris-HCl, pH 7, 2 m*M* EDTA, and 250 m*M* sucrose using a Potter homogenizer. At this stage the protein concentration is determined using the detergent compatible (DC) protein assay (Bio-Rad, Hercules, CA). Aliquots can be frozen in liquid nitrogen until use.

For SDS–PAGE, aliquots are denatured after addition of 0.25 volumes of 5 × sample buffer (60 m*M* Tris-HCl, pH 6.8, 10% SDS, 25% glycerol, 5% 2-mercaptoethanol, 0.1% bromphenol blue) at 60° for 15 min and sonicated in a sonication water bath (47 kHz; Bransonic, Aasoest, Holland) for another 15 min. Proteins are separated on 8% (w/v) SDS–polyacrylamide gels and transferred to nitrocellulose filters (0.45-μm pore size; Millipore, Bedford, MA) using a semidry blotting apparatus (Pharmacia-LKB, Uppsala, Sweden). For the immunodetection of ATPGP1, the blots were blocked in TBS [20 m*M* Tris-HCl, pH 7.5, 137 m*M* NaCl, 0.1% (v/v) Tween 20] containing 7.5% (w/v) nonfat milk powder and incubated in a 1:1000 dilution in TBS of anti-ATPGP1 fusion protein antiserum or Myc1-9E10 monoclonal antibody (purchased from BAbCO, Richmond, CA).[24] After washing in TBS and incubating with horseradish peroxidase-conjugated secondary antibody [goat anti-rabbit and rabbit anti-mouse immunoglo-

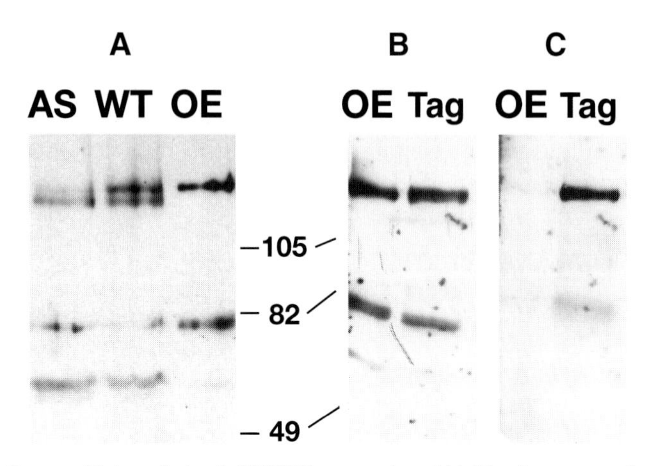

FIG. 4. Immunoblot analysis of ATPGP1 expression. (A) Membrane proteins extracted from wild-type plants (WT) and transgenic plants that express at high levels the *atpgp1* gene in sense (OE) and antisense (AS) orientation were separated by SDS–PAGE and blotted onto nitrocellulose. Samples contained 100 μg (WT and AS) or 4 μg (OE) protein. The gel blot was incubated with antifusion protein antiserum and developed using the ECL detection system (Amersham). (B) Immunoblot analysis of c-MYC epitope-tagged ATPGP1 expression. Samples containing 30 μg protein extracted from transgenic plants constitutively expressing ATPGP1 (OE) and c-MYC-tagged ATPGP1 (Tag) were separated by SDS–PAGE and blotted. The blot was incubated with antifusion protein antiserum and developed as described for part A. (C) The blot shown in part B was stripped from bound antibody and reprobed with Myc1-9E10 monoclonal antibody and developed again using the ECL system.

bulia G (IgG), respectively], the blots are developed using the ECL (enhanced chemiluminescence) immunodetection system (Amersham International, UK). For reprobing Western blots with different antibodies, blots were stripped from bound antibody by incubation in 100 m*M* 2-mercaptoethanol, 2% (w/v) SDS, 62.5 m*M* Tris-HCl, pH 6.7, for 30 min at 50°.

ATPGP1 Expression in Wild-Type and Transgenic Plants

Figure 4A shows a Western blot with protein extracts from wild-type plants and from transgenic plants constitutively and ectopically expressing the wild-type *atpgp1* gene in either sense or antisense orientation. The construction and the phenotype of the transgenic plants are described elsewhere.[25] As is evident from the figure, the anti-ATPGP1 fusion protein antibody detects two bands of slightly different sizes in the expected range

[25] M. Sidler, B. Vogel, and R. Dudler, submitted.

of about 150 kDA with extracts from wild-type plants. This suggests that the antibody recognizes not only ATPGP1, but also cross-reacts with at least another protein of similar size, presumably another member of the ABC superfamily. This would not be particularly surprising because the fusion protein contained the conserved ATP-binding domain characterizing the family.

That this is indeed the case is evident from the analysis of the transgenic plants. In extracts from plants constitutively expressing the ATPGP1 to high levels, only the larger of the two bands is visible (lane OE; note that 25 times less protein was loaded than in lanes WT and AS). Thus, the larger band represents ATPGP1. In contrast, with extracts from transgenic plants expressing the *atpgp1* antisense construct (lane AS), the lower band is much stronger than the upper one, indicating that the antisense expression results in the suppression of ATPGP1 synthesis, and thus ATPGP1 is hardly detectable. The additional band at 82 kDA is a derivative of ATPGP1 since it is also visible in the OE lane, where 25 times less protein was loaded than in the other lanes, while the cross-reacting band located between the 82- and 49-kDa size markers is not.

A protein blot with extracts (30 μg) from transgenic plants[25] constitutively expressing wild-type ATPGP1 or c-MYC-tagged ATPGP1 probed with antifusion protein antiserum shows that both proteins are of the same size and abundance (Fig. 4B). When the same blot is stripped from bound antibody and reprobed with monoclonal Myc1-9E10 antibody, only tagged ATPGP1 is recognizable (Fig. 4C). This indicates that the addition of the tag does not alter the stability of ATPGP1. The plants expressing c-MYC-tagged ATPGP1 should be useful for immunohistochemical localization studies. Preliminary results with these plants indicate that ATPGP1 is localized in the plasmalemma. However, the tag approach can also have its pitfalls. This is exemplified by our observation that transgenic tobacco plants harboring the gene encoding tagged ATPGP1 accumulate a protein of a size indistinguishable from that of type ATPGP1 that is recognized by the antifusion protein antiserum. However, it is not recognized by the monoclonal antibody Myc1-9E10. The most likely interpretation is that in tobacco, but not in *Arabidopsis,* the C terminus of the tagged ATPGP1 is proteolytically processed and the tag thereby removed.

Acknowledgments

We would like to thank Susanne Graf for excellent technical assistance. This work was supported by the Kanton Zürich and the Swiss National Science Foundation (grant 31-37145.93).

[12] Functional Analysis of *pfmdr1* Gene of *Plasmodium falciparum*

By SARAH VOLKMAN and DYANN WIRTH

Introduction

Drug resistance in microorganisms is emerging as an important obstacle to effective treatment of infectious diseases, yet molecular mechanisms underlying this resistance remain to be elucidated. Multidrug resistance (*mdr*) gene homologs have been implicated in mediating drug resistance in parasitic protozoans including *Plasmodium* and *Leishmania* and our work has focused on the functional characterization of these *mdr* genes and their role in drug resistance. In *Plasmodium falciparum,* a candidate drug resistance gene, *pfmdr1,* has been identified and overexpression of *pfmdr1* is strongly correlated with mefloquine resistance. We have taken two experimental approaches, including heterologous expression of *pfmdr1* in yeast and subcellular localization studies in *P. falciparum,* to address the functional nature of Pgh1 in a mechanism of drug resistance. Expression of *pfmdr1* in yeast containing a null mutation in *ste6,* can complement STE6 function and restore mating to a level 1000-fold over background, demonstrating that Pgh1 functions as a transport molecule in this system. Furthermore, expression of the *pfmdr1* gene containing mutations associated with either chloroquine or mefloquine resistance in the wild abolishes this mating phenotype, suggesting that these mutations are important for the transport function of Pgh1.

Role of Drug Accumulation and Efflux in Resistance in *Plasmodium falciparum*

Two important observations were made concerning differences in chloroquine accumulation between drug-sensitive and drug-resistant parasites. First, it was demonstrated that verapamil, a calcium channel blocker, can reverse chloroquine resistance at the same concentration at which it reverses resistance in multidrug-resistant (MDR) cultured neoplastic cells.[1] Second, it was shown that chloroquine-resistant parasites release drug 40 to 50 times more rapidly than a susceptible parasite.[2] These data suggested that

[1] S. K. Martin, A. M. J. Oduola, and W. K. Milhous, *Science* **235,** 899 (1987).

[2] D. J. Krogstad, I. Y. Gluzman, D. E. Kyle, A. M. J. Oduola, S. K. Martin, W. K. Milhous, and P. H. Schlesinger, *Science* **238,** 1283 (1987).

METHODS IN ENZYMOLOGY, VOL. 292 0076-6879/98 $25.00

chloroquine resistance is mediated by an efflux mechanism that results in decreased accumulation of chloroquine in the resistant parasite, and directed investigation into a mechanism of drug resistance that would involve a P-glycoprotein-like molecule. Additionally, the presence of a similar mechanism in *P. falciparum* may account for the cross-resistance patterns among antimalarials observed in the field, as P-glycoprotein mediates resistance to many structurally unrelated drugs. Note that the ability of this rapid efflux to account for differences in chloroquine accumulation between a sensitive and a resistant parasite is controversial and other studies have found differences in drug influx between drug-sensitive and -resistant parasites. Note also that efflux studies have not been done for mefloquine since it is a highly lipophilic drug, but reversal of mefloquine resistance has been demonstrated with compounds such as penfluridol.[3]

Identification of *mdr*-like Genes in *Plasmodium falciparum*

The phenotypic relationship of drug efflux and resistance reversal between mammalian MDR and chloroquine resistance was exploited to identify the *P. falciparum* multiple drug resistance or *pfmdr1* gene from the parasite.[4,5] Using P-glycoprotein-mediated MDR as a model (see Ref. 6) the mechanism of drug resistance in malaria was predicted to involve overexpression of *pfmdr1* to increase efflux of drug from the parasite, effectively making the parasite resistant to the effects of the drug. The protein encoded by the *pfmdr1* gene, Pgh1, is a 160-kDa protein localized to the digestive vacuole of the parasite.[7] This subcellular localization implicates Pgh1 in the modulation of intracellular chloroquine concentrations to reduce net accumulation in the food vacuole, the site of chloroquine's target, rather than reduce total chloroquine in the parasite.[8] Other reports suggest that Pgh1 expression, although mainly to the digestive vacuole membrane, may have some expression on other membrane sites in the parasitized red blood cell. Still, overexpression of a drug pump on the digestive vacuole membrane would allow the parasite to separate the drug from the target for the drug, which for chloroquine is presumably within the food vacuole. Despite the identification of an *mdr* homolog in *P. falciparum*, its role in chloroquine

[3] A. M. J. Oduola, G. O. Omitowoju, L. Gerena, D. E. Kyle, W. K. Milhous, A. Sowunmi, and L. A. Salako, *Trans R. Soc. Trop. Med. Hyg.* **87,** 81 (1992).

[4] C. M. Wilson, A. E. Serrano, A. Wasley, M. P. Bogenschutz, A. H. Shankar, and D. F. Wirth, *Science* **244,** 1184 (1989).

[5] S. J. Foote, J. K. Thompson, A. F. Cowman, and D. J. Kemp, *Cell* **57,** 921 (1989).

[6] M. M. Gottesman and I. Pastan, *Ann. Rev. Biochem.* **62,** 385 (1993).

[7] A. F. Cowman, S. Karcz, D. Galatis, and J. G. Culvenor, *J. Cell Biol.* **113,** 1033 (1991).

[8] A. F. Cowman, *Parasit. Today* **7,** 70 (1991).

resistance is not clearly established. Overexpression of *pfmdr1* does not account for chloroquine resistance, since a correlation between *pfmdr1* copy number and drug resistance does not hold for all strains of parasites resistant to chloroquine,[5] nor does simple overexpression of Pgh1 correlate with chloroquine resistance.[7] The story is different for mefloquine resistance, where overexpression of *pfmdr1* correlates with mefloquine resistance in both laboratory and field isolates. The observed differences of *pfmdr1* expression between chloroquine- and mefloquine-resistant parasites may be indicative of distinct roles for Pgh1 in mechanisms of resistance to these drugs.

These findings argued that, at least for chloroquine, the simplest model of increased drug efflux mediated by increased expression of the *pfmdr1* gene could not explain the resistance. Thus, other mechanisms involving a role for *pfmdr1* were explored. Sequence analysis identified genetic polymorphisms within the *pfmdr1* gene between drug-sensitive and -resistant parasites, which were used to predict correctly the chloroquine phenotype of 34 out of 36 isolates in a double-blind study,[9] suggesting that these mutations were important for the chloroquine-resistant phenotype in these field isolates. However, analysis of a genetic cross between a chloroquine-sensitive and a chloroquine-resistant strain[10] did not support the linkage of the parental *pfmdr1* gene identified by a restriction fragment length polymorphism to the chloroquine resistance or sensitivity phenotype and drug efflux characteristics of 16 progeny from this cross. Subsequent studies have demonstrated that these genetic changes in *pfmdr1* are not sufficient for chloroquine resistance,[11] since chloroquine-resistant isolates that do not contain these mutations have been identified. Nevertheless, the importance of these mutations, which were selected in the wild under natural drug pressures, has not been eliminated because subsequent analysis of these mutations in *pfmdr1* expressed in heterologous systems argues for their significance in a transport function for Pgh1.

More recent evidence suggests a direct relationship between *pfmdr1* copy number and both chloroquine and mefloquine drug pressure, arguing that Pgh1 is involved in a mechanism of drug resistance.[12] The first of these studies placed chloroquine-resistant isolates under increasing chloroquine

[9] S. J. Foote, D. E. Kyle, R. K. Martin, A. M. J. Oduola, K. Forsyth, D. J. Kemp, and A. F. Cowman, *Nature* **345,** 255 (1990).

[10] T. E. Wellems, L. J. Panton, I. Y. Gluzman, V. E. do Rosario, R. W. Gwadz, A. Walker-Jonah, and D. J. Krogstad, *Nature* **345,** 253 (1990).

[11] C. M. Wilson, S. K. Volkman, S. Thaithong, R. K. Martin, D. E. Kyle, W. K. Milhous, and D. F. Wirth, *Mol. Biochem. Parasit.* **57,** 151 (1993).

[12] D. A. Barnes, S. J. Foote, D. Galatis, D. J. Kemp, and A. F. Cowman, *EMBO J.* **11,** 3067 (1992).

pressure and found that as chloroquine resistance levels increased in these isolates, deamplification of *pfmdr1* and increased sensitivity to mefloquine occurred. These indicate an inverse relationship between chloroquine and mefloquine resistance: as parasites became more resistant to chloroquine, they became more sensitive to mefloquine. They also indicated that Pgh1 may function differently in response to these two drugs: as the parasite becomes more resistant to chloroquine it reduces its *pfmdr1* copy number. A similar correlation has been observed by several groups,[4,13,14] which placed laboratory strains under mefloquine pressure to generate mefloquine-resistant parasites. In concert with this acquisition of mefloquine resistance, *pfmdr1* copy number increased while chloroquine sensitivity increased; additionally, as several of these strains became resistant to mefloquine, they became cross-resistant to either quinine or halofantrine. Importantly, these laboratory studies are supported by the demonstration that mefloquine-resistant field isolates have increased *pfmdr1* copy number and expression,[11] as well as by long-standing epidemiological observations in nature that sensitivity to chloroquine returns to an area of high mefloquine use, coincident with the increase in mefloquine resistance. In addition, the characteristics of cross-resistance, associated with mefloquine resistance in studies of laboratory isolates, have also been observed in the field, suggesting that Pgh1 expression changes differently in response to either chloroquine or mefloquine resistance in field isolates.

Functional Analysis of Pgh1 in Heterologous Systems

It therefore seemed possible that Pgh1 might mediate both chloroquine susceptibility and mefloquine resistance. Chloroquine sensitivity mediated by Pgh1 expression was further demonstrated by expression of the *pfmdr1* gene product in Chinese hamster ovary (CHO) cells resulting in a two- to threefold increase in sensitivity to chloroquine as compared to CHO cells transfected with the expression plasmid alone.[15] Pgh1 expression was localized to the lysosomal membrane and presumably acts to increase the amount of chloroquine in the lysosome resulting in the observed increased sensitivity to chloroquine. This increased accumulation could either be the result of chloroquine transport, or of changes in the pH gradient to alter chloroquine uptake; as pH changes were detected, the authors suggested that the function of Pgh1 was to alter the pH gradient across the lysosome. Interestingly,

[13] S. A. Peel, S. C. Merritt, J. Handy, and R. S. Baric, *Am. J. Trop. Med. Hyg.* **48,** 385 (1993).

[14] A. F. Cowman, D. Galatis, and J. K. Thompson, *Proc. Natl. Acad. Sci. U.S.A.* **91,** 1143 (1994).

[15] H. H. G. van Es, S. Karcz, F. Chu, A. F. Cowman, S. Vidal, P. Gros, and E. Schurr, *Mol. Cell. Biol.* **14,** 2419 (1994).

expression of a Pgh1 molecule containing two amino acid changes associated with chloroquine-resistant field isolates[9] abolished the increased sensitivity conferred by Pgh1 in this system, arguing that these mutations, which result in a serine to cysteine residue change at amino acid position 1034 and an asparagine to aspartic acid residue change at amino acid position 1042, are important for this hypersensitivity effect. Both of these mutations are predicted to reside in transmembrane domain 11 of Pgh1[9,16] where homologous amino acid substitutions in the mouse *mdr1a* and *mdr1b* genes dramatically alter the level of resistance to adriamycin, colchicine,, and vinblastine.[17–20] Together these data suggest that these residues are important for determining either levels of drug resistance or specificity of substrates transported by members of this gene family.

Functional Role for Pgh1 as Transporter

Does Pgh1 therefore function as a transporter or as a channel? We have demonstrated that Pgh1 can function to export a small peptide, the yeast pheromone **a**-factor, arguing that Pgh1 functions as a carrier rather than an ionophore. We expressed *pfmdr1* in *Saccharomyces cerevisiae* deficient for STE6 expression,[21] a protein whose functional role has been defined as a transport molecule that exports mating pheromone **a**-factor.[22–24] Since **a**-factor is required for mating,[25] the frequency of mating reflects the efficiency of **a**-factor transport. Expression of Pgh1 in this system demonstrates that *pfmdr1* expression complements the *ste6* mutation and restores mating to a level 1000-fold above background.[21] Interestingly, the same mutations in *pfmdr1* previously identified in chloroquine resistance isolates[9] abolish restoration of a mating phenotype. Normally, most of the STE6 is localized to intracellular membranous structures consistent with the Golgi, and only a very small percentage of the protein is located at the cell surface where it is functional. These data argue that Pgh1 can function

[16] M. Raymond, P. Gros, M. Whiteway, and D. Y. Thomas, *Science* **256,** 232 (1992).

[17] S. I-H. Hsu, D. Cohen, L. S. Kirschner, L. Lothstein, M. Hartstein, and S. B. Horwitz, *Mol. Cell. Biol.* **10,** 3596 (1990).

[18] P. Gros, R. Dhir, J. Croop, and F. Talbot, *Proc. Natl. Acad. Sci. U.S.A.* **88,** 7289 (1991).

[19] R. Dhir, K. Grizzuti, S. Kajiji, and P. Gros, *Biochemistry* **32,** 9492 (1993).

[20] S. Kajiji, F. Talbot, K. Grizzuti, V. Van Dyke-Phillips, M. Agresti, A. R. Safa, and P. Gros, *Biochemistry* **32,** 4185 (1993).

[21] S. K. Volkman, A. F. Cowman, and D. F. Wirth, *Proc. Natl. Acad. Sci. U.S.A.* **92,** 8921 (1995).

[22] J. P. McGrath and A. Varshavsky, *Nature* **340,** 400 (1989).

[23] K. Kuchler and J. Thorner, *Proc. Natl. Acad. Sci. U.S.A.* **89,** 2302 (1992).

[24] S. Michaelis, *Sem. Cell Biol.* **4,** 17 (1993).

[25] S. Michaelis and I. Herskowitz, *Mol. Cell. Biol.* **8,** 1309 (1988).

as a transporter, and suggests that localization of expression may give us important clues to the functional role of Pgh1 in drug resistance.

How can we reconcile the findings that overexpression of *pfmdr1* correlates with both mefloquine resistance and chloroquine hypersensitivity? One explanation may be that chloroquine and the other quinine derivatives like mefloquine have distinct targets, or these targets are localized differently in the parasite. Drug pressure studies with chloroquine and expression of *pfmdr1* in CHO cells suggest that Pgh1 mediates chloroquine uptake into the parasite food vacuole, allowing more drug to enter the site of drug target and thus rendering the parasite more sensitive to its effects. If Pgh1 functions to drive mefloquine transport into the food vacuole, how does overexpression provide a mechanism of resistance? This might occur if the target for mefloquine is not located inside the food vacuole; then, sequestration of mefloquine into the food vacuole would provide a mechanism of protection or resistance to the effects of the drug. However, there is good evidence to suggest that both chloroquine and mefloquine act in the food vacuole to inhibit the heme polymerization process, consistent with ultrastructural data that suggests the same. Nevertheless, other drugs such as halofantrine and artemisinin and derivative compounds have not been shown to inhibit heme polymerization and therefore the targets for these compounds may be located outside the food vacuole. If this is the case, then sequestration of drug in the digestive vacuole may prevent access to its target, effectively resulting in a resistant phenotype. Although one cannot rule out the possibility of separate targets for chloroquine and the other quinoline derivatives as an explanation for the different levels of Pgh1 expression associated with these drug phenotypes, it seems an unlikely explanation at least for chloroquine and mefloquine.

Role of Subcellular Localization in Drug Resistance

As an alternative explanation we suggest that differential localization of Pgh1 can resolve the apparent discrepancy between Pgh1 expression and resistance to chloroquine or mefloquine. Assuming that drug resistance is accomplished by separating drug from its target of action, perhaps Pgh1 expression at different sites in the infected red blood cell provides both chloroquine uptake into the food vacuole and mefloquine release from the parasite. Expression of Pgh1 at the food vacuole transports chloroquine into this organelle, whereas overexpression of Pgh1 allows for transport of mefloquine away from the food vacuole, for example, out of the parasite or out of the surrounding host erythrocyte. Evidence suggesting that Pgh1 can mediate the transport of two distinct molecules into or across distinct membrane compartments comes from Pgh1 expression in yeast deficient for STE6. We have demonstrated that this expression allows for export of **a**-factor from the yeast as shown by complementation of a mating phenotype. In addition, we have recently found these yeast expressing Pgh1 to be hypersensitive to the antifungal agent FK520, presumably by allowing drug to accumulate within intracellular membranous compartments. If we look at the expression of STE6, which is an unstable molecule with a very short half-life in yeast, most of the protein is localized to internal membranes with only a small amount at the plasma membrane.[26] This small amount of expression at the cell surface appears sufficient for efficient export of **a**-factor from the cells, suggesting that location of STE6 expression is more important than quantity of expression in mediating this function, whereas FK520 is not being eliminated from the yeast since they are hypersensitive to its effects, arguing that for this substrate transport by Pgh1 results in the concentration of drug in the yeast. Pgh1 may have a functional role at the plasma membrane that allows for drug extrusion of mefloquine. Topologically, a transport molecule that directs transport into the food vacuole of one substrate could mediate the transport of another substrate out of the cell through the process of membrane recycling by the parasite.

How is this differential expression or localization achieved? Genetic variability in the malaria genome may account for polymorphic expression of a particular gene product; such genetic variability has been specifically identified for *pfmdr1* at both the DNA and RNA levels. Differential expression of a gene may involve many levels, including mutations in the gene, changes in targeting sequences, or regulation of stage of expression. Mutations have been identified in *pfmdr1* within the coding region, and a polymorphism has been identified in the 3′ flanking region, associated with chloroquine resistance. Genetic polymorphisms in malaria have recently been associated with antigenic variation; yet, neither the mechanisms of how the expression of these variants nor of how the targeting of these variants is regulated have been defined. Also at the DNA level is the presence of multiple copies of the *pfmdr1* gene in mefloquine-resistant parasites. Possibly these copies are genetically distinct in a manner that can account for differential localization in the parasite. Therefore, genetic polymorphism plays a possible yet undetermined role in the differential expression or localization of Pgh1, which may help determine the drug sensitivity or resistance phenotype in a given parasite.

In terms of stage-specific expression, we have identified two distinct mRNA transcripts for the *pfmdr1* gene[27] that are developmentally regulated. These two transcripts are significantly larger than the predicted coding

[26] R. Kolling and C. P. Hollenberg, *EMBO J.* **13,** 3261 (1994).

[27] S. K. Volkman, C. M. Wilson, and D. F. Wirth, *Mol. Biochem. Parasit.* **57,** 203 (1993).

region for *pfmdr1,* suggesting that there is information in the flanking regions that is important for the regulation of expression of Pgh1. These two transcripts were found to be collinear with the DNA across the coding region; however, just 3′ to the predicted stop codon, across the region of the previously identified DNA polymorphism,[9] the RNA transcripts diverge from the DNA. Perhaps the genetic variability evident at the 3′ end of the *pfmdr1* gene is important for differential expression or localization in the parasite. The functional nature of these transcripts and the possibility that there are distinct products or localization signals encoded by these transcripts is yet to be defined, but it is intriguing why there are two distinct transcripts for this gene and suggestive of an as yet undefined functional relationship to the expression of *pfmdr1.*

Another possible mechanism of how Pgh1 functions comes from our understanding of how members of the ABC gene family regulate transport in other cell systems. A great deal has been learned about the bifunctional nature of the proteins encoded by members of this gene family. Although they provide a transport functon, many also provide alteration of ion movements across a membrane, either as transporters of ions themselves, such as the cystic fibrosis transmembrane regulator, or as regulators of ion channels, such as P-glycoprotein. Since ion movements, especially proton gradients, are thought to be important for antimalarial accumulation in *P. falciparum,* the ability of Pgh1 to regulate ion channels may explain the effects of differential Pgh1 expression on chloroquine and mefloquine resistance. Possibly expression of Pgh1 at the food vacuole membrane and other membranes external to the parasite, say, the erythrocyte plasma membrane, results in the regulation of endogenous ion channels that are distinct to these membrane sites. If, for example, Pgh1 at the food vacuole regulates the proton pump, pH gradients that drive chloroquine accumulation may be regulated in one manner, whereas Pgh1 at the erythrocyte plasma membrane may regulate the chloride ion channel, which may be important for mefloquine uptake at this membrane site. Thus, the regulation of different endogenous channels may account for how Pgh1 expression can confer susceptibility to chloroquine while conferring resistance to mefloquine.

[13] Amplification of ABC Transporter Gene *pgpA* and of Other Heavy Metal Resistance Genes in *Leishmania tarentolae* and Their Study by Gene Transfection and Gene Disruption

By MARC OUELLETTE, ANASS HAIMEUR, KATHERINE GRONDIN, DANIELLE LÉGARÉ, and BARBARA PAPADOPOULOU

Introduction

Leishmania are protozoan parasites distributed worldwide that affect more than 12 million people. The clinical manifestations vary, depending on the species, from self-healing cutaneous lesions to disseminated visceral infections. The only effective way to control leishmaniasis is chemotherapy and the first-line drug against all forms of *Leishmania* infections rely on antiquated pentavalent antimony [Sb(V)]. Although still effective, longer courses of therapy with larger drug regimens are required to achieve a cure, suggesting that the parasite is becoming less susceptible to the drug. Indeed, clinical failures using Sb(V) drugs are now encountered and in several cases this is due to resistant parasites isolated from patients who did not respond to therapy.[1]

Mode(s) of action as well as the resistance mechanism(s) of antimonials in *Leishmania* are poorly understood but are thought to be multifactorial.[2] A detailed analysis of resistant mutants should be instructive and, using a step-by-step protocol, *Leishmania* cells of various species were selected for resistance to Sb(+5) (Pentostam and Glucantime). Sb(+5) is probably converted to Sb(+3) *in vivo,* which is likely to be the active form of the metal. Cell lines were therefore also selected for resistance to Sb(+3) (antimony potassium tartrate). The chemical properties of antimonials and arsenicals and their biological effects are similar. Cells were therefore also selected for As(+3) (sodium arsenite) resistance, and since radioactive ^{73}As As is available, arsenite-resistant mutants are often used as a paradigm to study oxyanion resistance mechanisms in *Leishmania.* The relatedness between Sb(+5), Sb(+3), and As(+3) is illustrated by cross-resistance

[1] M. Ouellette and B. Papadopoulou, *Parasitol. Today* **9,** 150 (1993).
[2] P. Borst and M. Ouellette, *Annu. Rev. Microbiol.* **49,** 427 (1995).

METHODS IN ENZYMOLOGY, VOL. 292
0076-6879/98 $25.00

between the different metals and by the active extrusion of arsenite in antimony resistant cell lines.[3]

Several ABC transporters have been found in protozoan parasites, and several of these transporters are linked with drug resistance.[2] At least one P-glycoprotein homolog is present in *Leishmania*[4] and its overexpression leads to resistance to vinblastine and other drugs that are part of the mammalian multidrug resistance phenotype.[5] The antibody C-219, specific for mammalian P-glycoproteins,[6] was found to cross-react with overproduced *Leishmania* protein in two different drug mutants[7,8] one of which was selected for Sb(+5) resistance. These cross-reacting bands have not been identified, however. Other ABC transporters have also been isolated in *Leishmania.* The first one is PgpA[9] an ABC transporter more closely related to MRP[10] than to P-glycoproteins. The *pgpA* gene is part of a genomic locus (the H locus) frequently amplified after selection with a number of different drugs including As(+3).[11,12] In addition to *pgpA,* other loci are also amplified in oxyanion-resistant cells. For example, a locus first observed as part of a linear amplicon[13] is derived from an 800-kb chromosome and is present in all *Leishmania tarentolae* arsenite-resistant mutants.

Leishmania, therefore, often respond to drug selection by amplifying specific portion of its genome, and the amplicon formed can be found as extrachromosomal circles or linear elements.[1,14] In this article we describe how amplicons can be detected and isolated and how by gene transfection and gene disruption we can study the function of heavy metal resistance genes in *Leishmania.*

[3] S. Dey, B. Papadopoulou, G. Roy, K. Grondin, D. Dou, B. P. Rosen, and M. Ouellette, *Mol. Biochem. Parasitol.* **67,** 49 (1994).

[4] D. M. Henderson, C. D. Sifri, M. Rodgers, D. F. Wirth, N. Hendrickson, and B. Ullman, *Mol. Cell. Biol.* **12,** 2855 (1992).

[5] L. M. C. Chow, A. K. C. Wong, B. Ullman, and D. F. Wirth, *Mol. Biochem. Parasitol.* **60,** 195 (1993).

[6] E. Georges, G. Bradley, J. Gariepy, and V. Ling, *Proc. Natl. Acad. Sci. U.S.A.* **87,** 152 (1990).

[7] F. Gamarro, M. V. Amador, M. J. Chiquero, D. Légaré, M. Ouellette, and S. Castanys, *Biochem. Pharmacol.* **47,** 1939 (1994).

[8] M. Grogl, R. K. Martin, A. M. J. Odula, W. K. Milhous, and D. E. Kyle, *Am. J. Trop. Med. Hyg.* **45,** 98 (1991).

[9] M. Ouellette, F. Fase-Fowler, and P. Borst, *EMBO J.* **9,** 1027 (1990).

[10] S. P. C. Cole, G. Bhardwaj, J. H. Gerlach, J. E. Mackie, C. E. Grant, K. C. Almquist, A. J. Stewart, E. U. Kurz, A. M. V. Duncan, and R. G. Deely, *Science* **258,** 1650 (1992).

[11] S. Detke, K. Katakura, and K.-P. Chang, *Exp. Cell Res.* **180,** 161 (1989).

[12] M. Ouellette, E. Hettema, D. Wust, F. Fase-Fowler, and P. Borst, *EMBO J.* **10,** 1009 (1991).

[13] K. Grondin, B. Papadopoulou, and M. Ouellette, *Nucl. Acids Res.* **21,** 1895 (1993).

[14] S. M. Beverley, *Annu. Rev. Microbiol.* **45,** 417 (1991).

Detection of Gene Amplification

The complexity of the *Leishmania* genome is small enough (36 Mbp) so that it is possible to detect gene amplification (as small as a 5- to 10-fold increase in copy number) by simple examination of digested DNAs of wild-type and resistant mutants run in parallel in an agarose gel.[15,16] Digestion with enzymes recognizing AT-rich sequences (hence cutting rarely *Leishmania* DNA with its 65% GC content) such as *Hin*dIII and *Eco*RI usually permit us to visualize the amplified band more easily. An example is shown in Fig. 1A. Total DNA of *Leishmania* is prepared as described[17] with some modifications. Five milliliters of cell culture are harvested at 3000 rpm for 5 min. Cells are washed once with HEPES buffer (21 m*M* HEPES, pH 7.05, 137 m*M* NaCl, 5 m*M* KCl, 0.7 m*M* Na_2HPO_4, 6 m*M* glucose) and then resuspended in 5 ml of resuspension buffer (100 m*M* NaCl, 100 m*M* EDTA, 10 m*M* Tris, pH 8.0). Cells are lysed by the addition of sodium dodecyl sulfate (SDS final concentration 1.0%) and digested with proteinase K (50 μg/ml) for at least 30 min at 37°. Total DNA is extracted twice with an equal volume of phenol, with gentle shaking for 15 min. DNA is precipitated with 2 volumes of ethanol (99%) and collected using a tissue culture 10-μl inoculation loop. The DNA is then resuspended in TE (10 m*M* Tris, 1 m*M* EDTA, pH 7.4) treated with RNaseA (20 μg/ml, for 30 min at 37°) and kept at 4°. We also use an alternative method to prepare genomic DNA that has been described.[18] The latter technique is more rapid but in our hands gives lower yields than with the first method described.

Five micrograms of total DNAs are digested with a restriction enzyme according to the manufacturer recommendation and an equivalent amount of DNAs from the wild-type and a mutant are run on a 0.7% agarose gel in 1X TBE buffer (90 m*M* Tris–borate, pH 8.0, 2 m*M* EDTA) as 35 V/cm for 16 hr. Gels are stained with ethidium bromide (0.5 μg/ml) and examined under the UV lamp at 300 nm. Amplified bands can easily be detected as more prominent stained bands in mutants (see arrows in Fig. 1A, lane 2) when compared to wild-type DNA (Fig. 1A, lane 1). The amplified bands can be isolated using one of several means (electroelution, low melting point agarose, commercial gel purification kits) and either cloned in plasmid vectors or used directly as a probe for further investigation of the role of

[15] J. A. Coderre, S. M. Beverley, R. T. Schimke, and D. V. Santi, *Proc. Natl. Acad. Sci. U.S.A.* **80,** 2132 (1983).

[16] T. C. White, F. Fase-Fowler, H. van Luenen, J. Calafat, and P. Borst, *J. Biol. Chem.* **263,** 16,977 (1988).

[17] A. Bernards, L. H. T. van der Ploeg, A. C. C. Frasch, P. Borst, J. C. Boothroyd, S. Coleman, and G. A. M. Cross, *Cell* **27,** 497 (1981).

[18] E. Medina-Acosta and G. A. M. Cross, *Mol. Biochem. Parasitol.* **59,** 327 (1993).

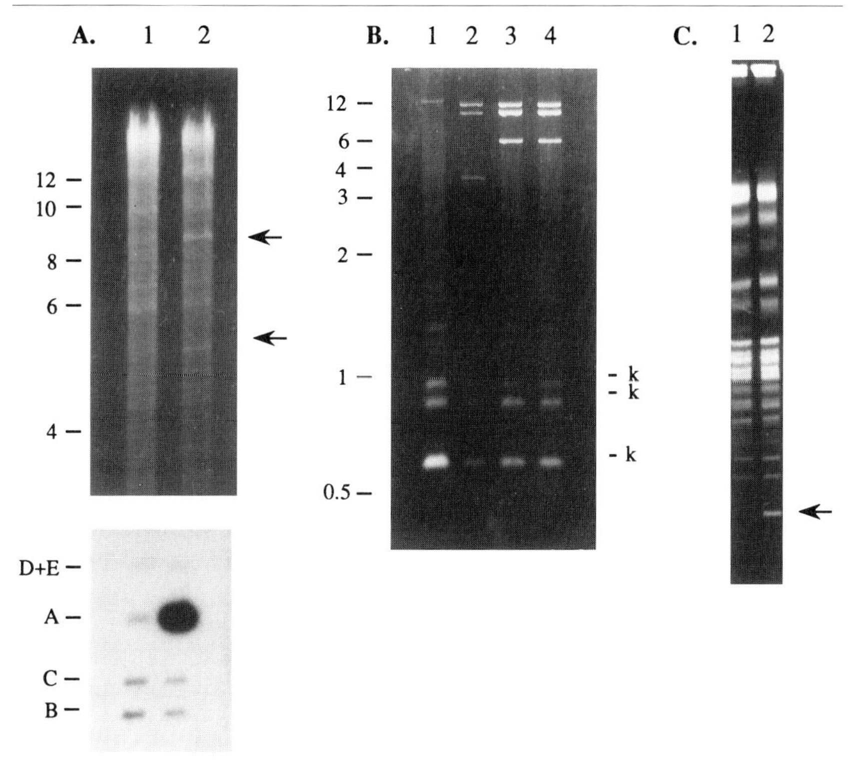

FIG. 1. Isolation and characterization of amplified DNA in drug-resistant *L. tarentolae.*[12,16] (A) Detection of amplified DNA by comparison of total DNAs of wild-type and mutant digested with *Hin*dIII, electrophoresed in agarose gels, and stained with ethidium bromide. Lane 1, TarII WT; lane 2, TarVIa MTX 1000. The arrows point to some of the amplified bands. The gel was blotted and hybridized with a probe spanning the nucleotide-binding site of *pgpA* to reveal a large gene family *pgpA, B, C, D,* and *E.* (B) Isolation of circular DNA by rapid circle minipreparation using alkaline lysis (lanes 1, 2, 3) or Hirt extraction (lane 4), which was digested with *Hin*dIII and electrophoresed in an 0.7% agarose gel. Lane 1, TarII WT; lane 2, TarVIa MTX 1000; lanes 3 and 4, TarIIAs 20.3 (C) Chromosome separation using TAFE electrophoresis of TarII WT (lane 1) and TarIIAs 20.2 (lane 2). Cells were embedded in agarose, lysed *in situ,* and electrophoresed in a TAFE apparatus in 0.25X TBE at 15° at 350 mA for 36 hr with the following conditions (12 hr) with 30-sec pulses, 12 hr with 1.5-min pulses, and 12 hr with 2.5-min pulses). The arrow indicates the 50-kb amplified linear amplicon.

the amplicons in resistance. Amplification of several *Leishmania* loci were observed in this manner. The H locus was found amplified in several mutants (Fig. 1A, lane 2) and since this region was found amplified in several species selected for a number of unrelated drugs[1,14] we tested whether a P-glycoprotein homolog was present on the amplified DNA. Hybridization using an heterologous human *mdr1* probe on Southern blots of genomic DNA, despite gene amplification, gave inconclusive results, however.

Isolation of Circular Amplicons

It is possible to isolate circular amplicons of *Leishmania* by using procedures enriching for circular episomes. Two methods have been applied successfully to isolate large circular amplicons (>100 kb) of *Leishmania* free from the bulk of chromosomal DNA. The first one is based on the Hirt extraction.[19] Five milliliters of cells are spun down at 3000 rpm for 5 min and washed in HEPES buffer. Cells are resuspended in 3 ml of TE (100 m*M*, 10 m*M*, pH 8) and lysed with 0.75 ml of 10% SDS followed by the addition of 1.25 ml of 5 *M* NaCl. Chromosomal DNA is then precipitated in high salt overnight at 4°, and spun at 17,000 rpm for 1 hr. The supernatant is carefully collected and phenol extracted twice. Circular DNA is precipitated in ethanol overnight at −20° and spun at 7500 rpm for 25 min and resuspended gently in 10 m*M* Tris, pH 8, 1 m*M* EDTA with 10 μg of RNase A/ml. The second method that we use on a more routine basis to prepare episomes from *Leishmania* is a standard alkaline lysis preparation used to prepare plasmids from bacterial cells.[20] Because the circles are large, special care is taken to collect the supernatant with cut tips with wider opening, to perform the phenol extractions slowly for 10 min, and to resuspend the pellet without pipetting. DNAs obtained using any of these two procedures are relatively free of contaminant chromosomal DNA (Fig. 1B). The smaller molecular weight bands also observed in wild-type cells (some of which are labeled with a *k* in Fig. 1B) correspond to the minicircles of kinetoplast DNA (the mitochondrial DNA made of complex network of maxi and minicircles[21]). Although no circles were observed in wild-type cells (Fig. 1B, lane 1) they were isolated in both a methotrexate and arsenite-resistant mutants (Fig. 1B, lanes 2 and 3, respectively).

When the H locus, amplified as part of an extrachromosomal circle, was isolated by Hirt extraction, digested with restriction enzymes (see Fig. 1B, lane 2), and blotted on filters, a band hybridized weakly but consistently with the human *mdr1* probe (not shown). Characterization of the smallest region still hybridizing weakly to the *mdr1* probe permitted us to eventually isolate and characterize *pgpA*. The level of homology between *pgpA* and *mdr1* is so low that we were only able to observe the hybridizing signal because we were working with several copies of the target sequence by having isolated the plasmid. Indeed, at the time of its characterization PgpA was found to be the most divergent P-glycoprotein.[9] With other sequences

[19] B. Hirt, *J. Mol. Biol.* **26,** 365 (1967).

[20] H. C. Birnboim and J. Doly, *Nucl. Acids Res.* **7,** 1513 (1979).

[21] T. A. Shapiro and P. T. Englund, *Annu. Rev. Microbiol.* **49,** 117 (1995).

that eventually became available, PgpA was found to have branched out from the main P-glycoprotein family and, along with MRP and other ABC transporters, they now constitute their own subgroup. When a region spanning the nucleotide-binding site of *pgpA* was used as a probe and hybridized to a Southern blot of *Leishmania* genomic DNA, several bands hybridized and *pgpA* was found to be part of a gene family (Fig. 1A). In some mutants only *pgpA* is amplified, but in others, *pgpA, pgpB,* and *pgpC* are coamplified showing that some of the genes are linked.[12] Amplification of *pgpD* and *pgpE,* linked on a large chromosome,[22] has never been observed although an increase in *pgpE* RNA was observed in one drug-resistant mutant.[7] In several arsenite-resistant mutants, including the one shown in Fig. 1B, lane 3, *pgpA* is amplified as part of the small extrachromosomal circles.

Isolation of Linear Amplicons

In ethidium bromide-stained gels of digested DNA similar to the one shown in Fig. 1A, some mutants clearly show DNA amplification but circles cannot be isolated. This may be due either to the large size of circular amplicons or to the fact that the amplified DNA is not circular. Linear amplicons have been described in *Leishmania* cells[14] and they can be visualized easily in chromosome sized gels, such as TAFE and CHEF, especially if they are smaller than the smallest *Leishmania* chromosome. *Leishmania* has 36 chromosomes ranging in size from 300 to 3150 kb (Fig. 1C, lane 1). The migration of the linear amplicons is pulse dependent and they hybridize to telomeric probes.[13,14,23] In addition to the *pgpA*-containing circular amplicon, a 50-kb linear amplicon (indicated by an arrow in Fig. 1C, lane 2) is frequently observed in *L.* tarentolae cells selected for resistance to arsenite.[13] Linear amplicons smaller than 200 kb are easily gel purified with the Sephaglas kit (Pharmacia, Piscataway, NJ) according to the manufacturer's recommendations. The linear DNA isolated in such a manner remains intact; we have succeeded in retransfecting large linear amplicons into wild-type cells.[23] Linear amplicons of any size can also be isolated using agarase (Boehringer Mannheim, Germany). The band of interest is cut and 0.04 volume of 25X agarose buffer is added (750 m*M* Bis-Tris, 250 m*M* EDTA, pH 6.0). The solution is incubated for 15 min at 65° until the agarose is completely molten. The molten agarose is cooled to 45° and one unit of agarase per 100 mg of agarose is added and incubated for 1 hr. Oligosaccharides are precipitated with 0.1 volume of 3 *M* sodium acetate, pH 5.5, for 15 min on ice and spun for 15 min at 4°. The supernatant is collected and DNA is precipitated with 3 volumes of ice-cold ethanol (99%). The isolated DNA from the linear amplicon can then be transfected, or used as a

[22] D. Légaré, E. Hettema, and M. Ouellette, *Mol. Biochem. Parasitol.* **68,** 81 (1994).
[23] B. Papadopoulou, G. Roy, and Ouellette M., *Nucl. Acids Res.* **21,** 4305 (1993).

probe, or digested for cloning experiments to further study its role in resistance.

Role of Amplified Genes in Resistance by Gene Transfection and Gene Disruption

The *pgpA* gene is frequently amplified in arsenite-resistant mutants, and when those mutants are grown in the absence of the drug they loose the *pgpA*-containing amplicon and part of their resistance. The role of PgpA was studied more directly using gene transfection, a technique available since the early 1990s for Kinetoplastidae. Constructs containing genomic inserts encoding *pgpA* (there are no introns in protein coding genes of Kinetoplastidae) and a dominant selectable marker [neomycin phosphotransferase (*neo*) and hygromycin phosphotransferase (*hyg*) conferring resistance to G418 and to hygromycin B, respectively, are the two most commonly used ones, but others are also available[2]] are introduced into wild-type cells by electroporation. Each transfection requires 5×10^7 cells of late log-phase promastigotes. Cells are harvested by centrifugation at 3000 rpm for 5 min, and washed once in 10 ml of HEPES buffer and resuspended at 1×10^8 cells/ml. Cells can be kept on ice for up to 2 hr. Five micrograms of DNA is mixed with 0.5 ml of the cell suspension in a sterile tube and then transferred to 0.2-cm electroporation cuvettes. If high transfection yields are required, increasing the concentration of DNA may help. However, the amount of DNA to saturate transfection varies with *Leishmania* species: 10 μg is enough for *L. tarentolae* but the efficiency of transfection is linear with up to 80 μg for *L. major*.[24] DNA is introduced into the cells using a Bio-Rad (Richmond, CA) gene pulser apparatus (voltage at 0.45 kV and the capacitance at 500 μF). Cells are immediately transferred to 5 ml of SDM-79 medium[25] and are incubated for 24 hr. Following overnight incubation, 5 ml of fresh SDM-79 medium is added as well as the appropriate drug, and cells are incubated for an additional 24 hr. The selective pressure applied is dependent on the species but should be around 100 μg/ml for hygromycin and 40 μg/ml for G418. Instead of G418, we can also use paromomycin[26] (400 μg/ml), which is much cheaper and in our hands often turns out to be a better selective drug. The next day, 300 μl of the culture is transferred to 5 ml of fresh SDM-79 medium with the appropriate drug selection. Transfectants are usually growing after 4–6 days. Other transfection protocols for *Leishmania* are available.[24,27]

[24] G. M. Kapler, C. Coburn, and S. M. Beverley, *Mol. Cell. Biol.* **10,** 1084 (1990).

[25] R. Brun and M. Schônenberger, *Acta Trop.* **36,** 289 (1979).

[26] F. J. Gueiros-Filho and S. M. Beverley, *Exp. Parasitol.* **78,** 425 (1994).

[27] A. Laban, J. Finbarr Tobin, M. A. Curotta de Lafaille, and D. F. Wirth, *Nature* **343,** 572 (1990).

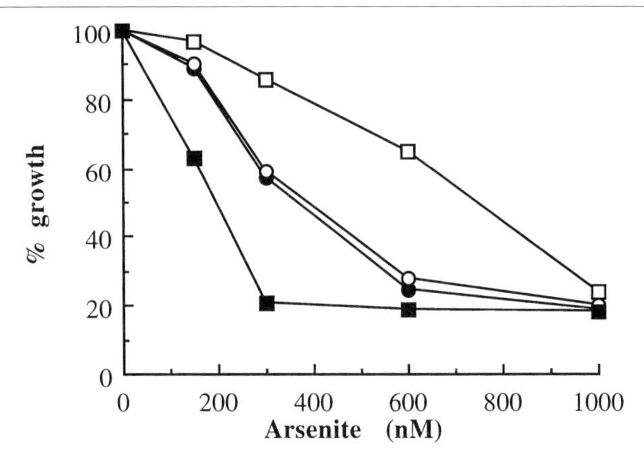

FIG. 2. Arsenite sensitivity of *L. tarentolae* TarII wild-type and transfectants. ●, TarIIWT; □, TarII transfected with *pgpA;* ■, TarII *pgpA* null mutant; ○, TarII transfected with the 50-kb linear amplicon. Cells were grown in SDM-79 in the presence of various concentrations of arsenite, and optical density was measured at 600 nm after 72 hr of growth using an automated microplate reader (Organon Teknika, Microwell System).

The copy number of *pgpA* is increased in transfectants, which is correlated with increased *pgpA* RNA[28] and increased gene product.[29] Growth curves indicated that PgpA is involved in resistance to arsenite and antimony but resistance observed in *L. tarentolae* were invariably lower than the ones observed in *L. major.*[28,30] Transfection of *pgpA* in *L. tarentolae* only gave a twofold increase in resistance (Fig. 2) and resistance was not directly linked to its copy number.[28]

We proposed that PgpA does not recognize the free metal but the metal linked to glutathione or another thiol molecule.[2] Therefore, to achieve resistance, cells need, in addition to a pump, a high level of the thiol molecules to be conjugated to the metal and possibly increased enzymatic activity conjugating the thiol to the metal. Transfection of *pgpA* alone may therefore not lead to high resistance, and different species of *Leishmania* may differ in their thiol metabolism, explaining the differences in resistance levels observed among different species.

PgpA as Thiol–Metal Plasma Membrane Pump

By transport experiments using ^{73}As we have shown that every mutant selected for oxyanion resistance has an increased efflux of arsenite.[3] The

[28] B. Papadopoulou, G. Roy, S. Dey, B. P. Rosen, and M. Ouellette, *J. Biol. Chem.* **269,** 11,980 (1994).

[29] D. Légaré, B. Papadopoulou, G. Roy, R. Mukhopadhyay, A. Haimeur, S. Dey, K. Grondin, B. P. Rosen, and M. Ouellette, *Exp. Parasitol.* **87,** 275 (1997).

[30] H. L. Callahan and S. M. Beverley, *J. Biol. Chem.* **266,** 18,427 (1991).

efflux system was characterized further by preparing everted plasma membrane vesicles of *L. tarentolae.* The metal was not a substrate for the pump but when it was first conjugated to glutathione or to other thiols, there was an ATP-dependent uptake of the conjugate.[31] The activity of the pump, however, was not increased in vesicles prepared from mutants. Thus the ATP requiring pump is not rate limiting for resistance and a step prior to the pump such as synthesis of thiols or conjugation to the metals may be rate limiting.

Amplification of *pgpA* did not correlate with increased extrusion,[3] but because the activity of the pump is not increased in vesicles prepared from resistant mutants we tested whether PgpA corresponds to this thiol metal pump by gene disruption experiments. In *Leishmania,* DNA integrates exclusively by homologous recombination. However, as the parasite is diploid and no sexual cycle has yet been described, two successive round of targeting are required with two different markers. The *pgpA* gene was disrupted in a wild-type strain and in an arsenite-resistant mutant.[32] Expression cassettes encoding resistance to G418 or hygromycin are cloned into the gene to be disrupted and at least 2 μg of the linearized construct with usually at least 1 kb of homologous sequences on each side of the marker are electroporated inside cells. Transfectants are selected as described earlier for episomal vectors except that between each targeting round, the transfectants are cloned on SDM-79 plates. Disruption of onc allele is verified by Southern blot and a second round of targeting is done on a clone with another marker to disrupt the remaining intact allele. The wild-type cells with no *pgpA* present became more sensitive to arsenite (Fig. 2). An increase in sensitivity was also observed in arsenite-resistant mutants with no PgpA but cells were still much more resistant than wild-type (Fig. 3) indicating that PgpA is dispensable for high-level oxyanion resistance. Plasma membrane vesicles prepared from a *pgpA* null mutant had the same rate of uptake of conjugated-metal inside vesicles as wild-type cells,[31] further indicating that PgpA is not the main plasma membrane GS-X pump.

Possible Resistance Mechanism for PgpA

Transport studies indicated, alternatively, that PgpA does not change the steady-state accumulation of metals,[28] that it reduces active drug up-

[31] S. Dey, M. Ouellette, J. Lightbody, B. Papadopoulou, and B. P. Rosen. *Proc. Natl. Acad. Sci. U.S.A.* **93,** 2192 (1996).

[32] B. Papadopoulou, G. Roy, S. Dey, B. P. Rosen, M. Olivier, and M. Ouellette, *Biochem. Biophys. Res. Commun.* **224,** 772 (1996).

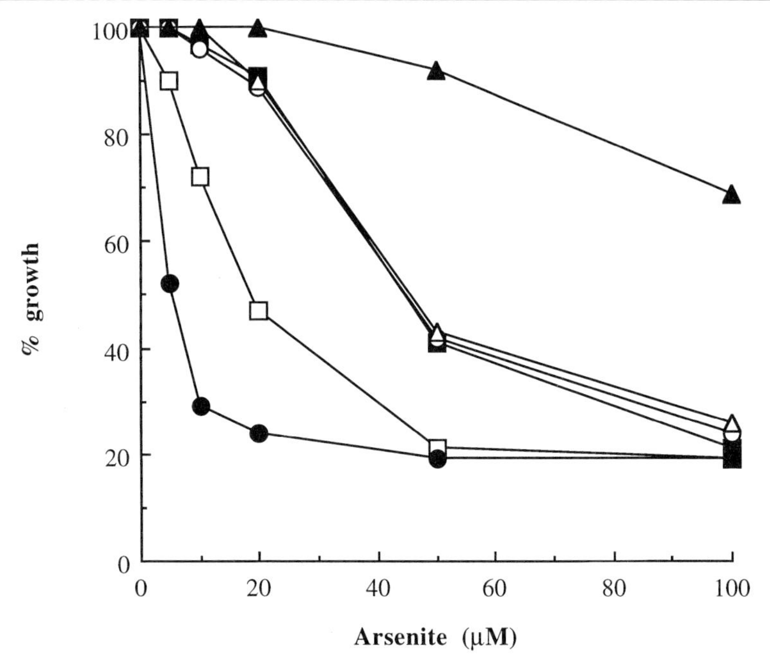

FIG. 3. Arsenite sensitivity of the *L. tarentolae* arsenite-resistant mutant TarIIAs20.3 and of its revertant and transfectants. ■, TarIIAs20.3; □, TarIIAs20.3 rev (mutant grown without arsenite); △, TarIIAs20.3 rev transfected with *pgpA;* ●, TarIIAs20.3 rev *pgpA* null mutant; ○, TarIIAs20.3 new transfected with the linear amplicon; ▲, TarIIAs20.3 rev transfected with *pgpA* and the linear amplicon. Cells were grown in SDM-79 at various concentrations of arsenite, and optical density was measured at 600 nm after 72 hr of growth.

take[33] or is associated with active efflux.[34] The mechanism by which PgpA confers resistance is therefore unclear but our analysis indicated that PgpA is probably not a plasma membrane efflux system. Our proposal that PgpA is a GS-X pump[2] provides a plausible explanation for our inability to reproduce the high level of resistance observed in stepwise selected mutants.[28] The high-level resistance obtained by selection of *Leishmania* with arsenite could be due to simultaneous selection for increased GSH biosynthesis, increased GST activity and increased export of GSH–arsenite conjugates. If transfection only provides high levels of the pump (PgpA), only low-level resistance may result. We propose further that PgpA confers resistance by transporting the drug conjugated to a thiol into a subcellular compartment, preventing it from reaching its cellular target. Several ABC

[33] H. L. Callahan, W. L. Roberts, P. M. Rainey, and S. M. Beverley, *Mol. Biochem. Parasitol.* **68,** 145 (1994).

[34] A. K. Singh, H. Y. Liu, and S. T. Lee, *Mol. Biochem. Parasitol.* **66,** 161 (1994).

transporters are located in the membrane of the vacuole, some of which are recognizing metals linked to glutathione or related molecules.[35] Subcellular localization of PgpA is currently under way with anti-PgpA antibodies and immunogold labeling electron microscopy.

A candidate gene cooperating with PgpA to confer resistance might reside in a region found in a 50-kb linear amplicon unrelated to *pgpA*[13] (Fig. 1C, lane 2). Mutants grown in the absence of the drug concomitantly lost this amplicon and part of their resistance, resulting in partial revertants (Fig. 3). Transfection of the linear amplicon into wild-type cells did not lead to an increase in oxyanion resistance (Fig. 2), but transfection into the partial revertant restored resistance to the level of the parent mutant (Fig. 3). To transfect the linear amplicon, a dominant marker (e.g., *neo*) is first integrated into a noncoding region of the linear amplicon by transfection and homologous recombination. The *neo*-containing linear amplicon is isolated as described earlier and transfected in cells. The preceding result suggests that the gene product encoded by the linear amplicon requires another mutation absent in the wild-type cells but present in the partial revertant to confer resistance. When *pgpA* was cotransfected with the linear amplicon in the partial revertant, resistance levels were higher than expected if resistance was due to the simple addition of the contributions of PgpA and the linear amplicon (Fig. 3). The gene present on the 50-kb linear amplicon was recently characterized and the basis for its synergism with PgpA was unraveled.[36]

Conclusions

Resistance to oxyanions in *Leishmania* is a complex phenomenon. The ease of detection of gene amplification has permitted the isolation of genes involved in resistance including one encoding for the ABC transporter PgpA and for a gene collaborating with it and possibly involved in thiol metabolism. Several other amplified genes in *L. tarentolae* resistant to oxyanions have been isolated using the methodology described in this report.[36] Studies of drug resistance mechanisms in parasites have long been correlative, but now with our ability to introduce genes by electroporation and to disrupt genes by homologous recombination we can test directly the role of genes in resistance. With our increased understanding of resistance mechanisms in laboratory strains we should soon be in a position to test the resistance mechanisms prevailing in field strains isolated from patients who are unresponsive to pentavalent antimony treat-

[35] D. F. Ortiz, T. Ruscitti, K. F. McCue, and D. W. Ow, *J. Biol. Chem.* **270,** 4721 (1995).

[36] K. Grondin, A. Haimeur, R. Mukhopadhyay, B. P. Rosen, and M. Ouellette, *EMBO J.* **16,** 3057 (1997).

ment. This should be beneficial for the millions of people suffering from *Leishmania* infections.

Acknowledgments

Work in the laboratory on oxyanion resistance is supported by the Medical Research Council. MO is supported by a FRSQ scholarship and a Burroughs Wellcome Fund New Investigator Award in molecular parasitology, BP by an MRC scholarship, and KG and DL by FCAR studentships.

[14] Functional Assays for Analysis of Yeast *ste6* Mutants

By GABY L. NIJBROEK and SUSAN MICHAELIS

Introduction

Ste6p is an ABC transporter of *Saccharomyces cerevisiae* that mediates the export of the **a**-factor mating pheromone. Mating in *S. cerevisiae* results from the fusion of two haploid cells of opposite mating type, designated *MAT***a** and *MAT*α (here referred to as **a** and α). The mating pheromones secreted by these haploid cells, **a**-factor and α-factor, respectively, interact with receptors on cells of the opposite mating type to initiate a signaling cascade that ultimately leads to cell fusion. The ability of cells to mate thus depends on their proficiency to produce and respond to mating pheromones. Correspondingly, cells that are unable to mate are referred to as sterile (*ste*) mutants. The *STE6* gene was identified on the basis of the **a** cell-specific sterile phenotype of a *ste6* mutant[1,2] and this mutant was shown to synthesize, but not secrete **a**-factor.[3,4] Two lines of evidence suggest that the sole function of Ste6p is in **a**-factor export: (1) The only phenotypes of a *MAT***a** Δ*ste6* strain are lack of **a**-factor export and sterility[5,6] and (2) *STE6,* along with the **a**-factor genes (*MFA1* and *MFA2*), are coregulated at the transcriptional level; these three genes are coordinately highly expressed in **a** cells and are tightly shut off in α cells.[1,7]

[1] K. L. Wilson and I. Herskowitz, *Mol. Cell. Biol.* **4,** 2420 (1984).
[2] J. D. Rine, PhD thesis, University of Oregon, Eugene (1979).
[3] K. Kuchler, R. E. Sterne, and J. Thorner, *EMBO J.* **8,** 3973 (1989).
[4] J. P. McGrath and A. Varshavsky, *Nature* **340,** 400 (1989).
[5] S. Michaelis, *Semin. Cell Biol.* **4,** 17 (1993).
[6] C. Berkower and S. Michaelis, *EMBO J.* **10,** 3777 (1991).
[7] S. Michaelis and I. Herskowitz, *Mol. Cell. Biol.* **8,** 1309 (1988).

0076-6879/98 $25.00

The **a**-factor pheromone is synthesized as a precursor (36 or 38 amino acids long) that undergoes extensive C- and N-terminal modifications. The earliest processing events occur at the C-terminal CAAX box of **a**-factor, and are followed by two N-terminal proteolytic cleavage steps.[8] The fully processed form is referred to as mature **a**-factor. It is 12 amino acids long and contains a C-terminal cysteine that is derivatized by a thioether-linked farnesyl and a carboxylmethyl ester, making **a**-factor extremely hydrophobic. It is this fully mature form of **a**-factor that is exported from the cell by the ABC transporter Ste6p. The export of **a**-factor via Ste6p represents a "nonclassical" export mechanism. In terms of both processing and secretion, **a**-factor contrasts with α-factor, which is synthesized as a precursor with a typical N-terminal signal sequence and is processed and secreted using the "classical" secretory pathway.

Ste6p is one of only a handful of eukaryotic ABC transporters whose true physiological substrate is known. Therefore, Ste6p serves as an excellent model to dissect the structure and function of an ABC protein at the molecular and the biochemical level. Little is known about the life cycle and intracellular trafficking of most ABC proteins; presumably these processes are complex and involve a variety of cellular components. By studying mutant and wild-type *STE6* proteins and by making use of the powerful genetic and molecular maneuvers possible in yeast, we anticipate that the identity of these components will be revealed. The yeast system also allows the functional analysis of mammalian genes that are related to *STE6*. When Ste6p is compared to members of the ABC family, the mammalian multidrug resistance (MDR) proteins (which also transport hydrophobic compounds) exhibit the strongest homology. Interestingly, mouse *MDR3* and plasmodium MDR (pf MDR) have been shown to functionally complement *ste6*Δ for mating.[9]

In this chapter, we discuss assays for measuring *STE6* function and for analyzing its intracellular trafficking. In addition, we mention insights that have been gained from *STE6* mutagenesis studies and comment on the use of a Δ*ste6* strain to assay heterologously expressed mammalian and microbial ABC proteins in yeast.

Assays to Measure STE6 Function

Ste6p function can be indirectly measured by determining the amount of **a**-factor exported from **a** cells. Several biological and immunochemical assays are useful to ascertain *STE6* function in terms of **a**-factor export.[10]

[8] P. Chen, S. S. Sapperstein, J. C. Choi, and S. Michaelis, *J. Cell Biol.* **136,** 251 (1997).

[9] M. Raymond, P. Gros, M. Whiteway, and D. Y. Thomas, *Science* **256,** 232 (1992).

[10] C. Berkower and S. Michaelis, *in* "Membrane Protein Transport" (S. Rothman, ed.), Vol. 3, pp. 231–277, JAI Press, Greenwich, CT, 1996.

These assays vary in sensitivity and technical difficulty and each has distinct advantages and disadvantages. Presently, direct biochemical assays to examine the utilization of ATP by Ste6p and to assess the interactions of Ste6p with **a**-factor are still lacking; however, the development of such a biochemical assay is a major focus of studies on Ste6p. In the following paragraphs, we discuss the *in vivo* biological assays that are available to assess Ste6p export function.

Functional Assays; Mating and Halo Tests

Mating Assays. The mating assay is used to determine the efficiency of diploid formation resulting from the mating of auxotrophic **a** and α cells. For example, this procedure can be used to measure the mating efficiency of *ste6* mutants, or to measure the ability of heterologous ABC transporters (such as *mMDR* and *pfMDR*) to complement a Δ*ste6* defect. The mating test is the assay of choice and can be sensitive, quantitative, and easily done. The procedure allows one to calculate the mating efficiency of either slightly or severely defective *ste6* mutants with a range from 100 to 0.001% (i.e., 6 logs) mating, as compared to wild-type (WT). A Δ*ste6* mutant will not mate under any condition and thus exhibits a mating efficiency of $<0.001\%$.

In a typical mating experiment, approximately 20–40% of wild-type **a** and α cells undergo mating[11] (Berkower, unpublished observation, 1990). In wild-type cells, *STE6* appears to be present in excess, because overexpression of *STE6* does not lead to an increase in mating or **a**-factor export (Berkower, unpublished observation, 1990). It is unclear how much Ste6p activity needs to be lowered to have an effect on mating. Thus, the assay is most informative when quantitating the mating efficiency of *ste6* mutants that are noticeably impaired for function (i.e., mutants that exhibit $<100\%$ mating as compared to a wild-type *STE6* strain), since in these mutants *STE6* is clearly rate limiting. In contrast to our uncertainty about the minimum Ste6p activity needed for 100% mating, more is known about the levels of **a**-factor that are required for optimal mating. Browne *et al.*[12] have shown that only 20% of the wild-type amount of extracellular **a**-factor is sufficient for maximal mating at 30° suggesting that the levels of the pheromones and the Ste6p transporter are typically in vast excess. Note also that wild-type yeast cells are slightly temperature sensitive (Ts-) for mating, thus the mating process occurs less efficiently at 37° than at 30°. At 37°, the production and secretion of **a**-factor are not affected, suggesting that components other than **a**-factor or Ste6p are Ts- for mating.

[11] C. Guthrie and G. R. Fink, "Guide to Yeast Genetics and Molecular Biology." Academic Press, San Diego, 1991.

[12] B. L. Browne, V. McClendon, and D. M. Bedwell, *J. Bacteriol.* **178,** 1712 (1996).

Two protocols that allow for a quantitative evaluation of the mating efficiency are the filter mating assay[13] and the plate mating assay.[6] The filter mating assay is the more reproducible of the two since **a** and α cells, which are concentrated on a filter, are allowed to form mating pairs for a fixed period of time. In the quantitative plate mating assay, on the other hand, the time of mating is not fixed; this assay is highly dependent on, and may be influenced by the physiologic conditions of the plate, but yields reproducible results when carefully performed.[6,10] For the following protocols, all the media and plates are made as described in the *Yeast Genetics Course Manual.*[13] The media is defined as follows: SD (synthetic deplete) media is lacking all amino acids and SC (synthetic complete) or YEPD (yeast extract peptone dextrose), which are rich media, contain all amino acids. SC dropout media lacks specific amino acids and can be used to maintain plasmids selectively.

PROCEDURE FOR QUANTITATIVE FILTER MATING ASSAY. The filter mating assay involves challenging a specific number of auxotrophic **a** and α cells to mate over a fixed amount of time (in general 4 hr), then assaying the efficiency of diploid formation by counting prototrophs that grow on selective media and comparing this number to wild-type. Various dilutions of the mating mix are plated to facilitate obtaining an accurate count of diploids. The assay is carried out using a modified version of the previously described procedure.[13] Mating partners are grown in rich or minimal media (i.e., YEPD or SC dropout, respectively) at 30° to midexponential phase (OD_{600} ~ 0.7), and an equal number of **a** and α cells (~0.25 OD_{600} units of each) are mixed in 5 ml of sterile H_2O. We mate our **a** strains (which contain some combination of the markers *trp1, leu2, ura3,* and *his4*) to the wild-type α partner SM1068 (*lys1*$^-$),[7,14] but other complementary auxotrophic strains may also be used. The cells are gently vortexed and concentrated on a sterile 0.45-μm nitrocellulose filter (Millipore, Bedford, MA) using a filter manifold (Hoeffer Scientific Instruments, San Francisco, CA) that has been previously sterilized by rinsing with 95% ethanol. Using sterilized forceps, the filter is placed (cell side up) on a prewarmed 30° YEPD plate (which is permissive for mating) and incubated at 30° for 4–6 hr. The filter is vortexed in 5 ml of sterile H_2O to resuspend the cells, and several dilutions of the cell suspension are made in YEPD. A log phase culture of cells at OD_{600} ~ 0.6 grown in selective media (SC dropout) contains ~1×10^6 cell/ml while a log phase culture grown in rich media contains ~3×10^6 cell/ml. To calculate the number of viable cells for each strain, 0.1 ml of a

[13] C. Kaiser, S. Michaelis, and A. Mitchell, "Methods in Yeast Genetics, A Cold Spring Harbor Course Manual" 1994 ed. Cold Spring Harbor Laboratory Press, New York, 1994.

[14] K. Fujimura-Kamada, F. J. Nouvet, and S. Michaelis, *J. Cell Biol.* **136,** 271 (1997).

10^{-3} dilution (~100 cells) of the culture is plated onto YEPD so that the number of colonies can be easily counted. For the wild-type control strain, dilutions of 10^{-3}–10^{-5} are typically spread onto SD plates. The mating frequency of each strain is estimated by plating out 0.1 ml of 10^{-1}–10^{-4} dilutions of the filter-mated cells onto SD media, which selects for diploids. All plates are incubated for 2–5 days at 30°. Diploids are counted and normalized to the total number of viable cells plated. The resulting ratio is the frequency of diploid formation for each mutant, which is then expressed as a percentage relative to the wild-type ratio.

PROCEDURE FOR QUANTITATIVE PLATE MATING ASSAY. The quantitative plate mating procedure is carried out as described by Berkower and Michaelis.[6] Wild-type and *ste6* mutant **a** cells are grown to stationary phase in SC dropout medium, and serially diluted into fresh YEPD liquid media in 10-fold increments (10^{-1}–10^{-4} dilutions). The number of viable cells for each strain is determined as described in the previous paragraph. A 0.1-ml aliquot of each dilution is mixed with a vast excess of the α mating tester (~10^7 cells in 0.1 ml of fresh YEPD) and spread onto an SD plate. The presence of fresh YEPD allows the cells to form microcolonies, which occupy a larger surface area on the plate than single cells, and consequently optimizes their chances of finding mating partners.[14] Thus, the stringency of the plate mating assay can be increased or decreased by using less or more YEPD, respectively. Mating plates are incubated at 30° and the resulting diploids are counted after 3–5 days. The number of diploids is normalized to the total number of viable **a** cells and the mating efficiency (as compared to wild-type) is calculated as previously described.

PROCEDURE FOR QUALITATIVE PATCH MATING ASSAY. The qualitative patch mating assay is a rapid and convenient assay that involves the mating of a patch of auxotrophic **a** cells to a lawn of α cells on plates that are selective for prototrophic diploids. Although the patch assay is not a quantitative procedure, one can compare the mating of an unknown *ste6* mutant to that of a known, quantitated *ste6* mutant to get an approximate value for the mating efficiency (see Fig. 1).

To perform this assay, **a** cells are patched onto the appropriate selective plates and incubated at 30° for 24–48 hr to generate a patch "master plate." The patch master is then replica-plated (using velvets) onto an SD plate spread with a lawn of the α mating tester and incubated at 30° for 2–3 days. The lawn of α cells is made by diluting a clump of cells directly from a petri plate into fresh YEPD (until turbid), then spreading 0.3 ml onto an SD plate and allowing the plate to dry for ~5 min. The mating assay can be varied in terms of its stringency by altering the conditions as discussed earlier for the quantitative plate mating. For example, the α mating tester can be resuspended in a mixture of YEPD and water, and/or the mating

A

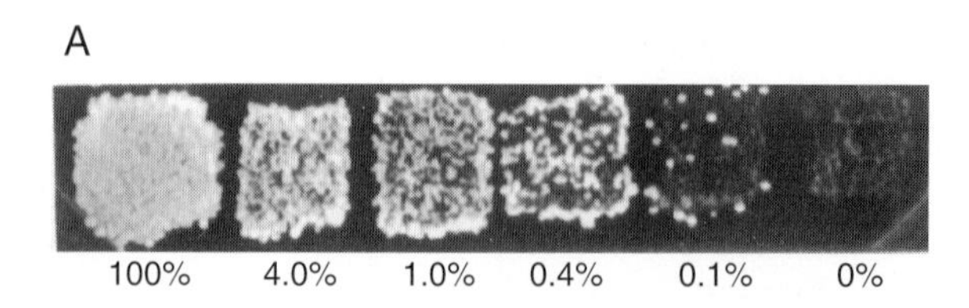

B

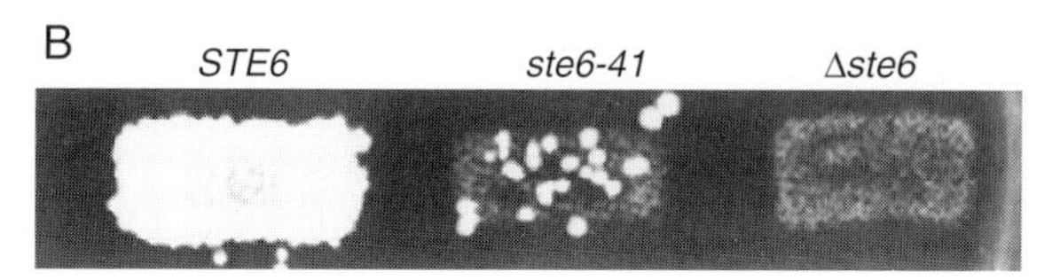

FIG. 1. Qualitative mating assays of wild-type and *ste6* strains; correlation with quantitative mating tests. (A) Patch mating of wild-type and mutant **a** strains to the α mating tester, SM1068. Each patch shown represents a different strain. The mating efficiency for each strain has been quantitatively determined (indicated by percentage) as compared to wild type (100%) at left. (B) Qualitative patch mating assay of wild-type, *ste6*Δ, and a *ste6* point mutant *ste6-41*. Quantitative mating indicates that *ste6-41* has a mating efficiency of ~0.1% (G. L. Nijbroek, unpublished observation, 1996).

plates can be placed at 37°. Both of these conditions increase the stringency. Another way to vary the stringency is to use a mating tester that is defective for mating, such as a *far1* tester.[15,16] Only prototrophic diploids are able to form colonies on the SD plate. Efficient maters form a confluent patch, whereas poor maters exhibit only a few colonies per patch (see Fig. 1B).

Halo Assays to Measure Extracellular **a**-*Factor*. The halo assay relies on the fact that the interaction of extracellular **a**-factor with its receptor on α cells causes a G_1 growth arrest. The **a**-factor halo assay is made profoundly more sensitive by using a supersensitive α *sst2* mating partner for which the growth arrest is irreversible, for instance, RC757, also called SM1086.[7] A patch of **a** cells replica-plated onto a lawn of α *sst2* cells results in a clear zone or halo (surrounding the patch) where growth of the lawn has been inhibited. The assay is so sensitive that even a *ste6* deletion mutant may exhibit a low level of **a**-factor halo activity, presumably due to the release of intracellular **a**-factor from a small number of lysing cells on the plate. Therefore, the halo assay is rarely used when evaluating severe *ste6* mutants since one cannot easily discriminate between severely defective *ste6* mutants and a Δ*ste6* strain. For less severe *ste6* mutants, however, the

[15] N. Valtz, P. Matthias, and I. Herskowitz, *J. Cell. Biol.* **131,** 863 (1995).
[16] N. Adames, K. Blundell, M. N. Ashby, and C. Boone, *Science* **270,** 464 (1995).

halo test can indicate a general trend in terms of **a**-factor export (for instance, see Fig. 2A). The halo assay can be made quantitative by directly spotting serial dilutions of concentrated extracellular **a**-factor onto a lawn of supersensitive α *sst2* cells (see Fig. 2B). Several protocols that use a halo test to measure the amount of **a**-factor secreted by **a** cells are described next.

"SPOT" HALO ASSAY: QUANTITATION OF EXTRACELLULAR **a**-FACTOR ADHERING TO TUBE WALL. The **a**-factor spot test indirectly measures the

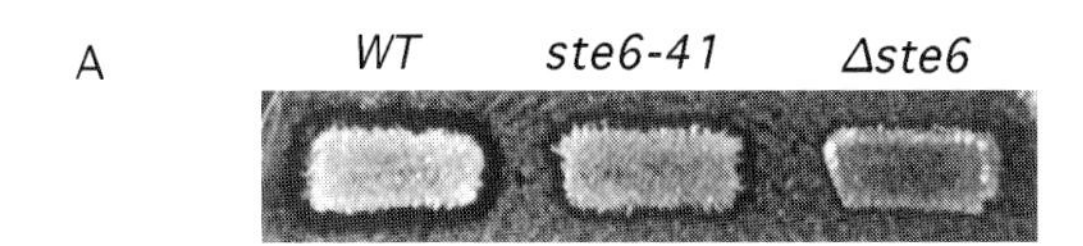

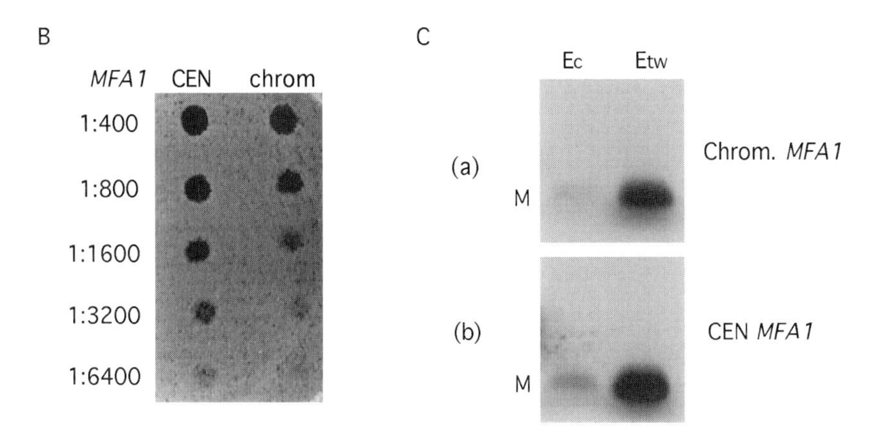

FIG. 2. Qualitative and quantitative halo assays. (A) Qualitative patch halo assays of wild-type and *ste6* strains were performed as described in the text. Note the residual halo for the Δ*ste6* mutant, as discussed. (B) Quantitative **a**-factor spot halo assay of a wild-type (chromosomal *MFA1*) and *CEN MFA1* strain. Concentrated tube-washed **a**-factor was prepared as described and spotted at the indicated dilutions on a lawn of α cells (SM1086). *CEN MFA1* is present in 1–3 copies per cell and correspondingly has a higher end point than chromosomal *MFA1* (1 : 3200 versus 1 : 1600). This result supports the notion that Ste6p is not rate limiting for **a**-factor export in wild-type cells. (C) Immunoprecipitation of **a**-factor to determine the fraction of extracellular **a**-factor that sticks to the walls of polystyrene tubes. The same strains as in (B) were labeled with [^{35}S]cysteine for 30 min. Extracellular culture fluid (Ec) and extracellular tube-washed (Etw) fractions were prepared and immunoprecipitated with **a**-factor antiserum as described in the text. Mature **a**-factor (M) is labeled. Bands were quantified by PhosphorImager (Molecular Dynamics, Sunnyvale, CA), and the portion of **a**-factor adhering to the culture tube (Etw) was calculated to be >85% of the total extracellular **a**-factor pool.

amount of **a**-factor that has been secreted into the culture fluid by Ste6p. A striking feature of **a**-factor is that it is extremely hydrophobic and sticks to the walls of glass, polystyrene, or polypropylene tubes.[3,8] On washing the tube with methanol, this bound **a**-factor becomes soluble and can be concentrated. In our hands, the amount of **a**-factor recovered from the sides of the culture tube represents at least 85% of the total pool of extracellular **a**-factor. An experiment comparing the amount of **a**-factor present in the culture fluid versus on the sides of the culture tube is shown in Fig. 2C.

To prepare extracellular **a**-factor, a 5-ml culture of **a** cells is grown until saturation (usually overnight) in a 15-ml polystyrene tube in SC dropout medium. Note that YEPD medium contains compounds that appear to compete with **a**-factor for binding to tube walls. The culture medium is discarded, the tubes are thoroughly rinsed with distilled water, and the remaining drops of water are removed with a Kimwipe. The sides of the tube are rinsed with enough methanol to cover the walls (~400 μl) and vortexed well. The contents are then transferred to an Eppendorf tube and dried under vacuum at high heat for ~40 min. The dried sample is resuspended in 50 μl methanol and a dilution series (2- to 10-fold increments) is prepared. The **a**-factor concentrate is assayed by preparing serial dilutions (in 2-fold or higher increments) in YEPD plus 250 μg/ml azide-free bovine serum albumin (BSA) Pentax Fraction V (Miles Scientific, Naperville, IL) on Parafilm, and spotting 2 μl onto a lawn of supersensitive *α sst2* cells. BSA is present to saturate nonspecific binding sites for **a**-factor. **a**-Factor will stick to pipette tips, so for each spot a fresh tip is used, which is first exposed to the YEPD–BSA solution. The lawn of supersensitive *α* cells is prepared by resuspending a clump of *α sst2* cells (SM1086) into YEPD until just turbid. The plates are incubated at 30° for 20 hr and the end point is determined. The highest dilution for which a clear halo is observed is called the end point, as shown in Fig. 2B. By comparing the end points from different cultures to wild type, the relative amount of **a**-factor secreted by these strains can be determined. Samples are stored at −20° and will maintain their activity for at least 6 months.

"SPOT" HALO ASSAY: MEASURING **a**-FACTOR PRESENT IN CULTURE FLUID. This assay measures the amount of **a**-factor present in the culture fluid per se (and not on the tube walls) and involves concentrating extracellular **a**-factor by ultrafiltration, generating serial dilutions of the concentrated preparation, and determining the maximal dilution at which a halo is still detected on the lawn of *α* cells. To use this method, it is best to grow cells in YEPD, in which case the adherence of **a**-factor to the tube walls is minimized. Generally, the **a**-factor remaining in the YEPD culture fluid represents ~10–25% of total secreted **a**-factor; the remainder is bound to the tube walls.

Wild-type and *ste6* strains bearing plasmids containing alleles of *STE6* or *MFA1* are grown in 5 ml YEPD or SC dropout media at 30° until saturated. The saturated overnight culture (0.5 ml) is diluted into 4.5 ml of YEPD to OD_{600} 1.0 and grown overnight to saturation ($OD_{600} \sim 25$). This period of growing out in YEPD for four to five generations ensures that plasmid loss is minimal and improves the yield of **a**-factor that remains in the culture fluid.

Preparing Centricons. The Centricon ultrafiltration units (Centricon 30, Amicon Corp., Danvers, MA) are assembled for filtration and 0.4 mg of azide-free BSA Pentax Fraction V from a 0.2 mg/ml stock is added to each to saturate nonspecific binding sites for **a**-factor. The Centricons are centrifuged in a JA-20 rotor at 3000*g* for 30 min at 4°. The retentate cups are rinsed with the BSA filtrate, then the top of the Centricon is capped with the retentate cup and inverted. The Centricons are centrifuged again in a JA-20 rotor at 800*g* for 2 min at 4° and emptied. Centricons can be reused by first rinsing with distilled water, soaking overnight in 70% ethanol, rinsing again with distilled water and 95% (v/v) ethanol, and storing in ethanol.

Preparing supernates for concentration. The cells are removed from the 5-ml culture by centrifugation at 1000*g* in a TJ-6 centrifuge (Beckman, Fullerton, CA). The supernatant is transferred to a new tube and recentrifuged (this step is repeated two more times). Two milliliters of the final supernatant is added to each Centricon reservoir and spun at 3000*g* in a JA-20 rotor for 30 min at 4° (save 50–100 μl for use as a preconcentrated control). The retentate cup is attached to the Centricon unit, inverted and spun at 800*g* in a JA-20 rotor for 2 min at 4°. The **a**-factor concentrate is assayed by spot dilution as before and the remaining samples are stored at 4° or −20°.

PROCEDURE FOR QUALITATIVE PATCH HALO. The patch halo assay allows one to visualize the growth arrest of supersensitive α cells in the form of a halo around a patch of cells on a plate. A patch master of **a** cells is generated on either a YEPD or SC dropout plate (as mentioned before), incubated at 30° for 24–48 hr, and then replica plated onto YEPD plates spread with an α halo tester (as described earlier). By visually comparing the width of the clear zone of a *ste6* mutant to a wild-type and Δ*ste6* strain, one is able to determine the approximate level of **a**-factor activity and thus indirectly the transport activity of Ste6p (see Fig. 2C). However, based on this assay, one can only conclude that a particular *ste6* mutant has either a moderate or severe defect.

Immunochemical Assay; Immunoprecipitation of Radiolabeled **a**-*Factor*

Another direct way of measuring Ste6p function is to determine the amount of **a**-factor exported from **a** cells immunochemically. This procedure

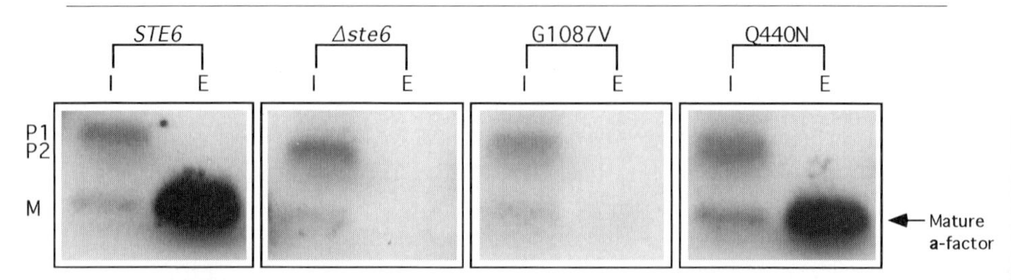

FIG. 3. Immunoprecipitation of **a**-factor in strains expressing wild-type and mutant forms of STE6. Cells were labeled with [^{35}S]cysteine. Intracellular (I) and extracellular (E) fractions were prepared, immunoprecipitated with **a**-factor antiserum, and analyzed by SDS–PAGE, as described [P. Chen, S. S. Sapperstein, J. C. Choi, and S. Michaelis, *J. Cell Biol.* **136,** 16 (1997); and S. Sapperstein, C. Berkower, and S. Michaelis, *Mol. Cell. Biol.* **14,** 1438 (1994)]. Intracellular species of **a**-factor precursors (P1 and P2) and mature (M) forms are labeled. All of the strains used have a chromosomal deletion of *STE6,* and contain one of the following *CEN* plasmids: *STE6* (WT), no insert (Δ*ste6*), *STE6* with an NBF1 point mutation (Q440N), or *STE6* with an NBF2 point mutation (G1087V). Results from quantitative mating assays indicate that these mutants have a mating efficiency of 0.3% (G1087V) and 87% (Q440N).

involves metabolically labeling **a**-factor, separating the intracellular from the extracellular fractions, pooling the extracellular culture fluid and tube wall fractions, and then immunoprecipitating **a**-factor.[3,6,8,17] The amount of **a**-factor exported from cells by Ste6p is determined quantitatively and is generally comparable to the amount of mating observed by the biological assay described earlier.[6] This method is very accurate, but the drawback is that the immunoprecipitation of **a**-factor is technically more difficult than the mating and halo assays and requires the possession of **a**-factor antibodies. An example of a typical immunoprecipitation experiment is shown in Fig. 3. The procedures for metabolic labeling and immunoprecipitation of **a**-factor are described in detail elsewhere.[8]

Summary

In summary, several assays can be used to indirectly measure Ste6p function. These are the mating assay, the halo assay, and **a**-factor immunoprecipitation. These assays can be informative about the extent of a *ste6* defect. The mating assay is the assay of choice because it is technically easy and particularly useful for assaying the complementation of the Δ*ste6* defect by other ABC transporters. The immunoprecipitation of **a**-factor is useful for correlating the amount of mating with the amount of **a**-factor exported. However, this assay is technically more difficult and labor-intensive than the mating assay. Finally, the halo assay is useful in combination with the mating assay to identify gross transport defects.

[17] C. Hrycyna, S. Sapperstein, S. Clarke, and S. Michaelis, *EMBO J.* **10,** 1699 (1991).

Assays for STE6 Trafficking and Stability

Ste6p is a 145-kDa protein that is metabolically unstable, with a half-life of ~30 min at 30°.[18,19] On reaching the plasma membrane, Ste6p is rapidly and constitutively endocytosed and degraded in the vacuole in a *PEP4*-dependent manner. In a *pep4* mutant, undegraded Ste6p accumulates in the vacuole and, correspondingly, the half-life of Ste6p increases more than fivefold.[18,19] Similarly, in mutants defective for endocytosis, Ste6p half-life and surface accumulation are increased.[18] Since Ste6p is rapidly and constitutively endocytosed, immunofluorescence of Ste6p in a wild-type strain does not show plasma membrane staining; rather, Ste6p colocalizes with the Golgi marker, Kex2p.[19a] An explanation for this observation is that the Golgi represents a slow point in Ste6p trafficking. It is not known whether Ste6p is functional intracellularly or if Ste6p is only functional at the plasma membrane. Furthermore, Ste6p has been shown to be ubiquitinated and has a threefold longer half-life in a *ubc4 ubc5* (ubiquitin-defective) double mutant,[19] suggesting that conjugation of ubiquitin to Ste6p is critical for its normal cellular trafficking. The importance of ubiquitination in the trafficking of a membrane protein was recently demonstrated by Hicke *et al.*[20] who showed that the ubiquitination of Ste2p (the α-factor receptor) is required for its ligand-stimulated endocytosis and vacuolar degradation. Similar studies with Ste3p (the **a**-factor receptor) indicate that it also must be ubiquitinated prior to endocytosis.[21,22] It is likely that ubiquintination is also critical for the endocytosis of Ste6p.[19a]

Several assays are available for examining Ste6p localization and stability. We routinely follow Ste6p trafficking by immunofluorescence (IF) and determine its stability by metabolic labeling and immunoprecipitation (IP) or by Western analysis. Due to difficulty in generating Ste6p antisera that are optimal for IF, we epitope-tagged Ste6p to enable its detection. The advantage of using this approach is that highly specific monoclonal antibodies are commercially available. For example, our laboratory makes use of the triply iterated hemagglutinin (HA) epitope from the *influenzae* virus,[23] the *myc* epitope,[24] and the green fluorescent protein (GFP) from the jellyfish *Aequoria victoria.*[25] Using these tags, we have constructed several different

[18] C. Berkower, D. Loayza, and S. Michaelis, *Mol. Biol. Cell* **5,** 1185 (1994).
[19] R. Kölling and C. P. Hollenberg, *EMBO J.* **13,** 3261 (1994).
[19a] D. Loayza and S. Michaelis, *Mol. Cell Biol.* **18,** 779 (1998).
[20] L. Hicke and H. Riezman, *Cell* **84,** 277 (1996).
[21] N. G. Davis, J. L. Horecka, and G. F. Sprague Jr., *J. Cell Biol.* **122,** 53 (1993).
[22] A. F. Roth and N. G. Davis, *J. Cell Biol.* **134,** 661 (1996).
[23] M. Tyers, G. Tokiwa, R. Nash, and B. Futcher, *EMBO J.* **11,** 1773 (1992).
[24] G. Evan, G. Lewis, G. Ramsay, and J. Bishop, *Mol. Cell Biol.* **5,** 3610 (1985).
[25] D. C. Prasher, V. K. Eckenrode, W. W. Ward, F. G. Prendergast, and M. J. Cormier, *Gene* **111,** 229 (1992).

N- and C-terminally and internally tagged *STE6* constructs. The *STE6* constructs have the HA or *myc* epitope inserted at either codon 7, at codon 1290 directly in front of the *STE6* termination codon, or in a predicted extracellular loop between the first and second membrane spans at codon 70. Similarly, we have generated *STE6-GFP* constructs by inserting a 238-residue portion of GFP directly in front of the termination codon of *STE6*. All of the epitope-tagged *STE6* constructs complement a Δ*ste6* mutant for mating and appear to be wild-type in all respects. It is worthwhile to point out that *MDR* has also been shown to be flexible in terms of tolerating insertion of an epitope (adenosine deaminase) at its C terminus,[26] suggesting that the C terminus may be an ideal location for epitope tagging of ABC proteins in general.

In the following paragraphs, we discuss in detail protocols for the IP, IF, and Western analyses of *STE6*. Representative IP and IF experiments are shown in Figs. 4 and 5.

PROCEDURE FOR STE6 PULSE-CHASE METABOLIC LABELING AND IMMUNOPRECIPITATION. Metabolic labeling followed by IP is performed to determine the stability of wild-type and mutant Ste6p (see Fig. 4) and may give important clues as to the nature of a *ste6* defect when combined with the immunofluorescence data. For example, we have observed that several ER-retained *ste6* mutants have an increased half-life, presumably because they fail to reach the plasma membrane and thus cannot undergo subsequent endocytosis and degradation, while other ER-retained *ste6* mutants are hyperunstable and undergo rapid ubiquitin-dependent degradation (Loayza and Michaelis, manuscript submitted, 1998). Most *ste6* mutants exhibit normal stability and localization, but are impaired for function (G. L. Nijbroek, manuscript in preparation, 1998).

Pulse-chase metabolic labeling of cells for IP is performed as described in Berkower *et al.*[18] Strains are grown for 2 days to saturation at 30° in 5 ml SC dropout media. The cultures are diluted 1 : 20 in SC dropout media, and grown for one to three generations to OD_{600} 0.4–1.0. Cells are grown in this manner because the master vacuolar protease, Pep4p, is expressed during stationary phase and remains present for several generations after dilution into fresh media.[27,28] Then 2.5 OD_{600} units of cell (per time point) are harvested in screw cap tubes by centrifuging in a Beckman TJ-6 centrifuge at 1000*g* for 5 min, and resuspended in 0.5 ml SD media containing

[26] U. A. Germann, K.-V. Chin, I. Pastan, and M. M. Gottesman, *FASEB J.* **4,** 1501 (1990).

[27] G. Ammerer, C. P. Hunter, J. H. Rothman, G. C. Saari, V. A. Valls, and T. H. Stevens, *Mol. Cell Biol.* **6,** 2490 (1986).

[28] C. A. Woolford, L. B. Daniels, F. J. Park, E. W. Jones, J. N. Van Arsdell, and M. A. Innis, *Mol. Cell Biol.* **6,** 2500 (1986).

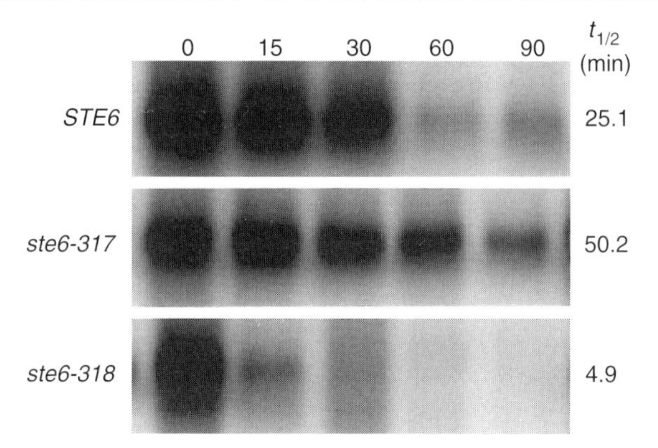

FIG. 4. Pulse-chase analysis of HA-tagged wild-type *STE6* and two *ste6* mutants. Strains that express *STE6* from a high copy number (2μ) plasmid were pulse-labeled for 10 min and the label was chased for the indicated times (in minutes). Samples were immunoprecipitated with HA antiserum and analyzed by SDS–PAGE as described. *STE6* counts were quantified by PhosphorImager analysis and the half-life ($t_{1/2}$) was determined (indicated to the right of each gel). One mutant (Ste6-317p) is more stable than wild-type Ste6p while the other mutant (Ste6-318p) is highly unstable.

required supplements. Cells are labeled with 75 μCi of Expre^{35}S^{35}S (New England Nuclear, Boston MA) for 10 min and chased for 0, 15, 30, 60, and 90 min by adding 10 μl of 50× chase mix (1 *M* cysteine, 1 *M* methionine) per time point. The chase reactions are terminated by transferring the cells to an Eppendorf tube on ice containing 0.5 ml of 2× azide stop mix (40 m*M* cysteine, 40 m*M* methionine, 20 m*M* NaN_3, 500 μg/ml BSA (Miles Laboratories, Elkhart, IN)).

Processing samples. The cells are microfuged at 4° for 30 sec, the supernatant is aspirated, and the cell pellets are resuspended in 1 ml cold distilled water. The cells are lysed by adding 150 μl of fresh NaOH/2-mercaptoethanol (1.14 *M* 2-mercaptoethanol in 1.0 ml 2 *M* NaOH), mixed well, and left on ice for 15 min. Intracellular Ste6p is precipitated for 15 min on ice with 150 μl of freshly made 50% trichloroacetic acid (J. T. Baker, Phillipsburg, NJ). The sample is centrifuged for 5 min at 4°, the supernatant is aspirated, and the pellet is resuspended in 30 μl of TCA sample buffer (80 m*M* Tris-HCl, pH 8.0, 8 m*M* EDTA, pH 8.0, 120 m*M* DTT, 3.5% SDS (w/v), 0.15% glycerol, 0.08% Tris base (w/v), 0.01% bromophenol blue (w/v)). Resuspension of the pellet is facilitated by grinding the pellet with a sealed Pasteur pipette. The sample is neutralized with 1–2 μl of 2 *M* Tris base when necessary (if yellow) and heated at 37° for 30 min. Note that samples containing Ste6p should never be boiled to prevent aggregation.

Immunoprecipitation. Samples are resuspended in 1.5 ml of buffer A (1% Triton X-100, 150 m*M* NaCl, 5 m*M* EDTA, 50 m*M* Tris-HCl, pH 7.5) containing 0.5% aprotinin (Sigma, St. Louis, MO) and 1 m*M* phenylmethylsulfonyl fluoride (PMSF) (in methanol) and left on ice for 1 hr. The samples are vortexed, centrifuged for 1 min at room temperature to remove insoluble debris, and the supernatant transferred to new Eppendorf tubes (avoiding the pellet). This step is repeated before immunoprecipitating with 1 μl of HA antiserum 12CA5 (Boehringer Mannheim, Indianapolis, IN) overnight at 4° on a rotating wheel. To collect immunoprecipitates, 45 μl of protein A-Sepharose CL-4B beads (Pharmacia, Piscataway, NJ) suspended in buffer A (1:3 beads: total volume ratio) is added to each sample and the tubes are continuously mixed for 90 min at 4°. The beads are pulse pelleted (10 sec) at 13,600*g* and washed with 1 ml buffer B (0.1% Triton X-100, 0.02% SDS, 150 m*M* NaCl, 50 m*M* Tris-HCl, pH 7.5, 5 m*M* EDTA) containing 1 m*M* PMSF and 0.5% aprotinin. This step is repeated three more times, then the samples are washed once with 1 ml wash buffer C (150 m*M* NaCl, 50 m*M* Tris-Cl, pH 7.5, 5 m*M* EDTA) containing 1 m*M* PMSF and 0.5% aprotinin. Bound immune complexes are released from the beads by the addition of 30 μl of 2× Laemmli buffer (20% glycerol, 10% 2-mercaptoethanol, 4% SDS, 125 m*M* Tris-HCl, pH 6.8, 0.02% bromophenol blue). Samples are incubated at 37° for 30 min and run at 35 mA on 8% SDS–polyacrylamide gels for $2\frac{1}{2}$ hr. Gels are dried under vacuum and analyzed by PhosphorImager. The half-life is determined by taking the slope of the line, which is obtained when plotting quantitated Ste6p counts (in %) versus chase times using CA-Cricket Graph III (Computer Associates, Islandia, NY).

PROCEDURE FOR Ste6p-HA INDIRECT IMMUNOFLUORESCENCE. Immunofluorescence (IF) is performed to determine the localization of wild-type and mutant forms of Ste6p within the cell (see Fig. 5). To detect wild-type and mutant Ste6p-HA, we make use of HA antiserum 12CA5 (Boehringer Mannheim). Coimmunofluorescence with well-defined subcellular markers, such as Kar2p (ER)[29] and Kex2p (Golgi),[30] or vital dyes, DAPI (nucleus) and FM4-64 (vacuole),[31] is important in determining the exact location of a *ste6* mutant protein. The IF assay is useful when looking for *ste6* mutants, which lead to altered localization. For example, we screened a large collection of mating defective *ste6* mutants by IF and found several that were ER-retained (Loayza, Ph.D. Thesis, Johns Hopkins University, Baltimore, MD, 1998; G. L. Nijbroek, unpublished observation, 1996). Determining the localization of *ste6* mutants within the cell by IF is extremely valuable;

[29] M. D. Rose, L. M. Misra, and J. P. Vogel, *Cell* **57,** 1211 (1989).
[30] R. S. Fuller, A. J. Brake, and J. Thorner, *Science* **246,** 482 (1989).
[31] T. A. Vida and S. D. Emr, *J. Cell. Biol.* **128,** 779 (1995).

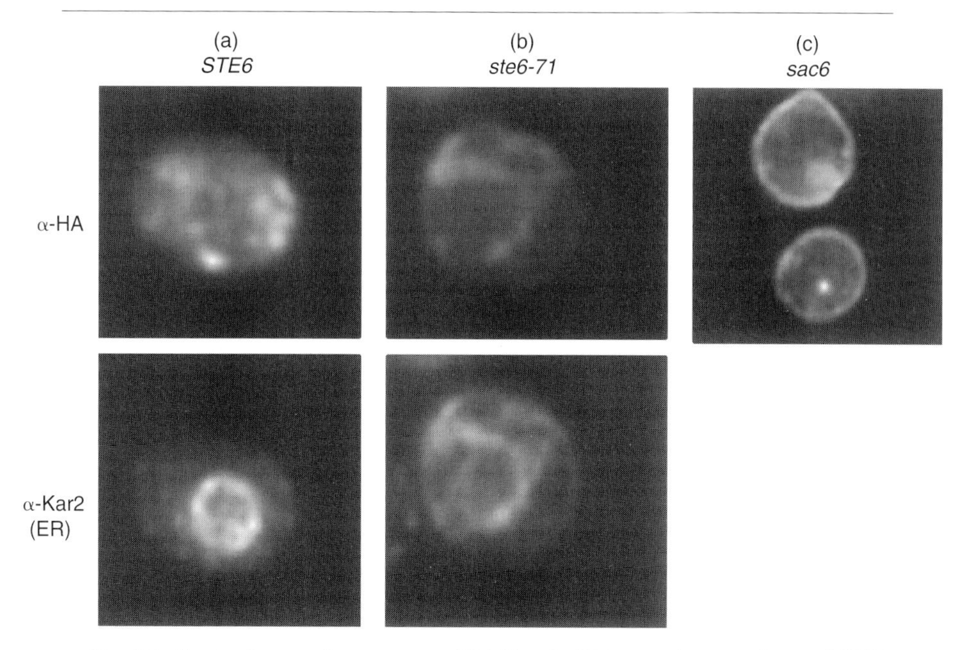

FIG. 5. Indirect coimmunofluorescence of *KAR2* and wild-type and mutant forms of *STE6*. Indirect coimmunofluorescence of an ER marker, Kar2p, with Ste6p-HA (a) and the Ste6-71p-HA mutant (b) in Δ*ste6* cells as described in the text. The cell surface staining pattern of *STE6* in a mutant defective for endocytosis (*sac6*) is shown in (c). Wild-type Ste6p exhibits a punctate, Golgi staining pattern (a) while the mutant Ste6-71p is ER-retained as shown by its coimmunofluorescence with Kar2P (b). Wild-type *STE6* accumulates at the plasma membrane in a mutant defective for endocytosis (c).

however, the procedure can be technically demanding. High copy (2μ or *Gal-STE6*) constructs may be used for IF, since low copy (*CEN*) levels of Ste6p are difficult to detect.

Preparation of yeast cells for IF is performed essentially as described in Rose *et al.*[32] Cells are grown to OD_{600} 0.5–1.0. In the case of temperature-sensitive mutants, the log phase cultures are grown at restrictive temperature for 40 min prior to fixation. For fixation, 5 OD_{600} units of mid-log cells are harvested at 1000*g* in a Beckman TJ-6 centrifuge and resuspended in 5 ml KP_i buffer (0.1 *M* potassium phosphate, pH 6.5). Then 0.6 ml of 37% formaldehyde is slowly added while gently vortexing, and the cells are incubated at 30° (or at restrictive temperature) for 40 min with gentle agitation. Cells are washed twice with 5 ml KP_i and once with 5 ml KPS

[32] M. D. Rose, F. Winston, and P. Hieter, "Methods in Yeast Genetics, A Laboratory Course Manual." Cold Spring Harbor Laboratory Press, New York, 1990.

(0.1 *M* potassium phosphate, pH 6.5, 1.2 *M* sorbitol), then resuspended in 1 ml KPS. To remove the cell wall, 142 m*M* 2-mercaptoethanol and 0.1 mg zymolyase 100T (ICN, Costa Mesa, CA) are added, and the cells are incubated at 30° for 10 min with mixing (different batches of zymolyase may require altering this time). The resultant spheroplasts are washed once with 5 ml KPS, then resuspended in 1 ml KPST (0.1 *M* potassium phosphate, pH 6.5, 1.2 *M* sorbitol, 0.1% Tween 20) by gently pipetting up and down. Spheroplasts are very fragile and should never be vortexed. Fixed and permeabilized spheroplasts (15 μl) are adsorbed to polylysine-coated multiwell slides in a humid chamber for 15 min and then incubated in 15 μl of PBST (0.04 *M* K_2HPO_4, 0.01 *M* NaCl, 0.1% Tween 20, 10 mg/ml BSA, 0.1% NaN_3) for 15 min. Care must be taken from this point on to ensure that the cells do not dry out. Polylysine-coated multiwell slides are prepared by coating slides for 15 min with 15 μl of 0.1% polylysine (Sigma) and washing four times with distilled water. The adsorbed cells are incubated with 15 μl of primary antibody diluted into PBST (2.15 μg/ml for HA monoclonal antiserum 12CA5 (Boehringer Mannheim) at room temperature overnight. The cells are then washed four times with 15 μl of PBST and incubated with 15 μl of secondary antibody diluted in PBST (1:500 goat anti-mouse rhodamine, Boehringer Mannheim) for 2–4 hr in the dark to avoid photobleaching. Instead of using goat anti-mouse rhodamine 12, the cells may be incubated with a different secondary antibody; for instance, 15 μl of goat anti-mouse Cy3 (Jackson ImmunoResearch, West Grove, PA). This Cy3 antibody is brighter than rhodamine and is useful for visualizing *CEN* levels of Ste6p-HA. The cells are again washed four times with 15 μl PBST. At this point, the cell nuclei can be stained with DAPI by first washing once with 15 μl PBS (0.04 *M* K_2HPO_4, 0.01 *M* NaCl, 0.1% NaN_3) and then incubating for 5 min with 1 μg/mol DAPI. The cells are washed once more with 15 μl PBS, and the slides are allowed to dry in the dark (e.g., under aluminum foil, in a cabinet). Mounting medium[32] is added to each well, and coverslips are mounted with nail polish. Slides can then be viewed immediately or can be stored at −20°.

Alternatively, direct (and not indirect) IF of Ste6p-HA can be performed by incubating the cells with 15 μl of 8 μg/ml HA-TRITC conjugated mouse monoclonal antiserum (Boehringer Mannheim) diluted into PBST. The results from direct IF of Ste6p-HA are comparable to indirect IF of Ste6p. Direct IF has the advantage of being slightly less time consuming, since the secondary antibody step is not carried out.

PROCEDURE FOR WESTERN ANALYSIS OF Ste6p-HA. Nonradioactive Ste6p-HA Western blots are carried out to determine the amount of total protein produced by wild-type and *ste6* mutants. The Ste6p-HA Western blot is performed essentially as described in *Molecular Cloning: A Lab*

Manual.[33] As before, the samples are not boiled but heated at 37°. Ste6p-HA is detected by chemiluminescence (kit and procedure according to Boehringer Mannheim).

In summary, several assays can be used to assess Ste6p localization and stability, including immunoprecipitation, immunofluorescence, and Western blots. An additional assay that we did not discuss here is the subcellular fractionation of Ste6p[34] (Schmidt, submitted, 1998). These assays allow one to obtain information about the nature of the *ste6* defect. Immunoprecipitations are carried out to determine the half-life of mating-defective *ste6* mutants and may allow one to distinguish among loss of function, stability, or trafficking mutants. For instance, a *ste6* mutant with an increased half-life may be retained in a compartment in the cell (such as the ER or vacuole), while a rapidly degraded protein will have a shortened half-life. Generally, the stability of a *ste6* mutant is determined before the IF experiments are carried out. The reasons for this are twofold: (1) Highly unstable mutants are difficult to detect due to the low level of protein and (2) IF experiments are technically more difficult than IP experiments. Ste6p Western blots are performed to detect altered steady-state levels of *ste6* protein and to identify potential promoter or trafficking mutations.

Structure–Function Analysis of *STE6*

STE6 is comprised of two homologous halves, each half containing a membrane-spanning domain (MSD), comprised of six predicted membrane spans, and a nucleotide-binding domain (NBD). To dissect *STE6* at the molecular level, we and others have undertaken several approaches involving site-directed and random mutagenesis of the NBDs and full-length *STE6*. In addition, we constructed partial *STE6* molecules that do not contain a full set of MSD or NBD modules. These studies are summarized here.

Partial Molecules of STE6 to Reconstitute Functional Transporter

It was reasonable to expect that coexpression of separately encoded half-molecules of *STE6* could reconstitute a functional protein since certain ABC half-transporters, for example, *TAP1* and *TAP2,* heterodimerize to form a functional full-length transporter. We severed the *STE6* gene into separately encoded half-molecules, N-1/2 (MSD1-NBD1) and C-1/2

[33] J. Sambrook, E. F. Fritsch, and T. Maniatis, "Molecular Cloning, A Laboratory Manual." Cold Spring Harbor Laboratory Press, New York, 1989.

[34] R. Kolling and C. P. Hollenberg, *EMBO J.* **13,** 3261 (1994).

(MSD2-NBD2) and showed that while individually expressed molecules could not promote **a**-factor export or mating, coexpression of both half-molecules in the same strain restored *STE6* function.[6] Similarly, these separately expressed portions of *STE6* were shown to physically associate by coimmunoprecipitation.[35] N-1/4 and C-3/4 molecules were also shown to interact and restore *STE6* function. Taken together, these and other results suggest that the interaction between domains of Ste6p occurs via its MSDs instead of its NBDs.

In addition, we have demonstrated that a Ste6p half-molecule, which on its own is completely inactive, when coexpressed with a defective full-length *ste6* molecule, is capable of "rescuing" the mutant *ste6* protein to restore function to the nearly wild-type level.[35] This result suggests that a productive interaction between the half-molecule and the full-length protein is possible.

Site-Directed and Random Mutagenesis of STE6

The high degree of sequence conservation in the NBDs among all ABC proteins suggested that these regions contained critically important residues that are necessary for function. We and others have created mutations in the NBDs of *STE6* analogous to those found in *CFTR*, to determine the significance of these residues in **a**-factor export.[6,12,35] We found that every mutation made in conserved residues of the Walker A and B motifs had an effect on Ste6p function, measured as a reduction in mating efficiency (0.3–26%) relative to wild type. However, none of the center region mutations had any significant effect on the ability of Ste6p to transport **a**-factor, including a mutation analogous to the ΔF508 mutation of *CFTR*. These results lend support to the notion that the center region may not be directly involved in ATP binding or hydrolysis, but rather may play a role in substrate specificity or in interactions between the NBDs and the MSDs.[6]

We have also employed a random mutagenesis strategy to generate missense mutations to uncover regions within *STE6*, other than the NBDs, that were important for function, localization, or stability. Among the collection of *ste6* mutants, we identified several that were ER-retained (D. Loayza, Ph.D. Thesis, Johns Hopkins University, Baltimore, MD, 1998; G. L. Nijbroek, unpublished observations, 1996). Since the ER quality control machinery that is reponsible for retaining misfolded proteins has not yet been identified, these mutants are expected to aid in the identification of these components.

In addition, we also identified several *ste6* mutants containing premature

[35] C. Berkower, D. Taglicht, and S. Michaelis, *J. Biol. Chem.* **271,** 22983 (1996).

termination codons that have low, but substantial amounts of *STE6* activity, suggestive of nonsense suppression and read-through (G. L. Nijbroek, unpublished observations, 1996). Interestingly, Fearon *et al.*[36] have shown that a nonsense mutation in *STE6* NBD1 led to an extremely high level of read-through; this phenomenon was shown to be highly dependent on the local sequence content.[37]

Other Types of ste6 Mutants

In a search for sterile mutants with specific defects in cell fusion, which is a late step of mating, several *ste6* alleles were identified with only partially impaired ability to transport **a**-factor but with severe fusion defects.[38] While these results initially suggested a specific role for *STE6* in cell fusion, we and others[39] hypothesized that Ste6p is not directly involved in cell fusion, but that instead cell fusion requires an extremely high level of extracellular **a**-factor. Accordingly, we showed that mutations in **a**-factor processing genes, such as *ram1* and *axl1,* also lead to cell fusion defects. Furthermore, by placing *MFA1* under control of a repressable promoter, we showed that cell fusion was promoted when high levels of **a**-factor were present, but was reduced when levels of **a**-factor were decreased.[39] Thus, the cell-fusion defect associated with certain *ste6* alleles is likely to be due to reduced **a**-factor export, resulting in lowered extracellular **a**-factor levels, rather than to a direct role for Ste6p in promoting cell fusion.

Assaying Heterologous ABC Proteins Using *ste6* Deletion in Yeast

An advantage of using yeast is the ability to express and assay nonyeast ABC proteins by measuring **a**-factor export in a strain that is deleted for *STE6* (Δ*ste6*) using the mating assay. A growing number of ABC proteins have been expressed in yeast and have been shown to complement a Δ*ste6* strain. These include mouse *MDR3,*[9] *Plasmodium falciparum pfMDR1,*[40] *Candida albicans HST6,*[40a] and human *MRP.*[41] Levels of complementation as measured by mating efficiency (as described previously) varied from

[36] K. Fearon, V. McClendon, B. Bonetti, and D. M. Bedwell, *J. Biol. Chem.* **269,** 17802 (1994).

[37] B. Bonetti, L. Fu, J. Moon, and D. Bedwell, *J. Mol. Biol.* **251,** 334 (1995).

[38] L. Elia and L. Marsh, *J. Cell Biol.* **135,** 741 (1996).

[39] V. Brizzio, E. Gammie, G. L. Nijbroek, S. Michaelis, and M. Rose, *J. Cell Biol.* **135,** 1727 (1996).

[40] S. K. Volkman, A. F. Cowman, and D. F. Wirth, *Proc. Natl. Acad. Sci. U.S.A.* **92,** 8921 (1995).

[40a] M. Raymond, D. Dignard, A. M. Alarco, N. Mainville, B. B. Magee, and D. Y. Thomas, *Mol. Microbiol.* **27,** 587 (1998).

[41] S. Ruetz, U. Delling, M. Brault, E. Schurr, and P. Gros, *Proc. Natl. Acad. Sci. U.S.A.* **93,** 9942 (1996).

about 0.5% (mouse *MDR3*) to 13% (human *MRP*) compared to a wild-type *STE6* strain (100%). The ability of nonyeast proteins to complement the **a**-factor export defect suggests that structural and functional determinants have been evolutionarily conserved in these (in some instances) very distant members of the ABC superfamily.

STE6 can also be a powerful tool when used to create chimeric molecules with other ABC genes. Teem *et al.*[42] constructed a *STE6–CFTR* hybrid gene and identified revertants of the ER-retained *CFTR–ΔF508* mutation in yeast using the mating assay. These revertants partially suppressed the chloride channel defects in mammalian cells due to the lack of functional CFTR. Thus, the use of *STE6* chimeras may provide a system to identify potential intragenic suppressors in other related ABC proteins. It should be noted, however, that while the *STE6–CFTR ΔF508* chimera appears to be a good model for *CFTR–ΔF508* at the genetic level for functional studies, it is not an accurate model for protein localization, since unlike the *CFTR–ΔF508* mutant in mammalian cells, the mutant *STE6–CFTR* chimera is not ER-retained in yeast.[43]

Summary

As a member of the ABC superfamily, *STE6* is unique in that it has a well-characterized substrate, **a**-factor, and can be easily manipulated in the yeast system. Functional assays have been extensively used, and methods to examine trafficking and stability of *STE6* are well established. In addition, *STE6* chimeras and *ste6* deletion strains are useful for the analysis of many nonyeast ABC proteins. Continuing studies of *STE6* are expected to aid in the identification of novel cellular components involved in the trafficking and functioning of not only *STE6*, but of other members of the ABC superfamily as well.

Acknowledgments

Many past and present members of our laboratory contributed to the development of these protocols. We are grateful to Walter Schmidt, Amy Tam, and Diego Loayza for comments on this manuscript, and to Carol Berkower and Amy Tam who generated the **a**-factor immunoprecipitation and *sac6* immunofluorescence data, shown in Figs. 3 and 5, respectively. This work was supported by grants GM51508 and DK48977 from the National Institutes of Health to S. M. and by a Predoctoral Minority Fellowship from the NIGMS to G. L. N.

[42] J. L. Teem, H. A. Berger, L. S. Ostedgaard, D. P. Rich, L.-C. Tsui, and M. J. Welsh, *Cell* **73,** 335 (1993).

[43] C. Paddon, D. Loayza, L. Vangelista, R. Solari, and S. Michaelis, *Mol. Micro.* **19,** 1007 (1996).

[15] ABC Transporters Involved in Transport of Eye Pigment Precursors in *Drosophila melanogaster*

By GARY D. EWART and ANTHONY J. HOWELLS

Introduction

The usual red-brown color of the eyes of the vinegar fly *Drosophila melanogaster* is due to the presence of two light-screening pigments: *xanthommatin*, which is brown, and a class of pigments known as *drosopterins*, which are red. These pigments are deposited in membrane-bound granules in specialized pigment cells in each ommatidium of the compound eye. They serve not in light detection, but to isolate optically each ommatidium from its neighbor and thus enhance visual acuity. The biology of *Drosophila* eye pigmentation has been reviewed by Summers *et al.*[1]

Investigation of *Drosophila* eye color mutants has a long tradition, starting as far back as the 1930s[2] and, consequently, the molecular genetics and biochemistry of pigment synthesis are well characterized.[1] Xanthommatin is derived biosynthetically from tryptophan and drosopterins are derived from guanosine triphosphate (GTP). The reaction pathways leading to pigment biosynthesis within cells are not discussed further here. Instead, the subject of this article is the characterization of the genes and proteins involved in transport of the pigment precursor molecules from the haemolymph into the fly eye pigment cells.

Studies of eye color mutant flies revealed three genetic loci (*white*, *brown*, and *scarlet*) in which mutations do not alter levels of the enzymes involved in pigment biosynthesis, but instead interfere with the ability of cells to take up pigment precursors. The genes were named for the eye color phenotype of null mutants: Null mutations at the *white* locus (w^0) deposit neither xanthommatin nor drosopterins in the pigment cells and consequently have white eyes. The red eyes of flies with null mutations at *scarlet* (st^0) are due to normal levels of drosopterins but the absence of xanthommatin. Conversely, *brown* null flies (bw^0) do not have red pigments in their eyes, while the xanthommatin level is wild type. The biochemical studies of Sullivan *et al.*[3–5] indicate that the *w* and *st* genes are involved in

[1] K. M. Summers, A. J. Howells, and N. A. Pyliotis, *Adv. Insect. Physiol.* **16,** 119 (1982).
[2] B. Ephrussi, *in* "Cold Spring Harbor Symposium on Quantitative Biology," Vol. 10, pp. 40–48. Cold Spring Harbor Laboratory Press, New York, 1942.
[3] D. T. Sullivan, L. A. Bell, D. R. Paton, and M. C. Sullivan, *Biochem. Genet.* **18,** 1109 (1980).

0076-6879/98 $25.00

the uptake of tryptophan and two intermediates of the xanthommatin pathway (kynurenine and 3-hydroxy-kynurenine), while the *w* and *bw* genes are involved in the uptake of the drosopterin precursor guanine.

Cloning of *w*, *st*, and *bw* Genes

The first gene of the transporter trio to be cloned was the *white* gene,[6–8] reviewed by Hazelrigg.[9] A genomic clone was sequenced[10] and the intron–exon junctions were deduced with only minor modifications being required when a cDNA clone was subsequently isolated.[11] The putative White protein contains 687 amino acids, and database searches identified homology between a region in the White protein and the nucleotide-binding domain observed in the "rapidly growing family of proteins" now known as the ABC transporters (or traffic ATPases).[12]

In 1988 a section of the *Drosophila* polytene chromosome containing the *brown* gene was cloned and used to isolate a cDNA from which was deduced the 675 residue amino acid sequence of the Brown protein.[13] Database searching revealed the White protein as the best match with Brown at 29% overall residue identity. Again homology with the nucleotide-binding domain of other ABC transporters was detected.

Finally, in 1989 the cloning and partial sequencing of the *scarlet* gene was reported[14] and since then the full-length cDNA sequence has been determined.[15,16] The deduced amino acid sequence of Scarlet is 666 residues long, contains a nucleotide-binding domain, and alignment with White and Brown shows that Scarlet shares about 35 and 32% identical residues with those two proteins, respectively.[15]

---

[4] D. T. Sullivan and M. C. Sullivan, *Biochem. Genet.* **13,** 603 (1975).

[5] D. T. Sullivan, L. A. Bell, and D. R. Paton, *Biochem. Genet.* **17,** 565 (1979).

[6] P. Bingham, R. Levis, and G. Rubin, *Cell* **25,** 693 (1981).

[7] M. Goldberg, R. Paro, and W. J. Gehring, *EMBO J.* **1,** 93 (1982).

[8] V. Pirrotta, C. Hadfield, and G. Pretorius, *EMBO J* **2,** 927 (1983).

[9] T. Hazelrigg, *Trends in Genet.* **3,** 43 (1987).

[10] K. O'Hare, C. Murphy, R. Levis, and G. M. Rubin, *J. Mol. Biol.* **180,** 437 (1984).

[11] M. Pepling and S. Mount, *Nucleic Acids Res.* **18,** 1633 (1990).

[12] S. M. Mount, *Nature* **325,** 487 (1987).

[13] T. D. Dreesen, D. H. Johnson, and S. Henikoff, *Mol. Cell Biol.* **8,** 5206 (1988).

[14] R. G. Tearle, J. M. Belote, M. McKeown, B. S. Baker, and A. J. Howells, *Genetics* **122,** 595 (1989).

[15] J. F. M. ten Have, M. M. Green, and A. J. Howells, *Mol. Gen. Genet.* **249,** 673 (1995).

[16] J. F. M. ten Have, "Molecular characterisation of spontaneous mutations at the *scarlet* locus of *Drosophila melanogaster.*" Master's Thesis, Australian National University (1993).

The comparisons of the White, Brown, and Scarlet amino acid sequences (Fig. 1A) clearly show that they are related proteins and that they belong to the ABC transporter superfamily. Each of the proteins contains a single nucleotide-binding domain, in the N-terminal half of their sequences, and a hydrophobic C-terminal half that hydropathy plots (Fig. 1B) predict to contain six membrane-spanning α helices.

Because the structural model for general ABC transporters indicates that the functional unit contains two nucleotide-binding domains and two transmembrane domains, each of the *Drosophila* proteins is capable of comprising half of a minimal ABC transporter complex. This prediction fits well with the physiologic studies of genetic mutants by Sullivan's group,[3–5] which demonstrate that both *white* and *brown* are required for transport of drosopterin precursors and that both *white* and *scarlet* are required for the transport of xanthommatin precursors. Therefore, the currently accepted working model, as depicted in Fig. 2, is that the drosopterin precursor transporter is a heterodimer containing White and Brown subunits, and the xanthommatin precursor transporter is a heterodimer containing White and Scarlet subunits (first discussed by Dreesen *et al.*[13]).

Drosophila Transporters as Model Eukaryotic ABC Transporters

The *Drosophila* pigment precursor transporters represent a powerful experimental system for the investigation of eukaryotic ABC transporters, allowing an integrated approach combining genetics, biochemical, and recombinant DNA technologies: None of the genes is essential and mutant phenotypes are easily scored. A large number of eye color mutant strains, with the mutations localized in the *white*, *brown*, or *scarlet* genes, have already been catalogued and are maintained in various strain collections around the world (information can be obtained on the Internet via the Flybase link[17]). More eye color mutants can be readily generated because protocols for chemical mutagenesis of *Drosophila* are well established.[18] Chemical mutagenesis is a particularly appropriate approach for identification of functionally significant residues because of the high proportion of single base changes generated by this method, which produce point mutations in the amino acid sequence as opposed to insertion or deletion which ablate the function of the proteins.

PCR (polymerase chain reaction) technology and the usually small size of *Drosophila* introns make it possible to rapidly amplify, clone, and se-

[17] "Flybase" internet address: http://flybase.bio.indiana.edu:82/1/stocks/.

[18] L. R. Lacy, M. T. Eisenberg, and C. J. Osgood, *Mutation Res.* **162,** 47 (1986).

A

```
   Scarlet  ........msdsdskridve.........aperveqhel.qvmpvgstie
     White  mgqedqellirggskhpsaehlnngdsgaasqscinqgfgqaknygtllp
     Brown  ..................................................
 Consensus  --------------------------------------------------

   Scarlet  vpsldstpklSkrnsserslplrsyskWspt.eQgat....lvwrdlcvy
     White  psppedsgsgSgqlaenltyawhnmdiFgav.nQpgsgwrqlvnrtrglf
     Brown  ...mqesggsSgqggpslclewkqlnyYvpdqeQsnysfwnecrkkrel.
 Consensus  ----------S----------------------Q----------------

   Scarlet  tnvggsgqrmkriinnstGaiqpGtLmAlMGsSGsGKTTLMstlafRqpa
     White  cnerhipaprkhllknvcGvaypGeLlAvMGsSGaGKTTLLnalafRspq
     Brown  ...........rilqdasGhmktGdLiAiLGgSGaGKTTLLaaisqRlrg
 Consensus  ------------------G----G-L-A--G-SG-GKTTL------R---
                                          Walker A

   Scarlet  gtvv..qGdiliNGrrigp.fMhrnhgYvyQdDlflgsvsvlEHLnFmah
     White  giqvspsGmrllNGqpvdakeMqarcaYvqQdDlfigsltarEHLiFqam
     Brown  nl....tGdvvlNGmamerhqMtrissFlpQfEinvktftayEHLyFmsh
 Consensus  -------G----NG-------M--------Q-----------EHL-F---

   Scarlet  lrLdRrvskeerrliIkellertgLlsaaqTrIgsgddkkvLSGGERKRL
     White  vrMpRhltyrqrvarVdqviqelsLskcqhTiIgvpgrvkgLSGGERKRL
     Brown  fkMhRrttkaekrqrVadlllavgLrdaahTrI......qqLSGGERKRL
 Consensus  ----R-------------------L-----T-I--------LSGGERKRL
                                                    "Loop3"

   Scarlet  aFAvEllnnPviLfCDEPTtGLDSYsAqqlVatLyeLaqk..........
     White  aFAsEaltdPplLiCDEPTsGLDSFtAhsvVqvLkkLsqk..........
     Brown  sLAeElitdPifLfCDEPTtGLDSFsAysvIktLrhLctrrriakhslnq
 Consensus  --A-E----P--L-CDEPT-GLDS--A------L--L-------------
                       Walker B

   Scarlet  ..................................................
     White  ..................................................
     Brown  vygedsfetpsgessasgsgsksiemevvaeshesllqtmrelpalgvls
 Consensus  --------------------------------------------------

   Scarlet  .......gttilctIHQPsSqlFDnFnnVmLLadGRVaFtGspqhAlsFF
     White  .......gktviltIHQPsSelFElFdkIlLMaeGRVaFlGtpseAvdFF
     Brown  nspngthkkaaicsIHQPtSdiFElFthIiLMdgGRIvYqGrteqAakFF
 Consensus  --------------IHQP-S--F--F----L---GR----G----A--FF
```

FIG. 1. (A) Alignment of the White, Scarlet, and Brown amino acid sequences. The GCG program PILEUP was used with a gap weight of 3.0 and gap length weight of 0.10. The

```
  Scarlet  anhGyycPeaYNPADFligvLAtdpGyEqasqrsaqhlcdqfavssaakq
    White  syvGaqcPtnYNPADFyvqvLAvvpGrEiesrdriakicdnfaiskvard
    Brown  tdlGyelPlnCNPADFylktLAdkeGkEna.........gavlrakyehe
Consensus  ---G---P---NPADF----LA---G-E----------------------

  Scarlet  rDmLv...nLeihMaqsgn.fpFdtevesfrgvaWykrfhvvwlRaivtl
    White  mEqLlatknLekpLeqpengytYka........tWfmqfravlwRswlsv
    Brown  tDgLysgswL...Larsysg.dYlkhvqnfkkirWiyqvyllmvRfmted
Consensus  ---L-----L------------------------W---------R-----

  Scarlet  LrdptiqwlrFiqkiamAfiigacFaGtteps QlgVqaVqGalFiMisen
    White  LkepllvkvrLiqttmvAiligliFlG.qqltQvgVmnInGaiFlFltnm
    Brown  LrnirsgliaFgffmitAvtlslmYsGigqltQrtVqdVgGsiFmLsnem
Consensus  L----------------A--------G-----Q--V----G--F------
                                TMH1

  Scarlet  tYhpmYsvlnlFpqgFPlfmREtrsglYstggYYaanilallPgmiiepl
    White  tFqnvFatinvFtseLPvfmREarsrlYrcdtYFlgktiaelPlfltvpl
    Brown  iFtfsYgvtyiFpaaLPiirREvgegtYslsaYYvalvlsfvPvaffkgy
Consensus  -----------F----P---RE-----Y----Y---------P-------
           TMH2                                    TMH3

  Scarlet  IFviIcYwltglrstfyaFgvtamcvvLvmnvataCGcFfStafnSvpLA
    White  VFtaIaYpmiglragvlhFfnclalvtLvanvstsFGyLiScassStsMA
    Brown  VFlsViYasiyytrgfllYlsmgflmsLsavaavgYGvFlSslfeSdkMA
Consensus  -F----Y--------------------L--------G---S----S---A
                                   TMH4

  Scarlet  maylvPldyiFMitsGiFiqvnslPvafwwtqFLSwmlYaNEaMtaaqWs
    White  lsvgpPviipFLlfgGfFlnsgsvPvylkwlsYLSwfrYaNEgLlinqWa
    Brown  secaaPfdliFLifgGtYmnvdtvPg....lkYLSlffYsNEaLmykfWi
Consensus  -----P----F----G--------P--------LS---Y-NE------W-
              TMH5

  Scarlet  gVqnitcfqesadlpCfhtGqdVLdkytFnesn..vyrnllaMvglyfgF
    White  dVepgeisctssnttCpssGkvILetlnFsaad..lpldyvgLailivsF
    Brown  dIdnidc.pvnedhpCiktGveVLqqgsYrnadytywldcfsLvvvaviF
Consensus  ---------------C---G---L-------------------------F
                                                   TMH6

  Scarlet  hllgYyclWrrarkl....
    White  rvlaYlalRlrarrke*..
    Brown  hivsFglvRryihrsgyy*
Consensus  -------------------
```

Walker A and B motifs and the "loop 3'" region characteristic of ABC transporter nucleotide-binding domains are underlined and labeled. The amino acids predicted to be in transmembrane helices (TMH) are in bold text. (B) Hydropathy plots of the three subunits with the TMH regions underlined.

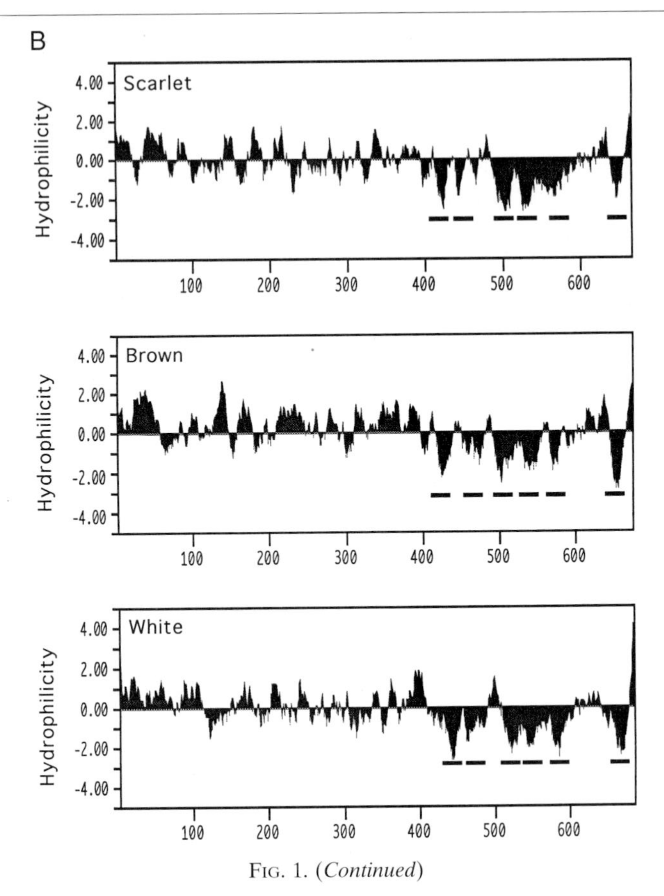

FIG. 1. (*Continued*)

quence mutant alleles directly from genomic DNA. The P-element mediated germline transduction system for *D. melanogaster*[19] makes it possible to test *in vivo* functioning of proteins specifically altered by site-directed mutagenesis. Finally, biochemical assays have been developed to measure individual pigment levels[20] and also to measure the function of the transporters *in situ*.[5]

Despite the apparent utility of this system only a very few groups have been involved in the molecular characterization of the *w*, *st* and *bw* eye color genes and even fewer are working on the encoded proteins (this is no doubt in part due to the lack of medical significance attached to these particular ABC transporters). Consequently, understanding of the struc-

[19] D. B. Roberts, ed., "*Drosophila*: A Practical Approach." IRL Press, Washington, D. C., 1986.
[20] D. J. Nolte, *J. Genet.* **52,** 127 (1954).

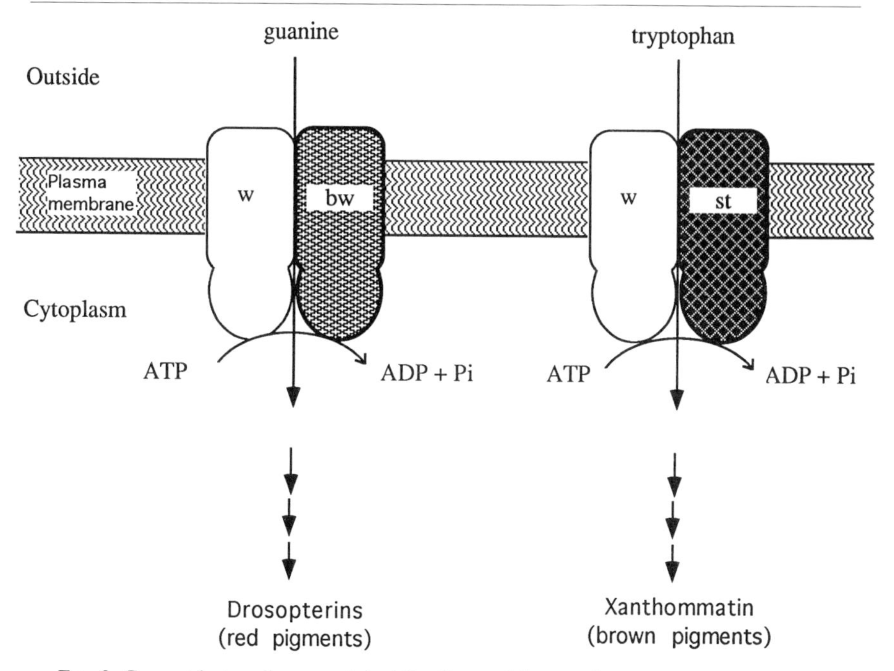

FIG. 2. Current heterodimer model of the *Drosophila* eye pigment precursor ABC transporters.

ture–function relationship in these transporters is at a very early stage compared to, say, CFTR or the P-glycoprotein. However, a number of mutations in the amino acid sequence of White and Brown have been characterized by our group[21,22] that both support and extend the low-resolution structural model depicted in Fig. 2. Discussion of these mutations and future directions of the research are the topics of the rest of this article.

Identification of Functionally Significant Residues by Characterization of Mutants

ATP Binding Domain of White

The *white* minigene, in which the majority of the 3-kb intron 1 is deleted, has been used for some time as a marker in P-element transformation

[21] G. D. Ewart, D. Cannell, G. B. Cox, and A. J. Howells, *J. Biol. Chem.* **269,** 10,370 (1994).
[22] G. D. Ewart, unpublished results and A. J. Howells (1997).

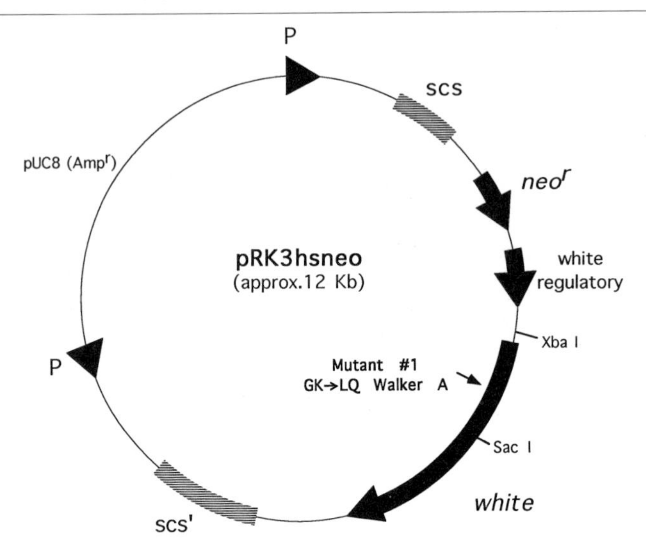

FIG. 3. Shuttle vector pRK3hsneo used in P-element mediated germline transduction experiment to introduce mutated *white* genes into transgenic fly strains. Constructed as described in Ewart *et al.*[21]

experiments.[23] This system is ideal for assessment of the effects of mutations on White protein function. We constructed the vector pRK3hsneo (Fig. 3) and introduced site-directed mutations to the *white* minigene to change the codons for the invariant GlyLys amino acid doublet in the Walker A motif. We then injected pRK3hsneo[*w*mut], containing the mutated gene, into *Drosophila* embryos to create transgenic flies.[21] The neomycin resistance gene was used to select transformants, and a PCR test was designed to confirm the presence of the mutated gene. The resultant transgenic flies had white eyes indicating that the mutated *white* gene was nonfunctional. This result indicates that an intact Walker A motif in White is essential to function of both of the eye pigment precursor transporters, and, as a corollary, that the nucleotide-binding domains of Brown or Scarlet are insufficient to support transport activity. This result is consistent with the effects of similar mutations in other members of the ABC transporter family.[24–27]

[23] R. Kellum and P. Schedl, *Cell* **64,** 941 (1991).

[24] D. Parsonage, S. Wilke-Mounts, and A. E. Senior, *J. Biol. Chem.* **262,** 8022 (1987).

[25] M. Azzaria, E. Schurr, and P. Gros, *Mol. Cell. Biol.* **9,** 5289 (1989).

[26] G. B. Cox, D. Webb, and H. Rosenberg, *J. Bacteriol.* **171,** 1531 (1989).

[27] C. Berkower and S. Michaelis, *EMBO J.* **10,** 3777 (1991).

Mutations Defining Regions of Interaction between White and Brown

As discussed earlier, the low-resolution structural model predicts that White and Brown subunit dimerization is required for formation of an active pteridine precursor transporter. This notion is strongly supported by a pair of interacting alleles of *white* (w^{co2}) and *brown* (bw^{6}) that were first characterized genetically by Farmer and Fairbanks.[28,29] w^{co2} and bw^{6} are both recessive alleles which interact such that drosopterin levels in the eyes appear normal unless there is no other functional copy of w^{+} or bw^{+} in the fly. In terms of the heterodimer model for the encoded transporter subunits, this means that, in the presence of bw^{+}, the w^{co2} alleles produce a White protein that dimerizes normally with Brown to make an active transporter. Similarly, in the presence of w^{+}, bw^{6} produces a functionally competent Brown subunit. However, w^{co2}-encoded White subunits and bw^{6}-encoded Brown subunits are unable to form a complex that is active in transport of drosopterin precursors. We have cloned and sequenced both of these interacting alleles from genomic DNA[21] and identified single amino acid substitutions in the transmembrane domains of each protein: In w^{co2} glycine-588 is changed to serine; in bw^{6} asparagine-638 is changed to threonine.

Taking into account the genetic evidence, it can be concluded that neither the serine nor the threonine substitution dramatically alters the global conformation of their respective subunits, nor does either mutation alone prevent White–Brown dimerization (since they can each be incorporated into an active transporter in complex with the complementary wild-type subunit). However, when the mutations are present together in the same transporter the combined effect of the two small changes prevents formation of an active complex. In terms of hydropathy-based prediction of the transmembrane helices, the residue affected by the w^{co2} mutation is near the C-terminal side of the fifth putative transmembrane helix of White and the bw^{6} mutation is near the N-terminal end of the sixth transmembrane helix of Brown. This places both mutations close to the external surface of the membrane. The most likely interpretation of the interacting nature of these two mutations is that the fifth transmembrane helix of White and the sixth transmembrane helix of Brown are in contact in the functional dimer. Since both mutations introduce hydroxylated residues, it is possible that the new hydrogen bonding potentials may be responsible for nonfunctional conformations in the interacting regions.

[28] J. L. Farmer, *Heredity* **39,** 297 (1977).

[29] J. L. Farmer and D. J. Fairbanks, *Drosophila Info. Service* **63,** 50 (1986).

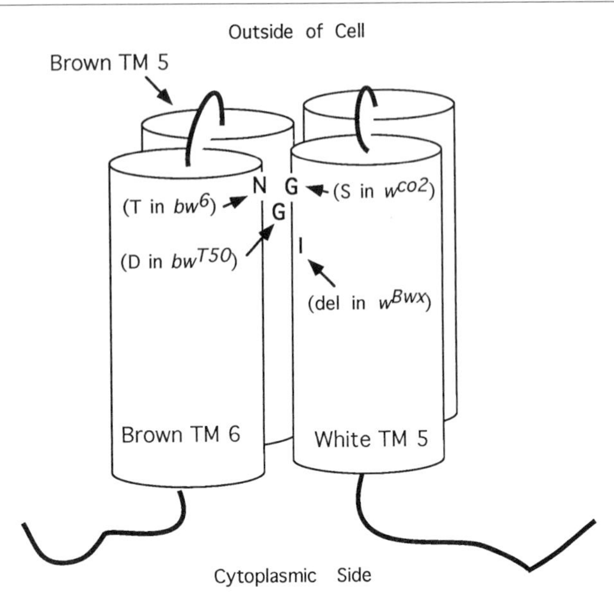

FIG. 4. Putative model of the interaction between TMH5 of White and TMHs 5 and 6 of Brown based on the mutant alleles described in the text.

Another w^{co2}-interacting *brown* allele (bw^{T50}) was identified and isolated as a result of genetic crosses between flies homozygous for w^{co2} and 36 wild-type laboratory strains of *D. melanogaster*. This gene also contained a single base substitution (G to A) compared to wild type, which resulted in the change of glycine-578 to aspartate. Interestingly, glycine-578 is predicted to be near the C-terminal side of the fifth transmembrane helix of Brown, which, like bw^6 and w^{co2}, places this residue near the external surface of the membrane. Using the same argument given earlier, this result also clearly points to a physical interaction between the fifth transmembrane helix of Brown and transmembrane helix five of White. Taken together, these three alleles allow prediction of the packing of the relevant helices, as depicted in Fig. 4.

The w^{Bwx} allele was chosen for characterization because it is another allele that produces a partial eye color phenotype suggestive that it might be due to a point mutation in the amino acid sequence of White. In addition, it is a semidominant allele such that w^{Bwx}/w^+ heterozygous flies also exhibit partial reduction in eye drosopterin levels,[30] whereas for the majority of mutant alleles, one copy of a fully functional gene is sufficient to produce

[30] D. L. Lindsay and E. H. Grell, "Genetic Variation in *Drosophila melanogaster*." Carnegie Institute, Washington, D.C., 1968.

wild-type eye color. Xanthommatin levels are unaffected by this allele, which indicates that the White subunit it encodes is synthesized and is capable of interacting with the Scarlet subunit to make a functional tryptophan transporter.

After PCR and sequencing, the mutation in w^{Bwx} was identified as the deletion of the triplet codon encoding isoleucine at residue 581. This residue is predicted to be in the middle of transmembrane helix five of White. Clearly, deletion of an amino acid from the middle of an α helix would result in alteration of the faces of that helix and change its interactions with the helices it is in contact with in the packed bundle of the native complex. In that regard it is noteworthy that Ile-581 is only two turns of a helix away from Gly-588 that is mutated in w^{co2}. Again this supports the model that there are specific interactions between transmembrane helix 5 of White and the Brown subunit, which are involved in the functional drosopterin precursor transporter.

Attempting to conceptualize the semidominant phenotype of w^{Bwx} raises the possibility that the native transporter may be a higher order multimer than a simple dimer: As mentioned, for most recessive *white* alleles the level of expression from a single functional gene in heterozygotes is sufficient to produce enough transport activity for wild-type pigmentation. If the transporter were active as a dimer, the formation of nonfunctional dimers from w^{Bwx}-encoded and bw^{+}-encoded subunits should not interfere with the activity of the wild-type dimers, which would be present at levels sufficient to allow accumulation of wild-type levels of drosopterins in the eyes. Therefore, the w^{Bwx}-encoded subunit must somehow interfere with the assembly of the wild-type subunit into functional transporters. One way for this to occur is if the presence of a w^{Bwx}-encoded subunit in a higher order multimer is sufficient to abolish activity. Then the residual activity resulting in partial pigmentation could be explained by the statistically determined small proportion of complexes that are present containing all wild-type subunits.

Implications for White–Scarlet Interactions

The observation that xanthommatin levels are unaffected by the w^{Bwx} allele indicates that the interaction between White and Scarlet subunits is of a different nature to those of the White and Brown subunits. Although the White subunit is involved in both transporters the deletion of Ile581 in the White subunit has no significant effect on the function of the xanthommatin precursor transporter, while it is clearly deleterious for the drosopterin precursor transporter. Indirect support for this idea also comes from the fact that no w^{co2} interacting alleles of *scarlet* have been reported even though the genetic screening experiments used would have easily detected such alleles.

Future Directions

A number of genetically characterized alleles of *white* have been reported that affect the level of one of the eye pigments while leaving the other pigment unaffected.[31] Such mutants are prime candidates for containing informative point mutations in the White subunit that would be useful in further defining the functional regions of these transporters. Therefore, cloning and sequencing of these alleles is a high priority. P-element transduction can be used to confirm the causative link between the eye color phenotype and the mutations detected in defective alleles.

Chemical mutagenesis to produce more mutant alleles of *white*, *brown*, and *scarlet* genes will also prove fruitful in identification of important amino acids.

The transmembrane-folding models of the three subunits predict a number of charged residues and prolines to be buried in the hydrophobic environment of the membrane. These should be targeted for site-directed mutagenesis and their functional significance assessed by P-element transduction. Data from such experiments will be used to refine the structural models and contribute to construction of a mechanistic model.

An important fundamental question concerning the mechanism of ABC transporters is how the energy of ATP hydrolysis is coupled to the pumping of substrates. A detailed model for this process has been proposed by Webb *et al.*[32] for the bacterial phosphate transporter. Their model implicates movement of a proline-containing kinked helical hairpin as the mechanical means by which a conformational change is transmitted across a membrane. In that regard, mutations to the prolines in the transmembrane helices of the subunits may be particularly informative.

Ultimately, direct structural information will be necessary to understand fully the structure–function relationship in these transporters. To that end, we have begun attempts to express the *Drosophila* proteins in bacteria and insect Sf9 (*Spodoptera frugiperda* fall armyworm ovary) cells to allow isolation of enough material for structural studies.

[31] Z. Zachar and P. M. Bingham, *Cell* **30,** 529 (1982).
[32] D. C. Webb, H. Rosenberg, and G. B. Cox, *J. Biol. Chem.* **267,** 24,661 (1992).

[16] Isolation of Altered-Function Mutants and Genetic Suppressor Elements of Multidrug Transporter P-Glycoprotein by Expression Selection from Retroviral Libraries

By IGOR B. RONINSON, DONALD ZUHN, ADAM RUTH, and DAVID DE GRAAF*

Introduction

Most of the known mutational changes in different proteins are either neutral or detrimental to the protein function. Phenotypic analysis of such mutations helps to identify the essential amino acid residues, but it does not indicate the specific functions of the mutated protein regions. Two relatively rare types of mutations, however, are apt to provide mechanistic information on the structure–functional organization of the proteins. The first type of mutations are those that do not abolish the protein function but alter it in a phenotypically identifiable and biochemically analyzable manner. The phenotypes of such mutations give a clear indication for the specific function of the altered residues. The second type are dominant negative mutations, which not only abolish the protein function but also allow the mutant protein to inhibit the activity of its normal counterpart. A dominant negative effect indicates that the mutant protein participates in some essential molecular interactions and competes with the normal protein for binding to some of its physiologic partner molecules. Dominant negative phenotype frequently results from truncation or mutational inactivation of one or more functional domains of a protein.[1] The smallest of the known dominant negative mutants are protein fragments of 25–100 amino acids, which are likely to represent a single protein domain capable of functional interactions. Functionally active cDNA fragments that encode either such peptides or short antisense RNA molecules inhibiting gene expression are termed *genetic suppressor elements* (GSEs).[2,3]

The multidrug transporter P-glycoprotein (Pgp), the product of the *MDR1* (multidrug resistance) gene, acts as an ATPase efflux pump for

* Present address: MIT/Whitehead Institute Center for Genome Research, Functional Genomics, Cambridge, Massachusetts 02139.

[1] I. Herskowitz, *Nature* **329,** 219 (1987).

[2] T. A. Holzmayer, D. G. Pestov, and I. B. Roninson, *Nucleic Acids Res.* **20,** 711 (1992).

[3] I. B. Roninson, A. V. Gudkov, T. A. Holzmayer, D. J. Kirschling, A. R. Kazarov, C. R. Zelnick, I. A. Mazo, S. Axenovich, and R. Thimmapaya, *Cancer Res.* **55,** 4023 (1995).

0076-6879/98 $25.00

various hydrophobic substrates, including different cytotoxic drugs.[4,5] Pgp studies have provided some of the most striking examples of altered-function mutants, which differ in the levels of resistance that they confer to different drugs. The importance of such mutants can be illustrated on the example of the first identified mutation of this type, the 185 Gly → Val substitution, which occurred in the *MDR1* gene of human KB cells selected for resistance to colchicine. This substitution increases the level of Pgp-mediated colchicine resistance by approximately three-fold, while decreasing the level of resistance to some other Pgp-transported drugs, such as vinblastine.[6] In accordance with this change in resistance, cells expressing the mutant Pgp accumulate less colchicine and more vinblastine than the cells expressing the same amount of the wild-type protein.[7] The mutant protein also changes the kinetics of Pgp-mediated drug transport and the sensitivity of Pgp to different inhibitors.[8,9] Biochemical analysis showed that the 185 Gly → Val substitution increases the ability of colchicine and decreases the ability of vinblastine to stimulate the ATPase activity of Pgp.[10] On the other hand, the mutation decreases Pgp binding of a photoactive analog of colchicine and increases the binding of a vinblastine analog.[7] These data not only indicate a region of Pgp that is involved in determining its substrate specificity, but also provide a link between the disassociation of bound substrates from Pgp and substrate stimulation of the Pgp ATPase activity.

Unfortunately, spontaneous mutations of Pgp are very rare among drug-selected Pgp-overexpressing cell lines,[11] with only one published example (apart from the 185 Gly → Val substitution) known to us.[12] A few mutations that arose as cloning artifacts or were generated through site-directed mutagenesis of known or suspected functional regions of Pgp were also found to change the relative levels of Pgp-mediated resistance to different

[4] I. B. Roninson (ed.), "Molecular and Cellular Biology of Multidrug Resistance in Tumor Cells." Plenum Press, New York, 1991.

[5] M. M. Gottesman and I. Pastan, *Annu. Rev. Biochem.* **62,** 385 (1993).

[6] K. Choi, C.-J. Chen, M. Kriegler, and I. B. Roninson, *Cell* **53,** 519 (1988).

[7] A. R. Safa, R. K. Stern, K. Choi, M. Agresti, I. Tamai, N. D. Mehta, and I. B. Roninson, *Proc. Natl. Acad. Sci. U.S.A.* **87,** 7225 (1990).

[8] W. D. Stein, C. Cardarelli, I. Pastan, and M. M. Gottesman, *Mol. Pharmacol* **45,** 763 (1994).

[9] C. O. Cardarelli, I. Aksentijevich, I. Pastan, and M. M. Gottesman, *Cancer Res.* **55,** 1086 (1995).

[10] M. Müller, E. Bakos, E. Welker, A. Varadi, U. A. Germann, M. M. Gottesman, B. S. Morse, I. B. Roninson, and B. Sarkadi, *J. Biol. Chem.* **271,** 1877 (1996).

[11] C.-J. Chen and I. B. Roninson, unpublished data (1989).

[12] S. E. Devine, V. Ling, and P. W. Melera, *Proc. Natl. Acad. Sci. U.S.A.* **89,** 4564 (1992).

drugs,[13–15] but the number of such mutations is very small. Our laboratory has therefore undertaken a search for additional altered-function Pgp mutants, through random mutagenesis of MDR1 cDNA cloned in a mammalian expression vector, followed by transduction of the mutagenized vector into Pgp-negative recipient cells and expression selection of mutants with the desired phenotype. We have also used the general strategy for the isolation of GSEs from a library of randomly fragmented cDNA[2,16] to generate dominant negative mutants of Pgp (no such mutants had been previously isolated). In contrast to altered-function Pgp mutants, which are selected for the desired function using Pgp-negative recipient cells, dominant negative mutants (including GSEs) are selected in Pgp-positive recipients using selection strategies that are based on the inhibition of the Pgp function.

A pivotal feature of our protocols for the isolation of altered-function mutants or GSEs in mammalian cells is the use of retroviral expression vectors to construct the libraries of mutagenized or fragmented cDNA of the target gene. The main advantage of retroviral infection over DNA transfection-based methods of gene transfer is a very high efficiency of stable transduction, with 50–100% transduction rates readily achievable for many recipient cell lines. On the other hand, recombinant retroviruses can be generated from plasmid vectors much more easily than other types of recombinant viruses used for gene transfer in mammalian cells (such as adeno-associated viruses or adenoviruses). Recombinant retroviruses are recovered as tissue culture supernatants after a single step of transient transfection into packaging cell lines that express the proteins of the retroviral particle. The ease of this procedure makes it possible to convert even very complex plasmid libraries into the corresponding retroviral populations without significant loss of sequence representation.[16,17] The general scheme of expression selection from retroviral libraries is illustrated in Fig. 1. This strategy includes the steps of library construction, transfection of packaging cells, infection of recipient cells, selection of the infected cells for the desired phenotype, recovery of integrated proviruses from the

[13] P. Gros, R. Dhir, J. Croop, and F. Talbot, *Proc. Natl. Acad. Sci. U.S.A.* **88,** 7289 (1991).

[14] T. Hoof, A. Demmer, M. R. Hadam, J. R. Riordan, and B. Tummler, *J. Biol. Chem.* **269,** 20,575 (1994).

[15] S. J. Currier, S. E. Kane, M. C. Willingham, C. O. Cardarelli, I. Pastan, and M. M. Gottesman, *J. Biol. Chem.* **267,** 25,153 (1992).

[16] A. V. Gudkov, C. Zelnick, A. R. Kazarov, R. Thimmapaya, D. P. Suttle, W. T. Beck, and I. B. Roninson, *Proc. Natl. Acad. Sci. U.S.A.* **90,** 3231 (1993).

[17] A. V. Gudkov, A. R. Kazarov, R. Thimmapaya, S. Axenovich, I. Mazo, and I. B. Roninson, *Proc. Natl. Acad. Sci. U.S.A.* **91,** 3744 (1994).

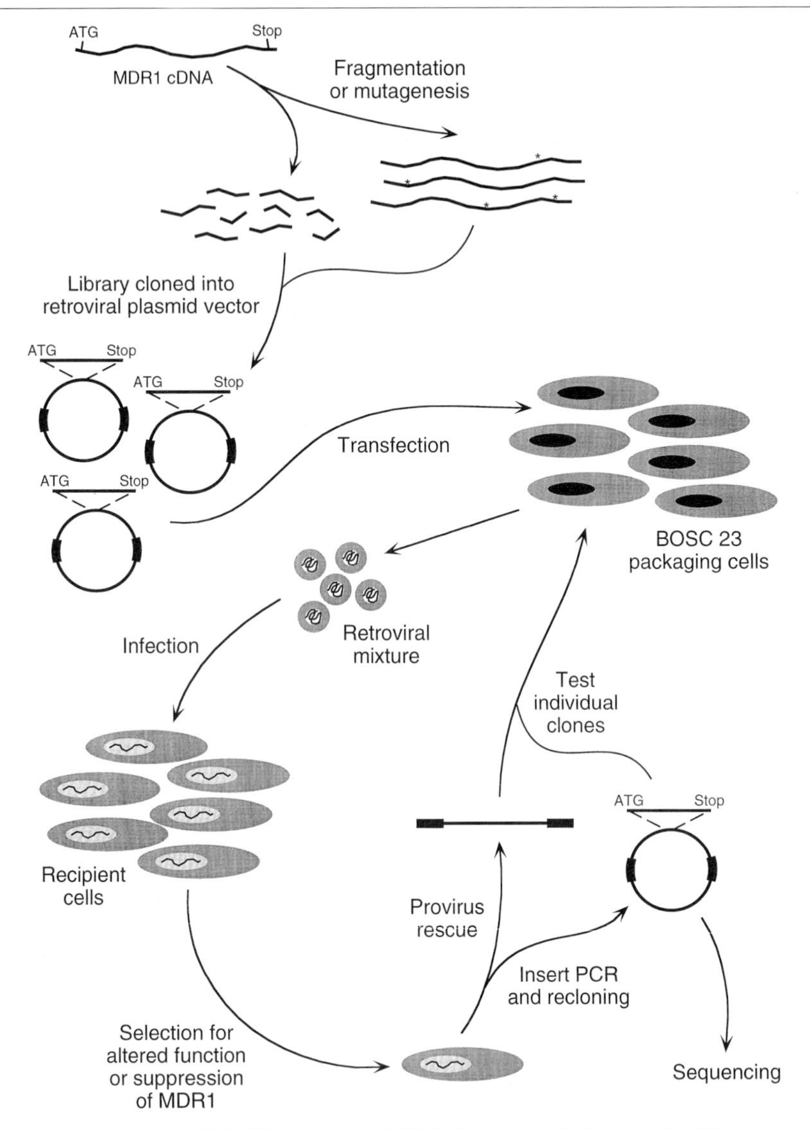

FIG. 1. Isolation of MDR1 mutants and GSEs from retroviral expression libraries.

selected cells, additional rounds of transduction and selection (if needed), and functional and sequence analysis of the retroviral clones enriched after selection. These methods, which are described here for the human *MDR1* gene, should be applicable (with appropriate modifications) to the analysis of other ABC transporters.

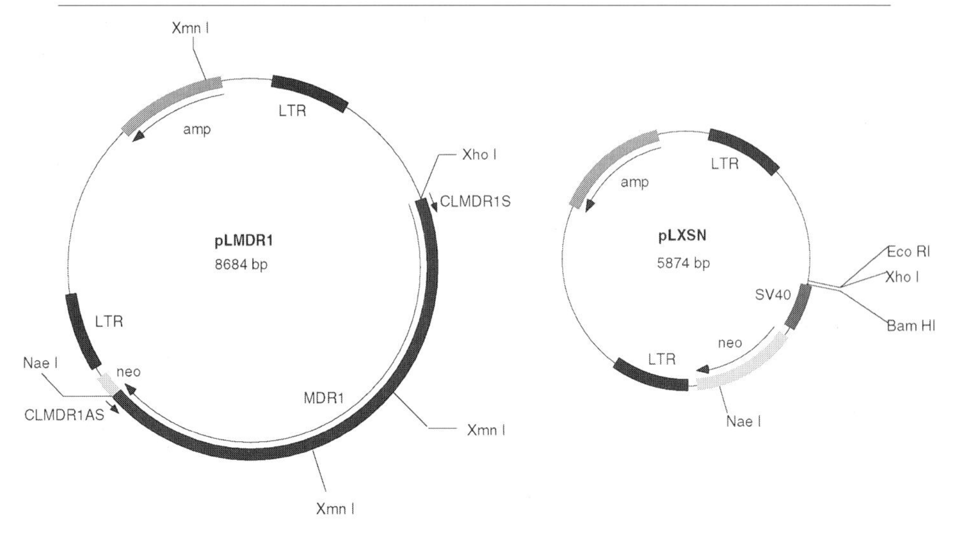

FIG. 2. Maps of pLMDR1 and pLXSN plasmids.

Isolation of Altered-Function P-Glycoprotein Mutants

We are using expression selection to isolate altered-function mutants of the human MDR1 Pgp. Examples of selectable mutant phenotypes include a higher level of resistance to specific Pgp-transported drugs relative to the wild-type Pgp, resistance to combinations of Pgp-transported drugs with specific Pgp inhibitors, altered reactivity with specific anti-Pgp antibodies, or altered uptake of Pgp-transported fluorescent dyes. The latter approach can also be used to isolate dominant negative mutants of Pgp, as described in the next section. Our mutation-selection strategy is based on the use of a retroviral vector pLMDR1[18] (Fig. 2), which was derived by inserting the coding sequence of the *MDR1* gene (without the 5′ or 3′ untranslated sequences) into the backbone of Moloney virus-based vector LXSN[19] (Fig. 2). A randomly mutated plasmid DNA preparation of pLMDR1 is generated by propagating the plasmid in a mutator strain of *Escherichia coli*. This strain, MutD5, is deficient in DNA polymerase proofreading activity and provides a high frequency of base substitutions.[20] The conditions of MutD5 growth have been selected to achieve the mutation fre-

[18] D. De Graaf, R. C. Sharma, E. B. Mechetner, R. T. Schimke, and I. B. Roninson, *Proc. Natl. Acad. Sci. U.S.A.* **93,** 1238 (1996).

[19] A. D. Miller and G. J. Rosman, *Biotechniques* **7,** 980 (1989).

[20] R. M. Schaaper, *Proc. Natl. Acad. Sci. U.S.A.* **85,** 8126 (1988).

quency of approximately 1/1000 bp. A library of mutant LMDR1 retroviruses is generated by transfecting the mutagenized plasmid preparation into a retroviral packaging cell line BOSC 23,[21] which produces a high titer of recombinant retroviruses upon transient transfection. The mutant retroviral library, contained in the supernatant of the transfected BOSC 23 cells, is used to infect recipient cells (e.g., mouse NIH 3T3 fibroblasts), which are then selected for drug resistance under the conditions that are too stringent to allow for the survival of cells infected with an unmutagenized pLMDR1 retrovirus (these conditions are determined from preliminary experiments). The *MDR1* cDNA sequence, contained in the integrated proviruses of cells that survive the selection, is recovered by polymerase chain reaction (PCR) and recloned into the same pLXSN vector. Individual retroviral clones are then tested for functional activity; if necessary, the mixture of the recovered clones is used for one or more additional rounds of transduction and selection. Once a mutant retroviral clone conferring the desired phenotype is identified, the specific mutations present in this clone are identified by sequence analysis. If more than one sense-altering mutation is found, individual mutations can be tested for functionality by site-directed mutagenesis of the wild-type retroviral plasmid vector.

Procedure 1: Preparation of Mutagenized Plasmid DNA

Solutions

Luria broth (LB)
SOC (2% Bacto-tryptone; 0.5% yeast extract; 10 m*M* NaCl; 2.5 m*M* KCl; 10 m*M* $MgCl_2$; 10 m*M* $MgSO_4$; 20 m*M* glucose)
LB agar
MacConkey agar with 1% w/v maltose
Ampicillin (Sigma, St. Louis, MO) 100 mg/ml stock

Equipment and Materials

Plasmid Maxi Kit (Qiagen, Valencia, CA)
Bacterial electroporation apparatus: Gene Pulser and Pulse Controller (Bio-Rad, Richmond, CA)

The *E. coli* MutD5 mutator strain was a gift of Dr. B. Wanner (Purdue University, Indiana). A similar mutator strain (*mutDmutS* genotype) is

[21] W. S. Pear, G. P. Nolan, M. L. Scott, and D. Baltimore, *Proc. Natl. Acad. Sci. U.S.A.* **90,** 8392 (1993).

commercially available from Stratagene (La Jolla, CA), but we have not evaluated that strain for its applicability to our mutation-selection strategy. As a surrogate marker for the plasmid mutation frequency, we use the mutation rate of bacterial genes involved in maltose metabolism. This rate is evaluated by growing bacteria on MacConkey media with maltose and scoring the incidence of mutant colonies, deficient in maltose metabolism. These colonies appear white, as opposed to the red wild-type colonies. In our experience, the desired mutation frequency in the pLMDR1 plasmid (approximately 1/1000 bp) corresponds to the bacterial mutation rate yielding 65–75% white colonies in this assay.

1. Streak *E. coli* MutD5 on MacConkey plates with 1% w/v maltose and incubate at 37° overnight. Pick 12–24 red colonies and grow them individually in 5 ml of LB overnight. Streak the cultures on MacConkey plates with 1% maltose and grow at 37° overnight to determine the mutation frequency. Select and use a culture yielding 20–30% white (including partly white) colonies.
2. The bacterial culture is made electroporation competent (other methods of transformation are inefficient with MutD5). Grow 1 liter of bacterial culture in LB to an A_{600} of approximately 0.5, chill cells on ice 30 min, centrifuge 15 min at 4° and 4000g, resuspend in 1 liter cold deionized water, centrifuge, resuspend in 0.5 liter cold deionized water, centrifuge, resuspend in 20 ml cold 10% glycerol, centrifuge, resuspend in 3 ml cold 10% glycerol, freeze in aliquots on dry ice, store at −70°.
3. Transform MutD5 with pLMDR1 plasmid DNA by electroporation according to manufacturer's protocol (Bio-Rad). Grow in 1 ml SOC for 2 hr at 37°, shaking at 225 rpm. Plate cells on ten 150- × 15-mm LB agar plates containing 50 μg/ml ampicillin; grow for 18 hr at 37°. As a control, transform the same plasmid into the wild-type (not hypermutable) INVαF′ strain of *E. coli* (Invitrogen, Carlsbad, CA).
4. Pool transformed colonies and grow in 1 liter LB (without ampicillin) for 6 hr. To test the mutation rate, plate small aliquots of both MutD5 and wild-type transformant cultures on three to five 150- × 15-mm MacConkey plates with 1% maltose. Grow at 37° overnight and determine the percentage of white colonies. The MutD5 should have approximately 65–75% white colonies, while the value for the wild-type transformants should be 0.05–0.1%.
5. Isolate plasmid from the bulk of the MutD5 transformant culture using the Plasmid Maxi kit according to manufacturer's instructions. Determine DNA concentration by spectrophotometry and check plasmid quality by electrophoresis in 0.8% agarose gel.

Procedure 2: Transfection of Packaging Cells and Infection of Recipient Cells

This procedure is modified from Pear *et al.*[21]

Solutions

Media: Dulbecco's modified Eagle's (DME) medium (Sigma) containing 10% FetalClone II (FCII) serum (HyClone) supplemented with L-glutamine (Mediatech, Herndon, VA) to a concentration of 2 m*M* and with penicillin–streptomycin (Mediatech) to a concentration of 100 U/ml penicillin and 100 μg/ml streptomycin
Phosphate-buffered saline (PBS; Sigma)
Chloroquine (Sigma) (25 m*M*): dissolve in deionized H_2O, sterilize by passing through 0.2-μm filter, and freeze in aliquots at $-20°$
Salmon sperm DNA (Sigma) (5 mg/ml): dissolve in deionized H_2O at a concentration of 10 mg/ml, shake 2–4 hr, adjust to 0.1 *M* NaCl, pass through 17-gauge needle 12 times, extract with phenol once and phenol : chloroform once, precipitate with two volumes ethanol, dissolve in deionized H_2O to 10 mg/ml, boil 10 min, adjust concentration to 5 mg/ml, store in aliquots at $-20°$
2 *M* sodium acetate, pH 4.5
2 *M* $CaCl_2$
2× HBS: 50 m*M* HEPES, pH 7.05, 10 m*M* KCl, 12 m*M* dextrose, 280 m*M* NaCl, 1.5 m*M* Na_2HPO_4; store in aliquots at $-20°$
4 mg/ml Polybrene (hexadimethrine bromide; Sigma): dissolve in deionized H_2O, sterilize by passing through 0.2-μm filter, and store in aliquots at $-20°$

1. Plate 2–3 × 10^6 BOSC 23 cells per 60- × 15-mm plate (25–30 plates total) in 4 ml media, 18–24 hr prior to transfection.
2. Precipitate 5 μg pLMDR1 plasmid DNA (mutagenized and unmutagenized control) with 15 μg of salmon sperm DNA carrier using 0.1 volumes sodium acetate and 2 volumes ice-cold ethanol. Pellet DNA, wash with 70% (v/v) ethanol, and dry the pellet in tissue culture hood.
3. Immediately prior to transfection, change media on BOSC 23 cells to 4 ml media containing 25 μM chloroquine.
4. Resuspend DNA pellet in deionized H_2O, add 62 μl 2 M $CaCl_2$, bring volume of DNA/$CaCl_2$ up to 500 μl with deionized H_2O.
5. Add 500 μl 2× HBS dropwise to DNA/$CaCl_2$ with bubbling to form precipitate. Within 1 to 2 min, add the precipitate dropwise to BOSC 23 cells. Gently agitate plates to mix and incubate at 37°.
6. After 6 hr, wash cells twice with PBS and renew with 4 ml media.

7. Plate 5 × 10^5 recipient cells (NIH 3T3) per 100- × 20-mm plate (25–30 plates total), 18–24 hr prior to infection.

8. Virus-containing media may be harvested from BOSC 23 cells at 24, 48, or 72 hr following transfection. Use 10 ml media from one plate of BOSC 23 cells to infect a single plate of recipient cells. Prior to infection, remove contaminating packaging cells from viral supernatant by passing through a 0.2-μm Spin X II filter (Costar, Acton, MA). Add polybrene to a final concentration of 4 μg/ml and place on recipient cells for 4–16 hr. To increase the transduction rate, cells may be repeatedly infected with the supernatant collected from BOSC 23 cells on two or three consecutive days.

9. Determine the transduction rate (see Procedure 3) 24–48 hr after infection to ensure that a total of at least 8 × 10^6 cells have been infected. Begin selection of recipient cells 48–72 hr following infection.

Procedure 3: Determination of MDR1 Transduction Rate by Indirect Immunofluorescence and Flow Cytometry

Solutions

PBS, PBS^- (Sigma)
EDTA (Sigma); 2 m*M* in PBS^-
Murine anti-human Pgp monoclonal antibody UIC2 (IgG_{2a}), (commercially available from Immunotech, Miami, FL); 20 μg/ml in PBS
Control murine monoclonal antibody UPC10 (mouse IgG_{2a}) (Sigma); 20 μg/ml in PBS
Secondary antibody: sheep anti-mouse IgG, *R*-phycoerythrin conjugate (Sigma); 20 μg/ml in PBS
Propidium iodide (Sigma); 1 mg/ml in deionized H_2O

Equipment

Fluorescence-activated cell sorter (FACS) (FACSort, Becton Dickinson, Franklin Lakes, NJ); the procedure may require some modifications for use with other FACS models.

The transduction rate of the LMDR1 retrovirus is estimated as the percentage of infected cells that stain with a monoclonal antibody UIC2 specific for the human MDR1 Pgp.[22] This percentage is determined by indirect immunofluorescence using FACS. An unrelated antibody of the same isotype (UPC10) is used as a negative control.

1. Remove cells from plates with 2 m*M* EDTA and wash once with PBS. Place 10^6 cells into each of two centrifuge tubes. Pellet cells by centrifugation and wash the pellet twice with PBS.

[22] E. B. Mechetner and I. B. Roninson, *Proc. Natl. Acad. Sci. U.S.A.* **89,** 5824 (1992).

2. Resuspend one pellet in 100 μl of UIC2 and the other in 100 μl UPC10 (20 μg/ml of each antibody in PBS). Incubate on ice for 45 min. Following incubation, add 5 ml PBS and spin down cells.

3. Resuspend each pellet in 100 μl of *R*-phycoerythrin-conjugated secondary antibody (20 μg/ml) and incubate on ice for 45 min. Following incubation, add 5 ml PBS and spin down cells.

4. Resuspend cells in 1 ml of DME media (without serum) containing 15 μg/ml of propidium iodide; transfer to 6-ml tubes (Falcon; needed for use with FACSort). Keep the tubes on ice until FACS analysis.

5. Carry out FACS analysis using the instrument manufacturer's protocol. Determine the percentage of UIC2-staining cells by comparison with the UPC10-treated sample. The expected transduction rates for NIH 3T3 cells are 90–100% (for the unmutagenized pLMDR1) or 50–75% (for the mutagenized vector).

Procedure 4: Drug Selection and Recovery of MDR1 cDNA Inserts

Solutions

Media: same as Procedure 2
PBS (Sigma)
Etoposide (Sigma); 1 mg/ml in dimethyl sulfoxide (DMSO)
Crystal violet (Sigma), 9% (w/v) in 10% methanol
*Xho*I and restriction buffer (New England Biolabs, Beverly, MA)
*Nae*I and restriction buffer (New England Biolabs)
Taq DNA polymerase and *Taq* DNA polymerase buffer (Promega, Madison, WI)
Taq Extender and *Taq* Extender buffer (Stratagene, La Jolla, CA)
25 m*M* $MgCl_2$
dNTP mix (Pharmacia, Piscataway, NJ): 2.5m*M* each of dATP, dGTP, dCTP, dTTP
Primers (Oligos Etc., Wilsonville, OR): CLMDR1s, AATTCGTTAACTCGAGATGGATCTTGAAGGGGACGCAATGGA; CLMDR1as
GGCCACTGGCGCTTTGTTCCAGCCTGGACACTGA; 1 mg/ml in deionized H_2O
E. coli INVαF′ transformation-competent cells (Invitrogen)

Equipment and Materials

QIAamp Tissue kit for DNA extraction (Qiagen)
QIAquick PCR purification kit (Qiagen)
QIAquick Gel Extraction kit (Qiagen)
Plasmid Maxi kit (Qiagen)

QIA Prep Spin plasmid miniprep kit (Qiagen)
Cetus Thermal Cycler (Perkin-Elmer, Norwalk, CT)
TKO 100 minifluorometer (Hoefer Scientific Instruments, San Francisco, CA): quantitation based on fluorescence resulting from the binding of bisbenzimidazole (Hoechst 33258)

The following procedure is used for the selection of mutants conferring an elevated level of etoposide resistance relative to the wild-type Pgp. For other types of selection, preliminary experiments with unmutagenized LMDR1 are carried out to determine the selective conditions where the wild-type *MDR1* sequence provides little or no resistance.

1. Plate 1.2×10^7 NIH 3T3 cells transduced with the mutagenized LMDR1 on twelve 150- × 25-mm dishes at 2.5×10^5 cells per plate. As a control, plate cells transduced with unmutagenized LMDR1 on two plates, under the same conditions.

2. The following day wash cells once with PBS and add media containing 2 μg/ml of etoposide. Change the media (with etoposide) every 3 days for 10 days. The control plates at the end of this period should contain no colonies (some cells would still be attached to the plate but they will not be growing), while multiple colonies should be growing on plates with cells transduced with the mutagenized virus (Fig. 3).

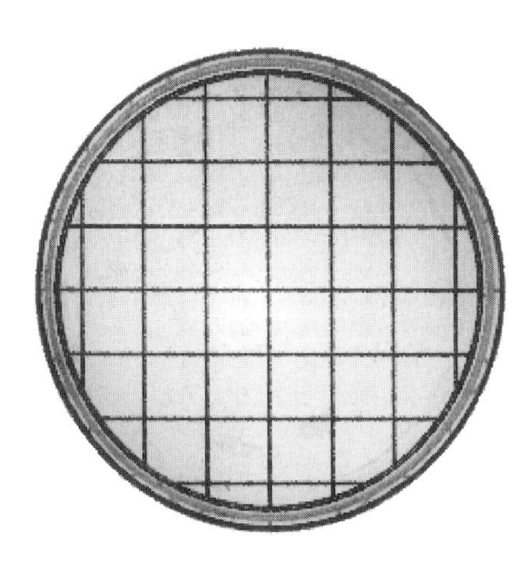

NIH 3T3 infected with unmutagenized
pLMDR1 (97% infection rate)

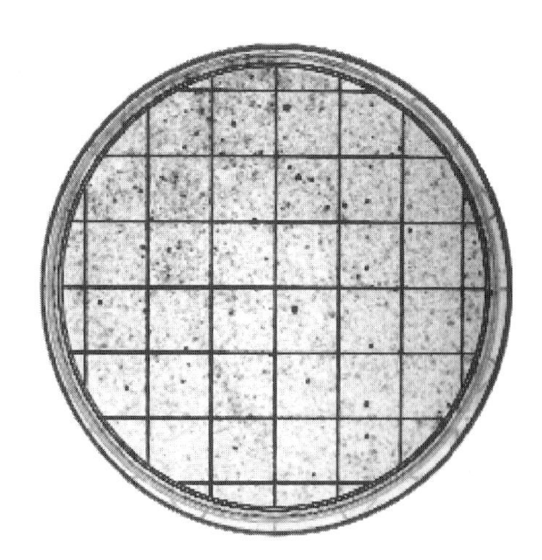

NIH 3T3 infected with mutagenized
pLMDR1 (67% infection rate)

FIG. 3. Survival of NIH 3T3 cells transduced with unmutagenized or mutagenized LMDR1 retrovirus in 2 μg/ml etoposide.

3. Pool the surviving colonies from the experimental plates. Expand this population for genomic DNA extraction or for additional steps of selection at higher drug concentrations (if desired).

4. For DNA isolation, collect 5×10^6 cells in 200 μl of PBS. Isolate genomic DNA using QIAamp Tissue kit.

5. Use PCR amplification (under the conditions for "long-range" PCR) to recover MDR1 cDNA inserts from the proviruses integrated in the selected cells. The CLMDR1s and CLMDR1as primers flank the MDR1 sequence. For recloning into pLXSN (Fig. 2), these primers contain the *Xho*I restriction site and a partial *Nae*I site (for blunt-end ligation with *Nae*I-digested vector). Reaction mixtures (50 μl volume) include 1 μg of genomic DNA, 2 units of *Taq* DNA polymerase, 2 units of *Taq* Extender, 5 units of *Taq* Extender buffer, 300 ng each of primers CLMDR1s and CLMDR1as and 4 μl of dNTP mix. PCR conditions are one cycle of 94° for 3 min, 65° for 1 min, 72° for 5 min, followed by 27 cycles of 94° for 1 min, 65° for 1 min, 72° for 2.5 min, and one final cycle of 94° for 1 min, 65° for 1 min, and 72° for 5 min.

6. Purify the PCR products using QIAquick PCR purification kit. Quantitate the product yield with TKO 100 Minifluorometer, using manufacturer's instructions. The size of the PCR products (3.85 kb) is verified by agarose gel electrophoresis.

7. For recloning, digest PCR-amplified MDR1 cDNA at 37° with 20 U/μg of *Xho*I for 1 hr in manufacturer-supplied restriction buffer. Purify the digested products using QIAquick PCR purification kit.

8. Digest retroviral plasmid vector pLXSN (a gift of A. D. Miller, Fred Hutchison Cancer Center) with 5 U/μg of *Xho*I for 1 hr and with 10 U/μg of *Nae*I for 16 hr. Separate the digested fragments of pLXSN by electrophoresis in a 0.8% agarose gel. Excise the larger (4.8-kb) fragment from the gel and purify using the QIAquick Gel Extraction kit. Determine DNA concentration by fluorometry.

9. Set up ligation in a 10 μl volume, using 200 ng total DNA per reaction with an insert to vector ratio of 1:1, in T4 DNA ligase buffer with 6 Weiss units of T4 DNA ligase. Incubate the ligation mixture for 16 hr at 16°.

10. Transform *E. coli* INVαF′ competent cells according to the manufacturer's protocol and plate on agar plates with 100 μg/ml ampicillin. Pick 8–12 individual bacterial colonies for each ligation mixture, grow them overnight in 5 ml of LB.

11. Identify plasmids containing MDR1 cDNA inserts by using 5 μl of each bacterial culture for a PCR reaction (no DNA extraction necessary) with any pair of MDR1-specific primers [we usually use the primers and PCR conditions developed for reverse transcriptase (RT)-PCR analysis of

MDR1 mRNA[23]]. The positive clones can be checked for the absence of gross rearrangements in MDR1 cDNA by plasmid minipreparation isolation (using the QIAPrep Spin Miniprep kit) and restriction digestion with *Xmn*I (Fig. 2).

Procedure 5: Identification of Altered-Function Mutants

Solutions

Media: see Procedure 2
PBS (Sigma)
Etoposide (Sigma); 1 mg/ml stock in DMSO
Methanol (Sigma)
Crystal violet (Sigma); 9% (w/vol) in 10% (vol/vol) methanol
MDR1-specific primers for DNA sequencing (as needed)

Equipment and Materials

Sequitherm Cycle Sequencing kit (Epicentre Technologies, Madison, WI)
DNA sequencing equipment

The reconstituted plasmid clones, generated in Procedure 4, may be individually tested for functional activity. If the selection was carried out under low-stringency conditions (where unmutagenized pLMDR1 produces substantial survival), it would be advantageous to repeat the selection under higher stringency conditions, using a population of clones isolated after the first round of selection. Individual clones will then be isolated and analyzed only after the second round of selection. The following assay procedure is used to identify clones conferring increased resistance to etoposide.

1. Using the reconstituted retroviral plasmid clones, carry out retroviral transduction of NIH 3T3 cells and FACS analysis, as described in Procedures 2 and 3. Ensure that the tested and the control (LMDR1) retroviruses produce similar infection rates.
2. Plate cell populations transduced with individual retroviruses at 200 cells per well in 10 six-well tissue culture plates (Falcon). The next day, wash cells once with PBS and add media containing etoposide in the following concentrations: 0, 0.5, 1, 2, 4, 6, 8, 10, 15, 20 μg/ml (three wells per each concentration). Carefully replace drug-containing media every 3 days (without disturbing the attached cells), for a total of 10 days (or until colonies are visible).

[23] K. E. Noonan, C. Beck, T. A. Holzmayer, J. E. Chin, J. S. Wunder, I. L. Andrulis, A. F. Gazdar, C. L. Willman, B. Griffith, D. D. Von Hoff, and I. B. Roninson, *Proc. Natl. Acad. Sci. U.S.A.* **87,** 7160 (1990).

3. At the end of incubation, stain plates with crystal violet and count the number of colonies in each well. Plot the number of colonies arising at each concentration of etoposide as a percentage of the number of colonies formed in the absence of drug. Identify retroviral clones that would produce increased resistance to etoposide relative to LMDR1.

4. Sequence MDR1 cDNA of the clones that confer elevated etoposide resistance, using standard dideoxy sequencing techniques. Use a series of MDR1-specific sequencing primers spanning the MDR1 cDNA sequence and spaced closely enough to obtain overlapping sequence readouts from the adjacent primers. Identify mutated bases and confirm by sequencing the opposite strand. Once mutations are identified, their functionality can be verified by introducing individual mutations *de novo* into the wild-type pLMDR1 plasmid, followed by functional testing.

Isolation of GSEs from MDR1 cDNA

While dominant negative mutants of Pgp (arising from point mutations) can in principle be isolated from the above-described mutagenized retroviral population, GSE selection from a retroviral library of randomly fragmented MDR1 cDNA allows for rapid delineation of the functional domains of the protein that act as dominant negative mutants when expressed in isolation. To obtain random fragments, we digest cloned cDNA with DNase I in the presence of Mn^{2+}; under these conditions, the enzyme preferentially generates fragments with blunt rather than staggered ends. The random fragments are then supplied with linkers that provide translation initiation and termination codons and cloned as a library ($3–10 \times 10^3$ recombinant clones per 1 kb of cDNA) in a retroviral plasmid vector. The library is transfected into packaging cells and the resulting retroviral population is used to infect a Pgp-expressing recipient cell line. The recipient cell line that we have utilized for this purpose is KB-8-5, a derivative of human KB-3-1 carcinoma cells, isolated by selection for approximately a four-fold increase in colchicine resistance.[24]

MDR1-derived GSEs (or other types of dominant negative mutants) are selected on the basis of their ability to decrease the Pgp function. Several different selection strategies may be used for this purpose, including decreased Pgp expression on the cell surface or selection for cell death or growth arrest in the presence of low doses of Pgp-transported drugs. The protocol with which we have the most experience, however, is based on the ability of Pgp to decrease intracellular accumulation of a fluorescent

[24] S. Akiyama, A. Fojo, J. A. Hanover, I. Pastan, and M. M. Gottesman, *Somat. Cell Mol. Genet.* **11,** 117 (1985).

vital dye rhodamine 123 (Rh123). Cells with decreased Pgp function show brighter staining with Rh123[25] and can therefore be isolated using a FACS. After FACS selection of Rh123-bright cells, the cDNA inserts from the integrated proviruses are recovered by PCR, recloned into a retroviral vector, and tested for Pgp inhibitory activity either immediately or after one or more additional rounds of transduction and selection.

Functional testing of GSEs for the ability to decrease Pgp function in KB-8-5 cells is not straightforward. As originally discovered by Hanchett *et al.*,[26] who analyzed Pgp inhibition in KB-8-5 cells transfected with a vector expressing an antisense RNA complementary to the 5′ portion of MDR1 cDNA, only a minority of transfected cells respond to the inhibitory effect of antisense RNA. The factors determining the heterogeneity of this response are as yet unknown. We have observed that such heterogeneous response occurs not only with antisense-oriented but also with sense-oriented (peptide-encoding) GSEs, introduced into KB-8-5 cells by retroviral infection. It is conceivable that some other Pgp-expressing cell lines may be more responsive to GSE inhibition than KB-8-5. Despite the relatively poor response to Pgp inhibition, an objective effect of GSEs in KB-8-5 cells can be demonstrated by a microscopic assay for colony fluorescence, designed by Hanchett *et al.*[26] Functional GSEs are identified by a statistically significant increase in the number of Rh123-bright colonies, relative to control cells infected with an insert-free retroviral vector. Sequence analysis of such GSEs is then used to determine whether they are sense- or antisense-oriented relative to the *MDR1* gene, and which segments of the protein are encoded by the sense-oriented GSEs.

Procedure 6: Construction of Random Fragment Library

This procedure is modified from Holzmayer *et al.*[2] and Gudkov *et al.*[16]

Solutions

10× DNase I buffer: 500 m*M* Tris-HCl, pH 7.5, 100 m*M* $MnCl_2$, 10 mg/ml bovine serum albumin (BSA); prepare fresh for each use

DNase I (Sigma): dissolve in 0.01 *N* HCl at a concentration of 1 mg/ml, store in aliquots at −70°

50 m*M* EDTA, pH 8.0

3000 U/ml T4 DNA polymerase (New England BioLabs)

5000 U/ml Klenow fragment (New England BioLabs)

5× T4 DNA polymerase buffer: 165 m*M* Tris–acetate, pH 8.0, 33 m*M*

[25] A. A. Neyfakh, *Exp. Cell Res.* **174,** 168 (1988).

[26] L. A. Hanchett, R. M. Baker, and B. J. Dolnick, *Somat. Cell Mol. Genet.* **20,** 463 (1995).

potassium acetate, 5 m*M* magnesium acetate, 2.5 m*M* dithiothreitol (DTT), 0.5 mg/ml BSA; sterilize by passing through a 0.2-μm filter and store at −20°

10× T4 DNA ligase buffer (New England BioLabs)

10× *Taq* DNA polymerase buffer (Promega); reactions are supplemented with $MgCl_2$ (Promega) to a final concentration of 15 m*M*

5,000 U/ml *Taq* DNA Polymerase (Promega)

dNTP mix (Pharmacia): 2.5 m*M* each of dATP, dGTP, dCTP, dTTP

Primers (Oligos, Etc.); SRH: TACCGAATTCAAGCTTATGGATGGATG; ASRH: CATCCATCCATAAGCTTGAATTC; SCB: TGAGTGAGTGAATCGATGGATCC; ASCB: TATAGGATCCATCGATTCACTCACTCA; dissolve in deionized H_2O at 1 mg/ml

Linkers: mix 50 μg of appropriate primers (SRH and ASRH; SCB and ASCB), heat 5 min at 80°, slowly cool to room temperature in a water bath to anneal

10 mg/ml BSA (New England BioLabs)

Electroporation-competent *E. coli* HB101 or InvαF′: prepare as described in Procedure 1, step 2; store at −70°

Equipment and Materials

QIAquick PCR purification kit (Qiagen)

QIAquick Gel Extraction kit (Qiagen)

Plasmid Maxi kit (Qiagen)

TKO 100 Minifluorometer (Hoefer Scientific Instruments)

Bacterial electroporation apparatus: Gene Pulser and Pulser Controller (Bio-Rad)

1. Isolate *MDR1* cDNA from a plasmid vector (such as pLMDR1, Fig. 2) by digestion with appropriate restriction endonucleases (e.g., *Xho*I and *Nae*I). Recover *MDR1* cDNA insert by agarose gel electrophoresis followed by purification using QIAquick Gel Extraction kit.

2. Digest 500 ng of target cDNA in 1× DNase I buffer (30 μl reaction volume) with several different concentrations of DNase I (0.5–5 ng per reaction recommended). Digest 10 min at 16°.

3. Add 3 μl 50 m*M* EDTA to stop further DNase I digestion. Remove DNase I using QIAquick PCR purification kit.

4. Analyze one-fourth of purified cDNA digests by agarose gel electrophoresis to identify a digest with fragment sizes ranging from approximately 100 to 700 bp (using 123-bp ladder [GIBCO-BRL, Gaithersburg, MD] as electrophoretic markers).

5. To blunt fragment ends of the remainder of cDNA digest, add 10 units T4 DNA polymerase and 2 μl dNTP mix to cDNA digest in 1× T4 DNA polymerase buffer; total reaction volume of 50 μl. Incubate 30 min at 37°. Add 10 units Klenow fragment and incubate 15 min at 37°. Purify PCR products using QIAquick PCR purification kit.

6. Ligate 100 ng of cDNA fragments with 1 μg each of the SRH–ASRH linker (containing translation initiation codons in three reading frames) and the SCB-ASCB linker (containing translation termination codons in three reading frames), in 1× T4 DNA ligase buffer supplemented with 2.5 μl BSA and 12 Weiss units T4 DNA ligase; total reaction volume of 25 μl. A control mixture is prepared without ligase. Incubate ligation mixtures overnight at room temperature.

7. Analyze ligation efficiency by PCR using linker primers SRH and ASCB. Reactions are assembled in 1× PCR buffer with 2 μl dNTP mix, 1 μl of ligation mixtures, 250 ng of each primer, and 1 unit *Taq* DNA polymerase in a total volume of 25 μl. Reactions are performed under the following conditions: 35 cycles of 94° for 30 sec, 55° for 45 sec, and 72° for 1.5 min. Only ligase-containing mixtures should generate smears of appropriately sized fragments.

8. Size-select PCR-amplified ligation products (100–700 bp or some other chosen size range) by agarose gel electrophoresis and purify using QIAquick gel extraction kit.

9. Prepare several different dilutions of the size-selected fragments and analyze by PCR amplification as described earlier. Select a dilution giving the best yield and distribution of PCR products and perform large-scale amplification of fragments (10- to 50-μl reactions).

10. Following purification with QIAquick PCR purification kit, quantitate the yield of the PCR-amplified fragments by fluorometry.

11. Prepare fragments for cloning into the chosen retroviral vector, such as LXSN (Fig. 2) or LNCX[19], by digesting with appropriate restriction endonucleases. The linkers contain *Eco*RI and *Hin*dIII sites (SRH–ASRH linker) or *Bam*HI and *Cla*I sites (SCB–ASCB linker). Digest fragments for 1 hr with 20 U/μg of the corresponding restriction endonuclease (*Eco*RI and *Bam*HI for cloning into LXSN) in the appropriate buffer and under the conditions specified by the manufacturer. Digest vector with 5 U/μg of each restriction endonuclease.

12. Separate fragments and vector from undesired digestion products by agarose gel electrophoresis and purify with QIAquick Gel Extraction kit. Quantify vector and fragments by fluorometry.

13. Ligate cDNA fragments to vector. Several different insert : vector ratios (e.g., 1:3, 1:5 and 1:10 molar ratios) should be used with 200 ng

total DNA per reaction. Reactions are performed in 1× T4 DNA ligase buffer with 1 μl BSA and 6 Weiss units T4 DNA ligase; total reaction volume is 10 μl. Incubate at 16° for 16–18 hr.

14. Transform *E. coli* by electroporation using the protocol supplied by the electroporation device manufacturer and plate cells on several 150- × 15-mm agar plates with 100 μg/ml ampicillin.

15. Select at least 20 individual colonies to determine the frequency of recombinants by PCR analysis using the linker primers SRH and ASCB. Perform PCR analysis using 5 μl of overnight cultures prepared from selected colonies.

16. Pool remaining colonies and isolate plasmid DNA from the random fragment library using Plasmid Maxi kit.

Procedure 7: FACS Selection for Decreased Rhodamine 123 Efflux

This procedure is modified from Chaudhary and Roninson.[27]

Solutions

Media: see Procedure 2
PBS, PBS^- (Sigma)
2 m*M* EDTA in PBS^-
1 mg/ml rhodamine 123 (Rh123; Sigma): dissolve in 70% methanol, sterilize by passing through 0.2-μm filter, and store at 4° in light-protected containers
1 mg/ml propidium iodide (PI; Sigma): dissolve in deionized H_2O, sterilize by passing through 0.2-μm filter and freeze in aliquots at −20°

Equipment and Materials

Fluorescence activated cell sorter (FACS; Coulter Epics Coherent); some steps in the procedure may need to be modified depending on the FACS model to be used

1. Carry out retroviral transduction of KB-8-5 cells with the random fragment library, as described in Procedure 2. To use BOSC 23 ecotropic packaging cells, KB-8-5 cells should be made infectable with ecotropic retrovirus by stable transfection with the plasmid pJET[28] (a gift of J. M. Cunningham, Brigham and Women's Hospital, Boston, MA), expressing the murine retroviral receptor. Otherwise, amphotropic packaging cells

[27] P. M. Chaudhary and I. B. Roninson, *Cell* **66,** 85 (1991).
[28] L. M. Albritton, L. Tseng, D. Scadden, and J. M. Cunningham, *Cell* **57,** 659 (1989).

(such as BING[29]) should be used. Infected cells are selected with 400 μg/ml G418 (GIBCO-BRL).

2. Remove cells from plates with 2 m*M* EDTA and wash once with media.

3. Pellet 1×10^7 cells and resuspend in 25 ml ice-cold media containing 0.5 μg/ml Rh123. Media used for this and all subsequent steps should be passed through a 0.45-μm cellulose acetate filter (Nalgen, Rochester, NY) to remove particulate matter that may obstruct FACS machine lines during sort. Place cells on ice for 15 min.

4. Pellet and wash with 25 ml ice-cold media.

5. Pellet and resuspend in 25 ml media. Incubate cells at 37° for 45–120 min to allow for Rh123 efflux.

6. Sterilize six scintillation vials by washing twice with PBS and twice with ethanol; allow to dry in tissue culture hood and add 25 μl of PI. Place vials on ice.

7. Condition 4–12 well plates with 1 ml DME–20% FCII per well at 37° for 2–4 hr.

8. Prepare 4–15 ml recovery tubes containing 14 ml DME–50% FCII and place on ice.

9. Pellet cells following efflux (step 5) and wash with 10 ml ice-cold media.

10. Pellet cells and resuspend in 3.5 ml ice-cold media. Transfer 0.5-ml aliquots to vials (step 6), mix, and place on ice.

11. Carry out FACS analysis and sorting using the instrument manufacturer's protocol. Pgp-inhibited cells are characterized by brighter Rh123 staining (Fig. 4). Isolate the top 5% Rh123-brightest cells.

12. Collect sorted cells in recovery tubes (step 8), pellet, and resuspend in 12 ml DME–20% FCII. Transfer 0.5-ml aliquots to conditioned 12 well plates (step 7).

13. To ensure that samples were not contaminated during the sorting process, allow cells to grow for at least 4–6 days before renewing with fresh media. Expand cells and repeat the Rh123 efflux assay, to ensure that the sorted cells accumulate more Rh123 than the original population (Fig. 4).

Procedure 8: Recovery of cDNA Fragments

Solutions

dNTP mix (Pharmacia): 2.5 m*M* each of dATP, dGTP, dCTP, dTTP
Primers (oligonucleotides, etc.); SRH: TACCGAATTCAAGCTTAT-

[29] W. S. Pear, M. L. Scott, and G. P. Nolan, In: P. Robbins (ed.), *Gene Therapy Protocols*, Humana Press, Totowa, NJ, p. 41 (1996).

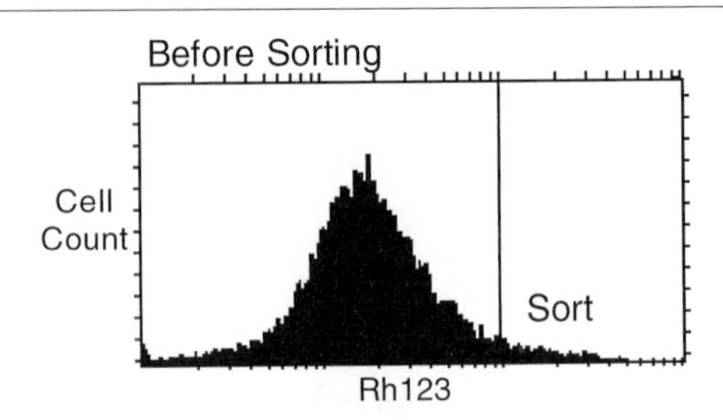

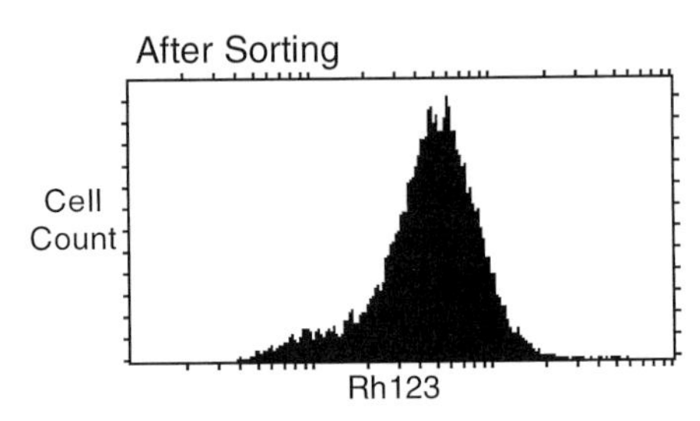

FIG. 4. Flow cytometric profiles of Rh123 efflux in KB-8-5 cells transduced with MDR1 GSE library, before and after sorting. Rh123 fluorescence (*X* axis) is plotted as channel numbers on logarithmic scale.

GGATGGATG; ASCB: TATAGGATCCATCGATTCACTCAC-TCA; dissolve in deionized H_2O at 1 mg/ml

10× *Taq* DNA polymerase buffer (Promega); reactions are supplemented with $MgCl_2$ to a final concentration of 15 m*M*

5000 U/ml *Taq* DNA polymerase (Promega)

Elution buffer: 0.5 *M* ammonium acetate, 10 m*M* magnesium acetate, 1 m*M* EDTA, pH 8.0, 0.1% SDS

1 *M* TE: 10 m*M* Tris-HCl, pH 7.5, 1 m*M* EDTA

2 *M* sodium acetate, pH 4.5

Equipment and Materials

QIAamp Tissue kit (Qiagen)

QIAquick PCR Purification kit (Qiagen)

TKO 100 Minifluorometer (Hoefer Scientific Instruments)

Bacterial electroporation apparatus: Gene Pulser and Pulse Controller (Bio-Rad)

1. Isolate genomic DNA from cells using QIAamp Tissue kit.
2. Recover inserts for cloning by PCR amplification using linker primers SRH and ASCB. Reactions consist of 1 μg genomic DNA, 1 unit *Taq*

DNA polymerase, 2 μl dNTP mix, and 250 ng of each primer in 1× *Taq* DNA polymerase buffer with a reaction volume of 25 μl. Reactions are performed under the following conditions: 1 cycle of 94° for 2 min, 55° for 45 sec, and 72° for 5 min; 30 cycles of 94° for 30 sec, 55° for 45 sec, and 72° for 1.5 min; 1 cycle of 94° for 30 sec, 55° for 45 sec, and 72° for 5 min.

3. PCR products are purifed using QIAquick columns and quantitated by fluorometry.

4. Digest PCR products for 1 hr with 20 U/μg of appropriate restriction endonucleases (*Eco*RI and *Bam*HI for LXSN) under the conditions specified by the manufacturer. Digest vector with 5 U/μg of each restriction endonuclease.

5. Separate digested PCR reactions by polyacrylamide gel electrophoresis. Excise desired inserts for recloning.

6. DNA is purified from the gel by the procedure described in Sambrook *et al.*[30] Transfer gel slice to a microfuge tube and crush slice with a pipette tip. Add 500 μl of elution buffer and incubate 3–4 hr at 37° on a rotating wheel.

7. Centrifuge sample 1 min at 12,000*g* and 4° in a microfuge and transfer supernatant to a fresh microfuge tube. Add an additional 250 μl of elution buffer to polyacrylamide pellet, vortex briefly, and centrifuge. Combine recovered supernatants.

8. Remove remaining fragments of polyacrylamide from pooled supernatant by passing through a Millex GV 0.22-μm syringe filter (Millipore, Bedford, MA). Add 2 volumes ethanol and place on ice for 30 min.

9. Recover DNA by centrifugation at 12,000*g* for 10 min at 4° in a microfuge.

10. Dissolve pellet in 200 μl TE and precipitate again using 0.1 volume sodium acetate and 2 volumes ethanol. Centrifuge and wash with 70% ethanol. Centrifuge briefly and resuspend in 25 μl TE.

11. Quantify inserts by fluorometry and ligate to vector using 200 ng total DNA per reaction with an insert : vector ratio of 1 : 1 or 1 : 3. Reactions are performed in 1× T4 DNA ligase buffer with 1 μl BSA and 6 Weiss units T4 DNA ligase with a total reaction volume of 10 μl. Incubate at 16° for 16–18 hr.

12. Transform *E. coli* by bacterial electroporation and plate on selective media.

13. Analyze recloned inserts from individual bacterial clones by PCR using the linker primers SRH and ASCB. Perform PCR analysis using 5 μl from overnight cultures prepared from selected colonies. In addition, PCR

[30] J. Sambrook, E. F. Fritsch, and T. Maniatis, *in* "Molecular Cloning. A Laboratory Manual," pp. 6.46–6.48. Cold Spring Harbor Laboratory Press, Cold Spring Harbor, New York, 1989.

products may be further analyzed by digesting products with frequently cutting restriction endonucleases (i.e., *Alu*I) to generate distinct restriction fragment profiles, followed by polyacrylamide gel electrophoresis.

Procedure 9: Testing of Selected Clones by Colony Assay for Rhodamine-123 Efflux

This procedure is modified from Hanchett *et al.*[26]

Solutions

Media: see Procedure 2
PBS (Sigma)
1 mg/ml Rh123: see Procedure 7
FluroSave (Calbiochem)

Equipment and Materials

Fluorescent microscope (Leitz Wetzlar); some steps may need to be modified to conform to the specific microscope

1. Sterilize 22- × 22-mm coverslips with 70% ethanol and dry in tissue culture hood. Place coverslips in 60- × 15-mm plates and cover with 4 ml media.
2. Add 1.5×10^4 cells and grow 6–7 days until colonies of at least 200 cells in size have formed.
3. Rinse colonies with 3 ml PBS and add 3 ml media containing 5 μg/ml Rh123. Place colonies in dark at room temperature for 30 min.
4. Rinse colonies with 3 ml 37° PBS and add 4 ml dye-free media. Incubate at 37° for 60 min.
5. Rinse colonies with 3 ml ice-cold PBS. Place coverslips on a drop of FluroSave reagent on slides.
6. Analyze colonies by fluorescent microscopy to determine the colony phenotype. Four colony phenotypes can be observed: uniform bright, rhodamine-dull, heterogeneous, and sectored (Fig. 5). In uniform bright colonies, all cells in the colony retain high levels of Rh123. For rhodamine-dull colonies, none of the cells retain a detectable amount of Rh123. Most of the cells in a heterogeneous colony will retain some detectable level of Rh123; however, individual cells will display different levels of Rh123 retention. Sectored colonies contain contiguous sections that retain high levels of Rh123 and sections that do not. Score the phenotypes of at least 100 colonies for each sample; use χ^2 test to evaluate statistically significant differences between controls and tested cell populations. For vector controls, between 0 and 2.5% of the colonies should be bright or sectored,

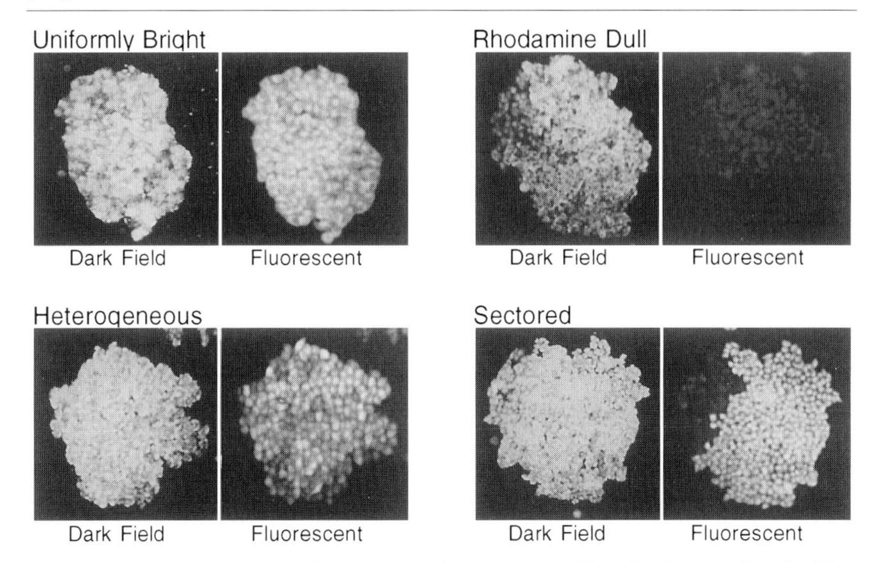

FIG. 5. Different types of Rh123 staining in colonies of KB-8-5 cells transduced with a MDR1-derived GSE.

while GSE-transduced samples should yield between 4 and 12% bright or sectored colonies.

Concluding Remarks

The above-described expression–selection procedures can be used to identify the amino acid residues and protein domains responsible for different intra- and intermolecular interactions of the multidrug transporter, and for various other purposes, such as developing drug-tailored versions of the *MDR1* gene for gene therapy applications, analyzing the mechanism of Pgp interaction with different resistance-reversing agents, isolating second-site revertants of Pgp mutants, or identifying the epitopes of different anti-Pgp antibodies. The same experimental strategies should also be applicable to other mammalian ABC transporters, but the design of the cellular selection protocols would depend on the properties of the specific transporter. Some of the more general selection strategies may involve FACS selection, either on the basis of altered immunoreactivity or transporter-dependent cellular staining with fluorescent dyes. The multifunctional nature of this class of enzymes and their complex patterns of molecular interactions virtually guarantee the finding of interesting and unexpected mutant phenotypes, analysis of which should prove intellectually and scientifically satisfying.

Acknowledgments

The procedures described in this chapter were developed with pivotal contributions from several past and present members of our laboratory, including T. A. Holzmayer, A. V. Gudkov, E. B. Mechetner, P. M. Chaudhary, B.-D. Chang, K. E. Noonan, B. Schott, K. Choi, and B. S. Morse. We are grateful to Drs. A. D. Miller, W. Pear, D. Baltimore, M. M. Gottesman, J. M. Cunningham, and B. Wanner for various clones and cell lines used in this study. This work was supported by grants R37 CA40333, R01 CA56736, and R01 CA62099 from the National Cancer Institute.

[17] Selection and Maintenance of Multidrug-Resistant Cells

By MICHAEL M. GOTTESMAN, CAROL CARDARELLI, SARAH GOLDENBERG, THOMAS LICHT, and IRA PASTAN

Introduction

The identification and cloning of two of the best studied ATP-binding cassette (ABC) transporters, P-glycoprotein (the product of the *MDR1* gene, a direct transporter of multiple hydrophobic cationic or neutral drugs) and MRP (the product of the *MRP* gene, a transporter of anionic hydrophobic drugs and drugs conjugated or cotransported with anionic moieties such as glutathione and sulfate), depended on the selection of cell lines in tissue culture that were multidrug resistant. This article details means for selection and maintenance of such multidrug-resistant cells, including specialized techniques such as transfection and transduction of multidrug resistance genes and introduction of multidrug resistance genes into bone marrow to form drug-resistant bone marrow colony-forming units (cfus).

General Principles

Multidrug-resistant cells can be selected as individual clones or as mass populations. In the former approach, a cell line is carefully cloned so that all the cells in the population derive from a single cell, and these cloned cells are exposed to a single cytotoxic agent (or combination of agents) in doses that kill the great majority of the cells in the population, but not all. This approach results in single-step resistance to the agent, and individual surviving clones may be screened for cross-resistance to other drugs. If higher level resistance is desired (for example when it might be necessary

0076-6879/98 $25.00

to clone a gene that is amplified in a multidrug-resistant cell or to detect differentially expressed mRNA), the cloned resistant cell population can be expanded and reselected under the original selection conditions, but at higher stringency (i.e., higher drug concentrations). Surviving clones from this second step can be screened for the pattern of multidrug resistance found in the original resistant clone. This approach was used to isolate the original KB cell lines from which the human *MDR1* gene was cloned based on amplification of the *MDR1* gene.[1] The advantage of this general strategy is that it is much more likely to yield multidrug-resistant populations in which a single gene is responsible for the drug resistance. Similar strategies using mutagenesis at the first step, either by chemical agents or insertions of activating or inactivating retrotransposons, can also be devised so that only a single gene will be activated (or inactivated), resulting in a multidrug resistance phenotype.

A second approach is to expose an uncloned cell population to a selective drug in gradually increasing doses, allowing the population to die back and reexpand after each increase in the stringency of selection. This approach may result in highly multidrug-resistant populations in which the mechanisms of multidrug resistance may be relatively unique, particularly if a potent mechanism was responsible for resistance at an early stage of selection and the derivatives of the original resistant cell have overgrown the population. But more likely it will result in a heterogeneous population of resistant cells in which each cell expresses multiple, different mechanisms of resistance. This kind of resistance has been called *multifactorial multidrug resistance* and is thought by some workers to more closely represent the situation in human tumors exposed over long periods of time *in vivo* to multiple rounds of cytotoxic drug treatment. Although individual multidrug-resistant cells can be cloned from such populations, even single clones are likely to have multiple mechanisms for their drug resistance acquired during the multiple steps of selection used to obtain the resistant populations. The advantage of this approach is that it casts a wide net for many different possible kinds of drug resistance. The disadvantage is that it is difficult to clone drug resistance genes from such mixed populations, since any one drug resistance gene may be expressed at relatively low levels, and it is very hard to sort out which mechanisms are significant contributors to the overall multidrug resistance of the population. The MCF-7 doxorubicin resistant human breast cancer cell populations that have been studied in many laboratories are an example of multidrug-resistant cells selected as

[1] S.-I. Akiyama, A. Fojo, J. A. Hanover, I. Pastan, and M. M. Gottesman, *Somat. Cell Mol. Genet.* **11,** 117 (1985).

mass populations in which many different mechanisms of resistance are expressed.[2]

Another important consideration is the choice of the starting cell line used for selection. Some drug resistance genes may not be expressed in certain cell types; other cell types are already quite drug resistant and selection may yield a set of resistance genes different from those selected in drug-sensitive cells. We began our studies on the human *MDR1* gene by choosing a cell line (KB) that was very drug sensitive (and, hence, as it turned out, expressed the *MDR1* gene only at very low levels), cloned with high efficiency (allowing the rescue of a high percentage of drug-resistant clones), and which, under conditions of appropriate stringency of selection, yielded reasonable numbers of multidrug-resistant clones.[1] By studying tissues and cancers in which the *MDR1* gene was expressed,[3] it became apparent that expression of the *MDR1* gene in lung cancer cell lines, many of which were quite multidrug resistant, was rare. Cole and Deeley and their collaborators used a lung cancer cell line to select for high-level expression of the *MRP* gene, allowing the cloning of this gene because expression of *MDR1* in this cell system did not occur.[4] This strategy illustrates the importance of choosing a starting cell line with care.

Selection of Multidrug-Resistant Cells from Cloned Populations

1. Once a cell line has been chosen, the cell line is cloned by plating 100 cells on a 100-mm tissue culture dish and by picking one or more resulting clones. We do this by simply removing the medium from the dish, circling the clones with a magic marker on the bottom of the dish, and gently scraping the cells from a single clone with a 1-ml plastic pipette, aspirating the cells with a small amount of medium, and introducing them into the wells of a 24-well dish. Cells that are sensitive to mechanical shear can be removed from the dishes with a cloning ring into which trypsin solution has been added. Once the population of cloned cells has expanded, the cloning efficiency of the cell line is determined. This can be accomplished by plating a predetermined number of cells (usually 200–500 on a 100-mm tissue culture dish), allowing the cells to grow into clones, and staining with 0.5% methylene blue in 50% ethanol for 30 min. The blue colonies can be

[2] C. R. Fairchild, S. P. Ivy, C. S. Kao-Shan, J. Whang-Peng, N. Rosen, M. A. Israel, P. W. Melera, K. H. Lowan, and M. E. Goldsmith, *Cancer Res.* **47,** 5141 (1987).

[3] L. J. Goldstein, H. Galski, A. Fojo, M. Willingham, S-L. Lai, A. Gazdar, R. Pirker, A. Green, W. Crist, G. Brodeur, C. Grant, M. Lieber, J. Cossman, M. M. Gottesman, and I. Pastan, *J. Natl. Cancer Inst.* **81,** 116 (1989).

[4] S. P. C. Cole, G. Bhardwaj, J. H. Gerlach, J. E. Mackie, C. E. Grant, and R. Deeley, *Science* **258,** 1650 (1992).

easily counted after the dishes are rinsed to remove the stain, and the colony-forming efficiency calculated. If the colony-forming efficiency is less than 10%, the cloned population may not be satisfactory for selection of multidrug-resistant clones.

2. Some attention should be paid to the medium used for growth and selection of drug-resistant clones. A commercial medium containing all appropriate nutrients, as well as sodium pyruvate (0.11 mg/ml), should be used to cultivate the cell line. This medium should be supplemented with 5 m*M* glutamine (GIBCO, Grand Island, NY), 50 units/ml penicillin (GIBCO), 50 μg/ml streptomycin (GIBCO), and carefully chosen serum. We routinely screen large lots of serum by determining cloning efficiency in 2, 5, and 10% serum. (For toxic sera, cloning efficiency will drop with increasing serum concentration; for nontoxic sera, higher concentrations of serum usually result in higher cloning efficiency.) A toxic serum may itself serve as a means of selecting for resistant cell populations, thereby confusing the selection process and confounding the mechanism of drug resistance. A serum that has inadequate growth stimulatory properties may result in selection of clones that grow better in the inadequate serum and, hence, are not really drug resistant in selective medium, but only grow faster. This can result in a kind of pseudo-multidrug resistance, especially in early steps of selection.

3. To determine the appropriate concentration of a drug to use for selection, it is necessary to screen the clone of interest for its cloning efficiency in varying concentrations of the drug. We do this by generating killing curves essentially by measuring the cloning efficiency of 300 cells in 60-mm tissue culture dishes containing different amounts of cytotoxic drug to which the clones are continually exposed. Depending on the drug and the sensitivity of the cell line, cells are plated without drug and in increasing concentrations of drug (usually beginning at 1 ng/ml) with approximately three-fold increments in drug concentration. Ten concentration points will cover four logs of drug concentration (1, 3, 10 ng/ml, etc., to 10 μg/ml), and the concentration increments conveniently give equal spacing on the x axis when the result are plotted on semilog paper. A good starting point for selection is the minimum concentration of drug that reduces the cloning efficiency below 1%. Because drug resistance in some cell populations is dependent on the density of cells in the dish in which they are selected, it is necessary to confirm that the chosen drug concentration is effective in killing most of the cells under the conditions of selection. Normally, we apply the selection to no more than 1–2 million cells in a 100-mm dish.

4. Drugs used for selection should be in sterile solutions and stored under conditions that maintain their potency. Most of the drugs used to select *MDR1* or *MRP*-expressing cells are hydrophobic and hence can be

dissolved in 100% dimethyl sulfoxide (DMSO, Sigma, St. Louis, MO). Solutions of 10–100 mg/ml can be easily prepared and stored in the dark at −20°. Dilutions into selective medium should be made so that the final concentration of DMSO is below 0.5%. If aqueous solutions are made, they should be filter sterilized before use.

5. Once drug-resistant clones are obtained, they can be expanded in selective medium, initially by growth in multiwell dishes, and eventually in T-25 or T-75 flasks (Costar, Cambridge, MA). Several vials of each of the clones of interest should be stored in liquid nitrogen as soon as is possible, to preserve the phenotype of these populations. Continued selection, even under stable selection conditions, results in phenotypic and/or genetic drift of selected populations. This can be manifested as changes in copy number of amplified genes, integration or excision from the chromosome of amplified genes,[5] or changes in the predominant mechanism of drug resistance in populations in which multiple mechanisms of resistance have occurred. We freeze our cells in DMSO as follows: exponentially growing (subconfluent) cultures are harvested from a T-75 flask by trypsinization. Initially, the culture is checked under a phase microscope to determine the condition of the cells before removing and discarding the growth medium; 1 ml of trypsin–ethylenediaminetetraacetic acid (EDTA, Sigma) is added to the flask for the minimal time needed to release the cells from their monolayer. Depending on the cell line, this process can take anywhere from 1 to 6 min and can be verified by checking the flask under a binocular microscope. To nullify the action of the trypsin, 9 ml of growth medium containing 10% fetal bovine serum (FBS) is added to the 1 ml of trypsinized cells; this mixture is gently pipetted up and down several times in the flask to disperse the cells. The total cell number is then determined by counting with either a hemocytometer or, more routinely, by removing 0.2 ml of the cell suspension from the T-75 flask, adding it to a dilution vial containing 9.8 ml of Isoton II (Coulter Corporation, Miami, FL) (1 : 50 dilution), and counting it on a Coulter counter. After the cells have been counted, the remaining cell suspension is pipetted from the T-75 flask into a 15-ml conical capped centrifuge tube. This tube is placed into an IEC tabletop centrifuge and the cells are spun down at 1500 rpm for 5 min. After centrifugation, the supernatant is removed and discarded. The cell pellet is resuspended in an appropriate volume of growth medium containing 10% DMSO (minus drug, even if the cell line has previously been kept under selection), which will allow for the dispensing of 10^6 cells per milliliter per Nunc (Daigger and Company, Inc., Lincolnshire, IL) cryovial. This procedure needs to be done quickly in order to prevent the cells from clumping or remaining in

[5] P. V. Schoenlein, D.-W. Shen, J. T. Barrett, I. Pastan, and M. M. Gottesman, *Mol. Biol. Cell* **3,** 507 (1992).

prolonged contact with the DMSO. The filled cryovials are placed on wet ice until they can be transferred directly into a −70° freezer. An alternate freeze-down method that may improve viability for some cell lines when the cells are defrosted is to place a cryovial in the −70° freezer in a Nalgene (Daigger and Company, Inc., Lincolnshire, IL) freezing container containing 250 ml of isopropyl alcohol for 24 hr before transferring it into a liquid nitrogen freezer. The cells remain in a −70° freezer for at least 24 hr before they are transferred into a vapor-phase liquid nitrogen freezer. Freezer storage temperatures are kept below −50° to prevent any physical damage to the cells that might be caused by the formation of ice crystals. Cells maintained under these conditions should remain viable indefinitely.

Defrosting cells for reculturing after they have been in liquid nitrogen storage can be accomplished by using the following method: the cryovial containing the desired cells is removed from the freezer, and its contents are rapidly thawed by hand swirling in a 37° waterbath. The cryovial containing the thawed cells is placed into a sterile hood where it is wiped down with 70% ethanol. Its contents, 1 ml of cells in 10% DMSO, are immediately transferred into a 15-ml conical centrifuge tube containing 9 ml of 10% FBS growth medium. The cells are spun down at 1500 rpm for 5 min, and the supernatant containing the DMSO is removed and discarded. The cell pellet is resuspended in 2 ml of fresh growth medium with 1 ml of this suspension being dispensed into a T-25 flask containing growth medium minus drug (i.e., a drug-sensitive parental cell line or in the case of a multidrug-resistant cell line, a "safe flask"). A second T-25 flask also receives 1 ml of the thawed cells into growth medium containing the appropriate concentration of selection drug needed to culture a drug-resistant cell line.

After being frozen, the multidrug-resistant cell lines, especially those derived from the human KB epidermoid carcinoma cells, can be slow to regain their growth vigor. If this is the case, it is advisable to postpone splitting the T-25 flask under drug selection until after the cells begin to proliferate. When this occurs, transfer the entire T-25 cell population into a T-75 flask and let it expand again before cautiously splitting it 1:4 or 1:10. As soon as the cells are established in culture, freeze down multiple copies of each line to establish a healthy cell inventory.

6. Individual drug-resistant clones, either derived directly by selection or by cloning out of mixed populations of resistant clones, should be screened for cross-resistance by generating killing curves using a variety of different drugs of interest. Frequently it is desirable to determine whether the drug resistance phenotype is reversed by relatively nontoxic drugs. We do these experiments by first determining the maximal tolerated concentration of the reversing agent. (In this case, we usually start at 1–100 ng/ml

and go up to 100 μg/ml in the killing curve.) We choose a concentration that results in no significant loss of cloning efficiency of the cell line and then add this concentration to all dishes used to generate a killing curve for a cytotoxic drug. The killing curve in the presence and absence of the reversing agent can then be determined. By comparing the LD_{50} (or LD_{10}) in the presence or absence of the reversing agent, a reversing potency can be determined. It is also possible to use this approach to select for cells that are multidrug resistant in the presence of an agent that inhibits a specific mechanism of drug resistance. By using inhibitors of P-glycoprotein at nontoxic concentrations (such as verapamil or cyclosporin A), it is possible to select for cell lines that are multidrug-resistant based on non-*MDR1* mechanisms, or that express mutant P-glycoproteins.[6]

7. To maintain drug-resistant cell populations, we expose the cells continuously to the selecting drug. For most high-level drug-resistant cell lines, resistance will be lost in time under nonselective conditions. At low levels of resistance, where unstable amplification mechanisms and/or high levels of gene expression are not required, cells may be stable for long periods of time in the absence of selective medium. In either case, it is essential after 2–3 months of growth of cell populations in cell culture to return to the freezer to defrost originally selected cell populations.

Transfection and Transduction of Multidrug Resistance Genes

Now that several drug resistance genes have been cloned, it is useful to be able to introduce such genes into drug-sensitive cell lines. Standard transfection and transduction methods, using calcium phosphate or cationic lipid preparations and retroviral supernatants, may be used. Chapter 40 (Baudard) describes methodology for transfection of an *MDR1* cDNA into drug-sensitive cells, thereby making them drug resistant. The same principles used for selection of drug-resistant clones are used to determine conditions for selection of transfectants. It is worth pointing out that human cells in tissue culture tend to be more sensitive to most cytotoxic agents than rodent (mouse, rat, hamster) cells. Hence, we usually find that selection of human cells for most cytotoxic drugs occurs at approximately one log lower concentrations (frequently 1–10 ng/ml for many anticancer drugs) than for rodent cells (20–100 ng/ml; sometimes higher for Chinese hamster ovary cells).

A more efficient way to introduce multidrug resistance genes into cells is by retroviral transduction. Retroviral packaging cell lines have been

[6] C. O. Cardarelli, I. Aksentijevich, I. Pastan, and M. M. Gottesman, *Cancer Res.* **55,** 1086 (1995).

prepared for many *MDR1*-based vectors.[7] We use the following protocol for *MDR1* retroviral transduction: pHaMDR1/A, a retroviral vector that carries the human *MDR1* cDNA, was used to transfect mouse PA-12 cells that already had the ability to package viral transcripts into infectious viral particles by virtue of the amphotropic, defective, murine leukemia virus genome they harbored. The now multidrug-resistant PA-12 cells that could produce encapsidated virus and grow in medium containing 60 ng/ml colchicine were renamed PA12MDR1/A-1 cells.

The supernatant medium in which the packaging cells are grown, and which contains the virus, has been used to transduce many different cell types. A typical protocol for mouse lymphoma L5178Y cells is as follows: the supernatant from PA12MDR1/A cells is filtered, diluted 1:5 with McCoy's 5A medium plus 10% horse serum, and added to the L5178Y cells along with polybrene at a concentration of 2 μg/ml. After 2 days of growth in McCoy's, the lymphoma cells are fed fresh McCoy's medium. After killing curves are done to find the tolerable colchicine dose, the cells are then continually grown under selection in colchicine to maintain their multidrug resistance.[8]

Introduction of Multidrug Resistance Genes into Bone Marrow

Because of the potential for use of multidrug resistance genes as selectable markers for human gene therapy, techniques have been developed for *ex vivo* retroviral transduction of mouse bone marrow cells and determination of the drug-resistance of cfus in methyl cellulose derived from these transduced cells.[9] We use the following protocol to obtain mouse bone marrow, transduce it with *MDR1* retroviral supernatants, and determine the number of hematopoietic cfus that are drug resistant.

To stimulate mouse bone marrow stem cells to produce more progeny, 5-FU is administered IV (150 mg/kg) to 6- to 10-week-old mice to kill the more mature progenitor cells. This drug is administered 2–4 days before the harvest of bone marrow from sacrificed mice is to take place.[10] Bone marrow from tibias and femurs is removed, subjected to repeated aspirations with a 20-gauge needle, and filtered to remove clumps of cells and small bone chips in order to obtain a single-cell suspension of marrow cells.

[7] M. M. Gottesman, C. A. Hrycyna, P. V. Schoenlein, U. A. Germann, and I. Pastan, *Annu. Rev. Genet.* **29,** 607 (1995).

[8] P. Chiba, G. Ecker, D. Schhid, J. Drach, B. Tell, S. Goldenberg, and V. Gekeler, *Mol. Pharmacol.* **49,** 122 (1996).

[9] J. R. McLachlin, M. A. Eglitis, K. Ueda, P. W. Kantoff, I. Pastan, W. R. Anderson, and M. M. Gottesman, *J. Natl. Cancer Inst.* **82,** 1260 (1990).

[10] T. Licht, I. Aksentijevich, M. M. Gottesman, and I. Pastan, *Blood* **86,** 111 (1995).

Next, a Ficoll separation is performed to eliminate from the bone marrow cells (BMC) as many nonmononuclear cells as possible. To do this, BMC are washed twice in PBS without Ca^{2+} and Mg^{2+} and resuspended in fresh PBS in volumes ranging from 3 to 15 ml depending on the volume of cells. Ficoll-Paque (4–5 ml) is placed into 15-ml conical tubes, and 2 ml of BMC cell suspension [in phosphate-buffered solution (PBS)] is very carefully overlaid on the Ficoll to avoid mixing. Additional PBS is added to bring the total volume in the tube to 13 ml. After 20–30 min of centrifugation at 1000 rpm at 4° without a break, the tubes will have two distinct layers and a narrow interphase band. Five milliliters of the upper layer is removed very carefully and discarded. The interphase layer is then removed carefully to avoid removing Ficoll in the bottom layer. The cells in the interphase are the desired mononuclears, which are then washed twice in medium + FBS.

Immunomagnetic microbeads and MACS columns (Miltenyi, Sunnyvale, CA) are used to deplete the mononuclear BMC of Lin^+ and MHC II^+ cells, and this is followed by another magnetic sorting that yields the positive isolation of Sca-1^+ cells which are now Lin^- MHC II^- Sca-1^+ and therefore likely to contain a large proportion of stem cells. These cells are seeded to tissue culture flasks containing growth medium and cytokines for 18–24 hr at 37° in order to stimulate cell division, which in turn facilitates stable retroviral integration into the cells' genome.[10]

The next step is to expose the BMC to virus producer cells previously seeded to tissue culture flasks and then irradiated (30 Gy). The virus producer cells carry retroviral vectors which contain human *MDR1* cDNA. When encapsidated, the viruses are extruded into the producer cells' supernatant medium. Bone marrow cells at a concentration of 10^6 cells per milliliter of medium are cocultured with subconfluent virus producer cells for 6–8 days in flasks containing Iscove's medium (MDM), polybrene at 4 μg/ml, 20% FBS, 0.5 μmol/L 2-mercaptoethanol, and a combination of cytokines as follows: murine IL-1β at 100 ng/ml (Peprotech, Rochky Hill, NJ); human recombinant erythropoietin at 1 U/ml (Sigma); murine IL-3 at 200 U/ml; human IL-6 at 200 U/ml (both from Becton Dickinson Labware, Bedford, MA); and rat stem cell factor at 100 ng/ml (Amgen, Thousand Oaks, CA). [Note: Recombinant mouse stem cell factor is now available (R & D Systems, Minneapolis, MN), but its dose relative to rat stem cell factor has not been determined in our laboratory.]

The BMC which grow in suspension are carefully harvested from the coculture flasks containing virus producer cells attached to the flask bottoms. The BMC are placed in fresh flasks to be grown for 24 hr in medium containing the growth factors listed earlier after which they are selected for 48–96 hr or longer in the presence of drugs.[10]

The drug-resistant cells are now ready to be used in one or more different ways:

1. Colony assays of hematopoietic cells may be done to evaluate the cells' drug resistance.
2. The cells may be used to transplant recipient mice irradiated (900 R) on the same day the transplant is carried out. The transplant consists of 10^7 donor cells suspended in 250 μl of PBS. The bone marrow of these mice may be harvested 8–10 weeks later to be used for colony assays to test the viability of their drug resistance.[10]
3. The resistant BMC may be analyzed for P-glycoprotein activity with fluorescence-activated cell sorting (FACS) using monoclonal antibody or rhodamine-123 efflux study.[11]

Colony Assays

All of the mature cells in blood and bone marrow are derived from pluripotent stem cells, but they have become differentiated and can therefore be distinguished from each other. However, stem cells occur in very low numbers—perhaps one in 10^5 total cells. They can best be identified using *in vitro* colony assays during which pluripotent stem cells can grow and differentiate.

The growth medium most often used in this lab is a ready-made methylcellulose-based mixture containing recombinant cytokines, which is intended for colony assays of murine BMC—MethoCult GF M3434 (Stem Cell Technologies Inc., Vancouver, BC, Canada). Its components are: 0.9% methylcellulose in Iscove's MDM; 15% FBS; 1% bovine serum albumin; 10 μg/ml bovine pancreatic insulin; 200 μg/ml human transferrin (iron-saturated); 10^{-4} *M* 2-mercaptoethanol; 2 m*M* L-glutamine; 10 ng/ml rm IL-3; 10 ng/ml rh IL-6; 50 ng/ml rm SCF; and 3 units/ml rh erythropoietin. This combination of growth factors is suitable to stimulate cfu-GEMM (granulocytes-erythrocytes-megakaryocytes-macrophages). Other combinations lead to colonies of more differentiated precursor cells.

The methylcellulose serves to keep the differentiating progeny of each stem cell together—they form a colony made visible with the use of an inverted microscope.

The MethoCult GF growth medium can be obtained already dispensed into tubes—3 ml per tube. The tubes are stored frozen at −20° and must therefore be warmed at 37° to melt the contents and then vortexed briefly to mix. What results is a slow-flowing substance with the consistency of sludge. The air bubbles trapped in the medium and the frothy cap of bubbles on the top require that the tubes be kept for at least 10 min until some of these bubbles break up. Once melted, however, the methylcellulose medium is quite stable with regard to its ability to avoid gelling and stay workable.

[11] T. Licht, I. Pastan, M. M. Gottesman, and F. Hermann, *Ann. Hematol.* **69,** 159 (1994).

A series of 20× drug concentrations to be added to each of the tubes is prepared. Each drug concentration is added to one tube in a volume of 0.15 ml. For example, for a colchicine killing curve, final concentrations of 0–70 ng/ml of the drug are used. Dilutions are prepared using a stock concentration of 1 mg/ml in DMSO, but medium without serum is used to dilute the stock. The tubes are vortexed well to get the drugs evenly distributed.

A suspension of bone marrow cells is prepared such that the concentration is 10^6 cells/ml. When 0.15 ml of this is added to each tube, the final concentration is 5×10^4 cells/ml. Vortexting must now be done briefly and at a relatively low speed to avoid damage to the cells.

One milliliter of each tube's contents is slowly added to each of two or three 35-mm dishes, depending on the amount of air bubbles remaining in each tube. Each dish must be gently rotated to distribute its sample evenly. Moderate-sized bubbles can be gently punctured with a pipette tip, and small bubbles will disappear during incubation.

If there are two duplicate 35-mm dishes, they are to be placed with lids closed in a 100-mm dish along with an open 35-mm dish filled with 2 ml sterile water. The lid of the 100-mm dish is closed. If three replicate 35-mm dishes have been filled, two are dealt with as above, except that the third dish is placed in a 100-mm dish along with a water dish. The sample dishes can dry up quite readily if they are not incubated with a water dish.

Our assays run for 1 week, at the end of which colonies are counted using a stereoscopic microscope with the help of a grid located on the bottom of a 100-mm dish (Licht and Goldenberg, unpublished data, 1995).

[18] Monoclonal Antibodies to P-Glycoprotein: Preparation and Applications to Basic and Clinical Research

By MIKIHIKO NAITO and TAKASHI TSURUO

Introduction

One of the major problems in cancer chemotherapy is the development of drug resistance during treatment. The nature of the drug resistance is complex, but it is now generally accepted that P-glycoprotein, a member of the ABC transporter family protein, plays an important role in cellular drug resistance.[1–3]

[1] J. A. Endicott and V. Ling, *Annu. Rev. Biochem.* **58,** 137 (1989).

0076-6879/98 $25.00

P-Glycoprotein is a multidrug transporter originally found on the surface of multidrug-resistant (MDR) tumor cells.[4–6] P-Glycoprotein can transport various compounds concomitant with the consumption of adenosine triphosphate (ATP).[7,8] These include anticancer drugs[9–13] (*Vinca* alkaloids, anthracyclines, etoposide, taxol), calcium channel blockers[14,15] (verapamil, diltiazem, azidopine), immunomodulators[16–18] (cyclosporin A, FK-506), cardiac glycoside[19] (digoxin), fluorescent dyes[20] (rhodamine 123, fluo-3), and steroids[21] (cortisol, aldosterone). MDR tumor cells transport antitumor drugs outside the cells, thus reducing their cellular accumulation. This represents the cellular mechanism of MDR. Besides MDR tumor cells, P-glycoprotein is expressed on normal cells of secretory tissues, such as kidney, liver, intestine, and adrenal gland as well as capillary endothelial cells in brain and hematopoietic progenitor cells in bone marrow.[22–24]

---

[2] M. M. Gottesman and I. Pastan, *Annu. Rev. Biochem.* **62,** 385 (1993).

[3] T. Tsuruo, *Jpn. J. Cancer Res.* **79,** 285 (1988).

[4] R. L. Juliano and V. Ling, *Biochim. Biophys. Acta* **455,** 152 (1976).

[5] W. T. Beck, T. J. Mueller, and L. R. Tanzer, *Cancer Res.* **39,** 2070 (1979).

[6] D. W. Shen, C. Cardarelli, J. Hwang, M. Cornwell, N. Richert, S. Ishii, I. Pastan, and M. Gottesmann, *J. Biol. Chem.* **261,** 7762 (1986).

[7] M. Naito, H. Hamada, and T. Tsuruo, *J. Biol. Chem.* **263,** 11887 (1988).

[8] M. Horio, M. M. Gottesmann, and I. Pastan, *Proc. Natl. Acad. Sci. U.S.A.* **85,** 3580 (1988).

[9] A. Fojo, S. Akiyama, M. M. Gottesmann, and I. Pastan, *Cancer Res.* **45,** 3002 (1985).

[10] M. Inaba and R. K. Johnson, *Biochem. Pharmacol.* **27,** 2123 (1978).

[11] K. Dano, *Biochim. Biophys. Acta* **323,** 466 (1973).

[12] T. Tsuruo, H. Iida-Saito, H. Kawabata, T. Oh-hara, H. Hamada, and T. Utakoji, *Jpn. J. Cancer Res.* **77,** 682 (1986).

[13] T. Tatsuta, M. Naito, T. Oh-hara, I. Sugawara, and T. Tsuruo, *J. Biol. Chem.* **267,** 20383 (1992).

[14] K. Yusa and T. Tsuruo, *Cancer Res.* **49,** 5002 (1989).

[15] T. Saeki, K. Ueda, Y. Tanigawara, R. Hori, and T. Komano, *FEBS Lett.* **324,** 99 (1993).

[16] M. Naito, H. Tsuge, C. Kuroko, T. Koyama, A. Tomida, T. Tatsuta, Y. Heike, and T. Tsuruo, *J. Natl. Cancer Inst.* **85,** 311 (1993).

[17] T. Saeki, K. Ueda, Y. Tanigawara, R. Hori, and T. Komano, *J. Biol. Chem.* **268,** 6077 (1993).

[18] A. Shirai, M. Naito, T. Tatsuta, J. Dong, K. Hanaoka, K. Mikami, T. Oh-hara, and T. Tsuruo, *Biochim. Biophys. Acta* **1222,** 400 (1994).

[19] Y. Tanigawara, N. Okamura, M. Hirai, M. Yasuhara, K. Ueda, N. Kioka, T. Komano, and R. Hori, *J. Pharmacol. Exp. Ther.* **263,** 840 (1992).

[20] D. M. Wall, X. F. Hu, J. R. Zalcberg, and J. D. Parkin, *J. Natl. Cancer Inst.* **83,** 206 (1991).

[21] K. Ueda, N. Okamura, M. Hirai, Y. Tanigawara, T. Saeki, N. Kioka, T. Komano, and R. Hori, *J. Biol. Chem.* **267,** 24248 (1992).

[22] F. Thiebaut, T. Tsuruo, H. Hamada, M. M. Gottesman, I. Pastan, and M. C. Willingham, *Proc. Natl. Acad. Sci. U.S.A.* **84,** 7735 (1987).

[23] C. Cordon-Cardo, J. P. O'Brien, D. Casals, L. Rittman-Grauer, J. L. Biedler, M. R. Melamed, and J. R. Bertino, *Proc. Natl. Acad. Sci. U.S.A.* **86,** 695 (1989).

[24] P. M. Chaudhary and I. B. Roninson, *Cell* **66,** 85 (1991).

P-Glycoprotein in such normal tissues can excrete toxic compounds, thereby protecting the cells and tissues from harmful effects of the compounds, and can secrete steroids.

In the study of MDR, considerable numbers of monoclonal antibodies to P-glycoprotein have been developed, and they significantly contributed to our understanding of the molecular and cellular functions of P-glycoprotein. The purpose of this article is to give an overview of the monoclonal antibodies to P-glycoprotein and their applications to basic and clinical research.

Monoclonal Antibodies to P-Glycoprotein

In 1985, Kartner *et al.*[25] developed C219 monoclonal antibody against Chinese hamster P-glycoprotein. C219 recognizes a conserved epitope of P-glycoprotein, and the antibody can bind to all classes of P-glycoprotein from rodents (*mdr1a, mdr1b,* and *mdr2*) and humans (*MDR1* and *MDR2*).[26] Western blot analysis and immunohistochemical analysis have been carried out with the C219 antibody to detect P-glycoprotein in MDR cells. However, since the epitope for C219 is located in the cytoplasmic domain of P-glycoprotein,[26] this antibody cannot be applied to living cells for flow cytometric analysis.

A monoclonal antibody applicable to flow cytometric analysis on living cells was developed in our laboratory in 1986.[27] The antibody, MRK-16, was developed after repeated immunization with intact K562/ADM cells expressing large amounts of P-glycoprotein. As a result, MRK-16 can bind to the extracellular domain of P-glycoprotein. The epitope for MRK-16 located on fourth loop of the extracellular domain[27,28] is unique to the human *MDR1* P-glycoprotein. Therefore, MRK-16 can only bind to human *MDR1* P-glycoprotein but cannot bind to the human *MDR2* P-glycoprotein or P-glycoprotein of other species. The only exception so far is bovine brain endothelial cells that expressed P-glycoprotein and were immunostained with MRK-16.[29]

MRK-16 can bind to natural, but not denatured, human *MDR1* P-glycoprotein, and thus, MRK-16 is not applicable to the Western blot technique. Solubilization with most detergents also affects the binding ability of MRK-16, but immunoprecipitation is successfully performed after

[25] N. Kartner, D. Everden-Porelle, G. Bradley, and V. Ling, *Nature* **316,** 820 (1985).

[26] E. Georges, G. Bradley, J. Gariepy, and V. Ling, *Proc. Natl. Acad. Sci. U.S.A.* **87,** 152 (1990).

[27] E. Georges, T. Tsuruo, and V. Ling, *J. Biol. Chem.* **268,** 1792 (1993).

[28] M.-C. Chevallier-Multon, H. Hamada, T. Tsuruo, J.-B. Le Pecq, and M. Lipinski, *J. Cell. Pharmacol.* **2,** 165 (1991).

[29] A. Tsuji, T. Terasaki, Y. Takabatake, Y. Tenda, I. Tamai, T. Yamashima, S. Moritani, T. Tsuruo, and J. Yamashita, *Life Sci.* **51,** 1427 (1992).

solubilization of P-glycoprotein with 1% CHAPS, 0.5% deoxycholate, and 1% glycocholate.[30–32] These observations indicate that MRK-16 recognizes the higher structure of human *MDR1* P-glycoprotein.

In addition to C219 and MRK-16, many monoclonal antibodies to P-glycoprotein have been developed (Table I). These antibodies have been used to study molecular and cellular functions of P-glycoprotein.

Preparation of Monoclonal Antibody That Recognizes Extracellular Epitope of P-Glycoprotein

BALB/c mice are immunized weekly for 7 weeks with intraperitoneal injection of 10^7 intact K562/ADM cells that express large amounts of P-glycoprotein. Spleen cells from immune mice are fused with P3-X63-Ag8-653 myeloma cells, and hybridomas are selected in hypoxanthine/aminopterin/thymidine medium in the usual manner. The culture supernatants from growing hybridomas are tested by cellular ELISA using the intact cells fixed on 96-well microtiter plates pretreated with 0.001% poly(L-lysine),[33] and the hybridomas producing antibodies that react with K562/ADM cells but not with the parental K562 cells are selected.

The antigens recognized by the monoclonal antibodies are visualized on Western blot analysis or immunoprecipitation after cell surface proteins of K562/ADM are biotinylated by incubating the cells with 100 μg/ml of NHS-LC biotin (Pierce, Rockford, IL) in 0.1 *M* HEPES (pH 8.0)–saline for 40 min with occasional agitation at room temperature. The labeled cells are solubilized in lysis buffer (50 m*M* Tris, 100 m*M* NaCl, 2 m*M* $CaCl_2$, 2 m*M* $MgCl_2$, 1% CHAPS) containing protease inhibitors [50 m*M* iodoacetamide, 1 m*M* phenylmethylsulfonyl fluoride (PMSF), 1 mg/ml soybean trypsin inhibitor, 0.1 U/ml aprotinin] for 30 min at 4°, and immunoprecipitated with the antibodies. The precipitated proteins are subjected to SDS–PAGE, transferred onto nitrocellulose membrane, reacted with peroxidase-conjugated avidin, and detected with an enhanced chemiluminescence mixture (Amersham) using X-ray films. When monoclonal antibodies cannot bind to denatured or solubilized antigens, antibodies bound to natural proteins on the cell surface are chemically cross-linked to the antigens[34] by treating the antibody-bound cells with 1 m*M* DSP (Pierce) in phosphate-buffered saline for 30 min at 25°. Then, solubilization and immunoprecipita-

[30] H. Hamada and T. Tsuruo, *Proc. Natl. Acad. Sci. U.S.A.* **83,** 7785 (1986).

[31] H. Hamada and T. Tsuruo, *J. Biol. Chem.* **263,** 1454 (1988).

[32] M. Naito and T. Tsuruo, *J. Cancer Res. Clin. Oncol.* **121,** 582 (1995).

[33] R. H. Kennet, *in* "Monoclonal Antibodies" (R. H. Kennet, T. J. McKearn, and K. B. Bechtol, eds.), p. 376. Plenum, New York, 1980.

[34] H. Hamada and T. Tsuruo, *Anal. Biochem.* **160,** 483 (1987).

TABLE I
MONOCLONAL ANTIBODIES TO P-GLYCOPROTEIN

Antibodies	Immunoglobulin class	Epitope in P-glycoprotein	Refs.
C219	IgG_{2a}	Cytoplasmic	a
JSB-1	IgG_1	Cytoplasmic	b
MRK-16	IgG_{2a}	Extracellular	c
MRK-17	IgG_1	Extracellular	c
HYB-241	IgG_1	Extracellular	d
UIC2	IgG_{2a}	Extracellular	e
4E3	IgG_{2a}	Extracellular	f
7G4	IgG_{2a}	Extracellular	g
17F9	IgG_{2b}	Extracellular	h
MAb57	IgG_{2a}	Extracellular	i
F4	IgG_1	Extracellular	j

[a] N. Kartner, D. Everden-Porelle, G. Bradley, and V. Ling, *Nature* **316,** 820 (1985).
[b] R. J. Scheper, J. W. M. Bulte, B. J. G. P., J. J. Quak, E. Van der Schoot, A. J. M. Balm, C. J. L. M. Meijer, H. J. Broxterman, C. M. Kuiper, J. Lankelma, and H. M. Pinedo, *Int. J. Cancer* **42,** 389 (1988).
[c] H. Hamada and T. Tsuruo, *Proc. Natl. Acad. Sci. U.S.A.* **83,** 7785 (1986).
[d] M. B. Meyers, L. Rittman-Graner, J. P. O'Brien, and A. R. Safa, *Cancer Res.* **49,** 3209 (1989).
[e] E. B. Merchetner and I. B. Roninson, *Proc. Natl. Acad. Sci. U.S.A.* **89,** 5824 (1992).
[f] R. J. Arceci, K. Stieglitz, B. J., J. Schinkel, F. Baas, and J. Croop, *Cancer Res.* **53,** 310 (1993).
[g] A. H. Schinkel, R. J. Arceci, J. J. M. Smit, E. Wagenaar, F. Baas, M. Dolle, T. Tsuruo, E. B. Mechetner, I. B. Roninson, and P. Borst, *Int. J. Cancer* **55,** 478 (1993).
[h] M. Aihara, Y. Aihara, G. Schmidt-Wolf, I. Schmidt-Wolf, B. I. Sikic, K. G. Blume, and N. J. Chao, *Blood* **77,** 2079 (1991).
[i] C. Cenciarelli, S. J. Currier, M. C. Willingham, F. Thiebaut, U. A. Germann, A. V. Rutherford, M. M. Gottesmann, S. Barca, M. Tombesi, S. Morrone, A. Santoni, M. Mariani, C. Romani, M. L. Dupuis, and M. Cianfriglia, *Int. J. Cancer* **47,** 533 (1991).
[j] T. M. Chu, E. Kawinski, and T. H. Lin, *Hybridoma* **12,** 417 (1993).

tion are performed as usual, and the precipitated cross-linked complex is boiled in Laemmli sample buffer containing 5% 2-mercaptoethanol for 5 min. The disulfide bonds of the antigen–antibody complex introduced by DSP is cleaved under these reducing conditions. These samples are subjected to SDS–PAGE, and the proteins are visualized by staining the cells or after blotting to nitrocellulose membranes.

Antibodies to cell surface molecules overexpressed in K562/ADM cells compared with the parental K562 cells were generated by these methods.[30]

Because P-glycoprotein is not the only protein overexpressed in MDR tumor cells, antibodies to other proteins, such as 85-kDa protein (CD36)[35–37] and 300-kDa protein,[38] were obtained together with the antibodies to P-glycoprotein.

Research Application of Monoclonal Antibodies to P-Glycoprotein

Gene Cloning

The C219 monoclonal antibody was used to isolate P-glycoprotein cDNA clones from library in λgt11 expression vector.[39] The cloned cDNA was confirmed to encode P-glycoprotein by reactivity to the gene product of other monoclonal antibodies that could bind to the independent epitope of P-glycoprotein.[25,26] The cloned cDNA was also used to isolate family genes, and *pgp1, pgp2,* and *pgp3* (alternatively *mdr1a, mdr1b,* and *mdr2,* respectively) genes were isolated from Chinese hamster ovary (CHO) cells.[40]

Purification of P-Glycoprotein

Membrane proteins of K562/ADM cells were solubilized with 1% CHAPS, and P-glycoprotein was purified to a homogenous band by immunoaffinity chromatography with MRK-16.[31] The purified protein has an ATPase activity, which is elevated by the addition of verapamil, a substrate transported by P-glycoprotein. The ATPase activity was also demonstrated with the immunoprecipitated P-glycoprotein.[41] Reconstitution of the purified P-glycoprotein into liposomes was also carried out, and drug transport that depended on ATP was significantly demonstrated.[32]

Modulation of Transport Functions

Several monoclonal antibodies were reported to modulate the transport function of P-glycoprotein when antibodies were added to the culture me-

[35] H. Hamada, E. Okochi, M. Watanabe, T. Oh-hara, Y. Sugimoto, H. Kawabata, and T. Tsuruo, *Cancer Res.* **48,** 7082 (1988).

[36] K. Noguchi, M. Naito, K. Tezuka, S. Ishii, H. Seimiya, Y. Sugimoto, E. Amman, and T. Tsuruo, *Biochem. Biophys. Res. Commun.* **192,** 88 (1993).

[37] Y. Sugimoto, H. Hamada, S. Tsukahara, K. Noguchi, K. Yamaguchi, M. Sato, and T. Tsuruo, *Cancer Res.* **53,** 2538 (1993).

[38] Y. Sugimoto, E. Okochi, H. Hamada, T. Oh-hara, and T. Tsuruo, *Biochem. Biophys. Res. Commun.* **169,** 686 (1990).

[39] J. R. Riordan, K. Deuchars, N. Kartner, N. Alon, J. Trent, and V. Ling, *Nature* **316,** 817 (1985).

[40] W. F. Ng, F. Sarangi, R. L. Zastawny, L. Veinot-Drebot, and V. Ling, *Mol. Cell. Biol.* **9,** 1224 (1989).

[41] H. Hamada and T. Tsuruo, *Cancer Res.* **48,** 4926 (1989).

dium. MRK-16 increased cellular accumulation of vincristine, actinomycin D, and cyclosporin A in MDR tumor cells.[16,30] HYB-241 also enhanced cellular accumulation of vincristine, vinblastine, and actinomycin D in *MDR1*-transfected cells.[42,43] Consequently, these monoclonal antibodies enhanced cellular sensitivity to such drugs. The potentiation of the cellular sensitivity to drugs of the P-glycoprotein substrate was also demonstrated with UIC2 monoclonal antibody.[44]

Detection of P-Glycoprotein

Monoclonal antibodies to P-glycoprotein have been used to detect and quantitate the expression of P-glycoprotein. Usual immunological methods, including Western blot analysis, immunoprecipitation, flow cytometry, and immunohistochemical analysis were carried out. The results indicate that P-glycoprotein expression correlates well with the degree of drug resistance in experimental MDR tumor systems[45,46] and also in clinical samples.[47–50] These observations imply a clinical usefulness for the monoclonal antibodies to P-glycoprotein in detecting the emergence of drug-resistant tumors.

Selective Killing of MDR Cells

Since the expression of P-glycoprotein is a clinically relevant mechanism of MDR,[51] selective killing of tumors that express P-glycoprotein is an effective way to overcome drug resistance. For this purpose, MRK-16 was conjugated to *Pseudomonas* exotoxin, and the conjugate shows selective toxicity to P-glycoprotein expressing tumors, depending on their expression level.[52] Bispecific antibodies (one half of the molecule recognizes

[42] M. B. Meyers, L. Rittman-Graner, J. P. O'Brien, and A. R. Safa, *Cancer Res.* **49,** 3209 (1989).

[43] L. S. Rittmann-Grauer, M. A. Yong, V. Sanders, and D. G. Mackensen, *Cancer Res.* **52,** 1810 (1992).

[44] E. B. Merchetner and I. B. Roninson, *Proc. Natl. Acad. Sci. U.S.A.* **89,** 5824 (1992).

[45] H. G. Keizer, G. J. Schuurhius, H. J. Broxterman, J. Lankelma, W. G. E. J. Schoonen, J. van Rijn, H. M. Pinedo, and H. Joenje, *Cancer Res.* **49,** 2988 (1989).

[46] J. Dong, M. Naito, T. Tatsuta, H. Seimiya, O. Johdo, and T. Tsuruo, *Oncol. Res.* **7,** 245 (1995).

[47] W. S. Dalton, T. M. Grogan, P. S. Meltzer, R. J. Scheper, B. G. M. Durie, C. W. Taylor, T. P. Miller, and S. E. Salmon, *J. Clin. Oncol.* **7,** 415 (1989).

[48] P. Verrelle, F. Meissonnier, Y. Fonck, V. Feillel, C. Dionet, F. Kwiatkowski, R. Plagne, and J. Chassagne, *J. Natl. Cancer Inst.* **83,** 111 (1991).

[49] S. E. Salmon, T. M. Grogan, T. Miller, R. Scheper, and W. S. Dalton, *J. Natl. Cancer Inst.* **81,** 696 (1989).

[50] H. S. Chan, P. S. Thorner, G. Haddad, and V. Ling, *J. Clin. Oncol.* **8,** 689 (1990).

[51] P. R. Twentyman, *J. Natl. Cancer Inst.* **84,** 1458 (1992).

[52] D. J. FitzGerald, M. C. Willingham, C. O. Cardarelli, H. Hamada, T. Tsuruo, M. M. Gottesman, and I. Pastan, *Proc. Natl. Acad. Sci. U.S.A.* **84,** 4288 (1987).

P-glycoprotein and the other half CD3) can cross-link P-glycoprotein expressing tumors to cytotoxic T lymphocytes, resulting in selective killing of the MDR tumors that express P-glycoprotein.[53,54] Anti-P-glycoprotein monoclonal antibodies were also used to eliminate MDR tumors contaminating bone marrow cells before autologous bone marrow transplantation. Magnetic affinity cell sorting using MRK-16 coupled to magnetic particles has been developed as a means of removing MDR tumors that express P-glycoprotein.[55]

Summary

P-Glycoprotein, a member of the ABC transporter group, plays an important role in anticancer drug resistance. Many monoclonal antibodies to P-glycoprotein have been developed and used to study the molecular and cellular functions of P-glycoprotein. These monoclonal antibodies were used for gene cloning, purification and reconstitution of P-glycoprotein, analysis of transport functions of P-glycoprotein, immunological detection and selective killing of tumors that express P-glycoprotein for diagnostic, and therapeutic purposes.

[53] J. van Duk, T. Tsuruo, D. M. Segal, R. L. H. Bolhuis, R. Colognola, R. J. van de Griend, G. J. Fleuren, and S. O. Warnaar, *Int. J. Cancer* **44,** 738 (1989).

[54] Y. Heike, K. Okumura, and T. Tsuruo, *Jpn. J. Cancer Res.* **83,** 366 (1992).

[55] R. Padmanabhan, T. Tsuruo, S. E. Kane, M. C. Willingham, B. H. Howard, M. M. Gottesman, and I. Pastan, *J. Natl. Cancer Inst.* **83,** 565 (1991).

[19] Topology of P-Glycoproteins

By WILLIAM R. SKACH

Introduction

Protein topology in biologic systems refers to the location of a protein or region of protein with respect to a membrane-bound compartment. In recent years, significant efforts have been aimed at defining protein topology, particularly for complex membrane proteins such as ABC transporters where compartmental localization of different peptide loops contributes to important aspects of function. Using a variety of techniques several investigators have attempted to map directly cytosolic and extracytosolic peptide regions of P-glycoprotein (Pgp) at the plasma membrane. These

0076-6879/98 $25.00

methods, which have largely supported conventional hydropathy-based topologic models, include (1) protease digestion,[1] (2) cell surface labeling with epitope specific antibodies,[2–4] and (3) cross-linking with sulfhydryl-specific reagents.[5] Using an alternate approach, other investigators have attempted to define Pgp topology by examining folding pathways through which Pgp and other ABC transporters are assembled into cellular membranes.[6–10] In this article, the focus is on these latter techniques, which have generated new insights and questions regarding Pgp folding, topology, and general mechanisms of polytopic protein biogenesis.

For most eukaryotic polytopic proteins, transmembrane topology is established in the endoplasmic reticulum (ER) membrane through a series of programmed folding events.[11] These events are directed through interactions between specific sequence determinants encoded in the nascent chain (topogenic sequences) and their cognate cytosolic and membrane-bound receptors.[12,13] Such topogenic sequences may include (1) signal or signal anchor sequences, which direct ER membrane targeting and translocation, as well as (2) stop transfer sequences, which terminate translocation and integrate the chain into the lipid bilayer.[14–18] One method used to study complex protein topology, therefore, is to define topogenic sequences and the translocation events they direct at the ER membrane. In this manner, the process of establishing complex transmembrane topology (i.e., topogenesis) may be dissected into a series of distinct folding steps. Through such studies,

[1] A. Yoshimura, Y. Kuwazuru, T. Sumizawa, M. Ichikawa, S. Ikeda, T. Uda, and S. Akitama, *J. Biol. Chem.* **264,** 16282 (1989).

[2] C. Kast, V. Canfield, R. Levenson, and P. Gros, *J. Biol. Chem.* **271,** 9240 (1996).

[3] E. Georges, T. Tsuro, and V. Ling, *J. Biol. Chem.* **268,** 1792 (1993).

[4] A. Schinkel, R. Arceci, J. Smit, E. Wagenaar, F. Baas, M. Dolle, T. Tsururo, E. Mechetner, I. Foninson, and P. Borst, *Int. J. Cancer* **55,** 478 (1993).

[5] T. Loo and D. Clarke, *J. Biol. Chem.* **270,** 843 (1995).

[6] W. Skach, M. C. Calayag, and V. Lingappa, *J. Biol. Chem.* **268,** 6903 (1993).

[7] W. Skach and V. Lingappa, *Cancer Res.* **54,** 3202 (1994).

[8] J.-T. Zhang and V. Ling, *J. Biol. Chem.* **266,** 18224 (1991).

[9] J.-T. Zhang, M. Duthie, and V. Ling, *J. Biol. Chem.* **268,** 15,101 (1993).

[10] O. Beja and E. Bibi, *J. Biol. Chem.* **270,** 12351 (1995).

[11] G. Blobel, *Proc. Natl. Acad. Sci. U.S.A.* **77,** 1496 (1980).

[12] P. Walter and V. R. Lingappa, *Ann. Rev. Cell Biol.* **2,** 499 (1986).

[13] W. Skach and V. Lingappa, *in* "Mechanisms of Intracellular Trafficking and Processing of Proproteins" (Y. P. Loh, ed.), p. 19. CRC Press, Boca Raton, Florida, 1993.

[14] M. Friedlander and G. Blobel, *Nature* **318,** 338 (1985).

[15] Y. Audigier, M. Friedlander, and G. Blobel, *Proc. Nat. Acad. Sci. U.S.A.* **84,** 5783 (1987).

[16] J. Lipp, N. Flint, M. T. Haeuptle, and B. Dobberstein, *J. Cell Biol.* **109,** 2013 (1989).

[17] R. E. Rothman, D. W. Andrews, M. C. Calayag, and V. R. Lingappa, *J. Biol. Chem.* **263,** 10470 (1988).

[18] H. Wessels and M. Spies, *Cell* **55,** 61 (1988).

it has recently become clear that not all polytopic proteins utilize conventional, cotranslational assembly pathways predicted by early biogenesis models.[19] For example, transmembrane topology and membrane integration of certain peptide regions within Pgp are directed through redundant and/or cooperative translocation mechanisms.[7,20] For other regions of Pgp, topogenic sequences have been identified that direct the chain into multiple alternate transmembrane orientations in the ER membrane.[6,8,9,21] Thus a thorough understanding of Pgp topology will require knowledge of transmembrane assembly mechanisms as well as final transmembrane orientation in different cellular compartments.

General Approach to Polytopic Protein Topogenesis

Several complementary techniques have been used to study Pgp biogenesis and topology in the ER membrane. These techniques exploit the physical integrity and/or asymmetric enzymatic activities of the ER membrane, which include protease protection of translocated peptide domains,[22] cleavage of the chain by signal peptidase complex,[23] and covalent addition of N-linked high mannose oligosaccharides.[24] Three separate experimental steps are required for this approach: (1) generation of cDNA expression vectors encoding specific regions of Pgp for which topologic information is desired, (2) protein expression in cell-free or whole-cell systems that faithfully carry out topogenesis events at the ER membrane, and (3) determination of nascent chain topology in right-side-out ER-derived microsomes. Because of the complex topology of Pgp, it is difficult to define the topology of all peptide regions simultaneously. For this reason, individual cDNA constructs are designed to define the topology of a particular peptide region or site. By examining a series of such constructs the topology for larger regions of protein may then be determined.

C Terminus Translocation Reporter

C Terminus translocation reporters have been widely used to study protein topology.[25] In this approach a series of fusion proteins is generated

[19] W. A. Braell and H. F. Lodish, *Cell* **28,** 23 (1982).

[20] W. Skach and V. Lingappa, *J. Biol. Chem.* **268,** 23552 (1993).

[21] J. Zhang and V. Ling, *Biochemistry* **34,** 9159 (1995).

[22] G. Blobel and B. Dobberstein, *J. Cell Biol.* **67,** 852 (1975).

[23] G. Blobel and B. Dobberstein, *J. Cell Biol.* **67,** 835 (1975).

[24] C. Hirshberg and M. Sneider, *Ann. Rev. Biochem.* **56,** 63 (1987).

[25] D. Boyd, C. Manoil, S. Froshauer, J. Millan, N. Green, K. McGovern, C. Lee, and J. Beckwith, *in* "Protein Folding: Deciphering the Second Half of the Genetic Code" (L. Gierasch and J. King, eds.), p. 314. American Association for the Advancement of Science, Washington, D.C., 1990.

that contain a heterologous reporter domain engineered onto sequential sites in the coding sequence. Topology of the reporter (e.g., cytosolic versus extracytosolic) is determined for each fusion protein and used to infer topology at the fusion site. In bacterial systems, reporter domains typically exhibit enzymatic activity that is dependent on cytoplasmic versus periplasmic location.[25] In eukaryotic expression systems, reporter topology is determined in right-side-out ER-derived microsomal membranes by protection of the reporter from exogenously added protease.[6,7,26] A potential limitation of this technique is that reporters usually reflect protein topology that is directed in a linear manner, that is, where N terminus topogenic determinants direct translocation of C terminus peptide regions. However, C terminus reporters have been remarkably successful in mapping the topology of diverse polytopic proteins, particularly in prokaryotes. In addition, this technique has provided important insight regarding cooperative interactions between topogenic determinants that direct polytopic protein biogenesis.[27]

Reporter domains should ideally (1) lack intrinsic topogenic activity, (2) passively and efficiently follow topogenic information presented in the nascent chain, (3) be intrinsically sensitive to protease digestion, and (4) be readily identifiable (e.g., be of sufficient size and antigenicity for easy recovery and visualization). One such reporter is derived from a 142-residue C terminus fragment of bovine prolactin.[17,28] This reporter (P) has been extensively studied in a variety of secretory, bitopic, and polytopic contexts and fulfills the criteria given. High-affinity antisera reactive against the P domain is readily available (ICN, Costa Mesa, CA), and a variety of vectors now exist that encode this domain in a convenient context for cloning heterologous cDNA fragments.

Generating Fusion Proteins

Fusion sites are chosen to test predictions of a specific topologic model, usually based on hydropathy analysis and/or homology with related proteins as illustrated in Fig. 1. At least 15 amino acid residues should be left between the predicted boundary of a transmembrane (TM) segment and the reporter fusion site to avoid removing residues involved in directing topology.[29] Where the boundaries of TM segments are ambiguous, several closely spaced fusion sites should be used. Fusion proteins are engineered by amplifying template cDNA using the polymerase chain reaction (PCR) and

[26] L.-B. Shi, W. Skach, V. Lingappa, and A. Verkman, *Biochemistry* **34,** 8250 (1995).
[27] J. Calamia and C. Manoil, *J. Mol. Biol.* **224,** 539 (1992).
[28] D. W. Andrews, E. Perara, C. Lesser, and V. R. Lingappa, *J. Biol. Chem.* **263,** 15791 (1988).
[29] E. Hartman, T. A. Rapoport, and H. F. Lodish, *Proc. Natl. Acad. Sci. U.S.A.* **86,** 5786 (1989).

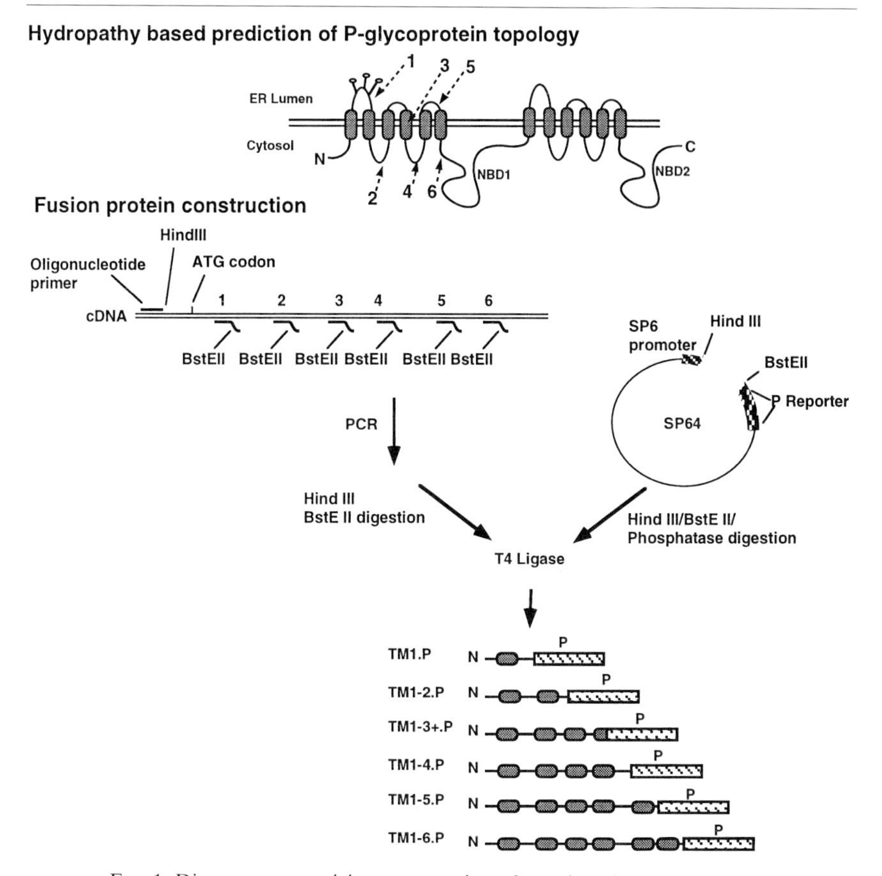

FIG. 1. Diagram summarizing construction of a series of fusion proteins.

ligating these fragments into an appropriate plasmid expression vector (Fig. 1). An in-frame *Bst*EII restriction site is engineered into antisense oligonucleotides and used for ligating the PCR fragment onto the P reporter coding sequence.

Ten cycles of PCR are performed in a 100-μl volume containing 100 pmol of sense and antisense oligonucleotides, 0.1 μg template DNA, 50 m*M* KCl, 1.5 m*M* $MgCl_2$, 0.01% gelatin, 1 m*M* each of dATP, dTTP, dGTP, and dCTP, 10 m*M* Tris, pH 7.8, and 2.5 units *Taq* DNA polymerase (Boehringer Mannheim, Germany). Plasmid DNA encoding Pgp (0.1 μg) (or desired protein) is added as a template. PCR fragments are digested with appropriate restriction enzymes (*Hin*dIII and *Bst*EII), separated in low melting point agarose, and extracted with phenol/chloroform and ethanol precipitation. Fragments are ligated (T4 DNA ligase, New England Biolabs,

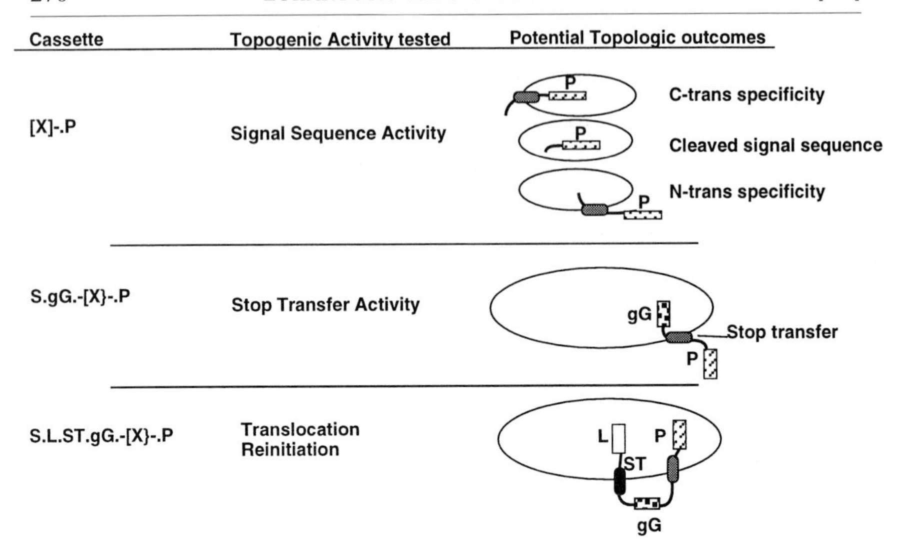

FIG. 2. Schematic diagram of chimeric cassettes used for defining specificity of topogenic determinants.

Beverly, MA) into a previously digested (*Hin*dIII and *Bst*EII) and phosphatase-treated (calf intestinal phosphatase, New England Biolabs) expression vector. Typically, 10–50% of products recovered from a PCR reaction are ligated to 0.1–0.2 μg of vector DNA for 2–6 hr at 18°. One convenient vector for these studies is derived from pSP64 (Promega, Madison, WI), which encodes the SP6 promoter for *in vitro* transcription. An untranslated region from *Xenopus* globin cDNA is located immediately 3′ to the promoter and serves to increase translation efficiency.[30]

Characterization of P-Glycoprotein Topogenic Determinants

The hallmark of topogenic sequences is that they faithfully direct events of topogenesis when engineered into appropriate heterologous contexts.[17,31–33] Individual topogenic determinants that direct translocation and membrane integration events may thus be evaluated using a series of defined chimeric cassettes that encode proteins of known topology.[17] Examples of these cassettes are illustrated in Fig. 2. *De novo* signal sequence activity is

[30] D. A. Meltin, P. A. Krieg, M. R. Bebagliati, T. Maniatis, K. Zinn, and M. R. Green, *Nucleic Acids Res.* **12,** 7035 (1984).

[31] V. R. Lingappa, J. Chaidez, C. S. Yost, and J. Hedgpeth, *Proc. Natl. Acad. Sci. U.S.A.* **81,** 456 (1984).

[32] E. Perara and V. R. Lingappa, *J. Cell. Biol.* **101,** 2292 (1985).

[33] C. S. Yost, J. Hedgpeth, and V. R. Lingappa, *Cell* **34,** 759 (1983).

identified using the chimera [X]-.P, where [X] represents a putative topogenic sequence (e.g., TM segments together with flanking residues), and P represents the prolactin-derived reporter. Signal and signal anchor activities are identified by testing whether [X] functions to translocate the chimeric protein (or regions of the protein) into the ER lumen. In a similar manner, stop transfer (ST) activity is tested using the cassette S.gG.-[X]-.P, where "S" and "gG" represent an N terminus signal sequence and a modified globin passenger domain, respectively. ER targeting and translocation are initiated by the signal sequence, and in the absence of other topogenic information the chain is fully secretory. If [X] exhibits ST activity, then it will terminate translocation and direct chains into an N-trans (type I) transmembrane topology with the gG domain in the ER lumen and the P domain in the cytosol. Additional topogenic activities may be tested using increasingly complex cassettes. For example, the ability of Pgp determinants to reinitate translocation (as opposed to *de novo* translocation) may be determined using the cassette S.L.ST.gG.-[X]-.P, where L represents a passenger domain derived from β-lactamase and ST represents the stop transfer sequence from transmembrane immunoglobulin M (IgM). This chimera mimics the native presentation of internal Pgp topogenic sequences that emerge from the ribosome on a transmembrane chain that has been pretargeted and integrated into the ER membrane.

Chimeric proteins are generated by amplifying Pgp cDNA using the PCR and ligating these fragments into appropriate vectors with engineered restriction sites as described in Fig. 1. Oligonucleotide primers are designed to flank coding sequences of potential topogenic determinants, which usually include predicted membrane-spanning segments along with at least 15 flanking residues.

Although topological analysis of individual chimeric proteins is relatively straightforward, extrapolation of these results to understand folding pathways and topology of Pgp (or other proteins) may be difficult. In some cases, individual determinants may exhibit topogenic activities different from those predicted by theoretical topologic models.[6] Alternatively, topogenic activities may support the proposed topology but identify novel pathways by which that topology is achieved.[7,20] It is therefore important to corroborate topologic predictions derived from individual topogenic sequences with additional independent studies.

Protein Expression Systems

Several expression systems have been used to study Pgp topology and biogenesis in the ER membrane. The most common include microinjected *Xenopus* oocytes and rabbit reticulocyte lysate (RRL) supplemented with canine pancrease microsomes. Both systems provide a versatile, rapid expression system, which reconstitutes early trafficking and folding events at the ER membrane.[34] Oocytes provide the added advantage of generating proteins in living cells and, in our hands, are usually more efficient at establishing polytopic protein topology than reconstituted cell-free expression systems.

Preparation of Reticulocyte Lysate

Although several commercial preparations of RRL and canine pancrease microsomal membranes are available that give satisfactory results, it is often more economical or convenient to prepare these reagents. These procedures are performed as follows. For a more detailed description the reader is referred to previous volumes.[35,36]

Healthy adult (2.5–4 kg) New Zealand White rabbits are injected subcutaneously for 3 consecutive days with 50 mg acetylphenylhydrazine (2.5% in 20% ethanol/KOH, pH 7). On day 7 rabbits are heparinized with 2000 U heparin and anesthetized with 25–40 mg/kg pentobarbital. Thoracotomy is performed; the heart is cannulated with a 16-gauge needle, and approximately 80–120 ml of blood is obtained and collected on ice. Blood is centrifuged at 3000 rpm (Sorval GSA rotor) for 10 min, and packed cells are washed three times in 1 volume 0.14 *M* NaCl, 0.05 *M* KCl, and 5 m*M* $MgCl_2$, each time resuspending the cells by gentle swirling. Cells are then lysed in 1 volume distilled water (RNase free) with vigorous mixing for 5 min at 4° and the mixture centrifuged at 15,000*g* for 15 min. The supernatant is collected and 1 m*M* hemin is added (40 μl/ml lysate). Lysate is then aliquoted, frozen in liquid nitrogen, and stored at −80°. Hemin solution is mixed in the following order: 6.44 mg hemin (bovine crystalline type I, Sigma, St. Louis, MO), 0.25 ml 1 *N* KOH, 0.5 ml 0.2 *M* Tris-HCl (pH 7–8), 8.5 ml ethylene glycol, 0.19 ml 1 *N* HCl, 0.5 ml H_2O. Adjust final pH between 7 and 8.

Preparation of Microsomal Membranes

The pancreas is surgically removed from a freshly euthanized dog and placed in ice-cold buffer A [50 m*M* potassium acetate, 6 m*M* magnesium acetate, 1 m*M* EDTA, 0.25 *M* sucrose, 1 m*M* dithiothreitol (DTT), 0.5 m*M*

[34] V. R. Lingappa, J. R. Lingappa, R. Prasad, K. E. Ebner, and G. Blobel, *Proc. Natl. Acad. Sci. U.S.A.* **75,** 2338 (1978).

[35] R. Jackson and T. Hunt, *Methods Enzymol.* **96,** 50 (1983).

[36] P. Walter and G. Blobel, *Methods Enzymol.* **96,** 84 (1983).

phenylmethylsulfonyl fluoride (PMSF), 50 m*M* triethanolamine (TEA), pH 7.5]. When cool, it is weighed and then gently ground by hand in a tissue grinder (food grinders are adequate) until a pulpy consistency is reached. Ice-cold buffer A (4 ml/g starting weight) is added, and the tissue is homogenized by three to four passes through a coarse Teflon–glass homogenizer connected to a high-speed motor (hand drill) and then by two to three passes with a fine Dounce homogenizer until a smooth consistency is achieved. Homogenate is centrifuged 600*g* for 10 min and then again at 10,000*g* for 10 min and the pellets discarded. The supernatant is layered onto a 5–8 ml sucrose cushion (50 m*M* potassium acetate, 6 m*M* magnesium acetate, 1 m*M* EDTA, 1.3 *M* sucrose, 50 m*M* TEA, pH 7.5, and centrifuged at 150,000*g* for 3 hr. The fluffy pellet is removed and resuspended in 0.5 ml/g starting material of 0.25 *M* sucrose, 1 m*M* DTT, 50 m*M* TEA, pH 7.5, using a hand-held Teflon–glass homogenizer. This mixture is diluted to a final OD_{280} of 50, frozen in liquid nitrogen and stored at −80°. All steps are performed on ice.

Nuclease Digestion

Prior to use, native RNA in RRL and microsomal membrane preparations is digested by addition of $CaCl_2$ (1/100 volume of 0.1 *M* $CaCl_2$) and staphylococcal nuclease (150 U/ml, Pharmacia, Piscataway, NJ) and incubation at 25° for 10 min. Nuclease is inactivated by addition of 1/50 volume EGTA (2 m*M* final concentration). Samples are aliquoted, frozen in liquid nitrogen, and stored at −80° for up to 3 months.

Care and Harvesting of Xenopus laevis Oocytes

Mature, female *Xenopus laevis* (NASCO, Fort Atkinson, WI) are kept in 20-gal aquariums with distilled water containing 0.4–0.6% rock salt. Water temperature is kept below 22° (18° is optimal) and day length is adjusted to 12 hr. Ovaries are surgically removed from anesthetized frogs (placed in 0.15% tricaine for 30 min) through a 1-cm paramedian abdominal incision. Oocytes are placed directly in modified Barth's solution (MBSH) containing 88 m*M* NaCl, 1.0 m*M* KCl, 2.4 m*M* $NaHCO_3$, 0.82 m*M* $MgSO_4$, 0.33 m*M* $Ca(NO_3)_2$, 0.41 m*M* $CaCl_2$, 10 m*M* sodium *N*-2-hydroxyethylpiperazine-*N'*-2-ethanesulfonic acid (HEPES) containing 100 units/ml penicillin, 100 μg/ml streptomycin and 50 μg/ml gentamicin and kept at 18°. Mature oocytes are harvested for microinjection by manual dissection from ovary tissue and stored for no more than 5 days after removal from frogs. Damaged or discolored oocytes are discarded.

In Vivo and *in Vitro* Protein Expression

RNA is transcribed *in vitro* for 1 hr at 40° in 10-μl reactions containing 2–4 μg plasmid DNA, 40 m*M* Tris, pH 7.5, 6.0 m*M* magnesium acetate, 2 m*M* spermidine, 0.5 m*M* each of ATP, CTP, UTP, 0.1 m*M* GTP, 0.5 m*M* GpppG, 10 m*M* DTT, 0.2 mg/ml bovine calf tRNA (Sigma), 0.75 U/ml RNase inhibitor (Promega, Madison, WI), and 0.4 U/ml SP6 RNA polymerase (New England Biolabs).

In vitro translation in RRL is carried out at 25° for 1–1.5 hr in 10- to 40-μl volumes. Each 10-μl reaction volume contains 2 μl transcription mixture, 4 μl nuclease-treated reticulocyte lysate, 2 μl Emix5×, 0.5 μl CB20×, 0.1 μl creatine kinase (4 mg/ml in 50% glycerol, 10 m*M* Tris, pH 7.5), 0.1 μl tRNA (10 mg/ml), 0.1 μl RNase inhibitor (20 U/μl), 0.2 μl H_2O, and 1.0 μl nuclease-treated microsomal membranes (or H_2O as a control). Emix5× contains 5 m*M* ATP, 5 m*M* GTP, 60 m*M* creatine phosphate, 0.2 m*M* each of 19 essential amino acids, except methionine (Sigma) and 5 μCi/μl [^{35}S]methionine (Tran^{35}S-label, ICN), CB20× contains 2 *M* potassium acetate, 16 m*M* magnesium acetate, and 40 m*M* Tris, pH 7.5.

In vivo translation in oocytes. Transcription mixture (2.0 μl) is added to 50 μCi of [^{35}S]methionine [0.5 μl of a 10× concentrated Tran^{35}S-label (ICN)] and loaded into a micropipette (25 μm in diameter) attached to a micromanipulator. Oocytes are positioned on nylon mesh on an ice-cold stage and injected with 50 nl of labeled transcription mixture. Following injection, oocytes are incubated in MBSH at 18°. Two to four hours after injection oocytes are homogenized in a 1.5-ml Eppendorf tube using 3 volumes (3 μl/oocyte) of ice-cold homogenization buffer (0.25 *M* sucrose, 50 m*M* Tris, pH 7.5, 50 m*M* potassium acetate, 5 m*M* magnesium acetate, and 1 m*M* DTT) using a hand-held Teflon homogenizer.

Assay of Protein Topology

N-Linked Glycosylation

N-linked carbohydrates are covalently added to nascent chains by oligosaccharyltransferase exclusively in the ER lumen.[24] However, not all potential N-linked glycosylation consensus sites are utilized, particularly sites located within 15 residues of TM segment boundaries of sites located in different peptide loops within a single polytopic chain.[37,38] Thus, while N-linked glycosylation of peptide domains provides powerful evidence for

[37] I. Nilsson and G. von Heijne, *J. Biol. Chem.* **268,** 5798 (1993).

[38] C. Landolt-Marticorena and R. Reithmeier, *Biochem. J.* **302,** 253 (1994).

translocation of the acceptor site, lack of glycosylation at any given site provides little information on protein topology.

N-linked glycosylation of chains in the ER lumen is usually detected by a shift in mobility corresponding to acquisition of high mannose, core carbohydrates using either of two methods: (1) digestion with endoglycosidase H or (2) inhibition of oligosaccharyltransferase. Endoglycosidase H digestion is best performed immediately following immunoprecipitation of translation products from either RRL or oocytes by adding 50 μl of 0.1% SDS, 0.1 *M* sodium citrate, pH 5.5, to protein A beads. The sample is heated to 100° for 1 min, and supernatant is removed and divided into two aliquots. To one aliquot is added 0.2 U endoglycosidase H (New England Biolabs) and both samples are incubated at 37° for 4–6 hr prior to analysis by SDS–PAGE. Alternatively, oligosaccharyltransferase may be inhibited by addition of a tripeptide, AcAsn-Try-Thr, to the assembled RRL translation mixture (0.2 m*M* final concentration). In both cases, glycosylation is confirmed by demonstrating a shift in migration (3 kDa per oligosaccharide chain) when comparing control and treated chains by SDS–PAGE.

Protease Digestion Assay

One of the most widely used techniques to define protein topology in cell membranes involves digestion of peptide regions with a membrane-impermeable protease such as proteinase K (PK, Boehringer Mannheim, Indianapolis, IN). Because the intact ER membrane defines a closed physical space, PK has access to only cytosolically exposed domains. Thus following PK digestion, cytosolic proteins are completely degraded; secretory proteins remain fully protected; and transmembrane proteins generate protease protected fragments (diagramed in Fig. 3). In the presence of nondenaturing detergent, membrane integrity is disrupted, and PK gains access to all protein domains.

Proteolysis is performed on ice following completion of translation in RRL or following homogenization of oocytes. One-tenth volume of 0.1 *M* $CaCl_2$ is added, and each reaction is divided into three separate aliquots (5-μl minimum volume per tube). One sample serves as a control; to a second sample is added one-tenth volume 2.0 m*M* PK stock solution; to a third sample is added one-tenth volume PK stock solution and one-tenth volume 10% Triton X-100. Samples are incubated on ice for 1 hr. Residual protease is inactivated by addition of one-tenth volume 0.1 *M* PMSF in dimethyl sulfoxide (DMSO) and rapid transfer of the entire sample into 10 volumes of 1% SDS, 0.1 *M* Tris, pH 8.0, preheated to 100° in a boiling bath. Samples are incubated at 100° for at least 5 min prior to subsequent analysis. A known secretory control protein should be tested to determine the efficiency of translocation and PK protection by ER membrane under

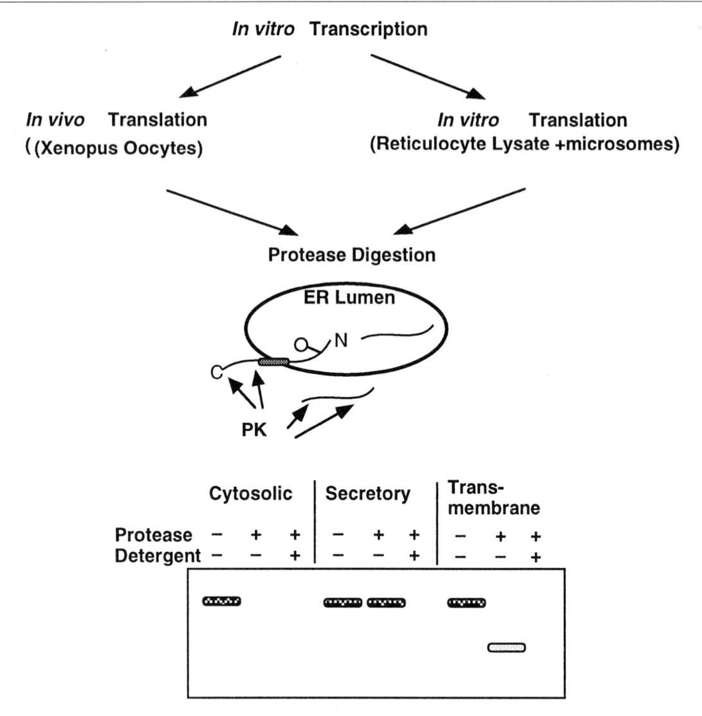

FIG. 3. Protease digestion of right-side-out ER-derived microsomal membranes. Protease digestion results for different topological orientations are schematized.

specific conditions used. To prepare PK stock solution, PK is dissolved in H_2O (2 mg/ml), incubated at 37° for 30 min, and stored in aliquots at −80°.

Following protease inactivation, RRL translation products may be visualized directly by SDS–PAGE and autoradiography. Oocyte products are immunoprecipitated prior to analysis by dilution into 10 volumes of 0.1 *M* NaCl, 0.1 *M* Tris, pH 8.0 10 m*M* EDTA, and 1% Triton X-100. This solution is incubated at 4° for at least 2 hr and centrifuged at 16,000*g* for 10 min at 4°. Supernatants are removed, combined with 5 μl of Affi-Gel protein A beads (Bio-Rad, Hercules, CA) and 1 μl of appropriate antisera and mixed at 4° for 4–10 hr. Beads are washed three times with 0.5 ml of 0.1 *M* NaCl, 0.1 *M* Tris, pH 8.0, 10 m*M* EDTA, and 1% Triton X-100 and twice with 0.1 *M* NaCl, 0.1 *M* Tris, pH 8.0, prior to SDS–PAGE an autoradiography. Typically, material from two or three oocytes or 0.5–1.0 μl of RRL translation mixture is loaded per lane.

Proteolysis should be performed using three parallel samples as described. Only peptide fragments protected in the absence but not presence of detergent are scored as translocated. The efficiency of translocation for any given peptide fragment is determined by comparing the intensity of protein bands that represent full-length chains to bands that represent protease-protected fragments and correcting for relative methionine content.

Mapping P-Glyoprotein Topology

Figure 4 illustrates one analysis of Pgp-derived fusion proteins using techniques described earlier. These proteins were constructed by engineering the P reporter after each of the first six TM segments in the N terminus hydrophobic domain. Proteins were expressed by RNA microinjection in *Xenopus* oocytes, and oocyte homogenates were digested with PK and immunoprecipitated with anti-prolactin antisera prior to SDS–PAGE. Each fusion protein generated P-reactive chains n the absence of protease (lanes 1, 4, 7, 10, 12, Fig. 4). Multiple protein bands in these lanes resulted from variable glycosylation of chains at N-linked consensus sites located between TM1 and TM2 and/or cleavage by signal peptidase complex during translocation of the reporter into the ER lumen (lanes 2, 8, Fig. 4). Following PK digestion in the absence of detergent, protected P-reactive fragments

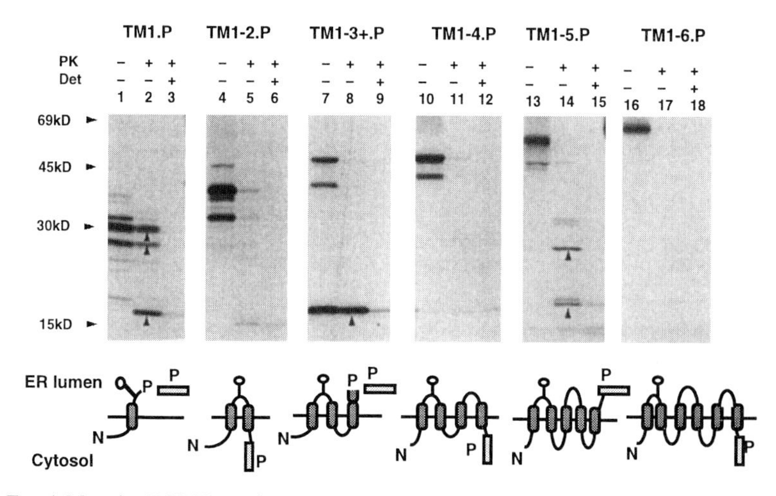

FIG. 4. Mapping PGP N terminus topology with a translocation reporter. Plasmids encoding the P reporter engineered at six sequential sites in the PGP coding sequence (as in Fig. 1) were expressed in Xenopus ocytes, digested with PK, immunoprecipitated with anti-prolactin antisera, and analyzed by SDS–PAGE. Upward arrows (lanes 2, 8, 14) indicate protease-protected, translocated P-reactive fragments. Topology of each chain deduced from these data is indicated. [Adapted with permission from W. Skach and V. Lingappa, *Cancer Res.* **54,** 3202 (1994)].

were recovered from plasmids TM1.P, TM1-3^{+}.P and TM1-5.P, while little or no protection was observed when the reporter was located downstream of TM2, TM4, or TM6. These results suggest that TM1, TM3, and TM5 function as signal sequences to initiate translocation of the reporter into the ER lumen, while TM2, TM4, and TM6 function as ST sequences to terminate translocation and orient downstream peptide loops in the cytosol. In this manner, the process of Pgp topogenesis may be reconstructed by following cytosolic versus lumenal topology of the reporter as a function of nascent chain elongation.

It is important to reemphasize that techniques described in this review are designed to investigate the mechanism of Pgp folding and assembly in the ER membrane. Because, transmembrane topology of proteins at the ER membrane has traditionally been considered to reflect topology throughout the cell, these techniques provide important information for understanding structural aspects of multispanning proteins. However, recent studies have suggested that transmembrane topology may be more dynamic than previously thought. For certain proteins (e.g., colicin, SecA, SecG) transmembrane topology may be transiently influenced by transmembrane voltage protential or ATP hydrolysis.[39–41] Alternatively, during prion protein biogenesis, cytosolic factors appear to be involved in directing nascent chains into two different topologic fates.[42] Studies of Pgp topogenesis have suggested that specific peptide regions may also exhibit more than one transmembrane structure in the ER membrane.[6,8,9] For this reason, topological analyses using these as well as other techniques should be interpreted cautiously, because the topology of Pgp (and possibly other proteins) within a given cellular membrane of intracellular compartment may reflect only a subset of TM structures attained during the lifetime of a given molecule.

[39] S. Slatin, X.-Q. Qui, K. Jakes, and A. Finkelstein, *Nature* **371,** 158 (1994).

[40] A. Economou and W. Wickner, *Cell* **78,** 835 (1994).

[41] K. Nishiyama, T. Suzuki, and H. Tokuda, *Cell* **85,** 71 (1996).

[42] C. D. Lopez, C. S. Yost, S. B. Prusiner, R. M. Myers, and V. R. Lingappa, *Science* **248,** 226 (1990).

[20] Use of Cell-Free Systems to Determine P-Glycoprotein Transmembrane Topology

By JIAN-TING ZHANG

Introduction

P-Glycoprotein (Pgp) is a polytopic plasma membrane protein consisting of 12 putative transmembrane (TM) segments.[1-4] It belongs to the ATP-binding cassette (ABC) transporter superfamily. Knowledge of its detailed membrane folding is essential to understand its function. However, deduction of the topology of a membrane protein from its sequence is a difficult task that is often based only on hydropathy plots.[5] Such prediction may generate a misleading topology.[6,7] In this chapter, the application of a cell-free expression system combined with cDNA engineering to dissect the detailed topological folding of Pgp is described. The methods described here also help to determine the topogenic sequences in Pgp and to identify other factors that control the Pgp topology. This method can also be used to determine the topological folding of other simple membrane proteins.[8]

There are two well-characterized and frequently used cell-free expression systems, rabbit reticulocyte lysate (RRL) and wheat germ extract (WGE) both supplemented with microsomal membranes (RM) from dog pancreatic endoplasmic reticulum (ER). Both systems have been widely used in the characterization of topogenic sequences in secretory and membrane proteins and for the identification of components of the insertion machinery. Use of these systems to determine topology is based on the belief that most transmembrane proteins in eukaryotic cells acquire their final membrane topologies during or immediately after synthesis on the rough ER.[9-12] The cell-free system is fast, convenient, and easy to manipu-

[1] J. A. Endicott and V. Ling, *Annu. Rev. Biochem.* **58,** 137 (1989).
[2] C. Higgins, *Annu. Rev. Cell Biol.* **8,** 67 (1992).
[3] M. M. Gottesman and I. Pastan, *Annu. Rev. Biochem.* **62,** 385 (1993).
[4] S. Childs and V. Ling, in "Important Advances in Oncology," (V. T. DeVita, S. Hellman, and S. A. Rosenberg, eds.), p. 21. Lippincott Co., Philadelphia, 1994.
[5] J. Kyte and R. F. Doolittle, *J. Mol. Biol.* **157,** 105 (1982).
[6] R. J. Leonard, C. G. Labarca, P. Charnet, P. N. Davidson, and H. A. Lester, *Science* **242,** 1578 (1988).
[7] G. Yellen, M. Jurman, T. Abramson, and R. MacKinnon, *Science* **251,** 9393 (1991).
[8] H. P. Wessels, J. P. Beltzer, and M. Spiess, *Methods Cell Biol.* **34,** 287 (1991).
[9] G. Goldman and G. Blobel, *J. Cell Biol.* **90,** 236 (1981).
[10] W. A. Braell and H. F. Lodish, *Cell* **28,** 23 (1982).

0076-6879/98 $25.00

late. It avoids complications imposed by other processes such as protein trafficking. The protein of interest is the only protein newly synthesized in the cell-free system, eliminating interferences by other membrane proteins in the analysis of the results.

In Vitro Transcription, Cell-Free Translation, and Membrane Translocation

In Vitro Transcription Using SP6 RNA Polymerase

Materials. SP6 RNA polymerase; RNasin (RNase inhibitor); RNA cap analog (^{7}mGpppG, 300 A_{260} units/ml), rNTPs (2.5 m*M* each of ATP, CTP, GTP, UTP); dithiothreitol (DTT) (0.1 *M*); RQ1 DNase; 5× transcription buffer (200 m*M* Tris-HCl, pH 7.5, 30 m*M* $MgCl_2$, 10 m*M* spermidine, 100 m*M* NaCl). We normally purchase the cap analog from Pharmacia (Piscataway, NJ) and other reagents from Promega (Madison, WI).

Methods. cDNA is cloned into pGEM-4z in both orientations. One orientation is used to make the sense and the other one for making the antisense RNA transcript. The recombinant DNA is linearized using the appropriate restriction endonuclease downstream of the SP6 promoter, ethanol precipitated, and redissolved in H_2O.

The following is mixed in a test tube: 10 μl 5× transcription buffer, 5 μl DTT, 10 μl rNTPs, 50 units of RNasin, 3 μg linearized DNA, 2 μl RNA cap analog, and 10–20 units SP6 RNA polymerase. DEPC-treated H_2O is added to a final volume of 50 μl. After 1–1.5 hr incubation at 40°, 20 units RNasin and 2.5 units RQ1 DNase are added to the reaction and the DNA templates are digested for 15 min at 37°. The reaction is then extracted with phenol/chloroform and chloroform once each. The RNA transcripts are finally precipitated by addition of 5 μl 3 *M* sodium acetate (pH 5.2) and 125 μl ethanol at −70° for ≥20 min. After microcentrifugation at 14,000 rpm for 10 min, the pellet is dried and redissolved in 40 μl diethyl pyrocarbonate (DEPC)-treated H_2O containing 1 unit/μl RNasin. Typically, a yield of ~5 μg RNA is obtained. If T7 RNA polymerase is used, incubation at 37° is performed. It is worth noting that mixing the reaction is best carried out at room temperature because the spermidine in the reaction buffer may precipitate DNA on ice.

[11] D. Brown and R. Simoni, *Proc. Natl. Acad. Sci. U.S.A.* **81,** 1674 (1984).
[12] H. Wessels and M. Spiess, *Cell* **55,** 61 (1988).

In Vitro Translation and Membrane Insertion

Materials. Nuclease-treated RRL, WGE, canine pancreatic microsomal membranes (RM), [^{35}S]methionine, amino acid mixture (without methionine), RNasin, and 10% Triton X-100. Both RRL and WGE translation systems are commercially available. We normally purchase them from Promega. The amino acid mixtures come with RRL or WGE. [^{35}S]Methionine is normally purchased from New England Nuclear (Boston, MA).

Methods. To perform a coupled translation–translocation reaction in RRL, mix the following in a test tube: 17.5 μl RRL, 0.5 μl RNasin, 0.5 μl amino acid mixture (minus methionine), 2.5 μl [^{35}S]methionine, 1.8 μl RM, and 3 μl RNA transcript. The mixture is incubated at 30° for 1–1.5 hr. Aliquots of reactions are then processed for the membrane insertion and topology analyses of the newly synthesized proteins. To achieve high translation efficiency, it is advisable that the concentration of potassium acetate (especially for the WGE translation system) and RNA transcripts be optimized. To perform a control translation in the absence of RM, 0.1–0.5% Triton X-100 should be included in the reaction to maximize the translation efficiency.[13,14] The translation efficiency of polytopic membrane proteins is normally low in the absence of RM. The reason for this is not known. However, it is possible that the RRL contains a large amount of signal recognition particle (SRP), which arrests translation of membrane and secretory proteins.[15] This translation arrest can be released by inclusion of RM due to the interaction between SRP and its receptors on the RM. The action of Triton X-100 to release the translation arrest in the absence of RM is still poorly understood. Figure 1 shows an example of translation of full-length Pgp molecules. A protein of 150 kDa was produced from the sense transcript of Pgp (lane 3, Fig. 1), but not from the antisense transcript (lane 4, Fig. 1) in the RRL cell-free system.

Determining Transmembrane Topology of P-Glycoprotein

Membrane Insertion and Integration

To investigate the membrane topology of Pgp, nascent Pgp molecules have to be shown to be inserted and integrated in the membrane. To achieve this goal, the translation mixture is treated with an alkaline buffer (0.1 *M* Na_2CO_3, pH 11.5). Under such strong alkaline condition, biological mem-

[13] R. Gregory, S. Cheng, D. Rich, J. Marshall, S. Paul, K. Hehir, L. Ostedgaard, K. Klinger, M. Welsh, and A. Smith, *Nature* **347,** 382 (1990).

[14] J. T. Zhang, M. Duthie, and V. Ling, *J. Biol. Chem.* **268,** 15101 (1993).

[15] R. Jagus, *Methods Enzymol.* **152,** 267 (1987).

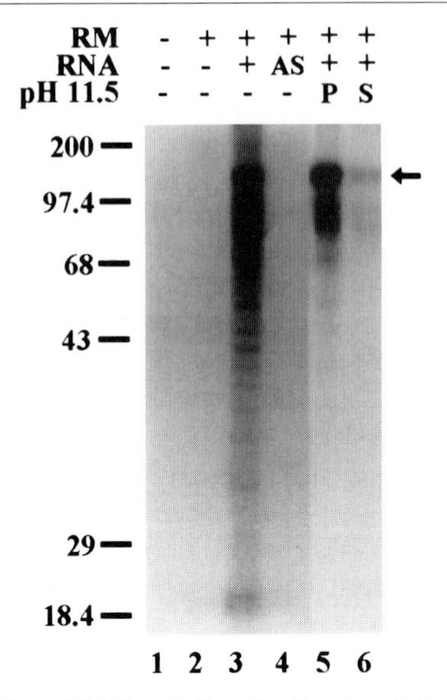

FIG. 1. *In vitro* translation of full-length Pgp. *In vitro* translation in RRL was directed by sense (lane 3) or antisense (lane 4) Pgp transcripts. Lanes 1 and 2 depict control reactions in the absence of RNA transcripts. After alkaline (pH 11.5) extraction to remove peripheral proteins, the nascent Pgp product was mainly found in the membrane pellet (lane 5) and little was observed in the supernatant (lane 6). Data first appeared in Zhang and Ling[17] and are reproduced with permission. The arrow indicates the full-length Pgp product.

branes are disrupted to open sheets and stripped of peripheral proteins.[16] This method has been proven to be a useful empirical procedure to distinguish between integral membrane proteins from peripheral proteins and proteins translocated into the RM lumen. Translation mixture (2 μl) is diluted in 20 μl 0.1 *M* $NaCO_3$, pH 11.5, incubated on ice for 20 min, and followed by centrifugation for 15 min. The supernatant is carefully removed without disrupting the membrane pellet. Both pellet and supernant fractions are solubilized in SDS–PAGE sample buffer and analyzed by gel electrophoresis and fluorography. As shown in Fig. 1, the majority of the nascent Pgp molecules are resistant to alkaline extraction, indicating that they are integrated in the membrane (compare lanes 5 and 6, Fig. 1).

Engineering Reporter at cDNA Level

To analyze the topology of Pgp, a reporter peptide was used to indicate membrane sidedness of specific Pgp domains. A reporter needs to have one or more potential N-linked glycosylation sites and a desirable antibody

[16] K. Howell and G. Palade, *J. Cell Biol.* **92,** 822 (1982).

epitope. It should not contain any potential membrane-spanning sequences that may affect the membrane translocation. The correct reading frame after linking the reporter sequence should be confirmed by double-strand DNA sequencing. The reporter used in dissecting Pgp topology is derived from the first ATP-binding fold of Chinese hamster *pgp1* Pgp.[14] This reporter contains a potential N-linked glycosylation site, an epitope for the monoclonal antibody C219, and a desirable molecular mass for SDS–PAGE analysis. It does not contain any potential membrane-spanning sequences. Figure 2A shows the schematic diagram of a fusion construct, pGPGP-N4.

Glycosylation/Deglycosylation

The most obvious modification occurring in the RM is N-linked glycosylation of asparagine residues in the consensus sequence Asn-X-Ser/Thr (where X is any amino acid but Pro) of a nascent protein. Many but not all of these sequences, when translocated and exposed to the RM lumen, are modified in one step by a glycosyltransferase located in the RM lumen. Addition of each high mannose core oligosaccharide chain results in an increase of 2–3 kDa in the apparent molecular mass on an SDS–PAGE. Addition of more oligosaccharide chains has additive effects on the apparent size of the protein. Comparing translation products in the absence and presence of RM will show a difference in size if the products are glycosylated.

The N-linked glycosylation can be easily confirmed by enzymatic digestion with peptide *N*-glycosidase F (PNGase F), which specifically cleaves between the innermost *N,N'*-diacetylchitobiose and the asparagine to which the oligosaccharide is linked. The removal of oligosaccharide chains can be controlled using limited conditions and the intermediate protein products containing different number of oligosaccharide chains can be observed. Therefore, a glycosylated reporter will indicate the lumenal location of the Pgp domain to which the reporter is attached and the number of oligosaccharide chains in a nascent protein can be determined.

Normally, membranes from an 8-μl translation mixture are pelleted, resuspended in 40 μl 0.1 *M* sodium phosphate, pH 7.1, and divided into four aliquots. To each aliquot, the following is added: 1 μl 10% 2-mercaptoethanol, 2.5 μl 10% Nonidet P-40 (NP-40), 1 μl 100 m*M* phenylmethylsulfonyl fluoride (PMSF), and 4.9 μl H_2O. Then, a different amount of PNGase F in 0.8 μl is added to each tube followed by addition of 0.8 μl 5% sodium dodecyl sulfate (SDS). The reaction is performed at either room temperature for 2 min or 37° for 1 hr, then stopped by addition of SDS–PAGE sample buffer, and analyzed on an SDS–PAGE.

Figure 2C shows a deglycosylation study of pGPGP-N4 protein. Translation of pGPGP-N4 transcript in the presence of RM generated two major

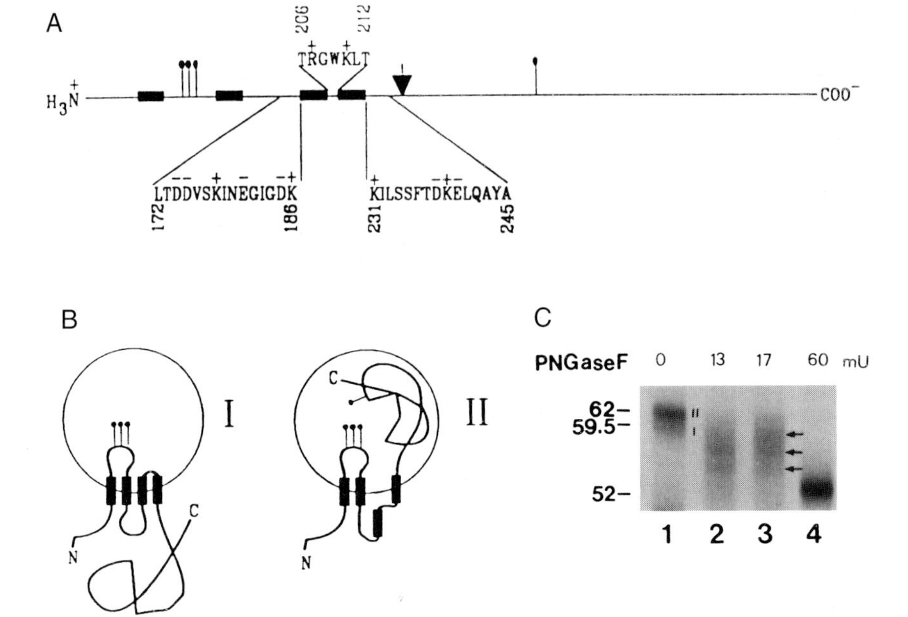

FIG. 2. Coupled *in vitro* translation of pGPGP-N4. (A) Schematic linear structure of pGPGP-N4 protein. The pGPGP-N4 protein is derived from hamster *pgp1* Pgp and consists of 550 amino acids, 4 TM segments, and 4 consensus N-linked glycosylation sites (⫯). The arrow indicates the fusion site between the N-terminal TM domain and the C-terminal ATP-binding domain reporter. Amino acids surrounding TM3 and TM4 are shown in single-letter code with charged amino acids marked by (+) or (−). (B) Two observed membrane orientations of pGPGP-N4 protein in RM. In model I, both N and C termini are on the cytoplasmic side (outside of RM) and all four TM segments (solid bars) are in the membrane. In model II, only three TM segments are in the membrane and the C terminus is in the RM lumen with an oligosaccharide chain (⫯) attached. (C) *In vitro* translation and endoglycosidase treatments of pGPGP-N4 proteins. Two mature products of 62 kDa (representing model II) and 59.5 kDa (representing model I) were translated from the *in vitro* transcripts in the presence of RM (lane 1). Lanes 2–3 show limited endoglycosidase PNGase F treatment performed at room temperature for 2 min. The partially deglycosylated peptide is indicated by arrows (lanes 2 and 3) and the fully deglycosylated peptide is 52 kDa (lane 4). (Data first appeared in Zhang *et al.*[27] and are reprinted with permission.)

proteins of 62 and 59.5 kDa (lane 1, Fig. 2C). Both proteins were shifted to 52 kDa after complete endoglycosidase PNGase F digestion (37°, 1 hr) (lane 4, Fig. 2C), suggesting that they are glycosylated forms of the same peptide backbone. Limited PNGase F treatment (room temperature, 2 min) showed that the 62- and the 59.5-kDa proteins contain four and three oligosaccharide chains, respectively (lanes 2 and 3, Fig. 2C). The two differentially glycosylated proteins (62 and 59.5 kDa) correspond to the molecules with two different topological orientations as shown in Fig. 2B.

Protease Digestion/Membrane Protection

Because RM are closed vesicles with a cytoplasmic-side-out orientation, protein products or their domains located cytoplasmically are exposed and can be digested by proteases, whereas the ones located in the lumen (extracellular) are protected from digestion by the membranes. The membrane protection can be confirmed by permeabilizing the membrane with 1% Triton X-100 prior to the digestion.

To perform a digestion, membranes from an 8-μl translation reaction are collected and resuspended in 60 μl STBS (0.25 *M* sucrose, 10 m*M* Tris-HCl, pH 7.4, 150 m*M* NaCl), and then divided into three aliquots. To one aliquot, 1 μl 20% Triton X-100 and 1 μl 1 mg/ml proteinase K is added. To the second aliquot, 1 μl proteinase K only is added. The third aliquot is used as an untreated control. The reaction is performed on ice for 30 min and stopped by addition of 2 μl of 100 m*M* PMSF. The membrane fraction is then collected by centrifugation and immediately solubilized in sample buffer for SDS–PAGE. If desired, the membrane-protected fragments can be subjected to endoglycosidase treatment as described earlier to confirm the lumenal location of oligosaccharide chains attached to the membrane-protected fragments. To identify the protease-resistant reporter, monoclonal antibody C219, which has an epitope in the reporter, can be used to immunoprecipitate the membrane-protected peptides.

Figure 3D shows a digestion profile of the translation products of pGPGP-N4. Proteins of 42, 39, and 19 kDa were resistant to proteinase K digestion (lane 1, Fig. 3D). Treatment of these fragments with endoglycosidase PNGase F reduced the size of all three protease-resistant fragments (lane 2, Fig. 3D), indicating that they all are glycosylated. The 19-kDa peptides presumably represent the fragment containing the TM1, TM2, and the glycosylated loop (see Fig. 2B). The 42-kDa peptide represents the glycosylated reporter located in the RM lumen as shown for the model II structure (Fig. 2B). The 39-kDa peptide was shown to be an internally initiated protein product that has been translocated into the RM lumen.[14]

Control of Topological Folding of P-Glycoprotein

Using the method just described, an alternate topology of Pgp was observed (Fig. 4). To determine what controls the membrane folding of a Pgp sequence, we investigated whether other cytoplasmic factors and charged amino acids in Pgp involve controlling the Pgp topology. In the pGPGP-N4 protein, two positive charges between the TM3 and TM4 were mutated to neutral or negatively charged residues (Fig. 3A). The effects of these mutations on the membrane folding of the TM3 and TM4 were investigated using the cell-free system. As shown in lane 1 of Fig. 3B, the

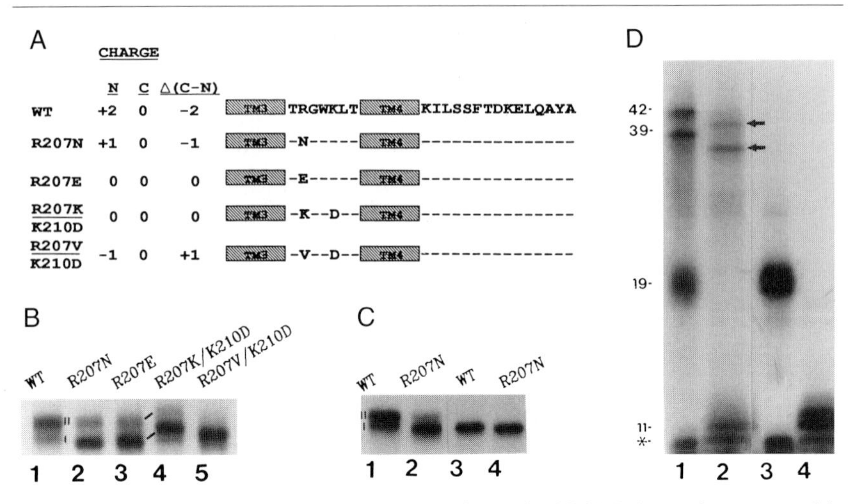

FIG. 3. Topology of mutant pGPGP-N4 proteins and pGPGP-N4 proteins generated in WGE. (A) Amino-acid sequences flanking TM4 of wild-type and mutant pGPGP-N4 proteins. The 7 amino acids at N terminus and 15 amino acids at C terminus of TM4 (hatched bar) are shown. The dashed lines denote sequence identity of mutants with the wild-type molecule. The mutations at N terminus to TM4 are shown at specific positions. Net charge of N-terminal (N) and C-terminal (C) sequences of TM4 and the total net charge [Δ(C–N)] flanking TM4 are shown on the left with the name of each construct. (B) Expression of wild-type and mutant pGPGP-N4 proteins. Wild-type and mutant pGPGP-N4 proteins were translated in the presence of RM and the membrane fraction was analyzed by SDS–PAGE. I and II denote molecules with model I and II orientations, as shown in Fig. 2B. (C) *In vitro* translation of wild-type and mutant pGPGP-N4 proteins in WGE. Wild-type (lanes 1 and 3) and mutant (lanes 2 and 4) pGPGP-N4 were translated in RRL (lanes 1 and 2) or WGE (lanes 3 and 4) in the presence of RM. I and II indicate the model I (62-kDa) and II (59.5-kDa) structures, as shown in Fig. 2B, respectively. (D) Proteinase K treatment of pGPGP-N4 proteins. Wild-type pGPGP-N4 proteins translated in RRL (lanes 1 and 2) and WGE (lanes 3 and 4) were treated by proteinase K (lanes 1 and 3), and proteinase K followed by PNGase F (lanes 2 and 4). The protected C-terminal reporter was observed only in the RRL system. (Data were published in Zhang *et al.*[27] and Zhang and Ling[30] and are reproduced with permission. Copyright 1995 American Chemical Society.)

wild-type sequence generated more 62-kDa protein (model II topology) than the 59.5-kDa protein (model I topology) (see Fig. 2B for the topology). However, mutations of the Arg-207 alone reduced the production of the model II topology (lanes 2 and 3, Fig. 3B). Mutation of both Arg-207 and Lys-210 to neutral or negatively charged amino acids eliminated the generation of the model II topology (lane 5, Fig. 3B). Thus, generation of the model II topology of pGPGP-N4 proteins was apparently due to the presence of the two positive charges between the TM3 and TM4.

When the wild-type pGPGP-N4 transcript was used to direct translation

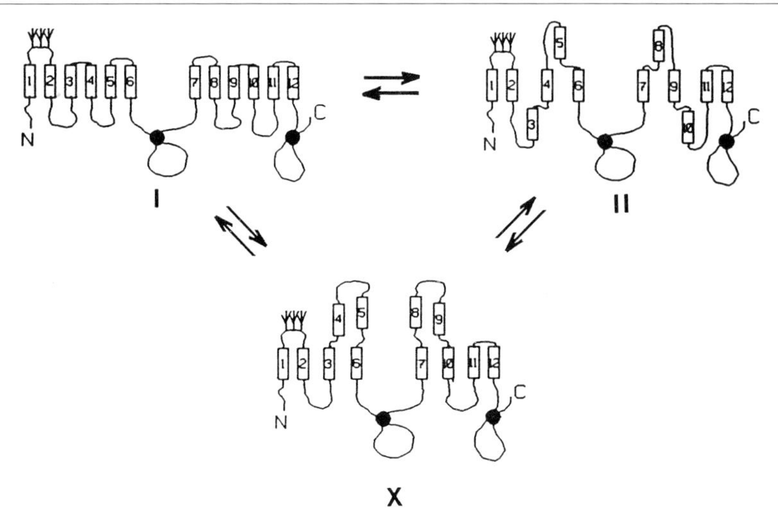

FIG. 4. Topological model of P-glycoprotein. The model I structure is predicted based on hydropathy plot analysis of amino acid sequence. The model II structure is the alternate topology observed with Pgp expressed in cell-free expression systems. The model denoted by X is a potential intermediate topology during the conformational change. The C-terminal half in the model X has been observed by Skach *et al.*[18] using frog oocyte expression system. The branched symbols indicate oligosaccharide chains. The solid circle represents the ATP-binding domain. Conformational changes of Pgp may be involved in transporting drug substrates.

in the WGE expression system, only the 59.5-kDa protein (model I topology) was produced, whereas both 59.9- and 62-kDa proteins were produced in RRL (compare lane 1 with lane 3, in Fig. 3C). However, the model II molecule was produced from the R207N mutant in the WGE system, suggesting that the charges are also important in the WGE system. Further study showed that the WGE system contains a protein factor, which is missing in RRL, that may affect the membrane folding of Pgp sequences.

Discussion

Using the method described, the topology of Pgp has been determined. Interestingly, an alternate topology of Pgp was observed.[14,17] An alternate topology for Pgp has also been observed using *Xenopus* oocytes and bacteria expression systems.[18–20] The finding of a possible alternate topology of Pgp

[17] J. T. Zhang and V. Ling, *J. Biol. Chem.* **266,** 18224 (1991).
[18] W. R. Skach, M. C. Calayag, and V. R. Lingappa, *J. Biol. Chem.* **268,** 6903 (1993).
[19] E. Bibi and O. Béjà, *J. Biol. Chem.* **269,** 19910 (1994).
[20] O. Béjà and E. Bibi, *J. Biol. Chem.* **270,** 12351 (1995).

is important for our understanding of the molecular basis of drug transport by Pgp. A topological change may be involved during drug transport (Fig. 4). Thus an alternate topology for Pgp may represent a different functional state of the molecule. The cytoplasmically located hydrophobic segments TM3, TM10, and their flanking sequences may be involved in binding drugs in the cytoplasm. Through conformational change, the TM3 and TM10 may help translocate the drug substrate from cytoplasm to the outside of cells where the extracellularly located hydrophobic segments TM5, TM8, and their flanking sequences help release drugs. Energy from ATP hydrolysis may be used to support this topological conversion. Recently, the gating of colicin Ia, a voltage-sensitive ion channel, has been shown to involve a massive change in membrane topology between the opening and closed states.[21] At least 31 amino acids of this colicin appear to be translocated from the *cis* side of the membrane to the *trans* side when the channel changes from the closed to the open state.

The fidelity and quality control of the cell-free systems have been questioned.[22] It was shown that misfolded mutant Glut1 glucose transporters were generated in the RRL, whereas only the correctly folded proteins were detected in a frog oocyte expression system. A Cys-less mutant human *MDR1* Pgp was also found to have only the predicted topology when expressed in mammalian cells.[23] However, *MDR1* Pgp molecules with an alternate topology in the C-terminal half were detected in a frog oocyte expression system.[18] The alternate topology of Pgp observed in cell-free systems has also been demonstrated in the plasma membrane of MDR cells using site-specific antibodies[24] or epitope insertion.[25] Kast *et al.*[26] also mapped the topology of Pgp using epitope insertion. Unfortunately, the important loops linking TM3, TM4, and TM5 were not addressed by these authors due to functional disruption. Furthermore, it has been shown that the information for an alternate topology of Pgp was encoded in its amino acid sequence.[27] A study of CFTR using a cell-free system[28] generated only the predicted topology and it is the same as that expressed in Chinese Hamster Ovary (CHO) cells.[29] Observation of the predicted and the alter-

[21] S. L. Slatin, X-Q. Qiu, K. S. Jakes, and A. Finkelstein, *Nature* **371,** 158 (1995).
[22] R. C. Hresko, M. Kruse, M. Strube, and M. Mueckler, *J. Biol. Chem.* **269,** 20482 (1994).
[23] T. W. Loo and D. M. Clarke, *J. Biol. Chem.* **270,** 843 (1995).
[24] M. Zhang, G. Wang, A. Shapiro, and J. T. Zhang, *Biochemistry* **35,** 9728 (1996).
[25] A. Bragin, and W. R. Skach, *Proc. AACR* **37,** 327a (1996).
[26] C. Kast, V. Canfield, R. Levenson, and P. Gros, *Biochemistry* **34,** 4402 (1995).
[27] J. T. Zhang, C.-H. Lee, M. Duthie, and V. Ling, *J. Biol. Chem.* **270,** 1742 (1995).
[28] M. A. Chen and J. T. Zhang, *Mol. Memb. Biol.* **13,** 33 (1996).
[29] X.-B. Chang, Y.-X. Hou, T. J. Jensen, and J. R. Riordan, *J. Biol. Chem.* **269,** 18572 (1994).
[30] J. T. Zhang and V. Ling, *Biochemistry* **34,** 9159 (1995).

nate topologies of Pgp in different laboratories may reflect "two sides of the same coin." Other unstable intermediate states or topologies may also exist (Fig. 4). Further investigation will elucidate whether there is a relationship between topology and drug transport by Pgp.

Acknowledgment

The author would like to thank Dr. Victor Ling and Dr. Luis Reuss for their support of the P-glycoprotein topology study. Critical comments on this manuscript by Dr. Luis Reuss are greatly appreciated. This work was supported by grant CA64539, National Institutes of Health.

[21] Photoaffinity Labels for Characterizing Drug Interaction Sites of P-Glycoprotein

By AHMAD R. SAFA

Introduction

The demonstration by several investigators that the membranes of multidrug-resistant rodent and human cells contained a 150- to 180-kDa phosphoglycoprotein termed P-glycoprotein[1] suggested that this protein might be involved in mediating multidrug resistance (MDR). Analysis of the MDR cell lines and drug uptake experiments have revealed that reduced drug accumulation is responsible for MDR. Decreased drug accumulation is partially the result of an increased rate of drug efflux. Much evidence indicates that overexpression of P-glycoprotein is responsible for increased energy-dependent drug efflux from MDR cells.[2] We[3,4] and others[5] have shown by transport experiments using plasma membrane vesicles obtained from MDR cells and apical plasma membranes of bile canaliculi possessing P-glycoprotein that vinblastine, vincristine, daunorubicin, colchicine, verapamil, and cyclosporin A can be transported into inside-out membrane vesicles.[3–5] This transport is temperature and adenosine triphosphate (ATP) dependent, occurs against concentration gradients, and follows Michaelis–Menten kinetics.

[1] J. A. Endicott and V. Ling, *Annu. Rev. Biochem.* **58,** 137 (1989).

[2] M. M. Gottesman and I. Pastan, *Annu. Rev. Biochem.* **62,** 385 (1993).

[3] I. Tamai and A. R. Safa, *J. Biol. Chem.* **265,** 16509 (1990).

[4] F. A. Sinicrope, P. K. Dudeja, B. M. Bissonnette, A. R. Safa, and T. A. Brasitus, *J. Biol. Chem.* **267,** 24995 (1992).

[5] M. Horio, M. M. Gottesman, and I. Pastan, *Proc. Natl. Acad. Sci. U.S.A.* **85,** 3580 (1990).

0076-6879/98 $25.00

P-Glycoprotein is encoded by a family of two closely linked genes (MDR1 and MDR2, also called MDR3) on chromosome 7 in humans, three genes in Chinese hamsters (*pgp1, pgp2,* and *pgp3*), and three genes in mice (*mdr1, mdr2,* and *mdr3,* or *mdr1b, mdr2,* and *mdr1a,* respectively).[2,6] Transfection experiments using gene-specific complementary DNA have shown conclusively that the human MDR1 gene or the corresponding Chinese hamster (*pgp1* and *pgp2*) or mouse genes (*mdr1a* and *mdr1b*) are sufficient to confer the MDR phenotype in sensitive cells. However, transfection experiments with full-length human MDR2 (MDR3), *pgp2,* or *mdr2* cDNA have failed to confer MDR phenotype in sensitive cells.[2,6,7]

The putative amino acid sequence shows that human and rodent P-glycoprotein are composed of 1280 and 1276 amino acids, respectively, and have glycosylation sites in the first extracytoplasmic loop.[2,7,8] A striking structural feature of P-glycoprotein is an internal duplication that includes approximately 500 amino acids. Each duplicate segment of the molecule has a consensus ATP-binding site and six hydrophobic transmembrane domains with significant homology to bacterial transport proteins.[7,8] The protein binds to ATP[9] and its pure form displays Mg^{2+}-dependent ATPase activity.[10]

Delineating the architecture of the drug-binding sites of P-glycoprotein for both cytotoxic agents and the MDR modulators will be invaluable not only to understand how drugs interact with this protein, but also to design more useful and specific inhibitors of P-glycoprotein function. One way to identify and characterize P-glycoprotein drug-binding sites is to use photoaffinity analogs of both MDR-related agents and MDR modulators. Therefore, the synthesis and use of photoaffinity analogs in identifying, characterizing, and localizing the drug-binding sites of P-glycoprotein structure are reviewed in this article.

The identification of the P-glycoprotein drug-binding sites by photoaffinity labeling is based on the assumption that a reversible complex occurs between the photoactive drug analog and P-glycoprotein binding sites that specifically recognizes the structural characteristics of the parent drug. Ultraviolet irradiation then converts the photoactive analog into a reactive

[6] J. L. Biedler, *Cancer* (*Supp.*) **70,** 1799 (1992).

[7] I. B. Roninson, I. Pastan, and M. M. Gottesman, *in* "Molecular and Cellular Biology of Multidrug Resistance in Tumor Cells" (I. B. Roninson, ed.), pp. 91–104. Plenum Press, New York and London, 1991.

[8] P. Gros, M. Ramond, and D. Housman, *in* "Molecular and Cellular Biology of Multidrug Resistance in Tumor Cells" (I. B. Roninson, ed.), pp. 73–88. Plenum Press, New York and London, 1991.

[9] M. M. Cornwell, T. Tsuruo, M. M. Gottesman, and I. Pastan, *FASEB J.* **1,** 51 (1987).

[10] H. Hamada and T. Tsuruo, *J. Biol. Chem.* **263,** 1454 (1988).

nitrene intermediate, which then covalently attaches at or near the drug-binding sites. To identify P-glycoprotein as a specific drug-binding protein, we synthesized a number of photoaffinity analogs of cytotoxic agents and MDR modulators. To synthesize these photoaffinity analogs, we employed two approaches: (1) In most cases, we synthesized primary amine derivatives of the parent drugs and then reacted them with *N*-hydroxysuccinimidyl-4-azido[3,5-^{3}H]benzoate([^{3}H]NAB) or *N*-hydroxysuccinimidyl-4-azido[3-^{125}I]salicylate([^{125}I]NAS), or (2) the nonradioactive photoaffinity analogs were synthesized and the final compound was then radioiodinated.

Synthesis and Use of Photoaffinity Analogs of MDR-Related Drugs and MDR Modulators

We originally synthesized and used photoaffinity analogs of vinblastine, *N*-(*p*-azido[3-^{125}I]salicyl)-*N*′-β-aminoethylvindesine ([^{125}I]NASV) (Fig. 1) and *N*-(*p*-azido[3,5-^{3}H]benzoyl)-*N*′-β-aminoethylvindesine[11,12] ([^{3}H]NABV) to identify and characterize the drug-binding sites of P-glycoprotein. For the synthesis of [^{125}I]NASV, a two-step reaction is carried out. In the first step, 1.67 nmol of *N*-hydroxysuccinimidyl-4-azidosalicylate (NAS) (Pierce Chemical Co., Rockford, IL) are dissolved in 15 μl of acetonitrile. Five micromoles of 0.5 *M* sodium phosphate, pH 7, and 5 mCi Na^{125}I (specific activity 2200 Ci/mmol) in 10 μl of 0.1 *M* NaOH (Amersham, Arlington Heights, IL) are added. Then 2.5 nmol of chloramine-T in 10 μl of a mixture of acetonitrile and dimethylformamide (1 : 1, v/v) are added, and the mixture is kept for 2 min at room temperature. Then 300 μl of 10% NaCl is added and the reaction mixture is extracted with 300 μl of ethyl acetate. Two microliters of the reaction mixture is chromotographed on a silica gel G thin layer using benzene, chloroform, ethyl acetate, and acetic acid (1 : 1 : 1 : 0.1, v/v).[13] As revealed by autoradiography, the product [^{125}I]NAS gives one main spot of radioactivity (R_f 0.4).

In the second step, [^{125}I]NAS is dissolved in chloroform and 0.5 μmol of *N*-β-aminoethylvindesine[12] is added. The mixture is kept at 4° for 48–72 hr. The [^{125}I]NASV is purified by either reversed-phase high-performance liquid chromatography (HPLC) or silica gel chromatography.[11,12] The HPLC method consists of a 30-min gradient of 30–80% methanol in distilled H_2O containing 0.19% diethylamine at a flow rate of 1.5 ml/min. In this system, [^{125}I]NASV has a retention time of 22.6 min. [^{125}I]NASV gives a

[11] A. R. Safa, C. J. Glover, M. B. Meyers, J. L. Biedler, and R. L. Felsted, *J. Biol. Chem.* **261,** 6137 (1986).

[12] A. R. Safa and R. L. Felsted, *J. Biol. Chem.* **262,** 1261 (1987).

[13] I. Ji, J. Shin and T. H. Ji, *Anal. Biochem.* **151,** 348 (1985).

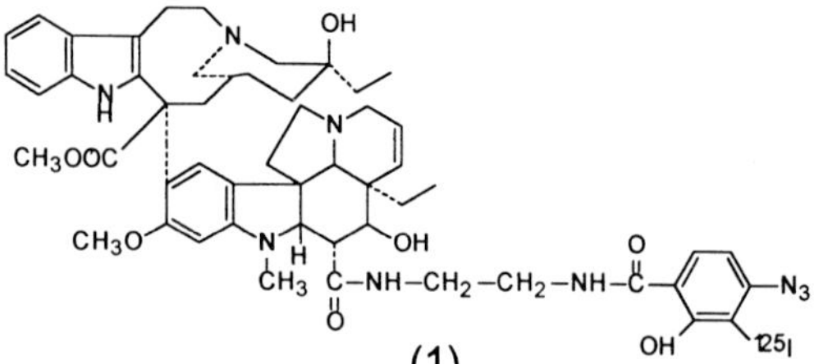

(1)

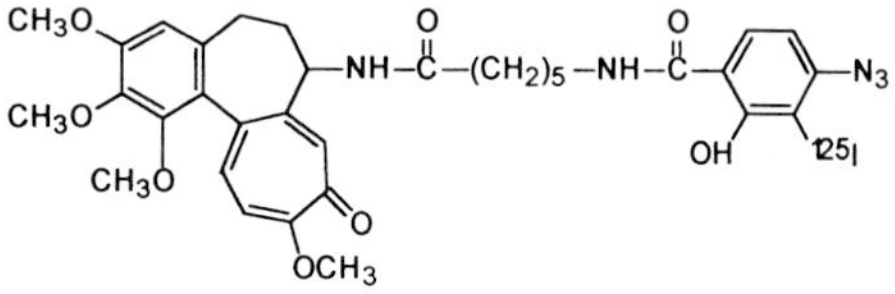

(2)

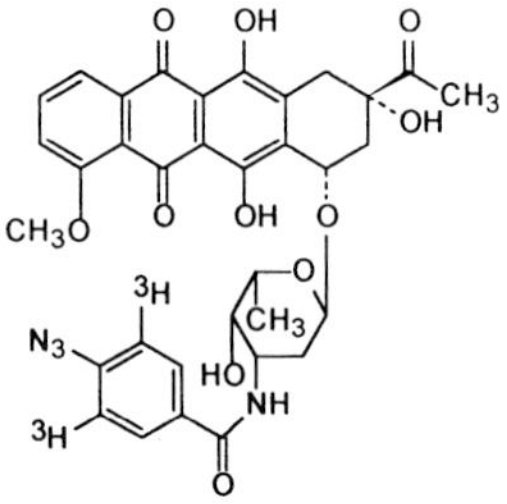

(3)

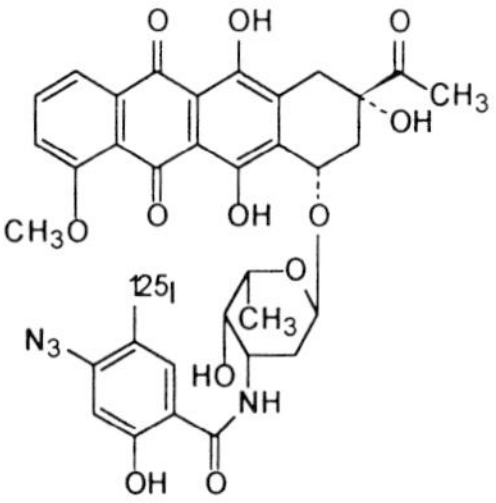

(4)

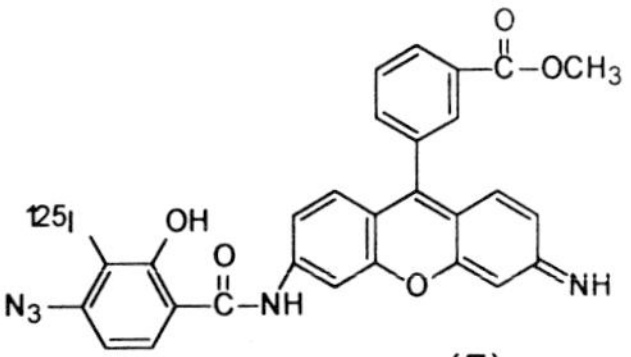

(5)

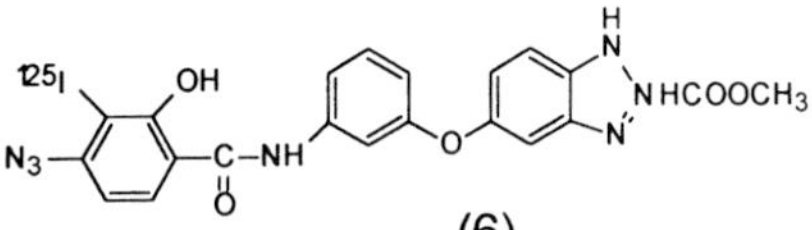

(6)

FIG. 1. Structures of photoaffinity analogs of MDR-related drugs known to covalently label P-glycoprotein: (1) [^{125}I]NASV, (2) [^{125}I]NASC, (3) [^{3}H]NAB–DNR, (4) [^{125}I]NAS–DNR, (5) [^{125}I]ASA–Rh 123, and (6) [^{125}I]ASA–BZ.

single radioactive spot in silica gel thin-layer chromatography (TLC) with fluorescent indicator (Analtech, Newark, DE), with R_f of 0.66 for solvent I (chloroform: methanol: H_2O, 80:20:2, v/v) and R_f of 0.78 for solvent II (chloroform: methanol: 40% aqueous methylamine, 80:20:4, v/v) as revealed by autoradiography.

For purification using silica gel chromatography, the reaction mixture is absorbed to a silica gel column (0.4 × 5 cm) equilibrated in chloroform to remove impurities. The column is then washed with 1% (v/v) methanol in chloroform and the [^{125}I]NASV is removed by 3% (v/v) methanol in chloroform. The purified [^{125}I]NASV is dried with a stream of nitrogen gas. The product purified by this method is 90% pure as indicated by TLC using solvents I and II.

For the synthesis of [^{3}H]NABV, 4.5 nmol of *N*-hydroxysuccinimidyl-4-azido[3,5-^{3}H]benzoate (50.6 Ci/mmol) in 0.05 ml 2-propanol (New England Nuclear, Boston, MA) are added to 0.45 ml (0.5–0.7 μmol) solution of β-aminoethylvindesine[12] in chloroform, and the mixture is maintained at 4° for 48–72 hr. The product is purified by reversed-phase HPLC on a μBondapak C_{18} column (Waters Associates, Milford, MA) using a 30-min gradient of 60–90% (v/v) methanol in distilled H_2O containing 0.19% diethylamine at a flow rate of 1.5 ml/min, with an additional 10 min isocratic 90% (v/v) methanol at the end of the gradient. In this system, vinblastine and [^{3}H]NABV have retention times of 9.43 and 16.48 min, respectively. [^{3}H]NABV gives a single radioactive spot on silica gel TLC with R_f of 0.72 using solvent I.[12]

Since the initial synthesis of [^{3}H]NABV and [^{125}I]NASV, we[14–17] and others[18–26] have made a number of P-glycoprotein-specific photoaffinity

[14] A. R. Safa, *Proc. Natl. Acad. Sci. U.S.A.* **85,** 7187 (1988).

[15] A. R. Safa, N. D. Mehta, and M. Agresti, *Biochem. Biophys. Res. Commun.* **162,** 1402 (1989).

[16] W. Priebe, T. Przewloka, I. Fokt, R. Perez-Soler, and A. R. Safa, *Proc. Am. Assoc. Cancer Res.* **36,** 334 (1995).

[17] A. R. Safa, M. Agresti, D. Bryk, and I. Tamai, *Biochemistry* **33,** 256 (1994).

[18] W. T. Beck and X.-D. Qian, *Biochem. Pharmacol.* **43,** 89 (1992).

[19] S.-I. Akiyama, A. Yoshimura, H. Kikuchi, T. Sumizawa, M. Kuwano, and T. Tahara, *Mol. Pharmacol.* **36,** 730 (1989).

[20] B. M. J. Foxwell, A. Mackie, V. Ling and B. Ryffel, *Mol. Pharmacol.* **36,** 543 (1989).

[21] D. I. Morris, L. A. Speicher, A. E. Ruoho, K. D. Tew, and K. B. Seamon, *Biochemistry* **30,** 8371 (1991).

[22] N. Baker, R. K. Prichard, and E. Georges, *Mol. Pharmacol.* **45,** 1145 (1994).

[23] B. Nare, Z. Liu, R. K. Prichard, and E. Georges, *Biochem. Pharmacol.* **48,** 2115 (1994).

[24] C. Borchers, W.-R. Ulrich, K. Klemm, W. Ise, V. Gekeler, S. Haas, A. Schodl, J. Conrad, M. Przybylski, and R. Boer, *Mol. Pharmacol.* **48,** 21 (1995).

[25] L. A. Speicher, N. Laing, L. R. Barone, J. D. Robbins, K. B. Seamon, and K. D. Tew, *Mol. Pharmacol.* **46,** 866 (1994).

[26] L. A. Speicher, L. R. Barone, A. E. Chapman, G. R. Hudes, N. Laing, C. D. Smith, and K. D. Tew, *J. Natl. Cancer Inst.* **86,** 688 (1994).

analogs. We prepared the photoaffinity analogs of colchicine, *N*-(*p*-azido[3,5-^{3}H]benzoyl)aminohexanoyl deacetylcolchicine ([^{3}H]NABC) and *N*-(*p*-azido[3-^{125}I]salicyl)aminohexanoyl deacetylcolchicine ([^{125}I]NASC) (Fig. 1), by procedures similar to those described for [^{3}H]NABV and [^{125}I]NASV using aminohexanoyl deacetylcolchicine trifluoroacetate, [^{3}H]NAB and [^{125}I]NAS, respectively.[15]

We[16,27] and others[18] have synthesized photoaffinity analogs of daunorubicin (DNR) and doxorubicin (DOX). A photoaffinity analog of DOX, *N*-(*p*-azido[3,5-^{3}H]benzoyl)doxorubicin ([^{3}H]NAB–DOX) (Fig. 1) is synthesized by using 100 μl of 1 mg/ml DOX in methanol mixed with 50 μl of [^{3}H]NAB. After keeping the mixture at room temperature for 72–96 hr, the reaction mixture is applied to silica gel TLC, which is developed in solvent III (chloroform : methanol : H_2O, 80 : 20 : 3, v/v). The silica gel TLC is scraped in 0.5-cm increments and the radioactivity is determined by scintillation counter. The product, [^{3}H]NAB–DOX, gives a spot on TLC with R_f of 0.68. [^{3}H]NAB–DOX can be purified from the silica gel by scraping the radioactive band from the plate. [^{3}H]NAB–DOX can be extracted from silica gel by 10% methanol in chloroform, dried by nitrogen gas, and kept in dimethyl sulfoxide (DMSO) for photolabeling experiments. A similar procedure has been used by Beck and Qian[18] to prepare [^{3}H]NAB–DNR (Fig. 1). Detailed alternative procedures for synthesizing these photoaffinity probes have also been reported.[28] A radioiodinated photoaffinity analog of DNR, *N*-(*p*-azido[3-^{125}I]salicyl)daunorubicin ([^{125}I]NAS–DNR or [^{125}I]WP622) (Fig. 1) can be synthesized by preparing the nonradioactive NAS–DNR by azidosalicylation of DNR with nonradioactive NAS[16] (Pierce Chemical Co., Rockford, IL) and then radioiodinating NAS–DNR as described earlier for preparing [^{125}I]NAS.

Using procedures similar to those already described, the syntheses of two more photoaffinity analogs of MDR-related cytotoxic agents, rhodamine 123 ([^{125}I]azidosalicylic (ASA) Rh 123) and benzimidazole ([^{125}I]ASA–BZ),[22,23] have also been reported.

Among MDR modulators, we initially reported binding of a dihydropyridine photoaffinity analog, [^{3}H]azidopine[28] (Fig. 2). Subsequently, we reported the synthesis and use of two photoaffinity analogs of verapamil, *N*-(*p*-azido[3-^{125}I]salicyl)aminomethylverapamil ([^{125}I]NAS–VP) (Fig. 2) and *N*-(*p*-azido[3,5-^{3}H]benzoyl)aminomethylverapamil ([^{3}H]NAB–VP).[14] For preparing these photoprobes, a methylamine derivative of verapamil,

[27] A. R. Safa, M. Agresti, and S. Kajiji, *Proc. 9th NCI–EORTC Symp. New Drugs Cancer Therapy,* p. 119 (Abstract 419), 1996.

[28] A. R. Safa, C. J. Glover, J. L. Sewell, M. B. Meyers, J. L. Biedler, and R. L. Felsted, *J. Biol. Chem.* **262,** 7884 (1987).

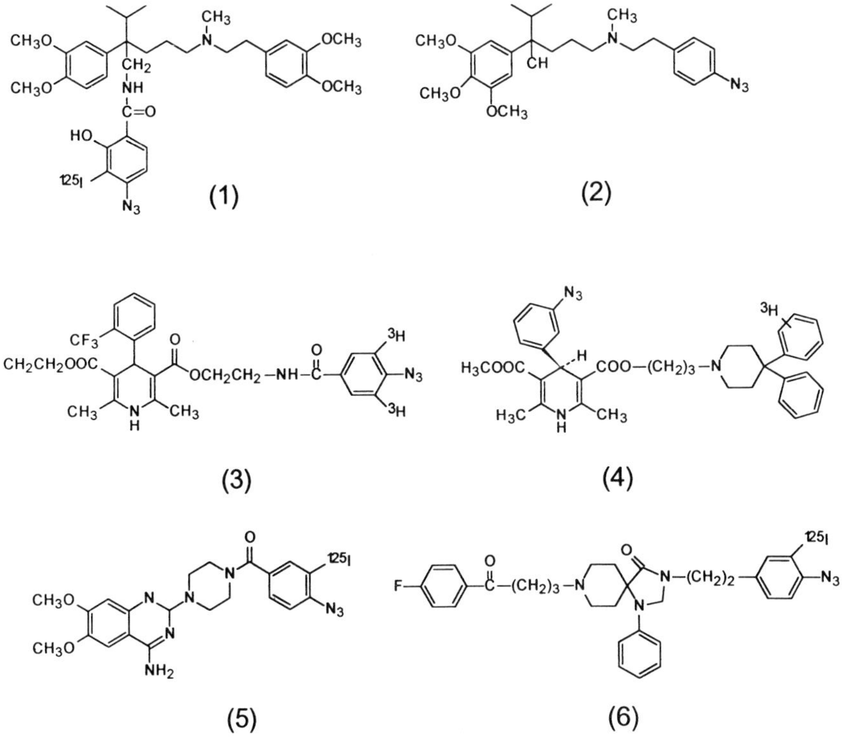

(1) (2)

(3) (4)

(5) (6)

FIG. 2. Structures of photoaffinity analogs of MDR modulators known to covalently label P-glycoprotein: (1) [^{125}I]NAS-VP, (2) LU-49888, (3) [^{3}H]azidopine, (4) [^{3}H]B9209-005, (5) [^{125}I]AAP, and (6) [^{125}I]NAPS.

(±)-5-[(3,4-dimethoxyphenethyl)methylamino]-2-(3,4-dimethoxyphenyl)-2-isopropylpentamine (DMDI)[29] is synthesized and reacted with [^{3}H]NAB or [^{125}I]NAS as described earlier for the preparation of [^{3}H]NABV and [^{125}I]NASV. The products are purified by a silica gel column as described for the preparation of the photoaffinity analogs of vinblastine. [^{3}H]NAB–VP gives a radioactive spot on silica gel TLC plates with fluorescent indicator (Analtech, Newark, DE) (solvent I, R_f of 0.76). [^{125}I]NAS–VP gave an R_f of 0.68 using solvent II. [^{3}H]NAB–VP can also be purified from silica gel TLC plates using solvent IV (chloroform : methanol : acetic acid, 100 : 5 : 2.5, v/v). In this latter solvent, [^{3}H]NAB–VP has an R_f of 0.24, DMDI an R_f of 0.04, an unreacted NAB an R_f of 0.67.

[29] L. J. Theodore, W. L. Nelson, R. H. Zobrist, K. M. Giacomini, and J. C. Giacomini, *J. Med. Chem.* **29,** 1784 (1986).

Alternatively, [^{125}I]NAB–VP can be prepared using nonradioactive NAS–VP by dissolving NAS (10 μmol) (Pierce Chemical Co., Rockford, IL) and DMDI (5 μmol) in 1 ml of tetrahydrofuran and incubating overnight at 4°. The product, NAS–VP, is purified by silica gel TLC as described earlier and then radioiodinated as described for the iodination of NAS. Since the initial synthesis of verapamil photoaffinity probes, several photoaffinity analogs of MDR modulators have been synthesized (Figs. 2 and 3) and used to characterize the drug-binding sites of P-glycoprotein.

Photoaffinity Labeling

The photoaffinity mixture contains 1–100 mM ^{125}I-labeled photoaffinity analog (specific activity 40–2200 Ci/mmol) or 50 nM [^{3}H]azidopine (50 Ci/

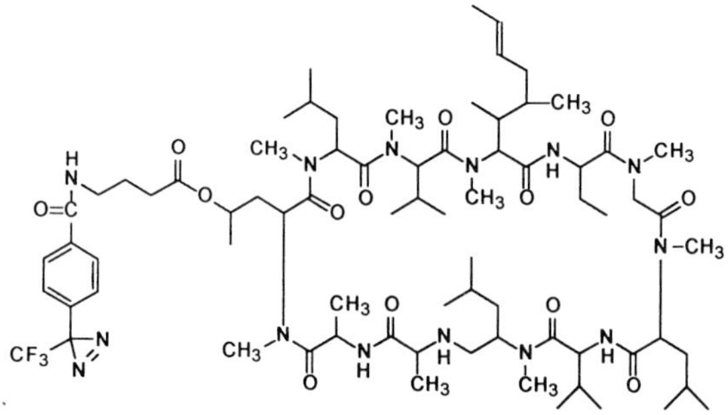

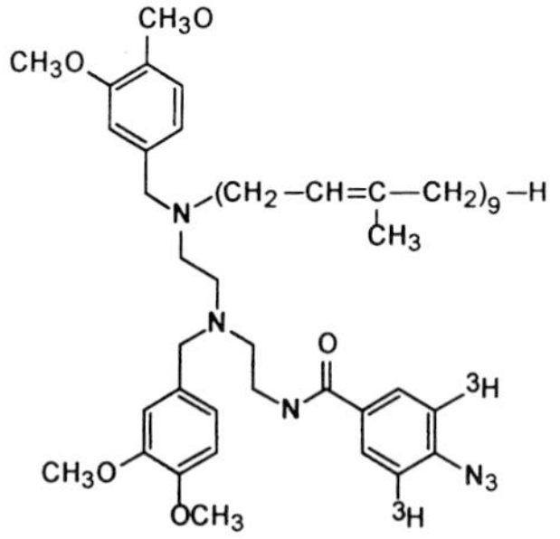

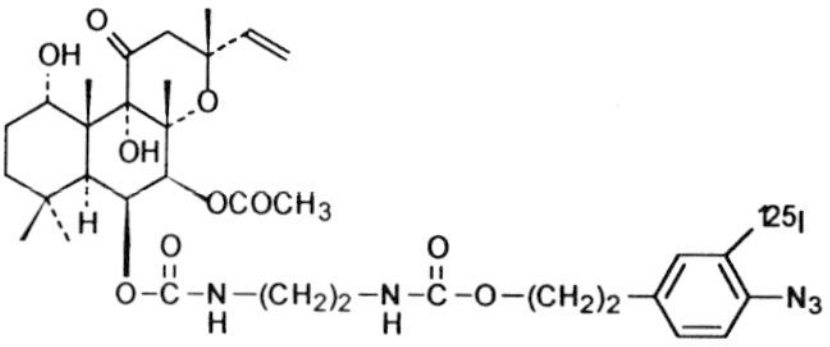

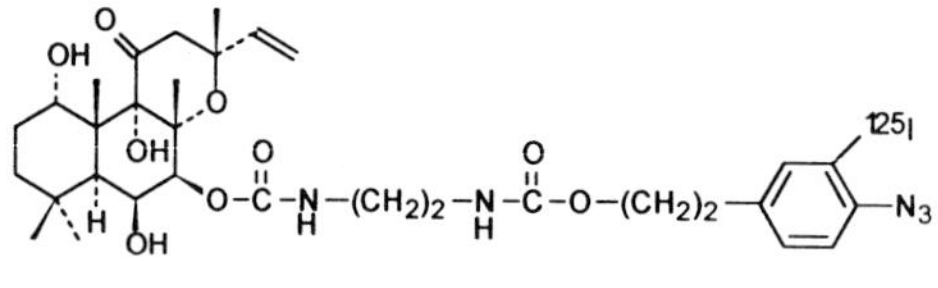

Fig. 3. Structures of photoaffinity analogs of MDR modulators known to covalently label P-glycoprotein: (1) cyclosporin aziridine, (2) synthetic isoprenoid, *N*-solanesyl-*N'*,*N'*-bis(3,4-dimethoxybenzoyl)ethylenediamine, (3) and (4) forskolin analogs, and (5) estramustine photoaffinity analog.

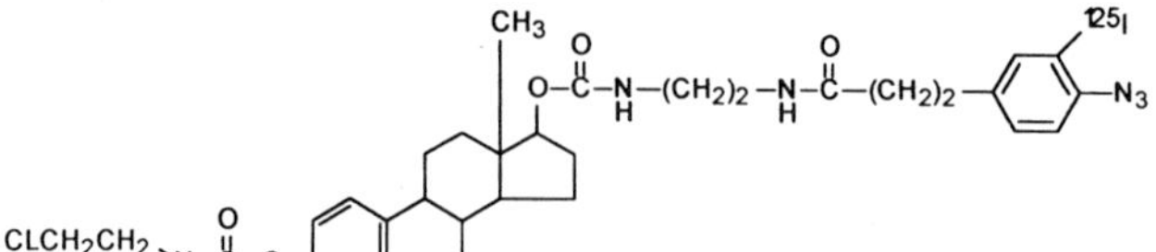

mmol), 40 mM potassium phosphate buffer (pH 7.0)[11] or 10 mM Tris-HCl (pH 7.4), 250 mM sucrose, 10 mM NaCl (buffer A) containing 4% dimethyl sulfoxide and 10–100 μg plasma membranes in a final volume of 0.05 ml in polyvinyl chloride V-microtiter wells (Dynatech Laboratories, Alexandria, VA). Photolabeling with intact viable cells ($0.5–2 \times 10^6$ cells) is carried out in Hanks' balanced salt solution (HBSS). Samples in the presence or absence of competitors are incubated for 15–30 min to equilibrate and photoactivation is accomplished by 10–20 min of irradiation with a UV lamp equipped with two 15-W self-filtering 302- or 306-nm lamps (Model xx-15, Ultra-Violet Products, San Gabriel, CA) suspended 3.4 cm above the photolabeling mixtures in a water bath at room temperature. After UV irradiation, samples are processed for sodium dodecyl sulfate–polyacrylamide gel electrophoresis (SDS–PAGE), fluorography, or autoradiography.

Immunoprecipitation

For immunoprecipitation of P-glycoprotein, 1×10^7 cells in 200 μl HBSS and/or membrane vesicles in 200 μl buffer A (0.2–0.5 mg/assay) photoaffinity labeled with the radioactive probes are solubilized for 1 hr at 4° with 200 μl radioimmune precipitating (RIPA) buffer [1% sodium deoxycholate, 1% Triton X-100, 0.1% SDS, 2% aprotinin, 1 mM phenylmethylsulfonyl fluoride (PMSF), 1 mM EDTA, and 0.15 M NaCl in 50 mM Tris-HCl, pH 7.0]. The mixture is then incubated with 20 μl of anti-P-glycoprotein rabbit polyclonal antisera or anti-P-glycoprotein monoclonal antibody for 1 hr at 4°. Nonimmune rabbit sera or isotype-specific IgG can be used as controls. Then 100 μl of protein A-Sepharose is added and samples are incubated for 1 hr at 4°. Immune complexes are recovered at 13,000g at 4° after washing once with RIPA buffer and twice with 50 mM Tris-HCl buffer (pH 7.0) containing 150 mM NaCl, 1 mM EDTA, and 1 mM PMSF (buffer B). The immune complexes are then eluted with SDS-sample buffer without heating before separation of solubilized antigen by SDS–PAGE.

SDS–PAGE

Photolabeled samples are processed by mixing an equal volume of SDS sample buffer (4% SDS, 8 M urea, 0.1 M dithiothreitol, 20% glycerol, 0.04% bromophenol blue and 0.08 M Tris-HCl, pH 6.8), and the suspensions are transferred to 0.5-ml microcentrifuge tubes and heated for 5 min at 100° in a heating block. Aliquots of 75 μl are applied to separate slots of a polyacrylamide slab gel. SDS–PAGE is run with a slab gel apparatus (Model 220, Bio-Rad Laboratories, Richmond, CA) on 1.5-mm-thick gels made with a 15-slot sample comb. Electrophoresis is carried out as previously described[30] using a linear polyacrylamide gradient of 5% polyacrylamide/4.5 *M* urea to 15% polyacrylamide/4.5 *M* urea, with a 4% polyacrylamide/4.5 *M* urea stacking gel. Electrophoresis is performed at 8 mA/gel for 16 hr. The gel is fixed and stained in 0.25% Coomassie Brilliant Blue in 45% (v/v) methanol and 10% (v/v) acetic acid, and destained by diffusion against 20% (v/v) methanol and 10% (v/v) acetic acid.

Low and high molecular weight standards (Bio-Rad Laboratories, Richmond, CA) ranging from 14.4 to 200 kDa are included in all runs. The gels are then dried using a Savant gel drier on Whatman paper. Autoradiography is performed using Kodak X-Omat AR film and a DuPont intensifying screen at −70°. Fluorography is performed by the procedure of Bonner and Laskey[31] prior to drying and exposing the film. The data are quantitated by either (1) excising the gel slices corresponding to the particular bands on the fluorogram or autoradiogram and determining the radioactivity, or (2) after autoradiography or fluorography, a densitometer is used to determine the amount of radioactivity corresponding to P-glycoprotein or its fragments.

Identification and Characterization of Drug-Binding Sites of P-Glycoprotein

Sucrose density gradient purified plasma membranes, mixed membrane vesicles, or intact cells were incubated with [^{125}I]NASV at 25° for 15 min and then irradiated at 302 nm (for [^{3}H]NABV) or 366 nm (for [^{125}I]NASV) for 20 min to activate the analog. Polyacrylamide gel electrophoresis of solubilized photoaffinity labeled samples followed by autoradiography showed that [^{125}I]NASV specifically bound only to a 150- to 170-kDa protein in Chinese hamster cells expressing MDR (Fig. 4A) and in MDR variants of KB-3-1 human epidermoid carcinoma cells, KB-V and KB-C4. A polyclonal antibody recognizing P-glycoprotein immunoprecipitated the 150- to 170-kDa [^{125}I]NASV-bound protein.[11] Immunoprecipitation of [^{125}I]NASV photoaffinity labeled P-glycoprotein using transfectant cell lines expressing either the wild-type Gly-185 or mutant Val-185 P-glycoprotein (KB-GRC1 or KB-VSV1 cell lines, respectively), and the anti-P-glycoprotein monoclonal antibody C219[32] (Fig. 4B) revealed that [^{125}I]NASV binds to P-

[30] U. K. Laemmli, *Nature (London)* **227,** 680 (1970).
[31] W. M. Bonner and A. Laskey, *Eur. J. Biochem.* **46,** 83 (1974).
[32] N. Kartner, D. Evernden-Parelle, G. Bradley, and V. Ling, *Nature (London)* **316,** 820 (1985).

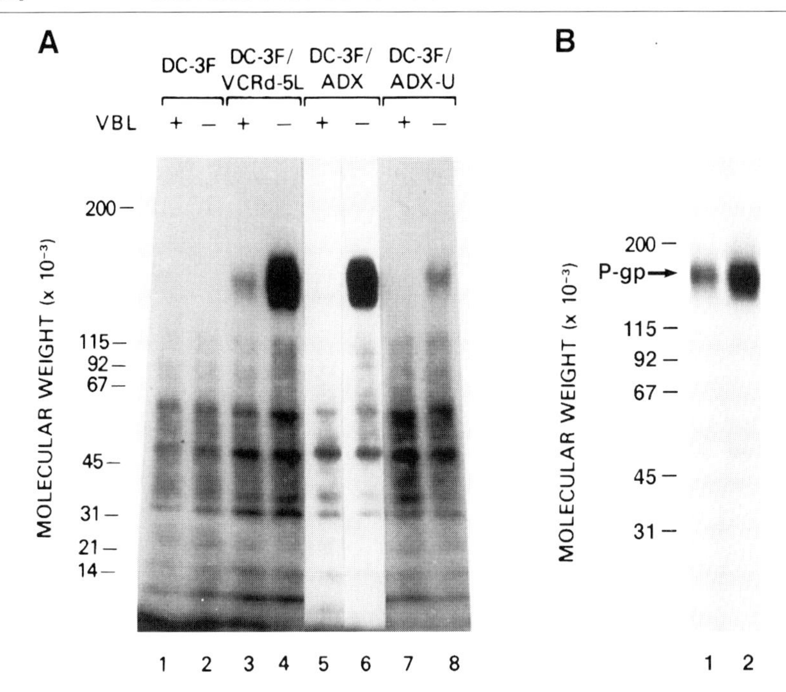

FIG. 4. (A) SDS–PAGE autoradiography of [^{125}I]NASV photoaffinity labeled sucrose density gradient purified plasma membranes (15 μg of protein) from drug-sensitive Chinese hamster DC-3F cells (lanes 1 and 2), drug-resistant DC-3F/VCRd-5L cells (lanes 3 and 4), drug-resistant DC-3F/ADX cells (lanes 5 and 5), and revertant DC-3F/ADX-U cells (lanes 7 and 8). Photoaffinity labeling was carried out in the presence (lanes 1, 3, 5, and 7) or absence (lanes 2, 4, 6, and 8) of 10 μM vinblastine. VBL, vinblastine. (B) SDS–PAGE autoradiography of immunoprecipitates of [^{125}I]NASV photoaffinity labeled detergent solubilized KB-GRC1 or KB-VSV1 cells (8 $\times$ 10^6 cells) obtained with anti-P-glycoprotein monoclonal antibody C219 as described in the text. [Reproduced with permission from Safa *et al. J. Biol. Chem.* **261,** 6137–6140 (1986). Copyright 1986 The American Society for Biochemistry and Molecular Biology.]

glycoprotein and that it binds more effectively to the mutant Val-185 protein.[33]

[^{125}I]NASV photolabeling of P-glycoprotein in membrane vesicles of the MDR Chinese hamster DC-3F/VCRd-5L and human MDR KB-V1 carcinoma cell lines was inhibited in a dose-dependent manner by vinblastine with half-maximal inhibition at 0.5–1 μM.[34] Evidence for the multidrug

[33] A. R. Safa, R. Kaplan-Stern, K. Choi, M. Agresti, I. Tamai, N. D. Mehta, and I. B. Roninson, *Proc. Natl. Acad. Sci. U.S.A.* **87,** 7225 (1990).

[34] M. M. Cornwell, A. R. Safa, R. L. Felsted, M. M. Gottesman, and I. Pastan, *Proc. Natl. Acad. Sci. U.S.A.* **83,** 3847 (1986).

specificity of the P-glycoprotein *Vinca* alkaloid-binding site was also seen, and while vinblastine, actinomycin D, doxorubicin, or daunorubicin inhibited [^{125}I]NASV photolabeling of P-glycoprotein to various degrees, colchicine did not block the binding, suggesting either that colchicine binds to a separate site or that it has lower affinity for the vinblastine binding site of P-glycoprotein.[11,34]

There has been major interest in developing strategies to circumvent MDR *in vitro* and ultimately in resistant tumor cells in patients. It is known that many diverse lipophilic agents enhance drug cytotoxicity in MDR cells. These agents possess many structural and chemical features common to chemotherapeutic drugs that bind to P-glycoprotein,[18,35] and it is known that modulating agents may reverse MDR by competitively inhibiting chemotherapeutic drug binding to P-glycoprotein.[3–5] We[3] and others[5] have clearly demonstrated that many of these agents were able to compete with vinblastine transport and photolabeling of P-glycoprotein.[3,5,34,35] It has been suggested that a common pharmacophore, the minimum set of structural and functional features required for compounds modulating MDR, includes lipophilicity, a basic nitrogen atom, and two planar aromatic domains.[36] Using a series of reserpine and yohimbine analogs, we[37] demonstrated that not only are these domains necessary for interaction with P-glycoprotein and reversal of MDR, but also that they must exist in a well-defined conformation. Moreover, using these analogs, a direct correlation was found between inhibition of [^{125}I]NASV photolabeling and reversal of vinblastine resistance with a correlation coefficient of 0.94.[37] These data suggest that the relative disposition of aromatic rings and the basic nitrogen atom is important for the interaction of a modulator with P-glycoprotein and reversing MDR. Although these structural features may be required for many agents reversing MDR, cyclosporins, which are potent inhibitors of P-glycoprotein function, lack such determinants. One common feature among reversing agents is that all are lipophilic.

Colchicine is known to be transported by P-glycoprotein. To identify P-glycoprotein as a specific acceptor for colchicine, we synthesized and used photoactive analogs of colchicine (Fig. 1).[15] Both ^{3}H- and ^{125}I-labeled photoaffinity analogs bound specifically to P-glycoprotein.[26] Photolabeling of P-glycoprotein by [^{125}I]NASC was inhibited by other compounds to which MDR cells are resistant with the following order: vinblastine > vincristine > doxorubicin > actinomycin D > colchicine.[15] These results

[35] A. R. Safa, *Cancer Invest.* **11,** 46 (1993).

[36] J. M. Zamora, H. L. Pearce, and W. T. Beck, *Mol. Pharmacol.* **33,** 454 (1988).

[37] H. L. Pearce, A. R. Safa, N. J. Bach, M. A. Winter, M. C. Cirtain, and W. T. Beck, *Proc. Natl. Acad. Sci. U.S.A.* **86,** 5128 (1989).

and previous photoaffinity labeling experiments with [^{125}I]NASV[34,35] suggest that vinblastine and colchicine may bind to separate sites or that P-glycoprotein may have a common drug–acceptor site displaying high affinity for *Vinca* alkaloids and lower affinity for colchicine.

A high specific activity ^{125}I-labeled Bolton–Hunter derivative of daunorubicin, iodomycin, which like daunorubicin is an inherently photoactive compound, can bind specifically to P-glycoprotein. The binding of [^{125}I]iodomycin was inhibited by vinblastine > daunorubicin > colchicine.[34] We[35] and others[18] have also used an ^{3}H-labeled azidobenzoyl derivative of daunorubicin, *N*-(*p*-azido[3,5-^{3}H]benzoyl)daunorubicin ([^{3}H]NABD) (Fig. 1) and showed that the analog can specifically bind to P-glycoprotein. Anthracyclines were shown to be poor inhibitors of P-glycoprotein photoaffinity labeling with the vinblastine photoaffinity analog, suggesting that anthracyclines and *Vinca* alkaloids either have different binding affinities or separate binding sites on P-glycoprotein. As discussed earlier, we have prepared a tritiated photoaffinity analog of doxorubicin, *N*-(*p*-azido-[3,5-^{3}H]benzoyl)-doxorubicin ([^{3}H]NAB–DOX) and a radioiodinated photoaffinity analog of daunorubicin, *N*-(*p*-azido[3-^{125}I]salicyl)daunorubicin ([^{125}I]NAS–DNR) (Fig. 1), and found that these analogs bind specifically to P-glycoprotein.[16] Photolabeling of P-glycoprotein with 0.5 μM [^{3}H]NAB–DOX was inhibited by cyclosporin A, verapamil, vinblastine, NAB–DOX, and doxorubicin with 50% inhibition concentrations of 2.6, 20, 30, 32, and 300 μM, respectively. Photoaffinity labeling with the radioiodinated probes was carried out in SH-SY56/VCR human neuroblastoma MDR cells using 1 nM [^{125}I]NAS–DNR. Daunorubicin and doxorubicin at 250 μM inhibited [^{125}I]NAS–DNR photolabeling of P-glycoprotein by 60 and 77%, respectively.

We have provided the first experimental confirmation that an amine group at C-3′ is an important structural element of anthracyclines recognized by P-glycoprotein and that the hydroxyl group at C-14 might significantly affect the binding affinities of anthracyclines to P-glycoprotein.[16]

Photoactive analogs of the MDR-related drugs rhodamine 123 (Rh 123), ^{125}I-labeled azidosalicyclic acid (ASA)–Rh 123 ([^{125}I]ADA–Rh 123), and benzimidazole (BZ) ([^{125}I]ASA–BZ) have also been shown to specifically photolabel P-glycoprotein.[22,23] Interestingly, vinblastine and verapamil, but not colchicine, inhibited the binding of these photoaffinity drugs to P-glycoprotein.[22,23]

Interaction of MDR Modulators with P-Glycoprotein

To examine whether compounds known to reverse the MDR phenotype bind directly to P-glycoprotein and inhibit efflux of the cytotoxic agents, we initially used a photoaffinity analog of dihydropyridine, [^{3}H]azidopine[28] (Fig. 2), and provided the first experimental evidence that [^{3}H]azidopine binds to P-glycoprotein specifically and that this dihydropyridine calcium channel blocker effectively competed for the binding.[28] In addition, [^{3}H]azidopine photolabeling was partially blocked by verapamil (Fig. 2) and diltiazem, but was stimulated by excess prenylamine and bepridil.[28] The stimulatory effect caused by prenylamine and bepridil may be a positive allosteric effect of these agents on [^{3}H]azidopine labeling by binding to a separate site on P-glycoprotein and increasing the reversible binding of azidopine and its covalent coupling. In contrast, both prenylamine and bepridil inhibited [^{125}I]NASV and [^{125}I]NAS–VP photolabeling of P-glycoprotein, suggesting that these agents may have a negative allosteric effect on vinblastine and verapamil binding (unpublished data). The [^{3}H]azidopine photolabeling of P-glycoprotein also was inhibited by excess vinblastine > actinomycin D > doxorubicin, demonstrating a broad P-glycoprotein drug recognition capacity.[28] These data suggest a direct function for P-glycoprotein in the MDR phenotype and an important role in its circumvention by MDR modulators.

Cornwell *et al.*[34] originally described that the calcium channel blocker verapamil, a known MDR modulator, could inhibit photoaffinity labeling of P-glycoprotein by [^{125}I]NASV.[34] This result suggested that MDR reversal by verapamil may be due to competition of the reversing agent for the cytotoxic drug-binding site of P-glycoprotein. To demonstrate directly that P-glycoprotein is a target for verapamil, we initially synthesized ^{3}H- and ^{125}I-labeled photoaffinity analogs of verapamil and showed that they bind specifically to this protein.[14] The structures of the radioiodinated photoaffinity analogs of verapamil, *N*-(*p*-azido-[3,5-^{3}H]benzoyl)aminomethylverapamil ([^{3}H]NAB–VP) and *N*-(*p*-azido-[3-^{125}I]salicyl)aminomethyl verapamil ([^{125}I]NAS–VP), are shown in Fig. 2. Photoaffinity labeling of P-glycoprotein was inhibited in a dose-dependent manner by verapamil and vinblastine[14] and by other calcium channel blockers. Photolabeling was also partially inhibited by two of the drugs to which these cells are cross-resistant, doxorubicin and actinomycin D.[14] Subsequently, Yusa and Tsuruo[38] using [^{3}H]NAB–VP and Beck and Qian[18] using an optically pure photoaffinity analog of verapamil, LU-49888 (Fig. 2), reported similar results. These results indicate the multidrug specificity of the verapamil-binding site of P-glycoprotein.

Among MDR reversing agents, phenothiazines (PTZs) and related compounds may sensitize MDR by interacting with a specific binding site(s) on P-glycoprotein and by other mechanisms. To identify a binding site for

[38] K. Yusa and T. Tsuruo, *Cancer Res.* **49,** 5002 (1989).

PTZs and related compounds on P-glycoprotein and to examine whether these compounds and other MDR modulators bind to the same domains of P-glycoprotein, we synthesized and used a butyrophenone D_2-dopamine receptor photoaffinity probe, *N*-(*p*-azido-3-[^{125}I]iodophenethyl)spiperone ([^{125}I]NAPS)[17] (Fig. 2). [^{125}I]NAPS was actively effluxed from vincristine-resistant SH-SY5Y/VCR human neuroblastoma cells and nonradioactive I–NAPS was a potent chemosensitizing agent. After photolabeling, the probe bound specifically and with high efficiency to P-glycoprotein. The efficiency of [^{125}I]NAPS binding to P-glycoprotein was five- to six-fold more than that of [^{3}H]azidopine and [^{125}I]arylazidoprazosin ([^{125}I]AAP) (Fig. 2), known photoaffinity analogs for P-glycoprotein.[39,40] [^{125}I]NAPS photolabeling of P-glycoprotein was preferentially competed by MDR-related drugs, with vinblastine > vincristine > colchicine > doxorubicin > actinomycin D. Many drugs that are known to reverse MDR were potent inhibitors of [^{125}I]NAPS binding to P-glycoprotein. While PTZs and related compounds were potent inhibitors of [^{125}I]NAPS binding to P-glycoprotein, most of them enhanced the binding of [^{125}I]AAP significantly.[17] *cis*-Flupentixol increased the binding of [^{125}I]AAP to P-glycoprotein nine-fold more than did *trans*-flupentixol, but both were potent inhibitors of [^{125}I]NAPS binding, suggesting their stereoselective effect on the [^{125}I]AAP-binding site.

In addition to [^{3}H]azidopine, [^{125}I]NAS–VP, [^{3}H]NAB–VP, [^{125}I]AAP, and [^{125}I]NAPS, other investigators[18–26,28] have used or synthesized several photoaffinity analogs of MDR modulators (Figs. 2 and 3) and have shown that they specifically bind to P-glycoprotein. These photoactive analogs are synthetic isoprenoid,[19] cyclosporin A,[20] forskolin[21] and estramustine.[25,26] Progesterone,[41] the synthetic progestin R5020 (promegestone),[35] and corticosterone[42] are also inherently photoactivatable and have been shown to bind to P-glycoprotein. Whether these compounds label similar or different sites on P-glycoprotein and whether they share closely related or separate sites with cytotoxic agents remain to be determined.

A photoaffinity analog of ATP, 8-azido-ATP, also affinity labels P-glycoprotein, and ATP and GTP can compete for its binding[9] but this labeling is not competed with other P-glycoprotein substrates, indicating that these agents do not bind to the nucleotide-binding sites of P-glycoprotein.

[39] A. R. Safa, M. Agresti, I. Tamai, N. D. Mehta, and S. Vahabi, *Biochem. Biophys. Res. Commun.* **166,** 259 (1990).

[40] L. M. Greenberger, C.-P. H. Yang, E. Gindin, and S. B. Horwitz, *J. Biol. Chem.* **265,** 4394 (1990).

[41] X.-D. Qian and W. T. Beck, *J. Biol. Chem.* **265,** 18753 (1990).

[42] D. C. Wolf and S. B. Horwitz, *Int. J. Cancer* **52,** 141 (1992).

Although a vast number of lipophilic agents can reverse MDR and inhibit P-glycoprotein-mediated drug efflux,[18,35] the molecular mechanisms of MDR reversal by these agents and the exact location of their binding site(s) within P-glycoprotein remain to be found. It is important to understand how cytotoxic agents and MDR modulators interact with each other at the drug binding site(s) of P-glycoprotein in order to design effective MDR modulators.

Our kinetic analysis of the binding of vinblastine and azidopine[43] to P-glycoprotein provides evidence that these agents bind to separate sites, and cyclosporin A interacts competitively with vinblastine but noncompetitively with the azidopine binding site. Ferry *et al.*[44] have also confirmed these data and suggested that not only azidopine, but other dihydropyridine calcium channel blockers, bind to a site on P-glycoprotein different from the vinblastine binding site. Interestingly, we (unpublished data) have also used kinetic analysis to show that taxol also binds to a different site from that of vinblastine. However, whether taxol and dihydropyridine calcium channel blockers bind to the same or different sites remains to be found.

Location of Drug-Binding Site(s) of P-Glycoprotein

The amino acid sequences of the drug-binding sites of P-glycoprotein have not been identified. To identify the drug-binding domains of P-glycoprotein, two approaches have been employed: (1) site-directed mutagenesis[33,45,46] and (2) identification of the photoaffinity labeled regions of P-glycoprotein by peptide mapping and using site-directed anti-peptide antibodies to immunoprecipitate the drug-bound fragments after protease digestion.[47–49] Site-directed mutagenesis studies have demonstrated that a single amino acid substitution in transmembrane domains (TMDs) 6, 11, or 12 modulates P-glycoprotein activity and substrate specificity.[45,46,50,51] Some progress has been made in identifying the site in P-glycoprotein labeled by various photoaffinity labels.[47–49] By proteolytic digestion and

[43] I. Tamai and A. R. Safa, *J. Biol. Chem.* **266,** 16796 (1991).

[44] D. R. Ferry, M. A. Russell, and M. H. Cullen, *Biochem. Biophys. Res. Commun.* **188,** 440 (1992).

[45] S. Kajiji, F. Talbot, K. Grizzuti, V. Van Dyke-Phillips, M. Agresti, A. R. Safa, and P. Gros, *Biochemistry* **32,** 4185 (1993).

[46] W. T. Loo and D. M. Clarke, *Biochemistry* **33,** 14049 (1994).

[47] L. M. Greenberger, *J. Biol. Chem.* **268,** 11417 (1993).

[48] D. I. Morris, L. M. Greenberger, E. P. Bruggemann, C. Cardarelli, M. M. Gottesman, I. Pastan, and K. B. Seamon, *Mol. Pharmacol.* **46,** 329 (1994).

[49] A. R. Safa and M. Agresti, *Am. Assoc. Cancer Res.* **37,** 327 (Abst. 2226), (1996).

[50] S. E. Devine, V. Ling and P. W. Malera, *Proc. Natl. Acad. Sci. U.S.A.* **89,** 4564 (1992).

[51] W. T. Loo and D. M. Clarke, *J. Biol. Chem.* **268,** 19965 (1993).

using antibodies to specific fragments of P-glycoprotein, [^{3}H]azidopine-, [^{125}I]AAP-, and [^{125}I]azidoforskolin-binding regions were localized to TMD 5 or 6 or to the cytoplasmic region immediately following TMD 6, and a region within TMD 12 or the cytoplasmic domain immediately following it.[47,48] We have found that maximum digestion of [^{125}I]NAS–VP photolabeled human P-glycoprotein with *Staphylococcus aureus* V8 protease resulted in two major-6 and 11-kDA and one minor 12-kDa [^{125}I]NAS–VP-bound peptides. To identify the location of these peptides within P-glycoprotein, we produced site-directed anti-peptide antisera to several domains of P-glycoprotein including peptide sequences located amino terminal or carboxy terminal to TMDs 6, 11, and 12, respectively. The 6-kDa [^{125}I]NAS–VP-bound domain of P-glycoprotein was immunoprecipitated by antisera raised against amino acid residues carboxy terminal to TMD 6, indicating that this 6-kDa verapamil-binding domain is located within or immediately carboxy terminal to TMD 6. The location(s) of the 11- and 12-kDa [^{125}I]NAS–VP-bound domains remain to be found. Interestingly, kinetic data showed a single high affinity [^{125}I]NAS–VP-binding site in P-glycoprotein, suggesting that spatially these three peptides may form a single binding site for [^{125}I]NAS–VP.[49] Similarly, we have used kinetic analysis to show that [^{3}H]azidopine binds to a single site within the P-glycoprotein structure.[43]

To identify the specific [^{125}I]NAPS-binding domain(s) of P-glycoprotein and to determine if the photolabeled site(s) for this probe is identical to the [^{125}I]AAP-binding domain(s) of P-glycoprotein, we performed peptide mapping by *S. aureus* V8 protease digestion of this protein. Our results revealed that [^{125}I]NAPS binds to two major and a number of minor fragments of P-glycoprotein. The major [^{125}I]NAPS-bound fragments were 6 and 8 kDa while [^{125}I]AAP bound primarily to an 8-kDa fragment of P-glycoprotein.[17] While both [^{125}I]NAPS and [^{125}I]AAP bind to a common 8-kDa fragment of P-glycoprotein, the former binds to the other domains as well. The 6-kDa [^{125}I]NAPS-binding fragment was immunoprecipitated with an antibody that recognized the carboxy-terminal region of TMD 6. These results suggest several possibilities: (1) P-glycoprotein may have several binding sites, and [^{125}I]NAPS binds to a number of these sites; (2) the drug-bound fragments may form a single drug-binding site for [^{125}I]NAPS, which partially overlaps with the binding site domain of [^{125}I]AAP; or (3) both photoaffinity analogs bind to adjacent amino acid residues of a common domain, while [^{125}I]NAPS also binds to other drug-binding sites. Further support that [^{125}I]NAPS binds to separate sites from [^{125}I]AAP was provided by competition experiments with other MDR modulators. For instance, while many PTZs and related compounds inhibited the binding of [^{125}I]NAPS, they significantly increased the binding of

[^{125}I]AAP to P-glycoprotein. It is possible that these compounds competitively inhibit the binding of [^{125}I]NAPS to P-glycoprotein while allosterically affecting the [^{125}I]AAP-binding site and increasing the binding of the latter photoaffinity probe to P-glycoprotein. As previously discussed, we have shown the presence of a single class of binding sites on P-glycoprotein for [^{3}H]azidopine that is distinct from that of *Vinca* alkaloids and cyclosporin A.[43] It has been shown that *trans*-flupentixol reverses MDR more effectively than *cis*-flupentixol, but both compounds equally inhibited [^{3}H]azidopine binding to P-glycoprotein.[52] While our results also showed that these stereoisomers equally inhibited the binding of [^{125}I]NAPS[17] and [^{3}H]azidopine to P-glycoprotein, *cis*-flupentixol increased the binding of [^{125}I]AAP to P-glycoprotein 9- to 10-fold more than did *trans*-flupentixol. This is the first indication that P-glycoprotein may interact with its substrates stereoselectively.

The specific amino acid residues involved in drug recognition by P-glycoprotein have not yet been identified. Nevertheless, analysis of P-glycoprotein mutations that express altered patterns of drug resistance relative to the normal protein has permitted identification of particular functional domains of P-glycoprotein that may be involved in drug recognition and transport. Sequences of the drug-bound peptides, site-directed mutagenesis studies and molecular modeling analysis will be necessary to delineate the three-dimensional structure of the drug-binding sites of P-glycoprotein.

Conclusions

Although a vast amount of data on the structure-activity of P-glycoprotein is available, the molecular architecture of its drug-binding sites, their exact locations in the P-glycoprotein molecule, and the total number of drug-binding sites remain to be found. Our P-glycoprotein drug-binding data suggest the existence of multiple drug-binding sites and overlapping sites within the P-glycoprotein structure. Our kinetic results suggest the presence of a binding site for vinblastine, verapamil, cyclosporin A, and probably phenothiazines and related agents; a second binding site for dihydropyridine calcium channel blockers; a third site for the calcium channel blockers bepridil and prenylamine, and megestrol acetate, a synthetic congener of progesterone[53]; and a fourth binding site for I-AAP. Therefore, the substrates for P-glycoprotein may interact competitively as well as allosterically for binding to this protein.

[52] J. M. Ford, E. P. Bruggeman, I. Pastan, M. M. Gottesman, and W. N. Hait, *Cancer Res.* **50,** 1748 (1990).

[53] G. F. Fleming, J. M. Amato, M. Agresti, and A. R. Safa, *Cancer Chemother. Pharmacol.* **29,** 445 (1992).

Acknowledgments

I am indebted to Dr. Mary D. McCauley for excellent editorial assistance and to Mr. Young Edward Choi for preparing the chemical structures. The work in the author's laboratory was supported by National Cancer Institute grants CA-47652 and CA-56078 and American Cancer Society grants DHP-100 and DHP-100A.

[22] Identification of Drug Interaction Sites in P-Glycoprotein

By LEE M. GREENBERGER

Introduction

P-Glycoprotein behaves as an energy-dependent efflux pump capable of transporting structurally diverse anticancer drugs such as taxol, vinblastine, doxorubicin, and etoposide.[1] When the protein is overexpressed, as it is in drug-selected cells, or in some tumors, multiple drug resistance (MDR) can occur. Beyond this, agents that bind to P-glycoprotein and block the transport of anticancer drugs can resensitize MDR cells to chemotherapy and may have clinical utility.[2]

The molecular mechanisms governing P-glycoprotein function are not well known. Information is needed (1) to understand how the protein can bind structurally diverse anticancer drugs, (2) to define the localization of these sites and understand how they are related to the chemosensitizing binding sites, and (3) to determine how ATP binding and hydrolysis are linked to drug transport. Answers to these questions will ultimately require a three-dimensional analysis. Nevertheless, mapping of the drug-binding site(s) in P-glycoprotein using relatively simplistic approaches have already provided important insights. The methods used to identify drug interaction sites in P-glycoprotein are discussed here.

Before discussing these methods, it must be appreciated that P-glycoprotein is composed of two cassettes. Each cassette contains six putative transmembrane (TM) domains (based on hydropathy plot analysis), followed by an adenosine triphosphate (ATP)-binding motif. The exact number and position of TM domains has been questioned based on recent

[1] M. M. Gottesman and I. Pastan, *Annu. Rev. Biochem.* **62,** 385 (1993).

[2] L. M. Greenberger, D. Cohen, and S. B. Horwitz, *in* "In Vitro Models of Multiple Drug Resistance" (L. J. Goldstein and R. F. Ozols, eds.), pp. 69–106. Kluwer Academic Publishers, Norwell, Massachusetts, 1994.

0076-6879/98 $25.00

biochemical analyses.[3,4] P-glycoprotein belongs to a large family, known as ATP-binding cassette (ABC) transporters, which in mammalian systems contains either the two cassettes linked in tandem (as in P-glycoprotein) or one cassette that is likely to operate as a dimer.[5] In bacteria, many ABC proteins contain elements of the cassette that are present on subunits that work together. Collectively, ABC transporters allow the movement of small molecules including peptides, ions, and drugs through the membrane. Since structural conservation exists, it seems likely that ABC transporters will share common mechanistic properties with specialization for individual ligands, and methods described here will have utility beyond P-glycoprotein.

Photoaffinity Labeling of P-Glycoprotein

Choice of Photoaffinity Probe

Biochemical analysis of the drug-binding sites in P-glycoprotein relies heavily on the use of radiolabeled photoactive molecules that, when activated by ultraviolet light, covalently bind to P-glycoprotein. A variety of radiolabeled probes have been reviewed.[6] Examples of photoactive probes include analogs of anticancer drugs (vinblastine, colchicine, taxol, doxorubicin) and photoactive analogs of chemosensitizers (verapamil, forskolin, cyclosporin A, prazosin, and azidopine). Almost all of the analogs contain tritiated or iodinated phenyl-substituted arylazide groups (see Fig. 1). Exceptions are a phenyldiazarine containing cyclosporine A analog,[7] and progesterone, which is inherently photoactive.[8] On ultraviolet irradiation, the azido group becomes a reactive intermediate nitrene species that ultimately forms a covalent bond with proteins in its immediate proximity.[9] It must be appreciated that these probes impose certain limitations that could have an important bearing on mechanistic inferences. First, the addition of the photoactive moiety changes the structure of the drug itself. Therefore, it is important to show that P-glycoprotein has resistance to the chemically modified molecule, or such a drug resensitizes drug-resistant cells to anticancer agents. Second, the reactive domain of the molecule may be a distance from the actual site of drug contact. Finally, if the reactive interme-

[3] O. Beja and E. Bibi, *J. Biol. Chem.* **270,** 12351 (1995).
[4] E. Bibi and O. Beja, *J. Biol. Chem.* **269,** 19910 (1994).
[5] C. F. Higgins, *Annu. Rev. Cell Biol.* **8,** 67 (1992).
[6] W. T. Beck and X. Qian, *Biochem. Pharmacol.* **43,** 89 (1992).
[7] B. M. J. Foxwell, A. Mackie, V. Ling, and B. Ryffel, *Mol. Pharmacol.* **36,** 543 (1989).
[8] X.-D. Qian and W. T. Beck, *J. Biol. Chem.* **365,** 8753 (1990).
[9] J. Brunner, *Annu. Rev. Biochem.* **62,** 483 (1993).

diate has a long half-life, the molecule can diffuse from the true drug-binding site at the time of irradiation.

For initial methodological descriptions of drug localization studies in P-glycoprotein, the probes can be discussed interchangeably, since some, if not all of these structurally diverse probes bind to common, if not identical regions in the protein.[10,11] Consistent with this, the binding of diverse photoactive probes were specifically and preferentially inhibited by vinblastine > doxorubicin > colchicine, regardless of the photoaffinity probe used.[2] This is in contrast to studies using equilibrium binding analysis where competitive and noncompetitive binding with molecules that interact with P-glycoprotein have been described.[12–16] The latter suggests that ligands for P-glycoprotein may have distinct binding domains, and photoactive binding sites selectively identify some of these sites.

Application with [^{125}I]Iodoarylazidoprazosin or [^{3}H]Azidopine

For drug mapping studies, we relied heavily on a 1,4-dihydropyridine derivative, [^{3}H]azidopine, the prazosin derivative [^{125}I]iodoarylazidoprazosin ([^{125}I]IAAP), and a photoactive, iodinated derivative of forskolin, [^{125}I]-6-O-[[2-[3-(4-azido-3-[^{125}I]iodophenyl)propionamido]ethyl]carbamyl]forskolin ([^{125}I]-6-AIPP-forskolin) (Fig. 1). These photoactive compounds are known to reverse drug resistance and bind to P-glycoprotein. Experimentation with [^{125}I]IAAP is easiest because it is iodinated, commercially available, and stable for at least 2 months (2200 Ci/mmol; DuPont NEN, Boston, MA). Labeling with [^{125}I]-6-AIPP-forskolin appears to be 10-fold more efficient than [^{125}I]IAAP,[10] but the forskolin analog is not commercially available. For the general procedure, 200 μg of membrane protein is brought to a volume of 121 μl with 50 mM Tris, pH 7.4. The material is placed in a 96-well tissue culture dish. If desired, 2.6 μl of a 1 mM candidate inhibitor is added (final concentration 20 μM), mixed by pipetting repeatedly, and incubated for 5 min at room temperature.

[10] D. I. Morris, L. M. Greenberger, E. P. Bruggemann, C. Cardarelli, M. M. Gottesman, I. Pastan, and K. B. Seamon, *Mol. Pharmacol.* **46,** 329 (1994).

[11] L. M. Greenberger, C.-P. H. Yang, E. Gindin, and S. B. Horwitz, *J. Biol. Chem.* **265,** 4394 (1990).

[12] D. R. Ferry, P. J. Malkhandi, M. A. Russell, and D. J. Kerr, *Biochem. Pharmacol.* **49,** 1851 (1995).

[13] D. R. Ferry, M. A. Russell, and M. H. Cullen, *Biochem. Biophys. Res. Commun.* **188,** 440 (1992).

[14] E. Spoelstra, H. Westerhoff, H. Pinedo, H. Dekker, and J. Lankelma, *Eur. J. Biochem.* **221,** 363 (1994).

[15] I. Tamai and A. R. Safa, *J. Biol. Chem.* **266,** 16796 (1991).

[16] I. Tamai and A. R. Safa, *J. Biol. Chem.* **265,** 16509 (1990).

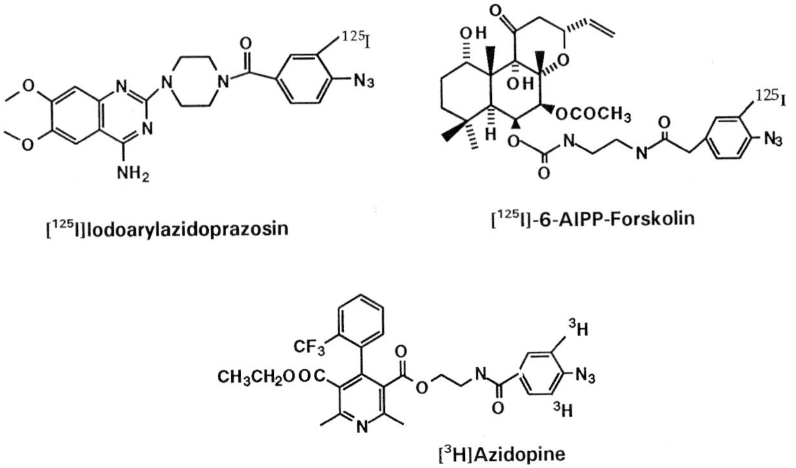

[^{125}I]Iodoarylazidoprazosin

[^{125}I]-6-AIPP-Forskolin

[^{3}H]Azidopine

FIG. 1. Structure of [^{3}H]azidopine, [^{125}I]iodoarylazidoprazosin and a [^{125}I]-6-AIPP-forskolin.

The photolabel is now prepared. To do this, [^{125}I]IAAP is diluted (in the presence of dim light to avoid activating the molecule) to approximately 40–80 n*M* in 50 m*M* Tris, pH 7.4, and 6.65 μl of diluted material is added to the wells. (The final [^{125}I]IAAP concentration, 2–4 n*M*, is 5000-fold lower than the candidate inhibitor.) After mixing by pipetting, the sample is split into two wells, the 96-well plate is wrapped in aluminum foil, and incubated for 1 hr at room temperature. The plate is then unwrapped, the cover removed, and samples irradiated approximately 3 cm from an ultraviolet source for 5 min on ice (Ultra-Violet Products, San Gabriel, CA, model R52G; output 1250 μW). The reaction is terminated by the addition of 5× Laemmli sample buffer. The resultant sample is subjected to SDS–PAGE on a 7% gel. If the sample is not pelleted at 100,000*g* (Beckman Airfuge, Palo Alto, CA: 30 min, 22 psi at 4°) prior to the addition of Laemmli sample buffer, but is instead run directly on a gel, great care is taken when discarding the lower chamber buffer since it contains free [^{125}I]-IAAP.

After electrophoresis, the gel is stained (15 min with 0.2% Coomassie Brilliant Blue R-250 followed by destaining for more than 4 hr) to confirm equal protein loading, dried, and the radioactive image detected on XAR-5 film (Kodak, Rochester, NY). Exposure time when using membranes from S1-B1-20 cells is approximately 20–60 min. Quantitative analysis of the radioactive bands is done by cutting the radioactive band from the gel and counting with a gamma counter or analyzing the image captured on film with an image scanner (Molecular Devices, Menlo Park, CA). If azidopine is used for labeling, the analysis is done with 50–100 n*M* azidopine, irradiation is for 15 min, and the gel is enhanced prior to exposure to X-ray film.

Considerations

The procedure works well when materials are obtained from relatively pure membranes, crude membranes, or whole-cell suspensions (for the latter, see Ref. 17). Best results are obtained if comparative analysis is done with drug-sensitive and drug-resistant cells that contain no or high levels of P-glycoprotein, respectively. Cells that are highly drug resistant (>1000-fold resistance to antitumor drugs) are preferred, because it is difficult to detect photolabeled P-glycoprotein from cells that have low levels of P-glycoprotein. This is probably due to the poor efficiency of photolabeling. Highly resistant MDR cell lines that have been used for domain mapping include KB-V1, S1-B1-20, and J7.V1-1. Relatively pure membrane preparations are obtained by density-gradient centrifugation according to Roy and Horwitz[18] with the following modifications: (1) Cell pellets are disrupted by sonication for 30 sec and (2) a 20–40% step gradient replaces the 20/34/40% step gradient in the original method. The final pellet is resuspended in 10 m*M* Tris, pH 7.4. This procedure requires overnight centrifugation. Alternatively, a more crude and quickly obtained membrane preparation can be derived by hypotonic lysis, douncing, and differential centrifugation as described by Morris *et al.*[10]

In some cases, it is important to obtain vesicles. Vesicles are prepared by nitrogen cavitation and sucrose gradient centrifugation according to Lever.[19] In all subcellular fraction procedures, after the initial disruption of cells in a protease inhibitor cocktail containing 1 m*M* phenylmethylsulfonyl fluoride (PMSF), 100 U/ml aprotonin, 30 μM leupeptin, and 1 μg/ml pepstatin, only 100 U/ml aprotinin is maintained for the rest of the purification. P-Glycoprotein is stable under these conditions for greater than 1 yr at −80°. The concentration of photoaffinity label probe in the reaction, and its relative concentration compared to a candidate inhibitor of labeling, should be noted. We have not reached saturating amounts of [^{125}I]IAAP,[11] whereas half-maximal labeling is achieved with approximately 100 n*M* [^{3}H]-azidopine.[20]

[17] E. P. Bruggemann, U. Germann, M. M. Gottesman, and I. Pastan, *J. Biol. Chem.* **264,** 15483 (1989).

[18] S. N. Roy and S. B. Horwitz, *Cancer Res.* **45,** 3856 (1985).

[19] J. E. Lever, *J. Biol. Chem.* **252,** 1990 (1977).

[20] A. R. Safa, C. J. Glover, J. L. Sewell, M. B. Meyers, J. L. Biedler, and R. L. Felsted, *J. Biol. Chem.* **262,** 7884 (1987).

Because P-glycoprotein interacts with diverse chemical structures, it is important to determine if photoaffinity labeled drugs bind to common sites. This can be addressed initially by digesting photoaffinity labeled P-glycoprotein and comparing the peptide map. Mapping is easily achieved using in-gel enzyme digestion according to published methods.[21] Under these conditions, the peptide maps for [^{125}I]IAAP and [^{3}H]azidopine, as well as [^{125}I]-6-AIPP-forskolin and [^{125}I]IAPP, are identical.[10,11] Therefore, these photoaffinity probes bind to common domains in P-glycoprotein. Based on these digestions, two small molecular weight species are identified after *Staphylococcus aureus* V8 protease or chymotrypsin digestion. With both enzymes, major species are found at approximately 5–12 kDa. This suggests that the photoaffinity binding domains are restricted to a small epitope in the 170-kDa intact protein.

Immunological Mapping

Two approaches can be used to localize photoaffinity drug bindings sites in P-glycoprotein. Immunological mapping has proven the most productive. However, direct amino acid sequence analyses of peptide digests of photoaffinity labeled P-glycoprotein have been attempted because others have had success with other heavily embedded membrane proteins.[22] Initial efforts toward direct mapping of binding sites in P-glycoprotein failed since the resultant radiolabeled fragment (purified by high-performance liquid chromatography) did not contain sufficient amounts of purified, radiolabeled material for subsequent amino acid sequence analysis.

For immunological mapping, a battery of site-directed antibodies to P-glycoprotein is used (Fig. 2). The initial strategy is to make antibodies to synthetic peptides (14–20 oligomers) that mimic each hydrophilic loop of P-glycoprotein. Peptides are conjugated to keyhole limpet hemocyanin, and antisera are generated in rabbits according to published methods.[23] Because P-glycoprotein has sustantial homology between its two cassettes, the synthetic peptides are chosen so that homology with the second half of P-glycoprotein is less than 15%. High titer serum to many but not all epitopes is obtained. Specificity to each epitope is good, because the ability of the antibody to detect P-glycoprotein on immunoblot is eliminated if the peptide is preincubated with corresponding antiserum prior to incuba-

[21] D. W. Cleveland, S. G. Fischer, M. W. Kirschner, and U. K. Laemmli, *J. Biol. Chem.* **252,** 102 (1977).

[22] S. Regulla, T. Schneider, W. Nastaincyzk, H. E. Meyer, and F. Hofmann, *EMBO J.* **10,** 45 (1991).

[23] E. Harlow and D. Lane, *in* "Antibodies: A Laboratory Manual," pp. 53–137. Cold Spring Harbor Laboratory Press, New York, 1988.

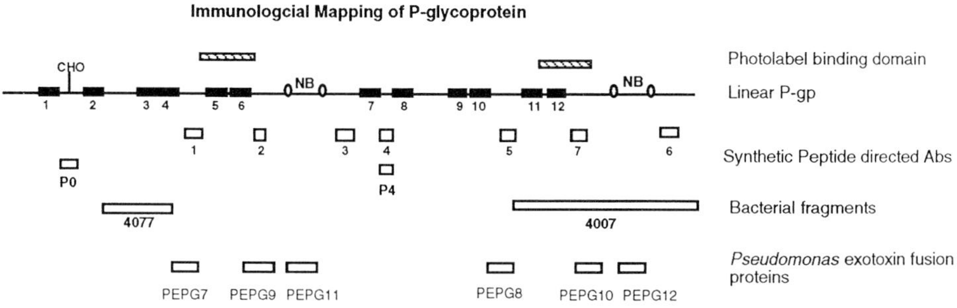

FIG. 2. Antibodies used for the mapping of photoaffinity binding domains in P-glycoprotein. Position of epitopes with respect to the linear sequence of P-glycoprotein is indicated. Predicted photoaffinity labeling domains denoted by hatched bar. Transmembrane domains 1–12, glycosylation (CHO), and nucleotide-binding (NB) domain are indicated in the linear sequence. The two circles in the linear sequence indicate the Walker A and B motifs. Antiserum was generated to synthetic peptides 1–7,[26,27] P0, and P4,[17] bacterial fragments,[25] and *Pseudomonas* exotoxin fusion proteins.[24]

tion with the blot. Furthermore, where available, peptide from the corresponding second half of P-glycoprotein does not compete the immunoreactivity of the appropriate antibody with P-glycoprotein. Antibodies to larger hydrophilic regions of P-glycoprotein, which are made by others, are also invaluable. In this case, antisera raised against fusion proteins between *Pseudomonas* exotoxin and hydrophilic loops of P-glycoprotein (70–95 amino acids in length),[24] or antisera raised against fragments of P-glycoprotein expressed in *Escherichia coli* are generated.[25]

Our immunological strategy for mapping is to go from low to high resolution. Therefore, to get large fragments of P-glycoprotein, the digestion is restricted to P-glycoprotein in vesicles. In vesicles, because the protein is embedded in the membrane, proteolytic enzymes have access to far fewer sites than when the protein is denatured and removed from the membrane.

Application: Low-Resolution Localization

After photoaffinity labeling 50 μg of vesicles from highly resistant cells with 20 nM [^{125}I]IAAP, the irradiated material is centrifuged for 15 min

[24] E. P. Bruggemann, V. Chaudhary, M. M. Gottesman, and I. Pastan, *BioTechniques* **10,** 202 (1991).

[25] S. Tanaka, S. J. Currier, E. P. Bruggemann, K. Ueda, U. A. Germann, I. Pastan, and M. M. Gottesman, *Biochem. Biophys. Res. Commun.* **166,** 180 (1990).

[26] L. M. Greenberger, *J. Biol. Chem.* **268,** 11417 (1993).

[27] L. M. Greenberger, C. J. Lisanti, J. T. Silva, and S. B. Horwitz, *J. Biol. Chem.* **266,** 20744 (1991).

at 4° in a Beckman Airfuge. The resultant pellet is resuspended in 10 μl of 10 mM Tris, pH 7.4, containing 250 mM sucrose and incubated with 0–2500 μg/ml of tosylphenylalanylchloromethyl ketone-treated trypsin or chymotrypsin for 1 hr at 37° in a total volume of 20 μl. The digestion is stopped by the addition of aprotinin, phenylmethylsulfonyl fluoride, and soybean trypsin inhibitor to attain a final concentration of 0.25 μg/ml, 3.5 μM, and 1.0 μg/ml, respectively. After a 5-min incubation at 4°, the samples are prepared for immunoprecipitation by solubilizing in an equal volume of 2% sodium dodecyl sulfate (SDS), containing 2 mg/ml bovine serum albumin (BSA) in 50 mM Tris, pH 7.4. The samples are then diluted five-fold in a solution containing 1.25% Triton, 190 mM NaCl, and 50 mM Tris, pH 7.4. Then, immunoprecipitation is done as previously described. Briefly, samples are incubated with antiserum overnight at 4°, followed by incubation with a 100-μl slurry of protein A-agarose (1 mg powder/100 μl 50 mM Tris, pH 7.4). After rocking for 2 hr, samples are washed, and eluted into Laemmli sample buffer. The resultant fragments are resolved on 10% gels.

Based on this analysis, two major [^{125}I]IAAP drug-binding domains, one in each cassette within P-glycoprotein, were identified. Similar conclusions have been reached with [^{3}H]azidopine[28] and [^{125}I]-6-AIPP-forskolin.[10] In the N-terminal half of P-glycoprotein, a 95-kDa glycosylated [^{125}I]IAAP-labeled fragment was detected with antiserum 1. (Antiserum 2 is not used since the titer was too low to be useful.) In the C-terminal half of P-glycoprotein, a 40-kDa [^{125}I]IAAP-labeled fragment was detected with antiserum 6 but not antiserum 5. If the 40-kDa fragment was subjected to in-gel digestion with V8 protease, a 6-kDa fragment was obtained. Therefore, this suggests that a 6-kDa photoaffinity binding site is located between TM11 and the C terminus of P-glycoprotein.

Application: High-Resolution Immunologic Mapping

P-Glycoprotein in vesicles was highly resistant to trypsin and digestion was not done to completion. Complete digestion is important, since it allows for quantitative assessment of putative binding domains. Therefore, an alternative strategy was developed. In this case, photoaffinity labeled P-glycoprotein is denatured and extracted from the membrane prior to enzymatic digestion. The resultant material is cleaved proteolytically and then immunopurified with site-specific antibodies.

To conduct the experiment, P-glycoprotein in membrane is photoaffinity labeled. After the material is pelleted (15 min at 4° in a Beckman Airfuge),

[28] E. P. Bruggemann, S. J. Currier, M. M. Gottesman, and I. Pastan, *J. Biol. Chem.* **267,** 21020 (1992).

it is extracted into 1% SDS containing 10 m*M* Tris, pH 7.4, to completely denature the protein. To make the preparation compatible with trypsin digestion, this material is then diluted into 10 m*M* Tris, pH 7.4, containing 1.25% Triton X-100 to obtain the final concentration of 0.2% SDS and 1.0% Triton X-100. Even under these conditions, large amounts of trypsin are needed. However, complete digestion is achieved and the solutions are compatible with immunoprecipitation. After digestion for 1 hr at 37° with 0.01–1 mg/ml trypsin, the reaction is stopped with protease inhibitors, antiserum is added and then precipitated with protein A-agarose.

The resultant fragments were low molecular weight and superior resolution was achieved with 10% Tris-tricine low molecular weight gels as described in Schagger and von Jagow.[29] Under these conditions, antiserum 7, but not antisera 1, 5, or 6, immunoprecipitated a 4-kDa photoaffinity labeled fragment. Therefore, the photoaffinity drug-binding epitope was immediately C-terminal to TM11 and extended to the N-terminal of the Walker A motif in this cassette. Similarly, antiserum PEPG9 immunoprecipitated a 5-kDa fragment. Therefore, in the N-terminal half of P-glycoprotein, the epitope could be localized between TM5 and the C-terminal region of the Walker A motif. Subsequently, Morris *et al.*,[10] who used [^{125}I]-6-AIPP-forskolin, showed that a 6-kDa photolabeled tryptic fragment was detected by antiserum to PEPG9 but not to antisera PEPG7 or PEPG11. Furthermore, a 17-amino-acid synthetic peptide, which mimicked the extreme N-terminal region of the PEPG9 epitope, inhibited immunoprecipitation of this 6-kDa fragment, while the next C-terminal most amino acid synthetic peptide did not inhibit immunoprecipitation. Therefore, the position of the photoaffinity binding domain in the N-terminal half of the molecule was likely to span putative TM5 to TM6. In all of these studies, the predicted photolabeling domain was based on (1) the size of the photolabeled fragment, (2) the antibody used to immunoprecipitate the fragment, and (3) the predicted site for enzyme digestion based on the known sequence. If sufficient radiolabeled fragments could be purified, amino acid sequence analysis would help refine the positioning.

Molecular Biology Approaches

It is essential to determine if the predicted photoaffinity drug-binding domains are functionally important. To evaluate this, the effect of mutations in putative drug-binding sites on drug resistance has been examined. A variety of mutant P-glycoproteins, many of which were made prior to knowing the photolabel binding domains, have been created by making

[29] H. Schagger and G. von Jagow, *Anal. Biochem.* **166,** 368 (1987).

point mutations, deletions, insertions, and chimeric molecules composed of P-glycoprotein family members. In general, these data indicate that alteration in the putative photoaffinity labeling domains can alter drug resistance. However, in many cases resistance profiles were selectively altered for only one or more anticancer agents. For example, if individual point mutations were made in each of the 31 phenylalanine residues in human P-glycoprotein encoded by MDR1, it was found that only two mutations, located in putative TM 6 or TM 12, drastically altered the drug resistance phenotype. Furthermore, alterations in photoaffinity labeling were observed with the F335A mutants.[30] This does not exclude the fact that other regions in P-glycoprotein play a functional role in drug binding or transport. For example, mutations outside the putative photolabeling regions, particularly in transmembrane domains, can alter resistance profiles for some substrates for P-glycoprotein. In particular, point mutations in G185V in human P-glycoprotein (immediately before TM3)[31] and S941F or S939F in TM11 of mouse P-glycoprotein encoded by *mdr1* and *mdr3*, respectively, altered drug resistance profiles[32] and the ability to photolabel the mutant proteins.[33,34] In summary, these data suggest that different drugs have overlapping but distinct binding sites, and some of the drug-binding sites may exist outside the photolabeling regions.

We have used a more general strategy to test whether functionally important regions in P-glycoprotein reside within or near the putative photoaffinity labeling sites.[35] Our initial analysis was restricted to the C-terminal half of P-glycoprotein. To do this, we evaluated the function of chimeric P-glycoproteins composed of human MDR1 and MDR3. This is logical since human MDR3 has 77% amino acid identity with MDR1, but unlike MDR1, does not mediate MDR.[11] In the initial experiment, the region of TM10 to the C terminus of MDR1 was replaced with the corresponding region of MDR3. The resultant chimeric molecule was not functional. By progressively decreasing the replacement size, it was found that replacement of only TM12 markedly impaired resistance to actinomycin D, vincristine, and doxorubicin (but not colchicine). The phenotype was associated with greater than a 75% reduction in the ability to photoaffinity label the chimeric protein with [^{125}I]IAAP. In contrast, the drug

[30] T. W. Loo and D. M. Clarke, *J. Biol. Chem.* **268,** 19965 (1993).

[31] K. Choi, C.-J. Chen, M. Kriegler, and I. B. Roninson, *Cell* **53,** 519 (1988).

[32] P. Gros, R. Dhir, and F. Talbot, *Proc. Natl. Acad. Sci. U.S.A.* **88,** 7289 (1991).

[33] A. R. Safa, R. K. Stern, K. Choi, M. Agresti, I. Tamai, N. D. Mehta, and I. B. Roninson, *Proc. Natl. Acad. Sci. U.S.A.* **87,** 7225 (1990).

[34] S. Kajiji, F. Talbot, K. Grizzuti, V. Van Dyke-Phillips, M. Agresti, A. R. Safa, and P. Gros, *Biochemistry* **32,** 4185 (1993).

[35] X. P. Zhang, K. I. Collins, and L. M. Greenberger, *J. Biol. Chem.* **270,** 5441 (1995).

resistance profile and photolabeling efficiency were not altered if TM11 was replaced. Surprisingly, replacement of the loop between TM11 and TM12 appeared to create a more efficient form of P-glycoprotein, since protein levels were reduced compared to wild-type P-glycoprotein, yet resistance to actinomycin D, colchicine, and doxorubicin was elevated two- to three-fold. In this case, the ability to photolabel the protein was reduced further than expected when normalized to the amount of P-glycoprotein expressed. These data suggest that the loop between TM11 and TM12 as well as TM12 itself are in proximity to domains that come in contact with substrates for P-glycoprotein.

Conclusions

The photoaffinity labeling data and genetic studies suggest that homologous transmembrane domains of each cassette of P-glycoprotein contain drug-binding sites. Such regions are likely to interact with each other. However, it is highly likely that many additional hydrophobic amino acids come in contact with chemotherapeutic agents and chemosensitizers. The use of photoaffinity labeled drugs that contain photoactive groups in different locations in the same or related molecules may help identify other residues involved in drug binding and determine the spatial arrangement of such sites.[36] Some of these domains may be common to multiple anticancer drugs and chemosensitizers, while others may be uniquely related to a particular drug. Beyond this, in the future, it will be necessary to distinguish between sites involved in drug binding versus those involved in drug translocation. The data and methods described here will help form the foundation for a more complete understanding of drug-binding pores in P-glycoprotein, which will ultimately require structural resolution in three dimensions.

Acknowledgment

I thank Dr. Robert Mallon for critical review of this manuscript.

[36] C. Borchers, W.-R. Ulrich, K. Klemm, W. Ise, V. Gekeler, S. Haas, A. Schodl, J. Conrad, M. Przybylski, and R. Boer, *Mol. Pharmacol.* **48,** 1 (1995).

[23] Photoaffinity Labeling of Human P-Glycoprotein: Effect of Modulator Interaction and ATP Hydrolysis on Substrate Binding

By SAIBAL DEY, MURALIDHARA RAMACHANDRA, IRA PASTAN, MICHAEL M. GOTTESMAN, and SURESH V. AMBUDKAR

Introduction

The human multidrug transporter, P-glycoprotein (Pgp), is a 1280-amino-acid integral membrane protein that confers resistance to a number of structurally diverse anticancer drugs by adenosine triphosphate (ATP)-dependent extrusion across the plasma membrane.[1] P-Glycoprotein shows internal homology of amino acid sequence between its N- and C-terminal halves.[2] Each half of the protein contains six putative transmembrane regions (TMs) followed by a consensus nucleotide-binding domain (NBD). Direct involvement of the TM domains in substrate recognition and the role of NBDs in ATP binding/hydrolysis are well established. Directed mutagenesis of certain amino acid residues, in either of the two NBDs, which abolishes ATPase activity of Pgp, results in impaired drug transport, suggesting a tight coupling between ATP hydrolysis and drug translocation.[3,4] However, substrate recognition and ATP binding are independent of each other.

Four major steps appear to be involved in drug transport: (1) substrate recognition, (2) ATP binding, (3) ATP hydrolysis, and (4) coupling of ATP hydrolysis to drug translocation. The scheme for catalytic cycle of Pgp, originally put forward by Senior *et al.*,[5] and the modified scheme by Dey *et al.*,[6] propose a conformational change in the drug recognition domain on hydrolysis of ATP. We have demonstrated that stabilizing a transition state conformation of Pgp, during ATP hydrolysis, significantly reduces the affinity of the drug-binding site for photoactivatable analogs of substrates.[6,7]

[1] M. M. Gottesman and I. Pastan, *Annu. Rev. Biochem.* **62,** 385 (1993).

[2] C.-J. Chen, J. E. Chin, K. Ueda, D. P. Clark, I. Pastan, M. M. Gottesman, and I. B. Roninson, *Cell* **47,** 381 (1986).

[3] M. Azzaria, E. Schurr, and P. Gros, *Mol. Cell. Biol.* **9,** 5289 (1989).

[4] T. W. Loo and D. M. Clarke, *J. Biol. Chem.* **270,** 21449 (1995).

[5] A. E. Senior, M. K. al-Shawi, and I. L. Urbatsch, *FEBS Lett.* **377,** 285 (1995).

[6] S. Dey, M. Ramachandra, I. Pastan, M. M. Gottesman and S. V. Ambudkar, *Proc. Natl. Acad. Sci. U.S.A.* **94,** 10594 (1997).

[7] M. Ramachandra, S. V. Ambudkar, D. Chen, C. A. Hrycyna, S. Dey, M. M. Gottesman, and I. Pastan, *Biochemistry,* in press (1998).

METHODS IN ENZYMOLOGY, VOL. 292
0076-6879/98 $25.00

This change in affinity for the substrates can be used as an effective assay to measure the extent of coupling between ATP hydrolysis and substrate binding.

Photoaffinity labeling of Pgp with analogs of substrates and modulators has revealed two major areas, one on each homologous half of the protein, as primary sites of drug interaction.[8,9] The presence of a highly susceptible trypsin cleavage site at the linker region between the N- and C-terminal halves of the protein has allowed us to study the substrate-binding properties of the two sites separately.[6] Work from this laboratory has provided evidence that the two major photolabeled areas of Pgp represent nonidentical drug interaction sites.[6]

In this article, we discuss methods that can be used to evaluate the differential effect of ATP hydrolysis and modulator interaction on substrate binding to these two sites. Although results presented in this article deal with Pgp in isolated membranes from baculovirus-infected insect cells, these assays can be successfully used for studies with Pgp expressed in human and murine cells as well as with the purified protein on its reconstitution into proteoliposomes.

Methods

Expression and Purification of $(His)_6$-Tagged Pgp by Metal Affinity Chromatography

For efficient expression of Pgp and its purification and reconstitution with relative ease, insect cells were infected with recombinant baculovirus containing the human *MDR1* gene with a $(His)_6$ tag at the C-terminal end of the protein. Construction of the recombinant baculovirus with wild-type human *MDR1* gene for expression in insect cells has been described in [31].[10a,b] Therefore the detailed protocol for construction of the virus is not discussed here. $(His)_6$-Tagged Pgp when expressed in High Five (Invitrogen, Carlsbad, CA) insect cells exhibited properties comparable to that of the wild-type protein.[7] Isolation of membranes from the insect cells and rapid purification of the $(His)_6$-tagged Pgp by metal-affinity chromatography are discussed in the following sections.

[8] E. P. Bruggemann, U. A. Germann, M. M. Gottesman, and I. Pastan, *J. Biol. Chem.* **264,** 15483 (1989).

[9] L. M. Greenberger, C. J. Lisanti, J. T. Silva, and S. B. Horwitz, *J. Biol. Chem.* **266,** 20744 (1991).

[10a] U. A. Germann, M. C. Willingham, I. Pastan, and M. M. Gottesman, *Biochemistry* **29,** 2295 (1990).

[10b] U. A. Germann, *Methods Enzymol.* **292,** [31], 1998 (this volume).

I. Strategy for Generating Recombinant Baculovirus Encoding $(His)_6$-Tagged Pgp. Standard recombinant DNA procedures were used for the construction of transfer vector pBacPAK-MDR1(H_6) encoding $(His)_6$-tagged human *MDR1*. Using this transfer vector recombinant baculovirus was generated and purified as described in [32].[10b]

II. Preparation of Insect Cell Membranes. The isolated crude membrane fraction from baculovirus-infected High Five cells contains high levels of Pgp in an active form suitable for biochemical characterization as well as for purification of the protein.[7,10,11]

Protocol

1. Grow High Five insect cells in serum-free Ex-Cell 400 medium (JRH Biosciences, Lenexa, KS) as monolayer cultures in 165-cm^2 tissue culture flasks.
2. Infect 2×10^7 cells/flask with recombinant baculovirus BV-MDR1(H_6) at a multiplicity of infection of 10 for 2 hr in a volume of 4 ml of serum-free Ex-Cell 400 medium.
3. After 2 hr, feed cells with another 16 ml of the same medium and incubate at 27° for 72 hr.
4. Scrape off insect cells from the bottom of the flask and harvest by centrifugation at 500*g* for 10 min.
5. Wash cells twice with ice cold Dulbecco's phosphate buffered saline (PBS) (without $CaCl_2$ and $MgCl_2$; Gibco-BRL, Gaithersburg, MD) supplemented with 1% aprotinin (Sigma, St. Louis, MO, 100% stock).
6. Resuspend cells in 2 ml/T-162 flask of the lysis buffer [10 m*M* Tris-HCl, pH 7.5, 10 m*M* NaCl, 1 m*M* $MgCl_2$, 1 m*M* dithiothreitol (DTT, ICN, Costa Mesa, CA), 1 m*M* 4-(2-aminoethyl)-benzenesulfonyl fluoride hydrochloride (AEBSF, ICN), and 1% aprotinin and leave on dry ice for 10 min. (Cells could be stored at −70° for future use at this point.)
7. Thaw frozen cells on water at room temperature and let swell on ice for 40 min.
8. Disrupt thawed cells by repeated strokes of a Dounce homogenizer 40 times with loose fitting (pestle A), and 40 times with tight fitting (pestle B), partially submerged in ice water.
9. Following lysis, remove undisrupted cells and nuclei by centrifugation at 500*g* for 15 min.

[11] G. L. Evans, B. Ni, C. A. Hrycyna, D. Chen, S. V. Ambudkar, I. Pastan, U. A. Germann, and M. M. Gottesman, *J. Bioenerg. Biomembr.* **27,** 43 (1995).

10. Dilute supernatant with resuspension buffer (10 m*M* Tris-HCl, pH 7.5, 50 m*M* NaCl, 300 m*M* mannitol, 1 m*M* DTT, 1 m*M* AEBSF, and 1% aprotinin) and centrifuge at 100,000*g* for 1 hr.
11. Resuspend pellet in the same buffer and wash it once by centrifuging at 100,000*g* for 1 hr.
12. Resuspend the pellet in a resuspension buffer containing 10% (v/v) glycerol by passing through a hypodermic needle (gauge size 19 and then gauge size 23).
13. Store membranes in aliquots at −80°. Determine protein content by modified Lowry method[12] using bovine serum albumin (BSA) as a standard.

III. Purification of Pgp by Metal Affinity Chromatography. Insect cell membranes are solubilized using octyl β-D-glucopyranoside (octylglucoside from Calbiochem, San Diego, CA) and $(His)_6$-tagged Pgp is purified using metal-affinity chromatography as described.[7]

Protocol

1. Resuspend at 2 mg/ml of crude membranes in 20 ml of 20 m*M* Tris-HCl, pH 8.0, 20% glycerol, 150 m*M* NaCl, 2 m*M* 2-mercaptoethanol (Sigma), 0.4% lipid mixture containing *E. coli* bulk phospholipid, phosphatidylcholine, phosphatidylserine, and cholesterol (all from Avanti Polar Lipids, Alabaster, AL) at 60 : 17.5 : 10 : 12.5 (w/w),[13] respectively, 2.0% octylglucoside, 1.5 m*M* $MgCl_2$, 1 m*M* AEBSF, 2 μg/ml pepstatin (Sigma), 2 μg/ml leupeptin (Sigma), and 1% aprotinin, and incubate on ice for 20 min.
2. After 20 min, remove insoluble material by centrifugation at 100,000*g* for 1 hr and store the supernatant containing solubilized protein on ice.
3. Prewash the supplied Talon resin (Clontech, Palo Alto, CA) once with buffer A containing 20 m*M* Tris-HCl, pH 8.0, 100 m*M* NaCl, 20% glycerol, 2.5 m*M* 2-mercaptoethanol, 0.1% lipid mixture (same as above) and 1.25% octylglucoside, 1 m*M* $MgCl_2$, 1 m*M* AEBSF, 2 μg/ml pepstatin, 2 μg/ml leupeptin, and 1% aprotinin.
4. Add imidazole (Sigma) to a final concentration of 2 m*M* from a stock adjusted to pH 8.0 to the detergent extract (10 mg protein) and incubate for 30 min at 4° on a rotary shaker with 0.5 ml of prewashed Talon metal-affinity resin.

[12] J. L. Bailey, *in* "Techniques in Protein Chemistry," pp. 340–341. Elsevier Publishing Co., New York, 1967.

[13] S. V. Ambudkar, I. H. Lelong, J. Zhang, C. O. Cardarelli, M. M. Gottesman, and I. Pastan, *Proc. Natl. Acad. Sci. U.S.A.* **89,** 8472 (1992).

5. Pellet the metal-affinity beads by centrifugation for 5 min at 500*g* and wash twice by resuspending and incubating in 10 ml buffer A at 4° for 10 min on a rotary shaker.
6. Resuspend beads in 1 ml buffer A and apply on to a 4-ml disposable column (Bio-Rad, Hercules, CA).
7. Wash twice with 5 ml of buffer A containing 500 m*M* KCl.
8. Elute proteins stepwise in 2 ml each of buffer B (same as buffer A except with 20 m*M* Tris-HCl, pH 6.8 instead of pH 8.0) containing 10, 100, and 200 m*M* imidazole.
9. Concentrate fractions eluted from the column using Centriprep-50 (Amicon, Beverly, MA), and store in aliquots at −70°.

IV. Photoaffinity Labeling of the N- and C-Terminal Halves of Pgp with [^{125}I]IAAP. Loo and Clarke[14] have shown that a close association between the N- and C-terminal halves of Pgp is essential for drug interaction and drug-stimulated ATPase activity. On the other hand, separation of the two halves is required to study the drug-binding properties of the two photoaffinity labeled sites. To find a way around this problem, we have taken advantage of a highly susceptible trypsin cleavage site at the linker region between the two homologous halves of Pgp.[8,15] A mild trypsin treatment cleaves Pgp at the linker region without altering its ability to interact with substrates (Fig. 1). Photoaffinity labeling of the cleaved molecule with substrate analog can be modulated by Pgp reversing agents (or modulators), much the same way as in the intact molecule (Fig. 1). As depicted in Fig. 2, the cleaved Pgp molecule also retains its ability to hydrolyze ATP. Therefore quantitation of [^{125}]IAPP labeling of the two halves gives similar results irrespective of whether the trypsin digestion is carried out before or after the photoaffinity labeling (Fig. 1). This is due to the fact that even after trypsin treatment, both the fragments are tightly associated with each other and can be separated only after denaturation.

Protocol

1. Incubate Pgp containing membranes (10 μg protein) in 10 m*M* Tris-HCl, pH 7.5, 50 m*M* NaCl, 300 m*M* mannitol, and 1% aprotinin (labeling buffer) in a final volume of 100 μl, for 10 min at 37° with 7.5–10 μg of TPCK treated trypsin (Sigma Chemicals), freshly prepared in 1 m*M* HCl.
2. Following incubation with trypsin, add 37.5 μg of soybean trypsin inhibitor (Sigma Chemicals), dissolved in H_2O, to the tube and incubate for 2 min at room temperature to stop the reaction.

[14] T. W. Loo and D. M. Clarke, *J. Biol. Chem.* **269,** 7750 (1994).
[15] E. Georges, J. T. Zhang, and V. Ling, *J. Cell. Physiol.* **148,** 479 (1991).

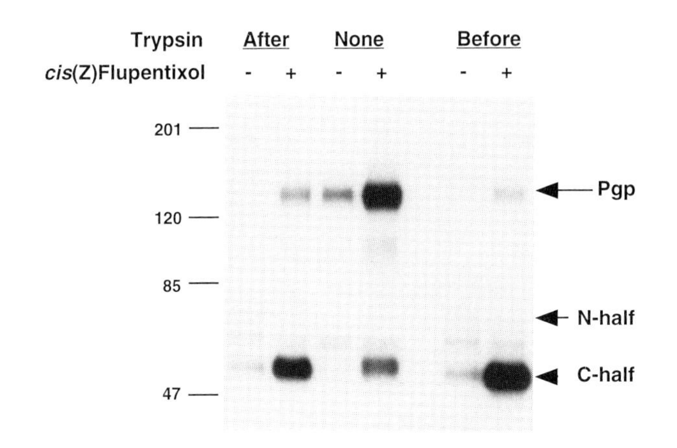

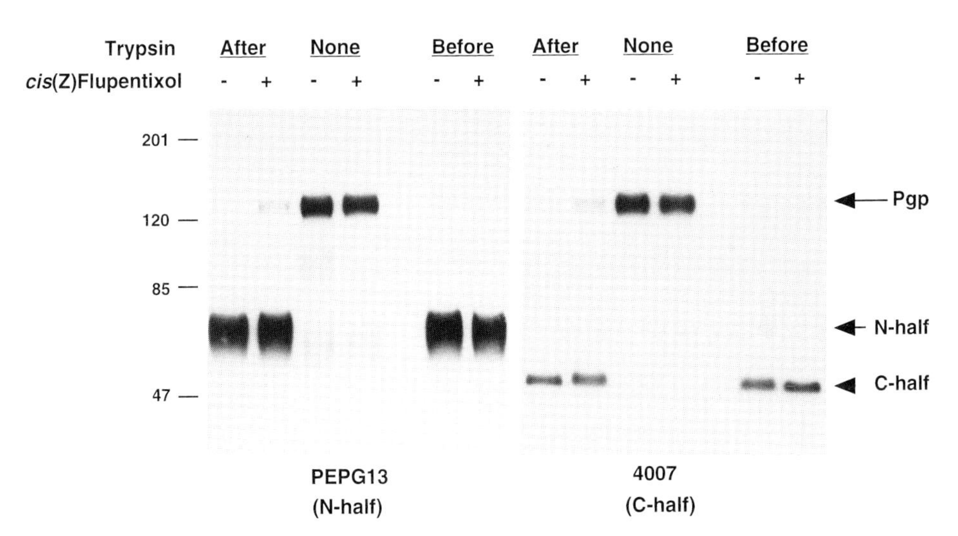

FIG. 1. [^{125}I]IAAP labeling of the N- and C-terminal halves of Pgp. (A) 10 μg of insect cell membranes containing recombinant Pgp were photoaffinity labeled with 2 n*M* [^{125}I]IAAP in the presence and absence of 50 μM *cis*(Z)flupentixol. Membranes were incubated with 7.5 μg of trypsin for 15 min at 37° either before or after photoaffinity labeling. Digestion with trypsin was stopped by adding a five-fold excess of soybean trypsin inhibitor. 2 μg of each sample was run on an 8% SDS–PAGE, dried, and exposed to X-ray film. (B) The identity of the two halves of the Pgp molecule was determined by immunoblot analysis using polyclonal antibodies PEPG 13 and 4007, specific for the N- and C-terminal halves of Pgp, respectively. The arrow with long tail, the arrowhead with short tail, and the arrowhead show the position of Pgp (140 kDa), N-half (75 kDa), and C-half (60 kDa), respectively.

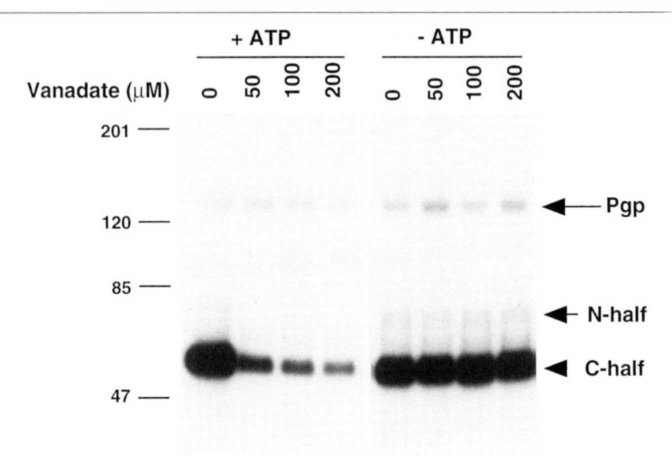

FIG. 2. Effect of vanadate trapping on [^{125}I]IAAP photoaffinity labeling of the N- and C-terminal halves of Pgp. Pgp containing insect cell membranes were subjected to limited trypsin digestion as described in the legend to Fig. 1. Trypsinized membranes were incubated at 37° for 10 min with 4 n*M* [^{125}I]IAAP, 10 μ*M* *cis*(Z)flupentixol, 5 m*M* $MgCl_2$, and varying concentrations (0–200 μ*M*) of vanadate, in the presence and absence of 2.5 m*M* ATP, prior to UV exposure for 10 min. After photoaffinity labeling with [^{125}I]IAAP the N- and C-terminal halves of Pgp were separated by SDS–PAGE as mentioned. Dried gels were exposed to X-ray film for autoradiography.

3. Incubate trypsinized membranes for 10 min with 2–5 n*M* [^{125}]iodoarylazidoprazosin ([^{125}I]IAAP) (2200 Ci/mmol, DuPont-NEN, Boston, MA), at room temperature under subdued light.
4. Where appropriate, preincubate membranes for 3–5 min with indicated concentrations of substrates or modulators of Pgp prior to the addition of [^{125}I]IAAP.
5. Following incubation, illuminate membranes with a UV lamp (366 nm, General Electric, Plainville, CT) for 10 min, placing the tubes on water at room temperature.
6. After illumination, add 25 μl of 5× SDS–PAGE sample buffer to the reaction mixture and keep at 21–23° for another 30 min before analyzing by SDS–PAGE.
7. At the end of the incubation period vortex the samples and load 2–4 μg of protein to each lane of an 8% Tris-glycine gel (Novex, San Diego, CA) and separate by SDS–PAGE.
8. Fix gels for 15 min in a solution containing 30% methanol and 10% acetic acid.
9. Treat with Novex gel drying solution for another 10 min and air dry for 16–18 hr between Dryerase minicellophane (Novex).

10. Expose dried gels to Kodak (Rochester, NY) Biomax-MR X-ray films at −80°.
11. Cut out the radioactive bands from the gel, and soak in 1 ml of tissue solubilizer (Solvable from Packard) at room temperature for 16 hr.
12. Dilute samples to 15 ml with Biosafe II scintillation fluid (RPI, Mount Propsect, NY) and determine radioactivity associated with each band in a scintillation counter.
13. Alternatively, expose the dried gel to a phosphorimager screen for 3–5 hr and quantify the radioactive bands using the Storm 860 system (Molecular Dynamics, Sunnyvale, CA).

V. Effect of Vanadate Trapping on Photoaffinity Labeling of the N- and C-Terminal Halves of Pgp. Inhibition of Pgp ATPase activity by sodium orthovanadate results from trapping of a MgADP·Vi complex (analog of MgADP·Pi) at the catalytic site. Stabilization of this transition state conformation of Pgp during its catalytic cycle greatly reduces the affinity of the drug-binding sites for photoaffinity analogs of the substrates.[6,7] Using crude membranes as well as purified and reconstituted Pgp, we have demonstrated that ATP hydrolysis is essential for this phenomenon. Here we show that vanadate trapping and its effect on [^{125}I]IAAP labeling is unaffected by cleaving Pgp at its linker region with mild trypsin treatment (Fig. 3).

Protocol

1. Trypsinize Pgp containing membranes (10 μg protein), as mentioned earlier, in 10 m*M* Tris-HCl, pH 7.5, 50 m*M* NaCl, 300 m*M* mannitol, and 1% aprotinin (labeling buffer) in a final volume of 100 μl, as described.
2. Incubate trypsinized membranes at 37° for 10 min with 5 n*M* [^{125}I]IAAP, 10 μM *cis*(Z)flupentixol (optional), 10 m*M* MgC12, 5 m*M* ATP, and varying concentrations (0–200 μM) of sodium orthovanadate (Sigma).
3. Follow steps 5 through 12 of Section IV (under Methods) Protocol.

Note: Preparation of sodium orthovanadate: Dissolve 9.4 mg of sodium orthovanadate (Sigma) in 1 ml of H_2O. Boil the solution for 3 min at 100° and slowly cool down to room temperature. Measure absorbance at 268 nm and calculate concentration using an extinction coefficient of OD 3.6 = 1 m*M*. Dilute to a stock solution of 10 m*M* in water. To obtain maximum inhibition of ATPase activity of Pgp, use only freshly prepared vanadate solution.

VI. Studies with Purified and Reconstituted Pgp. Effect of vanadate trapping and modulator interaction on [^{125}I]IAAP labeling can be assessed

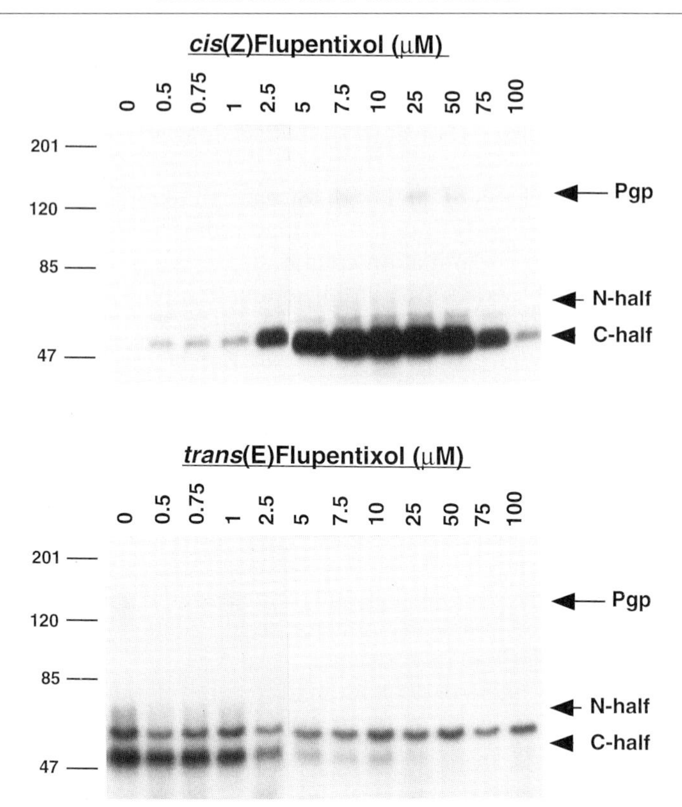

FIG. 3. Effect of *cis*(Z)flupentixol and *trans*(E)flupentixol on [^{125}I]IAAP labeling of the N- and C-terminal sites of Pgp. Photoaffinity labeling of Pgp containing crude insect cell membranes was carried out with 2 n*M* of [^{125}I]IAAP in the presence of varying concentrations (0–100 μM) of *cis*(Z)flupentixol (*upper*) or *trans*(E)flupentixol (*lower*). Following labeling, membranes were incubated with trypsin and then with trypsin inhibitor as mentioned in Fig. 1. N- and C-terminal halves of Pgp were separated in an 8% gel by SDS–PAGE and [^{125}I]IAAP labeling was quantified by measuring radioactivity associated with each fragment, as described in the text.

using purified and reconstituted Pgp.[7] Purified Pgp is reconstituted into phospholipid vesicles by rapid dilution of detergent.

Protocol

1. Dilute (20-fold in a final volume of 50 μl) purified Pgp preparation (~1 μg protein) containing 1.25% octylglucoside, and 0.1% lipid mixture in a reaction mixture containing 50 m*M* Tris-HCl, pH 7.0, 125 m*M* KCl, and 1 m*M* DTT and incubate at room temperature for 5 min.

2. Add 4 nM [^{125}I]IAAP and incubate samples in dark at 37° for 10 min.
3. After 10 min of incubation, expose samples to a UV lamp (366 nm, General Electric) on ice for 30 min.
4. Follow steps 6 through 12 of Section IV (under Methods) Protocol.

Note: For vanadate-induced inhibition include 10 mM $MgCl_2$, 5 mM ATP, and varying concentrations (0–400 μM) of sodium orthovanadate prior to the addition of [^{125}I]IAAP.

Notes and Conclusions

Over the years, photoaffinity labeling has evolved as one of the most widely used techniques for the identification and characterization of substrate and inhibitor binding sites of Pgp. This aspect of photoaffinity labeling has been discussed by Safa[15a] and Greenberger[15b] in this volume. Commercial availability, high specific activity, and minimum deviation in its structure from the parent compound (prazosin) provides an advantage to [^{125}I]IAAP over other photoaffinity analogs of Pgp substrates. Figure 1A shows distribution of [^{125}I]IAAP labeling between the N- and C-terminal halves of Pgp, which are separated by SDS–PAGE following limited trypsin digestion at the linker region. The identity of the two halves was determined by immunoblot analysis with specific antibodies against the N- (PEPG13)[16] and C-terminal (4007)[17] halves of Pgp (Fig. 1B). Pgp retained its ability to bind [^{125}I]IAAP even when trypsin treatment was carried out before photoaffinity labeling (Fig. 1). Under a similar condition the stimulatory effect of *cis*(Z)flupentixol on [^{125}I]IAAP labeling of the C-terminal half remained unaltered, suggesting that the effect of modulator interaction on substrate binding of the two sites of Pgp can be studied efficiently by this method (Figs. 1 and 3).

Inhibition of [^{125}I]IAAP labeling of Pgp results from trapping of ADP · Vi at the catalytic site. When Pgp was cleaved by trypsin treatment at the linker region, before the start of the reaction, vanadate trapping had the same inhibitory effect on photoaffinity labeling with [^{125}I]IAAP as was observed for intact Pgp (Fig. 2). Lack of inhibition of [^{125}I]IAAP labeling in the absence of ATP shows the specificity of the reaction and indirectly demonstrates that the cleaved molecule is able to hydrolyze ATP (Fig. 2).

[15a] A. R. Safa, *Methods Enzymol.* **292,** [21], 1998 (this volume).
[15b] L. M. Greenberger, *Methods Enzymol.* **292,** [22], 1998 (this volume).
[16] E. P. Bruggemann, V. Chaudhary, M. M. Gottesman, and I. Pastan, *BioTechniques* **10,** 202 (1991).
[17] S. Tanaka, S. J. Currier, E. P. Bruggemann, K. Ueda, U. A. Germann, I. Pastan, and M. M. Gottesman, *Biochem. Biophys. Res. Commun.* **166,** 180 (1990).

Taken together, these results suggest that vanadate trapping provides an efficient assay to study the extent of coupling between the substrate binding site(s) and the catalytic domain. In addition, single tryptic cleavage at the linker region allows one to evaluate the effect of vanadate trapping on the two [^{125}I]IAAP binding sites of Pgp separately. The effect of allosteric modulators, such as flupentixol, can also be studied on both the substrate-binding sites using the same strategy. Thus, the drug-binding properties of the two sites comprised of the two halves of Pgp can be determined without affecting the structural and functional integrity of the protein.

[24] Identification of Phosphorylation Sites in Human MDR1 P-Glycoprotein

By TIMOTHY C. CHAMBERS

Introduction

Overexpression of P-glycoprotein (Pgp) is the best characterized experimental mechanism of multidrug resistance (MDR) and has been implicated in several studies as an important factor in clinical resistance to chemotherapeutic drugs.[1] P-glycoprotein undergoes covalent modification by phosphorylation as first demonstrated by Carlsen, Ling and co-workers.[2] A large body of evidence based on pharmacologic, biochemical, and molecular approaches has suggested that phosphorylation of Pgp plays a modulatory role in drug transport.[3] To complement these studies and more rigorously demonstrate the role of phosphorylation in Pgp function by site-directed mutagenesis, we embarked on a project to identify the sites phosphorylated. A short communication describing this work has appeared.[4] In the following chapter the methodology utilized is described in more detail. The procedure represents an amalgam of previously described protocols for protein and peptide purification, adapted to suit the particular characteristics of Pgp. In particular, the methods for enzymatic digestion and initial peptide purification follow those described by Stone *et al.*, and the interested researcher

[1] M. M. Gottesman and I. Pastan, *Annu. Rev. Biochem.* **62,** 385 (1993).
[2] S. A. Carlsen, J. E. Till, and V. Ling, *Biochim. Biophys. Acta* **467,** 238 (1977).
[3] U. A. Germann, T. C. Chambers, S. V. Ambudkar, I. Pastan, and M. M. Gottesman, *J. Bioenerg. Biomembr.* **27,** 53 (1995).
[4] T. C. Chambers, J. Pohl, R. L. Raynor, and J. F. Kuo, *J. Biol. Chem.* **268,** 4592 (1993).

METHODS IN ENZYMOLOGY, VOL. 292
0076-6879/98 $25.00

is referred to their excellent article.[5] With appropriate modification, the general strategy and procedure should be applicable to the identification of phosphorylation sites in any reasonably abundant protein, including other ABC transporters.

Rationale

It was shown previously that tryptic phosphopeptide map patterns of Pgp, isolated either from intact ^{32}P-labeled control or phorbol ester-stimulated KB-V1 cells, or from reaction mixtures of Pgp phosphorylated by protein kinase C (PKC), were essentially identical, with three distinct phosphopeptides in each.[6] This suggested that PKC or a closely related kinase was responsible for phosphorylation *in vivo,* and that the sites phosphorylated by PKC *in vitro* were the physiologically relevant sites. A strategy was therefore developed to identify the *in vitro* sites, anticipating such a correspondence, and reasoning that identification of the *in vivo* sites would require ^{32}P-labeling of cells on an impractically large scale. Membrane-associated Pgp is utilized as substrate rather than purified Pgp, in order to retain Pgp in a more native conformation and not expose irrelevant sites to the exogenous kinase, and also to avoid solubility problems.

Strategy

The most important initial consideration is deciding on the scale of the procedure. With about 50 pmol of each purified phosphopeptide needed for sequence analysis, and assuming a loss of 95% during digestion and peptide purification, a requirement for about 1 nmol (150–170 μg) of purified phosphorylated Pgp was determined. It was estimated by gel scanning that Pgp represents 5% of the protein in purified plasma membrane vesicles from KB-V1 cells. Again assuming significant losses during Pgp purification, this translated to a requirement for about 18 mg of membrane vesicle protein, containing an estimated 900 μg Pgp, as starting material for each experiment. The strategy adopted is as follows: Prepare plasma membrane vesicles from KB-V1 cells; phosphorylate Pgp to a maximum extent with highly purified PKC; purify ^{32}P-labeled Pgp by preparative SDS–PAGE; reduce and carboxamidomethylate Pgp; digest Pgp with endoproteinase Lys-C or trypsin; partially purify resultant phosphopeptides by reversed-

[5] K. L. Stone, M. B. LoPresti, J. M. Crawford, R. DeAngelis, and K. R. Williams, *in* "A Practical Guide to Protein and Peptide Purification for Microsequencing" (P. T. Matsudaira, ed.), p. 31. Academic Press, San Diego, 1989.

[6] T. C. Chambers, B. Zheng, and J. F. Kuo, *Mol. Pharmacol.* **41,** 1008 (1992).

phase high-performance liquid chromatography (RP-HPLC); isolate and further purify individual phosphopeptides; determine amino acid sequence; identify the sites by comparison of the sequences obtained with the published predicted sequence of human MDR1.[7]

Procedure

Preparation of Membrane Vesicles and PKC

Approximately 10^9 KB-V1 cells are harvested from sixteen 24- × 24-cm Nunc tissue culture dishes and membrane vesicles prepared using the nitrogen cavitation method as previously described.[8] Vesicles are suspended at 10 mg/ml protein in 10 m*M* Tris-HCl, pH 7.5, 0.25 *M* sucrose, and 1 m*M* phenylmethylsulfonyl fluoride (PMSF), and stored in aliquots of 1 mg at −70°. Typical yields are 15–20 mg protein. PKC, comprising a mixture of isozymes but mainly the calcium-dependent forms, is purified from pig brain as described.[9] Commercially available preparations (Calbiochem, La Jolla, CA) can substitute.

Phosphorylation of P-Glycoprotein

Multiple (six or eight) phosphorylation reaction mixtures (1 ml) each contain 2–3 mg membrane vesicles, PKC (sufficient to incorporate 0.5 nmol phosphate/min into histone H1 under similar conditions), 20 m*M* Tris-HCl, pH 7.5, 0.25 *M* sucrose, 5 m*M* $MgCl_2$, 0.2 m*M* $CaCl_2$, 50 μg/ml phosphatidylserine, 0.2 μM okadaic acid, 1 m*M* ouabain, 1 m*M* sodium orthovanadate, 1 m*M* PMSF, and 1 m*M* [γ-^{32}P]ATP (specific activity 500 cpm/pmol). Reactions are carried out for 60 min at 30° and terminated by the addition of 0.25 ml 5 × sodium dodecyl sulfate (SDS) sample buffer[10] without boiling. In pilot experiments, these conditions were found to result in maximum phosphorylation of Pgp by PKC. Ouabain, vanadate, and okadaic acid are essential to inhibit membrane-associated ATPases and phosphatases, and phosphate incorporation is also improved with relatively high concentrations of ATP. Under optimal conditions, a phosphorylation stoichiometry of about 1 mol phosphate/mol of Pgp was the maximum

[7] C-J. Chen, J. E. Chin, K. Ueda, D. P. Clark, I. Pastan, M. M. Gottesman, and I. B. Roninson, *Cell* **47,** 381 (1986).

[8] M. M. Cornwell, M. M. Gottesman, and I. H. Pastan, *J. Biol. Chem.* **261,** 7921 (1986).

[9] P. R. Girard, G. J. Mazzei, and J. F. Kuo, *J. Biol. Chem.* **261,** 370 (1986).

[10] U. K. Laemmli, *Nature* **227,** 680 (1970).

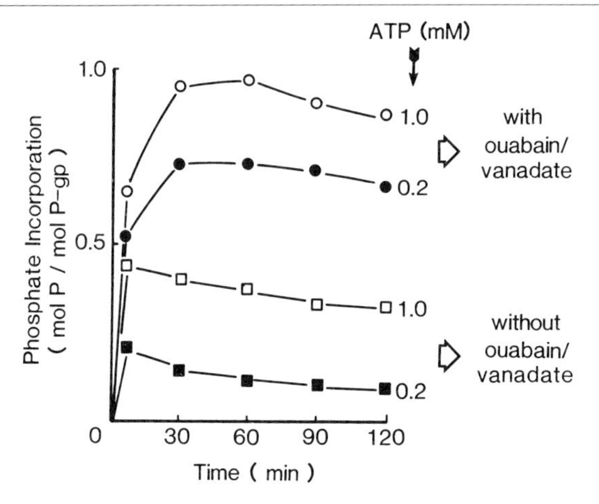

FIG. 1. Stoichiometry of Pgp phosphorylation under different conditions. Membrane vesicles from KB-V1 cells were incubated with PKC under standard conditions in the presence (○, ●) or absence (□, ■) of 1 mM ouabain and 1 mM sodium orthovanadate, and with either 0.2 mM ATP (specific activity 1000 cpm/pmol) or 1 mM ATP (200 cpm/pmol), as indicated. Aliquots were removed at the times indicated, Pgp was resolved by SDS–PAGE, and phosphate incorporation was determined by scintillation counting of gel bands. Pgp was estimated to represent 5% of the membrane vesicle protein in the calculation of phosphorylation stoichiometry.

obtained (Fig. 1), even though at least three sites were phosphorylated (see below). In retrospect, phosphorylation by the exogenous PKC was probably limited to those Pgp molecules with their linker regions exposed on the outer surface of the vesicle, while molecules in the opposite orientation have no access to PKC and ATP. Phosphorylation of Pgp by endogenous membrane kinase(s) was negligible under these conditions, in part due to inactivation on freezing/thawing of vesicles.

Purification of Phosphorylated P-Glycoprotein

Each terminated reaction mixture is subjected to SDS–PAGE with a 6% acrylamide resolving gel of 0.15 × 15 × 15 cm. A preparative comb is utilized with a single large slot for the sample and a small slot in which prestained molecular weight standards are run. The upper buffer contains 0.1 mM sodium thioglycolate to scavenge free radicals and oxidants to prevent damage to susceptiple amino acids.[11] Electrophoresis is carried out at 150 V for 1 hr and then at 250 V for another 4 hr or so until the dye-

[11] M. W. Hunkapiller, E. Lujan, F. Ostrander, and L. E. Hood, *Methods Enzymol.* **91,** 227 (1983).

front has completely electrophoresed out of the gel. The lower buffer contains excess [γ-^{32}P]ATP; we routinely collect and store the buffer for 6 months until the radioactivity has decayed. Each gel is removed from its plates and 1 cm of the bottom edge removed to ensure no residual [γ-^{32}P]ATP is present. The gel is then carefully wrapped in plastic wrap, with one side covered with a single, wrinkle-free layer. It is essential to use the dry gel without any kind of washing since Pgp recovery is extremely poor from previously wetted gels. The wrapped gel is then taped to a square of 3MM paper and several orientation spots made on the paper with "radioactive ink" (0.5 μl of [γ-^{32}P]ATP/SDS sample buffer, 1:10). Autoradiography with Kodak XAR5 film or equivalent is performed without a screen at room temperature for 2–3 hr. Phosphorylated Pgp, the major phosphorylated species visible on the developed film, is evident as a strip of radioactivity at 150–170 kDa. ^{32}P-Labeled Pgp is located on the gel and the gel strips are minced with a scalpel into 1-mm cubes and divided into two 50-ml plastic screw-capped tubes. Three volumes (about 30 ml) of freshly made 0.1 *M* NH_4HCO_3, pH 8.0, are added to each tube and the contents are mixed end-over-end at 30° or room temperature overnight. The gel pieces are removed by centrifugation (2000*g*, 15 min), and the eluants are pooled and concentrated to a volume of 2 ml by ultrafiltration with a YM10 membrane (Amicon, Danvers, MA). The sample is then dialyzed overnight at 4° against 2 liters of 5 m*M* NH_4HCO_3 and 0.05% SDS to remove salts originally present in the unwashed gel pieces. The dialyzate, clarified by centrifugation if necessary, is divided between two microfuge tubes and dried by rotary evaporation in a Speed-Vac. The samples are dissolved, quite readily due to the presence of SDS, in a total combined volume of 100 μl H_2O. The remainder of the procedure is carried out in the same tube, thus minimizing losses due to transfer. The protein is precipitated by the addition of 0.9 ml acetone containing 1 m*M* HCl and incubation at −20° for 3 hr. The precipitated protein is recovered by centrifugation (10,000*g*, 5 min), washed twice in 100 μl cold acetone to ensure complete removal of any residual SDS, air dried, and dissolved in 50 μl 8 *M* urea and 0.4 *M* NH_4HCO_3. At this point it is convenient to determine protein concentration and total radioactivity, from which the stoichiometry of phosphorylation can be calculated. The purity can also be checked by SDS–PAGE using a minigel. A 2-μl aliquot, diluted to 20 μl with H_2O, is generally sufficient for all measurements.

Reduction, Carboxamidomethylation, and Enzymatic Digestion

Carboxamidomethylation of the protein is first performed to improve subsequent proteolytic digestion.[5] Dithiothreitol (5 μl of 45 m*M*) is added to the ~50-μl sample of [^{32}P]Pgp. After incubation at 50° for 15 min and cooling to room temperature, 5 μl 100 m*M* iodoacetamide is added and incubation at room temperature carried out for 15 min. The urea concentration is reduced to 2 *M* by the addition of 140 μl H_2O. Enzymatic digestion is carried out at 37° for 24 hr with endoproteinase Lys-C (3 μg per 100 μg Pgp) or trypsin (5 μg per 100 μg Pgp). We used sequencing grade Lys-C or trypsin from Sigma (St. Louis, MO), but enzymes of equal quality are available from other sources. Other proteases may be used, but many are not compatible with the high concentration of urea required to maintain Pgp in solution.

RP-HPLC

Buffer A is 0.06% trifluoroacetic acid (TFA)/H_2O and buffer B is 0.05% TFA/80% acetonitrile; the absorbances at 218 n*M* are matched by the addition of more TFA to buffer A if required. The digests are centrifuged briefly to remove insoluble material and injected directly onto an analytical (4.6- × 250-mm) Vydac C_{18} RP-HPLC column equilibrated in buffer A. The column is washed for 10 min with buffer A at a flow rate of 0.5 ml/min and then a gradient is run from 0 to 80% acetonitrile with an increase of 0.73% acetonitrile/min (Lys-C digest) or 0.5% acetonitrile/min (trypsin digest). The absorbance at 218 n*M* is recorded with a flow cell. Fractions of 0.5–0.6 ml are collected into microfuge tubes and radioactivity determined by Cerenkov counting.

Purification of Phosphopeptide Generated by Lys-C Digestion

A typical elution pattern of peptides generated by digestion of Pgp with Lys-C is shown in Fig. 2. As indicated by the absorbance profile, a large number of peptides are bound and resolve well during elution in the acetonitrile gradient. Under optimal conditions, essentially all of the radioactivity is found in a single asymmetric peak eluting at 28–32% acetonitrile (not shown, but indicated by the bar in Fig. 2).[4] The peak of radioactivity corresponds to a phosphopeptide having an apparent molecular mass of 9.3 kDa (Fig. 2). Fractions containing the phosphopeptide are pooled and the solvent removed by rotary evaporation. The pooled fractions span a region of the chromatogram containing several peaks, and the presence of multiple peptides can be confirmed by analyzing a portion of the sample by high-resolution 16.5% acrylamide Tris/Tricine/SDS gel electrophoresis[12] and silver staining (Fig. 3). The 9.3-kDa peptide is well resolved from contaminating peptides, suggesting the use of such a gel in final purification.

[12] H. Schagger and G. von Jagow, *Anal. Biochem.* **166,** 368 (1987).

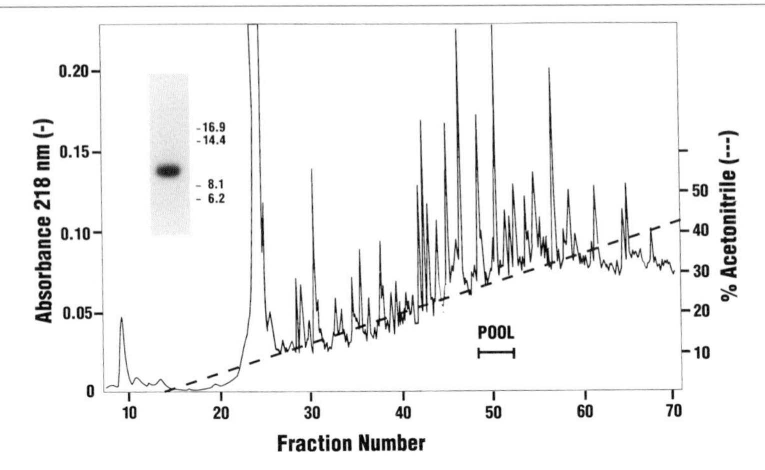

FIG. 2. RP-HPLC of products of Lys-C digestion of purified phosphorylated Pgp. HPLC was performed with a 4.6- × 250-mm Vydac C_{18} column as described in the text. The absorbance profile at 218 n*M* and acetonitrile gradient are presented. The bar indicates the fractions pooled. These contained the major phosphorylated product, a phosphopeptide of apparent M_r 9300, which is shown on the left analyzed by 16.5% acrylamide high-resolution SDS–PAGE and autoradiography. Molecular mass standards (in kilodaltons) are indicated.

The bulk of the sample is subjected to electrophoresis in a 0.1- × 15- × 15-cm Tris/Tricine gel, and the peptides transferred to a polyvinylidene difluoride (PVDF) membrane. Electrophoretic transfer is performed at 4° for 16 hr in a buffer system (pH 8.3) of 12 m*M* Tris, 80 m*M* glycine, 0.05% SDS, and 15% (v/v) methanol. The 9.3-kDa phosphopeptide is located on the membrane by autoradiography (Fig. 3), the relevant section is excised, and radioactivity is determined by Cerenkov counting. An estimate of the molar quantity of peptide can then be made from the known phosphorylation stoichiometry.

Amino Acid Sequence of Lys-C Phosphopeptide

For amino acid sequence analysis from PVDF membrane, a ProBlott reaction cartridge is utilized. Automated Edman degradation is performed in a model 477A protein sequencer online connected with a phenylthiohydantoin (PTH) analyzer (model 120A), both from Applied Biosystems, Inc. (Foster City, CA). The amino acid sequence analysis of the 9.3-kDa Lys-C phosphopeptide is shown in Table I. The first 19 amino acids are identified, the limitation being progressive losses of material at each cycle as expected. The sequence corresponds uniquely to residues 625–643 of human MDR1 and indicates that the peptide is derived from the linker region.[7] The

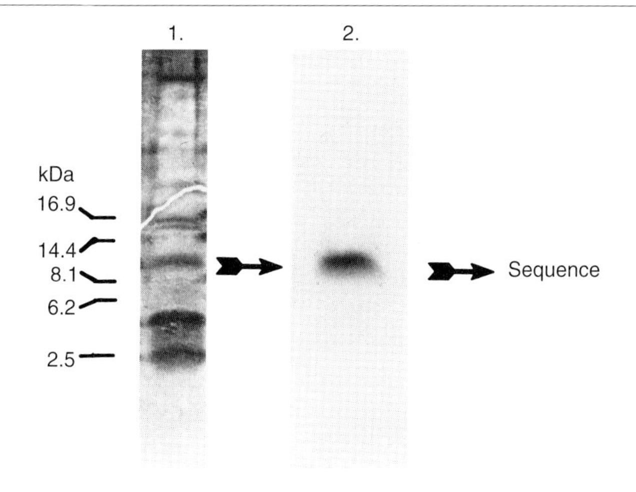

FIG. 3. Final purification of 9.3-kDa Lys-C phosphopeptide. Solvent was removed from the pooled fractions (Fig. 2) and the sample taken up in SDS buffer. A portion (5%) of the sample was analyzed by 16.5% acrylamide SDS–PAGE and silver staining as shown in lane 1. The bulk of the sample was similarly resolved by gel fractionation and the peptides transferred to PVDF membrane. An autoradigraph of the membrane is shown in lane 2. The radioactive band was cut out and subjected to amino acid sequence analysis.

carboxy terminus is tentatively assigned to lysine-702 based on the predicted molecular mass of the peptide and the specificity of Lys-C.[4] However, termination at an alternative lysine residue, for example, lysine-685, is not excluded since the actual molecular mass of the peptide is unknown. Phosphoamino acid analysis indicates that only serine residues are phosphorylated.[4] Inspection of the sequence predicted to be encompassed by the Lys-C peptide indicates the presence of several serine residues occurring within PKC consensus motifs of the type $(R/K_{1-3}X_{0-2})$-S-$(X_{0-2}R/K_{1-3})$.[13] These included Ser-661, Ser-667, Ser-671, and Ser-683; Ser-675 represented a site with minimal potential for PKC recognition.[4]

HPLC followed by high-resolution gel electrophoresis proved to be a powerful combination for phosphopeptide purification that may well be applicable to the identification of phosphorylated regions in other proteins. To identify the actual sites of phosphorylation, however, it became clear

[13] P. J. Kennelly and E. G. Krebs, *J. Biol. Chem.* **266,** 15555 (1991).

TABLE I
SEQUENCE ANALYSIS OF LYS-C PHOSPHOPEPTIDE[a]

Cycle number	Amino acid identified	Yield of PTH-amino acid (pmol)
1	L	36.1
2	V	23.9
3	T	12.4
4	M	15.5
5	Q	20.7
6	T	8.3
7	A	17.2
8	G	18.9
9	N	11.1
10	E	7.9
11	V	7.0
12	E	6.1
13	L	7.6
14	E	3.7
15	N	4.8
16	A	5.7
17	A	6.5
18	D	3.9
19	E	1.3
20+	n.d.	—

[a] An estimated 50 pmol was subjected to sequencing; n.d., not determined.

that isolation of smaller fragments from [^{32}P]Pgp would be necessary, so methods for the purification of tryptic phosphopeptides were developed.

Separation of Tryptic Phosphopeptides

A typical HPLC of trypsin-digested [^{32}P]Pgp is presented in Fig. 4. Four peaks of radioactivity are reproducibly resolved. Peak 1 (Fig. 4) is completely unretarded and peak 2 is retarded but unbound under initial conditions, properties consistent with relatively small and/or hydrophilic phosphopeptides. Peak 3 and peak 4 (Fig. 4) are bound to the column and eluted at 6 and 12% acetonitrile, respectively. Peaks 1, 2, and 4 (Fig. 4) appear to represent authentic peptides since they each resolved as single, discrete species when subjected to two-dimensional cellulose thin-layer separation. Peak 3 (Fig. 4), on the other hand, remained at the origin, and on further examination failed to resolve in SDS gels suggesting a molecular mass of <2 kDa. Peak 3 therefore likely represents a phosphopeptide derived from incomplete digestion. Peak 1 proved extremely difficult to

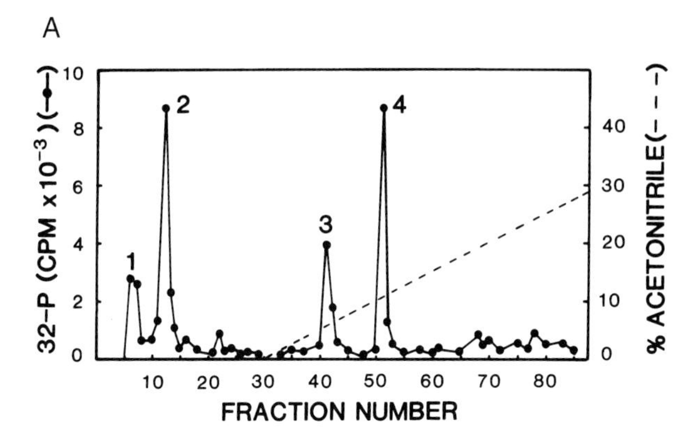

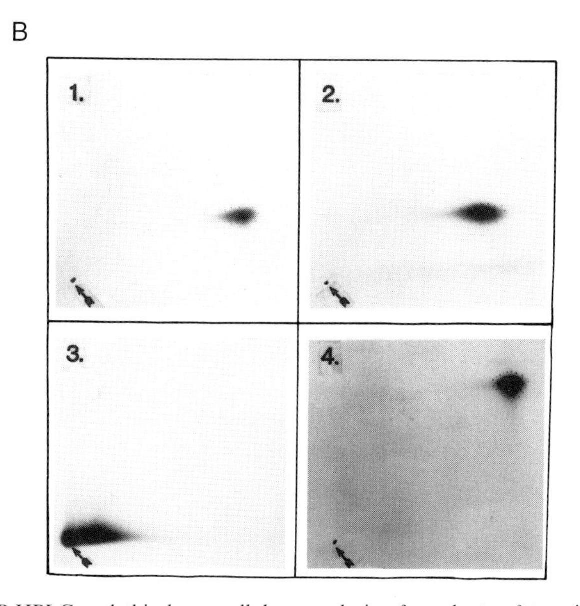

FIG. 4. RP-HPLC and thin-layer cellulose analysis of products of trypsin digestion of purified phosphorylated Pgp. (A) HPLC was performed with a 4.6- × 250-mm Vydac C_{18} column as described in the text. The major radioactive peaks are labeled 1–4 in order of elution. (B) An aliquot (0.1 ml) of each peak was analyzed by cellulose thin-layer with high-voltage electrophoresis at pH 3.7 from left to right in the first dimension and by ascending chromatography in the second dimension. Autoradiographs are shown. Arrows indicate origins. [Reprinted with permission from T. C. Chambers, J. Pohl, R. L. Raynor, and J. F. Kuo, *J. Biol. Chem.* **268,** 4592 (1993)].

purify, mainly because it would not bind to any type of C_{18} column regardless of initial conditions, and because the flow-through fractions contained many impurities. Purification and sequence analysis of the other two tryptic phosphopeptides is described below.

Purification and Sequence Analysis of Tryptic Phosphopeptide 2

Tryptic phosphopeptide 2 (TP2) is purified by preparative two-dimensional electrophoresis/chromatography on cellulose thin-layer followed by narrow-bore HPLC as follows. The relevant HPLC fractions are pooled (in this case, the single peak fraction in Fig. 4 is selected) and the solvent removed by rotary evaporation. The sample is dissolved in 20 μl 5% (v/v) acetic acid and 0.5% (v/v) pyridine and applied to a cellulose thin-layer plate (Eastman-Kodak, Rochester, NY) as described.[14] Electrophoresis is performed toward the cathode for 2.5 hr at 450 V. After the plate has completely dried, second dimension chromatography is performed with a solvent system of *n*-butanol/pyridine/glacial acetic acid/water (50 : 33 : 1 : 40). The plate is exposed to X-ray film and the phosphopeptide spot located. The relevant area of the cellulose is scraped away from the plastic backing with a spatula and collected. This is most conveniently performed by vacuuming the sample into the wide end of a Pasteur pipette plugged at the neck with glass wool and attached via tubing to a vacuum outlet. The pipette is then disconnected and turned right side up to form a column with the cellulose packed on the glass wool. The phosphopeptide is eluted from the cellulose with 1 ml of 50% (v/v) pyridine, which is then removed by rotary evaporation. The sample is dissolved in 0.25 ml of 0.06% heptafluorobutyric acid (HFBA) and injected onto a narrow-bore (2.1 × 150-mm) Vydac C_{18} column equilibrated in the same solvent. The use of HFBA instead of TFA as hydrophobic counterion has been shown to increase the retention time of peptides during RP-HPLC.[15] It was reasoned that TP2, which failed to bind to the C_{18} column with TFA as counterion (Fig. 4), might bind in the presence of HFBA, thereby providing the necessary final purification step. The column is eluted with a linear gradient of 0–50% acetonitrile (0.5%/min at 0.2 ml/min) and fractions of 0.2 ml are collected and radioactivity determined by Cerenkov counting. TP2 indeed bound to the column under these conditions and elutes as a single peak of radioactivity at around 4% acetonitrile (Fig. 5). Radioactive fractions are pooled, the volume reduced to 50 μl by rotary evaporation, and the sample subjected to amino acid sequence analysis.

Amino acid sequence analysis indicated that TP2 was the tripeptide

[14] R. W. Gracy, *Methods Enzymol.* **47,** 195 (1977).

[15] H. P. J. Bennett, C. A. Browne, and S. Solomon, *J. Liquid Chromatogr.* **3,** 1353 (1980).

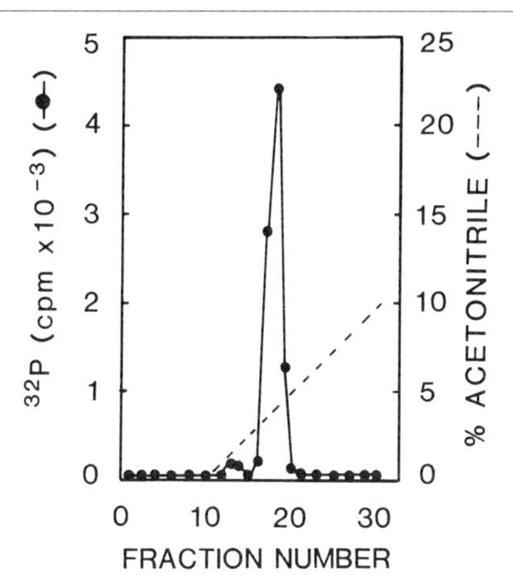

FIG. 5. Final purification of tryptic phosphopeptide 2. The elution of the phosphopeptide in an acetonitrile gradient from a 2.1- × 150-mm Vydac C_{18} column equilibrated in heptafluorobutyric acid is shown. See text for other details.

SVR; no PTH-amino acids were recovered in cycles 4–10 (Table II). This sequence occurs only once in human MDR1 (amino acids 671–673)[7] and also occurs within the portion of Pgp predicted to be encompassed by the Lys-C phosphopeptide.[4] These results therefore unambiguously identified Ser-671 as one of the PKC phosphorylation sites. Further evidence that the lone serine residue was indeed modified by phosphorylation was suggested by the presence in the first cycle of a species termed Ser′ (Table II). During Edman degradation, phosphoserine undergoes β-elimination to form dehydroalanine. After cleavage, an addition reaction occurs with dithiothreitol (DTT), and an adduct of PTH-dehydroalanine with DTT

TABLE II
SEQUENCE ANALYSIS OF TRYPTIC PHOSPHOPEPTIDE 2[a]

Cycle number	Amino acid identified	Yield of PTH-amino acid (pmol)
1	S(P)[b]	7 (Ser), 7 (Ser′)[c]
2	V	36.0
3	R	10.1
4–10	—	—

[a] An estimated 40 pmol was subjected to sequencing.
[b] S(P) represents phosphoserine.
[c] Ser′ represents an adduct of PTH-dehydroalanine with dithiothreitol, typically formed more readily from phosphoserine than from serine.

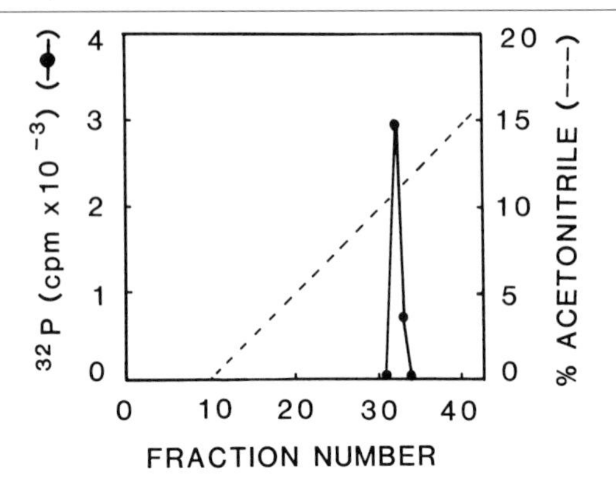

FIG. 6. Final purification of tryptic phosphopeptide 4. The elution of the phosphopeptide in an acetonitrile gradient from a 2.1- × 150-mm Vydac C_{18} column equilibrated in TFA is shown. See text for other details.

(Ser′) is formed and detected in the chromatogram.[16] This adduct is also formed from unphosphorylated serine, but to a much lesser extent. The presence in a cycle of significant amounts of Ser′ versus Ser is strongly indicative of the presence of phosphoserine.

Purification and Sequence Analysis of Tryptic Phosphopeptide 4

The peak fraction in Fig. 4 is selected and the solvent removed by rotary evaporation. The sample is then subjected to two-dimensional cellulose thin-layer separation and narrow-bore HPLC as described earlier for TP2 except that TFA is used in the solvents for HPLC. The phosphopeptide elutes from the narrow-bore column at 11–12% acetonitrile as expected (Fig. 6). Amino acid analysis of the purified peptide is shown in Table III. Five amino acids were identified. The yields of Ser versus Ser′ in the first two cycles strongly suggests that the second serine and not the first is phosphorylated. The sequence SSLIR occurs only once in MDR1 (amino acids 660–664) and identifies and confirms Ser-661 as a PKC site that, like Ser-671, is present as a PKC consensus site within the Lys-C phosphopeptide.

Confirmation of Phosphorylation Site Identification

Sequence analysis of purified phosphopeptides from Pgp phosphorylated by PKC enabled the identification of two specific phosphorylation

[16] H. E. Meyer, E. Hoffman-Posorske, and L. M. G. Heilmeyer, Jr., *Methods Enzymol.* **201,** 169 (1991).

TABLE III
SEQUENCE ANALYSIS OF TRYPTIC PHOSPHOPEPTIDE 4[a]

Cycle number	Amino acid identified	Yield of PTH-amino acid (pmol)
1	S	13 (Ser), 6 (Ser′)
2	S(P)	3 (Ser), 9 (Ser′)
3	L	14.4
4	I	15.2
5	R	2.0
6–10	—	—

[a] An estimated 28 pmol was subjected to sequencing.

sites, Ser-661 and Ser-671. TP1 appeared to contain a third distinct site with an uncertain identity. As with any phosphoprotein, it is important to confirm the phosphorylation site identification with independent methods, especially in this case where the analysis was incomplete. The phosphorylation of a chemically synthesized peptide, corresponding to residues 656–689 of human MDR1 and termed PG-2, was therefore examined to confirm that Ser-661 and Ser-671 were phosphorylation sites and to determine the origin of TP1.[17] PG-2 proved to be one of the most effective PKC substrates documented to date, with an apparent K_m of 1.3 μM and a V_{max}/K_m value superior to histone H1. The peptide was phosphorylated by PKC to a maximum stoichiometry of 3 mol phosphate/mol peptide and three phosphorylation sites, Ser-661, Ser-667, and Ser-671, were unambiguously identified. The electrophoretic and chromatographic properties of a tryptic subfragment of PG-2 containing Ser-667, with the sequence STR, were identical to those of TP1 obtained from Pgp. This indicated that TP1 indeed represented a unique phosphopeptide, and not one derived from incomplete digestion, and identified Ser-667 as the third PKC site in the protein.

Ultimately, the validity of phosphorylation site identification must be tested by site-directed mutagenesis and analysis of mutants expressed in intact cells. To this end, MDR1 mutants were constructed in which serines 661, 667, 671, 675, and 683 were replaced with either alanine or aspartic acid.[18] The mutants were expressed in NIH 3T3 fibroblasts and characterized. In contrast to wild-type Pgp, the mutants exhibited no detectable phosphorylation *in vivo* following metabolic labeling of cells with $[^{32}P]P_i$. This evidence can be regarded as proof that the basal phosphorylation

[17] T. C. Chambers, J. Pohl, D. B. Glass, and J. F. Kuo, *Biochem. J.* **299,** 309 (1994).
[18] U. A. Germann, T. C. Chambers, S. V. Ambudkar, T. Licht, C. O. Cardarelli, I. Pastan, and M. M. Gottesman, *J. Biol. Chem.* **271,** 1708 (1996).

sites of human MDR1 Pgp are indeed among the specific serine residues identified by sequencing and targeted for substitution. In addition, the phosphorylation-defective mutants failed to serve as substrates *in vitro* for PKC, PKA, or a novel Pgp kinase isolated from KB-V1 cells.[18] A more recent independent study also utilized a mutagenesis approach to show that the PKC sites are located in the linker region of human MDR1.[19]

Acknowledgments

I am indebted to Dr. J. F. Kuo for his support, encouragement, and interest throughout the course of this project. I thank Dr. Jan Pohl for sequence analysis, peptide synthesis, and invaluable advice on peptide purification. I thank Robert Raynor for providing PKC and Dr. Michael Gottesman for the KB-V1 cell line. This work was supported in part by grant CH-513 from the American Cancer Society.

[19] H. R. Goodfellow, A. Sardini, S. Ruetz, R. Callaghan, P. Gros, P. A. McNaughton, and C. F. Higgins, *J. Biol. Chem.* **271,** 13668 (1996).

[25] Identification of *in Vivo* Phosphorylation Sites for Basic-Directed Kinases in Murine *mdr1b* P-Glycoprotein by Combination of Mass Spectrometry and Site-Directed Mutagenesis

By J. S. Glavy, M. Wolfson, E. Nieves, E.-K. Han, C.-P. H. Yang, S. B. Horwitz, and G. A. Orr

Introduction

The reversible phosphorylation/dephosphorylation of proteins on serine, threonine, and tyrosine residues constitutes an important mechanism for regulating the activity of proteins in eukaryotic cells.[1,2] It is known that the multidrug resistant (MDR) transporter, P-glycoprotein, is phosphorylated *in vivo* but the relevance of this posttranslational modification to transporter function is poorly understood.[3–7] The approach that we have

[1] T. Hunter, *Cell* **80,** 225 (1995).
[2] P. Cohen, *Trends Biochem. Sci.* **17,** 408 (1992).
[3] W. Marsh and M. S. Center, *Biochem. Pharmacol.* **34,** 4180 (1985).
[4] S. N. Roy and S. B. Horwitz, *Cancer Res.* **45,** 3856 (1985).
[5] H. Hamada, K. Hagiwara, T. Nakajima, and T. Tsuruo, *Cancer Res.* **47,** 2860 (1987).
[6] T. C. Chambers, E. M. McAvoy, J. W. Jacobs, and G. Eilon, *J. Biol. Chem.* **265,** 7679 (1990).
[7] L. D. Ma, D. Marquardt, L. Takemoto, and M. S. Center, *J. Biol. Chem.* **266,** 5593 (1991).

0076-6879/98 $25.00

employed to investigate the physiologic relevance of P-glycoprotein phosphorylation is first to map the *in vivo* sites of phosphorylation and then to mutate the identified sites, singly or in combination, to nonphosphorylated residues, for example, a serine to alanine substitution. The creation of stably transfected cell lines carrying the desired mutations in P-glycoprotein allows dissection of the role of phosphorylation in P-glycoprotein function. Our studies have been performed on the murine *mdr1b* P-glycoprotein.

Identification of Major *in vivo* Phosphorylation Domain in Murine *mdr1b* P-Glycoprotein

To identify the precise location of all the *in vivo* phosphorylation sites in a 1250-amino-acid membrane-associated glycoprotein may seem daunting. However, studies performed on the regulation of the cystic fibrosis transmembrane conductance regulator (CFTR), another member of the ATP-binding cassette (ABC) family,[8] proved to be successful and informative. CFTR is a chloride channel, the activity of which is positively regulated by phosphorylation.[9,10] CFTR has a transmembrane-spanning topology similar to that of P-glycoprotein and the major phosphorylation domain in the channel was shown to be the R domain.[11-13] This domain, which serves to link the two homologous halves of the glycoprotein, is 241 amino acids in length and contains 24% charged residues in a repeating acidic and basic side-chain motif.[14] Phosphorylation of serine/threonine residues by protein kinase A (PKA) has been shown to be directly involved in regulation of the channel opening. In addition, phosphorylation of CFTR by protein kinase C (PKC) has been implicated in the turnover of CFTR.[15]

The region of P-glycoprotein analogous to the R domain of CFTR has been termed the *linker region* (amino acids 631–687). Although smaller than the R domain, the linker region of all mammalian P-glycoproteins is

[8] C. F. Higgins, *Annu. Rev. Cell Biol.* **8,** 67 (1992).

[9] P. M. Quinton, *Science* **245,** 1066 (1989).

[10] M. J. Welsh, *FASEB J.* **4,** 2718 (1990).

[11] X. B. Change, J. A. Tabcharani, Y. X. Hou, T. J. Jensen, N. Kartner, N. Alon, J. W. Hanrahan, and J. R. Riordan, *J. Biol. Chem.* **268,** 11304 (1993).

[12] S. H. Cheng, D. P. Rich, J. Marshall, R. J. Gregory, M. J. Welsh, and A. E. Smith, *Cell* **66,** 1027 (1991).

[13] M. R. Picciotto, J. A. Cohn, G. Bertuzzi, P. Greengard, and A. C. Nairn, *J. Biol. Chem.* **267,** 12742 (1992).

[14] J. R. Riordan, J. M. Rommens, B. Kerem, N. Alon, R. Rozmahel, Z. Grzelczak, J. Zielenski, S. Lok, N. Plavisic, J. L. Chou, M. L. Drumm, M. C. Iannuzzi, F. S. Collins, and L. C. Tsui, *Science* **245,** 1066 (1989).

[15] W. Breuer, H. Glickstein, N. Kartner, J. R. Riordan, D. A. Ausiello, and I. Z. Cabantchik, *J. Biol. Chem.* **268,** 13935 (1993).

```
         - -     -   + +-- -   -    +-          +++  ++          - -++    +-  ---
(626) TAGNEIELGNEACKSKDEIDNLDMSSKDSGSSLIRRRSTRKSICGPHDQDRKLSTKEALDEDVPP (690)  mdr1a
      : ::::: :: :  :      :       :  : : :::: :     :        :: :   : ::: :::::
(629) TRGNEIEPGNNAYGSQSDTDASELTSEESKSPLIRR-SIYRSVHRKQDQERRLSMKEAVDEDVPL (692)  mdr1b
        +  - -             - -  -    -- +      ++     +  +++ - -++    +-  ---
```

FIG. 1. Deduced amino acid sequences of the linker region of murine *mdr1a* and *mdr1b* P-glycoproteins. The predicted sequences are given in the single-letter code. Basic and acidic amino acids are designated + and −, respectively. Serine and threonine residues are in italics and colons indicate identical residues.

also highly charged and contains clusters of consensus phosphorylation sites for both acidic- and basic-directed kinases (see Fig. 1). We have used a combination of CNBr digestion and immunoblot analysis to determine whether this region of the murine *mdr1b* P-glycoprotein is a major phosphorylation domain after labeling of J7.V1-1, a murine vinblastine-resistant cell line, with $[^{32}P]P_1$ or labeling of membranes with $[\gamma\text{-}^{32}P]ATP$ and either PKA or PKC.[16] Inspection of the deduced amino acid sequence of the *mdr1b* P-glycoprotein indicates that CNBr digestion should generate a peptide fragment (amino acids 627–682), which contains the majority of the linker region including all of the potential phosphorylation sites.[17] A polyclonal antibody was raised against a synthetic peptide (amino acids 665–682) corresponding to a segment of the *mdr1b* linker region. Immunoblot analysis demonstrated that this antibody recognized specifically murine *mdr1b,* and not *mdr1a,* P-glycoprotein.[17]

Figure 2 shows that CNBr digestion of P-glycoprotein phosphorylated *in vivo* (Fig. 2A), *in vitro* with PKA (Fig. 2B), or *in vitro* with PCK (Fig. 2C) gave rise to a major 12-kDa ^{32}P-containing polypeptide that was specifically recognized by the *mdr1b* linker region-specific antibody. The minor, slower running ^{32}P-labeled peptides, which were also recognized by the antibody, were shown to be the result of incomplete CNBr digestion. The apparent size of this major ^{32}P-labeled CNBr fragment is approximately twice the predicted size for the CNBr fragment containing the linker region (5800 Da). However, a bacterially expressed murine *mdr1b* and linker region peptide (amino acids 621–687) also ran on sodium dodecyl sulfate–polyacrylamide gel electrophoresis (SDS–PAGE) at a position corresponding to almost twice its experimentally determined mass.[18] It is likely that the anomalous electrophoretic behavior of the linker region on SDS–PAGE is due to the large number of acidic residues in the peptide. These experi-

[16] G. A. Orr, E. K. Han, P. C. Browne, E. Nieves, B. M. O'Connor, C. P. Yang, and S. B. Horwitz, *J. Biol. Chem.* **268,** 25054 (1993).

[17] S. I. Hsu, L. Lothstein, and S. B. Horwitz, *J. Biol. Chem.* **264,** 12053 (1989).

[18] S. R. Juvvadi, J. S. Glavy, S. B. Horwitz, and G. A. Orr, *Biochem. Biophys. Res. Commun.* **230,** 442 (1997).

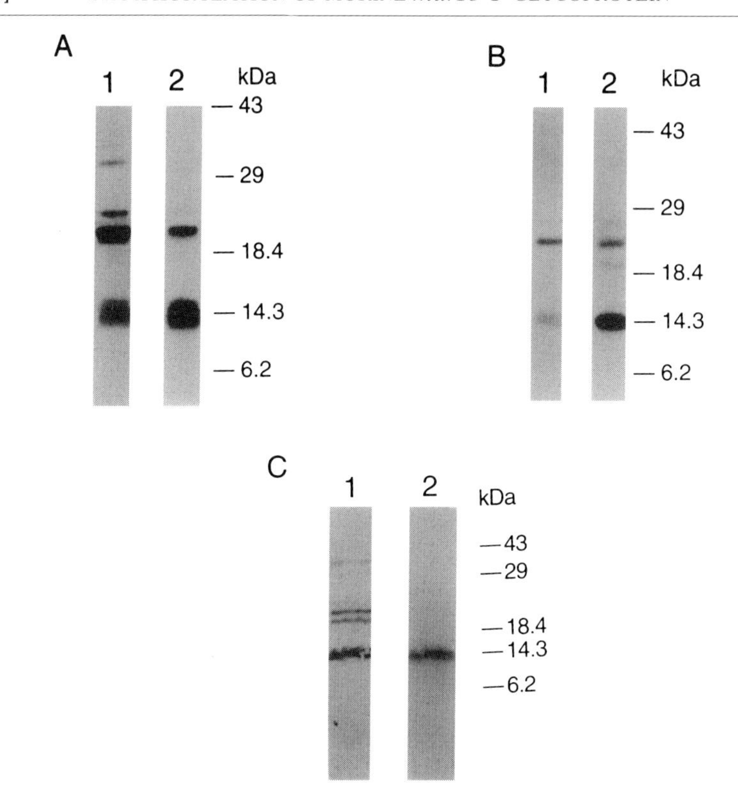

FIG. 2. Identification of the linker region as a major phosphorylation domain in murine *mdr1b* P-glycoprotein. *In vivo* and *in vitro* ^{32}P-labeled P-glycoproteins from J7.V1-1 cells were digested with CNBr, resolved by SDS–PAGE (15%) and transferred to nitrocellulose. ^{32}P-Labeled peptides were visualized by autoradiography and linker region-containing peptides by immunoblot analysis. (A) *In vivo* [^{32}P]P_i-labeled; (B) *in vitro* PKA-labeled; (C) *in vitro* PKC-labeled. In each panel, lane 1 is an immunoblot using an *mdr1b* linker region-specific polyclonal antibody; lane 2 is an autoradiogram of lane 1. In (A) and (B), immunoblotting was performed using the 2-chloronaphthol/H_2O_2 system. In (C), an enhanced chemiluminescence kit (ECL, Amersham) was employed. [Adapted with permission from G. A. Orr, E. K. Han, P. C. Browne, E. Nieves, B. M. O'Connor, C. P. Yang, and S. B. Horwitz, *J. Biol. Chem.* **268,** 25054 (1993).]

ments document unequivocally that the linker region of murine *mdr1b* P-glycoprotein is the major phosphorylation domain after *in vivo* labeling with [^{32}P]P_i and also after *in vitro* phosphorylation by either PKA or PKC.

Procedures

[^{32}P]P_i Labeling of Murine J7.V1-1 Cells. The vinblastine-resistant cell line, J7.V1-1, which is ~1000-fold resistant to vinblastine, is isolated by

stepwise selection from the drug-sensitive macrophage-like cell line, J774.2.[19] This drug-resistant cell line expresses predominantly the *mdr1b* P-glycoprotein.[17] For [^{32}P]P_i labeling studies, the cells are grown in 100-mm plates to 70–80% confluence and incubated for 4 hr in phosphate-free Dulbecco's modified Eagle's medium containing 100 μCi/ml [^{32}P]P_i. Membrane fractions are prepared as described previously[19] except that 50 n*M* calyculin A (Sigma, St. Louis, MO) is present in all solutions.

Phosphorylation of mdr1b P-Glycoprotein by PKA and PKC. PKA-catalyzed phosporylation J7.V1-1 membranes (100 μg), in a total volume of 100 μl, are suspended in 25 m*M* Tris-HCl, pH 7.5, containing 5 m*M* $MgCl_2$, 40 m*M* dithiothreitol (DTT), and the catalytic subunit of PKA (Sigma, St. Louis, MO, 5 units; 1 unit will transfer 1 pmol of phosphate from ATP to dephosphorylated casein/min).

PKC-catalyzed phosphorylation J7.V1-1 membranes (100 μg), in a total volume of 100 μl, are suspended in 20 m*M* Tris-HCl, pH 7.5, containing 200 m*M* sucrose, 5 m*M* $MgCl_2$, 0.2 m*M* $CaCl_2$, phosphatidylserine (6 μg), and PKC (Upstate Biotechnology, Lake Placid, NY; 0.1 unit, 1 unit will transfer 1 μmol of phosphate from ATP to histone H_1/min).

All phosphorylation reactions are performed at 37°, are initiated by the addition of 10 μM (γ-^{32}P]ATP (4–6 $\times$ 10^3 cpm/pmol), and are terminated after 30 min by the addition of 5$\times$ Laemmli sample application buffer.[20]

CNBr Digestion of ^{32}P-Labeled mdr1b P-Glycoprotein. ^{32}P-labeled membrane proteins prepared from J7.V1-1 cells after either *in vivo* or *in vitro* phosphorylation are separated by SDS–PAGE (6%) and electrotransferred to nitrocellulose paper. P-glycoprotein is visualized by autoradiography and the region of the blot containing P-glycoprotein is excised. The nitrocellulose strips are treated with CNBr (250 μl of a 100 mg/ml solution in 70% formic acid) at room temperature for 90 min.[21] Under these conditions >90% of the ^{32}P is released from the nitrocellulose. The supernatant is taken to dryness and the residue is dissolved and re-evaporated twice with H_2O to remove residual formic acid. Samples are dissolved in 1$\times$ sample application buffer, resolved by SDS–PAGE (15%), electrotransferred to nitrocellulose, and ^{32}P-labeled peptides visualized by autoradiography at −70° in the presence of an intensifying screen. After autoradiography, peptides containing the *mdr1b* linker region are visualized using the anti-*mdr1b* linker region polyclonal antibody. Immune complexes are visualized by sequential incubation with biotin-conjugated goat anti-rabbit IgG

[19] L. M. Greenberger, L. Lothstein, S. S. Williams, and S. B. Horwitz, *Proc. Natl. Acad. Sci. U.S.A.* **85,** 3762 (1988).

[20] U. K. Laemmli, *Nature* **227,** 680 (1970).

[21] K. X. Luo, T. R. Hurley, and B. M. Sefton, *Methods Enzymol.* **201,** 149 (1991).

and horseradish peroxidase-conjugated avidin followed by either the 2-chloronaphthol/H_2O_2 or the ECL (enhanced chemiluminescence) visualized method (Amersham, Arlington, IL).

Strategy for Identifying *in vivo* Sites of Phosphorylation in *mdr1b* P-Glycoprotein

The determination of the amino acid sequence of peptides containing phosphoserine and phosphothreonine residues is often problematic due to the lability of the phosphomonoester bonds under the conditions required for Edman degradation.[22,23] Both phosphoserine and phosphothreonine undergo base-catalyzed β-elimination of their phosphate groups. Indirect methods have therefore been developed for the assignment of phosphorylation sites. The best known involves the conversion of phosphoserine to *S*-ethylcysteine, prior to Edman degradation, by base-catalyzed β-elimination in the presence of ethanethiol.[22] *S*-Ethylcysteine forms a stable phenylthiohydantoin derivative. The reaction does not work, however, if phosphoserine is the *N*- or C-terminal amino acid. Furthermore, phosphothreonine does not undergo this modification. Methods based on the measurement of the released $[^{32}P]P_i$ are suspect due to inefficient extraction of inorganic phosphate into the organic phase.

The advent of modern mass spectrometers with their ability to measure accurately masses of both peptides and proteins with high sensitivity has caused a major revolution in protein structural determinations.[24,25] It affords one the opportunity to address questions that, only a few years ago, would have been impossible either as a consequence of sample limitations or due to the time, effort, and expense necessary to use alternative approaches. One powerful use of mass spectrometry is to detect posttranslational modifications to proteins and peptides. If the primary sequence of a protein is known, the accuracy of modern mass spectrometers is such that any deviation from the expected molecular mass is indicative of some type of posttranslational modification. In the case of phosphorylation, the expected mass of the protein increases by 80 Da for each phosphate introduced. Furthermore, chemical/proteolytic fragmentation of the protein combined with mass spectral analysis can also provide evidence for the site(s) of posttranslation modification.

[22] H. E. Meyer, P. E. Hoffmann, D. A. Donella, and H. Korte, *Methods Enzymol.* **201,** 206 (1991).
[23] P. J. Roach and Y. H. Wang, *Methods Enzymol.* **201,** 200 (1991).
[24] J. S. Andersen, B. Svensson, and P. Roepstorff, *Nat. Biotech.* **14,** 449 (1996).
[25] B. T. Chait and S. B. Kent, *Science* **257,** 1885 (1992).

Two mass spectrometry ionization methods used for protein characterization are electrospray ionization (ESI) and the matrix-assisted laser desorption/ionization (MALDI).[24,25] They serve different but complementary purposes. ESI-MS is more accurate than MALDI-MS but is intolerant to the presence of buffer components in the sample. This requirement means that the peptide/protein sample has to be purified by reverse-phase high-performance liquid chromatography (HPLC) prior to ESI-MS analysis. However, it is possible to introduce directly into the ESI mass spectrometer the effluent from HPLC, thus permitting on-line analysis of complex mixtures, for example, a total tryptic digest of a protein resolved by reverse-phase HPLC. In contrast, MALDI-MS, in some instances, allows the analysis of complex mixtures of peptides/proteins in the presence of buffer components. Although slightly less accurate than ESI-MS, this instrument is considerably more sensitive. With MALDI-MS, it is possible to obtain accurate masses with approximately $\leq$1 pmol of sample. ESI-MS requires 1–10 pmol of material.

There are 12 potential serine/threonine phosphorylation sites within the linker region of murine *mdr1b* P-glycoprotein. Our experimental strategy for determining the precise sites of *in vivo* phosphorylation in murine *mdr1b* P-glycoprotein is outlined schematically in Fig. 3. The method involves the direct comparison of the two-dimensional tryptic phosphopeptide maps derived from the *in vivo* [^{32}P]P$_i$-labeled *mdr1b* P-glycoprotein with those obtained from the recombinant *mdr1b* linker peptide phosphorylated *in vitro* by a variety of purified protein kinases. The sites of phosphorylation in the linker peptide are determined by ESI-MS after tryptic digestion. If a particular kinase phosphorylates more than one residue within the linker peptide, the tryptic phosphopeptides are resolved by reversed phase chromatography, and individual radiolabeled peaks are subjected to both ESI-MS and two-dimensional high-voltage electrophoresis/thin-layer chromatography (2D-HVE/TLC). It is important to stress that this method identifies only *in vivo* sites that are undergoing turnover during the labeling period with [^{32}P]P$_i$. Also the fact that a particular kinase phosphorylates a specific residue *in vitro* does not necessarily imply that the same kinase phosphorylates the site *in vivo.*

Phosphoamino acid analysis of *mdr1b* P-glycoprotein, immunoprecipitated from J7V1.1 cells after *in vivo* labeling with [^{32}P]P$_i$, indicated that only serine was phosphorylated.[16] The two-dimensional tryptic phosphopeptide map of *in vivo* ^{32}P-labeled *mdr1b* P-glycoprotein is shown in Fig. 4A. Six major phosphopeptides are clearly resolved and they have been designated 1–6. The tryptic phosphopeptide maps derived from the recombinant *mdr1b* linker peptide phosphorylated *in vitro* by either PKA or PKC are shown in Figs. 4B and 4C, respectively. A comixture of both *in vitro* phosphoryla-

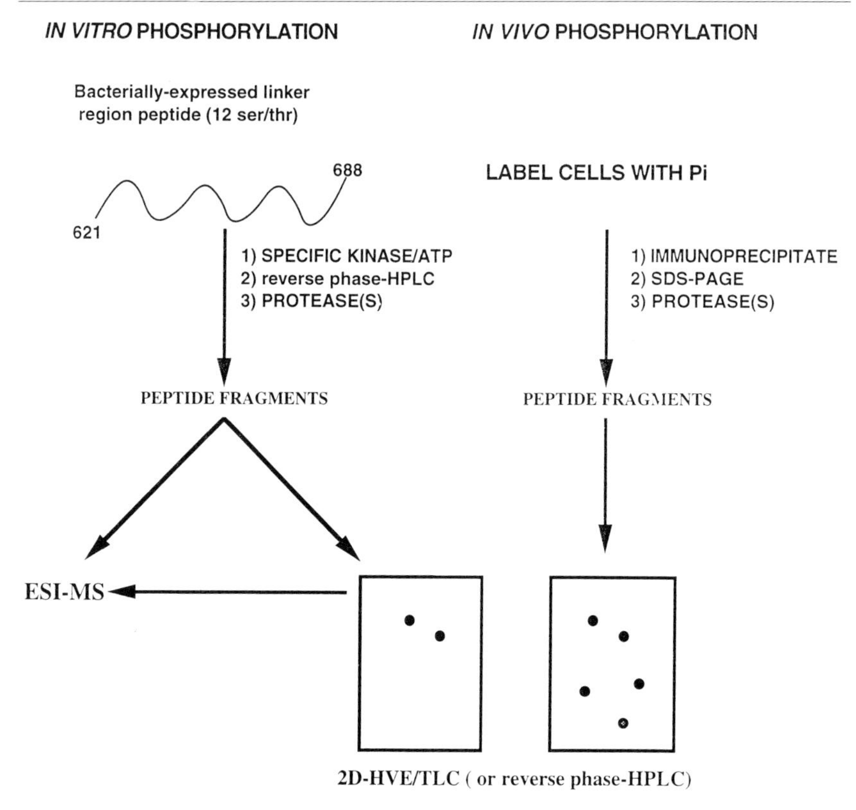

FIG. 3. Strategy for locating the *in vivo* phosphorylation sites in *mdr1b* P-glycoprotein.

tions is shown in Fig. 4D. It is apparent from these maps that the *in vitro* derived phosphopeptides 3–6 are similar to those generated by the combined *in vitro* PKA and PKC phosphorylations.

Inspection of the deduced amino acid sequence of *mdr1b* P-glycoprotein shows that tryptic digestion will generate a discrete peptide for each potential phosphorylation site within the linker region for basic-directed kinases.[16] These peptides are ^{659}SPLIR, ^{665}SIYR, ^{669}SVHR, and ^{681}LSMK. Each of these peptides can be clearly identified in the ESI-MS of the tryptic digest of the nonphosphorylated recombinant linker peptide (Fig. 5). Since the addition of a phosphate group will increase the mass of a peptide by 80 Da, comparison of the mass spectra of both nonphosphorylated and phosphorylated peptides can, in this case, lead to unambiguous determination of sites of phosphorylation. It is known, however, that the intensity of ions observed in ESI-MS is dependent on the chemical/physical properties of the individual peptides. This is illustrated in the ESI-MS analysis of the

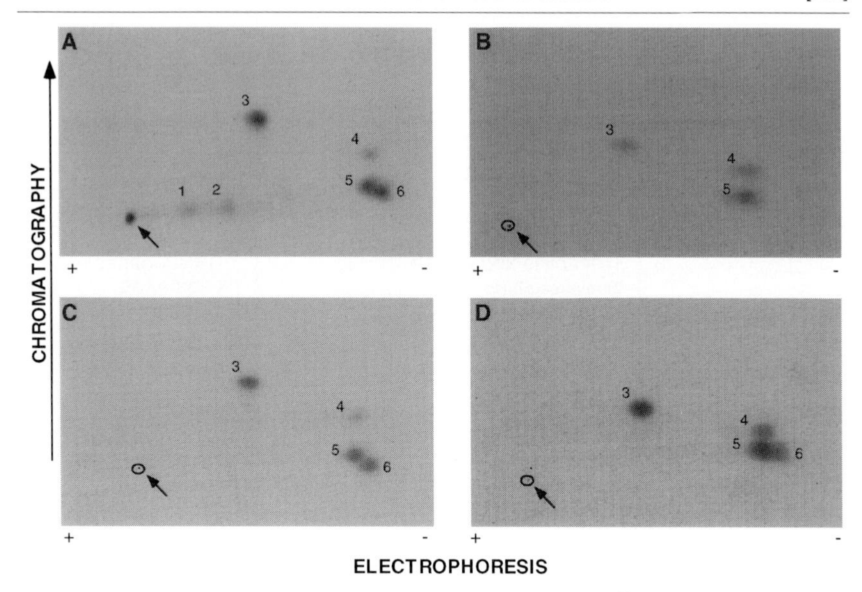

FIG. 4. Two-dimensional tryptic phosphopeptide maps of *in vivo* ^{32}P-labeled P-glycoprotein and *in vitro* ^{32}P-labeled recombinant linker peptide. Tryptic phosphopeptides were separated by electrophoresis at pH 1.9 in the first dimension (left, positive; right, negative) and by chromatography in the second dimension. Approximately 500 cpm were analyzed. (A) *In vivo* [^{32}P]P_i-labeled P-glycoprotein; (B) *in vitro* PKA-labeled recombinant linker peptide; (C) *in vitro* PKC-labeled recombinant linker peptide; (D) mixture of *in vitro* PKA/PKC-labeled recombinant linker peptide. The arrow (lower left) indicates the origin.

complete tryptic digest of the nonphosphorylated linker peptide in which the observed relative intensities for the four peptides of interest decrease in the order SIYR > SPLIR > SVHR > LSMK. For this reason it is more accurate to measure the change in relative intensities of the nonphosphorylated peptides when determining the order of addition of phosphate into the linker peptide.

In both the *in vitro* PKA and PKC phosphorylation reactions, conditions were chosen to maximize phosphate incorporation into the peptide. After PKA phosphorylation, both mono- and bisphosphorylated species could be identified by combined reverse-phase HPLC and ESI-MS. Although we were unable to purify the monophosphorylated form to homogeneity in these experiments, tryptic digestion followed by ESI-MS clearly established that serine-681 (LSMK) was the predominant site labeled. Note the complete absence of the ion corresponding to LSMK in the monophosphorylated linker peptide. Although LSMK is the predominant tryptic peptide containing serine-681 in the nonphosphorylated peptide, RLS(P)MK pre-

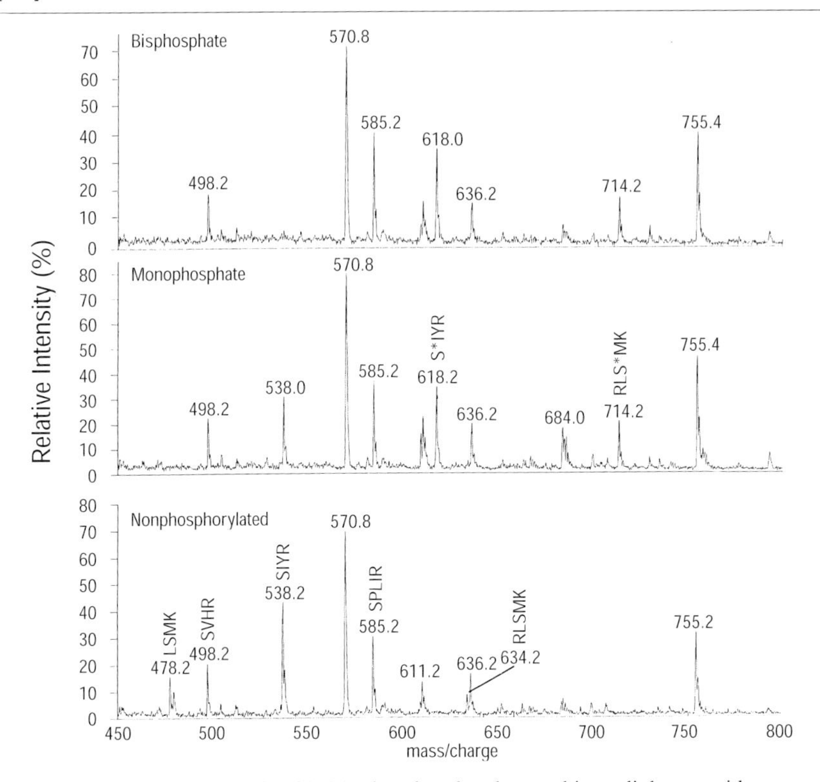

FIG. 5. ESI-MS analysis of PKA-phosphorylated recombinant linker peptide.

dominates in the phosphorylated peptides. It is known that trypsin does not cleave RXS(P) peptides efficiently.

In the bisphosphorylated peptide, the ion corresponding to SIYR is also missing indicating that serine-665 is also a PKA phosphorylation site. There is no evidence that either serine-659 (SPLIR) or serine-669 (SVHR) can be phosphorylated by this kinase. Individual phosphopeptides were resolved by reverse-phase HPLC and reanalyzed by both two-dimensional peptide mapping and ESI-MS. These studies allowed the identification of phosphopeptide 3 as S(P)IYR and phosphopeptide 5 as RLS(P)MK. No identifiable ions from phosphopeptide 4 were obtained. This phosphopeptide is probably a partial tryptic fragment of serine-681 since it, and phosphopeptide 6, were the only phosphopeptides absent from the *in vivo* phosphopeptide maps of the S681A P-glycoprotein mutant (data not shown).

ESI-MS analysis revealed that PKC introduced two phosphate groups into the linker peptide under the conditions used. Figure 6 shows the

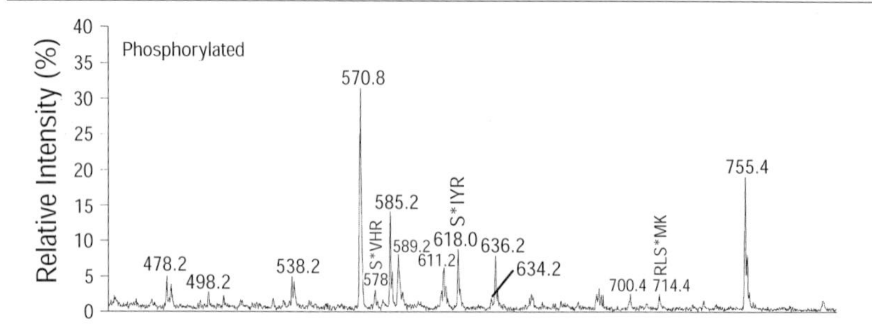

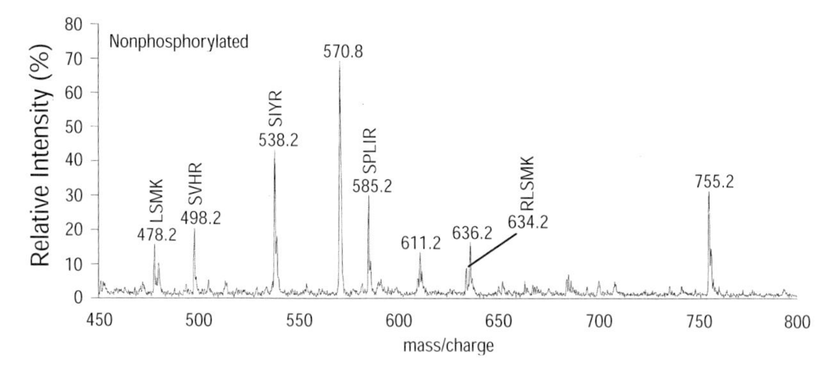

FIG. 6. ESI-MS analysis of PKC-phosphorylated recombinant linker peptide.

ESI-MS analysis of a tryptic digest of the PKC-phosphorylated recombinant linker peptide. The major PKC sites were identified as serine-665 and serine-669. Minor labeling of serine-681 was observed. In contrast, no labeling of serine-659 (SPLIR) was detected. ESI-MS analysis at earlier stages in the phosphorylation reaction showed that serine-669 was the preferred PKC site (data not shown). Analysis of the reverse-phase HPLC-resolved phosphopeptides led to the identification of phosphopeptide 6 as S(P)VHR.

These studies indicate that three of the four serine residues, serine-665 (phosphopeptide 3), serine-669 (phosphopeptide 6), and serine-681 (phosphopeptides 4/5), in the base domain of the linker region of murine *mdr1b* P-glycoprotein are utilized *in vivo.* More recent studies suggest that the two remaining tryptic phosphopeptides (1 and 2) are derived from the acidic domain of the linker region (see Fig. 1).[26] In human MDR1 P-glycoprotein serine-661, -667, and -671 were shown to be phosphorylated

[26] J. S. Glavy, S. B. Horwitz, and G. A. Orr, *J. Biol. Chem.* **272,** 5909 (1997).

in vitro by PKC whereas serine-667, -671, and -683 were phosphorylated by PKA.[27,28]

Procedures

Expression and Purification of Recombinant mdr1b Linker Region Peptide. The region corresponding to nucleotides 1970–2174 of the murine *mdr1b* P-glycoprotein cDNA, which encodes the linker region, is amplified by the polymerase chain reactin (PCR).[18] Restriction sites for *Sma*I and *Eco*RI are incorporated into the 5′ end of each primer. The PCR product, after purification, is subcloned into the pGEX-2T vector (Pharmacia, Piscataway, NJ), which expresses heterologous proteins fused to glutathione *S*-transferase (GST). Expression of recombinant protein is induced by isopropyl-β-thiogalactoside in *Escherichia coli* BL21-DE3 cells (Novagen, Madison, WI). The GST-linker peptide is fully soluble and is purified from crude bacterial lysates by affinity chromatography on glutathione-Sepharose. Elution of specifically bound proteins is accomplished with 5 m*M* glutathione. The linker peptide is cleaved from the GST fusion protein by thrombin cleavage at 37° for 60 min. The released peptide is purified by chromatography on Superdex 75 (2.6 × 60 cm) and eluted with 10 m*M* Tris-HCl, pH 7.4, and 140 m*M* NaCl. The authenticity of the bacterially expressed peptide is verified by both amino acid sequencing and ESI-MS.

Phosphorylation of Recombinant mdr1b Linker Region Peptide by Protein Kinases A and C

PKA CATALYZED PHOSPHORYLATION. The reaction mixture in a total volume of 100 μl contains 25 m*M*, Tris-HCl, pH 7.5, 5 m*M* $MgCl_2$, 25 μM DTT, 5 m*M* [γ-^{32}P]ATP (150 cpm/pmol), *mdr1b* linker peptide (450 pmol), and PKA (1 unit). Phosphorylations are performed at room temperature and the extent of ^{32}P incorporation is evaluated by the method of Roskoski[29] using Whatman (Clifton, NJ) P81 filters.

PKC CATALYZED PHOSPHORYLATION. The reaction mixture in a total volume of 100 μl contains 25 m*M* Tris-HCl, pH 7.5, 5 m*M* $MgCl_2$, 5 m*M* [γ-^{32}P]ATP (150 cpm/pmol), 2 m*M* DTT, 1 m*M* $CaCl_2$, 100 μg/ml phosphatidylserine, 2 μg/ml diolein, and *mdr1b* linker peptide (450 pmol). All other conditions are as described for PKC phosphorylation.

HPLC Purification of Phosphorylated Peptides and Tryptic Digestion. Phosphorylated peptides are purified by HPLC using an Aquapore RP300

[27] T. C. Chambers, J. Pohl, R. L. Raynor, and J. F. Kuo, *J. Biol. Chem.* **268,** 4592 (1993).

[28] T. C. Chambers, U. A. Germann, M. M. Gottesman, I. Pastan, J. F. Kuo, and S. V. Ambudkar, *Biochemistry* **34,** 14156 (1995).

[29] R. J. Roskoski, *Methods Enzymol.* **99,** 3 (1983).

reverse-phase column (2.1 × 200 mm) (Applied Biosystems, San Jose, CA) on a Hewlett-Packard 1090A system. Peptides are eluted with a linear gradient (0–40% over 40 min) of H_2O/acetonitrile containing 0.1% trifluoroacetic acid (TFA) at 0.2 ml/min and monitored at 214 nm. Peaks are collected manually, taken to dryness in a Speed-Vac, dissolved in 100 μl of 50 m*M* NH_4HCO_3/1 m*M* $CaCl_2$, and treated with TPCK-treated trypsin (1:100; trypsin:peptide; w/w) for 15 hr at room temperature. A second aliquot of trypsin is added and digestion continues for an additional 3 hr. The sample is taken to dryness in a Speed-Vac, dissolved and reevaporated three times with water. In some instances, the tryptic digests are rechromatographed on a C_8 reverse phase column using the conditions described to resolve individual phosphopeptides.

Immunoprecipitation of ^{32}P-Labeled P-Glycoprotein and Tryptic Digestion. ^{32}P-Labeled J7.V1-1 cells, from a single 100-mm cell culture dish, are suspended in 200 μl of buffer A (50 m*M* Tris-HCl, pH 7.4, containing 4 m*M* EDTA, 50 n*M* calyculin A, and 20 μg/ml each of aprotinin, leupeptin and pepstatin A) plus 1% SDS. After 5 min on ice, the sample is diluted with 800 μl of buffer A containing 2% Triton X-100 and centrifuged at 14,000 rpm for 5 min in an Eppendorf microfuge at 4° to remove particulate material. Immunoprecipitations are carried out for 15 hr using the anti-P-glycoprotein-specific antibody, R3 (1:50 dilution) at 4°. Immune complexes are isolated by incubation with protein A-Sepharose (100-μl packed resin) for 2 hr at 4° followed by four washes with buffer A containing 1 mg/ml bovine serum albumin. The beads are suspended in 1× Laemmli sample buffer,[20] subjected to SDS–PAGE (6%), and the resolved proteins electroblotted to nitrocellulose. ^{32}P-Labeled P-glycoprotein is visualized by autoradiography and the appropriate area of the blot carefully excised. The nitrocellulose strips are treated with 0.5% polyvinylpyrrolidone-30[21] in 0.1 *M* acetic acid for 30 min at room temperature, washed extensively with water, and digested with TPCK-treated trypsin (5 μg) in 50 m*M* NH_4HCO_3/1 m*M* $CaCl_2$ for 15 hr at 37°. A second aliquot of TPCK–trypsin (5 μg) is added and the digestion continued for an additional 3 hr. The nitrocellulose strips are removed and the buffer taken to dryness in a Speed-Vac, dissolved, and re-evaporated three times with water. Under these conditions greater than 90% of the radioactivity is released from the nitrocellulose.

Phosphoamino Acid Analysis and Two-Dimensional Phosphopeptide Mapping. ^{32}P-labeled P-glycoprotein is hydrolyzed in 6 *N* HCl at 110° for 90 min and the resulting phosphoamino acids separated in two dimensions on 0.1-mm-thick cellulose thin-layer plates (E. Merck, 20 × 20 cm) by electrophoresis at pH 1.9 for 20 min at 1.5 kV followed by electrophoresis in the second dimension at pH 3.5 for 20 min at 1.3 kV.[30] The ^{32}P-labeled

[30] W. J. Boyle, P. Van Der Geer, and T. Hunter, *Methods Enzymol.* **201,** 110 (1991).

amino acids are detected by autoradiography and phosphoamino acid markers by ninhydrin spray.

Tryptic phosphopeptides were separated in two dimensions on cellulose thin-layer plates by electrophoresis at pH 1.9 for 60 min at 0.65 kV followed by chromatography in the second dimension (1-butanol : pyridine : acetic acid, H_2O; 50 : 33 : 1 : 40, v/v).[30] A total of at least 500 cpm is routinely loaded onto each plate. Autoradiography is performed with Kodak XAR-5 film in the presence of an intensifying screen at $-70°$. All electrophoretic separations are performed on a HTLE 7000 (CBS Scientific, Del Mar, CA). Chromatographic separations are performed in tightly sealed glass tanks with freshly prepared buffer.

Electrospray Mass Spectrometry. ESI-MS analysis of total tryptic digests or HPLC-purified phosphopeptides is performed on a PE-SCIEX AP1-111 mass spectrometer using nitrogen as the nebulizer gas and an orifice voltage of 85 V. Mass spectral data are acquired from 400 to 1800 at steps of 0.2 amu with a 1-msec dwell time. Peptide masses are determined by analyzing plots of the relative abundance of each ion versus their mass-to-charge ratio (m/z) using software supplied by the manufacturer.

Site-Specific Mutagenesis of Basic-Directed Kinase Phosphorylation Sites in Linker Region of *mdr1b* P-Glycoprotein

To confirm the mass spectral analysis indicating that serine-669 and serine-681 were the principal *in vitro* phosphorylation sites for PKC and PKA, respectively, each serine was individually mutated to alanine. The serine to alanine substitutions were introduced by PCR-based methods and the wild-type and mutated P-glycoprotein cDNAs stably expressed in HeLa cells. As can be seen from Fig. 7, phosphorylation of the S681A mutant by PKA is reduced considerably compared to the wild-type P-glycoprotein. *In vitro* phosphorylation of this mutant by PKC is not affected. In contrast, the S669A mutant is a poor substrate for PKC (Fig. 8).

Expression of either the S669A, S681A, or a quadruple mutation in which all of the potential basic-directed kinase sites (serine-659, -665, -669, and -681) in murine *mdr1b* P-glycoprotein have been eliminated did not alter the ability of the stably transfected cell lines to induce the MDR phenotype. These data would suggest that phosphorylation of any of the four serines in the basic domain of the *mdr1b* linker region is not a direct on/off switch for drug transporter activity. Similar conclusions were reached on studies of the human MDR1 transporter.[31,32] Other phosphorylation

[31] U. A. Germann, T. C. Chambers, S. V. Ambudkar, T. Licht, C. O. Cardarelli, I. Pastan, and M. M. Gottesman, *J. Biol. Chem.* **271,** 1708 (1996).

[32] H. R. Goodfellow, A. Sardini, S. Ruetz, R. Callaghan, P. Gros, P. A. McNaughton, and C. F. Higgins, *J. Biol. Chem.* **271,** 13668 (1996).

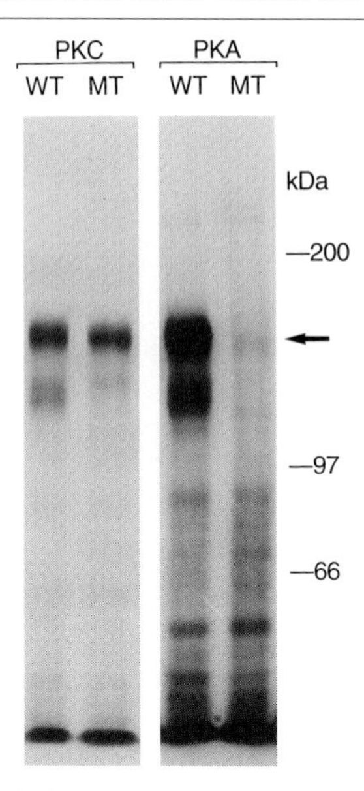

FIG. 7. *In vitro* PKA and PKC phosphorylation of membrane fractions from wild-type and S681A mutant *mdr1b* P-glycoprotein transfectants. Membrane fractions from each transfectant were phosphorylated using [γ-^{32}P]ATP in the presence of either PKA or PKC, resolved by SDS–PAGE (6%) and autoradiographed. Arrow indicates P-glycoprotein. [Reproduced with permission from G. A. Orr, E. K. Han, P. C. Browne, E. Nieves, B. M. O'Connor, C. P. Yang, and S. B. Horwitz, *J. Biol. Chem.* **268,** 25054 (1993).]

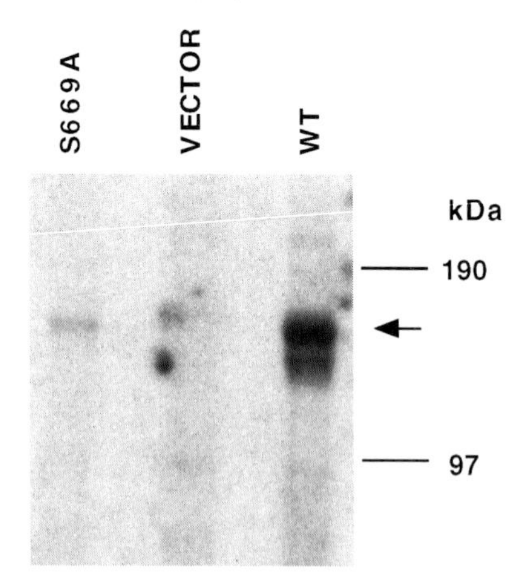

FIG. 8. *In vitro* PKC phosporylation of membrane fractions from wild-type and S669A mutant *mdr1b* P-glycoprotein transfectants. Membrane fractions from each transfectant were phosphorylated using [γ-^{32}P]ATP in the presence of PKC, resolved by SDS–PAGE (6%) and autoradiographed. Arrow indicates P-glycoprotein.

sites in murine *mdr1b* P-glycoprotein, however, remain to be identified. In addition, none of the mutagenesis/transfection studies reported have excluded the possibility that phosphorylation within the linker region is involved in other facets of P-glycoprotein activity including determination of transporter specificity, rates of biosynthesis, and turnover or intracellular trafficking.

In the past, efforts were made to demonstrate that the drug transport activity of P-glycoprotein was regulated by phosphorylation using protein kinase activators/inhibitors. For example, it was shown that protein kinase inhibitors can reduce P-glycoprotein phosphorylation *in vivo* with a concomitant increase in intracellular drug levels.[7,33,34] These studies have been questioned since many of the inhibitors employed were hydrophobic in nature and therefore could also block drug transport activity by directly interacting with the drug binding site(s) on P-glycoprotein.[35] More recent studies, however, have shown that two specific PKC inhibitors, safingol[36] and myristoylated peptides,[37] corresponding to the pseudosubstrate region of PKCα, can partially reverse multidrug resistance apparently without interacting with the transporter. The availability of phosphorylation-deficient P-glycoproteins will allow the dissection of the mechanisms whereby a diverse group of agents can modulate and/or reverse drug transport activity.

Procedures

Site-Directed Mutagenesis of Serine-669 and Serine-681 in mdr1b P-Glycoprotein cDNA. Site-directed mutagenesis is performed by the polymerase chain reaction (PCR). Both serine-669 and serine-681 are located between two *Bam*HI sites in the *mdr1b* P-glycoprotein cDNA. For each serine to alanine substitution, four primers are designed based on the nucleotide sequence at the two *Bam*HI sites [primer 1 (5′-ATAATGCTTATGGATCCCAGA) and primer 4 (5′-ATTTAGATTTAGGATCCGCCA)] and at either serine-669 or serine-681 (primers 2 and 3). For the S669A substitution, primers 2 and 3 are 5′-ATTTACAGAGCTGTCCACAGAAAGC and 5′-GCTTTCTGTGGACAGCTCTGTAAAT, respectively. For the S681A substitution, primers 2 and 3 are 5′-CTCTTTCATAGCAAGTCTTCT and 5′-AGAAGACTTGCTATGAAAGAG, respectively. Primers 2 and 3 con-

[33] T. C. Chambers, B. Zheng, and J. F. Kuo, *Mol. Pharmacol.* **41,** 1008 (1992).

[34] W. Sato, K. Yusa, M. Natio, and T. Tsuruo, *Biochem. Biophys. Res. Commun.* **173,** 1252 (1990).

[35] C. D. Smith and J. T. Zilfou, *J. Biol. Chem.* **270,** 28145 (1995).

[36] C. W. Sachs, A. R. Safa, S. D. Harrison, and R. L. Fine, *J. Biol. Chem.* **270,** 26639 (1995).

[37] K. P. Gupta, N. E. Ward, K. R. Gravitt, P. J. Bergman, and C. A. O'Brian, *J. Biol. Chem.* **271,** 2102 (1996).

tain the necessary base pair changes to achieve the desired serine to alanine substitutions. The first reaction contains primers 1 and 2 (40 pmol each), *Taq* polymerase (2.5 units), commercially available kit reagents (Amplitaq, Perkin-Elmer Cetus, Norwalk, CT) and template DNA in a total volume of 50 μl. Forty cycles, each including 1 min at 94°, 1 min at 52°, and 2 min at 72°, are carried out in a thermal cycler (Perkin-Elmer Cetus). The second reaction is performed in an identical manner except that primers 3 and 4 are used. The third reaction contains primers 1 and 4 plus 1 μl of each of the original reaction products. The final PCR reaction product is digested with *Bam*HI and ligated into pBluescript (Stratagene, La Jolla, CA) and the sequence of the mutated fragment determined using the dideoxynucleotide method. pC1.5, a full-length murine *mdr1b* P-glycoprotein cDNA, is ligated into the *Eco*RI site of pBluescript, from which the *Bam*HI site had been removed.[38] The vector containing pC1.5 is digested with *Bam*HI to release the wild-type *Bam*HI fragment, and the mutant sequence is then inserted. Orientation of the mutated insert is determined by DNA sequencing.

Stable Transfection of Wild-Type and Mutant mdr1b P-Glycoprotein cDNAs in HeLa Cells. pBluescript containing either the wild-type, S669A, or S681A *mdr1b* P-glycoprotein cDNAs is digested with *Hind*III and *Not*I, and the released fragments are ligated into the expression vector Rc/CMV (Invitrogen, San Diego, CA). Plasmid DNA is prepared by the cesium chloride method and used to transfect HeLa cells (5×10^5 cells in 100-mm plates) by either the calcium phosphate or the Lipofectin methods. After transfection, 500 cells are plated per 100-mm plates and subjected to selection using G418 (1.4 mg/ml). Individual G418-resistant clones are harvested 14 days after selection and further selected with vinblastine (1–3 n*M*).

Acknowledgments

This work was supported in part by U.S. Public Health Service grants CA39821 (to S.B.H.), CA56677 (to G.A.O.), and 5P30 CA13330. J.S.G. was supported by National Institutes of Health Training Program in Pharmacological Sciences grant 5T32 GM07260.

[38] S. Henikoff, *Methods Enzymol.* **155,** 156 (1987).

[26] P-Glycoprotein and Swelling-Activated Chloride Channels

By TAMARA D. BOND, CHRISTOPHER F. HIGGINS, and MIGUEL A. VALVERDE

Introduction

The multidrug resistance P-glycoprotein (Pgp) is an adenosine triphosphate (ATP)-dependent transporter that pumps hydrophobic molecules out of cells. In humans, Pgp is encoded by the *MDR1* gene. In rodents, the equivalent of the *MDR1* gene is duplicated: the products of the *mdr1a* and *mdr1b* genes share >80% sequence identity. When overexpressed, Pgp can confer resistance of cells and tumors to a wide variety of cytotoxic agents including chemotherapeutic drugs. In addition to its role as a transporter, Pgp also appears to modulate the activity of cell swelling-activated chloride channels, at least in some cell types, and may have a role in the regulation of cell volume. This latter role for Pgp has been the subject of some controversy that has arisen, at least in part, because of differences in methodology.

Cell Volume Regulation

Most animal cells regulate their volume in response to changes in the osmolality of the environment, or during bulk nutrient uptake or following hydroelectrolytic secretion.[1] Following exposure to hypoosmotic medium, and consequent cell swelling, regulatory volume decrease (RVD) is generally achieved by the loss of ions and other osmolytes from the cell and the concomitant loss of water. Of the various efflux systems that can contribute to RVD, swelling-activated chloride channels are present in most cell types studied.[2–4] These anion channels also appear to mediate efflux of taurine and other organic osmolytes.[4,5] Despite extensive electrophysiologic characterization of swelling-activated chloride conductances,[6] little is known about

[1] E. K. Hoffmann and L. O. Simonsen, *Physiol. Rev.* **69,** 315 (1989).

[2] M. D. Cahalan and R. S. Lewis, *in* "Current Topics in Membranes and Transport" (W. Guggino, ed.). Academic Press, San Diego, 1994.

[3] B. Nilius, J. Schrer, and G. Droogmans, *Br. J. Pharmacol.* **112,** 1049 (1994).

[4] K. Strange, F. Emma, and P. S. Jackson, *Am. J. Physiol.* **270,** C711 (1996).

[5] U. Banderali and G. Roy, *Am. J. Physiol.* **263,** C1200 (1992).

[6] M. A. Valverde, S. P. Hardy, and F. V. Sepúlveda, *FASEB J.* **9,** 505 (1995).

0076-6879/98 $25.00

the number of channel types, their molecular identities, or the mechanisms by which their activities are regulated.

P-Glycoprotein Modulates Swelling-Activated Chloride Channels

An association between Pgp expression and cell-swelling-activated chloride channels was first demonstrated in 1992.[7] Several subsequent and independent lines of evidence are consistent with this association: (1) expression of Pgp can influence the magnitude of cell-swelling-activated chloride currents[8–11]; (2) antibodies recognizing distinct epitopes of Pgp can inhibit these chloride conductances[12–14]; and (3) the activation of chloride conductances is reduced in zymogen granules from mice disrupted in the *mdr1a* gene.[15]

Two interpretations of these initial observations were possible: (1) that Pgp was itself a channel or (2) that Pgp influenced the activity of a heterologous channel protein(s).[7] It soon became apparent that swelling-activated chloride channel activity could be measured in several cell types that did not express significant levels of Pgp, arguing strongly against the hypothesis that Pgp is itself the channel protein.[16–22] Further evidence that Pgp is not

[7] M. A. Valverde, M. Diaz, F. V. Sepúlveda, D. R. Gill, S. C. Hyde, and C. F. Higgins, *Nature* **355,** 830 (1992).

[8] D. R. Gill, S. C. Hyde, C. F. Higgins, M. A. Valverde, G. M. Mintenig, and F. V. Sepúlveda, *Cell* **71,** 23 (1992).

[9] G. A. Altenberg, C. G. Vanoye, E. S. Han, J. W. Deitmer, and L. Reuss, *J. Biol. Chem.* **269,** 7145 (1994).

[10] D. B. Luckie, M. E. Krouse, K. L. Harper, T. C. Law, and J. J. Wine, *Am. J. Physiol.* **267,** C650 (1994).

[11] M. A. Valverde, T. D. Bond, S. P. Hardy, J. C. Taylor, C. F. Higgins, J. Altamirano, and F. J. Alvarez-Leefmans, *EMBO J.* **15,** 4460 (1996).

[12] F. Thévenod, I. Anderie, and I. Schulz, *J. Biol. Chem.* **269,** 24410 (1994).

[13] L. Reuss, G. A. Altenberg, C. G. Vanoye, and E. S. Han, *J. Physiol.* **489P,** 12S (1995).

[14] J. Wu, J. J. Zhang, H. Koppel, and T. J. C. Jacob, *J. Physiol.* **491.3,** 743 (1996).

[15] F. Thévenod, J-P. Hildebrandt, J. Striessnig, H. R. de Jonge, and I. Schulz, *J. Biol. Chem.* **271,** 3300 (1996).

[16] M. Diaz, M. A. Valverde, C. F. Higgins, C. Rucareaunu, and F. V. Sepúlveda, *Pflugers Arch.* **422,** 347 (1993).

[17] Y. J. Dong, G. Chen, G. E. Duran, K. Kouyama, A. C. Chao, B. I. Sikic, S. V. Gollapudi, S. Gupta, and P. Gardner, *Cancer Res.* **54,** 5029 (1994).

[18] A. Rasola, L. J. V. Galietta, D. C. Gruenert, and G. Romeo, *J. Biol. Chem.* **269,** 1432 (1994).

[19] G. R. Ehring, Y. V. Osipchuk, and M. D. Cahalan, *J. Gen. Physiol.* **104,** 1129 (1994).

[20] M. Tominaga, T. Tominaga, A. Miwa, and Y. Okada, *J. Biol. Chem.* **270,** 27887 (1995).

[21] S. P. Hardy, H. R. Goodfellow, M. A. Valverde, D. R. Gill, F. V. Sepúlveda, and C. F. Higgins, *EMBO J.* **14,** 68 (1995).

[22] F. Viana, K. Van Acker, D. De Greef, J. Eggermont, L. Raeymaekers, G. Droogmans, and B. Nilius, *J. Membr. Biol.* **145,** 87 (1995).

itself a channel came from studies on *Xenopus* oocytes: the endogenous swelling-activated chloride currents disappear several days after defolliculation yet subsequent expression of Pgp does not generate a new conductance.[23] Instead, direct evidence that Pgp can modulate channel activity has been presented[11,21] (see later discussion). The challenge is to ascertain how Pgp modulates channel activity and what significance, if any, this plays in cell physiology.

Measurement of Channel Activity

Swelling-activated chloride channels are generally activated by replacing an isoosmotic bathing solution with a hypoosmotic solution. Alternatively, channel activation can be achieved by replacing the intracellular solution (pipette solution in whole-cell patch-clamp studies) with a hypertonic solution.[19] Figure 1 shows the apparatus we have used to permit *rapid* transfer of cells from one bathing solution to another. Using this apparatus to deliver tetrodotoxin to block Na^+ channels demonstrates that cells are exposed to the new experimental conditions within 5 sec. As Pgp influences the kinetics of channel activation, rather than maximal activity which can be elicited (see later discussion), a rapid perfusion system is a crucial factor in assessing the role of Pgp in channel activation.

Characterization of Swelling-Activated Chloride Channel Modulated by P-Glycoprotein

Swelling-activated chloride channels/currents show different electrophysiologic and pharmacologic characteristics depending on the cell type under study. Most commonly these currents are outwardly rectifying and inactivate at positive potentials,[7,25] although in some cells the currents show no time-dependent inactivation at depolarizing voltages.[26] Inwardly rectifying[27] and ohmic[28] currents have also been reported. Moreover, single-channel conductances have been reported as 1–10 pS,[26,29] 20–40 pS,[25,30,31] or 300 pS.[28]

[23] X. K. Morin, T. D. Bond, T. W. Loo, D. M. Clarke, and C. E. Bear, *J. Physiol.* **486,** 707 (1995).
[24] S. Suzuki, M. Tachibana, and A. Kanebo, *J. Physiol.* **421,** 645. (1990).
[25] R. T. Worrell, A. G. Butt, W. H. Cliff, and R. A. Frizzell, *Am. J. Physiol.* **256,** C1111 (1989).
[26] R. S. Lewis, P. E. Ross, and M. D. Cahalan, *J. Gen. Physiol.* **101,** 801 (1993).
[27] S. Grunden, A. Thiemann, M. Pusch, and T. J. Jentsch, *Nature* **360,** 759 (1992).
[28] E. M. Schwiebert, J. W. Mills, and B. A. Stanton, *J. Biol. Chem.* **269,** 7081 (1994).
[29] O. Christensen and E. K. Hoffman, *J. Membr. Biol.* **129,** 13 (1992).
[30] C. K. Solc and J. J. Wine, *Am. J. Physiol.* **261,** C658 (1991).
[31] Y. Okada, C. C. H. Petersen, M. Kubo, S., Morishima, and M. Tominaga, *Jap. J. Physiol.* **44,** 403 (1994).

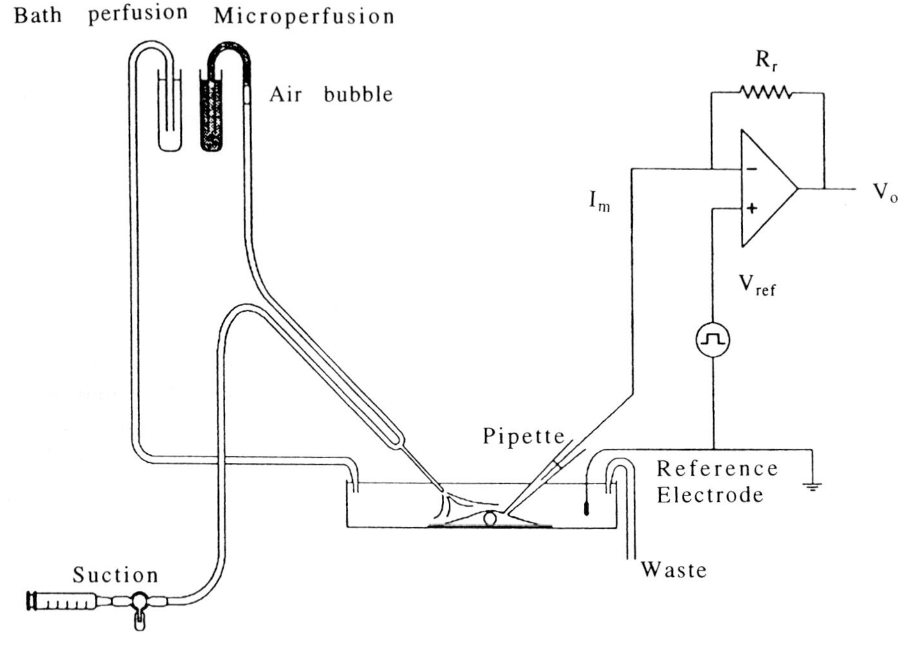

FIG. 1. Diagram of superfusion and electrical systems used to study swelling-activated chloride channels. The apparatus is a modification of that used by Suzuki *et al.*[24] and includes both a general superfusion system and a microperfusion system. The general superfusion system allows continuous renewal of the bathing solution throughout the experiment. The microperfusion system allows a narrow jet of solution to be directed at the cell or group of cells under study, permitting a very rapid change in the composition of the bathing solution. This microperfusion system works as follows. The 0.5-mm-diameter plastic capillary used to direct the solution is plugged into a looped region of a wider 3-mm plastic tube. The open end of the narrow tube is placed near the cell to allow fresh experimental solution to be projected at the cell. The 3-mm tube is filled with bath solution. Negative pressure (suction) is then applied using a syringe, introducing a small air bubble into the tube, after which the upper end is placed in the reservoir of experimental solution. Further suction introduces the new experimental solution into the tube (shaded), separated from the bath solution by the air bubble. Suction is applied until the air bubble crosses the junction with the 0.5-mm tube, at which time the experimental solution behind the air bubble flows through the 0.5-mm capillary and is squirted at the cell. Studies with channel blockers in the experimental solution show that this system exposes the cell to fresh solution with a delay of only 5 sec. I_m, Current across the cell membrane; R_r, feedback resistor; V_{ref}, reference voltage; V_o, output voltage.

The swelling-activated chloride currents modulated by Pgp are outwardly rectifying with a characteristic time-dependent decay at depolarizing voltages.[7] This type of current is typically observed in epithelial cells although it has also been observed in cells of different origins such as fibroblasts. Activation of this volume-sensitive chloride current has an absolute requirement for nonhydrolytic ATP binding, a characteristic we first described in 1992[8] and more recently discussed by Strange *et al.*[4] Identification of the single-channel conductance associated with this whole-cell current has been difficult, due to difficulties in activating the channels on cell-attached patches.[32] Four different strategies have been used to overcome this problem: (1) Noise analysis of swelling-activated, whole-cell chloride currents suggested a conductance of ~40 pS at depolarizing voltages[4]; (2) cell-attached patch-clamp recordings of single channels on cells previously swollen by hypoosmotic shock gave a conductance at +80 mV of 50 pS[30]; (3) simultaneous monitoring of whole-cell and cell-attached recordings of preswollen cells revealed a single-channel conductance of 25 pS[31]; and (4) whole-cell chloride currents were recorded following exposure to hypoosmotic solutions using the system described in Fig. 1. After activation of the currents (Fig. 2A), the pipette is pulled to form an excised outside-out patch, allowing the recording of single-channel events (Fig. 2B). The kinetics of channel inactivation/closing at positive potentials was found to be the same for whole-cell and single-channel currents (Figs. 2C and 2D). This type of analysis offers strong evidence that the 28-pS channel recorded under excised conditions (Fig. 2B) underlies the swelling-activated whole-cell currents regulated by Pgp in this cell type.

Pharmacologic studies of this outwardly rectifying, time-dependent current show a broad spectrum of compounds with blocking properties.[34] These blockers include (with IC_{50} ranging from 0.5 to 100 μM): (1) anthracene–carboxylic acid derivatives (e.g., NPPB); (2) stilbene disulfonate derivatives (e.g., DIDS); (3) multidrug resistance reversers (e.g., verapamil, dideoxyforskolin); (4) nonsteroidal antiestrogens (e.g., tamoxifen); and (5) fatty acids (e.g., arachidonic acid).

Intriguingly, many of the compounds that block the swelling-activated chloride channel, such as tamoxifen, verapamil, and dideoxyforskolin, also inhibit drug transport by Pgp. However, many of these compounds can also block the channel in the absence of Pgp.[16,35] The fact that drug transport and channel activity have similar (but distinct) pharmacologic profiles sug-

[32] Christensen and Hoffman[29] have shown the activation of a 5- to 7-pS chloride channel on cell-attached recordings following hypoosmotic shock, although its voltage dependence is not known and, unlike the Pgp-regulated channel, it is Ca^{2+} dependent. Also Banderali and Roy[33] reported the activation of a 60-pS chloride channel in cell-attached conditions, but its voltage-dependence differs (i.e., it lacks time-dependent inactivation at positive potentials).

[33] U. Banderali and G. Roy, *J. Membr. Biol.* **126,** 219 (1992).

[34] G. M. Mintenig, M. A. Valverde, F. V. Sepúlveda, D. R. Gill, S. C. Hyde, J. Kirk, and C. F. Higgins, *Receptors Channels* **1,** 305 (1993).

[35] M. A. Valverde, G. M. Mintenig, and F. V. Sepúlveda, *Pflugers Arch.* **525,** 552 (1993).

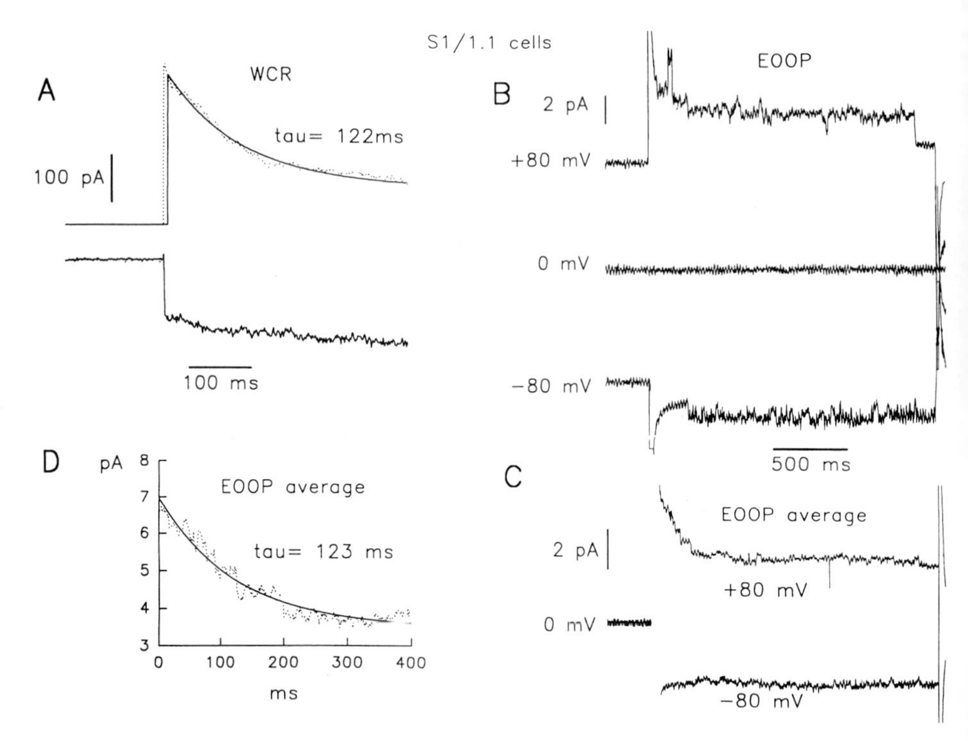

FIG. 2. Whole-cell and single-channel recordings of swelling-activated chloride channels. Chloride currents were activated by exposure to a 20% hypoosmotic solution in an S1/1.1 lung carcinoma cell, permanently transfected with the human MDR1 cDNA. (A) Whole-cell recording (WCR) obtained at +80 mV (top trace, dotted line) and −80 mV (lower trace). The decay in the current amplitude at +80 mV (top trace, solid line), reflecting a characteristic kinetic property of this current, was fitted by a single exponential function to give a time constant of 122 ms. (B) After recording whole-cell currents the pipette was pulled from the cell, generating an excised, outside-out patch (EOOP). In this configuration, stepwise single-channel currents were observed that disappeared by the end of the +80-mV pulse. In contrast, at −80 mV after an initial silent period, single-channel fluctuations were seen throughout the voltage pulse. The unitary conductance of this channel was 28 pS at +80 mV and 10 pS at −80 mV. (C) Average currents obtained from five voltage pulses to +80 mV and −80 mV after subtracting the capacitance component. The current obtained at the positive voltage showed a clear decay in amplitude during the initial 500 ms of the pulse, resembling the kinetics of the whole-cell current previously recorded in the same cell (A). (D) Exponential fitting of the average current obtained in (C). The decay in current amplitude could be fitted by an exponential curve giving a time constant of 123 ms, almost identical to the time constant measured for the decay of whole-cell currents (A). Together, these results suggest that this small-conductance (28 pS at +80 mV) chloride channel is the channel underlying the swelling-activated chloride currents recorded in several epithelial cell types and modulated by Pgp. The ionic compositions of the pipette and bath solutions were chosen to measure chloride currents. Pipette solution contained (in mM) 105 NMDGCl, 1.2 $MgCl_2$, 1 EGTA, 2 ATP, 10 HEPES, pH 7.4, 70 mannitol. Isoosmotic bathing solution contained (in mM) 105 NMDGCl, 1.3 $CaCl_2$, 0.5 $MgCl_2$, 10 HEPES, pH 7.4, 70 mannitol. The hypoosmotic bathing solution lacked mannitol.

gests either that Pgp and the channel have similar binding sites, or that a common subunit binds these compounds and can interact with both proteins. Note also that among cells expressing Pgp, dideoxyforskolin, verapamil, and tamoxifen block swelling-activated chloride currents in many cell types,[7,36,38] but not in all.[19,20,38] These differences suggest that there may be several channels with distinct pharmacologies that are differentially expressed, or that a single channel underlies all of these currents but has characteristics that are modified by accessory factors, with Pgp as a possible candidate. Both scenarios have been observed for inwardly rectifying K^+ channels.[39,40] The question of whether there are multiple swelling-activated chloride channel proteins will not be satisfactorily resolved until the channel protein(s) and its gene(s) are identified.

Channel Modulation by P-Glycoprotein

Two distinct, although perhaps related, effects of Pgp on swelling-activated chloride channels have been reported.

First, Pgp increases the rate of channel activation[11] (Fig. 3). Maximum channel activation is normally elicited over a period of several minutes after exposing cells to a hypoosmotic solution. The length of time taken to reach maximum activation in Chinese hamster ovary (CHO) cells depends on the degree of hypoosmotic shock (5–7 min at 40%; 10–12 min at 20%; 20–25 min at 8%). The rate at which maximum activation is achieved is greater in cells expressing mouse Pgp1a (LR73-mdr1a) than in non-Pgp-expressing cells (LR73) or in cells expressing mouse Pgp1b (LR73-mdr1b). Human Pgp in these cells causes a similar increase in the rate of channel activation to mouse Pgp1a. However, the maximum current that can be elicited does not differ between the cell types. This is best illustrated in Fig. 2B where cells have been exposed to a 20% hypoosmotic solution. At short time periods after exposure to hypoosmotic solution (e.g., 3 min), the current magnitude is considerably greater in Pgp1a-expressing cells than in nonexpressing cells. However, once the response is saturated, no difference in current magnitude is observed between the three cell types.

[36] J. J. Zhang, T. J. C. Jacob, M. A. Valverde, S. P. Hardy, G. M. Mintenig, D. R. Gill, S. C. Hyde, A. E. O. Trezise, and C. F. Higgins, *J. Clin. Invest.* **94,** 1690 (1994).

[37] C-Y. Chou, M-R. Shen, and S-N Wu, *Cancer Res.* **55,** 6077 (1995).

[38] J. L. Weaver, A. Aszalos, and L. McKinney, *Am. J. Physiol.* **270,** C1453 (1996).

[39] G. Krapivinsky, E. A. Gordon, K. Wickman, B. Velimirov, L. Krapivinsky, and D. E. Clapham, *Nature* **374,** 135 (1995).

[40] C. Ämmälä, A. Moorhouse, F. Gribble, R. Ashfield, P. Proks, P. A. Smith, H. Sakura, B. Coles, S. J. H. Ashcroft, and F. M. Ashcroft, *Nature* **379,** 545 (1996).

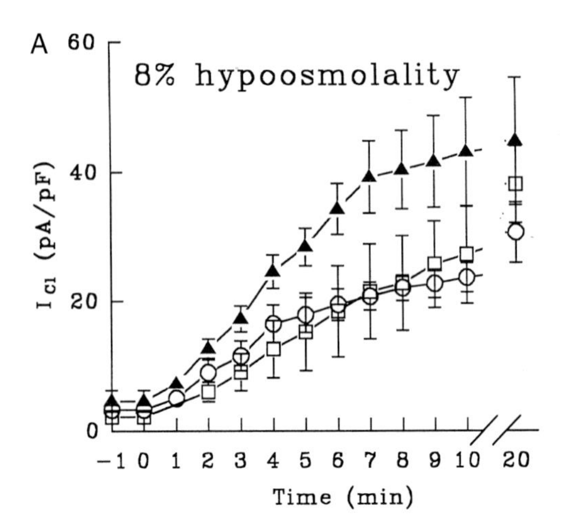

A
8% hypoosmolality
I_{Cl} (pA/pF)
Time (min)

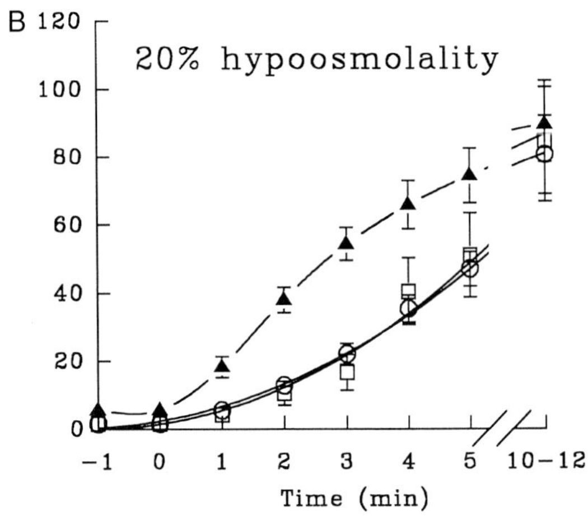

B
20% hypoosmolality
Time (min)

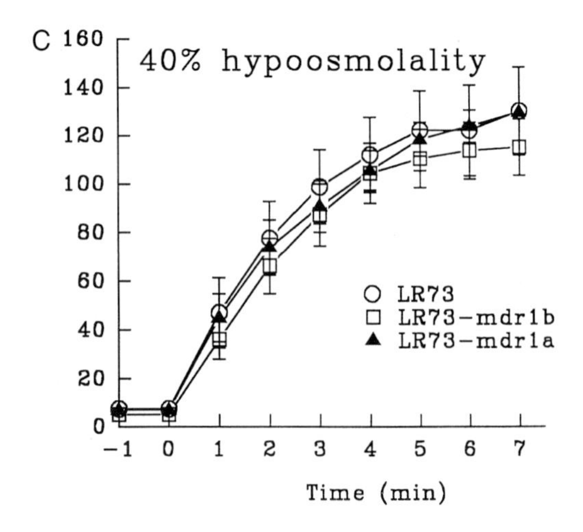

C
40% hypoosmolality
LR73
LR73-mdr1b
LR73-mdr1a
Time (min)

Second, Pgp expression increases the sensitivity of the channel to hypoosmotic solutions[10,11,19] (Fig. 3). That is, in cells expressing Pgp the channel can be more rapidly activated in response to small hypoosmotic gradients (e.g., 8 or 20% hypoosmolarity) than in non-Pgp-expressing cells (Figs. 3A and 3B). However, in response to larger hypoosmotic gradients (e.g., 40% hypoosmolarity) channel activation is similar in Pgp-expressing and nonexpressing cells (Fig. 3C).

Thus, in studies in which the maximum (saturating) response has been elicited, by measuring currents after an extended period of time or in cells exposed to a high osmotic gradient, the response will be saturated and any effect of Pgp on channel activation will fail to be detected.[18–20,38,41,42]

How Does P-Glycoprotein Modulate Channel Activity?

Little is known about the mechanism(s) by which Pgp influences channel activity. Because Pgp affects the rate of channel activation, rather than the maximum currents that can be elicited, it appears that Pgp affects the activation mechanisms. The effects of Pgp cannot simply be ascribed to high levels of protein in the membrane because equivalent expression of mouse *mdr1a*, but not *mdr1b*, alters the rate of channel activation.[11] Similarly, the effects cannot be ascribed to drug selection as the *mdr1a* and *mdr1b* cells are selected and grown in equivalent concentrations of drug. Regulation of channel activity by Pgp appears to be relatively specific. Although there have been reports linking Pgp to swelling-activated K^+ channels,[10,15] because of the close relationship between K^+ and Cl^- movement during RVD this may be an indirect consequence of its effects on Cl^- channel activity, which increases the driving force for K^+. In another

[41] C. De Greef, S. Van der Heyden, F. Viana, J. Eggermont, E. A. De Bruijn, L. Raeymaekers, G. Droogmans, and B. Nilius, *Pflugers Arch.* **430,** 296 (1995).

[42] K. Kunzelmann, I. N. Slotki, P. Klein, T. Koslowsky, D. A. Ausiello, R. Greger, and Z. I. Cabantchik, *J. Cell. Physiol.* **161,** 393 (1994).

FIG. 3. Pgp affects the rate of channel activation. The increases in chloride currents elicited by exposure to an (A) 8%, (B) 20%, or (C) 40% hypoosmotic challenge were monitored in Chinese hamster ovary cells (LR73) and derivatives permanently transfected with the mouse mdr1a (LR73-mdr1a) or mdr1b (LR73-mdr1b) genes. Cells were voltage clamped at 0 mV and 200-ms pulses to −80 mV and +80 mV were recorded each minute following hypoosmotic challenge until a steady-state current was reached. Values represent mean peak currents at +80 mV ± SEM obtained from LR73 (○; n = 7–10), LR73-1b (□; n = 7–8), and LR73-1a (▲; n = 9–11) cells. Experimental solutions used are similar to those outlined earlier, with appropriate modifications to NMDGCl and mannitol concentrations to achieve the necessary osmolarities.[11]

study,[11] we have demonstrated that K^+ channels in CHO cells studied under whole-cell conditions are not affected by expression of Pgp.

Modulation of channel activity by Pgp does not require Pgp-mediated transport because channel activity can be supported by nonhydrolyzable ATP, whereas drug transport requires ATP hydrolysis; mutations can be introduced into Pgp that severely impair its transporter activity yet have no significant effect on the channel activity[8]; and the pharmacology of the two activities, although related, is distinct.[34] However, substrates that are transported by Pgp, such as doxorubicin or vinblastine, reduce channel activity when added to the cytoplasmic face of the membrane.[3,8] This inhibition appears to depend on the presence of hydrolyzable ATP, suggesting that transport activity can, in some indirect manner, influence channel regulation.[8] The nature of the relationship between the transporter and channel regulatory activities of Pgp is not yet understood.

Phosphorylation of Pgp by PKC alters its ability to influence channel activation, implying a role for the *linker* region in channel activation. In contrast, phosphorylation of Pgp by PKC does not affect the rate of drug transport.[43,44] Thus, the principal function of Pgp phosphorylation may be in channel regulation. One scenario consistent with the available data is that Pgp interacts with the channel protein, either directly or indirectly through other protein components (e.g., the cytoskeleton).

The modulation of channel activity is also likely to be cell-type specific. Several proteins besides Pgp have been shown to influence swelling-activated chloride currents, including pI_{Cln}[45,46] and the band 3 anion exchanger.[47] As these proteins show cell-type specific expression, regulation will differ between cell types. Furthermore, other factors that play a role in regulation, such as PKC, differ from cell to cell. PKC activators have been found to influence channel activation in some cell types[21,48] but not in others.[20]

Physiological Significance

Most studies associating Pgp with swelling-activated chloride channels have been carried out on cultured cells. However, an association has been

[43] U. A. Germann, T. C. Chambers, S. V. Ambudkar, T. Licht, C. O. Cardarelli, I. Pastan, and M. M. Gottesman, *J. Biol. Chem.* **271,** 1708 (1996).

[44] H. R. Goodfellow, A. Sardini, S. Ruetz, R. Callaghan, P. Gros, P. A. McNaughton, and C. F. Higgins, *J. Biol. Chem.* **271,** 13668 (1996).

[45] M. Paulmichl, Y. Li, K. Wickman, M. Ackerman, E. Peralta, and D. Clapham, *Nature* **356,** 238 (1992).

[46] G. B. Krapivinsky, M. J. Ackerman, E. A. Gordon, L. D. Krapivinsky, and D. E. Clapham, *Cell* **76,** 439 (1994).

[47] B. Fievet, N. Gabillat, F. Borgese, and R. Motais, *EMBO J.* **14,** 5158 (1995).

[48] L. Reuss, G. S. Altenberg, C. G. Vanoye, and E. S. Han, *J. Physiol.* **489,** 12S (1995).

reported in several native Pgp-expressing cells including endothelia,[3] non-pigmented ciliary cells of the eye,[14] and pancreatic zymogen granules.[15]

Swelling-activated chloride channels are believed to play a role in the regulation of cell volume,[4] and agents that block these channel activities can inhibit RVD.[11,49] Significantly, in response to small hypoosmotic gradients (10–20%) where Pgp1a has a significant effect on the rate of channel activation, Pgp1a-expressing cells also exhibit more rapid volume recovery.[11] In contrast, in response to large osmotic gradients where Pgp has no detectable effect on channel activation, Pgp also has no significant effect on the rate of RVD. Thus, at least in the cultured cells tested, the effect of Pgp on the rate of channel activation is mirrored by an effect on cell volume recovery. Whether this reflects a role *in vivo* has yet to be ascertained. Interestingly, most non-Pgp-expressing cell types studied only show RVD in response to large (30–40%) osmotic gradients.[50] In contrast, native small-intestinal cells that express Pgp respond to small (5–10%) osmotic gradients.[51] Moreover, RVD can be inhibited by PKC activation[52] in native Pgp-expressing cells that exhibit typical swelling-activated chloride currents.[14] This may reflect the observation that PKC-mediated phosphorylation of Pgp reduces volume-activated channel activity in cultured cells.[21] Together, these findings are consistent with, but do not demonstrate, an *in vivo* role for Pgp in the modulation of RVD.

The modulation of channel activity by Pgp may not be unique to this ABC transporter.[53] Thus, for example, the cystic fibrosis gene product CFTR has been reported to influence the activity of the epithelial sodium channel,[54] an outwardly rectifying chloride channel,[55] and K^+ channels[56]; the sulfonylurea receptor appears to form a complex with a K^+ channel protein ($K_{ir}6.2$), imposing gibenclamide sensitivity on the channel[40,57]; and the *Escherichia coli* Sap peptide transporter interacts with a K^+ channel

[49] B. Sarkadi, R. Cheung, E. Mack, S. Grinstein, E. W. Gelfand, and A. Rothstein, *Am. J. Physiol.* **248,** C480 (1985).

[50] F. V. Alvarez-Leefmans, J. Altamirano, and W. E. Crowe, *Methods Neurosc.* **27,** 361 (1995).

[51] J. MacLeod in "Cellular and Molecular Physiology of Cell Volume Regulation" (K. Strange, ed.), p. 191. CRC Press, Boca Raton, Florida 1994.

[52] M. M. Civan, M. Coca-Prados, and K. Petersen-Yantorna, *Invest. Ophthalmol. Vis. Sci.* **35,** 2876 (1994).

[53] C. F. Higgins, *Cell* **82,** 693 (1995).

[54] M. J. Stutts, C. Canessa, J. C. Olsen, M. Hamrick, J. A. Cohn, B. Rossier, and R. C. Boucher, *Science* **269,** 847 (1995).

[55] M. Egan, T. Flotte, S. Afione, R. Solow, P. L. Zeitlin, B. J. Carter, and W. B. Guggino, *Nature* **358,** 581 (1992).

[56] M. A. Valverde, J. A. O'Brien, F. V. Sepúlveda, R. A. Ratcliff, M. J. Evans, and W. H. Colledge, *Proc. Natl. Acad. Sci. U.S.A.* **92,** 9038 (1995b).

[57] N. Inagaki, T. Gonoi, J. P. Clement, N. Namba, J. Inazawa, G. Gonzalez, L. Aguilar-Bryan, S. Seion, and J. Bryan, *Science* **270,** 1166 (1995).

pore to form the Trk K^+ channel.[58] Whether Pgp and these other ABC transporters modulate channel activity by related mechanisms remains to be determined.

Conclusions

The association of Pgp with swelling-activated chloride channels, while clearly demonstrated by some groups, has failed to be detected by others. This is, at least in part, because of methodological differences: (1) Channel properties vary between cell types, suggesting that there may be several distinct channels, not all of which are affected by Pgp; (2) factors that influence channel regulation (e.g., PKC isoforms) can differ between cell types; and (3) Pgp modulates the rate of channel activation such that, if studied under "saturating" conditions (i.e., high osmotic shock or after long periods of activation), any effect of Pgp will be missed.

There are hints that channel modulation by Pgp may play a role in cell volume regulation. The challenge now is to ascertain the *in vivo* significance of this role (if any) and to determine the mechanism by which Pgp modulates channel activation. This will, undoubtedly, require identification of the channel protein(s) and their genes.

Acknowledgments

We thank G. M. Mintenig for contributions to the single-channel recordings. This work was supported by the Imperial Cancer Research Campaign, the BBSRC, and the EU. C.F.H. is a Howard Hughes International Research Scholar.

[58] C. Parra-Lopez, R. Lin, A. Aspedon, and E. A. Groisman, *EMBO J.* **13,** 3964 (1994).

[27] Functional Expression of *mdr* and *mdr*-like cDNAs in *Escherichia coli*

By Eitan Bibi, Rotem Edgar, and Oded Béjà

Introduction

Although *Escherichia coli* has become a popular host for heterologous expression of eukaryotic soluble proteins, only a few attempts have been made to utilize it for functional expression of eukaryotic transport proteins. In this regard, bacterial expression systems may be invaluable also in facili-

0076-6879/98 $25.00

tating functional and structural studies of membrane transporters, using operations that can be accomplished considerably more efficiently in *E. coli* than in eukaryotes. The simple and rapid life cycle of the prokaryotic microorganism makes it attractive for large-scale screening of substrates and modulators, mutagenesis studies and selection of intragenic suppressors, topology studies using gene fusions, bioenergetic studies *in vivo* and in well-characterized vesicles, and finally for acquisition of purified proteins in quantities sufficient for biochemical studies.

This chapter describes the methodologies and tools used to examine the application of an *E. coli* expression system for study of the mouse multidrug resistance protein, Mdr1. This protein represents a growing number of ATP-binding cassette (ABC) proteins responsible for the simultaneous resistance to a large group of structurally unrelated cytotoxic compounds. More specifically, we describe strategies, expression vectors, and *E. coli* strains used for heterologous expression of Mdr1, for topology studies, and certain considerations regarding the reliability and legitimacy of the prokaryotic expression system.

General Considerations

In the past, only a few attempts to express eukaryotic transporters in *E. coli* have been described and none of the expressed transporters was characterized further, except for the mouse multidrug resistance protein Mdr1.[1] Accordingly, it is not known whether the methodology described here is also applicable to other transport systems. Moreover, in our experience and to our knowledge, heterologous expression of transport proteins is a tailor-made task, and different proteins may require different cloning and expression strategies. We would like to emphasize two major, general problems: One problem is that a variety of cytoplasmic, membrane, and periplasmic proteases of *E. coli* recognize and degrade foreign proteins. The use of certain protease-deficient *E. coli* strains for expression of the mouse Mdr1 is described later. Another common obstacle to successful expression of membrane proteins is *E. coli* is that sometimes even very low levels of expression may cause toxic effects. Although it is likely that the toxicity results primarily from membrane perturbation, it is also conceivable that the transport functions or the associated enzymatic activities of the heterologous transporters cause deleterious effects on the prokaryotic system. In either case the detrimental consequence of the expression of foreign membrane proteins in *E. coli* is probably membrane associated, suggesting that the eukaryotic protein is targeted and inserted into the prokaryotic

[1] E. Bibi, P. Gros, and H. R. Kaback, *Proc. Natl. Acad. Sci. U.S.A.* **90,** 9209 (1993).

plasma membrane. This proposal is in accordance with our primary assumption that biogenesis of membrane proteins in eukaryotes and prokaryotes follows universal signals and possibly also utilizes similar mechanisms. This assumption underlies a fundamental prerequisite for successful expression of functional transporters in *E. coli.* Another presumption is that posttranslational modification of Mdr proteins is probably not essential for its activity. This question has been studied in detail and it is now accepted that glycosylation and phosphorylation do not play a critical role in the function of Mdr.[2,3]

Expression System

Expression Plasmid

High copy number plasmids are not appropriate for expression of membrane proteins in *E. coli* probably because of the toxic effect of overexpression. Therefore, we chose a derivative of the medium copy number plasmid pBR322 (pT7-5), for expression of the mouse *mdr1* cDNA in *E. coli.* Full *mdr1* cDNA in the *Eco*RI site of pGEM7Zf (pGEMK4) is used as starting material, and the sites *Bam*HI and *Hin*dIII are introduced in the 5′ and the 3′ ends of the coding region, respectively. Using these sites, the *mdr1* gene is inserted into pT7-5(*lacY*)[1] instead of the *lacY* gene, under the control of the *lac* promoter–operator, to produce pT7-5/*mdr1*. This plasmid contains a modified *lac* promoter–operator that confers a lower level of expression of LacY. It is not known whether or not this modification is relevant to the successful expression of *mdr1* in *E. coli,* since the wild-type promoter–operator has never been analyzed in this system. Noteworthy, despite the weak promoter, after transformation with plasmid pT7-5/*mdr1* and 24 hr incubation at 30°, a small fraction of the colonies exhibits a different morphology (big colonies). These colonies probably do not contain the original plasmid (pT7-5/*mdr1*) and therefore only small colonies are used for further studies.

Escherichia coli Strains and Growth Conditions

Many attempts to express *mdr1* in a variety of *E. coli* strains were unsuccessful, as determined by immunoblotting with the anti-Mdr monoclonal antibody C219, and only low molecular weight products were frequently observed. Assuming that this may be due, in large part, to proteoly-

[2] N. D. Richert, L. Aldwin, D. Nitecki, M. M. Gottesman, and I. Pastan, *Biochemistry* **27,** 7607 (1988).

[3] U. A. Germann, T. C. Chambers, S. V. Ambudkar, T. Licht, C. O. Cardarelli, I. Pastan, and M. M. Gottesman, *J. Biol. Chem.* **271,** 1708 (1996).

sis, several *E. coli* strains deficient in cytoplasmic, inner, or outer membrane proteases were screened for *mdr1* expression. An immunoreactive band corresponding to intact, unglycosylated Mdr1 (ca. 140 kDa) was detected exclusively in membranes prepared from the *ompT* strain (UT5600 obtained from the *E. coli* Genetic Stock Center at Yale University as CGSC #7092) transformed with pT7-5/*mdr1*. In an attempt to compare the amount of Mdr1 in membranes from *mdr1*-transfected and overexpressing eukaryotic cells to the amount expressed in the prokaryotic system by immunoblotting, it was observed that the level of expression is comparable but lower in the heterologous system. The following growth conditions were selected for better expression of Mdr1 in *E. coli* UT5600. Growth is always conducted in LB containing 100 μg/ml ampicillin at 30°. Overnight cultures are diluted ≈1:20 and after 3 hr, isopropyl-1-thio-β-D-galactopyranoside (IPTG, 0.5 m*M* final concentration) is added for induction (4 hr). Other growth conditions such as 37° or different media yield lower levels of intact Mdr1.

Of importance is our unpublished observation that unlike *E. coli* UT5600, other *ompT* strains such as BL21[4] and SF110[5] did not produce detectable quantities of the full-length Mdr1 protein. Although this phenomenon has not been studied in detail, we would like to propose that another outer membrane protease may exist in these strains but not in UT5600. In fact, an OmpT homolog, OmpP, has been identified recently.[6] OmpP probably plays a role that is similar to OmpT in degradation of foreign proteins and unlike a number of *E. coli* K12 strains, UT5600 lacks both *ompT* and *ompP*.[6]

The involvement of other proteases in the degradation of heterologous Mdr1 is not known. Most strikingly however, we observed that an outer membrane permeability derivative of *E. coli* UT5600, UTL2 (see next section), expresses significantly higher quantities of Mdr1 (Fig. 1). A speculative explanation of this phenomenon is that the local operational concentration of periplasmic proteases in this mutant is decreased if they diffuse out through the outer membrane. It is most likely that periplasmic proteases such as DegP and its homologs contribute to the instability of heterologous Mdr1, as demonstrated for other foreign proteins.[7]

Functional Studies

A number of approaches are used to assess the function of heterologously expressed Mdr1: (1) photolabeling of the protein by photoactive

[4] J. Grodberg and J. J. Dunn, *J. Bacteriol.* **170,** 1245 (1988).

[5] F. Baneyx and G. Georgiou, *J. Bacteriol.* **172,** 491 (1990).

[6] A. Kaufmann, Y.-D. Stierhof, and U. Henning, *J. Bacteriol.* **176,** 359 (1994).

[7] K. L. Strauch and J. Beckwith, *Proc. Natl. Acad. Sci. U.S.A.* **85,** 1576 (1988).

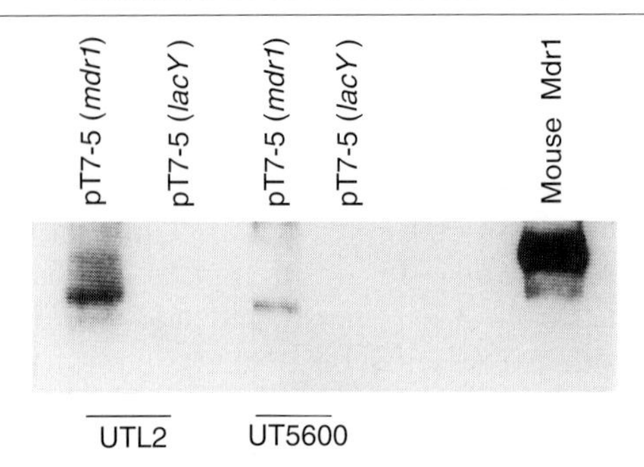

FIG. 1. Comparison of the levels of Mdr1 expression in *E. coli* UT5600, UTL2, or transfected hamster LR73 ovary cells. *Escherichia coli* UT5600 or UTL2 transformed with pT7-5(*mdr1*) or pT7-5(*lacY*) as control were induced with IPTG. Membranes were prepared and 15 μg of membrane proteins or 2.8 μg of membrane proteins from LR73 cells overexpressing Mdr1 were subjected to SDS–PAGE (6%), electroblotted, and the nitrocellulose paper was incubated with monoclonal antibodies C219. After incubation with horseradish peroxidase conjugated rabbit anti-mouse antibodies, followed by a short incubation with luminescent substrate (Amersham, Buckinghamshire, England), the nitrocellulose paper was exposed to film for 10 min. [Reprinted with permission from Béjà, O., and Bibi, E. *Proc. Natl. Acad. Sci. U.S.A.* **93,** 5969–5974 (1996).]

substrate analogs[1]; (2) *in vivo* resistance to the growth inhibiting effects of cytotoxic agents; and (3) direct transport of radiolabeled substrates. The main technical problem encountered in approaches 2 and 3 is the impermeability of the *E. coli* outer membrane to the hydrophobic cytotoxic drugs generally used as Mdr1 substrates. To overcome this problem, we use tetraphenylphosphonium (TPP^+) or tetraphenylarsonium (TPA^+), since both compounds are Mdr1 substrates[8] and are relatively soluble and membrane permeable. Another approach to solve the outer membrane permeability barrier, using mutagenesis, is described later.

Transport Assay

The most compelling data supporting the suggestion that the heterologous Mdr1 is functional has come from transport assays. To examine the activity of Mdr1, cells are harvested and washed once with ice-cold 0.1 *M* potassium phosphate buffer (pH 7.5). To enable better accessibility of the lipophilic substrates through the outer membrane, the cells are washed once in the same buffer containing 10 m*M* ethylenediaminetetraacetic acid

[8] P. Gros, F. Talbot, D. Tang-Wai, E. Bibi, and H. R. Kaback, *Biochemistry* **31,** 1992 (1992).

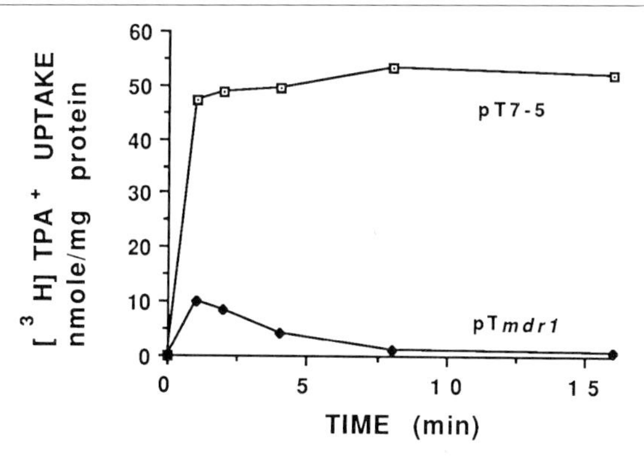

FIG. 2. Transport of TPA$^+$ by *E. coli* UT5600 harboring pT7-5 or pT*mdr1*. Cells were grown, induced with IPTG, and treated with EDTA as described in the text. Transport of [^{3}H]TPA$^+$ (23.8 mCi/mmol) at a final concentration of 0.4 m*M* was assayed at 34° by rapid filtration. [Reprinted with permission from Bibi, E., Gros, P., and Kaback, H. R. *Proc. Natl. Acad. Sci. U.S.A.* **90,** 9209–9213 (1993).]

(EDTA), centrifuged, and resuspended in 0.1 *M* potassium phosphate (pH 7.5) to an OD_{420} of 20 (ca. 2 mg protein/ml). EDTA permeabilizes the outer membrane without dissipating the electrical potential across the inner membrane. Transport of [^{3}H]TPA$^+$ is assayed at 34° by rapid filtration.[9] A typical transport experiment is shown in Fig. 2.

Outer Membrane Permeability Barrier

Functional expression of Mdr1 in *E. coli* might provide an invaluable tool to screen potential Mdr substrates and inhibitors. The major problem encountered in such studies, however, is the impermeability of the outer membrane of gram-negative bacteria, which protects microorganisms against the cytotoxic effects of many lipophilic cancer drugs and blocks accessibility of Mdr reversal agents. To overcome this problem, we constructed outer membrane permeability ("leaky") mutants of *E. coli* UT5600, thus allowing accessibility of lipophilic drugs to the cytoplasmic membrane. For this purpose, alkaline phosphatase (AP) is used as the reported enzyme. Cells expressing native AP grow as blue colonies on agar plates containing 5-bromo-4-chloro-3-indolyl phosphate (X-P). The blue hydrolysis product from X-P is insoluble and is retained within the colonies. However, if the outer membrane is permeable and the enzyme is able to

[9] W. R. Trumble, P. V. Viitanen, H. K. Sarkar, M. S. Poonian, and H. R. Kaback, *Biochem. Biophys. Res. Commun.* **119,** 860 (1984).

diffuse away from the colonies into the agar, the blue product generated appears as a halo around the colonies. To obtain leaky mutants, competent *E. coli* UT5600 cells are exposed to UV light (approximately 95% killing) and are immediately transformed with a plasmid encoding AP. Transformants are plated on agar containing 200 μg/ml X-P and 100 μg/ml ampicillin. After an 18-hr incubation at 37°, a number of colonies (approximately 3 out of 10,000) are able to form large blue halos, indicating that AP from these mutants is liberated into the medium.

Drug Susceptibility Tests

The drug sensitivity of permeable mutants with and without Mdr1 is tested using a semiquantitative disk assay on agar plates, or a quantitative growth inhibition assay in broth. When tested on solid media, overnight cultures are diluted to OD_{600} of 0.1 and grown at 30° for a few hours. Then 200 μl of cultures (OD_{600} of 0.5) are mixed with 3 ml of soft agar (LB with 0.8% agar) at 45° and poured over 1.5% agar LB square petri dishes. Fifteen minutes later, antibiotic filter disks are applied on each lawn and the appropriate amount of tested antibiotic material is carefully loaded on each filter disk. Inhibition zones are measured after overnight growth at 30°. When tested in liquid medium, overnight cultures are diluted into fresh LB broth containing ampicillin (100 μg/ml), and grown up to OD_{600} of 0.6. Cells are then diluted again and aliquoted (50 μl) into 96-well microplates containing 50 μl of various concentrations of the drugs. At the beginning of a typical experiment, the cell density in the wells (measured in a microplate autoreader) expressed as OD_{600} units is 0.03. Plates are incubated at 30° shaker and cell density is monitored by following the absorption at 600 nm. In experiments with chloroquine or quinidine the LB medium is supplemented with 60 m*M* Bis–Tris propane to maintain pH 7.4.

One outer membrane permeability mutant (UTL2[10]) was found to be significantly more sensitive to SDS and other hydrophobic compounds. Moreover, UTL2 is also more sensitive to the toxic effect of chloroquine, quinidine, daunomycin, puromycin, and rhodamine (Table I). High concentrations of verapamil are also more toxic to UTL2 than UT5600. Other drugs such as doxorubicin, vinblastine, vincristine, actinomycin D, and colchicine have no effect on either *E. coli* UT5600 or UTL2 at the highest concentrations that can be tested. When UTL2 cells harboring pT7-5/*mdr1* or vector without *mdr1* are exposed to various Mdr-related compounds in liquid medium, it is readily apparent that Mdr1 confers significant resistance against quinidine, chloroquine, puromycin, rhodamine, and daunomycin

[10] O. Béjà and E. Bibi, *Proc. Natl. Acad. Sci. U.S.A.* **93,** 5969 (1996).

TABLE I
DRUG RESISTANCE OF *E. coli* UT5600, UTL2, AND UTL2 EXPRESSING MDR1 OR LACTOSE PERMEASE[a]

Drug	D_{50} (μM)	
	UT5600	UTL2
EDTA	57	42
Erythromycin	6	7
SDS	267	26
Daunomycin	125	45
Chloroquine	136	67
Puromycin	115	32
Quinidine	475	66
TPA^+	240	50
Rhodamine 6G	200	27
	UTL2 pT7-5(*lacY*)	UTL2 pT7-5(*mdr1*)
Chloroquine	34	232
Puromycin	35	100
Quinidine	40	224
TPA^+	90	340
Rhodamine	20	75

[a] D_{50} is the concentration needed to inhibit growth by 50%. [Reprinted with permission from Béjà, O., and Bibi, E. *Proc. Natl. Acad. Sci. U.S.A.* **93,** 5969–5974 (1996).]

(Table I). Remarkably, Mdr1 is able to confer significant multidrug resistance, despite the high concentrations of drugs needed to inhibit growth of UTL2 (about 10–100 times the concentrations needed to inhibit growth of mammalian cells).

Reversal of Mdr1-Mediated Drug Resistance in E. coli

In addition to its effect on the susceptibility of *E. coli* to Mdr-related drugs, the outer membrane permeability barrier also prevents access of MDR reversal agents to the cytoplasmic membrane. The leaky UTL2 strain thus opens the possibility to examine inhibition of Mdr1 by various modulators *in vivo*. Two Mdr modulators have been studied so far: Reserpine, a known potent suppressor of the MDR phenotype, and doxorubicin, a known Mdr substrate, with which we have not been able to detect toxicity in *E. coli.* To examine the MDR-reversal phenomenon by growth inhibition assays. UTL2 cells harboring pT7-5(*mdr1*) are exposed to increasing concentrations of rhodamine, with or without the reversal agent. Both reserpine (25 μM) and doxorubicin (75 μM) completely abolish Mdr1-mediated resistance against rhodamine.[10]

Structural Studies

One advantage of the *E. coli* expression system for eukaryotic integral membrane proteins is the possibility of applying well-characterized genetic methods to analyze membrane protein topology.[11] Briefly, gene fusion encoding hybrid proteins composed of N-terminal fragments of the membrane protein attached to a cytoplasmic or a periplasmic reporter lacking its signal peptide are expressed in *E. coli.* A periplasmic reporter requires export to the periplasm in order to be enzymatically active, and acts as a sensor for periplasmic location of the protein sequence to which it is attached. We have used AP as the periplasmic reporter.[12] To facilitate rapid construction of Mdr1-AP hybrids, it is convenient to create a vector template in which the *phoA* gene (devoid of the leader peptide coding sequence) is fused in frame to the 3′ end of *mdr1*, with a unique restriction site in the junction. Using synthetic deoxyoligonucleotides for polymerase chain reaction (PCR), given fusions are constructed. Each fusion is made by PCR amplification of a 5′ region of the *mdr* gene encoding N-terminal Mdr1 polypeptide with the appropriate amino acid residue on its C terminus. The PCR amplified fragments are designed to contain the same unique site in the 3′ end and another unique site in the 5′ end. After digestion with these two restriction enzymes, fragments are ligated with the vector template digested with the same two enzymes. Positive transformants are usually detected by AP activity on X-P plates, and the DNA is analyzed by restriction enzyme digestion and by sequencing of the PCR amplified region and the *Nhe*I junction. The results of an extensive study of topology of certain transmembrane domains of the mouse Mdr1 support the idea that the N-terminal half of Mdr1 contains six transmembrane helices (TM) with minor modification regarding TM4.[13] In the C-terminal half of Mdr1 the gene fusion analysis led to a significant revision in the model of membrane topology and suggests that the homologous halves of Mdr are asymmetric in the membrane.[14] However, note that unlike methods used to examine the topology of a full-length Mdr, gene-fusion experiments provide information about the topology at the time it is created cotranslationally, during the biosynthesis of the protein and its insertion into the membrane.

[11] B. Traxler, D. Boyd, and J. Beckwith, *J. Membr. Biol.* **132,** 1 (1993).
[12] C. Manoil and J. Beckwith, *Proc. Natl. Acad. Sci. U.S.A.* **82,** 8129 (1985).
[13] E. Bibi and O. Béjà, *J. Biol. Chem.* **269,** 19910 (1994).
[14] O. Béjà and E. Bibi, *J. Biol. Chem.* **270,** 12351 (1995).

Thus, although we propose a new topology, it is possible that certain regions of Mdr are subjected to reversible conformational changes *in vivo,* imposed by intramolecular interactions, substrates, or other unknown effectors.

Topology of Ste6 from Saccharomyces cerevisiae in E. coli

The proposal that it may be valid to carry out structural analysis of eukaryotic membrane proteins in *E. coli* stems from the assumption that the endoplasmic reticulum (ER) membrane of eukaryotes and the cytoplasmic membrane of prokaryotes exhibit similar properties in regard to protein translocation, and that polytopic membrane proteins are assembled into their final membrane topology in both of these membranes. To explore this hypothesis, we have initiated a direct comparative examination of the topology of the N-terminal half of another member of the ABC superfamily, the yeast **a**-factor transporter, Ste6, both in its native system, *S. cerevisiae,* and in the heterologous prokaryotic system, *E. coli.*[15] The topogenic reporters, invertase in *S. cerevisiae* and alkaline phosphatase in *E. coli,* were fused to Ste6 at identical sites and the fusions were expressed in yeast and bacteria, respectively. When AP is used as a reporter for Ste6 in the heterologous *E. coli* expression system, the results clearly support a secondary structure model in which Ste6 contains six TMs in its N-terminal half (as predicted from the hydropathy profile). When invertase is used to probe the topology of Ste6 in the yeast, the results with all the hybrids except those in the loop between TM2 and TM3, support the same six-TM organization. Hybrids in loop 2 exhibit fewer unequivocal phenotypes, and a fraction of the invertase moieties can be detected in both orientations. Interestingly, and in line with these observations, previous gene-fusion studies with Mdr show that results obtained in the *E. coli* system could clarify uncertainties observed in analogous gene-fusion studies in eukaryotic expression systems. The possible explanation for the clear phenotypes obtained with gene fusions in *E. coli,* in contrast to the mixed phenotypes obtained in eukaryotic systems, is discussed elsewhere.[15] From this study and others we conclude that the topological determinants for membrane insertion of polytopic proteins in prokaryotic and in eukaryotic systems appear to be highly similar.

Bioenergetics of Heterologous Mdr1

The *E. coli* expression system is very powerful for studies of transport proteins *in vivo* and in membrane vesicles. In the case of heterologous

[15] D. Geller, D. Taglicht, R. Edgar, A. Tam, O. Pines, S. Michaelis, and E. Bibi, *J. Biol. Chem.* **271,** 13746 (1996).

Mdr1, however, the experimental situation is more complex. Substrates that we found useful for transport experiments with whole cells expressing Mdr1 are positively charged at neutral pH. Passive intracellular accumulation of these drugs is driven by the membrane electrical potential, and consequently changes in this parameter must alter the cytoplasmic concentration of the charged drug molecules, regardless of the function of Mdr1. Therefore, examination of the membrane potential (which is significantly higher in *E. coli* than in eukaryotic membranes) as a putative participant in the bioenergetics of Mdr1 *in vivo,* or in vesicles, is complicated. Moreover, in *E. coli,* ATP and the proton electrochemical potential are coupled via the F_1-F_0 ATP synthase. Experimental procedures, such as treatment with uncouplers, which affect the membrane electrical potential, or treatment with ATP depletion agents, are expected to affect also energy conversion via the F_1-F_0 ATP synthase. In these conditions it would be complicated to study the possible involvement of the membrane electrical potential or of ATP in Mdr1 function in *E. coli.* This problem can be resolved, at least to some extent, using mutants deficient in the major constituents of the proton ATPase (*unc* strains). Recently, we have used *E. coli* DK8 (*unc*) (kindly provided by Robert D. Simoni) and transferred the deletion by P1 transduction to *E. coli* UTL2 to produce UTL2*unc*. Very well characterized methodologies are currently being used to manipulate the intracellular levels of ATP and/or the membrane electrical potential in UTL2*unc*, in order to study various aspects of drug efflux by Mdr1 in *E. coli.* Regarding the function of heterologous Mdr1, it is relevant to note that a mutant of Mdr1, Mdr88, in which each of two Lys residues located in the conserved nucleotide-binding folds in Mdr1 (Lys-432 and Lys-1074) are replaced with Arg, retains efflux activity in *E. coli,*[1] whereas these mutations eliminate drug resistance in transfected mammalian cells.[16] Further examination of this mutant in *E. coli* UTL2*unc* may reveal the cause for this discrepancy. Although speculative, it is possible that the intrinsic differences between the mammalian and the prokaryotic cells (i.e., the membrane electrical potential or the local ATP concentrations) are responsible for the unexpected activity of Mdr88 in *E. coli.*

Expression of Other Eukaryotic ABC Proteins in *Escherichia coli*

Mouse Mdr2, Mdr3, and Human Mdr1

Preliminary attempts to use the expression system described earlier for the homologous mouse Mdr3 and human Mdr1 were unsuccessful. The

[16] M. Azzaria, E. Schurr, and P. Gros, *Mol. Cell. Biol.* **9,** 5289 (1989).

main complication stems from the toxic effect of these proteins even when expressed at a very low level, regulated by the relatively weak *lac*/ promoter–operator. Similar results were obtained with the mouse phosphatidylcholine transporter Mdr2. Functional expression of the human *MDR1* cDNA in *E. coli* has been reported.[17] In this study George *et al.* cloned the human Mdr1 coding sequence under the control of the lambda P_L promoter using a plasmid harboring also the lambda repressor (λcIts857) gene. The expression of the human Mdr1 was induced at 36–42°. As expected, induction of Mdr at 42° arrests growth due to the toxic effect of high levels of human Mdr1 in *E. coli.* For this reason, induction was carried out at lower temperatures for cytotoxicity assays that involve growth after induction.

Ste6 from S. cerevisiae

Unlike with human Mdr1 or mouse Mdr3, *E. coli* cells harboring *STE6* under the *lac*/promoter–operator grow normally, possibly because Ste6 is hardly expressed. One possibility is that Ste6 is not expressed in *E. coli* due to proteolysis (which is then very significant even in *E. coli* UT5600). Another speculative explanation is that there are very efficient intragenic ribosomal-binding sites followed by perfectly spaced translation start codons (ATG), which appear at high frequency throughout the Ste6 coding region. The consequences are apparent in immunoprecipitation experiments with cells expressing various Ste6–AP hybrids in the N-terminal half of the protein; in addition to the full-length hybrid protein, we also observed large quantities of rapidly migrating species. This phenomenon is even more prominent with fusions of AP in the C-terminal half of Ste6, and all the attempts to express full-length Ste6 or full-length Ste6–AP fusion protein were unsuccessful.

Concluding Remarks

In contrast to heterologous (soluble) enzymes, which can be isolated from *E. coli* extracts and assayed for activity *in vitro,* the proof that heterologous membrane transport proteins, such as Mdr1, are functional in bacteria, *in the same manner* as in eukaryotic cells, is much more complicated. The emphasis here is on "the same manner," but this may be impossible to demonstrate in practice. It is conceivable to assume that energy-driven eukaryotic membrane transporters (such as Mdr1) may behave differently in bacteria because of the inherent difference in the properties of prokary-

[17] A. M. George, M. W. Davey, and A. A. Mir, *Arch. Biochem. Biophys.* **333,** 66 (1996).

otic and eukaryotic cells, for example, the electrical potential, which is significantly higher in prokaryotes. Consequently, it is of interest and may be very informative even if heterologous Mdr1 exhibits "only" similar transport properties in both systems. It is, however, important to investigate the differences when pertinent, because they may shed light on important structural and functional aspects of mouse Mdr1 and possibly of other proteins from this family.

[28] Yeast Secretory Vesicle System for Expression and Functional Characterization of P-Glycoproteins

By STEPHAN RUETZ

Introduction

To investigate the proposed direct or indirect mechanism of P-glycoprotein (Pgp)-mediated drug resistance and transport,[1,2] mouse *mdr* genes[3] have been introduced and expressed in a heterologous yeast expression system suitable for structural and functional analysis. Initial experiments by Raymond *et al.*[4] have shown that expression of the mouse *mdr3* gene in the yeast strain JPY201 can partially complement a null mutation at the *ste6* locus and restore mating in this mutant. In the same yeast strain, the recombinant yeast Mdr3 protein is not only mediating **a**-pheromone transport, but also behaves as a fully functional drug transporter. It mediates ATP-dependent, temperature-sensitive vinblastine and colchicine transport into isolated, inside-out yeast plasma membrane vesicles.[5] This transport activity is inhibited by the known Pgp modulator verapamil and abrogated by prior incubation of the vesicles with the anti-Pgp antibody C219.[5] Finally, expression of Mdr3 in yeast confers resistance to growth inhibition by the antifungal agent and Pgp modulator FK520.[6]

Functional Mdr3 expression in JPY201 cells prompted us to take advantage of the mutant yeast strain *sec6-4*[7-9] to further characterize parameters

[1] M. M. Gottesman and I. Pastan, *Ann. Rev. Biochem.* **62,** 385 (1993).

[2] L. Beaudet and P. Gros, *Methods Enzymol.* **292,** [31], 1998 (this volume).

[3] P. Gros and E. Buschman, *Int. Rev. Cytol.* **137C,** 169 (1993).

[4] M. Raymond, P. Gros, M. Withway, and D. Y. Thomas, *Science* **256,** 232 (1992).

[5] S. Ruetz, M. Raymond, and P. Gros, *Proc. Natl. Acad. Sci. U.S.A.* **90,** 11588 (1993).

[6] M. Raymond, S. Ruetz, D. Thomas, and P. Gros, *Mol. Cell. Biol.* **14,** 277 (1994).

[7] P. Novick and R. Schekman, *Proc. Natl. Acad. Sci. U.S.A.* **76,** 1858 (1979).

[8] N. C. Walworth and P. J. Novick, *J. Cell Biol.* **105,** 163 (1987).

METHODS IN ENZYMOLOGY, VOL. 292
0076-6879/98 $25.00

of the underlying mechanism(s) of Pgp-mediated drug transport. This mutant is defective in the final step of the vesicular secretory pathway and accumulates at the nonpermissive temperature large amounts of unfused secretory vesicles (SVs). These SVs contain mainly newly synthesized membrane proteins commuting to the plasma membrane[8] including exogenous ones introduced as cloned cDNAs on expression plasmids. Using such mutant strains, Nakamoto *et al.*[9] originally produced and functionally characterized mutant forms of the yeast plasma membrane PMA1 H^+-ATPase. Based on these findings, we exploit the use of these yeast SVs as a powerful expression system for structural–functional analysis of Pgps.

In this article we describe the expression of the three mouse *mdr* gene products in SVs from *sec6-4* mutant strains. In addition, we describe two different functional assays that demonstrate the utility of these SVs to characterize membrane transport proteins. First, we discuss drug uptake assays in Mdr-expressing SVs used not only to show that Pgps actively mediate drug transport, but also to clarify several controversial aspects of the underlying mechanisms of Pgp action.[10] The second functional assay is a lipid flippase assay designed to test whether Pgps, in particular Mdr2, are actively translocating phosphatidylcholine within a membrane bilayer and, thus, function as phospholipid flippases.[11]

Yeast Secretory Vesicles

The major advantage of SVs over traditionally prepared vesicle fractions from plasma membranes is that SV preparations of high purity can be easily obtained in good yields. In addition, SVs remain fully intact during the isolation procedure and they are uniform with respect to their origin and size and, most importantly, to their polarity. From all that is known about the insertion of proteins into membranes during biogenesis, SVs are expected to be inside-out. Thus, membrane proteins inserted in the bilayer of these SVs are exposing their cytoplasmic domains to the medium. That means in the case of Pgps that the catalytic adenosine triphosphate (ATP)-binding domains are accessible from the outside and that Pgp-mediated vectorial transport of drug molecules occurs from the extravesicular medium into the lumen of SVs. Using this system we have shown that Mdr1 and Mdr3, but not Mdr2 expressed in SVs, are capable of significantly enhancing the accumulation of radiolabeled vinblastine and colchicine into

[9] R. Nakamoto, K. R. Rao, and C. W. Slayman, *J. Biol. Chem.* **266,** 7940 (1991).
[10] S. Ruetz and P. Gros, *J. Biol. Chem.* **269,** 12277 (1994).
[11] S. Ruetz and P. Gros, *Cell* **77,** 1071 (1994).

the lumen of SVs.[10] These transport activities are strictly ATP and temperature dependent and are completely abolished by the known Pgp modulators verapamil and vanadate.[10]

In addition, SVs contain substantial amounts of newly synthesized PMA1 ATPase trafficking to the plasma membrane. Already in SVs, this endogenous ATPase is fully functional as shown by Nakomoto *et al.*[9] and translocates H^+ into the lumen of these vesicles. In the presence of the membrane-impermeable anion gluconate in the buffer system (VB/Glu and TB/Glu buffer), this activity causes a strong pH gradient (inside acidic) and, consequently, the polarization of the luminal membrane surface (positively charged). We have determined a maximal electrochemical potential ($\Delta\mu_{H^+}$) of ~90 mV at an outside pH of 7.4, which can be maintained over several minutes. The electrical component ($\Delta\Psi$) of $\Delta\mu_{H^+}$ can be abolished by replacing in the buffer system gluconate with the membrane permeable anion nitrate (VB/NO_3 and TB/NO_3 buffer). The addition of nigericin dissipates the pH gradient. Thus, the SV expression system provides a powerful system not only to measure Pgp-mediated drug uptake, but also to determine the respective contribution of $\Delta\Psi$ and ΔpH to Pgp-mediated transport.

To assess whether Pgp-mediated drug transport is dependent or directly coupled to proton movement, the mouse *mdr3* gene is expressed in the *sec6-4* mutant strain SY4. This strain originates from the SY1 strain but the chromosomal *PMA1* gene is put under the control of the *GAL1* promoter. In galactose medium at 25°, the wild-type PMA1 ATPase is produced and supports normal growth. However, on switching the cells in glucose medium to 37°, the wild-type gene is turned off and SVs accumulate lacking endogenous H^+-ATPase but expressing recombinant Mdr3. Since in these SVs none of the normal intravesicular acidification or polarization of the membrane surface is detectable, they have turned out to be very suitable to study the contribution of a proton gradient or proton movement on Pgp-mediated transport.[10]

Expression of Three Mouse P-Glycoproteins in *sec6-4* Mutant Yeast Strains SY1 and SY4

Yeast Strains, Media, and Growth Conditions

Saccharomyces cerevisiae strains SY1 and SY4 used throughout this study were kindly provided by Dr. C. W. Slayman (Department of Cell Biology, Yale University, New Haven, CT) and have been previously described.[8,9] The relevant phenotype of SY1 cells in *MATα*, *ura3-52*, *leu2-3,112*, *his4-619*, and *sec6-4 GAL*, and for SY4 cells *MATα*, *ura3-52*, *leu2-*

3,112, his4-619, sec6-4 GAL, and *pma1::YIpGAL-PMA1*. They were grown at 25° in supplemented minimal medium (0.7% yeast nitrogen base without amino acids, 0.1 m*M* histidine, 0.2 m*M* leucine, and 0.2 m*M* uracil) containing 2% glucose (SY1) or 2% galactose (SY4).

Transformation of Yeast Cells

The high copy number plasmids pVT101-U (for SY1 cells) and pVT101-L (for SY4 cells) were used for the constitutive expression of the three mouse *mdr* genes from the alcohol dehydrogenase (ADH) promoter.[12] Full-length cDNAs for the mouse *mdr1*, *mdr2*, and *mdr3* genes were introduced by standard cloning techniques[13] into the unique *PvuII* site of the pVT101-U or pVT101-L expression vectors as described elsewhere.[6]

DNA transformations of SY1 and SY4 cells are performed by the lithium acetate method of Ito *et al.*,[14] which required the following modifications of the standard protocol. Cultures are grown to mid-logarithmic phase in YPD (1% yeast extract, 2% Bacto-peptone, 2% glucose), diluted to A_{600} of 0.8, and the cells are further incubated for 3 hr at 25° with constant shaking. Fifty-milliliter aliquots are harvested, washed once in lithium acetate buffer (LIAT buffer) consisting of 100 m*M* lithium acetate, 10 m*M* Tris-HCl, pH 8.0, and 1 m*M* EDTA, and resuspended in a final volume of 1 ml of LIAT buffer. In a 1.5-ml microfuge tube, 10–20 μg of plasmid DNA of the various constructs in 10 μl LIAT buffer are mixed with 40 μg (in 10 μl) of freshly prepared total yeast RNA and 150 μl of the yeast cell suspension. After incubation for 2 hr at 25°, the mixture is suspended in 0.7 ml of 36% polyethylene glycol, 100 m*M* lithium acetate, 10 m*M* Tris-HCl, pH 8.0, and incubated for an additional 2 hr at 25°. The samples are then heat-shocked for 5 min at 37°, the cells pelleted, washed once in 10 m*M* Tris-HCl, pH 8.0, 1 m*M* EDTA, and plated on synthetic medium lacking uracil (SD-Ura). Five to 10 transformants are picked, pooled, grown as a mass culture in SD-Ura medium, and followed by freezing at −80° in SD-Ura medium supplemented with 30% glycerol. All subsequent cultures are always started from these stock samples, and grown either in SD-Ura or YCG-medium (0.75% yeast nitrogen base without amino acids, 0.35% Bacto-casamino acids, 2% glucose).

SY1 transformants are selected for growth on supplemented minimal plates containing 2% glycose lacking uracil, whereas transformed SY4 cells are selected on plates lacking leucine and containing 2% galactose. Rou-

[12] T. Vernet, D. Dignard, and D. Y. Thomas, *Gene* **52,** 225 (1987).

[13] J. Sambrook, E. F. Fritsch and T. Maniatis, *in* "Molecular Cloning: A Laboratory Manual," 2nd Ed., Cold Spring Harbor Laboratory Press, New York, 1989.

[14] H. Ito, Y. Kukuda, K. Murata, and A. Kimura, *J. Bacteriol.* **153,** 163 (1983).

tinely, 5 to 10 SY1 transformants are picked and grown at 25° to a dense culture (A_{600} = ~4–5) in supplemented medium containing glucose and lacking uracil. From this stock, large cultures in YC (0.75% yeast nitrogen base without amino acids, 0.35% Bacto-casamino acids) containing 2% glucose are inoculated and grown at 25° to a mid-logarithmic culture (A_{600} = 1.5 ~ 2.0). To induce phenotypic expression of the *sec6-4* mutation and cellular accumulation of unfused SVs, the temperature of the cultures is shifted to 37° and growth continued for 2 hr. Growth is stopped by addition of 10 m*M* NaN_3 and cultures are immediately chilled in ice water. Cells are collected by centrifugation (8275*g* for 10 min), washed once in 10 m*M* Tris-HCl, pH 7.5, 5 m*M* NaN_3, and collected cell pellets stored finally at −70° in 3- to 5-g aliquots.

Similarly, SY4 transformants are picked, grown in supplemented medium containing galactose and lacking leucine, and used to inoculate large cultures in YP containing 2% galactose at 25° to mid-logarithmic culture (A_{600} = 1.5 ~ 2.0). The cells are then collected by centrifugation (8275*g* for 10 min), resuspended in 0.5 volume of prewarmed YC medium containing 2% glucose and incubated for 2 hr at 37°. Finally, the cells are collected and processed as described earlier for the SY1 transformants.

Isolation of Secretory Vesicles from sec6-4 Mutant Yeast Cells

To isolate Pgp expressing SVs from *sec6-4* mutant cells, a modification of the protocol originally described by Nakamoto *et al.*[9] is used. The procedure involves enzymatic digestion of the cell walls, stabilization of the plasma membranes with lectins (concanavalin A) for an effective removal by low-speed centrifugation, followed by osmotic lysis of the spheroplasts and purification of SVs by differential centrifugation steps. The lysate is first subjected to two consecutive low-speed centrifugations (10,000*g*) to remove unbroken cells, mitochondria, intact vacuoles, and nuclei. Under these conditions most of the endoplasmic reticulum (ER), Golgi, and concanavalin A-coated plasma membranes can be removed. In the following two high-speed centrifugation spins (100,000*g*) the remaining membranes are collected. This fraction is highly enriched with SVs sufficient for subsequent drug transport and phospholipid translocation assays. However, if pure SV populations are needed, the latter fraction can be further purified by gel-filtration chromatography as described elsewhere.[8,9]

Frozen yeast cells are thawed at room temperature, resuspended in 25 ml 100 m*M* Tris-SO_4, pH 9.4 (25°), collected by centrifugation (2000*g* for 10 min). They are converted to spheroplasts by incubating them with 1 mg zymolyase (Zymolyase-100T, ICN Biomedicals, Irvine, CA) per wet gram of yeast cells for 45 min at 37° in prewarmed SM-buffer consisting of 1.2 *M* sorbitol, 20 m*M* HEPES–KOH, pH 7.0, 2 m*M* EDTA supplemented with 10 m*M* NaN_3 and 40 m*M* 2-mercaptoethanol. The resulting spheroplasts are washed twice (5000*g* for 10 min) in SM buffer (25°), carefully resuspended in 10 ml SM buffer supplemented with 1 m*M* $CaCl_2$, 5 m*M* $MnSO_4$, and 1.5 mg concanavalin A (Con A, Pharmacia, Uppsala, Sweden) per gram wet cell. After an incubation period of 15–20 min on ice, the lectin-coated spheroplasts are washed again twice (5000*g* for 10 min) with ice-cold SM buffer and resuspended in cold hypotonic lysis buffer consisting of 0.6 *M* sorbitol, 20 m*M* HEPES–KOH, pH 7.0, 2 m*M* EDTA supplemented with protease inhibitors, phenylmethylsulfonyl fluoride (PMSF, 1 m*M*), pepstatin, leupeptin, and aprotinin (1 μg/ml each). The spheroplasts are kept for 10–20 min on ice, and further disrupted by Dounce homogenization (Wheaton Scientific, Millville, NJ) with 20–30 strokes using an A pestle. The lysate is centrifuged twice (10,000*g* for 10 min) to remove unbroken cells, membrane fragments, nuclei, and mitochondria. SVs are recovered from the supernatant by centrifugation (100,000*g* for 45 min), and the pellet resuspended in an appropriate vesicle buffer (VB buffer) containing 50 m*M* sucrose, 10 m*M* Tris–HEPES, pH 7.5, supplemented either with 100 m*M* potassium gluconate (VB/Glu buffer) or 100 m*M* potassium nitrate VB/NO_3 buffer. To the resuspended fraction EGTA is then added to a final concentration of 5 m*M* and the sample is centrifuged as before. The final pellet is resuspended in an appropriate volume (0.1–1 ml) of the same VB buffer with the aid of a syringe fitted with a 23-gauge needle and stored on ice until further use. Routinely, starting from 3–5 g wet yeast cells, the described isolation procedure results in a highly enriched SV fraction containing approximately 2.5–3 mg of protein.

We always use freshly prepared SV fractions for functional assays. In a pilot experiment we once tested SV fractions kept frozen at −70° prior to using them in drug transport assays. We found that the overall transport rates were significantly reduced. This loss of activity was probably not due to inactivation of Pgp but rather due to the fact that freezing and/or thawing makes SVs leaky as confirmed by electron microscopy.

Morphological and Biochemical Characterization of Secretory Vesicles

For morphological analysis of SVs preparations we have used negative stain electron microscopy using 2% (w/v) ammonium molybdate.[11] Examination of such micrographs has revealed that the isolated SV fractions consist mainly of a population of closed vesicles of homogeneous size with an average diameter of 75–100 nm. They appear to remain fully intact during the isolation procedure.

As a typical biochemical marker for SVs, Walworth and Novick[8] have originally used the enzyme invertase, both to demonstrate the enrichment of SVs in their purification and, at the same time, to monitor the integrity of the isolated vesicles. Accordingly, we have validated our isolation procedure by measuring invertase activities in SV fractions as described by Goldstein and Lampen.[15] However, since this enzyme is under hexose repression, invertase synthesis must be stimulated by growing the cells in growth medium containing low concentrations of glucose. Thus, *sec6-4* cells have to be shifted to 37° in YC medium containing 0.2% glucose and growth is continued for 2 hr. In the isolated SV fractions we have determined in the presence of 0.1% Triton X-100 an increase of the invertase activity over the total cell lysate by 22- to 26-fold. Little invertase activity (~30 times less) is observed in the absence of detergent, indicating that in the SV preparations the vesicles are mostly intact.

Another approach used[9,11] to analyze the integrity of the isolated SVs is to visualize the formation and dissipation of a transmembrane proton gradient due to the activity of the endogenous PMA1 H^+-ATPase (see earlier section) using the ΔpH-sensitive fluorescent dye acridine orange. This weak base accumulates in vesicles whenever intravesicular pH is acidic relative to the pH of the extravesicular buffer and, as a consequence, the strong fluorescent signal of the accumulated fluorophore is quenched. The degree of the fluorescence quenching (or decrease of absorption) is not only a measure of the magnitude of the ΔpH across the membrane bilayer but also indicates that the vesicle membrane is sufficient impermeable to maintain a proton gradient.

Finally, Pgp expression in SVs is shown by standard Western blotting techniques using the specific monoclonal anti-Pgp antibody C219.

Vesicular Drug Transport Studies

General Considerations

Radiolabeled MDR drug uptake measurements into SVs are determined by a rapid filtration technique using a filtration device (Glass Microanalysis, S.S. Support, 25 mm, Millipore Intertech, Bedford, MA). It allows filtration in a few seconds under mild vacuum of small volumes of liquids through a nitrocellulose filter. Thus, in a typical drug transport assay, vesicles incubated with a transport buffer containing radiolabeled drug molecules can be very efficiently separated from free labeled ligands. After additional washes of the trapped vesicles on the filter, vesicle-associated radioactivity

[15] A. Goldstein and J. O. Lampen, *Methods Enzymol.* **42,** 504 (1975).

is measured by dissolving the filters in an appropriate scintillation fluid and determining the radioactivity by liquid scintillation counting.

The transport buffer (TB buffer) should always be of the same osmolarity as the vesicle resuspension buffer (VB buffer) and contain the desired extravesicular concentrations of additives and ions (e.g., drugs, inhibitors, ATP, magnesium ions). For uptake measurements of a certain drug, the required radioactivity for a single time point is approximately 3×10^5 cpm. For example, for [^{3}H]FK506 transport as shown in Fig. 1, with the available specific activity of ~50 Ci/mmol for [^{3}H]FK506 ([^{3}H]dihydro-FK506, 50.3 Ci/1.86 TBq/mmol, Amersham, Buckinghamshire, England) the resulting tracer concentration in the TB buffer is about 0.1–0.25 μM assuming a 50% counting efficiency. Because in the transport assay a final concentration of 1 μM is used, unlabeled FK506 must be added.

The "cold stop" solution is used to terminate the reaction and should also be isosmotic to the final incubation. One of the most significant draw-

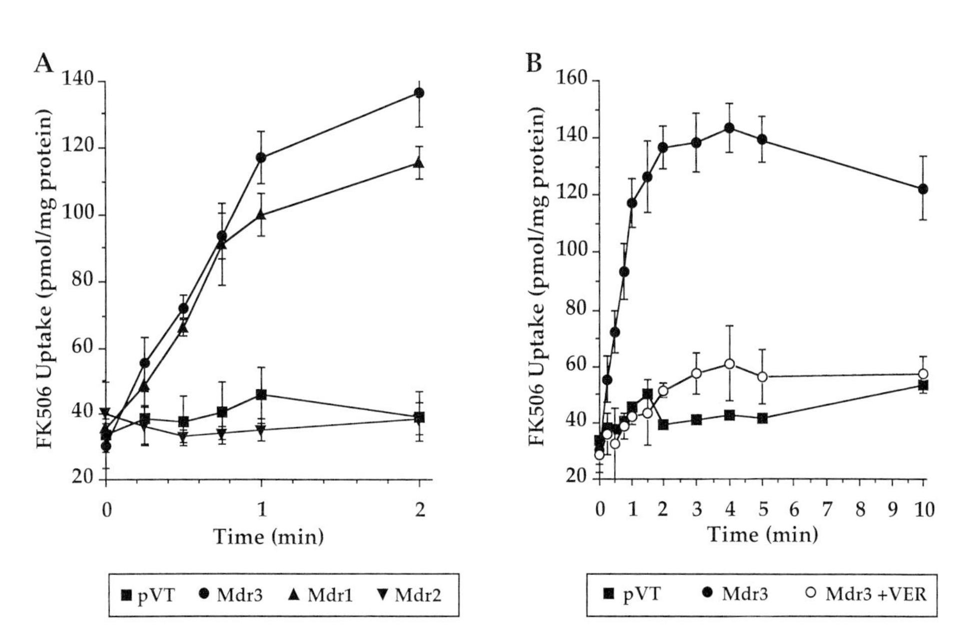

FIG. 1. Kinetics of FK506 accumulation in SVs from control and Mdr-expressing cells. SVs isolated from control (■), Mdr1- (▲), Mdr2- (▼), and Mdr3-expressing cells (● ○) are resuspended in VB buffer containing nitrate (VB/NO_3). Drug transport is initiated by diluting the SVs in transport buffer containing nitrate (VB/NO_3), 2.5 mM NaATP, 5 mM Mg^{2+}, [^{3}H]dihydro-FK506 (50.3 Ci/mmol) at a final concentration of 1 μM, and 10 μM nigericin. FK506 accumulation is determined in the absence (A) or in the presence of the known Pgp reversal agent verapamil (B). Data are the mean values ±SD of quadruplicates of a representative experiment.

backs in transport measurements of hydrophobic drugs is the binding problem, both the nonspecific binding to the filters and to the vesicle membranes. First, to quench nonspecific filter binding we routinely supplement the cold stop medium with 1–5 μM of the same drug as used in the transport assay; alternatively, we just add 1 μM vinblastine. In addition, it is advisable to prefilter this solution through a 0.45-μM fitler prior to use to guarantee a continuous flow during the filtration. Furthermore, the nonspecific binding to the filters is further reduced by soaking them in cold stop solution prior to starting the experiment. Also critical is determination of the optimal filtration rate prior to the experiment, which strongly influences the nonspecific filter binding. A stronger vacuum reduces the filtration time and, thus, can help to reduce the binding significantly. But forces that are too strong may cause vesicles to collapse during the filtration.

In addition, because of the high membrane-binding properties of most of these hydrophobic MDR drugs it is always critical to prepare appropriate vesicle/filter-binding blanks for all uptake series. We prepare this "zero time" blank by keeping 90 μl of radioactive transport buffer on ice and adding 10 μl of the cold SV suspension as a drop to the side of the tube. Immediately after mixing cold stop solution is added and the sample processed as described. The absolute amount of substrate taken up into the vesicles after a defined time interval is calculated using the following formula:

$$\text{pmol mg}^{-1} = [(cpm - B) \times \text{pmol substrate in assay}]/[cpm_{\text{tot}} \times \text{mg protein in assay}]$$

where cpm is the counted vesicle-associated radioactivity at a certain time point, B is the "zero time" blank, and cpm_{total} is the total amount of radioactivity used in the assay.

Practical Procedure

In this section we describe a detailed protocol for measuring radiolabeled FK506 accumulation into control, Mdr1-, Mdr2- and Mdr3-expressing SVs in the absence of a transmembrane electrochemical gradient ($\Delta\mu_{H^+}$). In Fig. 1, the data for such a typical experiment are shown. Similar experiments using [^{3}H]vinblastine, [^{3}H]colchicine, or [^{3}H]tetraphenylphosphonium bromide in the presence or absence of $\Delta\mu_{H^+}$ across the membrane bilayer have been published elsewhere.[10]

SV fractions resuspended in VB/NO_3 buffer are diluted to 10–15 mg protein/ml with the same buffer supplemented with creatine phosphokinase (3 μg/ml) and kept on ice until the start of the assay. All transport studies are carried out in a temperature-controlled water bath. Drug transport is initiated by diluting a 10-μl aliquot of the SV suspension in 90 μl of a prewarmed (37°) transport buffer (TB/NO_3 buffer) consisting of 10 m*M* magnesium nitrate, 100 m*M* potassium nitrate, 2.5 m*M* NaATP, 10 m*M* creatine phosphate, 50 m*M* sucrose, and 10 m*M* Tris–HEPES, pH 7. In addition, this TB buffer contains tracer amounts of radiolabeled drug supplemented to final concentrations of 1 m*M* with the corresponding cold drug. After the desired time period the reaction is terminated by adding 2 ml of an ice-cold stop solution (250 m*M* sucrose, 10 m*M* Tris-HCl pH 7.5). The vesicle-associated ligand is then separated from the free ligand by rapid filtration (~1 ml/sec) through a 0.65-μm nitrocellulose filter (type HA, Millipore, Bedford, MA) preequilibrated in stop solution. The filter is additionally washed twice with 2 ml of stop solution. The filters containing the trapped vesicles are then dissolved in an appropriate scintillation fluid and the SV-associated radioactivity determined by liquid scintillation counting. All measurements are usually carried out in triplicate and confirmed in several independent SV preparations. To perform radiolabeled drug uptake measurements in the presence of an electrochemical membrane potential, nitrate anions in the VB/NO_3 and TB/NO_3 buffers are isosmotically replaced by the membrane-impermeable anion gluconate.

NBD-Labeled Phosphatidylcholine Translocation Assay

The second part of this article describes the use of SVs fractions to demonstrate Pgp-mediated phosphatidylcholine (PC) translocation across the membrane bilayer. Since Mdr2 has been proposed to transport PC across the canalicular membrane from the cytoplasmic side to the outside domain of the hepatocyte,[16] a similar activity for recombinant Mdr2 expressed in SVs would be expected to mediate PC translocation from the outer to the inner leaflet of the membrane bilayer of these vesicles.

To measure lipid translocation activity within a bilayer, it is necessary to determine the asymmetric distribution of specific lipid molecules in the inner or outer leaflet. We have taken advantage of a fluorescent PC analog containing a 7-nitro-2,1,3-benzoxadiazol-4-yl group (NBD), which can be chemically reduced to the nonfluorescent 7-amino-2,1,3-benzoxadiazol-4-yl compound by dithionite.[17] Because the reducing agent is membrane impermeable, it will reduce in intact vesicles or liposomes only those NBD groups of labeled PC molecules (NBD-PC) which are present in the outer

[16] J. J. Smit, A. H. Schinkel, R. P. Oude Elferink, A. K. Groen, E. Wagenaar, L. van Deemter, C. A. A. Mol, R. Ottenhoff, N. M. van der Lugt, M. A. van Roon, M. A. van der Valk, G. J. Offerhaus, A. J. Berns, and P. Borst, *Cell* **75,** 451 (1993).

[17] J. C. McIntyre and R. G. Sleight, *Biochemistry* **30,** 11,819 (1991).

leaflet of the bilayer while leaving unaffected those either associated with the inner leaflet or present within the luminal space.[17] Only on detergent disruption of the SV membranes, will the fluorescent emission of the remaining NBD-labeled PC be destroyed, and this further reduction in fluorescence will directly reflect the amount of NBD-PC associated with the inner leaflet of the bilayer.

A schematic diagram of the experimental approach used is shown in Fig. 2. Fluorescent PC molecules (NBD-PC) are first transferred from donor liposomes (DL) into the outer leaflet of SV from control and Mdr-expressing yeast cells. Excess of DL is then completely removed and potential NBD-PC translocation activities within the bilayer are activated in the resulting labeled SV by incubation at 37°. To determine the relative distribution of NBD-PC molecules in the inner versus outer leaflet of the bilayer, dithionite is subsequently added in excess to destroy chemically

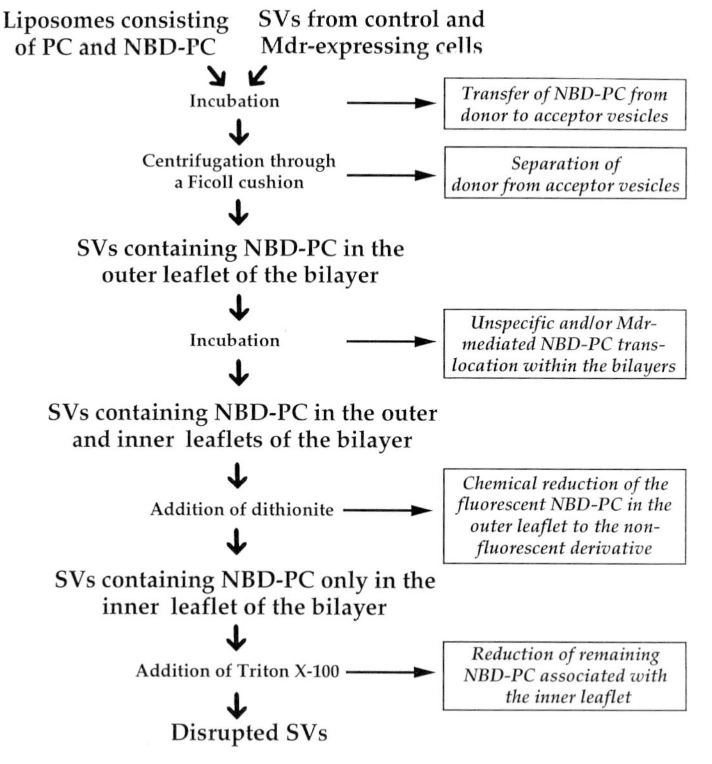

FIG. 2. Schematic diagram of the experimental approach used to measure the translocation of fluorescent phosphatidylcholine (NBD-PC) in the membrane bilayer of secretory vesicles (SVs).

the fluorescent groups of NBD-PC molecules associated with the outer leaflet. Once this first reaction is completed (i.e., no further decrease in fluorescence emission over time), the vesicles are disrupted with Triton X-100, and further loss of remaining fluorescence emission (second reaction) is measured, and used to calculate the fraction of NBD-PC molecules present in the inner leaflet of the bilayer. As shown in Fig. 3, using this approach we have monitored and compared the time-dependent changes in the distribution of NBD-PC within the two leaflets of the bilayer of SV from control (pVT) and Mdr2-expressing yeast cells.

Preparation of Secretory Vesicles Containing NBD-Labeled PC in Outer Leaflet of SV Membrane Bilayer

Secretory vesicles from Mdr1-, Mdr2-, or Mdr3-expressing cells were isolated as described earlier and resuspended in an appropriate volume (0.5–2 ml) of cold VB buffer. NBD-labeled PC is introduced into the outer leaflet of the SV membrane bilayer via small donor liposomes (DL). These liposomes contained 40 mol% NBD-PC and 60 mol% PC and are prepared by mixing appropriate amounts of the two lipid components (total amount 20 μmol/ml). This storage solvent is evaporated under nitrogen and remaining solvent tracers completely removed by a 4- to 6-hr vacuum desiccation. DL are prepared by ethanol injection using 10 μmol of lipid per milliliter of ethanol. Therefore, the dried lipids are accordingly dissolved in ethanol and injected at room temperature with a Hamilton syringe into 10 ml of a magnetically stirred vesicle buffer solution containing 50 m*M* sucrose, 100 m*M* potassium chloride, and 10 m*M* Tris–HEPES, pH 7.5, resulting in a solution with a final ethanol concentration of 7.5%. The resulting NBD-PC-containing DLs are subsequently dialyzed for 16 hr at 4° against the same buffer prior use.

In contrast to PC, the NBD-labeled lipid PC analog translocates with a much higher frequency between membranes, which is mainly due to its increased water solubility. We have taken advantage of this property and transferred NBD-PC from DLs by mixing a 100-μl aliquot of DLs containing 48 nmol NBD-PC with 2 ml of a SV suspension containing normally a total amount of 40–43 μmol of lipid, which corresponds to about 12.2 ± 0.45 μmol lipids per milligram/protein from either control, Mdr2-, or Mdr3-expressing cells. After an incubation period of 30 min on ice, the samples are then carefully applied on a 7.5% Ficoll 400 cushion 910 ml in VB buffer) and centrifuged in an SW40 (Beckmann, Palo Alto, CA) swing-out rotor at 90,000*g* for 30 min at 4°. Since DLs with an average diameter of ~30 nm are much smaller than isolated SVs, they cannot enter the Ficoll cushion, whereas NBD-PC labeled SVs are sedimented. Thus, by this procedure,

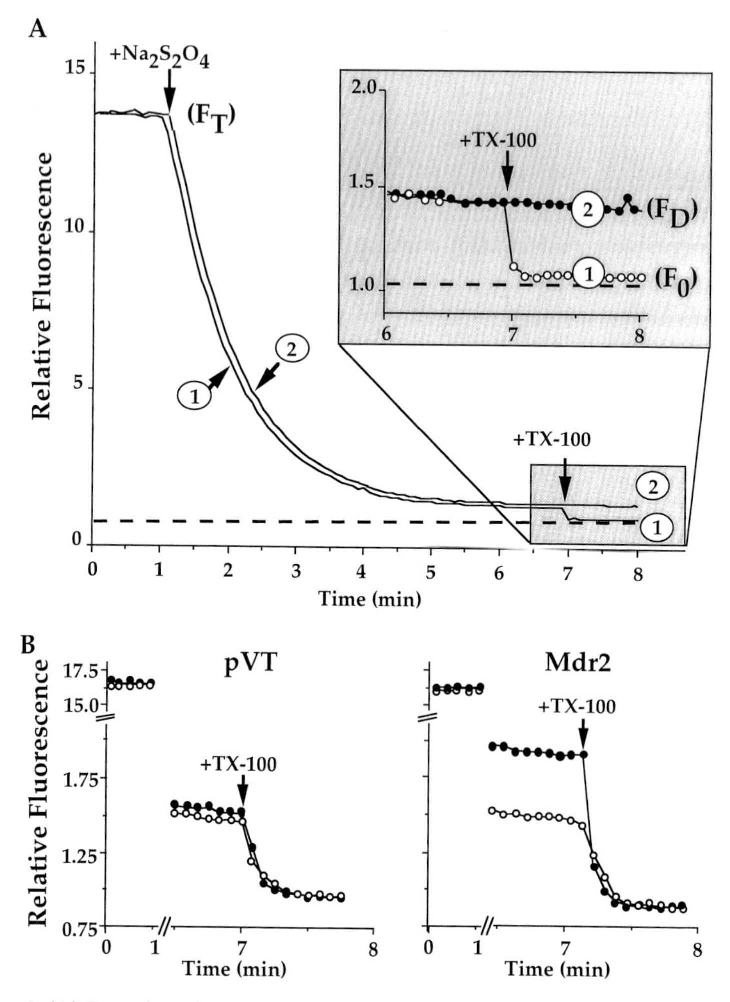

FIG. 3. (A) Detection of asymmetric distribution of NBD-PC molecules in the membrane bilayer of SVs. A 50-μl aliquot of SVs containing NBD-PC is diluted into 1.95 ml VB buffer and fluorescence ($\lambda_{ex} = 470$ nm, $\lambda_{em} = 540$ nm) is measured over time. After an initial constant baseline (F_T) is recorded, 25 μl of 1 *M* dithionite ($Na_2S_2O_4$) in 1 *M* Tris, pH 10, is added and the reduction of 7-nitro-2,1,3-benzoxadiazol-4yl-labeled PC (NBD) to 7-amino-2,1,3-benzoxadiazol-4-yl-labeled PC is monitored over time. After the reaction is completed with no further reduction in fluorescence (F_D, profile 2, *insert*), Triton X-100 (+Tx-100, profile 1, *insert*) is added to a final concentration of 1% (v/w) and a further loss of fluorescence emission followed, until a stable baseline is again obtained (F_0, *insert*). The dashed line represents the background fluorescence emission at 540 nm ($\lambda_{ex} = 470$ nm) of SV containing no NBD-PC. The fraction of NBD-PL in the outer leaflet can be calculated by the equation: percent (out) = 100 $[(F_T - F_D)/(F_T - F_0)]$ and the amount present in the inner leaflet of the bilayer

labeled SVs are very effectively separated from the excess of NBD-PC containing DL. It is important to emphasize that an efficient separation is critical for the subsequent PC translocation assay since incomplete removal of DLs would result in a too high background from labeled PC molecules associated with the luminal domain of DLs. We have also determined that under the established conditions approximately 97% of the transferred NBD-PC molecules are associated with the outer leaflet of the membrane bilayer, and only a neglectable portion of the fluorophore spontaneously flipped within the bilayer into the inner leaflet. Furthermore, we can estimate that ~8 nmol of fluorophore is transferred from DLs into SV membranes. This amount corresponds to about 0.025% of the total lipid content of SVs, which we have found does not affect the integrity of the SV membranes.[11]

Phospholipid Translocation Assay

The pelleted SV fractions containing incorporated NBD-PC are finally resuspended in 1.5–2 ml cold VB buffer and further diluted with the same buffer to obtain a relative fluorescence emission signal of about 15–20 OD in a 50-μl SV aliquot. To measure NBD-PC translocation, a 150 μl aliquot of NBD-PC containing SVs from control, Mdr2, or Mdr3-expressing cells in VB-buffer supplemented with creatine phosphokinase (3 μg/ml) is mixed with 150 μl cold VB buffer containing 5 m*M* NaATP, 10 m*M* magnesium gluconate, and 20 m*M* creatine phosphate. The assay is initiated by incubating the SV suspension at 37°. First, at predetermined time intevals, 100-μl aliquots are removed and immediately placed into a fluorescence cuvette containing 1.9 ml VB buffer (25°) supplemented with 2.5 m*M* EDTA and 10 μM verapamil. The fluorescence emission is recorded in a fluorescence spectrometer at a wavelength of 540 nm using an excitation wavelength of 470 nm. Solutions in the cuvettes are continuously mixed with a magnetic stirrer. An initial fluorescence emission of the SV samples is recorded over time (~1 min) to establish a constant and stable baseline (F_T). Second, 25 μl of a solution containing 1 *M* $Na_2S_2O_4$ in 1 *M* Tris, pH 10, is added to chemically reduce the fluorescent nitroaryl group of NBD associated

by: percent (in) = 100 $[(F_D - F_0)/(F_T - F_0)]$. (B) NBD-PC translocation in SVs from control (pVT) and Mdr2-expressing SVs. Translocation of NBD-PC from the outer to the inner leaflet in the membranes of SV from control (*left graph*) and Mdr2-expressing cells (*right graph*) is determined by measuring the specific asymmetric distribution of the labeled lipid in the SV bilayers (as described earlier) prior (○) and after a 10-min incubation (●) at 37° in the presence of ATP (2.5 m*M*), Mg^{2+} (5 m*M*), and an ATP regeneration system. [Reprinted with permission from S. Ruetz and P. Gros, *Cell* **77,** 1071 (1994). Copyright 1994 by Cell Press.]

with the outer leaflet of the SV bilayer, and the decrease of the fluorescence signal is further monitored for 5–6 min until a new stable emission baseline was recorded (F_D). Finally, NBD-PC fluorescence associated with the inner leaflet of the SV bilayer was revealed after disrupting the vesicles by addition of 80 μl of a 25% (w/v) Triton X-100 stock solution followed by further monitoring of the disappearance of the fluorescence signal.

Conclusion

Functional expression of Pgps in yeast, in particular, in SVs from the yeast mutant *sec6-4*, has resulted in a clearer understanding of the mechanism and function of the different Pgp isoforms. In this system, used previously to study mutants of the endogenous yeast plasma membrane PMA1 H^+-ATPase, accumulation of secretory vesicles can be induced. These post–Golgi vesicles contain newly synthesized proteins including recombinant ones introduced as cloned cDNAs on expression plasmids. The major advantages of these SVs are that they are all uniform in respect to their origin, size, and polarity. Membrane proteins are inserted in the bilayer of SVs in an "inside-out" orientation exposing their cytoplasmic domains to the medium side. This unique feature makes SVs extremely useful to study transport proteins that are functioning, for instance, in an intact cell as efflux pumps. In SVs, substrate transport can be determined in uptake measurements with the big advantages that the experimental conditions can be easily controlled from the outside. In addition, SVs are fully functional organelles in which the endogenous proton translocating PMA1 ATPase maintains a strong electrochemical gradient. Manipulation of $\Delta\mu_{H^+}$ allows the study of the contribution of the membrane potential and/or a pH gradient on substrate transport and, thus, the determination of specific parameters of the underlying mechanism(s) of action of a protein of interest. Furthermore, the abundance of SVs in these *sec6-4* cells and the ease of purifying them in good yields makes this system a powerful tool not only for expression and functional characterization but also for purification of membrane proteins.

[29] High-Level Expression of Mouse Mdr3 P-Glycoprotein in Yeast *Pichia pastoris* and Characterization of ATPase Activity

By LUCILLE BEAUDET, INA L. URBATSCH, and PHILIPPE GROS

Introduction

P-Glycoproteins (Pgps), which are eukaryotic members of the ATP-binding cassette (ABC) superfamily of transporters, cause cellular resistance to a wide variety of structurally unrelated cytotoxic drugs, including *Vinca* alkaloids, anthracyclins, colchicine, and ionophore peptides such as valinomycin and gramicidin D. The structural unit common to ABC transporters includes one membrane-associated region [typical six transmembrane domains (TMDs)] and one nucleotide-binding domain (NBD) of the Walker type.[1] In Pgps and most eukaryotic ABC transporters, this structural unit is duplicated, resulting in 12 TMDs and two NBDs.[2] Mammalian Pgps can be functionally divided into two groups: Class I Pgps (human MDR1, mouse Mdr1 and Mdr3, hamster Pgp1 and Pgp2) can convey multidrug resistance (MDR) to drug-sensitive cells, whereas class II Pgps (human MDR2, mouse Mdr2, and hamster Pgp3) apparently cannot. Human and mouse class II Pgps have been shown to function as phosphatidylcholine translocases in liver bile canaliculi.[3-5]

Pgp ATPase activity of the human MDR1[6-9] and the hamster Pgp1[10-14]

[1] J. E. Walker, M. Saraste, M. J. Runswick, and N. J. Gay *EMBO J.* **1,** 945 (1982).
[2] P. Gros and M. Hanna, *in* "Handbook of Biological Physics" (W. N. Konings, H. R. Kabak, and J. S. Lolkema, eds.), Vol. 2 p. 137. Elsevier, Amsterdam, 1996.
[3] S. Ruetz and P. Gros, *Cell* **77,** 1071 (1994).
[4] A. J. Smith, J. L. P. M. Timmermans-Hereijgers, B. Roelofsen, K. W. A. Wirtz, W. J. van Blitterswijk, J. J. M. Smit, A. H. Schinkel, and P. Borst, *FEBS Lett.* **354,** 263 (1994).
[5] A. T. Nies, Z. Gatmaitan, and I. M. Arias, *J. Lipids Res.* **37,** 1125 (1996).
[6] B. Sarkadi, E. M. Price, R. C. Boucher, U. A. Germann, and G. A. Scarborough, *J. Biol. Chem.* **267,** 4854 (1992).
[7] H. Hamada and T. Tsuruo, *J. Biol. Chem.* **263,** 1454 (1988).
[8] S. V. Ambudkar, I. H. Lelong, J. Zhang, C. O. Cardarelli, M. M. Gottesman, and I. Pastan, *Proc. Natl. Acad. Sci. U.S.A.* **89,** 8472 (1992).
[9] T. W. Loo and D. M. Clarke, *J. Biol. Chem.* **269,** 7750 (1994).
[10] C. A. Doige, X. Yu, and F. J. Sharom, *Biochim. Biophys. Acta* **1109,** 149 (1992).
[11] M. K. Al-Shawi and A. E. Senior, *J. Biol. Chem.* **268,** 4197 (1993).
[12] A. B. Shapiro and V. Ling, *J. Biol. Chem.* **269,** 3745 (1994).
[13] I. L. Urbatsch, M. K. Al-Shawi, and A. E. Senior, *Biochemistry* **33,** 7069 (1994).
[14] F. J. Sharom, X. Yu, J. W. K. Chu, and C. A. Doige, *Biochem. J.* **308,** 381 (1995).

0076-6879/98 $25.00

isoforms has been studied in membrane-enriched fractions[6,8–11] and in purified preparations of the protein.[7,12–15] Reported V_{max} values of the Pgp basal activity vary between 0.1 and 2 μmol/mg/min, with a K_m for MgATP between 0.5 and 1.4 m*M*. One of the striking features of the Pgp ATPase is that it can be stimulated by several MDR drugs and modulators (reviewed in Gros and Hanna[2]).

Two heterologous eukaryotic expression systems have been used for the characterization of the ATPase activity of wild-type and mutant variants of Pgp. The most commonly used is the baculovirus expression system.[6,16–18] Although baculovirus-infected *Spodoptera frugiperda* ovary (Sf9) cells produce large quantities of Pgp (3% of total membrane proteins[6]), it has been reported that the production of Pgp-carrying viruses was impaired in Sf9 insect cells, and that the integrity of the resulting virus stock had to be carefully ascertained due to possible recombinational events with endogenous genes.[18] A second expression system consists of the transient transfection of HEK 293 cells with a modified Pgp cDNA encoding an extra polyhistidine tag for affinity purification by nickel–chelate chromatography.[15] Although simple and fast, this procedure yields relatively low quantities of pure protein (6–12 μg) and its upscaling would be difficult. In general, expression and purification of Pgp from animal cell systems in which it is stably expressed involve time-consuming and expensive selection procedures that are difficult to adapt to large-scale purification of the protein, and to the simultaneous analysis of multiple mutants.

We have used yeast as a heterologous expression system to study wild-type and mutant variants of Pgps. Functional integrity of yeast-expressed Pgp has been demonstrated[19–24] with respect to the drug resistance characteristics of Mdr3 transformants (FK506, FK520, and valinomycin), the drug transport properties of Pgp in secretory vesicles, and by the capacity of *mdr3* to complement the endogenous *STE6* gene and restore mating in a Δ*ste6* strain. However, yeast has not been used so far to study Pgp ATPase activity, in part due to the relatively low level of Pgp expression achieved

[15] T. W. Loo and D. M. Clarke, *J. Biol. Chem.* **270,** 21449 (1995).
[16] U. S. Rao, *J. Biol. Chem.* **270,** 6686 (1995).
[17] M. Müller, E. Bakos, E. Welker, A. Váradi, U. A. Germann, M. M. Gottesman, B. S. Morse, I. B. Roninson, and B. Sarkadi, *J. Biol. Chem.* **271,** 1877 (1996).
[18] G. A. Scarborough, *J. Bioenerg. Biomembr.* **27,** 37 (1995).
[19] M. Raymond, P. Gros, M. Whiteway, and D. Y. Thomas. *Science* **256,** 232 (1992).
[20] M. Raymond, S. Ruetz, D. Y. Thomas, and P. Gros, *Mol. Cell. Biol.* **14,** 277 (1994).
[21] S. Ruetz and P. Gros, *J. Biol. Chem.* **269,** 12277 (1994).
[22] S. Ruetz, M. Raymond, and R. Gros, *Proc. Natl. Acad. Sci. U.S.A.* **90,** 11588 (1993).
[23] L. Beaudet and P. Gros, *J. Biol. Chem.* **270,** 17159 (1995).
[24] S. Ruetz, U. Delling, M. Brault, E. Schurr, and P. Gros, *Proc. Natl. Acad. Sci. U.S.A.* **93,** 9942 (1996).

in *Saccharomyces cerevisiae,* but also due to contamination of plasma membrane preparations with the endogenous H^+-ATPase (PMA-1).

Pichia pastoris is a methylotrophic yeast capable of metabolizing methanol as a sole carbon source. The first enzyme in the methanol utilization pathway is alcohol oxidase (AOX1), which converts methanol to formaldehyde.[25] When yeast cells are grown with methanol as a sole carbon source, the AOX1 enzyme is induced and accounts for 30% of total soluble protein.[25,26] An expression system based on the utilization of the strong *P. pastoris AOX1* promoter to direct the transcription of foreign genes has been developed[27] and is commercially available through InVitrogen (San Diego, CA). We used the *P. pastoris* expression system to overproduce the mouse Mdr3 Pgp isoform and to show that stable and high-level expression of intact Pgp can be achieved in the membrane fraction of *P. pastoris* cells. We have characterized the ATPase activity of this mouse Pgp following selective detergent solubilization and reconstitution in *Escherichia coli* lipids of Pgp-rich membranes fractionated on sucrose gradients.

Experimental Procedures

Plasmid Construction and Transformation

The expression system included *P. pastoris* yeast strain GS115 (*his4*) and expression vector pHILD2 is obtained from InVitrogen. A mouse 4.2-kb full-length *mdr3* cDNA including 57 base pairs of 5′ and 210 base pairs of 3′ untranslated sequences is excised from the pDR16 vector[28] using restriction enzymes *Kpn*I and *Cla*I (polylinker sites). The extremities of the *Kpn*I/*Cla*I fragment are repaired with T4 DNA polymerase and cloned into a pHILD2 vector that had been digested with *Eco*RI and repaired with T4 DNA polymerase. The resulting plasmid, named pHIL-*mdr3,* is propagated and purified from *E. coli* according to standard protocols.

pHIL-*mdr3* is digested with *Not*I and transformed into *P. pastoris* GS115 spheroplasts according to the manufacturer instructions. His^+ transformants are picked and streaked onto minimal methanol plates [MM: 1.34% (w/v) yeast nitrogen base (YNB), 0.04% (w/v) biotin, 0.5% (v/v) methanol, and 20% (w/v) agar] and minimal glycerol plates [MG: 1.34% (w/v) YNB, 0.04% (w/v) biotin, 1% (v/v) glycerol, and 20% (w/v) agar] to identify clones with

[25] I. J. van der Klei, W. Harder, and M. Veenhuis, *Yeast* **7,** 195 (1991).

[26] M. Veenhuis, J. P. Van Dijken, and W. Harder, *Adv. Microb. Physiol.* **24,** 1 (1983).

[27] K. Sreekrishna, R. H. Potenz, J. A. Cruze, W. R. McCombie, K. A. Parker, L. Nelles, P. K. Mazzaferro, K. A. Holden, R. G. Harrison, P. J. Wood, D. A. Phelps, C. E. Hubbard, and M. Fuke, *J. Basic Microbiol.* **28,** 265 (1988).

[28] A. Devault and P. Gros, *Mol. Cell. Biol.* **10,** 1652 (1990).

an impaired capacity to metabolize methanol (methanol utilizing slow or *mut*^{S}). Such colonies lost the endogenous *AOX1* gene after a successful homologous recombination event with pHIL-*mdr3*.

Membrane Preparation

Cell lysates and membrane fractions are prepared in ice-cold buffers supplemented with protease inhibitors [10 μg/ml leupeptin and pepstatin A, 100 m*M* ε-amino-*n*-caproic acid (EACA), 1 m*M* *p*-aminobenzamidine and 1 m*M* phenylmethylsulfonyl fluoride (PMSF)]. The initial screen for *P. pastoris* clones expressing Pgp is performed on *his*$^+$*mut*^{S} transformants as follows: Individual clones are inoculated in 50-ml polypropylene tubes containing 10 ml of MG medium and incubated overnight with strong agitation (250 rpm) at 30° up to an optical density (OD_{600}) of 1 to 2. Cells are recovered by centrifugation, resuspended in an identical volume of MM medium, and incubated at 30° for 48 hr to activate the *AOX1* promoter. Crude lysates of methanol-induced cells (5-ml equivalent) are prepared by vigorous vortexing of the cells with acide-washed beads (8 × 30-sec pulses at 4°) in 400 μl of a buffer containing 5% (v/v) glycerol, 50 m*M* sodium phosphate, pH 7.4, 1 m*M* ethylenediaminetetraacetic acid (EDTA), and protease inhibitors. Cell debris and glass beads are removed by centrifugation (10 min, 12,000*g*, 4°). Small-scale preparations of *P. pastoris* crude membranes are obtained from 100-ml cultures of induced cells according to a protocol we have previously described for Pgp expressed in *S. cerevisae*.[23]

For large-scale preparations of *P. pastoris* membranes, 1–3 liters of culture are obtained by consecutive growth and induction in MG and MM media, as described earlier, with the following modifications: 500-ml cultures are induced in 2-liter baffled flasks and methanol is replenished after 2 days. After 3 days of methanol induction, cells are pelleted (1500*g*) and resuspended in 1/20 volume of SM buffer (1.2 *M* sorbitol, 20 m*M* Tris-HCl, pH 7.0, and 5 m*M* EDTA) containing 20% glycerol and frozen at −80°. Crude membranes are prepared from these stocks according to the method of Perlin *et al.*[29] Briefly, frozen cells are rapidly thawed at 30° and concentrated by centrifugation to a density of 7.5 g/40 ml of ice-cold buffer A [0.33 *M* sucrose, 150 m*M* Tris-HCl, pH 7.4, 1 m*M* EGTA, 1 m*M* EDTA, 2 m*M* dithiothreitol (DTT) and protease inhibitors (Sigma, Oakville, Ontario, Canada)]. Cells are passed twice through a French press set at 20,000 psi, adding fresh PMSF to the samples every half-hour. Large cell debris and unbroken cells are removed by centrifugation (14,000*g*, 4°, 20 min) and crude membrane fractions are isolated from the supernatant by centrifuga-

[29] D. S. Perlin, S. L. Harris, D. Seto-Young, and J. E. Haber, *J. Biol. Chem.* **264,** 21857 (1989).

tion (200,000g, 4°, 90 min). Crude membrane pellets are resuspended in buffer B [10 m*M* Tris-HCl, pH 7.4, 1 m*M* EDTA, 10% (v/v) glycerol, and protease inhibitors], homogenized by three passages through a tight-fitting glass homogenizer, and concentrated by centrifugation (as described earlier). The washed crude membrane pellets were resuspended in 4 ml of ice-cold buffer B using a 23-gauge needle and layered on top of a discontinuous sucrose density gradient consisting of 16, 31, 43, and 53% (w/v) sucrose solutions[11,30] containing 10 m*M* Tris-HCl, pH 7.4, and 0.1 m*M* EDTA (TE), followed by centrifugation (53,000g, 4°, 16 hr). Membranes found at the various interfaces of the gradient are harvested, diluted to less then 0.25 *M* sucrose with TE, and recovered by centrifugation (200,000g, 4°, 3 hr). Membrane pellets are resuspended in buffer C (0.33 *M* sucrose, 40 m*M* Tris-HCl, pH 7.4, 1 m*M* EGTA, 1 m*M* EDTA, 2 m*M* DTT, and protease inhibitors) and kept frozen at −80°.

Protein concentration of samples is determined by the bicinchoninic acid protein assay[31] using bovine serum albumin (BSA) as a reference standard. Proteins are separated on a 7.5% gel by sodium dodecyl sulfate–polyacrylamide gel electrophoresis (SDS–PAGE) according to Laemmli.[32] After migration, gels are either stained with Coomassie Brilliant Blue R-250 or transferred to nitrocellulose. Immunodetection of Pgp is performed as previously described[23] using the anti-Pgp monoclonal antibody C219 (Signet Laboratories Inc., Dedham, MA) and immune complexes are revealed with the ECL (enhanced chemiluminescence) detection system (Amersham, Toronto, Ontario, Canada).

Photoaffinity Labeling with [^{125}I]Iodoarylazidoprazosin

For Pgp photoaffinity labeling experiments, crude membranes (250 μg/100 μl) are equilibrated with TE by passage through a Sephadex G-50 centrifuge column.[33] Membranes (2 μg) are incubated in the dark for 30 min with the drug analog [^{125}I]iodoarylazidoprazosin ([^{125}I]IAAP; 2200 Ci/mmol; Du Pont New England Nuclear, Boston, MA) at a concentration of 30 n*M*, in the absence or presence of excess vinblastine, colchicine, doxorubicin, or valinomycin. Membranes are then irradiated with UV for 1 min at 0°, as previously described.[34] Excess [^{125}I]IAAP is removed by

[30] J. R. Riordan and V. Ling, *J. Biol. Chem.* **254,** 12701 (1979).

[31] P. K. Smith, R. I. Krohn, G. T. Hermanson, A. K. Mallia, F. H. Gartner, M. D. Provenzano, E. K. Fujimoto, N. M. Goeke, B. J. Olson, and D. C. Klenk, *Anal. Biochem.* **150,** 76 (1985).

[32] U. K. Laemmli, *Nature* **227,** 680 (1970).

[33] H. S. Penefsky, *J. Biol. Chem.* **252,** 2891 (1977).

[34] S. Kajiji, F. Talbot, K. Grizzuti, V. Van Dyke-Phillips, M. Agresti, A. R. Safa, and P. Gros, *Biochemistry* **32,** 4185 (1993).

centrifugation (150,000g, 4°, 30 min) in a table-top ultracentrifuge [Beckman (Mississauga, Ontario, Canada); rotor TL-100]. Membrane pellets are dissolved in sample buffer [62.5 mM Tris-HCl, pH 6.8, 10% (v/v) glycerol, 5% (w/v) SDS, 5% (v/v) 2-mercaptoethanol, and 0.01% (w/v) bromphenol blue], incubated 30 min at 23°, and separated by SDS–PAGE on 7.5% gels. Gels are dried and exposed overnight at −60° to Kodak (Rochester, NY) X-AR films with one intensifying screen, or exposed for 4 hr to a phosphoimaging plate (type BASIII, Fuji) and analyzed with the Bio-Image Analyzer (Fuji, Mississauga, Ontario, Canada).

P-glycoprotein Solubilization and Reconstitution in Escherichia coli Lipids

P-glycoprotein is solubilized from the membrane fraction recovered at the interface of the 16% (w/v) and 31% (w/v) sucrose solutions of the step gradient (16/31 fraction) using deoxycholic acid sodium salt (DOC) by a modification of published protocols.[35] Briefly, 16/31 membranes are solubilized at a final concentration of 1 mg/ml in buffer D [0.33 M sucrose, 40 mM Tris-HCl, pH 7.4, 1 mM EGTA, 1 mM EDTA, 2 mM DTT, 1.4% (w/v) DOC, 20% (v/v) glycerol, 0.4% (w/v) *E. coli* lipids, and protease inhibitors] by gentle vortexing at room temperature. Once solubilized samples are ice-cooled and spun at 200,000g for 1 hr at 4° to remove insoluble material. *Escherichia coli* lipids, dissolved in buffer C, are then added to the solubilized fraction to a final concentration of 1.4% (w/v). Reconstitution is achieved by dialyzing the samples against 200 volumes of dialysis buffer (40 mM Tris-HCl, pH 7.4, 1 mM EGTA, 1 mM DTT, and 5 mM EACA) at 4° for 48 hr. Reconstituted samples are aliquoted and stored at −80°. The percentage of Pgp (w/w) present in the reconstituted samples is evaluated by laser densitometry of the samples separated on a 10% SDS–PAGE gel and stained with Coomassie Blue using the Sci-Scan 500 densitometer (United States Biochemicals, Cleveland, OH).

ATPase Assays

ATPase activity of the reconstituted fractions is estimated by measuring inorganic phosphate release by the colorimetric phosphate determination method of van Veldhoven and Mannaerts.[36] The basic ATPase cocktail consists of 40 mM Tris-HCl, pH 7.4, 10 mM adenosine triphosphate (ATP), and 10 mM $MgSO_4$. EGTA (0.1 mM) and sodium azide (10 mM) are added to eliminate contaminating Ca^{2+}-ATPase and F_1-ATPase activity, respectively. One to three micrograms of protein are added to 50 μl of

[35] D. Seto-Young, B. C. Monk, and D. S. Perlin, *Biochim Biophys. Acta* **1102,** 213 (1992).
[36] P. P. van Veldhoven and G. P. Mannaerts, *Anal. Biochem.* **161,** 45 (1987).

cocktail and incubated for 10–40 min at 37°. Both protein concentration and time of incubation are kept within the linear range of the ATPase reaction. For measurements of pH dependence of ATPase activity, the pHs of the cocktails are adjusted at 37° with 50 m*M* Tris–succinate (pH 4.7–8.5). P-Glycoprotein substrates or modulators are added from fresh stocks in dimethyl sulfoxide (DMSO).

Drugs and Chemicals

Most drugs and chemicals were purchased from Sigma (St. Louis, MO). Verapamil was obtained from ICN, valinomycin from Calbiochem (La Jolla, CA), and TPP from Aldrich (Madison, WI). Acetone/ether precipitated *E. coli* lipids were purchased from Avanti Polar Lipids (Birmingham, AL) and DOC from American Chemicals Ltd. (Montreal, Quebec, Canada).

Results

To express Pgp in *P. pastoris,* a full-length cDNA encoding the mouse Mdr3 Pgp isoform[28] was inserted into the expression vector pHILD2. This construct was designed to introduce the *mdr3* expression cassette at the *AOX1* chromosomal locus via a successful homologous recombination event. To favor this process, the pHILD2-*mdr3* plasmid was linearized with *Not*I. *Pichia pastoris* GS115 spheroplasts were transformed with the digested plasmid and his^+ transformants were selected. Individual his^+ clones were further screened to identify those with impaired growth in the presence of methanol as a sole carbon source (methanol utilizing slow or mut^S). These his^+mut^S clones should harbor a copy *mdr3* integrated at the *AOX1* locus and produce high levels of *mdr3* mRNA and of the Mdr3 Pgp isoform following methanol induction. Out of 50 his^+ clones initially selected, 11 were impaired for growth in methanol-containing medium (mut^S). Whole-cell extracts from methanol-induced cultures from these 11 mut^S clones were screened by Western blot analysis with the anti-Pgp monoclonal antibody C219. Six of these expressed an immunoreactive 140-kDa Pgp (data not shown). To confirm that this new protein corresponded to membrane-associated Pgp, crude membranes were prepared from the six positive clones (Fig. 1A, lanes 1–5, 7), from a control his^+mut^S pHILD2-*mdr3* transformant not expressing Pgp (Fig. 1A, lane 6), and from a his^+mut^S clone transformed with pHILD2 alone (Fig. 1A, lane 8). This analysis revealed the presence of an abundant 140-kDa species in the membrane fraction of the six positive clones that was readily visible by staining the gel with Coomassie blue (Fig. 1A, lanes 1–5, 7). Immunoblotting of this gel with either the mouse anti-Pgp monoclonal antibody C219 (Fig. 1B) or the rabbit isoform-specific anti-mouse Mdr3 polyclonal antibody B2037[28]

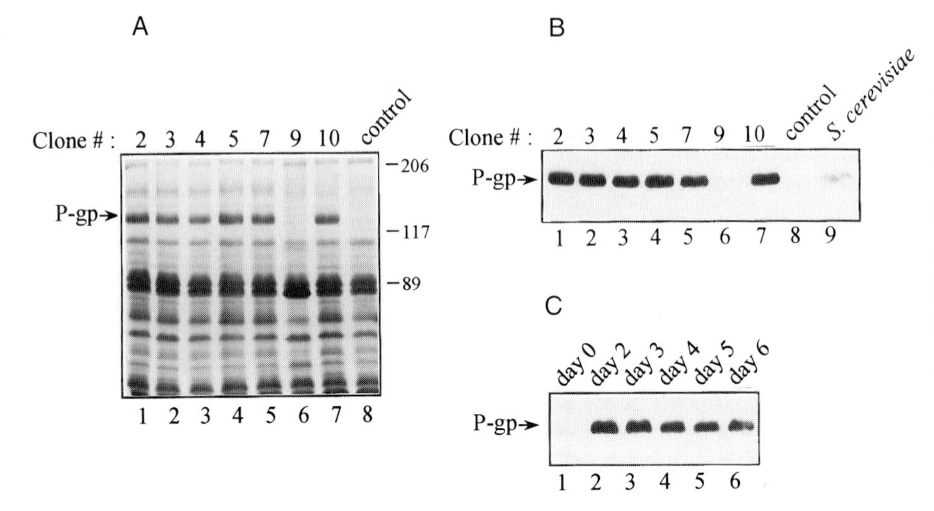

FIG. 1. Detection of mouse Mdr3 in membrane preparations from *P. pastoris.* (A) Crude membranes were prepared from pHIL-*mdr3 his*$^+$*mut*^{S} transformants (lanes 1–7), and from a control clone transformed with the pHILD2 vector alone (lane 8). Protein (15 μg) was loaded in each lane, separated by SDS–PAGE, and stained with Coomassie blue. (B) Western blot analysis of the samples shown in (A) (lanes 1 to 8) using the anti-Pgp monoclonal antibody C219. For comparison, crude membranes from a Pgp expressing *S. cerevisiae* clone (15 μg) were included (lane 9). (C) Time course of methanol induction of mouse Mdr3 expression (clone 5) as revealed by Western blot analysis. At day 0, a sample was taken before induction.

(data not shown) confirmed that the 140-kDa protein was indeed the mouse Mdr3 Pgp. The electrophoretic mobility of Mdr3 expressed in *P. pastoris* and in *S. cerevisiae* was very similar (Fig. 1B, lane 9), suggesting that full-length Mdr3 is expressed in *P. pastoris* and that it is probably not glycosylated, as opposed to Pgp expressed in mammalian cells.[37] A time course of methanol induction of Pgp expression in *P. pastoris* is shown in Fig. 1C: Induction of Pgp expression is completely dependent on the presence of methanol (undetected in glycerol-containing medium; Fig. 1C, lane 1), and is maximal after 2–3 days of methanol induction (Fig. 1C, lanes 2 and 3). Although significant expression of Pgp was detected after only 16 hr of methanol induction (not shown), we found that 2–3 days of methanol induction were necessary to get maximal yield of total membrane proteins.

Functional Pgp can be specifically labeled by a number of photoactivatable drug analogs.[38] including [^{3}H]azidopine and [^{125}I]IAAP. To verify that Pgp expressed in the membrane fraction of *P. pastoris* was inserted and

[37] E. Georges, J.-T. Zhang, and V. Ling, *J. Cell. Physiol.* **148,** 479 (1991).
[38] A. R. Safa, *Cancer Invest.* **10,** 295 (1992).

folded in a biologically active conformation, we determined if it could be photolabeled with [^{125}I]IAAP. For this, crude membrane fractions from Pgp expressing and control *P. pastoris* clones were cross-linked to [^{125}I]IAAP by UV light, the excess label was removed, and the pelleted membranes were analyzed by SDS–PAGE and autoradiography (Fig. 2A). These experiments showed that a polypeptide of 140 kDa present in Pgp-expressing clones (Fig. 2A, lane 1) and absent in control membranes (Fig. 2A, lane 6) was specifically labeled by [^{125}I]IAAP. [^{125}I]IAAP labeling of Pgp appeared specific since it was competed by excess of Pgp substrates vinblastine and valinomycin (Fig. 2A, lanes 2 and 5). Excess of MDR drugs colchicine (Fig.

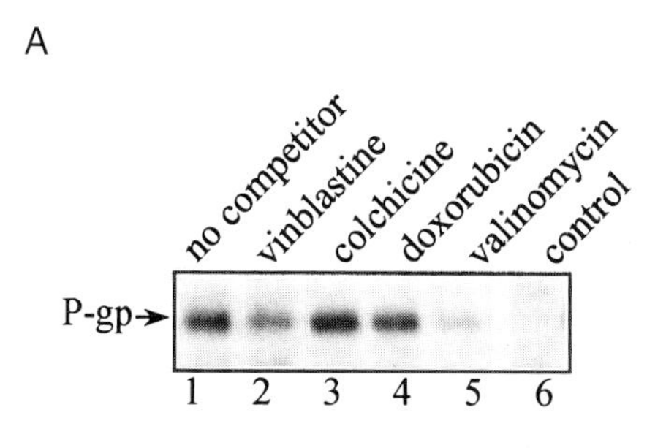

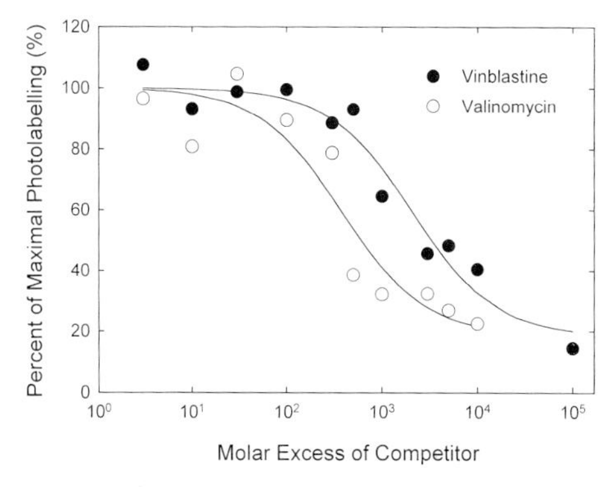

FIG. 2. Photolabeling by [^{125}I]IAAP of mouse Mdr3 in crude membranes from a pHIL-*mdr3* transformant (clone 5). (A) Photolabeling in the absence (lane 1) or presence of 1000-fold excess of vinblastine (lane 2), colchicine (lane 3), doxorubicin (lane 4), or valinomycin (lane 5). Photolabeling of crude membranes from a control clone transformed with the pHILD2 vector alone (lane 6). (B) Competition of photolabeling of mouse Mdr3 by increasing concentrations of vinblastine (●) or valinomycin (○). Photolabeling of mouse Mdr3 was quantified using the Bio-Image Analyzer (Fuji) after a 4-hr exposure of the dried SDS–PAGE gels to a phospho-imaging plate (Fuji imaging plate type BASIII).

2A, lane 3) and doxorubicin (Fig. 2A, lane 4) had little if any effect on [^{125}I]IAAP labeling of Pgp. A more detailed analysis of the competitive behavior of vinblastine and valinomycin on [^{125}I]IAAP labeling of Mdr3 is shown in Fig. 2B and suggests a 50% decrease (D_{50}) in binding caused by a 600 molar excess of valinomycin and a 3300 molar excess of vinblastine. The D_{50} measured for vinblastine is somewhat higher than that reported for Pgp expressed in mammalian cells membranes[39,40] or in purified plasma membranes from Pgp expressing *S. cerevisiae.*[20] Nevertheless, these results suggest that Mdr3 expressed in *P. pastoris* membranes is properly folded and capable of binding drug analogs with characteristics similar to those observed for Pgp expressed in other systems.

Specific properties of Pgp ATPase activity such as pH optimums and drug stimulation profiles are well established and were therefore used to monitor the enzymatic activity of the mouse Mdr3 Pgp isoform expressed in *P. pastoris.* In a first experiment, the total ATPase activity associated with crude membranes of control and Pgp expressing *P. pastoris* clones was measured and compared at different pH (data not shown). In this assay, the levels of total ATPase activity measured in Pgp-rich and control membranes were nearly identical. The optimum ATPase activity was around pH 6.0, suggesting that the yeast H^+-ATPase (PMA-1), which has optimal activity between pH 5.5 and 6.5 (reviewed in Serrano[41]), was the major ATPase present. At the reported optimal pH values for rodent Pgp ATPase (pH 7.5[11,13]), there was no difference between the Pgp-rich and control samples. To enrich the crude membrane preparations for Pgp and concomitantly reduce the level of contaminating endogenous yeast ATPases, crude membranes from control and Pgp-expressing cells were fractionated by centrifugation on a discontinuous sucrose gradient. Membrane fractions accumulating at the different interfaces were collected and analyzed for the presence of Pgp by immunoblotting. Under our conditions, Pgp was localized predominantly at the interface of the 16 and 31% sucrose solutions (16/31 interface) (Fig. 3A, lane 2) while very little was found in the other fractions. The total ATPase activity associated with membranes found at the 16/31 and 43/53 interfaces was measured at different pH. The Pgp positive 16/31 fraction showed a level of ATPase activity that was half of the maximal ATPase activity measured in the 43/53 membranes (0.7 mmol/mg/min versus 1.4 mmol/mg/min), but the two fractions still showed

[39] L. M. Greenberger, C. P. Yang, E. Gindin, and S. B. Horwitz, *J. Biol. Chem.* **265,** 4394 (1990).

[40] A. R. Safa, M. Agresti, I. Tamai, N. D. Mehta, and S. Vahabi, *Biochem. Biophys. Res. Commun.* **166,** 259 (1990).

[41] R. Serrano, *in* "The Molecular and Cellular Biology of the Yeast Saccharomyces: Genome Dynamics, Protein Synthesis, and Energetics" (J. R. Broach, J. R. Pringle, and E. W. Jones, eds.), p. 523. Cold Spring Harbor Laboratory Press, New York, 1991.

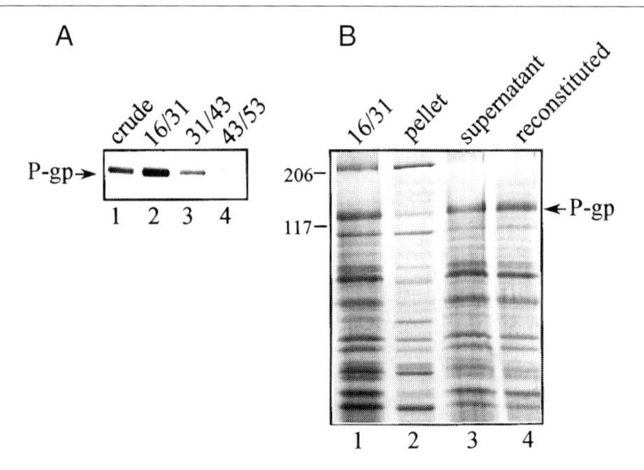

FIG. 3. (A) Fractionation of crude membranes from an Mdr3-expressing clone (lane 1) by density gradient centrifugation. Proteins (10 μg) from crude membranes and from the 16/31, 31/43, and 43/53 membrane fractions were separated by SDS–PAGE and analyzed by Western blot using the anti-Pgp antibody C219. (B) Solubilization and reconstitution of Mdr3-enriched membranes. Membranes from the 16/31 sucrose gradient interface (lane 1) were solubilized in 1.4% DOC, separated from the insoluble pellet (lane 2) by centrifugation, and the detergent extract (lane 3) was reconstituted in *E. coli* lipids by dialysis (lane 4). Proteins (10 μg) from the original 16/31 membrane fraction and the corresponding amount of protein from the insoluble, soluble, and reconstituted fractions were separated by SDS–PAGE and stained with Coomassie blue.

an optimum for ATPase activity around pH 6.0. Although previous studies had reported that the yeast PMA-1 was mostly associated with membranes sedimenting at the 43/53 interface,[42] the data suggest that significant levels of contaminating endogenous ATPases remain in the 16/31 fraction and interfere with the detection of the specific Pgp ATPase activity. Again, we did not detect any difference in ATPase activity between control and Pgp-expressing membranes in those fractions (data not shown). We also analyzed the Pgp-enriched 16/31 membrane fractions for the presence of ATPase activity inducible by verapamil and valinomycin. No such activity could be detected suggesting that contamination by endogenous ATPase was masking drug stimulation of the Pgp ATPase activity.

It has been previously reported that yeast plasma membrane H^+-ATPase (PMA-1) is not soluble in the detergent deoxycholate (DOC),[35] while Pgp is readily extracted from mammalian cells membranes by low concentration of this detergent.[7] We tried to take advantage of the different solubilization properties of the two enzymes to separate Pgp from PMA-1 in our membrane preparations. Membranes (16/31 interface) were extracted with 1.4%

[42] R. Serrano, *Methods Enzymol.* **157,** 533 (1988).

(w/v) DOC and the insoluble material was removed by centrifugation at 200,000*g*. Both soluble and insoluble fractions were analyzed by SDS–PAGE. Simple DOC solubilization was found to be very effective for extracting Pgp (Fig. 3B, lane 3) because very little of the 140-kDa Pgp remained associated within the insoluble pellet (Fig. 3B, lane 2). The solubilized proteins were then reconstituted in *E. coli* lipids by detergent dialysis (Fig. 3B, lane 4). Control membranes were solubilized and reconstituted by the same procedure and assayed for ATPase activity. The insoluble fraction of the control membranes retained characteristics (0.8 μmol/mg/min peaking at pH 6) similar to those measured in the crude membrane fractions, while the reconstituted soluble fraction showed very little remaining ATPase activity (data not shown). This indicated that the major contaminating ATPase activity of yeast membranes had been retained in the insoluble pellet. The ATPase activity of the reconstituted fractions from Pgp-expressing and control samples was then assayed in the presence of 100 μM verapamil (Fig. 4). Verapamil stimulated the reconstituted Pgp membranes' ATPase activity (Fig. 4) with a maximal specific activity of 160 nmol/min/mg observed at pH 7.7. The maximal specific activity achieved represents a 4.4-fold increase over background level. The pH optimum is very close to the value of pH 7.5 previously reported for the hamster isoform.[13] The absence of verapamil stimulation in control membranes that were solubilized and reconstituted by the same procedure (Fig. 4)

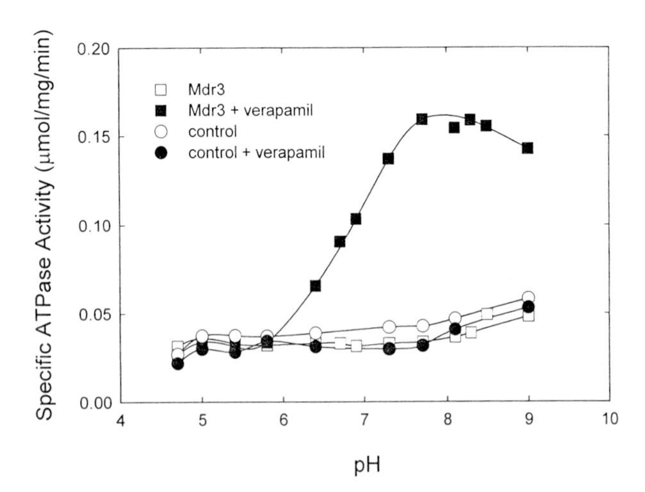

FIG. 4. ATPase activity of the reconstituted 16/31 membrane fractions from a Mdr3-expressing (□, ■) and from a control clone (○, ●) were measured at different pH in the absence (□, ○) or presence (■, ●) of 100 μM verapamil. Both membrane fractions were solubilized and reconstituted in *E. coli* lipids as outlined in Fig. 3.

demonstrated that Pgp was functional and indeed responsible for the observed drug-induced ATPase activity. In contrast to the human[6,8] and hamster[10,11,13] Pgp, we could not detect a basal level of Pgp ATPase activity in our reconstituted fraction, because non-drug-stimulated Pgp fractions showed the same background activity as control membranes, around 30 nmol/mg/min (Fig. 4). This could reflect (1) an intrinsic property of the mouse Mdr3 isoform, (2) an inhibitory effect of detergent and/or lipids used in our protocol, or (3) a level of basal activity too low to be measured over the background of residual yeast ATPases present in the reconstituted preparations.

The drug-stimulatable ATPase activity of the Mdr3 Pgp isoform reconstituted by the above-mentioned procedure was further analyzed for various MDR drugs and Pgp modulators. The patterns and magnitude of stimulation noted in these experiments were in good agreement with those described previously for the same compounds on other Pgp isoforms. The best activators (Fig. 5) were found to be valinomycin (5.4-fold at 100 μM), verapamil (3.8-fold at 100 μM), and vinblastine (2.6-fold at 30 μM). The characteristic activation at lower concentrations and inhibition at higher concentrations (bell-shaped curves) noted for verapamil and vinblastine had been previously observed for the hamster and human Pgps.[6,10,11,13,14] The plateau reached at high valinomycin concentrations has also been

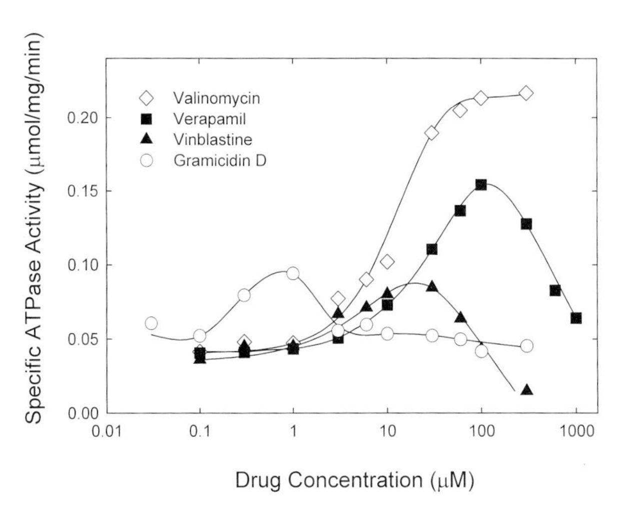

FIG. 5. Drug stimulation of the Mdr3-specific ATPase activity by valinomycin (◇), verapamil (■), vinblastine (▲), and gramicidin D (○). The 16/31 membrane fractions were solubilized and reconstituted in *E. coli* lipids as outlined in Fig. 3. Results are the mean value of independent experiments with two different preparations.

described for the hamster isoform.[43,44] Gramicidin D, in contrast to previously reported results describing its inhibitory effect on Pgp ATPase activity,[44] was stimulatory at low concentration, with a 2.3-fold stimulation at 1 μM (Fig. 5). Other known Pgp substrates such as tetraphenylphosphonium (TPP) (2.1-fold at 100 μM), diethylstilbestrol (DES) (2.6-fold at 300 μM), and FK506 (1.7 fold at 10 μM) also had a stimulatory effect on Pgp ATPase activity. Actinomycin D, doxorubicin, and colchicine did not stimulate Mdr3 ATPase activity in our system, in accordance with similar findings reported for hamster Pgp reconstituted in *E coli* lipids.[45]

A K_m for MgATP of 0.49 mM was measured for the mouse Mdr3 Pgp isoform in the presence of verapamil (data not shown). This low affinity for MgATP is similar to values determined for the human MDR1 and hamster Pgp1 isoforms, for which K_m values between 0.3 and 1.5 have been reported.[6,8,10–13] V_{max} of the verapamil-stimulated ATPase was 170 nmol/mg/min. Laser densitometry of Coomassie blue-stained SDS–PAGE gel revealed a Pgp content of 5% (w/w) in the reconstituted membranes, suggesting an extrapolated verapamil-induced specific activity for Mdr3 Pgp of 3.4 μmol/mg/min. This value is comparable to the verapamil-stimulated activity recently reported for purified hamster Pgp reconstituted in *E coli* lipids.[45] Similar results were obtained in the presence of valinomycin with a K_m for ATP of 0.44 mM and a V_{max} of 210 nmol/mg/min, which extrapolates to a specific activity of 4.2 μmol/mg/min for a pure Pgp preparation. Finally, the inhibitory effect of vanadate, which has been extensively characterized for the human MDR1[6,8] and hamster Pgp1[10–13,46,47] isoforms was measured for Mdr3 in our reconstituted fraction. We determined that 50% inhibition of the drug-induced ATPase activity of Mdr3 was achieved at 0.9 μM of vanadate (data not shown). This value compared well with that obtained for the purified hamster isoform by Shapiro and Ling[12] (0.8 μM).

Discussion

We report the high-level expression of the mouse Pgp isoform Mdr3 in the yeast *P. pastoris.* A simple protocol based on membrane fractionation on a discontinuous sucrose gradient, selective detergent solubilization, and reconstitution in *E. coli* lipids is presented for the rapid characterization of the ATPase activity of Pgp. Using this system, we demonstrated that

[43] F. J. Sharom, G. DiDiodato, X. Yu, and K. J. D. Ashbourne, *J. Biol. Chem.* **270,** 10334 (1995).
[44] M. J. Borgnia, G. D. Eytan, and Y. G. Assaraf, *J. Biol. Chem.* **271,** 3163 (1996).
[45] I. L. Urbatsch and A. E. Senior, *Arch. Biochem. Biophys.* **316,** 135 (1995).
[46] I. L. Ubratsch, B. Sankaran, J. Weber, and A. E. Senior, *J. Biol. Chem.* **270,** 19383 (1995).
[47] I. L. Urbatsch, B. Sankaran, S. Bhagat, and A. E. Senior, *J. Biol. Chem.* **270,** 26956 (1995).

the mouse Mdr3 Pgp isoform has an ATPase activity that is drug inducible and vanadate sensitive, and which has a relatively low K_m for MgATP (0.49 m*M*). Mouse Mdr3 ATPase activity is stimulated by several MDR drugs and Pgp modulators, including valinomycin, verapamil, vinblastine, gramicidin D (Fig. 5), as well as DES, TPP, and FK506. All the characteristics of the mouse Mdr3 ATPase are very similar to those observed for human MDR1[6,8] and hamster Pgp1.[10–14] More importantly, this rapid and inexpensive protocol should prove very useful for the high-throughput analysis of mutant Pgp variants generated by site-directed mutagenesis. This protocol is also easily amenable, on inclusion of additional purification steps, to the isolation in pure form of large amounts of Pgp for structural studies. Finally, this system should be applicable to parallel studies of other medically important members of the ABC superfamily of transporters such as MRP, the protein associated with resistance to chemotherapy in lung tumors,[48] Pgh-1, the product of the *pfmdr1* gene of *Plasmodium falcipartum* which is associated with resistance to 4-aminoquinoline drugs in the malarial parasites,[49] and the cystic fibrosis transmembrane conductance regulator (*CFTR*) gene product, mutations in which cause cystic fibrosis in humans.[50]

A major obstacle in measuring the ATPase activity of the *P. pastoris*-expressed Pgp was the presence of strong contaminating ATPases in the yeast membrane fractions. Experimental evidence, including the optimal pH of the contaminant ATPase and its resistance to deoxycholate extraction, suggests that the bulk of this activity was generated by endogenous PMA-1. Although sodium azide could readily inhibit mitochondrial F_1-ATPase at a concentration known not to affect Pgp ATPase (10 m*M*[11]), eliminating the PMA-1 activity by the inclusion of inhibitors in the assay reaction proved to be much more problematic. Indeed, *N*-ethylmaleimide (NEM), 7-chloro-4-nitrobenz-2-oxa-1,3,-diazole (NBD-chloride), and vanadate all inhibit PMA-1 in concentration ranges similar to those known to inhibit mammalian Pgp.[51–53] Certain combinations of PMA-1 inhibitors, for example, 100 μM *N*,*N*′-dicyclohexylcarbodiimide (DCCD) plus 30 μM

[48] G. J. R. Zaman and P. Borst, *in* "Multidrug Resistance in Cancer Cells: Molecular, Biochemical, and Biological Aspects" (S. Gupta and T. Tsuruo, eds.), p. 95. John Wiley & Sons Ltd., Chichester, 1996.

[49] P. Borst and M. Ouellette, *Annu. Rev. Microbiol.* **49,** 427 (1995).

[50] J. R. Riordan, J. M. Rommens, B.-S. Kerem, N. Alon, R. Rozmahel, Z. Grzelezak, J. Zielenski, S. Lok, N. Plavsic, J.-L. Chou, M. L. Drumm, M. C. Iannuzzi, F. S. Collins, and L.-C. Tsui, *Science* **245,** 1066 (1989).

[51] R. J. Brooker and C. W. Slayman, *J. Biol. Chem.* **257,** 12051 (1982).

[52] A. Wach and P. Graber, *Eur. J. Biochem.* **201,** 91 (1991).

[53] M. K. Al-Shawi, I. L. Urbatsch, and A. E. Senior, *J. Biol. Chem.* **269,** 8986 (1994).

DES,[54] seemed to reveal a Pgp-related ATPase activity in Pgp-expressing membranes but this ATPase activity was insensitive to MDR drugs or modulators, and therefore not truly representative for Pgp (data not shown). The reason for this behavior is unclear but could be linked to the effect of DES on Pgp ATPase activity: DES is a synthetic hormone that can stimulate Pgp ATPase activity (our unpublished results and Rao *et al.*[55]) but it also reduces the stimulatory effects of compounds such as verapamil (unpublished observations).

Our results clearly indicate that the more practicable strategy to circumvent the problem of contaminating ATPase is to perform a sucrose gradient enrichment of the crude membranes followed by a simple selective detergent extraction procedure. Yeast PMA-1 is known to sediment preferentially at high sucrose concentrations (43/53 interface[42]), while yeast-expressed mouse Mdr3 is recovered predominantly at the 16/31 interface (Fig. 3A). However, under our experimental conditions, the 16/31 membrane fraction still retained a high level of contaminating ATPase activity (data not shown), prompting us to attempt differential detergent solubilization of the two ATPase activities. Preliminary experiments had shown that octyl glucoside, dodecyl maltoside, and CHAPS were all ineffective to extract Pgp from the yeast 16/31 membranes, while bile acids such as cholate, DOC, and taurocholate readily extracted a large proportion of Pgp present in this membrane fraction (Fig. 3B, and unpublished observations). Because it was previously reported that PMA-1 is insoluble in DOC,[35] this bile acid was used for the selective solubilization of Pgp from the yeast 16/31 membrane fraction. After reconstitution in *E. coli* lipids, drug-stimulatable Pgp ATPase activity could be easily measured and characterized, without need for further purification of the enzyme.

The high levels of Pgp expression achieved in *P. pastoris* using the strong methanol-inducible *AOX1* gene promoter were similar for all clones tested (Fig. 1A), suggesting a fairly low degree of clonal variation in the amount of protein synthesized and expressed in the membrane. In reconstituted 16/31 membranes, Pgp was found to account for approximately 5% of the total membrane proteins (Fig. 3B); this number is comparable to the level of Pgp expressed in baculovirus-infected Sf9 cells (3%[6]), the most widely used system for the characterization of the ATPase activity of mutant Pgps and other ABC transporters. However, expression in *P. pastoris* is achieved at only a fraction of the cost of the baculovirus protocol. The levels of mouse Mdr3 expression achieved in the *P. pastoris* expression system seem to be vastly superior to those reached for the same protein

[54] R. Serrano, *Eur. J. Biochem.* **105,** 419 (1980).

[55] U. S. Rao, R. L. Fine, and G. A. Scarborough, *Biochem. Pharmacol.* **48,** 287 (1994).

in *S. cerevisiae*.[21] In our experience with *S. cerevisiae*, even the strong alcohol dehydrogenase gene promoter only allowed Mdr3 expression at the level of less than 0.5% of the total membrane protein fraction, which was too low to detect any Pgp-related ATPase activity.[21] In an independent study of the ABC transporter CFTR[56] expressed in *S. cerevisiae* under the transcriptional control of the *PMA-1* gene promoter, the heterologous protein was estimated to account for only 0.5% of total membrane protein.

In addition to high-level protein expression, a yeast heterologous expression system offers several advantages. Manipulating *P. pastoris* cells is simple and inexpensive and the cloning, transformation, and selection procedures are fast and well established. As eukaryotes, yeast cells usually perform posttranslational modifications and processing steps that increase the likelihood of functional expression of mammalian transport proteins in the membrane, which is an advantage over prokaryotic expression systems from which equivalent levels of expression could be achieved. In conclusion, the *P. pastoris* system is superior in several aspects to the currently available expression systems for production of Pgp and perhaps other ABC transporters. We are currently using this system to characterize a large series of Mdr3 mutants showing alterations in TMDs and NBDs. We are also initiating large-scale purification of Pgp from these cells for structural studies.

Acknowledgments

This work was supported by a grant (to P.G.) from the National Cancer Institute of Canada. L.B. was supported by a fellowship from the Medical Research Council of Canada (MRC) and P.G. by a senior Scientist Award from the MRC. The authors are indebted to Dr. R. MacGillivray (University of British Columbia) for suggesting the *P. pastoris* expression system, and to the department of Microbiology/Immunology (McGill University) for the use of their French press. The help of Greg Mowick for the manipulation of the French press was greatly appreciated.

[56] P. B. Huang, K. Stroffekova, J. Cuppoletti, S. K. Mahanty, and G. A. Scarborough, *Biochim. Biophys. Acta* **1281,** 80 (1996)

[30] Mutational Analysis of P-Glycoprotein in Yeast *Saccharomyces cerevisiae*

By LUCILLE BEAUDET and PHILIPPE GROS

Introduction

P-Glycoprotein (Pgp) expression in cultured cells and in tumor cells *in vivo* has been associated with the emergence of multidrug resistance or MDR.[1] In these cells, Pgp causes a reduction of intracellular drug accumulation through an enhanced drug efflux. So far, our understanding of Pgp has relied on immunological, biochemical, and genetic analyses of the protein performed in well-controlled experimental settings. *In vitro* genetic analysis using mutant and chimeric forms of Pgps has been used to define essential residues and protein domains involved in drug binding, ATP binding, and hydrolysis as well as to elucidate the membrane topology of the protein.[2–5]

Mutant Pgps have been studied in transfected mammalian host cells such as Chinese hamster ovary (CHO) cells, NIH 3T3 cells, and others.[6,7] Although such analysis is adequate for the study of a few discrete Pgp mutants, it is both an expensive and impractical approach for a systematic genetic analysis involving the characterization of a large number of mutants. Our laboratory has therefore turned to the yeast *Saccharomyces cerevisiae* to express and characterize wild-type and mutant variants of murine Pgps. All three mouse Pgp isoforms (Mdr1, Mdr2, and Mdr3) can be stably expressed in yeast, where they are properly targeted to the membrane. Yeast cells expressing wild-type Pgp acquire cellular resistance to well-known anticancer drugs such as Adriamycin[8] and actinomycin D[9]; to the antifungal drugs FK520 and FK506[6,10]; and to the ionophore valinomycin.[9,11]

[1] C. Shustik, W. Dalton, and P. Gros, *Molec. Aspects Med.* **16,** 1 (1995).
[2] X. Zhang, K. I. Collins, and L. M. Greenberger, *J. Biol. Chem.* **270,** 5441 (1995).
[3] M. Azzaria, E. Schurr, and P. Gros, *Mol. Cell. Biol.* **9,** 5289 (1989).
[4] C. Kast, V. Canfield, and R. Levenson, *Biochemistry* **34,** 4402 (1995).
[5] T. W. Loo and D. M. Clarke, *J. Biol. Chem.* **270,** 843 (1995).
[6] L. Beaudet and P. Gros, *J. Biol. Chem.* **270,** 17159 (1995).
[7] T. W. Loo and D. M. Clarke, *J. Biol. Chem.* **268,** 3143 (1993).
[8] S. Ruetz, M. Brault, C. Kast, C. Hemenway, J. Heitman, C. E. Grant, S. P. C. Cole, R. G. Deeley, and P. Gros, *J. Biol. Chem.* **271,** 4154 (1996).
[9] K. Ueda, A. M. Shimabuku, H. Konishi, Y. Fujii, S. Takebe, K. Nishi, M. Yoshida, T. Beppu, and T. Komano, *FEBS Lett.* **330,** 279 (1993).
[10] M. Raymond, S. Ruetz, D. Y. Thomas, and P. Gros, *Mol. Cell. Biol.* **14,** 277 (1994).
[11] K. Kuchler and J. Thorner, *Proc. Natl. Acad. Sci. U.S.A.* **89,** 2302 (1992).

METHODS IN ENZYMOLOGY, VOL. 292 0076-6879/98 $25.00

Moreover, structural similarity between Pgp[12] and its yeast homolog STE6 can translate into functional homology since Pgp (and also the ABC transporters pfMDR1[13] and MRP[8]) can partly complement the biological activity of the yeast transporter, by restoring mating in a *ste6* null mutant. Yeast-expressed Pgps can also be photolabeled by drugs[10] and ATP analogs (unpublished). We previously showed that mutations that are known to decrease drug transport by Pgp in CHO cells also impair in yeast both *ste6* complementation and drug resistance.[6,10] In our laboratory, we measure the effect of mutant Pgp expression on cellular drug resistance and complementation of a *ste6* deletion in *S. cerevisiae* cells.

P-Glycoprotein is composed of two homologous half proteins, with each half containing a highly conserved nucleotide-binding domain (NBD) of the Walker type[14] and a less conserved hydrophobic membrane-associated domain. In this article, we review the construction of a group of mutants and chimeras of the mouse Mdr3 Pgp isoform that were created to determine if the high degree of sequence identity between the two Pgp NBDs translates into functional homology, and to measure the ability of each NBD to function within the context of each half protein. In these mutants and chimeras, discrete Pgp segments and individual residues from the amino-terminal NBD (NBD1) were replaced with the corresponding ones found in the carboxy-terminal NBD (NBD2).[6] We present some of the results obtained with the NBD1 chimera series in yeast (drug resistance and mating efficiency) and compare them to those obtained from parallel expression studies performed in CHO cells with the same constructs. Using this approach, we identified a set of NBD1 residues that cannot be replaced with their NBD2 homologs without loss of function. In addition, mutations at these NBD positions affect the substrate specificity of the transporter.[6]

Mutagenesis Strategy

The following strategy was used for the construction and biological testing of the activity of mouse Mdr3 NBD1 chimeras and mutants: All NBD1 modifications were made within the limits of a "mutation cassette" defined by unique restriction sites artifically created by silent site-directed mutagenesis. This cassette was then used to reintroduce in a single step mutant and chimeric NBD1s back into the *mdr3* cDNA in expression vectors pEMC2b (generous gift from Dr. R. Kaufman, Yale University, New Haven CT) for studies in CHO cells (pEMC-*mdr3*), and pVT101-U[15]

[12] M. Raymond, P. Gros, M. Whiteway, and D. Y. Thomas, *Science* **256,** 232 (1992).
[13] S. K. Volkman, A. F. Cowman, and D. F. Wirth, *Proc. Natl. Acad. Sci. U.S.A.* **92,** 8921 (1995).
[14] J. E. Walker, M. Saraste, M. J. Runswick, and N. J. Gay, *EMBO J.* **1,** 945 (1982).
[15] T. Vernet, D. Dignard, and D. Y. Thomas, *Gene* **52,** 225 (1987).

for studies in yeast cells (pVT-*mdr3*). Chimeric NBD1 were generated by a three-step polymerase chain reaction (PCR) procedure while point mutations were created by site-directed mutagenesis using a commercially available kit (Amersham, Toronto, Ontario, Canada).

Mutation Cassettes

Mutation cassettes are gene segments encompassing domains to be mutated. They are flanked by unique restriction enzyme (RE) sites facilitating the replacement of a wild-type segment by its mutated versions in a single cloning step. Mutation cassettes should remain relatively short (the shorter the better), because each new modified cassette has to be sequenced to ensure that PCR or mutagenesis procedures did not introduce unwanted mutations. These cassettes are delineated by unique RE sites either naturally occurring in the cDNA or artificially introduced by mutagenesis. To allow a rigorous comparison with the wild-type protein, novel RE sites should not modify the amino acid sequence of the protein. For the construction of *mdr3* NBD1 mutants, a mutation cassette encompassing NBD1 was created by the introduction of two new RE sites by silent mutagenesis. In a first step, putative unique RE sites were selected from a list of enzymes that do not cut in *mdr3*, pVT101-U, or pEMC2b sequence. To determine which of these RE sites could be introduced without altering the protein sequence, each candidate site was decomposed into the three possible mRNA reading frames and a list of all possible amino acids for each codon was assembled. For example, in the case of an *Nru*I RE site (5′-TCGCGA-3′), the three reading frames are:

	Frame	Amino acids (one-letter code)
1	XXU CGC GAX	(F/S/Y/C/L/P/H/N/I/T/R/V/G/A/D, R, D/E)
2	XUC GCG AXX	(F/L/I/V, A, I/M/T/N/K/S/R)
3	UCG CGA	(S, R)

In reading frame 1 (XXU CGC GAX), the first amino acid has to be encoded by a codon that ends with a U (F/S/Y/C/L/P/H/N/I/T/R/V/G/A/D are all possible). The second amino acid is defined by the codon CGC (R) and the third is encoded by GAX codons (either D or E). In the second reading frame (XUC GCG AXX), the first amino acid codon should end with UC (F/L/I/V); the second amino acid, encoded by the GCG codon, is unambiguous and encodes an A; and the third amino acid is encoded by AXX codons (I/M/T/N/K/S/R). In the last reading frame, the only possibility is S (UCG) followed by R (CGA). A scan for all these potential amino acid combinations is then performed on the amino acid sequence of the protein area where the mutation cassette will be established. Once putative location and sites are identified, the nucleotide context of the new RE site should be checked in both DNA strands to ensure that it will not be part of a methylation-sensitive sequence, possibly interfering with subsequent RE cleavage.

In the case of *mdr3*, the sequence TCT CGA, encoding the dipeptide S-R, is found at position 1346, immediately upstream from the Walker A motif of NBD1. According to reading frame 3 for the *Nru*I RE site sequence (see earlier discussion), a single change from TCT to TCG generates the *Nru*I site without modifying the dipeptide sequence. The sequence TTG TCT ACC encoding the tripeptide L-S-T is found downstream from the Walker B motif (position 1906). There, a unique *Sal*I site (5′..G TCG AC..3′) is introduced by the simple change of the TCT codon to TCG while retaining the L-S-T sequence. Although the search within the amino acid sequence of a protein for such decomposable di- or tripeptides and the corresponding possible RE sites can be done manually, helpful listings of these sequences have been published (e.g., New England Biolabs, Beverly, MA).

Site-Directed Mutagenesis

A comprehensive review of methods for site-directed mutagenesis (M13 single-strand mutagenesis and megaprimer-mediated PCR) has been published,[16] and is not reviewed in detail here. In our laboratory, we routinely use the Sculptor *in vitro* mutagenesis kit (Amersham) to introduce point mutations. Although commercially available kits will usually give a high yield of the desired mutant clones, they are relatively expensive and only provide reagents for a limited number of mutagenesis reactions. When many mutations at the same codon or for neighboring codons are planned, individual mutant oligonucleotides can be pooled into the same reaction mixture, relying on DNA sequencing for the subsequent identification of individual mutations. Also, when a single template DNA has to undergo more than one mutation, two mutant oligonucleotides can be used in the same reaction, avoiding the extra step of mutating the template a second time.

For the introduction of *Nru*I and *Sal*I restriction sites into *mdr3*, a 2.2-kb *Eco*RI *mdr3* fragment (5′ half: polylinker to position 2249) was subcloned into phage vector M13mp18, and silent mutations creating *Nru*I (position 1346) and *Sal*I sites (position 1908) were introduced. The protocols for preparation of vector and oligonucleotides, and the conditions for annealing, elongation, ligation, and transformation, were exactly as

[16] I. G. Macara and W. H. Bronduk, *Methods Enzymol.* **257,** 107 (1995).

described in the manual provided with the mutagenesis kit. Screening for positive mutant inserts was done by looking for the appearance of novel RE sites in double-stranded DNA prepared from individual M13 clones. Using this approach, an average percentage of 60–95% of the M13 clones obtained was found to contain the desired mutations. The integrity of the *Bal*I to *Eco*RI *mdr3* fragment (positions 798–2249) containing the introduced *Nru*I and *Sal*I RE sites was verified by nucleotide sequencing prior to recloning into pEMC-*mdr3* and pVT-*mdr3* vectors. All subsequent NBD1 point mutations were made in the modified M13mp18 vector carrying the *Nru*I–*Sal*I cassette by the same mutagenesis procedure. Once sequenced, newly mutated cassettes were recloned as *Nru*I–*Sal*I fragments into pEMC-*mdr3* and pVT-*mdr3* vectors.

Construction of Chimeric mdr3 cDNAs

Generally speaking, chimeric genes are composed of gene fragments from different origins that are fused in frame to produce a recombinant protein. This strategy has been used extensively for the identification and study of "modular" functional domains in certain proteins or protein family members (such as membrane-bound or soluble receptors family members), but has also been used to introduce specific tags as proteolytic cleavage sites, epitopes, and others that facilitate biochemical studies (purification, localization) of a novel protein. In our study, we wished to exchange discrete homologous segments between the two highly conserved NBDs of Pgp to determine which, if any, residues of NBD1 could be substituted by its NBD2 counterpart without loss of function. Although preexisting RE sites have been used to exchange large Pgp domains between Pgp isoforms,[17,18] the more refined analysis of small homologous segments of each NBD requires an alternative approach that requires fusing segments at precise locations where such RE sites are not available.

The construction of chimeric *mdr3* cDNAs relied on the use of hybrid oligonucleotide primers and PCR amplification to fuse noncontiguous DNA segments (Fig. 1, fragments A, B, C) in three rounds of PCR. In our chimeric constructs, NBD1 fragments were replaced by the homologous segments of NBD2, while NBD2 was left intact. The strategy and methods were as follows (Fig. 1): First, individual NBD2 fragments to be duplicated (fragments B) were amplified by PCR using an *mdr3* 3′ half template and complementary oligonucleotide primers (primers 5 and 6). Similarly, fragments from NBD1 flanking individual fragments B on either side were amplified from an *mdr3* 5′ half template. The upstream 5′ fragments (frag-

[17] E. Buschman and P. Gros, *Mol. Cell. Biol.* **11(2),** 595 (1991).
[18] R. Dhir and P. Gros, *Biochemistry* **31,** 6103 (1992).

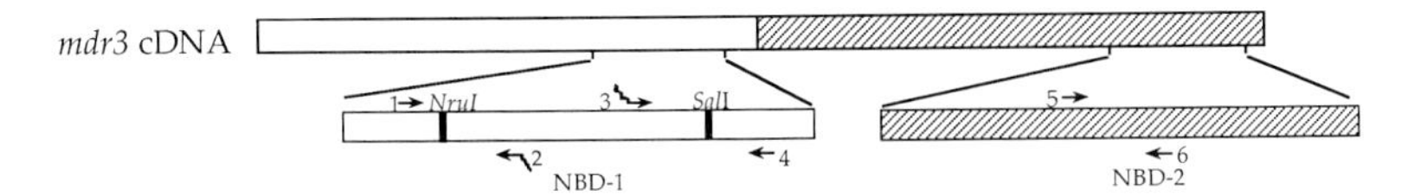

1st Step: Generation of Individual Fragments:

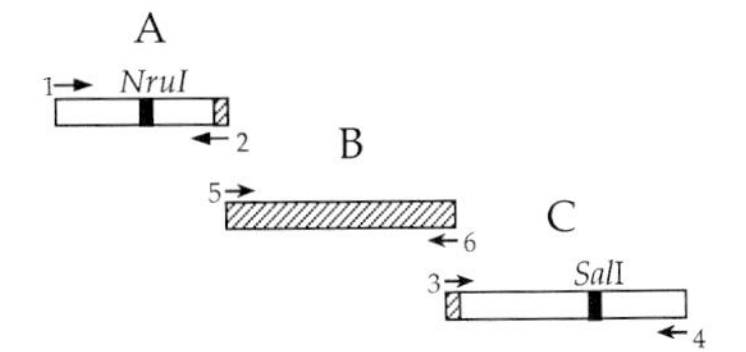

2d Step: Combination of Fragments

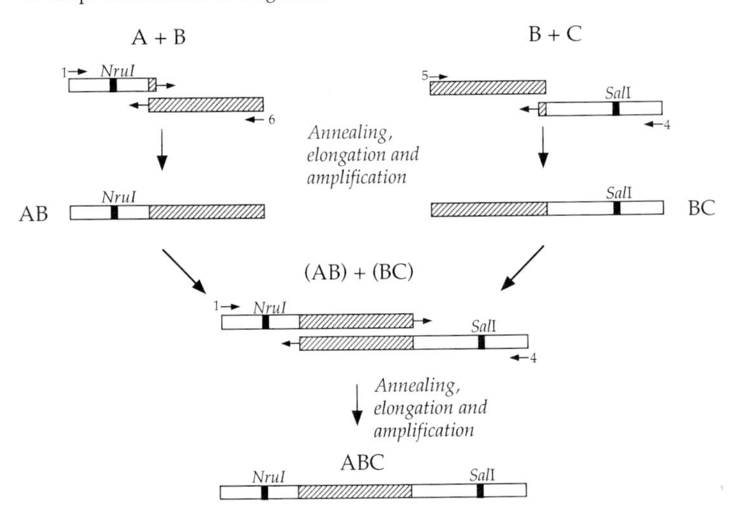

FIG. 1. Construction of chimeric *mdr3* cDNAs by a three-step PCR procedure. A full description of the procedure is given in the text.

ments A) were generated using one primer located 5′ of the *Nru*I site (primer 1) and an antisense chimeric primer in which 12 nucleotides at the 5′ end were complementary to the corresponding fragment B and 12 nucleotides at the 3′ end complementary to the adjacent NBD1 sequence (primer 2). Likewise, the downstream 3′ fragments (fragments C) were generated using one antisense primer located downstream of the *Sal*I site (primer 4), and a sense chimeric primer in which 12 nucleotides at the 5′

end were complementary to individual fragments B and the 12 at the 5′ end complementary to the adjacent NBD1 sequence (primer 3). To construct individual *mdr3* cDNAs with chimeric NBD1 segments, the NBD2-derived fragment B was mixed separately with either adjacent fragment A or C in a buffer consisting of 7 m*M* Tris (pH 7.5), 50 m*M* NaCl, and 7 m*M* $MgCl_2$. The DNA fragments were heat denatured (95°, 5 min), and complementary ends were allowed to reanneal (37° for 15 min) before filling with the Klenow fragment of DNA polymerase I and dNTPs (50 μM). Double-stranded hybrid molecules were recovered by phenol–chloroform extraction and ethanol precipitation, and the A–B and B–C chimeric molecules were amplified using either the proximal primer 1 and distal primer 6 (A–B hybrid) or proximal primer 5 and distal primer 4 (B-C hybrid). Finally, the A–B–C chimeric molecule was created by a similar strategy using gel purified A–B and B–C hybrid fragments mixed and amplified with the proximal primer from fragment A and the distal primer from fragment C (primers 1 and 4). It was then digested with the *Nru*I and *Sal*I enzymes, and the resulting cassette was gel-purified and cloned back into the expression vectors pVT-*mdr3* or pEMC-*mdr3*. Typical conditions for PCR amplification reactions were for 20 cycles (1 min at 94°, 1 min at 45°, and 1 min at 72°) in a buffer consisting of 20 m*M* Tris, pH 8.4, 1 mg/ml bovine serum albumin (BSA), 50 m*M* NaCl, 200 μM of each dNTP, $MgCl_2$ at concentrations ranging from 0.5 to 1.5 m*M*, 0.5 units of Vent polymerase (New England Biolabs, Canada), and 0.5 ng of template.

When using a PCR amplification approach to create chimeras, it is important to minimize the occurrence of mutations in the amplified products. For this, we use thermostable polymerases that carry proofreading ($3' \rightarrow 5'$-exonuclease) activity, such as $Vent_R$ and Deep $Vent_R$ DNA polymerase (New England Biolabs). Other experimental factors must be controlled to reduce misincorporation, in particular keeping the number of amplification cycles to a minimum. Primer length and sequence is also important and computer programs such as Amplify (WREngels@macc.wisc.edu) can be used to ensure that primers will not from internal secondary structures (hairpins), dimers, or bind elsewhere in the plasmid. To amplify DNA segments using primers with a perfect match, 15 base oligonucleotides in length are adequate, while chimeric primers must be considerably longer to maintain a good region of complementarity to each template. For amplification of *mdr3* fragments, 24-mer primers were used with 12 bases complementary to the NBD1 sequence and 12 bases complementary to the NBD2 sequence. If oligonucleotide synthesis is not a limitation, longer hybrid primers are preferable, allowing higher annealing temperature and increased specificity. Finally, for each pair of primers, amplification reactions should be optimized for annealing temperature and Mg^{2+} concentration. Enzyme concentration might also be critical when short primers are used since the exonuclease activity of $Vent_R$ enzymes can degrade primers from their 3′ end.

Assaying for Function

Two expression systems are used in our laboratory to assess the function of mutant Pgp molecules. We originally set up a mammalian CHO cell transfection assay, where the capacity of *mdr3* mutants to confer cellular resistance to structurally unrelated drugs allows the monitoring of possible alterations in the overall activity and substrate specificity of the encoded mutated Pgp. More recently, we have turned to the yeast *S. cerevisiae* as recipient cells, and analyzed the effect of *mdr3* expression on the appearance of cellular resistance to antifungal macrolites such as FK506 and FK520.A second functional assay takes advantage of the ability of *mdr3* to complement the yeast *ste6* gene and restore mating in an othewise sterile *ste6* null mutant.

Studies in Mammalian Cells

The CHO cell line LR73[19] is maintained in α-minimal essential medium supplemented with 10% fetal calf serum, penicillin (50 U/ml), streptomycin (50 μg/ml), and 2 m*M* glutamine (complete medium). Wild-type and mutant *mdr3* cDNAs cloned into expression plasmid pEMC2b are introduced in these cells by cotransfection with indicator plasmid pSV2*neo*[20] by a standard calcium phosphate precipitation method previously described.[3] pSV2*neo* cotransfectants are selected in complete medium supplemented with Geneticin (G418; Life Technologies, Inc./BRL, Montreal, Quebec, Canada) at a final concentration of 500 μg/ml and mass populations of $G418^R$ clones are harvested 10 days later. Further selection in complete medium containing a low concentration of the MDR drug vinblastine (VBL, 25 ng/ml) (Sigma, St. Louis, MO) ensures survival of stable *mdr3* transfectants only. Mass populations of VBL resistant (VBL^R) colonies are harvested 5–7 days later. VBL^R cells are harvested and frozen as several aliquots, which are thawed out before each experiment. P-Glycoprotein expression is evaluated by Western blotting of crude membrane fractions using the monoclonal antibody C219 (Centocor Corp., Philadelphia, PA) at 1 μg/ml or the isoform-specific anti-Mdr3 affinity purified polyclonal antibody (0.4 mg/ml).[21]

[19] J. W. Pollard and C. P. Stanners, *J. Cell Physiol.* **98,** 571 (1979).
[20] P. J. Southern and P. Berg, *J. Mol. Appl. Genet.* **1,** 327 (1982).
[21] D. F. Tang-Wai, A. Brossi, L. D. Arnold, and P. Gros, *Biochemistry* **32,** 6470 (1993).

Drug cytotoxicity assays are carried out using a method based on cell protein staining by sulforhodamine B.[22] In these experiments, a G418^R mass population of cells cotransfected with an *mdr3* cDNA cloned in pEMC2b in the antisense orientation is used as a negative control. Briefly, 5×10^3 cells from each VBL^R mass population are plated in duplicate in 96-well titer plates containing increasing concentrations of drugs. We routinely use VBL, colchicine (Sigma), doxorubicin hydrochloride (Sigma), actinomycin D (ACT) (Sigma), and gramicidin D (Sigma) as test substrates to establish substrate specificity of the Pgps analyzed. After a 3-day incubation at 37°, cells are fixed in 17% trichloroacetic acid, and stained with a solution of 0.4% sulforhodamine B (w/v) in 1% acetic acid (v/v). Excess stain is removed by five washes in 1% acetic acid and plates are air dried overnight. The stain is dissolved in 200 μl of Tris-Cl (10 mM, pH 9) and quantified with an enzyme-linked immunosorbent assay (ELISA) plate reader (Bio-Rad, Richmond, CA) set at 490 nm. The relative plating efficiency represents the ratio of the absorbance measured at a given drug concentration divided by the absorbance of the same population of cells plated in the absence of drug. The D_{50} is defined as the drug concentration required to reduce the relative plating efficiency of each cell population by 50%. This experiment is repeated three times for each mutant population. Figure 2b shows, as an example, ACT D_{50} values measured for the various NBD1 *mdr3* chimeras. A detailed description of the chimeric *mdr3* cDNA is published elsewhere.[6] Finally, should alterations in the level of overall activity or substrate specificity of the mutant Pgps be detected, further characterization of the biochemical basis for the observed mutant phenotype can be undertaken, performing drug transport experiments, photolabeling, and characterization of the ATPase activity.

Mutants showing complete loss of function are less easily studied in CHO transfectants, mainly because drug selection cannot be used to bring about high levels of protein expression. Nevertheless, individual G418^R clones can be individually picked, expanded, and analyzed for expression of the mutant proteins. The biochemical characteristics of the mutant Pgps can then be studied, using as controls individual G418^R clones expressing similar amounts of the wild-type protein and obtained according to the same selection procedure. We have previously applied this selection strategy to isolate and characterize cell clones expressing inactive mutant Pgps bearing discrete mutations in NBD1 and NBD2.[3] Other investigators have used fluorescence-activated cell sorting (FACS) analysis to isolate cell populations expressing the same amount of wild-type or mutant Pgps on their

[22] A. Devault and P. Gros, *Mol. Cell. Biol.* **10,** 1652 (1990).

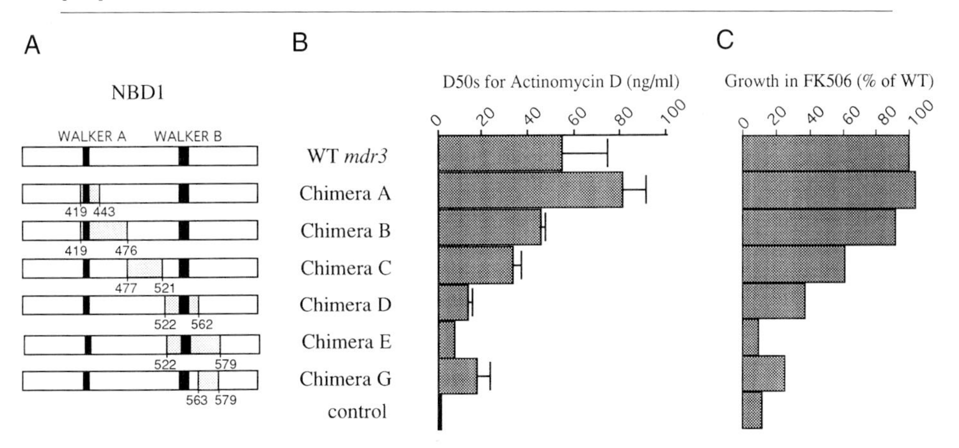

FIG. 2. Drug resistance phenotype of mammalian and yeast cells expressing wild-type or chimeric Mdr3 proteins. (a) Schematic representation of wild-type and chimeric NBD1, from the end of TM domain 6 to the beginning of the linker region. NBD1 segments replaced by the corresponding NBD2 sections are indicated (dotted boxes) with the amino acid position of their boundaries. Positions of the Walker A and B motifs are identified by black boxes. (b) D_{50} values of LR73 transfectants expressing wild-type or chimeric Pgp transfectants for actinomycin D. (c) Growth of yeast transformants expressing wild-type or chimeric *mdr3* cDNAs in medium containing FK506. Growth rates were determined by measuring the slope of the growth curves over a 4.5-hr period (between 16.5 and 21 hr) with growth rate of WT *mdr3* transformants set at 100%. [Adapted with permission from L. Beaudet and P. Gros, *J. Biol. Chem.* **270,** 17159 (1995).]

surface, irrespective of the level of activity of these Pgps.[2,23] This technique, which requires an antibody recognizing an external epitope of Pgp, has also been used to normalize the calculations of the degree of cellular drug resistance observed in mass populations of transfectants expressing different levels of distinct Pgp mutants.[23]

Studies in Yeast Cells

Mating and drug resistance assays are done using the yeast *S. cerevisiae* strain JPY201 (*MAT***a** *ste6Δ::HIS3 gal2 his3Δ200 leu2-3, 112 lys2-801 trpl ura3-52*) in which wild-type and mutant Pgps are expressed. Strain DC17 (*MATa his1*) is used as the tester strain in the mating assays. These two yeast strains were kindly provided by Dr. Martine Raymond (Institut de Recherches Cliniques, Montréal, Canada). JPY201 cells are transformed with the various *mdr3* mutants subcloned in the pVT101-U vector (pVT-

[23] T. Hoof, A. Demmer, M. R. Hadam, J. R. Riordan, and B. Tümmler, *J. Biol. Chem.* **269,** 20575 (1994).

mdr3) by the lithium acetate method. A step-by-step description of this technique is provided in the "Methods in Yeast Genetics."[24] Pools of ura$^+$ colonies are harvested, expanded in culture, and kept frozen until used in each experiment. A composition of YPD medium, minimal medium, and synthetic medium lacking uracil (SD-ura) is provided in a previous report.[25] The mating efficiency of pVT-*mdr3* yeast transformants is quantified by filter assay in the following way[26]: Approximately 1×10^6 exponentially growing JPY201 transformants are mixed with 5×10^6 DC17 tester cells and vacuum-filtered onto a 0.45-μm nitrocellulose filter (Millipore, Toronto, Ontario, Canada). Filters are deposited cell-side-up on well-dried YPD plates and incubated for 4 hr at 30° to allow mating to take place. Filters are then introduced into a 1.5-ml microtube and washed several times with 200 μl of sterile water. Various cell dilutions are plated onto minimal medium agar plates on which only fused diploid yeasts will grow. Control samples with no tester cells or with transformed yeasts that do not express Pgp are also included in the assay. The number of haploid JPY201 transformants used in the assay is determined by plating cell dilutions of the haploid transformant cultures onto SD-ura plates. Mating efficiency represents the ratio of the number of diploids obtained in a mating reaction to the number of haploid JPY201 transformants introduced in the mating reaction. Mating frequencies obtained for the mouse Mdr3 NBD1 chimeras are presented in Table I.

We estimate drug resistance of yeast transformants expressing either WT or mutant *mdr3* cDNAs by a growth inhibition assay using the antifungal peptide macrolite FK506. Stock solutions of FK506 (10 mg/ml) are prepared in ethanol and kept at −80° until use. The growth inhibition assay is carried out essentially as follows: Fresh stationary overnight cultures of pVT-*mdr3* transformants grown in SD-ura are diluted in duplicates to OD 0.005 (600 nm) in YPD medium, and added (50 μl) to an equal volume of YPD medium containing 100 μg/ml of FK506 (final concentration of 50 μg/ml) in a 96-well microwell titer plate. The plates are wrapped with Parafilm to prevent evaporation and incubated at 30° with constant agitation (250 rpm). Growth is monitored from 12 to 22 hr after seeding by optical density measurement every 90 min using an ELISA plate reader (Bio-Rad) set a 595 nm. Growth rate of the various *mdr3* transformants is calculated in the exponential growth phase over a 4.5-hr period and is expressed as

[24] C. Kaiser, S. Michaelis, and A. Mitchell, *in* "Methods in Yeast Genetics. A Cold Spring Harbor Laboratory Course Manual." Cold Spring Harbor Laboratory Press, New York, 1994.

[25] F. Sherman, *Methods Enzymol.* **194,** 3 (1991).

[26] G. F. J. Sprague, *Methods Enzymol.* **194,** 77 (1991).

TABLE I
MATING FREQUENCIES OF YEAST MASS POPULATIONS EXPRESSING WILD-TYPE OR CHIMERIC *mdr3* cDNAs

Construct[a]	Frequency[b]	WT frequency[c]
WT *mdr3*	$6.2 \times 10^{-3} \pm 3.6 \times 10^{-3}$	100
Chimera A	$4.2 \times 10^{-2} \pm 7.1 \times 10^{-3}$	680
Chimera B	$6.8 \times 10^{-3} \pm 2.7 \times 10^{-3}$	110
Chimera C	$9.2 \times 10^{-2} \pm 1.3 \times 10^{-2}$	1500
Chimera D	$3.2 \times 10^{-6} \pm 2.0 \times 10^{-6}$	0.05
Chimera E	$1.8 \times 10^{-6} \pm 8.7 \times 10^{-7}$	0.03
Chimera G	$3.9 \times 10^{-6} \pm 3.2 \times 10^{-6}$	0.06
pVT101-U	5.7×10^{-7}	0.01

[a] Schematic representation of constructs is provided in Fig. 2.
[b] Mating frequency represents the proportion of transformed **a**-type JPY201 cells that formed diploids on mating with α-type tester cells DC17 on minimal medium. Values are the average of at least three independent experiments. Standard deviations were not calculated for pVT101-U since diploid formation for this construct was observed only in one experiment.
[c] Percentage of WT (wild-type) frequency was calculated by setting WT value at 100%. Values lower than 0.2% were considered as background levels due to the extremely low number of diploid cells recovered from minimal plates.
Adapted from L. Beaudet and P. Gros, *J. Biol. Chem.* **270,** 17,159 (1995) with the permission of the publisher.

a percentage of control transformants expressing WT *mdr3*. Relative growth rates for our Mdr3 NBD1 chimeras are illustrated in Fig. 2c.

A quick comparison of (1) mating frequencies observed in *ste6* complementation assays (Table I), (2) the degree of FK506 resistance (Fig. 2c), and finally (3) the level of actinomycin D resistance conveyed by these same chimeras transfected in CHO cells (Fig. 2b) shows a very good correlation between all the assays. These results indicate that yeast is an appropriate host for the expression and rapid functional analysis of Pgp activity.

Isolation of Yeast Membranes and Identification of P-Glycoprotein

Membrane fractions in which the relative amount of mutant Pgps in different mass populations can be evaluated are isolated by a relatively simple small-scale procedure. Yeast membrane preparations are made from a 3-ml overnight culture grown in SD-ura that has reached stationary phase. The cells are pelleted in a microfuge tube (two consecutive spins of a 1.5-ml volume for 2 sec at maximum speed), washed once with 1 ml of sterile water, and resuspended in 300 μL of ice-cold lysis buffer consisting of 100 mM Tris-Cl (pH 8.0), 1 mM ethylenedia minetetra acetic acid (EDTA), 1 mM of dithiothreitol, 20% (v/v) glycerol and, protease inhibitors [1 mM of phenylmethyl sulfonyl fluoride (PMSF), 10 μg/ml of aprotinin, leupeptin, and pepstatin]. Approximately 450 mg of acid-washed glass beads (425–600 μm, Sigma) are added to the suspensions and cells are disrupted by vigorous vortexing (seven 30 sec bursts, with 30 sec on ice between vortexing). Cell debris and nuclei are removed by centrifugation (6000g, 5 min at 4°), the supernatant is recovered, and membranes are pelleted by centrifugation (100,000g, 30 min at 4°) in a table-top ultracentrifuge (Beckman, Mississauga, Ontario, Canada). Pellets are resuspended in 40 μl of lysis buffer and stored at −80°. Protein concentration is determined by the method of Bradford using a commercially available reagent (Bio-Rad), and a standard curve constructed using BSA, fraction V. Scaling up this procedure should yield enough material for photolabeling and peptide mapping studies. To determine the relative amount of Pgp expressed by different populations of yeast transformants or in individual yeast clones, Western blotting analysis can be performed using the mouse anti-Pgp monoclonal antibody C219, or the affinity-purified isoform specific anti-Mdr3 polyclonal antibody K2037 (see earlier discussion).[22]

Additional Measurements of Pgp Function in Yeast

Additional assays can be carried out in yeast cells to study further the biological activity of mutant variants of Pgp. In particular, secretory vesicles (SVs) from the yeast *sec6-4* mutant can be used as a vehicle to study the parameters of substrate transport. The various mouse Pgp isoforms can be expressed in the membrane of SVs, and these can be isolated in large numbers by a relatively simple two-step procedure. SVs are uniform in size and polarity, and tightly sealed such that vectorial drug transport can be easily studied in these preparations. We have used SVs to study the parameters of drug transport by Pgp,[27] and to demonstrate the lipid translocating properties of the Mdr2 isoform of Pgp.[28] The use of SVs to study Pgp function is described in detail in a separate article of this book. Drug-binding properties of wild-type and mutant variants of Pgp can also be directly studied in photolabeling experiments using photoactivatable drug analogs and in plasma membrane enriched fractions of yeast transformants.[10]

[27] S. Ruetz and P. Gros, *J. Biol. Chem.* **269,** 12277 (1994).

[28] S. Reutz and P. Gros, *Cell* **77,** 1071 (1994).

Conclusion

In conclusion, the yeast *S. cerevisiae* has proven to be a useful host to study the mechanism of action of Pgp and to identify structure–function relationships in this protein. Pgp expressed in yeast cells has conserved functional characteristics established for this protein in mammalian cells. Individual Pgp mutants can be expressed and characterized in yeast in a relatively rapid and inexpensive fashion, facilitating the high-throughput analysis of a large number of individual mutants. In addition, yeast can be used for additional genetic analysis of Pgp, including the selection of intragenic suppressors of null mutations in Pgp, an analysis virtually impossible to carry out in mammalian cells. Finally, yeast seems to be a suitable host for other members of the ABC transporters family, since we and others have recently been able to express functionally in yeast the human MRP protein,[8] and the *pfmdrl* gene product of *Plasmodium falciparum.*[13]

[31] Baculovirus-Mediated Expression of Human Multidrug Resistance cDNA in Insect Cells and Functional Analysis of Recombinant P-Glycoprotein

By URSULA A. GERMANN

Introduction

The availability of large quantities of biologically active protein is a prerequisite for studies of the structural, biochemical, and enzymatic properties of *MDR* gene products.[1,2] High-level expression of recombinant P-glycoprotein has been attempted with varying degrees of success in several heterologous expression systems, including bacteria, yeast, and insect cells.[3] Heterologous expression of *MDR* cDNAs in insect cells by recombinant baculovirus expression vectors has been quite successful and has contributed significantly to the recent progress in the study of the structure–function relationship of the human *MDR1* P-glycoprotein, its substrate interactions, and the role of its posttranslational modifications.[3–16]

[1] M. M. Gottesman and I. Pastan, *Annu. Rev. Biochem.* **62,** 385 (1993).
[2] U. A. Germann, *Eur. J. Cancer,* **32A,** 927 (1996).
[3] G. L. Evans, B. Ni, C. A. Hrycyna, D. Chen, S. V. Ambudkar, I. Pastan, U. A. Germann, and M. M. Gottesman, *J. Bioenerget. Biomembr.* **27,** 43 (1995).
[4] U. A. Germann, M. C. Willingham, I. Pastan, and M. M. Gottesman, *Biochemistry* **29,** 2295 (1990).

0076-6879/98 $25.00

The helper virus-independent baculovirus expression system developed by Smith, Summers, Miller, and colleagues is a powerful and versatile technology for the overexpression of eukaryotic cDNAs.[17–21] Some of the major advantages of this system are as follows:

1. The baculovirus expression system is safe due to the high species specificity of baculoviruses that infect insect cells exclusively.
2. Generally high-level expression of recombinant protein is achieved in cultured insect cells with reported levels ranging from 0.1% up to 50% of the total cellular protein.
3. Recombinant proteins produced in insect cells are usually targeted to the proper subcellular compartment, that is, membrane proteins will be anchored within the insect cell membrane.
4. Many posttranslational modifications, including *N*-linked glycosylation and phosphorylation, are performed in insect cells, hence, recombinant proteins produced in the baculovirus expression system structurally and functionally resemble their authentic counterpart.
5. Baculoviruses have the capacity to package and express relatively large foreign genes.

---

[5] B. Sarkadi, E. M. Price, R. C. Boucher, U. A. Germann, and G. A. Scarborough, *J. Biol. Chem.* **267,** 4854 (1992).

[6] L. Homolya, Z. Hollo, U. A. Germann, I. Pastan, M. M. Gottesman, and B. Sarkadi, *J. Biol. Chem.* **268,** 21493 (1993).

[7] B. Sarkadi, M. Müller, L. Homoloya, Z. Holló, J. Seprődi, U. A. Germann, M. M. Gottesman, E. M. Price, and R. C. Boucher, *FASEB J.* **8,** 766 (1994).

[8] U. S. Rao and G. A. Scarborough, *Mol. Pharmacol.* **45,** 773 (1994).

[9] U. S. Rao, R. L. Fine, and G. A. Scarborough, *Biochem. Pharmacol.* **48,** 287 (1994).

[10] V. V. Rao, M. L. Chiu, J. F. Kronauge, and D. Piwnica-Worms, *J. Nucl. Med.* **35,** 500 (1994).

[11] L. Zhang, C. W. Sachs, R. L. Fine, and P. Casey, *J. Biol. Chem.* **269,** 15973 (1994).

[12] T. W. Loo and D. M. Clarke, *J. Biol. Chem.* **269,** 7750 (1994).

[13] S. Ahmad, A. R. Safa, and R. I. Glazer, *Biochemistry* **33,** 10313 (1994).

[14] E. Welker, K. Szabó, Z. Holló, M. Müller, B. Sarkadi, and A. Váradi, *Biochem. Biophys. Res. Commun.* **216,** 602 (1995).

[15] G. A. Scarborough, *J. Bioenerget. Biomembr.* **27,** 37 (1995).

[16] M. Müller, E. Bakos, E. Welker, A. Váradi, U. A. Germann, M. M. Gottesman, B. S. Morse, I. B. Roninson, and B. Sarkadi, *J. Biol. Chem.* **271,** 1877 (1996).

[17] G. E. Smith, M. D. Summers and M. J. Fraser, *Mol. Cell. Biol.* **3,** 2156 (1983).

[18] L. K. Miller, *in* "Genetic Engineering in the Plant Sciences" (P. Panopoulos, ed.), p. 203. Praeger, New York, 1981.

[19] V. A. Luckow and M. D. Summers, *Bio/Technol.* **6,** 47 (1988).

[20] L. A. King and R. D. Possee, "The Baculovirus Expression System: a Laboratory Guide." Chapman and Hall, London, 1992.

[21] D. R. O'Reilly, L. K. Miller, and V. A. Luckow, "Baculovirus Expression Vectors: a Laboratory Manual." Oxford University Press, New York, 1994.

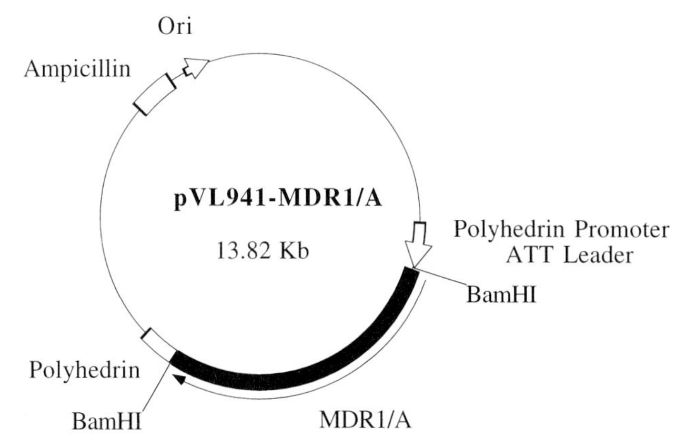

FIG. 1. Map of the *MDR1* baculovirus expression vector pVL941-MDR1/A. The pVL941-MDR1/A plasmid contains a 3.97-kb *MDR1* cDNA inserted as a *Bam*HI fragment under the control of the polyhedrin promoter. The direction of transcription is indicated by the arrows. The *MDR1* cDNA is flanked at both ends by parts of the AcNPV polyhedrin gene locus that are required for *MDR1* baculovirus production by homologous recombination with viral genomic DNA upon cotransfection into insect cells. The origin of replication (Ori) and the ampicillin resistance gene (Ampicillin) are derivatives of the pUC8 vector.

6. Simultaneous expression of multiple foreign genes within a single insect cell is possible.
7. The wide use of the baculovirus expression system has lead to the development and commercial availability of novel technologies and experimental kits that allow for reliable and rapid expression of heterologous gene products as full-length proteins and/or as fusion proteins.

The most extensively studied and most commonly used baculovirus strain is the *Autographa californica* nuclear polyhedrosis virus (AcNPV), a large double-stranded DNA virus, whose entire genome of 128 kbp has been mapped and partially sequenced.[22] Genetic disruption studies have led to the identification of several AcNPV genes that are nonessential for replication of the virus in cell culture and in insects.[19–21] Some of these nonessential genes, particularly the ones expressed very late in infection, are under the control of very strong promoters.[19–21] The replacement of such nonessential AcNPV genes by exogenous genes provides the basis of the baculovirus expression system. Many baculovirus expression vectors, including the one described here for expression of recombinant P-glycoprotein (Fig. 1), make use of the polyhedrin promoter and also

[22] J. M. Vlak and G. E. Smith, *J. Virol.* **41,** 1118 (1982).

contain flanking sequences of the polyhedrin gene, which are required for homologous recombination with AcNPV genomic DNA upon cotransfection into insect cells.[19–21]

The methods described in the following sections made possible baculovirus-mediated heterologous expression of a human *MDR1* cDNA in insect cells. High-level expression of recombinant P-glycoprotein is achieved, and the recombinant protein is structurally and functionally similar to native P-glycoprotein expressed in multidrug-resistant human cancer cells. Thus, the baculovirus expression system may be useful for the large-scale production and purification of recombinant P-glycoprotein for future studies of its molecular structure and mechanisms of action. The methods outlined here should also be suitable for the overexpression of P-glycoprotein-related proteins from other species, or of other types of ATP-binding cassette (ABC) transporters.

Plasmid Constructions

Baculovirus expression vectors containing an *MDR1* cDNA under control of the polyhedrin promoter have been constructed.[4,16] Alternatively, the p10 promoter has also been used to drive production of recombinant P-glycoprotein in insect cells.[14] Several polyhedrin-driven expression baculovirus vectors have been tested: pVL941[23] (in which a human *MDR1* cDNA is cloned in the single *Bam*HI cloning site), pVL1393[21] (human *MDR1* cDNA cloned as blunt-ended fragment in the single *Sma*I cloning site), and pAC373[24] (human *MDR1* cDNA cloned in the single *Bam*HI cloning site).[4,16] The pVL941-based *MDR1* baculovirus expression system gives rise to somewhat higher levels of recombinant P-glycoprotein expression in *Spodoptera frugiperda* (Sf9) insect ovary cells than the pVL1393- or the pAC373-based ones.[16]

The bacterial strain *Escherichia coli* DH5α was used as a host for human *MDR1*-containing subcloning and baculovirus expression vectors.[4,16] In the first report on baculovirus-mediated overexpression of P-glycoprotein,[4] a human *MDR1* cDNA insert was derived from the plasmid pMDRΔA, which contains a human *MDR1* cDNA with a deleted poly(A) addition site.[25] This *MDR1* cDNA was originally isolated from human KB-C1 multidrug-resistant cells and encodes a P-glycoprotein mutant

[23] V. A. Luckow and M. D. Summers, *Virology* **170,** 31 (1989).

[24] M. D. Summers and G. E. Smith, *Tex. Agric. Exp. Stn.* [*Bull.*] **1555,** (1987).

[25] I. Pastan, M. M. Gottesman, K. Ueda, E. Lovelace, A. V. Rutherford, and M. C. Willingham, *Proc. Natl. Acad. Sci. U.S.A.* **85,** 4486 (1988).

with a Gly-185 → Val substitution.[26,27] To introduce this *MDR1* cDNA into the pAC373 and later the pVL941 baculovirus expression vector, novel *Bam*HI restriction sites were introduced within the 5′ and 3′ untranslated regions of the *MDR1* cDNA at position −10 5′ of the translation start codon and at position +110 3′ of the translation stop codon (Fig. 1).[4,16]

Similar baculovirus expression vectors have been constructed for the heterologous expression of a wild-type *MDR1* cDNA isolated from KB-V1 cells, as well as of several *MDR1* mutants.[16] For expression of *MDR1*-related genes (e.g., a human *MDR2* cDNA[26] or *MDR* cDNAs from other species such as the mouse *mdr3*[28,29] and *mdr1*[30] cDNAs) the restriction sites required for cloning into the appropriate baculovirus expression vector can be easily introduced by the polymerase chain reaction (PCR) using specifically designed oligodeoxynucleotides.[31] Ideally, the 5′ end of an *MDR* cDNA insert should be chosen as close as possible to the translation start codon and unnecessary 3′ noncoding sequences should also be removed since the polyadenylation signals are usually provided by the baculovirus transfer plasmid. A pVL941-based expression plasmid similar to the pVL941-MDR1/A described earlier has also been used successfully for overexpression of a human *MDR2* cDNA in insect cells.[32]

Production of Recombinant *MDR1* Baculovirus

Insertion of an *MDR1* cDNA into the baculovirus genome is achieved by homologous recombination between the polyhedrin flanking sequences within the recombinant baculovirus expression vector (pVL941-MDR1/A, Fig. 1) and the polyhedrin gene within AcNPV viral DNA. A cotransfection is performed to introduce simultaneously the recombinant transfer plasmid and viral DNA into insect cells. The success of this procedure depends on the quality of both the plasmid and viral DNA. The pVL941-MDR1/A plasmid DNA is purified by two cesium chloride–ethidium bromide density gradient centrifugations. Baculovirus DNA is purified from wild-type AcNPV according to established protocols,[21] but it is highly recommended to use commercially available viral DNA, which is linearized within the target area for allelic replacement (e.g., PharMingen, San Diego, CA).

[26] C.-J. Chen, J. E. Chin, K. Ueda, D. P. Clark, I. Pastan, M. M. Gottesman, and I. B. Roninson, *Cell* **47,** 381 (1986).

[27] K. Choi, C.-J. Chen, M. Kriegler, and I. B. Roninson, *Cell* **53,** 519 (1989).

[28] S. I. Hsu, L. Lothstein, and S. B. Horwitz, *J. Biol. Chem.* **264,** 12053 (1989).

[29] A. Devault and P. Gros, *Mol. Cell. Biol.* **10,** 1652 (1990).

[30] P. Gros, J. Croop, and D. E. Housman, *Cell* **47,** 371 (1986).

[31] J. Sambrook, E. F. Fritsch, and T. Maniatis, "Molecular Cloning: a Laboratory Manual." Cold Spring Harbor Laboratory Press, New York, 1989.

[32] U. A. Germann and D. Shlyakhter, unpublished data, 1996.

The use of such linearized viral DNA decreases the proportion of single-crossover recombinants obtained during cotransfection and, most importantly, strongly reduces the background of nonrecombinant baculovirus.

Linearized AcNPV-derived DNA, termed BaculoGold DNA (PharMingen), contains a lethal 1.7-kb deletion downstream of the stop codon of the polyhedrin gene. This lethal deletion can be complemented only with an appropriate polyhedrin-based baculovirus expression transfer plasmid, thus permitting positive survival selection for recombinants. Moreover the BaculoGold baculovirus DNA contains a *lacZ* gene, which on homologous recombination with the recombinant pVL941-MDR1/A transfer plasmid is replaced by the *MDR1* gene. Consequently, recombinant *MDR1* baculovirus (usually 99%) will produce colorless plaques on 5-bromo-4-chloro-3-indoly-β-D-galactoside (X-Gal)-containing plates, while a small proportion of β-galactosidase-expressing nonrecombinant baculovirus produces blue plaques on X-Gal plates.

Sf9 insect cells are used as a host for cotransfection of the pVL941-MDR1/A baculovirus expression vector and the BaculoGold viral DNA by the calcium phosphate coprecipitation method. The Sf9 cells should be in logarithmic growth phase and cell viability measured by trypan blue exclusion [0.1 ml of 0.4% (w/v) trypan blue in phosphate-buffered saline (PBS) added to 1 ml of cells] should be greater than 97%. For each cotransfection 2×10^6 Sf9 cells are seeded in 4 ml of TNM-FH medium (i.e., Grace's medium supplemented with 3.3 g/liter yeastolate and 3.3 g/liter lactalbumin hydrolyzate) containing 10% fetal bovine serum in a 60-mm tissue culture plate and allowed to attach firmly at 27° for 60 min. Then 0.5 μg of BaculoGold virus DNA (0.1 μg/μl) and 2 μg of pVL941-MDR1/A plasmid DNA are coincubated in a sterile Eppendorf tube for 5 min at room temperature and mixed with 1 ml of transfection buffer B (125 m*M* HEPES, pH 7.1, 125 m*M* $CaCl_2$, 140 m*M* NaCl). Then the culture medium is removed from the insect cells and replaced with 1 ml of BaculoGold transfection buffer A (Grace's medium with 10% (v/v) fetal bovine serum). The transfection buffer B/DNA solution (1 ml) is added dropwise to the insect cells and after every 3–5 drops the tissue culture plate is moved gently back and forth on a plain and even surface for mixing. A fine calcium phosphate/DNA precipitate should form, which is incubated with the insect cells for 4 hr at 27°. Then the transfection solution is removed from the culture plate, 3 ml of TNM-FH medium is added, and the insect cells incubated at 27° for 5 days. The plate is checked for signs of infection (large floating cells with large nuclei), and the supernatant is collected and cleared of cells by centrifugation at 1000*g* and 25° for 10 min. This supernatant will contain recombinant (usually >99%) and a small amount of nonrecombinant baculovirus. The cotransfection supernatant is stored at 4° until use.

Recombinant *MDR1* baculovirus is purified by a plaque assay. To this end 2×10^6 logarithmically growing Sf9 cells are seeded in 4 ml of growth medium in several 60-mm tissue culture plates and allowed to attach for at least 1 hr. Two hours prior to infection the medium is replaced with 1 ml fresh TNM-FH supplemented with 10% (v/v) fetal bovine serum. Serial dilutions of the cotransfection supernatant are prepared (e.g., 10^{-2}, 10^{-3}, 10^{-4}, 10^{-5}) and 100 μl of diluted virus inoculum is added per plate. The plates are gently mixed to distribute the virus evenly and incubated at 27° for 1 hr. During this time a 0.8% low melting agarose (Seaplaque, FMC, Rockland, ME) overlay in 1 × TNM-FH/10% fetal bovine serum is prepared and equilibrated in a waterbath at 37–40°. For detection of nonrecombinant baculovirus 10 μl X-Gal stock solution (25 mg/ml X-Gal in dimethylformamide) is added per milliliter agarose overlay. At the end of the 1-hr incubation period the virus inoculum is removed from the plated insect cells and 4 ml agarose overlay is added slowly from one side of the plate and spread evenly. The plates are left undisturbed for 1 hr until the agarose is solidified. Then the plates are incubated in a humid atmosphere for 6 days at 27° until clearly visible plaques develop. From a plate in which individual plaques are well separated, recombinant (colorless) plaques are picked with a sterile Pasteur pipette and each is placed in 1 ml TNM-FH medium supplemented with 10% fetal bovine serum. The budded *MDR1* baculovirus particles are released from the agarose plug by vortexing and incubation at 4° overnight. Similarly, control baculovirus stocks may be prepared by picking a few blue, β-galactosidase-expressing plaques.

To amplify the recombinant baculovirus, 1×10^6 logarithmically growing Sf9 cells are seeded in 2 ml of growth medium on a 35-mm tissue culture plate. After the cells are attached to the plate, the growth medium is aspirated and 500 μl plaque pick stock is added. The cells are infected at room temperature for 1 hr with gentle rocking. Then 1.5 ml fresh TNM-FH/10% fetal bovine serum is added and the cells incubated for 4 days at 27°. The virus-containing supernatant is harvested by centrifugation for 10 min at 1000*g* and 25°. This is the designated passage-1 *MDR1* baculovirus stock. One milliliter of the passage-1 virus stock is diluted with 3 ml complete tissue culture medium (TNM-FH/10% fetal bovine serum). Four 10-cm tissue culture plates containing 5×10^6 Sf9 cells are infected with 1 ml diluted passage-1 virus stock for 1 hr at room temperature with gentle rocking, fed with 9 ml complete tissue culture medium and incubated at 27° for 4 days. The passage-2 *MDR1* baculovirus is harvested by centrifugation at 1000*g* and 25° for 10 min. Several 1-ml aliquots are stored at −80° as long-term backups. The remainder of the passage-2 *MDR1* baculovirus stock is used for titer determination by plaque assay or end point dilution,[21] as a source of DNA to verify the structure of the recombinant virus, and as inoculum for large-scale working stocks by infecting Sf9 cells at a multiplicity of infection (MOI) of 0.1. The MOI is calculated by dividing the number of plaque-forming units (pfu) by the cell number, that is, an MOI of 0.1 corresponds to 1 pfu/10 cells. High-titer *MDR1* baculovirus stocks ($\geq 10^8$ pfu/ml) should be obtained after one to two rounds of additional virus amplification.

Production of Recombinant P-Glycoprotein in Insect Cells

A time course experiment is performed for initial analysis of recombinant P-glycoprotein production in *MDR1* baculovirus-infected insect cells. To this end, 1×10^6 logarithmically growing Sf9 cells are seeded in 2 ml of growth medium into several 35-mm tissue culture plates and incubated at 27° for 30 min to 1 hr. One dish is needed per time point. Control experiments should be performed with a control baculovirus (e.g., β-galactosidase-expressing baculovirus or wild-type AcNPV), as well as noninfected insect cells. High-titer baculovirus stock is diluted in growth medium to a concentration of 1×10^7 pfu in 500 μl. After the cells are attached, the medium is aspirated, the diluted baculovirus stock added at an MOI of 10 and incubated for 1 hr at room temperature with gentle rocking. Then the virus inoculum is removed, 2 ml of fresh medium is added, and the cells incubated at 27°. Protein expression is analyzed at 24, 48, and 72 hr postinfection (pi). The culture medium is removed and the cells are harvested by vigorous pipetting into ice-cold DPBS-Ap [Dulbecco's phosphate-buffered saline (200 mg/liter KCl, 8000 mg/liter NaCl, 2160 mg/liter $Na_2HPO_4 \cdot 7H_2O$, 200 mg/liter KH_2PO_4), supplemented with 1% (v/v) aprotinin (Sigma, St. Louis, MO)]. The cells are washed twice with DPBS-Ap and lysed on ice for 15 min with ice-cold 1× RIPA [20 m*M* Tris-HCl, pH 7.2, 150 m*M* NaCl, 1% (v/v) Triton-X-100, 1% (w/v) deoxycholate, 0.1% (w/v) sodium dodecyl sulfate (SDS), 1 m*M* ethylenediaminetetraacetic acid (EDTA), 1% (v/v) aprotinin (Sigma)] and 1 μl DNase (Benzonase, EM Science, Gibbstown, NJ). The protein concentrations of the cell lysates are determined and adjusted by diluting more concentrated samples with 1× RIPA. After adding equal volumes of 2× SDS gel-loading buffer and heating the samples at 37° for 10 min, equal amounts of protein are analyzed on a 8% (w/v) SDS–polyacrylamide gel, electroblotted, and immunostained with a P-glycoprotein-specific antibody (e.g., C219 monoclonal antibody[33]).[34] Production of recombinant P-glycoprotein in *MDR1* baculovirus-infected Sf9 cells is expected to be maximal approximately 3 days after infection (Fig. 2). P-Glycoprotein produced in insect cells has

[33] N. Kartner, D. Evernden-Porelle, G. Bradley, and V. Ling, *Nature* **316,** 820 (1985).
[34] U. A. Germann, *Methods Mol. Biol.* **63,** 139 (1997).

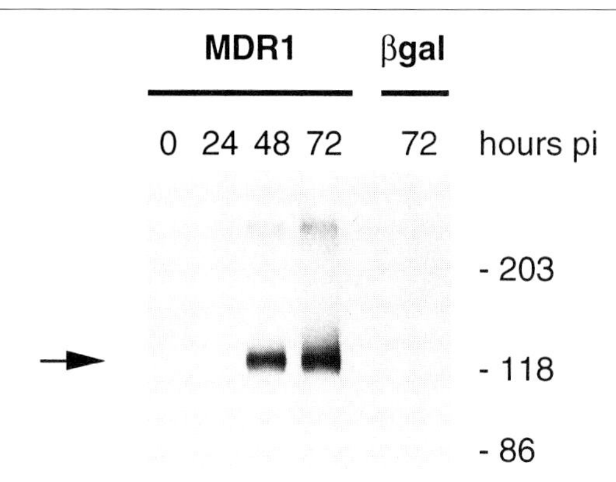

FIG. 2. Production of recombinant P-glycoprotein in *MDR1* baculovirus-infected Sf9 insect cells. Whole-cell extracts were prepared from *MDR1*- or β-galactosidase baculovirus-infected insect cells at different times postinfection (pi), resolved by SDS–polyacrylamide gel electrophoresis (SDS–PAGE), electrophoretically transferred to nitrocellulose, and immunostained using the P-glycoprotein-specific monoclonal antibody C219. On the right-hand side, molecular weight standards $\times 10^{-3}$ are indicated. The arrow on the left-hand side indicates recombinant P-glycoprotein.

a lower apparent molecular weight (120,000–140,000) than its authentic counterpart from multidrug-resistant human cells (160,000–180,000) because it is underglycosylated (Fig. 3).[4] However, glycosylation is not essential for the functional activity of the multidrug transporter.[1,2,35]

For large-scale production of recombinant P-glycoprotein, approximately 1.5×10^7 logarithmically growing Sf9 cells are plated in 30 ml of TNM-FH containing 10% fetal bovine serum into several 15-cm tissue culture dishes and incubated for 30 min to 1 hr at 27°. The attached insect cells are infected by aspirating the medium, adding high-titer *MDR1* baculovirus stock at an MOI of 10 in a minimum volume of 6 ml, and incubating at room temperature for 1 hr with gentle rocking. Then the virus inoculum is removed, 30 ml of fresh growth medium is added, and the cells are incubated at 27° for 3 days.

Other insect host cell lines may also be used for overexpression of *MDR* gene products. Time course experiments have shown that production of recombinant P-glycoprotein in Sf9, Sf21,[21] and High Five[21] cells is similar, but the yield in High Five cells is slightly higher.[32]

[35] A. H. Schinkel, S. Kemp, M. Dolle, G. Rudenko, and E. Wagenaar, *J. Biol. Chem.* **268,** 7474 (1993).

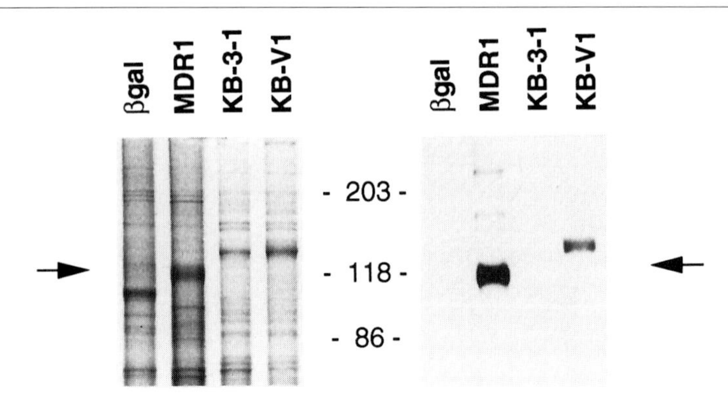

FIG. 3. Analysis of P-glycoprotein-enriched membranes from *MDR1* baculovirus-infected insect cells. *MDR1* and β-galactosidase baculovirus-infected insect cells, as well as drug-sensitive KB-3-1 and P-glycoprotein-expressing multidrug resistant KB-V1 human carcinoma cells, were fractionated by differential centrifugation. The light membrane fractions were resolved by SDS–PAGE and analyzed by staining with Coomassie blue (left-hand side) or subjected to a Western blot analysis using C219 P-glycoprotein-specific monoclonal antibody (right-hand side). Molecular weight standards ($\times 10^{-3}$) are indicated between the two gels. The arrows point to the signals from recombinant P-glycoprotein.

Preparation of P-Glycoprotein-Enriched Insect Cell Membranes

Immunofluorescence analyses and electron microscopy immunocytochemistry have demonstrated that recombinant P-glycoprotein produced in insect cells, similar to native P-glycoprotein,[36] is membrane associated and expressed at the cell surface, as well as in the Golgi apparatus.[4] However, recombinant P-glycoprotein is also present in the nuclear envelope of Sf9 cells and it has been hypothesized that this may reflect an inefficiency of the translocation machinery in processing the overexpressed foreign gene product.[4]

To facilitate functional analysis of the recombinant *MDR1* gene product, insect cell fractions enriched in P-glycoprotein are prepared. The *MDR1* baculovirus-infected cells are harvested, washed twice with ice-cold DPBS-Ap (see earlier description), and subjected to hypotonic lysis on ice in TMEP buffer [50 m*M* Tris-HCl, pH 7.0, 50 m*M* mannitol, 2 m*M* EGTA, 2 m*M* 2-mercaptoethanol, 0.5 m*M* phenylmethylsulfonyl fluoride (PMSF), 10 μg/ml leupeptin, and 1% (v/v) aprotinin (Sigma)]. The swollen cells are disrupted by 40 strokes in a tight-fitting Dounce homogenizer, and trypan blue staining is performed to ascertain that at least 90% of the insect cells

[36] M. C. Willingham, N. D. Richert, M. M. Cornwell, T. Tsuruo, H. Hamada, M. M. Gottesman, and I. H. Pastan, *J. Histochem. Cytochem.* **35,** 1451 (1987).

are ruptured. Intact cells and nuclear debris are removed by centrifugation at 500g and 4° for 10 min. The low-speed supernatant is centrifuged again at 100,000g and 4° for 1 hr and the high-speed pellet containing the light membranes is resuspended in TMEP buffer at a concentration of 2–4 mg/ml by use of a syringe with a 1.5-in.-long 18-gauge needle that is bent at a 90-degree angle. The insect cell membranes are stored in aliquots at −70° until use. A small sample of the cell membrane fraction is analyzed by SDS–PAGE and recombinant P-glycoprotein is visualized by Coomassie blue staining, or by immunostaining after electroblotting onto nitrocellulose (Fig. 3). The P-glycoprotein content in the insect cell membrane fractions is approximately 3% of the total protein based on densitometric analyses of Coomassie blue-stained gels.[5]

Photoaffinity Labeling of Recombinant P-Glycoprotein

Several photoaffinity labels have been shown to label P-glycoprotein specifically, such as photoaffinity analogs of cytotoxic agents and analogs of MDR reversing agents.[37] Commercially available P-glycoprotein-specific photoaffinity probes include [^{3}H]azidopine[38] (Amersham, Arlington Heights, IL) and [^{125}I]iodoarylazidoprazosin[39] (NEN, Boston, MA). Photolabeling of P-glycoprotein by [^{3}H]azidopine and [^{125}I]iodoarylazidoprazosin is inhibited in a dose-dependent manner by an excess of MDR drugs (e.g., vinblastine, daunomycin) and MDR modulators (e.g., verapamil).[37] Hence, competitive photolabeling experiments can be performed to analyze the drug-binding capacity of recombinant P-glycoprotein.[4] Similarly, the photoprobe [^{32}P]-8-azido-ATP (ICN, Costa Mesa, CA) and nonradioactive nucleotides can be used to study ATP binding by recombinant P-glycoprotein.[5]

Photolabeling experiments are performed by incubating cell membrane fractions containing 25 μg protein in 100 μl 10 mM Tris-HCl, pH 7.5, 10 mM NaCl, 1 mM $MgCl_2$ with 1 μCi [^{3}H]azidopine (approximately 0.25 μM) in the absence or presence of an inhibitory nonradioactive agent (usually 1 μl of a stock solution dissolved in dimethyl sulfoxide) for 1 hr at room temperature in the dark with gentle agitation. For UV cross-linking, the membrane-containing tubes are placed with open lids in a rack on ice and irradiated for 20 min at a distance of approximately 5 cm with a UV lamp using two 15-W self-filtering, long-wavelength UV tubes. The photolabeled membranes are diluted with an equal volume of 2× gel loading buffer,

[37] A. R. Safa, *Cancer Invest.* **11,** 46 (1993).

[38] E. P. Bruggemann, U. A. Germann, M. M. Gottesman, and I. Pastan, *J. Biol. Chem.* **264,** 15483 (1989).

[39] L. M. Greenberger, *J. Biol. Chem.* **268,** 11,417 (1993).

heated at 37° for 10 min, and analyzed by SDS–PAGE and fluorography. As shown in Fig. 4, recombinant P-glycoprotein is efficiently labeled by [^{3}H]azidopine and various substrates of the multidrug transporter (e.g., daunomycin > verapamil > vinblastine ~ vincristine) partially block photolabeling when added in excess, indicating that recombinant P-glycoprotein maintains the capacity to interact directly with structurally diverse MDR compounds and is functionally similar to the native human multidrug transporter.

Drug-Stimulated, Vanadate-Sensitive P-Glycoprotein ATPase Activity

Another simple assay for studying the functional interaction of various drugs with recombinant P-glycoprotein has been established after discovery of a high-capacity drug-stimulated ATPase activity associated with membranes prepared from *MDR1* baculovirus-infected insect cells.[5] Due to the shut-off of insect host cell protein synthesis upon infection with recombinant *MDR1* baculovirus, endogenous insect ATPase activities are lowered and measurement of the vanadate-sensitive P-glycoprotein ATPase activity and its dose-dependent modulation by MDR cytotoxics and MDR modulators is facilitated. This ATPase activity measured in membranes from *MDR1* baculovirus-infected Sf9 cells can be stimulated by a variety of MDR cytotoxics (e.g., vinblastine) and MDR modulators (e.g., verapamil) and correlates with the amount of recombinant P-glycoprotein synthesized, but is

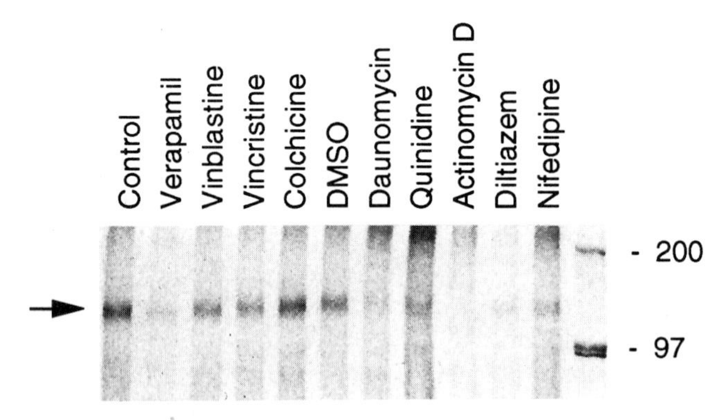

FIG. 4. Photolabeling of recombinant P-glycoprotein by [^{3}H]azidopine and inhibitory effects of MDR cytotoxics and MDR modulators. P-Glycoprotein-enriched membranes isolated from *MDR1* baculovirus-infected Sf9 insect were labeled with 1 μCi (0.25 μM) [^{3}H]azidopine in the absence or presence of a 400-fold excess of MDR cytotoxics or MDR modulators and analyzed by SDS–PAGE followed by fluorography. On the right-hand side, the ^{14}C-labeled molecular weight standards ($\times 10^{-3}$) are indicated. The arrow on the left-hand side points to the signals from recombinant P-glycoprotein.

not detectable in cell membranes prepared from uninfected or control baculovirus-infected Sf9 cells.[5,15,16] The P-glycoprotein ATPase activity is magnesium dependent, does not utilize ADP or AMP as substrates, and can be blocked by vanadate.[5] A K_m of 0.5 mM has been reported for MgATP.[5]

P-Glycoprotein ATPase assays are performed using 100-μg membranes from *MDR1* baculovirus-infected insect cells per test condition.[5] The insect cell membranes are suspended in 0.9-ml total volume containing 0.5 ml 2× ATPase assays buffer [100 mM Tris–4-morpholinoethanesulfonic acid, pH 6.8, 100 mM KCl, 20 mM $MgCl_2$, 4 mM dithiothreitol (DTT), 10 mM NaN_3 (inhibitor of F_1,F_0-ATPases), 4 mM EGTA (inhibitor of calcium-dependent ATPases), 2 mM ouabain (inhibitor of Na^+,K^+-ATPase)] in the absence or presence of 100 μM sodium orthovanadate (calculated for final assay volume of 1 ml), with or without a dilution series of a stimulant, for example, verapamil. Verapamil is added as 100× stock solutions in dimethyl sulfoxide (DMSO). Control assays are performed with DMSO alone. Duplicate samples are prewarmed for 10 min at 37° and reactions are initiated by addition of 100 μl 50 mM ATP sodium salt, pH 7.0. Aliquots of 100 μl are taken at various time points (e.g., 0, 6, 12, 20, 40, 60 min) and the reactions are stopped by adding an equal volume of 5% (w/v) SDS. The samples can be put on ice for a few minutes, but liberated inorganic phosphate (P_i) should be quantitated as soon as possible by use of a colorimetric method involving ammonium molybdate complexes.[40] The colorimetric reaction for detection of inorganic phosphate is initiated by adding 250 μl of reagent A [2.5 N H_2SO_4, 1% (w/v) ammonium molybdate, 0.014% (w/v) antimony potassium tartrate], 125 μl 1% (w/v) ascorbic acid, and 600 μl 20% (v/v) acetic acid and vortexing. After color development for 10–15 min at room temperature, the absorbance at 880 nm is determined and compared to the absorbance of a series of potassium phosphate standards (0.1–1 mM) reacted the same way. Results are expressed either as ATPase activity (μmol P_i/mg membrane protein) attributable to P-glycoprotein (i.e., ATPase activity measured in the absence of vanadate–ATPase measured in the presence of vanadate) or as vanadate-inhibitable verapamil-stimulated P-glycoprotein ATPase activity measured in comparison to control (samples without verapamil).

As shown in the time course experiment presented in Fig. 5, the P-glycoprotein ATPase activity increases linearly during at least 20 min. The basal vanadate-sensitive ATPase activity in membranes from *MDR1* baculovirus-infected insect cells is in the range of 30 nmol/mg membrane protein/min (Fig. 5), which is at least three-fold higher than the one measured in membranes from uninfected or control-infected insect cells.[5] Based

[40] B. Sarkadi, I. Szasz, A. Gerloczi, and G. Gardos, *Biochim. Biophys. Acta* **464,** 93 (1977).

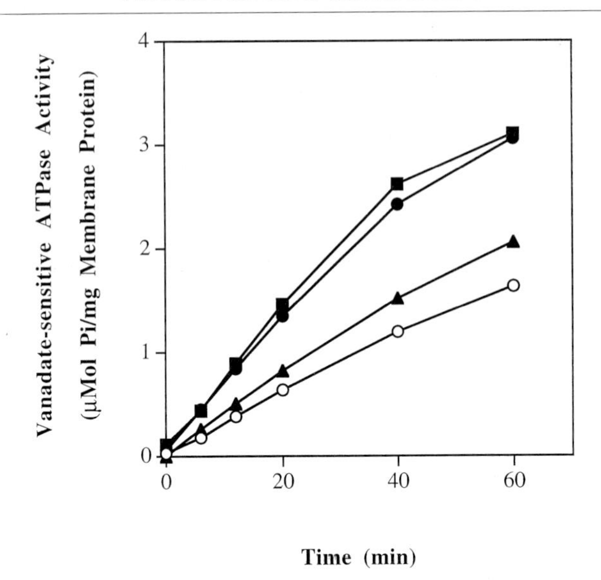

FIG. 5. Modulation of the vanadate-sensitive P-glycoprotein ATPase activity by verapamil. ATPase assays were performed with P-glycoprotein enriched membranes isolated from *MDR1* baculovirus-infected Sf9 insect cells. The vanadate-sensitive P-glycoprotein ATPase activity was determined as a function of time in the absence of verapamil (○) and in the presence of verapamil at 1 μM (▲), 10 μM (●), and 100 μM (■). Data represent means from duplicate samples.

on an estimated content of 3% recombinant *MDR1* gene product in the insect cell membrane fraction, the basal ATPase activity of P-glycoprotein is in the range of 1 μmol/mg P-glycoprotein/min (Fig. 5).[5] In the presence of verapamil, the P-glycoprotein ATPase activity is specifically increased in a dose-dependent manner with a slight stimulatory action by 1 μM verapamil and maximal activation by 10–100 μM verapamil (Fig. 5). At higher concentrations (e.g., 1 mM) stimulation of the P-glycoprotein ATPase activity is suboptimal.[5,14,16] Maximal activation of the vanadate-sensitive ATPase activity by verapamil is up to five-fold,[5] but varies up to two-fold between different batches of membranes, whereas the estimated drug concentration required for half-maximal activation is relatively constant.[32] To date, many P-glycoprotein-interacting agents have been demonstrated to stimulate the vanadate-sensitive ATPase activity in isolated membranes of *MDR1* baculovirus-infected Sf9 insect cells in a concentration-dependent manner, including MDR cytotoxics (e.g., vinblastine, vincristine, daunomycin, colchicine, etoposide), MDR modulators (e.g., verapamil, quinine, trifluoperazin, nifedipine, FK506, tamoxifen), steroids (e.g., progesterone, β-estradiol), prenylcysteine compounds derived from prenylated proteins, biologically active hydrophobic peptide derivatives (e.g., *N*-acetyleucylleucylnorleucinal, *N*-acetylleucylleucylmethional, calpeptin, leupeptin, formylmethionylleucylphenylalanine methyl ester, tyrosyl-D-alanylglycyl-*N*-methylphenylalanylglycinol, valinomycin, and gramicidin D), and hydrophobic acetoxymethyl ester (AM) derivatives of fluorescent cellular indicators [e.g., calcein-AM, Fura-2/AM, Fluo-3 AM, 2′,7′-bis(2-carboxyethyl)-5(6)-carboxyfluorescein (BCECF) AM].[5-9,11,15] Because no such drug stimulation of the vanadate-sensitive ATPase activity is observed in membranes isolated from uninfected or β-galactosidase expressing control Sf9, it has been suggested that this simple assay, in addition to MDR reversal assays performed in multidrug-resistant cancer cells, may be useful in the identification of novel chemosensitizers that interact with P-glycoprotein directly.[5,15]

Acknowledgments

The work of Balazs Sarkadi and his laboratory, Elmer M. Price, Richard C. Boucher, and Gene A. Scarborough in helping to characterize the vanadate-sensitive, drug-stimulatable ATPase activity of recombinant P-glycoprotein, and the advice and guidance of Michael M. Gottesman, Ira Pastan, and Mark C. Willingham are gratefully acknowledged. I also wish to thank Dina Shlyakhter for excellent technical assistance and Karl Münger for stimulating discussions and critical reading of this manuscript.

[32] Recombinant Vaccinia Virus Vectors for Functional Expression of P-Glycoprotein in Mammalian Cells

By MURALIDHARA RAMACHANDRA, MICHAEL M. GOTTESMAN, and IRA PASTAN

Introduction

High-level expression of a functional protein is an essential requirement in studies of P-glycoprotein (Pgp) to elucidate its structure–function relationships, membrane topology, posttranslational regulation, and bioenergetics of drug transport. Expression of Pgp in bacteria, in yeast, in insect cells using baculovirus, and in mammalian cells using transient and stable vectors has proven to be valuable for functional and biogenesis studies.[1-3]

[1] G. L. Evans, B. Ni, C. A. Hrycyna, D. Chen, S. V. Ambudkar, I. Pastan, U. A. Germann, and M. M. Gottesman, *J. Bioenerg. Biomembr.* **27,** 43 (1995).

[2] M. Ramachandra, S. V. Ambudkar, M. M. Gottesman, I. Pastan, and C. A. Hrycyna, *Mol. Biol. Cell* **7,** 1485 (1996).

0076-6879/98 $25.00

Expression in mammalian cells has the advantage that the cells used for expression resemble the cells in humans and other animals with regard to biosynthesis, posttranslational modification, sorting, assembly, and targeting of the recombinant proteins. Previously, high-level expression of Pgp in mammalian cells was achieved by selection of cells with multidrug resistance (MDR) substrates on introduction of *MDR1* cDNA.[3] However, this approach is time consuming. Also the common practice of selection of cells on introduction of *MDR1* cDNA with cytotoxic drugs to drive expression to high levels may lead to other cellular effects including activation of alternate endogenous drug resistance mechanisms and mutation of Pgp. Therefore, transient high-level mammalian expression systems in which properties of Pgp can be studied in the absence of prior drug selection offer clear advantages over systems that involve drug selection.

The recombinant vaccinia virus system has been one of the popular transient mammalian expression systems since 1982.[4,5] Vaccinia virus, a lytic DNA virus that belongs to the Poxviridae family, has been successfully used as a live vaccine to eradicate smallpox. The basis for using vaccinia virus as an expression vector is to introduce the gene of interest by targeted homologous recombination into a site of the vaccinia virus genome without considerably altering the ability of the virus to replicate and express proteins in the host cell (reviewed by Moss[6]). Several features of vaccinia viruses, including their ability to accept very large foreign DNA fragments of up to 25 kb, their ability to infect a variety of mammalian cells at nearly 100% efficiency, and the "appropriate" transport, secretion, processing, and posttranslational modifications of the protein they express, make the vaccinia virus expression system one of the systems of choice for expression of mammalian proteins.[6] One of the unique features of poxviruses including vaccinia virus is that they complete their life cycle in the cytoplasm of the infected cells. This property facilitates the integration of the foreign gene into the vaccinia virus genome and ensures that the transcription and processing of the viral DNA are carried out by viral enzymes that are packaged in the virus core. Thus, there is no dependence on nuclear transcriptional regulation and RNA processing, which generally complicate expression of viral genomes that are localized in the nucleus of the host cell. However,

[3] C. A. Hrycyna, S. Zhang, M. Ramachandra, B. Ni, I. Pastan, and M. Gottesman, *in* "Multidrug Resistance in Cancer Cells: Cellular, Biochemical, Molecular, and Biological Aspects" (S. Gupta and T. Tsuruo, eds.), p. 29. John Wiley and Sons, New York, 1996.

[4] D. Panicali and E. Paoletti, *Proc. Natl. Acad. Sci. U.S.A.* **79,** 4927 (1982).

[5] M. Mackett, G. L. Smith, and B. Moss, *Proc. Natl. Acad. Sci. U.S.A.* **79,** 7415 (1982).

[6] B. Moss, *Proc. Natl. Acad. Sci. U.S.A.* **93,** 11341 (1996).

because vaccinia virus lacks enzymes to splice out introns, only cDNAs should be used.

There are several variations of the vaccinia virus expression systems. In the most commonly used method, the desired gene is placed under the control of a vaccinia virus promoter and integrated into the viral genome.[7] In one of the popular variations, the bacteriophage T7 RNA polymerase gene is inserted into vaccinia genome such that a cDNA controlled by the bacteriophage T7 promoter, either in a transfected plasmid or a recombinant vaccinia virus, will be expressed in infected cells.[7,8] This recombinant vaccinia virus–bacteriophage T7 RNA polymerase hybrid expression (vaccinia–T7) system also utilizes the encephalomyocarditis virus internal ribosome entry site (IRES), which confers a high-affinity ribosome-binding site at the 5′ end of the RNA. The IRES is inserted between the T7 promoter and the gene of interest to improve the translatability of the largely uncapped vaccinia virus transcripts.[9] The presence of IRES has been shown to enhance the expression level of chloramphenicol acetyltransferase in the vaccinia–T7 system by 5- to 10-fold and to about 10% of the total cell protein[9] as estimated by densitometry of Coomassie blue-stained SDS–PAGE gel.

The vaccinia–T7 system offers some advantages over the other transient expression systems in that biological activities of mutants constructed in the expression/transfer plasmids can be assayed rapidly by transfection of cells infected with a vaccinia virus encoding T7 RNA polymerase (vTF 7-3),[8,10] without the necessity of generating a recombinant virus for each mutant. Detailed protocols for the use of this plasmid system for expression of Pgp are described elsewhere in this volume.[11] The amount of protein produced in this "infection–transfection" protocol is sufficiently high to carry out a number of biochemical analyses. For studies of human Pgp, the vaccinia–T7 expression from transfected plasmids is ideal to examine cell surface and total expression, binding and energy-dependent transport of drugs in intact cells, and to measure drug-stimulatable ATPase activity in crude membrane preparations.[2,11] However, the necessity of transfection limits the choice of host cell lines. Also synthesis of the recombinant protein

[7] B. Moss, *Science* **252,** 1662 (1991).

[8] T. R. Fuerst, E. G. Niles, F. W. Studier, and B. Moss, *Proc. Natl. Acad. Sci. U.S.A.* **83,** 8122 (1986).

[9] O. Elroy-Stein, T. R. Fuerst, and B. Moss, *Proc. Natl. Acad. Sci. U.S.A.* **86,** 6126 (1989).

[10] O. Elroy-Stein and B. Moss, *in* "Current Protocols in Molecular Biology" (F. M. Ausubel, R. Brent, R. E. Kinston, D. D. Moore, J. A. Smith, J. C. Seidman, and K. Struhl, eds.), p. 16.19.1. John Wiley and Sons, New York, 1991.

[11] C. A. Hrycyna, M. Ramachandra, I. Pastan, and M. M. Gottesman, *Methods Enzymol.* **292** [34] (1997) (this volume).

is achieved in only 70–80% of the cell population because of the efficiency of transfection. In the vaccinia–T7 system, instead of transfecting plasmids encoding the desired gene, generation and the use of recombinant viruses encoding the expression cassette (T7 promoter, IRES, cDNA of interest, and T7 terminator) would eliminate the limitation of transfection, increase the choice of host cells and permit scale-up of protein synthesis.

In this article, we describe methods to generate and use recombinant vaccinia virus encoding Pgp.

Strategy for Generation of Recombinant Vaccinia Virus

Construction of Transfer/Expression Plasmid

In the vaccinia–T7 system, as a first step, the desired gene is subcloned into a plasmid vector, pTM1,[8] which contains the IRES downstream of the T7 promoter. In this transfer vector, the *Nco*I site present at the 3′ end of the IRES contains the translation initiation codon and is used for insertion of the 5′ end of the protein-coding DNA fragment. The 3′ end of the DNA can be inserted into any downstream sites preceding the T7 terminator. Once the transfer/expression plasmid is constructed, protein expression can be achieved by transfection of cells that are infected with vTF 7-3. The amount of recombinant protein synthesized in the "infection–transfection" protocol is sufficient for a number of studies. For scale-up of protein synthesis, recombinant vaccinia viruses encoding the gene of interest can be generated using the same transfer/expression plasmid.

Generation and Isolation of Recombinant Vaccinia Viruses

Using segments of the vaccinia virus thymidine kinase (TK) gene flanking the expression cassette, the expression cassette in the chimeric plasmid is inserted into the TK locus of the viral genome by *in vivo* homologous recombination (Fig. 1). The recombination is achieved by infecting fibroblasts with the wild-type virus, followed by transfection with the recombinant plasmid containing the cDNA. The recombinant DNA molecules are then replicated and packaged into mature virions. The resulting recombinants are viable, because the insertion in the TK gene does not interrupt any essential function. The recombination occurs at a frequency of about 0.1% leading to inactivation of the viral TK gene, thus allowing isolation of TK^- plaques using human TK^- cells grown in the presence of a nucleoside analog, 5-bromodeoxyuridine (5-BrdU). Cells infected with the wild-type virus encoding the active TK gene product will phosphorylate and lethally incorporate the analog into DNA.

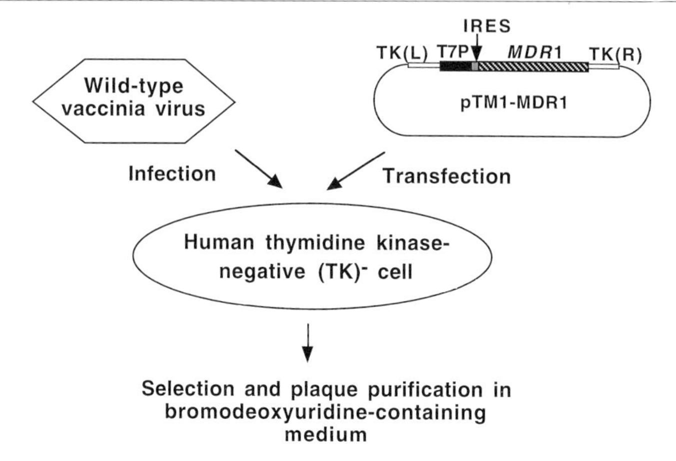

FIG. 1. An outline of the protocol used to generate vaccinia viruses encoding Pgp (see text for details).

Safety Issues

Poxviruses are infectious and can persist at ambient temperature when dried.[6] Parenteral inoculation, ingestion, and droplet or aerosol exposure of mucous membranes are the primary hazards to laboratory personnel. To prevent laboratory infection by vaccinia viruses, in the United States, the immunization practices advisory committee[12] recommends smallpox vaccination at 10-yr intervals for all personnel who work with recombinant vaccinia viruses. The vaccine (Wyeth Dryvax smallpox vaccine) can be obtained from the Centers for Disease Control and Prevention (Atlanta, GA) and should be given only under medical supervision. Eczema or an immunodeficiency disorder in a laboratory worker or a close contact may be a contraindication to vaccination. It is also recommended that the laboratory work with vaccinia virus should be carried out under biosafety level 2 conditions.[12] In addition, safety divisions or committees at the institute where research is being conducted usually must be notified for approval of vaccinia virus use. The low rate of virus-associated postimmunizing complications may be decreased with attenuated vaccinia virus strains (described later in this article). It has also been shown that TK$^-$ vaccinia viruses exhibit reduced virulence and thus may represent relatively safe recombinant vectors for laboratory research.[13] Nevertheless, precautions

[12] S. L. Katz and C. V. Broome, *Morb. Mortal. Wkly. Rep.* **40,** 1 (1991).
[13] R. M. L. Buller, G. L. Smith, K. Cremer, A. L. Notkins, and B. Moss, *Nature* (*London*) **317,** 813 (1985).

must be taken to prevent accidental exposure of personnel, to inactivate virus, and to sterilize solutions and equipment, which could contain recombinant vaccinia virus.

Techniques

The protocols described in this chapter are based on the methods described earlier,[10,14–16] which we have successfully adapted for generating vaccinia viruses encoding human Pgp.

Construction of the Transfer Vector

The wild-type or mutant cDNAs can be cloned into the vector pTM1 by following the strategy described before[2,11] using standard molecular biology techniques.[17] The vector pTM1-MDR1 was constructed by inserting the human *MDR1* cDNA at the 3′ end of the IRES downstream of the T7 promoter in pTM1.[2]

Preparation of Vaccinia Virus Stocks

To generate a recombinant virus, cells are infected with the wild-type vaccinia virus and then transfected with a pTM1-based transfer vector. In the following protocol, which describes the method to prepare the wild-type vaccinia virus stock, HeLa S3 cells are infected with the virus and after 3 days, cell lysates are prepared.

Materials

Confluent HeLa S3 (ATCC, Rockville, MD) cells grown as monolayer in 162-cm^2 tissue culture flask
Dulbecco's modified Eagle's medium (DMEM) (Quality Biological, Gaithersburg, MD) supplemented with 2.5% fetal bovine serum (FBS) (HyClone Laboratories, Logan, UT) (DMEM-2.5)
Wild-type vaccinia virus (Western Reserve strain, ATCC)
2.5% trypsin (GIBCO-BRL, Gaithersburg, MD)

[14] P. E. Earl, N. Cooper, and B. Moss, *in* "Current Protocols in Molecular Biology" (F. M. Ausubel, R. Brent, R. E. Kinston, D. D. Moore, J. A. Smith, J. C. Seidman, and K. Struhl, eds.), p. 16.16.1. John Wiley and Sons, New York, 1991.
[15] H. Chen and R. Padmanabhan, *Biotechniques* **17,** 41 (1994).
[16] P. E. Earl and B. Moss, *in* "Current Protocols in Molecular Biology" (F. M. Ausubel, R. Brent, R. E. Kinston, D. D. Moore, J. A. Smith, J. C. Seidman, and K. Struhl, eds.), p. 16.17.1. John Wiley and Sons, New York, 1991.
[17] T. Maniatis, E. F. Fritsch, and J. Sambrook, "Molecular Cloning: A Laboratory Manual." Cold Spring Harbor Laboratory Press, New York, 1982.

Method

1. Mix an equal volume of vaccinia virus stock with trypsin diluted 100-fold (or 0.025%) in 1 m*M* Tris (pH 8.0) and vortex vigorously. Incubate for 30 min at 37°, vortexing at 10- to 15-min intervals.

2. Dilute trypsinized virus in DMEM-2.5 to 2.5–7.5 plaque-forming units (pfu)/ml. Aspirate medium from HeLa monolayer cultures grown in 162-cm^2 flasks and add 4 ml trypsinized and diluted virus. Incubate for 2 hr in a CO_2 incubator at 37°, rocking the flask at every 30-min interval.

3. Add 25 ml DMEM-2.5 and incubate for 3 days in a CO_2 incubator at 37°.

4. Harvest infected cells by scraping, and pellet the cells by centrifuging at 1800*g* for 5 min at 4°.

5. Resuspend the cell pellet in DMEM-2.5 (2.0 ml for 162-cm^2 flask) by vortexing. Prepare the cell lysate by subjecting the cell suspension to three rounds of free–thaw cycles (5 min on a dry ice–ethanol bath, 5 min at 37° and vortex). Store the resulting virus stock in aliquots at −70°.

The titer of the virus stock can be determined by following the protocol described later in this article.

Transfection of Wild-Type Vaccinia Virus-Infected Cells with Vaccinia Transfer Vector

Monkey kidney (CV1) cells are first infected with the wild-type vaccinia virus and then transfected with the transfer vector. Recombination results in the generation of recombinant viruses.

Materials

CV1 cells (ATCC)
DMEM supplemented with no serum (serum-free DMEM), 2.5% FBS (DMEM-2.5), and 10% FBS (DMEM-10)
Wild-type vaccinia virus stock (Western Reserve strain)
0.025% trypsin (1:100 dilution of 2.5% trypsin in 1 m*M* Tris, pH 8.0)
Lipofectin (1 mg/ml)(GIBCO-BRL)

Method

1. Seed a 25-cm^2 flask with 1×10^6 CV1 cells in DMEM-10. Incubate overnight at 37° in a CO_2 incubator to grow to near confluency.

2. Mix an equal volume of the wild-type vaccinia virus stock and 0.025% trypsin and vortex vigorously. Incubate for 30 min in a 37° water bath, vortexing at 5- to 10-min intervals. Dilute the trypsinized virus in DMEM-2.5 to 1.5×10^5 pfu/ml.

3. Aspirate the old medium from the monolayer of CV1 cells and add 1 ml diluted viral solution (0.05 pfu/cell). Incubate for 2 hr at 37° in a CO_2 incubator, gently rocking at ~15-min intervals.

4. Approximately 45 min before the end of the infection period, add 17 μl lipofectin and 6 μg recombinant plasmid DNA to 1 ml serum-free DMEM in a 12- × 75-mm polystyrene tube, and gently mix. Leave at room temperature for 30–45 min.

5. Aspirate the virus inoculum from the monolayer of CV1 cells and add 1 ml DNA-lipofectin mixture. Incubate for 4 hr at 37° in a CO_2 incubator.

6. After a 4-hr incubation, add 4 ml of DMEM-10 and incubate for 2 days.

7. Dislodge the cells along with the medium by using a scraper. Collect the cell pellet by centrifugation at 1800*g* for 5 min at 4°.

8. Resuspend the cell pellet in 0.5 ml DMEM-2.5 and prepare the cell lysate by subjecting to three rounds of freeze–thaw cycles. Store the cell lysate at −70° until needed in the selection and screening procedure.

Isolation of Recombinant Virus by Plaque Purification

Lysate from infected-transfected cells (obtained in the previous step) is serially diluted and the dilutions are used to infect human TK^- cells grown in 96-well tissue culture plates in a medium containing BrdU. Well-isolated plaques are visualized by neutral red staining.

Materials

5 mg/ml BrdU (Aldrich, Milawaukee, WI) in water (filter sterilize and store at −20°)
DMEM containing 10% FBS and 50 μg/ml (1 : 100 dilution of the stock) BrdU (DMEM-10–BrdU)
DMEM containing 2.5% FBS and 50 μg/ml BrdU (DMEM-2.5–BrdU)
Human TK^-143 B osteosarcoma (TK^-) monolayer cells (ATCC) maintained in DMEM-10–BrdU
0.33% neutral red solution (GIBCO-BRL)
A cup sonicator such as Tekmar Sonic Disrupter (Tekmar Co, Cincinnati, OH)

Method

1. Trypsinize and seed TK^- cells in 96-well plates (3×10^6/plate) using DMEM-10–BrdU (100 μl/well). Incubate overnight at 37°.

2. Briefly sonicate (~30 sec) the infected-transfected cell lysate to break the clumps formed during storage using a cup sonicator and prepare serial dilutions (10 ml for each dilution) in DMEM-2.5–BrdU. Desired dilutions

for the first round of plaque purifications are 10^{-2}, 10^{-3}, and 10^{-4}. (*Note:* Occasionally, higher dilutions may be needed to obtain well-isolated plaques.)

3. Remove the medium from the wells of 96-well plates and add 100 μl of the diluted lysate to each well. Incubate at 37° for 2 days.

4. Remove the diluted infected–transfected cell lysate from the wells and gently (without disturbing the cell monolayer) add 100 μl DMEM-2.5–BrdU–neutral red medium (prepared by adding 227 μl of 0.33% neutral red to 30 ml DMEM-2.5–BrdU) to each well. Incubate at 37° for 4 hr to overnight.

5. Examine the plates on a light box and mark the wells containing single well-isolated plaques using a permanent marker. Incubate the plates containing marked wells at −70° for 1 hr to freeze the contents.

6. Thaw the plates and collect isolated plaques with the medium in separate sterile microcentrifuge tubes after scraping with the tip of a pipette. Subject the plaque isolates to two additional rounds of freeze–thaw cycles. Store the plaque isolates at −70° till further use.

7. Briefly sonicate (~30 sec) the plaque isolates using a cup sonicator and use 50 μl to infect confluent TK^- cells grown in 6-well plates. For infection, remove the medium, and add 50 μl lysate and 2 ml DMEM-2.5–BrdU to each well. Incubate for 2 days at 37°.

8. After 2 days, remove the medium, add 500 μl DMEM-2.5–BrdU to each well, and collect the cells with the medium after scraping with the back of a 1-ml pipette tip in separate microcentrifuge tubes. Prepare the cell lysates by subjecting to three rounds of freeze–thaw cycles as described before. Store the lysates at −70° until needed.

Screening of the Recombinant Vaccinia Virus Plaques

The plaques derived from vaccinia viruses encoding the gene of interest can be identified by methods such as DNA hybridization, polymerase chain reaction (PCR) screening, immunostaining, or immunoblot analysis. We routinely perform immunoblot analysis to identify recombinant vaccinia virus plaques encoding human *MDR1* as described next.

Materials

Near confluent HeLa or TK^- cells grown in 6-well 35-mm tissue culture dishes

DMEM-2.5 and DMEM-10

Cell lysis buffer containing 10 mM Tris (pH 8.0), 0.1% Triton X-100, 10 mM $MgSO_4$, 2 mM $CaCl_2$, 1 mM dithiothreitol (DTT) (freshly added), and 20 μg/ml DNase (Sigma) (freshly added)

Vaccinia virus encoding T7 RNA polymerase (vTF 7-3) (ATCC) stock

Method

1. Replace the medium from each well of 6-well dishes with 150 μl of the lysates from plaque isolates amplified in 6-well dishes (prepared in a previous step) and 350 μl DMEM-2.5 containing an appropriate volume of vTF 7-3 (at 2 pfu/cell; $\sim 1.0 \times 10^6$ cells in each well of a 6-well dish). Incubate at 37° for 1 hr in a CO_2 incubator with gentle rocking at ~15-min intervals.

2. Add 1.5 ml DMEM-10 and incubate for 24 hr in a CO_2 incubator at 37°.

3. Collect the infected cells after scraping either with the back of a sterile 1-ml pipette tip or plunger of 1-ml syringes. Spin in a microcentrifuge for 30 sec to pellet the cells. Aspirate and discard the medium.

4. Add 100 μl cell lysis buffer, mix well by vortexing, and subject to three rounds of freeze–thaw cycles.

5. Dilute 10 μl of the lysate (~40 μg total protein) with 2× Laemmli sample buffer[18] and subject to SDS–PAGE (8% gel) and immunoblot analysis.[2]

Second Round of Plaque Purification

Follow the steps described for the first round of plaque purification, but by using 10^{-4}, 10^{-5}, and 10^{-6} dilutions. After identifying the plaques that express Pgp by immunoblot analysis, amplify the isolated plaques in 6-well dishes and confirm protein expression as described before. Use the remaining lysates to infect TK^- cells grown in 75-cm^2 tissue culture flask in the presence of BrdU. Harvest cells after 2 days and prepare lysates in 500 μl DMEM without BrdU.

Large-Scale Amplification and Purification of Recombinant Virus

The cell lysates containing the plaque isolate are used to infect HeLa S3 monolayers and a large stock of virus is prepared. The same protocol is used to prepare vTF 7-3 stocks.

Materials

Confluent HeLa S3 monolayer cells grown in 162-cm^2 flasks (5–10 flasks for each virus)
DMEM-2.5
2.5% trypsin

[18] U. K. Laemmli, *Nature* **227,** 680 (1970).

10 m*M* Tris (pH 9.0) and 1 m*M* Tris (pH 9.0) (autoclaved or filter sterilized)
36% (w/v) sucrose in 10 m*M* Tris (pH 9.0) (autoclaved)

Method

1. Dilute the virus stock in 3 ml DMEM-2.5 and sonicate for 30 sec in a cup sonicator.

2. Aspirate the medium from 162-cm^2 flasks and add 4 ml DMEM-2.5 containing the diluted virus stock. If the virus stock is of known titer, use 1–3 pfu/cell. For a newly generated virus, it is not necessary to titer before large-scale amplification; the stock can be diluted depending on the number of flasks used. Leave for 2 hr in a CO_2 incubator at 37° with rocking at 20- to 30-min intervals.

3. Add 25 ml DMEM-2.5 and incubate for 3 days in a CO_2 incubator at 37°.

4. Detach the infected cells from the flask by scraping, and pellet the cells by centrifuging at 1800*g* for 5 min at 4°. Discard the supernatant.

5. Resuspend the cells in 10 m*M* Tris (pH 9.0). Use 2 ml for 2×10^7 cells (this is usually the number of cells from a 162-cm^2 tissue culture flask). Incubate on ice for 5 min.

6. Dounce homogenize to break the cells; 20–25 strokes with a tight fitting homogenizer should be sufficient. Check for complete cell breakage by microscopy. This and all subsequent manipulations should be done on ice.

7. Spin out the nuclei by centrifuging at 500*g* for 10 min at 4°. Resuspend the nuclear pellet in 5 ml 10 m*M* Tris (pH 9.0) and centrifuge again at 500*g* for 10 min. Combine the supernatants.

8. Add 0.01 volume of 2.5% trypsin and incubate at 37° for 30 min, vortexing every 10 min.

9. Layer onto an equal volume of 36% (w/v) sucrose in 10 m*M* Tris (pH 9.0) and centrifuge at 32,000*g* for 80 min at 4°.

10. Discard the supernatant. Resuspend the virus pellet in 1 m*M* Tris (pH 9.0) and store in aliquots at −70°.

Plaque Assay to Determine the Titer

CV1 monolayer cells are infected with serial dilutions of the virus stock. After 2 days, plaques are visualized by staining with crystal violet.

Materials

Confluent monolayer of CV1 cells
DMEM-10 and DMEM-2.5
0.1% (w/v) crystal violet (Sigma) in 20% (v/v) ethanol

Method

1. Seed wells of 6-well tissue culture dishes with CV1 cells in 2 ml DMEM-10 at a density of 5×10^5 cells per well. Incubate overnight in a CO_2 incubator at 37°.

2. Dilute the virus stock 100-fold in DMEM-2.5 and sonicate for 30 sec in a cup sonicator. Prepare further serial dilutions in the same medium to obtain 1 ml each of 10^{-8}, 10^{-9}, and 10^{-10}, using a fresh pipette for each dilution.

3. Aspirate the medium from 6-well tissue culture dishes containing confluent CV1 cells and infect cells in duplicate wells with 0.5 ml of the 10^{-8}, 10^{-9}, and 10^{-10} dilutions. Incubate the dishes at 37° in a CO_2 incubator, rocking at 15- to 30-min intervals.

4. Overlay each well with 1.5 ml DMEM-2.5 and place in a CO_2 incubator for 2 days at 37°.

5. Remove the medium and add 1.0 ml of 0.1% crystal violet solution to each well. Leave for 5 min at room temperature.

6. Aspirate the dye solution and allow wells to dry.

7. Determine the titer by counting plaques within the wells and multiplying by the dilution factor. Plaques appear as 1- to 2-mm-diameter clear areas.

Infection for Expression and Functional Analysis of Pgp

For expression of human Pgp, we have evaluated several mammalian cells including HeLa S3 (cervical epidermoid carcinoma), MCF-7 (breast adenocarcinoma), and human osteosarcoma cells. Cells are coinfected with vaccinia virus encoding *MDR1* (vvMDR1) and vTF 7-3 at a multiplicity of infection of 10 each, and harvested at 24 or 48 hr postinfection.

Materials

Near confluent HeLa S3 monolayer or other mammalian cells seeded in 75-cm^2 flasks (5–10 flasks for each virus) the day before
vTF 7-3 stock
DMEM-2.5 and 10

Method

1. Dilute the appropriate volume of vTF 7-3 and vvMDR1 (10 pfu/cell for each virus) in 3 ml DMEM-2.5.

2. Aspirate the medium from 75-cm^2 flask and replace with DMEM-2.5 containing the viruses. Incubate for 1 hr in a CO_2 incubator at 37°, gently rocking at 10- to 15-min intervals.

3. Add 12 ml DMEM-10 and incubate for 24–48 hr at 32° in a CO_2 incubator.

Analysis of Expression and Function of P-Glycoprotein

Essentially the same methods that are used for analyzing Pgp expressed in the "infection–transfection" protocol[2,11] can be used for characterizing Pgp synthesized in recombinant vaccinia virus-infected cells. These techniques include fluorescence-activated cell sorting (FACS) analysis of the intact cells after staining with surface epitope-specific human Pgp antibodies such as MRK-16[19] or UIC2[20] to determine cell-surface expression (Fig. 2A), immunoblot analysis with C219[21] (Fig. 2B) or other antibodies to estimate total protein expression, photoaffinity labeling with either whole cells or crude membrane preparations using [^{3}H]azidopine or [^{125}I]iodoarylazidoprazosin, photoaffinity labeling with [^{32}P]azido-ATP using crude membranes, measurement of drug-stimulated ATPase activity using crude membranes, and determination of drug transport using either fluorescent (Fig. 3) or radiolabeled compounds.

Recent Developments and Future Directions

The undesirable features of the traditional recombinant vaccinia virus systems, including the one described here, are the cytopathic effects of vaccinia virus and the safety precautions required for its use. It has been shown that the highly attenuated and avian host-range-restricted vaccinia virus known as modified vaccinia Ankara (MVA) can replace replication competent strains of vaccinia virus for employing the vaccinia–T7 system.[22] In addition, the host range restriction of MVA provides an added degree of personal safety in the laboratory. Plasmid vectors for construction of MVA recombinants[23,24] and the potential use of recombinant MVA as a safe vaccine candidate have been reported.[23] At the National Institutes of Health, work exclusively with MVA is permitted at biosafety level 1 without vaccination.[6]

Vaccinia virus infections even with the attenuated strains eventually lead to cell death after several days in culture. The time between the expression of the recombinant protein and lysis of the host cell limits the

[19] H. Hamada and T. Tsuruo, *Proc. Natl. Acad. Sci. U.S.A.* **83,** 7785 (1986).
[20] E. B. Mechetner and I. B. Roninson, *Proc. Natl. Acad. Sci. U.S.A.* **89,** 5824 (1992).
[21] E. Georges, G. Bradley, J. Gariepy, and V. Ling, *Proc. Natl. Acad. Sci. U.S.A.* **87,** 152 (1990).
[22] L. S. Wyatt, B. Moss, and S. Rozenblatt, *Virology* **210,** 202 (1995).
[23] G. Sutter, L. S. Wyatt, P. L. Foley, J. R. Bennink, and B. Moss, *Vaccine* **12,** 1032 (1994).
[24] M. W. Carroll and B. Moss, *Biotechniques* **19,** 352 (1995).

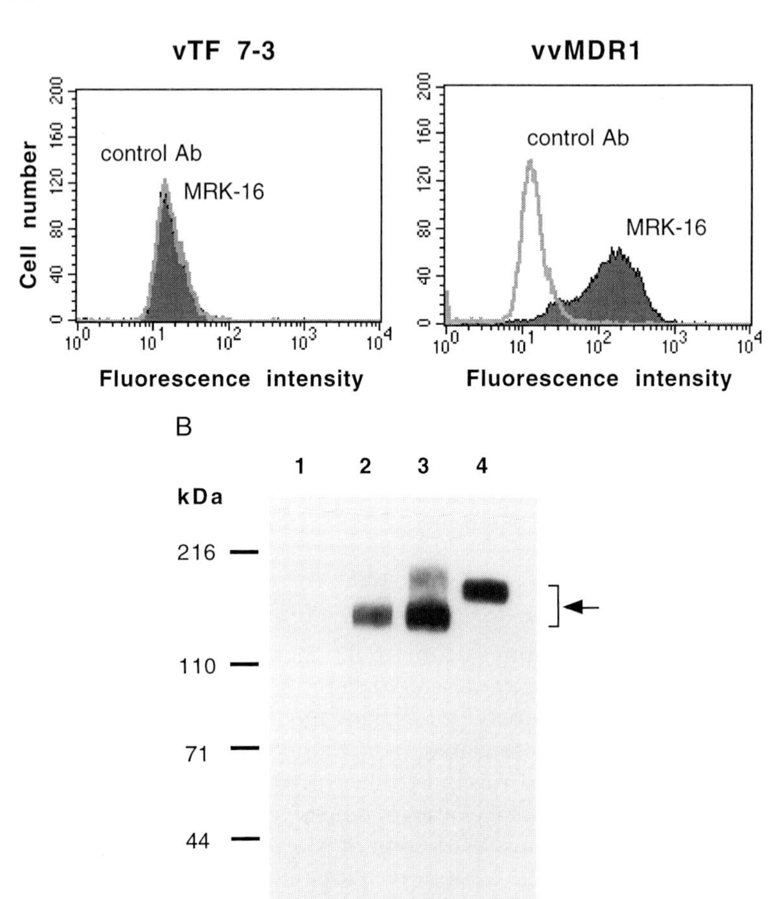

FIG. 2. Expression of Pgp using recombinant vaccinia viruses. HeLa cells were infected with vTF 7-3 alone (vTF 7-3) or coinfected with vTF 7-3 and vvMDR1 (vvMDR1) at a multiplicity of infection of 10 each. (A) Analysis of cell surface expression of Pgp. Twenty-four hours after infection, cells were subjected to FACS analysis after staining with human Pgp surface epitope-specific monoclonal antibody MRK-16 or mouse IgG_{2a} (as a negative control). (B) Immunoblot analysis of the membrane preparations from HeLa cells infected with vTF 7-3 (lane 1), HeLa cells infected with vTF 7-3 and transfected with pTM1-MDR1 (lane 2), HeLa cells coinfected with vTF 7-3 and vvMDR1 (lane 3), and *MDR1*-transfected and colchicine-selected NIH 3T3 cells [S. J. Currier, S. E. Kane, M. C. Willingham, C. O. Cardarelli, I. Pastan, and M. M. Gottesman, *J. Biol. Chem.* **267,** 25,153 (1992)] (lane 4). In each lane 0.75 μg of membrane protein was loaded. The position of Pgp is shown by an arrow. Compared to NIH 3T3 cells, the majority of the recombinant Pgp molecules synthesized in vaccinia virus-infected cells migrate faster because of either unglycosylation or underglycosylation for reasons not clearly known.

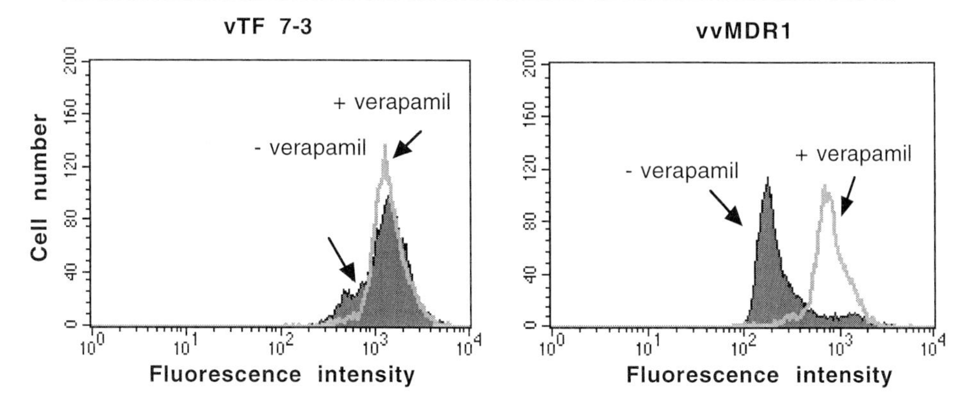

FIG. 3. Rhodamine 123 accumulation and the effect of verapamil on rhodamine 123 accumulation. Rhodamine 123 accumulation in HeLa cells infected with vTF 7-3 alone (vTF 7-3) (as a control) or coinfected with vTF 7-3 and vvMDR1 (vvMDR1) was determined by FACS in the presence or absence of 30 μM verapamil, a known reversing agent.

types of experiments that can be performed. In the case of Pgp expression using vaccinia viruses, it is not possible to perform cell proliferation assays to measure relative resistance to MDR drugs. It may be possible to overcome some of the shortcomings of the vaccinia virus expression system by cloning the desired gene into a conditonal-lethal mutant vaccinia virus,[25] or by performing the experiments under conditions in which virus DNA replication and late gene expression are inhibited (e.g., in the presence of cytosine arabinoside[26]). In the future, availability of a noncytopathic virus or modified conditions may allow one to use a recombinant vaccinia virus encoding Pgp in cytotoxicity assays to determine the relative resistance to drugs.

The major advantages of recombinant vaccinia viruses over other transient expression systems are that the protein of interest can be expressed in a variety of mammalian cell types at high efficiency. Thus, recombinant vaccinia viruses are useful in expression and functional analysis of Pgp and related transporters in virtually any desired mammalian cell line. The use of vaccinia–T7 system may be valuable to evaluate other reported[27] and postulated activities of Pgp, especially if these activities require cell-specific factors or posttranslational modifications.

[25] Y. Zhang and B. Moss, *Proc. Natl. Acad. Sci. U.S.A.* **88,** 1511 (1991).

[26] B. Moss, *in* "Recombinant Poxviruses" (M. M. Binns and G. L. Smith, eds.), p. 45. CRC Press, Boca Raton, Florida 1992.

[27] S. M. Simon and M. Schindler, *Proc. Natl. Acad. Sci. U.S.A.* **91,** 3497 (1994).

[33] Functional Expression of Human P-Glycoprotein from Plasmids using Vaccinia Virus–Bacteriophage T7 RNA Polymerase System

By CHRISTINE A. HRYCYNA, MURALIDHARA RAMACHANDRA, IRA PASTAN, and MICHAEL M. GOTTESMAN

Introduction

Transient expression of recombinant proteins in mammalian cells provides a rapid and convenient way to study the biochemistry of a protein of interest without the confounding effects that are normally associated with many of the stable expression systems.[1] However, many of the systems currently available do not allow for high-level expression of protein. A recent protocol designed to circumvent this problem uses HEK293 cells and histidine-tag affinity chromatography to purify partially and enrich for transiently expressed protein for use in *in vitro* biochemical analyses.[2,3] Transient viral transductants are known to produce larger amounts of recombinant protein within 24–48 hr; however, constructing the recombinant viruses is time consuming. A notable exception is the recombinant vaccinia-T7 RNA polymerase (vaccinia-T7) system developed by Bernard Moss and colleagues at the National Institutes of Health.[4–6] In this system, on infection with a recombinant vaccinia virus encoding the bacteriophage T7 RNA polymerase (vTF7-3), a cotransfected cDNA under control of the bacteriophage T7 promoter on a plasmid is expressed at high levels.[4,7]

For transient expression using the vaccinia-T7 system, virtually any mammalian cell can serve as the host and typically 70–80% of cells will express the protein of interest. Because it is believed that viral entry of the vaccinia virus apparently involves fusion of the viral envelope with the plasma membrane, this high transfection efficiency in the vaccinia-T7 system can be attributed to the fact that the cationic liposome-mediated transfec-

[1] C. A. Hrycyna, S. Zhang, M. Ramachandra, B. Ni, I. Pastan, and M. Gottesman, *in* "Multidrug Resistance in Cancer Cells: Cellular, Biochemical, Molecular, and Biological Aspects" (S. Gupta and T. Tsuruo, eds.), p. 29. John Wiley and Sons, New York, 1996.

[2] T. W. Loo and D. M. Clarke, *J. Biol. Chem.* **270,** 21449 (1995).

[3] T. W. Loo and D. M. Clarke, *Methods Enzymol.* **292,** [35], 1998 (this volume).

[4] T. R. Fuerst, E. G. Niles, F. W. Studier, and B. Moss, *Proc. Natl. Acad. Sci. U.S.A.* **83,** 8122 (1986).

[5] B. Moss, *Proc. Natl. Acad. Sci. U.S.A.* **93,** 11341 (1996).

[6] B. Moss, *Science* **252,** 1662 (1991).

[7] T. R. Fuerst, P. L. Earl, and B. Moss, *Mol. Cell Biol.* **7,** 2538 (1987).

METHODS IN ENZYMOLOGY, VOL. 292 0076-6879/98 $25.00

tion and infection steps are performed simultaneously allowing coentry of plasmid DNA and virus to each cell.[6] This procedure, however, is not completely efficient; although all of the cells are infected by the virus, only 70–80% contain plasmid DNA. Transfected plasmids are transcribed in the cytoplasm of vaccinia virus-infected cells, avoiding the requirement of plasmid transit to the nucleus, thus allowing virtually all transfected cells to express the protein of interest. Additionally, the presence of an encephalomyocarditis virus internal ribosomal entry site (IRES) sequence downstream from the T7 promoter on the expression plasmid makes the translation of the viral transcripts cap independent, significantly enhancing the expression level.[8] Alternatively, recombinant viruses encoding a similar expression cassette (T7 promoter-IRES-cDNA-T7 terminator) can also be generated.[9]

We have successfully expressed human P-glycoprotein (Pgp) in a variety of mammalian cells using this recombinant vaccinia-T7 RNA polymerase system and determined it to be a useful tool for rapidly analyzing the functional properties of Pgp.[10] The coinfection and transfection expression system is well suited for studying great numbers of mutant/chimeric constructs since large amounts of recombinant Pgp can be expressed rapidly and the cells remain intact allowing for whole-cell *in vivo* experiments to be performed (Fig. 1). Importantly, it is unnecessary to impose any drug selection on these cells, eliminating bias due to any possible pleiotropic cellular effects of drug exposure. We have also generated recombinant vaccinia viruses encoding Pgp[9]; however, although this method somewhat increases the amount of protein made and allows for virtually 100% of infected cells to express Pgp, this pursuit is time consuming and is not amenable to the study of numerous constructs. For mutational studies, there is no need to generate a new virus for each construct, because almost all functional aspects of Pgp can be studied upon protein expression using the infection–transfection procedure.

The protocols outlined in this article describe various methods used to express and functionally characterize human Pgp including the infection–transfection procedure, immunoblot analysis, photoaffinity labeling with nucleotide and substrate analogs, drug-stimulatable ATPase activity assays, fluorescent and radioactive substrate accumulation/efflux assays in intact cells, and determination of cell surface localization by flow cytometry. We

[8] O. Elroy-Stein, T. R. Fuerst, and B. Moss, *Proc. Natl. Acad. Sci. U.S.A.* **86,** 6126 (1989).

[9] M. Ramachandra, M. M. Gottesman, and I. Pastan, *Methods Enzymol.* **292,** [32], 1998 (this volume).

[10] M. Ramachandra, S. V. Ambudkar, M. M. Gottesman, I. Pastan, and C. A. Hrycyna, *Mol. Biol. Cell* **7,** 1485 (1996).

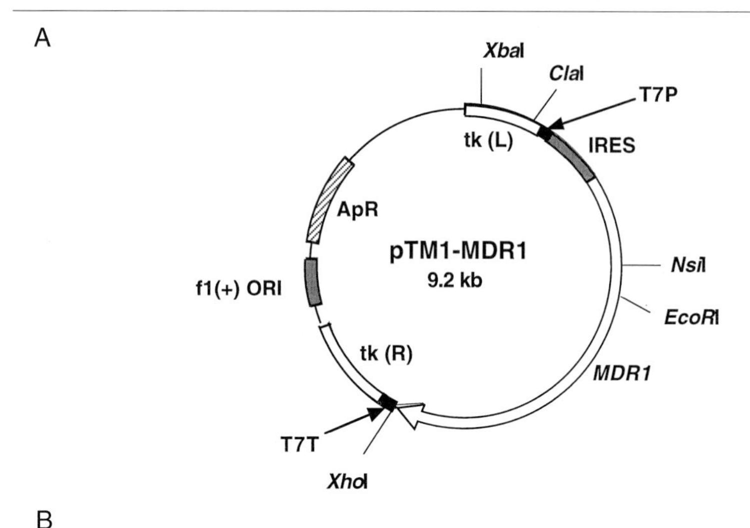

FIG. 1. Pgp-expression plasmid and schematic diagram of the infection–transfection protocol and functional assays. (A) The unique restriction sites and main features of the expression plasmid pTM1-*MDR1* are shown. In this vector, *MDR1* is controlled by the T7-promoter (T7P) and the presence of an IRES sequence between the T7-promoter and *MDR1* sequence facilitates cap-independent translation on transfection of cells infected with vTF 7-3, a recombinant vaccinia virus encoding T7 RNA polymerase. The T7 terminator (T7T) is present at the 3′-end of the *MDR1* cDNA. The expression cassette is flanked by segments of the vaccinia

have successfully adapted these assays for use with virally infected–transfected cells.[10]

Techniques and Protocols

Construction of Expression Vectors

The parental plasmid vector pTM1[11] is most commonly used to drive expression of cDNAs in this recombinant vaccinia virus-bacteriophage T7 RNA polymerase system. This plasmid, propagated in *Escherichia coli,* contains the T7 promoter, an IRES, a multiple cloning site for insertion of the cDNA of interest, and the T7 terminator (Fig. 1). It is ideal to place the cDNA of interest at the *Nco*I site present at the 3′-end of the IRES to maximize transcription efficiency. The 3′-end of the cDNA can be cloned into any site preceding the T7 terminator sequence. We used the following protocol to clone the human *MDR1* cDNA into the transfer vector pTM1 to obtain pTM1-*MDR1*.

Materials

Restriction enzymes (New England Biolabs, Beverly, MA)
Calf intestinal phosphatase (CIP) and polynucleotide kinase (New England Biolabs)
Competent DH5α *E. coli* (Life Technologies, Grand Island, NY)

Method

1. Obtain the DNA fragment (1183 bp) between the second codon and the *Eco*RI site in the coding sequence from human *MDR1* by polymerase chain reaction (PCR) using pMDRXS(WT) (Sugimoto, Gottesman, and Pastan, unpublished data) as a template.
2. Phosphorylate the 5′-end with polynucleotide kinase and digest with *Eco*RI and subclone the PCR fragment into the linearized vector

[11] O. Elroy-Stein and B. Moss, *in* "Current Protocols in Molecular Biology" (F. M. Ausubel, R. Brent, R. E. Kinston, D. D. Moore, J. A. Smith, J. C. Seidman, and K. Struhl, eds.), p. 16.19.1–16.19.9. John Wiley and Sons, New York, 1991.

virus thymidine kinase (tk) gene, which allows for the use of this construct to generate and isolate recombinant vaccinia virus. (B) Outline of the infection–transfection protocol and functional assays that can be performed in the infected–transfected cells. [Reproduced from *Molecular Biology of the Cell* (1996, Volume 7, 1485–1498), with permission of the American Society for Cell Biology.]

pTM1[11] that was prepared by digestion with *Nco*I, treatment with the Klenow fragment of *E. coli* DNA polymerase followed by digestion with *Eco*RI, to obtain pTM1-*MDR1*(NE).

3. Double-digest pTM1-*MDR1*(NE) with *Nsi*I and *Xho*I and ligate to the remaining coding sequence of *MDR1* isolated from pMDRXS(WT) by digesting with the same enzymes. The resultant expression vector is pTM1-*MDR1*.
4. Mutations can be readily introduced into pTM1-*MDR1* by cassette replacement of unique fragments made by sequence overlap PCR techniques.[12] All sequences should be subsequently verified by either manual or automated DNA sequencing (PRISM™ Ready Reaction DyeDeoxy™ Terminator Sequencing Kit, Perkin-Elmer Corporation, Norwalk, CT).

Expression of Pgp by an Infection–Transfection Protocol

The protocol[11] described next has been optimized for the amount of Pgp expression in human osteosarcoma and HeLa cells. These cell lines were chosen because they have negligible amounts of endogenous Pgp expression and low basal membrane-associated ATPase activity, and they are easily transfected. Similar conditions can also be used for other cell types although varying the time of expression as well as the temperature of incubation may be necessary. Cells are normally harvested at 24 hr for assays with intact cells and at 48 hr for isolating crude membranes after infection–transfection.

Materials

Minimum essential medium with Earle's salts (Life Technologies) supplemented with 4.5 g/liter glucose, 5 m*M* L-glutamine, 50 units/ml penicillin, 50 μg/ml streptomycin, and 10% fetal bovine serum (FBS) (HyClone Laboratories, Logan, UT) (EMEM). HeLa cells are grown in Dulbecco's modified Eagle's medium (DMEM) supplemented with 4.5 g/liter glucose, 5 m*M* L-glutamine, 50 units/ml penicillin, 50 μg/ml streptomycin, and 10% FBS (Quality Biological, Gaithersburg, MD)

Human osteosarcoma (HOS) American Type Culture Collection (ATCC) cells or HeLa cells were propagated as monolayer cultures at 37° in 5% CO_2 in their respective medium

[12] R. Higuchi, *in* "PCR Technology" (H. A. Erlich, ed.), pp. 61–70. W. H. Freeman and Company, New York, 1992.

Recombinant vaccinia virus encoding bacteriophage T7 RNA polymerase (vTF7-3) American Type Culture Collection (ATCC), with vTF7-3 propagated and purified as previously described[9–11,13]
Lipofectin® reagent (1 mg/mL) (Life Technologies)
Opti-MEM reduced serum medium (Life Technologies)
pTM1-*MDR1* plasmid DNA

Method

1. Split the cells the day before into 75-cm^2 tissue culture flasks such that they will be about 70–80% confluent on the day of infection–transfection.
2. For each transfection, place 3.0 mL Opti-MEM into a 15-mL polystyrene tube. Since DNA–Lipofectin complexes are known to bind to polypropylene, it is essential to use polystyrene tubes.
3. Vortex the Lipofectin solution, add 45 μl (45 μg) Lipofectin reagent to the Opti-MEM and vortex.
4. Add DNA to attain a 3:1 lipid:DNA ratio by adding 15 μg DNA (pTM1 as control or desired pTM1-*MDR1* construct) to the medium containing Lipofectin and gently mix by swirling. Do not vortex the solution after adding the DNA.
5. Incubate the tubes at room temperature for 20–30 min.
6. Five minutes before the end of the incubation period, add vTF7-3 to Opti-MEM and vortex 15–30 sec. Routinely, use 10 plaque forming units (pfu)/cell and use 0.5 mL of Opti-MEM per flask. If the virus is to be used for more than one flask, a master mix should be prepared. However, first vortex the virus in a small volume of Opti-MEM (0.5–3 mL), then dilute to the appropriate volume and vortex again to mix. If a nonpurified virus is being used, sonicate the virus in 0.5–3 mL Opti-MEM for 30 sec in a bath sonicator filled with 50% ice in water, dilute accordingly, and then vortex to mix.
7. Wash the cells once with 5–10 mL Opti-MEM medium. Add the 0.5 mL of the viral dilution to the flask and then gently place the DNA–Lipofectin mix directly on the cells. Mix the solutions well in the flask and incubate for 4 hr at 32° or 37°. For the first hour, gently swirl the solution in the flask every 15 min. The total volume of the infection–transfection is 3.5 mL.

[13] P. E. Earl, N. Cooper, and B. Moss, *in* "Current Protocols in Molecular Biology" (F. M. Ausubel, R. Brent, R. E. Kinston, D. D. Moore, J. A. Smith, J. C. Seidman, and K. Struhl, eds.), pp. 16.16.1–16.16.7. John Wiley and Sons, New York, 1991.

8. After 4 hr, add 12 ml complete medium and incubate for desired periods of time. Expression can be seen as early as 8 hr post–infection–transfection and maximum expression is usually achieved by 48 hr. Functional studies in intact cells are carried out at 20–24 hr post–infection–transfection.

Preparation of Crude Membranes

Crude membrane preparations from infected–transfected cells can be used for a number of applications including immunoblot analysis, photoaffinity nucleotide- and drug-binding studies, and drug-stimulatable ATPase activity assays.

Materials

Phosphate buffered saline without Ca^{2+} and Mg^{2+} (PBS)
Hypotonic lysis buffer: 10 m*M* Tris (pH 7.5), 10 m*M* NaCl, and 1 m*M* $MgCl_2$. Add the following reagents just prior to use: 1 m*M* dithiothreitol (DTT), 1% (v/v) aprotinin solution (0.1 U/mL) (Sigma, St. Louis, MO), and 2 m*M* 4-(2-aminoethyl)benzenesulfonyl fluoride hydrochloride (AEBSF) (ICN, Irvine, CA)
Micrococcal nuclease (*Staphylococccus aureus*) (Pharmacia Biotech, Piscataway, NJ)
Dounce homogenizer with pestle A
TSNa resuspension buffer: 10 m*M* Tris (pH 7.5), 50 m*M* NaCl, 250 m*M* sucrose. Add 1 m*M* DTT, 1% (v/v) aprotinin solution (0.1 U/mL), and 2 m*M* AEBSF just prior to use

Method

1. Harvest the cells by scraping and wash twice in PBS containing 1% (v/v) aprotinin solution (0.1 U/mL).
2. Resuspend the cell pellet in hypotonic lysis buffer in a volume of approximately 0.5 mL per 75-cm^2 tissue culture flask and freeze at −80°. Subsequently, thaw the cells and incubate on ice for 30–45 min.
3. Disrupt the cells using 50 strokes with a Dounce homogenizer using tight fitting pestle A only.
4. Dilute the homogenate two-fold with hypotonic lysis buffer and subsequently remove the undisrupted cells and nuclear debris by centrifugation at 500*g* for 10 min at 4°.
5. Transfer the low-speed supernatant to a new tube and incubate with micrococcal nuclease (50 U/mL) in the presence of 1 m*M* $CaCl_2$ for 20–30 min on ice.

6. Collect the membranes by centrifugation for 60 min at 100,000*g*, and resuspend in resuspension buffer containing 10% glycerol using a 1–3 ml syringe and a blunt-ended 23-gauge needle.
7. Store the resuspended membranes in small aliquots at −70° immediately and determine the protein content by a modified Lowry method[14] using bovine serum albumin as a standard.

Sodium Dodecyl Sulfate–Polyacrylamide Gel Electrophoresis (SDS–PAGE) and Immunoblot Analysis

Materials

SDS–PAGE gels
Laemmli SDS–PAGE sample loading buffer[15]
Running buffer: 25 m*M* Tris-glycine (pH 8.3) and 0.2% (w/v) SDS
Transfer buffer: 25 m*M* Tris-glycine (pH 8.3) and 0.2% (w/v) SDS + 20% (v/v) methanol (SDS optional)
Phosphate buffered saline without Ca^{2+} and Mg^{2+} with 0.05% (v/v) Tween 20 (PBST)
Nitrocellulose membranes (0.45-μm pore size, Schleicher and Schuell, Keene, NH)
Monoclonal anti-P-glycoprotein antibody C219 (Centocor, Malvern, PA)[16]
Secondary antibody: peroxidase-conjugated goat anti-mouse IgG (H + L) (Life Technologies)
Nonfat dry milk

Method

1. Incubate membrane preparations in an appropriate volume of Laemmli SDS–PAGE sample loading buffer (1× final concentration) at room temperature for 30 min.
2. Perform electrophoresis on 8.0% polyacrylamide gels at 170 V in running buffer.
3. Electroblot the proteins onto a nitrocellulose membrane in transfer buffer at 400 mA (constant amperage) for 60 min in the presence of an ice block.
4. After transfer, block the membrane for 30 min at room temperature in PSBT containing 20% (w/v) nonfat dry milk.

[14] J. L. Bailey, *in* "Techniques in Protein Chemistry," p. 340. Elsevier Publishing, New York, 1967.
[15] U. K. Laemmli, *Nature* **227,** 680 (1970).
[16] E. Georges, G. Bradley, J. Gariepy, and V. Ling, *Proc. Natl. Acad. Sci. U.S.A.* **87,** 152 (1990).

5. Incubate the blot in a 1:2000 dilution of C219 monoclonal antibody prepared in PBST containing 5% (w/v) nonfat dry milk for 2 hr to overnight at room temperature.
6. Wash the blot three times for 15 min in PSBT.
7. Incubate the blot for 60 min at a 1:4000 dilution of secondary antibody in PBST containing 5% (w/v) nonfat dry milk.
8. Wash three times for 15 min in PBST.
9. Develop the blot using an enhanced chemiluminescence kit (Amersham Life Science, Arlington Heights, IL) as per manufacturer's instructions.

[α-^{32}P]8-Azidoadenosine-5′-triphosphate Labeling of Pgp and Immunoprecipitation

Materials

[α-^{32}P]8-Azidoadenosine-5′-triphosphate ([α-^{32}P]8-azido-ATP) (specific activity 23 Ci/mmol; 2.0 mCi/mL) (ICN)

TSNa resuspension buffer: 10 m*M* Tris (pH 7.5), 50 m*M* NaCl, 250 m*M* sucrose. Add 1 m*M* DTT 1% (v/v) aprotinin solution (0.1 U/mL) (Sigma), and 2 m*M* AEBSF (ICN) just prior to use

2× Labeling buffer: 100 m*M* Tris-HCl (pH 7.5) with 2 m*M* DTT, 2% (v/v) aprotinin solution (0.2 U/mL) (Sigma), and 4 m*M* AEBSF (ICN) added just prior to use

1 *M* $MgCl_2$

1× RIPA buffer: 20 m*M* Tris-HCl (pH 7.2), 150 m*M* NaCl, 1% (v/v) Triton-X 100, 1% (w/v) sodium deoxycholate, 0.1% (w/v) sodium dodecyl sulfate (SDS) 1 m*M* ethylenediaminetetraacetic acid (EDTA)

Protein A agarose: 45–55% suspension in 20 m*M* sodium phosphate (pH 7.2), 0.15 *M* NaCl, 0.02% (w/v) merthiolate (Life Technologies)

Method

1. Dilute the membrane protein (50–100 μg total protein) in TSNa resuspension buffer or 1× labeling buffer containing 10 m*M* $MgCl_2$ (final concentration) to a final volume of 97.5 μl in a 1.5-ml microcentrifuge tube.
2. Incubate on ice for 3 min.
3. Add 5 μCi [α-^{32}P]8-azido-ATP per tube under subdued light conditions and incubate in the dark for 10 min on ice.
4. UV cross-link by exposing the reaction mix to a UV lamp at 365 nm (Black-Ray lamp, model XX-15, UVP, Upland, CA) for 15 min on ice.

5. Analyze the cross-linked products by immunoprecipitation as described in the following steps.
6. To each sample add 600 μL 1× cold RIPA buffer and antibody (e.g., 8 μL of anti-Pgp polyclonal antibody 4007.[17])
7. Incubate at 4° for 1 hr on a rotary shaker.
8. Add 25 μL protein A agarose beads and continue to incubate at 4° for an additional 2 hr on a rotary shaker.
9. Wash the beads three times with 1 mL 1× cold RIPA buffer. Pulse down beads at 14,000*g* for 10 sec, remove supernatant, and repeat twice.
10. Elute protein by incubation with 30 μL 2× SDS–PAGE sample buffer at room temperature for 30–60 min.
11. Load sample, including beads, and perform SDS–PAGE, fix the gel in a solution containing 30% (v/v) methanol and 10% (v/v) acetic acid for 30 min and subsequently dry the gel. Alternatively, transfer the proteins from the gel to a nitrocellulose membrane and subject the membrane to autoradiography.

Photoaffinity Labeling of Pgp with [^{3}H]Azidopine or [^{125}I]Iodoarylazidoprazosin

Materials

Iscove's modified Dulbecco's medium supplemented with 5% FBS or calf serum (CS) (IMEM) (Life Technologies)
Trypsin-EDTA: 0.25% (w/v) trypsin, 1 m*M* EDTA
Phosphate-buffered saline without Ca^{2+} and Mg^{2+} (PBS) with 1% (v/v) aprotinin solution (0.1 U/mL) (Sigma)
TD buffer: 10 m*M* Tris (pH 8.0), 0.1% (v/v) Triton X-100, 10 m*M* $MgSO_4$, 2 m*M* $CaCl_2$. Add 1% (v/v) aprotinin solution (0.1 U/mL) (Sigma) 2 m*M* AEBSF (ICN), 1 m*M* DTT, and 20 μg/ml deoxyribonuclease I (DNase) (Sigma) just prior to use. Alternatively, 50 U/mL micrococcal nuclease (*S. aureus*) (Pharmacia Biotech) in the presence of 1 m*M* $CaCl_2$ can be used in place of DNase
2× Labeling buffer: 100 m*M* Tris-HCl (pH 7.5) with 2 m*M* DTT, 2% (v/v) aprotinin solution (0.2 U/mL) (Sigma), and 4 m*M* AEBSF (ICN) added just prior to use
5× Laemmli SDS–PAGE sample loading buffer[15]
ENLIGHTNING® (DuPont-NEN, Boston, MA)

[17] S. Tanaka, S. J. Currier, E. P. Bruggemann, K. Ueda, U. A. Germann, I. Pastan, and M. M. Gottesman, *Biochem. Biophys. Res. Commun.* **166,** 180 (1990).

[^{3}H]Azidopine (specific activity, 46 Ci/mmol) (Amersham Life Science)

[^{125}I]Iodoarylazidoprazosin ([^{125}I]IAAP) (specific activity, 2200 Ci/mmol) (DuPont-NEN)

Labeling of Pgp in Intact Cells

Method

1. Harvest cells by trypsinization 24 hr after infection–transfection and wash once with 15 mL of IMEM.
2. Resuspend pellet in 1 mL IMEM, count cells, and aliquot 1×10^6 cells into a 1.5-mL microcentrifuge tube.
3. Centrifuge the cells for 30 sec at 14,000*g*.
4. Remove the supernatant and resuspend the cells in 98 μL PBS containing 2 m*M* AEBSF (ICN) added just prior to use.
5. Add 1 μCi of [^{3}H]azidopine or [^{125}I]IAAP (1–2 μL). For binding competition assays, 1 μL of 100× concentrated drugs dissolved in dimethyl sulfoxide (DMSO) are first added as the competitors (1% final concentration of DMSO) and incubated for 3–5 min at room temperature before adding the labeled drug.
6. Incubate the cells at room temperature on a rotary shaker in the dark for 60 min.
7. After incubation, directly expose the samples to a 365-nm UV lamp (Black-Ray lamp, UVP) on ice for 30 min.
8. After UV cross-linking, add 500 μL of cold PSB and centrifuge the cells for 30 sec at 14,000*g*
9. Resuspend the pellets in 100 μL TD buffer (1×10^6/100 μL TD buffer).
10. Subject the cells to three freeze–thaw cycles to lyse the cells. Freeze on dry ice for 5 min and thaw at 37° for 2–3 min each cycle and vortex between cycles.
11. Add 25 μL 5× Laemmli sample buffer to the resulting lysate and perform SDS–PAGE on an appropriate aliquot (50–60 μL). Load 5–10 μL of the sample on an SDS–PAGE gel run in parallel to compare sample loading by immunoblot analysis as described earlier.
12. Alternatively, the lysate can be subjected to immunoprecipitation as described earlier.
13. Fix gels in a solution containing 30% (v/v) methanol and 10% (v/v) acetic acid for 30 min.
14. Treat gels containing ^{3}H-labeled samples with ENLIGHTNING for 20 min, dry, and subject to autoradiography.
15. Immediately dry gels containing the ^{125}I-labeled samples without treatment with ENLIGHTNING and subject to autoradiography.

Labeling of Pgp in Membrane Preparations

Method

1. Dilute membrane preparations (25–50 μg) to 98 μL in 1× labeling buffer.
2. Add 1 μCi of [^{3}H]azidopine or [^{125}I]IAAP (1–2 μL). For binding competition assays, first add 1 μL of 100× concentrated drugs dissolved in DMSO (1% final concentration of DMSO to the membrane preparations and incubate at room temperature for 3 min. Subsequent to this incubation, add the [^{3}H]azidopine or [^{125}I]IAAP solution.
3. Incubate the sample at room temperature in the dark for 10 min.
4. After incubation, directly expose the samples to a 365-nm UV lamp (Black-Ray lamp, UVP) on ice for 15 min.
5. After UV cross-linking, add an appropriate volume of SDS–PAGE sample buffer (1× final concentration) and perform SDS–PAGE as described earlier. At this point, alternatively, the sample can be subjected to immunoprecipitation as described earlier.
6. Treat gels as described earlier.

Measurement of ATPase Activity

Pgp-associated substrate-stimulated ATPase activity is measured by determining the vanadate-sensitive release of inorganic phosphate from ATP with a colorimetric method as previously described[18] with some modifications. The vanadate-sensitive activities in the presence and absence of substrate are calculated as the differences between the ATPase activities obtained in the presence and absence of 300 μ*M* sodium orthovanadate. Briefly, membrane suspensions are first incubated at 37° for 5 min in the reaction mixture assay buffer and then substrates are added from stock solutions prepared in DMSO and the assay mixtures are then incubated for 3 min at 37°. The final concentration of DMSO in the assay medium is 1% (v/v), a concentration that does not exhibit any effect on the ATPase activity. The reactions are subsequently started by the addition of 5 m*M* ATP to the assay mixtures and incubated at 37° for desired periods of time (usually 20 min). Reactions are stopped by the addition of 5% (w/v) SDS solution and the amount of inorganic phosphate (P_i) released is measured by a colorimetric reaction.

[18] B. Sarkadi, E. M. Price, R. C. Boucher, U. A. Germann, and G. A. Scarborough, *J. Biol. Chem.* **267,** 4854 (1992).

Materials

Phosphate (P_i) reagents: 1% ammonium molybdate (or molybdic acid, ammonium salt) in 2.5 *N* sulfuric acid and 0.014% antimony potassium tartarate). Add 50 ml distilled water and 6.9 ml concentrated sulfuric acid (36.2 *N* solution) to a 250-ml glass beaker and then add 1 g ammonium molybdate powder and 14 mg of antimony potassium tartrate and stir for 20 min in a hood. Make up the final volume to 100 ml with distilled water and store the solution in a glass bottle covered with aluminum foil at room temperature

2× assay buffer (make fresh): 100 m*M* Tris (pH 7.5), 10 m*M* sodium azide, 4 m*M* ethylene-bis(oxyethylenenitrilo)tetraacetic acid (EGTA) (pH 7.0), 2 m*M* ouabain, 4 m*M* DTT, 100 m*M* KCl, and 20 m*M* $MgCl_2$

5% (w/v) SDS

1% (w/v) ascorbic acid prepared fresh and stored on ice

10 m*M* sodium orthovanadate: Prepare a fresh stock of 30–40 m*M* and incubate solution at 100° for 3 min prior to use. Read OD_{268} and adjust to a concentration of 10 m*M* (OD 3.6 = 1 m*M*)

100× Concentrated stocks of various drugs to be tested prepared in DMSO

100 m*M* Adenosine 5′-triphosphate (ATP) (disodium salt), pH 7.0. Store in aliquots at −80°

1 m*M* potassium phosphate. Store in aliquots at −20°

Method

1. Carry out the reaction in duplicate in either 13 × 100 mm or 12 × 75 mm glass tubes and use 10–20 μg membrane protein/reaction.
2. Mix the components as shown in Table I. Note that the volumes given are in microliters.
3. Incubate the components mixed as described in Table I at 37° for 3 min.
4. Add 5 μL 100 m*M* ATP.
5. Incubate at 37° for the desired time, add 100 μl 5% (w/v) SDS, and vortex.
6. To develop the color add 400 μl P_i reagent, vortex, and add 500 μl water. Add 200 μl 1% (w/v) ascorbic acid to each tube and vortex immediately. Incubate at room temperature for 10 min.
7. Read the optical density at 880 nm using ATP alone in assay buffer (tube V in Table I) as the blank.
8. Prepare the phosphate standard curve by taking 0, 10, 20, 30, 40, 50, 75, and 100 μl of 1 m*M* phosphate standard in a final volume of 100

TABLE I
SEQUENTIAL ADDITIONS[a] TO MEASURE SUBSTRATE-STIMULATED ATPASE ACTIVITY

Tube	I	II	III	IV	V
2× assay buffer	50	50	50	50	50
H_2O	(Adjust the final volume to 100 μl)				
10 mM sodium orthovanadate	—	3	—	3	—
Membranes	10–20 μg/tube				—
Incubate at 37° for 3 min and then add the following:					
Dimethyl sulfoxide (DMSO)	1	1	—	—	1
100× substrate stock	—	—	1	1	—

[a] Volumes are given in microliters.

μl and add 100 μl 5% (w/v) SDS, 400 μl P_i reagent, 500 μl water, and 200 μl 1% (w/v) ascorbic acid. Incubate the samples at room temperature for 10 min and read the optical density at 880 nm. The color is stable for 30 min. Usually, 1 OD corresponds to 58–62 nmol of phosphate.

Determination of Cell-Surface Expression of Pgp by MRK-16 Staining

Materials

Trypsin-EDTA: 0.25% (w/v) trypsin, 1 mM EDTA
Purified mouse IgG_{2a}, kappa (anti-TNP) (PharMingen, San Diego, CA)
MRK-16 monoclonal antibody[19] that recognizes an external epitope of human Pgp (1 mg/mL) (Hoechst, Japan)
FITC-labeled anti-mouse IgG_{2a} (PharMingen)
Iscove's modified Dulbecco's medium supplemented with 5% (v/v) FBS or CS (IMEM) (Life Technologies)
Phosphate-buffered saline without Ca^{2+} and Mg^{2+} (PBS)
Fluorescence-activated cell sort flow cytometer (Becton-Dickinson, San Jose, CA)
6-mL 12 × 75-mm polystyrene round-bottom tubes with caps (Falcon, Becton-Dickinson, Lincoln Park, NJ)

Method

1. Remove cells from flasks 24 hr after infection–transfection by trypsinization; wash once in IMEM.

[19] H. Hamada and T. Tsuruo, *Proc. Natl. Acad. Sci. U.S.A.* **83,** 7785 (1986).

2. Incubate 3.5–5 × 10^5 cells in a final volume of 200 μL of the same medium with 6 μg of MRK-16 or purified mouse IgG_{2a} (as an isotype control) in 6 mL polystyrene tubes with caps.
3. Incubate at 4° for 20–30 min.
4. Dilute cell suspensions to 4.5 ml in IMEM and centrifuge at 200*g* for 5 min.
5. Aspirate the wash medium and resuspend cell pellet in 200 μL IMEM containing 3 μg FITC-labeled anti-mouse IgG.
6. Incubate at 4° for 20–30 min under subdued light conditions.
7. Dilute cell suspensions to 4.5 ml in IMEM, centrifuge at 200*g* for 5 min.
8. Remove the medium and resuspend the cells in 500 μl of IMEM and centrifuge again at 200*g* for 5 min.
9. Remove the wash medium and resuspend the cell pellet in 350–500 μl PBS and analyze by fluorescence-activated cell sort flow cytometry (FACSort) (Becton-Dickinson FACS system)

Fluorescent Substrate Efflux and Accumulation Assays in Intact Cells

Materials

Phosphate-buffered saline without Ca^{2+} and Mg^{2+}
Trypsin-EDTA: 0.25% w/v trypsin, 1 m*M* EDTA
Iscove's modified Dulbecco's medium supplemented with 5% (v/v) FBS or CS (IMEM) (Life Technologies)
DMEM without glucose and phenol red (glucose-free DMEM) (Life Technologies)
Glucose-containing DMEM (Bio-Whittaker, Walkersville, MD)
2-Deoxy-glucose (Sigma)
Sodium azide (Sigma)
Rhodamine-123 (Eastman Kodak Co., Rochester, NY)
6-mL 12 × 75-mm polystyrene round-bottom tubes with caps (Falcon)
FACSort (Becton-Dickinson)

ATP-Dependent Rhodamine-123 Efflux Assay in Intact Cells

Method

1. Harvest and wash cells as described earlier for MRK-16 antibody staining.
2. To determine energy-dependent rhodamine-123 efflux, incubate 500,000 cells in either glucose-free DMEM containing 5 m*M* 2-deoxy-glucose and 10 m*M* sodium azide (to deplete energy) or 25 m*M*

glucose-containing DMEM for 20 min at room temperature in 6-mL polystyrene tubes with caps.

3. Pellet cells by centrifugation at 200*g* for 5 min and add 4.5 mL of their respective medium containing 0.5 μg/ml rhodamine 123 added from a stock of 1 mg/ml in ethanol.
4. Incubate at 37° for 40 min and pellet the cells by centrifugation at 200*g*.
5. Remove medium, resuspend cells in 0.5 mL of their respective media without rhodamine-123, and add an additional 4 mL of the same media.
6. Incubate for an additional 40 min at 37°.
7. Pellet cells and resuspend in 450 μl of ice-cold PBS and immediately analyze by FACS.

Fluorescent Substrate Accumulation Assay

Method

1. For fluorescent drug accumulation measurements, incubate 500,000 cells in 5 ml of IMEM (prewarmed to 37°) containing 5% FBS or CS and fluorescent substrate with or without reversing agent for 40 min at 37° in 6 mL polystyrene tubes with caps. The following concentration of fluorescent compounds can be used: rhodamine-123, 0.5 μg/mL; bodipy-verapamil, 0.5 μM (Molecular Probes, Eugene, OR); calcein AM, 0.5 μM (Molecular Probes); daunomycin, 4 μM (Calbiochem). Measurements of calcein AM accumulation are taken after a 10-min incubation at 37°.
2. Centrifuge cells at 200*g* for 5 min and remove medium.
3. Resuspend the cells in 300 μl ice-cold PBS and analyze by FACS.
4. For rhodamine-123 and daunomycin measurements, resuspend cells in 4.5 mL substrate-free medium with or without reversing agent.
5. Incubate an additional 40 min at 37°.
6. Pellet the cells by centrifugation at 200*g* for 5 min and remove the medium.
7. Resuspend the cells in 450 μl ice-cold PBS and analyze by FACS.

Drug Accumulation Assays Using Radiolabeled Compounds in Intact Cells

Drug accumulation assays in infected–transfected cells can be performed essentially as described by Stein *et al.*[20] However, for experimental

[20] W. D. Stein, C. O. Cardarelli, I. Pastan, and M. M. Gottesman, *Mol. Pharmacol.* **45,** 763 (1994).

ease and reproducibility of results, the modified protocol described in detail next has been developed allowing for the assay to be performed with cells in suspension.

Materials

Iscove's modified Dulbecco's medium (IMEM) (Life Technologies)
Trypsin-EDTA: 0.25% w/v trypsin, 1 m*M* EDTA
[^{3}H]Colchicine (74.0 Ci/mmol) (DuPont-NEN)
[^{3}H]Vinblastine (13.5 Ci/mmol) (Amersham Life Science)
[^{3}H]Daunomycin (4.4 Ci/mmol) (DuPont-NEN)
Formula-989 (NEF-989; Packard, Downers Grove, IL) liquid scintillation cocktail
Phosphate-buffered saline without Ca^{2+} and Mg^{2+} (PBS)
6-mL 12 × 75-mm polystyrene round-bottom tubes with caps (Falcon)

Method

1. Infect–transfect 70–80% confluent HOS or HeLa cells in 75-cm^2 flasks.
2. Remove cells from flasks 16–24 hr after infection–transfection by trypsinization and wash once in IMEM containing 5% FBS or CS.
3. To initiate the drug uptake, incubate 500,000 cells in 4 ml of IMEM containing 55 mM glucose and radioactive substrate for 50 min at 37° in 6-mL polystyrene tubes with caps. The following concentrations of radioactive substrates can be used: [^{3}H]colchicine (100 n*M*, 0.5 μCi/ml), [^{3}H]vinblastine (13.5 n*M*, 0.25 μCi/ml), or [^{3}H]daunomycin (113.5 n*M*, 0.5 μCi/ml). Other radioactively labeled compounds can also be used. To determine specificity or efficacy, assays should also be performed in the presence of a reversing agent or inhibitors. Assays should all be performed in duplicate or triplicate.
4. Gently resuspend cells in 500 μL ice-cold serum-free and glucose-free IMEM and add an additional 4 mL of the same medium to wash the cells.
5. Centrifuge cells at 200*g* for 5 min and remove medium.
6. For determining the nonspecific binding of drug to cells, add ice-cold IMEM containing the labeled drug and centrifuge immediately at 200*g* for 5 min. Remove the medium and wash the cells as described earlier keeping the cells on ice at all times.
7. Transfer cells with 400 μL of cold PBS to scintillation vials containing 15 ml of a high-flash point liquid scintillation cocktail such as Formula-989. Rinse the tube with an additional 400 μL PBS and count total radioactivity using a liquid scintillation counter.
8. Express the intracellular accumulation as picomoles of labeled drug per million cells.

Conclusions

The vaccinia-T7 expression system has proven to be ideal for functional studies of Pgp, the transporter associated with multidrug resistance (MDR). Previously, other heterologous expression systems have been explored but none have been completely satisfying.[1,21] Of those examined the baculovirus/insect cell system appears to be most promising for the large-scale synthesis of protein for biochemical and structural analysis. However, whole-cell functional assays are not feasible because the infection process appears to deplete the intracellular ATP levels and make the cell membrane leaky to drugs.[21] Expression of human Pgp in the yeast *Saccharomyces cerevisiae* and *E. coli* have met with limited success owing to low expression levels, toxicity, or intrinsically high ATPase levels, but neither has proven as versatile as the vaccinia-T7 system.[1,10,21,22] Higher level expression in the yeast *Pichia pastoris* appears to be feasible.[23] Historically, in order to express large amounts of protein in mammalian cells, it has been necessary to establish stably transfected cell lines selected stepwise in MDR drugs. Because this process can take weeks to months, these selection schemes are a constant source of debate because of the unknown pleiotropic effects of drug selection on cellular functions. The use of this vaccinia-T7 transient system that does not involve drug selection for Pgp expression eliminates the need to consider these possible complications in the interpretation of the observed phenotypes.

The major drawback of the vaccinia-T7 system is that the infected–transfected cells cannot be used to measure relative resistance to MDR drugs in cell proliferation assays since the infected cells are committed to virus-induced lysis. In the future, it may be possible to make use of a non- or reduced-cytopathic vaccinia virus (MVA-Pol)[24] that will allow for use in these types of assays. However, currently, changes in substrate specificity resulting from mutations can be examined with transport assays using fluorescent or radiolabeled compounds with vTF7-3 infected–transfected cells. Another system that appears to have some promise utilizes the canarypox virus, which has the advantage of not inhibiting host-cell replication[25,26] suggesting that these infected cells would remain targets for cytotoxic drugs.

[21] G. L. Evans, B. Ni, C. A. Hrycyna, D. Chen, S. V. Ambudkar, I. Pastan, U. A. Germann, and M. M. Gottesman, *J. Bioenerg. Biomembr.* **27,** 43 (1995).

[22] K. Kuchler and J. Thorner, *Proc. Natl. Acad. Sci. U.S.A.* **89,** 2302 (1992).

[23] L. Beaudet, I. L. Urbatsch, and P. Gros, *Methods Enzymol.* **292,** [29], 1997 (this volume).

[24] L. S. Wyatt, B. Moss, and S. Rozenblatt, *Virology* **210,** 202 (1995).

[25] J. Taylor, R. Weinberg, J. Tartaglia, C. Richardson, G. Alkhatib, D. Briedis, M. Appel, E. Norton, and E. Paoletti, *Virology* **187,** 321 (1992).

[26] J. Tartaglia, J. Taylor, W. I. Cox, J.-C. Audonnet, M. E. Perkus, and E. Paoletti, *in* "Aids Research Reviews," Vol. 3, p. 361. Marcel Dekker, New York, 1993.

[34] pHaMDR-DHFR Bicistronic Expression System for Mutational Analysis of P-Glycoprotein

By SHUDONG ZHANG, YOSHIKAZU SUGIMOTO, TZIPORA SHOSHANI, IRA PASTAN, and MICHAEL M. GOTTESMAN

Introduction

Mutagenesis is one of the most common approaches employed in the structure–function study of proteins both in somatic cells[1,2] and in germ cells.[3,4] There are many ways to introduce mutations into a molecule:

1. Making chimeras by changing functional domains between two proteins[5]
2. Site directed mutagenesis by doing polymerase chain (PCR) studies with either the upper primer or lower primer or both containing the expected sequence change[6]
3. PCR-based random mutagenesis by setting up PCR conditions under which one or more of the deoxynucleotides is limiting[7–9]
4. Targeted mutagenesis, usually by using a cre-loxP system for this purpose[10–12]

[1] D. Mendel, V. W. Cornish, and P. G. Schultz, *Annu. Rev. Biophys, Biomol. Struct.* **24,** 435 (1995).

[2] K. M. Ruppel, M. Lorenz, and J. A. Spudich, *Curr. Opin. Struct. Biol.* **5,** 181 (1995).

[3] V. L. Dellarco, R. P. Erickson, S. E. Lewis, and M. D. Shelby, *Environ. Mol. Mutagen.* **25,** 2 (1995).

[4] J. Rossant, C. Bernelot-Moens, and A. Nagy, *Phil. Trans. R. Soc. Lond. B Biol. Sci.* **339,** 207 (1993).

[5] S. M. Rybak, H. R. Hoogenboom, D. L. Newton, J. C. Raus, and R. J. Youle, *Cell Biophys.* **21,** 121 (1992).

[6] S. J. Currier, S. E. Kane, M. C. Willingham, C. O. Cardarelli, I. Pastan, and M. M. Gottesman, *J. Biol. Chem.* **267,** 25153 (1992).

[7] C. G. Lerner and M. Inouye, *Methods Mol. Biol.* **31,** 97 (1994).

[8] D. A. Resnick, A. D. Smith, A. Zhang, S. C. Geisler, E. Arnold, and G. F. Arnold, *AIDS Res. Hum. Retroviruses* **10,** S47 (1994).

[9] Y. Zhou, X. Zhang, and R. H. Ebright, *Nucleic Acids Res.* **19,** 6052 (1991).

[10] H. Gu, Y. R. Zou, and K. Rajewsky, *Cell* **73,** 1155 (1993).

[11] P. Hasty, R. Ramirez-Solis, R. Krumlauf, and A. Bradley, *Nature* **350,** 243 (1991).

[12] S. O'Gorman, D. T. Fox, and G. M. Wahl, *Science* **251,** 1351 (1991).

0076-6879/98 $25.00

5. Insertional mutagenesis with transposable elements in cultured cells *in vitro*[13–15] or in transgenic mice[16]
6. Radiation or chemical mutagenesis[17,18]

To analyze the function of the mutagenized gene, it is generally subcloned into an expression vector that usually has two different promoters for the gene of interest and the selective marker.[19,20] One of the disadvantages of this kind of vector is that the expression level of the gene of interest is sometimes not as high as desired. Frequently, the unselected gene (the gene of interest) is no longer expressed during passage of the transfectants. To address this problem, we have developed a bicistronic selection system in which expression of the gene of interest (*MDR1*) is tightly linked to selection of a second gene (DHFR).

Studies of the internal ribosomal entry site (IRES) have made it possible to develop a bicistronic or polycistronic vector in which a fusion transcript can be obtained from two or more genes linked together by one or more IRES sequences.[21,22] One of the unique features of this type of vector is that selection of the second gene downstream from the IRES guarantees the expression of the first gene. In this article we describe a bicistronic mammalian expression vector, pHaMDR-DHFR, that is being used in our laboratory to study the effect of mutations introduced into the human *MDR1* gene.

In previous studies of the multidrug transporter (*MDR1*), selection with MDR drugs such as vinblastine or colchicine after transfection was often employed, resulting in drug-resistant cell lines that might carry secondary mutations in either the *MDR1* gene or host. To avoid the need for selection with MDR substrates, we created a bicistronic vector system by using a mutant murine dihydrofolate reductase (DHFR) gene, which confers methotrexate resistance, as a dominant selectable marker that allows for expression of P-glycoprotein (Pgp) on the cell surface without the need for selection of resistance to MDR drugs.

[13] N. Amariglio and G. Rechavi, *Environ. Mol. Mutagen.* **21,** 212 (1993).
[14] M. Silverman, R. Showalter, and L. McCarter, *Methods Enzymol.* **204,** 515 (1991).
[15] R. P. Erickson and S. E. Lewis, *Environ. Mol. Mutagen.* **25,** 7 (1995).
[16] T. Rijkers, A. Peetz, and U. Ruther, *Transgenic Res.* **3,** 203 (1994).
[17] E. W. Vogel and A. T. Natarajan, *Mutat. Res.* **330,** 183 (1995).
[18] J. F. Ward, *Radiat. Res.* **142,** 362 (1995).
[19] T. Bick, G. P. Frick, D. Leonard, J. L. Leonard, and H. M. Goodman, *Proc. Soc. Exp. Biol. Med.* **206,** 185 (1994).
[20] D. B. Lowrie, R. E. Tascon, M. J. Colston, and C. L. Silva, *Vaccine* **12,** 1537 (1994).
[21] C. U. Hellen and E. Wimmer, *Curr. Top. Microbiol. Immunol.* **203,** 31 (1995).
[22] M. Schmid and E. Wimmer, *Arch. Virol.* **9,** 279 (1994).

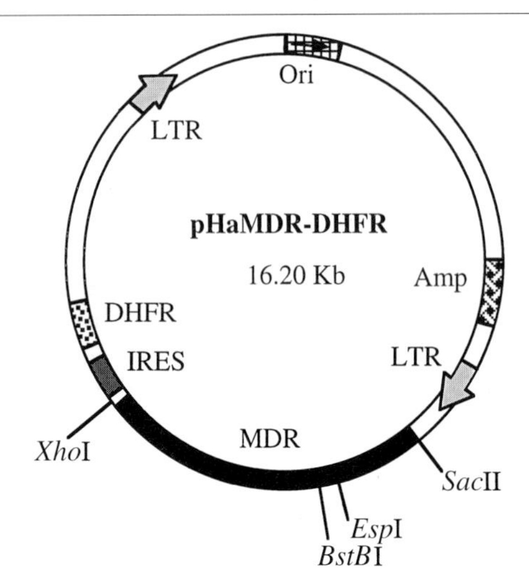

FIG. 1. The bicistronic expression vector pHaMDR-DHFR. Unique restriction sites flanking MDR and DHFR are labeled. LTR, long terminal repeat; *MDR,* human multidrug resistance gene; IRES, internal ribosomal entry site, DHFR, dihydrofolate reductase.

Construction of the pHaMDR-DHFR Bicistronic Vector System

A diagram depicting the construction of the mammalian expression vector pHaMDR-DHFR based on our retroviral vector pHaMDR[23] is shown in Fig. 1. The downstream open reading frame (ORF) after the IRES sequence derived from encephalomyocarditis virus was a mutant murine dihydrofolate reductase gene that served as a selectable marker. The substitution of an arginine for a leucine at codon 22 in DHFR makes it possible to select transfectants in DHFR$^+$ cells.[24] The upstream ORF, either wild-type or mutant human *MDR1,* was cloned before the IRES. Both ORFs were driven by the same promoter, in the long terminal repeat (LTR) from Harvey murine sarcoma virus. The mutant murine DHFR cDNA (576 bp) was obtained by PCR with primer 1 (5′ CCATGGTTC-GACCATTGAAC 3′) and primer 2 (5′ GAGAAGAAAGACTAACTC-GAGCAGGAAG 3′) from the vector pSVMAP1FLDHFR,[25] which contains DHFR cDNA as the template. To subclone the PCR product, *Nco*I and *Xho*I recognition sequences are included and underlined in primers 1 and 2, respectively. The ATG in primer 1 was used as the start codon for

[23] S. E. Kane, D. H. Reinhard, C. M. Fordis, I. Pastan, and M. M. Gottesman, *Gene* **84,** 439 (1989).

[24] C. C. Simonsen and A. D. Levinson, *Proc. Natl. Acad. Sci. U.S.A.* **80,** 2495 (1983).

[25] R. P. Ryseck, S. I. Hirai, M. Yaniv, and R. Bravo. *Nature* **334,** 535 (1988).

the second gene that is translated under control of the IRES. This PCR product (579 bp) was subcloned into the TA cloning vector (Invitrogen, San Diego, CA), sequenced and digested with *Nco*I and *Xho*I. The *Nco*I-*Xho*I fragment (566 bp) was then ligated in the vetor pSXLC,[26] which was digested with the same pair of restriction enzymes, resulting in pID. pID was cut with *Sac*II and *Xba*I and ligated to the wild-type *Esp*I-mutant human *MDR1* gene (3.9 kb) removed from the vector by *Sac*II-*Xba*I digestion, resulting in pMID. The *Sac*II-*Xho*I cassette in pMID was then cut out and ligated to the retrovirus backbone of pHaMDR (11.2 kb), which had been digested with the same pair of restriction enzymes, resulting in the bicistronic expression vector pHaMDR-DHFR. To introduce a unique *Esp*I restriction site before the TM5 transmembrane domain of Pgp in which we are interested, the *Bgl*II-*Bst*BI fragment (876 bp) from the wild-type *MDR1* cDNA in pSXbaMDR[26] was replaced with the recombinant 876-bp PCR DNA harboring an *Esp*I site.

Selection for Methotrexate Resistance Results in High-Level Expression of P-Glycoprotein

To characterize this vector system, we transfect pHaMDR-DHFR, which contains a wild-type *MDR1* cDNA, into KB-3-1 cells. After transfection, cells are grown in Dulbecco's modified Eagle's medium (DMEM) supplemented with 10% dialyzed bovine calf serum and selected in 30 ng/ml methotrexate for the expression of the second gene (DHFR) in the vector. As shown by fluorescence-activated cell sorting (FACS) analysis of Pgp expression on the cell surface (Fig. 2), most MTX-resistant cells after transfection express Pgp. This result is further confirmed by Western blot analysis, which measures total Pgp in the cells. Western blots also show that the average amount of Pgp in the transfectant cells could be increased further by selection in stepwise increments of methotrexate. Pgp expressed in the transfectants is functional, which is demonstrated with a fluorescent dye efflux assay and a direct drug cytotoxicity assay. In the fluorescent dye efflux assay, either rhodamine-123 or bodipy-verapamil, two known human Pgp substrates, is used to stain cells that are then analyzed by FACS. A reduction in the amount of these fluorescent dyes is shown for both dyes, indicating that Pgp on the cell surface is able to pump the dyes out of the transfected cells. Functional Pgp in the transfectants is also demonstrated, since some cells are resistant to 600 ng/ml vinblastine, whereas non-transfected control cells could not even survive 5–10 ng/ml of the drug.

[26] Y. Sugimoto, I. Aksentijevich, M. M. Gottesman, and I. Pastan, *Biotechnology* **12,** 694 (1994).

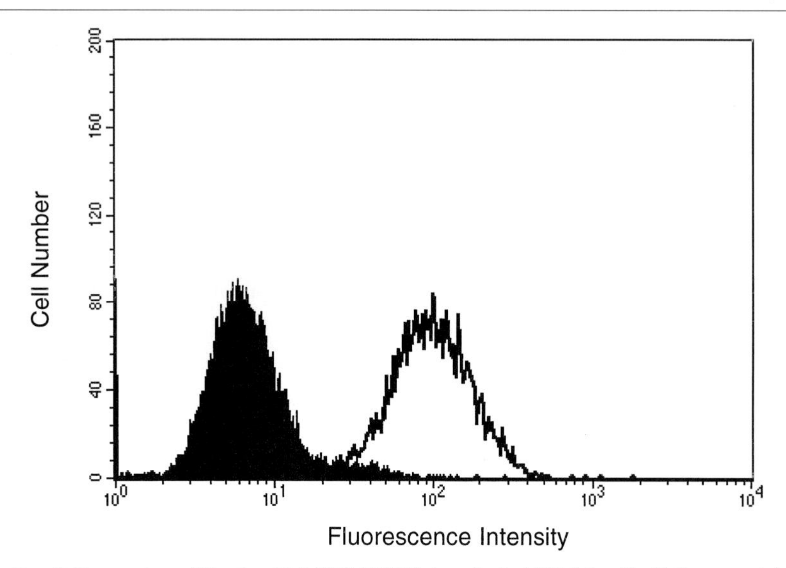

FIG. 2. Expression of Pgp in pHaMDR-DHFR transfected KB-3-1 cells. Cells were stained either with MRK16 antibody specific for human Pgp (empty) or with IgG antibody nonbinding control (filled).

Polymerase Chain Reaction-Based Mutagenesis of *MDR1*

Our initial structure–function studies of Pgp have been focused on transmembrane domains (TMs) 5,6 and 11,12 demonstrated previously by [^{3}H]azidopine photoaffinity labeling in our laboratory and elsewhere to be substrate interaction sites.[27,28] Mutagenesis studies further reveal that changes of several different amino acids throughout Pgp and in the predicted TMs 5,6, and 11,12 of Pgp change its drug resistance pattern.[29–33] So we chose the region from codon 283 to 383 (300-bp cDNA), including TMs 5 and 6 of human Pgp, as the target for our mutagenesis studies. To increase the likelihood of variation of the amino acid sequence in this region, we developed a protocol to mutate randomly the 300-bp *MDR1* cDNA fragment that we choose.

[27] E. P. Bruggemann, S. J. Currier, M. M. Gottesman, and I. Pastan, *J. Biol. Chem.* **267,** 21020 (1992).

[28] A. R. Safa, N. D. Mehta, and M. Agresti, *Biochem. Biophys. Res. Commun.* **162,** 1402 (1989).

[29] U. A. Germann, I. Pastan, and M. M. Gottesman, *Semin. Cell. Biol.* **4,** 63 (1993).

[30] X. Zhang, K. I. Collins, and L. M. Greenberger, *J. Biol. Chem.* **270,** 5441 (1995).

[31] T. W. Loo and D. M. Clarke, *Biochemistry* **33,** 14049 (1994).

[32] T. W. Loo and D. M. Clarke, *J. Biol. Chem.* **269,** 7243 (1994).

[33] T. W. Loo and D. M. Clarke, *J. Biol. Chem.* **268,** 19965 (1993).

Random Mutations Generated by Mut-PCR

Random mutagenesis has been described to probe the important regions in a given gene when functionally significant positions are not well known.[34–38] The fidelity of the Mut-PCR used for this purpose needs to be reduced to a certain level at which the error rate is sufficient to mutagenize a gene or a gene-size DNA fragment. This can be achieved by increasing the concentration of $MgCl_2$ and *Taq* polymerase, unbalancing the concentration of the four dNTPs, lowering the annealing temperature, and increasing the extension time. Among many different conditions we tried, the one described here consistently gave us 60% transformants that contained only one missense mutation; 20%, two or more mutations, and 20%, wild type. The Mut-PCR was conducted in a 100-μl reaction volume in PCR buffer from Perkin Elmer (Norwalk, CT) with 7 m*M* $MgCl_2$, 1 m*M* dGTP, 0.2 m*M* each of dATP, dCTP, and dTTP, 20 pmol of each primer, 8 ng plasmid DNA template, and 5 units *Taq* polymerase. Amplifications were done in a Perkin-Elmer automated DNA thermal cycler with a hot start at 94° for 3 min. The Mut-PCR was then run 35 cycles consisting of 1 min of denaturation at 94°, 1 min of annealing at 47°, and 1 min of extension at 72°, followed by a final extension for 7 min at 72°. The PCR DNA was then subcloned to a TA cloning vector. DNA was prepared from individual clones using a Wizard DNA minipreparation kit (Promega, Madison, WI), and the PCR DNA fragment inserted in the TA vector was sequenced to obtain information about the mutation rate. Mut-PCR DNA was digested with two unique restriction enzymes, *Esp*I and *Bst*BI, for cloning back into the pHaMDR-DHFR expression vector.

Mammalian Cell Transfection Library

DNA isolated from the bacterial libraries constructed using standard protocols (GIBCO-BRL, Gaithersburg, MD) was electroporated into KB-3-1 cells with an electroporator (Bio-Rad, Richmond, CA). KB cells were trypsinized and then washed once with normal growth medium and twice with phosphate-buffered solution (PBS) without Ca^{2+} or Mg^{2+}. For a 0.4-cm cuvette, 10^7 cells and 10–20 μg DNA resuspended in 0.8 ml of PBS were mixed gently, then incubated on ice for 10 min. Just prior to electroporation, cells were resuspended again with a pipette. In our hands,

[34] R. M. Myers, L. S. Lerman, and T. Maniatis, *Science* **229,** 242 (1985).

[35] M. D. Matteucci and H. L. Heyneker, *Nucleic Acids Res.* **24,** 5810 (1983).

[36] C. A. Hutchinson, S. K. Nordeen, K. Vogt, and M. H. Edgell, *Proc. Natl. Acad. Sci. U.S.A.* **83,** 710 (1986).

[37] K. M. Derbyshire, J. J. Salvo, and N. D. F. Grindley, *Gene* **46,** 145 (1986).

[38] J. D. Hermes, S. M. Parekh, S. C. Blacklow, H. Koster, and J. R. Knowles, *Gene* **84,** 143 (1989).

high transfection efficiency (about 10^{-4}/cell) was obtained by setting the voltage at 300–400 V and the electric field strength at 960 mF. The time constant was generally from 12 to 16 ms. After electroporation, cells were kept on ice for another 15 min and then plated to a 100-mm cell culture dish. Fifteen to 20 ml of DMEM was added to each dish, and dishes were then put back into an incubator at 37°, 5% CO_2, and 100% humidity. Twenty-four to 48 hr later, medium was replaced by DMEM containing methotrexate for selection as described earlier. Usually, slow-growing methotrexate-resistant colonies would appear on day 8–10, which could be isolated by week 2 or 3.

[35] Mutational Analysis of Human P-Glycoprotein

By TIP W. LOO and DAVID M. CLARKE

Introduction

The structure and mechanism of a membrane protein are often difficult to study due to low levels of expression in normal tissues and the technical difficulties associated with purifying enough protein for useful analyses. Molecular biology techniques, however, have made it possible to clone almost any cDNA coding for a membrane protein. This provides a powerful tool for the study of the structure–function relationships of the protein.

The cDNA is easily manipulated to introduce changes in the protein. These include changes to individual amino acids, deletion of segments of the molecule, insertion of antibody tags or reporter sequences into any region of the molecule, or expression of individual domains as separate polypeptides or as fusion proteins. Characterization of the mutant protein products can provide novel insights into the structure and function of the molecule. The usefulness of such an approach, however, depends on the availability of rapid and reproducible expression and assay systems. P-Glycoprotein has been particularly suitable for mutational analysis for four major reasons: (1) The protein is stable after transient or stable expression in a variety of heterologous expression systems such as in various mammalian cell lines and in yeast and insect cells (*Spodoptera frugiperda* fall armyworm ovary, Sf9).[1] (2) A growing number of assays are available for characterizing the protein products, such as drug-stimulated ATPase

[1] G. L. Evans, B. Ni, C. A. Hrycyna, D. Chen, S. V. Ambudkar, I. Pastan, U. A. Germann, and M. M. Gottesman, *J. Bioenerg. Biomembr.* **27,** 1 (1995).

0076-6879/98 $25.00

activity[2] or conferring of resistance to various cytotoxic drugs in transfected cells.[3,4] (3) Alterations to the structure of P-glycoprotein can be readily identified by checking the pattern of glycosylation. P-Glycoprotein is extensively glycosylated because it contains three N-linked glycosylation sites between predicted TM1 and TM2.[5] Many mutants with altered structure appear to be retained in the endoplasmic reticulum and are only core-glycosylated.[6] These mutants are sensitive to treatment with endoglycosidase H. (4) P-Glycoprotein can tolerate insertions and deletions and still retain function. For example, addition of a monoclonal antibody epitope tag,[7] or 10 tandem histidine residues at the COOH-terminal end of the molecule can be done with no apparent effect on function.[8] In addition, the two halves of the molecule can be expressed as separate polypeptides that can associate and function when expressed contemporaneously.[9] These properties make P-glycoprotein an ideal candidate for mutational analysis.

In this article, we outline some of the techniques used to construct, express, and characterize P-glycoprotein mutants.

Site-Directed Mutagenesis

The first step is to introduce the desired mutation into the cDNA of P-glycoprotein. The details of these techniques have been described elsewhere.[7] Briefly, a synthetic oligonucleotide(s) (18- to 21-mer) containing the desired mutation is required. These oligonucleotides can be synthesized by many commercial sources. A wide variety of approaches have been developed for site-directed mutagenesis. Many of these are available as a kit from many suppliers of molecular biology products. For the beginner, it is best to use a commercial kit because they are relatively inexpensive and the yield of positive mutants is claimed to be relatively high. One possible disadvantage of the commercial kit is that often the whole cDNA is used during mutagenesis, thus requiring that the full-length clone be sequenced to confirm that no extraneous mutations have been introduced.

[2] S. V. Ambudkar, I. H. Lelong, J. Zhang, C. O. Cardarelli, M. M. Gottesman, and I. Pastan, *Proc. Natl. Acad. Sci. U.S.A.* **89,** 8472 (1992).

[3] K. Ueda, C. Cardarelli, M. M. Gottesman, and I. Pastan, *Proc. Natl. Acad. Sci. U.S.A.* **84,** 3004 (1987).

[4] A. Devault and P. Gros, *Mol. Cell Biol.* **10,** 1652 (1990).

[5] A. F. Schinkel, S. Kemp, M. Dolle, G. Rudenko, and E. Wagenaar, *J. Biol. Chem.* **268,** 7474 (1993).

[6] T. W. Loo and D. M. Clarke, *J. Biol. Chem.* **268,** 19965 (1993).

[7] T. W. Loo and D. M. Clarke, *J. Biol. Chem.* **268,** 3143 (1993).

[8] T. W. Loo and D. M. Clarke, *J. Biol. Chem.* **270,** 21449 (1995).

[9] T. W. Loo and D. M. Clarke, *J. Biol. Chem.* **269,** 7750 (1994).

For the more experienced, mutagenesis by the Kunkel[10] method is often more satisfying. The yield of positive clones, however, is often lower than that claimed for commercial kits. One advantage of this method is that if suitable restriction sites are available, then one often confines the fragment of cDNA used for mutagenesis to the smallest size possible, subsequently reducing the amount of work involved in sequencing the fragment.

Expression of Mutants

Initially, we attempted to express human P-glycoprotein cDNA in yeast and bacteria, because of the potential to produce large amounts of P-glycoprotein in these expression systems. We followed the expression of epitope-tagged human P-glycoprotein, in *Pichia pastoris* (a methylotropic yeast) and in *Escherichia coli.* We were only able to detect degradation products of the full-length P-glycoprotein after expression in *Pichia pastoris* or *Escherichia coli.*[11] Human P-glycoprotein is extremely sensitive to proteolytic digestion in these expression systems. Therefore, expression of the wild-type and mutant P-glycoproteins required the development of other expression systems.

We now use three different expression systems to characterize P-glycoprotein mutants: transient expression in HEK (human embryonic kidney) 293 cells, stable expression in mouse. NIH 3T3 cells, and expression in insect Sf9 cells. The choice of expression system depends on the type of assay to be used.

Stable Expression

In previous studies, drug resistance assays were used to assess the effects of mutations on P-glycoprotein function. This assay requires the cells to be stably transfected with the mutant P-glycoprotein. Cell lines such as NIH 3T3[7] or chinese hamster ovary (CHO)[4] rather than HEK 293 cells are used for selection in the presence of cytotoxic agents. This is because HEK 293 cells are very easily dislodged from the tissue culture dish by any gentle shearing force such as simply squirting media onto the monolayer cells.

Numerous colonies are usually visible when selected in the presence of colchicine or vinblastine. Several of these clones are selected and expanded and then their growth characteristics in the presence of various cytotoxic agents can be compared to clones expressing the wild-type P-glycoprotein. This involves the seeding of equivalent number of cells in 96-well plates

[10] T. A. Kunkel, *Proc. Natl. Acad. Sci. U.S.A.* **82,** 488 (1985).

[11] T. W. Loo and D. M. Clarke, unpublished observations (1993).

and incubating the cells in the presence of increasing concentrations of cytotoxic agents. After 5–10 days of growth, cell viability is determined indirectly by a colorimetric assay that measures the ability of live cells to convert MTT [3-(4,5-dimethylthiazol-2-yl)-2,5-diphenyltetrazolium bromide] to formazan.[12] This is then standardized to the amount of P-glycoprotein expressed in an equivalent number of cells. It is often difficult to determine the concentration of cytotoxic agents to use during the initial selection of cells expressing a mutant P-glycoprotein. The mutation may drastically affect activity such that direct selection following transfection does not yield any drug-resistant colonies. One way to address this problem is to cotransfect the cells with a vector containing the neomycin-resistance gene together with the mutant cDNA and then select for G418 (geneticin)-resistant colonies. All the G418-resistant colonies are pooled and then reselected in the presence of a different concentration of cytotoxic agents. Another method of identifying clones expressing the mutant P-glycoprotein is to subject whole-cell extracts of individual G418-resistant colonies to immunoblot analysis. The major disadvantages of this type of drug-resistance profile study are that it is a very slow process (months to complete) and that it is only an indirect measure of P-glycoprotein activity. Other drug transporter(s) or alterations in the cell's physiology or potential recombination of the mutant P-glycoprotein with endogenous isoforms during the selection procedure could possibly affect the results. It is the best system, however, for studying temperature or glycerol sensitivities of biosynthetic processing mutants.[11,13]

Expression in Insect Cells

A second expression system is in Sf9 insect cells using a baculovirus vector.[9,14] This is really a transient system and generally cannot be used for measuring drug-resistant profiles because of the tendency of the infected cells to lyse 3–5 days after infection. This system, however, yields very high levels of underglycosylated P-glycoprotein that is still functional. The underglycosylated form of P-glycoprotein is suitable for crystallization studies. Another problem experienced when using this method for studying mutants is that it is tedious and time consuming. It can take months to generate the recombinant baculovirus containing the mutant P-glycoprotein and to obtain a sufficiently high titer virus stock. There is again the possibil-

[12] M. C. Alley, D. A. Scudiero, A. Monks, M. L. Hursey, M. J. Czerwinski, D. L. Fine, B. J. Abbott, J. G. Mayo, R. H. Shoemaker, and M. R. Boyd, *Cancer Res.* **48,** 589 (1988).

[13] T. W. Loo and D. M. Clarke, *J. Biol. Chem.* **269,** 28683 (1994).

[14] U. A. Germann, M. C. Willingham, I. Pastan, and M. M. Gottesman, *Biochemistry* **29,** 2295 (1990).

ity of recombination of the mutant P-glycoprotein cDNA with endogenous isoforms of insect cell P-glycoproteins.

Transient Expression in HEK 293 Cells

The most versatile expression system for studying P-glycoprotein mutants is transient expression in HEK 293 cells. This expression system has the following major advantages: (1) It is a simple and rapid system because P-glycoprotein mutants can be analyzed 36–48 hr after transfection. (2) Enough P-glycoprotein is usually produced (1–3 μg per 10-cm-diameter plate) such that it can be purified from a moderate number of transfected cells (ten to twenty 10-cm-diameter plates) for measurement of ATPase activity. (3) The P-glycoprotein is usually fully glycosylated and it is easy to distinguish between mature and core-glycosylated forms of the enzyme. (4) P-Glycoprotein and domains of P-glycoprotein can be expressed as separate polypeptides that are relatively stable in HEK 293 cells. The only disadvantage is that the cells are only loosely attached to the plates. This makes it rather difficult to do studies such as labeling of P-glycoprotein on the cell surface (e.g., biotin hydrazide)[13] directly on the monolayer cells. For these types of procedures, transient expression in COS-1 cells can be used.

Transient expression systems require the use of special expression vectors. Fortunately, these are readily available either from commercial sources (such as pcDNA3 or pCMV, Invitrogen, San Diego, CA) or from Dr. Randal Kaufman, Genetics Institute, Boston (p91023B, pMTs). All these vectors work well with P-glycoprotein[15] in either HEK 293 or COS-1 cells.

Transfection of HEK 293 Cells

Techniques such as purification of P-glycoprotein or labeling studies require relatively high expression levels of the enzyme. The procedure described below gives a relatively high yield of enzyme. The number of cells to be transfected depends on the assay. We routinely carry out transfections using 24-well plates if we only wish to see if the mutation affects processing of the enzyme. For purification purposes, we usually transfect twenty 10-cm-diameter plates of HEK 293 cells. The transfection procedure is based on the method described by Chen and Okayama[16] and is described next.

[15] T. W. Loo and D. M. Clarke, unpublished observations (1993).

[16] C. Chen and M. Okayama, *Mol. Cell Biol.* **7,** 2745 (1987).

Material and Solutions Required

Solution 1: 50 m*M* BES [*N,N*-bis(2-hydroxyethyl)-2-aminoethanesulfonic acid], 280 m*M* NaCl, 1.5 m*M* $Na_2HPO_4 \cdot 7H_2O$. Adjust pH to 6.96 with NaOH.

Solution 2: 2.5*M* $CaCl_2$. The quality of $CaCl_2$ seems important as we have most success with a tissue culture grade from Sigma (St. Louis, MO).

Media: Dulbecco's modified Eagle's medium (DMEM) supplemented with nonessential amino acids, 4 m*M* L-glutamine and 10% (v/v) fetal bovine serum (FBS).

Procedures

1. Two 75-cm^2 flasks of subconfluent HEK 293 cells are gently washed with phosphate-buffered saline (PBS). Incubate each monolayer with 3 ml of PBS for 10–20 min at room temperature.
2. While the cells are in PBS, the $CaCl_2$/DNA/BES mixture is prepared. In general, 270 μg of plasmid DNA (purified by CsCl-banding or on commerical DNA purification cartridges) is mixed with water to give a total volume of 7.65 ml. This will yield a final DNA concentration of 1.5 μg/ml when added to the cells. Add 0.85 ml of solution 2, mix gently, and then add 8.5 ml of solution 1. Let the mixture stand for 10 min at room temperature.

   The concentration of DNA required for optimal transfection will have to be determined empirically for each batch of solution 1. We generally prepare 4-liter batches of solution 1 and test its transfection efficiency with various concentrations of plasmid DNA (0.5–4 μg/ml). Solution 1 is then frozen at $-20°$ in 50-ml fractions until ready for use.
3. Aseptically aspirate the PBS from the flasks, add 3 ml of media to each flask, and then strike the flask against a solid object to dislodge the monolayer of HEK 293 cells. Pool the cells from both flasks, pipette up and down five times to break any clumps, and then adjust the volume to about 170 ml with media.
4. Add the DNA mixture from step 2, mix and immediately dispense 9 ml to each 10-cm-diameter plate. The cells are incubated at 37° for about 40 hr in 2–3% CO_2. Propagation of HEK 293 cells is done in 5% CO_2 and therefore requires a separate incubator.
5. For small-scale transfections, it is sometimes easier to plate the cells 1 day before transfection. The transfection procedure is the same as described except that complete media without cells is added to the DNA mixture before being added to the cells in the culture dishes.

Biosynthesis of P-Glycoprotein Mutants

Many single amino acid changes in P-glycoprotein cause a structural perturbation.[6,13,17] These mutants appear to be recognized by a cellular quality control mechanism such that they are retained in the endoplasmic reticulum as core-glycosylated biosynthetic intermediates. These mutants are readily identified by their sensitivity to endoglycosidase H. Mutants that leave the endoplasmic reticulum and pass through the Golgi apparatus become fully glycosylated and resistant to treatment with endoglycosidase H. These complex sugars, however, can be removed with PNGase F. These properties provide a convenient assay for identifying mutants that are defective in processing.

Procedures

1. Transfect a 24-well plate of HEK 293 cells with the desired mutant cDNAs as described earlier.
2. After 40 hr, gently wash the transfected cells twice with PBS. The cells are then solubilized by addition of 1 ml of denaturation buffer [0.5% (w/v) sodium dodecyl sulfate (SDS), 15 m*M* 2-mercaptoethanol, 10 m*M* EDTA, 50 μg/ml 4-(2-aminoethyl)benzenesulfonyl fluoride]. The sample can be passed through a Qiagen (Mississauga, Ontario, Canada) DNA minipreparation spin column to remove DNA.
3. Take 90 μl of the solubilized fraction and add 10 μl of 0.5 *M* sodium citrate, pH 5.5. Divide the mixture in two, and to one fraction add 1 μl of Endo H_f [NEB (Mississauga, Ontario, Canada) 1000 units] and to the other fraction add 1 μl water. Vortex the samples gently and incubate at room temperature for 15 min.
4. Stop the reaction by addition of 1 volume of 2× SDS sample buffer [0.25 *M* Tris-HCl, pH 6.8, 4% (w/v) SDS, 4% (v/v) 2-mercaptoethanol, and 20% (v/v) glycerol].
5. Analyze the samples by immunoblot analysis after SDS–PAGE on a 6.5% gel.
6. For digestion with PNGase F (NEB), mix 80 μl of the solubilized fraction with 10 μl of 10% (v/v) Nonidet P-40 (NP-40) and 10 μl of 0.5 *M* sodium phosphate, pH 7.5. Divide the mixture into two fractions and add 1 μl of PNGase F (1000 units) to one sample and 1 μl of water to the other. Incubate at 37° for 15 min. Analyze by immunoblot analysis as described earlier.

[17] T. W. Loo and D. M. Clarke, *J. Biol. Chem.* **269,** 7243 (1994).

Whole-cell extracts from HEK 293 cells transfected with human P-glycoprotein cDNA express two forms of P-glycoprotein with apparent masses of 170 and 150 kDa, respectively. The major product of 170 kDa corresponds to the mature form of the enzyme, whereas the minor 150-kDa product corresponds to the core-glycosylated intermediate. Treatment of the mature enzyme with PNGase F or the core-glycosylated form with either PNGase F or endoglycosidase H will yield the 140-kDa unglycosylated form of the enzyme. The mature form of the enzyme is resistant to endoglycosidase H.

Purification of P-Glycoprotein and Measurement of Drug-Stimulated ATPase Activity

P-Glycoprotein possesses ATPase activity that is stimulated by many drug substrates.[2,18–20] We have found that significant stimulation of ATPase activity occurs in the presence of a variety of compounds such as verapamil, vinblastine, colchicine, paclitaxel, thapsigargin, leupeptin, and rhodamine 123. The ATPase activity is also stimulated by compounds such as cyclosporin. There is a good correlation between a mutant's ability to confer resistance to vinblastine or colchicine and the level of stimulation of the ATPase activity by these compounds.[8] Therefore, measurement of drug-stimulated ATPase activity is a very useful assay for characterizing P-glycoprotein mutants. Unlike expression of P-glycoprotein in insect cells using a baculovirus vector, the membranes from transfected HEK 293 cells do not contain enough P-glycoprotein for one to measure ATPase activity on membrane preparations. Therefore, the protein must first be purified and then reconstituted into lipid for measurement of drug-stimulated ATPase activity. Purification of P-glycoprotein expressed in HEK 293 cells can readily be accomplished by using a histidine-tagged protein and nickel–chelate chromatography.[8] For purification purposes, we use a P-glycoprotein cDNA containing 10 tandem histidines at the COOH-terminal end of the molecule. We found that a tag containing 10 histidines was far superior than one using the conventional 6 histidines. The advantage of this method is that this is a fairly rapid purification procedure. Transfection, purification, and the ATPase assays can be completed within 48 hr. The ATPase assays are done immediately after purification since the purified enzyme loses about 20–40% of its activity after 24 hr at 4°.

[18] B. Sarkadi, E. M. Price, R. C. Boucher, U. A. Germann, and G. A. Scarborough, *J. Biol. Chem.* **267,** 4854 (1992).

[19] F. J. Sharom, X. Yu, and C. A. Doige, *J. Biol. Chem.* **268,** 24197 (1993).

[20] M. K. Al-Shawi and A. E. Senior, *J. Biol. Chem.* **268,** 4197 (1993).

Purification of Histidine-Tagged P-Glycoprotein

Materials and Solutions Required

Solution 1: 1 *M* Imidazole, pH 7.0.

Buffer A: 50 m*M* $NaPO_4$, 5 m*M* Tris-HCl, pH 8.0, 500 m*M* NaCl, 50 m*M* Imidazole, and 20% (v/v) glycerol.

Buffer B: Buffer A containing 1% (w/v) *n*-dodecyl-β-D-maltoside. (It is important to note that this detergent is poorly soluble at 4° and will precipitate when stored for long periods at this temperature.)

Buffer C: Buffer A containing 0.1% (w/v) *n*-dodecyl-β-D-maltoside.

Buffer D: 10 m*M* Tris-HCl, pH 7.5, 500 m*M* NaCl, 80 m*M* Imidazole, 0.1% (w/v) *n*-dodecyl-β-D-maltoside, and 20% (v/v) glycerol.

Buffer E: Buffer D but containing 300 m*M* Imidazole.

Buffer L: 10 m*M* Tris-HCl, pH 7.5, and 0.5 m*M* $MgCl_2$.

Buffer S: 10 m*M* Tris-HCl, pH 7.5, 0.5 *M* sucrose, and 0.3 *M* KCl.

Procedures. All solutions and steps are at 4°.

1. For each mutant, twenty 10-cm-diameter culture dishes of HEK 293 cells are transfected as described earlier.
2. Harvest cells 40 hr after transfection by scraping with a rubber policeman, wash twice with PBS, and suspend in 3 ml of buffer L. Incubate the cells on ice for 10 min and then homogenize (40 strokes) with a Dounce homogenizer.
3. Add 3 ml of buffer S, followed by another 20 strokes with the homogenizer.
4. Remove unbroken cells and nuclei by centrifugation at 4000*g* for 10 min. Transfer the supernatant to an ultracentrifuge tube containing 30 ml of 2× diluted buffer S and centrifuge at 100,000*g* for 1 hr.
5. Suspend the crude membranes in 0.3 ml of buffer A and solubilize by addition of 1 ml of buffer B. Leave on ice for 10 min.
6. Remove insoluble material by centrifugation at 16,000*g* for 15 min.
7. Load the supernatant onto a nickel spin column (Ni-NTA, Qiagen) which has been preequilibrated with buffer C, and then centrifuge at 50*g* for 5–10 min.
8. Wash the column three times with 0.6 ml of buffer D (250*g* for 3 min for each wash).
9. Elute histidine-tagged P-glycoprotein with 0.25 ml of buffer E at 50*g* for 3 min, then at 500*g* for 1 min.
10. Dilute the eluted material six-fold with buffer A (but containing no Imidazole) and reload the sample onto the same column that had been regenerated by washing with 1 *M* Imidazole containing 0.1% (w/v) *n*-dodecyl-β-D maltoside, followed by equilibration with buffer B.

11. Wash the column as described earlier and elute P-glycoprotein with 0.25 ml of buffer E.
12. Determine the yield of P-glycoprotein by measuring the amount of protein,[21] using bovine serum albumin as a standard, and subtracting the amount of protein obtained when the same number of control vector-transfected cells carried through the purification procedure. We often get between 6 and 12 μg of P-glycoprotein.

Measurement of Mg^{2+}-ATPase Activity

We have used a modified version of the method described by Chifflet *et al.*[22] Human P-glycoprotein purified by nickel–chelate chromatography exhibits little drug-stimulated ATPase activity without lipid. In the presence of lipid, however, high levels of ATPase activity are obtained in the presence of various substrates. For example, the ATPase activity of P-glycoprotein in the presence of 800 μM verapamil is about 1 μmol P_i/min/mg P-glycoprotein. This is 12- to 15-fold higher than the basal activity determined without drug substrates. Because P-glycoprotein in the presence of lipid and *n*-dodecyl-β-D-maltoside is not in a vesicular form, masking of ATP-binding sites is not a problem.

Materials and Solutions Required. Phosphatidylethanolamine [crude sheep brain, Sigma Type IIs commercial grade; composition: phosphatidylethanolamine (20–30%, w/w), phosphatidylcholine (18–20%, w/w), phosphatidylserine (10–12%, w/w), sphingomyelin (8–10%, w/w), and cerebrosides (3–4%, w/w)]: 100 mg is homogenized in 1 ml of TBS [10 m*M* Tris-HCl, pH 7.5, 150 m*M* NaCl, 2 m*M* dithiothreitol (DTT)] in a microcentrifuge tube and then diluted to 40 ml with TBS. The suspension is centrifuged at 12,000*g* for 30 min and the pellet suspended in a final volume of 1 ml of TBS. This is prepared immediately before use. Unwashed lipid contains significant amounts of inorganic phosphate that interferes with the colorimetric assay.

Solution A: 12% (w/v) SDS.
Solution B: 6% (w/v) L-Ascorbic acid in 1 *M* HCl (prepare fresh).
Solution D: 1% (w/v) Ammonium molybdate, 10% (w/v) SDS.
Solution BD: Mix 1 : 1 ratio of solutions B and D (prepare fresh).
Solution E: 2% (w/v) sodium citrate, 2% (w/v) sodium metaarsenite, 2% (w/v) acetic acid.
2× ATP solution: 100 m*M* Tris-HCl, pH 7.5, 100 m*M* NaCl, 10 m*M* $MgCl_2$, and 10 m*M* ATP.

[21] M. M. Bradford, *Anal. Biochem.* **72,** 248 (1976).
[22] S. Chifflet, A. Torriglia, R. Chiesa, and S. Tolosa, *Anal. Biochem.* **168,** 1 (1988).

20 mM verapamil-HCl (prepared in H_2O).
20 mM vinblastine (prepared in DMSO).
866 mM colchicine (prepared in DMSO).

Procedures

1. An equal volume of purified P-glycoprotein and lipid is mixed and sonicated briefly to clarify the mixture and then allowed to sit on ice for 10 min.
2. Eight microliters of the P-glycoprotein–lipid mixture is incubated with 2 μl of the various concentrations of drug substrates for 10 min at room temperature. (For human P-glycoprotein, maximal drug-stimulated ATPase activities are obtained in the presence of 800 μM verapamil, 100 μM vinblastine, or 3 mM colchicine.) If desired, an inhibitor of P-glycoprotein, such as 1 mM N-ethylmaleimide can be used to determine background or basal activity.
3. Start reaction by addition of an equal volume of 2× ATP solution and incubate the samples for 30–60 min at 37°. The reaction is linear as long as the final A_{850} <1.0.
4. Stop the reaction by addition of an equal volume of buffer A.
5. Add 40 μl of solution BD and incubate for 5 min at room temperature.
6. Add 60 μl of solution E and let the samples incubate for 10 min at 37°.
7. Measure absorbance at 850 nm using a 50-μl cuvette.
8. Determine the amount of inorganic phosphate released by using a standard curve of known amounts of phosphate.

An example of the types of results that can be obtained is shown for mutation of residue Leu305 of human P-glycoprotein. In an alanine-scanning mutagenesis approach combined with drug-resistance profiles to study the importance of residues in predicted transmembrane segments, we noted that mutation Leu-305 → Ala altered the drug resistance profile of the enzyme.[23] This mutant exhibits two- to three-fold increased ability to confer resistance to colchicine compared to wild-type enzyme, but its relative resistance to vinblastine was similar to that of wild-type enzyme. Similar results were obtained when Leu-305 was changed to serine or threonine (Fig. 1A). Mutation of Leu-305 → Phe, however, resulted in a mutant indistinguishable from wild-type enzyme, whereas mutations to charged residues (Lys or Arg) resulted in mutants that were unable to confer drug resistance. Obtaining drug resistance profiles for each mutant is tedious and time consuming. To determine if these drug resistance profiles correlate with the ATPase activities of the purified mutants, the mutants were puri-

[23] T. W. Loo and D. M. Clarke, unpublished observations (1996).

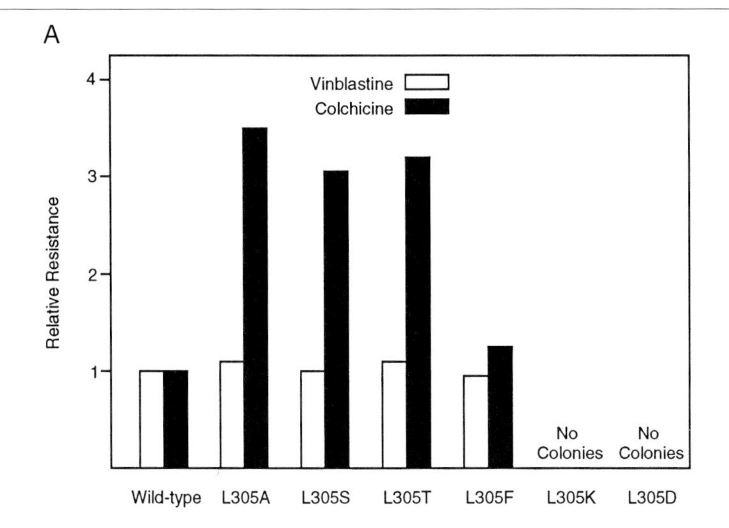

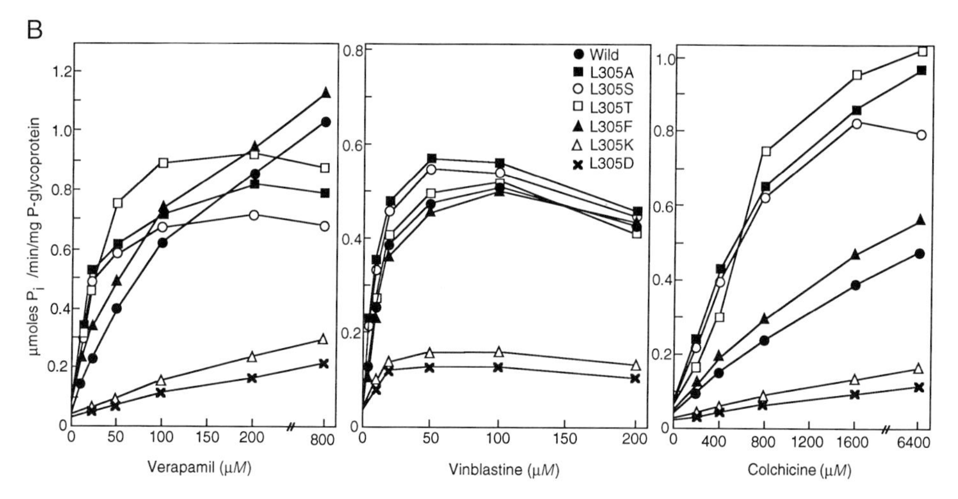

FIG. 1. Analysis of wild-type and Leu-305 mutants of human P-glycoprotein: drug resistance profiles and drug-stimulated ATPase activities. (A) NIH 3T3 cells were transfected with monoclonal antibody A52-tagged wild-type or mutant cDNAs, and drug-resistant colonies selected in the presence of 5 n*M* vinblastine or 45 n*M* colchicine. Drug-resistant colonies were selected, expanded, and the cell lines expressing comparable levels of P-glycoprotein–A52 were tested for their ability to confer resistance to vinblastine or colchicine, relative to cell lines expressing wild-type P-glycoprotein–A52 as described previously.[6] (B) P-glycoproteins–$(His)_{10}$ were expressed in HEK 293 cells, purified by nickel–chelate chromatography, and mixed with sheep brain lipid. ATPase activities were measured in the presence of various concentrations of verapamil, vinblastine, and colchicine.

fied. The cDNAs of these mutants were altered to encode for 10 tandem histidine residues at the COOH-terminal end of the molecules. The mutant cDNAs were expressed in HEK 293 cells, purified by nickel–chelate chromatography and reconstituted with phosphatidylethanolamine. The ATPase activities were then measured in the presence of various concentrations of verapamil, vinblastine, or colchicine. As shown in Fig. 1B, there is a good correlation between enhanced colchicine-resistance and colchicine-stimulated ATPase activities of mutants Leu-305 → Ala, Ser, or Thr. Similarly, mutant Leu-305 → Phe, which had a drug resistance profile that was indistinguishable from that of wild-type enzyme, also had similar ATPase activities compared to wild-type enzyme. Mutant Leu-305 → Lys or Asp was unable to confer drug resistance and showed drastically reduced drug-stimulated ATPase activities. Therefore, there is good correlation between the two assay systems. The ATPase assays, however, have the advantage that the results can be obtained in 2–3 days.

Acknowledgments

This work was supported by grant to David M. Clarke, as part of a group grant from the Medical Research Council of Canada. D.M.C. is a Scholar of the Medical Research Council of Canada.

[36] Purification and Reconstitution of Human P-Glycoprotein

By SURESH V. AMBUDKAR, ISABELLE H. LELONG, JIAPING ZHANG, and CAROL CARDARELLI

Introduction

It is widely known that certain tumors and cultured cells lines can develop simultaneous resistance to various types of cytotoxic, natural product, anticancer agents such as vinblastine, doxorubicin, paclitaxel, colchicine, and actinomycin D.[1-3] These drugs have little in common structurally,

[1] I. B. Roninson, "Molecular and Cellular Biology of Multidrug Resistance in Tumor Cells." Plenum Press, New York, 1991.

[2] M. M. Gottesman and I. Pastan, *Annu. Rev. Biochem.* **62,** 385 (1993).

[3] S. V. Ambudkar, I. Pastan, and M. M. Gottesman, *in* "Drug Transport in Antimicrobial and Anticancer Chemotherapy: Cellular and Biochemical Aspects of Multidrug Resistance" (N. H. Georgapapadakou, ed.), p. 525. Marcel Dekker, New York, 1995.

0076-6879/98 $25.00

except for hydrophobicity and in some cases a tendency to be positively charged at neutral pH.[3,4] The cells that are resistant to natural product cytotoxic agents accumulate much less drug than drug-sensitive cells due to the presence of an energy-dependent drug extrusion system. Cells with a high level of drug resistance express large quantities of a 150- to 170-kDa membrane phosphoglycoprotein, referred to as the P-glycoprotein (Pgp),[5] or the multidrug transporter.[2] In some instances, such as small cell lung carcinoma, cells overexpress a 190- to 200-kDa membrane protein known as multidrug resistance-associated protein, the glutathione conjugate transporter (MRP), which is encoded by the *MRP1* gene.[6,7]

Human *MDR1* cDNA encodes Pgp, a 1280-amino-acid integral membrane protein.[8] Hydropathy profiling of the deduced amino acid sequence has led to a secondary structure model of Pgp which contains 12 putative transmembrane segments with two ATP (nucleotide) binding/utilization domains on the cytoplasmic surface of the membrane.[8] Each half of the protein contains a hydrophobic region with six transmembrane helices and a hydrophilic region with one ATP binding/utilization site; the N-terminal half of the molecule exhibits 43% identity with the C-terminal half. The regions of greatest identity in the two halves of the Pgp molecule are also homologous to similar sequences in the ATP-binding domains of the ABC (ATP-binding cassette) superfamily of transporters.[2,9,10] This superfamily now includes more than 80 members.[11] However, Pgp alone is known to interact with and in many cases, transport a broad range of structurally heterogeneous substrates (e.g., anticancer drugs, reversing agents, hydrophobic peptides, steroids, and detergents).[2,3]

The work from various laboratories in recent years strongly suggests that Pgp catalyzes the ATP-dependent extrusion of natural cytotoxic drugs from multidrug-resistant cells. Pgp has also been proposed to function as

[4] M. M. Gottesman, I. Pastan, and S. V. Ambudkar, *Curr. Opin. Genet. Dev.* **6,** 610 (1996).

[5] R. L. Juliano, and V. Ling, *Biochim. Biophys. Acta* **455,** 152 (1976).

[6] S. P. C. Cole, S. G. Bhardwaj, J. H. Gerlach, J. E. Mackie, C. E. Grant, K. C. Almquist, A. J. Stewart, E. U. Kurz, A. M. V. Duncan, and R. G. Deeley, *Science* **258,** 1650 (1992).

[7] D. Lautier, Y. Cantrot, R. G. Deeley, and S. P. C. Cole, *Biochem. Pharmacol.* **52,** 967 (1996).

[8] C-J. Chen, C. E. Chin, K. Ueda, D. P. Clark, I. Pastan, M. M. Gottesman, and I. B. Roninson, *Cell* **47,** 381 (1986).

[9] C. F. Higgins, *Annu. Rev. Cell Biol.* **8,** 67 (1992).

[10] S. Childs and V. Ling, *in* "Important Advances in Oncology: The MDR Superfamily of Genes and Its Biological Implications" (V. T. DeVita, S. Hellman, and S. A. Rosenberg, eds.), p. 21. Lippincott Company, Philadelphia, 1994.

[11] H. Nikaido and J. A. Hall, *Methods Enzymol.* **292,** [1], 1998 (this volume); J. M. Croop, *Methods Enzymol.* **292,** [8], 1998 (this volume); R. Allikmets and M. Dean, *Methods Enzymol.* **292,** [9], 1998 (this volume).

a regulatory of volume-regulated chloride channel function[12] and as an ATP channel.[13] However, the channel activities or the regulatory role of Pgp have not yet been conclusively established. Moreover, some of the basic questions are as yet not well understood. For example, what defines the broad substrate specificity of the Pgp and how is the energy from ATP hydrolysis coupled to the drug transport? Similarly, very little is known about the mechanism underlying the drug efflux mediated by Pgp. The availability of purified biologically active Pgp and the development of an *in vitro* artificial membrane system will help to elucidate the mechanism of action and various other functions of this protein. Moreover, pure active protein is required to obtain the 2D and 3D crystal structures of Pgp. In this article, we describe the methods developed in our laboratory to obtain active pure Pgp for biochemical characterization of its ATP hydrolysis and drug transport activities.

Purification and Reconstitution of P-Glycoprotein from Human Multidrug-Resistant KB-V1 Cells

We have used the multidrug-resistant human carcinoma cell line KB-V1[14] for these studies, because these cells have been shown to overexpress only the *MDR1* gene.[15] In humans, another gene, *MDR2,* that is closely related to *MDR1* is also expressed and this gene product functions as an ATP-dependent phosphatidylcholine translocator.[16,17] By using KB-V1 cells as the starting material, a homogeneous preparation of purified Pgp can be obtained.

Isolation of Plasma Membrane Vesicles of KB-V1 Cells

The KB-V1 cells, a subclone of KB3-1, are grown in Dulbecco's modified Eagle's medium (DMEM) supplemented with 1 μg/ml vinblastine.[14] Cells are grown to confluence in 15-cm dishes (60 dishes/preparation) and scraped

[12] S. P. Hardy, H. R. Goodfellow, M. A. Valverde, D. R. Gill, F. V. Sepulveda, and C. F. Higgins, *EMBO J.* **14,** 68 (1995).

[13] E. H. Abraham, A. G. Prat, L. Gerweck, T. Seneveratne, R. J. Arceci, R. R. Kramer, G. Guidotti, and H. F. Cantiello, *Proc. Natl. Acad. Sci. U.S.A.* **90,** 312 (1992).

[14] D-w. Shen, A. Fojo, J. E. Chin, I. B. Roninson, N. Richert, I. Pastan, and M. M. Gottesman, *Science* **232,** 643 (1986).

[15] P. V. Schoenlein, D-w. Shen, J. T. Barrett, I. Pastan and M. M . Gottesman, *Mol. Cell. Biol.* **3,** 507 (1992).

[16] J. J. M. Smit, A. H. Schinkel, R. P. J. OudeElferink, A. K. Groen, E. Wagenaar, L. van Deemter, C. A. M. Mol, R. Ottenhoff, N. M. T. van der Lugt, M. A. van Roon, M. A. van der Valk, G. J. A. Offerhaus, A. J. M. Berns, and P. Borst, *Cell* **75,** 451 (1993).

[17] S. Ruetz and P. Gros, *Cell* **77,** 1071 (1994).

into ice-cold phosphate-buffered saline (PBS), pH 7.0, supplemented with 1% aprotinin (Sigma, St. Louis, MO), washed once with PBS and once with lysis buffer [10 m*M* Tris-HCl, pH 7.5, 50 m*M* NaCl, 0.25 *M* sucrose, 0.1 m*M* $CaCl_2$, 1% aprotinin, and 0.5 m*M* phenylmethylsulfonyl fluoride (PMSF) or 4-(aminoethyl)-benzenesulfonyl fluoride (AEBSF, ICN, Costa Mesa, CA)]. Plasma membrane vesicles are prepared by standard nitrogen cavitation followed by a sucrose density gradient procedure.[18] The yield of membrane vesicles is about 12 to 16 mg protein per 60 dishes of cells. These plasma membrane vesicles exhibit ATP-dependent [^{3}H]vinblastine transport.[18]

Solubilization of P-Glycoprotein

The protocol for the extraction of Pgp from membranes is based on our earlier studies with a variety of transport systems.[19–21] The purified KB-V1 plasma membrane vesicles (1.5–2 mg protein) are solubilized in a final volume of 1.2 ml with 1.5% octyl-β-D-glucopyranoside (octylglucoside; Calbiochem, La Jolla, CA) in the presence of 0.1 *M* 3-(*N*-morpholino)propanesulfonic acid (MOPS)/*N*-methyl-D-glucamine (NMDG, a substitute for monovalent cations such as Na^+ or K^+), pH 7.0, 20% (w/v) glycerol, 1.5 m*M* $MgCl_2$, 1 m*M* dithiothreitol (DTT), 1 m*M* AEBSF, 2 μg/ml pepstatin, 2 μg/ml leupeptin, 1% aprotinin, and 0.4% of a lipid mixture containing acetone–ether-washed *Escherichia coli* bulk phospholipid (70% phosphatidylethanolamine, 15% phosphatidylglycerol, and 15% cardiolipin), phosphatidylcholine, phosphatidylserine and cholesterol (60 : 17.5 : 10 : 12.5; all lipids are obtained from Avanti Polar Lipids, Alabaster, AL). The solubilization is initiated by addition of membrane vesicles to the mixture of all other components at 4° and incubated for 20 min on ice.[22] The solubilized protein (detergent extract) is separated from unextracted material by centrifugation at 152,000*g* for 1 hr at 4° in a Beckman-type 52.1 or 50 Ti rotor. This extract, which contains >90% of Pgp, is either used immediately for purification or quickly frozen in dry ice and stored at −80° until later use.

Purification Procedures

Initial screening of various column resins show that both DEAE-Sepharose CL-6B and agarose-linked wheat germ agglutinin (WGA) lectin

[18] M. Horio, M. M. Gottesman, and I. Pastan, *Proc. Natl. Acad. Sci. U.S.A.* **85,** 3580 (1988).
[19] S. V. Ambudkar and P. C. Maloney, *J. Biol. Chem.* **261,** 10079 (1986).
[20] S. V. Ambudkar and P. C. Maloney, *Methods Enzymol.* **125,** 558 (1986).
[21] P. C. Maloney and S. V. Ambudkar, *Arch. Biochem. Biophys.* **269,** 1 (1989).
[22] S. V. Ambudkar, I. H. Lelong, J. Zhang, C. O. Cardarelli, M. M. Gottesman, and I. Pastan, *Proc. Natl. Acad. Sci. U.S.A.* **89,** 8472 (1992).

column are most suitable for obtaining pure Pgp without significant loss of yield.[23] All steps are carried out at 4°. The DEAE-Sepharose CL-6B (Pharmacia, Piscataway, NJ) column (8-ml volume) is equilibrated with five volumes of buffer A [20 m*M* MOPS/NMDG, pH 7.4, 1.25% octylglucoside, 20% glycerol, 1 m*M* DTT, and protease inhibitors (1 m*M* AEBSF, 2 μg/ml pepstatin, 2 μg/ml leupeptin, and 1% aprotinin)]. Then 1.6 ml of octylglucoside extract containing 1.2–1.4 mg protein is loaded on each column. Four columns are run in parallel. The DEAE column is eluted with 20 ml of buffer A containing 0, 50, 100, 150, and 300 m*M* NaCl. Pgp is eluted with 0.1 *M* NaCl (Fig. 1A, lane 3).[22] These fractions are pooled and concentrated with Centriprep-100 concentrators (Amicon, Danvers, MA) by centrifugation as per the supplier's protocol. The concentrated fraction (2.5 ml; 200–300 μg protein) is loaded on the WGA column (EY Laboratories, San Mateo, CA, 2.5-ml volume) preequilibrated with 12.5 ml of buffer B (50 m*M* Tris-HCl, pH 8.0, 0.1 *M* NaCl, 1.25% octylglucoside, 0.1% lipid, 5% glycerol, and protease inhibitors). The protein is allowed to adsorb to the column for 30 min and then the column is eluted with 16 ml of buffer B containing 0, 100, and 250 m*M* *N*-acetyl-D-glucosamine (NAG, Sigma). The 250 m*M* NAG eluate containing most of the Pgp is concentrated to 1–1.5 ml (10–20 μg protein/ml) with Centriprep-100 concentrators. The concentrated 0.25 *M* NAG fraction in 0.3- to 0.5-ml aliquots was quickly frozen in dry ice and stored at −80°. The verapamil-stimulated ATPase activity of Pgp is stable for up to 8–10 months at −80°. Pgp in column fractions is detected with Western immunoblotting by using Pgp-specific mono- and polyclonal antisera.[22] With the DEAE-Sepharose column, 4- to 5-fold enrichment is observed and during the next step with the WGA column, an additional 10- to 12-fold purification is achieved (Fig. 1A, lanes 2–5). Pgp makes up approximately 3% of the protein in the detergent extract, as estimated by sodium dodecyl sulfate–polyacrylamide gel electrophoresis (SDS–PAGE). Thus, overall recovery is about 30–45% (see Table I).

Reconstitution

The reconstitution of pure Pgp into proteoliposomes is achieved by removal of detergent either rapidly by detergent dilution or by slow removal with detergent dialysis followed by Sephadex G-50 chromatography.[22,23] During reconstitution, the concentrated 0.25 *M* NAG fraction (7.5–10 μg protein) is mixed with excess (7.5–12.5 mg) lipid mixture (see section on solubilization) by adding bath-sonicated liposomes[20] in the presence of 1.1%

[23] S. V. Ambudkar, *J. Bioenerg. Biomembr.* **27,** 23 (1995).

octylglucoside in a final volume of 1–1.5 ml. Proteoliposomes are formed by 25-fold dilution at room temperature (22–23°) into a "loading" medium containing 25 m*M* MOPS/NMDG, pH 7.4, 150 m*M* NMDG chloride, and 2.5 m*M* $MgCl_2$.[20] After incubation for 20 min at 22–23°, the suspension is centrifuged at 152,000*g* for 1 hr in a Beckman-type 50.2 Ti rotor at 4°. The proteoliposomal pellet is washed once with the loading buffer and finally resuspended in 0.3–0.5 ml of loading buffer containing protease inhibitors. The proteoliposomes can be stored on ice for 24 hr without any significant loss in Pgp–ATPase activity. To reduce the lipid-to-protein ratio, proteoliposomes are also prepared by a detergent-dialysis method. In this case, the reconstitution mixture containing 5–10 μg protein (0.5–1 ml of 0.25 *M* NAG fraction) and 1.25–5 mg sonicated lipid is dialyzed against 1.5 L loading buffer for 18 hr at 4° and then applied to Sephadex G-50 (Pharmacia; Piscataway, NJ, 1 × 30 cm) column preequilibrated with the loading buffer. The proteoliposomes eluted in first 3–6 ml volume are collected by ultracentrifugation as described earlier. These proteoliposomes (or liposomes) have internal volume of 1 μl/mg phospholipid.[19] Typically, proteoliposomes prepared by the detergent-dilution method (liposomes) contain 14–16 mg phospholipid/ml, and this recovery (~70%) is independent of the protein concentration.

Other Assays

The protein content of membrane vesicles, detergent extract, column fractions, and proteoliposomes is measured with a modification of the method of Schaffner and Weissmann.[24] SDS–PAGE with 7.5, 8, or 10% acrylamide is performed as described by Laemmli[25] using a Bio-Rad (Richmond, CA) Mini-PROTEAN II apparatus. To prepare the samples of column fractions and proteoliposomes for electrophoresis, it is necessary to remove excess lipid by chloroform : methanol extraction, and this is done according to the method of Wessel and Flugge.[26] For Western blots, proteins are electrophoretically transferred to nitrocellulose membranes (Trans-Blot, Bio-Rad). The nitrocellulose blot is blocked for 1 hr in 3% gelatin in wash buffer and incubated with primary antibody [monoclonal C219[27] (0.5 μg/ml) or polyclonal PEPG-2 antiserum,[28] diluted 1 : 3000] in blocking solution for 1 hr at room temperature. After three washes in wash buffer,

[24] W. Schaffner and C. Weissmann, *Anal. Biochem.* **56,** 502 (1970).
[25] U. K. Laemmli, *Nature* (*London*) **227,** 680 (1970).
[26] D. Wessel and U. I. Flugge, *Anal. Biochem.* **138,** 141 (1984).
[27] E. Georges, G. Bradley, J. Gariepy, and V. Ling, *Proc. Natl. Acad. Sci. U.S.A.* **87,** 152 (1990).
[28] E. P. Bruggemann, V. Chaudhary, M. M. Gottesman, and I. Pastan, *Biotechniqes* **10,** 202 (1991).

the blots are incubated with secondary antibody (horseradish peroxidase-conjugated donkey antimouse IgG or goat anti-rabbit IgG) at 1:2500 dilution for 1 hr. The immunoblots are washed three times with wash buffer and developed by using an enhanced chemiluminescence (ECL) detection system (Amersham, Arlington Heights, IL) according to the instructions given by manufacturer.

Most of the Pgp from the WGA column is eluted with 0.25 *M* NAG and this fraction contains protein with maximum purity (>90%). The protein profile as visualized by silver stain is given in Fig. 1A. Pgp migrates as a diffuse band around 165–170 kDa (Fig. 1A). Under similar conditions, when the detergent extract of the parental cell line, KB-3-1, is chromatographed on DEAE and WGA columns, the 170-kDa band (Pgp) is not detectable on silver-stained 8% acrylamide gel (data not shown). The WGA eluate, when reconstituted into proteoliposomes by the detergent-dilution procedure (Fig. 1A, lane 5), shows two minor contaminant proteins with

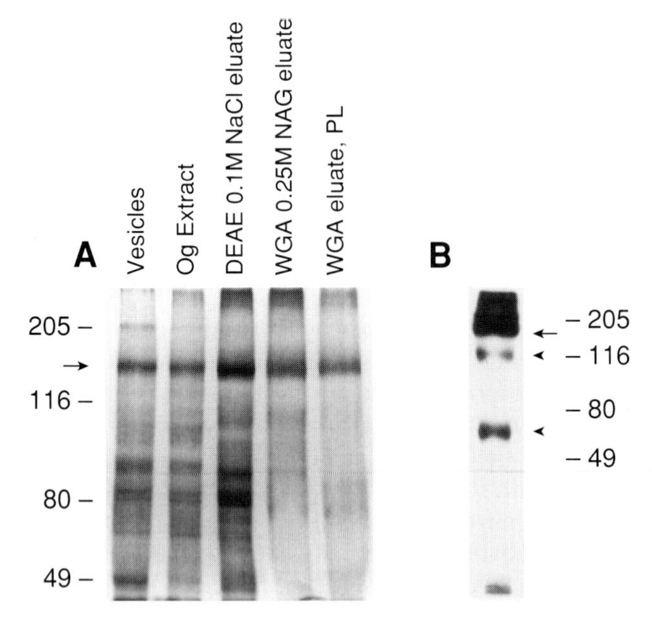

FIG. 1. Purification and reconstitution of P-glycoprotein by sequential chromatography on DEAE-Sepharose CL-6B, followed by wheat germ agglutinin column. (A) A silver-stained SDS–PAGE protein profile of column fractions. Lane 1, KB-V1 vesicles (2 μg); lane 2, octylglucoside extract (2 μg); lane 3, DEAE-Sepharose CL-6B, 0.1 *M* NaCl eluate pooled fractions (1 μg); lane 4, WGA column, 0.25 *M* NAG eluate pooled fractions (0.3 μg); and lane 5, proteoliposomes (PL) reconstituted with WGA 0.25 *M* NAG eluate (fraction in lane 4, 0.2 μg). (B) Western immunoblotting with polyclonal antiserum PEPG-2.[28] Proteoliposomes (0.2 μg protein) reconstituted with WGA 0.25 *M* NAG eluate (see A, lane 5). The arrow shows the position of Pgp and arrowheads show the Pgp fragments.

molecular weight of ~110 kDa (diffused band in Fig. 1A, lanes 4 and 5) and 55 kDa, respectively. Both these contaminants are fragments of Pgp. Based on the Western blotting with peptide specific polyclonal antibodies and the monoclonal antibody C219, it is clear that the 110-kDa fragment corresponds to the N-terminal and the 55-kDa fragment to the C-terminal region of Pgp and these fragments most probably are generated by a single cleavage in the region of residues 637 to 712 because both are recognized by the antisera PEPG2[28] (see Fig. 1B). Similar fragments are generated *in vitro* by the treatment of pure Pgp on ice for 5 min with trypsin (Pgp : trypsin, 1 : 2, data not shown).

Orientation of Purified P-Glycoprotein in Proteoliposomes

To assess fully the reconstitution of Pgp function, it is essential to determine whether the protein is reconstituted right-side out (i.e., same orientation as in a cell) or inside-out (i.e., cytoplasmic side facing outside). We assessed the orientation of Pgp in proteoliposomes by UV cross-linking with [α-^{32}P]ATP, followed by immunoprecipitation with Pgp specific monoclonal antibody, C219. The Pgp is reconstituted with predominantly (>90%) inside-out orientation. A similar orientation is observed when the Pgp is reconstituted either by detergent dilution or by detergent dialysis followed by Sephadex G-50 chromatography. This was also verified by lack of an increase in Pgp-associated ATPase activity in alamethicin-permeabilized proteoliposomes as compared to the intact ones (data not shown).

Drug-Stimulatable ATPase Activity of Purified and Reconstituted P-Glycoprotein

ATPase Assays

Proteoliposomes containing pure Pgp (1–2 μg protein/ml) are incubated with the given drug or equivalent volume of dimethyl sulfoxide (DMSO) in the presence or absence of 0.30 mM sodium orthovanadate for 5 min at 37° in 25 mM MOPS/NMDG, pH 7, 150 mM NMDGCl, 2 mM DTT, and 1.5 mM $MgCl_2$. The ATP hydrolysis is assayed at 37° by measuring the release of [^{32}P]Pi from [γ-^{32}P]ATP (0.6 mM Tris salt; 20 mCi/mmol).[22] At specified times (usually after 10, 20, and 30 min of incubation), 200 μl of the reaction mixture is mixed with 200 μl of 20% (w/v) trichloroacetic acid in a 16- × 125-mm glass test tube. To this, 0.7 ml Millipore HPLC grade water, 0.15 ml 0.15 M KCl, 0.2 ml 4.25% ammonium molybdate in

[28] E. P. Bruggemann, V. Chaudhary, M. M. Gottesman, and I. Pastan, *Biotechniqes* **10,** 202 (1991).

TABLE I
RECOVERY OF P-GLYCOPROTEIN-ASSOCIATED ATPASE ACTIVITY DURING PURIFICATION

Fraction[a]	Protein (mg)	Yield (%)	ATPase (μmol/min/mg protein)[b] Control	+30 μM Verapamil
Octylglucoside extract	5.00	100	0.41	0.43
DEAE-Sepharose CL-6B, 0.1 M NaCl eluate	0.85	17	1.40	5.40
Wheat germ agglutinin, 0.25-M NAG eluate	0.06	1.20	5.85	21.80
			1.70	6.40[c]

[a] Each fraction was reconstituted into proteoliposomes by the detergent-dilution method.
[b] The control (DMSO treated) and verapamil-stimulated ATPase activity in the presence and absence of 0.3 mM sodium orthovanadate was measured. Pgp-associated vanadate sensitive activity is shown.
[c] Note the significant loss of ATPase activity of Pgp when the NAG eluate was reconstituted by the detergent-dialysis followed by the Sephadex G-50 chromatography method.

0.85 N H_2SO_4, and 1 ml of isobutyl alcohol are added and the tubes vortexed for 30 sec. The reaction mixture is left at room temperature for 5 min to allow complete phase separation and then 0.2 ml of isobutanol (upper) layer in duplicate is used for the measurement of radioactivity by liquid scintillation counting.

The functional integrity of the purified Pgp in the reconstituted system is determined by assaying the drug-stimulatable ATPase activity in proteoliposomes. The recovery of Pgp–ATPase activity during purification is summarized in Table I. Proteoliposomes containing total membrane protein (octylglucoside extract) do not exhibit substrate-stimulatable ATPase; this may be due to a high background level of other ATPases or phosphatases. The degree of purification of Pgp based on a protein profile (Fig. 1A) and the enrichment in drug-stimulated ATPase level of WGA eluate strongly argues that the Pgp is obtained without significant loss of activity. The specific activity of Pgp–ATPase (5–20 μmol/min/mg protein) is similar to that of other ion-motive ATPases.[22,23,29] The ATPase activity of pure protein is stimulated by vinblastine, verapamil, daunorubicin, and colchicine (see Table II) but not by 50 μM camptothecin, which does not interact with Pgp.[22]

Transport of [³H]Vinblastine in Proteoliposomes Reconstituted with Purified P-Glycoprotein

Because the Pgp is reconstituted with inside-out orientation as discussed earlier, it is convenient to assay transport function by following ATP-

[29] P. L. Pedersen and L. M. Amzel, *J. Biol. Chem.* **268,** 9937 (1993).

TABLE II
EFFECT OF SELECTED DRUGS ON ATPASE ACTIVITY OF PURIFIED P-GLYCOPROTEIN[a]

Drug	Concentration required for half-maximal activation[b] (μM)	Stimulation
None (DMSO treated)	—	1.00
Vinblastine	5.0	2.95
Verapamil	15.0	4.20
Daunorubicin	20.0	1.50
Colchicine	100.0	1.40

[a] Purified Pgp (0.25 *M* NAG eluate) was reconstituted into proteoliposomes by the detergent-dilution method and the vanadate-sensitive ATPase activity in the presence of the indicated drug was measured.
[b] For each compound, six concentrations were used for the determination of these values.

dependent accumulation of [3H]vinblastine in the proteoliposomes (this is equivalent to the efflux of vinblastine from cells). In proteoliposomes prepared by the detergent-dilution method (lipid : protein, 1200–1500 : 1), although verapamil-stimulated ATPase activity can be measured (Table I), ATP-dependent vinblastine accumulation is not detected likely due to the high level of nonspecific binding of hydrophobic vinblastine to the phospholipid vesicles. To overcome this problem, we use proteoliposomes prepared by detergent dialysis followed by Sephadex G-50 chromatography (see earlier section). By this method lipid to protein ratio (200–400 : 1) in proteoliposomes is significantly reduced. The proteoliposomes containing pure Pgp (0.5–1 μg protein/ml) or an equivalent volume of liposomes (without protein) are incubated with 5–10 n*M* [^{3}H]vinblastine (0.2–0.5 μCi/ml), with or without 2 m*M* ATP, in a medium containing 25 m*M* MOPS/NMDG, pH 7.4, 150 m*M* NMDGCl, 2 m*M* DTT, and 5 m*M* $MgCl_2$. The transport is assayed by the filtration technique[30] using 0.22-μM pore size GSTF filters (Millipore, Bedford, MA). To reduce the nonspecific binding of [^{3}H]vinblastine, the filters are presoaked for 2 hr with 10% fetal calf serum in PBS. The data given in Fig. 2 show that proteoliposomes accumulate four- to fivefold more vinblastine in the presence of ATP and that this accumulation is totally blocked by vanadate, indicating that the hydrolysis of ATP is required for transport. In addition, omission of 5 m*M* $MgCl_2$ from the assay also blocked the vinblastine accumulation. The radioactivity associated with liposomes is not affected by ATP (data not shown). Maximum ATP-dependent drug accumulation is seen within 10 min. The vinblas-

[30] S. V. Ambudkar, V. Anantharam, and P. C. Maloney, *J. Biol. Chem.* **265,** 12287 (1990).

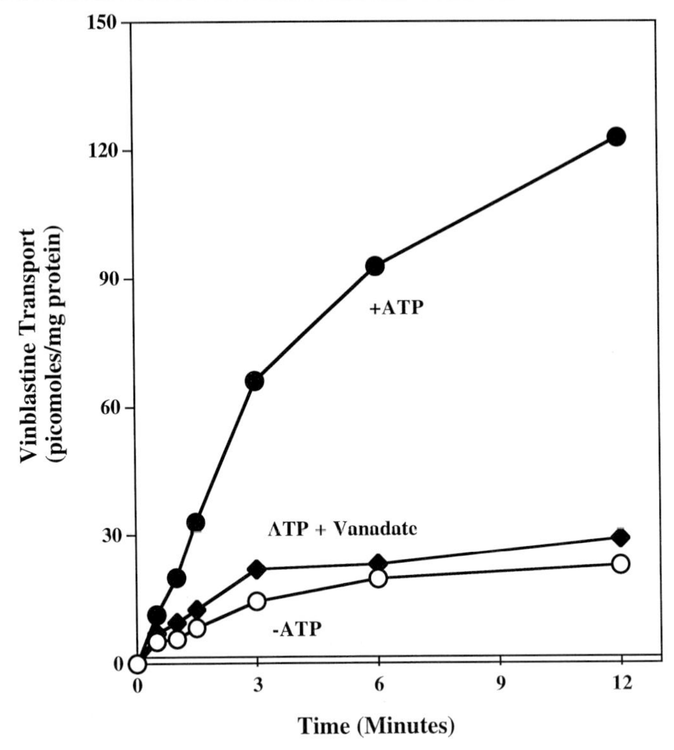

FIG. 2. Vinblastine transport by proteoliposomes reconstituted with purified Pgp. Proteoliposomes (0.5–1 μg protein/ml) were incubated with 5 nM [^{3}H]vinblastine (0.2 μCi/ml) in the presence or absence of 0.3 mM sodium orthovanadate for 5 min in a medium containing 25 mM MOPS/NMDG, pH 7.4, 150 mM NMDGCl, 2 mM DTT, and 5 mM $MgCl_2$ at room temperature. The assay was initiated by addition of 2 mM ATP (dosodium salt, pH adjusted to 7.4 with NMDG), and 200 μl of proteoliposomal suspension at the indicated time was applied under vacuum, to the center of a 0.22-μm GSTF Millipore filter. The filter was washed, under vacuum, with three 5-ml rinses with ice-cold buffer (25 mM MOPS/NMDG, pH 7.4, and 150 mM NMDGCl). The [^{3}H]vinblastine associated with proteoliposomes in the absence (open circles) or presence of ATP alone (filled circles) or both with ATP and vanadate (filled diamonds) is shown. The level of radioactivity bound to an equivalent volume of liposomes (without protein) with or without ATP was similar to that observed with proteoliposomes in the absence of ATP (data not shown).

tine accumulation is inhibited by 25 μM verapamil but not by camptothecin (50 μM; not shown). These proteoliposomes exhibit verapamil-stimulated ATPase activity at a much reduced level (Table I). This is probably due to loss of activity due to prolonged (18-hr) incubation during dialysis. Thus, it is possible to increase the vinblastine accumulation in these proteoliposomes significantly (four- to fivefold) by preventing the inactivation during dialysis.

Summary

Human Pgp from the vinblastine-resistant cell line, KB-V1, can be purified by sequential conventional chromatography on DEAE-sepharose CL-6B resin followed by a wheat germ agglutinin column. By including glycerol (osmolyte protectant) and lipid during the solubilization[19] and chromatography procedures[30] most of the biological activity of Pgp can be retained. The activity of Pgp in the detergent extract or in the concentrated column fractions is stable for at least 8–10 months when stored at −80°. However, repeated cycles of freezing and thawing of fractions result in considerable loss of activity. We have purified Pgp from KB-C1[14] (a subclone of KB 3-1 that is resistant to 1 μg/ml colchicine) by following the same protocol. When this method was used for purification of Pgp from *MDR1*-transfected NIH 3T3 transfectants (N3-V2400, grown in the presence of 2.4 μg/ml vinblastine),[31] the protein was eluted with 0.1 *M* NaCl from the DEAE-Sepharose CL-6B column as usual. However, during WGA lectin chromatography, the protein was eluted with a lower concentration of sugar (0.1 *M* instead of 0.25 *M* NAG). This altered elution pattern appears to be due to a difference in the glycosylation of human Pgp in mouse NIH 3T3 cells. This is consistent with the observation that human Pgp expressed in NIH 3T3 cells migrates faster compared to the protein from KB-V1 cells on 8–10% acrylamide gel.[31] Similarly, other workers have purified Chinese hamster Pgp either by a single-step chromatography on Reactive Red 120 agarose[32] or by a combination of anion exchange and immunoaffinity chromatography[33] (see the article by Senior *et al.*[34] for the purification and properties of ATPase activity of Chinese hamster Pgp). The high level of drug-stimulated ATP hydrolysis by Pgp (Table I), like other ion-transporting ATPases, indicates that this is a high-capacity pump that can function as an effective multidrug transporter. This is further supported by the qualitative demonstration of ATP-dependent vinblastine transport in proteoliposomes reconstituted with pure Pgp (see Fig. 2). Thus,

[31] U. A. Germann, T. C. Chambers, T. Licht, C. O. Cardarelli, I. Pastan, and M. M. Gottesman, *J. Biol. Chem.* **271,** 1708 (1996).

[32] I. L. Urbatsch, M. K. Al-Shawi, and A. E. Senior, *Biochemistry* **33,** 7069 (1994).

[33] A. B. Shapiro, and V. Ling, *J. Biol. Chem.* **270,** 16167 (1995).

[34] A. E. Senior, M. K. Al-Shawi, and I. L. Urbatsch, *Methods Enzymol.* **292,** [38], 1998 (this volume).

these experiments provide strong evidence that purified Pgp retains its activity and that it functions as an ATP-dependent drug transporter.

Acknowledgments

We are grateful to Drs. Michael M. Gottesman and Ira Pastan for advice and encouragement during the course of this work.

[37] Drug-Stimulatable ATPase Activity in Crude Membranes of Human *MDR1*-Transfected Mammalian Cells

By SURESH V. AMBUDKAR

Introduction

The increased expression of P-glycoprotein (Pgp), the *MDR1* gene product, has been linked to the development of resistance to multiple cytotoxic natural product anticancer drugs in certain cancers and cell lines derived from tumors.[1–3] Pgp, a member of the ATP-binding cassette (ABC) superfamily of transporters, functions as an ATP-dependent drug efflux pump with broad specificity for chemically unrelated lipophilic compounds.[4,5] Pgp, a 1280-amino-acid (160–170 kDa) membrane-associated protein, is comprised of 12 putative transmembrane segments with two ATP-binding/utilization sites. Each half of the molecule contains a lipophilic region with six transmembrane helices and a hydrophilic region with one ATP site. The amino-terminal half of the molecule is homologous (43% identity) with the carboxy-terminal half.[6]

Full-length cDNAs for the human and mouse multidrug resistance (MDR) genes (*MDR1, mdr1a,* and *mdr1b,* respectively) have been cloned

[1] M. M. Gottesman and I. Pastan, *Annu. Rev. Biochem.* **62,** 385 (1993).

[2] S. V. Ambudkar, I. Pastan, and M. M. Gottesman, *in* "Drug Transport in Antimicrobial and Anticancer Chemotherapy: Cellular and Biochemical Aspects of Multidrug Resistance" (N. H. Georgapapadakou, ed.), pp. 525–547. Marcel Dekker, New York, 1995.

[3] M. M. Gottesman, I. Pastan, and S. V. Ambudkar, *Curr. Opin. Genet. Dev.* **6,** 610 (1996).

[4] M. M. Gottesman, C. A. Hrycyna, P. V. Schoenlein, U. A. Germann, and I. Pastan, *Annu. Rev. Genet.* **29,** 607 (1995).

[5] U. A. Germann, *Eur. J. Cancer* **32A,** 927 (1996).

[6] C-J. Chen, C. E. Chin, K. Ueda, D. P. Clark, I. Pastan, M. M. Gottesman, and I. B. Roninson, *Cell* **47,** 381 (1986).

0076-6879/98 $25.00

into expression vectors, transfected into drug-sensitive cells, and shown to confer drug resistance.[4,5] Human (KB 3-1),[7,8] mouse (NIH 3T3),[9] and Chinese hamster (LR 73)[10] cell lines have been used for the stable expression of human and murine MDR genes. The site-directed mutagenesis approach is being used to elucidate the structure–function relationship of various domains of Pgp. The biochemical and molecular genetic studies in recent years have shown that both ATP sites can hydrolyze ATP and that proper functioning of both sites is required for ATP hydrolysis and subsequent drug efflux by Pgp.[3–5] Several mutants that are defective in drug transport function exhibit normal substrate and ATP binding but fail to catalyze drug-stimulated ATPase activity. Because the substrate-stimulated ATPase activity is tightly coupled to drug transport, this activity can be used as a diagnostic test to assess the effect of a particular mutation. Some of the cell lines such as human KB 3-1, HEK293, and Chinese hamster ovary LR-73, due to high level of other ATPases and/or alkaline phosphate, are not suitable for the measurement of Pgp-associated ATPase activity in crude membranes. In these instances, Pgp has to be at least partially purified to assay its ATPase activity.[11,12] Mouse NIH 3T3 fibroblasts are most suitable, due to the low level of background ATPases and other phosphatases activities, to assess the ATPase activity of Pgp even when the protein is expressed at low level. In this paper, the methods for assaying the drug-stimulatable ATPase activity of human Pgp expressed in murine NIH 3T3 transfectants are described.

Preparation of Crude Membranes from Rodent Cultured Cells by Hypotonic Lysis

Murine 3T3 fibroblasts are grown as monolayer cultures in Dulbecco's modified Eagle's medium (DMEM) supplemented with 4.5 g/liter glucose, 2 m*M* L-glutamine, 50 units/ml penicillin, 50 μg/ml streptomycin, and 10% (v/v) calf serum at 37° in 5% CO_2. NIH 3T3 cells are transfected with human *MDR1* (pHaMDR1/A-wt-CX construct), and mass populations of

[7] D.-w. Shen, A. Fojo, J. E. Chin, I. B. Roninson, N. Richert, I. Pastan, and M. M. Gottesman, *Science* **232,** 643 (1986).

[8] S. Zhang, Y. Sugimoto, T. Shoshani, I. Pastan, and M. M. Gottesman, *Methods Enzymol.* **292,** [34], 1998 (this volume).

[9] U. A. Germann, T. C. Chambers, S. V. Ambudkar, T. Licht, C. O. Cardarelli, I. Pastan, and M. M. Gottesman, *J. Biol Chem.* **271,** 1708 (1996).

[10] M. Hanna, M. Brault, T. Kwan, C. Kast, and P. Gros, *Biochemistry* **35,** 3625 (1996).

[11] T. W. Loo and D. M. Clarke, *Methods Enzymol.* **292,** [35], 1998 (this volume).

[12] S. V. Ambudkar, I. H. Lelong, J. Zhang, C. O. Cardarelli, M. M. Gottesman, and I. Pastan, *Proc. Natl. Acad. Sci. U.S.A.* **89,** 8472 (1992).

drug-resistant cells are selected by growing in increasing concentrations of vincristine ranging from 30 to 2400 ng/ml as described.[9,13] C3m, a colchicine-resistant subclone of NIH 3T3, is grown in the presence of 3 μg/ml colchicine. For crude membrane preparation, cells are grown to confluence in 15-cm dishes (3 to 10 dishes/preparation), scraped into ice-cold phosphate-buffered saline (PBS, without calcium and magnesium), pH 7.0, supplemented with 0.1 unit/ml aprotinin (Sigma Chemicals, St. Louis, MO), and then washed once with PBS and once with hypotonic lysis buffer (10 m*M* Tris-HCl, pH 7.5, 10 m*M* NaCl, 1 m*M* $MgCl_2$, and 0.1 unit/ml aprotinin; 15 ml buffer per dish) by centrifugation at 3500 rpm for 10 min at 4° in a Sorvall A6000 rotor (Sorvall, Newtown, CT). The washed cell pellet is resuspended in lysis buffer at 3 ml/dish and quickly frozen in dry ice. The frozen cell suspension is stored at −80° until later use. The cell suspension is kept at −80° for at least 1 hr prior to thawing to room temperature. The single freeze–thaw cycle results in optimal cell lysis. The thawed cell suspension is incubated on ice for 45 min and subsequently cells are disrupted by homogenization with 30 strokes of pestle A (loose fitting) and pestle B (tight fitting) in a handheld Dounce homogenizer. The cell lysate is diluted two- to three-fold with ice-cold isotonic TSNMDG buffer [10 m*M* Tris-HCl, pH 7.4, 250 m*M* sucrose, 50 m*M* *N*-methyl-D-glucamine (NMDG, Sigma) chloride, and 0.1 unit/ml aprotinin]. The unbroken cells and nuclei are removed from the lysate by centrifugation for 10 min at 2000 rpm at 4° in a Sorvall A6000 rotor. The low-speed supernatant is transferred to polycarbonate Beckman Type 45 Ti rotor tubes and the volume is brought to 60 ml with TSNMDG buffer. The crude membranes are collected by ultracentrifugation for 45 min at 4° in a Type 45 Ti rotor. The membrane pellet is resuspended in 5–7 ml of TSNMDG buffer using a 1- to 3-ml syringe fitted with a blunt-ended 23-gauge needle and washed once by ultracentrifugation. Finally, the washed membranes are resuspended in TSNMDG buffer (0.5 ml/15-cm dish), and small aliquots of 0.2–0.5 ml are quickly frozen in dry ice and stored at −80°. The protein content of crude membranes is determined by a modified Lowry method[14] using bovine serum albumin (Fraction V, Sigma) as a standard. Typically, the yield of crude membrane is 1.5–2.0 mg protein/15-cm dish). The verapamil-stimulated ATPase activity is stable for at least 8–12 months when membranes are stored at −80°. The exposure of membranes to repeated freeze–thaw cycles results in significant loss of Pgp-associated ATPase activity.

[13] M. M. Gottesman, C. O. Cardarelli, S. Goldenberg, T. Licht, and I. Pastan, *Methods Enzymol.* **292,** [17], 1998 (this volume).

[14] J. L. Bailey, *in* "Techniques in Protein Chemistry," pp. 340–341. Elsevier Publishing Co., New York, 1967.

ATPase Assays

Pgp-mediated ATP hydrolysis in crude membranes is measured by determining the release of inorganic phosphate from Mg-ATP in the presence of inhibitors of F- and P-type ATPases (i.e., sodium azide, EGTA, and ouabain)[15] with a colorimetric method as described previously,[16] with some modifications. Pgp-ATP activity is inhibited by incubation with low (0.3 m*M*) concentrations of sodium orthovanadate and, thus, ATP hydrolysis in the presence and absence of substrate and vanadate is measured to determine the vanadate-sensitive Pgp-specific activity. The ATPase assays are carried out at 37°. Crude membranes (10–30 μg protein) in a final volume of 0.1 ml are incubated with 0.3 m*M* sodium orthovanadate in the assay buffer containing 50 m*M* Tris-HCl, pH 7.5, 150 m*M* *N*-methyl-D-glucamine (NMDG)-Cl, 5 m*M* sodium azide, 1 m*M* ethylene glycol bis(β-aminoethyl ether)-*N,N,N′,N′*-tetraacetic acid (EGTA), 1 m*M* ouabain, 2 m*M* dithiothreitol (DTT), and 10 m*M* $MgCl_2$ for 5 min at 37°. Following the incubation with vanadate, either 1 μl of dimethyl sulfoxide (DMSO) or 1 μl of 3 m*M* verapamil prepared in DMSO is added and incubated for 3 min at 37°. The assay is initiated by addition of 10 μl of 50 m*M* disodium ATP (pH 7.0 adjusted with NMDG). Typically, the reaction is terminated after 20 min incubation by addition of 0.1 ml of 5% (w/v) sodium dodecyl sulfate (SDS). The ATP hydrolysis is linear in the presence and absence of substrate (verapamil) up to 30–35 and 50–60 min, respectively. To estimate the content of liberated inorganic phosphate (P_i), 0.5 ml high-performance liquid chromatography (HPLC) grade water, 0.4 ml P_i reagent (1% ammonium molybdate in 2.5 *N* sulfuric acid and 0.014% antimony potassium tartrate), and 0.2 ml 1% (w/v) ascorbic acid are added to each tube and mixed well by vortexing. After 10–12 min of incubation at room temperature, the blue color intensity is measured at 880 nm by using assay buffer containing ATP alone without membrane protein as a blank. The blue color is stable for at least an additional 10–15 min, after which time due to the ATP breakdown in acidic conditions, the intensity of color in blank and test samples increases with time.

Western Immunoblotting

Sodium dodecyl sulfate–polyacrylamide gel electrophoresis (SDS–PAGE) with 7.5 or 8% acrylamide is performed as described by Laemmli[17]

[15] P. L. Pedersen and L. M. Amzel, *J. Biol. Chem.* **268,** 9937 (1993).

[16] B. Sarkadi, E. M. Price, R. C. Boucher, U. A. Germann, and G. A. Scarborough, *J. Biol. Chem.* **267,** 4854 (1992).

[17] U. K. Laemmli, *Nature* (*Lond.*), **227,** 680 (1970).

using a Mini-PROTEAN II apparatus (Bio-Rad, Hercules, CA). For Western blots, proteins are electrophoretically transferred to 0.45 μm pure nitrocellulose membranes. The nitrocellulose blot was blocked for 1 hr in 2.5% gelatin in wash buffer [10 m*M* Tris-HCl, pH 7.4, 150 m*M* Nacl, 0.2% (v/v) Triton X-100, filtered through Whatman (Chifton, NJ) #3 filter paper] and incubated with primary antibody [monoclonal C219 (0.5 μg/ml)[18] or polyclonal PEPG-13 antiserum,[19] diluted 1:2000 (v/v)] in blocking solution containing 3% bovine serum albumin (BSA) in wash buffer for 1 hr at room temperature. After three washes in wash buffer, the blots are incubated with secondary antibody [horseradish peroxidase-conjugated donkey anti-mouse immunoglobulin G (IgG) or goat anti-rabbit IgG] at 1:2500 dilution for 1 hr. The immunoblots are washed three times with wash buffer and developed by using an enhanced chemiluminescence (ECL) detection system (Amersham, Arlington Heights, IL) by following the manufacturer's protocol.

Identification of Human Pgp Expressed in Murine NIH 3T3 Cells

The expression of human Pgp at the cell surface of NIH 3T3 *MDR1* transfectants selected in the presence of vincristine can be determined with fluorescence-activated cell sorting (FACS) after staining with human Pgp-specific monoclonal antibody MRK-16[20] or UIC2[21] that recognizes the extracellular region of the protein. Such analysis shows that the level of Pgp expression is dependent on the concentration of drug used for the selection of transfectants. As an alternative to the method of FACS analysis, we found that human Pgp expression in mouse NIH 3T3 cells can be determined by Western immunoblot using polyclonal antiserum PEPG-13. This antiserum was raised against 592- to 636-amino-acid residues of human Pgp overexpressed in *Escherichia coli* as a fusion protein with *Pseudomonas* exotoxin. As shown in Fig. 1A, PEPG-13 recognizes human Pgp expressed in vinblastine-resistant KB-V1 cells as well as in NIH 3T3 *MDR1* wild-type and phosphorylation-deficient mutant transfectants (5-Ala and 5-Asp). However, PEPG antiserum failed to recognize mouse Pgps overexpressed in C3m, a colchicine-resistant subclone of NIH 3T3 cells. Similarly, PEPG-13 did not recognize Chinese hamster Pgp in colchicine-resistant Chinese hamster ovary (CHO) cells (data not shown). On the other hand, both

[18] E. Georges, G. Bradely, J. Gariepy, and V. Ling, *Proc. Natl. Acad. Sci. U.S.A.* **87,** 152 (1990).
[19] E. P. Bruggemann, V. Chaudhary, M. M. Gottesman, and I. Pastan, *Biotechniques* **10,** 202 (1991).
[20] H. Hamada and T. Tsuruo, *Proc. Natl. Acad. Sci. U.S.A.* **83,** 7785 (1986).
[21] E. B. Mechetner and I. B. Roninson, *Proc. Natl. Acad. Sci. U.S.A.* **89,** 5824 (1992).

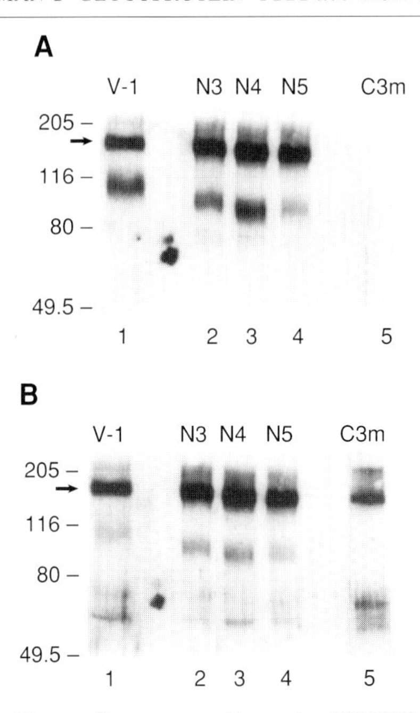

FIG. 1. Identification of human Pgp expressed in murine NIH 3T3 transfectants by Western immunoblotting. Crude membranes of NIH 3T3 *MDR1* transfectants selected in 2.4 μg/ml vincristine were prepared as described in the text and analyzed by Western blot using Pgp-specific polyclonal antiserum PEPG-13 (A) or monoclonal antibody C219 (B). The blot in panel A was stripped by incubating in buffer containing 10 m*M* sodium monobasic phosphate, pH 7.5, 2% SDS (w/v), and 2 m*M* 2-mercaptoethanol for 30 min (the stripping buffer was preheated to 75° and the buffer was changed four times at regular intervals during the 30-min incubation). The stripped blot was used for the immunodetection of Pgp with monoclonal antibody C219 as described in the text. Each lane contained 2 μg protein. Lane 1, plasma membrane vesicles of vinblastine-resistant human KB V-1 cells; lanes 2–4, NIH 3T3 *MDR1* transfectants selected at 2.4 μg/ml vincristine of wild-type (N3), and phosphorylation-deficient 5-Ala (N4) and 5-Asp (N5) mutants; lane 5, C3m, 3 μg/ml colchicine-resistant mouse NIH 3T3 subclone that overexpresses only mouse *mdr* gene(s). The arrow shows the position of Pgp.

human and mouse Pgps were detected with monoclonal antibody C219[18] (Fig. 1B), which recognized Pgps encoded by mammalian *MDR* genes.

Effect of Various Inhibitors on ATPase Activity in Crude Membranes of NIH 3T3 *MDR1* Transfectants Selected at 2.4 μg/ml Vincristine

Crude membrane preparation of NIH 3T3 *MDR1* transfectants is a mixture of various intracellular organelle membranes and plasma membranes. As a result, F-, V-, and P-type ATPases contribute to the activity measured in these membranes. To overcome this problem, a battery of inhibitors is used to inhibit these ATPases. To determine whether it is necessary to use multiple inhibitors, the effect of each inhibitor was tested. The F-type ATPase inhibitor, sodium azide, inhibited most of the ATPase activity (Fig. 2). Inclusion of EGTA and ouabain (an inhibitor of Na^+,K^+-ATPase) did not have any additional effect. The stimulation of Pgp-ATPase activity by verapamil was observed in the presence of sodium azide alone or along with EGTA and ouabain. More than 90% of the Pgp-independent activity was inhibited by 5–10 m*M* azide (Fig. 2). On the other hand, 1 m*M* EGTA did not have any effect (data not shown). Sodium azide, EGTA,

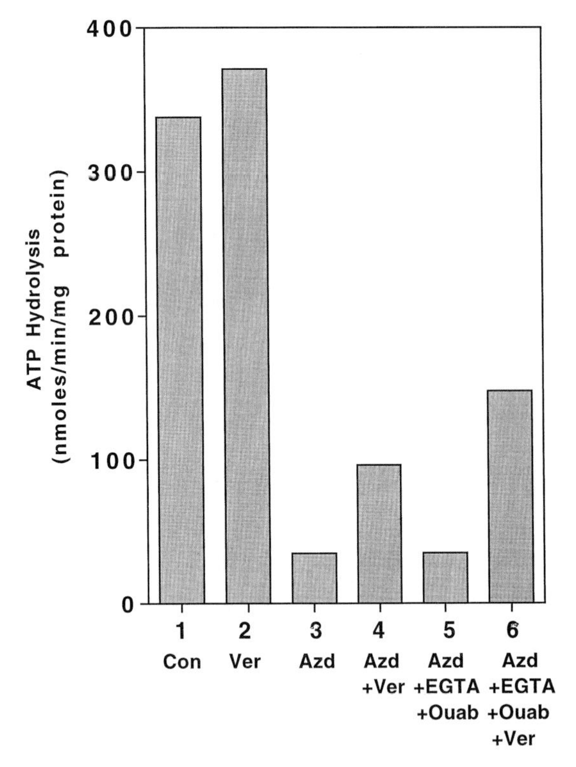

FIG. 2. Effect of various ATPase inhibitors on Pgp-ATPase in crude membranes of NIH 3T3 *MDR1* transfectant, N3-V2400. Crude membranes of N3-V2400 cells (100 μg protein/ml) were incubated with the indicated agent for 5 min at 37° prior to the determination of the ATPase activity. As indicated, tubes containing 30 μM verapamil were incubated for 3 min before the addition of ATP. 1, control without any inhibitor; 2, 30 μM verapamil; 3, 5 m*M* sodium azide; 4, 5 m*M* sodium azide and 30 μM verapamil; 5, 5 m*M* sodium azide, 1 m*M* EGTA, and 1 m*M* ouabain; 6, 5 m*M* sodium azide, 1 m*M* EGTA, 1 m*M* ouabain, and 30 μM verapamil.

and ouabain did not have any effect on the ATPase activity of partially purified Pgp in a reconstituted system.[12] However, 0.3 m*M* sodium orthovanadate inhibited >90% of basal and verapamil-stimulated Pgp-ATPase activity. Thus, F-type ATPase contributes most of the background activity in crude membranes of NIH 3T3 transfectants.

ATPase Activity of NIH 3T3 *MDR1* Transfectants Expressing Varying Levels of Human Pgp

To facilitate the measurement of ATPase activity of a given mutant Pgp in transfectants selected at low concentration, one should be able to detect the specific Pgp-associated activity above the background level. The selection of transfectants expressing high levels of Pgp is quite time consuming, thus, it is desirable to be able to assess the ATPase activity in transfectants expressing low levels of Pgp. For this reason, the ATPase activity of NIH 3T3 *MDR1* transfectants N3-V30, N3-V600, and N3-V2400 (selected with 30, 600, and 2400 ng/ml vincristine, respectively, as described[9]) was determined (Table I). The total ATPase activity in the absence of any ATPase inhibitor was more or less similar in all transfectants. The stimulation by verapamil as an indicator of Pgp-associated ATPase activity was seen in N3-V600 and N3-V2400. However, when the ATPase activity was assayed in the presence of azide, EGTA, and ouabain, the stimulation by

TABLE I
ATP HYDROLYSIS IN THE CRUDE MEMBRANES OF NIH 3T3 *MDR1* TRANSFECTANTS EXPRESSING VARYING LEVELS OF P-GLYCOPROTEIN

Transfectant	ATP hydrolysis[a] (nmol/min/mg protein) Control	+30 μM Verapamil	Concentration required for half-maximal stimulation: verapamil (μM)[b]	(-fold) Stimulation[b]
3T3 (untransfected)	319.30	312.00	—	1.00
N3-30 ng/ml vincristine[c]	233.30	239.00	1.00	3.50
N3-600 ng/ml vincristine	220.50	237.60	1.50	3.45
N3-2400 ng/ml vincristine	293.00	378.00	1.50	5.90

[a] ATPase assays were carried out in the absence of inhibitors (azide, EGTA, ouabain, and vanadate) as shown in Fig. 2.

[b] Verapamil-stimulated ATPase activity in the presence of azide, EGTA, and ouabain and in the presence and absence of 0.3 m*M* sodium orthovanadate was measured as described in the legend to Fig. 3. These values were calculated from the data shown in Fig. 3.

[c] NIH 3T3 *MDR1* transfectants were selected and grown in the presence of 30, 600, and 2400 ng/ml vincristine.[9]

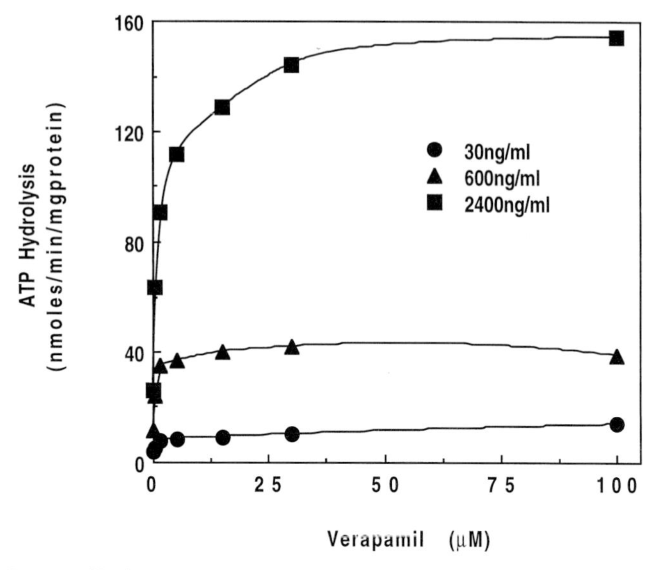

FIG. 3. Verapamil-stimulated ATPase activity in NIH 3T3 *MDR1* transfectants expressing varying levels of Pgp. Crude membranes (100–300 μg protein/ml) were incubated with the indicated concentration of verapamil in the assay buffer containing inhibitors of various ATPases as described in the text and in the presence and absence of 0.3 m*M* sodium orthovanadate. Only the vanadate-sensitive activities are shown. NIH 3T3 *MDR1* transfectants were grown in the presence of 30 ng/ml (N3-V30, ●), 600 ng/ml (N3-V600, ▲), and 2400 ng/ml (N3-V2400, ■) vincristine, respectively.[9]

verapamil in a concentration-dependent manner was seen even in N3V-30 membranes (Fig. 3). The properties of Pgp such as affinity for verapamil and the fold stimulation of ATPase activity was not affected by stepwise selection of N3-V600 and N3-V2400 in 20- and 80-fold concentration of vincristine compared to N3-V-30 (Table I).

ATPase Activity of Phosphorylation-Deficient Pgp Mutants Expressed in NIH 3T3 Cells

Human Pgp is phosphorylated at serine residue in the linker region at positions 661, 667, 671, 675, and 683 by a variety of kinases including protein kinase A, protein kinase C, and V-1 kinase.[22] Germann *et al.* demonstrated that replacement of serine at 661, 667, 671, 675, and 683 with either alanine (N4-V2400, 5-Ala) or aspartic acid (N5-V2400, 5-Asp) abolished the phos-

[22] U. A. Germann, T. C. Chambers, S. V. Ambudkar, I. Pastan, and M. M. Gottesman, *J. Bioenerg. Biomembr.* **27,** 53 (1995).

TABLE II
VERAPAMIL-STIMULATED Pgp–ATPase ACTIVITY IN CRUDE MEMBRANES OF PHOSPHORYLATION-DEFICIENT MUTANTS

Transfectant[a]	Concentration required for half-maximal stimulation[b] (μM)	Maximal stimulation (-fold)
N3-V2400 (wild type)	1.50	5.60
N4-V2400 (5-Ala)	2.00	5.70
N5-V2400 (5-Asp)	1.80	4.40

[a] Crude membranes of NIH 3T3 *MDR1* transfectants grown in the presence of 2.4 μg/ml vincristine were prepared as described in the text. N4-V2400 (5-Ala) and N5-V2400 (5-Asp) are phosphorylation-deficient mutant Pgps in which serine at positions 661, 667, 671, 675, and 683 was replaced with alanine (5-Ala) or aspartic acid (5-Asp), respectively.[9]

[b] Vanadate-sensitive ATPase activity in the presence of varying concentration of verapamil ranging from 0.05 to 300 μM was measured as described in the text for the determination of these values.

phorylation of Pgp both *in vivo* and *in vitro* by endogenous or externally added kinases.[9] The lack of phosphorylation did not affect the drug transport function or the ability of Pgp to confer drug resistance phenotype to sensitive cells. The verapamil-stimulated ATPase activity of these mutants in crude membranes was determined (Table II). The level of mutant Pgps in membranes was similar to that of wild type (Fig. 1). The affinity for verapamil and the extent of verapamil-stimulated activity in these mutants was similar to that of wild-type Pgp (Table II).

Conclusions

The data presented here clearly demonstrate that Pgp ATPase activity can be measured in the crude membranes of NIH 3T3 transfectants. The background activity of other ATPases can effectively be blocked with sodium azide alone. However, inclusion of other inhibitors such as EGTA and ouabain does not have any deleterious effect on Pgp activity. NIH 3T3 transfectants selected at low drug concentration are also suitable for characterization of Pgp ATPase activity. N3-V30 transfectants selected at 30 ng/ml vincristine exhibit four- to six-fold resistance to vincristine compared to untransfected NIH 3T3 cells. Thus, it should be feasible to measure ATPase activity of a given mutant Pgp expressed at similar level. Because NIH 3T3 cells are suitable for characterization of drug transport function[23]

[23] S. V. Ambudkar, C. O. Cardarelli, I. Pashinsky, and W. D. Stein, *J. Biol. Chem.* **272,** 21160 (1997).

and ATPase activity, they offer an ideal stable expression system to assess the structure–function relationship of various regions in Pgp.

Acknowledgments

I thank Drs. Ursula Germann, Ding-wu Shen, and Michael M. Gottesman for providing NIH 3T3 *MDR1* transfectants and colchicine-resistant C3m, a subclone of NIH 3T3 cells and Carol O. Cardarelli for growing NIH 3T3 transfectants.

[38] ATPase Activity of Chinese Hamster P-Glycoprotein

By ALAN E. SENIOR, MARWAN K. AL-SHAWI, and INA L. URBATSCH

Introduction

P-Glycoprotein (Pgp) is a plasma membrane glycoprotein that confers multidrug-resistant (MDR) phenotype on cells. It is of interest for a number of reasons: first, because it represents one mechanism of resistance to chemotherapy in human cancer[1,2]; second, because it occurs naturally in mammalian tissues and appears to play a role in protection from environmental poisons[3]; and third, because it is a prominent member of the rapidly expanding "ABC transporter" superfamily of transport proteins, many of which appear to be involved in human disease.[4]

The amino acid sequences of Pgp from human, mouse, and Chinese hamster suggested a structure containing two transmembrane domains (TMD) and two nucleotide-binding sites (NBS).[5,6] Each NBS contains the two consensus sequences (homology A and B) identified by Walker *et al.*[7] as diagnostic of nucleotide binding. Studies in whole cells have shown that Pgp excludes chemotherapeutic drugs from MDR cells in an ATP-dependent manner,[8] and subsequent studies in plasma membrane vesicles, proteoliposomes, and yeast secretory vesicles showed that drug transport

[1] M. M. Gottesman and I. Pastan, *Annu. Rev. Biochem.* **62,** 385 (1993).

[2] A. B. Shapiro and V. Ling, *J. Bioenerg. Biomemb.* **27,** 7 (1995).

[3] A. H. Schinkel, J. J. M. Smit, O. van Tellingen, J. H. Beijnen, E. Wagenaar, L. van Deempter, C. A. A. M. Mol, M. A. van der Valk, E. C. Robanus-Maandag, H. P. J. Te Riele, A. J. M. Burns and P. Borst, *Cell* **77,** 491 (1994).

[4] C. F. Higgins, *Annu. Rev. Cell Biol.* **8,** 67 (1992).

[5] P. Gros, J. Croop, and D. Housman, *Cell* **47,** 371 (1986).

[6] C. Chen, J. E. Chin, K. Ueda, D. P. Clark, I. Pastan, M. M. Gottesman, and I. G. Roninson, *Cell* **47,** 381 (1986).

[7] J. E. Walker, M. Saraste, M. J. Runswick, and N. J. Gay, *EMBO J.* **1,** 945 (1982).

[8] G. Bradley, P. F. Juranka, and V. Ling, *Biochim. Biophys. Acta* **948,** 87 (1988).

METHODS IN ENZYMOLOGY, VOL. 292
0076-6879/98 $25.00

was ATP-hydrolysis dependent.[9–12] The most widely accepted hypothesis is that Pgp functions as an ATP-driven drug export pump. Thus, it is important to determine molecular characteristics of ATP hydrolysis by Pgp. In this paper, we describe methods devised in our laboratory to obtain defined experimental systems for biochemical investigation of Pgp, and for characterization of Pgp ATPase activity.

CR1R12 Cell Line Overexpressing P-Glycoprotein

Chinese hamster ovary (CHO) cells, which are subjected to selection in media containing incrementally increasing concentrations of colchicine, overexpress Pgp and show an MDR phenotype. Here, the CH^RC5 cell line (which is partially MDR, and was originally derived from the AUXB1 cell line as described in Ref. 11) is rendered highly MDR by further selection as follows.

Cell lines are routinely passaged and grown on solid support in α-minimal essential medium (α-MEM) with L-glutamine, ribonucleosides, and deoxyribonucleosides (GIBCO, Grand Island, NY), 50 units/ml penicillin G, plus 50 μg/ml streptomycin sulfate (complete α-MEM), supplemented with 10% (v/v) fetal calf serum. CH^RC5 cells are grown on multiple 15-cm dishes, initially in the presence of 5 μg/ml colchicine, increasing the colchicine concentration incrementally by 5 μg/ml steps, and achieving stable growth at each new concentration. At a final concentration of 30 μg colchicine/ml, individual colonies are chosen and subjected to two purification cycles.

Nineteen derivative sublines were obtained in this manner, and assessed for MDR phenotype by colony-forming assays, content of P-glycoprotein by immunoblotting with C219 monoclonal antibody,[13] constitutive overexpression of Pgp, and vigorous growth in suspension culture. One cell line with desired characteristics was chosen and named CR1R12. In comparison to the parent, drug-sensitive AUXB1 cell line, CR1R12 cells show 1540-fold resistance to colchicine, 670-fold resistance to daunomycin, and 1200-fold resistance to vinblastine.[14] CR1R12 cells are routinely grown on plates in medium supplemented with 5 μg/ml colchicine to prevent reversion. Colchicine is added from a 100 mg/ml stock solution in sterile dimethyl sulfoxide (DMSO).

[9] S. V. Ambudkar, *J. Bioenerg. Biomembr.* **27,** 23 (1995).
[10] F. J. Sharom, *J. Bioenerg. Biomembr.* **27,** 15 (1995).
[11] V. Ling and L. H. Thompson, *J. Cell Physiol.* **83,** 103 (1974).
[12] S. Ruetz and P. Gros, *J. Biol. Chem.* **269,** 12277 (1994).
[13] E. Georges, G. Bradley, J. Gariepy, and V. Ling, *Proc. Natl. Acad. Sci. U.S.A.* **87,** 152 (1990).
[14] M. K. Al-Shawi and A. E. Senior, *J. Biol. Chem.* **268,** 4197 (1993).

Growth of CR1R12 Cells in Suspension Culture

CR1R12 cells are first grown on eighteen 150- × 25-mm plates as described earlier, with a final volume of 30 ml medium/plate. The cells are then removed by trypsinization and seeded into eight 1-liter or four 3-liter spinner flasks, containing 200 ml or 500 ml, respectively, of complete α-MEM supplemented with 10% Fetal Clone (HyClone, GIBCO, Grand Island, NY), without colchicine, so as to give initial cell density of 2.0–2.3 $\times 10^5$ cell/ml. Cells are incubated at 37° in a 5% (v/v) CO_2 incubator, and fed by doubling the medium volume every 24 hr. Cells are harvested at the end of log phase (96–120 hr; cell density ~5 $\times 10^5$/ml), spun down at 4° (Beckman JA10 rotor, 3000 rpm for 20 min), washed once, and resuspended at 2° in buffer A: 25 m*M* HEPES, pH 7.4, 10 m*M* $NaHCO_3$, 1 m*M* K_2HPO4, 1 m*M* $MgSO_4$, 1 m*M* $CaCl_2$, 125 m*M* NaCl, and 11 m*M* glucose. If cells are to be frozen at −70°, glycerol (20%, v/v) is added.

Preparation of Plasma Membranes Enriched in P-Glycoprotein

All operations are done at 4°. Around 3–5 $\times 10^9$ CR1R12 cells are spun down (JA10 rotor, 6000 rpm, 20 min) and resuspended in 30 ml of ice-cold buffer B: 0.25 *M* sucrose, 20 m*M* Tris-Cl, pH 7.4, 1 m*M* EGTA, 6.5 m*M* dithiothreitol (DTT), 15 m*M* *p*-aminobenzamidine, 0.5 m*M* phenylmethylsulfonyl fluoride (PMSF), 76 m*M* 6-aminohexanoic acid, 1 μM leupeptin, and 2 μM pepstatin A. The cell suspension is disrupted at 30 psi N_2 by nebulization in a Bio-Neb Cell Disruptor (Glas-Col Co., Terre Haute, IN). Disruption is monitored by microscopic examination and usually requires three to five passes.

The homogenate is centrifuged at 2400 rpm in a Beckman JA20 rotor for 10 min and the pellet resuspended in 5 ml buffer B ("nuclear fraction"). Surface lipids are removed from the supernatant, which is then centrifuged at 7100 rpm in a JA20 rotor for 10 min. The pelleted "mitochondrial fraction" is resuspended as described earlier. The supernatant is centrifuged in a Beckman 60Ti rotor at 22,000 rpm for 30 min. The pelleted "microsomal fraction" is resuspended in 2.5 ml buffer B, and is applied in two equal parts without freezing to two discontinuous sucrose gradients contained in Beckman SW28 rotor tubes. Each tube contains sucrose solutions as follows: 5 ml 60% (w/v), 7 ml 45% (w/v), 12 ml 31% (w/v), and 6 ml 16% (w/v). Tris-Cl (30 m*M*, pH 7.4) is present in all sucrose solutions. The gradients are spun 18 hr at 24,000 rpm, and allowed to decelerate without braking. Interface bands are removed by Pasteur pipette, diluted 1:10 in buffer B with 1 m*M* NaN_3, and centrifuged in the 60Ti rotor at 60,000 rpm for 3 hr. Final pellets are resuspended in 1 ml buffer B plus 1 m*M* NaN_3 and stored frozen at −70°.

The fraction accumulating at the 16/31% sucrose interface consists of purified plasma membranes. Protein assay by the bicinchoninic acid method[15] in the presence of 1% (w/v) sodium dodecyl sulfate (SDS) indicates a yield of 3–5 mg protein. The Pgp content, assayed by laser densitometry of Coomassie blue-stained SDS gels, ranges from 15 to 32% (w/w) of total membrane protein (average 21%). An example of such a stained gel is shown in Fig. 1, lanes 3 and 4. The CR1R12 cell plasma membranes show high, drug-stimulated ATPase activity, attributable to Pgp (specific activity in presence of 10 μM verapamil ~9 μmol/min/mg Pgp, equivalent to ~20 sec^{-1}, at 37°, pH 7.4). The Pgp ATPase activity is stable for several years at −70°. The Pgp isoform present is Pgp-1, which is the isoform equivalent to human MDR1.[16] There is low Na^+, K^+-ATPase and Ca^{2+}-ATPase activity present in the membranes, which may be suppressed by inclusion of 2 mM ouabain and 0.1 mM EGTA in assays. Ecto-ATPase and mitochondrial ATPase are absent. We have used these plasma membranes in a large number of diverse experiments designed to characterize the catalytic sites and ATP hydrolysis activity of Pgp (reviewed in Refs. 17 and 18). Obviously, there is a great deal of potential experimentation that can be done using this material, for example, identification of new inhibitors, modulators, and transport substrates of Pgp.

The fraction accumulating at the 31/45% sucrose interface contains 2–5% Pgp (w/w of total membrane protein) and is typically obtained in a yield of ~20 mg protein. It also shows significant drug-stimulated ATPase activity attributable to Pgp. Collected "nuclear" and "mitochondrial fractions" may be combined and subjected to disruption by three to five passes through the nebulizer, followed by differential centrifugation and sucrose density gradient fractionation as described earlier, yielding significant amounts of both 16/31% and 31/45% sucrose interface material, which contains Pgp in the same relative amounts as detailed earlier.

Purification of P-Glycoprotein to Homogeneity[19]

All steps are done at 4°. Plasma membranes (1–1.5 mg/ml final concentration) are solubilized for 15 min in buffer C: 50 mM Tris-Cl, pH 7.0, 6.5 mM DTT, 0.25 M sucrose, 1 mM EGTA, 5 mM 6-aminohexanoic acid, 1 mM NaN_3, 1 μM leupeptin, 2 μM pepstatin A, 20% (v/v) glycerol, 1.4%

[15] P. K. Smith, R. I. Krohn, G. T. Hermanson, A. K. Mallia, F. H. Gartner, M. D. Provenzano, E. K. Fujimoto, N. M. Goekie, B. J. Olson, and D. C. Klenk, *Anal. Biochem.* **150,** 76 (1985).

[16] M. K. Al-Shawi, I. L. Urbatsch, and A. E. Senior, *J. Biol. Chem.* **269,** 8986 (1994).

[17] A. E. Senior, M. K. Al-Shawi, and I. L. Urbatsch, *J. Bioenerg. Biomembr.* **27,** 31 (1995).

[18] A. E. Senior, M. K. Al-Shawi, and I. L. Urbatsch, *FEBS Lett.* **377,** 285 (1995).

[19] I. L. Urbatsch, M. K. Al-Shawi, and A. E. Senior, *Biochemistry* **33,** 7069 (1994).

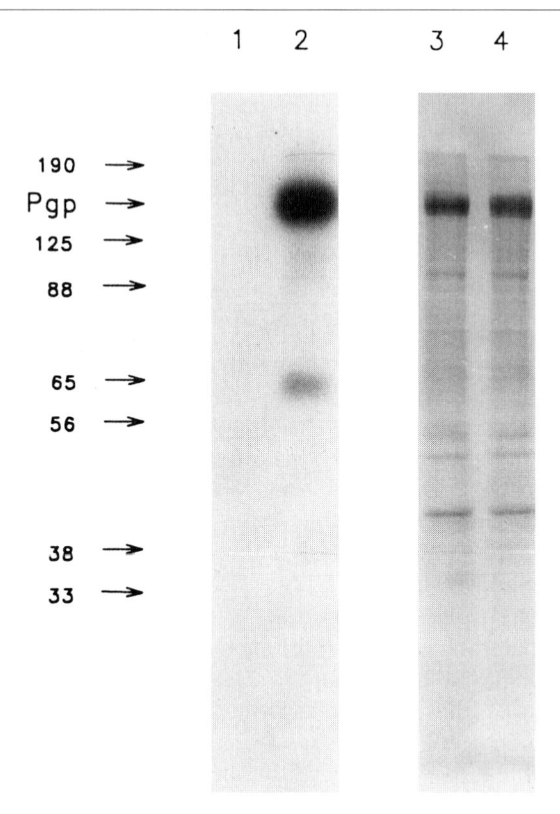

FIG. 1. Vanadate trapping of 8-azido-ATP and specific photolabeling of Pgp. Plasma membranes from CR1R12 cells were preincubated wtih 200 μM 8-azido[α-^{32}P]ATP, 3 mM $MgSO_4$, 40 mM Tris-Cl, pH 7.4, and 0.1 mM EGTA in the absence (lanes 1 and 3) or presence (lanes 2 and 4) of 200 μM vanadate. Unbound ligands were removed by passage through centrifuge columns. The degree of ATPase inhibition was 80% (with vanadate) or zero (without vanadate). The eluates were placed on ice and irradiated for 5 min (λ = 254 nm, 5.5 mW/cm^2). Samples were subjected to SDS–PAGE, then to autoradiography (lanes 1 and 2) or staining with Coomassie blue (lanes 3 and 4). [Reproduced with permission from I. L. Urbatsch, B. Sankaran, S. Bhagat, and A. E. Senior, *J. Biol. Chem.* **270,** 19387 (1995)].

(w/v) *n*-octyl-β-D-glucopyranoside (octylglucoside, Calbiochem, La Jolla, CA), 0.4% (w/v) acetone/ether precipitated *Escherichia coli* lipid (Avanti Corp., Alabaster, AL), 50 mM $MgCl_2$, 50 mM NaATP. [*E. coli* lipids are dried from $CHCl_3$ solution under N_2, resuspended at 5% (w/v) in buffer C by sonication on ice, and stored in light-protected aliquots at $-20°$ under N_2.]

The suspension is spun for 60 min at 50,000 rpm in a Beckman 50Ti rotor. The supernatant contains $\geq$90% of the Pgp and is applied to a Reactive Red 120 agarose column (Sigma, St. Louis, MO, Type 3000-Cl).

The column (5 ml of resin in a 10-ml syringe) is washed with 25 ml of buffer C, which contains only 1.2% (w/v) octylglucoside and 0.2% (w/v) acetone/ether precipitated *E. coli* lipid. Ten milligrams total protein is applied to the column at 0.4 ml/min flow rate, and the column is washed with five volumes of buffer C at the same flow rate. The adsorbed Pgp is eluted with a gradient (50-ml total volume) of 0–500 m*M* NaCl in buffer C, at a flow rate of 0.4 ml/min, collecting 3-ml fractions. The column is then washed with 5 ml buffer C plus 500 m*M* NaCl, and 5 ml buffer C plus 1 *M* NaCl. Pure Pgp elutes at 50–450 m*M* NaCl and those fractions are pooled. Subsequent NaCl gradient fractions, and the 0.5 *M* and 1 *M* NaCl eluates, contain highly enriched Pgp.

The Pgp is reconstituted into proteoliposomes by dialysis of pooled fractions for 16 hr against 20 volumes of buffer containing degassed, N_2-saturated 50 m*M* Tris-Cl, pH 7.4, 1 m*M* DTT, 1 m*M* EGTA, and 5 m*M* 6-aminohexanoic acid, with change in buffer at 3 hr. The proteoliposomes were harvested by centrifugation in the 60Ti rotor at 60,000 rpm for 3 hr, resuspended in fresh dialysis buffer at 0.5–1 mg protein/ml, and stored at −70°. Before use, proteoliposomes are exposed to a single freeze–thaw cycle.

The yield of pure Pgp from 10-mg plasma membranes is ~1 mg. The specific ATPase activity (in the presence of 50 μM verapamil) is 4.2 μmol/min/mg, equivalent to 9.2 sec^{-1}, at 37°, pH 7.4. The ATPase activity is stable for years at −70°.

The "31/45% sucrose interface" membrane fraction (see earlier discussion) can be solubilized and chromatographically fractionated on Reactive Red 120 agarose by the same procedure, yielding slightly impure Pgp. Further purification of this material on a Cibacron Blue column yields highly pure Pgp[19] at a yield of ~400 μg from 20 mg of starting membranes. The Cibacron Blue column (Sigma, Type 3000-Cl) is prepared as for Reactive Red 120 above (1 ml packed resin per 2 mg protein). The protein is applied, the column is washed with five volumes of buffer C, then the column is eluted with 1.5 volumes of buffer C containing 1 *M* NaCl. The eluted material is reconstituted into proteoliposomes as was done earlier, and is seen on SDS gels to contain pure Pgp.

Steady-State Assay of ATP Hydrolysis

Malachite Green Assay

Nucleotide hydrolysis by Pgp is allowed to proceed, the reaction is stopped, and P_i released is estimated by the method of Van Veldhoven and Mannaerts.[20] The advantages are that P_i released is measured sensitively (down to 0.5 nmol), that the procedure may be used for assays at different temperature, pH, ionic strength, etc., using different nucleotides, and that it allows investigation of effects of drugs, modulators, activators, and inhibitors directly on Pgp.

The assay medium consists of 40 m*M* Tris-Cl, pH 7.4, 0.1 m*M* EGTA, 2 m*M* ouabain, 15 m*M* $MgSO_4$, and 10 m*M* NaATP at 37°. The assay is carried out in 50 μl prewarmed medium, starting the reaction with 0.5–1 μg of plasma membranes or 0.25 μg pure Pgp. Assays are stopped by addition of 1 ml ice-cold 20 m*M* H_2SO_4, vortexed, and kept on ice. Typical assay times are 0.5, 10.5, 20.5, and 30.5 min, and the linear rate of P_i release is calculated. Since ATP is slowly hydrolyzed under acidic conditions, we perform color development as follows: 200 μl of reagent A[20] is added to the ice-cold tube, which is vortexed and incubated at 23° for 10 min. Then 200 μl of reagent C[20] is added and absorbance at 610 nm is read after a 25-min incubation at 23°.

Ouabain is added from a 400 m*M* stock solution in DMSO. It is not needed with pure Pgp, which contains no Na^+,K^+-ATPase. Drugs and modulators are added in carrier solvents DMSO or ethanol, keeping the final concentration of the solvent to $<1\%$ (v/v).

Coupled Enzyme Assay of ATP Hydrolysis

The assay utilizes an ATP-regenerating system composed of phosphoenol pyruvate and pyruvate kinase, linked to oxidation of NADH by pyruvate and lactate dehydrogenase. It has the advantages of conservation of enzyme and direct generation of time courses.

The assay medium consists of 40 m*M* Tris-Cl, pH 7.4, 0.1 m*M* EGTA, 10 m*M* NaATP, 12 m*M* $MgSO_4$, 3 m*M* phosphoenolpyruvate, 0.33 m*M* NADH, 10 units lactate hydrogenase (Boehringer-Mannheim, Indianapolis, IN), and 10 units pyruvate kinase (Boehringer-Mannheim, Indianapolis, IN). Ouabain and drugs or modulators are added as required (see earlier discussion). Typically, 1 ml of assay medium at 37° in a cuvette is continuously monitored at 340 nm, and the reaction is started by addition of 1–2 μg pure Pgp or 5–10 μg of plasma membranes. A second or third addition of NADH is necessary during long assays.

Brief Description of Catalytic Properties of Plasma Membrane and Purified Reconstituted P-Glycoprotein

The major points to be made are that the plasma membrane Pgp and purified reconstituted Pgp both show substantial ATPase activity, compara-

[20] P. P. Van Veldhoven and G. P. Mannaerts, *Anal. Biochem.* **161,** 45 (1987).

ble in rate to that of other transport ATPases, and that, in general, the catalytic properties of the two preparations are very similar. Thus, the kinetic parameters k_{cat}, K_m, and k_{cat}/K_m, the pH dependence of ATPase, cation dependence and sensitivity, and specificities for nucleotide are all similar.[17] Both preparations are covalently inactivated by *N*-ethylmaleimide (NEM) and 7-chloro-4-nitrobenzo-2-oxa-1,3-diazole (NBD-Cl), with MgATP affording protection against either reagent. NEM labels by reacting with a Cys residue in the P-loop of each of the two NBS, whereas NBD-Cl labels Cys or Lys residues. 8-Azido-ATP is a good substrate, which on UV illumination inactivates and photolabels. The maximal stoichiometry of photolabeling is 2 mol/mol Pgp, with each NBS being labeled. Pgp ATPase activity is not inhibited by NaN_3 ouabain, fluorescein isothiocyanate (FITC), dicyclohexylcarbodiimide (DCCD), or EGTA, but is potently inhibited by vanadate.[14,16,19]

The degree and spectrum of drug stimulation of the ATPase activity of purified reconstituted Pgp was found to depend on the lipids used for reconstitution.[21] Whereas plasma membrane Pgp ATPase is stimulated by colchicine, vinblastine, daunomycin, and verapamil, only the latter stimulated the ATPase of pure Pgp reconstituted in *E. coli* lipids. However, when bovine liver or sheep brain lipids were used, all of the drugs were seen to stimulate. The ATPase activity of pure Pgp reconstituted in *E. coli* lipids is stimulated by a wide range of other compounds, however, including nifedipine, trifluoperazine, rhodamine 123, Fura-2AM, Indo-1AM, progesterone, dexamethasone, hydrocortisone, leupeptin, pepstatin A, quinidine, and triphenylphosphonium ion.[19]

Vanadate Trapping of Nucleotide in Catalytic Sites of Pgp

We found that preincubation of plasma membrane or pure reconstituted Pgp with orthovanadate (Vi) and MgATP or MgADP led to stable inhibition of ATPase activity.[22] In a typical experiment with plasma membranes, 10 μg of protein is incubated for 20 min at 37° in a 100-μl volume containing 200 μM sodium orthovanadate, 1 mM NaATP or NaADP, 3 mM $MgSO_4$, 0.1 mM EGTA, and 40 mM Tris-Cl, pH 7.4, then passed through a centrifuge column consisting of 1 ml of Sephadex G-50 (fine) topped with a 10-mm layer of Dowex AG1-X8 (Bio-Rad, Richmond, CA) preequilibrated in 40 mM Tris-Cl, pH 7.4, and 0.1 mM EGTA. The centrifuge columns remove all unbound ligand. The eluted Pgp is found to be ~90% inhibited in terms of ATPase activity, and to contain 1 mol ADP/mol Pgp. Inhibition and nucleotide-trapping are dependent on the presence of Vi, divalent cation

[21] I. L. Urbatsch and A. E. Senior, *Arch. Biochem. Biophys.* **316,** 135 (1995).

[22] I. L. Urbatsch, B. Sankaran, J. Weber, and A. E. Senior, *J. Biol. Chem.* **270,** 19383 (1995).

(Mg, Mn, or Co), and nucleotide in the preincubation. The reactivation of ATPase occurs slowly ($t_{1/2}$ = 84 min at 37°, 311 min at 23°) and correlates with release of trapped nucleotide.

The vanadate-trapping procedure has proved valuable in a number of ways.[22,23] One example is shown in Fig. 1. In this experiment, 200 μM 8-azido [α-^{32}P]ATP was preincubated with CR1R12 plasma membranes, Vi, and $MgSO_4$ as described earlier, and eluted through centrifuge columns. The eluates were then UV-irradiated and subjected to SDS–PAGE. Autoradiograms are shown in lanes 1 and 2. In lane 1, Vi was omitted from the preincubation, and there was no covalent labeling of Pgp. In lane 2 the preincubation contained nucleotide, Vi, and $MgSO_4$, and specific covalent labeling of Pgp by 8-azido-ATP was achieved. Immunoblotting experiments demonstrated that the minor labeled bands in lane 2, of molecular size 65 and 100 kDa, both reacted with C219 anti-Pgp monoclonal antibody, and correspond to proteolytic fragments of Pgp.[24] Lanes 3 and 4 of Fig. 1 show Coomassie blue-stained SDS gels of the samples in lanes 1 and 2, respectively. It is evident from this example that vanadate trapping of nucleotide offers an excellent approach to achieving specific photolabeling of Pgp.

In further applications of the vanadate-trapping technique it was shown[22,23] that (1) both of the NBSs in Pgp are active in nucleotide hydrolysis; (2) a minimal scheme for Vi trapping of nucleotide, deduced from studies of the characteristics of Vi inhibition, predicts also a minimal scheme for the normal ATP hydrolysis catalytic pathway (shown in Ref. 18); and (3) Vi-induced trapping of nucleotide at either catalytic site completely inhibits steady-state catalysis. This last observation, taken together with the fact that the stably inhibited Pgp · MgADP · Vi species likely mimics the catalytic reaction transition state, has prompted us to postulate a catalytic cycle of Pgp ATP hydrolysis in which the two sites mandatorily alternate in catalysis.[18] In this cycle, drug transport is postulated to be coupled to relaxation of a high-energy catalytic site conformation that is generated by the hydrolysis step.

Summary

We have developed two defined experimental systems for biochemical investigation of P-glycoprotein, namely, plasma membranes highly enriched in Pgp, obtained from the CR1R12 Chinese hamster ovary cell line, and pure, reconstituted Pgp, obtained by solubilization of Pgp from CR1R12

[23] I. L. Urbatsch, B. Sankaran, S. Bhagat, and A. E. Senior, *J. Biol. Chem.* **270,** 26596 (1995).
[24] E. Georges, J.-T. Zhang, and V. Ling, *J. Cell Physiol.* **148,** 479 (1991).

plasma membranes, Reactive Red 120 chromatography, and reconstitution in liposomes.

Studies of the ATPase catalytic mechanism by kinetic methods and covalent inactivation have been greatly facilitated by the availability of these experimental systems. The technique of vanadate trapping of nucleotide has been particularly useful. As a result of these studies, we now have explicit, testable, proposals for (1) the normal catalytic pathway of ATP hydrolysis, (2) a postulated alternating catalytic site cycle, and (3) coupling of ATP hydrolysis to drug transport. The experimental methods described here should prove valuable for future studies of Pgp and of ABC transporters in general.

Acknowledgments

This work was supported by American Cancer Society grant BE2 (1990–1992) and by NIH grant GM50156 (1994–1996). I.L.U. gratefully acknowledges support from DFG grant Ur45/1-1. We thank Sumedha Bhagat for excellent technical assistance, and Drs. Joachim Weber, Banumathi Sankaran, and James Rebbeor for valuable comments. Figure 1 is reproduced with permission from Ref. 22.

[39] Construction of *MDR1* Vectors for Gene Therapy

By YOSHIKAZU SUGIMOTO, MICHAEL M. GOTTESMAN, IRA PASTAN, and TAKASHI TSURUO

Introduction

Gene therapy is a promising strategy for the treatment of cancer and many inherited diseases. Transduction of hematopoietic precursor cells *ex vivo* using amphotropic retrovirus vectors has been studied extensively to introduce foreign genes into blood cells of patients. Two important issues need to be considered for the success of this type of gene therapy: (1) efficient transduction of early hematopoietic precursors is essential, and (2) even though target cells are efficiently transduced with retrovirus vectors, the transduced cells will be mixed *in vivo* with a large number of uncorrected resident marrow cells (dilution effect). Unless the therapeutic gene itself confers a selective growth advantage, as may be the case in the gene therapy of adenosine deaminase deficiency, low transduction rates together with this *in vivo* dilution effect may limit the success of this form of gene therapy. The use of a dominant drug-selectable marker *in vitro*

0076-6879/98 $25.00

and *in vivo* should allow for the selection and enrichment of cells expressing the transduced gene and would solve these problems.

As a positive drug-selectable marker, we used a human multidrug resistance gene, *MDR1*.[1] The *MDR1* gene encodes the plasma membrane P-glycoprotein with a molecular mass of 170,000. Cells that overexpress the *MDR1* gene show resistance to multiple drugs such as *Vinca* alkaloids, anthracyclines, epipodophyllotoxins, and taxol. P-Glycoprotein acts as an ATP-dependent efflux pump for various structurally unrelated natural product antitumor agents. The *MDR1* gene was shown to confer multidrug resistance when the gene was introduced into drug-sensitive cells such as bone marrow cells.[2,3] Therefore, if the *MDR1* gene can be introduced into the hematopoietic progenitor cells of cancer patients, it may be possible to treat patients with higher doses of anticancer agents without fear of leukopenia and life-threatening infection developing. If the *MDR1* gene can be introduced with a therapeutic gene to bone marrow cells of patients with genetic diseases, it may be possible to use the *MDR1* gene as a dominant selectable marker *in vivo* to coexpress otherwise nonselectable genes.[4,5]

Another problem is that it is likely with current vector systems that cells other than the target cells will be transduced. For example, we have proposed the use of *MDR1* retroviral vectors to protect normal bone marrow cells of cancer patients from intensive chemotherapy. However, if the bone marrow of a patient contains contaminating cancer cells and they receive the *MDR1* retrovirus, multidrug-resistant cancer cells may result. To eliminate such unintentionally transduced cells, coexpression of a negative drug-selectable marker that confers hypersensitivity to a specific drug would be useful.

As a negative drug-selectable marker, we used the herpes simplex virus thymidine kinase (HSV-TK) gene that acts both as a selectable marker in thymidine kinase (TK)-deficient cells and as a suicide gene both *in vitro* and *in vivo*.[6] Cells that express HSV-TK show hypersensitivity to nucleoside

[1] M. M. Gottesman, C. Hrycyna, P. V. Schoenlein, U. A. Germann, and I. Pastan, *Annu. Rev. Genet.* **29,** 607 (1995).

[2] I. Pastan, M. M. Gottesman, K. Ueda, E. Lovelace, A. V. Rotherford, and M. C. Willingham, *Proc. Natl. Acad. Sci. U.S.A.* **85,** 4486 (1988).

[3] J. R. McLachlin, M. A. Eglitis, K. Ueda, P. W. Kantoff, I. H. Pastan, W. F. Anderson, and M. M. Gottesman, *J. Natl. Cancer Inst.* **82,** 1260 (1990).

[4] M. M. Gottesman, S. V. Ambudkar, B. Ni, J. M. Aran, Y. Sugimoto, C. O. Cardarelli, and I. Pastan, *Cold Spring Harbor Symp. Quant. Biol.* **59,** 677 (1994).

[5] M. M. Gottesman, U. A. Germann, I. Aksentijevich, Y. Sugimoto, C. O. Cardarelli, and I. Pastan, *Ann. N.Y. Acad. Sci.* **716,** 126 (1994).

[6] K. W. Culver, Z. Ram, S. Wallbridge, H. Ishii, E. H. Oldfield, and R. M. Blaese, *Science* **256,** 1550 (1992).

analogs such as ganciclovir, because HSV-TK can phosphorylate ganciclovir more efficiently than the endogenous TK of mammalian cells. Therefore, if the HSV-TK gene can be introduced with a therapeutic gene, it may be possible to eliminate unintentionally tranduced cells using ganciclovir.

To coexpress drug-selectable markers with therapeutic genes, we have developed a new retroviral vector system, pSXLC/pHa, that utilizes a putative internal ribosome entry site (IRES) isolated from encephalomyocarditis (EMC) virus.[7] In this construct, a single mRNA is transcribed under control of an upstream promoter, and two gene products are translated independently from a bicistronic mRNA. One of these gene products is translated in a cap-dependent manner and the other is translated under control of the IRES. Here we describe the construction of *MDR1* bicistronic vectors with two different purposes.

The first set of vectors that we made was comprised of Ha-αGal-IRES-MDR and Ha-MDR-IRES-αGal for the coexpression of the *MDR1* gene and human α-galactosidase A gene.[8] One vector, Ha-αGal-IRES-MDR, has an α-galactosidase A cDNA upstream from the IRES and the *MDR1* cDNA under control of the IRES, and the other vector, Ha-MDR-IRES-αGal, has the *MDR1* cDNA upstream from the IRES and the α-galactosidase A cDNA under control of the IRES. These vectors could be used for the gene therapy of Fabry disease. Fabry disease results from mutations in the α-galactosidase A gene that cause progressive glycosphingolipid deposition leading to early demise from renal, cardiac, or cerebrovascular disease.[9] Enzyme replacement therapy may be relatively ineffective for the treatment of Fabry disease because the half-life of injected α-galactosidase A, at least in the circulation, is very short. No effective treatment protocol has been established for this disease so far. Coexpression of the *MDR1* gene with α-galactosidase A may be effective for treatment of Fabry disease for the following reasons: (1) Expression of a normal cDNA for the single-chain enzyme α-galactosidase A should cure the disease. (2) The disease is a result of the ineffectiveness of macrophages to digest glycosphingolipids. Therefore, correction of the enzyme deficiency in macrophages and other cells by the introduction of the α-galactosidase A cDNA into bone marrow cells may result in the normal catabolism of ceramide trihexoside. (3) This disease leads to premature death in affected males and there are no other effective treatments for the disease. The cells in

[7] Y. Sugimoto, I. Aksentijevich, I. Pastan, and M. M. Gottesman, *Bio/Technol.* **12,** 694 (1994).

[8] Y. Sugimoto, I. Aksentijevich, G. J. Murray, R. O. Brady, I. Pastan, and M. M. Gottesman, *Human Gene Ther.* **6,** 905 (1996).

[9] R. J. Desnick and D. F. Bishop, *in* "The Metabolic Basis of Inherited Disease" (C. R. Scriber, A. L. Beaded, W. S. Sly, and D. Vale, eds.), 6th Ed., p. 1751. McGraw-Hill, New York, 1989.

which these vectors have been introduced are multidrug resistant with high expression of human P-glycoprotein. Most of the drug-resistant cells showed higher activity of α-galactosidase A than the parental cells, with levels dependent on whether the *MDR1* cDNA was upstream or downstream from the IRES. High titers of amphotropic retrovirus capable of transducing cultured cells were obtained with these vectors.[8] These results suggest the usefulness of the *MDR1* gene as a dominant selectable marker in gene therapy of Fabry disease.

Another *MDR1* bicistronic vector is Ha-MDR-IRES-TK, which allows for coexpression of the cap-dependent *MDR1* gene and the IRES-dependent HSV-TK gene.[10] As described earlier, when *MDR1* retroviral vectors are used to introduce the *MDR1* gene into normal bone marrow cells of cancer patients for intensive chemotherapy, if the bone marrow of a patient contains contaminating cancer cells and they receive the *MDR1* retrovirus, multidrug-resistant cancer cells may result. The vector Ha-MDR-IRES-TK should be better than vectors with the *MDR1* gene alone because it is possible to eliminate cancer cells that have been unintentionally transduced with the *MDR1* retrovirus. The cells in which these vectors have been introduced are multidrug resistant with high expression of human P-glycoprotein. Most of the drug-resistant cells transduced with Ha-MDR-IRES-TK showed hypersensitivity to ganciclovir.[10]

Construction of *MDR1* Bicistronic Retrovirus Vectors

Construction of IRES Subcloning Vectors

Our laboratory uses a pHa retroviral vector in which the promoter of Harvey murine sarcoma virus LTR drives the expression of the *MDR1* gene in mammalian cells.[11] The HaMDR vector, which carries the wild-type *MDR1* cDNA in the pHa vector, is used as a control *MDR1* vector throughout the study. The pHa vector has only two unique restriction sites, *Sac*II and *Xho*I, for cloning of cDNAs that are to be expressed. To make the cloning of two cDNAs into the pHa vector easier, we designed a series of subcloning vectors. The basic subcloning vector is pSX, which is derived from pGEM2 (Promega, Madison, WI). The pSX vector has only two restriction sites, *Sac*II and *Xho*I, between the Sp6 and T7 promoters of pGEM2, and all the other multicloning sites of pGEM2 are deleted.

[10] Y. Sugimoto, C. A. Hrycyna, I. Aksentijevich, I. Pastan, and M. M. Gottesman, *Clin. Cancer Res.* **1,** 447 (1995).

[11] R. A. Sokolic, S. Sekhsaria, Y. Sugimoto, N. Whiting-Theobald, G. F. Linton, F. Li, M. M. Gottesman, and H. L. Malech, *Blood* **87,** 42 (1996).

First we insert a new multicloning site *Sac*II–*Bam*HI–*Bgl*II–*Sac*I–*Xba*I–*Sal*I–*Ase*I–*Nco*I–*Xho*I between the *Sac*II and *Xho*I sites of pSX, and the resulting plasmid is termed pSXL.[7] All sites except the *Ase*I site are unique in pSXL. Next we clone an IRES fragment (0.5 kb) of EMC virus between the *Ase*I and *Nco*I sites of pSXL. This IRES-containing plasmid pSXLC has a single *Nco*I restriction site (CCATGG) at the 3′-end of the IRES sequence for the cloning of the downstream cDNA that is translated under control of the IRES. The ATG sequence within the *Nco*I site must be used as the first ATG codon of the downstream cDNA because the IRES-dependent translation starts from the ATG codon of the *Nco*I site. The multicloning site *Sac*II–*Bam*HI–*Bgl*II–*Sac*I–*Xba*I–*Sal*I upstream from the IRES is unique in pSXLC and can be used to insert the upstream cDNA that is translated in a cap-dependent manner. Since this pSXLC plasmid does not have a promoter for transcription in mammalian cells, the whole insert should be transferred to the pHa retroviral expression vector for expression. After cloning of two cDNAs, two restriction sites, *Sac*II and *Xho*I, at both ends are used to transfer the whole insert to the pHa vector.[7]

To insert the *MDR1* cDNA downstream from the IRES of pSXLC, we create a new *Nco*I end at the first ATG codon of the *MDR1* cDNA and ligate this cDNA with *Nco*I, *Xho*I, *Xho*I-digested pSXLC. For this purpose, a synthetic adapter (52-mer/48-mer) from −1 (newly created *Nco*I end) to +51 (unique *Dra*I site) of the *MDR1* cDNA is ligated with the *Dra*I–*Xho*I fragment of the wild-type *MDR1* cDNA without polyadenylation signals, and the resulting full-length *MDR1* cDNA is subcloned between the *Nco*I and *Xho*I sites of pSXLC (pSXLC-MDR). To express the IRES-dependent *MDR1* cDNA in mammalian cells, the *Sac*II–*Xho*I fragment of pSXLC-MDR is transferred into the pHa vector (pHa-MCS-IRES-MDR).[7]

To insert HSV-TK downstream from the IRES of pSXLC, we create a new *Nco*I site at the first ATG codon of the HSV-TK open reading frame and ligate the modified HSV-TK DNA with *Nco*I, *Xho*I-digested pSXLC. For this purpose, first we delete a polyadenylation site in the 3′-noncoding region of the HSV-TK gene and introduce a *Xho*I site. Next a synthetic adapter (27-mer/27-mer) from −1 (newly created *Nco*I end) to +26 (unique *Mlu*I site) of HSV-TK DNA is ligated with the *Mlu*I–*Xho*I fragment of HSV-TK DNA. The resulting *Nco*I–*Xho*I fragment of full-length HSV-TK DNA is subcloned into pSXLC. To express the IRES-dependent HSV-TK in mammalian cells, the *Sac*II–*Xho*I fragment of pSXLC-TK is transferred into the pHa vector (pHa-MCS-IRES-TK).[7]

For the cloning of cDNAs that have internal *Xho*I sites, we construct another subcloning vector pSSLC. pSSLC is a derivative of pSXLC. In pSSLC, the *Sal*I site of pSXLC is deleted and the *Xho*I site of pSXLC is changed to a *Sal*I site. The pSSLC vector has a multicloning site of *Sac*II–*Bam*HI–*Bgl*II–*Sac*I–*Xba*I upstream from the IRES for the subcloning on cap-dependent cDNA. The IRES-dependent cDNA is inserted between the *Nco*I and *Sal*I sites downstream from the IRES. After cloning of the two cDNAs, two restriction sites, *Sac*II and *Sal*I, at both ends can be used to transfer the whole insert between the *Sac*II and *Xho*I sites of the pHa vector.

Construction of Bicistronic Vectors

To illustrate the construction of *MDR1*-bicistronic vectors, we show our work on the coexpression of P-glycoprotein and α-galactosidase A.[8] We construct two types of retroviral vectors for this purpose. The cloning strategy is summarized in Fig. 1.

One construct is pHa-αGal-IRES-MDR, which has the α-galactosidase A cDNA upstream from the IRES. The plasmid pSXLC-MDR has the entire open reading frame of the *MDR1* cDNA downstream from the IRES sequence and 5 unique sites (*Sac*II, *Bam*HI, *Sac*I, *Xba*I, and *Sal*I) for the cloning of another gene upstream from the IRES.

The human α-galactosidase A cDNA has two potential polyadenylation signals (^{1255}AATACA and ^{1272}ATTAAA) within the open reading frame. To inactivate these two polyadenylation sites, we make a synthetic adapter (60-mer/67-mer), which has the sequence of α-galactosidase A cDNA from 1231 (*Alw*NI site) to 1288 (three bases after the termination codon) and introduces nonsense mutations in each polyadenylation signal (^{1255}AACACA and ^{1272}TTTAAA) as well as a 3′-*Xho*I site at the 3′-end of the cDNA. The adapter is ligated to the *Alw*NI-digested 3′-fragment (804–1231) of the α-galactosidase A cDNA. The cDNA fragment is digested with *Bgl*II (nucleotide 825) and *Xho*I, and ligated with the rest of α-galactosidase A cDNA in the *Sma*I site of pGEM7 (Promega). The resulting plasmid with the full-length α-galactosidase A cDNA is termed pGEM7/αGal. The nucleotide sequence of the α-galactosidase A cDNA is confirmed using an automatic sequencer (373A, Applied Biosystems, Foster City, CA). We then clone the *Bam*HI, *Xba*I fragment of the α-galactosidase A cDNA into the *Bam*HI, *Xba*I-digested pSXLC-MDR (pSXLC-αGal-MDR). The pSXLC-αGal-MDR insert is isolated after *Sac*II, *Xho*I digestion and transferred into the pHa retroviral vector. The resulting retrovirus construct is termed pHa-αGal-IRES-MDR.

Another construct we make is pHa-MDR-IRES-αGal, which has the α-galactosidase A cDNA downstream from the IRES. To insert the α-galactosidase A cDNA between the *Nco*I and *Xho*I sites of pSXLC, we modified the 5′-end of the cDNA. The IRES uses the ATG sequence within

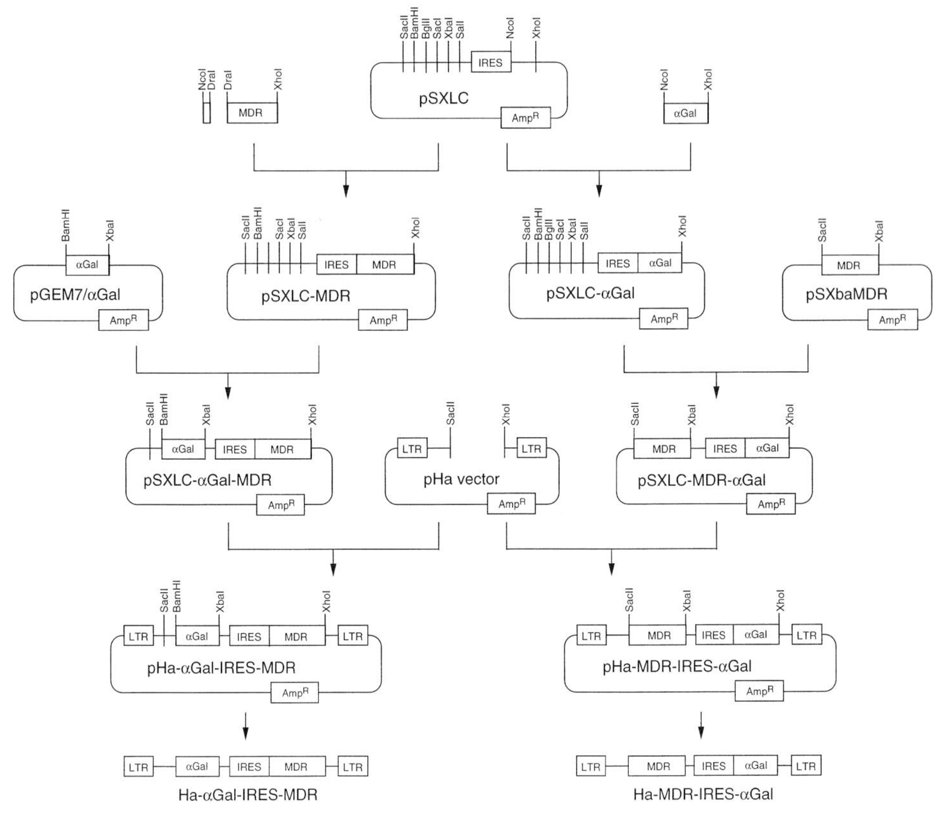

FIG. 1. Strategies for the construction of *MDR1*–α-galactosidase A bicistronic retrovirus vectors. All the restriction sites indicated in this figure are unique sites in each plasmid. The drawing is not to scale. MDR, human multidrug resistance gene *MDR1*; αGal, human α-galactosidase A cDNA; IRES, internal ribosome entry site; Amp^R, ampicillin resistance gene; LTR, long terminal repeat of Harvey murine sarcoma virus.

the single *Nco*I site (CCATGG) at the 3′-end as the first codon. Since the second G nucleotide may be important for efficient translation, we changed the +4 nucleotide C of α-galactosidase A cDNA to G. This C $\rightarrow$ G change at the +4 nucleotide causes mutation of the second amino acid sequence of human α-galactosidase A. For this purpose we use two synthetic adapters from nucleotide −1 (newly created *Nco*I end) to +35 (unique *Bss*HII site) of α-galactosidase A cDNA. The original amino acid sequence of α-galactosidase A at its N terminus is Met–Gln–Leu–Arg. The 5′-nucleotide sequence of the first adapter (36-mer) is CATGGCCCTGAGG, which changes the second amino acid from Gln to Ala (Met–Ala–Leu–Arg). The

5′-sequence of the second adapter (42-mer) is CATGGCCATGCAGC-TGAGG, which adds Met–Ala at the N terminus of the signal peptide (Met–Ala–Met–Gln–Leu–Arg). These adapters are ligated with the *Bss*HII–*Xho*I fragment of α-galactosidase A cDNA, and the full-length α-galactosidase A cDNAs are subcloned into pSXLC. These plasmids, pSXLC-αGal(1) with the 35-mer adapter and pSXLC-αGal(2) with the 41-mer adapter, have four unique sites (*Sac*II, *Bam*HI, *Xba*I, and *Sal*I) for cloning of the upstream gene. To insert the *MDR1* cDNA upstream from the IRES of pSXLC, we first make pSXbaMDR, which contains *MDR1* cDNA with 5′-*Sac*II and 3′-*Xba*I sites in the pSX backbone. Next we subclone the *Sac*II, *Xba*I-digested *MDR1* cDNA from pSXbaMDR between the *Sac*II and *Xba*I sites of pSXLC-αGal(1) and pSXLC-αGal(2). The inserts of these plasmids are isolated after *Sac*II, *Xho*I digestion and transferred into the pHa retroviral vector. These retrovirus expression constructs are termed pHa-MDR-IRES-αGal(1) and pHa-MDR-IRES-αGal(2), respectively. pHa-MDR-IRES-αGal is the representative name of these two vectors.

The strategy for the construction of pHa-MDR-IRES-TK, which allows for coexpression of the cap-dependent *MDR1* gene and the IRES-dependent HSV-TK gene, is as follows:[7,10] The plasmid pSXLC-TK has the entire open reading frame of the HSV-TK gene downstream from the IRES sequence and six unique sites (*Sac*II, *Bam*HI, *Bgl*II, *Sac*I, *Xga*I, and *Sal*I) for the cloning of another gene upstream from the IRES. We insert a *Sac*II, *Xba*I-digested *MDR1* cDNA fragment from pSXbaMDR between the *Sac*II and *Xba*I sites of pSXLC-TK (pSXLC-MDR-TK). The pSXLC-MDR-TK insert is isolated after *Sac*II, *Xho*I digestion and transferred into the pHa retroviral vector (pHa-MDR-IRES-TK). As a control TK vector, pHa-MCS-IRES-TK, which carries the insert of pSXLC-TK, is also constructed.

Cell Culture and Assay of Drug Sensitivity

The ecotropic retrovirus packaging cell line ψ-CRE and the mouse fibroblast cell line NIH 3T3 are cultured in Dulbecco's modified Eagle's medium (DMEM) supplemented with 10% calf serum. The amphotropic retrovirus packaging cell line PA317 and mouse thymidine kinase-deficient cell line Ltk^- are grown in DMEM supplemented with 10% fetal bovine serum. The sensitivity of cultured cell lines to vincristine is evaluated by inhibition of cell growth after incubation at 37° for 6 days in the presence of various concentrations of the drug. Cell numbers are determined in a Coulter counter, and the concentration of drug that inhibits cell growth by 50% (IC_{50}) is determined.

DNA Transfection

Transfection is carried out using the high-efficiency calcium phosphate coprecipitation method. Recipient cells are plated at 5×10^5 cells per 100-mm dish on day 1 and transfected with 20 μg of the expression plasmid DNA on day 2. Cells are exposed to the DNA precipitate until day 3 when the medium is aspirated and fresh medium added. On day 4, the cells are split at 1:10 or 1:100. The cells are selected with 25–35 ng/ml vincristine when *MDR1* expression plasmids are used for transfection.

Retrovirus Transduction

For the vincristine-resistant colony formation assay of *MDR1* retroviruses, the retrovirus-producing cells (PA317 origin) and the recipient cells (NIH 3T3) are plated at 5×10^6 cells and 3×10^4 cells per 100-mm dish, respectively, on day 1. On day 2, the retrovirus-containing supernatant is collected, passed through a 0.45-μm pore filter to remove cells and debris, and added to each dish of recipient cells in the presence of 4–6 μg/ml Polybrene (Aldrich, Milwaukee, WI). On day 4, medium is removed, and fresh medium containing 25 ng/ml vincristine is added. On days 10–12, the medium is removed and the colonies are stained with 0.5% methylene blue dissolved in 50% methanol.

FACS Analysis

To examine the expression of human P-glycoprotein on the surface of cultured cell lines transduced with *MDR1* retroviruses, fluorescent-activated cell sorting (FACS) analysis is used in which cells are reacted with human P-glycoprotein-specific monoclonal antibody MARK16.[12,13] For the detection of P-glycoprotein expressed in cultured cells of nonhematopoietic origin, cells (10^5–10^6 in 50 μl) harvested after trypsinization are incubated with the $F(ab')_2$ fragment of MRK16 (100 μg/ml, 5 μg antibody per reaction), washed, and incubated with fluorescein-conjugated $F(ab')_2$ fragment of goat anti-mouse IgC $F(ab')_2$ (1:10 diluted, Cappel, Durham, NC). For the detection of P-glycoprotein expressed in murine or human white blood cells, cells (10^5–10^6 in 50 μl) are incubated with the biotinylated $F(ab')_2$ fragment of MRK16 (100 μg/ml, 5 μg antibody per reaction), washed, and incubated with *R*-phycoerythrin-conjugated streptavidin (25 μg/ml, GIBCO, Grand Island, NY). The fluorescence straining level was analyzed using FACSort (Becton-Dickinson, San Jose, CA).

[12] H. Hamada and T. Tsuruo, *Proc. Natl. Acad. Sci. U.S.A.* **83,** 7785 (1986).

[13] E. Okochi, T. Iwahashi, K. Ariyoshi, H. Watabe, T. Tsuruo, and K. Ono, *J. Immunol. Methods* **187,** 127 (1995).

α-Galactosidase Enzyme Activity Assay

Cell pellets in microtubes are resuspended in 10 volumes of extraction buffer containing 50 m*M* phosphate buffer, pH 6.0, 0.2% Triton X-100, and 20 μg/ml aprotinin (Sigma, St. Louis, MO) and sonicated on ice for 90 sec. Lysates are centrifuged at 14,000 rpm for 30 sec at 4°. The resulting supernatants are used as cell extracts. Protein concentrations of the cell extracts are determined by the Protein Assay Kit (Bio-Rad, Hercules, CA). The α-galactosidase A activity in the cell lysate is determined using 3.5 m*M* 4-methylumbelliferyl-α-D-galactopyranoside as a substrate in the presence of 100 m*M* *N*-acetylgalactosamine. Data (percent of activity in parental cells) are expressed as means of four determinations.

Transduction and Transplantation of Murine and Human Bone Marrow Cells

Seven-week old male C57BL/6 mice are treated intravenously with 150 mg/kg 5-fluorouracil. After 6 days, the bone marrow is harvested from both femurs, and mononuclear cells are isolated. The resulting cells are cultured in Iscove's modified Dulbecco's medium (IMDM) supplemented with 20% fetal bovine serum (FBS), 50 μ*M* 2-mercaptoethanol, 100 ng/ml rat stem cell factor (SCF; Amgen, Thousand Oaks, CA), 20 ng/ml mouse interleukin 3 (IL-3; KIRIN Brewery Co. Ltd., Tokyo), and 100 ng/ml human interleukin 6 (IL-6; Serono Japan Ltd., Tokyo). On the next day, retrovirus-containing supernatant is added to the bone marrow cells with 8 μg/ml Polybrene. Retrovirus-containing supernatant is changed once or twice a day for 3–4 days. After the transduction, the cells are washed and implanted into lethally irradiated female C57BL/6 mice. The mice are then treated once or twice a month with vinblastine (6 or 12 mg/kg, intravenously). The expression of human P-glycoprotein in peripheral blood cells of the mice is examined by FACS with MRK16. A portion of the transduced bone marrow cells is further cultured *in vitro* and treated with vincristine.

Human CD34-positive cells are purified from frozen bone marrow cells of cancer patients using the MACS system (Miltenyi Biotec GmbH) and cultured in IMDM supplemented with 20% (v/v) fetal bovine serum, 50 μ*M* 2-mercaptoethanol, 100 ng/ml human stem cell factor (SCF; Amgen), 20 ng/ml human interleukin 3 (IL-3; KIRIN Brewery Co. Ltd.), and 100 ng/ml IL-6. On the next day, retrovirus-containing supernatant is added to the bone marrow cells with 6 μg/ml Polybrene. Retrovirus-containing supernatant is changed once or twice a day for 3 days. The expression of human P-glycoprotein in human CD34 cells is examined by FACS with MRK16.

Coexpression of Two Genes from Bicistronic Retroviruses

DNA Transfection

The retroviral expression constructs, pHaMDR, pHa-MCS-IRES-MDR, pHa-αGal-IRES-MDR, pHa-MDR-IRES-αGal(1), and pHa-MDR-IRES-αGal(2) are transfected into ψ-CRE, a murine retrovirus-packaging cell line. After 10 days, approximately 30 vincristine-resistant colonies per dish are observed in pHa-αGal-IRES-MDR-transfected cells when transfected cells are split 1:100 on day 4 (transfection efficiency is 1.1×10^{-3}). This transfection efficiency is three-fold less than that of pHa-MCS-IRES-MDR, a control *MDR1* vector whose *MDR1* gene is also controlled by the IRES, and seven-fold less than that of pHaMDR. In contrast, two pHa-MDR-IRES-αGal constructs conferred multidrug resistance similar to pHaMDR (transfection efficiency was 6×10^{-3}).

Retrovirus Production

The retrovirus titers of PA317 cells (mixed populations) producing Ha-αGal-IRES-MDR or Ha-MDR-IRES-αGal retroviruses are examined using NIH 3T3 cells as recipient cells. Retrovirus titers of pHa-αGal-IRES-MDR are 4- to 10-fold less than those of HaMDR. Retrovirus titers of Ha-MDR-IRES-αGal are slightly less than those of HaMDR. To obtain retrovirus-producing cells with higher titers, we isolate single PA317 clones transduced with Ha-αGal-IRES-MDR or Ha-MDR-IRES-αGal retrovirus. The highest titer of Ha-αGal-IRES-MDR retrovirus we obtained from 13 clones of PA317 was 3.0×10^4/ml, and the highest titers of Ha-MDR-IRES-αGal(1) and Ha-MDR-IRES-αGal(2) retrovirus from 8 clones each of PA317 were 8.4×10^4/ml and 1.2×10^5/ml, respectively.

Coexpression of P-Glycoprotein and α-Galactosidase A After Retrovirus Transduction

To determine whether the *MDR1* and α-galactosidase A genes are coexpressed in clonal cell lines, we randomly isolate NIH 3T3 clones transduced with the *MDR1*-α-galactosidase A retroviruses and determine the α-galactosidase A activity, P-glycoprotein expression, and vincristine sensitivity. The α-galactosidase A activity (x-fold increase in activity as compared to the activity of NIH 3T3 cells), the mean channel of the FACS fluorogram for each population, and the IC_{50} values for vincristine are shown in Fig. 2. Cells transduced with Ha-αGal-IRES-MDR and Ha-MDR-IRES-αGal show high α-galactosidase A activity and P-glycoprotein expression when

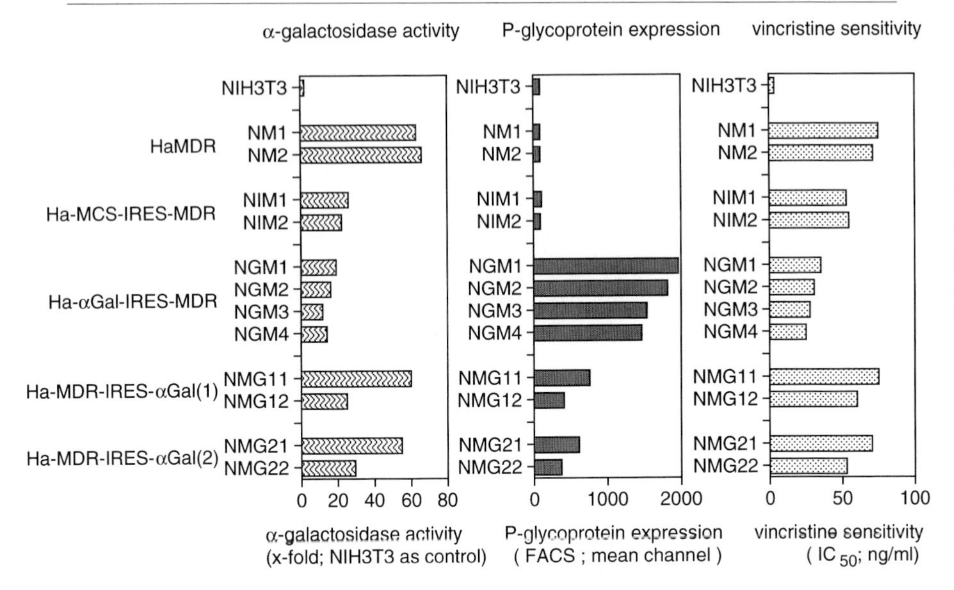

FIG. 2. α-Galactosidase activity, P-glycoprotein expression, and sensitivity to vincristine of NIH 3T3 clones transduced with the vectors. The α-galactosidase A activity in the cell lysate was determined using 3.5 m*M* 4-methylumbelliferyl-α-D-galactopyranoside as a substrate in the presence of 100 m*M* *N*-acetylgalactosamine. Data (relative value when the activity of NIH 3T3 was assigned as 1) are expressed as means for four determinations. To examine the expression of human P-glycoprotein on the cell surface, cells (10^5–10^6 in 50 μl) harvested after trypsinization were incubated with MRK16 antibody (100 μg/ml, 5 μg antibody per reaction), washed, and incubated with fluorescein-conjugated $F(ab')_2$ fragment of goat anti-mouse IgG $F(ab')_2$ (1:10 diluted, Cappel, Durham, NC). The fluorescence staining level was analyzed using FACSort (Becton-Dickinson, San Jose, CA). The sensitivity of cultured cell lines to vincristine was evaluated by inhibition of cell growth after incubation at 37° for 6 days in the presence of various concentrations of the drug. Cell numbers were determined in a Coulter counter, and the concentration of drug that inhibits cell growth by 50% (IC_{50}) was determined.

compared to the parental NIH 3T3 cells. As shown in Fig. 2, all four clones (NGM1–NGM4) transduced with Ha-αGal-IRES-MDR show 15- to 20-fold higher α-galactosidase A activity than the parental NIH 3T3 cells. These transductants also show 6- to 9-fold higher resistance to vincristine than the parental cells. The expression of human P-glycoprotein in the transduced clones is confirmed by FACS analysis. This result clearly demonstrates that the transferred retrovirus results in coexpression of two genes in the transduced cells. All four clones (NMG11, NMG12, NMG21, and NMG22) transduced with Ha-MDR-IRES-αGal show higher α-galactosidase A activity than the parental NIH 3T3 cells. Cells transduced with

Ha-MDR-IRES-αGal showed higher levels of drug resistance (13- to 19-fold) and lower levels of α-galactosidase A activity (4- to 8-fold) than those with Ha-αGal-IRES-MDR. In summary, we observe higher levels of expression of the genes inserted upstream from the IRES than downstream from the IRES. Therefore, it seems reasonable to conclude that translation of the genes under the control of the IRES is less efficient than cap-dependent translation. This vector system provides the flexibility to express different ratios of drug resistance to nonselectable gene expression depending on which gene is upstream from the IRES and which is downstream.

Among Ha-MDR-IRES-αGal-transduced clones, NMG11 and NMG21 show higher levels of α-galactosidase A activity, P-glycoprotein expression, and vincristine resistance than NMG12 and NMG22. All the vincristine-selected Ha-αGal-IRES-MDR-transduced clones showed similar levels of α-galactosidase A activity, P-glycoprotein expression, and vincristine resistance. Therefore, by using similar polycistronic constructs, it should be possible to control the expression level of the foreign gene in the transduced cells so that all the cells express an appropriate amount of the gene. This is a desirable feature for a vector system if the expression of the transduced gene at an undesired level is toxic to cells. The parallel expression of the gene of interest with cell surface P-glycoprotein enables the removal of cells expressing undesirable amounts of the gene of interest using cell sorting or panning methods *ex vivo* with monoclonal antibodies to P-glycoprotein. This approach could also be used to select *ex vivo* for cells expressing high levels of P-glycoprotein and a second gene of interest, possibly obviating or eliminating the need for selection with toxic drugs *in vivo*. Polycistronic vector systems that employ *MDR1* as a selectable marker appear to be flexible and efficient vehicles for introduction of nonselectable genes into cells.

Coexpression of P-Glycoprotein and HSV-TK

We examined the probability of the coexpression of two genes using Ha-MDR-IRES-TK. In this construct, both the *MDR1* gene and the HSV-TK gene can be used as drug-selectable markers when a TK-deficient cell line Ltk$^-$ is used as the recipient. The pHa-MDR-IRES-TK-transfected Ltk$^-$ cells are selected with either vincristine (30 ng/ml) or HAT medium, and 20 vincristine-resistant clones and 20 HAT-resistant clones are isolated. Growth of each clone is examined in the presence of vincristine (15–30 ng/ml) or HAT medium. All the HAT-resistant clones (20/20) show more resistance to vincristine than the parental Ltk$^-$, even though some clones show only marginal levels of resistance. On the other hand, 19/20 vincristine-resistant clones show HAT resistance. Only one clone shows HAT sensitivity. The pHa-MDR-IRES-TK clones selected with vincristine appear to have higher levels of vincristine resistance than the transfectants selected with HAT medium. In separate experiments, the overall frequencies of ganciclovir-resistant cells in vincristine-resistant, Ha-MDR-IRES-TK-transduced cells were 1% or less when clonal retrovirus-producer cells were used as the source of retrovirus. Therefore, it is possible to conclude that almost 100% coexpression of two genes could be achieved with bicistronic retrovirus vectors.

Transduction of Bone Marrow Cells with Bicistronic Retrovirus

Mouse bone marrow cells are harvested from male C57BL/6 mice pretreated with 150 mg/kg 5-fluorouracil. The cells are cultured in IMDM supplemented with 20% FBS, 50 μM 2-mercaptoethanol, 100 ng/ml rat SCF, 100 ng/ml human IL-6, and 20 ng/ml mouse IL-3. The cells are transduced with Ha-MDR-IRES-TK retrovirus for 3 days (days 1, 2, and 3). The expression of cell surface P glycoprotein was analyzed on day 4 by FACS using the biotinylated $F(ab')_2$ fragment of MRK16 antibody (Fig. 3). Approximately 26% of cells are successfully transduced with Ha-MDR-IRES-TK (Fig. 3B). When the Ha-MDR-IRES-TK-transduced cells are treated with vincristine and ganciclovir, all the cells die (data not shown). The purpose of this study is to treat tumor cells unintentionally transduced with *MDR1* vectors. We do not intend to eliminate *MDR1*-transduced bone

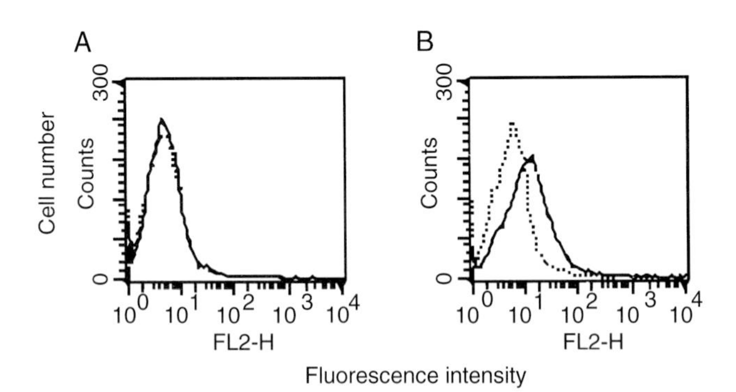

FIG. 3. Expression of cell surface P-glycoprotein in murine bone marrow cells transduced with Ha-MDR-IRES-TK. Cells (10^6) were incubated with the biotinylated $F(ab')_2$ fragment of MRK16 (100 μg/ml), washed, and incubated with *R*-phycoerythrin-conjugated streptavidin (25 μg/ml). The fluorescence staining level was analyzed using FACSort (solid lines). As controls, the cells were strained with *R*-phycoerythrin-conjugated streptavidin (25 μg/ml) without MRK16 staining and analyzed (dotted lines). (A) Nontransduced cells; (B) Ha-MDR-IRES-TK-transduced cells.

marrow cells with ganciclovir. The present experiment, however, proves that it is possible to transduce bone marrow cells with our bicistronic vectors.

Other Genes Successfully Expressed in Bicistronic Retrovirus Vectors

We previously reported the construction of a bicistronic retroviral vector in a somewhat less flexible system for the coexpression of cap-dependent *MDR1* gene and IRES-dependent human glucocerebrosidase gene and demonstrated the colchicine-selected coexpression of glucocerebrosidase in NIH 3T3 cells transfected with the vector.[14]

In another study, we create a bicistronic vector for the gene therapy of the inherited hematopoietic disorder, X-linked chronic granulomatous disease (X-CGD). The chronic granulomatous diseases are a group of four inherited disorders with a common phenotype characterized by a failure or blood phagocytic cells to generate superoxide anion, leading to recurrent infections and granuloma formation. The X-CGD is caused by the mutation of gp91-phox, a transmembrane large subunit of NADPH oxidase responsible for superoxide generation. For the gene therapy of X-CGD, we construct a bicistronic retrovirus vector MFG-gp91-IRES-MDR, which allows for coexpression of cap-dependent gp91 gene and IRES-dependent *MDR1* gene in MFG retrovirus vector backbone.[11] Epstein–Barr (EB) virus-transformed B cells established from X-CGD patients are successfully transduced with MFG-gp91-IRES-MDR and the vincristine-selected cells show high superoxide-producing activity.[11] *MDR1*-bicistronic retrovirus vectors for the correction of p47-deficient CGD and p67-deficient CGD are constructed (unpublished results of authors, 1997).

These bicistronic vectors would enable the selection of genetically modified cells both *in vitro* and *in vivo,* and would increase the effectiveness of somatic gene therapy.

[14] J. M. Aran, M. M. Gottesman, and I. Pastan, *Proc. Natl. Acad. Sci. U.S.A.* **91,** 3176 (1994).

[40] Construction of *MDR1* Adeno-Associated Virus Vectors for Gene Therapy

By MARION BAUDARD

Introduction

The multidrug resistance gene (*MDR1*) encodes a membrane protein, P-glycoprotein (Pgp), able to extrude many endogenous substrates or xenobiotic compounds out of cells. Transfer studies indicate that *MDR1* transgene expression protects cells transfected or transduced with *MDR1* containing vectors from cytotoxic Pgp substrates.[1] Ongoing clinical protocols are already evaluating whether transfer of the *MDR1* gene into bone marrow cells from patients administered chemotherapy for solid tumors will limit the myelosuppressive side effect of cytotoxic agents and allow treatment dose intensification.[2,3] Another potential of *MDR1* transfer for gene therapy in humans is that *MDR1* expression allows the selection of cells coexpressing an otherwise nonselectable gene.[4,5] Efficient gene therapy of genetic diseases is often limited by the low proportion of successfully transduced or transfected cells and the usual impossibility of selecting the targeted cells based on transgene expression.

Retrovirus-mediated gene therapy has been favored by a number of groups. Retroviral vectors efficiently integrate into the host cell genome with resulting long-term expression. However, retroviruses contain transcriptional regulation elements and the random integration of retroviral sequences may lead to activation of a deleterious gene (e.g., protooncogene). In addition, transduction of quiescent cells, such as hematopoietic stem cells, is difficult since at least one cycle of cell division is necessary for provirus integration to occur.[6]

[1] M. M. Gottesman and I. Pastan, *Annu. Rev. Biochem.* **62,** 385 (1993).

[2] J. A. O'Shaughnessy, K. H. Cowan, A. W. Nienhuis, K. T. McDonagh, B. P. Sorrentino, C. E. Dunbar, Y. Chiang, W. Wilson, B. Goldspiel, D. Kohler, M. Cottler-Fox, S. F. Leltman, M. M. Gottesman, I. Pastan, A. Denicoff, M. Noone, and R. Gress, *Hum. Gene Ther.* **5,** 891 (1994).

[3] C. Hesdorffer, K. Antman, A. Bank, M. Fetell, G. Mears, and M. Begg, *Hum. Gene Ther.* **5,** 1151 (1994).

[4] J. M. Aran, M. M. Gottesman, and I. Pastan, *Proc. Natl. Acad. Sci. U.S.A.* **91,** 3176 (1994).

[5] Y. Sugimoto, I. Aksentijevich, G. J. Murray, R. O. Brady, I. Pastan, and M. M. Gottesman, *Hum. Gene Ther.* **6,** 905 (1995).

[6] A. D. Miller, *Blood* **76,** 271 (1990).

0076-6879/98 $25.00

The adeno-associated virus (AAV2) is an attractive alternative to retroviral vectors. This parvovirus was initially isolated as a tissue culture contaminant and was found as a nonpathogenic coinfecting agent during an adenovirus outbreak in children. Efficient replication and lytic growth of wild-type AAV2 depend on coinfection by a helper virus (adenovirus or herpes simplex virus). In the absence of a helper virus, the AAV virion can absorb to the cell and penetrate to the nucleus where the AAV genome is uncoated and stably integrates into the host cell chromosome.[7,8] Studies of human cell lines with stable integrated wild-type AAV sequences show that in 60–70% of the cases, integration occurs at a specific site (AAVS1) on chromosome 19 (19q13-qter) with no resulting insertional mutagenesis. The remaining integration events occur at a limited number of chromosomal sites.[9,10] The wide host cell range of AAV, its potentials for site-specific integration, and transduction of noncycling cells[11] are other attractive features of AAV vectors for transfer of the *MDR1* gene.

Recombinant *MDR1*-AAV Vectors

The wild type AAV2 genome is a 4680-nucleotide-long, single-stranded linear DNA molecule constituted of two main genes, *rep* and *cap*, flanked by characteristic 145-base-pair (bp) inverted terminal repeats (ITR).[12] The ITRs are self-complementary and the hairpin structure formed by folding over the first 125 nucleotides at the 3′-end of the parental strand serves as the primer for the initiation of DNA synthesis. The ITR sequences contain all of the *cis*-acting sequences required for the AAV replication origin, for encapsidation of the genome into particles, and for efficient integration into the host cell chromosome. Thus, all AAV vectors that are to be used as AAV-transducing vectors must have the ITR sequence at each end of the packaged vector genome. The *rep* gene encodes at least four multifunctional, overlapping, nonstructural proteins, expressed by differential splicing from two promoters, p5 and p19 (named according to their approximate map position). Rep proteins are required for efficient AAV DNA replication and production of single-stranded progeny genomes for encapsidation. Capsid proteins are encoded by the *cap* gene controlled by the p40 promoter.

[7] B. J. Carter, *Curr. Opin. Biotechnol.* **3,** 533 (1992).
[8] K. I. Berns, *Microbiol. Rev.* **54,** 316 (1990).
[9] R. M. Kotin, J. C. Menninger, D. C. Ward, and K. I. Berns, *Genomics* **10,** 831 (1991).
[10] R. J. Samulski, *Curr. Opin. Genet. Develop.* **3,** 74 (1993).
[11] T. R. Flotte, S. A. Afione, and P. L. Zeitlin, *Am. J. Resp. Cell. Mol. Biol.* **11,** 517 (1994).
[12] N. Muzyczka, *Curr. Topics Microbiol. Immunol.* **158,** 97 (1992).

In current recombinant AAV vectors, most of the AAV coding sequences are deleted and replaced by a promoter–transgene cassette; therefore the *rep* and *cap* functions have to be provided *in trans* during the production of virions. When the size of the rAAV genome is 120% or more of the size of the wild-type AAV genome, packaging efficiency decreases drastically.[12] Selection of a small and efficient promoter is critical for AAV vectors designed to transfer genes such as *MDR1* cDNA (3.8 kb) whose size approaches the packaging limit. Construction of an AAV-based *MDR1*-containing vector requires very compact transcription regulatory elements. We analyzed the suitability of the AAV-p5 promoter for expression because this promoter forms a convenient cassette with the 5′ ITR and is one of the smallest promoters available.[13] We designed an AAV-based plasmid, named ITR-P5-*MDR1,* in which the human *MDR1* cDNA was placed downstream from the p5 promoter (Fig. 1). The p5 promoter showed an activity similar in strength to that of the retroviral Harvey murine sarcoma virus LTR (long terminal repeat). Flotte *et al.*[11] demonstrated that ITR sequences are also able to drive expression of the CFTR (cystic fibrosis transmembrane conductance regulator) cDNA. We confirmed the promoter activity of ITR sequences.[13] We showed that a cassette containing the 5′ ITR and 90 additional base pairs efficiently drove *MDR1* expression, using a vector in which viral DNA sequence requirement was reduced to a total of 380 bp. A possible alternative approach to increase the capacity of AAV vectors for *MDR1* insertion would be to use mutations that delete regions of the Pgp protein. Until now attempts to shorten *MDR1* coding sequences did not result in the production of a functional multidrug transporter.[1]

Methods: Production of Recombinant Adeno-Associated Virus Particles, Titration of Stocks, and Transduction Procedures

Production of Recombinant Virions

Recombinant AAV (rAAV) vectors are packaged into infectious virions on transfection of the vector-containing plasmid into helper virus-infected cells.[14,15] When an AAV-based plasmid is transfected into human cells in the presence of helper virus, the AAV sequences are rescued from the plasmid and a normal lytic cycle ensues. AAV vector may be encapsidated

[13] M. Baudard, T. R. Flotte, J. M. Aran, A. R. Thierry, I. Pastan, M. G. Pang, W. G. Kearns, and M. M. Gottesman, *Hum. Gene Ther.* **7,** 1309 (1996).

[14] T. R. Flotte, S. A. Afione, R. Solow, M. L. Drumm, D. Markakis, W. B. Guggino, P. L. Zeitlin, and B. J. Carter, *J. Biol. Chem.* **268,** 3781 (1993).

[15] S. Chatterjee and K. K. Wong, *Methods* **5,** 51 (1993).

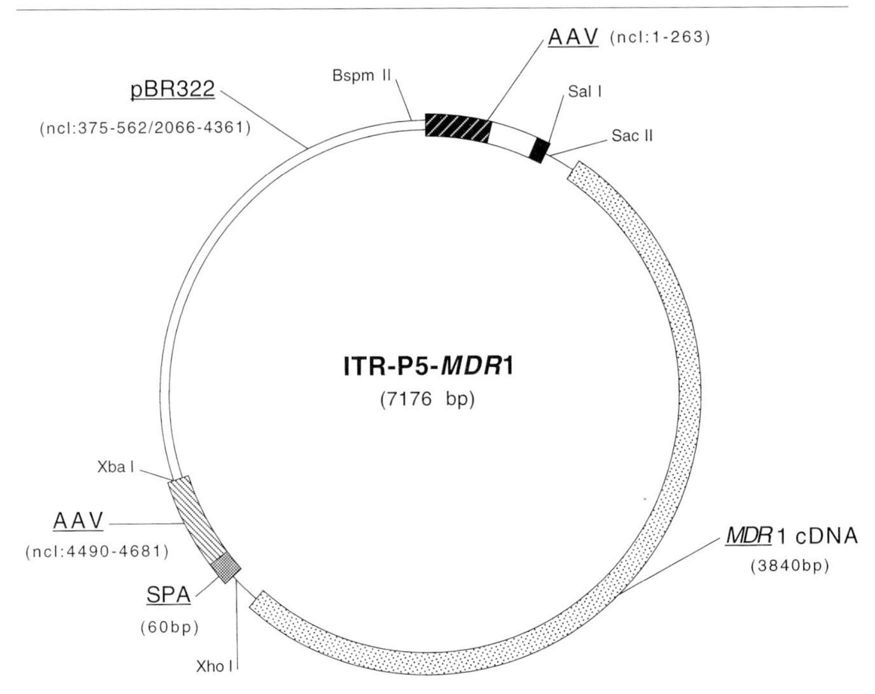

FIG. 1. Map of the ITR-P5-*MDR1* AAV expression vector. The human *MDR1* cDNA is placed downstream from a cassette constituted of the first 263 bp of the wild-type (wt) AAV genome in this pBR322-derived plasmid. ▨, 5′ AAV ITR, wt-AAV nucleotides 1–145; □, wt-AAV nucleotides 146–254; ■, AAV-p5 promoter TATA box (ncl: 255–261); ▨, human *MDR1* cDNA; ▩, SPA (synthetic polyadenylation site); ▧, 3′ AAV ITR (ncl: 4490–4681). Drawing is not to scale. Relevant restriction endonuclease sites are indicated.

in a variety of cell lines, including HeLa, KB, or 293 cells. The 293 cells, originally derived from human embryonic kidney cells transformed with the adenovirus type 5E1A and E1B genes, constitute the cell line of choice. The adenoviruses Ad2 or Ad5 and the plasmid pAAV/Ad are the most extensively used helper viruses and helper plasmid. pAAV/Ad contains functional AAV *rep* and *cap* genes flanked by the Ad5 terminal repeats, this plasmid is deleted for the AAV replication origins and therefore cannot replicate.[16]

The following is a brief description of a commonly used protocol for production of AAV particles.

1. Some 293 cells are plated onto 100-mm tissue culture dishes and grown at 37° in 5% CO_2 in Eagle's minimal essential medium supple-

[16] R. J. Samulski, L.-S. Chang, and T. Shenk, *J. Virol.* **63,** 3822 (1989).

mented with 2 mmol/L L-glutamine, penicillin–streptomycin antibiotic mix and 10% fetal bovine serum, to 60–70% confluence.

2. Medium is replaced 2 hr before transfection and cells are infected with Ad2 or Ad5 at a multiplicity of infection of 1–5, 1 hr before transfection.
3. Typically 10–15 μg of the AAV plasmid to be packaged is cotransfected with 2–4 μg of the helper plasmid per 100-mm dish, using standard calcium phosphate coprecipitation methods.[17]

Reagents

$CaCl_2$, 2.5 *M*
Sterile H_2O
HBS(2×): 50 m*M* *N*-2-Hydroxyethylpiperazine-*N*′-2-ethanesulfonic acid (HEPES), pH 7.1, 280 m*M* NaCl, 1.5 m*M* Na_2HPO_4

Protocol

3-1. Set up two tubes: tube 1—DNA (AAV vector and packaging vector), 25 μl 2.5 *M* $CaCl_2$, H_2O to 250 μl; tube 2—250 μl 2× HBS.
3-2. Using a 1-ml pipette, bubble air into the solution of 2× HBS; at the same time, add DNA solution dropwise to the HBS. Continue bubbling air through the solution for 5 sec after all the DNA is added. Let the mixture sit at room temperature for 30 min to allow precipitate to form.
3-3. Add DNA/HBS mix to cells in complete medium, dropwise with gentle swirling. No glycerol shock is necessary since the entry of the plasmid DNA into cells is promoted by the helper virus function. The medium is changed 4 hr after transfection.

4. Cell lysate and cesium chloride gradient are prepared. Three to 4 days after transfection, cells showing complete Ad2-induced cytopathic effect are harvested by gently scraping, pelleted by low-speed (4000 rpm) centrifugation, resuspended in 10 m*M* Tris-HCl buffer, pH 8.0, then lysed by three freezing–thawing cycles in a dry ice/37° water bath. The cell lysate is treated with DNase I (10 μg/ml) at 37° for 15 min to degrade nonvirion DNA. A brief treatment with a final concentration of 0.025% trypsin for 30 min at 37°, followed by the addition of an equal amount of serum to neutralize the trypsin, frequently facilitates the release of the encapsidated vectors. Based on differences in virion density between Ad (1.35 g/cm^3) and AAV (wild-type AAV density is 1.41 g/cm^3; the density of rAAV depends

[17] C. M. Gorman, G. T. Merlino, M. C. Willingham, I. Pastan, and B. H. Howard, *Proc. Natl. Acad. Sci. U.S.A.* **79,** 6777 (1982).

on the size of the DNA encapsidated) and on the relative resistance to heat of AAV compared to Ad, rAAV particles can then be separated from contaminant helper virus by CsCl density gradient ultracentrifugation (refractive index: 1.37, 35000 rpm at 15° overnight without brake) and heat treatment (65° for 15 min). Fractions to be used for transduction assays are then dialyzed against 1× SSC (sodium saline citrate buffer) three times for 1 hr at room temperature. Crude cell lysate can be employed in analytic experiments after a 2- to 7-day culture with indicator cells to ascertain complete helper virus inactivation.

Vector Titration

For vectors encoding a drug resistance gene such as *MDR1,* the titer of the transducing vector stocks can be determined by drug-resistant colony formation assays.[15,16] This method estimates the actual number of biologically active transducing virions. DNA dot–blot hybridization can also be used for titration of the vector preparation. Briefly, an aliquot of the virus stock is treated with DNase (50 μg/ml of crude virus lysate; for 1 hr at 37°), which then is inactivated (95° for 15 min). The subsequent steps consist consecutively of proteinase K treatment (1 mg/ml, 37° for 1 hr), phenol extraction, denaturation (with 0.2 *N* NaOH), and a 10-min boiling step. Then 10 volumes of cold 10× SSC are added and the resulting mixture is applied to a nitrocellulose sheet with a dot–blot apparatus and probed with a ^{32}P-labeled DNA probe. The intensities of dots are compared with those of an AAV DNA preparation of known concentration. This procedure allows estimation of the amount of vector DNA in the stock; it does not measure the actual biologically active transducing titer of a vector stock.

Cell Transduction and Selection of MDR1 Expressing Cells

Vector transductions are performed by the direct addition of desired amounts of vectors to cells in serum-free medium in minimal volumes. After an adsorption period of approximately 45 min at 37°, medium containing serum may be added and the cells cultured as usual.

MDR1 expression can be selected by growth of transduced cells in the presence of cytotoxic drugs (for example, colchicine and vincristine are two extensively used cytotoxic substrates of the multidrug transporter). The concentration of a given cytotoxic agent to be used for the initial selection step can vary significantly from one cell type to another and needs to be determined in advance so that only cells that take up and express *MDR1* will survive. Cell-sorting technology using a monoclonal antibody directed against an external epitope of the multidrug transporter may also be used.[18]

[18] S. E. Kane and M. M. Gottesman, *Methods Enzymol.* **217,** 34 (1993).

Integration of *rep* Negative Adeno-Associated Virus Vectors

In transduction protocols and with current rAAV, transduction efficiencies reach 50–95% when high multiplicites of infection are used. However, the proportion of long-term expressing cells is much lower, suggesting that viral uptake and initial gene expression might occur more efficiently than the integration step that leads to persistence of the rAAV genome and stable expression.[19,20] Wild-type AAV demonstrates efficient site-specific integration for established human cell lines but this property might be lost by *rep* negative rAAV. This is of particular importance for *MDR1* gene therapy since the packaging limit of the AAV capsid necessitates the removal of the *rep* and *cap* genes to accommodate the *MDR1* cDNA. Rep proteins may facilitate site-specific integration of AAV. A binding site for Rep78 and Rep68 has been identified near the AAVS1 site, and it has been demonstrated that Rep proteins can mediate the formation of a complex between the ends of the AAV genome and the chromosome 19 integration locus.[21] Integration of rAAV containing the ITRs but no coding sequences may occur.[13,22,23] In transduction experiments a possible role for Rep proteins present in the rAAV virions cannot be ruled out. We transfected an *MDR1*-AAV vector lacking *rep* sequences using the $CaPO_4$ coprecipitation procedure; integration of the recombinant genome was assessed by fluorescent *in situ* hybridization (FISH) analysis.[13] Whether integration was mediated by the ITR could not be determined. Vectors containing ITRs appear to integrate more efficiently than vectors without ITRs, and the ITRs may mediate integration of *rep* negative rAAV vectors via a different mechanism than that used by wild-type AAV.[10] If this is the case, then *rep* negative AAV vectors may not integrate randomly but at a selected number of chromosomal locations with reduced risk of insertional mutagenesis. Another mechanism to reduce concerns of insertional mutagenesis is to develop a strategy for targeting rAAV integration, such as the addition of *rep* sequences *in trans.*

Recent studies have demonstrated that the rAAV genome frequently remains episomal, although gene expression may persist for several cell divisions. As mentioned earlier, FISH analysis of cells that we transfected with *MDR1*-AAV vectors showed integration of rAAV sequences into

[19] S. Goodman, X. Xiao, R. E. Donahue, A. Moulton, J. Miller, C. Walsh, N. S. Young, R. J. Samulski, and A. W. Nienhuis, *Blood* **84,** 1492 (1994).

[20] J. L. Miller, C. E. Walsh, P. A. Ney, R. J. Samulski, and A. W. Nienhuis, *Blood* **82,** 1900 (1993).

[21] R. A. Owens, M. D. Weitzman, S. R. M. Kyostio, and B. Carter, *Virology* **67,** 997 (1993).

[22] C. E. Walsh, J. M. Liu, X. Xiao, N. S. Young, A. W. Nienhuis, and R. J. Samulski, *Proc. Natl. Acad. Sci. U.S.A.* **89,** 7257 (1992).

[23] A. N. Shelling and M. G. Smith, *Gene Ther.* **1,** 165 (1994).

the host cell genome.[13] Southern analysis of low-molecular-weight DNA extracted from these drug-resistant cells demonstrated the persistence of nonintegrated episomal forms of the AAV-*MDR1* plasmid whose transcription might have contributed to Pgp expression. As long as selection pressure was maintained *MDR1* expression was stable, but withdrawal of the selecting agent from the culture medium resulted in loss of Pgp expression subsequent to the loss of nonintegrated AAV vector forms as well as the extinction of integrated *MDR1* cDNA expression.

The use of *MDR1* as a selectable marker might allow both sustained persistence of episomal form of *MDR1*-containing AAV vectors and selection of cells coexpressing an otherwise nonselectable transgene. The size of the *MDR1* coding region is close to the packaging limit for AAV transduction vectors. Therefore, for cotransfer protocols the advantages of AAV-based *MDR1*-containing vectors might be combined with those of nonviral gene delivery systems such as cationic liposomes, for example. Cationic liposomes spontaneously complex with plasmid DNA and the fusion of the complex with the cell membrane results in uptake and expression of the DNA.[24] There is apparently no size limit for the plasmid for successful complex formation and this system might allow insertion of a larger cassette in AAV-based *MDR1*-containing vectors.

Conclusion

Because of the large size of the *MDR1* cDNA, the AAV coding sequences must be deleted prior to the insertion of *MDR1* between the ITRs of rAAV vectors meant for transduction protocols. Whether all *rep* negative rAAV vectors are capable of stable integration, the mechanisms that contribute to site-specific integration still need to be investigated. The fact that rAAVs frequently remain episomal and may be diluted and lost with cell division may limit the use of AAV vectors when long-term expression is expected. However, with selectable markers such as *MDR1* it may be possible to maintain nonintegrated episomal rAAV in proliferating cells; since only cells carrying such episomal AAV-*MDR1* would survive the selection, this advantage should be useful for gene therapy with AAV-*MDR1* vectors *in vivo* as well as *in vitro*.

[24] J. K. Rose, L. Buonocore, and M. A. Whitt, *BioTechniques* **10,** 520 (1991).

[41] Retroviral Transfer of Multidrug Transporter to Murine Hematopoietic Stem Cells

By THOMAS LICHT, MICHAEL M. GOTTESMAN, and IRA PASTAN

Introduction

The multidrug transporter, P-glycoprotein, is constitutively expressed in hematopoietic stem cells[1] and subpopulations of lymphocytes,[2,3] whereas physiologic expression levels are very low in mature granulocytes, monocytes, and their lineage-determined precursor cells.[4,5] Low expression of P-glycoprotein is associated with high vulnerability of these cells to the cytotoxic effects of anticancer chemotherapy. Thus leukocytopenia, resulting in bacterial and fungal infections, is a common adverse event of drug treatment of malignant tumors. It has been suggested that transfer and expression of chemoresistance genes like *MDR1* in hematopoietic tissues may protect them from the toxicity of antineoplastic drugs.[6,7] This has first been demonstrated in a transgenic mouse model in which a full-length *MDR1* cDNA was expressed under control of a chicken β-actin promotor.[8] Virtually all bone marrow cells of these animals expressed P-glycoprotein, and toxic substrates of the multidrug transporter such as taxol and daunomycin could be administered at dose levels that were lethal for animals of the respective background strains.[9] This protection could be transferred to lethally irradiated recipient mice by transplantation of bone marrow ob-

[1] P. M. Chaudhary and I. B. Roninson, *Cell* **66,** 85 (1991).

[2] A. A. Neyfakh, A. S. Serpinskaya, A. V. Chervonsky, S. G. Apasov, and A. R. Kazarov, *Exp. Cell Res.* **185,** 496 (1989).

[3] P. M. Chaudhary, E. B. Mechetner, and I. B. Roninson, *Blood* **80,** 2729 (1992).

[4] D. Drach, S. Zhao, J. Drach, R. Mahadevia, C. Gattringer, H. Huber, and M. Andreeff, *Blood* **80,** 2735 (1992).

[5] W. T. Klimecki, B. W. Futscher, T. M. Grogan, and W. S. Dalton, *Blood* **83,** 2451 (1994).

[6] D. Banerjee, S. C. Zhao, M.-X. Li, B. I. Schweitzer, S. Mineishi, and J. R. Bertino, *Stem Cells* **12,** 387 (1994).

[7] M. M. Gottesman, U. A. Germann, I. Aksentijevich, Y. Sugimoto, C. O. Cardarelli, and I. Pastan, *Ann. N.Y. Acad. Sci.* **716,** 126 (1994).

[8] H. Galski, M. Sullivan, M. C. Willingham, K.-V. Chin, M. M. Gottesman, I. Pastan, and G. T. Merlino, *Mol. Cell Biol.* **9,** 4357 (1989).

[9] G. H. Mickisch, T. Licht, G. T. Merlino, M. M. Gottesman, and I. Pastan, *Cancer Res.* **51,** 5417 (1991).

METHODS IN ENZYMOLOGY, VOL. 292
0076-6879/98 $25.00

tained from transgenic mice.[10] More recently, procedures have been established facilitating retroviral *MDR1* gene transfer to murine bone marrow cells and its expression *in vivo*.[11,12] Based on these and other studies, clinical trials have been initiated that were aimed at investigating transduction of the *MDR1* gene into bone marrow cells of cancer patients.[13-15]

Transfer of an *MDR1* cDNA to bone marrow cells of cancer patients may bear the risk of transduction of this cDNA into previously undetected micrometastases, thereby rendering tumor cells chemoresistant. This risk should be substantially reduced if isolated cell preparations of hematopoietic progenitor or stem cells rather then whole bone marrow are used. In addition, only stem cells have the capacity of long-term repopulation of bone marrow because they have the capacity of self-renewal. More mature progenitor cells become more differentiated with each step of cell duplication, and the life span of clones generated by these progenitors is limited.[16] Metamyelocytes and their progeny are terminally differentiated and eventually undergo programmed cell death. Transduction of such cells does not contribute to stable expression of transgenes in bone marrow *in vivo*. Thus, fractionation of bone marrow cells may be advantageous to increase the multiplicity of infection (MOI) to the preferred target cells.

Various methods have been developed to isolate hematopoietic stem cells that have high proliferative capacity *in vivo* and the potential to generate granulocytes, lymphocytes, and red blood cell precursors. Although the stem cell compartment appears to be heterogeneous, it is generally accepted that pluripotent stem cells of mice express no lineage-specific antigens (Lin^-), except for the c-kit protein (CD117), the Sca-1 antigen (also referred to as Ly-6.A/E) and low levels of CD90 (Thy-1).[17-20] A

[10] G. H. Mickisch, I. Aksentijevich, P. V. Schoenlein, L. J. Goldstein, H. Galski, C. Staehle, D. H. Sachs, and I. Pastan, *Blood* **79,** 1087 (1992).

[11] S. Podda, M. Ward, A. Himelstein, C. Richardson, E. delaFlor-Weiss, L. Smith, M. Gottesman, I. Pastan, and A. Bank, *Proc. Natl. Acad. Sci. U.S.A.* **89,** 9676 (1992).

[12] B. P. Sorrentino, S. J. Brandt, D. Bodine, M. M. Gottesman, I. Pastan, A. Cline, and A. W. Nienhuis, *Science* **257,** 99 (1992).

[13] J. A. O'Shaughnessy, K. H. Cowan, A. W. Nienhuis, K. T. McDonagh, B. P. Sorrentino, C. E. Dunbar, Y. Chiang, W. Wilson, B. Goldspiel, D. Kohler, M. Cottler-Fox, S. F. Leitman, M. M. Gottesman, I. Pastan, A. Denicoff, M. Noone, and R. Gress, *Hum. Gene Ther.* **5,** 891 (1994).

[14] A. B. Deisseroth, J. Kavanagh, and R. Champlin, *Hum. Gene Ther.* **5,** 1507 (1994).

[15] C. Hesdorffer, K. Antman, A. Bank, M. Fetell, G. Mears, and M. Begg, *Hum. Gene Ther.* **5,** 1151 (1994).

[16] J. R. Lemischka, *Semin. Immunol.* **3,** 349 (1991).

[17] G. J. Spangrude, S. Heimfeld, and I. L. Weissman, *Science* **241,** 58 (1988).

[18] S. J. Szilvassy, P. M. Lansdorp, R. K. Humphries, A. C. Eaves, and C. J. Eaves, *Blood* **74,** 930 (1989).

method described by Spangrude *et al.*[17] is widely used for isolation of $Lin^{-}Thy\text{-}1^{low}Sca\text{-}1^{+}$ cells by fluorescent-activated cell sorting (FACS).[17] In the human hematopoietic system the most primitive stem cells do not express the major histocompatibility complex (MHC) class II antigen, HLA-DR.[21] Since cell sorting by FACS may cause impaired cell viability, we have developed a method for immunomagnetic separation of murine bone marrow cells that express Sca-1 but fail to display lineage or MHC class II-associated antigens. This population has stem cell properties *in vivo* as previously reported.[22] In this article we describe a cell separation procedure and a protocol for transfer of an *MDR1* cDNA that permits expression of functional human P-glycoprotein in mouse stem cells and their progeny in the bone marrow.

Immunomagnetic Cell Fractionation

Six to 12-week-old C57BL/6 donor mice, obtained from NCI Frederick Cancer Research and Development Center, Frederick, MD, are treated IV with 150 mg/kg 5-fluorouracil (5-FU) 2–6 days before bone marrow cells are harvested. Pretreatment with 5-FU activates quiescent stem cells and thereby increases the efficiency of retroviral gene transduction,[23,24] because retroviral sequences integrate only in the genome of dividing cells.[25]

An outline of the fractionation procedure is depicted in Fig. 1A. Bone marrow is harvested from both femurs and tibias, and a single-cell suspensions is prepared by repeated flushing through 20-gauge needles. Low-density mononuclear cells are isolated from buffy coats by density-gradient centrifugation over Ficoll-Paque (Pharmacia, Piscataway, NJ). After two washings with Iscove's modified Dulbecco's medium (IMDM; Gibco/BRL, Gaithersburg, MD), hematopoietic stem cells are enriched by a two-step fractionation procedure. First, cells are incubated for 20 min at 4° with sheep anti-rat immunoglobulin (Ig)-coated Dynabeads (Dynal, Grand Neck, NY) after incubation with a mixture of commercially available monoclonal antibodies including Gr-1, MAC-1, TER119 (10 μg each), 5 μg anti-CD8a, 10 μg B220 (anti-CD45R) (all antibodies from Pharmingen, San Diego,

[19] S. Okada, H. Nakuchi, K. Nagayoshi, S.-I. Nishikawa, M. Yasusada, and T. Suda, *Blood* **80,** 3044 (1992).

[20] S. J. Szilvassy and S. Cory, *Blood* **81,** 2310 (1993).

[21] E. F. Srour, J. F. Brandt, R. A. Briddell, T. Leemhuis, K. van Besien, and R. Hoffman, *Blood Cells* **17,** 287 (1991).

[22] T. Licht, I. Aksentijevich, M. M. Gottesman, and I. Pastan, *Blood* **86,** 111 (1995).

[23] R. Wieder, K. Cornetta, S. W. Kessler, and W. F. Anderson, *Blood* **77,** 448 (1991).

[24] D. M. Bodine, K. T. McDonagh, N. E. Seidel, and A. W. Nienhuis, *Exp. Hematol.* **19,** 206 (1991).

[25] D. G. Miller, M. A. Adam, and A. D. Miller, *Mol. Cell Biol.* **10,** 4239 (1992).

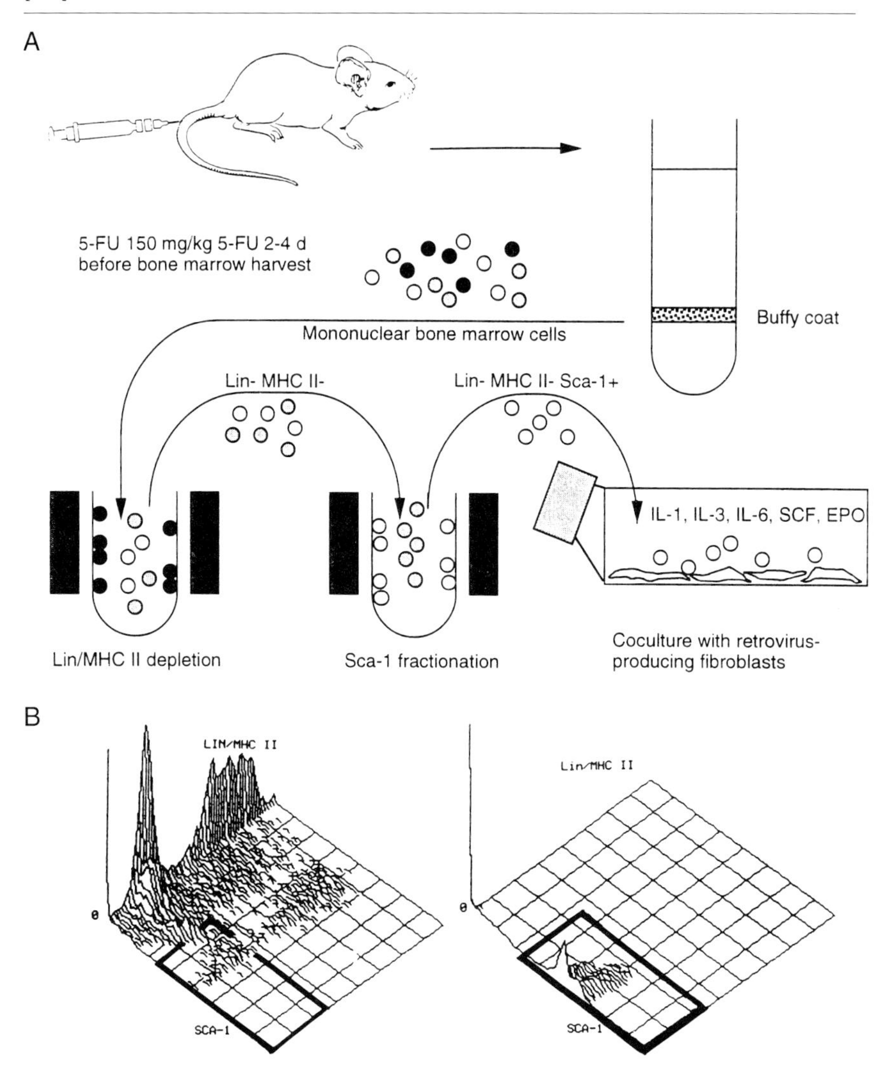

FIG. 1. Immunomagnetic fractionation of murine stem cells. (A) Schematic diagram of the procedure. Two to 4 days after IV administration of 150 mg/kg 5-FU to donor mice, bone marrow was harvested. Mononuclear cells were prepared by density gradient centrifugation. Cells expressing lineage or MHC class II antigens were depleted, followed by positive separation of cells expressing Sca-1. Sorted cells were transferred to tissue culture flasks containing pregrown virus-producing fibroblasts. (B) Expression of lineage/MHC class II and Sca-I antigens in normal marrow and in sorted cells. Cells were stained with a mixture of PE-labeled antibodies Gr-1, MAC-1, TER119, anti-CD3, anti-CD4, anti-CD8a, anti-CD45R, and anti-I-a, and with Sca-1 conjugated to FITC. Shown are plots of unseparated marrow (left) and fractionated cells (right).

CA), anti-CD3 (GIBCO, Grand Island, NY), and 10 μg of an antibody to the MHC class II-associated I-a antigen (Pharmingen). Cells expressing these antigens are depleted using an MPC-1 magnetic device (Dynal). Cells that are not bound are washed with 10 ml of medium, followed by incubation with Dynabeads that are labeled with monoclonal antibody Sca-1 (Pharmingen). After 20 min cells attached to the Sca-1-coupled Dynabeads are magnetically isolated.

Analysis of enrichment by this procedure is performed employing antibodies labeled with fluorescent compounds. Sca-1 is conjugated to fluorescein isothiocyanate (FITC). All other antibodies are conjugated to phycoerythrin (PE). Thereby it is possible to analyze directly the fractionated cell population without restaining of the cells. The enriched $Lin^-MHC\ II^-Sca\text{-}1^+$ population is compared to whole normal bone marrow stained with the respective fluorescent antibodies bound to Dynabeads. Isotype-identical nonbinding antibodies (Pharmingen) are attached to Dynabeads and used as negative controls for analysis of unfractionated cells.

The populations are analyzed using a FACSort cytometer with CELL-Quest or LYSIS II software (Becton-Dickinson, San Jose, CA). By gating in the forward- and side-scatter plot, 99.96% of the beads are excluded due to their small size (4.50 μm). For two-color analysis of $Lin^-MHC\ II^-Sca\text{-}1^+$ cells, the percentage of cells nonspecifically stained with isotype-identical nonbinding control antibodies in the respective quadrant (which is less than 1% in all samples) is subtracted from the positive populations. Additional experiments revealed that the use of monoclonal antibodies coupled to Dynabeads did not significantly alter the percentages of labeled cells as compared to cells that were directly stained with the respective labeled antibodies.

We found that the vast majority of cells in bone marrow expressed lineage-specific or MHC class II antigens (Fig. 1B). Only 0.35% of cells from bone marrow treated with 5-FU displayed the $Lin^-MHC\ II^-Sca\text{-}1^+$ phenotype. After immunomagnetic sorting this percentage was increased to 56 to 68%. Cells expressing lineage markers or the I-a antigen were almost completely removed (less than 0.3%); the remaining cells had a $Lin^-MHC\ II^-Sca\text{-}1^-$ phenotype. It has been shown that $Lin^-Sca\text{-}1^-$ cells contain short-lived B-cell precursors, but no cells with the capacity of repopulating cells of myelomonocytoid lineages.[26]

Culture of Hematopoietic Stem Cells and Retroviral *MDR1* Gene Transfer

Vector pHaMDR1/A, which contains a full-length *MDR1* cDNA promoted by Harvey sarcoma virus long terminal repeats, is used to generate

[26] N. Uchida, H. L. Aguila, W. F. Fleming, L. Jerabek, and I. Weissman, *Blood* **83,** 3758 (1990).

retroviruses for transduction of the multidrug transporter. An ecotropic producer cell line is engineered from GP+E86 packaging cells by transfection with pHaMDR1/A as described.[11] Retroviral titers of 5×10^5 to 3×10^6 colony-forming units (cfu)/ml are obtained by repeated subcloning of the producer cells and selecting them in colchicine at 60 ng/ml. Retroviruses that contained a neomycin-resistance (*neo*^{R}) gene are used as a negative control for transduction of the *MDR1* cDNA. We also examine other packaging cell lines. In general, transduction is more efficient with the use of ecotropic as compared to amphotropic viruses. Investigations have shown that very few amphotropic receptors are present on hematopoietic stem cells.[27,28]

Fractionated hematopoietic cells are transferred to tissue culture flasks in which virus producers have been pregrown to 30–40% confluency. Alternatively, virus producers can be grown to confluency and irradiated (30 Gy) before addition of hematopoietic cells. The latter approach ensures that no viable virus-producing fibroblasts are contained in cells transplanted into recipient animals but may result in detachment of parts of the fibroblast layer after several days. Cocultures are performed in IMDM supplemented with 20% fetal bovine serum (HyClone, Logan, UT), 50 U/ml penicillin, 50 μg/ml streptomycin, and 0.5 m*M* 2-mercaptoethanol. Serum is critical for the experiment, and extensive testing had to be performed in order to identify a suitable lot. Polybrene (Sigma, St. Louis, MO) at 4 μg/ml was added to increase the binding of virus particles to cells. Cells are stimulated with a combination of recombinant growth factors, including rat stem cell factor (SCF), also referred to as c-kit ligand or Steel factor, at 100 ng/ml (kindly provided by Amgen, Thousand Oaks, CA), murine interleukin (IL)-1β at 100 ng/ml (Peprotech, Rocky Hill, NJ), human erythropoietin at 1 U/ml (Sigma, St. Louis, MO), murine IL-3 at 200 U/ml, and human IL-6 at 200 U/ml (both from Becton-Dickinson Labware, Bedford, MA).

Nonadherent cells are harvested from cocultures after 6–8 days. Cells loosely attached to fibroblasts are detached by addition of PBS containing Ca^{2+} and Mg^{2+} and gentle swirling of tissue culture flasks. Great care is taken to avoid contamination of hematopoietic cells with fibroblasts. By flow cytometry analysis the virus producers displayed higher side-scatter and forward-scatter signals than the hematopoietic progenitor cells. In addition, due to continuous selection in colchicine at 60 ng/ml P-glycoprotein expression levels of the producer cells are much higher than levels observed in transduced but unselected cells (not shown). Gating for the combined forward- and side-scatter signals and for P-glycoprotein

[27] C. Richardson, M. Ward, S. Podda, and A. Bank, *Blood* **84,** 433 (1994).

[28] D. Orlic, L. J. Girard, C. T. Jordan, S. M. Anderson, A. P. Cline, and D. M. Bodine, *Proc. Natl. Acad. Sci. U.S.A.* **93,** 11097 (1996).

expression reveal that the contamination of harvested cells with virus producers is less than 0.3% in a typical experiment. Cultures in which the virus producers came off the flasks and contaminated hematopoietic cells are discarded.

Preservation of Immature $Lin^-MHC\ II^-Sca\text{-}1^+$ Phenotype in Cultured Hematopoietic Cells

The hematopoietic cells are expanded while being cocultured with virus-producing cells in the presence of growth factors. In a typical experiment their number increases from 3×10^4 to 2.8×10^6 cells after 8 days. Not only growth factors but also the presence of fibroblasts is beneficial to enhance expansion of $Lin^-MHC\ II^-Sca^-1^+$ cells. Most beads detach from the hematopoietic cells during coculturing. In some experiments the culture period is extended to examine preservation of the stem cell phenotype or its loss by differentiation of hematopoietic cells. Cells are analyzed by flow cytometry after staining with monoclonal antibody Sca-1 conjugated to FITC and a mixture of PE-coupled antibodies to lineage antigens (Gr-1, MAC-1, TER119, anti-CD45R, anti-CD3, anti-CD4, anti-CD8a, and anti-I-a).

As shown in Fig. 2A, the percentage of cells displaying the $Lin^-MHC\ II^-Sca\text{-}1^+$ phenotype decreases from 56–68% to 12% after a 6-day culture period. Due to the expansion by activation of stem cells, however, the absolute number of these cells is apparently not reduced. After prolongation of the culture period a shift of the cell population from $Lin^-MHC\ II^-Sca\text{-}1^+$ to lineage-committed cells is observed. The majority of these cells react with the myelomonocytoid marker, MAC-1, but B and T lymphocytes are not detectable in cultured cells (Fig. 2B). Interestingly, many cells display both the Sca-1 and lineage or MHC class II antigens. In contrast to conditions *in vitro,* in normal bone marrow most $Lin^+Sca\text{-}1^+$ cells are T cells.[29]

Differentiation of stem cells in culture depends on stimulation with growth factors. In many protocols for gene transfer into hematopoietic cells SCF, IL-3 and IL-6 are combined. These factors have synergistic effects on early progenitors and stem cells.[30,31] Addition of IL-1 and erythropoietin to this combination has been proposed for expansion of human stem cells

[29] M. van de Rijn, S. Heimfeld, G. J. Spangrude, and I. L. Weissman, *Proc. Natl. Acad. Sci. U.S.A.* **86,** 4634 (1989).

[30] K. Ikebuchi, G. G. Wong, S. C. Clark, J. N. Ihle, Y. Hirai, and M. Ogawa, *Proc. Natl. Acad. Sci. U.S.A.* **84,** 9035 (1987).

[31] M. A. S. Moore, *Blood* **78,** 1 (1991).

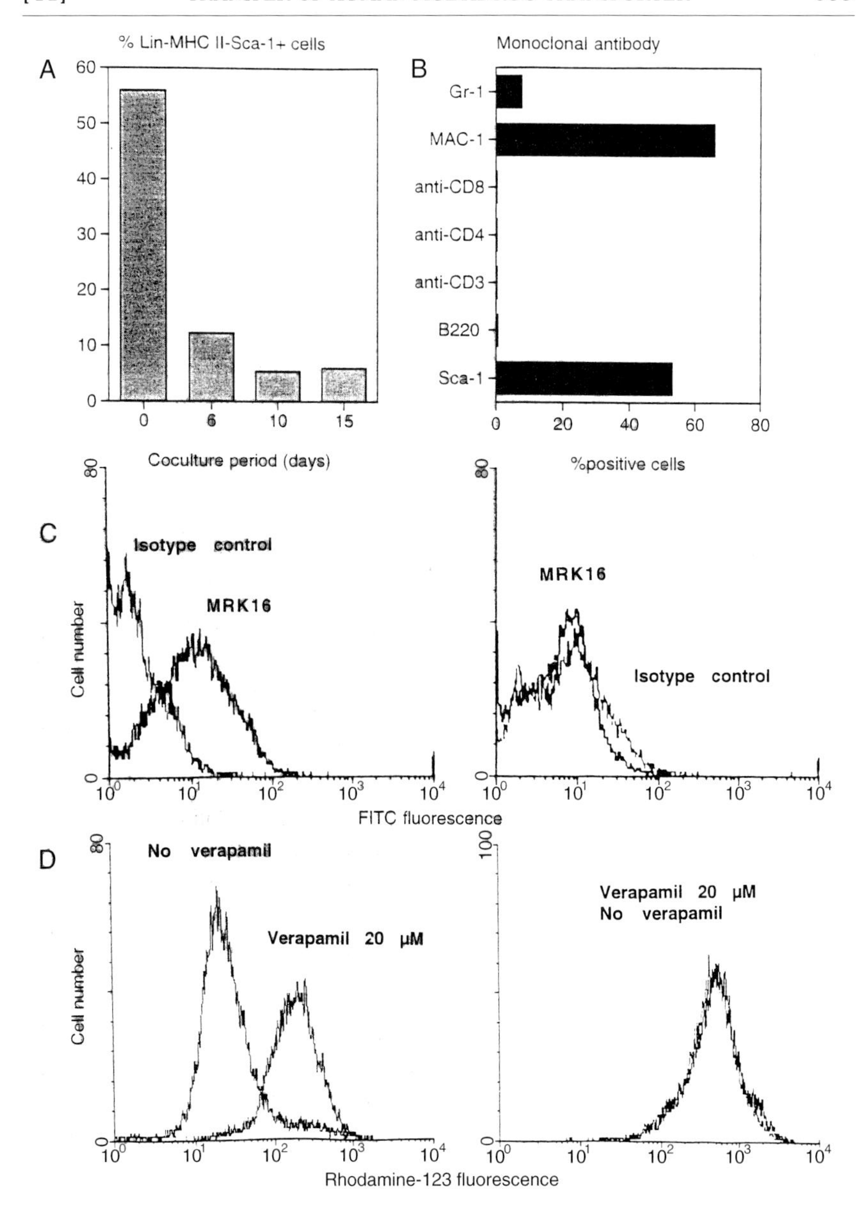

FIG. 2. Culture and transduction of Lin⁻MHC II⁻Sca-1⁺. (A) Preservation of Lin⁻MHC II⁻Sca-1⁺ phenotype in coculture with GP+E86 MDR1/A virus-producing fibroblasts. Cells were harvested from cocultures at indicated times and analyzed as described in the legend to Fig. 1B. (B) Differentiation in culture: Cells were harvested after 9 days in coculture and stained with PE-labeled antibodies to lineage-specific markers Gr-1 (granulocytes), MAC-1

apheresed from the peripheral blood.[32] In our hands, a combination of these growth factors has been useful for culture and transduction of murine stem cells. Factors that promote mainly differentiation of stem cells are avoided. Other investigators have reported successful transduction of hematopoietic stem cells with IL-3, IL-6, SCF, and leukemia inhibitory factor (LIF),[33] or with IL-3, SCF, erythropoietin, and granulocyte-macrophage colony-stimulating factor (GM-CSF).[34] It has been reported that IL-1 and IL-3 may inhibit development of lymphocytes from stem cells.[35] Thus, the optimal combination of growth factors remains to be defined; other factors such as flt-3 ligand may be beneficial.

Expression of Functional P-Glycoprotein in Transduced LIN$^-$MHC II$^-$SCA-1$^+$ Bone Marrow Cells

Expression of P-glycoprotein is analyzed using monoclonal antibody MRK16, kindly provided by Hoechst Japan Ltd. This antibody is specific for the protein encoded by the human *MDR1* gene and does not cross-react with murine P-glycoproteins.[36] To minimize nonspecific binding, samples containing 2.5×10^5 cells are first incubated with 1 μg of a blocking Fc-receptor antibody (Pharmingen) in PBS with 0.1% BSA. Five minutes later 5 μg of MRK16 conjugated to FITC was added. After a 20-min incubation at 4° cells are washed twice with phosphate-buffered saline (PBS). Fluorescence intensity is assessed by flow cytometry; all samples are compared to cells prepared under identical conditions which are stained with a FITC-conjugated nonbinding mouse IgG_{2a}-control antibody (Becton-Dickinson). P-glycoprotein expression is found in greater than 60% of fractionated bone marrow cells after coculturing for 6 days (Fig. 2C). The

[32] W. Brugger, W. Mocklin, S. Heimfeld, R. J. Berenson, R. Mertelsmann, and L. Kanz, *Blood* **81,** 2579 (1993).

[33] S. J. Szilvassy and S. Cory, *Blood* **84,** 74 (1994).

[34] L. Lu, M. Xiao, D. W. Clapp, Z.-H. Li, and H. E. Broxmeyer, *J. Exp. Med.* **178,** 2089 (1993).

[35] F. Hirayama, and M. Ogawa. *Blood* **86,** 4527 (1995).

[36] H. Hamada and T. Tsuruo, *Proc. Natl. Acad. Sci. U.S.A.* **83,** 7785 (1986).

(myelomonocytic precursors), anti-CD3 (pan-T cell), anti-CD4 (T-helper cells), anti-CD8a (cytotoxic T cells), B220 (pre-B and B cells), or with Sca-1. (C) P-glycoprotein expression of cells transduced with vector pHaMDR1/A (left) or, as a negative control, a *neo*^{R} vector (right), Lin$^-$MHC II$^-$Sca-1$^+$ cells were cocultured for 6 days. Staining was performed with monoclonal antibody MRK16 as described in the text. (D) Rhodamine-123 efflux from Lin$^-$MHC II$^-$Sca-1$^+$ cells after a 6-day coculture with pHaMDR1/A (left) or a *neo*^{R} vector (right). The experimental procedure is described in the text.

levels of expression are similar to those observed with colchicine-selected KB-8-5 cells (data not shown). Previous studies had revealed that KB-8-5 cells express *MDR1* at levels comparable to drug-resistant tumors in patients.[37]

The transport function of P-glycoprotein is assessed by efflux of a fluorescent substrate in the presence or absence of an inhibitor. To this end, cells are first incubated for 15 min at 37° with rhodamine 123 (1 μg/ml) in medium without phenol red containing 5% serum, then spun down at 400*g*, and resuspended in medium with 5% serum and distributed to two tubes. In one tube verapamil (20–25 μM) is added. After 120 min at 37°, cells are pelleted and supernatants are removed. Tubes are transferred to ice, and cell fluorescence is immediately analyzed by FACS. Active efflux of rhodamine 123 is detected in the majority of hematopoietic cells harvested from cocultures (Fig. 2D).

To exclude the possibility that only differentiating cells are transduced in coculture, resorting experiments are performed. Using the described mixture of monoclonal antibodies, cells harvested from cocultures that displayed the original $Lin^-MHC\ II^-Sca\text{-}1^+$ phenotype are isolated by immunomagnetic sorting with microbeads and MiniMACS separation columns (Miltenyi, Sunnyvale, CA). Staining with antibody MRK16 and rhodamine 123 efflux experiments revealed that at least 40% of the primitive $Lin^-MHC\ II^-Sca\text{-}1^+$ cells expressed functional human P-glycoprotein (data not shown).

P-Glycoprotein Expression *in vivo* After Transplantation of *MDR1*-Transduced Stem Cells

To investigate gene expression *in vivo,* fractionated, *ex vivo* transduced $Lin^-MHC\ II^-Sca\text{-}1^+$ cells from C57BL/6 donors are transplanted into sublethally irradiated (2.5–3.5 Gy) SCID mice. At least 4×10^4, but in most experiments 2.5×10^5 cells, are injected into the lateral tail veins of recipient mice of the strains C57BL/6-SCID-SzJ (Jackson Laboratories, Bar Harbor, ME) or CB.17-SCID (Charles River Laboratories, Wilmington, MA). The presence of human *MDR1* cDNA in peripheral blood is assessed by polymerase chain reaction (PCR) with primers 5′-AGATCAACTCGTAGGAGTGTC-3′ (forward) and 5′-GTTTCTGTATGGTACCTGCAA-3′ (reverse), amplifying a human *MDR1* fragment from nucleotide positions 1996–2790. DNA from 70 μl

[37] L. Goldstein, H. Galski, A. Fojo, M. Willingham, S.-I. Lai, A. Gadzar, R. Pirker, A. Green, W. Crist, G. M. Brodeur, M. Lieber, J. Cossman, M. M. Gottesman, and I. Pastan, *J. Natl. Cancer Inst.* **81,** 116 (1989).

of blood is amplified in 35 cycles (92° for 1 min, 52° for 2 min, 72° for 2 min). PCR products are visualized by ethidium bromide staining on a 1.4% (w/v) agarose gel. Five to 7 weeks after transplantation the *MDR1* cDNA is detected in peripheral blood of 14 of 17 mice (78%).

Nine weeks after transplantation mice are sacrificed, and mononuclear bone marrow cells are assayed by flow cytometry with antibody MRK16. In bone marrow from five of seven mice (71%) P-glycoprotein is detectable. As shown in Fig. 3, in some mice the fluorescence intensity of the whole-cell population appears to be shifted toward higher levels compared to the nonbinding isotype control antibody. No nonspecific staining is found under the conditions described in mononuclear cells of C57BL/6 mice transplanted with *neo*^{R}-transduced stem cells (Fig. 3). We should mention that polynucleated granulocytes from mice of several strains reacted with this antibody in a nonspecific fashion.

The product of the transgene, P-glycoprotein, is functional in bone marrow of mice transplanted with *MDR1*-transduced Lin^{-1}MHC II$^-$Sca-1$^+$ cells, but not with *neo*^{R}-transduced stem cells as assessed with rhodamine 123 efflux experiments (data not shown).

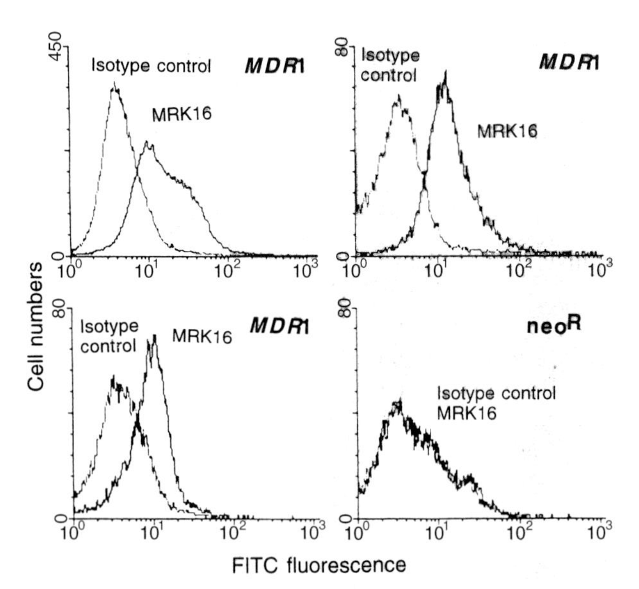

FIG. 3. Expression of P-glycoprotein *in vivo.* 2.5 × 10^5 *MDR1*- or *neo*^{R}-transduced Lin$^-$MHC II$^-$Sca-1$^+$ cells were transplanted into sublethally irradiated (2.5 Gy) recipient SCID mice. Seven to 9 weeks later, mononuclear bone marrow cells of recipient mice were analyzed with antibody MRK16 as described in text.

Conclusions

The protocol described here facilitates expression of functional human P-glycoprotein in a high proportion of bone marrow cells after retroviral transfer to purified hematopoietic stem cells. High numbers of hematopoietic progenitor cells with the capacity of engrafting bone marrow of recipient SCID mice are generated in coculture by stimulation with growth factors. Transfer of the multidrug transporter to hematopoietic stem cells may be useful to ameliorate the adverse effects of anticancer drug treatment. In addition, polycistronic vectors have been constructed in which an *MDR1* cDNA is combined with genes that correct genetically determined diseases.[38–40] It has been suggested that selection with cytotoxic substrates of the multidrug transporter may permit increased expression in bone marrow of both *MDR1* and the therapeutic gene. Transduction with such vectors might improve low transgene expression, which has hampered clinical gene therapy of hematopoietic disorders.[41] The method described here may be helpful to establish principles and applications of this treatment in animal studies.

[38] J. M. Aran, M. M. Gottesman, and I. Pastan, *Proc. Natl. Acad. Sci. U.S.A.* **91,** 3176 (1994).
[39] Y. Sugimoto, I. Aksentijevich, G. Murray, R. O. Brady, and I. Pastan, *Hum. Gene Ther.* **6,** 905 (1995).
[40] R. A. Sokolic, S. Sekhsaria, Y. Sugimoto, N. Whiting-Theobald, G. F. Linton, F. Li, M. M. Gottesman, and H. L. Malech, *Blood* **87,** 42 (1996).
[41] T. Licht, I. Pastan, M. M. Gottesman, and F. Herrmann, *Ann. Hematol.* **72,** 184 (1996).

[42] Retroviral Transfer of Human *MDR1* Gene into Human T Lymphocytes

By CAROLINE G. L. LEE, IRA PASTAN, and MICHAEL M. GOTTESMAN

Introduction

The *MDR1* multidrug transporter has potential usefulness as a clinically relevant dominant selectable marker for the gene therapy of various inherited and acquired disease.[1] Nonetheless, targeting gene delivery to the appropriate cell type is important for successful gene therapy. For the

[1] M. M. Gottesman, C. A. Hrycyna, P. V. Schoenlein, U. A. Germann, and I. Pastan, *Annu. Rev. Genet.* **29,** 607 (1995).

0076-6879/98 $25.00

treatment of adenosine deaminase (ADA) deficiency and acquired immune deficiency syndrome (AIDS), which affect T lymphocytes, it would be useful to target T cells. The transfer of the *MDR1* gene into T lymphocytes may also be useful for the study of possible involvement of the *MDR1* transporter in various T-cell functions.

Unfortunately, unlike other cell types, T cells are rather refractory to most nonviral and viral gene transfer methods. Electroporation[2] and retroviral transduction (e.g., Refs. 3 and 4) provide feasible means for the stable introduction of genes into these lymphocytes. However, electroporation of T cells often results in low viability (<10%) and very slow recovery of the surviving cells. Likewise, retroviral transduction suffers from low levels of gene transfer.[3,5] Even if gene transfer is successful, expression of the gene product in T cells is often low and dependent on the viral promoters and other transcriptional control elements.[6]

Recently, various methods have been developed that have resulted in improved retroviral transduction of human T lymphocytes. The use of the retroviral packaging cell line, PG13, rather than conventional amphotropic packaging lines like PA317, was found to greatly improve gene transfer into human T cells.[7–9] In these PG13 packaging cells, the *gag* and *pol* genes are derived from the Moloney murine leukemia virus while the *env* gene is derived from the gibbon ape leukemia virus (GALV). The cell surface receptor (*glvr-1*) for GALV was shown to be a sodium-dependent phosphate symporter and it was reported that depleting cells of extracellular phosphate resulted in increased receptor gene expression.[10]

[2] C. G. L. Lee, K. T. Jeang, M. Martin, I. Pastan, and M. M. Gottesman, *Antisense Nucleic Acids Dev.* **7,** 511 (1997).

[3] K. Culver, K. Cornetta, R. Morgan, S. Morecki, P. Aebersold, A. Kasid, M. Lotze, S. A. Rosenberg, W. F. Anderson, and R. M. Blaese, *Proc. Natl. Acad. Sci. U.S.A.* **88,** 3155 (1991).

[4] K. W. Culver, R. A. Morgan, W. R. Osborne, R. T. Lee, D. Lenschow, C. Able, K. Cornetta, W. F. Anderson, and R. M. Blaese, *Hum. Gene Ther.* **1,** 399 (1990).

[5] A. Kasid, S. Morecki, P. Aebersold, K. Cornetta, K. Culver, S. Freeman, E. Director, M. T. Lotze, R. M. Blaese, W. F. Anderson, and S. A. Rosenberg, *Proc. Natl. Acad. Sci. U.S.A.* **87,** 473 (1990).

[6] I. Riviere, K. Brose, and R. C. Mulligan, *Proc. Natl. Acad. Sci. U.S.A.* **92,** 6733 (1995).

[7] B. A. Bunnell, L. M. Muul, R. E. Donahue, R. M. Blaese, and R. A. Morgan, *Proc. Natl. Acad. Sci. U.S.A.* **92,** 7739 (1995).

[8] J. S. Lam, M. E. Reeves, R. Cowherd, S. A. Rosenberg, and P. Hwu, *Hum. Gene Ther.* **7,** 1415 (1996).

[9] C. von Kalle, H. P. Kiem, S. Goehle, B. Darovsky, S. Heimfeld, B. Torok-Storb, R. Storb, and F. G. Schuening, *Blood* **84,** 2890 (1994).

[10] M. P. Kavanaugh, D. G. Miller, W. Zhang, W. Law, S. L. Kozak, D. Kabat, and A. D. Miller, *Proc. Natl. Acad. Sci. U.S.A.* **91,** 7071 (1994).

This was found to translate into higher transduction efficiency in phosphate-deprived cells.[7] It was also found that transduction efficiency can be further improved by incubating the T cells with the viruses from the PG13 producer cell line at 32° instead of 37° presumably by prolonging the half-life of the retroviruses.[7,11] To facilitate retroviral transduction, polycations such as polybrene and protamine sulfate are commonly used. These polycations act by neutralizing the anionic charge on the cell surface. The use of the cationic lipid, DOSPA/DOPE (lipofectamine) (GIBCO-BRL, Gaithersburg, MD), in place of polybrene was demonstrated to improve gene delivery.[12,13] The reason for this improvement remains unclear. It was suggested that perhaps, in addition to the cationic properties of this liposome, its fusogenic activity may somehow further help the virus–receptor–liposome complex cross the cell membrane.[13] Another major barrier to achieving high retroviral transduction efficiency is the combination of the relatively short half-life of the viruses and the physical constraints imposed by Brownian motion of the viral particles resulting in low effective contact between the virus and the target cells.[14,15] This limitation has been overcome by the use of flow-through transduction methods that direct the movement of the virus toward the target cells via gravity by passing the virus solution through a porous membrane supporting the target cells.[14,15] Another way of colocalizing the viruses with the target cells would be to centrifuge the target cells in the viral media at 32° for 60 min.[7] However, this method of transduction often results in less viable target cells.

In this article, we describe the employment of an optimized combination of the above recently established procedures that improve retroviral gene delivery to efficiently transduce the *MDR1* multidrug transporter into a human 12D7 T-cell line. The protocols outlined in this article include the generation of high-titer recombinant retroviral PG13 producer cell lines, assays to quickly screen and identify recombinant producer clones with high vector titer, and an optimized method of transducing *MDR1*-expressing retrovirus into the 12D7 cell line.

[11] R. M. Kotin, M. Siniscalco, R. J. Samulski, X. D. Zhu, L. Hunter, C. A. Laughlin, S. McLaughlin, N. Muzyczka, M. Rocchi, and K. I. Berns, *Proc. Natl. Acad. Sci. U.S.A.* **87,** 2211 (1990).

[12] C. P. Hodgson and F. Solaiman, *Nature Biotechnol.* **14,** 339 (1996).

[13] J. Dybing, C. M. Lynch, P. Hara, L. Jurus, H.-P. Kiem, and P. Anklesaria, *Hum. Gene Ther.* **8,** 1685 (1997).

[14] A. S. Chuck and B. O. Palsson, *Hum. Gene Ther.* **7,** 743 (1996).

[15] A. S. Chuck, M. F. Clarke, and B. O. Palsson, *Hum. Gene Ther.* **7,** 1527 (1996).

Protocols

A. Cells and Culture Conditions

KB-3-1, a drug-sensitive clone of the human KB epidermal carcinoma cell line[16]; PA317, a murine amphotropic packaging cell line[17] (ATCC: CRL-9078); and PG13, a gibbon ape leukemia virus packaging line[18] (ATCC CRL-10686) are propagated as monolayer cultures at 37° with 5% CO_2 in complete DMEM [Dulbecco's modified Eagle's medium (DMEM) in high glucose (4.5 g/liter) containing 5 m*M* L-glutamine, 50 U/ml penicillin, 50 μg/ml streptomycin, and 10% (v/v) fetal bovine serum (FBS)]. It is desirable to add HAT selective media (30 μM hypoxanthine, 1 μM amethopterin, and 20 μM thymidine)[17,18] to both PA317 and PG13 producer cell lines to select against loss of plasmids conferring the packaging functions, especially after prolonged passage of these cells.[17,18] The human T-cell lymphoblastic leukemia line, 12D7, a derivative of CEM (ATCC CCL 119), is maintained as a suspension culture at 37° with 5% CO_2 in complete RPMI media (RPMI media 1640 supplemented with 10% FBS, 5 m*M* L-glutamine, 50 U/ml penicillin and 50 μg/ml streptomycin).

B. Generation of a Stable MDR1 Expressing Retroviral Producer Cell Line

A high titer of recombinant retrovirus is crucial for successful transduction of human T lymphocytes. Recombinant retroviruses obtained via infection are more desirable since a provirus resulting from infection is usually more stable, containing only one vector integrant compared to those derived from transfection, which can lead to rearrangement of tandemly integrated plasmids.[19] It has also been reported that vector titer can be increased by the infection of a producer cell line with another recombinant packaging cell line of a different host range as this facilitates multiple rounds of infection ("ping-pong infection") resulting in amplification of vector copy number and titer.[20,21] Packaging cell lines of a different host range have to

[16] S. Akiyama, A. Fojo, J. A. Hanover, I. Pastan, and M. M. Gottesman, *Somat. Cell Mol. Genet.* **11,** 117 (1985).

[17] A. D. Miller and C. Buttimore, *Mol. Cell Biol.* **6,** 2895 (1986).

[18] A. D. Miller, J. V. Garcia, N. von Suhr, C. M. Lynch, C. Wilson, and M. V. Eiden, *J. Virol.* **65,** 2220 (1991).

[19] A. D. Miller, D. G. Miller, J. V. Garcia, and C. M. Lynch, *Methods Enzymol.* **217,** 581 (1993).

[20] C. M. Lynch and A. D. Miller, *J. Virol.* **65,** 3887 (1991).

[21] A. D. Miller and G. J. Rosman, *BioTechniques* **7,** 980 (1989).

be used for infecting another producer line because the recipient cell surface receptors would already have been occupied by the retroviral envelope of the endogenous retrovirus, hence blocking further infection by viruses utilizing the same cell surface receptors. On the other hand, infection by retroviruses utilizing different cell surface receptors for entry is unaffected.[19] Hence, the establishment of a stable PG13 producer cell line that produces high titer of retroviruses expressing *MDR1* involves several steps (Fig. 1). The first step is the introduction of plasmid DNA containing *MDR1* retroviral vector into a heterologous producer cell line like PA317, an amphotropic packaging line. Calcium phosphate-mediated transfection[22,23] would be the method of choice since it is highly efficient and results in high expression of the introduced gene. The transfected cells are then subjected to selection and the surviving cells pooled. Viruses produced from these pooled drug-resistant PA317 are subsequently used to infect the PG13 packaging line. These infected cells are then placed in selective medium and, thereafter, clones isolated from the surviving cells are assayed for vector titer. We have obtained PG13 producer cells with *MDR1* titers of $\sim 6 \times 10^5$ cfu/ml using this protocol.[2] A further increase in titer may be achievable by first transfecting into an ecotropic producer line, then infecting an amphotropic packaging line with viruses from the ecotropic line before finally infecting the PG13 producer cells with viruses obtained from the amphotropic packaging line.

1. Transfection of pHaMDR1 Retroviral Construct into the Amphotropic Producer Cell Line, PA317

Materials and reagents

pHa*MDR1*, a retroviral construct encoding the *MDR1* multidrug transporter driven by the Harvey sarcoma retroviral long terminal repeat (LTR). The quality of the DNA is important for achieving high-efficiency transfection of the producer cell line. Plasmid DNA preparation using either cesium chloride purification or Qiagen EndoFree Plasmid Maxi Kit (Qiagen, Chatsworth, CA) gives good-quality DNA suitable for transfection

PA317 producer cells

Trypsin solution: 0.25% trypsin, 1 m*M* ethylenediaminetetraacetic acid (EDTA, pH 7.2)

[22] C. Chen and H. Okayama, *Mol. Cell. Biol.* **7,** 2745 (1987).

[23] C. M. Gorman, G. T. Merlino, M. C. Willingham, I. Pastan, and B. H. Howard, *Proc. Natl. Acad. Sci. U.S.A.* **79,** 6777 (1982).

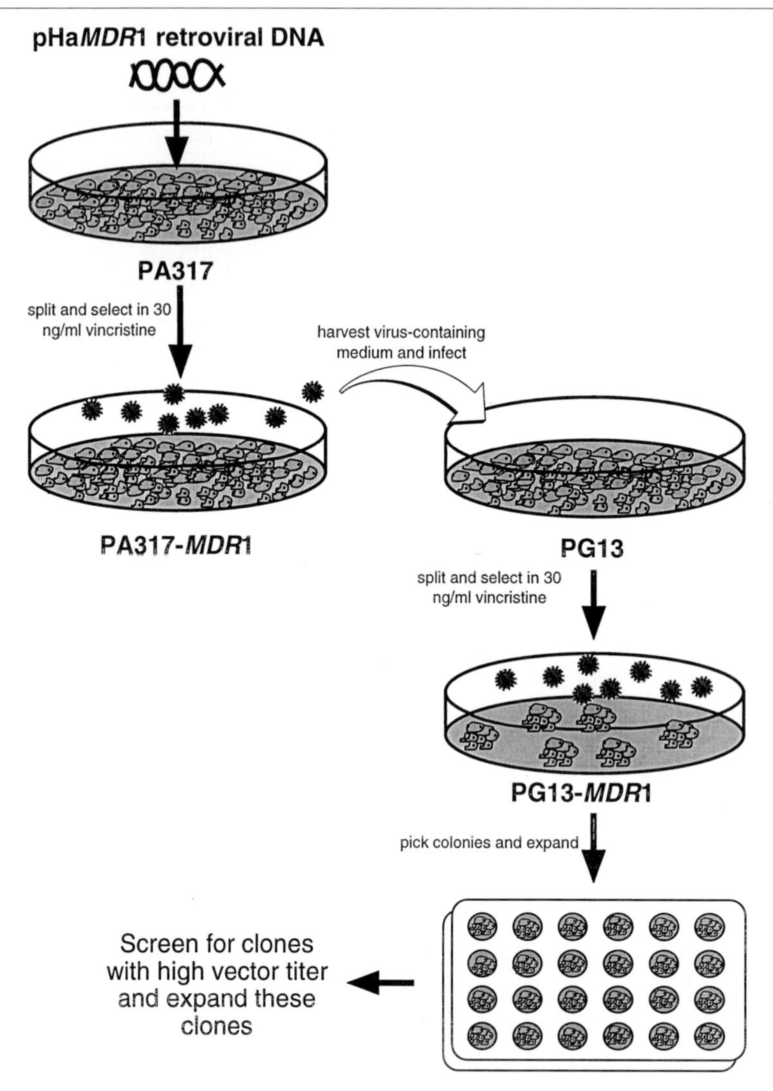

FIG. 1. Schematic flowchart for the generation of stable *MDR1*-expressing PG13 producer cell line.

Reagents for transfection[24]: 2× HEPES-buffered saline (2 × HBS): 280 mM NaCl, 10 mM KCl, 1.5 mM $Na_2HPO_4 \cdot 2H_2O$, 12 mM dextrose, and 50 mM HEPES (pH 7.05), and 2.5 M $CaCl_2$

100 μg/ml stock vincristine (Sigma, St. Louis, MO) solution dissolved in dimethyl sulfoxide (DMSO)

Method

Day 1: Harvest exponentially growing PA317 cells and replate them at ~5 × 10^6 cells into each 100-mm dish.

Day 2: On the day of transfection, feed the cells with 10 ml complete DMEM 0.5–3 hr prior to the addition of the DNA–$CaPO_4$ complex. Prepare two clear tubes (e.g., 15-ml polystyrene tubes) for each 100-mm dish: In one tube, slowly add 30 μl of 2.5 M $CaCl_2$ to 270 μl of DNA solution containing 15–20 μg pHa*MDR1* plasmid. In the other tube add 300 μl 2 × HBS solution. DNA–$CaPO_4$ precipitate is formed by gently vortexing (setting Vortex 2 of the Daigger Vortex Genie 2) (Lincolnshire, IL) the tube containing 2 × HBS while slowly adding (dropwise) the DNA–$CaCl_2$ solution. Incubate this mixture for 20–30 min at room temperature. A fine precipitate should form. Large, clumpy precipitates will result in lower transfection efficiencies. Gently transfer the DNA–$CaPO_4$ suspension mixture dropwise into the medium above the cell monolayer. Rock the dish gently to mix. Incubate these cells in a humidified incubator at 37°, 5% CO_2, overnight.

Day 3: Wash the cells once with phosphate-buffered saline (PBS) and refeed with complete DMEM.

Day 4: Dilute the 100 μg/ml vincristine stock solution to 30 ng/ml in complete DMEM. Trypsinize and seed the cells at 1:6 to 1:10 dilutions into 100-mm dishes containing complete DMEM plus 30 ng/ml vincristine. We found that this concentration of vincristine is sufficient to kill >90% of PA317 cells that do not express the *MDR1* gene.

Days 11–17: A few hundred/thousand colonies will be obtained in dishes containing PA317 cells transfected with pHa*MDR1,* while in dishes containing mock transfected cells, few or no colonies should be formed. The drug-resistant colonies from the recombinant PA317-*MDR1* producer cells are pooled by trypsinizing these colonies and replating them onto 100-mm dishes/T-75 flasks for immediate or later infection of the PG13 producer cells. It is important at this

[24] J. Sambrook, E. F. Fritsch, and T. Manniatis, *in* "Molecular Cloning, A Laboratory Manual." Cold Spring Harbor Laboratory Press, New York, 1989.

point to store frozen aliquots of some of these recombinant PA317-*MDR1* producer cells.

2. Transduction of PG13 Packaging Cell Line with Viruses from Pooled Recombinant PA317-MDR1 Producer Cells

Method

Day 1: Seed exponentially growing recombinant PA317-*MDR1* producer cells at a density of 3×10^6 cells per 100-mm dish in 10 ml complete DMEM *without* drug selection. At the same time, the target PG13 producer cells are seeded at 5×10^4 cells per 100-mm dish in complete DMEM containing 4 μg/ml polybrene (Sigma).

Day 2: Remove the virus-containing medium from PA317-*MDR1* producer cells and pass is through a low protein-binding 0.45-μm filter (e.g., Sterile Acrodisc, 0.45 μm, Gelman, Ann Arbor, MI) to remove producer cells and cell debris. Replace the medium from the PG13 producer cells with filtered PA317-*MDR1* virus-containing medium in the presence of 4 μg/ml polybrene. About 8 hr later, these infected PG13 cells are rinsed once with PBS and refed with complete DMEM.

Day 4: Trypsinize the infected PG13 cells and seed them at 1:10 to 1:50 dilutions into 100-mm dishes containing complete DMEM plus 30 ng/ml vincristine. This drug concentration is sufficient to kill >90% of parental PG13 cells.

Days 11–17: Drug-resistant colonies will be formed in PG13 cells infected with PA317-*MDR1* but not in mock infected PG13 cells. Clones of drug-resistant cells can be obtained either by single-cell cloning via limiting dilution, which is more time consuming since it involves a second step of cloning, or by directly picking well-separated single colonies.

Isolation of colonies. It is important that only well-separated colonies are isolated; hence, cells should be split at higher dilutions before drug selection. A simple method of picking colonies is as follows. Warm the trypsin solution to 37°. Add 1 ml complete DMEM medium to the wells of a 24-well plate. Rinse each dish, from which colonies are to be picked with PBS once and remove all liquid. Pipette up 4 μl of trypsin using an ART® 10P or REACH tip (Molecular Bio-Products Inc., San Diego, CA). Add the trypsin solution onto isolated colonies by depressing and holding the Pipetman at the depressed position against the isolated colonies for 10–15 sec. Thereafter, pipette up and down to separate the cells in the colony and pipette the separated cells into one well of the 24-cell plate. Rinse the colony with media from that well once. Repeat for the rest of the colonies. It is important not to allow the dish to be left dry for too long; hence, after about 7–10 min, add fresh complete DMEM to that dish and move on to the next dish. After some practice, it is possible to pick at least 12 colonies within 10 min. When all the wells are inoculated with cells, incubate these cells at 37°, 5% CO_2, in a humidified chamber. Depending on the size of the colony that is picked, the cells within the wells should be confluent by about 4–7 days. At this point, one can quickly estimate the vector titer by one of the two methods described later before a more definitive determination of the titer of the few promising clones is performed.

3. Quick Screening for PG13 Producer Clones with High Vector Titer

The two methods described here do not accurately determine the vector titer of the producer clone but do provide a rapid way to identify useful clones from a large number of producer clones.

a. VECTOR TITER SCREENING BY SEMIQUANTITATIVE RNA DOT BLOT

Materials and reagents

8 *M* LiCl
70% Ethanol (ETOH)
15% Formaldehyde (v/v)
Nylon membrane (e.g., Nylon N^+, Amersham, Arlington Heights, IL)
Manifold dot-blot apparatus (Schleicher and Schuell, Keene, NH)
20× SSC (3 *M* sodium chloride, 0.3 *M* sodium citrate, pH 7.0)
^{32}P-labeled *MDR1* probe (specific activity ~1×10^8 cpm/ug)
Rapid-Hyb buffer (Amersham)

Method

1. Collect 0.75 ml of medium from the producer clones that have been fed the day before into 1.5-ml RNAse-free Eppendorf tubes.
2. Centrifuge the medium at 3000*g* for 5 min at 4° to remove the producer cells and cell debris. Transfer the medium to a new RNAse-free Eppendorf tube.
3. Add an equal volume of 8 *M* LiCl and place on dry ice for about 15 min.
4. Pellet the RNA at 14,000*g* for 30 min at 4° and wash the pellet with 70% ETOH. Air dry the pellet briefly. The pellet obtained can either be used for reverse transcriptase–polymerase chain reaction (RT-PCR) or for RNA dot-blot analysis. We found that RT-PCR, although useful in identifying clones that express *MDR1,* is less indicative of the actual titer of the clone than RNA dot-blot analyses unless semiquantitative PCR (which is more tedious) is performed.
5. Denature the RNA by resuspending the pellet in 15% formaldehyde solution. Add 150 μl 20× SSC to each tube and mix. Incubate the

samples at 50° for 5 min and quickly chill the tubes on ice. Recently, the use of 180 μl of the viral supernatant directly for RNA dot-blot screening was reported.[25]

6. Prepare a Nylon$^+$ membrane by soaking it in 10× SSC and place it onto the Manifold dot-blot apparatus. Load the samples onto the wells and turn on the vacuum. Rinse wells with 10× SSC.
7. Remove the membrane from the dot-blot apparatus and cross-link the RNA to the membrane using the UV Stratalinker 1800 (Stratagene, La Jolla, CA). Prehybridize and hybridize using ^{32}P-labeled *MDR1* probe and Rapid-Hyb (Amersham) per manufacturer's instructions. Results can be obtained the same day using this hybridization protocol by quantitating the dot densities with the STORM 860 phosphorimaging system (Molecular Dynamics, Sunnyvale, CA). Darker dots correlate with clones of higher vector titer. Nonradioactive methods of hybridization can also be used.

b. VECTOR TITER SCREENING VIA DRUG-RESISTANT COLONY-FORMATION ASSAY AFTER TRANSDUCTION OF KB-3-1 CELLS

Method. Although this method of determining the titer takes slightly longer than the hybridization method just described, it is less difficult and gives approximate transducing titers.

1. Seed exponentially growing KB-3-1 cells at 1×10^4 cells per well in six-well plates in 2 ml of complete DMEM containing 4 μg/ml polybrene the day before performing this assay.
2. On the day of assay, add to the KB-3-1 cells, 1 μl of medium from the PG13 producer clones that were fed 1 day earlier with medium that does not contain any selective drugs.
3. Two days later, remove media from the KB-3-1 cells and add 2.5 ng/ml vincristine. This concentration of vincristine was found to kill >95% of drug-sensitive parental KB-3-1 cells.
4. About 7 days after selection, colonies will be formed and these colonies can be scored by removing the media, staining the cells with 0.5% methylene blue in 50% methanol, and counting the colonies manually or automatically using the AlphaImager IS-1000 Digital Imaging System (Alpha Innotech Corp., San Leandro, CA) and its associated software. Producer clones with higher vector titer will have more resistant KB-3-1 colonies.

Once promising PG13-*MDR1* producer clones have been identified, the vector titer can then be determined on these clones using the method described in Section B,2 (under Protocols) with the following modifications:

[25] M. Onodera, A. Yachie, D. M. Nelson, H. Welchlin, R. A. Morgan, and R. M. Blaese, *Hum. Gene Ther.* **8,** 1189 (1997).

1. The producer cells in this case will be the different clones of PG13-*MDR1*. The target cells used will be KB-3-1 instead of PG13 producer cells and they are seeded at 2×10^4 cell per 100-mm dish.
2. Instead of using all the virus-containing medium from the producer cells for infection, serial dilutions of the virus-containing medium from each of the PG13-*MDR1* producer clones will be added to each 100-mm dish with the KB-3-1 target cells. It is best to have at least three to four different dilutions that differ by 10-fold each to obtain a countable number of colonies for at least two of the dilutions so that one can confirm that the titer is reproducible.
3. The target KB-3-1 cells are not trypsinized or plated at different dilutions before selection. The concentration of vincristine used for selection is 2.5 ng/ml vincristine instead of 30 ng/ml vincristine as is the case for PG13 cells.
4. Once colonies are formed they are stained and scored as described in Section B,3,b, step 4 (under Protocols). The vector titer is expressed as colony-forming units per milliliter (cfu/ml) and is defined by the number of colonies formed on infection with 1 ml of virus supernatant after selection with 2.5 ng/ml vincristine. It is important to take the dilution factor into account when calculating the vector titer.

C. Transduction of 12D7, A Human T-Leukemic Cell Line, With PG13-MDR1 Retrovirus

The protocol for the infection of the 12D7 T-cell line with medium containing PG13-*MDR1* retrovirus includes the use of flow-through transduction over 3–4 days at 32° with 5% CO_2 with lipofectamine to facilitate transduction (Fig. 2).

Materials and reagents

0.4-μ Pore size Transwell-COL 6 well plates (Costar, Cambridge, MA)
Phosphate-free RPMI 1640 (GIBCO-BRL): sodium phosphate is omitted and 10% dialyzed FBS is added
DOSPA/DOPE polycationic lipid (LipofectAMINE™, GIBCO-BRL). Stock cationic lipid concentration: 2 mg/ml

Methods

1. The day before infection, seed exponentially growing recombinant PG13-*MDR1* producer cells at a density of 3×10^6 cells per 100-mm dish in 5 ml complete RPMI 1640 without drug selection in a humidified incubator at 37° with 5% CO_2.

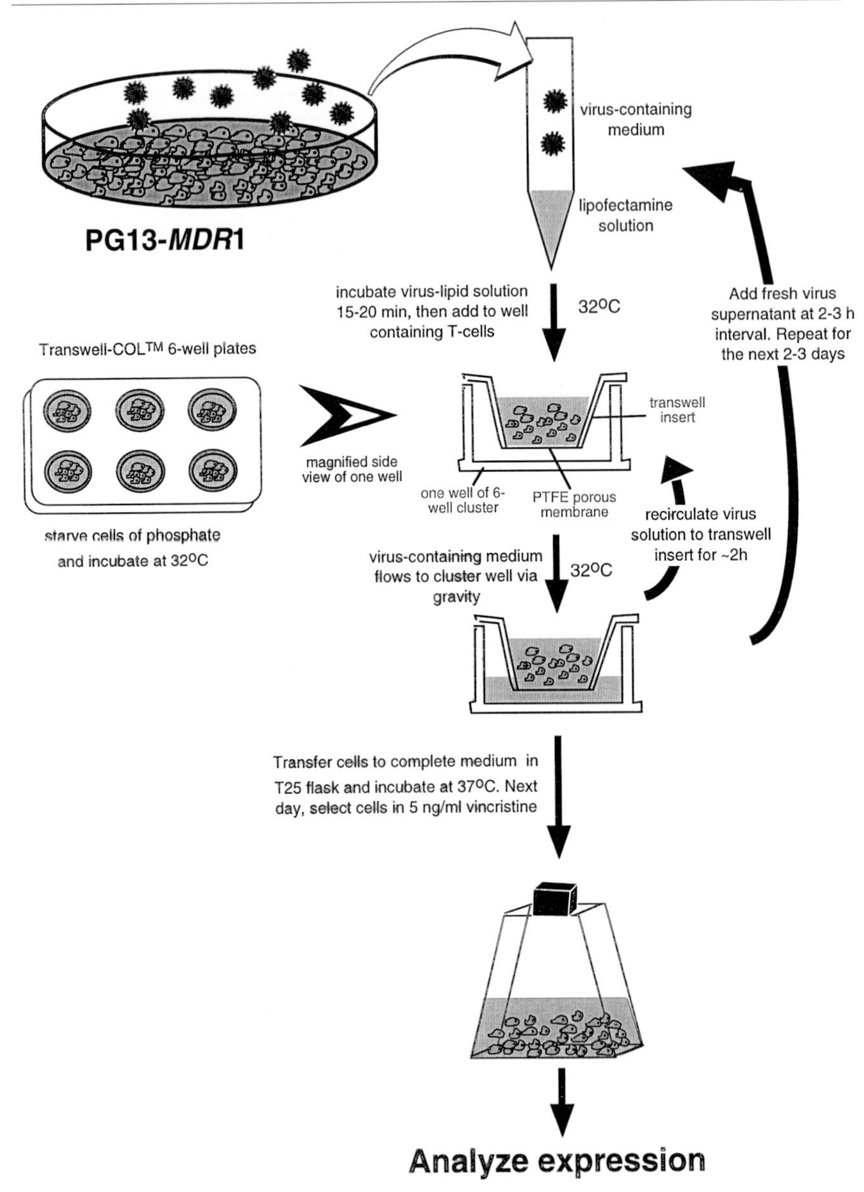

FIG. 2. Schematic flowchart for the transduction of 12D7, a human T-cell line with viruses from PG13-*MDR1* producer cells.

2. On the day of infection, equilibrate the 0.4 μm Transwell polytetrafluoroethylene (PTFE) membrane at 32° by adding complete RPMI into the Transwell insert and the cluster plate (see Fig. 2).
3. After an hour or two, remove the media and seed exponentially growing 12D7 cells in phosphate-free RPMI 1640 media at 1×10^5 cells/well onto the PTFE membrane in the Transwell insert. The medium but not the cells from the insert will flow through the membranes onto the cluster well. Ensure that the insert is not completely dry at any time. Incubate at 32° in a humidified chamber with 5% CO_2 for an hour.
4. In the meantime, 15 μl LipofectAMINE is added to 1 ml serum-free RPMI 1640 in a polystyrene tube. One milliliter of the virus-containing complete medium from step 1 is then combined with the lipid-containing medium and the mixture is incubated at room temperature for 15–20 min.
5. Remove the virus-containing medium from the cluster well of the Transwell system and allow more of the media to flow from the Transwell insert. Repeat until the cells on the insert are relatively dry. (This is to prevent further dilution of the virus solution.) Immediately add the virus–lipid complex onto the Transwell insert and allow the viral solution to flow through the cells on the membrane via gravity. Incubate at 32°, 5% CO_2, in a humidified chamber. Depending on the density of cells on the membrane, the flow will take about 10–30 min before the media in the insert and cluster well equilibrate. During the first 2 hr, when the most of the viruses are still viable, constantly recirculate the viral-containing media that gravitationally flow to the cluster well back to the Transwell insert and allow the process to repeat.
6. Remove the old viral media and repeat steps 4 and 5 once or twice at 2- to 3-hr intervals.
7. Repeat steps 4–6 for 12–16 hr daily for the next 2–3 days.
8. Thereafter, feed the infected 12D7 cells with complete RPMI 1640 medium and transfer these cells to T-25 flasks and incubate them at 37°, 5% CO_2, in a humidified chamber, overnight.
9. Although vector DNA and transcripts can be detected at this juncture, the expression is not sufficiently high to be detectable by fluorescence-activated cell sorting (FACS) analyses (described later). Hence, to select for *MDR1* expressing 12D7 cells, 5 ng/ml vincristine is added to the cells. This concentration of vincristine was found to kill 90% of the parental 12D7 cells. Our experience has been that different T cells as well as different derivative clones of the same T-cell line often express different amounts of the *MDR1* multidrug

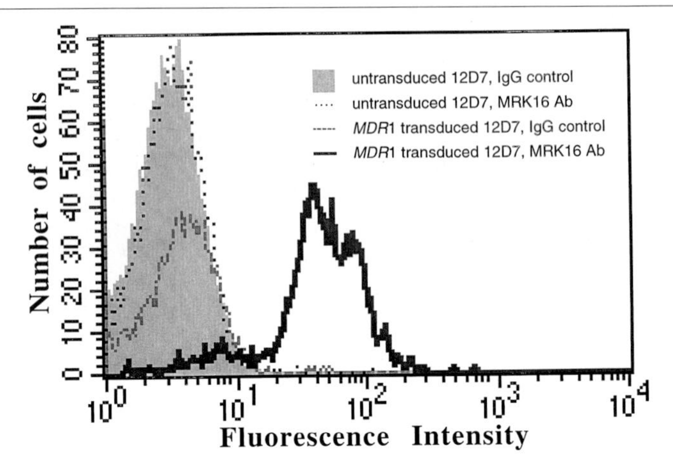

FIG. 3. Expression of *MDR1* in retrovirally transduced 12D7 as determined by FACS analysis using human P-glycoprotein specific monoclonal antibody, MRK16 (see Chapter [33] by Hrycyna *et al.*)

transporter. Hence it is important to determine the drug concentration that will kill ≥90% of parental cells before transduction and selection of these cells are performed. The procedure for this determination has been reported.[2] After about 2–3 weeks under selection, most of the population of the PG13-*MDR1* infected 12D7 cells should express *MDR1* while the mock (PG13) infected 12D7 cells should not survive the selection. At this point, one can proceed to obtain single clones of 12D7-*MDR1* cells by limiting dilution. However, we found that nearly all the cells express *MDR1* on selection although some may express more *MDR1* than others (Fig. 3). If a homogenous population of *MDR1*-expressing cells is desired, FACS can be performed using MRK16, a monoclonal antibody recognizing a human P-glycoprotein surface epitope.[26]

D. Analyses of Transduced Cells

Various assays can be performed to determine if the transduction and selection procedure is successful. A quick way to determine if the gene is transferred to the T cell is to perform PCR on lysed and neutralized transduced T cells using one primer that resides in the vector and the other within the *MDR1* gene. This prevents amplification and detection of the

[26] C. A. Hrycyna, M. Ramachandra, I. Pastan, and M. M. Gottesman, *Methods Enzymol.* **292,** [33], 1998 (this volume).

endogenous *MDR1* gene. Southern blot analysis[24] can then be done to determine if the transduced gene is integrated. Gene expression can be determined either with RT-PCR, Northern,[24] Western,[26] or FACS analyses.[26]

Conclusion

Long-term stable expression (~1 yr) of the transduced *MDR1* gene in 12D7 and CEM T-cell lines has been achieved using this protocol. Upon selection, it is possible to achieve >90% positive cells. Since the initial population of cells does not express much P-glycoprotein, it is important to carefully determine the concentration of drug to be used. The appropriate concentration of drug would be one that would kill 90% of untransduced cells. Too low a concentration of drug will result in a high background of untransduced cells. On the other hand, an unduly high concentration of drug will kill too many cells (especially cells that are expressing low amounts of P-glycoprotein) and it would be very difficult to expand this limited population of *MDR1*-expressing cells. If high expression of *MDR1* is desired, higher concentration of drug can be added later.

Although the flow-through method of transduction is rather involved and time consuming, it should be amenable to automation where one can device a pump system to recirculate or continually add, at a desired flow rate, fresh viruses onto cells that are resting on the membrane. A plausible alternative to flow-through transduction would be to coculture the T lymphocytes with the producer cells. This should allow continuous transduction of freshly produced recombinant viruses from the producer cells. However, this procedure often results in cross-contamination of the T cells with the producer cells. Another drawback of this procedure is that the T lymphocytes tend to settle onto the producer cells, making recovery of the infected T cells difficult. A different approach to coculture involving the use of Transwell-COL plates similar to the ones used for flow-through transduction has also been reported.[27] In this procedure, the producer cells are seeded on the membrane in the Transwell insert while the target cells are maintained in the cluster well below, thus separating the producer cells from the target cells.

Finally, since the cell lines used here are of human origin and the producer cell lines are making defective retroviruses capable of transducing human cells, it is important to observe safety procedures. All of these cells should be handled carefully and experiments should be performed under

[27] W. T. Germeraad, N. Asami, S. Fujimoto, O. Mazda, and Y. Katsura, *Blood* **84,** 780 (1994).

Biosafety level 2 (BL-2) containment. One should also check for producer cell contamination in the transduced target cells.

Acknowledgments

We would like to thank Thomas Licht and Zhou Yi for helpful discussions and comments.

[43] Construction and Analysis of Multidrug Resistance Transgenic Mice

By GREGORY L. EVANS

Introduction

Transgenic mice derived by pronuclear injection have proven a useful tool for *in vivo* study of multidrug resistance (MDR) mediated by the *MDR1* gene product, human P-glycoprotein (HPgp). That HPgp-associated MDR is one barrier to the successful chemotherapeutic treatment of certain human cancers, especially hematologic ones, is well documented.[1] Another barrier to successful chemotherapy with MDR drugs, such as taxol, vinblastine, and bisantrene, is dose-limiting bone marrow suppression. To circumvent these obstacles, it is desirable to develop animal models that can be used to screen potential MDR reversal agents. In addition, it is desirable to use animal models to measure the antitumor effects of MDR drug dose escalation, and in the long term to evaluate anticancer therapies directed against multidrug-resistant tumors of various histology. Further potential long-term uses might be to confirm *in vivo* the differing cross-resistance patterns reported *in vitro* for different variants of mammalian P-glycoproteins (Pgps),[2] which may have implications for gene therapy protocols involving protection of specific tissues from specific drugs, and to assess HPgp-mediated protection of specific tissues from environmental carcinogens.

Various types of MDR animal models, each with their own advantages and disadvantages, have been utilized to address these broad goals.[3] What

[1] W. T. Bellamy and W. S. Dalton, *Adv. Clin. Chem.* **31,** 1 (1994).

[2] M. M. Gottesman, C. A. Hrycyna, P. V. Schoenlein, U. A. Germann, and I. Pastan, *Annu. Rev. Genet.* **29,** 607 (1995).

[3] M. M. Gottesman, G. H. Mickisch, and I. Pastan, *in* "Anticancer Drug Resistance: Advances in Molecular and Clinical Research" (L. J. Goldstein and R. F. Ozols, eds.), p. 107. Kluwer Academic, Norwell, MA, 1994.

0076-6879/98 $25.00

follows in this article is a technical discussion of the generation of transgenic mice expressing human *MDR1* from two different transgenes in hematopoietic cells.[4,5] Transgenic mouse models in which HPgp is expressed in blood cells have the advantages of well-defined ectopic expression at physiologic levels, a single mechanism of drug resistance, convenient and rapid assays, and simple determination of reversing agent selectivity, that is, preferential chemosensitization in HPgp-positive cells. They have the disadvantages in this context of not bearing a tumor, and of the difficulty of data interpretation in the presence of mouse *mdr* genes. The recent construction and analysis of *mdr1* knockout mice has validated the original hypothesis of Gottesman and Pastan that the normal function of *MDR* genes (here defined as human *MDR*, murine *mdr*, and hamster *pgp* genes) is to provide xenobiotic protection for privileged sites such as the brain.[6]

While the original β-actin-MDR transgenic mice (actin mice) exhibited an MDR phenotype in bone marrow (BM) for the first few generations,[4,7–11] after several generations of inbreeding the MDR phenotype was lost, due to the loss of transgene expression. In an attempt to replace these mice, the second-generation Eμ-MDR transgenic mice (Eμ mice) were constructed,[5] with which most of the protocols described here were developed or applied. Although the Eμ-MDR mice proved not to be useful for drug screening, the methodology used to develop them can be generalized.

Construction of MDR Transgenic Mice

Initial Considerations: Expression Strategies, Vector Construction, and Mouse Strain Background

The most important considerations in preparing to develop MDR transgenic mice are which tissues to target (and which dominant control

[4] H. Galski, M. Sullivan, M. C. Willingham, K.-V. Chin, M. M. Gottesman, I. Pastan, and G. T. Merlino, *Mol. Cell. Biol.* **9,** 4357 (1989).

[5] G. L. Evans, T. Licht, C. Jhappan, G. T. Merlino, M. M. Gottesman, and I. Pastan, unpublished results (1995).

[6] M. M. Gottesman and I. Pastan, *J. Biol. Chem.* **263,** 12163 (1988).

[7] G. H. Mickisch, G. T. Merlino, H. Galski, M. M. Gottesman, and I. Pastan, *Proc. Natl. Acad. Sci. U.S.A.* **88,** 547 (1991).

[8] G. H. Mickisch, T. Licht, G. T. Merlino, M. M. Gottesman, and I. Pastan, *Cancer Res.* **51,** 5417 (1991).

[9] G. H. Mickisch, G. T. Merlino, P. M. Aiken, M. M. Gottesman, and I. Pastan, *J. Urol.* **146,** 447 (1991).

[10] G. H. Mickisch, L. H. Pai, M. M. Gottesman, and I. Pastan, *Cancer Res.* **52,** 4427 (1992).

[11] G. H. Mickisch, I. Aksentijevich, P. V. Schoenlein, L. J. Goldstein, H. Galski, C. Stahle, D. H. Sachs, I. Pastan, and M. M. Gottesman, *Blood* **79,** 1087 (1992).

elements to use), which form of HPgp to use (i.e., wild type, or any of a number of mutant forms with distinct MDR drug cross-resistance patterns *in vitro*[2]), and, secondarily, which mouse background to use. The choice of tissue to target will naturally affect the type and ease of assay one might use to screen MDR reversal agents, but it also will affect the types of MDR tumors one could potentially study in double transgenic mice created by breeding, if this is a secondary goal. The choice of which HPgp isoform to use is important not only for physiologic relevance (the wild type should mimic the human situation the best), but also because it is now clear that various human and murine Pgp isoforms are differentially inhibited by different MDR reversal agents.[12,13]

The initial strategy undertaken was to express human *MDR1* in a wide range of mouse tissues. To this end, a minimal chicken β-actin promoter, predicted to be active in most cell types, was fused using standard cloning methods,[14] to the human *MDR1* cDNA bearing its endogenous polyadenylation [poly(A)] signal, and encoding the 1280-amino-acid Val^{185} mutant form of HPgp (Fig. 1). This form of HPgp confers increased relative resistance to colchicine and etoposide *in vitro*. Importantly, all of the cloning steps for both transgenes discussed here were carried out in the pGEM2 (Promega, Madison, WI) promoterless bacterial plasmid in bacterial strain DH5 (Life Technologies), to reduce potential host toxicity associated with human *MDR1* expression in bacteria.[15] Even so, bacterial colonies bearing the various *MDR1* plasmid vectors were noticeably and reproducibly smaller than *MDR1*-free controls. Also, in planning the cloning strategy, it was important to leave rare restriction sites at the ends of the transgene to facilitate separation from the plasmid vector (see later discussion).

Surprisingly, one founder mouse carrying the actin-MDR transgene expressed it only in the BM and spleen (see later discussion), for unknown reasons, but perhaps due to serendipitous transgene integration near a hematopoietic enhancer element. These mice proved very useful since they allowed for simple measurement of drug resistance *in vivo* (resistance to leukopenia) by following peripheral white blood cell counts, which reflected toxic insults to proliferating bone marrow precursors (see later discussion). However, because the useful lifetime of the actin mice was limited by the

[12] C. O. Cardarelli, I. Aksentijevich, I. Pastan, and M. M. Gottesman, *Cancer Res.* **55,** 1086 (1995).

[13] D. F. Tang-Wai, S. Kajiji, F. DiCapua, D. de Graaf, I. B. Roninson, and P. Gros, *Biochemistry* **34,** 32 (1995).

[14] J. Sambrook, E. F. Fritsch, and T. Maniatis, *in* "Molecular Cloning," 2nd Ed., Cold Spring Harbor Laboratory, New York, 1989.

[15] G. L. Evans, B. Ni, C. A. Hrycyna, D. Chen, S. V. Ambudkar, I. Pastan, U. A. Germann, and M. M. Gottesman, *J. Bioenerg. Biomembr.* **27,** 43 (1995).

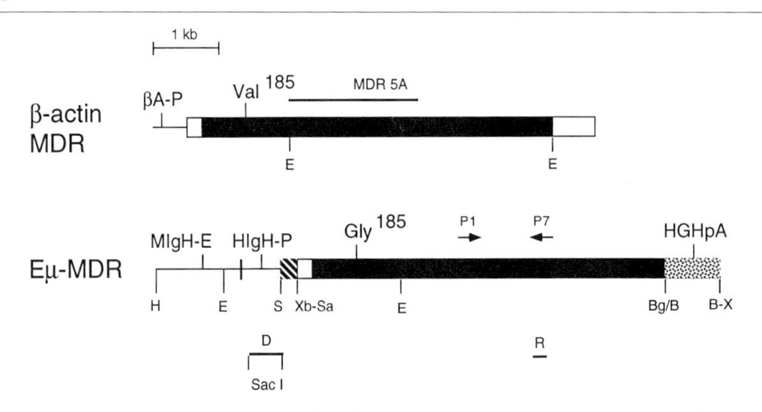

FIG. 1. Schematic diagram of human *MDR1* transgenes used for the generation of transgenic mice by pronuclear injection. *Top*, the first-generation β-actin-MDR construct (4.7 kb) consisted of a minimal 0.3-kb chicken β-actin promoter (βA-P) fused to the 4.4-kb human *MDR1* cDNA, with the endogenous poly(A) addition signal, encoding a mutant form of P-glycoprotein with valine at amino acid position 185. The bar indicates the position of the *MDR1* 5A probe used for Southern and RNA slot blots. *Bottom*, the second-generation Eμ-MDR construct (6.4 kb) contained the following elements (see text for references): the intronic murine immunoglobulin heavy chain enhancer (MIgH-E; Eμ; 0.9 kb), human immunoglobulin heavy chain promoter (HIgH-P; Pμ; 0.6 kb), rat insulin II intron A (0.2 kb), human growth hormone poly(A) addition signal (HGHpA; 0.6 kb), and the 4.1-kb human *MDR1* cDNA [without the endogenous poly(A) signal] encoding wild-type P-glycoprotein with a glycine residue at amino acid position 185. Bars denote the locations of probes for Southern blots (D; bounded by *Sac*I restriction sites), and RNase protection (R); the arrows denote polymerase chain reaction (PCR) primers (P1 and P7). The sequence of the PCR primers, which amplify *MDR1* nucleotides 1996–2790, is as follows: P1 forward (21-mer), 5′-AGATCAACTCGTAGGAGT GTC-3′; P7 reverse (21-mer), 5′-GTTTCTGTATGGTACCTGCAA-3′. Symbols: filled boxes, human *MDR1* translated region; open boxes, human *MDR1* untranslated sequences; cross-hatched box, rat insulin II intron A; stippled box, human growth hormone poly(A) addition sequence. Restriction enzymes: E, *Eco*RI; H, *Hind*III; S, *Sal*I; Xb, *Xba*I; Sa, *Sac*II; Bg, *Bgl*II; B, *Bam*HI; X, *Xho* I. Hyphens indicate adjacent restriction sites; forward slash indicates the fusion of two sites. [β-actin-MDR schematic adapted from H. Galski, M. Sullivan, M. C. Willingham, K.-V. Chin, M. M. Gottesman, I. Pastan, and G. T. Merlino, *Mol. Cell Biol.* **9,** 4357 (1989).]

extinction of transgene expression, it was desirable to develop a second line of mice with the properties of the first. To do this, we used the well-studied lymphoid-specific enhancer Eμ (mouse) and promoter Pμ (human), derived from the immunoglobulin (Ig) heavy chain loci. These elements are known to direct transgene expression primarily to pre-B and mature B cells, but also somewhat to T cells in numerous lines of transgenic mice.[16] The Eμ transgene consisted of Eμ, Pμ, rat insulin II intron A, the human

[16] J. M. Adams and S. Cory, *Biochim. Biophys. Acta,* **1072,** 9 (1991).

MDR1 wild-type (Gly^{185}) cDNA without its endogenous polyadenylation signal, and the human growth hormone (HGH) polyadenylation signal (Fig. 1). The 5′-intron[17] and HGH[18] sequences were included so as to improve RNA processing and translation, and were reported in the transgenic literature to work well with other heterologous genes. This transgene vector was constructed in multiple cloning steps, and the DNA sequence was determined around cloning sites and *MDR*1 Gly^{185}. The actin and Eμ transgenes were prepared for microinjection by separation from bacterial plasmid sequences with restriction digestion and purification by either electrophoresis,[4] or sucrose gradient centrifugation,[19] followed by dialysis against very high quality water.

As far as the genetic background for transgenesis is concerned, it is generally best to use inbred strains, since then one can be sure that sibling mice, and all transgenic mice generated by inbreeding will be syngeneic. This is desirable for reproducibility of results, as well as for histocompatibility, if any sort of transplant procedures (i.e., MDR donor to nontransgenic sibling recipient) are envisioned in the long run. One such very well characterized strain that we used for the Eμ mice is FVB/N.[20] In considering potential genetic backgrounds, it is also wise to consider the properties of prospective strains, to the extent that they are known, in the context of your goals. For example, in retrospect, a strain other than FVB/N may have been better, since we learned that in this strain B cells make up only approximately 10% of PB white cells, compared to about 30% in C57Bl/6 (see later discussion).[5] For one who hopes to target *MDR1* to B cells and detect resistance to leukopenia by simple white counts, the higher this number is, the better. For the actin mice, the genetic background was C57Bl/6 $\times$ SJL F$_1$, which when inbred over many generations yields allogeneic progeny due to independent segregation of genetic markers. Whether this mixture of genetic backgrounds contributed to the loss of actin-MDR transgene expression with inbreeding is not clear. The isolation of fertilized mouse embryos, microinjection of transgene DNA, and reimplantation into CR NIH or CD1 pseudopregnant foster mothers were carried out using standard methods.[21]

[17] P. T. Lomedico, N. Rosenthal, A. Efstratiadis, W. Gilbert, R. Kolodner, and R. Tizard, *Cell* **18,** 545 (1979).

[18] M. J. Low, R. E. Hammer, R. H. Goodman, J. F. Habener, R. D. Palmiter, and R. L. Brinster, *Cell* **41,** 211 (1985).

[19] C. Jhappan, C. Stahle, R. N. Harkins, N. Fausto, G. H. Smith, and G. T. Merlino, *Cell* **61,** 1137 (1990).

[20] M. Taketo, A. C. Schroeder, L. E. Mobraaten, K. B. Gunning, G. Hanten, R. R. Fox, T. H. Roderick, C. L. Stewart, F. Lilly, C. T. Hansen, and P. A. Overbeek, *Proc. Natl. Acad. Sci. U.S.A.* **88,** 2065 (1991).

[21] B. Hogan, R. Beddington, F. Costantini, and E. Lacy, *in* "Manipulating the Mouse Embryo," 2nd Ed., Cold Spring Harbor Laboratory, New York, 1994.

DNA Methods for Screening Founder Mice

The DNA methods we use to screen the founder generations of MDR transgenic mice are unique only in the choice of probes and conditions that result in transgene-specific signals. The actin founder mice are screened by Southern blot of tail DNA, prepared by standard methods,[21] digested with *Eco*RI, using the 1.3-kilobase (kb) human *MDR1* 5A cDNA probe (Fig. 1), which contains sequences highly homologous to the mouse *mdr* genes. With low stringency washing at 50°, both the mouse and human genes are detected; whereas at 65°, only the human *MDR1* transgene (3.1 kb) is detected. The results are that five transgene-positive founders are identified, and it is estimated, by including known amounts of standard DNAs on Southern blots, or on DNA slot blots, that the actin founder mouse with bone marrow and spleen transgene expression carries about three copies of the transgene.

Because the number of Eμ founder mice is large (55), they are initially screened via PCR (polymerase chain reaction) on a tail DNA template, using primers derived from the end of exon 16 and beginning of exon 23 of the human *MDR1* cDNA (Fig. 1).[22] These primers, which on amplification of transgenic tail DNA (1–2 μg DNA, annealing temperature of 53°, 30 cycles) give a product of 0.8 kb from a human *MDR1* cDNA template, are chosen because they would amplify a very large 18.6-kb product from homologous sequences in mouse DNA. This product does not interfere with the assay for transgene *MDR1* DNA sequences. However, a disadvantage of this primer pair is that it gives the same product with any *MDR1* plasmid template as with transgenic tail DNA, so contaminating *MDR1* cDNA can give false positives. With this in mind, primer pairs that generate a band specific to the transgene (e.g., one primer from the promoter and one from the gene body) may be ideal. Using the internal *MDR1* primer assay, including many independent negative controls, and counting even faint bands as positive, 14 transgene-positive mice are identified. To confirm that the transgene is present and intact, a Southern blot (capillary transfer) assay is used for a second round of screening. The probe is a 0.4-kb *Sac*I–*Sac*I human Pμ probe derived from the transgene (Fig. 1), and labeled with ^{32}P by random priming. Tail DNAs (15–30 μg) are digested with *Eco*RI, and the blots washed at high stringency. Transgene-positive mice give a 2.1-kb band (shown schematically in Fig. 1). Of the 14 PCR-positive founders, 11 are positive by Southern. Transgene copy numbers, estimated as above, ranged from 1 to 300, with the founder showing the best transgene expression, described later, having about 10 copies. In addition, for both

[22] C.-J. Chen, D. Clark, K. Ueda, I. Pastan, M. M. Gottesman, and I. B. Roninson, *J. Biol. Chem.* **265,** 506 (1990).

the actin and Eμ transgenic mouse lines described later, based on Southern blots with various restriction digestions and on pedigree analysis, the *MDR1* transgenes are integrated in tandem into a single site on a single autosome and transmitted through the germ line.

Methods for Identification of Homozygous Transgenic Mice

Having identified positive founder mice, one then ideally screens the founders by some sort of preliminary expression assay to select a single founder for more detailed analysis. Left with a single hemizygous F_0 founder of interest then, it is desirable to generate homozygotes, primarily because they are a rich resource for generating large numbers of hemizygotes for analysis, without screening, by backcrossing to normal mice. Secondarily, one is often interested in analyzing dosage effects by studying phenotypic differences between hemizygotes and homozygotes. To obtain homozygotes, one first generates an F_1 hemizygote by backcrossing the F_0 founder with the parent nontransgenic mouse strain. Here, for screening one can safely resort to a single rapid screening method, with good controls, since the founders have already been carefully identified. Conventional or unconventional[23] PCR methods, vacuum or pressure Southern blots (Stratagene, La Jolla, CA), or interphase FISH (fluorescence *in situ* hybridization; see later discussion) on peripheral blood (PB) cells, can all be completed in 1–4 days. Next, transgene hemizygotes are inbred to generate homozygotes, the progeny from which should be 25% homozygotes. The most reliable conventional method (backcross method) for identifying them is by first doing a careful semiquantitative Southern blot, perhaps with two probes, one for the transgene and one for normalization to a nonvarying gene. This eliminates the transgene-negative mice and provides a few candidates for homozygotes which have an apparently higher amount of transgene DNA. These candidates are then definitively tested by backcrossing and looking for litters with 100% transgene-positive pups (by Southern blot). The problem with this method is obviously that it is very slow, requiring 4–7 weeks, depending on whether sacrifice of the pups tested is desired or not, from the point of beginning the analysis of the offspring of the two hemizygotes, to the result. Ideally, it is desirable to establish a homozygous line of mice, so one needs to identify one homozygous male and female, which can add even more time to the process.

To speed up this process, we developed a rapid (2-day) and reliable protocol to assess transgene zygosity in the PB white cells of the Eμ transgenic mice by interphase two-color FISH (a "colored Southern blot

[23] K. Reue and S. Rehnmark, *BioTechniques* **17,** 252 (1994).

on a slide"). Because this method is very similar to one developed independently by Dinchuk et al.,[24] and because it is not directly relevant to the study of *MDR* genes, only the differences in the two methods are highlighted here. Briefly, on the first day, 100 μl of PB is collected from the periorbital sinus of a mouse in heparinized microhematocrit capillary tubes (Fisher, Pittsburgh, PA), the red cells are lysed, the white cells swelled in hypotonic conditions, and then lysed, and their nuclei fixed and dropped onto microscope slides. From this point on, now-standard methods for *in situ* cytogenetic hybridization are used to detect the transgene (red color) and a reference homozygous locus (green color). The transgene probe is prepared by nick translation of gel-isolated 6.4-kb Eμ-MDR transgene in the presence of digoxigenin-labeled dUTP; the control probe is prepared the same way with biotinylated dUTP and a whole cosmid template bearing an insert (about 30 kb) derived from mouse chromosome 17. After *in situ* hybridization of the probe mixture on the slides overnight, on the second day, the slides are washed, and then processed for detection with sandwich fluorochromes. Lastly, enough nuclei (typically 60–75) are visualized in a fluorescence microscope to identify 50 nuclei with two green control dots, and the number of red transgene dots in each is counted.

In contrast to the published report,[24] we carried out a large pilot trial of this technique using 10 each of Eμ-MDR homozygotes, hemizygotes, and FVB/N normal mice. In a blinded study, one investigator provided the 30 PB samples from Eμ mice of known zygosity (determined by the conventional backcross method), 10 at a time; two other investigators performed the analysis and counted the results independently, so that 1000 nuclei were counted for each of the three classes of mice. The results can be summarized as follows, expressed as the percentage of 1000 nuclei from each zygosity class having the indicated number of transgene spots: homozygotes, 2 spots (92.2%); hemizygotes, 1 spot (93.8%); and FVB/N normal mice (nullizygotes), 0 spots (100%).[25] Based on these results, this technique provides a viable and very rapid (2-day) method not only for identifying homozygotes, but also, because the background is so low, for routine screening of postfounder transgenic animals. In addition to the assumption of integration of tandem copies of transgene at a single locus, one critical factor for the general utility of this type of method is the total contiguous transgene insert size (monomer size multiplied by number of tandem copies), here 64 kb, which should be roughly greater than 20 kb for best results without additional amplification steps or imaging methods.

[24] J. E. Dinchuk, K. Kelley, and A. L. Boyle, *BioTechniques* **17,** 954 (1994).

[25] B. Wolf-Ledbetter, G. L. Evans, S. Pack, C. Jhappan, G. T. Merlino, I. Pastan, M. M. Gottesman, and D. H. Ledbetter, unpublished results (1994).

In addition to our method, and the previously cited published method,[24] two other variations of this method, which are useful for very large transgene inserts, have appeared in the literature.[26,27]

Analysis of MDR Transgenic Mice

RNA and Protein Methods for Analysis of MDR1 Transgene Expression

Ideally, one hopes that the transgenic mice of interest will express the transgene at a high level independent of the positions of transgene integration in the different founders. In reality this is not the case, and it is desirable to have a quick and simple assay to pick out the founder with the best expression in the tissue of interest. With a human *MDR1* transgene, this assay should ideally measure transgene protein levels or function at the cell surface. For the actin mice, expression is expected in many tissues, and no quick assay for transgene expression, short of a complete analysis of transgene RNA in transgenic tissues, is available. Total RNA is prepared from 18 different tissues using the standard guanidine isothiocyanate method,[28] and detected by slot blot using the *MDR1* 5A Southern probe (high stringency), and a human γ-actin probe (low stringency) for normalization. The results indicate significant expression in transgenic BM and spleen, at RNA levels similar to those in human KB-8-5 cells, which are 3- to 6-fold resistant to MDR drugs in colony assays *in vitro*.[4] Weaker expression is seen in skeletal muscle and ovary. Importantly, by varying the stringency of the RNA slot blot washes after low stringency hybridization with the *MDR1* probe, it is possible to estimate that there is about 20-fold more human *MDR1* RNA than RNA from homologous mouse *mdr* genes in the BM of the actin mice. The presence of transgene RNA is also confirmed by Northern blot and primer extension assays.

The expression of *MDR1* protein at the cell surface is confirmed in most BM cells by immunofluorescence with HPgp-specific monoclonal antibody MRK16[29] (a gift of Hoechst Japan Ltd, Kawagoe City, Saitama, Japan). In addition, later it was possible to show by differential white counts obtained with blood smears treated with Wright's stain (Accustain; Sigma), combined with total white counts, that upon (MDR) drug challenge of the actin mice (see later discussion), PB granulocytes, lymphocytes, and monocytes are resistant to the effects of the drugs seen in normal mice.[7] This indicated that

[26] R. R. Swiger, J. D. Tucker, and J. A. Heddle, *BioTechniques* **18,** 952 (1995).
[27] H. Nishino, J. F. Herath, R. B. Jenkins, and S. S. Sommer, *BioTechniques* **19,** 587 (1995).
[28] P. Chomczynski and N. Sacchi, *Anal. Biochem.* **162,** 156 (1987).
[29] H. Hamada and T. Tsuruo, *Proc. Natl. Acad. Sci. U.S.A.* **83,** 7785 (1986).

all of the major BM precursors expressed HPgp. As mentioned previously, approximately 4 years after the actin mice were made, after extensive inbreeding with C57Bl/6 X SJL F_1 mice, transgene expression was undetectable by RNA slot–blot, while transgene DNA was still present.[30] This silencing of transgene expression, which could have occurred by distal genetic, or epigenetic mechanisms, has not been further investigated.

The eleven $E\mu$ founder mice are initially screened for expression by indirect immunostaining of PB mononuclear cells (PBMC) with MRK16, and FACS analysis. PBMC are isolated by centrifugation through Ficoll. Approximately 200 μl of blood is collected as above, and layered onto 2 ml of calcium- and magnesium-free phosphate-buffered saline (PBS), on top of 5 ml of lymphocyte separation medium (LSM; Organon Teknika, Durham, NC) in a 15-ml screw-cap centrifuge tube (Sarstedt, Newton, NC). The tubes are centrifuged at 400g for 30 min at 4°, and the buffy coat on top of the LSM layer is removed with a 1-ml pipette to a 6-ml open-top FACS tube (Falcon). The PBMC are washed twice, by filling the tubes with PBS plus 0.1% (w/v) bovine serum albumin (BSA), covering the tubes with parafilm, inverting several times, and pelleting the cells at 400g for 7 min at 4°. After cell counting in a hemacytometer, 1×10^6 cells are transferred to each of two fresh FACS tubes per test sample. Volumes are adjusted to about 200 μl of PBS, and 5 μg of primary MRK16 antibody, or of an isotype-matched IgG_{2a} control antibody (Pharmingen) is added and incubated at 4° for 30 min. After washing the cells three times as described earlier, secondary goat anti-mouse IgG–FITC (fluorescein isothiocyanate) antibody (Jackson Immunoresearch), used at a 1 : 5000 dilution is added and incubated at 4° for 30 min. Cells are washed three times, and analyzed by FACS (Becton-Dickinson FACSort instrument with LYSIS II software). This analysis is typically carried out with control cell lines (KB-3-1, KB-8-5, and KB-V1) that express none, a low, and a high amount, respectively, of HPgp RNA[31] and surface protein. The results (not shown) indicate weak HPgp surface expression in PBMC from 4 out of 11 founders, with one (DM3) clearly higher than the other three.

Transgene RNA is measured in 19 tissues of F_1 hemizygotes derived from DM3 by backcrossing, via semiquantitative RNase protection analysis, which is generally more sensitive and specific than blotting methods. Mice are sacrificed by cervical dislocation, or with carbon dioxide, tissues dissected, and total RNA prepared as above for the actin mice, except for the addition of a step at the end for treatment with RQ1 RNase-

[30] M. Siegsmund, I. Aksentijevich, M. M. Gottesman, and I Pastan, unpublished results (1992).

[31] A. Fojo, K. Ueda, D. Slamon, D. Poplack, M. M. Gottesman, and I. Pastan, *Proc. Natl. Acad. Sci. U.S.A.* **84,** 265 (1987).

free DNase (Promega), as described.[19] The antisense riboprobe consists of a 153-nucleotide (nt) RNA transcript, which protected a 113-nt fragment (see Fig. 1) from human *MDR1* nucleotides 2671–2784 (corresponding to amino acids 890–927, between the predicted transmembrane domains 10 and 11). This riboprobe is developed by first subcloning from the Eμ-MDR transgene plasmid a 0.85-kb *Pvu*II–*Asp*718 *MDR1* DNA fragment containing the desired probe sequences into the riboprobe vector pSP72 (Promega). The resulting plasmid, containing *Bfa*I sites in the *MDR1* insert, and in the vector, is digested with *Bfa*I to release a 0.9-kb band containing the phage T7 promoter and the desired *MDR1* sequences. This band is isolated out of an agarose gel using GENECLEAN II (Bio 101) and used as the template for the synthesis of a uniformly radiolabeled riboprobe via run-off *in vitro* transcription using the Riboprobe Gemini System II (Promega) with [α-^{32}P]GTP (400 Ci/mmol; Amersham). The resulting 153-nt human *MDR1* riboprobe is hybridized overnight to 50 μg of tissue RNA, yeast tRNA, or RNA from human KB-8-5 cells at 55°, digested with (controls without) RNases A and T1 at 30° for up to 2 hr, treated with sodium dodecyl sulfate (SDS) and proteinase K, and then extracted with phenol/chloroform/isoamyl alcohol, all essentially as described in the Promega technical bulletin. The expected 113-nt product protected from digestion by test RNA is visualized by electrophoresis on 8 *M* urea, 4% polyacrylamide gels (30 × 40 cm, 0.4-mm thick), run with radiolabeled (fill-in) of pBR322/*Msp*I-digested markers (New England Biolabs, Beverly, MA).

The results (Fig. 2) demonstrate significant expression in the BM and spleen, as expected, at levels quite similar to that seen in human KB-8-5 cells in the same assay. Expression is also seen in the small intestine (presumably lymphocyte-rich Peyer's patches) and stomach. Weaker expression is reproducibly seen in the thymus, colon, and lymph nodes. Surprising, PB RNA is negative, for unknown reasons. The BM, spleen, lymph nodes, and PB (lowest) should contain the largest concentrations of B cells, and the thymus, spleen, and lymph nodes the largest concentrations of T cells. No expression is seen when the same analysis is carried out with normal FVB/N mice, or with F_1 hemizygotes derived from two of the other founders apparently positive in the initial FACS expression screen. Expression of the mouse *mdr* and human *MDR1* genes is not compared here, though this is possible, either with two probes (one mouse and one human) of similar specific activity but slightly different sizes, or perhaps with one probe that is protected to different extents by the two RNAs. The expression of a 4.4-kb transgene transcript is confirmed by Northern blot (not shown), but the signal is much weaker than seen in the RNase protection assay, owing to lower sensitivity. Taken together, these data suggest that the strongest Eμ-

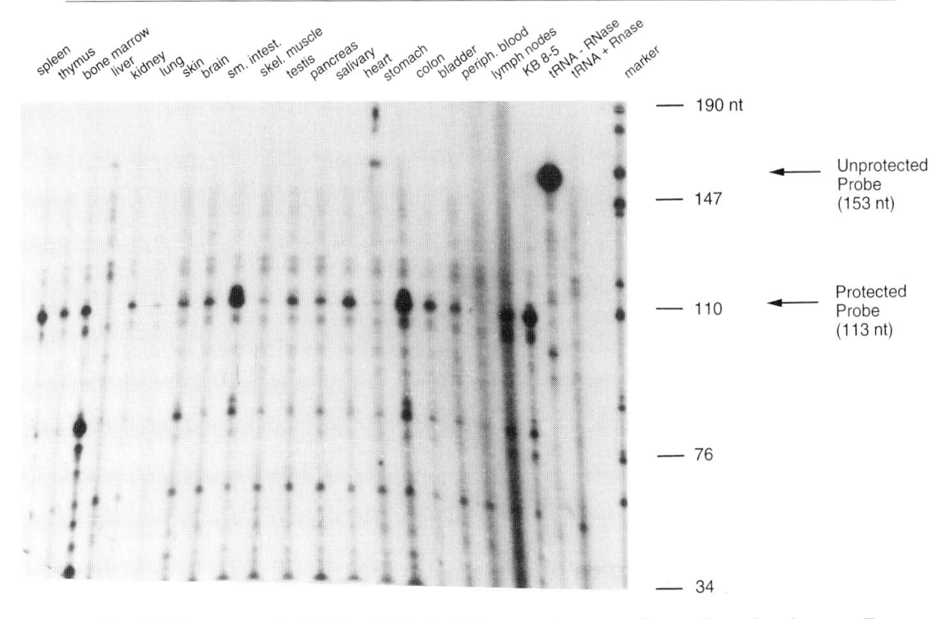

FIG. 2. Measurement of HPgp RNA in 19 transgenic mouse tissues from hemizygous Eμ-MDR mice, and in human KB-8-5 cells. Arrows indicate the position of migration of the 153-nt unprotected riboprobe (minus RNase), and of the smaller 113-nt riboprobe fragment protected from RNase degradation by total RNA from the various transgenic tissues, or from human KB-8-5 cells. Yeast transfer RNA (tRNA) serves as a negative control.

MDR transgene RNA expression is occurring in B lineage cells found in the BM and spleen, as expected.

In contrast to the RNA measurements, the results of attempts at detection of transgenic Eμ-MDR protein (HPgp) in these mice are less convincing and less reproducible, either owing to inefficient expression of preexisting mRNA, or just to the lower sensitivity and reproducibility of the protein assays, or both. Initially, we tried to measure surface HPgp on PBMC from several generations (F_1–F_4) of Eμ mice by direct, one-color FACS with MRK16-FITC as described for the initial expression assay above. Conjugation of MRK16 antibody to FITC for direct staining is carried out as described.[32] The results are quite variable, with some mice showing expression like that of the founder, some showing no expression at all, and some showing expression in the absence of transgene DNA. Attempts to demonstrate HPgp expression on the surface of B cells via two-color FACS with the CD45R/B220 pan-B cell antibody (Pharmingen),

[32] T. Licht, M. M. Gottesman, and I. Pastan, *Methods Enzymol.* **292,** [41], 1998 (this volume).

along with MRK16, even on BM mononuclear cells (BMMC) and spleen mononuclear cells (SMC) from homozygous Eμ-MDR mice, meet with partial success. Both direct and indirect immunofluorescence staining are tried with FACS, but the theoretically more sensitive indirect approach, using a mouse primary antibody (MRK16), is hindered by the presence of surface immunoglobulin on B cells, which is bound by the secondary antibody, giving rise to a large background of nonspecific binding. The results are that the direct MRK16-FITC approach, when using anti-FcγII/III receptor (CD32/16) antibody (Pharmingen) to block Fc receptors on leukocytes, gives less background but no convincing results for HPgp expression. Clearly, none of the FACS assays, as constituted with these antibodies and this low level of transgene expression, is reliable enough to reproducibly demonstrate HPgp expression.

Given the limited utility of the FACS assays in this context, a more biochemical approach can be taken, namely a Western blot on membrane protein derived from homozygote transgenic, and normal FVB/N BMMC and SMC, or from human KB-8-5 cells using either the human-specific anti-Pgp polyclonal antibody PEPG13,[33] or the pan-Pgp monoclonal antibody C219,[34] which should detect both mouse and human Pgps. Immune complexes are detected with either donkey anti-rabbit IgG (PEPG13), or goat anti-mouse IgG (C219) secondary antibodies, both coupled to horseradish peroxidase, and visualized with an enhanced chemiluminescence kit (ECL; all from Amersham). The results (not shown) with C219 antibody are uninformative, likely due again to interference by cross-reaction of the anti-mouse secondary antibody to abundant membrane-associated immunoglobulin. However, the results with PEPG13 antibody confirm the presence in the transgenic samples, but not the normal, of a low level of HPgp, again similar to KB-8-5 cells, with an apparent molecular mass of 120 kilodaltons (kDa), rather than the expected 140 kDa, both smaller than the 170-kDa band seen with KB-8-5 human cells. This reduction in the apparent molecular mass of human Pgp when expressed in mouse cells has also been seen in tissue culture systems.[35]

Measurement of Functional HPGP by Rhodamine Efflux in Lymphoid Tissues

To document the expression of functional HPgp in B lymphoid cells in the Eμ mice, verapamil-sensitive efflux of rhodamine 123 is assayed

[33] E. P. Bruggemann, V. Chaudhary, M. M. Gottesman, and I. Pastan, *BioTechniques* **10,** 202 (1991).

[34] N. Kartner, D. Evernden-Porelle, G. Bradley, and V. Ling, *Nature* **316,** 820 (1985).

[35] S. V. Ambudkar, unpublished results (1995).

in MC from hematologic tissues. Typically no more than 4 mice (<3 weeks of age is best for thymic MC yield) are analyzed per day. The protocol described here is a modification of another described in this volume.[32] To begin, PB (300–600 μl) is collected as above from normal and homozygote transgenic mice, the mice are sacrificed, and the femur and tibia (for BM), spleen, and thymus removed. The long bones are placed in a 10-cm tissue culture dish with 5 ml of PBS, and BM cells are flushed from the bones by forcing the PBS through the bones using a 25-gauge needle and a 3-ml disposable syringe; clumps are broken by several passes through the needle. The spleen is similarly placed in a dish with PBS and teased apart using sharp forceps, and then clumps are broken using a 19-gauge needle and a 6-ml syringe. BMMC, SMC, and PBMC are then prepared from the BM and spleen cell suspensions, and the whole blood. The cells are washed twice and then counted, as described above for the initial screening of the Eμ founders. Typically, about 1×10^7 SMC, 2×10^6 BMMC cells, 1×10^6 thymic MC, and 8×10^5 PBMC are recovered in this way. If desired, cardiac puncture instead of periorbital bleeding can be used to increase the volume of PB collected and PBMC yield. Depending on the yield, $1–5 \times 10^5$ cells of each type are then transferred to each of two fresh FACS tubes (one for verapamil, one without; see later discussion) and the tubes filled with Iscove's modified Dulbecco's medium (IMDM) with glucose, and without phenol red (Life Technologies), plus 5% (v/v) fetal bovine serum (Hyclone, Logan, UT) previously filtered through a large-pore Whatman (Clifton, NJ) prefilter, and then a 0.2-μm disposable filter, to reduce background fluorescence), plus 0.4 μg/ml rhodamine 123 (see next section for preparation of stock). The cells are then incubated in the dark at 37° for 15 min to allow cellular loading of rhodamine, and then pelleted as for washes, being very careful not to aspirate any of the cells along with the supernatant. The cells are resuspended (tubes filled to the top) in rhodamine-free IMDM (plus serum, as before), plus or minus 25 μM RS-verapamil (see later discussion for stock preparation), by briefly holding a small section of parafilm on top of one tube at a time, and inverting three times. The cells are then placed at 37° for 2.5 hr to allow efflux of rhodamine from the cells, pelleted, leaving about 300 μl of supernatant after aspiration, and resuspended by scraping the tube across a plastic pronged rack. They are immediately analyzed for green rhodamine fluorescence (FL1 channel) by FACS in gated lymphoid cells, first identified by forward- and side-scatter properties on dot plots.

The results of these assays clearly indicate that there is no detectable Pgp function in thymic MC from normal or transgenic mice (not shown). In contrast, low-level rhodamine efflux activity is seen in a major fraction of transgenic but not normal lymphoid-gated BMMC (Fig. 3), at least half of which are $B220^+$, IgD^- pre-B cells (see later discussion). A 2- to 3-fold reduction in rhodamine fluorescence is seen in the absence, compared to the presence of verapamil. The rhodamine efflux assay does not differentiate between mouse and human Pgps, but taken together with other data presented in this chapter, these data provide evidence for the presence of functional HPgp, albeit at a low level, in transgenic BM. Intermediate results are seen in normal lymphoid-gated SMC and PBMC, where Pgp activity is detected in a minor fraction (20–35%) of cells, consistent with published reports of Pgp activity in subsets of normal murine $CD8^+$ T cells, macrophages, and NK cells.[36] In the transgenic counterpart tissues, there are in relative terms reproducibly 10–20% more $rhodamine^{dull}$ cells, presumably due to activity in splenic and PB B cells superimposed on this significant normal T-cell Pgp background. Lastly, in about 50% of the transgenic mice, and none of the normal mice treated three times with bisantrene, which selects for *MDR1* expressing cells, over the course of 50 days (see later discussion; two independent experiments), a significant increase in rhodamine efflux activity, providing at least a suggestion of selection of HPgp-positive cells, is seen in SMC and PBMC, but not BMMC. The rhodamine efflux assay does not include Pgp reversing agents more potent than verapamil (e.g., PSC 833), which may increase sensitivity, but even as constituted with verapamil, it is very sensitive and reproducible.

MDR Drugs for Use in Mice: Sources, Preparation, and Dose

With the exception of rhodamine 123 and verapamil, all drug solutions are made fresh on the day of use and discarded after use. The cytotoxic MDR drugs (except rhodamine 123) are listed in order of decreasing resistance of NIH 3T3 cells expressing wild-type HPgp, relative to unmodified NIH 3T3 cells.[37] The suggested sublethal drug doses, which vary with different strains of mice, and routes (IP, intraperitoneal; IV, intravenous), are those producing significant drops (50–75%) in white blood counts in normal mice after 5–7 days. IP and IV injections were typically given in volumes less than 400 μl. Reversing agents were typically given 15 min before the administration of cytotoxic drugs.

MDR Drugs

Bisantrene: Reagent grade (Lederle Labs, Pearl River, NY), solution of 1–10 mg/ml in sterile PBS, pH 6.6; heat in 65° water for 5–10 min to completely dissolve drug [dose 30–60 mg/kg, IV bolus]

[36] A. A. Neyfakh, A. S. Serpinskaya, A. V. Chervonsky, S. G. Apasov, and A. R. Kazorov, *Exp. Cell Res.* **185,** 496 (1989).

[37] C. O. Cardarelli, M. M. Gottesman, and I. Pastan, unpublished results (1994).

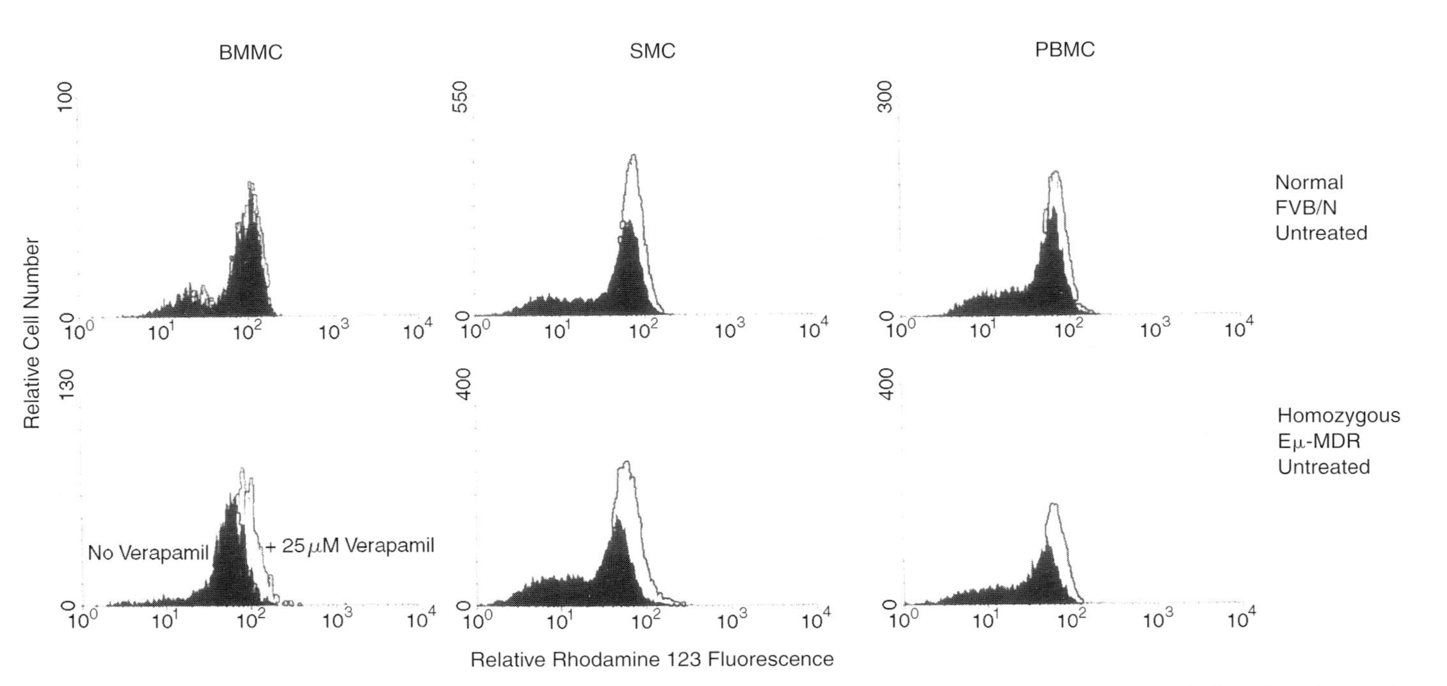

FIG. 3. Measurement of Pgp function in lymphoid-gated BMMC, SMC, and PBMC from homozygous Eμ-MDR transgenic (*lower*), and normal FVB/N (*upper*) mice, in the absence of *in vivo* drug treatment, by verapamil-sensitive rhodamine 123 efflux. Histograms showing cellular rhodamine fluorescence in the absence (shaded black) or presence (unshaded) of 25 μM RS-verapamil are shown.

Taxol: Reagent grade [Developmental Therapeutics Branch, National Cancer Institute (DTB/NCI); Sigma]; clinical grade drug (best) may be obtained from Bristol-Myers Squibb (Syracuse, NY); solution of 0.5–3 mg/ml. Method 1: dissolve weighed drug in 1/10 of final volume ethanol, add 5/10 final volume of dimethyl sulfoxide (DMSO), add remaining volume as sterile water. Method 2: dissolve weighed drug in 1 ml of 1:1 cremophor EL (Sigma):ethanol, sonicate at room temperature until solution becomes clear, dilute 10-fold with sterile PBS (5% cremophor, 5% ethanol final)[38] (1–25 mg/kg IP or IV).

Actinomycin D, Vincristine, Vinblastine, Doxorubicin, and Daunomycin: Reagent grade (Sigma), solution of 0.5 mg/ml (act D) or 1 mg/ml (rest) in sterile PBS (actD: 1–2 mg/kg, IP; rest: 5–10 mg/kg, IP)

Etoposide: Clinical grade (Bristol-Myers Squibb, Syracuse, NY); dissolve in 1/10 final volume ethanol, add 3/10 final volume DMSO, add remaining volume as sterile water (10 mg/kg IP)

Rhodamine 123: Reagent grade (Sigma); solution of 1 mg/ml in ethanol (stock for dye efflux studies *ex vivo*; not for use in animals). This solution is light sensitive, and should be stored in a dark bottle at 4°.

MDR Reversal Agents

R-Verapamil, *RS*-Verapamil, Quinine, and Quinidine: Clinical grade (*R*-verapamil) (BASF Bioresearch Corp, Cambridge, MA); reagent grade (Sigma), dissolve weighed drug in sterile PBS to give 1–5 mg/ml solution. Heat as necessary, up to 100°, for high concentrations (1–50 mg/kg, quinidine 25–150 mg/kg; IP). For rhodamine efflux studies *in vitro,* a 20 m*M* stock solution of *RS*-verapamil was prepared in DMSO. This solution is light sensitive, and frozen aliquots should be stored in the dark.

PSC 833: Clinical grade (Sandoz, East Hanover, NJ); for oral administration, dissolve weighed drug in 2/10 final volume ethanol, add 4/10 final volume Labrafil (Sandoz), and rest of volume as corn oil to give 5–10 mg/ml solution (25–100 mg/kg *per os*).[39]

Cyclosporin A: Clinical grade (Sandoz AG, Basel, Switzerland; or Pfizer Central Research, Groton, CT); solution of 1 mg/ml in sterile PBS (5 mg/kg, IP).

Non-MDR Drugs

Cisplatin: Reagent grade (Sigma); solution of 1 mg/ml in sterile PBS, stir for up to 2 hr at room temperature to completely dissolve drug (1–10 mg/kg, IP).

[38] Y. Sugimoto and T. Tsuruo, unpublished results (1994).

[39] T. Watanabe, H. Tsuge, T. Oh-Hara, M. Naito, and T. Tsuruo, *Acta Oncologica* **34,** 235 (1995).

Methotrexate: Reagent grade (Sigma); solution of 10 mg/ml, prepared as with cisplatin (50–100 mg/kg, IP).

5-Fluorouracil: Reagent grade (Fluka, Ronkonkoma, NY); solution of 10 mg/ml in PBS, heat at 55° for about 2 hr to completely dissolve drug (100–300 mg/kg, IP).

Drug Resistance Measurements in White Blood Cells

Using both the actin and Eμ mice, HPgp-mediated drug resistance was measured *in vivo* by resistance to MDR drug-induced leukopenia, as described in this section. Because the actin mice apparently expressed HPgp in all hematopoietic cell lineages, leukopenia is measured by total white blood cell count (TWBCC). In contrast, HPgp expression is more restricted in the Eμ mice, occurring primarily in the B lymphoid lineage, so with those mice leukopenia is assayed by measurement of B cells in the blood. Experiments with the actin mice include at least three animals per group (normal and hemizygous MDR transgenic), and at a minimum TWBCCs are determined both on day 0, before the administration of drug and/or reversing agent, and on day 5. The doses of MDR drugs (see previous section) resulting in significant drops (50–75%) of normal TWBCCs are determined empirically. For each TWBCC, 50–75 μl of blood is collected as above and placed on a microscope slide. Then 10 μl of blood is added to 190 μl of 3% acetic acid in a microfuge tube for red blood cell lysis, and refractile PB white cells are counted in a hemacytometer. An example is shown in Fig. 4A, where it is apparent that 8 mg/kg daunomycin causes a 70% decrease in the TWBCC of normal mice by day 5, but does not significantly reduce the TWBCC of actin-MDR transgenic mice. However, when 150 mg/kg *R*-verapamil, a known *MDR1* inhibitor which itself is not toxic in this model (Fig. 4B), is given 15 min prior to daunomycin, resistance to leukopenia is not observed. The precision in these experiments is very high, with less than 10% variation in TWBCCs between animals and between experiments. The results are that this resistance to leukopenia is seen with many of the classical MDR drugs,[4,7] and had a magnitude, based on maximum tolerated doses for taxol, compared to normal mice, of approximately 10-fold.[8] In addition, this resistance can be transferred to drug-sensitive mice by bone marrow transplant,[11] and can be reversed by many known, and some unknown, specific MDR inhibitors.[7,9] Were it not for the loss of transgene (HPgp) expression seen in this mice over time (see earlier discussion), this would have been a very useful and convenient system for screening MDR reversal agents.

The assay used with the Eμ mice is quite similar, except the data readout is peripheral B cell number instead of TWBCC. We wanted to maximize the sensitivity of the assay, so we chose to use the MDR chemotherapeutic

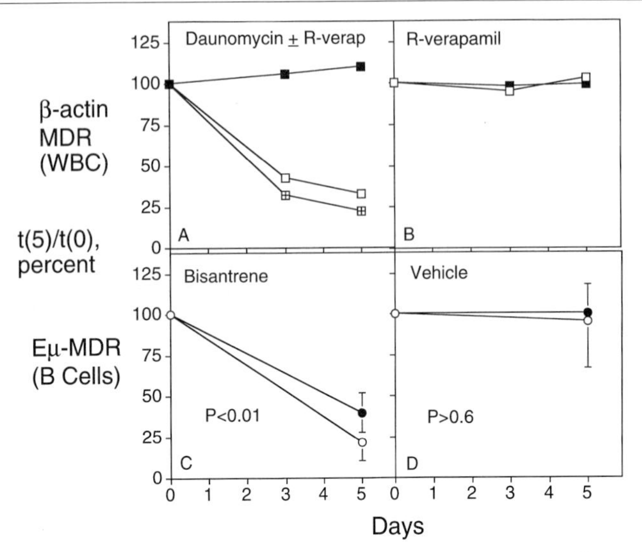

FIG. 4. Resistance to MDR drug-induced leukopenia in β-actin–MDR and Eμ-MDR transgenic mice. (A) and (B). On day 0, hemizygous β-actin–MDR mice (filled or hatched boxes), and control normal mice (open boxes) were injected intraperitoneally with 8 mg/kg daunomycin alone, 8 mg/kg daunomycin and 150 mg/kg *R*-verapamil (hatched boxes), or 150 mg/kg *R*-verapamil alone (B). Total white blood cell counts (WBC) are expressed as average counts at day 3 or 5 for each group, relative to day 0. (C) and (D). On day 0, homozygous Eμ-MDR mice (filled circles) and control normal mice (open circles) were injected intravenously with 45 mg/kg of bisantrene hydrochloride (C), or with vehicle (D). The results from one representative experiment are expressed as the average B cell count at day 5 for each animal, relative to day 0. Error bars represent standard deviations, and *P* values were determined using the unpaired two-tailed Student's *t* test for independent samples, assuming unknown but equal variances in the two groups of mice. P values <0.05 were considered to be statistically significant. [(A) and (B) adapted from G. H. Mickisch, G. T. Merlino, H. Galski, M. M. Gottesman, and I. Pastan, *Proc. Natl. Acad. Sci. U.S.A.* **88,** 547 (1991).]

drug bisantrene, a topoisomerase II poison to which cells expressing HPgp have recently been shown to be highly resistant,[40] and for which the dose-limiting toxicity for chemotherapy is BM suppression. Furthermore, tissue culture cells expressing HPgp are more resistant to this drug than most other MDR drugs,[37] and normal C57Bl/6 mice, transplanted with BM cells previously transduced *in vitro* with an *MDR1* retroviral vector, were characterized by resistance to bisantrene-induced leukope-

[40] X. P. Zhang, M. K. Ritke, J. C. Yalowich, M. L. Slovak, J. Pelkey Ho, K. I. Collins, T. Annable, R. J. Arceci, F. E. Durr, and L. M. Greenberger, *Oncol. Res.* **6,** 291 (1994).

nia.[41] Importantly, in normal control mice but not the *MDR1* transplant recipients, bisantrene treatment was especially toxic for B cells and macrophages, but not for T cells and granulocytes.

Having decided to use bisantrene to evaluate drug resistance in the Eμ mice, a dose–response analysis measuring the effect of intravenous drug on TWBCC and peripheral blood B cell count (PBBCC) in normal FVB/N mice is carried out. PBBCCs are obtained by combining total white counts with FACS analysis of blood cells stained with CD45R/B220 antibody. The entire procedure is as follows: bisantrene is prepared as described in the previous section and administered IV on day 0 at doses of 10–60 mg/kg. On days 0 (before drug administration), 3, 5, and 7, 75–150 μl of PB, collected as above, is placed in a microfuge tube (1.5 ml) containing just a few dry crystals of EDTA (ethylenediaminetetraacetic acid, disodium salt, dihydrate) to prevent clotting, and to minimize dilution artifacts. Then 10 μl of blood is saved for the determination of TWBCC, as above except that red blood cells are lysed by 20-fold dilution in Becton-Dickinson FACS lysing solution (BDFLS), which gives cleaner cell preparations than 3% acetic acid. Absolute TWBCCs average about 8000/μl in normal untreated mice. The balance of the blood (60–120 μl; 0.5–1.0 $\times$ 10^6 white cells) is split in half and transferred to FACS tubes for determination of the percentage of B lineage cells in the white blood cell population, using procedures similar to those described above for the initial analysis of Eμ-MDR transgene expression. The PB cells are spun, washed once, and left with about 200 μl of supernatant. Then 0.2 μg of CD45R/B220-PE (R-phycoerythrin), or IgG_{2a}-PE isotype matched control antibody (both from Pharmingen) is added to the two tubes, the cells are resuspended and mixed with the antibody by pipetting up and down, and the tubes are incubated at 4° in the dark for >30 min. The cells are then washed once, red cells lysed by the addition of 2 ml BDFLS by slow pipetting up and down, and the released hemoglobin removed by washing the white cells three times, or until the supernatant is clear. The cells are then either immediately analyzed by FACS, or kept in the dark for 1–2 hr at room temperature until FACS analysis can be initiated. The $B220^+$ B cell population is identified by histogram, and its relative size in 10,000 total events determined. The B-cell percentage determined in this way ranges from 6 to 15% in untreated normal or transgenic FVB/N mice, compared to 50–60% for T cells in the same mice, determined with CD3-FITC antibody. The B-cell percentages are combined with the TWBCCs to give absolute B-cell numbers.

[41] I. Aksentijevich, C. O. Cardarelli, I. Pastan, and M. M. Gottesman, unpublished results (1995).

The results for the dose–response analysis for intravenous bisantrene in normal mice (not shown) are analyzed by comparing the B-cell counts in the same animal on day 0 and day 5, for many drug treated animals (5 per dose group). They indicated that a dose of 45 mg/kg bisantrene at the 5 day nadir does not significantly reduce the TWBCCs (30–40% reduction), but at the same time B cells are down by 80%. The 60 mg/kg dose causes an 80% drop in TWBCC and is lethal in most mice by 3 days. These conditions then are used to evaluate drug resistance in the Eμ mice compared to normal mice, with four groups of 8–10 animals each: transgenic homozygote/drug; transgenic homozygote/vehicle; normal/drug; and normal/vehicle. The results from a representative experiment are shown in Figs. 4C and 4D. Reproducible, but very low level, resistance to bisantrene is seen in the Eμ-MDR mice, with transgenic B cells dropping by about 60% compared to 80% for normals. The vehicle does not cause a significant drop in the B cells of any mice. This difference is statistically significant ($P < 0.01$) when the data are analyzed on a per-animal basis, but is not so when the data are analyzed on a per-group basis, pointing to the small magnitude of resistance, the variation in baseline PBBCCs between each animal, which is sometimes significant, and/or the difficulty in accurately measuring TWBCCs with <10% variation. This analysis does not include the effects *in vivo* of MDR reversal agents on this weak B-cell resistance to bisantrene.

In summary, there is no significant difference in TWBCCs between the population of bisantrene-treated normal and transgenic mice. In untreated normal FVB/N mice the predominant PB white cell type is the $CD3^+$ T cell (50–60%). While the exact B lineage cell type targeted by bisantrene is not identified, using FACS analysis of BMMC and SMC stained with antibodies binding a B lineage marker (B220-PE), and surface immunoglobulin (IgD-FITC;Igh-C^a haplotype; Pharmingen), it is possible to differentiate effects on pre-B ($B220^+$, IgD^-) versus mature B ($B220^+$, IgD^+) cells. BM and spleen cell suspensions are prepared as described in the rhodamine efflux section. The results are that in vehicle-treated normal or transgenic BM, the primary site of B-cell production, pre-B cells are predominant (6:1 pre-B:mature B), while in the spleen mature B cells were more abundant (3:1). On day 2, the percentage of pre-B cells in the total BM B cell population of treated normal mice drops by 60% compared to vehicle controls; mature BM B cells increased by three-fold. In contrast, for homozygous transgenic mice the changes are smaller but in the same direction: pre-B cells drops by 30%, and mature B increases by nearly two-fold. Similar, but smaller changes are seen in the spleen on day 2, and at later times (days 5 and 9), there is no significant difference between treated normal and transgenic BM or spleen. These data suggest that bisantrene is inhibiting the proliferation of at least one type of cell (i.e., the pre-B cell) expressing functional HPgp (see section on rhodamine efflux). The possibility that the bisantrene-resistant B-cell phenotype is selectable is investigated by treating transgenic and normal mice three times with bisantrene (45 mg/kg), with 21 days between bolus injections. (B-cell counts reduced by bisantrene under these conditions recovered to normal levels in this time span.) The results are that, compared to the 5-day assay, the magnitude of B-cell resistance to bisantrene is not significantly increased after round 3 of drug treatment; the transgenic mice still exhibits a B-cell drop of 50–60%. Other MDR drugs, including daunomycin and taxol, were tested less exhaustively in transgenic hemizygotes in similar assays, with no greater resistance seen.

Perspectives and Conclusion

It is unfortunate that neither of the two lines of MDR transgenic mice described here are currently useful for screening MDR reversal agents. The actin mice no longer express HPgp, and the Eμ mice do not express functional protein at high enough levels, and/or express it in sufficiently permissive cell type(s), to provide a simple and reproducible quantitative assay. Notably, though total BM from both lines of mice apparently expressed HPgp RNA at levels similar to drug-resistant human KB-8-5 cells (independent semiquantitative assays), suggesting in fact a higher RNA level per cell in the Eμ mice since expression there was more lineage restricted, the Eμ mice expressed a much lower level of drug resistance. These three sources (actin mice, Eμ mice, and KB-8-5 cells) of HPgp-positive cells were not directly and quantitatively compared for surface protein level and rhodamine efflux activity, which are more meaningful assays. The generation of any future lines of MDR transgenic mice should ideally await the identification of better dominant transcriptional control elements which are resistant to positional silencing, to facilitate the identification of and extend the useful lifetime of lines of interest. These lines might express HPgp minigenes bearing endogenous *MDR1* introns, and be made in, or crossed into the murine *mdr* knockout genetic background, so as to subtract the effects of mouse *mdr* genes. The construction of MDR transgenic mice, and especially of MDR tumor models, for example, by making double transgenic mice, is still in principle potentially very useful, but is in practice a slow process fraught with difficulties.

Acknowledgments

This work was supported by a postdoctoral fellowship from the American Cancer Society (PF-3990) and by the National Cancer Institute. Special thanks goes to Thomas Licht for extremely generous assistance with hematological procedures, expression and rhodamine efflux analysis by FACS, and for the initial screening of the Eμ founder mice. I also thank Chammelli Jhappan for microinjections and help with transgenic mouse procedures; Glenn Merlino for help with dissection of mouse tissues and RNase protection analysis; Betty Wolf-Ledbetter and Svetlana Pack for very generous help with interphase FISH procedures; Emmett Schmidt and William LaRochelle for providing plasmid DNAs; Lee Greenberger and Lederle Labs for the gift of bisantrene; Ivan Aksentijevich for help with the preparation of bisantrene solutions; Yoshikazu Sugimoto for helpful discussions and for taxol preparation method 2; Carol Cardarelli for maintaining cell lines and for *in vitro* drug resistance profiles; Joyce Sharrar, Ann Schombert, and Jennie Evans for administrative assistance; and Ira Pastan and Michael M. Gottesman for intellectual guidance and support.

[44] Cloning, Transfer, and Characterization of Multidrug Resistance Protein

By CAROLINE E. GRANT, GABU BHARDWAJ, SUSAN P. C. COLE, and ROGER G. DEELEY

Introduction

Experimentally, multidrug resistance is most often studied in tumor cell lines that have been selected in only a single natural product cytotoxic agent but which display cross-resistance to structurally unrelated drugs.[1,2] The *in vitro* multidrug resistance phenotype differs from that seen in patients in that it normally does not include resistance to alkylating agents, antimetabolites, or platinum-containing drugs. Typically, the spectrum of drugs to which multidrug-resistant cell lines are cross-resistant includes several classes of natural product drugs and their synthetic congeners, including anthracyclines, *Vinca* alkaloids, and epipodophyllotoxins. It is characteristic of the multidrug resistance phenotype that it includes drugs with different intracellular targets and mechanisms of cytotoxicity.

In vitro, overexpression of either of two proteins, the 170-kDa P-glycoprotein (Pgp) or the 190-kDa multidrug resistance protein (MRP), has been shown to confer the multidrug resistance phenotype.[3,4] Both proteins are integral membrane proteins belonging to the ATP-binding cassette (ABC) superfamily of transport proteins. Cell lines that overexpress either Pgp or

[1] S. P. C. Cole, *in* "Lung Cancer: Principles and Practice" (H. I. Pass, J. B. Mitchell, D. H. Johnson, and A. T. Turrisi, eds.), p. 169. Lippincott-Raven, Philadelphia, 1996.

[2] M. M. Gottesman and I. Pastan, *Annu. Rev. Biochem.* **62,** 385 (1993).

[3] R. G. Deeley and S. P. C. Cole, *in* "Molecular Genetics of Drug Resistance, Modern Genetics," Vol. 3 (J. D. Hayes and C. R. Wolf, eds.), p. 247. Harwood Academic Press, Langhorne, PA, 1997.

[4] K.-V. Chin, I. Pastan, and M. M. Gottesman, *Adv. Cancer Res.* **60,** 157 (1993).

0076-6879/98 $25.00

MRP have been isolated using similar drug selection protocols. In addition, overexpression of either protein confers resistance to a similar, but not identical, profile of drugs. However, at the level of primary structure, the similarity between Pgp and MRP is very low with only 15% amino acid identity. Although much is now known about Pgp and MRP, details of the mechanisms by which these two proteins mediate drug resistance and what determines their substrate specificities are just beginning to be understood. To this end, stably-transfected cell populations overexpressing Pgp or MRP have been an invaluable resource. This chapter describes the initial cloning of MRP and the production of two stably transfected cell populations that have been used in its characterization.

Background

Prior to 1992, Pgp was the only protein known to be capable of conferring a multidrug resistance phenotype when overexpressed in previously drug-sensitive cells. In addition, several drug-resistant human tumors and many multidrug-resistant cell lines had been shown to overexpress Pgp, providing evidence that its overexpression may be an important factor in clinical multidrug resistance.[5] However, many tumors and cell lines that clearly displayed multidrug resistance were found not to overexpress Pgp which suggested that mechanisms other than overexpression of Pgp could produce a similar phenotype. One of the most extensively characterized of the non-Pgp multidrug-resistant cell lines, H69AR, was derived from the small cell lung cancer cell line NCI-H69 by selection in doxorubicin.[6,7] Many changes in gene expression had been detected in H69AR cells, including reduced topoisomerase II, and altered levels of GSH-associated drug detoxification enzymes.[7] However, these changes failed to account fully for the range of drugs to which H69AR cells were resistant. In addition, prolonged growth of H69AR cells in the absence of drug resulted in a cell line, H69PR, that had markedly reduced resistance to the multidrug resistance-associated drugs.[8] This concommitant loss of resistance to a range of drugs suggested that the multidrug-resistant phenotype of H69AR cells resulted from changes in the expression of one or a small number of genes, rather than a complex multifactorial process. It also suggested that the proteins responsible for the multidrug resistance phenotype and their corresponding mRNAs may be overexpressed in H69AR cells relative to the H69 and H69PR cell lines.

[5] I. Pastan and M. M. Gottesman, *Annu. Rev. Med.* **42,** 277 (1991).
[6] S. E. L. Mirski, J. H. Gerlach, and S. P. C. Cole, *Cancer Res.* **47,** 2594 (1987).
[7] S. P. C. Cole, *Can. J. Physiol. Pharmacol.* **70,** 313 (1992).
[8] S. P. C. Cole, M. J. Pinkoski, G. Bhardwaj, and R. G. Deeley, *Br. J. Cancer* **65,** 498 (1992).

Strategy

A number of approaches can be used to clone mRNAs on the basis of differences in expression levels between cell lines or tumor types. These include the polymerase chain reaction (PCR)-based differential display, subtractive hybridization, and differential screening.[9–13] In differential display, which is the most recently developed of these methods, subsets of mRNAs are reverse transcribed from their 3′ termini and then amplified by PCR using a set of degenerate anchored oligo(dT) primers and an arbitrary decamer. The products generated with the same primer sets from two or more mRNA samples are then compared using high-resolution polyacrylamide gel electrophoresis. Those products that appear to be amplified differentially from the different mRNA populations can be recovered, and cloned or used directly to probe a cDNA library. In subtractive hybridization, a cDNA prepared from one sample of mRNA is depleted of sequences common to a second sample of mRNA by controlled hybridization under conditions of RNA excess, followed by subsequent removal of RNA/DNA duplexes. There are many variations of this technique in which the subtracted cDNA can be used either directly as a probe to screen a complete cDNA library, or cloned to produce a subtracted library for random screening. In principal, differential hybridization is the simplest of the approaches. In this case, a cDNA library prepared from the cell or tissue of interest is screened both with a total cDNA probe synthesized from the mRNA population used to generate the library and with a similar probe prepared from the mRNA population used for comparison. Comparison of the results of the two screenings serves to identify cDNA clones derived from mRNAs whose relative levels of expression differ between the two mRNA populations.

The relative advantages and limitations of each approach have been discussed in depth elsewhere.[14,15] In general, both differential display and subtractive hybridization require a higher level of proficiency in the isolation and handling of RNA samples than does differential hybridization. Differ-

[9] P. Liang and A. B. Pardee, *Science* **257,** 967 (1992).

[10] P. Liang, L. Averboukh, and A. B. Pardee, *Nucl. Acids Res.* **21,** 3269 (1993).

[11] C. R. King, D. S. Udell, and R. G. Deeley, *J. Biol. Chem.* **254,** 6781 (1979).

[12] G. Scherer, J. Telford, C. Baldari, and V. Pirrotta, *Dev. Biol.* **86,** 438 (1981).

[13] P. Masiakowski, R. Breathnach, J. Bloch, F. Gannon, A. Krust, and P. Chambon, *Nucl. Acids Res.* **10,** 7895 (1982).

[14] J. Sambrook, E. F. Fritsch, and T. Maniatis, eds., "Molecular Cloning: A Laboratory Manual," 2nd Ed. Cold Spring Harbor Laboratory, New York, 1989.

[15] F. M. Ausubel, R. Brent, R. E. Kingston, D. D. Moore, J. G. Seidmon, J. A. Smith, and K. Struhl, eds., "Current Protocols in Molecular Biology." Wiley Interscience, New York, 1987.

ential display is particularly sensitive to the quality of RNA template since small variations in the efficiency of cDNA synthesis can produce an abundance of false-positive signals. Both differential display and subtractive hybridization techniques work best when there are only a small number of mRNA species that differ in relative abundance between the cell lines or tissues of interest. They are particularly efficient techniques when there are qualitative differences between two mRNA populations. In differential hybridization, the cDNA probe represents all mRNA species in the cell line or tissue used, and each species should be represented in the probe in proportion to its relative abundance in the total mRNA population. Consequently, clones derived from rare mRNA species will be essentially undetectable, as will small differences in the relative concentration of a particular mRNA species in the two populations. Because of these limitations, differential hybridization is most reliable when the mRNAs of interest are moderately or highly abundant with at least a five-fold difference in expression between mRNA populations.

H69AR cells display moderate to high levels of resistance to a range of drugs when compared to the parental H69 cell line.[7] This suggested that differences in the relative levels of multidrug resistance-associated mRNAs might be detectable by a simple differential screening approach. However, to increase the probability of detecting such mRNAs, and to reduce the likelihood of selecting false positives, we incorporated a prehybridization step with poly(A)-enriched H69 RNA during preparation of the probes.[16] This step provides some of the advantages of using a subtracted cDNA probe for screening and was introduced to suppress the signal from cDNAs corresponding to RNAs present at similar concentrations in both resistant and sensitive cells. Total H69AR and H69 cDNA preparations were hybridized with a modest molar excess of poly(A)-enriched RNA from H69 cells. The conditions used were designed to result in cDNAs from most of the highly or moderately abundant mRNAs common to both resistant and sensitive cells forming RNA/cDNA hybrids. Although no physical separation of free and duplexed cDNAs was carried out, the latter are effectively depleted from the probe when it is used subsequently for screening the library. This has the additional advantage of diminishing the strong background signal from cDNA clones derived from residual rRNA, which can constitute a significant component of libraries generated by random priming of poly(A)-enriched RNA preparations. In theory, the prehybridization step might be expected to obviate the need for duplicate screening of the library; however, in practice, the duplicate screening provides a useful control that facilitates identification of true positives.

[16] R. Wiskocil, P. Goldman, and R. G. Deeley, *J. Biol. Chem.* **256,** 9662 (1981).

Screening Methods

cDNA Library

There are many commercially available kits for making λgt11 or similar cDNA bacteriophage libraries. Therefore, only the points relevant to screening of the library are discussed here.

Although it has been customary to isolate poly(A)-enriched RNA from cells and tissues using oligo(dT)-cellulose chromatography of total RNA,[17] a number of less labor-intensive kits are now commercially available. We have found that methods using biotinylated oligo(dT) and streptavidin-linked magnetic beads yield preparations with relatively low levels of rRNA contamination, especially if the amount of total RNA used approaches the maximum amount suggested by the manufacturer. The number of phage plaques to be screened is influenced by the amount of ribosomal cDNA contamination of the library. This can be assessed by probing several thousand plaques from the library with a cloned fragment of rRNA. Generally, 5×10^5 to 2×10^6 phage particles of a cDNA library are screened depending on the level of contamination by ribosomal cDNAs and the abundance of the target cDNA.

Library Probes

The quality of poly(A)-enriched RNA preparations used for synthesis of cDNA can be asssessed both visually by ethidium bromide staining of RNA separated on a formaldehyde-agarose denaturing gel, and by hybridization of the corresponding Northern blot to identify a mRNA species of known size and abundance.[14,15] We also routinely measure the total and trichloroacetic acid-precipitable radioactivities[14] in the cDNA synthesis reactions to ensure similar incorporation of radioactivity into probes used for subtractive and differential hybridization.

For synthesis of each cDNA, 120 μCi of [α^{32}P]dATP (3000 Ci/mmol, 10 μCi/μl) is dried in a siliconized microcentrifuge tube. In separate tubes, 2 μg of H69AR or H69 poly(A)-enriched RNA in 10 μl of RNase-free water is denatured by heating at 65° for 5 min then quickly chilled on ice. The RNA solutions are used to resuspend the dried radiolabeled dATP, and a reaction mix is added so that the final volume of 25 μl contains reverse transcriptase (RT) buffer (50 m*M* Tris-HCl, pH 8.3, at 42°, 8 m*M* $MgCl_2$, 50 m*M* KCl), 20 U of placental RNase inhibitor (Gibco-BRL, Burlington, Ontario, Canada), 4 m*M* dithiothreitol (DTT), 2.5 μg of synthetic random nonanucleotide primers, 800 μ*M* each of dTTP, dCTP, and

[17] H. Aviv and P. Leder, *Proc. Natl. Acad. Sci. U.S.A.* **69,** 1408 (1972).

dGTP, 4.8 μM dATP, and 20 U of avian myeloblastosis virus (AMV) reverse transcriptase (Life Sciences, St. Petersburg, FL). The reaction is incubated at 42° for 1 hr. The solutions are adjusted to 0.1 *M* NaOH, 10 m*M* EDTA, and 0.1% sodium dodecyl sulfate (SDS), and incubated at 70° for 1 hr to remove RNA. The radiolabeled single-stranded cDNA is then separated from the unincorporated dNTPs by chromatography on, or centrifugation through, a small Sephadex G-50 column. Normally we recover 50–100× 10^6 cpm or 15–30% of the available radiolabeled nucleotide from each reaction. For both subtracted and differential hybridization probes, 2× 10^6 cpm of labeled cDNA is used per 137-mm filter. The radiolabeled single-stranded cDNA is used directly in differential hybridization.

For "subtracted" probes, we find it advantageous to eliminate any risk of RNase contamination of the cDNA immediately prior to the hybridization step. The appropriate amount of cDNA is precipitated in 0.2 volumes of 5 *M* ammonium acetate and 2.5 volumes of ethanol. Following centrifugation, the cDNA is resuspended in 100 μl of RNase-free water and a 10-fold excess of H69 poly(A)-enriched RNA (20 μg), proteinase K buffer, 0.5 *M* NaCl, and 15 μg of proteinase K added. The cDNA/RNA mixture is incubated at 37° for 1 hr, phenol/chloroform extracted twice, chloroform extracted twice, and precipitated with ethanol. For hybridization, the pelleted cDNA/RNA mix is resuspended in 5 μl of hybridization buffer (10 m*M* *N*-2-hydroxyethylpiperazine-*N'*-2-ethanesulfonic acid [HEPES], 0.6 *M* NaCl, 2 m*M* EDTA; pH 7.0). The mixture is drawn into a siliconized 20-μl glass capillary tube and both ends of the tube are flame-sealed. The sealed tubes are placed in boiling water for 2 min and then transferred to a waterbath at 68° for 24 hr. The "subtracted" probes are used immediately without separating the cDNA/RNA hybrids from the remaining single-stranded cDNA.

Screening

Detailed descriptions for the screening of phage libraries are available elsewhere.[14,15,18] Although cDNA libraries may be screened at densities ranging from 25,000 to 50,000 plaque-forming units (pfu/150-mm petri dish, we suggest much lower densities (10,000 to 15,000/dish) for subtractive and differential hybridization. Phage seeded at lower densities produces larger plaques, resulting in stronger, less ambiguous autoradiographic signals. It is essential that equivalent amounts of phage DNA be transferred to each of the duplicate filters. This is done by exposing the filters to the phage

[18] T. Yamamoto and Y. Kadowaki, *Methods Enzymol.* **254,** 169 (1995).

plaques for different time periods. The time required can be determined empirically using a cloned phage stock and hybridization of the filters with the same probe. In our hands, a 1-min contact with the first filter and a 5-min contact with the second filter gives approximately equal amounts of DNA on each. An additional useful precaution during the screening is to strip and rehybridize the duplicate filters with the alternate probes.

In screening for MRP, the filters are prehybridized for 4–5 hr at 42° with 5 filters/bag and 25 ml of hybridization solution [50% formamide, 5× SSPE (1× = 0.15 *M* NaCl, 1 m*M* NaH_2PO4, 1 m*M* EDTA, pH 7.4), 0.5% SDS, 4× Denhardt's solution [50× = 1% each of bovine serum albumin (BSA), polyvinylpyrrolidone, and Ficoll], and 100 μg/ml denatured, sheared herring testis DNA]. Fresh solution is used for hybridizing overnight at 42°. The filters are washed twice for 10 min at room temperature with 2× SSC (1× = 0.15 *M* NaCl, 15 m*M* sodium citrate; pH 7) and 0.1% SDS, three times for 30 min at 52° in large volumes of 0.1× SSC, 0.1% SDS, and exposed to X-ray film at −70° with an intensifying screen for one to several days.

Potential positive clones are rescreened with fresh probe at a density of 1000 to 3000 pfu/plate depending on the density of phage used in the primary screen. Plaque purified clones that still appear positive can be conveniently compared in the final screening by spotting 2 μl from a plate lysate of each clone in a grid pattern on a bacterial lawn several hours old. Phage DNA from such a gridded array was transferred to duplicate filters for hybridization to either H69AR or H69 cDNA probes.

Analysis of Positive Clones

Clones that gave strong differential signals were immediately used as probes to analyze Northern blots of poly(A)-enriched RNA isolated from H69, H69AR, and H69PR cells. The expression levels of a mRNA recognized by one clone, mrp-10, correlated well with the drug resistance levels.[19] The mrp-10 cDNA hybridized to a mRNA of approximately 6.5 kb that was overexpressed approximately 100-fold and 5-fold in H69AR cells and H69PR cells, respectively, when compared with H69 cells (Fig. 1). This mRNA was also overexpressed in a multidrug-resistant HeLa cell line that does not overexpress Pgp, suggesting that it was a good candidate for a multidrug resistance-associated mRNA.

[19] S. P. C. Cole, G. Bhardwaj, J. H. Gerlach, J. E. Mackie, C. E. Grant, K. C. Almquist, A. J. Stewart, E. U. Kurz, A. M. V. Duncan, and R. G. Deeley, *Science* **258,** 1650 (1992).

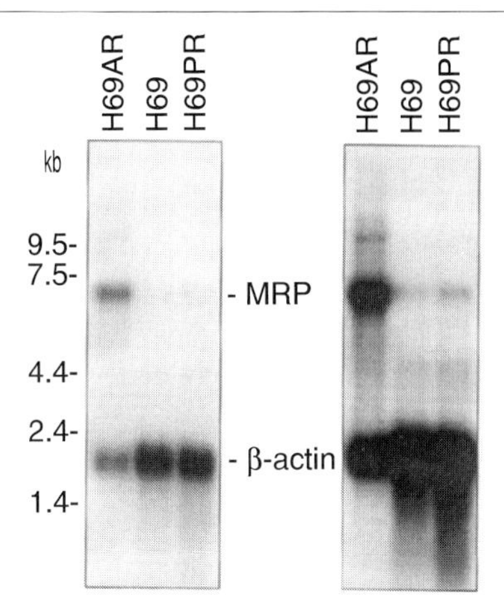

FIG. 1. To assess the relative abundance of MRP in H69, H69AR, and H69PR cells, a Northern blot of poly(A)-enriched RNA from each cell line was hybridized with a 1.8-kb cDNA fragment of mrp-10 (mrp-10.2). Identical results were obtained when the remaining portion of mrp-10 cDNA (mrp-10.1) was used as a probe. Variations in the amount of RNA loaded in each lane were assessed by probing the blot with a human β-actin cDNA. The autoradiograph shown on the right-hand side is a prolonged exposure of that shown on the left.

Isolation of a Full-Length cDNA

The 2.8-kb mrp-10 cDNA yielded two fragments (10.1 and 10.2) when digested with the restriction enzyme *Eco*RI. Both fragments were subcloned into a pGEM3Zf$^+$ plasmid vector (Stratagene, La Jolla, CA) and sequenced by the dideoxy chain termination method.[20] The combined sequences revealed an extended open reading frame (ORF) plus a short stretch of what appeared to be a 3′ untranslated region (UTR). A comparison of the putative amino acid sequence encoded by the ORF with the translated GenBank and SwissProt databases revealed conserved motifs characteristic of the ABC transporter superfamily. To obtain the rest of the coding sequence, the λgt11 cDNA library was rescreened using 5′ proximal fragments or oligonucleotide probes. Of the clones isolated, four contained small deletions of 100–300 bp within the putative coding region of the MRP mRNA. To determine which was the correct sequence for the MRP mRNA, we used PCR to compare the size of these MRP cDNA fragments to those generated from random-primed H69 or H69AR cDNAs (RT-PCR). For all the primer sets tested, a number of MRP cDNA fragments are amplified. Since total cellular poly(A)-enriched RNA is used to make the λgt11 library,

[20] F. Sanger, S. Nicklen, and A. R. Coulson, *Proc. Natl. Acad. Sci. U.S.A.* **74,** 5463 (1977).

it is probable that the variant cDNA fragments and clones are derived from alternatively spliced nuclear transcripts. Therefore, the correctly spliced transcript is assumed to be represented by the most abundant cDNA fragment amplified by RT-PCR. The size of each cloned fragment was subsequently compared to that generated from an equivalent region of the MRP mRNA by RT-PCR. Ultimately, four cDNA clones of between 1.6 and 4 kb and with a minimum of 400 bp of overlapping sequence were used to compile most of the MRP mRNA sequence. An estimate of the length of the 5′ UTR was obtained from an additional seven clones complementary to the 5′ proximal portion of the MRP mRNA. Three of these contained 5′ UTRs of 175 bases, whereas the remaining four contained 193 or 196 bases. Thus, a mRNA sequence was compiled of 5011 nucleotides containing 196 nucleotides of 5′ UTR, 219 nucleotides of 3′ UTR, and an ORF encoding a protein of 1531 amino acids.

Functional Tests

Southern analysis indicates that the MRP gene is amplified 40- to 50-fold in H69AR cells when compared with H69 or H69PR cells.[19,21] This raised the possibility that the MRP gene did not confer the drug-resistant phenotype but was simply coamplified with the gene that did. In addition, it was possible that MRP overexpression was only one of many components necessary to generate the phenotype. Therefore, it was important to determine if overexpression of MRP was both necessary and sufficient to confer resistance. To do this, a full-length MRP cDNA was constructed and cloned into a eukaryotic expression vector, which was then transfected into a drug-sensitive cell line. The transfected cell lines were analyzed for MRP mRNA and protein expression and, subsequently, the pharmacological phenotype resulting from MRP overexpression was determined.

Vector Construction and Transfection

A number of cDNA fragments and PCR products were used to assemble the full-length MRP cDNA in a plasmid vector. The first construct contains 86 nucleotides of MRP mRNA 5′ UTR and is designated MRP1. Since the MRP mRNA 5′ UTR is very GC-rich and could affect the translational efficiency of vector-encoded MRP mRNA, a second construct eliminating this sequence and adding a consensus Kozak sequence[22] at the translational initiation site was also made (MRP2). When the fidelity of the sequences

[21] M. L. Slovak, J. P. Ho, G. Bhardwaj, E. U. Kurz, R. G. Deeley, and S. P. C. Cole, *Cancer Res.* **53,** 3221 (1993).

[22] M. Kozak, *Nucl. Acids Res.* **15,** 8125 (1987).

had been confirmed using both restriction mapping and sequencing of the PCR-generated segments, the MRP cassettes were transferred into expression vectors suitable for use in human cell lines. We have used two vectors, both of which express MRP under the control of the enhancer/promoter sequence from the immediate early gene of the human cytomegalovirus (CMV). One vector, pRc/CMV (Invitrogen, La Jolla, CA), randomly integrates in the genome of recipient cells and contains a selectable marker for neomycin (G418) resistance. The other vector, pCEBV7,[23] replicates episomally in human cells and confers resistance to hygromycin B. With both vectors, the use of chemotherapeutic drugs to select an MRP-expressing population is avoided to eliminate the risk of introducing secondary changes that might affect the drug resistance characteristics of the cells. MRP has now been introduced into several different cell lines using a variety of transfection techniques. In general, the episomal vector is found to provide somewhat higher and more uniform levels of expression. However, some care has to be exercised in propagating the transfected cell population since recombination of the episomal vector, resulting in the loss of MRP expression, has occurred in some instances. Contrary to our expectations, there did not appear to be any difference in the level of MRP mRNA or protein expressed from the MRP1 and MRP2 constructs.

For all vectors, supercoiled DNA is isolated using Qiagen-tips (Qiagen Inc., Chatsworth, CA). Both the pRc/CMV vector and the vector containing the MRP construct (pRc/CMV-MRP1) were transfected into HeLa cells using a standard calcium phosphate transfection protocol.[14] The episomal vector, pCEBV7, and the corresponding MRP expression vector, pCEBV7-MRP1, were transfected into HeLa cells with lipofectin (Gibco-BRL) as directed by the manufacturer. After 48 hr, either G418 (Sigma, St. Louis, MO; 200 μg/ml final concentration) or hygromycin B (Sigma, 100 μg/ml final concentration) is added to the growth medium to select for transfected cells. The concentration of G418 or hygromycin B required for optimal selection is very cell-type dependent. We have found empirically that a concentration that kills 100% of nontransfected cells over 5–7 days works well for selection purposes. The cells are typically exposed to concentrations ranging from 50 to 400 μg/ml. Usually there are a sufficient number of cells to begin characterization 2–3 weeks following transfection. In our analyses, we used cell populations rather than cloned cell lines.

Analysis of Transfected Populations

The MRP-transfected populations are analyzed by Northern and Western blotting[14,15] to determine the MRP mRNA and protein expression

[23] G. M. Wilson and R. G. Deeley, *Plasmid* **33,** 198 (1995).

levels. The MRP mRNA produced from the expression vectors is designed to be approximately 1 kb smaller than the endogenous MRP mRNA by deleting portions of the 5′ and 3′ UTRs so that vector-encoded and endogenous mRNAs can be easily distinguished. MRP protein expression levels are determined using total cell lysates and membrane-enriched fractions.[24–26] The MRP protein is detected on Western blots by an affinity-purified, rabbit polyclonal antibody raised against a synthetic peptide (12 amino acids) corresponding to a unique portion of the predicted MRP amino acid sequence[27] or with monoclonal antibodies generated from mice immunized with H69AR cell membranes.[26]

Chemosensitivity Testing

The relative drug resistance of the transfected cell populations is tested using a colorimetric assay for cell survival.[28] After cells have been treated with different concentrations of cytotoxic drugs for a period of time, they are exposed to the tetrazolium salt 3-(4,5-dimethylthiazol-2-yl)-2,5-diphenyltetrazolium bromide (MTT). The mitochondrial dehydrogenase present in viable cells reduces the MTT to form a blue formazan product, which is measured spectrophotometrically. For HeLa cells, 10,000 cells are plated in each well of a 96-well tissue culture plate in 100 μl of growth medium. The cells are allowed to grow for 24 hr, then 100 μl of growth medium containing two times the required concentration of drug is added to each well. The drug concentrations required for the dose–response curve are determined empirically for each drug and cell type but usually span three orders of magnitude. Each drug concentration is tested in quadruplicate. After 72 hr of drug exposure, 100 μl of medium is removed, 25 μl of a 2 mg/ml MTT solution is added to each well, and the plate is incubated for 3 hr at 37°. Then 2-propanol : 1*N* HCl (24 : 1) solution (100 μl) is added to the wells, the contents mixed, and the absorbance measured at 570 nm using a mictrotiter plate reader. The absorbance measured for the drug-treated cells is expressed as a percentage of the absorbance measured for the control cells that received no drug. For statistical purposes, each drug should be tested in at least three independent experiments. Resistance factors are expressed as the ratio of the IC_{50} of the cell line transfected

[24] C. E. Grant, G. Valdimarsson, D. R. Hipfner, K. C. Almquist, S. P. C. Cole, and R. G. Deeley, *Cancer Res.* **54,** 357 (1994).

[25] S. P. C. Cole, K. E. Sparks, K. Fraser, D. W. Loe, C. E. Grant, G. M. Wilson, and R. G. Deeley, *Cancer Res.* **54,** 5902 (1994).

[26] D. R. Hipfner, S. D. Gauldie, R. G. Deeley, and S. P. C. Cole, *Cancer Res.* **54,** 5788 (1994).

[27] K. C. Almquist, D. W. Loe, D. R. Hipfner, J. E. Mackie, S. P. C. Cole, and R. G. Deeley, *Cancer Res.* **55,** 102 (1995).

[28] S. P. C. Cole, *Cancer Chemother. Pharmacol.* **26,** 250 (1990).

TABLE I
RESISTANCE OF MRP-TRANSFECTED HELA CELLS TO CHEMOTHERAPEUTIC DRUGS

Drug	IC_{50} (μM)[a] C1	IC_{50} (μM)[a] T5	Resistance factor[b]
Doxorubicin	0.28 ± 0.10 (n = 15)	1.58 ± 0.90[c] (n = 15)	5.6
Daunorubicin	0.09 ± 0.05 (n = 3)	0.46 ± 0.32[c] (n = 3)	5.1
MX2[d]	3.53 ± 2.44 (n = 3)	5.83 ± 4.25 (n = 3)	1.7
Mitoxantrone	0.51 ± 0.25 (n = 3)	0.60 ± 0.28 (n = 3)	1.2
Vincristine	6.14 ± 1.21 nM (n = 9)	51.33 ± 20.74 nM[c] (n = 9)	8.4
Vinblastine	0.76 nM (0.90, 0.62)	3.3 nM (5.2, 1.4)	4.3
VP-16	4.6 (4, 5.2)	54 (75, 32)	11.6
Taxol	3.6 ng/ml (3.9, 3.2)	7.1 ng/ml (8.0, 6.1)	2.0
Cisplatin	1.70 ± 0.75 (n = 3)	1.60 ± 0.30 (n = 3)	0.9
Carboplatin	48 ± 28 (n = 6)	54 ± 18 (n = 6)	1.1

[a] The IC_{50} of each drug was determined using the 3-(4,5-dimethylthiazol-2-yl)-2,5-diphenyltetrazolium bromide assay and is expressed as μM concentration unless otherwise indicated.

[b] The resistance factors were calculated as the ratio of the IC_{50} of HeLa cells transfected with the pRc/CMV-MRP1 (T5) expression vector to the IC_{50} of the HeLa cells transfected with the corresponding pRc/CMV (C1) vector alone.

[c] Values obtained with the resistant cell line, T5, are significantly different from those obtained with the corresponding sensitive cell line, C1 ($P < 0.05$).

[d] 3′-Deamino-3′-morpholino-13-deoxo-10-hydroxycarminomycin.

with the MRP expression vector for a given drug relative to that of the cell line transfected with the vector alone. Data obtained for one series of experiments are shown in Table I, and demonstrate that HeLa cells transfected with the integrating MRP expression vector, pRc/CMV-MRP1, are resistant to multiple chemotherapeutic drugs.

Drug Accumulation and Efflux

Drug accumulation and efflux in intact cells are usually measured using either a radiolabeled drug or, in the case of the anthracyclines, by measuring cellular fluorescence. The choice of drug is often influenced by the pattern of resistance of the cells and its availability in a radiolabeled form. Accumulation and efflux of radiolabeled vincristine (VCR) in MRP-transfected HeLa cells are measured as follows.[29]

The transfected HeLa cells at a concentration of 5×10^6 cells/ml are incubated with 1 μM [^{3}H]vincristine (specific activity, 4.5 Ci/mmol; 1 μCi/ml; Amersham, Oakville, Ontario) in growth medium supplemented with 5 mM HEPES buffer at 37°. To measure drug accumulation, aliquots of the cell suspension are removed at intervals for up to 2 hr. An aliquot is taken at time zero to measure nonspecific cell-associated radioactivity. Aliquots are diluted in ice-cold 1% bovine serum albumin (BSA) in phosphate-buffered saline (PBS) or ice-cold PBS to stop drug accumulation. Following centrifugation, the cell pellets are washed twice and then solubilized in 1% SDS. Cell-associated radioactivity is determined by liquid scintillation counting in an appropriate scintillation fluid. Results are expressed as picomoles cell-associated VCR/10^6 cells after subtraction of the zero time point versus time. Under these conditions, steady levels of VCR accumulation in the MRP-transfected HeLa cells were approximately 45% less than in cells transfected with the vector alone.[25]

To measure drug efflux, cells must first be loaded with drug to steady-state levels of accumulation, washed free of the drug, resuspended in drug-free medium, and the amount of drug remaining associated with the cells monitored over time at 37°. Steady-state concentrations of VCR are reached in the transfected HeLa cells by 60 min.[25] Consequently, to measure drug efflux, the cells are loaded with 1 μM [^{3}H]VCR for 60 min at 37°, washed in ice-cold medium, then incubated in prewarmed (37°) drug-free medium for up to 2 hr. Aliquots of the cell suspension are removed at various times and processed to determine the [^{3}H]VCR that remains associated with the cells as described earlier for drug accumulation. Data are expressed as percent of radioactivity remaining versus time. Investigation of the influence of MRP overexpression on drug efflux has yielded variable results. The variation observed may be attributable to cell type, level of MRP expression, or its subcellular distribution. Our MRP transfectants consistently show a modest but significant increase in the rate of [^{3}H]VCR efflux.[25]

The energy dependence of drug transport in intact cells can be demonstrated by comparing drug accumulation and efflux in cells depleted of ATP. This can be accomplished by preincubation of cells for 20 min in glucose-free medium supplemented with 10 mM deoxyglucose (Sigma) and 100 nM rotenone (Sigma). Under these conditions, intracellular ATP levels

[29] S. P. C. Cole, E. R. Chanda, F. P. Dicke, J. H. Gerlach, and S. E. L. Mirski, *Cancer Res.* **51,** 3345 (1991).

in the transfected cells were reduced more than 95% as measured by a luciferase-based somatic cell assay kit (Sigma).[25] VCR accumulation is measured as described earlier. Control cells are preincubated in medium containing 10 m*M* glucose. Depletion of ATP restored levels of VCR accumulation in the MRP-transfected cells to those found in the vector-transfected cells.[25]

[45] Transport Function and Substrate Specificity of Multidrug Resistance Protein

By DIETRICH KEPPLER, GABRIELE JEDLITSCHKY, and INKA LEIER

Introduction

The multidrug resistance protein (MRP1) has been cloned and sequenced from multidrug-resistant human lung cancer cells and identified as an integral membrane glycoprotein of about 190 kDa belonging to the superfamily of ATP-binding cassette (ABC) transporters.[1] MRP1 is expressed in most tissues and cell types and may be overexpressed in tumor cells.[1–6] Primary active, unidirectional, ATP-dependent transport of amphiphilic anions, particularly of conjugates of lipophilic substances with glutathione, glucuronate, or sulfate, has been recognized as the function of MRP1.[7–11] An isoform of MRP1 with a related sequence and a similar

[1] S. P. C. Cole, G. Bhardwaj, J. H. Gerlach, J. E. Mackie, C. R. Grant, K. C. Almquist, A. J. Stewart, E. U. Kurz, A. M. Duncan, and R. G. Deeley, *Science* **258,** 1650 (1992).

[2] G. J. R. Zaman, C. H. M. Versantvoort, J. J. M. Smit, E. W. H. M. Eijdems, M. de Haas, A. J. Smith, H. J. Broxterman, N. H. Mulder, E. G. E. de Vries, F. Baas, and P. Borst, *Cancer Res.* **53,** 1747 (1993).

[3] N. Krishnamachary and M. S. Center, *Cancer Res.* **53,** 3658 (1993).

[4] G. D. Kruh, K. T. Gaughan, A. Godwin, and A. Chan, *J. Natl. Cancer Inst.* **87,** 1256 (1995).

[5] L. Pulaski, G. Jedlitschky, I. Leier, U. Buchholz, and D. Keppler, *Eur. J. Biochem.* **241,** 644 (1996).

[6] M. J. Flens, G. J. R. Zaman, P. van der Valk, M. A. Izquierdo, A. B. Schroeijers, G. L. Scheffer, P. van der Groep, M. de Haas, C. J. L. M. Meijer, and R. J. Scheper, *Am. J. Pathol.* **148,** 1237 (1996).

[7] G. Jedlitschky, I. Leier, U. Buchholz, M. Center, and D. Keppler, *Cancer Res.* **54,** 4833 (1994).

[8] I. Leier, G. Jedlitschky, U. Buchholz, S. P. C. Cole, R. G. Deeley, and D. Keppler, *J. Biol. Chem.* **269,** 27807 (1994).

[9] M. Müller, C. Meijer, G. J. R. Zaman, P. Borst, R. J. Scheper, N. H. Mulder, E. G. E. de Vries, and P. L. M. Jansen, *Proc. Natl. Acad. Sci. U.S.A.* **91,** 13033 (1994).

[10] G. Jedlitschky, I. Leier, U. Buchholz, K. Barnouin, G. Kurz, and D. Keppler, *Cancer Res.* **56,** 988 (1996).

0076-6879/98 $25.00

function has been cloned from human and rat and localized predominantly to the hepatocyte canalicular membrane.[12,13] This isoform has been termed MRP2, or canalicular MRP (cMRP),[12] or canalicular multispecific organic anion transporter (cMOAT).[13] Canalicular (apical) MRP (i.e., MRP2) is deficient in human Dubin–Johnson syndrome[14] and in two mutant rat strains lacking the ATP-dependent transport of anionic conjugates across the hepatocyte canalicular membrane.[12,13] From a functional point of view, MRP1 and the isoform MRP2 may be termed conjugate export pumps with a broad substrate specificity.[7–12]

Elucidation of the function of MRP1 was closely linked to the characterization of the membrane proteins mediating the ATP-dependent transport of the endogenous glutathione *S*-conjugate leukotriene C_4 (LTC_4).[15,16] LTC_4 has a high affinity for MRP1, with a K_m value of 97 nM,[8] and its photolabile conjugated triene structure has enabled direct photoaffinity labeling of MRP1.[7,8,16,17] The identification of LTC_4 and other cysteinyl leukotrienes as MRP1 substrates suggested the use of leukotriene structural analogs as inhibitors of MRP1-mediated transport.[7,8,16] A considerable number of leukotriene (particularly LTD_4) receptor antagonists, developed for the treatment of human asthma,[18–20] have been recognized as potent inhibitors of ATP-dependent transport mediated by MRP1[7,8,10,11,21] and MRP2.[12] Accordingly, some leukotriene receptor antagonists are useful for modulating the MRP1-mediated multidrug resistance of tumor cell lines.[22]

---

[11] I. Leier, G. Jedlitschky, U. Buchholz, M. Center, S. P. C. Cole, R. G. Deeley, and D. Keppler, *Biochem. J.* **314,** 433 (1996).

[12] M. Büchler, J. König, M. Brom, J. Kartenbeck, H. Spring, T. Horie, and D. Keppler, *J. Biol. Chem.* **271,** 15091 (1996).

[13] C. C. Paulusma, P. J. Bosma, G. J. Zaman, C. T. Bakker, G. L. Scheffer, R. J. Scheper, P. Borst, and R. P. Oude Elferink, *Science* **271,** 1126 (1996).

[14] J. Kartenbeck, U. Leuschner, R. Mayer, and D. Keppler, *Hepatology* **23,** 1061 (1996).

[15] D. Keppler, *Rev. Physiol. Biochem. Pharmacol.* **121,** 1 (1992).

[16] I. Leier, G. Jedlitschky, U. Buchholz, and D. Keppler, *Eur. J. Biochem.* **220,** 599 (1994).

[17] D. W. Loe, K. C. Almquist, R. G. Deeley, and S. P. C. Cole, *J. Biol. Chem.* **271,** 9675 (1996).

[18] J. Augstein, J. B. Farmer, T. B. Lee, P. Sheard, and M. L. Tattersall, *Nature New Biol.* **245,** 215 (1973).

[19] T. R. Jones, R. Zamboni, M. Belley, E. Champion, L. Charette, A. W. Ford-Hutchinson, R. Frenette, J. Y. Gauthier, S. Leger, P. Masson, S. McFarlane, H. Piechuta, J. Rokach, H. Williams, R. M. Young, R. N. DeHaven, and S. S. Pong, *Can. J. Physiol. Pharmacol.* **67,** 17 (1989).

[20] N. J. Cuthbert, S. R. Tudhope, P. J. Gardiner, T. S. Abram, A. M. Thompson, R. J. Maxey, and M. A. Jennings, *Ann. N.Y. Acad. Sci.* **629,** 402 (1991).

[21] T. Schaub, T. Ishikawa, and D. Keppler, *FEBS Lett.* **279,** 83 (1991).

[22] V. Gekeler, W. Ise, K. H. Sanders, W.-R. Ulrich, and J. Beck, *Biochem. Biophys. Res. Commun.* **208,** 345 (1995).

The substrate specificity of MRP1 and additional MRP isoforms should be defined in inside-out-oriented membrane vesicles where substrate concentrations can be adjusted and metabolism or complex formation of the substrate, as it may occur in intact cells, can be controlled. Moreover, the ATP dependency of MRP1-mediated transport cannot be analyzed in intact cells where a selective modulation of the ATP concentration is impossible because of its rapid equilibrium with other nucleotides. We describe here the determination of the kinetic properties, substrate specificity, and inhibitors of MRP1-mediated transport in membrane vesicles from MRP1-overexpressing drug-selected cells,[7] cells transfected with a MRP1 expression vector,[8,10] or membranes rich in the apical isoform of MRP1 (i.e., MRP2).[12]

Experimental Procedures

Preparation of Plasma Membrane Vesicles

Selection of the cell line or tissue type influences not only the technique for membrane vesicle preparation but also the resulting percentage of inside-out-oriented plasma membrane vesicles. A sufficient amount of inside-out-oriented vesicles is essential, since only this fraction, with the ATP-binding domains oriented to the outer surface, mediates ATP-dependent transport of a labeled substrate into the vesicle. Some cell lines and membrane preparations may contain MRP1[1] together with the apical isoform MRP2 (cMRP/cMOAT).[12,13] Because of the similar transport function of the MRP isoforms it may be necessary to discriminate between them by immunoblotting or immunofluorescence microscopy.[12]

Method. This procedure is based on Refs. 7, 8, 10, 16, and 21

1. Harvest cells (about 3×10^9) from the cell culture by centrifugation (1200*g*, 10 min, 4°) and wash twice in ice-cold phosphate-buffered saline (PBS: 0.15 *M* NaCl, 50 m*M* KH_2PO_4, pH 7.4).
2. Hypotonic lysis of the cell pellet (about 5 ml) is induced by 40-fold dilution with hypotonic buffer (0.5 m*M* sodium phosphate, pH 7.0, 0.1 m*M* EDTA) supplemented with proteinase inhibitors [0.1 m*M* phenylmethylsulfonyl fluoride (PMSF), 2.8 μM E64 [*trans*-epoxysuccinyl-L-leucylamido(4-guanidino)butane], 1 μM leupeptin, and 0.3 μM aprotinin]. After gentle stirring on ice for 1.5 hr, the cell lysate is centrifuged at 100,000*g* for 40 min at 4°.
3. Resuspend the pellet after centrifugation in 20 ml of hypotonic buffer and homogenize with a Potter–Elvehjem homogenizer (500 rpm, 2 strokes/min, 30 strokes, 4°). Dilute the homogenate with incubation buffer (0.25 *M* sucrose, 10 m*M* Tris-HCl, pH 7.4) and centrifuge for

10 min at 12,000g at 4°. Store the resulting postnuclear supernatant on ice and resuspend the corresponding pellet in 20 ml of incubation buffer supplemented with proteinase inhibitors. Homogenize again and centrifuge as described earlier. Combine both postnuclear supernatants and centrifuge at 100,000g for 40 min at 4°. Resuspend the pellet in 20 ml of incubation buffer and homogenize manually by 50 strokes with a tightly fitting Dounce B (glass/glass) homogenizer (Fisher Scientific, Pittsburgh, PA) on ice.

4. Dilute homogenate by addition of 10 ml of incubation buffer and layer this crude membrane fraction on top of a 38% (w/v) sucrose solution in 5 mM HEPES–KOH, pH 7.4. Centrifuge at 280,000g for 2 hr at 4° in a swing-out rotor.
5. Collect the turbid layer at the interface, dilute in 20 ml of incubation buffer, resuspend, homogenize by 20 strokes with a Dounce B homogenizer on ice, and wash by centrifugation at 100,000g for 40 min at 4°. The resulting pellet is diluted in 1 ml of incubation buffer.
6. Vesicles are formed by passing the suspension 20 times through a 27-gauge needle with a syringe. Aliquots of the membrane vesicle suspension are frozen and stored in liquid nitrogen.
7. Determine the protein concentration of the membrane vesicle preparation by the Lowry method. Measurement of the Na^+,K^+-ATPase activity[23] may serve to assess the enrichment of plasma membranes relative to the original homogenate. This enrichment should be 15- to 30-fold.[16]

The sidedness of the membrane vesicle preparation may be estimated by the activity of an ectoenzyme, nucleotide pyrophosphatase (EC 3.6.1.9), in the presence or absence of Triton X-100 for solubilization of the membrane vesicle.[24] The percentage of inside-out-oriented plasma membrane vesicles from HL60 cells,[7] mastocytoma cells,[16] or HeLa cells[8] should range from 30 to 40%.

The preparation of plasma membrane vesicles from the liver (hepatocyte) canalicular membrane is highly enriched with the canalicular (apical) isoform MRP2.[12] Methods for the isolation of hepatocyte canalicular membranes have been described in detail.[24,25] The percentage of inside-out-oriented vesicles in these preparations amounts to 32% ± 5%.[24] The techniques for determination of MRP2-mediated transport, for example, of

[23] B. F. Scharschmidt, E. B. Keeffe, N. M. Blankenship, and R. K. Ockner, *J. Lab. Med.* **93,** 790 (1979).

[24] M. Böhme, M. Müller, I. Leier, G. Jedlitschky, and D. Keppler, *Gastroenterology* **107,** 255 (1994).

[25] P. J. Meier and J. L. Boyer, *Methods Enzymol.* **192,** 534 (1990).

$[^3H]LTC_4$,[12] are the same as described later for ATP-dependent transport mediated by MRP1.[7,8]

Measurement of ATP-Dependent Transport of Leukotriene C_4

Various labeled substrates for the conjugate export pumps encoded by the *MRP1* and *MRP2* genes may serve to assess the transport rate in membrane vesicle preparations.[7–12,17,21] However, because of its high affinity, specificity for MRP1[8,10,17] and MRP2,[12] and commercial availability, $[^3H]LTC_4$ has become the preferred substrate for transport measurements. Stability of LTC_4 must be ascertained because the arachidonate backbone of the molecule may undergo oxidative degradation and photoisomerization.[26] Moreover, the glutathione moiety of LTC_4 may be rapidly degraded by the plasma membrane enzymes γ-glutamyltransferase (EC 2.3.2.2) and LTD_4 dipeptidase (EC 3.4.13.19). This degradation yields the cysteinylglycine derivative LTD_4 and the cysteine derivative LTE_4, which are less efficient substrates for MRP1.[8] Separation of the membrane vesicles from the extravesicular labeled substrate terminates the transport assay. This is achieved with most substrates by rapid filtration through nitrocellulose membranes.[7–11,16,27,28] However, with substrates more hydrophobic than LTC_4 which bind strongly to the filters and to the membrane vesicles, small Sephadex G-50 columns[10,28] or glass filters[17] should be employed. Nonspecific time-dependent binding to membrane vesicles, particularly with lipophilic labeled substrates, is a source of considerable error. Depending on the substrate, it is necessary to include blanks in which ATP is replaced by an equimolar concentration of 5′-AMP[7,8,16,28] or by nonhydrolyzable ATP analogs such as adenosine 5′-[β,γ-methylene]triphosphate.[11,29] Addition of reduce glutathione (5 m*M*) strongly reduces the binding of LTC_4 to membrane-bound glutathione *S*-transferase (EC 2.5.1.18).

Method. This method is based on Refs. 16, 28, and 29.

1. Soak nitrocellulose filters (0.2-μm pore size, 25-mm diameter) in incubation buffer (0.25 *M* sucrose, 10 m*M* Tris-HCl, pH 7.4). Prepare rapid filtration apparatus (Millipore, Bedford, MA).
2. Preincubate at 37° for 1 min in a final volume of 110 μl, at pH 7.4, the following transport assay mixture:
   4 m*M* ATP (potassium salt)
   10 m*M* creatine phosphate (Tris salt)

[26] E. Falk, M. Müller, M. Huber, D. Keppler, and G. Kurz, *Eur. J. Biochem.* **186,** 741 (1989).
[27] T. Ishikawa, M. Müller, C. Klünemann, T. Schaub, and D. Keppler, *J. Biol. Chem.* **265,** 19279 (1990).
[28] M. Böhme, M. Büchler, M. Müller, and D. Keppler, *FEBS Lett.* **333,** 193 (1993).
[29] M. Büchler, M. Böhme, H. Ortlepp, and D. Keppler, *Eur. J. Biochem.* **224,** 345 (1994).

10 mM $MgCl_2$
10 mM Tris-HCl, pH 7.4
5 mM glutathione (reduced); recommended for [^{3}H]LTC_4 transport by canalicular membrane vesicles
0.25 M sucrose
100 μg/ml creatine phosphokinase (2 units/110 μl)
50 nM [^{3}H]LTC_4 (50 nCi/110 μl)

3. Prepare blanks by replacing ATP by 5′-AMP; otherwise same treatment as for transport assays.
4. Quickly thaw membrane vesicle suspension at 37° and store on ice for about 40 min before use.
5. Initiate transport assay at 37° by addition of membrane vesicle suspension (30 μg/110 μl).
6. Remove 20-μl aliquots, at 30- or 60-sec intervals, dilute immediately with 1 ml of ice-cold incubation buffer, and filter immediately through nitrocellulose membrane using the vacuum of the filtration apparatus (200 mbar). Rinse twice with 5 ml of cold incubation buffer.
7. Dissolve filter membranes in 10 ml of scintillation fluid and count for radioactivity.
8. Calculate transport rate of [^{3}H]LTC_4 into the membrane vesicle, corresponding to ATP-dependent export across the plasma membrane of intact cells, by subtracting the corresponding values in the presence of 5′-AMP from those in the presence of ATP (Fig. 1).

Alternatively, the rapid filtration through nitrocellulose filters described above can be replaced by centrifugation of the vesicles through a gel matrix.[10,28,29] For this procedure, prepare NICK spin columns (Pharmacia, Uppsala, Sweden) (0.2 g Sephadex G-50/3.3 ml), rinse them with the incubation buffer, and centrifuge at 400g for 4 min at 4°. Dilute aliquots of the incubations in 80 μl of ice-cold incubation buffer and load immediately onto the columns. After rinsing with 100 μl of incubation buffer, the columns are centrifuged at 400g for 4 min at 4°. Collect the effluents and count for radioactivity. This method is more suitable for MRP1 or MRP2 substrates, which exhibit a high background due to binding of the labeled substrate to the filters and to membrane vesicles. This is the case for transport measurements with several cytotoxic drugs and their conjugates including glucuronosyl[^{3}H]etoposide[10] and glutathione *S*-conjugates of the alkylating agents melphalan[10] and chlorambucil.

Substrate Specificity and Kinetic Properties of MRP1

The substrate specificity of recombinant human MRP1 determined in plasma membrane vesicles[8,10] from cells transfected with an MRP1 expres-

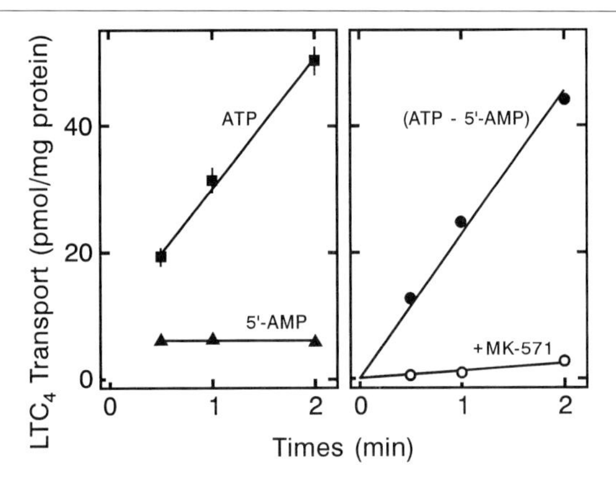

FIG. 1. Transport of [^{3}H]LTC$_4$ (50 nM) into plasma membrane vesicles prepared from MRP1-overexpressing HL60/ADR cells.[7] Apparent transport in the presence of 4 mM ATP or 4 mM 5′-AMP is shown on the left-hand side. Determination of net ATP-dependent [^{3}H]LTC$_4$ transport, as shown on the right-hand side, requires subtraction of blank values in the presence of 5′-AMP. The quinoline-based inhibitor MK-571 (5 μM) is a structural analog of LTD$_4$[19] and a potent inhibitor of MRP1-mediated transport.[7,8,10,11]

sion vector[30] is indistinguishable from the one determined in membrane vesicles from drug-selected MRP1-overexpressing cells, such as the HL60/ADR line.[3,7] The glutathione *S*-conjugate LTC$_4$ is the substrate with the highest affinity known at present, whereas ATP-dependent transport of oxidized glutathione (GSSG) has a relatively low affinity with a K_m value of 93 μM.[11] Reduced glutathione (GSH) itself is not a substrate for MRP1,[11] however, it may undergo complex formation or serve as a cosubstrate with cationic substances. This is indicated by the ATP-dependent transport of vincristine in the presence of GSH[17] and is suggested by the increased release of GSH from MRP1-overexpressing cells in the presence of arsenite.[31] The glutathione moiety is not a structural determinant of the substrate properties of MRP1 as indicated by the efficient transport of the *N*-acetylcysteine *S*-conjugate *N*-acetyl-LTE$_4$,[8] and of the glucuronate conjugates 17β-glucuronosylestradiol,[10,32] monoglucuronosylbilirubin,[33] and bis-

[30] C. E. Grant, G. Valdimarsson, D. R. Hipfner, K. C. Almquist, S. P. C. Cole, and R. G. Deeley, *Cancer Res.* **54,** 357 (1994).

[31] G. J. R. Zaman, J. Lankelma, O. van Tellingen, J. Beijnen, H. Dekker, C. Paulusma, R. P. J. Oude Elferink, F. Baas, and P. Borst, *Proc. Natl. Acad. Sci. U.S.A.* **92,** 7690 (1995).

[32] D. W. Loe, K. C. Almquist, S. P. C. Cole, and R. G. Deeley, *J. Biol. Chem.* **271,** 9683 (1996).

[33] G. Jedlitschky, I. Leier, U. Buchholz, J. Hummel-Eisenbeiss, B. Burchell, and D. Keppler, *Biochem. J.* **327,** 305 (1997).

TABLE I
SUBSTRATE SPECIFICITY OF MRP1[a]

Substrate	Concentration (μM)	Velocity (pmol × min^{-1} × mg^{-1})
LTC_4[b]	0.05	55
LTD_4	0.05	15
LTE_4	0.05	8
N-Acetyl-LTE_4	0.05	3
Dinitrophenyl-SG[c]	0.20	25
GSSG[d]	0.20	<0.5
GSSG	100	230
17β-Glucuronosylestradiol	0.20	3
Monoglucuronosylbilirubin	0.5	5
Bisglucuronosylbilirubin	0.5	3
3α-Sulfatolithocholyltaurine	5	27
Folate	5	1.4
Methotrexate	5	1.5

[a] ATP-dependent transport in membrane vesicles from MRP1-transfected HeLa T5 cells.[8,10,33]
[b] LTC_4, Leukotriene C_4.
[c] SG, Glutathione residue.
[d] GSSG, Glutathione disulfide.

glucuronosylbilirubin[33] (Table I). In addition, the 3α-sulfate conjugate of the bile salt lithocholyltaurine is a substrate for MRP1.[10] These results indicate that the term *glutathione S-conjugate* (GS-X) pump is not appropriate for description of the function of MRP1. On the other hand, many organic anions, including monoanionic bile salts, are *not* substrates transported by MRP1. Therefore, the term *multispecific organic anion transporter* (MOAT) suggests a substrate specificity much broader than that of MRP1. A ranking of the substrates for MRP1 according to the V_{max}/K_m ratios (Table II) is as follows[10]: LTC_4 > LTD_4 > *S*-(2,4-dinitrophenyl)glutathione > 17β-glucuronosylestradiol > monoglucuronosylbilirubin > 3α-sulfatolithocholyltaurine > GSSG.

Note that a similar ranking of substrates describes the specificity of the apical isoform MRP2 (cMRP/cMOAT), in spite of major differences in the amino acid sequence of both ABC transporters with a 49% overall amino acid sequence identity.[12] It will be of interest to identify the substrate binding sites and the common structural motifs determining the substrate specificity of these distinct members of the MRP family.

Inhibitors of MRP1-Mediated Transport

The most potent inhibitors for the *MRP1* gene-encoded conjugate export pump known at present are structural analogs of the cysteinyl leuko-

TABLE II
EFFICIENCY OF DIFFERENT SUBSTRATES FOR MRP1-MEDIATED TRANSPORT[a]

Substrate	K_m (μM)	V_{max}/K_m ($\mu l \times mg^{-1} \times min^{-1}$)
LTC_4[b]	0.1	1031
Dinitrophenyl-SG[c]	3.6	114
17β-Glucuronosylestradiol	1.5	42
GSSG[d]	93	7

[a] Transport into inside-out membrane vesicles from MRP1-transfected HeLa T5 cells.[10,11]
[b] LTC_4, Leukotriene C_4.
[c] SG, Glutathione residue.
[d] GSSG, Glutathione disulfide.

trienes (Table III). It is in line with our previous studies on the inhibition of ATP-dependent LTC_4 transport in mastocytoma cell membranes[16,21] and hepatocyte canalicular membranes[12] that LTD_4 receptor antagonists act as potent competitive inhibitors of MRP1 and MRP2.[8,12] *S*-Decylglutathione, which is a structural analog of LTC_4, also acts as an inhibitor of MRP1-mediated transport of LTC_4 with a K_i value of 116 nM.[17] The LTD_4 analog MK-571, a monoanionic quinoline derivative developed as receptor antagonist for the treatment of allergy and asthma,[19] inhibits MRP1-mediated membrane transport of LTC_4, 17β-glucuronosylestradiol, and GSSG with nanomolar K_i and K_m/K_i ratios of 0.16, 0.46, and 155, respectively.[10] These

TABLE III
INHIBITORS OF MRP1-MEDIATED ATP-DEPENDENT TRANSPORT OF LEUKOTRIENE C_4[a]

Inhibitor	K_i (μM)
Leukotriene D_4 receptor antagonists	
MK-571[b]	0.6
FPL-55712[c]	0.4
BAY-u9773[d]	0.7
Cyclosporin A	5
Cyclosporin PSC-833	27

[a] K_i values for inhibition of LTC_4 transport in membrane vesicles from MRP1-transfected HeLa T5 cells.[8]
[b] Jones *et al.* (1989).[19]
[c] Augstein *et al.* (1973).[18]
[d] Cuthbert *et al.* (1991).[20]

competitive inhibitors are less active on intact cells and *in vivo* than on inside-out-oriented membrane vesicles, since their binding competes with the substrate (LTC_4) on the cytosolic side of the transporter as suggested by competitive photoaffinity labeling.[8] Potent inhibitors of the ATP-dependent transport mediated by MDR1–P-glycoprotein, such as cyclosporin A and the nonimmunosuppressive cyclosporin derivative PSC 833, are only weak inhibitors of MRP1-mediated transport, with K_i values of 5 and 27 μM respectively.[8] The inhibitors listed in Table III clearly discriminate, therefore, between the ATP-dependent export pumps encoded by the *MDR1* gene and the *MRP1* gene.

Acknowledgments

This work was supported in parts by the Deutsche Forschungsgemeinschaft, Bonn (SFB 352/B3, SFB 601/A2), and the Fonds der Chemischen Industrie, Frankfurt, Germany.

[46] Heterologous Expression Systems for Study of Cystic Fibrosis Transmembrane Conductance Regulator

By Xiu-Bao Chang, Norbert Kartner, Fabian S. Seibert, Andrei A. Aleksandrov, Andrew W. Kloser, Gretchen L. Kiser, and John R. Riordan

Introduction

This article emphasizes the special needs for, limitations to, and requirements of vectors and hosts used for the expression of the cystic fibrosis transmembrane conductance regulator (CFTR), rather than providing a comprehensive technical description of all the heterologous systems that have been employed for this purpose. At the outset, it is important to state that while heterologous expression has become an integral part of the study of the protein product of virtually all cloned genes, it was absolutely essential for investigations of the structure, function, and biosynthesis of CFTR. This was not only for the obvious reason that the CFTR gene was identified and characterized before the protein was detected but also because it is present endogenously at very low copy number (less than 1000 copies per cell). Hence, no natural rich source of the protein exists.

Initial attempts to express CFTR heterologously were thwarted by difficulties in propagating plasmids containing the full-length CFTR cDNA in

METHODS IN ENZYMOLOGY, VOL. 292
0076-6879/98 $25.00

bacteria.[1–3] When the partial cDNAs that were originally isolated[4] were ligated together in bacterial plasmids such as pBlueScript (Stratagene, La Jolla, CA), growth was slow, yielding low numbers of small colonies. Inserts of the appropriate size for the full-length cDNA could be excised from these plasmids but sequencing invariably revealed rearrangement in the exon 6 coding region. These were usually small changes but sufficient to disable a putative prokaryotic promoter sequence. In one case, we found that the AG dinucleotide at positions 921 and 922 immediately 5′ of the first member of a pair of 13-nucleotide direct repeat sequences had been deleted.[3] To reduce the probability of rearrangements involving this repeat, we made silent nucleotide changes by PCR (polymerase chain reaction) mutagenesis at the third positions of three codons in the first member of the repeat, making its sequence different from the second member.[3] This construct could be propagated, albeit with low efficiency, without sequence rearrangements occurring. Other investigators made similar manipulations to solve this problem including the utilization of low copy plasmids to further reduce the frequency of sequence rearrangements. We encountered similar problems when assembling an expressible full-length cDNA coding for the CFTR homolog from *Squalus acanthias*.[5,6] Both mutagenesis of sequences within a putative bacterial promoter and employment of the low copy pACYC plasmid were necessary for stable propagation. It is assumed that some property of the CFTR sequence downstream of exon 6 is extremely toxic to bacteria. However, there is no definitive evidence as to whether this toxicity is related to the channel-forming capacity, to some other property of the protein, or indeed if the detrimental effect occurs at the level of the protein.

Historically, full-length CFTR cDNA was first expressed in epithelial cells from CF patients using retroviral[1] and plasmid[2] vectors to show that it restored cAMP-dependent Cl^- permeability. Following that, expression

[1] M. L. Drumm, H. A. Pope, W. H. Cliff, J. M. Rommens, S. A. Marvin, L. C. Tsui, F. S. Collins, R. A. Frizzell, and J. M. Wilson, *Cell* **62,** 1227 (1990).

[2] D. P. Rich, M. P. Anderson, R. J. Gregory, S. H. Cheng, S. Paul, D. M. Jefferson, J. D. McCann, K. W. Klinger, A. E. Smith, and M. J. Welsh, *Nature* **347,** 358 (1990).

[3] C. E. Bear, F. Duguay, A. L. Naismith, N. Kartner, J. W. Hanrahan, and J. R. Riordan, *J. Biol. Chem.* **266,** 19142 (1991).

[4] J. R. Riordan, J. M. Rommens, B. Kerem, N. Alon, R. Rozmahel, Z. Grzelczak, J. Zielenski, S. Lok, N. Plavsic, J. L. Chou, M. L. Drumm, M. C. Iannuzzi, F. S. Collins, and L.-C. Tsui, *Science* **245,** 1066 (1989).

[5] J. W. Hanrahan, F. Duguay, S. Sansom, N. Alon, T. Jensen, and J. R. Riordan, *MDIBL* **32,** 48 (1993).

[6] J. R. Riordan, B. Forbush III, and J. W. Hanrahan, *J. Exp. Biol.* **196,** 405 (1994).

in nonepithelial cells[7–9] including even nonvertebrate Sf9 (*Spodoptera frugiperda* fall armyworm ovary) insect cells[7] provided suggestive evidence that the CFTR gene might code for a cAMP-regulated Cl^- channel. Subsequently, expression in a wide variety of transient and stable mammalian cell systems, in *Xenopus* oocytes,[3] in insect cells,[7] in the yeast *Saccharomyces cerevisiae,*[10,11] and in transgenic mice[12] has been employed for studies of CFTR. Several of these systems have been developed in this laboratory with stable mammalian cell lines [CHO (chinese hamster ovary) and BHK (baby hamster kidney)], *Xenopus* oocytes, baculovirus-infected caterpillar cells, and more recently yeast[11] providing the greatest utility.

Stable CFTR Expression in CHO and BHK Cells

Because CHO cells provide one of the most robust and versatile mammalian cell culture systems, we have established permanent CFTR-expressing CHO cell lines.[13,14] Because we were already aware that there were difficulties in expressing CFTR stably at high levels, we have employed the amplifiable pNUT plasmid (Fig. 1).[15] Selection of colonies surviving the presence of the dihydrofolate reductase (DHFR) inhibiting drug, methotrexate ensures the presence of multiple copies of the plasmid DNA, which contains the sequence coding for DHFR. Selection at methotrexate concentrations between 10 and 100 μM provides cell lines expressing increasing amounts of CFTR, as monitored by immunoblots and cAMP-stimulated chloride channel activity.[13,14] The plasmid contains a strong, inducible metallathionein promoter,[15] enabling further control of expression level with heavy metals and glucocorticoids. As with many commonly used expression

[7] N. Kartner, J. W. Hanrahan, T. J. Jensen, A. L. Naismith, S. Sun, C. A. Ackerley, E. F. Reyes, L.-C. Tsui, J. M. Rommens, C. E. Bear, and J. R. Riordan, *Cell* **64,** 681 (1991).

[8] M. P. Anderson, D. P. Rich, R. J. Gregory, A. E. Smith, and M. J. Welsh, *Science* **251,** 679 (1991).

[9] J. M. Rommens, S. Dho, C. E. Bear, N. Kartner, D. Kennedy, J. R. Riordan, L.-C. Tsui, and J. K. Foskett, *Proc. Natl. Acad. Sci. U.S.A.* **88,** 7500 (1991).

[10] P. Huang, K. Stroffekova, J. Cuppoletti, S. K. Mahanty, and G. A. Scarborough, *Biochim. Biophys. Acta* **1281,** 80 (1996).

[11] G. L. Kiser, A. W. Kloser, E. Belzai, A. Goffeau, and J. R. Riordan, unpublished observations (1997).

[12] J. N. Snouwaert, K. K. Brigman, A. M. Latour, N. N. Malouf, R. C. Boucher, O. Smithies, and B. H. Keller, *Science* **257,** 1083 (1992).

[13] J. A. Tabcharani, X.-B. Chang, J. R. Riordan, and J. W. Hanrahan, *Nature* **352,** 628 (1991).

[14] X.-B. Chang, J. A. Tabcharani, Y.-X. Hou, T. J. Jensen, N. Kartner, N. Alon, J. W. Hanrahan, and J. R. Riordan, *J. Biol. Chem.* **268,** 11304 (1993).

[15] R. D. Palmiter, R. R. Behringer, C. J. Quaife, F. Maxwell, I. H. Maxwell, and R. L. Brinste, *Cell* **50,** 435 (1987).

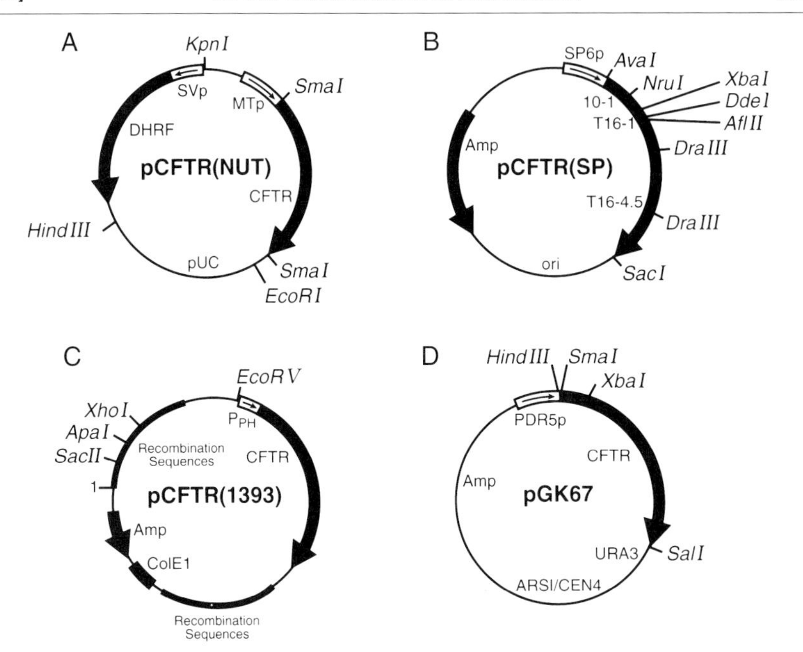

FIG. 1. Maps of CFTR expression plasmids. (A) Full-length CFTR cDNA cloned into *Sma*I site of pNUT plasmid[13] so that its transcription is driven by the mouse metallothionein I promoter (Mtp). The mutant dihydrofolate reductase-coding sequence under control of the SV40 (simian virus 40) early promoter (Svp) is also indicated. This enzyme provides for amplification of the plasmid sequence in cells under selection with methotrexate. Both CHO and BHK cells were transfected with this plasmid by the calcium phosphate method of Chen and Okayama,[54] grown in the presence of methotrexate for approximately 10 days when individual colonies were picked and expanded. Levels of CFTR expression were monitored by immunoblots of cell lysates or isolated membranes (see Fig. 3) and by assays of cAMP-stimulated Cl^- channel activity (see Fig. 2). These cell lines could then be maintained stably by growth in methotrexate-containing medium and as frozen stocks. (B) CFTR cDNA cloned at *Ava*I and *Sac*I sites of pSP64-poly(A) (Promega, Madison, WI) transcription plasmid. cRNA is generated from the SP6 RNA polymerase promoter (SP6p). CFTR partial cDNAs 10-1, T16-1, and T16-4.s originally isolated[4] and used to assemble the full-length coding sequence are indicated as are the XbaI, DdeI, and AflII sites employed during mutagenesis of repeat sequences responsible for unwanted sequence rearrangements.[3] Injection of cRNA obtained using this plasmid causes robust CFTR expression in *Xenopus* occytes.[3] (C) Baculovirus transfer plasmid pVL1393[33] containing CFTR coding sequence under control of the polyhedron promoter (P_{PH}). Sequences involved in recombination with wild-type baculovirus sequence are also indicated. (D) Yeast shuttle vector, pGK67, was constructed by insertion of the CFTR sequence downstream of the pleitotropic drug resistance promoter (PDR5p).[40] Positions of the origin of replication (ARSI) and the contromere sequence (CEN4) are indicated. Transformation of yeast containing a mutant *pdr1-3* gene enabled PDR5 promoter activity and CFTR expression.

plasmids, transcription of pNUT inserts is strongly stimulated (10–100×) by sodium butyrate, and this treatment can further elevate the level of CFTR protein present in these permanent cells.[16] However, even with this amplification of integrated CFTR cDNA copy number and strong stimulation of transcription, the amount of CFTR remains at less than 0.1% of the membrane protein,[13,14,16] and is not visible in SDS–PAGE by usual staining procedures above the background of other membrane proteins. This is in drastic contrast to the amount of the structurally related P-glycoprotein, which accumulates in the plasma membranes of multidrug-resistant CHO cells (i.e., >10%[17,18]), thus emphasizing that accumulation of CFTR in mammalian cells is limited, as also observed by other investigators.[19,20] As a consequence, CHO cells, which can be grown in large quantities in suspension and provide an excellent source for the purification of many recombinant mammalian proteins, have proven of limited use for purification of CFTR in quantity.[21]

On the other hand, CFTR-expressing CHO cell lines have proven ideal for studies of the ion channel function[13,14,22] and biosynthetic processing of CFTR.[23–25] When grown firmly attached to a substrate, cAMP-stimulated halide efflux can be readily followed by measuring the appearance of either a radioactive isotope ($^{125}I^-$ or $^{36}Cl^-$) in extracellular fluid or I^-, potentiometrically with an ion-sensitive electrode (Fig. 2).[14] Cl^--sensitive electrodes do not provide adequate sensitivity. Direct measurements of charge movement through the CFTR channel are also conveniently made using these cells because they are ideal for patch-clamping (Fig. 2).[26] We have established

[16] F. S. Seibert, J. A. Tabcharani, X.-B. Chang, A. M. Dulhanty, C. Mathews, J. W. Hanrahan, and J. R. Riordan, *J. Biol. Chem.* **270,** 2158 (1995).

[17] N. Kartner, D. Evernden-Porelle, G. Bradley, and V. Ling, *Nature* **316,** 820 (1985).

[18] M. K. Al-Shawi and A. E. Senior, *J. Biol. Chem.* **268,** 4197 (1993).

[19] R. J. Gregory, S. H. Cheng, D. P. Rich, J. Marshall, P. Sucharita, K. Hehir, L. Ostedgaard, K. W. Klinger, M. J. Welsh, and A. E. Smith, *Nature* **347,** 382 (1990).

[20] D. D. Koerberl, C. L. Halbert, A. Krumm, and A. D. Miller, *Hum. Gene Ther.* **6,** 469 (1995).

[21] C. R. O'Riordan, A. Erickson, C. Bears, C. Li, P. Manavalan, K. X. Wang, J. Marshall, R. K. Scheule, J. M. McPherson, S. H. Cheng, and A. E. Smith, *J. Biol. Chem.* **270,** 17033 (1995).

[22] J. A. Tabcharani, J. M. Rommens, Y.-X. Hou, J. R. Riordan, and J. W. Hanrahan, *Nature* **366,** 79 (1993).

[23] S. Pind, J. R. Riordan, and D. B. Williams, *J. Biol. Chem.* **269,** 12784 (1994).

[24] G. L. Lukacs, A. Mohamed, N. Kartner, X.-B. Chang, J. R. Riordan, and S. Grinstein, *EMBO J.* **13,** 6076 (1994).

[25] T. J. Jensen, M. A. Loo, S. Pind, D. B. Williams, A. L. Goldberg, and J. R. Riordan, *Cell* **83,** 129 (1995).

[26] J. W. Hanrahan, J. A. Tabcharani, F. Becq, C. J. Mathews, O. Augustinas, T. J. Jensen, X.-B. Chang, and J. R. Riordan, *in* "Ion Channels and Genetic Diseases" (D. C. Dawson and R. A. Frizzell, eds.), Soc. Gen. Physiologists 48th Annu. Symp., p. 125. The Rockefeller University Press, New York, 1995.

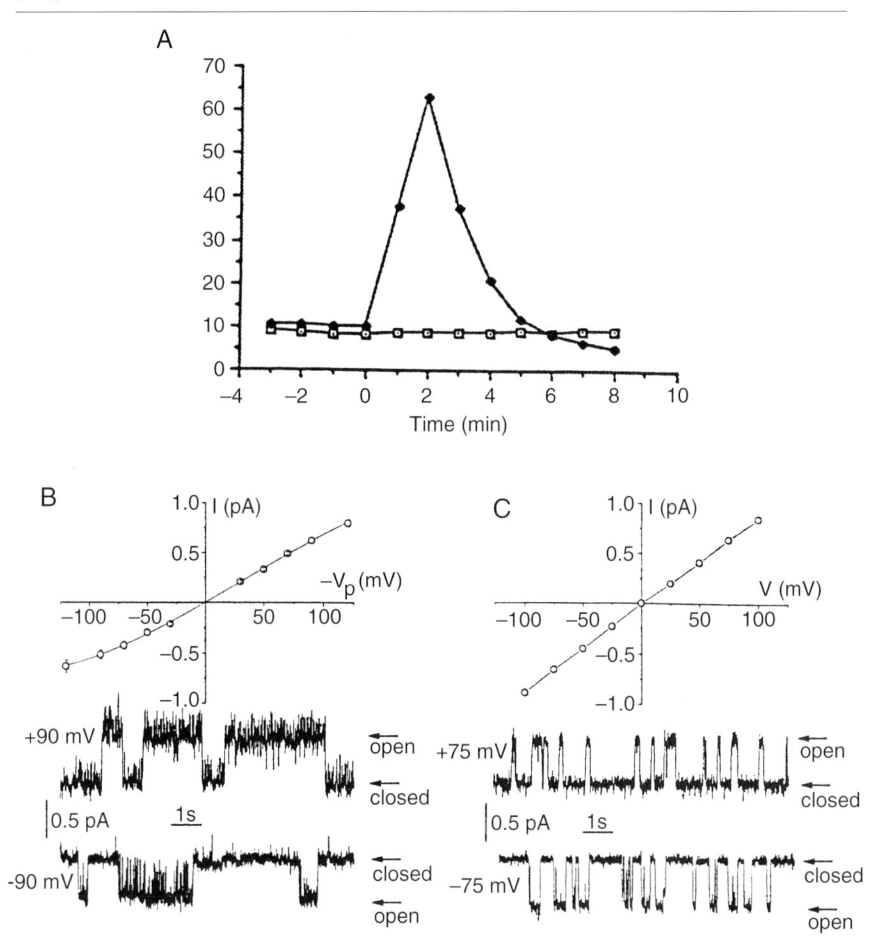

FIG. 2. Assays of CFTR chloride channel activity in CHO cells stably expressing CFTR. (A) Cells were loaded with 135 m*M* NaI and amounts of I^- released at the indicated times after stimulation with cAMP (time 0) determined potentiometrically.[14] (B) Single chloride channel activity detected in the same cells by whole-cell patch clamping. Single-channel tracings at two of the voltages used to determine the current–voltage relationship are shown below it. Pipette and bath solutions contained 150 m*M* Cl^- and single-channel conductance was estimated at ~7 pS. A low-pass filter with a cut off frequency of 100 Hz was employed. (C) Similar single-channel analysis obtained in a planar lipid bilayer on fusion of membranes isolated from the same cells.[14] They were phosphorylated with PKA, which was also present with ATP (1 m*M*) on the *cis* side of the bilayer. Symmetrical solutions containing 150 m*M* Cl^- were used. Single-channel conductance was ~8 pS. Currents were filtered at 50 Hz.

lines expressing different amounts of CFTR by selection in different methotrexate concentrations such that either low or high numbers of channels may be observed in a patch. The former is advantageous for quantitative characterization of single-channel properties and the latter can provide large current signals that readily reveal regulatory responses to phosphorylation and nucleotides, for example. In addition to their utility for ion-flux measurements and patch clamping, CFTR-expressing CHO cells have been used to isolate CFTR-containing membrane vesicles that can be fused with planar lipid bilayers. Figure 2 illustrates the properties of a single CFTR channel examined in this way.

Because of the novel behavior of wild-type CFTR in the early secretory pathway[25] and the fact that the ΔF508 mutant protein, which most patients with cystic fibrosis possess, does not proceed beyond the endoplasmic reticulum (ER),[27] studies of biosynthetic processing have been a major focus of CF research. Pulse-chase experiments have shown that the kinetics of the conversion of the immature core-glycosylated CFTR to the mature molecule with complex N-linked oligosaccharide chains is the same in CHO cell lines as in the epithelial cells in which CFTR is expressed endogenously.[24] Hence, it has been possible to study the details of the maturation and proteolysis of wild-type and mutant CFTR, including the involvement of molecular chaperones,[23] ubiquitination, and proteasomal degradation.[25]

Using the same expression plasmid and methods of selection, recombinant CFTRs have also been expressed in BHK cells deficient in dihydrofolate reductase,[15] allowing a higher and more uniform level of amplification and therefore more CFTR protein production. Although these cells are not as convenient for multiple applications as CHO, they can be grown in large quantities as a source of protein purification. We obtain the highest level of CFTR expression in mammalian cells when the CFTR sequence is maximally amplified and transcription strongly stimulated by sodium butyrate (2 m*M* for 24–48 hr). Recently, we inserted the 10 histidine-residue tag into the pNUT-CFTR plasmid and were able to purify the fully glycosylated protein in reasonable yield from detergent-solubilized membranes of butyrate-treated BHK cells.[28]

The principal disadvantage of permanent mammalian cell lines expressing recombinant CFTR is the limited amount of the protein that can be accumulated in their membranes. Major advantages are the consistency of the cultures, which can be passaged nearly indefinitely in the presence of

[27] S. H. Cheng, R. J. Gregory, J. Marshall, S. Paul, D. W. Souza, G. A. White, C. R. O'Riordan, and A. E. Smith, *Cell* **63,** 827 (1990).

[28] L. A. Aleksandrov, Y.-X. Hou, T. J. Jensen, L. Cui, A. A. Aleksandrov, X.-B. Chang, and J. R. Riordan, submitted (1998).

selective drug (methotrexate) and the fact that all of the clonally derived cells express the protein. This is a special advantage for patch-clamp experiments in which all cells can be patched and also for high-throughput assays to screen for agents influencing anion permeation through CFTR. The cells are healthy and metabolically stable in contrast to cells that are transiently transfected with plasmids or infected with viral vectors (see later discussion).

Transient CFTR Expression in Mammalian Cells

Although stable cell lines have many advantages for extensive mechanistic studies as outlined earlier, their development is laborious and time consuming. Transient expression is preferable from both of these points of view. Furthermore, it provides expression, which is independent of the site of gene incorporation so that within an experiment, expression levels of mutant and wild-type CFTR are well matched. To screen rapidly the effects of multiple mutations, for example, we transfer the CFTR coding sequence to the expression plasmid pcDNA3 (InVitrogen, Carlsbad, CA), which utilizes the strong cytomegalovirus (CMV) promoter. Forty-eight hours after calcium phosphate transfection into 293 or COS-1 cells, both immature and mature CFTR bands are readily detectable in immunoblots, and cAMP-stimulated, iodide efflux from the COS-1 cells reflects the ion channel activity.[29] The 293 cells are also suitable for patch clamping although the proportion of cells expressing channels varies with efficiency of transfection. We find the confluency of the cells to be the most important factor influencing this efficiency; it is essential to carry out the transfections at a subconfluent stage of growth.

Injection of *Xenopus* Oocytes with CFTR cRNA

Once initial studies indicated that CFTR was itself a chloride channel, it seemed likely that *Xenopus* oocytes would provide a useful expression system for studies of its electrogenic properties. We have confirmed that this was the case,[3] observing very large Cl^--carrying whole-cell currents on stimulation by cAMP or the catalytic subunit of protein kinase A, 1–4 days after injection of the cRNA generated using a pSP64 transcription plasmid (Fig. 1). These CFTR currents are distinct from and do not influence those due to endogenous Ca^{2+}-dependent Cl^- channels. Concomitantly, immu-

[29] F. S. Seibert, P. Linsdell, T. W. Loo, J. W. Hanrahan, D. M. Clarke, and J. R. Riordan, *J. Biol. Chem.* **271,** 15139 (1996).

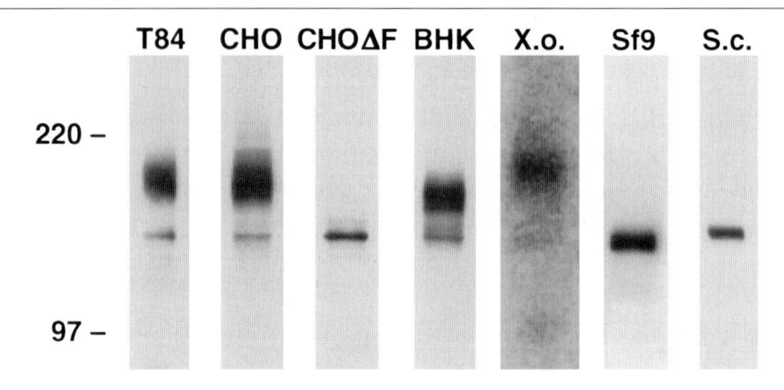

FIG. 3. CFTR protein bands detected in Western blots probed with the monoclonal antibody, M3A7. Lysates of each of the cell types indicated were subjected to SDS–PAGE (6% acrylamide) and transferred to nitrocellulose, which, after incubation with the primary antibody and a goat anti-mouse secondary antibody, was developed by ECL (enhanced chemiluminescence, Amersham, Arlington Heights, IL). T-84, colon carcinoma cell line; CHO, a CFTR-expressing CHO cell line obtained from the parent K1 line as described; CHOΔF, a similarly derived line employing the ΔF508 CFTR sequence instead of the wild type; BHK, a CFTR-expressing BHK cell line established from parental BHK cells in the same way; X.o., *Xenopus* oocytes 48 hr after injection with CFTR cRNA; Sf9, insect Sf9 cells 2 days after infection with the recombinant baculovirus containing the wild-type CFTR sequence; S.c., *Saccharomyces cerivisiae* containing a mutant *pdr1-3* gene expressing CFTR under control for the PDR5 promotor.

noblots detect a high-molecular-weight CFTR protein (Fig. 3), consistent with the presence of complex oligosaccharide chains.[3] Single-channel analysis by patch clamping of these oocytes after removal of the vitelin membrane reveals properties very similar to those observed in epithelial cells and other mammalian and insect cells expressing the protein heterologously. Other investigators have subsequently exploited the robust expression of CFTR in *Xenopus* oocytes to study the properties of wild-type and mutant CFTR channels.[30] Notably, expression in oocytes is more permissive than mammalian cells for processing of CFTR mutants,[30] and even fragments[31] of the protein can generate chloride channel activity. This may be at least partially due to the lower temperature (~20°) at which oocytes are maintained or may reflect other differences in the machinery involved in biosynthetic processing.

The usefulness of *Xenopus* oocytes is limited almost exclusively to electrophysiologic studies of the CFTR channel. They are less convenient and informative for biochemical experiments, however, compared to the

[30] M. L. Drumm, D. J. Wilkinson, L. S. Smit, R. T. Worrell, T. V. Strong, R. A. Frizzell, D. C. Dawson, and F. S. Collins, *Science* **254,** 1797 (1991).

[31] T. P. Carroll, M. M. Morales, S. B. Fulmer, S. S. Allen, T. R. Flotte, G. R. Cutting, and W. B. Guggino, *J. Biol. Chem.* **270,** 11941 (1995).

other expression systems. In addition to being very revealing of CFTR channel function, CFTR-expressing oocytes are useful reporters of the activities of receptors that couple via G proteins to adenylate cyclase.[32] The extreme sensitivity of CFTR to activation by the product of adenylate cyclase, cAMP, can provide a quantitative readout of receptor activity when its cRNA has been injected.

Baculovirus-Mediated Expression in Insect Cells

Because of the limited amounts of the protein produced in mammalian cells, we expressed CFTR in Sf9 insect cells employing the powerful, late-acting polyhedron gene promoter of *Autographa californica* nuclear polyhedrosis virus.[33] The CFTR sequence is cloned into the baculovirus transfer vector, pVL1393 (Fig. 1).[7] SF9 cells are cotransfected with this DNA and that of the wild-type virus. Plaques arising from recombinant CFTR-containing viruses are detected by the absence of polyhedron-containing occlusion particles and confirmed by probing plaque lifts with a CFTR DNA sequence. Infection of cells with the thus-identified recombinant virus results in the synthesis of abundant amounts of protein with chloride channel properties similar to those in other CFTR-expressing cells, as judged by I^- efflux assays, whole-cell anion current measurements, and single chloride channel analysis. This provides adequate starting material for CFTR purification; approximately 0.5 mg can be obtained from a 1 liter culture.[34] This can be reconstituted into phospholipid vesicles that were fused with planar lipid bilayers in which single chloride channels stimulated by ATP and protein kinase A are observed.[34] Their voltage dependence and conductance are similar to those observed by patch clamping the plasma membrane of CFTR-expressing cells. Notably, this provides formal evidence that the CFTR polypeptide is sufficient to generate a regulated low-conductance chloride channel.

Insect cells are known to be incapable of maturation of N-linked oligosaccharide chains,[35] and no mature CFTR brand containing such chains was observed in SDS–PAGE (Fig. 3), consistent with other evidence that oligosaccharides do not influence CFTR channel function. As in *Xenopus* oocytes, some mutant forms of CFTR, including ΔF508, which are mispro-

[32] Y. Uezono, J. Bradley, C. Min, N. A. McCarty, M. Quick, J. R. Riordan, C. Chavkin, K. Zinn, H. A. Lester, and N. Davidson, *Recep. Channels* **1,** 233 (1993).

[33] V. A. Lucknow and M. D. Summers, *Virology* **170,** 31 (1989).

[34] C. E. Bear, C. Li, N. Kartner, R. J. Bridges, T. J. Jensen, M. Ramjeesingh, and J. R. Riordan, *Cell* **68,** 809 (1992).

[35] J. Vialard, M. Lalumiere, T. Vernet, D. Briedis, G. Alkhatib, D. Henning, D. Levin, and C. Richardson, *J. Virol.* **64,** 37 (1990).

cessed and not transported beyond the ER of mammalian cells, do reach the surface of insect cells and exhibit chloride channel activity there.[36] This has been interpreted as reflecting just the lower temperature (~27°) of cell growth compared to mammalian cells. However, other molecular differences between the cell types may also be responsible. It is also important to recognize that mechanisms of ER transport are likely to be damaged late after viral infection when proteins are produced due to the action of the late-acting polyhedron promoter. This may be advantageous in the sense of resulting in less stringent ER quality control of mutant proteins, but it also means that this expression system is not useful for studies of mechanisms of ER processing and transport. This important caveat also applies to other viral expression systems such as *Vaccinia,*[37] widely used in mammalian cells. The fact that infection at an appropriate multiplicity of infection for the correct time must be performed repeatedly makes the baculovirus and other viral systems less convenient than stable mammalian cell lines. However, the baculovirus-infected insect cells are useful and informative for studies of channel function and are by far the richest source of CFTR for work with the purified protein.

CFTR Expression in *Saccharomyces Cerevisiae*

There are at least two major reasons to express CFTR in budding yeast. The first is the potential to provide a large-scale, low-cost source of the protein for purification. Huang and Scarborough[38] have made some progress toward this objective, but CFTR production is only 5–10% of that in Sf9 cells relative to total membrane protein.[10] We have undertaken an alternative yeast expression system in order to exploit the second major advantage it offers: a means of elucidating molecular mechanisms of ER processing and transport. These mechanisms have been extensively studied for several secretory and membranes in yeast owing to the existence of a large number of mutations in different steps in these pathways.[39]

Utilizing a yeast shuttle vector, we have expressed human CFTR under the transcriptional control of a yeast plasma membrane drug transporter, Pdr5p,[40] which is structurally related to CFTR. Activity of the PDR5 promo-

[36] C. Li, M. Ramjeesingh, E. Reyes, T. Jensen, X.-B. Chang, J. M. Rommens, and C. E. Bear, *Nat. Genet.* **3,** 311 (1993).

[37] L. S. Wyatt, B. Moss, and S. Rozenblatt, *Virology* **210,** 202 (1995).

[38] P. Huang and G. A. Scarborough, unpublished observations (1997).

[39] N. K. Pryer, L. J. Wuestehube, and R. Schekman, *Annu. Rev. Biochem.* **61,** 471 (1992).

[40] E. Balzi, M. Wang, S. Leterme, L. Van Dyck, and A. Goffeau, *J. Biol. Chem.* **269,** 2206 (1994).

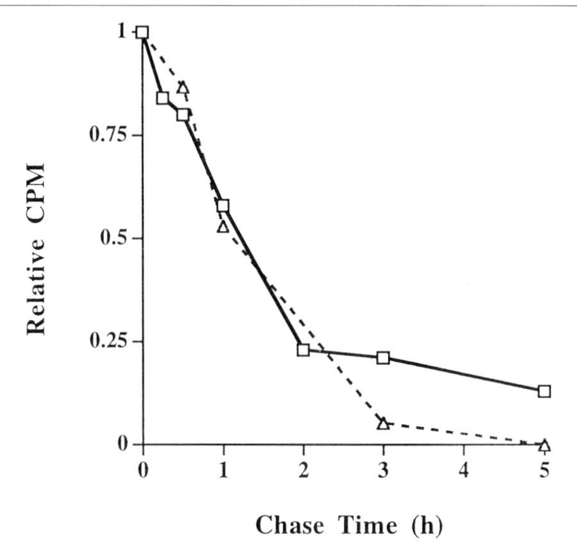

FIG. 4. Pulse-chase experiments indicating CFTR turnover in a stable CHO cell line (△) and in *S. cerevisiae* (□). After pulse-labeling for 15 min with [^{35}S]methionine, cells were transferred to methionine-replete media for the times indicated, lysed in RIPA buffer, and immunoprecipitated using antibody M3A7 and protein G agarose beads.[25] Following SDS–PAGE, radioactivity associated with 140-kDa CFTR bands was determined directly using a Packard Instant Imager (Downers Grove, IL). In the case of CHO cells, this band represents only the core-glycosylated ER form of the protein; in yeast sufficient oligosaccharide is not added to allow distinction of immature and mature CFTR bands; radioactivity associated with the single band observed is indicated.

tor is partly controlled by the transcriptional regulator, Pdr1p.[41] CFTR expression is detected only in *pdr1-3* gain-of-function mutant strains. Subcellular fractionation showed colocalization with the plasma membrane H^+-ATPase (encoded by PMA1).[42] Pulse-chase experiments reveal biphasic kinetics of turnover similar to that in mammalian cells (Fig. 4). Therefore, it should be possible to analyze intracellular processing and transport using the many available mutant strains defective at various steps in the secretory pathway,[39] in chaperones and their cofactors,[43–46] and in proteasomal[47] and

[41] T. M. McGuire, E. Carvajal, D. Katzmann, M. Wagner, W. S. Moye-Rowley, A. Goffeau, and J. Golin, *Gene* **167,** 151 (1995).
[42] R. Rao and C. W. Slayman, *Biophys. J.* **62,** 228 (1992).
[43] D. M. Cyr, X. Lu, and M. G. Douglas, *J. Biol. Chem.* **267,** 20927 (1992).
[44] F. U. Hartl, *Nature* **381,** 571 (1996).
[45] O. Kandror, L. Busconi, M. Sherman, and A. L. Goldberg, *J. Biol. Chem.* **269,** 23575 (1994).
[46] E. A. Craig, B. D. Gambill, and R. J. Nelson, *Microbiol. Rev.* **57,** 402 (1993).
[47] L. Hicke and H. Riezman, *Cell* **84,** 277 (1996).

vacuolar[48] degradation. In addition to this primary pursuit of biosynthetic processing mechanisms, we have used another yeast mutation, *sec6-4* to facilitate functional studies of CFTR in yeast. The temperature-sensitive *sec6-4* mutation blocks fusion of secretory vesicles with the plasma membrane.[49] Hence, the vesicles containing proteins destined for the plasma membrane accumulate within the cells at the restrictive temperature. Studies of the H^+-ATPase in these vesicles[42,50] showed that the vesicles were tightly sealed and of homogeneous orientation, enabling measurement of H^+ translocation. Since these vesicles are osmotically intact, maintain a large membrane potential, and have the regulatory cytoplasmic domains of CFTR accessible on their exterior surfaces, they are useful for assays of Cl^- channel activity either by ion flux measurements or after fusion with planar lipid bilayers.

Conclusions

Despite the apparent toxicity of CFTR to prokaryotes and severe limitations on the amount of CFTR protein that can accumulate in mammalian cells, heterologous expression is possible in all eukaryotic systems that have been attempted. Indeed, heterologous expression is absolutely essential for most CFTR studies because of its low abundance in all native sources. The limitation on the amount of CFTR protein that can be expressed in mammalian cells is not fully understood, although down-regulation appears to occur at several levels from transcription[20] to the rapid turnover of the nascent chain at the ER.[27] These means of attenuating the amount of the polypeptide may be part of the overall regulatory scheme that contributes to the control of net chloride channel activity in response to different demands of the epithelia tissues in which it normally functions.

No single heterologous expression system is sufficient for every avenue of CFTR investigation required to gain an understanding of its structure, function, and biosynthetic processing. As we have pointed out, each system has advantages and disadvantages for different types of experiments. Although additional specific vectors and hosts have been successfully employed by other investigators, the systems described here are representative of most of the approaches available. We have not attempted the use of stable cell lines in which CFTR is driven by a very tightly controlled unducible promoter such as in the tetracycline[51] system. Expression in transgenic

[48] D. J. Klionsky, P. K. Herman, and S. D. Emr. *Microbiol. Rev.* **54,** 266 (1990).
[49] N. C. Walworth and P. J. Novick, *J. Cell Biol.* **105,** 163 (1987).
[50] R. K. Nakamoto, R. Rao, and C. W. Slayman, *J. Biol. Chem.* **266,** 7940 (1991).
[51] M. Gossen and H. Bujard, *Proc. Nat. Acad. Sci. U.S.A.* **89,** 5547 (1992).

mice, which is beyond the scope of this article, has many potential advantages, including the detection of tissue-specific effects on CFTR synthesis and function. The use of promoters that directed expression to lung epithelial cells[52] showed that overexpression was without obvious detrimental effects and, in intestinal epithelial cells, was capable of restoring chloride and fluid secretion to animals in which the endogenous CFTR gene had been inactivated.[53]

[52] S. Yei, N. Mittereder, S. Wert, J. A. Whitsett, R. W. Wilmott, and B. C. Trapnell, *Hum. Gene Ther.* **5,** 731 (1994).

[53] L. Zhou, C. R. Dey, S. E. Wert, M. D. DuVall, R. A. Frizzell, and J. A. Whitsett, *Science* **266,** 1705 (1994).

[54] C. Chen and M. Okayama, *Mol. Cell Biol.* **7,** 2745 (1987).

[47] Characterization of Polyclonal and Monoclonal Antibodies to Cystic Fibrosis Transmembrane Conductance Regulator

By NORBERT KARTNER and JOHN R. RIORDAN

Introduction

Interest in the detection and quantitation of the cystic fibrosis (CF) gene product, the cystic fibrosis transmembrane conductance regulator (CFTR), followed immediately on the heels of the gene cloning.[1–3] This came at a time when almost nothing was known about CFTR protein structure, except what was predicted from cDNA sequence data,[2] and little was known about its normal localization and expression. How does one develop reagents that are sensitive and specific for a protein that is hypothetical, insofar as it has never before been detected, let alone purified, its function is largely unknown, and its primary sequence and structure have only been deduced from DNA sequences? Specific antibodies are the logical choice of reagents,

[1] J. M. Rommens, M. C. Iannuzzi, B.-S. Kerem, M. L. Drumm, G. Melmer, M. Dean, R. Rozmahel, J. L. Cole, D. Kennedy, N. Hidaka, M. Zsiga, M. Buchwald, J. R. Riordan, L.-C. Tsui, and F. S. Collins, *Science* **245,** 1059 (1989).

[2] J. R. Riordan, J. M. Rommens, B.-S. Kerem, N. Alon, R. Rozmahel, Z. Grzelczak, J. Zielenski, S. Lok, N. Plavsic, J.-L. Chou, M. L. Drumm, M. C. Ianuzzi, F. S. Collins, and L.-C. Tsui, *Science* **245,** 1066 (1989).

[3] B.-S. Kerem, J. M. Rommens, J. A. Buchannan, D. Markiewicz, T. K. Cox, A. Chakravarti, M. Buchwald, and L.-C. Tsui, *Science* **245,** 1073 (1989).

0076-6879/98 $25.00

and such reagents were in the process of being developed by the present authors, and others, even before any direct information on endogenous localization or any rich sources of CFTR were available. Indeed, the localization and quantitation of endogenous protein expression awaited the development of such reagents.

Antibody development was made possible by applying solid–phase oligopeptide synthesis and recombinant DNA technology to produce immunogens corresponding to parts of the CFTR molecule. Immunization of rabbits to produce polyclonal antisera or of mice to produce monoclonal antibody (MAb)-secreting hybridoma cell lines ultimately provided the CFTR-specific reagents that have made rapid advances in the understanding of CFTR structure and function possible. This article describes the procedures used to generate useful antibody reagents, how they were characterized, and a critical evaluation of the methods and of the usefulness of the reagents that were obtained. Most of the applications of such reagents have been published elsewhere, and will not be reproduced here except by way of reference; rather, we try to establish some general guidelines for successfully producing sensitive and specific reagents, and to illustrate some of the pitfalls, not just for CFTR, but for any similar protein that may have begun its existence as a "predicted translation product" from a string of DNA sequence.

Immunogen Preparation and Immunization

When CFTR was first cloned,[1–3] neither native protein nor expressed fragments were yet at hand and, consequently, the development of anti-oligopeptide antibodies was pursued at the outset. A second strategy for antibody development made use of cloned cDNA fragments representing part of the CFTR coding sequence.[1,2] These fragments were cloned into the convenient pGEX vector for bacterial fusion protein expression.[4–6] CFTR polypeptide fragments produced in this system were used in the immunization of animals for antibody production. This latter approach was most successful, and it will be described in detail; however, a disadvantage compared with the synthetic oligopeptide approach was that it necessitated the time-consuming optimization of the expression system and CFTR fragment purification, and so it followed some months after our use of synthetic oligopeptide antigens.

[4] D. G. Smith and K. S. Johnson, *Gene* **67,** 31 (1988).

[5] N. Kartner, J. W. Hanrahan, T. J. Jensen, A. L. Naismith, S. Sun, C. A. Ackerley, E. F. Reyes, L.-C. Tsui, J. M. Rommens, C. E. Bear, and J. R. Riordan, *Cell* **64,** 681 (1991).

[6] N. Kartner, O. Augustinas, T. J. Jensen, A. L. Naismith, and J. R. Riordan, *Nature Genet.* **1,** 321 (1992).

TABLE I
SYNTHETIC PEPTIDE AND FUSION PROTEIN IMMUNOGENS

Designation	Location in CFTR[a]	Oligopeptide sequence[b]
P1	D58–C75	DRELASKKNPKLINALRRC
P2	Y28–S45	CYRQRLELSDIYQIPSVDS
P3	E725–E746	CEEDSDEPLERRLSLVPDSEQGE
P4	R104–R117	CRIIASYDPDNKEER
P6	E279–K294	CEAMEKMIENLRQTELK
P7	D373–E384	CDFLQKQEYKTLE
P9	R1066–K1080	CRAFGRQPYFETLFHK
P10	A1465–L1480	CAALKEETEEEVQDTRL
FspRI	M348–K698	GST–fusion protein[c]
XH13	F712–T756	GST–fusion protein
RC1-1	N1197–L1480	GST–fusion protein

[a] Location within CFTR of the N-terminal and C-terminal amino acid position for each oligopeptide or fusion protein is indicated by the conventional single-letter amino acid code for that position, followed by its published numerical sequence position (see Introduction).

[b] Single-letter amino acid codes are given for sequences as synthesized. Except for P1, which has an endogenous cysteine, N-terminal cysteine residues were added to facilitate coupling to carrier protein.

[c] GST is *S. japonicum* glutathione transferase; fusion proteins are described in Fig. 1.

Preparation of Immunogens

Oligopeptides are synthesized with N-terminal cysteine residues to facilitate coupling to a carrier protein and are obtained, purified and reduced, in C-terminal amide form.[7] Oligopeptides are kept under argon, and all buffers used for oligopeptide work are vacuum degassed and purged with nitrogen before use. The sequences of the oligopeptide antigens that are used to immunize rabbits and mice for anti-CFTR antibody production are shown in Table I and their locations are indicated in Fig. 1, relative to a linear map of CFTR that shows its domain organization.[2] Oligopeptides are coupled to a carrier protein, keyhole limpet hemocyanin (KLH), using *m*-maleimidobenzoyl-*N*-hydroxysuccinimide ester (MBS) essentially according to the method of Marcel *et al.*[8] In a typical coupling, 4 mg of synthetic oligopeptide coupled to 7 mg KLH is obtained in 0.5 ml of phosphate-buffered saline (PBS; 150 m*M* NaCl, 20 m*M* sodium phosphate, pH 7.3, throughout this work). This mixture is used directly, after mixing with

[7] Multiple Peptide Systems, San Diego, California.

[8] Y. L. Marcel, T. L. Innerarity, C. Spilman, R. W. Mahley, A. A. Protter, and R. W. Milne, *Arteriosclerosis* **7,** 166 (1987).

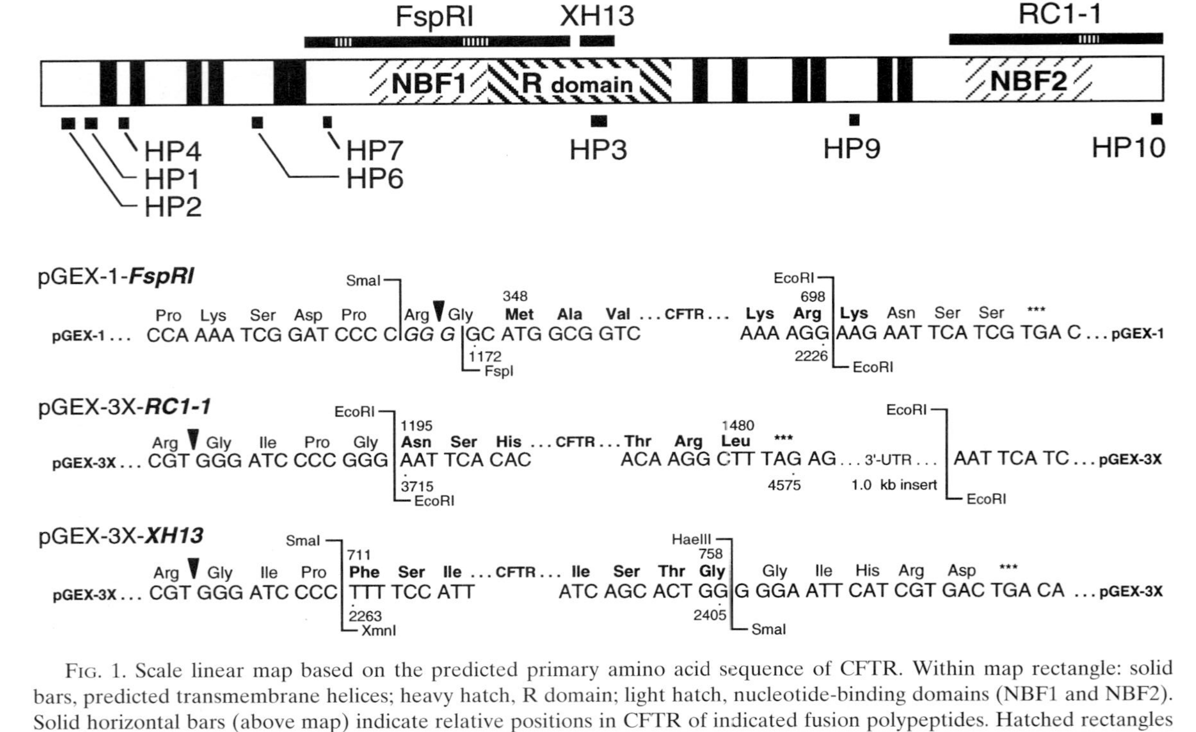

FIG. 1. Scale linear map based on the predicted primary amino acid sequence of CFTR. Within map rectangle: solid bars, predicted transmembrane helices; heavy hatch, R domain; light hatch, nucleotide-binding domains (NBF1 and NBF2). Solid horizontal bars (above map) indicate relative positions in CFTR of indicated fusion polypeptides. Hatched rectangles within FspRI bar denote known limits of the region within which some MAbs bind; L12B4 (left hatched rectangle, in pre-NBF1 region), and L3C4 (right hatched rectangle, at C-terminal end of NBF1). Hatched rectangle within RC1-1 bar denotes known limits of region (C-terminal end of NBF2) within which the MAb, M3A7, binds. Short solid bars (below map) indicate relative positions in CFTR of indicated oligopeptides. Beneath the map are shown the constructs used to generate CFTR fusion proteins in pGEX vectors. Bold amino acids represent CFTR coding sequences, which are preceded in the expressed fusion protein by *Schistosoma japonicum* glutathione transferase (GST) and short vector sequences. The CFTR coding sequence is followed by a short vector sequence for FspRI, but for RC1-1 the CFTR termination codon is used. Numbers

adjuvant, for the immunization of animals (see section on Immunization Schedules).

We use two large fragments of CFTR cDNA to generate fusion proteins, designated FspRI and RC1-1. FspRI is derived from the cDNA clone T16-4.5,[1,9] where *Fsp*I and *Eco*RI sites yielded a fragment including nucleotides 1172 to 2226, according to the published sequence.[2] This fragment is cloned into the *Sma*I site of the pGEX-1 vector.[4,6] RC1-1 is derived from cDNA clone C1-1/5,[2,10] from an *Eco*RI site at nucleotide 3175 to the end of the cDNA fragment,[1,2,6] beyond the translation stop codon, and is cloned into the *Eco*RI site of the pGEX-3X vector.[4,6] A third, small fragment designated XH13, derived from cDNA clone T16-4.5,[2,9] is also cloned into pGEX-3X, this comprising the cDNA sequence from the *Xmn*I site at nucleotide 2263 to the *Hae*III site at nucleotide 2405.[2,6] The CFTR fragments produced as fusion proteins are summarized in Table I, and the constructs in pGEX vectors and the location of resulting protein fragments within the primary sequence of CFTR are shown in Fig. 1. All three fragments are expressed in bacteria, as fusions with the pGEX vector-encoded *Schistosoma japonicum* glutathione *S*-transferase (GST), as described.[4,6]

The fusion protein XH13 is soluble and is purified on glutathione agarose beads, as described.[4] The fusion proteins FspRI and RC1-1 are not soluble. This means that one of the main advantages of using the pGEX vector system, rapid affinity purification of the product from bacterial lysates, is not possible. Insolubility of fusion proteins is a common problem, and during the time that it took to establish a method for purifying FspRI and RC1-1, we took the unconventional shortcut of beginning the immunization schedule of mice using a mixture of whole bacteria that expressed the fusion

[9] Accession No. 61150; American Type Culture Collection, Rockville, Maryland.

[10] Accession No. 61160; American Type Culture Collection, Rockville, Maryland.

above amino acids indicate positions in published CFTR amino acid sequence; numbers below nucleotide bases indicate positions in published CFTR cDNA sequence (see text). Arrowheads indicate Factor X_a cleavage sites. pGEX-1 has no engineered Factor X_a site, but in pGEX-1-FspRI a cloning artifact introduced bases GGG (italics) between the vector *Sma*I site and the insert FspRI site, thereby introducing Arg–Gly at the N terminus of the insert polypeptide, and fortuitously providing a Factor X_a cleavage site. MAbs L11E8 and M3A7 were determined to be IgG_1, and L12B4 and L3C4 were IgG_{2a}. Mapping of the epitopes for L3C4, L12B4, and M3A7 was done by preparing synthetic oligonucleotide primers incorporating restriction-site ends to produce overlapping halves of the original cDNA used to prepare fusion proteins, and cloning the halves into the pGEX vector. Colonies were screened with MAb, and the cDNA half giving a positive reaction was iteratively subdivided and rescreened. By this means, the epitopes were crudely localized to within ~30 amino acids, as follows: L3C4, A559 to K593; L12B4, N386 to A412; and M3A7, K1365 to C1395.

proteins, sonicated with adjuvant. Because this approach was successful, and can considerably reduce the amount of time and effort required for hybridoma production, preparation of the bacteria for immunization and for fusion protein purification is described in detail.

For bacteria transformed with pGEX constructs of both FspRI and RC1-1, lysates are prepared by first growing 50-ml bacteria cultures to saturation overnight in Luria–Bertani (LB) medium containing 50 μg/ml ampicillin. This starter culture is transferred to 500 ml of fresh medium in a 2-liter shaker flask, and induction is initiated after 1 h of growth in a 37° shaker by adding 12 mg of solid isopropyl-β-D-thiogalactopyranoside (IPTG). After 4 hr of further growth, the bacteria are pelleted at 4° in 250-ml bottles at 10,000*g* for 10 min and washed once by dispersing the pellet into 30 ml PBS and centrifuging at 10,000*g* for 5 min in a tared 50-ml tube. Excess buffer is drained and the pellet wet weight determined (typically 1–2 g). The pellet is dispersed into 10% dimethyl sulfoxide (DMSO) in PBS at 100 mg/ml (wet w/v) and stored at −80° until further use (see section on Immunization Schedules). Preparation of an inclusion body fraction does not significantly enrich for these particular fusion proteins, but approximately 3–5% of total bacterial protein is CFTR fusion protein, as determined from Coomassie Blue R250 staining after sodium dodecyl sulfate–polyacrylamide gel electrophoresis (SDS–PAGE) gels.[11]

Fusion protein purification is required for immunization of rabbits and is used for the final injections in mice and for hybridoma screening. Bacteria stored in 10% DMSO (see above) are sonicated to disrupt the cells and centrifuged at 12,000*g* for 5 min. The supernatant is discarded and the pellet solubilized in sample solubilization buffer for SDS–PAGE,[11] and is then separated on 1.5-mm preparative 10% polyacrylamide gels. Proteins are visualized by soaking gels in ice-cold 0.25 *M* KCl and viewing them under low-angle lighting, over a dark background. Fusion protein bands can be identified by comparison with a control lysate lane from bacteria expressing GST alone and M_r markers. M_r of the fusion protein bands is determined in advance by Coomassie Blue R250 staining of an analytical gel. The fusion protein bands are excised, and the protein is electroeluted in a Model 422 Electro-Eluter.[12] Eluates are concentrated to 5–10 mg/ml protein in Centricon 30 centrifugal microconcentrators,[13] and the SDS concentration is greatly reduced by cold precipitation at 0°, overnight, in 12-mm-diameter scored glass tubes. Crystallized SDS is pelleted by centrifugation in a refrigerated microfuge at 0°, and the supernatant is retained

[11] U. K. Laemmli, *Nature* **227,** 680 (1970).

[12] Bio-Rad Laboratories, Richmond, California.

[13] Amicon, Beverly, Massachusetts.

and used directly as a source of purified fusion protein. Yields are typically on the order of 100 μg per gel (approximately 1 mg per gram of bacterial pellet wet weight). Protein determinations are done using a modified method of Lowry *et al.*[14] that is insensitive to detergents and reducing agents,[15] and purity is judged using SDS–PAGE and Coomassie Blue R250 staining.[11]

For analytical purposes, for injection into rabbits, and for final intravenous (IV) injections into mice, fusion proteins are cleaved with activated plasma coagulation Factor X protease (Factor X_a)[16] in 50 m*M* Tris-HCl, pH 8.0, 100 m*M* NaCl, 1 m*M* $CaCl_2$, as described.[4] Typically, 50 μg fusion protein and 500 ng Factor X_a are incubated for 3 hr at 25° in a volume of 100 μl. The products are then separated on 12% polyacrylamide gels for FspRI and RC1-1 and 22% gels for XH13 and purified by band excision and electroelution as described earlier for fusion proteins. Cleavage gives predicted fragment sizes by removing the GST portion of the fusion protein, and in the case of RC1-1, gives three CFTR-derived bands, one being the intact fragment (uncut, internally by Factor X_a), and the other two being due to a predicted internal Factor X_a cleavage site between R1245 and T1246, according to the published sequence.[2] For RC1-1, all three bands are excised, electroeluted, and pooled for use in immunization of animals.

Immunization Schedules

Rabbits are bled for preimmune sera and then immunized every 4 weeks with approximately 1 mg equivalent of oligopeptide, as the KLH-conjugate, or 100 μg of the CFTR portion of purified and cleaved fusion protein (see above). Inoculum is prepared by bringing the volume of the protein solution to 0.25 ml in PBS and adding an equal volume of Freund's adjuvant and emulsifying the mixture by passage between syringes. The first inoculation is done subcutaneously at multiple interscapular sites (0.1 ml per site) with complete adjuvant; subsequent inoculations are in incomplete adjuvant. Bleeds are taken 10 days after immunization to determine specific antibody titer. Useful titers are generally achieved after the 4th to 6th bleeds, and after 8 to 10 bleeds the rabbits are exsanguinated and sera with acceptable titers are pooled.

BALB/c mice, 12–16 weeks old are immunized with the equivalent of 400 μg of oligopeptide per intraperitoneal (IP) injection. The conjugated oligopeptide in PBS is mixed 1:1 with Freund's adjuvant (complete for first injection, incomplete for all others), and 100 μl is injected IP. A final

[14] O. H. Lowry, N. J. Rosebrough, A. L. Farr, and R. J. Randall, *J. Biol. Chem.* **193,** 265 (1951).
[15] G. L. Peterson, *Methods Enzymol.* **91,** 95 (1983).
[16] Sigma Chemical Co., St. Louis, Missouri.

IV injection is done with the equivalent of 200 μg of oligopeptide in 100 μl of PBS. Injections are IP on days 0, 14, 42, 56, and 63, with the final IV injection on day 63. In the case of fusion protein antigens, 20 μl (2 mg wet weight) of thawed and resuspended bacteria from frozen stock in 10% DMSO medium (see section on Preparation of Synthetic Antigens) is transferred to a 1.5-ml microcentrifuge tube, 5 μl of 10% (w/v) SDS is added to this, and the mixture is heated in a 95° heating block for 5 min, then 225 μl PBS and 250 μl of Freund's adjuvant (complete for first injection, incomplete for second injection) is added to the cooled sample. The mixture is sonicated, using a microprobe sonicator[17] with a 2.3-mm probe to produce a stable emulsion. Injections are 100 μl IP on days 0 and 14. Further injections are done with fusion protein (purified as described) by preparing inocula using 25 μl of purified protein (at 10 mg/ml protein concentration), to which is added 225 μl of PBS and 250 μl of Freund's incomplete adjuvant, this mixture being sonicated with a microprobe sonicator[17] to form a stable emulsion. Then 100 μl of this is injected into mice IP to hyperimmunize the animals with 100 μg of fusion protein on days 42, 56, and 63. Additionally, 100 μg of fusion protein is diluted to 100 μl in PBS, sterile filtered by centrifugation through a 0.22-μm pore-size microcentrifuge filter unit, and injected IV via the tail vein, on day 63.

Antibody Production and Purification

Preparation and affinity purification of polyclonal rabbit antibodies is done according to routine methods (see, e.g., Ref. 18). An exception arises when rabbit antisera are used for identification of CFTR R-domain fusion protein in Western blots of bacterial lysates,[19] where neither crude nor affinity purified immunoglobulins gives sufficiently clean background. In this instance, sera are preabsorbed with an acetone powder prepared from bacteria expressing the control vector, pGEX-3X (nonfusion) alone, then ammonium sulfate precipitated, and the immunoglobulins further purified by FPLC[20] on Protein A Superose 12.[19–21] This method of preabsorption yields results in Western blots of bacterial extracts that are far superior to those obtained using affinity purified immunoglobulins (Igs).

We decided at the outset that the use of MAbs for Western blotting

[17] Micro Ultrasonic Disrupter (50 watt output); Kontes, Vineland, New Jersey.

[18] E. Harlow and D. P. Lane, *in* "Antibodies: A Laboratory Manual." Cold Spring Harbor Laboratory, New York, 1988.

[19] A. M. Dulhanty and J. R. Riordan, *Biochemistry* **33,** 4072 (1994).

[20] Pharmacia Biotech, Uppsala, Sweden.

[21] J. Sambrook, E. F. Fritsch, and T. Maniatis, *in* "Molecular Cloning: A Laboratory Manual," 2nd Ed., p. 18.15. Cold Spring Harbor Laboratory Press, New York, 1989.

had highest priority and, consequently, a nitrocellulose dot-blot assay is used for hybridoma screening where fusion protein is used as antigen. In the case of synthetic oligopeptides, enzyme-linked immunosorbent assays (ELISA) are used, where plates are coated with the synthetic oligopeptide that is used as immunogen. An enormous amount of literature exists that describes the use of ELISA for screening (see, e.g., Ref. 18); however, while this latter approach is widely used, it proved unsuccessful for the isolation of useful hybridomas in our hands, and so it is not described further here. Instead, we will describe in detail the use of the successful dot-blot screening assay.

Hybridoma Production and Screening

On day 67 (see section on Immunization Schedules) spleen cells from immunized mice are mixed 3:1 with Sp2/0-Ag14 murine myeloma fusion partner cells[22,23] and fused according to an adaptation of standard polyethylene glycol (PEG) cell–cell fusion methods.[18,24,25] Typically six 96-well plates are seeded per spleen and when wells with good growth in selective medium appear to be about one-quarter confluent, spent medium from a feeding is collected and screened for specific antibody production. Candidate hybridomas are selected on the basis of their stable secretion of antibodies with specificity toward the purified oligopeptide antigen used for immunization or, where fusion protein antigens are used, candidate hybridomas are selected on the basis of their stable secretion of antibodies with specificity for the CFTR fusion protein used as immunogen, in a differential dot-blot assay in a first round of screening (see Fig. 2),[6] and Western blot detection of the full-length recombinant CFTR protein produced in a baculovirus expression vector system in a second round of screening.[5,6] The dot-blot assay is performed by marking an array of 5- × 25-mm rectangles on sheets of nitrocellulose, numbering the right end of each rectangle and spotting the left end with 1 μl of fusion protein antigen at 0.1 mg/ml in 1% SDS solution, containing a trace of Pyronin Y so as to leave a faint pink marker spot. A second, nonoverlapping spot bearing the control nonfusion protein (GST) is spotted to the right of the antigen spot. The spots are dried and the nitrocellulose blots blocked with 3% bovine serum albumin (BSA) in Tris-buffered saline (TBS), according to the Western blotting protocol of Towbin *et al.*[26] Then the rectangles are cut from the wet sheets of nitrocellu-

[22] Accession No. CRL1581; American Type Culture Collection, Rockville, Maryland.
[23] M. Shulman, C. D. Wilden, and G. Köhler, *Nature* **276,** 269 (1978).
[24] M. L. Gefter, D. H. Margulies, and M. D. Scharff, *Somat. Cell Genet.* **3,** 231 (1977).
[25] R. H. Kennett, *Methods Enzymol.* **58,** 345 (1979).
[26] H. Towbin, T. Staehlin, and J. Gordon, *Proc. Natl. Acad. Sci. U.S.A.* **76,** 4350 (1979).

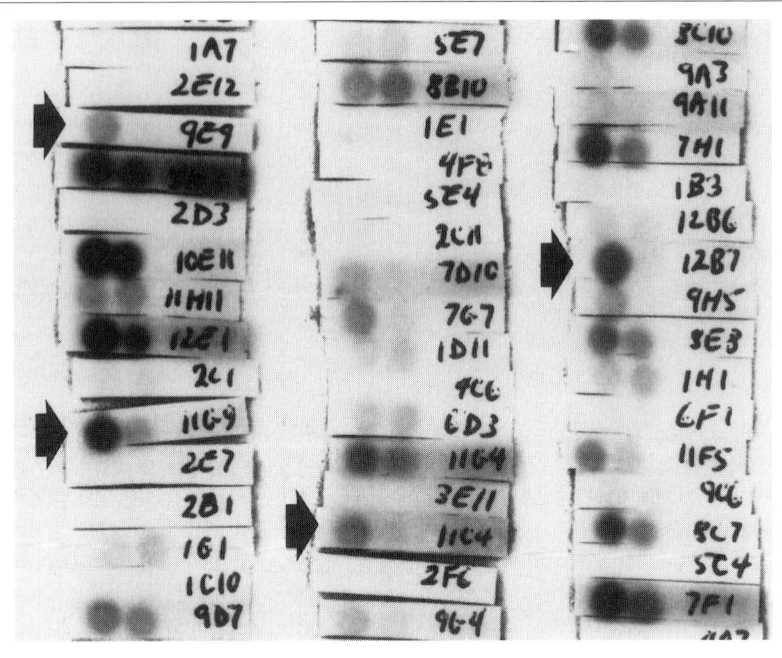

FIG. 2. Hybridoma screening with a dot-blot assay. Typically, 600–900 wells were positive for growth out of 1200 wells seeded from a pooled pair of spleens. Use of standard microtiter plate technology makes it easier to screen all wells than to pick those that are positive for growth, so 1200 test strips would normally be processed in a hybridoma screening. Colonies suitable for screening usually formed in microtiter plates within 9–12 days and were screened using nitrocellulose dot blots of purified GST-CFTR fragment fusion proteins and control GST. Nitrocellulose rectangles are labeled with the microtiter plate well identification number at their right end and spotted with fusion protein at their left end (1-μl spots, each containing 0.1 μg protein). To the right of each fusion protein spot is a control GST spot. After probing with hybridoma supernatants and radioactive second antibody, test strips were taped to plastic sheets, photocopied, then exposed to film. Shown here is a sandwich of the photocopy and processed film which shows the test strip outlines and identification numbers and the superimposed results of the screening. Arrows indicate some candidate hybridomas, where differential staining is as expected for CFTR-specific antibody. The rightmost arrow shows a relatively strong positive for fusion protein with no staining for the GST control, the ideal signal for a candidate CFTR-specific hybridoma. Double positives indicate GST-specific reactivity, which is not of interest. The microtiter plate well of origin is easily read from the in-register photocopy (in this example "12B7" for the clone indicated by the rightmost arrow).

lose with a sharp scalpel and inserted individually into the wells of 96 deep-well microtiter plates.[27] where each well had previously been loaded with a 100-μl aliquot of test hybridoma supernatant diluted to 1 ml with 3% BSA/TBS. Incubation is allowed for 4 hr, then the strips are removed and

[27] Beckman Instruments, Fullerton, California.

processed in batch in a plastic tub. This consists of washing out the primary antibody, incubating with ^{125}I-labeled second antibody (goat anti-mouse IgG, at 10^6 cpm/ml),[28,29] washing out the second antibody, arranging the wet strips in rows on stiff vinyl sheets (cut from 4-mil clear vinyl page protectors to form a 20- $\times$ 25-cm "booklet"), blotting the strips dry, taping their edges, and then exposing the assembled "booklet" to film overnight (see Fig. 2). Candidate wells are recloned twice by limiting dilution.[18]

Monoclonal Antibody Purification

Hybridoma supernatants containing up to 10 μg/ml of MAb can be used for further characterization of MAbs, but larger scale production and purification is required for more critical applications. MAbs developed toward fusion proteins are described further here and listed in Table II, and larger quantities of them are obtained by inoculating pristane-primed BALB/c mice with 2.5×10^6 viable hybridoma cells and harvesting ascites fluid (2–5 ml per mouse) when there was noticeable abdominal distension (1–4 weeks). Purification is by ammonium sulfate precipitation of immunoglobulins followed by FPLC (fast protein liquid chromatography), sequentially by affinity desorption from protein A-Superose 12, or protein G, and then anion exchange on Mono Q.[20] IgG_1 MAbs are passed directly over a protein G column, the bound fraction is eluted with 50 m*M* MES in 1% (v/v) acetic acid, and the eluate is neutralized immediately with 2 *M* Tris base (approximately 1/10 vol.). The eluate is then exchanged into 100 m*M* Tris-HCl, pH 8.0, on a desalting column and passed over a protein A column to remove contaminating IgG isotypes. For IgG_{2a} antibodies the immunoglobulin fraction is loaded onto a protein A column in 100 m*M* Tris-HCl, pH 8.0, and eluted with a linear gradient starting in loading buffer and ending in 50 m*M* MES in 1% (v/v) acetic acid. The specific antibody peak is collected and neutralized quickly with 2 *M* Tris-base. This method allows separation of isotypes and is crucial to obtaining clean backgrounds, especially in subsequent immunohistochemical methods with L12B4. MAbs are further desalted into 20 m*M* Tris-HCl, pH 7.6, and then loaded onto a Mono Q column, which is subsequently washed in the same buffer. The Mono Q column is eluted with a linear gradient ending in 0.3 *M* NaCl and 20 m*M* Tris-HCl, pH 7.6. The antibody peak is collected, adjusted to 2 mg/ml protein in 150 m*M* NaCl, 50 m*M* Tris-HCl, pH 7.5, 0.04% sodium azide, and an equal volume of glycerol is added to the solution for storage at $-20°$. In this form, the MAbs are stable for at least 2 years.

[28] W. M. Hunter and F. C. Greenwood, *Nature* **194,** 495 (1962).
[29] S. Sonoda and M. Schlamowitz, *Immunochemistry* **7,** 885 (1970).

TABLE II
POLYCLONAL AND MONOCLONAL ANTIBODY CHARACTERIZATION

Immunogens		Antibodies		Antibody characterization[a]						
				Rabbit polyclonal			Murine monoclonal			
Antigen	Location in CFTR[b]	Polyclonal	MAbs (total number)[c]	WB	IHC	IP	EIA	WB	IHC	IP
P1	D58–C75	HP-1	P1-1,3,6 (3)	±	−	−	++	−	−	−
P2	Y28–S45	HP-2	P2-2,5 (2)	±	−	−	++	−	−	−
P3	E725–E746	HP-3/PAb6[d]		±	−	+[e]				
P4	R104–R117	HP-4	K11C5 (1)	±	−	−	+	−	−	−
P6	E279–K294	HP-6		±	−	−				
P7	D373–E384	HP-7	P7-9,14 (2)	−	−	−	++	−	−	−
P9	R1066–K1080	HP-9		±	−	−				
P10	A1465–L1480	HP-10		+	−	±				
FspRI	M348–K698	αFspRI	L3C4, L11E8,[e] L12B4 (12)	±	±	±		++	+	++
XH13	F712–T756	αXH13	J14H10 (2)	±[e]	−	−		±	+	+
RC1-1	N1197–L1480	αRC1-1	M3A7 (10)	±	±	±		++	++	++

[a] EIA, Enzyme-linked immunosorbent assay (ELISA); IHC, immunohistochemistry; IP, immunoprecipitation; WB, Western blot. −, negative; ±, weak positive; +, positive; ++, strong positive; blank, not done.

[b] Location is indicated by single-letter amino acid codes, and sequence positions, of the peptide N-terminal and C-terminal amino acids, according to the published sequence of CFTR (see Introduction and Table I).

[c] Antibodies discussed in the text are listed, and designations of different antibodies raised toward a common immunogen are shown. Numbers in parentheses indicate total positive clones identified in primary hybridoma screening.

[d] An acetone powder preabsorbed fraction of HP-3 has been designated PAb6 (see section on Antibody Production and Purification).

[e] Antibodies localizing to the R domain of CFTR were reactive toward the original immunogen (see Table I) and toward isolated R domain, which could be immunoprecipitated, but were not reactive toward full-length CFTR.

Antibody Characterization: Applications and Limitations

Although our primary concern was to obtain antibodies useful for Western blots, we also screened the available panels of polyclonal antibodies and MAbs for immunoprecipitation and immunohistochemistry applications. With this characterization, our findings were that the MAbs developed toward fusion proteins were superior for all applications; however, it is instructive to compare results for the other antibodies.

Western Blotting

We have previously published the use of both polyclonal rabbit antibodies raised against synthetic oligopeptides and fusion proteins, and murine MAbs raised against fusion proteins.[5,6] These results, and some previously unpublished work, are summarized in Table II. A panel of antibodies with epitopes ranging across the full length of CFTR is used in Western blots of the Sf9 (*Spodoptera frugiperda* fall armyworm ovary cells) baculovirus expression system to show that full-length CFTR is being produced in those cells, concomitant with expression of a cAMP-stimulated anion channel, this being the first evidence that CFTR is indeed a regulated chloride channel.[5] Affinity-purified rabbit antisera, and especially the MAbs L12B4 and M3A7, raised against large fusion protein fragments of CFTR, yield strong bands at approximately M_r 140,000 in Western blots of CFTR-expressing Sf9 cells, and none in control β-galactosidase-expressing Sf9 cells. The affinity-purified rabbit polyclonal antibody HP-10 is used in Western blotting of lysates from human colon carcinoma T84 cells[30] and CFTR-expressing Sf9 cells to show that *N*-glycanase treatment alters the M_r of CFTR expressed in mammalian cells, but not that expressed in Sf9 cells.[5] The shift in M_r of mammalian-expressed CFTR from a diffuse band at approximately M_r 170,000 to a sharper band at approximately M_r 140,000 on *N*-glycanase treatment is consistent with the complex glycosylation of CFTR expected in mammalian cells.[31] That the *N*-glycanase treated mammalian-expressed CFTR gives a band of apparent M_r identical to that of the SF9-expressed protein is also consistent with the expected presence of only core glycosylation in the insect system, which is unable to process N-linked carbohydrate to a complex form.[32] Similar results are obtained when probing blots with M3A7 and L12B4, and it has been shown that these

[30] Accession No. CCL-248; American Type Culture Collection, Rockville, Maryland.

[31] S. H. Cheng, D. P. Rich, J. Marshall, R. J. Gregory, M. J. Welsh, and A. E. Smith, *Cell* **66,** 1027 (1991).

[32] J. Vialard, M. Lalumiere, T. Venet, D. Briedis, G. Alkhatib, D. Henning, D. Levin, and C. Richardson, *J. Virol.* **64,** 37 (1990).

antibodies detected both the wild-type and the mutant ΔF508 CFTR with equal facility.[6]

Another heterologous expression system is also developed in the Chinese hamster ovary CHO-K1 cell line,[33] resulting in the transfectant cell lines, BQ2, expressing the wild-type human CFTR, and BQΔF, expressing the mutant ΔF508 CFTR.[34,35] When we compare Western blots of detergent extracts from the CHO cell lines, we observe only poor staining of a putative CFTR band with rabbit antisera prepared against fusion protein and none with those prepared against synthetic oligopeptides, except HP-10, which gives a weak band. The putative CFTR band is seen only in the overexpressing BQ2 cells, as a diffuse band at approximately M_r 170,000. No such band is seen in controls and none is seen in the BQΔF cells expressing the ΔF508 CFTR, which undergoes biosynthetic arrest as a result of a processing defect and is seen only in its core-glycosylated form at approximately M_r 140,000.[31] Furthermore, the protein degradation that accompanies the biosynthetic arrest of ΔF508 CTFR reduces the amount detectable compared with wild-type CFTR levels.[36] MAbs prepared against synthetic oligopeptides are unable to detect any bands in the CHO system. In contrast, the MAbs L3C4, L12B4, and M3A7 that are prepared against fusion protein give strong diffuse bands at M_r 170,000 in CFTR-expressing BQ2 cells, and none in the parental cell line. In BQΔF cells, sharp paired bands are seen at approximately M_r 140,000. The sharpness of the bands is likely because the protein has not been processed to complex glycosylation, and the bands may be paired because of the use of alternate translation initiation sites.[37,37a] Blotting in the BQΔF system indicates a high degree of sensitivity and specificity. Interestingly, the MAbs L11E8 and JI4H10, which proved useful for other purposes, do not detect full-length CFTR in Western blots, even though they detect the partial polypeptides, *Fsp*RI and isolated R domain, by Western blotting with no difficulty. We have partially characterized eight different MAbs with specificity toward R domain and found the same phenomenon to hold true for all of them.[38] This suggests

[33] Accession No. CCL-61; American Type Culture Collection, Rockville, Maryland.

[34] X.-B. Chang, J. A. Tabcharani, Y.-X. Hou, T. J. Jensen, N. Kartner, N. Alon, J. W. Hanrahan, and J. R. Riordan, *J. Biol. Chem.* **268,** 11304 (1993).

[35] J. A. Tabcharani, X.-B. Chang, J. R. Riordan, and J. W. Hanrahan, *Nature* **352,** 628 (1991).

[36] G. L. Lukacs, A. Mohamed, N. Kartner, X. B. Chang, J. R. Riordan, and S. Grinstein, *EMBO J.* **13,** 6076 (1994).

[37] T. P. Carroll, M. M. Morales, S. B. Fulmer, S. S. Allen, T. R. Flotte, G. R. Cutting, and W. B. Guggino, *J. Biol. Chem.* **270,** 11941 (1995).

[37a] S. Pind, A. Mohamed, X.-B. Chang, Y.-X. Hou, T. J. Jensen, D. B. Williams, and J. R. Riordan, unpublished data.

[38] N. Kartner, A. M. Dulhanty, and J. R. Riordan, unpublished data (1993).

that there may be some conformational sequestration of sites within the R domain even after SDS-denaturation of full-length CFTR. In this regard, it is interesting that the predicted M_r of CFTR is 168,000 (unglycosylated), compared with an observed value of approximately 140,000. It seems likely that some conformation is preserved, even under the denaturing conditions of SDS–PAGE, that causes anomalous migration and may account for the inability of anti-R domain antibodies to detect R domain in the environment of the full-length protein.

In the baculovirus system CFTR is grossly overexpressed, and the CHO system is also a heterologous overexpression system. We were interested in comparing results in tissue culture models representing more realistic, relatively low-expression systems. Western blots of detergent extracts of human epithelial cell lines having endogenous functional expression of CFTR were thus probed. T84,[30] Caco-2,[39] and HT-29[40] are human colon carcinoma epithelial cell lines that express CFTR, and Calu-3[41] is a human lung adenocarcinoma cell line that expresses CFTR. PANC-1[42] is a human colon carcinoma epithelial cell line that has a very low level of CFTR expression detectable only by RT-PCR, and HeLa[43] is a human epitheloid cell line with no detectable CFTR expression. Rabbit polyclonal antibodies prepared against fusion protein show weak M_r 170,000 bands in T84 and HT-29 cells that appear absent in PANC-1 cells, but nonspecific background was high. With polyclonal antibodies prepared against synthetic oligopeptides, no difference between T84, HT-29, and PANC-1 can be discerned except with HP-10, which detects a diffuse band at approximately M_r 170,000 in T84 and HT-29 and none in PANC-1. The MAbs prepared against fusion proteins, L3C4, L12B4, and M3A7 (not the R domain antibodies L11E8 or J14H10) detect a discreet, clear, but somewhat diffuse band in T84, Caco-2, HT-29, and Calu-3 cells and no bands are seen in PANC-1 or HeLa. Astonishingly, all of the MAbs prepared against synthetic oligopeptides yield clear and very sharp bands at approximately M_r 170,000 that are of equal intensity in both T84 and PANC-1 cells. This band clearly is not CFTR, and no evidence of the characteristic diffuse band seen in all other mammalian CFTR-expressing cell lines can be discerned. Some of the salient features of Western blotting with MAbs are summarized in Fig. 3.

The MAb, M3A7, has proved particularly useful for Western blotting. It has been used in direct detection of CFTR expression in *Xenopus laevis*

[39] Accession No. HTB-37; American Type Culture Collection, Rockville, Maryland.
[40] Accession No. HTB-38; American Type Culture Collection, Rockville, Maryland.
[41] Accession No. HTB-55; American Type Culture Collection, Rockville, Maryland.
[42] Accession No. CRL-1469; American Type Culture Collection, Rockville, Maryland.
[43] Accession No. CCL-2; American Type Culture Collection, Rockville, Maryland.

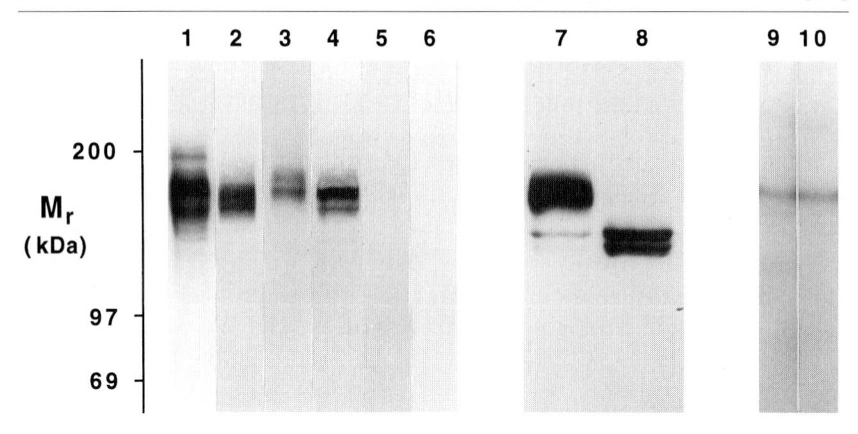

FIG. 3. Western blotting of test cell lines. Shown here are blots of 6% polyacrylamide SDS gels. Lanes 1–8 contained ~20 μg of protein as RIPA extract of whole cells and were probed with M3A7. Chemiluminescence film detection was adjusted to optimize resolution of the complexity of the banding pattern for CFTR in each lane, so this figure does not represent a quantitative comparison of cell lines; rather, we illustrate the diffuse and complex nature of the band for CFTR, which falls at ~M_r 170,000. Lane 1, Calu-3; lane 2, T84; lane 3, Caco-2; lane 4, HT-29; lane 5, PANC-1; lane 6, HeLa; lane 7, BQ2; lane 8, BQΔF. Note the diffuse and multiple banding of CFTR (lanes 1–4). Transillumination of the original chemiluminescence films reveals additional banding within the dense regions of lanes 1 and 2, which is difficult to reproduce here. Differences in mean M_r depending on cell line are also evident (e.g., compare lane 3 with lanes 2 and 4). Note the absence of staining in negative controls (lanes 5 and 6). A shift in M_r is observed for ΔF508 CFTR to ~140,000 (lane 8) compared with wild-type CFTR (lane 7). In lane 7, the large amount of heterologously expressed CFTR yields a heavy, mature glycosylated band (known as "band C") and faint immature CFTR bands (known as "band B"). Variations in CFTR band M_r and complexity are likely due to differences in glycosylation in clonally established tumor cell lines. The CFTR band hererogeneity is likely contributed to by multiple phosphorylation sites, complex glycosylation, alternate start codons for translation, and possibly steady-state intracellular degradation (extracts were prepared on ice with a cocktail of protease inhibitors). Differential extraction of bands could also be observed with different detergent mixtures (not shown). The complexity of the CFTR bands may be obscured in heavily loaded gels, and the diffuse nature of the bands may be less apparent in gels of inappropriately high polyacrylamide concentration. Lanes 9 and 10 are T84 and PANC-1 (50 μg protein each), respectively, stained with a representative MAb produced against oligopeptide antigen (P1-3; see Table II). Note identical staining in negative control (PANC-1, lane 10) and T84 (lane 9), and relative sharpness of bands (compared with T84 in lane 2) which are also ~170,000 (see text for discussion). All lanes shown here are to identical scale in the vertical dimension and were prepared on the same electrophoresis apparatus, although lanes 7 and 8 were run with wide-tooth combs and lanes 9 and 10 were run as preparative gels, where strips were cut for secondary hybridoma screening.

oocytes after CFTR cRNA microinjection,[44] and it has been used to monitor CFTR-containing fractions by both one- and two-dimensional Western blots in the purification and reconstitution scheme that was used to demonstrate directly that CFTR itself is a regulated anion channel.[45] M3A7 has been used to monitor regulation of CFTR expression in intestinal epithelial cells after differentiation,[46] after stimulation of intracellular cAMP levels,[47] and after stimulation of protein kinase C activity.[48] M3A7 has also proved a valuable tool in Western blotting for the study of membrane topology,[49] the regulation, biosynthesis, stability, and turnover of both wild-type and mutant forms of CFTR,[34,36,50,51] and to monitor expression of engineered mutant forms of CFTR in structure and function studies.[52,53] Interestingly, although MAbs raised against synthetic oligopeptides detected the individual oligopeptides used as immunogens both in ELISA assays and when immobilized on nitrocellulose in dot-blot assays with excellent sensitivity and specificity, they did not detect a band in Western blots of Sf9 cells expressing large amounts of full-length CFTR from the baculovirus expression vector system, or appropriate fusion protein fragment of CFTR, and were unsuccessful at detecting full-length CFTR in any heterologous or endogenous CFTR expression system.

Immunoprecipitation and Site-Directed Perturbation of CFTR Function

Affinity purified rabbit polyclonal antibodies give poor results with a relatively high background in immunoprecipitation experiments. In contrast, the MAbs raised against fusion proteins that are useful for Western blotting are also able to cleanly immunoprecipitate CFTR. Pind *et al.*[54]

[44] C. E. Bear, F. Duguay, A. L. Naismith, N. Kartner, J. W. Hanrahan, and J. R. Riordan, *J. Biol. Chem.* **266,** 19142 (1991).
[45] C. E. Bear, C. Li, N. Kartner, R. J. Bridges, T. J. Jensen, M. Ramjeesingh, and J. R. Riordan, *Cell* **68,** 809 (1992).
[46] R. Sood, C. Bear, W. Auerbach, E. Reyes, T. Jensen, N. Kartner, J. R. Riordan, and M. Buchwald, *EMBO J.* **11,** 2487 (1992).
[47] W. Breuer. N. Kartner, J. R. Riordan, and Z. I. Cabantchik, *J. Biol. Chem.* **267,** 10465 (1992).
[48] W. Breuer, H. Glickstein, N. Kartner, J. R. Riordan, D. A. Ausiello, and I. Z. Cabantchik, *J. Biol. Chem.* **268,** 13935 (1993).
[49] X.-B. Chang, Y. X. Hou, T. J. Jensen, and J. R. Riordan, *J. Biol. Chem.* **269,** 18572 (1994).
[50] G. L. Lukacs, X.-B. Chang, C. Bear, N. Kartner, A. Mohamed, J. R. Riordan, and S. Grinstein, *J. Biol. Chem.* **268,** 21592 (1993).
[51] T. J. Jensen, M. A. Loo, S. Pind, D. B. Williams, A. L. Goldberg, and J. R. Riordan, *Cell* **83,** 129 (1995).
[52] F. S. Seibert, J. A. Tabcharani, X.-B. Chang, A. M. Dulhanty, C. Matthews, J. W. Hanrahan, and J. R. Riordan, *J. Biol. Chem.* **270,** 2158 (1995).
[53] F. Becq, T. J. Jensen, X.-B. Chang, A. Savoia, J. H. Rommens, L.-C. Tsui, M. Buchwald, J. R. Riordan, and J. W. Hanrahan, *Proc. Natl. Acad. Sci. U.S.A.* **91,** 9160 (1994).
[54] S. Pind, J. R. Riordan, and D. B. Williams, *J. Biol. Chem.* **269,** 12784 (1994).

used a cocktail of M3A7 and L12B4 (2 μg/ml each) to coprecipitate CFTR and calnexin, a chaperone that binds CFTR and may play a role in the retention of ΔF508 CFTR in the endoplasmic reticulum. The immune complexes are recovered from detergent extracts on protein G agarose beads and CFTR recovery is quantitative when both antibodies are used, but not when either one is used alone.[55] CFTR has also been immunoprecipitated for *in vivo* and *in vitro* phosphorylation studies using M3A7,[34,52] and to monitor its state of phosphorylation in the presence of phosphatases and phosphatase inhibitors.[53] After cyanogen bromide cleavage of the protein, fragments of R domain are successfully recovered by immunoprecipitation with L11E8, and also with the polyclonal rabbit antibody PAb6, an acetone powder-absorbed fraction of HP-3.[19,52] It is of interest that none of the MAbs described here that were raised against synthetic oligopeptides is able to immunoprecipitate full-length CFTR from detergent extracts.

In studies of CFTR function in isolated endosomes, Lukacs *et al.*[56] showed that dissipation of a pH gradient by stimulation of CFTR channel opening could be completely inhibited with 50 n*M* M3A7. Similarly, inhibition of cell surface CFTR channel activity in fused respiratory epithelial cells by microinjection of M3A7 has been demonstrated.[57] Functional perturbation of cAMP-activated anion currents due to CFTR cRNA expression in *Xenopus laevis* oocytes was also shown with M3A7, where microinjection to an estimated concentration of 200 n*M* abolished stimulated channel activity. Interestingly, a similar concentration of L12B4 appeared to stimulate channel activity in this system, even in the absence of cAMP-mediated activation.[58] Further work to determine the significance of these effects with regard to structural perturbation of NBF2 and the pre-NBF1 regions of CFTR, with M3A7 and L12B4, respectively, could provide valuable insights into the molecular mechanism of CFTR function.

Immunohistochemistry

A survey of tissue staining has been done with all of the antibodies described here to determine their usefulness for immunohistochemistry. Our previously reported observations regarding immunolocalization of CFTR are based largely on the use of the MAbs, M3A7 and L12B4.[6] While other MAbs developed toward fusion proteins corroborated our findings with the latter two, they display neither the same specificity, nor the sensitiv-

[55] S. Pind, N. Kartner, and J. R. Riordan, unpublished data (1993).

[56] G. L. Lukacs, X.-B. Chang, N. Kartner, O. D. Rotstein, J. R. Riordan, and S. Grinstein, *J. Biol. Chem.* **267,** 14568 (1992).

[57] U. H. Schroder and E. Fromter, *Eur. J. Physiol.* **430,** 257 (1995).

[58] N. Kartner, T. Bond, C. E. Bear, and J. R. Riordan, unpublished data (1993).

ity. MAbs developed against synthetic oligopeptides show no CFTR-specific staining whatever, but in some cases appear to be reactive toward the extracellular matrix. Rabbit polyclonal antibodies, whether or not they are affinity purified, give significant background that made interpretation difficult, and none has sufficient sensitivity to detect CFTR in normal sweat duct.

Using L12B4 or M3A7, immunolocalization of CFTR in epithelial tissues is observed as a strong staining at the apical membranes of intercalated duct cells in human pancreas, but not in rat pancreas, in the apical membranes of striated duct cells of the rat parotid, and in the apical membranes of lower crypt cells in the human and rat gut. More importantly, though somewhat weaker, CFTR-specific staining is also seen at the apical membranes of the human sweat duct luminal cells, and to a lesser extent at the basolateral membranes of the same cells. In skin biopsies from ΔF508 homozygote patients no apical staining is seen in the lumen of the reabsorptive duct.[6] This corroborates *in vitro* studies showing that a processing defect is associated with this mutation.[59,60] Furthermore, normal localization to the apical membrane is observed with another mutation, G551D, which is not subject to a processing defect.[6,59] Localization of CFTR in lung tissue is inconclusive. While staining of submucosal glands of bronchi is observed with M3A7, perhaps the most sensitive and specific of our MAbs, similar staining is observed with a nonspecific, isotype-matched (IgG_1) control MAb.[61] This equivocal observation is disappointing, because of the importance of CFTR to lung pathology in CF, but difficulty in detection and low levels of expression in the lung have been reported by others.[62–64] Overall, our observations are consistent with those of other published works using well-characterized antibodies, *in situ* hybridization, and functional evidence of CFTR localization.[64–66]

[59] S. H. Cheng, R. J. Gregory, J. Marshall, S. Paul, D. Souza, G. A. White, C. R. O'Riordan, and A. E. Smith, *Cell* **63,** 827 (1990).

[60] R. J. Gregory, D. P. Rich, S. H. Cheng, D. W. Souza, S. Paul, P. Manavalan, M. P. Anderson, M. J. Welsh, and A. E. Smith, *Mol. Cell. Biol.* **11,** 3886 (1991).

[61] Pharmingen, San Diego, California.

[62] J. F. Engelhardt, M. Zepeda, J. A. Cohn, J. R. Yankaskas, and J. M. Wilson, *J. Clin. Invest.* **93,** 737 (1994).

[63] M. A. Rosenfeld, K. Yoshimura, B. C. Trapnell, K. Yoneyama, E. R. Rosenthal, W. Dalemans, M. Fukayama, J. Bargon, L. E. Stier, L. Stratford-Perricaudet, M. Perricaudet, W. B. Guggino, A. Pavirani, J. P. Lecocq, and R. G. Crystal, *Cell* **68,** 143 (1992).

[64] A. E. O. Trezise and M. Buchwald, *Nature* **353,** 434 (1991).

[65] C. R. Marino, L. M. Matovcik, F. S. Gorelick, and J. A. Cohn, *J. Clin. Invest.* **88,** 712 (1991).

[66] J. A. Cohn, O. Melhus, L. J. Page, K. L. Dittrich, and S. A. Vigna, *Biochem. Biophys. Res. Comm.* **181,** 36 (1991).

Cross-Reactivity

With the exception of M3A7, significant cross-reactivities are observed for all antibodies described here (see Discussion). In Western blots, all polyclonal antibodies give considerable background banding in epithelial systems where CFTR is expressed at relatively low levels. L12B4 also shows cross-reactive bands in some systems, but L12B4 and M3A7 can be used as a cocktail to obtain clean, quantitative recovery of CFTR in immunoprecipitation from detergent-solubilized cells. For reasons that are unclear, antibodies localizing to the R domain do not detect full-length CFTR in Western blots, even though they have no difficulty staining isolated R domain. MAbs developed against synthetic oligonucleotides seem to stain specifically non-CFTR protein bands in Western blots, with no apparent specificity for CFTR. This is the case even though the same MAbs easily stained dot blots of their respective immunogenic oligopeptides, derived from CFTR sequences, with no apparent cross-reactivity among oligopeptides used to produce other MAbs. In immunostaining, L12B4, M3A7, and J14H10 show clean staining of apical membranes in human pancreatic intercalated duct cells, but the MAbs L11E8 and L3C4 show cross-reactivity to other structures, making them less suitable for CFTR tissue localization. None of the MAbs developed toward synthetic oligopeptides stain any cellular structure. Most staining that is judged to be cross-reactive can be significantly reduced by competition with appropriate fusion protein or synthetic oligopeptide immunogen, indicating that the cross-reactive binding is likely via the variable region, antigen-binding portion of the antibody molecule and not via the Fc region or structural domains.

Discussion

Two important factors have a major impact on the outcome of attempts to produce specific antibody reagents. One is the nature of the antigen used for immunization, and the other is the method employed to screen for the specific antibodies that are sought, this being of primary importance in selecting useful hybridomas. In our hands, the use of synthetic oligopeptides as immunogens for the purpose of raising antibodies toward a full-length protein was not successful. While limited results were obtained for polyclonal sera, the MAbs generated by this approach were of little use. In contrast, the use of large fragments of CFTR as fusion proteins proved to be a successful approach. In the case of MAb development we were able to use unpurified bacterial lysates in the initial immunization of mice, because in the screening process a single clone with the desired specificity can be isolated. We have also done screening of recloned hybridomas with dot blots of bacterial lysates with little or no background in control lysates. This suggests that immunization and screening for hybridoma production could be done without purification of fusion protein, provided it is expressed at reasonable levels in the bacterial cells. This could prove to be a rapid and cost-effective general approach to hybridoma production where cloned cDNA fragments are available.

For the production of MAbs, we have found that the method employed for screening hybridomas should be designed around the application that has highest priority. That this is wise is especially evident in our efforts to produce MAbs specific for CFTR in Western blotting and immunohistochemical applications by screening hybridomas with oligopeptide immunogens in an ELISA assay. In spite of success in isolating positive clones that secreted MAbs with good specificity and sensitivity toward the CFTR-derived oligopeptides, these MAbs failed to identify full-length CFTR in their intended applications, Furthermore, these antibodies identified specific bands in Western blots that could easily have been misinterpreted as CFTR had we not screened a wide selection of cell lines, both CFTR-expressing and negative controls. Employing a screening assay using larger fragments of CFTR, or full-length CFTR, in a dot-blot or Western blot assay may have yielded more useful MAbs. We have concluded that the approach of using oligopeptide antigens as immunogens, combined with an ELISA screening assay to produce MAbs reactive toward full-length CFTR in Western blots is not a good one. That this approach is fraught with difficulties is evident from reports by other workers, as well.[67] On the other hand, the use of oligopeptide immunogens to produce polyclonal sera had some limited success in our hands, and has been highly successful where an oligopeptide from the C terminus of CFTR was used by another laboratory.[62,66] Our most successful approach, the use of large fragments or full-length protein and a dot-blot screening assay of hybridomas, has been used successfully by us to produce useful antibody reagents to both CFTR and the structurally related P-glycoprotein.[6,68]

A comparison of CFTR antibody reagents that have appeared in the literature is beyond the scope of the present article. As is inevitable in a new area of research, much of the interpretation of the CFTR literature since the cloning of the gene is confounded by reference to antibodies that in retrospect clearly cannot be specific for CFTR. In this regard, reflecting on what constitutes reasonable criteria for judging whether antibodies are specific for a target protein, in this case CFTR, and how to critically assess

[67] J. Walker, J. Watson, C. Holmes, A. Edelman, and G. Banting, *J. Cell Sci.* **108,** 2433 (1995).
[68] N. Kartner, D. Evernden-Porelle, G. Bradley, and V. Ling, *Nature* **316,** 820 (1985).

potential pitfalls of antibody use seem relevant here. We present some points for discussion:

1. CFTR is a glycoprotein with complex glycosylation in mammalian cells,[59] alternate translational start sites,[37,37a] and an anomalous M_r in SDS–PAGE.[5,6,59] This means that CFTR is detected as a diffuse band, or as a complex banding pattern in the vicinity of M_r 170,000 (see Fig. 3). Staining of a single sharp band in a test cell line should be taken as a strong indication that the antibody is not detecting CFTR, but some other protein that is not necessarily related in any way to CFTR (see point 4).

2. CFTR is expressed endogenously at detectable levels in the commonly available human epitheloid cell lines, Calu-3, T84, HT-29, and Caco-2 (Fig. 3). CFTR is not expressed at detectable levels in the commonly available epitheloid cell lines, HeLa and PANC-1 (Fig. 3). As is apparent from our experience with MAbs developed toward synthetic oligopeptide antigens, it is essential to test negative controls, and to reject antibodies that give apparent "CFTR" bands in extracts from cell lines that do not express CFTR. A good test would consist of side-by-side immunoblot lanes containing 50 μg protein from RIPA[18] extracts of the positive and negative epithelial pair, T84 and PANC-1, or any pair consisting of a well-characterized CFTR-expressing transfectant and its nonexpressing parental or mock-transfected partner. We should stress, however, that heterologous overexpression systems may yield results for poor antibodies of low affinity, and that the T84 and PANC-1 pair constitutes a more rigorous test. The phenomenon of seeing sharp bands at approximately M_r 170,000 in both positive and negative cell lines, as was the case for all of our anti-oligopeptide MAbs, seems to defy explanation, yet other workers have observed this phenomenon.[67] The bands are not CFTR; their sharpness suggests a soluble protein that is likely of cytoplasmic origin, and speculation that the bands are related in any way to CFTR may be unfounded (see point 4).

3. ΔF508 CFTR in heterologous expression systems is misprocessed and degraded,[36,59,60] as is the case for endogenous expression.[6] Comparison of wild-type and mutant CFTR from such cell lines should show obvious differences, including a shift to approximately M_r 140,000, a reduction in the complexity of the banding (a sharper band, or bands), and a relative decrease in protein expression level, compared with wild-type CFTR (Fig. 3), even though mRNA levels are similar to those in the cells expressing wild-type CFTR. In the CHO system that we have used,[34,35] a doublet band is seen for ΔF508 CFTR (Fig. 3), which likely is the result of alternate translational start sites.[37,37a]

4. It is a commonly held myth that affinity purified antibodies and MAbs are monospecific. No antibody is monospecific! The finite complexity of a

given system may be such that an antibody behaves within its confines as though it were monospecific, but whenever one ventures into new applications for the antibody, or applies it to a new cell line or tissue, one must be aware that cross-reactivity with proteins that are not the target protein is a real possibility. Obtaining a band of about the right size in an immunoblot of a one-dimensional SDS–PAGE gel is not proof of identity of the target protein, and neither is competition for antibody binding with antigen. This latter point is particularly important in view of the fact that many workers show no other "proof" of specificity than successful competition with antigen, usually a short synthetic oligopeptide. In our hands, all cross-reactivities of anti-CFTR antibodies with clearly non-CFTR proteins were mediated by antibody binding-site interactions that could be competed with antigen. This could be because the cross-reactive proteins have short stretches of primary sequence identity with CFTR, and not necessarily any overall homology. On the other hand, antibodies reactive toward short, contiguous amino acid sequences make use of only a fraction of the binding-site "footprint," which is approximately 750 $Å^2$ in area.[69,70] Regions of alternate binding specificity may exist over that area and overlap with the binding region that the immunogen of selection occupies. This means that competition with oligopeptide antigen may sterically abolish binding to cross-reactive proteins that are completely unrelated in amino acid sequence to the competing antigen. Consequently, competition with antigen not only is not proof of identity, it is not even proof of similarity. Ultimately, the only real proof of identity is a careful characterization of the putative target protein. This may mean having to purify enough of it to perform peptide mapping and peptide microsequencing, a daunting task. The present authors are aware of only a single example in the CF field of such a thorough, critical evaluation of a "CFTR candidate," and in that case the protein turned out to be unrelated to CFTR.[71]

It is hoped that thoughtful consideration of the preceding points will assist the reader in critically assessing the validity of conclusions based on antibody studies in the literature of the CF field, and other areas of research.

Conclusion

We have produced a number of antibody reagents, both polyclonal rabbit sera and MAbs, which have been useful for detecting CFTR in a

[69] Y. Patterson, S. W. Englander, and H. Roder, *Science* **249,** 755 (1990).

[70] W. G. Laver, G. M. Air, R. G. Webster, and S. S. Smith-Gill, *Cell* **61,** 553 (1990).

[71] R. C. Boucher, University of North Carolina, reported at the Cystic Fibrosis Foundation Conference: The CF Gene—Nine Months Later, Williamsburg, Virginia (1990).

variety of systems. As an approach toward producing specific antibodies, we found that immunizing animals with synthetic oligopeptide antigens was less useful than using large fragments of CFTR expressed as fusion proteins. This was especially true in the production of hybridomas, where the combination of immunization with oligopeptides and an ELISA-based screening assay led to the production of reagents that not only were not specific for CFTR, but gave misleading cross-reactivities. On the other hand, the combination of immunization with fusion protein antigens and a dot-blot screening assay proved highly successful. Our results suggest also that employing crude bacterial lysates from a fusion protein expression system for both immunization and screening may be a rapid and cost-effective method for developing useful MAb reagents. One of the MAbs that we have developed, M3A7, has proved to be a sensitive and specific reagent for the detection of CFTR in Western blotting, immunoprecipitation and tissue immunolocalization, as well as for the functional inhibition of CFTR in cells and organelles.

[48] Identification of Cystic Fibrosis Transmembrane Conductance Regulator in Renal Endosomes

By ISABELLE T. CRAWFORD and PETER C. MALONEY

Introduction

The disease, cystic fibrosis (CF), arises from mutations in the gene encoding a protein known as CFTR.[1,2] This disease is manifested as a defect in epithelial chloride (Cl) secretion in the airways and pancreas, resulting from failure to activate via a cAMP-mediated pathway one or more of the Cl channels required for fluid secretion.[3–5] Study of the purified CFTR protein[4] and the finding of abnormal Cl channel activity in natural or

[1] B. Kerem, J. M. Rommens, J. A. Buchanan, D. Markiewicz, T. K. Cox, A. Chakravarti, M. Buchwald, and L. C. Tsui, *Science* **245,** 1073 (1989).

[2] J. M. Rommens, M. C. Iannuzzi, B-S. Kerem, M. L. Drumm, G. Melmer, M. Dean, R. Rozmahel, J. L. Cole, D. Kennedy, N. Hidaka, M. Zsiga, M. Buchwald, J. R. Riordan, L.-C. Tsui, and F. S. Collins, *Science* **245,** 1059 (1989).

[3] M. P. Anderson, D. P. Rich, R. J. Gregory, A. E. Smith, and M. J. Welsh, *Science* **251,** 679 (1991).

[4] C. E. Bear, C. Li, N. Kartner, R. J. Bridges, T. J. Jensen, M. Ramjeesingh, and J. R. Riordan, *Cell* **68,** 809 (1992).

[5] S. H. Cheng, R. J. Gregory, J. Marshall, S. Paul, D. W. Souza, G. A. White, C. R. O'Riordan, and A. E. Smith, *Cell* **63,** 827 (1990).

0076-6879/98 $25.00

engineered CFTR mutants[6,7] suggest that CFTR is itself a Cl channel active in fluid secretion. However, since transfection of mutant cell lines by normal CFTR cDNA results in the appearance of both the CFTR Cl channel and an additional Cl channel,[8] regulation of fluid secretion may require an ensemble of events involving several CFTR-related functions.[9]

Although the pathologies associated with CF are centered in the airways and pancreas, CFTR itself appears to have a broader distribution, including the apical regions of cells in proximal and distal kidney tubules.[10] Because new findings suggest CFTR may have roles distinct from its function in fluid secretion,[11–13] we thought that study of renal CFTR would be worthwhile. Here, we describe localization of CFTR in endosomal membrane vesicles isolated from the kidney and outline factors important in the solubilization of renal CFTR and the initial steps of its purification.

Materials and Methods

Renal Membrane Isolation

Following sodium pentobarbital-induced anesthesia (150 mg/kg body weight), the renal cortex is removed from kidneys of male New Zealand White rabbits (NZR Ventures, Manchester, MD). Cortical tissue is finely minced (on ice), suspended in 4° buffer A (300 m*M* mannitol, 12 m*M* HEPES–Tris, pH 7.5) at 10 ml/g tissue, and dispersed with 20 strokes of a Potter–Elvehjem Teflon–glass homogenizer operating at 2000 rpm. The suspension was immediately supplemented with 5 m*M* $MgCl_2$, 5 m*M* EGTA, 1 m*M* dithiothreitol (DTT), and protease inhibitors [1 m*M* *p*-aminobenzoic acid, *p*-chloromercuribenzoate, iodoacetamide, and phenanthroline, 0.5 m*M* phenylmethylsulfonyl fluoride (PMSF), 10 μg/ml chymo-

[6] M. P. Anderson, R. J. Gregory, S. Thompson, D. W. Souza, S. Paul, R. C. Mulligan, A. E. Smith, and M. J. Welsh, *Science* **253,** 202 (1991).

[7] G. M. Denning, M. P. Anderson, J. F. Amara, J. Marshall, A. E. Smith, and M. J. Welsh, *Nature* **358,** 761 (1992).

[8] M. Egan, T. Flotte, S. Afione, R. Solow, P. L. Zeitlin, B. J. Carter, and W. B. Guggino, *Nature* **358,** 581 (1992).

[9] E. M. Schwiebert, M. E. Egan, T. H. Hwang, S. B. Fulmer, S. S. Allen, G. R. Cutting, and W. B. Guggino, *Cell* **81,** 1063 (1995).

[10] I. Crawford, P. C. Maloney, P. L. Zeitlin, W. B. Guggino, S. C. Hyde, H. Turley, K. C. Gatter, A. Harris, and C. F. Higgins, *Proc. Nat. Acad. Sci. U.S.A.* **88,** 9262 (1991).

[11] J. Barasch, B. Kiss, A. Prince, L. Saiman, D. C. Gruenert, and Q. Al-Awqati, *Nature* **352,** 70 (1991).

[12] J. Biwersi and A. S. Verkman, *Am. J. Physiol.* **266,** C149 (1994).

[13] G. L. Lukacs, X.-B. Chang, N. Kartner, O. D. Rotstein, J. R. Riordan, and S. Grinstein, *J. Biol. Chem.* **267,** 14568 (1992).

statin and α-macroglobulin, and 5 μg/ml pepstatin]. After removing nuclei, debris and unbroken cells are removed by two low-speed centrifugations (noted as P1 and P2 in Fig. 1), and brush border (apical) membranes are isolated by centrifugation after magnesium aggregation[14,15]; mitochondria are found in the underlying solid pellet. On collection of brush border and mitochondrial membranes, magnesium in the remaining supernatant is elevated to 30 m*M*, and after 1 hr at 4°, endosomes are isolated by centrifugation[16] and suspended in buffer A with protease inhibitors. A final high-speed centrifugation is used to isolate residual membranes; this material is noted as P3 in Fig. 1. Basolateral membranes are purified by percoll density gradient centrifugation[17] of a separate aliquot of the parent supernatant. Marker enzyme analysis, performed within 4 hr of fractionation, uses membranes at ca. 10 mg protein/ml in gluconate-based buffer B (100 m*M* mannitol, 100 m*M* potassium gluconate, 5 m*M* magnesium gluconate, 5 m*M* HEPES–Tris, pH 7.0).

Marker Enzyme Analysis

Membrane fractions are identified as follows. Brush border membranes are characterized by alkaline phosphatase (APase) (Sigma, St. Louis, MO, Technical Bulletin #104); basolateral membranes are identified by K^+-sensitive *p*-nitrophenyl phosphatase (PNPase)[18]; cytochrome-*c* oxidase is used to mark mitochondria,[19] and lysosomal, endoplasmic reticulum, and Golgi membranes are identified, respectively, by acid phosphatase (AcidPase) (Sigma, Technical Bulletin 10), glucose-6-phosphatase (G6Pase),[20] and thiamin pyrophosphatase (TPPase).[21] Protein assays[22] are standardized with bovine serum albumin (BSA).

Endosomal Acidification

ATP-dependent acidification of endosomes[23] is measured for samples (5–20 μl, 50–200 μg protein) suspended in 3 ml Cl- or gluconate-based

[14] P. S. Aronson, *J. Membr. Biol.* **42,** 81 (1978).
[15] A. G. Booth and A. J. Kenny, *Biochem. J.* **142,** 575 (1974).
[16] S. A. Hilden, C. A. Johns, and N. E. Madias, *Am. J. Physiol.* (*Renal*) **255,** F885 (1988).
[17] B. Sacktor, I. Rosenbloom, C. T. Liang, and L. Cheng, *J. Membr. Biol.* **60,** 63 (1981).
[18] K. Ahmed and J. D. Judah, *Biochim. Biophys. Acta* **93,** 603 (1964).
[19] A. Tzagoloff, P. C. Yang, D. C. Wharton, and J. S. Rieske, *Biochim. Biophys. Acta* **96,** 1 (1965).
[20] W. J. Arion, B. K. Walkin, W. Carlson, and A. J. Lange, *J. Biol. Chem.* **247,** 2558 (1972).
[21] J. Meldolesi, J. D. Jamieson, and G. E. Palade, *J. Cell Biol.* **49,** 109 (1971).
[22] W. Schaffner and C. Weissman, *Anal. Biochem.* **56,** 502 (1973).
[23] G. M. Denning, L. S. Ostedgaard, S. H. Cheng, A. E. Smith, and M. J. Welsh, *J. Clin. Invest.* **89,** 339 (1992).

buffer B. Following addition of 3 μM acridine orange, the sample is placed in the cuvette of an SLM-Aminco model SPF500C fluorescence spectrophotometer. Baseline level of fluorescence is recorded (excitation 490 nm, emission 520 nm), and on adding 0.25 mM Tris–ATP, formation of a pH gradient (acid interior) is measured by fluorescence quenching.[24]

Antibodies

A peptide-directed polyclonal antibody noted as No. 181 is most often used as the routine probe. This antibody, prepared against a CFTR r.415-427 (human CFTR), reacts with CFTR in a variety of cell lines or tissue preparations from mammalian species, including mouse and rat.[10,25,26] The monoclonal antibody (Genzyme Corp., Framingham, MA) is directed against an R domain fusion protein (human CFTR).

Gel Electrophoresis and Immunoblotting

Protein (5–50 μg/lane) is separated by SDS–PAGE at 25°, usually with 7% SDS polyacrylamide gels.[27] For Western blot analysis, protein is transferred to nitrocellulose at 4° (125 V, 200 mA).[28] The nitrocellulose support is subsequently blocked (30–60 min) by preincubation with gelatin (3% gelatin, 150 mM NaCl, 0.5% Triton X-100, 10 mM Tris-HCl, pH 7.4) and then probed with appropriately diluted antibody (1 : 1000 to 1 : 10,000) in buffer of the same composition except that 2.5% bovine serum albumin (BSA) replaces 3% gelatin. Anti-CFTR antibody binding is visualized by enhanced chemiluminescence (Amersham), using titrations of primary and secondary antibodies that give a maximum signal-to-noise (S/N) ratio after a 15- to 60-sec exposure. Excess lipid (e.g., in solubilized extracts) is removed by chloroform/methanol extraction prior to electrophoresis.

Solubilization and Partial Purification of CFTR

Endosomes are initially exposed (20 min at 4°) to 1.5% 3-[(3-cholamidopropyl)dimethylammonio]-1-propane sulfonate (CHAPS) in the presence of 0.4% *Escherichia coli* phospholipid,[29] 20 mM MOPS/K, 1 mM DTT, 0.75

[24] M. P. D'Souza, S. V. Ambudkar, J. T. August, and P. C. Maloney, *Proc. Natl. Acad. Sci. U.S.A.* **84,** 6980 (1987).

[25] A. E. O. Trezise and M. Buchwald, *Nature* **353,** 434 (1991).

[26] P. L. Zeitlin, I. Crawford, L. Lu, S. Woel, M. E. Cohen, M. Donowitz, M. H. Montrose, A. Hamosh, G. Cutting, D. C. Gruenert, R. Huganir, P. C. Maloney, and W. B. Guggino, *Proc. Natl. Acad. Sci. U.S.A.* **89,** 344 (1992).

[27] U. K. Laemmli, *Nature* **227,** 680 (1970).

[28] H. Towbin, T. Staehlin, and J. Gordon, *Proc. Natl. Acad. Sci. U.S.A.* **76,** 4350 (1979).

[29] S. V. Ambudkar and P. C. Maloney, *J. Biol. Chem.* **261,** 10079 (1986).

TABLE I
MARKER ENZYME ACTIVITIES IN MEMBRANE FRACTIONS OF RENAL CORTEX[a]

Marker	Membrane fraction			
	Homogenate	Brush border	Basolateral	Endosome
APase	46 ± 4 (100)	520 ± 11 (32)	111 ± 6 (2)	163 ± 6 (12)
PNPase	12 ± 2 (100)	39 ± 4 (1)	208 ± 12 (12)	ND[b]
Protein	(100)	(1.4)	(0.9)	(2.4)

[a] Mean values ± SEM of specific activities (nmol/min per mg protein) for the indicated marker enzymes as determined in three to five separate experiments. Values in parentheses give total activity (nmol/min) or protein content (mg/ml) relative to the parent homogenate.
[b] ND, Not determined.

m*M* phenylmethylsulfonyl fluoride (PMSF), and 20% (v/v) glycerol, pH 7. Following centrifugation at 37,000*g* for 60 min, CFTR-immunoreactive material (antibody No. 181) is found entirely in the insoluble pellet. After resuspension to its original volume, the insoluble material is treated with 1.5% octyl-β-D-glucoside (octylglucoside) (other conditions as above), resulting in solubilization of most (ca. 75%) of the immunoreactive material. Solubilized protein is then applied to a 20-ml DEAE-Sepharose CL-6B column equilibrated at 25° or 4° with buffer C (20 m*M* MES, 50 m*M* NaCl, 1.25% octylglucoside, 0.1% *E. coli* lipid, 1 m*M* DTT, 0.75 m*M* PMSF, and 20% (v/v) glycerol, pH 6). Thirteen 10-ml fractions are collected by stepwise elution with increasing salt (50 m*M* per step) using buffer C containing 50–400 m*M* NaCl. Fractions 10–12 (240–280 m*M* NaCl) contain all the input CFTR as measured by antibody reactivity.

Results

Subcellular Localization of CFTR

CFTR is known to reside in the apical membranes of several secretory cells,[7,10,25,30] but preliminary work indicated a more complex pattern in the kidney.[10] Consequently, to localize renal CFTR more precisely, we used renal cortex to prepare three membrane fractions (brush border, basolateral, and endosomal membranes) identifiable by marker enzyme analysis (Table I). Alkaline phosphatase was enriched 11-fold in brush border membranes; K^+-sensitive *p*-nitrophenylphosphatase showed 17-fold enrichment

[30] C. R. Marino, L. M. Matovcik, F. S. Gorelick, and J. A. Cohn, *J. Clin. Invest.* **88,** 712 (1991).

in basolateral membrane preparations; whereas endosomes, isolated by extended magnesium precipitation,[16] showed the expected ATP-dependent acidification (below). The specific activities of glucose-6-phosphatase, thiamin pyrophosphatase, acid phosphatase, and cytochrome oxidase were not elevated in these preparations (not shown), indicating the absence of significant contamination by endoplasmic reticulum, Golgi, lysosomal, and mitochondrial components, respectively.

The distribution of CFTR-immunoreactive material in these fractions was monitored with a monoclonal antibody directed to an epitope in the CFTR R domain (Fig. 1). Western blots showed prominent CFTR immunoreactivity at positions corresponding to ca. 165 and 195 kDa for endosome membranes (Fig. 1, lane E). Brush border membranes also gave positive signals (Fig. 1, lane B), but less strongly so than endosomes. By contrast, there was little or no reactivity in basolateral membranes, or in tests with a mitochondrial fraction or with material discarded with the initial, low-speed pellets (Fig. 1B, lanes BL, M, and P1, respectively); a minor portion of CFTR immunoreactivity was found in vesicles isolated by high-speed centrifugation of the supernatant from which endosomes had been isolated (Fig. 1, lane P3); this is presumed to reflect residual endosomes. A similar distribution of reactivity (endosomes > brush border membranes > homogenate) was found using the polyclonal antibody No. 181 (not shown here; see below) which had identified CFTR in renal tubules by cytochemistry.[10]

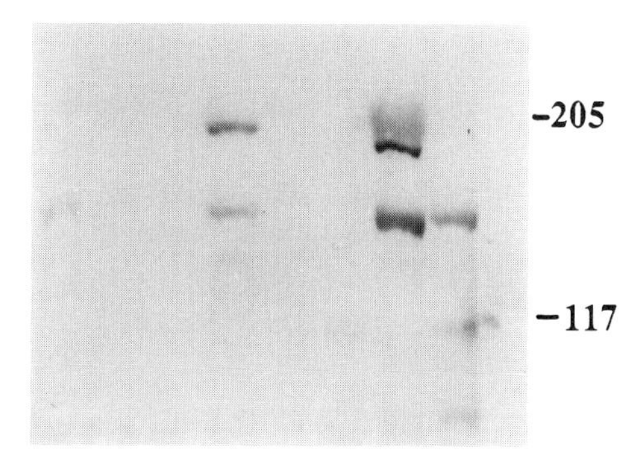

FIG. 1. Distribution of CFTR immunoreactivity in membrane fractions derived from renal cortex. A 7% SDS–PAGE gel was probed with the anti-CFTR monoclonal antibody (1:1000 dilution). Lane designations: H, homogenate; P1 and P2, low-speed pellets; B, brush border membranes; BL, basolateral membranes; M, mitochondrial pellet; E, endosomal membranes; P3, final high-speed pellet; S, final supernate. The mobilities of molecular weight markers are indicated.

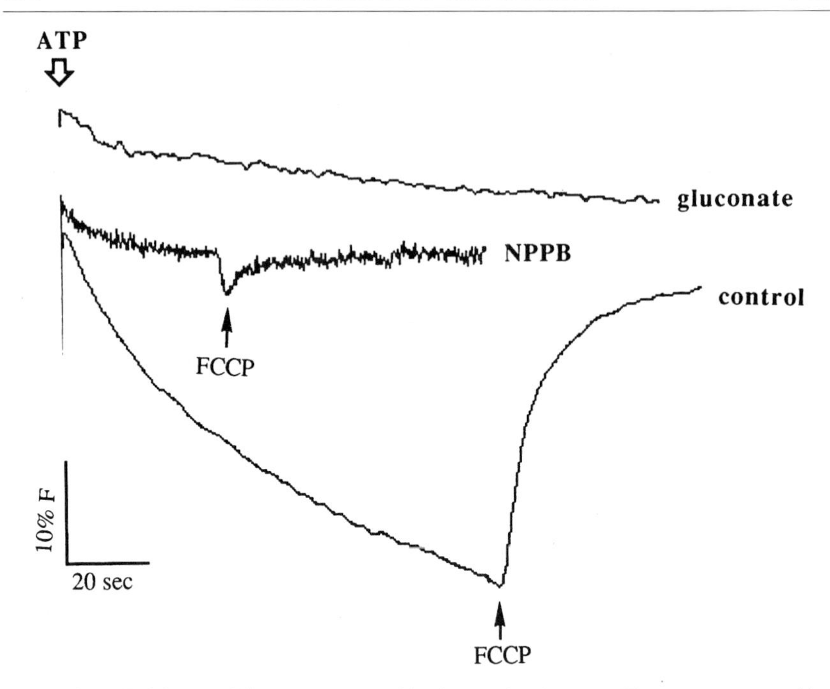

FIG. 2. ATP-driven acridine orange quenching by renal endosomes. Fluorescence quenching was monitored in Cl-based buffer with 0.25 m*M* Tris–ATP, as described in the Methods section. The downward arrow shows addition of ATP; the upward arrows show addition of the protonophore, FCCP (1 μM). The control and NPPB sample contained endosomes suspended in a Cl-based buffer; gluconate replaced Cl in the remaining sample, as indicated. NPPB was used at 100 μM.

Additional trials with the polyclonal antibody gave positive reactions with endosomes from human and bovine kidney as well as those from rabbit renal medulla (not shown). It therefore seemed likely that the CFTR immunoreactivity detected by immunocytochemistry of kidney tubules[10] arises in renal endosomes.

Dependence of Endosomal Acidification on Cl

To assess the ATP-dependent acidification expected of renal endosomes,[31,32] membranes were placed in a Cl-containing medium with the weak fluorescent base, acridine orange. Addition of ATP led to a quenching of fluorescence (Fig. 2, control), indicating acidification of the endosomal

[31] I. Mellman, R. Fuchs, and A. Helenius, *Ann. Rev. Biochem.* **55,** 663 (1986).
[32] G. Rudnick, *Annu. Rev. Physiol.* **48,** 403 (1986).

lumen; we could also detect an equivalent acidification of vesicles in the parent homogenate, although this required increased (≥10-fold) amounts of protein (not shown). Endosomal acidification was not observed when Cl in the medium was replaced by gluconate, nor was there acidification after pretreatment of endosomes with a Cl channel blocker, NPPB (Fig. 2). In this latter case it was possible to show that endosomes retained their permeability barrier, since adding 1 μM valinomycin in a low potassium medium gave acidification due to a proton entry driven by the potassium diffusion potential (negative inside) (not given) (see Ref. 24). This proved that NPPB did not damage renal endosomes and that acridine orange was capable of reporting acidification in the presence of the drug. Taken together, then, these findings suggest that CFTR or some other Cl channel is required for endosomal acidification.

Solubilization and Partial Purification of Endosomal CFTR

The preceding studies led us to attribute CFTR immunoreactivity in renal endosomes to interactions with protein(s) migrating at ca. 165 and/or 195 kDa. Earlier work with antibody No. 181 had identified CFTR as a 165-kDa protein appearing after transfection of tissue culture cells with CFTR-cDNA or after injection of *Xenopus* oocytes with CFTR-mRNA.[8] But because we had no equivalent information regarding the 195-kDa species, which may reflect a more highly glycosylated form of renal CFTR,[33] or a reactivity not directly related to CFTR, we directed subsequent biochemical work to solubilization of the 165-kDa species.

We surveyed several agents for their effectiveness in extracting CFTR from renal endosomes, using a protocol[29] that included endosomal membranes (1 mg protein), excess lipid (0.4%), glycerol (20%), and detergents (CHAPS, octylglucoside, octyl-β-D-maltoside, dodecyl-β-D-maltoside, Triton X-100, each at 0.5–2%). Using antibody No. 181 as a probe, we then followed recovery of the 165-kDa reactive species in the solubilized extract or its depletion in the remaining pellets. When it became apparent that no one of these detergents solubilized more than about 20% of endosomal CFTR, we moved to explore sequential extractions with pairs of detergents. This strategy proved satisfactory, and after further preliminary trials, we settled on an extraction using CHAPS and octylglucoside as the most effective method.

Use of the double-detergent protocol is illustrated by the experiment of Fig. 3. In that case, an initial solubilization with 1.5% CHAPS released

[33] L. S. Prince, R. B. Workman, Jr., and R. B. Marchase, *Proc. Natl. Acad. Sci. U.S.A.* **91,** 5192 (1994).

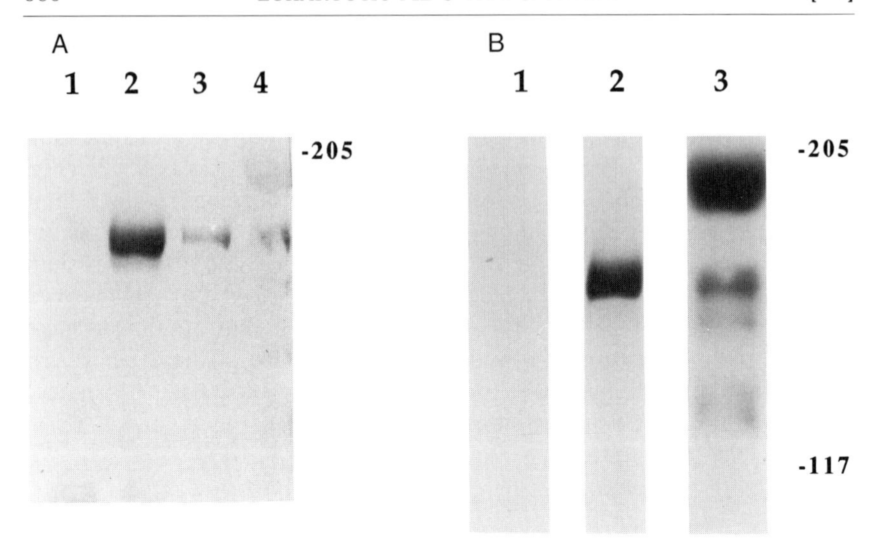

FIG. 3. Solubilization of CFTR from renal endosomes. (A) Endosomes (2 mg protein) were extracted with 1.5% CHAPS. Unextracted material was then twice reextracted with 1.5% octylglucoside. Solubilization of the CFTR immunoreactive material was monitored using antibody No. 181 (1:1000 dilution). Lane designations: lane 1, CHAPS-solubilized material; lanes 2 and 3, octylglucoside-solubilized material, first and second extractions; lane 4, pellet after second octylglucoside extraction. (B) In a separate experiment endosomes were solubilized at pH 6 (lane 1), pH 7 (lane 2), or pH 8 (lane 3) and CFTR immunoreactive material was identified using the Genzyme monoclonal antibody (1:1000 dilution).

about 75% of total protein, but little or none of the 165-kDa CFTR immunoreactivity (Fig. 3A); instead, immunoreactivity remained in the insoluble pellet (not shown). Subsequent exposure of this pellet to 1.5% octylglucoside released half of the residual endosomal protein and most of the 165-kDa immunoreactivity; the remaining 165-kDa reactivity was extractable by further exposure to octylglucoside (Fig. 3A), although this tertiary step was not normally used. Using this stepwise extraction method, we were able to establish two additional points. First, we showed that the solubilized material could be incorporated into proteoliposomes made by a subsequent detergent dilution[29] (not shown). This finding, along with the observed resistance to CHAPS solubilization, argues against the idea that CFTR immunoreactivity reflects a peripheral membrane protein released on solubilization of endosomes. In further work, we also found that extraction of CFTR was importantly influenced by pH (Fig. 3B). For example, when endosomes were extracted at pH 6, neither the 165-kDa nor the 195-kDa reactive material was released by the secondary extraction with octylglucoside; at pH 7, only the 165-kDa material was solubilized, while at pH 8, both could be solubilized (with reduced efficiency in extraction of the 165-kDa species) (not given) (Fig. 3B). For this reason, during partial purification of CFTR (below), we exploited solubilization conditions (pH 7) that maximized yield of the 165-kDa species.

Partial purification of CFTR immunoreactive material was effected by anion-exchange chromatography (Fig. 4). An endosomal detergent extract, prepared by stepwise detergent extraction at pH 7 (above), was placed on DEAE-Sepharose, and elution of retained material was followed as salt in the running buffer increased from 50 to 400 m*M*. The major endosomal protein(s) of 120–200 kDa eluted at about 75–150 m*M* NaCl (fractions 2–5), while CFTR immunoreactive material eluted at 240–280 m*M* NaCl (Fig. 4). Coomassie staining of this latter fraction showed 15 or more proteins (Fig. 4B), one of which migrated with the size (ca. 165 kDa) expected of CFTR. This fraction (240–280 m*M* NaCl) contained most of the starting CFTR immunoreactivity but only a small part (ca. 1–2%) of the input protein, reflecting a considerable purification of CFTR. However, the low absolute recovery of protein has precluded further purification at this time.

Discussion

Microscopic techniques[7,10,25,30,34] show that CFTR is predominantly targeted to the apical membranes of epithelia involved in fluid transport, so that the earlier finding of CFTR immunoreactivity in the human kidney[10] led us to expect that CFTR would reside in the renal brush border membranes, as it does in the gut.[25] Although we did find CFTR immunoreactivity in renal brush border, the major portion of this material copurified with endosomes (Fig. 1), as judged by both the absence of other markers and by the parallel enrichment of ATP-dependent acidification. It is unlikely that this distribution reflects contamination of endosomes by brush border, since the brush border marker (alkaline phosphatase) showed the opposite distribution. On the other hand, the positive reaction by brush border could arise from contamination by endosomes, or because fusion and retrieval mechanisms involving endosome and plasma membranes have established a steady-state distribution of CFTR between the two compartments.[33,35] Either of these latter possibilities is compatible with other findings. Thus, restriction of CFTR to an intracellular localization would be consistent with the absence of a cAMP-stimulated component to net fluid transport

[34] E. Puchelle, D. Gaillard, D. Ploton, J. Hinnrasky, C. Fuchey, M. Boutterin, J. Jacquot, D. Dreyer, A. Pavirani, and W. Dalemans, *Am. J. Respir. Cell Mol. Biol.* **7,** 485 (1992).

[35] P. Webster, L. Vanacore, A. C. Nairn, and C. R. Marino, *Am. J. Physiol.* **267,** C340 (1994).

A

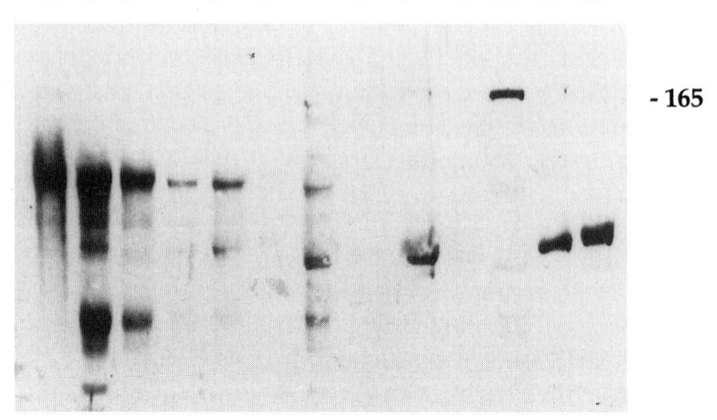

B

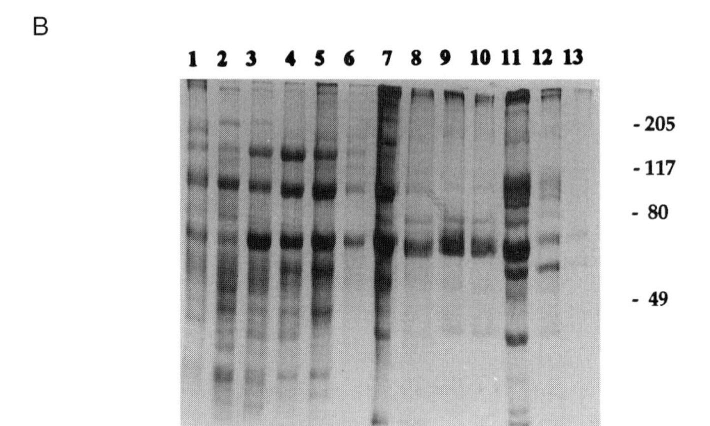

FIG. 4. Partial purification of CFTR. An endosomal (2 mg protein) detergent extract prepared at pH 7 was applied to DEAE-Sepharose. Material retained on the column was recovered by stepwise elution with a NaCl gradient (50–400 mM), concentrated during extraction of lipid,[36] and processed for protein assay and SDS–PAGE. (A) Western blot of a 10% SDS–PAGE gel probed with anti-CFTR antibody No. 181 (1:1000 dilution). Lane 1, crude extract; lanes 2–13, fractions 2–13. The ca. 165-kDa region is indicated. (B) A duplicate SDS–PAGE gel was stained with Coomassie brilliant blue to visualize protein.

in the kidney, but the presence of protein kinase A-mediated increases of Cl permeability in renal endosomes.[37] Alternatively, CFTR-containing endosomes may contribute to cell membrane Cl conductance by recycling to the plasma membrane in a cAMP-independent fashion.

Our experiments were also aimed at using renal tissue to develop techniques for solubilization and purification of CFTR from its native environment. Bear *et al.*[4] solubilized CFTR using sodium dodecylsulfate and found that the extracted material displayed Cl channel activity in planar lipid bilayers,[4] but it should not be assumed that CFTR retains all aspects of its native function under these relatively harsh conditions. For this reason, we adopted techniques known to optimize recovery of activity of other membrane transport systems,[29] including several related to CFTR.[38,39] The results of our initial efforts are encouraging. Endosomal protein represents ca. 2% of kidney protein (Table I), and CFTR is sufficiently enriched in renal endosomes to enable simple chromatographic procedures. In a first step, we used anion-exchange chromatography to separate the full-length protein from major endosomal proteins of similar size, yielding a further enrichment estimated as ca. 100-fold.

Acknowledgments

We thank Dr. William B. Guggino for helpful discussions and for use of antibodies No. 181 and Dr. Jean Beck for advice on isolation of renal membrane fractions. We are also grateful to Dr. Marshall Montrose and Ms. Jennifer Lewis for help during measurements of acridine orange fluorescence, and we acknowledge the technical assistance of Mr. Christopher Beardmore. This work was supported by a grant from the U.S. Public Health Service (DK44015).

[36] Z.-S. Ruan, V. Anantharam, I. T. Crawford, S. V. Ambudkar, S. Y. Rhee, M. J. Allison, and P. C. Maloney, *J. Biol. Chem.* **267,** 10537 (1992).

[37] W. R. Reenstra, I. Sabolic, H.-R. Bae, and A. S. Verkman, *Biochemistry* **31,** 175 (1992).

[38] L. Bishop, R. Agbayani, Jr., S. V. Ambudkar, P. C. Maloney, and G. F.-L. Ames, *Proc. Natl. Acad. Sci. U.S.A.* **86,** 6953 (1989).

[39] S. V. Ambudkar, I. H. Lelong, J. Zhang, C. O. Cardarelli, M. M. Gottesman, and I. Pastan, *Proc. Natl. Acad. Sci. U.S.A.* **89,** 8472 (1992).

[49] Assays of Dynamics, Mechanisms, and Regulation of ATP Transport and Release: Implications for Study of ABC Transporter Function

By ERIK M. SCHWIEBERT, MARIE E. EGAN, and WILLIAM B. GUGGINO

Introduction

Adenosine 5′-triphosphate (ATP) is the primary energy source within cells required for countless enzymatic reactions. In addition, ATP is also an autocrine or paracrine agonist that has bioactivity within a wide variety of tissues.[1] Classical examples of this agonist behavior include thrombin-induced release of ATP and its metabolite, adenosine 5′-diphosphate (ADP), for platelet self-aggregation, release of ATP from purinergic nerves in the brain and in dorsal root ganglia of the spinal cord for pain perception or nociception, and release of ATP along with catecholamines, histamine, and acetylcholine from autonomic nerve terminals, adrenal chromaffin cells, and mast cells. Once released, ATP is free to bind to a growing family of P_2 purinergic ATP receptors that have two subfamilies: a seven transmembrane-spanning P_{2Y} purinergic receptor subclass (up to five members have been cloned), which signals through G proteins, and a two transmembrane-spanning "ionotropic" P_{2X} purinergic receptor subclass (up to seven members have been cloned), which signals by allowing the influx of Ca^{2+} from extracellular stores.[2] Once released, ATP can bind to these receptors and/or be subject to degradation in all extracellular spaces, especially in vascular beds such as the lung,[3] by ecto-ATPases, ecto-ADPases, and ecto-nucleotidases.[4] ADP also has affinity that exceeds that of ATP for a specific subset of P_{2Y} purinergic receptors. Adenosine, a metabolite of ATP, is also known for potent biological activity as an agonist again as a local mediator within tissues and has its own family of P_1 purinergic adenosine receptors.[5]

[1] For review of ATP agonist biology, *see* J. L. Gordon, *Biochem. J.* **233,** 310 (1986).

[2] For reviews of ATP agonist biology and P_2 purinergic receptor biology, *see* M. P. Abbracchio and G. Burnstock, *Pharmacol. Ther.* **64,** 445 (1994); E. A. Barnard, G. Burnstock, and T. E. Webb, *Trends Physiol. Sci.* **15,** 67 (1994); B. B. Fredholm *et al.*, *Pharmacol. Rev.* **46,** 143 (1994).

[3] See Gordon (1986) in footnote 1.

[4] For review of ATP degradative enzymes, *see* U. S. Ryan, *Annu. Rev. Physiol.* **44,** 223 (1982).

[5] For review of adenosine agonist biology and P_1 adenosine receptor biology, *see* G. L. Stiles, *J. Biol. Chem.* **267**(10), 6451 (1992); M. G. Collins and S. M. O. Hourani, *Trends Pharmacol. Sci.* **14,** 360 (1993); H. H. Dalziel and D. P. Westfall, *Pharmacol. Rev.* **46**(4), 449 (1994).

METHODS IN ENZYMOLOGY, VOL. 292
0076-6879/98 $25.00

With that background in mind, the task of this article is to describe the development and use of assays designed to study the release of ATP that precedes its or its metabolites' extracellular agonist functions. ATP-binding cassette (ABC) transporters such as cystic fibrosis transmembrane conductance regulator (CFTR) and multidrug resistance (*mdr*) have been implicated in conductive transport of ATP, and these molecules as well as the sulfonylurea receptor (SUR) in pancreatic β cells may act as ATP sensors within cells.[6] It is important to emphasize initially that a concentration gradient exists in all cells for efflux or release of ATP; however, specific pathways such as those governed by ABC transporters are required to move negatively charged ATP across the lipid bilayer. Intracellular concentrations of ATP range from 3 to 5 mM (depending on the metabolic state of a given cell or tissue) and extracellular concentrations are negligible due to the eventual degradation of the released ATP. By analogy, this gradient for ATP efflux is equivalent to or greater than Ca^{2+} influx from extracellular stores (1–2 mM to 50–100 nM). The mechanisms of ATP release can be grouped into three categories: (1) facilitated transport of ATP through a specific transporter or "flippase" down a favorable concentration gradient, (2) conductive transport through an ATP-specific or anion-selective channel with the favorable gradient, and (3) release of ATP by exocytosis as exemplified by thrombin-induced platelet ATP/ADP release and ATP release from presynaptic nerve terminals as a neurotransmitter. Nucleoside transporters exist for the transport of adenosine 5′-monophosphate (5′-AMP) and for adenosine[7]; therefore, they may exist for ATP. Exocytic release of ATP is well documented[8]; however, the mechanism of how ATP is loaded into these vesicles by specialized cells is unknown. CFTR and *mdr* have been implicated in conductive transport of ATP; however, they may also facilitate or regulate conductive transport by another class of ATP-selective channels. Moreover, CFTR has been implicated as a regulator or facilitator of cAMP-dependent vesicle trafficking. CFTR was also hypothesized to be a "flippase" transporter soon after its identification. Thus, CFTR, *mdr*, or other ABC transporters may govern multiple mechanisms of ATP release. Figure 1 illustrates a model of the cellular mechanisms of ATP release, degradation, binding, and signaling.

[6] For ABC transporter functions, *see* E. M. Schwiebert et al., *Cell* **81,** 1063 (1995); M. J. Stutts et al., *Science* **269,** 847 (1995); I. L. Reisin et al., *J. Biol. Chem.* **269,** 20584 (1994); E. H. Abraham et al., *Proc. Natl. Acad. Sci. U.S.A.* **90,** 312 (1993); D. R. Gill et al., *Cell* **71,** 23 (1992); C. F. Higgins, *Cell* **82,** 693 (1995); L. Aguilar-Bryan et al., *Science* **268,** 423 (1995); N. Inagaki et al., *Science* **270,** 1166 (1995).

[7] For review of nucleoside transporters, *see* J. Deckert, P. F. Morgan, and P. J. Marangos, *Life Sci.* **42,** 1331 (1988).

[8] See Gordon (1986) in footnote 1.

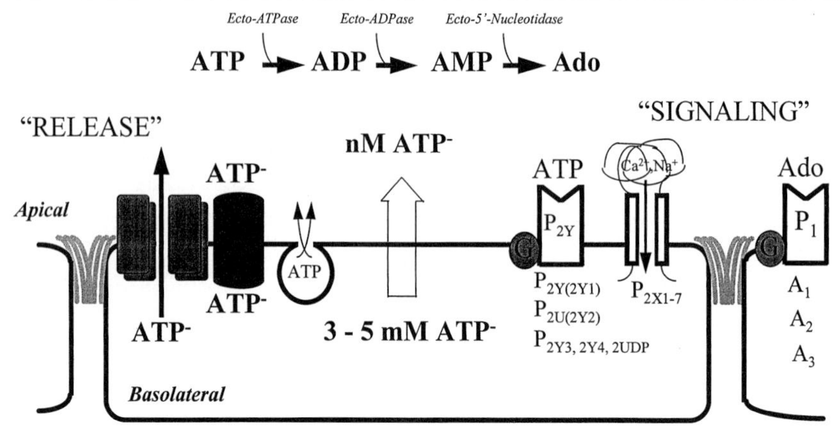

FIG. 1. Dynamics of ATP release, degradation, and signaling.

We have developed three different assays to study the role of ABC transporters or other pathways in ATP transport and release. The first is a radiolabeled [γ-^{32}P]ATP release assay in which the transport and release of loaded radiolabeled ATP is measured and trapped as ^{32}P-labeled glucose 6-phosphate with the help of the ATP scavenging enzyme, hexokinase. The second is a nonradioactive bioassay measuring released ATP as luminescence from the luciferase–luciferin reaction. Each ATP that is released creates a photon of light that is collected by a luminometer. The third is single-channel patch-clamp analysis of excised membrane patches of cells expressing CFTR in which Cl^- conduction versus ATP^- conduction can be measured. The assays will be discussed in this order.

Assays of ATP Release from Cultures of Epithelial and Nonepithelial Cells

Preparation of Cultured Cells

Epithelial cells are grown to confluence on collagen-coated tissue culture flasks until seeded into 6-well plates for [γ-^{32}P]ATP release/trapping assays or into 35-mm dishes for ATP bioluminescence assays. Because ATP release is dependent on cell culture conditions and degree of cellular differentiation, each cell type must be experimented with to find the correct culture conditions to ensure reproducibility. A given cell type can show considerable variability in ATP release depending on the feeding schedule, cell density, or the extracellular matrix on which the cells are grown. In the case of the

IB3-1 CF bronchial epithelial cells,[9] the cells appear epithelial-like only when grown on collagen; otherwise, they appear undifferentiated. In general, each cell type is fed the day before the assay, grown to confluence by the day of assay, and grown on collagen [Vitrogen 100; Collagen Corp., Palo Alto, CA; diluted 1:15 with phosphate-buffered saline (PBS); excess solution after coating removed by aspiration]. To compare fairly across cell types, all cells should be standardized with respect to culture conditions.

[γ-^{32}P]ATP Release/Trapping Assay

[γ-^{32}P]ATP (3000 Ci/mmol specific activity; 0.25 mCi per 6-well plate; NEN-Dupont, Boston, MA) is loaded into IB3-1 CF cells grown on collagen and transfected transiently[10] at 50% confluence via a three-step protocol at room temperature: (1) 15-min permeabilization with streptolysin O (1 U/ml; Sigma, St. Louis, MO) in a buffer containing (in mM): NaCl, 140; PIPES, 100; glucose, 5; KCl, 2.7; EGTA, 2.7, Na_2-ATP, 1 (pH 7.4) with 0.1% bovine serum albumin (BSA) and 0.25 mCi of [γ-^{32}P]ATP; (2) three vigorous washes with a standard Ringer's solution containing (in mM): NaCl, 140; $CaCl_2$, 1.5; $MgCl_2$, 1; HEPES, 5; glucose, 5;K_2HPO_4/KH_2PO_4, 5 (pH 7.4) to remove the streptolysin O and (3) a 2-hr incubation in the same Ringer's solution with another dose of 0.25 mCi of [γ-^{32}P]ATP to allow the cells to recover from the permeabilization and to accumulate more labeled ATP. Before starting the release assay time course, cells are washed five times with Ringer's solution. A key part to the design of this assay is to include hexokinase (Sigma, 1 U/ml, Type IV isolated from bakers yeast), an ATP scavenger, in the Ringer's solution. In contrast to other ATP scavengers such as the ATPase/ADPase apyrase, hexokinase traps released [γ-^{32}P]ATP as glucose 6-[^{32}P]phosphate before the labeled ATP is subject to degradation by the cultured cell themselves (see Fig. 1). Ringer's solution plus hexokinase is added at time "0 min" and extracted at four time points between 0 and 10 min with or without cyclic AMP agonists in an unpaired manner. The four samples are pooled after counting and are normalized as counts referable to generation of glucose 6-[^{32}P]phosphate.

Biochemical analysis can be utilized to determine the percentage of the counts that are reflective of free ^{32}P or ^{32}P incorporated into phospholipids versus glucose 6-[^{32}P]phosphate, reflective of intact [γ-^{32}P]ATP released from cells. To differentiate between free phosphate or labeled phospholipids and glucose 6-[^{32}P]phosphate, free phosphate and labeled phospholipids are trapped from samples with an equal volume (1 ml) of 5% (w/v) ammonium molybdate in 0.85 N sulfuric acid, extracted with an equal volume of

[9] P. L. Zeitlin et al., *Am. J. Respir. Cell Mol. Biol.* **4,** 313 (1991).

[10] E. M. Schwiebert et al., *Cell* **81,** 1063 (1995).

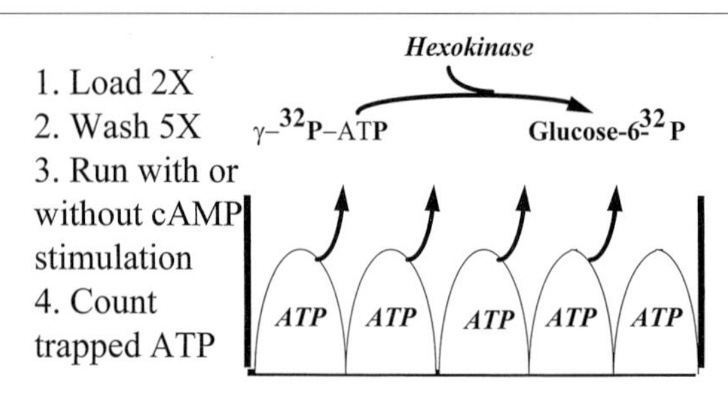

FIG. 2. Schematic of radioactive ATP release/trapping assay.

butanol, and the aqueous phase counted and compared to unextracted samples.[11] Reproducibly, 75–80% of the counts after extraction remain as glucose 6-phosphate in our experiments. Alternatively, a separate series of experiments can be performed where cells are loaded [γ-^{32}P]ATP but with [^{3}H]glucose in the bath. After the experiment, extracted samples are compared to unextracted samples as above.[11] Each sample is loaded on an anion-exchange column containing anion-exchange resin AG 1-X2 (Bio-Rad, Richmond, CA). The column is washed five times with water to remove free labeled glucose. Alkaline phosphatase (1 U/ml, Sigma, St. Louis, MO) is added to the column to cleave off labeled glucose bound to the column as [^{3}H]glucose 6-[^{32}P]phosphate. Bound labeled ^{32}P is eluted from the column with increasing NaCl concentrations (0–500 mM). After correcting for specific activity of [^{3}H]glucose and [γ-^{32}P]ATP, extracted samples have an approximate 1:1 ratio of [^{3}H]glucose : ^{32}P, indicative of "pure" glucose 6-phosphate in the extracted samples. In unextracted samples, the ratio is 0.75 : 1 [^{3}H]glucose : ^{32}P, which agrees with the extraction values shown earlier, and suggests that 75% of ^{32}P is complexed with glucose as glucose 6-phosphate and 25% is free ^{32}P or labeled phospholipid. Counts can be corrected based on this biochemical analysis; however, the assay is relatively specific for release of [γ-^{32}P]ATP and trapping of glucose 6-[^{32}P]phosphate with hexokinase. Figure 2 includes a schematic diagram of the assay. Data generated by this assay from parental IB3-1 cells and IB3-1 cells transfected transiently with wild-type CFTR have been published.[12] Advantages and disadvantages of this assay and the bioluminescence assay will be discussed later.

ATP Bioluminescence Assays

Figure 3 includes a schematic diagram of this assay (Fig. 3A) and some representative data from ATP bioluminescence assays performed on epithe-

[11] S. V. Ambudkar and P. C. Maloney, *J. Biol. Chem.* **259,** 12576 (1984).

[12] See Schwiebert et al. (1995) in footnote 6.

lial cells (Figs. 3B, 3C, and 3D). Cells are grown as above. The remainder of the assay is performed in the absence of direct light (only enough to take notes, handle the dish/reagents, and operate the luminometer). Only sterile, cell culture tested serum-free medium (OptiMEM-I, Gibco-BRL, Gaithersburg, MD) or filtered Ringer's solution (see above) can be used. Bacterial contamination could alter results. Cells are washed two times in sterile phosphate-buffered saline (PBS). Luciferase–luciferin reagent (Sigma, 50 mg, L1761, isolated from the firefly *Photinus pyralis*) is diluted 1 : 20 in OptiMEM-I medium (Gibco-BRL; contains no endogenous ATP) and 0.7–1 ml of the reagent in medium is added to the dish of confluent cells. This reagent and medium or Ringer's solution gives the best results; other reagents purchased from other sources do not provide good results for these bioassays because they are designed for lysis of cells for *luc* reporter gene assays. The cells are then transferred immediately into a 35-mm dish adapter and placed inside a Turner TD 20/20 luminometer (Turner/Promega, Madison, WI) for study. The luminometer can be set at different sensitivities (40% sensitivity in our experiments) and 30-sec time points are taken. In our experiments, a 15-sec delay followed by a 15-sec collection of photons is performed for each time point.

The essence of the assay exploits the requirement of the luciferase–luciferin reaction on ATP as a catalyst. In this assay, endogenous ATP produced and released by the cells themselves is being measured; 1 ATP molecule is approximately equivalent to 1 photon created by the luciferase–luciferin reaction. Data are expressed as luminescence in arbitrary units; however, values obtained from cells under basal and stimulated conditions can be correlated with a standard curve of ATP (see Table I) performed with serial dilutions of a 1 *M* stock of ATP-Mg^{2+} from 10^{-3} *M* (saturates the luminometer) to 10^{-12} through 10^{-15} *M* (values negligible and equivalent). The effective range of measurement of ATP under these strict conditions and a constant dilution of reagent in medium is 10^{-5} *M* through 10^{-10} *M*. ATP scavengers such as hexokinase and apyrase will eliminate the luminescence signal to background levels (see Table I). Background luminescence with these reagents is negligible (<1 arbitrary unit) (see also Table I).

Time controls and vehicle controls should be performed for each cell studied (see Fig. 3B for examples). ATP bioluminescence under basal conditions is a measure of ATP release and accumulation versus degradation; therefore, each cell will behave differently in the assay depending on the relative ability to release ATP (number and activity of release mechanisms) and to degrade ATP (number and activity of ecto-ATPases). Vehicle controls are also necessary for each cell. Study of the effects of intracellular cAMP (addition of forskolin) or calcium (addition of ionomycin) requires the addition of an agonist into the medium bathing the

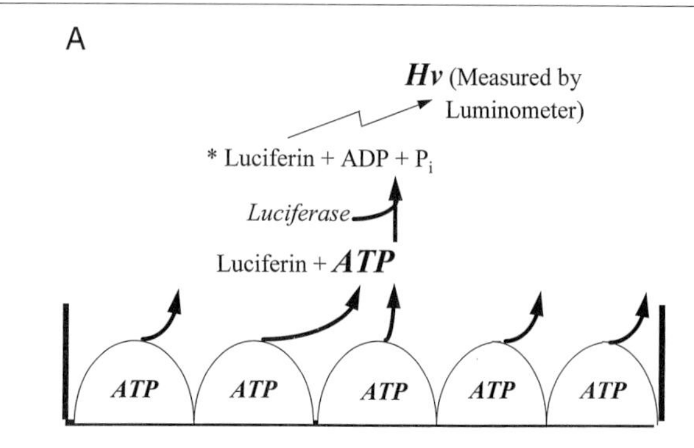

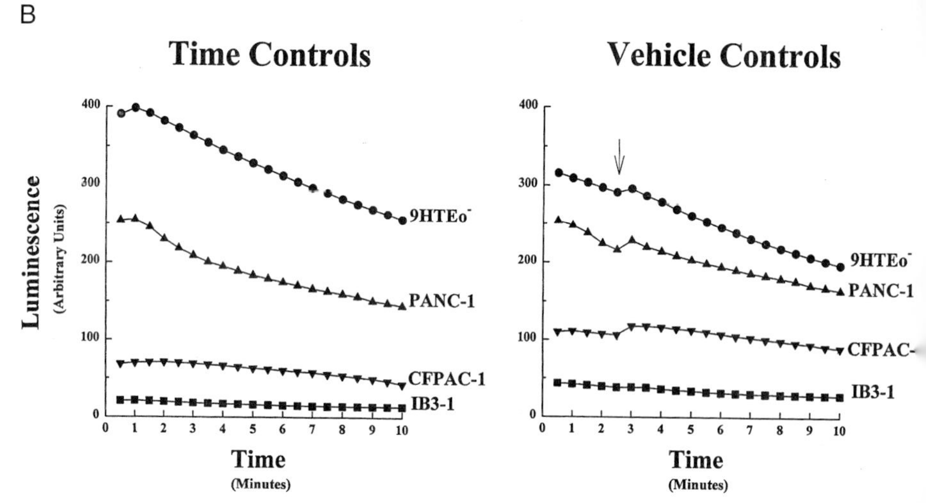

FIG. 3. ATP bioluminescence assay. (A) Schematic of the assay. This reaction occurs in a sealed chamber in the Turner luminometer where a dish of confluent cells is placed inside a holder that is lowered into the chamber. (B) Representative time and vehicle control experiments expressed as real luminescence values in arbitrary units versus time. (C) Representative cyclic AMP-induced ATP release experiments expressed as real luminescence values versus time. Note that cells expressing wild-type CFTR ($9HTEo^-$, PANC-1) had higher basal luminescence and responded to forskolin, whereas cells expressing ΔF508-CFTR (IB3-1, CFPAC-1) had much lower basal luminescence and failed to respond to forskolin. (D) Representative Ca_i^{2+} and swelling-induced ATP release experiments expressed as real luminescence values versus time. Note that different cells show different types of responses that are transient or sustained in nature.

cells; thus, simulation of this operation (removing the dish, tipping the dish, adding a small volume of vehicle, swirling the medium three times, and reinserting the dish) is very important. Vehicle controls are important because ATP on the periphery of the dish or in unstirred layers may

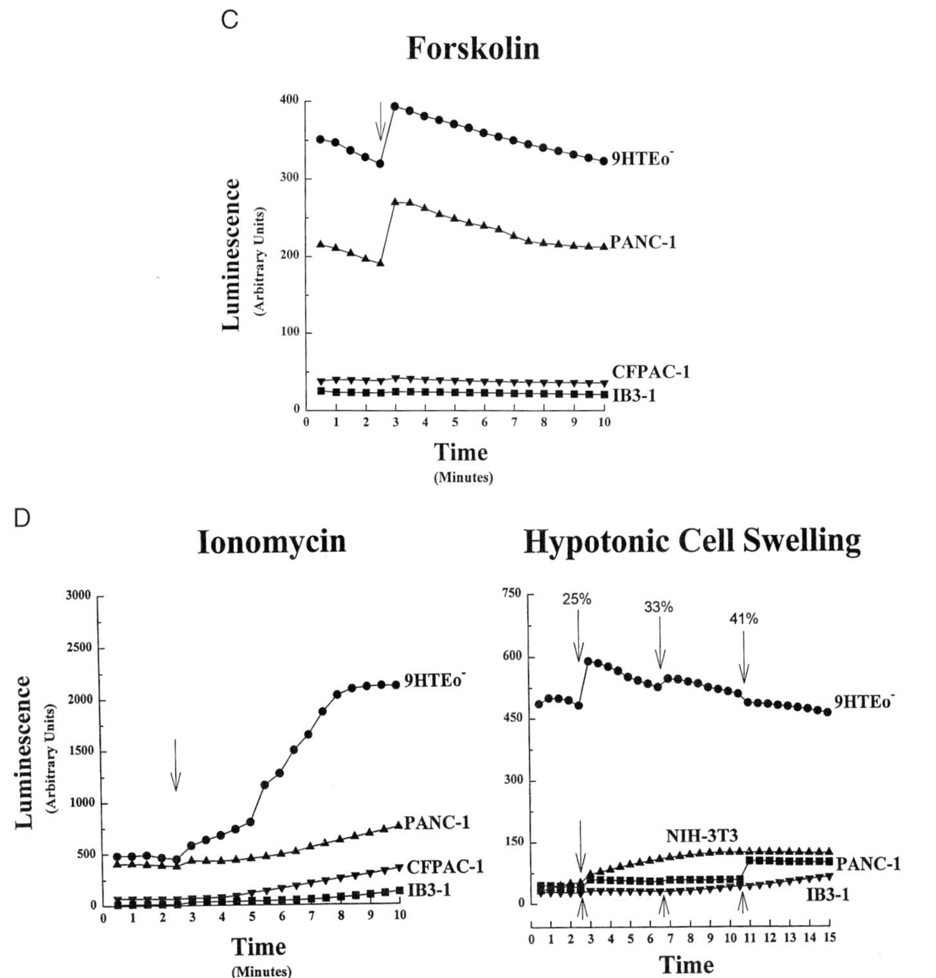

FIG. 3. (*continued*)

be revealed due merely to mixing of the solution and can enhance the luminescence signal. This is more significant if the basal rate of ATP release is high or if the cell type of interest grows as islands in culture rather than as a homogenous layer. Once these controls are performed, addition of agonists such as forskolin (Fig. 3C) or ionomycin (Fig. 3D) or a stimulus such as hypotonic cell swelling (dilution of the medium osmolality with sterile deionized water; Fig. 3D) can be investigated. Inhibitors of this regulation or of hypothesized mechanisms of ATP release can be tested

TABLE I
DETECTION OF ATP WITH LUCIFERASE–LUCIFERIN REAGENT DILUTED IN SERUM-FREE MEDIUM[a]

Dose of ATP-Mg^{2+}	Luminescence	Data values
0 ATP	1.017 ± 0.795 (7)	0.335 ± 0.155 (6) (HK Block)
10^{-15} ATP	0.835 ± .881 (3)	
10^{-14} ATP	0.937 ± 1.000 (3)	
10^{-13} ATP	1.063 ± 1.103 (3)	
10^{-12} ATP	1.142 ± 1.208 (3)	
10^{-11} ATP	1.200 ± 1.263 (3)	
10^{-10} ATP	1.434 ± 1.557 (3)	
10^{-9} ATP	2.733 ± 1.242 (3)	
10^{-8} ATP	10.45 ± 3.249 (3)	
10^{-7} ATP	84.67 ± 32.82 (3)	
		438.8 ± 39.67 (36) (9HT Basal)
		527.9 ± 49.39 (9) (9HT cAMP)
10^{-6} ATP	934.1 ± 200.5 (3)	
		4348 ± 1026 (6) (9HT Ca^{2+})
10^{-5} ATP	7570 ± 1934 (3)	
10^{-4} ATP	>9999	
10^{-3} ATP	>9999	

[a] Luminescence (in arbitrary units) is shown for a standard curve made up of serial dilutions of ATP-Mg^{2+} mixed in luciferase–luciferin reagent in OptiMEM-I medium. Values are shown as mean ± SEM for luminescence signal observed from experiments with 9HTEo^{-} normal human tracheal epithelial cells[19] including basal luminescence, cAMP- and Ca^{2+}-stimulated luminescence, and hexokinase blockade of the luminescence signal in the presence of 9HTEo^{-} cells.

subsequently on basal or stimulated ATP bioluminescence. ATP scavengers like hexokinase or apyrase can be used to abolish the ATP bioluminescence signal. This assay is being used to study many normal and CF epithelial cell culture models derived from human airway, rat trachea, human pancreas, human intestinal, and mouse and human kidney as well as heterologous mammalian cell expression systems and even *Xenopus* oocytes overexpressing CFTR, CFTR mutants, *mdr*, and *mdr* mutants.

Comparison of Radioactive and Nonradioactive ATP Release Assays

Advantages of both the [γ-^{32}P]ATP release/trapping assay and the ATP bioluminescence assay lie in the measurement of the release of intact ATP from cells. Advantages of the ATP bioluminescence assay over the radioactive assay are cost, safety, the generation of time courses, paired analysis of the effect of an agonist on ATP release, study of the dynamics of ATP release versus degradation, and possible elucidation of the mechanism of ATP release (inhibition of CFTR with glibenclamide; inhibition of exocytic release with brefeldin A, a vesicle trafficking inhibitor, or nocodazole, a microtubule stabilizing agent, etc.). Moreover, the radioactive assay is dependent on adequate loading of ATP; streptolysin O permeabilization is necessary to enhance our loading but is not an ideal maneuver when studying physiologic systems. Despite all of these differences, both assays show no cAMP-induced ATP release in parental IB3-1 CF cells and a similar one-fold increase in ATP release in response to cAMP in wild-type CFTR-transfected IB3-1 cells.[13] Finally, it may be possible to study ATP transport and release using the bioluminescence assay from samples harvested from apical and basolateral sides of epithelial cells grown on permeable supports, from tissues mounted on permeable supports, or from single or groups of *Xenopus* oocytes injected with cRNA for CFTR, *mdr*, or variants or mutants of each ABC transporter, although modifications will have to made concerning the handling of cells, oocytes, and samples.

Single-Channel Patch-Clamp Recordings of ATP-Selective Currents

As alluded to earlier, CFTR as well as *mdr* have been implicated in conductive transport of ATP in epithelial and heterologous cells.[14] To investigate whether expression of CFTR can result in conduction of ATP^- in addition to Cl^-, single-channel patch-clamp recordings of IB3-1 CF human airway epithelial cells transduced with and expressing wild-type CFTR as well as *Xenopus* oocytes injected with wild-type and mutant CFTR cRNAs can be performed to assess relative conduction of Cl^- and ATP.[15] Epithelial cells are seeded onto collagen-coated glass coverslips for single-channel recording. Oocytes are placed in a hypertonic solution (in m*M*): 220 *N*-methylglucamine, 220 aspartic acid, 2 $MgCl_2$, 10 EGTA, 10 HEPES, pH 7.4, at room temperature, and allowed to shrink for 2–5 min to aid removal of the vitelline membrane. Immediately following removal of this membrane, the oocyte is transferred to a chamber for patch-clamp recording and exposed to the solutions outlined below. For single-channel recordings of ATP^- versus Cl^- currents, standard 141 m*M* NaCl-containing bath solu-

[13] E. M. Schwiebert and W. B. Guggino, unpublished observations.

[14] For ABC transporter functions, *see* E. M. Schwiebert et al., *Cell* **81,** 1063 (1995); M. J. Stutts et al., *Science* **269,** 847 (1995); I. L. Reisin et al., *J. Biol. Chem.* **269,** 20584 (1994); E. H. Abraham et al., *Proc. Natl. Acad. Sci. U.S.A.* **90,** 312 (1993); D. R. Gill et al., *Cell* **71,** 23 (1992); C. F. Higgins, *Cell* **82,** 693 (1995); L. Aguilar-Bryan et al., *Science* **268,** 423 (1995); N. Inagaki et al., *Science* **270,** 1166 (1995).

[15] M. E. Egan, E. M. Schwiebert, and W. B. Guggino, unpublished observations.

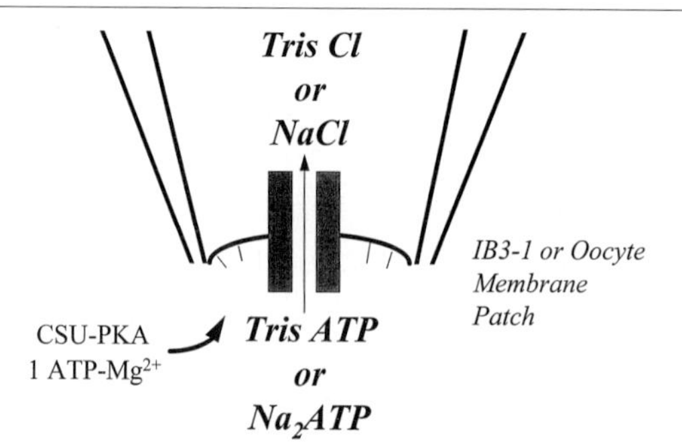

FIG. 4. Single-channel patch-clamp analysis of ATP ionic currents.

tion is exchanged for a nominally Cl^--free Na_2-ATP solution, which contained (in mM): sodium gluconate, 40; Na_2-ATP, 50; calcium gluconate, 0.25; EGTA, 1; HEPES, 5 (pH 7.4). For complete elimination of Cl^- in pipette and bath solutions, the pipette solution is (in mM): sodium gluconate, 140; calcium gluconate, 0.5; HEPES, 5 (pH 7.4). For elimination of Na^+ from the solutions, 145 mM Tris-Cl or 140 mM Tris–ATP-containing solution is used (all other constituents the same). Figure 4 presents a schematic diagram of the single-channel patch-clamp assay for both cells and oocytes.

In IB3-1 human airway cell membrane patches expressing CFTR, single-channel Cl^- conductance is 9 picosiemens (pS) consistent with our and other published results.[16] In patches from IB3-1 cells, single-channel ATP conductance is 4–5 pS, indicating activity by CFTR itself or a closely associated channel that conducts the ATP^- anions.[17] The open probability, a measure of the activity of a given conductance, is ~0.5 for Cl^- conduction and ~0.2 for ATP^- conduction in IB3-1 cell membrane patches. In oocyte membrane patches expressing wild-type CFTR, Cl^- conductive activity is similar (~0.5) but ATP^- conductive activity is much reduced (<0.1), suggesting that cofactors or regulators necessary for producing this CFTR-associated ATP conducting activity are more abundant in the CF human airway epithelial cell expression system.[18] We have not studied *mdr* in these expression systems.

[16] See Schwiebert et al. (1995) in footnote 6.

[17] M. E. Egan, unpublished observations.

[18] M. E. Egan, unpublished observations.

[19] D. C. Gruenert et al., *Proc. Natl. Acad. Sci. U.S.A.* **85,** 5951 (1988).

Conclusion

Utilization of these assays in concert is a powerful approach to studying the role of ABC transporters, other channel or transporter molecules, and the regulatory machinery of a given cell that regulates these molecules or regulated the bioavailability of ATP (i.e., release mechanisms versus degradative mechanisms). There is little doubt of the importance of ATP as an agonist within tissues or in the circulation within vascular beds. The next challenge will be to elucidate the molecules important for the transport of ATP, the regulators of those ABC transporters, and the biological significance of ATP agonists for a given cell or tissue.

Acknowledgments

We thank Dr. Peter Maloney and Dr. Peter Pedersen of the Johns Hopkins CF SCOR Center for advice with the design of the [γ-^{32}P]ATP release/trapping assay. We thank Dr. David McCoy of Dartmouth Medical School for discussions concerning ATP and adenosine receptor biology and nucleoside transporter biology. This work was supported by a National Research Service Award to E.M.S. (HL 08832-01-04), an NIH grant (HL 03023) and a Cystic Fibrosis Foundation grant to M.E.E., and Cystic Fibrosis Foundation and NIH grants (HL 47122, DR 32753, DK 48977) to W.B.G.

[50] Overexpression, Purification, and Function of First Nucleotide-Binding Fold of Cystic Fibrosis Transmembrane Conductance Regulator

By YOUNG HEE KO and PETER L. PEDERSEN

Introduction

Cystic fibrosis (CF) is the most common lethal autosomal recessive genetic disease in the Caucasian population. The disease results from mutations in the gene that codes for the cystic fibrosis transmembrane conductance regulator (CFTR) protein that is comprised of 1480 amino acid residues.[1–3] Early investigators predicted that CFTR embrances two

[1] J. M. Rommens, M. C. Iannuzzi, B. Kerem, M. L. Drumm, G. Melmer, M. Dean, R. Rozmahel, J. L. Cole, D. Kennedy, N. Hidaka, M. Zsiga, M. Buchwald, J. R. Riordan, L. Tsui, and F. S. Collins, *Science* **245,** 1059 (1989).

[2] J. R. Riordan, J. M. Rommens, B. Karem, N. Alon, R. Rozmahel, Z. Grzelczak, J. Zielenski, S. Lok, J. Plavsik, J. Chou, M. L. Drumm, M. C. Iannuzzi, F. S. Collins, and L. Tsui, *Science* **245,** 1066 (1989).

0076-6879/98 $25.00

transmembrane domains, each having six putative membrane-spanning segments; two nucleotide-binding folds, NBF1 and NBF2, each possessing a Walker A and B nucleotide-binding consensus sequence[4]; and a regulatory (R) domain having numerous acidic and basic amino acid residues as well as serine and threonine residues that can be phosphorylated by some cAMP-dependent protein kinases.[5,6]

Mutations in CFTR occur within or near the two NBFs. Significantly, most patients with cystic fibrosis have at least one allele in which the codon for phenylalanine-508 in NBF1 is deleted.[7–9] It has been established that CFTR functions as a chloride ion channel regulated by cAMP-dependent phosphorylations.[10–16] Studies indicate that interaction of intracellular ATP with NBF1 (or NBF2) may be essential for chloride channel activation.[17,18] Thus, the roles of the NBFs are involved in the elucidation of the molecular basis of CF.

Our laboratory found that a synthetic peptide embracing the Walker A consensus sequence and the region surrounding F508 contains less β-sheet

[3] B. Kerem, J. M. Rommens, J. A. Buchanan, D. Markiewicz, T. K. Cox, A. Chakravarti, M. Buchwald, and L. Tsui, *Science* **245,** 1073 (1989).

[4] J. E. Walker, M. Saraste, J. Runswick, and N. J. Gay, *EMBO J.* **1,** 945 (1982).

[5] S. H. Cheng, D. P. Rich, J. Marshall, R. J. Gregory, M. J. Welch, and A. E. Smith, *Cell* **66,** 1027 (1991).

[6] M. R. Picciotto, J. Cohn, G. Bertuzzi, P. Greengard, and A. C. Nairn, *J. Biol. Chem.* **267,** 12742 (1992).

[7] G. R. Cutting, L. M. Kasch, B. J. Rosenstein, J. Zielenski, L. Tsui, S. E. Antonarakis, and H. H. Karazian, Jr., *Nature* **346,** 366 (1990).

[8] B. Kerem, J. Zielenski, D. Markiewicz, D. Bozon, E. Gazit, J. Yahav, D. Kennedy, J. R. Riordan, F. S. Collins, J. M. Rommens, and L. Tsui, *Proc. Natl. Acad. Sci. U.S.A.* **87,** 8447 (1990).

[9] J. P. Cheadle, A. L. Meredith, and N. Aljader, *Hum. Mol. Genet.* **1,** 123 (1992).

[10] M. P. Anderson, R. J. Gregory, S. Thompson, A. E. Souza, P. Sucharita, R. C. Mulligan, A. E. Smith, and M. J. Welsh, *Science* **253,** 202 (1991).

[11] D. P. Rich, R. J. Gregory, M. P. Anderson, P. Manavalan, A. E. Smith, and M. J. Welsh, *Science* **253,** 205 (1991).

[12] T. V. McDonald, P. T. Nghiem, P. Gardner, and C. L. Martens, *J. Biol. Chem.* **267,** 3242 (1992).

[13] C. E. Bear, R. J. Bridges, T. J. Jensen, M. Ramjeesingh, and J. R. Riordan, *Cell* **68,** 809 (1992).

[14] B. C. Tilly, M. C. Winter, L. S. Ostedgaard, C. O'Riordan, A. E. Smith, and M. J. Welsh, *Biol. Chem.* **267,** 9470 (1992).

[15] M. Egan, T. Flotte, A. Afoine, R. Solow, P. L. Zeitlin, B. J. Carter, and W. B. Guggino, *Nature* **358,** 581 (1992).

[16] M. L. Drumm, D. J. Wilkinson, L. S. Smit, R. T. Worrell, T. V. Strong, R. A. Frizzell, D. C. Dawson, and F. S. Collins, *Science* **254,** 1797 (1991).

[17] M. P. Anderson, H. A. Berger, D. P. Rich, R. J. Gregory, A. E. Smith, and M. J. Welsh, *Cell* **67,** 775 (1991).

[18] P. M. Quinton and M. M. Reddy, *Nature* **360,** 79 (1992).

structure in the ΔF508 mutant than in the wild-type peptide.[19,20] The peptide studies demonstrated also that this structural change is due to destabilization of the mutant peptide structure, indicating that the ΔF508 mutation results in a protein folding defect.[19–21] *In vivo* studies conducted in other laboratories support this finding in that lowering the temperature of cells harboring the ΔF508 mutant protein that reaches the plasma membrane increases the amount of functional protein at the plasma membrane.[22,23] The functional mutant protein exhibits reduced open channel probabilities relative to the wild-type protein.[16,22] The structural basis of the altered channel function of the mutant protein in the membrane is currently unknown. Thus, the production of both purified wild-type and ΔF508 mutant forms of NBF1 in a state amenable to biochemical and structural studies is necessary to define the structural change in the mutant protein that leads to altered function.

With the above in mind, we rationalized that fusion of NBF1 and its ΔF508 mutant with a highly soluble protein such as maltose-binding protein, whose crystalline structure at high resolution had already been established,[24] might facilitate purification and structural characterization. Precedence for this approach is evident from the recent successful elucidation of the three-dimensional structure of actin by first forming a complex with DNaseI.[25] We report herein the result of studies covering the fusion of both the wild-type and ΔF508 mutant NBF1 domains with the maltose-binding protein along with the characterization of the products.

Experimental Methods

Materials

Clones T16-1 and Cl-1/5 are obtained from American Type Culture Collection (ATCC, Rockville, MD). Primers used in the polymerase chain reaction (PCR) procedures are synthesized in the Johns Hopkins University Protein/Peptide/DNA Synthesis Facility (Baltimore, MD) using an Applied

[19] P. J. Thomas, P. Shenbagamurthi, X. Ysern, and P. L. Pedersen, *Science* **251,** 555 (1991).

[20] P. J. Thomas, P. Shenbagamurthi, J. M. Hullihen, and P. L. Pedersen, *J. Biol. Chem.* **267,** 5727 (1992).

[21] P. J. Thomas, Y. H. Ko, and P. L. Pedersen, *FEBS Lett.* **312,** 7 (1992).

[22] G. M. Denning, M. P. Anderson, J. F. Amara, J. Marshall, A. E. Smith, and M. J. Welsh, *Nature* **358,** 761 (1992).

[23] C. Li, M. Ramjeesingh, E. Reyes, T. Jensen, X. Change, J. M. Rommens, and C. E. Bear, *Nature Genet.* **3,** 311 (1993).

[24] W. Kabsch, H. G. Mannherz, D. Suck, E. F. Pai, and K. C. Holmes, *Nature* **347,** 37 (1990).

[25] J. C. Spurlino, G. Y. Lu, and F. A. Quiocho, *J. Biol. Chem.* **266,** 5202 (1991).

BioSystems Model 380B Synthesizer. The pMAL-cR1 expression vector, the purification system, Factor X_a, restriction enzymes, and T4 DNA ligase are obtained from New England BioLabs. Trinitrophenyl-nucleotides are obtained from Molecular Probes. DNA polymerase and the Plasmid Quick DNA Purification Kit are purchased from Stratagene (La Jolla, CA). The DNA sequencing kit (Sequenase Version 2.0 DNA) is from United States Biochemical (Cleveland, OH), and the maltose-binding protein (MBP) from New England BioLabs. Waters-Millipore (Bedford, MA) supplies the Protein PAK 300 SW HPLC gel filtration column (7.5-mm inside diameter × 30-mm length) and uranyl acetate is purchased from Ted Pella. All other reagents are of the highest quality commercially obtainable.

Methods

Construction of an Expression Vector Containing CFTR NBF1. An outline of the method employed for synthesizing the construct is presented in Fig. 1. NBF1 of CFTR is synthesized by the expression cassette polymerase

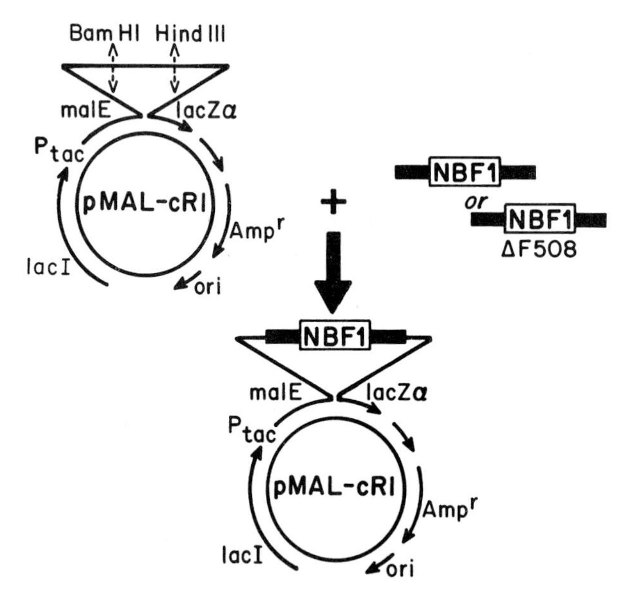

FIG. 1. Construction of the expression vector containing NBF1 of CFTR. The cDNA of NBF1 or ΔF508,NBF1 is ligated into *Bam*HI/*Hin*dIII sites of pMAL-cR1 in frame with the MBP (*malE* gene product). The insertion of the NBF1 cDNA into this plasmid results in disruption of the *lacZα* gene and allows blue to white colony selection when bacterial cells are grown on plates containing IPTG and X-Gal. The pMAL-cR1 is 6133 base pairs long and the NBF1 and its mutant are, respectively, 471 and 468 base pairs long. For experimental details see text. [Reproduced with permission from Y. H. Ko, P. J. Thomas, M. R. Delannoy, and P. L. Pedersen, *J. Biol. Chem.* **268,** 24330 (1993).]

chain reaction (ECPCR).[26,27] Restriction enzyme sites (*Bgl*II and *Hin*dIII), GC clamps, and a stop codon are incorporated into the primers, including a *Bgl*II site in the forward primer and a *Hin*dIII site in the reverse primer. The sequence of the primers used to amplify NBF1 from F433 (base 1424) to S589 (base 1899) follows:

Forward primer: 5′-C G C G C A G A T C T T T C T C A C T T C T
(35 bases) T G G T A C T C C T G T C-3′
Reverse primer: 5′-G C G C A A G C T T T T A G C T T T C A A
(36 bases) A T A T T T C T T T T T C T G-3′

The ECPCR is conducted in a 100-μl reaction volume comprising 20 mM Tris-Cl, pH 8.8, 2.5 units Pfu polymerase, 0.5 μM primers, 50 ng of template cDNA, 200 μM dNTPs, 1.5 mM $MgCl_2$, 10 mM KCl, 6 mM $(NH_4)_2SO_4$, and 0.1% Triton X-100. The conditions for each PCR cycle follow: denaturation, 94° for 43 sec; annealing at 43° for 1 min; and elongation at 73° for 1 min, 15 sec.

Clones T16-1 and Cl-1/5 used as templates to amplify the wild-type NBF1 and the mutant ΔF508,NBF1, respectively, are prepared for the PCR by using a Plasmid Quick Midi column (Stratagene). PCR amplified products (ca. 468–471 base pairs expected) are purified by using Prep-A-Gene DNA purification matrix kit (BioRad, Richmond, CA) and are subsequently digested with *Bgl*II and *Hin*dIII restriction enzymes. The digested PCR products are subsequently subjected to agarose gel electrophoresis, further purified by using Prep-A-Gene DNA Purification Matrix kit (BioRad), and finally ligated into the *Bam*HI/*Hin*DIII cut pMAL-cR1 expression vector[28,29] in frame with the MBP. The DNA sequence encoding for the protein sequence -I-E-G-R-, which is recognized by Factor X_a, is present between the DNA sequences for MBP and NBF1. The insertion of NBF1 into the vector disrupts the expression of *lacZ*α DNA and provides a rapid selection method for the correct recombinant. White colonies instead of blue colonies are formed on X-Gal (5-bromo-4-chloro-3-indolyl-β-galactoside) in an α-complementing host such as TB1[30] or JM107.[31]

[26] K. D. MacFerrin, M. P. Terranova, S. L. Schreiber, and G. L. Verdine, *Proc. Natl. Acad. Sci. U.S.A.* **87,** 1937 (1990).

[27] R. K. Saiki, S. Scharf, F. Faloona, K. B. Mullis, G. T. Horn, H. A. Erlich, and N. Arnheim, *Science* **230,** 1350 (1985).

[28] C. Guan, P. Li, and P. D. Riggs, and H. Inouye, *Gene* **67,** 21 (1987).

[29] C. V. Maina, P. D. Riggs, A. G. Granadea III, B. E. Slatko, L. S. Moran, J. A. Tagliamonte, L. A. McReynolds, and C. Guan, *Gene* **74,** 365 (1988).

[30] T. C. Johnston, R. B. Thompson, and T. O. Baldwin, *J. Biol. Chem.* **261,** 4805 (1986).

[31] C. Yanisch-Perron, J. Vieira, and J. Messing, *Gene* **33,** 103 (1985).

Competent TB1 *Escherichia coli* are transformed with the above ligation mixture and grown on LB/agar/ampicillin plates. Plasmids purified from the ampicillin-resistant white colonies are sequenced by Sanger's method[32,33] to confirm the fidelity of the PCR step and formation of the proper construct. All other molecular biological methods employed are conducted by standard procedures.[34]

Overexpression and Purification of MBP–NBF1 Fusion Proteins. TB1 *E. coli* harboring the appropriate recombinant plasmids are grown at 37° in LB media containing 0.2% glucose and ampicillin (0.1 mg/ml) until absorbance at 600 nm reaches about 0.5 units ($\sim 2 \times 10^{10}$ cells/ml). Cells are then induced with 0.3 m*M* isopropylthiogalactoside (ITPG). Incubation continues for an additional 3 hr, after which the cells are harvested by centrifugation at 4000*g* for 20 min. A combination of lysozyme and sodium deoxycholate is used to lyse the cells.[35,36] To remove the broken cell walls and debris, the lysate is centrifuged at 4°, 28,000 rpm, for 1 hr using a Beckman ultracentrifuge L7-55 fitted with a SW 28 rotor. The supernatant is then incubated with 55% saturated ammonium sulfate at 4° for 30 min, and centrifuged at 12,000*g* for 40 min at 4° in a Sorvall centrifuge, Model RC 2B, equipped with a SS 34 rotor (radius of 4.25 in.). To obtain the protein at a concentration of ca. 2.5 mg/ml, the pellet is resuspended in 40 ml of the column buffer (10 m*M* sodium phosphate, 0.5 *M* NaCl, and 1.0 m*M* EGTA, pH 7.2).

Purification is carried out by loading the resuspended sample onto a column containing amylose resin that has been previously equilibrated with the column buffer. The binding capacity of the column is 2 mg fusion protein per milliliter of bed volume. The column is washed with three column volumes of column buffer containing 0.25% Tween 20 and finally with column buffer without the surfactant. Elution of the bound fusion protein is accomplished at a flow rate of 1 ml/min with 20 m*M* maltose dissolved in the column buffer. Fractions containing the fusion proteins, detected by UV absorbance at 280 nm, are pooled and concentrated using Centri-Prep 30 Concentrator (Amicon, Denvers, MA).

SDS–PAGE is carried out as described by Laemmli.[37] Gels are stained with Coomassie dye to determine the degree of overexpression of the recombinant protein and to estimate its purity. Gas-phase sequencing of

[32] F. Sanger, S. Niklen, and A. R. Coulson, *Proc. Natl. Acad. Sci. U.S.A.* **74,** 5463 (1977).
[33] S. Tabor and C. C. Richardson, *Proc. Natl. Acad. Sci. U.S.A.* **84,** 4767 (1987).
[34] J. Sambrook, E. F. Fritsch, and T. Maniatis, *in* "Molecular Cloning: A Laboratory Manual," 2nd Ed. Cold Spring Harbor Laboratory Press, New York, 1989.
[35] F. A. O. Marston, *Methods Enzymol.* **3,** 59–88 (1987).
[36] B. Withold, H. Van Heerikhuizen, and L. De Leij, *Biochim. Biophys. Acta* **443,** 534 (1976).
[37] U. K. Laemmli, *Nature* **277,** 680 (1970).

the N terminus[38,39] of the purified protein, as well as an amino acid analysis, establishes that the product isolated is the desired fusion protein.

TNP–ATP Binding. Fluorescence enhancement of the ATP analog, TNP–ATP[40–42] is used to study the nucleotide-binding characteristics of the fusion proteins in the presence or absence of urea. The procedure used has been described previously.[19] Samples, 4 μM protein in 20 mM HEPES, pH 7.4, and at concentrations of TNP–ATP indicted, are excited at 410 nm (slit width, 1 mm). The emission is then measured at 550 nm (slit width, 2 mm) using an Aminco spectrofluorometer (Model SPF-125) (Silver Spring, MD). Finally, ATP is added after TNP–ATP binding to demonstrate that ATP can displace bound TNP–ATP.

Circular Dichroism Spectroscopy. To assess the secondary structures of the fusion proteins, circular dichroism (CD) spectra are obtained by the method previously described[20] at 23.2° in a 2-mm path length cuvette using an AVIV spectropolarimeter. Wild-type and mutant fusion proteins are present at a concentration of 1.3 μM in a 0.4-ml system containing 10 mM Tris-Cl buffer, pH 7.6.

The MBP–NBF1 fusion protein is treated with 0.08% SDS, incubated at 23° for 16 hr, and then subjected to the action of Factor X_a at a protease to fusion protein ratio of 1:100 for 14 hr at 4°. The degree of cleavage is monitored by SDS–PAGE.

Results

Overexpression and Purification of MBP–NBF1 and MBP–(ΔF508)NBF1

The correct orientation and cDNA sequence of CFTR NBF1 and its ΔF508 mutant in the pMAL-cR1 expression vector in frame with the MBP was first confirmed by sequencing the recombinant plasmids. The entire codon for phenylalanine in the mutant was absent. As shown in Fig. 2, the overexpression of fusion proteins under the control of the *tac* promoter after induction with IPTG is evident when the induced expression level is compared with the uninduced cell lysate. Because the MBP exhibits a high affinity for amylose, the purification of the fusion proteins by amylose column affinity chromatography was simple and rapid. A yield of approxi-

[38] P. Edman, *Acta. Chem. Scandinavica* **4,** 283 (1950).
[39] M. W. Hunkapiller and L. E. Hood, *Science* **219,** 650 (1983).
[40] T. Hiratsuka and K. Uchida, *Biochim. Biophys. Acta* **320,** 635 (1973).
[41] T. Hiratsuka, *Biochim. Biophys. Acta* **453,** 293 (1976).
[42] T. Hiratsuka, *Biochim. Biophys. Acta* **719,** 509 (1982).

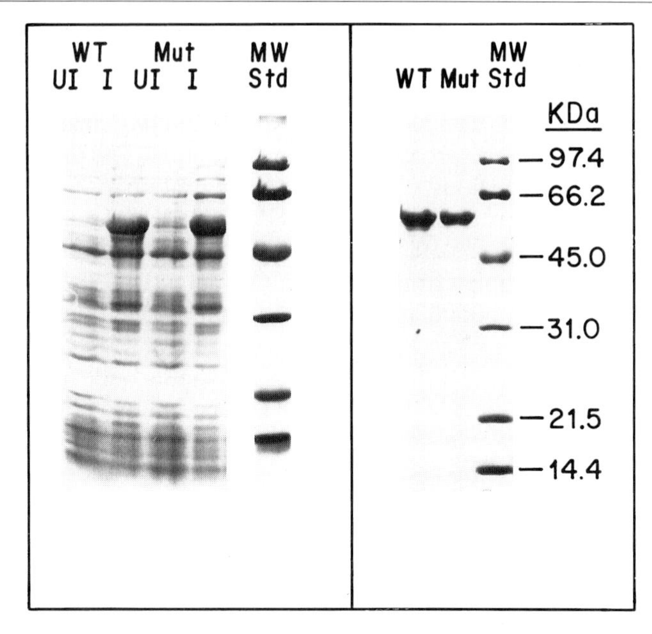

FIG. 2. Analysis by SDS–PAGE of overexpressed and purified wild-type and mutant MBP–NBF1 fusion proteins. *Left:* To demonstrate the overexpression of the MBP–NBF1 fusion proteins, the whole-cell lysate (300 μl, 2×10^8 cells/ml) before IPTG induction was solubilized in SDS sample buffer and loaded onto the gel designated as "uninduced cells" (UI). SDS–PAGE was then carried out by the method by Laemmli [U. K. Laemmli, *Nature* **277,** 680 (1970)]. The *E. coli* cell lysate (75 μl, stationary phase cells) after IPTG induction was treated in the same manner and designated "induced cells" (I). Molecular weight markers are from top to bottom, phosphorylase b (97.4 kDa), bovine albumin (66.2 kDa), ovalbumin (45 kDa), carbonic anhydrase (31 kDa), trypsin inhibitor (21.5 kDa), and lysozyme (14.4 kDa). *Right:* The fusion proteins (25 μg) purified by amylose affinity chromatography were subjected to SDS–PAGE. Both proteins exhibit an approximate molecular weight of 60 kDa. [Reproduced with permission from Y. H. Ko, P. J. Thomas, M. R. Delannoy, and P. L. Pedersen, *J. Biol. Chem.* **268,** 24330 (1993).]

mately 25 mg of purified proteins per liter of cell culture was achieved. The resultant proteins appear >95% pure as judged by SDS–PAGE (Fig. 2) and have the expected apparent molecular weights of 60,000. Moreover, amino acid analysis gave experimental to theoretical ratios very close to one in most cases. The ratio of picomoles of phenylalanine to the number of picomoles of other amino acids, obtained from amino acids analysis, is consistently lower for the ΔF508 mutant fusion protein than for the wild type, confirming that the number of phenylalanine residues is less in the mutant. In experiments not presented here, N-terminal sequence analysis (10 amino acids) confirmed the identity and purity of the fusion proteins.

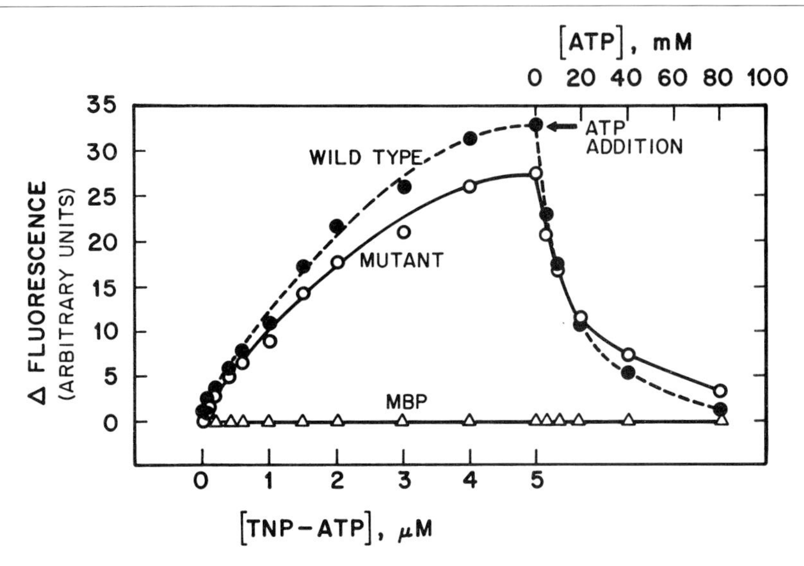

FIG. 3. Nucleotide-binding function of MBP–NBF1 fusion proteins. The enhanced extrinsic fluorescence of TNP–ATP on binding to the fusion proteins was observed at a 550-nm emission wavelength after exciting the samples at 410 nm. When ATP was added to the sample to compete off the bound TNP–ATP the fluorescence of the TNP–ATP rapidly decreased. See the text for details. [Reproduced with permission from Y. H. Ko, P. J. Thomas, M. R. Delannoy, and P. L. Pedersen, *J. Biol. Chem.* **268,** 24330 (1993).]

TNP–ATP Binding Characteristics

TNP–ATP was found to bind similarly to both the wild-type and mutant fusion proteins as depicted in Fig. 3. ATP effectively displaces all of the bound TNP–ATP indicating that the site involved is capable of binding the physiologic substrate. The apparent K_D values determined according to a method previously reported[43] for binding of TNP–ATP to the wild-type and mutant fusion proteins are 1.8 and 2.3 μM, respectively. The correct K_D values for ATP binding, determined by competition analysis between TNP–ATP and ATP, are 1.8 and 1.9 mM, respectively. The stoichiometry of TNP–ATP binding is approximately 1 for both types of fusion proteins under the conditions employed. In the control experiment, MBP alone did not bind TNP–ATP even at concentrations exceeding 5 μM. This important control emphasizes that the nucleotide-binding properties of the fusion proteins, as expected, are due to interaction of TNP–ATP or ATP with NBF1 or ΔF508,NBF1 per se and not with the MBP.

[43] G. P. Mullen, P. Shenbagamurthi, and A. S. Mildvan, *J. Biol. Chem.* **264,** 19637 (1989).

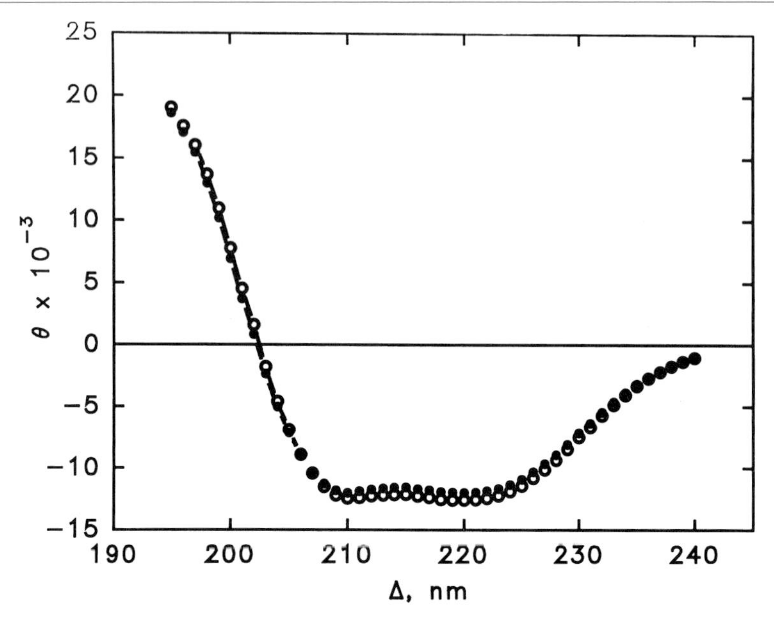

FIG. 4. Circular dichroism spectra of the MBP–NBF_1 fusion proteins. The molar mean residue ellipticity of the wild-type or mutant protein as a function of wavelength is depicted. Fusion proteins (1.3 μM) both dissolved in 10 mM Tris-Cl, pH 7.6, were subjected to CD spectroscopic studies using a 2-mm path length cuvette as described in the text. [Reproduced with permission from Y. H. Ko, P. J. Thomas, M. R. Delannoy, and P. L. Pedersen. *J. Biol. Chem.* **268,** 24330 (1993).]

Secondary Structure as Assessed by Circular Dichroism Spectroscopy

The overall secondary structures of the wild-type and ΔF508 mutant fusion proteins were estimated by CD spectropolarimetry. As shown in Fig. 4, the resultant spectra, which summarize the mean residue molar ellipticity as a function of wavelength, reveal no obvious differences between the wild-type and ΔF508 mutant fusion proteins. In both cases, the mean residue molar ellipticity of the fusion proteins exhibits a maximum at 197 nm and two minima, one at 209 nm and the other at 223 nm. Deconvolution of the spectra using the Prosec program[44] predicts that both wild-type and ΔF508 mutant fusion proteins contain nearly identical secondary structural elements distributed as approximately 40% α helix, 24% β strand, 13% turns, and 24% random coil. Thus, both fusion proteins are highly structured as assessed by this technique.

[44] C. T. Chang, C.-S. C. Wu, and J. T. Yang, *Anal. Biochem.* **91,** 13 (1978).

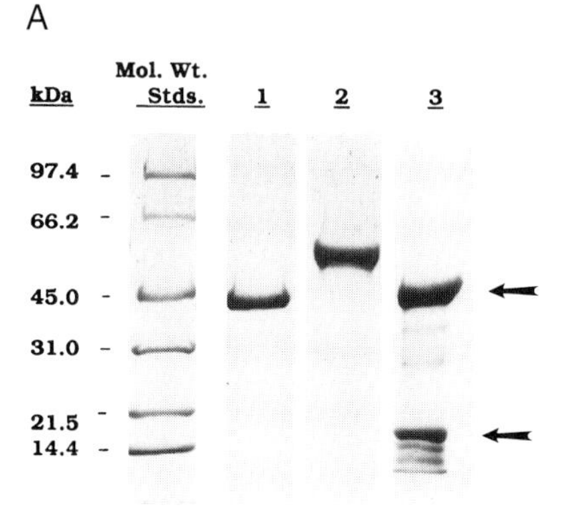

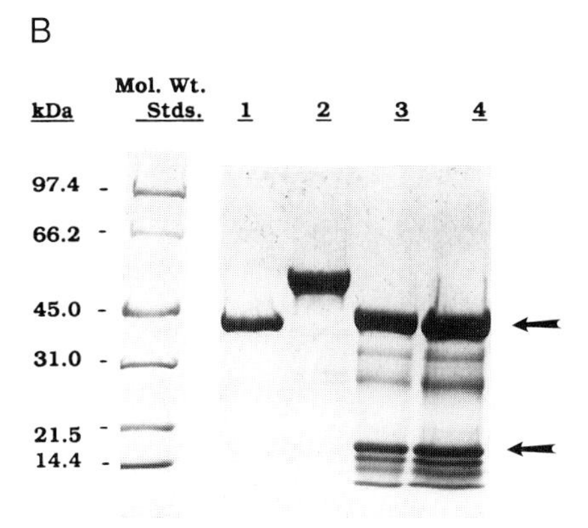

FIG. 5. Factor X_a cleavage of the wild-type MBP–NBF_1 fusion protein. (A) The wild-type fusion protein (MBP–NBF1) was treated with Factor X_a exactly as described in the text and then subjected to SDS–PAGE. In all cases 5 μg protein was loaded onto the gel, Lane 1, untreated MBP control; lane 2, untreated MBP-NBF1 fusion protein control; lane 3, MBP–NBF1 treated with Factor X_a. (B) Separation of NBF1 from the Factor X_a cleavage mixture was attempted by binding the MBP component to an amylose resin. To do this, the product from the cleavage reaction was mixed with the resin, incubated at room temperature for 10 min, and then centrifuged at 4000*g* for 10 min. The resultant pellet was washed with 20 m*M* HEPES, pH 7.5, and centrifuged again for 10 min at 4000*g*. Then, the proteins bound to the resin were eluted with 20 m*M* maltose containing 20 m*M* HEPES, 100 m*M* NaCl, and 2 m*M*, $CaCl_2$, pH 7.5 Both the bound and unbound fractions (5 μg) were then analyzed by SDS–PAGE. Lane 1, untreated MBP control; lane 2, untreated MBP–NBF1 fusion protein control; lane 3, the amylose unbound fraction; lane 4, the amylose bound fraction. In both (*A*) and (*B*) the upper arrow indicates the position of the MBP and the lower arrow the position of the cleaved NBF1. The identity of the NBF1 was confirmed by N-terminal sequencing of the band. Molecular weight markers are the same as those indicated in the legend to Fig. 2. See text for explanation of multiple bands in A, lane 3, and B, lanes 3 and 4. [Reproduced with permission from Y. H. Ko, P. J. Thomas, M. R. Delannoy, and P. L. Pedersen. *J. Biol. Chem.* **268,** 24330 (1993)].

Proteolytic Cleavage of MBP–NBF1 Fusion Protein Followed by Chromatography on Amylose Column

The fusion junction between the MBP and NBF_1 contains a recognition site for protease Factor X_a. Attempts to cleave this site using a variety of conditions in which ionic strength, various detergents, solvents, pH, and temperature were tested proved unsuccessful suggesting a strong interaction between the MPB and NBF1 within the fusion protein that masked access of the protease to the cleavage site. However, at a low concentration of SDS employed in the proteolytic incubation medium, and as shown in the subsequent SDS–PAGE profiles (Fig. 5A), NBF_1 is now almost completely cleaved from the fusion protein. If the cleaved fusion protein is first subjected to chromatography on an amylose column, the NBF1 remains associated with the MBP and coelutes with it on addition of the maltose containing buffer (Fig. 5B). The identity of the cleaved NBF1 was confirmed by sequencing the N terminus of the protein band designated by the lower arrow in Figure 5B. (Note that in Fig. 5A, lane 3, and in Fig. 5B, lanes 3 and 4, that Factor X_a causes some nonspecific cleavage of the maltose-binding protein and the NBF_1 as indicated by several additional bands.)

These results indicate that within the MBP–NBF_1 fusion complex a tight association occurs between the two noncovalently bound fusion partners that is retained even after cleavage of the fusion junction.

Acknowledgments

The authors are grateful to Dr. Wu-Schyong Liu for amino acid and N-terminal sequence analysis, to Dr. Mario Bianchet for help in crystallizing the fusion proteins, to Jody Franklin for preparation of oligonucleotides, and to Jackie Seidl for processing this manuscript for publication.

[51] Recombinant Synthesis of Cystic Fibrosis Transmembrane Conductance Regulator and Functional Nucleotide-Binding Domains

By SCOTT A. KING and ERIC J. SORSCHER

Introduction

Large yields of purified, recombinant proteins are often required for functional and structural studies. Prokaryotic and eukaryotic expression systems now allow sufficient yields of many proteins, including complex

0076-6879/98 $25.00

integral membrane proteins such as those in the ATP-binding cassette (ABC) gene family, for experiments that otherwise might not be possible due to the endogenous (low) cellular levels of these polypeptides. The cystic fibrosis transmembrane conductance regulator (CFTR) is an integral membrane protein that is expressed at very low endogenous levels in mammalian epithelial cells (in some cases, as little as 1 mRNA molecule/cell).[1,2] The quantities of CFTR polypeptides necessary for structural and functional measurements led us to develop new synthetic protocols that take advantage of recent advances in recombinant protein expression and purification. With these protocols we have been able to produce CFTR domains for (1) antibody production, (2) circular dichroism (CD) spectroscopy, (3) nucleotide-binding studies and (4) analysis of the interactions between the CFTR first nucleotide binding domain (NBD1) and other cellular proteins.

Cystic Fibrosis Transmembrane Conductance Regulator

Cystic fibrosis is the most common lethal genetic disease among Caucasians, with an incidence of approximately 1 in 2000 to 3000 live births. Clinically, the disease is characterized by chronic obstructive lung disease with recurrent respiratory infections, intestinal obstruction, elevated sweat electrolytes, male infertility, and pancreatic insufficiency. The disease results from mutations in the CFTR protein, a member of the ABC family of transporter proteins.[1,2] While specific functions have been attributed to CFTR, the physiologic role of CFTR is not completely understood. Besides chloride conductance, there is evidence supporting a role for CFTR in ATP transport, acidification of lysosomes, and *Pseudomonas* internalization. CFTR also regulates the activity of other epithelial Cl^- channel proteins and epithelial Na^+ channels. The range of these roles ascribed to CFTR confounds simple models of CFTR function.[3,4]

The CFTR is a 150- to 170-kDa protein including 1480 amino acids.[1,3,4] As an integral membrane protein, CFTR is translated on the rough endoplasmic reticulum (RER) and inserted into the ER membrane. The trans-

[1] J. R. Riordan, R. M. Rommens, B. S. Kerem, N. Alon, R. Rozmahel, Z. Grzelcsak, J. Zielenski, S. Lok, N. Plavsik, J. Chou, M. Drumm, M. Iannuzi, F. S. Collins, and L.-C. Tsui, *Science* **245,** 1066 (1989).

[2] B. C. Trapnell, C. S. Chu, P. K. Paakko, T. C. Banks, K. Yoshimura, V. J. Ferrans, M. S. Chernick, and R. G. Crystal, *Proc. Natl. Acad. Science U.S.A.* **88,** 6565 (1991).

[3] M. J. Welsh, L.-C. Tsui, T. F. Boat, and A. L. Beaudet, *in* "The Metabolic and Molecular Bases of Inherited Disease: Membrane Transport Systems" (Cr. Scriver, A. L. Beaudet, W. S. Sly, and D. Valle, eds.), Vol. 3. McGraw-Hill New York, 1995.

[4] T. Jilling and K. Kirk, *Int. Rev. Cytol.* **172,** 193 (1997).

lated product (known as band A, approximately 135 kDa) becomes glycosylated (band B, approximately 150 kDa) and trafficked to the Golgi complex where the glycoprotein is further modified into a mature form (band C, approximately 170 kDa). This mature CFTR inserts into the apical membranes of polarized epithelial cells. The topology of CFTR is believed to resemble other ABC transporters. The protein is composed of two symmetric halves, each containing six predicted membrane-spanning regions and an NBD. Bridging the two halves is a segment termed the regulatory (R) domain, consisting of multiple consensus phosphorylation sites for both protein kinase A and protein kinase C, and containing a high proportion of charged amino acids. This R domain is unusual among the ABC family of proteins.

More than 400 clinically important mutations have been described in the CFTR gene. However, a small subset of these accounts for the vast majority of mutated alleles in the CF population. Distribution of mutations over the gene is nonrandom with apparent "hot spots." For example, the CFTR first nucleotide binding domain (NBD1) contains a disproportional abundance of mutations that cause disease. The most common defective CF allele results from the loss of three bases encoding a phenylalanine residue at CFTR position 508 (ΔF508). This mutation accounts for approximately 70% of defective alleles in CF patients, and is present in at least one allele in more than 90% of the CF individuals in the United States.

The preponderance of mutations in NBD1 has focused attention on functional and structural studies of this region. NBD1 consists of amino acids 433–586 of CFTR, and contains consensus motifs (Walker A and B sequences) for ATP binding and hydrolysis. The molecular defects associated with different NBD1 mutations are heterogeneous. Many of the mutant proteins, including a glycine → aspartic acid replacement at CFTR postion 551 (G551D), have suppressed chloride conductance.[3–5] Although ΔF508 CFTR may have nearly normal anion conductance properties, disease associated with this allele is believed to result from a failure to process the ΔF508 protein beyond the ER to the Golgi apparatus; instead the protein is degraded very rapidly through a mechanism that appears to involve the ubiquitin/proteasome pathway.[3,4] Intrinsic properties of wild-type and mutant NBD1 and related CFTR peptides have been studied in an attempt to understand the effects of regional mutations on CFTR function and biogenesis.

[5] J. Logan, D. Heistand, P. Daram, Z. Huang, D. D. Muccio, B. Haley, W. Cook, and E. J. Sorscher, *J. Clin. Invest.* **94,** 228 (1994).

Methods for Protein Expression and Purification

Two protocols for peptide expression and purification are described next. In general, higher yields of smaller CFTR polypeptides can be obtained by prokaryotic expression, although CFTR proteins expressed in this fashion are packaged within insoluble inclusion bodies. The eukaryotic (baculovirus) system may retain a larger fraction of CFTR proteins in soluble form, although the overall yield is generally lower, and the process requires construction and isolation of more complex viral vectors. Both procedures express the recombinant CFTR proteins as fusions with glutathione *S*-transferase (GST). GST fusion proteins are suitable for single-step purification on glutathione affinity agarose. In addition, GST appears to stabilize CFTR polypeptides, improving the yield in some cases (unpublished observations, 1992).

Steps for Protein Expression and Purification in Escherichia coli

CFTR NBD1 yield by the protocol described below is 100–200 μg of isolated, purified NBD1 from 200 ml of *E. coli* starting material.

NBD1 DNA Cloned into the pGEX-2T Vector System. The NBD1 region of CFTR [amino acids, (a.a.) 433–586] is amplified by polymerase chain reaction (PCR) with flanking *Eco*RI sites. The PCR product and the pGEX-2T vector (Pharmacia, Piscataway, NJ) are digested with *Eco*RI and the gel purified products ligated.[6–8] The cloning is designed to derive a polypeptide in which GST and NBD1 are in-frame and separated by an engineered thrombin cleavage site. Correct recombinants are identified by restriction mapping and by DNA sequencing of the full-length NBD1 inserts.

GST–NBD1 Fusion Protein Expressed in E. coli. DH5α *E. coli* transformed with pGEX-2T:NBD1 are grown overnight at 37° in NCZYM broth. The culture is diluted 1:9, grown another 2 hr (to obtain log phase growth) and cooled to room temperature. Isopropyl-β-D-galactopyranoside (IPTG) is added (1 m*M*) to induce transcription from the pGEX-2T *tac* promoter. After 4 hr, the culture is pelleted by centrifugation, resuspended in 1 ml phosphate-buffered saline (PBS)/0.1% Triton X-100 (150 m*M* NaCl, 16 m*M* NaH_2PO_4, 4 m*M* Na_2HPO_4, pH 7.3, and 0.1% Triton X-100) per

[6] J. Hartman, Z. Huang, T. A. Rado, S. Peng, T. Jilling, D. D. Muccio, and E. J. Sorscher, *J. Biol. Chem.* **267,** 6455 (1992).

[7] J. Hartman, R. A. Frizzell, T. A. Rado, D. J. Benos, and E. J. Sorscher, *Biotechnol. Bioeng.* **39,** 828 (1992).

[8] N. Arispe, E. Rojas, J. Hartman, E. J. Sorscher, and H. Pollard, *Proc. Natl. Acad. Sci. U.S.A.* **89,** 1539 (1992).

50 ml *E. coli,* and sonicated (five 1-sec pulses, followed by two continuous pulses of 10 sec each on ice with a 3-mm microtip probe).

Isolation and Solubilization of Inclusion Bodies. The cell lysate after sonication is centrifuged at 2000*g* for 5 min at 4° to isolate inclusion bodies. The pellet is then solubilized in 8 *M* urea. Because the majority of expressed protein is found in inclusion bodies, further purification of CFTR domains is performed using these structures rather than from the soluble fraction.

Removal of Urea and Thrombin Cleavage of the Fusion Protein. The solubilized pellet is dialyzed against a 1-liter bath of 50 m*M* Tris, pH 7.4, and the Tris buffer changed five times. Then 150 m*M* NaCl and 2.5 m*M* $CaCl_2$ are added to adjust the buffer for thrombin cleavage. Human thrombin (Sigma, St. Louis, MO) is added at a concentration of 20 units/10 mg of fusion protein. The reaction was left at 25° for 3 hr.

Gel Purification of NBD Peptide. The reaction mixture is separated on a preparative 12% acrylamide gel (Bio-Rad, Richmond, CA). Fractions containing the 21-kDa NBD1 band are determined by SDS–PAGE and these fractions are pooled.

Dialysis of NBD1 Peptide Into Urea for Storage. The isolated peptide is dialyzed into 6 *M* urea with 0.1% Dowex resin to absorb residual SDS (6 × 200-ml dialysis changes for 5–10 mg NBD1 protein). The purified, recombinant NBD-1 appears stable at 4° in urea for up to 3 months when isolated by this technique.

Dialysis of NBD1 Into 10 mM Tris for Structural and Functional Studies. Prior to studies of NBD1, the protein is dialyzed into 10 m*M* Tris, pH 7.4 (5 dialysis changes of 1 liter each). This dialysis is performed over a 24- or 48-hour period to ensure removal of urea.

Alternate Purification Protocol by Glutathione Agarose. GST–NBD1 fusion proteins that remain soluble after recombinant overexpression can be purified by loading cellular lysates directly onto glutathione agarose beads (Sigma). The solubility of highly aggregated GST–NBD1 can be enhanced by induction of protein synthesis and growth at 25°, allowing single-step affinity purification.[7] Yields from this approach tend to be lower than yields from the urea solubilization method described earlier. When levels of expressed GST–NBD1 protein are compared to levels of the NBD1 peptide produced without GST fusion, the fusion protein is expressed at much higher levels (data not shown). This higher yield may be due to the ability of GST to stabilize the overall polypeptide.

Eukaryotic Protein Expression and Purification

Autographa californica nuclear polyhedrosis virus (AcNPV) is often used for expression of recombinant proteins, including complex integral membrane proteins. The late viral genes, including p10 or polyhedrin, are not required for replication and therefore are suitable locations for recombinant gene substitution. The baculoviral genome is relatively large (120–130 kb) and requires specialized techniques to engineer recombinant virus, including a homologous recombination step with viral genomic DNA. These recombination events occur at a frequency of only about 0.1–0.2% after cotransfection of genomic DNA with a recombinant transfer vector. A fluorescence-activated cell sorting (FACS) step during viral purification simplifies recombinant viral purification. Single-step purification of recombinant GST–CFTR fusion proteins by glutathione agarose following insect cell expression can also be used to facilitate CFTR protein recovery.[9,10] Yield from this eukaryotic expression system was 1–5 μg per 10^6 cells, although much higher yields of CFTR using recombinant insect cells have been reported.[11]

Steps for Protein Production from Eukaryotic Spodoptera frugiperda (Sf9) Moth Ovary Cells

Production of Plasmid Vectors. PCR primers are designed to amplify (1) GST–thrombin cleavage site cDNA or (2) GST–Factor X_a cleavage site cDNA from the prokaryotic expression vectors pGEX-2T and pGEX-3T (Pharmacia, Piscataway NJ). The products are engineered to contain 5′ *Spe*-1 and 3′ *Nhe*-1 sites. The PCR (polymerase chain reaction) product is cloned into the *Nhe*-1 site of the baculovirus transfer vector pJVETLZ, a vector expressing β-galactosidase for blue-white screening (gift of Dr. Chris Richardson, Biotechnology Research Institute, Montreal, Canada). Foreign genes of interest, such as the full-length CFTR or relevant domains, are cloned into the 3′ *Nhe*-1 site to construct GST–CFTR fusions. The final products G-BAC-1 (X_a cleavage site) and G-BAC-2 (thrombin cleavage site) are suitable for (1) homologous recombination into the AcNPV genome, (2) FACS and plaque purification of recombinant virus, (3) glutathione agarose purification of recombinant fusion protein, and (4) cleavage of the fusion protein, in order to yield the CFTR polypeptide of interest.

Transfection of Spodoptera frugiperda (Sf9) Cells with AcNPV Genome and Transfer Plasmids. Sf9 cells are seeded at 2×10^6 cells per T25 flask and grown overnight at 27° in Grace's insect medium containing 10% (v/v)

[9] S. Peng, M. A. Sommerfelt, G. Berta, A. K. Berry, K. Kirk, E. Hunter, and E. J. Sorscher, *BioTechniques* **14,** 274 (1993).

[10] S. Peng, M. A. Sommerfelt, J. Logan, Z. Huang, T. Jilling, K. Kirk, E. Hunter and E. J. Sorscher, *Protein Express. Purific.* **4,** 95 (1993).

[11] C. E. Bear, C. Li, N. Kartner, R. J. Bridges, T. J. Jensen, M. Ramjeesingh, and J. R. Riordan, *Cell* **68,** 809 (1992).

fetal bovine serum (FBS). The next day the cells are washed with serum-free Grace's insect medium and then maintained in 1 ml serum-free medium for the transfection. Then 100 ng purified wild-type AcNPV DNA and 1 μg G-BAC containing CFTR or CFTR domains are combined in 12 μl doubly distilled H_2O. The DNA is mixed with 8 μg lipofectin in 12 μl doubly distilled H_2O. The DNA/lipid mixture is incubated at room temperature for 15 min and then added dropwise to Sf9 cells. Five to 24 hr later, 1 ml Grace's insect medium plus 10% FBS is applied to the cells.

Isolation of Cells Producing Recombinant Virus by Fluorescence-Activated Cell Sorting. Seven days after transfection, media are removed from cells and 33 μM ImagGene alkyl-derivatized fluorescein di-β-galactopyranoside (Molecular Probes, Eugene, OR) in serum-free media is added. This substrate is cleaved by β-galactosidase expressed from the G-BAC vector. After 15 min at 27°, the cells are scraped and sorted by FACS. The highest β-galactosidase expressing population of Sf9 cells (comprising about 0.5% of the harvested culture) is isolated for further purification.

Further Purification of Sorted Cells by One of Two Protocols

PROTOCOL A. Forty-five highly fluorescent cells are plated on a background of 2×10^6 wild-type Sf9 cells. Seventy-two hours later, the medium is collected and centrifuged to remove cells and debris. The clarified supernatant is then added to 2×10^6 wild-type Sf9 cells per well of a 6-well tray. Then 0.4 ml of this medium at dilutions (in fresh medium) of 10^0, 10^{-1}, 10^{-2}, and 10^{-3} is added to the wells for 1 hr at 27°. The inoculum is removed and each well overlaid with 4 ml of molten agar containing β-galactosidase substrate, 5-bromo-4-chloro-3-indolyl-β-D-galactosidase (X-Gal). After the agar has solidifed, 2 ml of Grace's insect medium with 10% FBS is added. Seventy-two hours later, blue plaques are visible and recombinant viral particles are removed by inserting a sterile pipette tip into the plaque and excising an agar plug.

PROTOCOL B. Twenty, 200, or 2000 FACS-positive cells are cocultured with 2×10^6 wild-type Sf9 cells in 6-well trays. After 4 hr, the medium is replaced with 4 ml of molten agar containing X-Gal and, on agar solidification, 2 ml of Grace's insect medium plus 10% FBS was overlaid. After an additional 72 hr at 27°, blue plaques are visible and the viral particles are removed by inserting a pipette tip into the plaque and excising an agar plug.

Excised agar plus are transferred to sterile vials with 1 ml Grace's insect medium. Overnight incubation at 4° is used to elute the viral particles. Additional purification steps using standard techniques[12] are performed by

[12] M. D. Summers and G. E. Smith, *in* "A Manual of Methods of Baculovirus Vectors and Insect Cell Culture Procedures," Bulletin No. 1555. Texas Agricultural Experiment Station, Texas A&M University, College Station, 1987.

administering serial dilutions of virus to 6-well trays of Sf9 cells. In many cases, the use of FACS expedited recovery of recombinant baculovirus by several weeks.

Verification of Recombinant Protein Expression. To verify expression of CFTR fusion proteins, Sf9 cells are infected with recombinant virus or wild-type virus at a multiplicity of infection (MOI) of 5 or 10. After 3 days, the cells are lysed in Laemmli lysis buffer. A small amount of the lysate is run on a 12% polyacrylamide gel and stained with Coomassie Brilliant Blue. A strong 31-kDa protein band representing the overexpression of polyhedrin(p31) is visible in wild-type virus-infected cells but absent in cells infected with the recombinant virus.

Production of Recombinant Fusion Protein in Sf9 Cells by Inoculation with Purified Recombinant AcNPV. Sf9 cells are inoculated with recombinant AcNPV (purified after one of the above protocols) at an MOI of 5–10 and harvested by scraping 3 days later. The cells are pelleted by centrifugation at 1000*g* for 5 min at 4° and suspended in PBS/1% Triton X-100 (150 m*M* NaCl, 16 m*M* NaH_2PO_4, 4 m*M* Na_2HPO_4, pH 7.3, and 1% Triton X-100) at 1 ml per 4×10^7 cells. The cell slurry is sonicated on ice by five pulses of 1 sec each followed by two continuous pulses of 10 sec at an output of 125 W with a 3-mm microtip probe. The insoluble fraction is removed by centrifugation at 10,000*g* for 5 min at 4°.

Affinity Purification of Recombinant Fusion Protein on Glutathione Agarose. Reduced glutathione agarose (GSH-Ag) at 12 μM reduced glutathione per milliliter hydrated agarose (Sigma) is stored at 4°. GSH-Ag is equilibrated with PBS/1% Triton X-100 prior to use. The soluble fraction of cell lysate is added to GSH-Ag (50 μl GSH-Ag/ml cell lysate) and the mixture rotated overnight at 4°. The following morning, the GSH-Ag is washed five times with PBS. Thrombin cleavage and liberation of fusion proteins affixed to the glutathione agarose can be achieved in a cutting buffer containing 50 m*M* Tris, pH 7.4, 150 m*M* NaCl, and 2.5 m*M* $CaCl_2$ with 20 units human thrombin/mg fusion protein at 25° for 3 hr. Alternatively, the fusion proteins are eluted intact from the GSH-Ag by overnight incubation at 4° with five volumes of 50 m*M* Tris–HCl, pH 8.0, containing 5 m*M* reduced glutathione. The fusion protein can also be eluted by Laemmli sample buffer and then loaded directly on a polyacrylamide gel.

Characterization of CFTR Polypeptides Synthesized and Immunoprecipitated

Purification of CFTR NBD1. GST–NBD1 fusion protein is synthesized in *E. coli,* and the GST segment removed by thrombin cleavage as described earlier. The cleaved products are electrophoresed and fractionated through a preparative acrylamide gel (Bio-Rad) (see Fig. 1). Examples of fractions

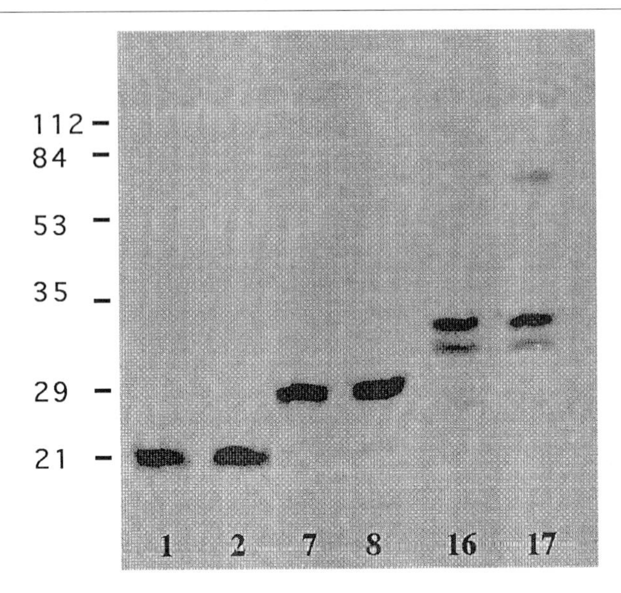

FIG. 1. Protein purification of CFTR NBD1.

containing the NBD1 (as verified by amino terminal sequencing and amino acid compositional analysis) are shown in fractions 7 and 8. Examples of samples containing lower molecular weight proteins (fractions 1 and 2) and higher molecular weight proteins (fractions 16 and 17) are also shown. The numerical values indicate the locations of the molecular mass standards (in kilodaltons).

Recombinant CFTR NBD-1 Binds Nucleotide Analogs. Figure 2 demonstrates a binding study with CFTR NBDs and the nucleotide analog trinitrophenyl-ATP (TNP–ATP). Binding affinity to several nucleotides is very similar for the wild-type and ΔF508 NBD1,[5,6] but certain point mutations in the domain markedly disrupt nucleotide binding. For example, a glycine → aspartic acid replacement at CFTR position 551 within a highly conserved ABC protein motif (LSG<u>G</u>QXQR, glycine corresponding to position 551 underlined) disrupts the nucleotide interaction.[5] The same appears to be true when a critical lysine residue in the Walker A site of NBD1 is substituted for methionine (K464M.)

CD Spectroscopy of CFTR NBD1 After Overexpression and Purification from E. coli. Far-UV circular dichroism (CD) spectra (Fig. 3) demonstrate very similar structures for refolded wild-type and ΔF508 NBD1. The NBD1 structures are found to be high in β-sheet composition, and much lower in α helix and β turn.

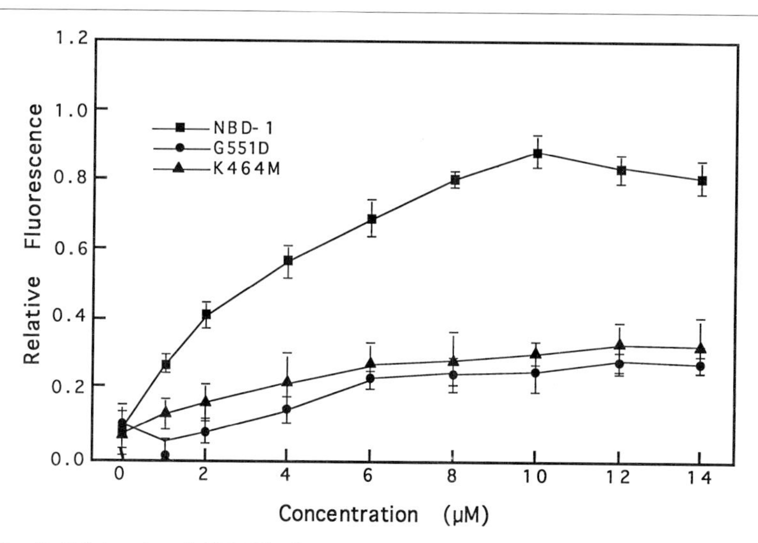

FIG. 2. Trinitrophenyl ATP Binding to Recombinant CFTR NBD1: 250 μg/ml purified NBD1 in 10 mM Tris, pH 7.4, was incubated at 4° with increasing concentrations of trinitrophenyl ATP (0–14 μM TNP–ATP). Maximal binding for the wild-type NBD1 was approximately four times that of the G551D and K464M domains. Maximal binding of the ΔF508 NBD1 was similar to wild type.[5,6] Binding of wild-type NBD1 could be inhibited in the presence of 5 mM Mg_2 ATP (with 50% inhibition at approximately 5 mM Mg_2 ATP, data not shown).

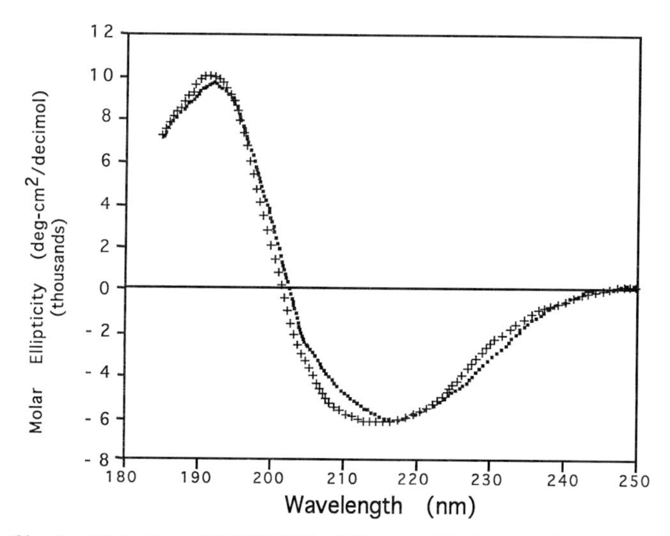

FIG. 3. Circular Dichroism of CFTR NBDs. Wild-type CFTR NBD1(−−−−−) and ΔF508 NBD1(++++) spectra after refolding, suggesting very similar structural content.[5,6]

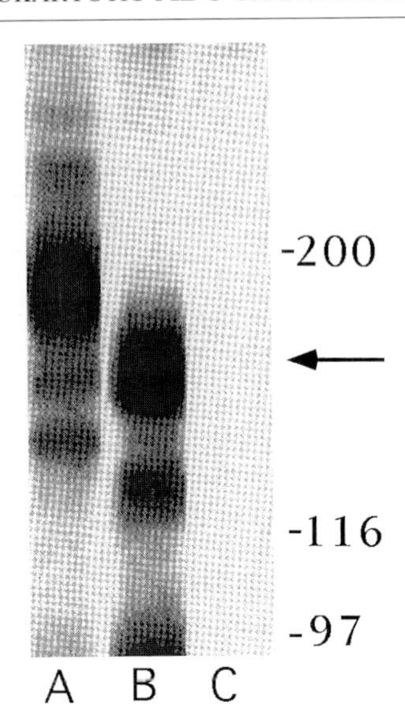

FIG. 4. Immunoprecipitation of CFTR following insect cell overexpression. A polyclonal antibody raised against CFTR NBD1 (Fig. 1) was used to immunoprecipitate full-length CFTR from SF9 cells (lane B) or T84 colonic carcinoma cells (lane A). The CFTR was phosphorylated *in vitro* according to a published protocol.

Antibody Production. A polyclonal antibody to wild-type NBD1 efficiently immunoprecipitates CFTR and has also been used for histochemistry and characterization of CF mice.[10,13] Figure 4 shows an immunoprecipitation of the CFTR synthesized in insect cells and prepared as described above (lane B, Fig. 4). Preimmune serum control is shown in lane C (Fig. 4). Lane A (Fig. 4) shows immunoprecipitation of nonrecombinant CFTR from a human colonic carcinoma cell line (T84). The difference in molecular mass after insect cell expression has been attributed by several laboratories to different glycosylation patterns in the two cell types. An arrow indicates the mature form of CFTR after insect cell overexpression.

Conclusion

Our laboratory has developed structural information concerning wild-type, ΔF508, and other NBD1 polypeptides by CD spectroscopy, and has

[13] P. Hasty, W. K. O'Neal, Q. Liu, A. P. Morris, Z. Bebok, G. B. Shumyatsky, T. Jilling, E. J. Sorscher, A. Bradley, and A. L. Beaudet, *Som. Cell Mol. Genet.* **21,** 177 (1995).

characterized nucleotide binding of these polypeptides. Useful amounts of purified proteins could be obtained by either of two mechanisms. Both the eukaryotic and prokaryotic systems utilize GST fusion with the expressed protein, and allow purification on glutathione agarose beads. The prokaryotic system provides higher yields of smaller polypeptides, although in many cases the products must be extricated from inclusion bodies during the purification process. Larger CFTR fragments and whole CFTR may require the use of insect cell expression in order to optimize yields.

Acknowledgments

The authors wish to thank Dr. Donald Muccio for preparing the circular dichroism analysis used in this project; Zhen Huang, M.D., Shenyi Peng, M.D., and Hui Wen, B.S., for excellent technical assistance; and Mrs. Bonnie Parrott for administrative assistance. We are also indebted to Dr. Maja Sommerfelt for help establishing the baculovirus expression systems.

[52] Cationic Lipid Formulations for Intracellular Gene Delivery of Cystic Fibrosis Transmembrane Conductance Regulator to Airway Epithelia

By SENG HING CHENG, JOHN MARSHALL, RONALD K. SCHEULE, and ALAN E. SMITH

Introduction

The development of gene delivery vehicles that are safe and efficacious is an important prerequisite for the successful treatment of acquired and inherited diseases by gene therapy. Gene transfer vectors that are being developed include recombinant viral vectors and nonviral, synthetic self-assembling systems such as those formed using polycationic molecular conjugates and cationic lipids.[1] Synthetic cationic lipids have been used successfully for delivery of a number of genes to a variety of different cell types *in vitro* and *in vivo.*[1,2] However, the efficiency of gene transfer, at least of the early generation of cationic lipids, particularly *in vivo,* has been much lower than could be attained with viral-based vectors. Despite this impediment, cationic lipids continue to gain increasing attention as vehicles for gene delivery primarily because of their simplicity, ease of production, their

[1] P. L. Felgner, T. R. Gadek, M. Holm, R. Roman, H. W. Chan, M. Wenz, J. P. Northrop, G. M. Ringold, and M. Danielson, *Proc. Natl. Acad. Sci. U.S.A.* **84,** 7413 (1987).
[2] X. Gao and L. Huang, *Gene Ther.* **2,** 710 (1995).

0076-6879/98 $25.00

ability to package and deliver very large transgenes, and because they present a very different safety profile than viral vectors. Additionally, because cationic lipid : plasmid DNA (pDNA) complexes, unlike viral vectors, are nonproteinaceous, they may not elicit a host immune response that could limit persistence of transgene expression and reduced efficacy following repeated administrations of the vector.

Cationic lipids that have been described to date are invariably composed of a hydrophobic lipid anchor group such as cholesterol or diacyl chains, linked to a positively charged headgroup containing amines or polyamines using a variety of linkers of different lengths and compositions.[2] For maximal gene transfer activity, most cationic lipids require coformulation with a neutral lipid such as dioleoylphosphatidylethanolamine (DOPE) or cholesterol. The mechanism by which these neutral lipids enhance transfection is unclear but one hypothesis is that they may facilitate escape of the pDNA from endosomal compartments. The gene transfer vehicle is assembled by combining the cationic lipid : DOPE liposomes with pDNA containing the transgene of interest. The mixing of the oppositely charged liposomes and DNA results in condensation of both moieties into so-called lipid : pDNA complexes that tend to be somewhat heterogeneous in size and composition. The activity of these complexes for any given target cell is greatly influenced by the relative proportions of the three constituents and maximal activity requires extensive optimization of the ratios of cationic lipid : neutral lipid : pDNA.

Since the mechanism(s) of cationic lipid-mediated gene transfer has not been determined, much of the recent development of new cationic lipid structures has been empirical, requiring systematic screening and testing for optimal activity in their respective target cells.[3–5] Using this strategy, we have synthesized and evaluated more than 150 different cationic lipid structures as candidate vehicles for gene therapy of cystic fibrosis (CF). In this article, we describe our approach to screen and optimize formulations of these compounds for high-efficiency gene delivery to airway epithelial cells in culture and into mouse lungs *in vivo.* Our strategy involved first identifying the active compounds and their respective formulations with a primary high-throughput screen using cells in culture, followed by determining their gene transduction activity in mouse lungs. Using this approach, we have been able (1) to identify novel cationic lipid structures that are

[3] J. P. Behr, B. Demeneix, J. P. Loeffler, and J. P. Mutul, *Proc. Natl. Acad. Sci. U.S.A.* **86,** 6982 (1989).

[4] J. H. Felgner, R. Kumar, C. N. Sridhar, C. J. Wheeler, Y. J. Tsai, R. Border, P. Ramsey, M. Martin, and P. L. Felgner, *J. Biol. Chem.* **269,** 2550 (1994).

[5] I. Solodin, C. S. Brown, M. S. Bruno, C. Y. Chow, E. Jang, R. J. Debs, and T. D. Heath, *Biochemistry* **34,** 13537 (1995).

effective for transfection of airway cells *in vitro* and *in vivo,* (2) to define the ratios of cationic lipid, colipid, DNA, and excipients for optimal activity, (3) to rank the relative performance of the different cationic lipids in relation to each other in terms of activity, toxicity, and potency, and (4) to show that although the optimal formulations of cationic lipids determined *in vitro* are preserved *in vivo,* their activity *in vitro* is generally not predictive of *in vivo* activity.

Optimization of Cationic Lipid : pDNA Formulations for *in vitro* Transfection

Many factors have been shown to influence the transfection activity of cationic lipids *in vitro.* For example, optimal activity is critically dependent on using a formulation containing the appropriate ratios and concentrations of cationic lipid : neutral lipid : pDNA.[4,6] The performance of a cationic lipid formulation is also influenced by the excipient used, the presence of serum and salt in the transfection milieu, the density and type of target cells, the duration of contact with the cells, and the pH. Because the cationic lipids that have been described are structurally very diverse, it is likely that they will all behave differently under the varied conditions and therefore optimization will require an empirical survey of the known variables. For high-throughput evaluation of these transfection parameters, we have adapted the 96-well microtiter plate assay described by Felgner *et al.*[4] Since our objective is delivery of the CFTR (cystic fibrosis transmembrane conductance regulator) cDNA to the airways of CF patients, our analysis was performed using an immortalized CF airway epithelial cell line, CFT1.[7]

Procedure

The lipid mixtures are generally prepared by mixing chloroform solutions of the cationic lipid and neutral lipid in the appropriate ratios. The organic solvent is then removed by evaporation on a rotavapor to produce dried lipid films. The vials of lipid films are placed *in vacuo* overnight to remove residual solvent and then stored at −70° under argon until required for use. Although it has been suggested that some cationic lipid formulations are stable over a period of up to several months following reconstitution, we have detected a reduction in performance and a change in size of the liposomes on storage in the liquid form. Furthermore, DOPE, an unsaturated lipid present in these formulations, is more likely to undergo hy-

[6] A. J. Fasbender, J. Zabner, and M. J. Welsh, *Am. J. Physiol.* **269,** L45 (1995).

[7] J. R. Yankaskas, J. E. Haizlip, M. Conrad, D. Koval, E. Lazarowski, A. M. Paradiso, C. A. Rinehart, B. Sarkadi, R. Schlegel, and R. C. Boucher, *Am. J. Physiol.* **264,** C1219 (1993).

drolytic and oxidative decomposition[8] under these conditions than when stored as a dried film. These resultant hydrolysis and oxidation products may have unwanted biological activities. Therefore, following solvation, we normally use the reconstituted liposomes within 24 hr.

For the studies here, CFT1 cells are seeded onto 96-well microtiter plates at 7500 cells per well and incubated in a humidified, 10% CO_2 atmosphere at 37° for up to 7 days or until they reach confluence. The cells are used at these high densities since they are more capable of tolerating the toxicity associated with the cationic lipids and because they may be a better model of an intact airway epithelium *in vivo*. Just prior to use, the dried films of cationic lipid:DOPE are hydrated with 1 ml water for 10 min and then vortexed for 2 min to generate a monodisperse suspension. The cationic lipid formulation (diluted to 670 μM cationic lipid with OptiMEM) and a plasmid expression vector for β-galactosidase (pCF1-βGal) diluted with OptiMEM to 960 μM (based on average molecular weight of nucleotide of 330 daltons) are two-fold serially diluted with OptiMEM into two separate microtiter plates. The lipid is two fold serially diluted with OptiMEM across one 96-well plate eight times to generate concentrations of cationic lipid from 335 to 2.6 μM and the pDNA is similarly diluted in a perpendicular direction in the other 96-well plate to generate nucleotide concentrations from 480 to 3.8 μM. An equal volume from the lipid dilution plate is then transferred to the corresponding wells of the DNA dilution plate to generate an array of 64 different formulations of lipid:pDNA complexes. After the complexes are allowed to form for 15 min at room temperature, 100 μl from each well is transferred into the corresponding well on a microtiter plate of CFT1 cells (media removed prior to addition) and the complexes allowed to transfect for 6 hr at 10% (v/v) CO_2 atmosphere and 37°. The cells are supplemented with 50 μl 30% fetal bovine serum (FBS, diluted with OptiMEM) at 6 hr and a further 100 μl 10% FBS at 24 hr posttransfection. The extent of cationic lipid-mediated transfection is then determined 48 hr later by measuring the levels of β-galactosidase in the wells.

To assay for β-galactosidase activity, the culture media is first aspirated, the cells washed once with phosphate-buffered saline (PBS), and then incubated with 50 μl of lysis buffer (250 mM Tris-HCl, pH 8, 0.15% Triton X-100). After 30 min at room temperature, the plates are gently vortexed for 10 sec to provide mixing and a 5-μl aliquot is removed for determination of total protein content using the Coomassie Plus protein assay (Pierce, Rockford, IL). The levels of protein are measured to provide an estimate of the number of surviving cells as a result of the toxicity associated with

[8] H. Farhood, R. Bottega, R. M. Epand, and L. Huang, *Biochim. Biophys. Acta* **1111,** 239 (1992).

the cationic lipid formulations. A low level of residual protein in the wells is indicative of a highly toxic lipid : pDNA formulation. β-Galactosidase activity in the lysates is determined using either CPRG (chlorophenol red galactopyranoside) or ONPG (*o*-nitrophenol-β-D-galactopyranoside) as substrates. Routinely, ONPG is used but CPRG can be used if an increase in sensitivity of detection is required. For assay using the ONPG substrate, 150 μl of ONPG buffer (0.1 *M* sodium phosphate buffer, pH 7.5, 0.011 *M* $MgCl_2$, 1 mg/ml ONPG, 0.045 *M* 2-mercaptoethanol) is added to each well. After a yellow color develops, the optical density of the reactions are read with a microtiter plate reader at 405 nm. The amount of β-galactosidase is determined using a series of β-galactosidase standards (Sigma St. Louis, MO) included on every plate. The assay using CPRG is similar except that 1 mg/ml of CPRG is used instead of ONPG in the buffer. The red color that develops is quantitated at 570 nm.

Results

For purposes of illustration, the four cationic lipids shown in Fig. 1 are selected. The cationic lipid DOTAP is commercially available from Avanti Polar Lipids (Birmingham, AL) coformulated with DOPE (1 : 1, w/w) or from Boehringer Mannheim (Indianapolis, IN) without any neutral lipid. DC-Chol[2] (gift from Dr. L. Huang, University of Pittsburgh) represents a different type of cationic lipid containing a cholesterol instead of dialkyl chains as the lipid anchor and the lipids #48 and #53 are compounds that we have prepared that contain a cholesterol anchor linked to polyamine headgroups. Figure 2 shows examples of results of transfection activities obtained with these cationic lipids. Using the assay grid, the ratios of cationic lipid : neutral lipid and cationic lipid : pDNA for high-efficiency transfection of the CFT1 cells can be readily determined. No transfection activity is observed when the cells are treated with pCF1-βGal alone in the absence of complexed cationic lipid.

Most cationic lipids exhibit low transfection activity when used in the absence of the neutral colipid DOPE. Inclusion of DOPE in the lipid formulation increases gene transduction activity, the extent of which varies for different cationic lipids. The optimal amount of DOPE to include for maximal transfection activity is also dependent on the cationic lipid. For example, in the case of #48 and #53, maximal activity is attained when the cationic lipids are formulated with DOPE in a molar ratio of 1 : 1 (Fig. 2). Formulations that contain no DOPE (1 : 0) or which contain cationic lipid : DOPE in a molar ratio of 1 : 2 display significantly reduced performance. Formulations of DC-Chol are judged to be optimal when coformulated with DOPE at a molar ratio of DC-Chol : DOPE of 3 : 2. The cationic

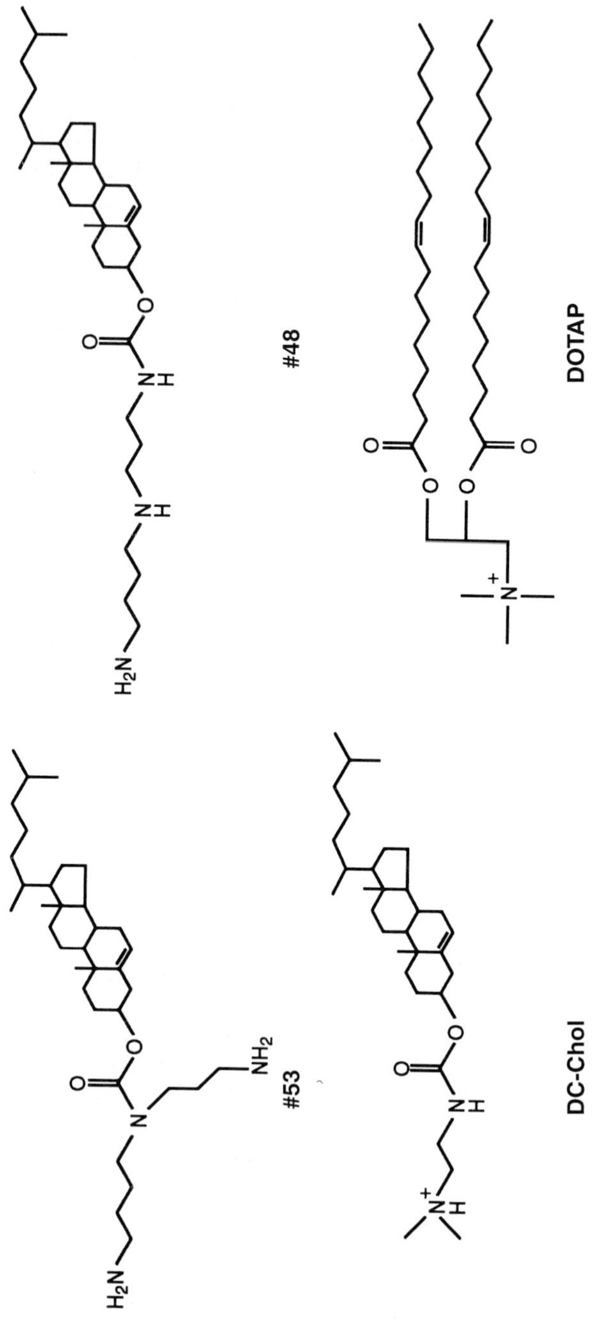

#48

DOTAP

#53

DC-Chol

FIG. 1. Structures of cationic lipids.

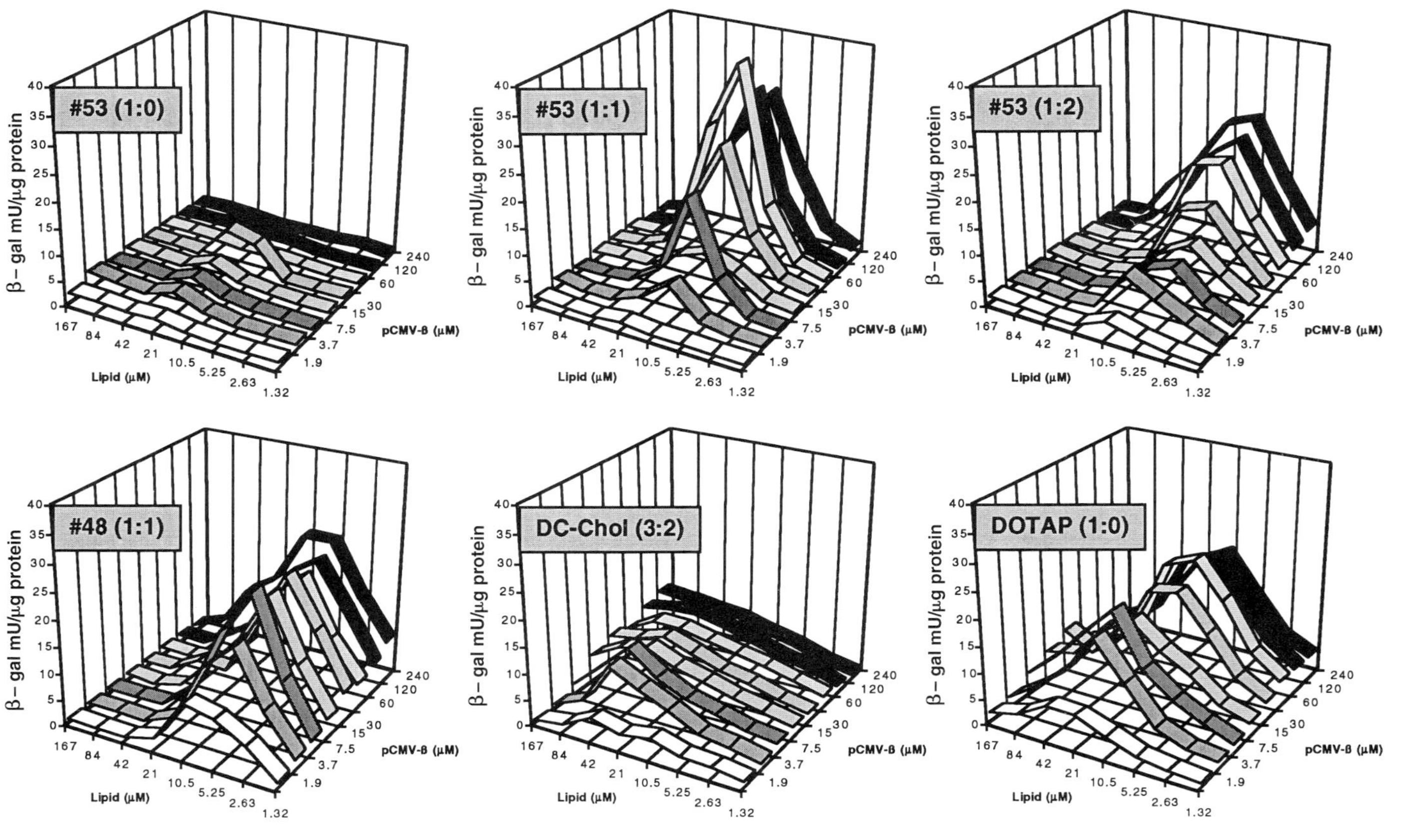

FIG. 2. Transfection activity of cationic lipid : DOPE : pDNA formulations in CFT1 cells. CFT1 cells in a 96-well microtiter plate were transfected with an array of different formulations of cationic lipid : DOPE : pCF1-βGal complexes. At 48 hr posttransfection, the extent of gene transfer was determined by measuring the levels of β-galactosidase in the wells. The numbers in brackets represent the molar ratio of cationic lipid : DOPE.

lipid DOTAP represents an exception and performs equally effectively when formulated in the presence or absence of a neutral colipid (Fig. 2). Other cationic lipids that reportedly also perform equally effectively in the absence of DOPE are DOGS[3] and βAE-DMRIE.[9] Cationic lipids that are active in the absence of a neutral colipid have the advantage that such formulations are less complex and also in principle more stable, since DOPE can undergo hydrolytic and oxidative decomposition. The basis for these observed variations for the requirement of the neutral lipid among the different cationic lipids is unclear. Previous studies have rationalized the need for DOPE as a requirement for transition between the bilayer and hexagonal $H_{\|}$ phase states. That DOTAP, DOGS, and βAE-DMRIE are able to function in the absence of DOPE therefore suggests that perhaps the mechanism(s) of gene transduction are very different for the various compounds. However, with the limited data available, there is no obvious correlation that one can discern between any particular class of cationic lipid structures and their requirement for DOPE.

The presentation of the data in the manner shown in Fig. 2 also allows selection of the optimal ratios of cationic lipid : pDNA for transfection of the CFT1 cells. Most cationic lipids are active over a broad range of cationic lipid to pDNA molar ratios. For example, lipids #48 and #53 are most active over the range of molar ratios of cationic lipid : pDNA of between 1 : 2 to 1 : 8 (Fig. 2). It is interesting to note that all the cationic lipid : pDNA complexes formed at the higher ratios contain a molar excess of pDNA and, in consequence, may exhibit a net negative surface zeta (ζ) potential. This observation contrasts with the proposed mechanism for cationic lipid-mediated gene transfer.[1–4] It has been suggested that an excess of positively charged cationic lipids is required for facilitating contact with the negatively charged cell membranes and it is this interaction that facilitates gene transfer. Clearly, the studies with the cationic lipids used here indicate that a strict requirement for a positively charged complex is not essential and indeed is much less optimal than for negatively charged complexes.

In general, at a fixed cationic lipid concentration, a DNA dose-dependent increase in expression is observed (Fig. 2). However, at a fixed DNA concentration, the cationic lipid dose–response curves are invariably bell shaped as a consequence of the toxicity associated with high concentrations of the lipids. In addition to determining the optimal conditions for use, the assays shown in Fig. 2 also allow a determination of the relative rank order of performance of the different cationic lipids. For example, based on the results shown in Fig. 2, the optimal formulation of cationic lipid #53 is more

[9] C. Wheeler, L. Sukhu, G. Yang, Y. Tsai, C. Bustamente, P. Felgner, J. Norman, and M. Manthorpe, *Biochim. Biophys. Acta* **1280,** 1 (1996).

active than #48, which in turn is more effective than DC-Chol or DOTAP (Fig. 3). The absolute values of β-galactosidase activity detected in these assays reflect the efficiency of gene transfer of the respective cationic lipids. A similar rank order is obtained if the number of transfected cells are quantitated following histochemical staining for the enzyme.

By analyzing a large number of analogs or derivatized structures in this way, it is possible to discern cationic lipid structure–activity relationships. For example, a comparison of the cationic lipids #53, #48, and DC-Chol (Fig. 1), which differ only in their headgroup composition, suggests that at least in the context of a cholesterol anchor, headgroups that are potentially multivalent are more active than those that are potentially monovalent. Furthermore, that #53 and #48 exhibit different activities also suggests that the specific configuration of the same spermidine headgroup on these lipids also influences activity. Moreover, these assays also allow for determination of the relative potency of the different cationic lipids. For example, lipids #48 and #53 are judged to be more potent than either DC-Chol or DOTAP by virtue of the fact that the former lipids are able to attain maximal transfection at much lower amounts of cationic lipid (Fig. 2).

Factors Affecting Efficiency of Gene Transfer *in vitro*

The transfection efficiency of any given cationic lipid is dictated primarily by the structure of the cationic lipid. However, in addition to structure, other factors have been recognized that can influence the efficacy of cationic lipid-mediated transfection. These factors can be divided into two categories: (1) those that affect the formulation or physical characteristics of the complex and (2) those that are influenced by the type or nature of the specific target cells into which gene transfer is desired. Examples of factors that can affect formulation include the dose concentration of the cationic lipid : pDNA complex, the excipients used in their preparation, the size and stability of the lipid : pDNA complexes that are formed, the order of addition of the components and time after complex formation, and the net charge (zeta potential) realized at the surface of these complexes. Efficacy of transfection is also influenced by cell type, whether they are primary or immortalized, their state of confluence, the duration of contact of the complexes with the cells, and the choice of promoters used for directing expression of the transgene.

1. For many cationic lipids, one of the principal limitations to their effective use is their ability to be prepared in a stable form at high concentrations. Most complexes formed at high concentrations have a tendency to aggregate and precipitate out of suspension. This deficiency has obvious

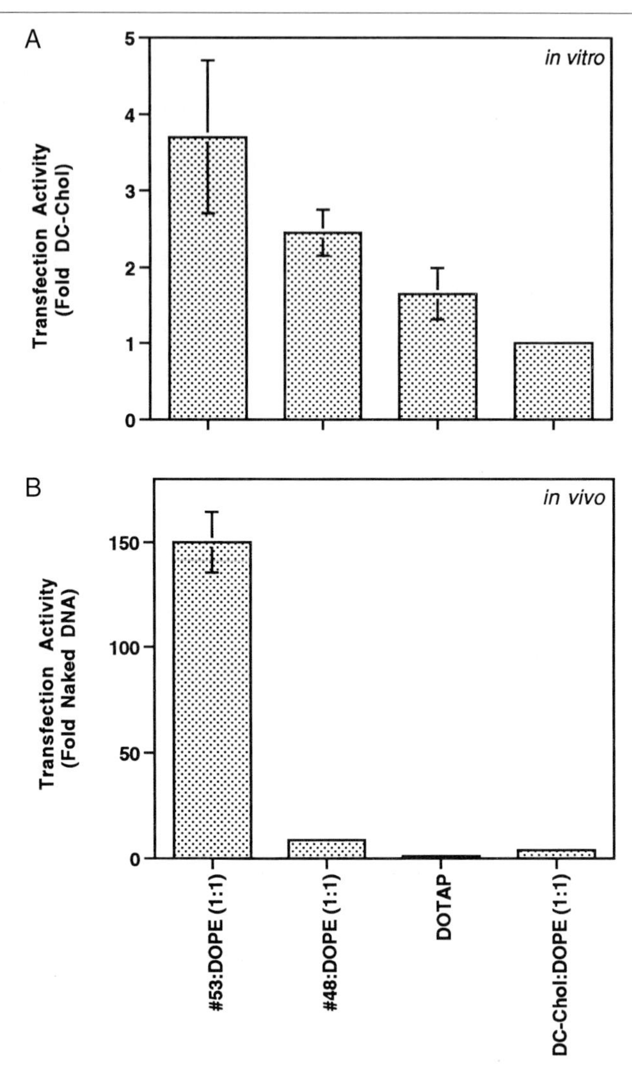

FIG. 3. Rank order of cationic lipids *in vitro* and *in vivo*. (A) The activities of the different cationic lipids assayed using the 96-well screen are presented as fold DC-Chol. The values represent the peak expression levels observed for each of the different cationic lipids. The data shown were from four separate experiments, each performed in triplicate. The data are expressed as mean ± SEM ($n = 12$). (B) Expression obtained following intranasal instillation of BALB/c mice with 100 μl of 1 mM of the respective cationic lipid, the corresponding amount of DOPE (shown in parentheses), and 4 mM pCF1-CAT. In the case of DC-Chol, 1.3 mM of the cationic lipid and 0.9 mM of pCF1-CAT were used. The data were from three separate experiments with each formulation tested in five BALB/c mice. The data are expressed as mean ± SEM ($n = 15$).

implications for their effective use, especially in *in vivo* settings. However, it is not unreasonable to predict that this restriction can be rectified, at least for some cationic lipids, by formulation development. Indeed it has been demonstrated that with DC-Chol formulations, precipitation of the complexes can be minimized by increasing the pH during complex formation.[10] We have also observed that at least in the case of #53, inclusion of salt can increase the stability of the complexes at high concentrations. Another intervention we have found that minimizes precipitation of the complexes involves warming the components to 30° prior to complex formation.

The size of the complexes that are formed also appears to be an important variable. Studies using the cationic lipid DMRIE have shown that larger vesicles generated by simple vortexing tend to be more active than smaller vesicles generated by sonication.[4] However, the opposite has been shown to apply for DC-Chol liposomes, where smaller vesicles are reportedly superior.[2] Again, the basis for these observed differences among the different cationic lipids is not well understood. Of note, however, is the observation that freshly prepared cationic lipid : pDNA complexes tend to be very heterogenous in size and, as yet, it is unclear which subfraction of these complexes may be the most active and responsible for the observed gene transfer activity. Studies to define further the active fraction of the lipid : pDNA complex need to be pursued to resolve these discrepancies.

The choice of excipient used during transfection also impacts the efficacy of gene delivery. Most *in vitro* transfections reported to date are performed in a serum-free environment such as that provided by OptiMEM (Gibco-BRL, Gaithersburg, MD). We have observed that several cationic lipids exhibit reduced activity when used in the presence of serum. Similar observations have been reported for DMRIE.[6] It has been proposed that nucleases present in serum may be responsible for the observed loss in activity. Alternatively, the lipid : pDNA complexes by virtue of their surface charge may be rendered inactive by the adsorption of serum proteins. If so, then cationic lipids that bind DNA with greater avidity and formulations that are formed at a higher lipid : pDNA ratio (to provide protection from nucleases), and which are closer to charge neutrality, should be more efficacious when used in the presence of serum. Indeed we have observed that not all cationic lipids are affected equally by serum, suggesting that certain cationic lipids or their formulations are more proficient at facilitating protection from nuclease digestion or the absorption of serum proteins. Although these effects may not be important for *in vitro* applications, because *in vivo* gene transfer applications will necessarily require contact with serum,

[10] N. J. Caplen, E. Kinrade, F. Sorgi, X. Gao, D. Gruenert, D. Geddes, C. Coutelle, L. Huang, E. W. F. W. Alton, and R. Williamson, *Gene Ther.* **2,** 603 (1995).

formulations of cationic lipids that are immune to the inhibitory effects of serum seem desirable.

2. The optimal formulation for transfection of different cell lines can vary. These variations mainly reflect the relative ability of the different cell types to tolerate the toxicity associated with the cationic lipids. For this reason, the optimal concentrations of cationic lipid and pDNA for transfection of primary cells and cells that have been grown to confluence are invariably much higher than for immortalized or nonconfluent cells.[6] The optimal ratios of the components in the complex for transfection are usually fairly well conserved for a number of different cell lines but not all. For example we have observed that the optimal formulation of #53 : pDNA for transfection of CFT1, 293, and HeLa cells is similar but differs from those for C2C12 and 3T3 cells. Similar findings have been reported from a study of DMRIE and βAE-DMRIE with 10 different cell lines.[9] The rank order of performance of a series of cationic lipids tested on a particular cell line may remain unaltered when analyzed on different cells but this again does not apply to all cell types.[9] Thus depending on the cell lines used, evaluation of the relative activities of different cationic lipids on one cell line can be but is not necessarily predictive of their performance in a different cell type.

When analyzed under conditions of low toxicity, the efficacy of transfection increases as the concentration of the cationic lipid : pDNA complex is increased or as the duration of exposure of the cells to the complexes is prolonged.[6] Significant activity can be detected after a 30-min incubation and efficacy increases as the time of contact is lengthened up to 20 hr. These results suggest that especially for *in vivo* applications, interventions aimed at prolonging the contact time between the lipid : pDNA complexes and the target cells or that increase the concentration of the formulations will improve the efficacy of gene transfer.

Yet another variable that impacts efficacy is the selection of appropriate transcriptional units for expression of the transgene following its delivery to the target cells. The judicious use of appropriate promoters and the selection of regulatory elements (enhancers, introns, polyadenylation signal sequences) that allows for maximal and persistent expression in the intended target cells is desirable. In general, for high-level ubiquitous expression, strong viral promoters such as those from cytomegalovirus (CMV), Rous sarcoma virus (RSV), or simian virus 40 (SV40) are used. However, some of these viral promoters, at least under some conditions, can be subject to transcriptional inactivation. Restriction of expression to select cell types can be achieved by use of tissue-specific promoters. Examples of such promoters include the Clara cell CC10 or surfactant SPC promoters for expression in the airway. To promote greater persistence in expression, constitutively active housekeeping-type promoters such as phosphoglycerate kinase and glyceraldehyde-3-phosphate dehydrogenase have been used. Therefore, selection of the promoter for use will depend on the particular application and the intended magnitude and longevity of expression required.

Optimization of Cationic Lipid Formulations for *in vivo* Gene Delivery

Efficient gene delivery *in vivo* presents several additional challenges to those already outlined above for *in vitro* transfection. For the lung, these include (1) methods for delivering the cationic lipid : pDNA complexes to the appropriate airway epithelial cells, (2) the presence of mucus and biological fluids lining the epithelial cells that may inactivate the complex, and (3) mucociliary action that may clear or reduce the contact time of the complex with the target cells. Although the initial *in vitro* screening used an airway epithelial cell line, because a fully differentiated airway epithelium is very dissimilar to cells grown in tissue culture, it is possible that the performance of the optimized cationic lipid formulations determined *in vitro* may not be predictive of their activity *in vivo*.[5] To test this, the gene transfer activity of the various lipid : pDNA complexes is examined following intranasal instillation into the lungs of BALB/c mice.

Procedure

To facilitate rapid screening in a large number of animals, the lipid : pDNA complexes are instilled intranasally into mice. Gene delivery to the lung is assessed using pCF1-CAT, a vector containing the chloramphenicol acetyltransferase reporter gene. Female BALB/c mice (4–6 weeks old) are first anesthetized with metaphane by placing the animals into a container containing gauze treated with the anesthetic. Soon after the animals are unconscious (less than 1 min), they are withdrawn from the box, placed upright with their noses up and pressure is applied to the lower mandible to immobilize the tongue. This maneuver is performed to minimize loss of the instilled complexes by swallowing. The complexes are applied dropwise (25 μl) onto the nares of the mice using a pipetteman and are internalized by natural breathing. Up to 100 μl of the complex can be administered in 1 min. For screening purposes, the animals are sacrificed 48 hr postinstillation by intraperitoneal injection of 50 μl of sodium pentobarbital. The lungs are harvested and frozen on dry ice until ready for analysis. Routinely, each formulation of lipid : pDNA is tested for five BALB/c mice and this repeated on at least three separate days. Although most of our testing has been performed using BALB/c mice, similar results are obtained with C57BL/6, DBA/1, and BALB/c nu/nu mice.

The CAT enzymatic activity in the lungs is assayed using the method of Gorman *et al.*[11] Mouse lungs are first weighed, then homogenized for 15 sec in two volumes of homogenization buffer (250 mM Tris-HCl, pH 7.8 containing 5 mM EDTA) using a handheld tissue tearer (Biospec Products, VWR, Bridgeport, NJ). The cells are lysed by three freeze–thaw cycles and the cleared supernatant then heated at 65° for 20 min to inactivate mammalian deacetylases. Up to 10 μl of the extract is added to 10 μl of 10 mM acetyl-CoA, 14 μl of ^{14}C-labeled chloramphenicol (NEN, Boston, MA) and homogenization buffer to a total volume of 176 μl and the mixture incubated at 37° for 15–60 min. Longer incubations may be required if the CAT activity is low, in which case the reaction mixture should be supplemented with fresh acetyl-CoA. At the end of the reaction period, the chloramphenicol and acetylated chloramphenicol are extracted with 1 ml of ethyl acetate. Following vortexing, the reaction mixture is microcentrifuged for 1 min and the upper solvent layer collected into clean tubes. The ethyl acetate is removed from the samples in a Speed-Vac evaporator (Integrated Separation Systems, Natick, MA) for about 60 min. The sample is resuspended in 20 μl ethyl acetate, spotted (5 μl at a time) onto plastic-backed thin-layer chromatography (TLC) plates (Whatman, Clifton, NJ), and then developed in a chromatography tank equilibrated with freshly prepared chloroform : methanol (19 : 1). The levels of CAT in the lysates are determined by quantitating the amount of acetylated chloramphenicol using an Instant Imager (Packard, Downer's Grove, IL) and comparing these values to those obtained using known CAT standards.

Results

Since the number of target cells in the lung is larger than in the *in vitro* experiments, a proportionately greater amount of the complex is used for the *in vivo* experiments. However, because the volume that can be instilled into the lung is limited, the concentration of the complex used is necessarily higher. Routinely, 100 μl of complex at a concentration of between 0.5 and 1.5 mM cationic lipid and between 2 and 8 mM pDNA is administered to each mouse. Figure 4 shows an example of results obtained using cationic lipid #53 and the plasmid expression vector pCF1-CAT. As might be expected, lungs of animals that are instilled with #53 : DOPE alone do not exhibit any CAT activity (Fig. 4, lanes 1 and 2). Interestingly, in contrast to the *in vitro* data, lungs treated with pCF1-CAT alone contain some CAT activity, indicating that naked DNA is able to mediate gene transfer into the airway cells *in vivo* (Fig. 4, lanes 3 and 4). However, when a complex

[11] C. M. Gorman, L. F. Moffat, and B. H. Howard, *Mol. Cell Biol.* **2,** 1044 (1982).

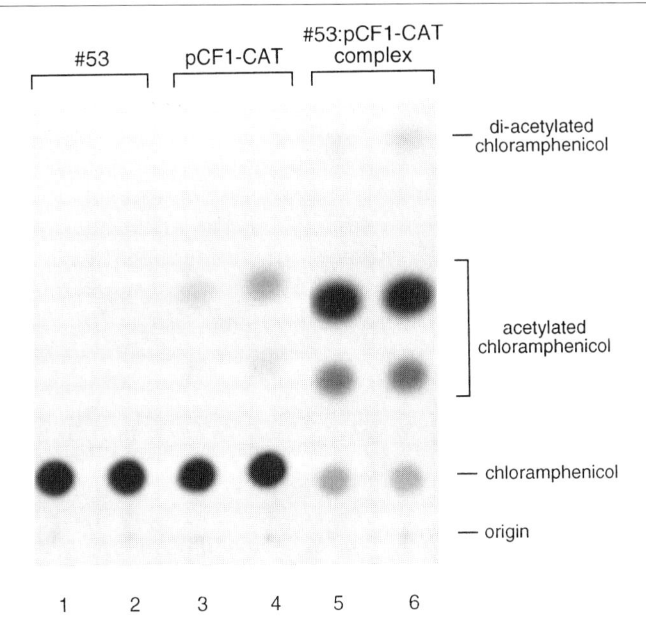

FIG. 4. *In vivo* gene transfer activity of naked DNA and #53 : pDNA complexes. Autoradiograph of TLC plate showing CAT activity in extracts from two different BALC/c mouse lungs instilled with 100 μl of either #53 : DOPE (lanes 1 and 2), pCF1-CAT (lanes 3 and 4), or #53 : pCF1-CAT (lanes 5 and 6). Autoradiography for lanes 1 to 4 was for 24 hr, and that for lanes 5 and 6 was for 4 hr.

of #53 : pCF1-CAT is used, gene transfer activity is markedly enhanced over that observed with naked DNA (Fig. 4, lanes 5 and 6). With optimization of the #53 : pCF1-CAT formulation, gene transfer activity has been obtained that is 150-fold higher than that observed with naked DNA. Analysis of #48, DOTAP and DC-Chol formulations under similar conditions give transfection activities that are 8-, 1-, and 3.8-fold, respectively, greater than that obtained with naked DNA (Fig. 3B). The observation that naked DNA has activity *in vivo* suggests that attempts to demonstrate efficacy of a particular cationic lipid must be compared against the background expression level obtained in the absence of the cationic lipid.

Routinely, cationic lipids judged to be active from the *in vitro* screens are further evaluated *in vivo*. Normally, these cationic lipids are tested using the optimal ratios determined from the *in vitro* analysis, but at a higher concentration. The formulations of lipid : pDNA to be tested *in vivo* are limited only by their toxicity and their ability to be formulated stably at higher concentrations. Since BALB/c mice can tolerate a 100-μl volume of most cationic lipids up to a concentration of 1 m*M*, testing is generally performed using 1 m*M* cationic lipid and the corresponding amount of pDNA based on the *in vitro* determined optimum values. Following this analysis, those cationic lipids determined to be particularly effective (transfection activity greater than 50-fold over that obtained with naked DNA) are then selected for further optimization studies. Optimization of the cationic lipid formulation for gene transduction *in vivo* is performed using a screening grid similar to that used in the *in vitro* assays albeit on a smaller scale. Typically, we screen 20 different formulations comprising 0.5–1.5 m*M* cationic lipid (in 0.25-m*M* increments) and between 2 and 8 m*M* DNA (in 2-m*M* increments). In general, we have found that the optimal ratio of cationic lipid : neutral lipid : pDNA for mediating gene transfer *in vivo* is similar to that determined from the *in vitro* screens. For example, in the case of lipid #53, formulations composed of #53 : DOPE in a molar ratio of 1 : 1, and #53 : pDNA in molar ratios ranging between 1 : 2 to 1 : 8 are most effective *in vivo*. These ratios are similar to those determined to be optimal for *in vitro* transfection (Fig. 2).

Although the active formulations determined from the *in vitro* transfection screens are fairly predictive of their optimal *in vivo* formulation, the relative performance of the different cationic lipids obtained using the *in vitro* analysis is not always reflective of their performance *in vivo*.[5,12,13] For example, the cationic lipids #48, DC-Chol, and DOTAP performed much less effectively *in vivo* than would have been predicted based on the *in vitro* analysis (Fig. 3). In general, based on our analysis of more than 150 compounds, although the correlation between *in vitro* and *in vivo* activity is not absolute, cationic lipids that transfect well *in vivo* are usually among the best performing lipids *in vitro*. Also, cationic lipids that perform poorly *in vitro* are invariably ineffective *in vivo*. Hence, the *in vitro* assay is useful for identifying ineffectual lipids and for prioritizing the cationic lipids for evaluation *in vivo*. The observed lack of correlation may reflect in part the sensitivity of some of the cationic lipid : pDNA complexes to serum inactivation.

Factors Affecting Efficacy of Gene Transfer *in vivo*

Besides the ratio of components that make up the complex, many other variables can affect the efficacy of cationic lipid-mediated gene transfer *in vivo*. The more important of these factors include the purity of the DNA,

[12] E. R. Lee, J. Marshall, C. S. Siegel, C. Jiang, N. S. Yew, M. R. Nichols, J. B. Nietupski, R. J. Ziegler, M. Lane, K. X. Wang, N. C. Wan, R. K. Scheule, D. J. Harris, A. E. Smith, and S. H. Cheng, *Hum. Gene Ther.* **7,** 1701 (1996).

[13] C. J. Wheeler, P. L. Felgner, Y. J. Tsai, J. Marshall, L. Sukhu, G. Doh, J. Hartikka, J. Nietupski, M. Manthorpe, M. Nichols, M. Plewe, X. Liang, J. Norman, A. E. Smith, and S. H. Cheng, *Proc. Natl. Acad. Sci. U.S.A.* **93,** 11454 (1996).

the toxicity associated with the cationic lipids, the ability of the cationic lipids to be formulated stably at high concentrations, and the ability of the cationic lipids to resist inactivation by serum.

Most cationic lipids elicit an inflammatory response when introduced into the lungs of BALB/c mice that is dose-dependent but which resolves over time. Toxicity therefore limits the amount of cationic lipid that can be administered to the animals. Because gene transduction activity invariably increases with increasing amounts of cationic lipid, and because toxicity is primarily cationic lipid mediated, structures that are active but exhibit low toxicity are expected to be more efficacious. No correlation is apparent however between the toxicity of the cationic lipid and its ability to mediate gene transfer. Presently, the basis for the observed lipid-mediated toxicity is not clear, but at high concentrations may involve a detergent-like effect. Toxicity may also be a manifestation of the inability of transfected cells to metabolize and secrete the cationic lipids. The resultant accumulation of large amounts of undigestable lipids over a period of time in the transfected cells may lead to lysosomal storage disease-like abnormalities. If this is the case, then the design of cationic lipids that are predicted to be biodegradable may alleviate this problem.

Since the relative efficiency of cationic lipid-mediated gene transfer is lower than that of virus-based vectors, a proportionately larger mass of DNA has to be delivered to achieve the same biological endpoint.[12,13] Therefore, yet another potential limitation for the use of cationic lipids *in vivo* is their ability to be formulated as highly concentrated suspensions. We have observed that a number of cationic lipids can be formulated stably in complexes that contain up to 1.5 m*M* cationic lipid. However, based on our studies at this concentration, tens of milliliters of the complex may have to be delivered to the human lung to achieve a biological effect. Ideally, an improvement in formulation that will allow a reduction of this volume by at least 5- to 10-fold is desirable for practical use in the clinic.

The quality of the pDNA is also an important parameter affecting *in vivo* performance. Care should be taken to ensure that the pDNA is free of bacterial endotoxins since its presence can affect the efficacy of gene transfer significantly and can contribute to the inflammatory response *in vivo.* A direct comparison has shown that when tested *in vivo* endotoxin-containing pDNA can be four-fold less effective than pDNA that is free of contaminating endotoxins. Furthermore, since transfection efficiency varies with the lipid:pDNA ratio, errors in quantitating DNA will also affect the activity of the complex. Additionally, since bacterial RNA and chromosomal DNA can bind cationic lipids and therefore compete for pDNA binding, care should be taken to ensure removal of these contaminants from pDNA preparations.

Cationic Lipid-Mediated Transfection of CFTR cDNA Into Airway Epithelial Cells

CFTR, the product of the gene associated with cystic fibrosis, is a cAMP-stimulated chloride (Cl^-) channel. Many approaches have been developed for detection of CFTR functional activity following transfection *in vitro* and *in vivo*. These include electrophysiological (patch-clamp, transepithelial electrolyte transport in an Ussing chamber (University of Iowa, Iowa City, IA) or planar lipid bilayer, nasal potential difference measurements)[14–17] and biochemical (SPQ [6-methoxy-*N*-(3-sulfopropyl)quinolinium] fluorescence assay, fluid transport measurements)[17–19] assays. For rapid determination of the efficacy of cationic lipid-mediated transfection of CFTR cDNA, we routinely used fluorescence digital imaging microscopy (FDIM) in conjunction with the halide-sensitive fluorophore SPQ.[20]

Procedure

CFT1 or Fischer rat thyroid (FRT)[12] epithelial cells that lack CFTR cAMP-stimulated Cl^- channel activity are plated onto collagen-coated glass coverslips 24 hr prior to transfection. Following cationic lipid-mediated transfection with plasmids encoding CFTR, the cells are incubated at 37° for a further 48 hr prior to analysis. Cells are loaded with SPQ by either including 10 m*M* SPQ in the growth media for 12–18 hr or after hypotonic shock (with 50% v/v water) for 4 min at room temperature. The SPQ fluorescence is initially quenched by incubating the cells for 25–45 min in a buffer containing 135 m*M* NaI, 2.4 m*M* K_2HPO_4, 0.6 m*M* KH_2PO_4, 1 m*M* $MgSO_4$, 1 m*M* $CaSO_4$, 10 m*M* dextrose, and 10 m*M* *N*-2-hydroxyethylpiperazine-*N*′-2-ethanesulfonic acid (HEPES), pH 7.4. After measuring baseline fluorescence for 2 min, the 135 m*M* NaI solution is replaced with one containing 135 m*M* $NaNO_3$, and fluorescence is measured for another 16 min. SPQ fluorescence is quenched by I^- but not by NO_3^-. Forskolin (20 μM) and 3-isobutyl-1-methylxanthine (IBMX) (100 μM) are added 5 min

[14] M. P. Anderson, D. P. Rich, R. J. Gregory, A. E. Smith, and M. J. Welsh, *Science* **251,** 679 (1991).

[15] C. E. Bear, C. Li, N. Kartner, R. J. Bridges, T. J. Jensen, M. Ramjeesingh, and J. R. Riordan, *Cell* **68,** 809 (1992).

[16] B. R. Grubb, R. J. Pickles, H. Ye, J. R. Yankaskas, R. N. Vick, J. F. Engelhardt, J. M. Wilson, L. G. Johnson, and R. C. Boucher, *Nature* **371,** 802 (1994).

[17] C. Jiang, S. P. O'Connor, D. Armentano, P. B. Berthelette, S. C. Schiavi, D. M. Jefferson, A. E. Smith, S. C. Wadsworth, and S. H. Cheng, *Am. J. Physiol.* **271,** L527 (1996).

[18] J. J. Smith and M. J. Welsh, *J. Clin. Invest.* **91,** 1590 (1993).

[19] S. H. Cheng, D. P. Rich, J. Marshall, R. J. Gregory, M. J. Welsh, and A. E. Smith, *Cell* **66,** 1027 (1991).

[20] N. P. Illsley and A. S. Verkman, *Biochemistry* **26,** 1215 (1987).

after anion substitution to increase intracellular cAMP. In this assay, an increase in halide permeability results in a more rapid increase in SPQ fluorescence. It is the rate of change rather than the absolute change in signal that is the important variable in evaluating anion permeability. Differences in absolute levels reflect quantitative differences between groups in SPQ loading, size of cells, or number of cells studied.[20]

Fluorescence of SPQ in single cells is measured with a Nikon inverted microscope (Nikon, Melville, NY), a digital imaging system from Universal Imaging (West Chester, PA), and a Hamamatsu ICCD camera (Hamamatsu Photonics, Japan). Excitation is at 350 nm and emission is at >400 nm. Cells are chosen for analysis without prior knowledge of the rate of change in fluorescence. In each experiment, up to five microscopic fields of between 90 and 100 cells are examined on a given day, and studies under each condition are repeated on at least three different days. Since expression of CFTR is heterogenous, the data shown are for the 20% of cells in each field exhibiting the greatest response. Data are presented as mean ± S.E. of fluorescence at time t (F_t) minus the baseline fluorescence (F_0, the average fluorescence measured in the presence of I^- for 2 min prior to ion substitution).

Results

Figure 5 shows an example of results obtained following transfection of CFT1 cells with cationic lipid : pDNA complexes. The assay takes advantage of the fact that SPQ fluorescence is more effectively quenched by iodide than by chloride, and that cAMP-stimulated Cl^- channels are permeable to iodide whereas several other Cl^- transport processes are not. Parental CFT1 cells and cells that had been transfected with #53 : pCF1-βGal complexes do not display cAMP-stimulated Cl^- channel activity. In contrast, CFT1 cells that have been transfected with #53 : pCF1-CFTR complexes display a rapid increase in SPQ fluorescence on stimulation with cAMP agonists, indicating an increased anion permeability (Fig. 5). Measurable cAMP-stimulated Cl^- channel activity can routinely be detected in 30% of the cells analyzed. This frequency however is dependent on the cationic lipid species with more effective cationic lipids in general giving higher frequencies of a measurable signal. Hence the SPQ fluorescence assay can be used as a guide to determine the relative activities of different cationic lipid formulations. The rate of change in SPQ fluorescence following stimulation with cAMP agonists can also be used as an indicator of the number of CFTR molecules in the cell. Cells containing a greater number of CFTR molecules effect a greater rate of change in SPQ fluorescence. This assay has also been demonstrated to be able to discriminate between brushed

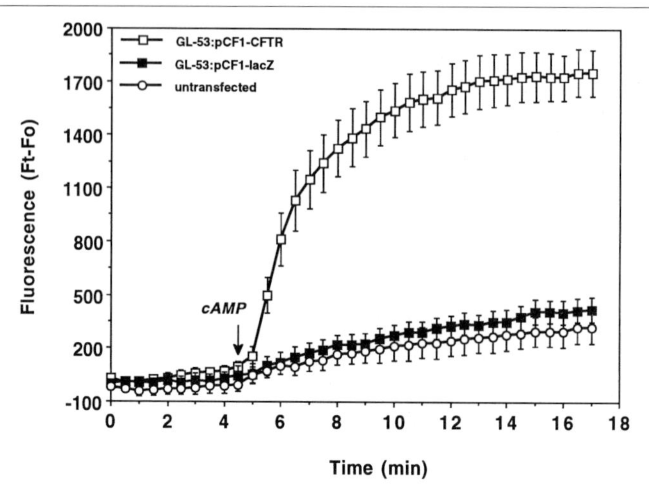

FIG. 5. SPQ halide efflux assay of CFT1 cells. Change in SPQ fluorescence is shown for CFT1 cells ($n = 20$; n is the number of cells) and CFT1 cells that had either been transfected with #53 : pCF1-CFTR ($n = 25$) or #53 : pCF1-lacZ ($n = 20$) complexes. NO_3^- was substituted for I^- in the bathing solution at 0 min. Four minutes later (arrow), cells were stimulated with 20 μM forskolin and 100 μM IBMX to increase intracellular cAMP.

nasal epithelial cells from CF and non-CF subjects.[21] Therefore, in addition to evaluating the efficacy of cationic lipid-mediated gene transfer *in vitro,* the SPQ fluorescence assay should also be useful for assessing the efficacy of CFTR gene transfer in human clinical studies.

Concluding Remarks

Much progress has been made in the recent past in terms of developing new cationic lipid structures with increased potency and activity. However, despite these advances, it is clear that much improvement in efficiency is still necessary for this technology to be of widespread benefit in the clinic. It is to be hoped that as our understanding of cationic lipid structure–activity relationships and of cellular barriers to gene delivery increases, yet more potent cationic lipid structures and formulations will be developed.

In the absence of a detailed understanding of the mechanism of cationic lipid-mediated gene delivery, it is apparent from our studies that an extensive and essentially empirical optimization of the formulations is necessary

[21] M. Stern, F. M. Munkonge, N. J. Caplen, F. Sorgi, L. Huang, D. M. Geddes, and E. W. F. W. Alton, *Gene Ther.* **2,** 766 (1995).

for maximal transfection activity. It is also likely that the results of such optimization studies for a particular cell type or target organ cannot be applied universally to different cells or organs. Therefore, caution should also be used when attempting to predict the performance of different cationic lipids in different applications.

Acknowledgments

The authors thank Dr. Craig Siegel and David Harris for the cationic lipids #48 and #53; Samantha Rudginsky, Margaret Nichols, and Jennifer Nietupski for technical assistance; and Dr. Nelson Yew for critical reading of the manuscript.

[53] Adeno-Associated Virus Vectors for Gene Therapy of Cystic Fibrosis

By TERENCE R. FLOTTE and BARRIE J. CARTER

Gene Therapy for Cystic Fibrosis Lung Disease

Cystic fibrosis (CF) is the most common genetic disease in North America, with an estimated incidence of 1 in 2750 live births, and a carrier frequency of approximately 4%.[1] Although CF can affect vitually every organ in which there is epithelial ion transport, the primary clinical manifestations of the disease are due to exocrine pancreatic insufficiency and obstructive lung disease.[2] While pancreatic disease is readily compensated for in most patients by use of oral pancreatic enzyme supplementation, CF lung disease continues to be the leading cause of morbidity and mortality among CF patients.

In 1989, the gene that is defective in CF was identified and cloned and a detailed characterization of the structure and function of its protein product, the CF transmembrane conductance regulator (CFTR), showed that it was a new member of the ATP-binding cassette (ABC) superfamily of transmembrane proteins.[3] The discovery of the CF gene also prompted

[1] S. C. Fitzsimmons, 1993 Annual Data Report, CF Foundation Patient Registry, 1994.

[2] T. F. Boat, M. J. Welsh, and A. L. Beaudet, *in* "The metabolic basis of inherited disease" (C. L. Scriver, A. L. Beaudet, W. S. Sly, and D. Valle, eds.), 6th Ed., pp. 2649–2680. McGraw-Hill, New York, 1989.

[3] J. R. Riordan, J. M. Rommens, B. S. Kerem, N. Alon, R. Rozmahel, Z. Grzelczak, J. Zielenski, S. Lok, N. Plavsic, J.-L. Chou, M. L. Drumm, M. C. Iannuzzi, F. S. Collins, and L.-C. Tsui, *Science* **245,** 1066 (1989).

0076-6879/98 $25.00

speculation that specific therapies, such as gene augmentation, would lead to rapid changes in the clinical care of CF patients. However, this task is not as simple as it may seem. The tissue that must be corrected is the epithelium of the conducting airways of the lungs, a complex tissue with a large surface area (~10 m^2) and a very specific histologic organization. Furthermore, pathology due to the disease itself creates additional obstacles to the efficacy and safety of gene transfer. Because of these factors, a fundamental understanding of the normal histology of the airway and the pathology of CF lung disease is essential for a complete appreciation of the CF gene therapy problem.

Airway Epithelium

The epithelial lining of the airways is a complex tissue, which differs in its organization in the more proximal large airways as compared with the more distal small airways. The proximal intrathoracic airways include the trachea, with its C-rings of cartilage, and the eight generations of branching bronchi, with partial rings and plates of cartilage.[4] These proximal airways are lined by a pseudostratified columnar epithelium composed of at least 15 distinct cell types. The predominant cell types are ciliated cells, secretory cells, and basal cells. Ciliated cells abut both the basement membrane and the lumenal surface and are involved in both mucociliary clearance and the production of aqueous secretions that comprise the periciliary fluid layer. Secretory cells are of two types: serous cells and mucous or goblet cells. Goblet cells contain granules filled with mucous glycoproteins, which, after they are secreted, contribute to the blanket of gelatinous mucus that lies atop the periciliary sol layer. Basal cells are smaller polygonal cells that sit on the basement membrane and may serve as progenitors for other epithelial cell types. Less abundant cell types within the epithelium include intermediate cells, brush cells, and neuroendocrine or Kultschitsky cells. The epithelium of the proximal airways also contains complex submucosal gland structures, which also function in the production of serous and mucous glycoprotein components of the airway mucus. Interestingly, these glands contain serous cells that are natural "hot spots" of CFTR expression.[5]

The distal conducting airways include an additional eight generations of branching noncartilaginous bronchioles, which are lined by a low columnar to cuboidal epithelium that is simpler in organization than that of bronchi. The last generation of bronchioles to contain smooth muscle are known as terminal bronchioles. Distal to the terminal bronchioles are the

[4] J. F. Murray, *in* "The Normal Lung," 2nd Ed. W.B. Saunders Co., Philadelphia, 1986.

[5] J. F. Engelhardt, J. R. Yankaskas, S. A. Ernst, Y. Yang, C. R. Marino, R. C. Boucher, J. A. Cohn, and J. M. Wilson, *Nature Genet.* **2,** 240 (1992).

gas exchanging or respiratory units, which anatomically consist of respiratory branchioles, alveolar ducts, alveolar sacs, and alveoli. Within the bronchiolar epithelium, several cell types can be recognized, including ciliated cells, secretory cells, and Clara cells. Clara cells have multiple functions in the bronchiolar epithelium, but they may also serve as progenitor cells for other cell types.[4]

The airway epithelium is active in electrolyte transport from a very early stage in prenatal development.[4] The primary pumps and channels involved in this process are the following (Fig. 1): On the basolateral surface, there are the oubain-sensitive Na^+,K^+-ATPase and the furosemide-sensitive NaCl cotransporter. On the apical surface are the amiloride-sensitive epithelial Na^+ channel (ENaC) and several chloride channels including the DPPC-sensitive cystic fibrosis transmembrane conductance regulator (CFTR), the DIDS-sensitive outwardly rectifying chloride channel

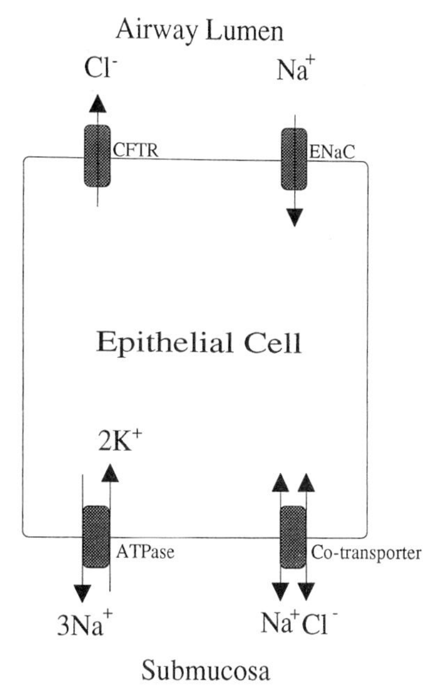

FIG. 1. Ion channels participating in fluid absorption and secretion at the airway surface. A typical airway epithelial cell is depicted with the airway lumen or apical surface shown at the top and the submucosal or basolateral side shown at the bottom. The primary pumps and channels involved in regulating ion and water transport are depicted, including the oubain-sensitive Na^+,K^+-ATPase, the furosemide-sensitive cotransporter, the amiloride-sensitive epithelial Na channel (ENaC), and CFTR. CFTR serves as both a chloride channel and a regulator of other chloride channels, such as the outwardly rectifying chloride channel (ORCC), which is not shown in this simplified diagram.

(ORCC), and perhaps some other "alternative" chloride channels, such as the volume-regulated ClC-2 channel and a calcium-regulated chloride channel. In an oversimplified view of the transport process, we may envision that the Na^+,K^+-ATPase generates an electrochemical gradient, which may drive either Na^+ absorption through ENaC or chloride secretion or both, depending on the relative conductance of these pathways on the apical membrane. The counterion can then be passively transferred through the basolateral cotransporter. The movement of water is likewise passively dictated by whether the predominant movement of ions is inward or outward.

Generally speaking, chloride secretion is greater in prenatal life, when the lung is fluid filled, than in postnatal life.[4] In postnatal life, choride secretion may still predominate in the more distal airways, where it provides a medium for the mucociliary transport of trapped particles and other unwanted materials to be mobilized upward. As the fluid is moved upward it is gradually reabsorbed as sodium absorption predominates in the upper airways. CFTR is the key to regulating this process. In CF, the absence of CFTR leads to an imbalance such that chloride transport at the apical surface is markedly deficient and sodium absorption is overactive. This is due not only to CFTR's intrinsic chloride channel activity but to interactions with other channels, such as ORCC and ENaC.[6,7] Interestingly, prenatal epithelial chloride transport is preserved in CF, suggesting that alternative pathways of chloride channel regulation are operative in that setting.

Pathophysiology of Cystic Fibrosis Lung Disease

CF lung disease involves a disorder of insufficient hydration of the conducting airways as one of the primary abnormalities, but several other specific changes occur at an early stage in diseased CF airways. Bronchoscopy studies[8] show that even in early stages of the disease process, many CF patients display two other hallmarks of this disease: lower respiratory tract colonization with a specific bacterial flora, which includes *Staphylococcus aureus* and *Pseudomonas aeruginosa,* and a form of airway inflammation in which infiltration by polymorphonuclear neutrophilic leukocytes (PMNs) predominate. The causal and mechanistic relationships between the abnormalities in ion transport, bacterial colonization, and airway inflammation are not known with certainty.

[6] M. Egan, T. Flotte, S. Afione, R. Solow, P. L. Zeitlin, B. J. Carter, and W. B. Guggino, *Nature* **358,** 581 (1992).

[7] E. Schwiebert, T. R. Flotte, G. Cutting, and W. B. Guggino, *Am. J. Physiol.* **266** (*Cell Physiol.* **35**), C1464 (1994).

[8] M. W. Konstan, K. A. Hilliard, T. M. Norvell, and M. Berger, *Am. J. Respir. Crit. Care Med.* **150,** 448 (1994).

One theory that has been widely held is that insufficient chloride and water transport[9] leads directly to an abnormality of the periciliary fluid and an underhydration of the superficial mucous layer, thus creating a milieu that is uniquely suitable for *S. aureus, P. aeruginosa,* and related organisms to colonize and persist in the lower airways. This colonization would then lead to the release of proinflammatory cytokines, such as interleukin 6 (IL-6), IL-8, tumor necrosis factor α (TNF-α), and IL-1β, which in turn set the processes of PMN infiltration and inflammatory injury into motion.[10] This theory is appealing since it simply and directly links the most obvious implications of the genetic defect with the primary aspects of the pathophysiology. One deficiency of this theory is that it does not explain why *S. aureus* and *P. aeruginosa* should be the favored pathogens in this model instead of the more usual respiratory pathogens, such as *Streptococcus pneumoniae.*

According to another theory,[11] the link between ion transport and the specific bacterial pathogens in CF airways is the PMN itself. This recently proposed model postulates that PMN function is impaired in the local airway surface environment of CF airways. Alteration of PMN function might lead to a specific susceptibility to organisms such as *Pseudomonas, Staphylococcus,* and *Aspergillus fumigatus* (a fungal colonizer in CF) for which PMN phagocytosis is the primary defense. This theory is supported by the finding that these same organisms are the major pathogens seen in primary PMN defects such as chronic granulomatous disease (CGD), a single-gene defect in oxidative killing of bacteria by PMNs. Abnormal ionic composition of airway surface fluid might also affect epithelial defensin function. A related theory holds that defective internalization of *Pseudomonas* by CFTR-defective airway epithelial cells leads to impaired clearance of these organisms.[11]

Another theory holds that *S. aureus* and *P. aeruginosa* bind preferentially to CF-defective bronchial epithelial cells as compared with non-CF cells. A recent study[12] indicated that both the abundance of the G_{M1}-ganglioside on the surface of CF bronchial epithelial cells and the binding of *P. aeruginosa* were increased compared with CFTR-corrected cells. Presumably, the alteration in the G_{M1}-ganglioside content on glycosylated

[9] M. L. Drumm, H. A. Pope, W. H. Cliff, J. M. Rommens, S. A. Sheila, L.-C. Tsui, F. S. Collins, R. A. Frizzell, and J. M. Wilson, *Cell* **62,** 1227 (1990).

[10] G. Kronborg, M. B. Hansen, M. Svenson, A. Fomsgaard, N. Hoiby, and K. Bendtzen, *Ped. Pulmonol.* **15,** 292 (1993).

[11] G. B. Pier, M. Grout, T. S. Zaidi, J. C. Olsen, L. G. Johnson, J. R. Yankaskas, J. B. Goldberg, *Science* **271,** 64 (1996).

[12] L. Imundo, J. Barasch, A. Prince, Q. Al-Awqatic, *Proc. Natl. Acad. Sci. U.S.A.* **92,** 3019 (1995).

proteins would be due to intracellular effects of CFTR such as regulation of membrane turnover[13] or endosomal pH.[14]

The validity of each of these theories could have important implications for CF gene therapy. Cell-mixing studies performed by Johnson *et al.*[15] showed that if as few as 5–10% of airway epithelial cells in a monolayer were complemented, the net chloride flux (as measured by short-circuit current) would approach the normal range. This would suggest that if ionic composition and hydration of airway lining fluid is the key to airway pathology in CF, then gene therapy could be successful if only a minority of cells were successfully transduced. If the pathogen-binding theory is correct, however, a much higher proportion of cells might need to be corrected in order to significantly affect disease progression.

Vectors Available for Cystic Fibrosis Gene Therapy

Given the current understanding of the pathophysiology of the disease, the current goal of gene therapy would be to express CFTR protein in as many cells as possible within the epithelium of the conducting airways. The ideal vector would accomplish this with minimal or no toxicity and have a very long duration of action, so that the frequency of administration can be minimized. None of the currently available technologies for gene transfer has been proven to accomplish all of these goals.

Several gene therapy vectors are based on viruses that have evolved specific mechanisms for inserting nucleic acid into the nucleus of a target cell so that it can be transcribed and translated into the desired protein product. Those which have been developed for CF gene therapy include the Moloney murine leukemia virus (MMLV), the group C adenoviruses (Ad2 and Ad5), and adeno-associated virus type 2 (AAV). Nonviral vectors for gene transfer, including liposomes and DNA–protein conjugates, have also been developed and are in preclinical or early clinical trials.[16,17]

Viral Vectors

MMLV is a murine retrovirus that integrates into the host cell genome to establish persistent infections. The retrovirus virion is composed of a

[13] N. A. Bradbury, T. Jilling, G. Berta, E. J. Sorscher, R. I. Bridges, K. L. Kirk, *Science* **256,** 530 (1992).

[14] G. L. Lukacs, X. B. Chang, N. Kartner, O. D. Rotstein, J. R. Riordan, S. Grinstein, *J. Biol. Chem.* **267,** 14568 (1992).

[15] L. G. Johnson, J. C. Olsen, B. Sarkadi, K. L. Moore, R. Swanstrom, R. C. Boucher, *Nature Genet.* **2,** 21 (1992).

[16] N. J. Caplen, E. W. F. W. Alton, P. G. Middleton, J. R. Dorin, B. J. Stevenson, X. Gao, S. R. Durham, P. K. Jeffery, M. D. Hodson, C. Coutelle, L. Huang, D. J. Porteous, R. Williamson, and D. M. Geddes, *Nature Med.* **1,** 39 (1995).

[17] T. Ferkol, C. S. Kaetzel, and P. B. Davis, *J. Clin. Invest.* **92,** 2394 (1993).

lipid envelope surrounding a nucleocapsid containing two single-stranded 8-kb RNA copies of the genome as well as several proteins, including reverse transcriptase (RT). After the retrovirus enters the cell by membrane fusion, RT converts the RNA genome to DNA, which can then migrate to the nucleus and integrate into the host cell genome if the cell is mitotically active.[18] The retrovirus genome consists of two genes, *gag-pol* and *env*, flanked by long terminal repeat (LTR) sequences. Just downstream from the left-hand LTR is the *psi* sequence, which serves as a signal for packaging. Vectors are produced by replacing *gag-pol* and *env* with the foreign gene of interest, and then supplying these gene functions *in trans* from a second *psi*$^-$ construct within the packaging cell. Recombinant retrovirus particles are assembled within the packaging cell and released into the medium by budding from the cell membrane. Vectors produced in this way have been used to transfer a wide variety of genes into human cells *in vitro* and have been used for clinical *ex vivo* gene transfer experiments in human subjects.

In the context of CF gene therapy, retrovirus vectors were used in early experiments to complement the CF defect in cells in culture. Drumm *et al.*[19] used an MMLV-based construct to insert a normal copy of CFTR into the CF defective pancreatic adenocarcinoma cell line, CFPAC-1. This resulted in restoration of cAMP-mediated activation of chloride conductance, which is the hallmark of CFTR function. Despite this early success *in vitro,* there are inherent problems with a retrovirus-based approach to CF gene therapy *in vivo.* First, the cells of the bronchial epithelium are not well suited to removal, manipulation *ex vivo*, and reimplantation as has been done with cell targets for other clinical trials of retroviral gene transfer. Second, the majority of target cells in the bronchial epithelium are nondividing and so would not be permissive for retrovirus integration and expression. Third, the titers of retrovirus vectors are low, which would translate into a very large volume of vector stock being required to achieve the desired multiplicity of infection.

Ad vectors can overcome those particular problems since they can be produced at high titers, can infect nondividing cells, and are effective for *in vivo* gene transfer.[20–22] Ad is a large (36 kilobase-pair) DNA virus that

[18] D. G. Miller, M. A. Adam, and A. D. Miller, *Mol. Cell Biol.* **10,** 4239 (1990).

[19] M. L. Drumm, H. A. Pope, W. H. Cliff, J. M. Rommens, S. A. Sheila, L.-C. Tsui, F. S. Collins, R. A. Frizzell, and J. M. Wilson, *Cell* **62,** 1227 (1990).

[20] M. A. Rosenfeld, W. Siegfried, K. Yoshimura, K. Yoneyama, M. Fukayama, L. E. Stier, P. K. Paakko, P. Gilardi, L. D. Stratford-Perricaudet, M. Perricaudet, S. Jaalat, A. Pavirani, J.-P. Lecocq, and R. G. Crystal, *Science* **252,** 431 (1991).

[21] M. A. Rosenfeld, K. Yoshimura, B. C. Trapnell, K. Yoneyama, E. R. Rosenthal, W. Dalemans, M. Fukayama, J. Bargon, L. E. Stier, L. Stratford-Perricaudet, M. Perricaudet, W. B. Guggino, J.-P. Lecocq, and R. G. Crystal, *Cell* **68,** 143 (1992).

naturally infects the respiratory and gastrointestinal tracts of humans, causing mild inflammatory illnesses such as rhinitis, conjuctivitis, pharyngitis, and diarhea. Ad infection in human cells is generally lytic, that is, the virus infects a cell, produces progeny, and then lyses the cell to release the progeny virions. In this highly organized process, the immediate early genes are activated first, followed by the early genes, and finally the late genes. The immediate early genes (E1a and E1b) regulate transcription of viral and host cell genes. These, in turn, activate the early genes (E2a, E3, E4) that are required for viral DNA synthesis. After DNA synthesis has occurred, the late genes (L1, L2, L3) are activated to produce the structural proteins that make up the capsids of the progeny virions. First-generation Ad vectors were produced by deleting E1a and E1b (and sometimes E3) from the viral DNA and inserting the foreign gene of interest in their place. These vectors were then packaged in cell lines such as 293, which constitutively express E1a and E1b.

Ad vectors have been shown to transfer genes efficiently to the airway epithelium *in vivo* in cotton rats and primates and have been used in several clinical trials in CF patients. In three of these trials,[23–25] transient gene transfer (for 1–2 weeks) was detected. In two of these three trials, however, dose-related airway inflammation was also noted, which resulted in clinical toxicity. Ad vector-related inflammation may relate to direct induction of IL-6 and IL-8 cytokines with subsequent neutrophil recruitment, or to cell-mediated immune responses to viral proteins, which are produced at low levels in the target cells. Efforts have been made to circumvent Ad vector toxicity by developing vectors in which additional early gene deletions have been made, but the safety of these constructs has yet to be confirmed in clinical trials. Interestingly, these second- and third-generation Ad vectors have also had a prolonged duration of expression in animals, suggesting that the transient nature of Ad vector expression *in vivo* may be due to immune-mediated elimination of transduced cells.[26,27]

[22] J. F. Engelhardt, R. H. Simon, Y. Yang, M. Zepeda, S. Weber-Pendelton, B. Doranz, M. Grossman, and J. M. Wilson, *Hum. Gene Ther.* **4,** 759 (1993).

[23] R. G. Crystal, N. G. McElvaney, M. A. Rosenfeld, C.-S. Chu, A. A. Mastrangeli, G. H. Jogn, S. L. Brody, H. A. Jaffe, N. T. Eissa, and C. Daniel, *Nature Genet.* **8,** 42 (1994).

[24] L. Zabner, L. A. Couture, R. J. Gregory, S. M. Graham, A. E. Smith, and M. J. Welsh, *Cell* **75,** 207 (1993).

[25] M. R. Knowles, K. W. Hohneker, Z. Zhou, J. C. Olsen, T. L. Noah, P. C. Hu, M. W. Leigh, J. F. Engelhardt, L. J. Edwards, K. R. Jones, J. M. Wilson, R. C. Boucher, *N. Engl. J. Med.* **333,** 823 (1995).

[26] Y. Yang, F. A. Nunes, K. Berensci, E. F. Furht, E. Gönzcöl, and J. M. Wilson, *Proc. Natl. Acad. Sci. U.S.A.* **91,** 4407 (1994).

[27] R. H. Simon, J. F. Engelhardt, Y. Yang, M. Zepeda, S. Weber-Pendleton, M. Grossman, and J. M. Wilson, *Hum. Gene Ther.* **4,** 771 (1993).

Adeno-Associated Virus

Adeno-associated virus is an alternative DNA virus that could avoid the inflammatory toxicity of Ad. AAV is a nonpathogenic human parvovirus, which commonly infects humans, but is not associated with any disease state.[28,29] AAV generally requires helper virus infection (usually adenovirus or herpesvirus) in order to replicate itself in a productive life cycle.[30–32] In the absence of helper virus infection, AAV latency is established.[33–35] During latency tandem copies of AAV DNA are usually integrated within one particular region of chromosome 19, the AAVS1 site.[36–41] Latency is a complex process, and episomal copies may also be detected for prolonged periods of time.[34,35] It is not clear whether these episomal copies can be replicated directly, or perhaps exist in a dynamic equilibrium with integrated copies. If cells are later infected with helper virus, the integrated proviral DNA can be efficiently rescued and enter the lytic phase of the life cycle.

The 20-mm AAV particle consists of a single-stranded 4.7-kb DNA molecule encapsidated in an icosahedral protein coat. The AAV genome consists of two genes, *rep* and *cap,* flanked by palindromic inverted terminal repeat (ITR) sequences (Fig. 2).[42] The *rep* gene codes for functions required for replication and is transcribed from two promoters, p_5 and p_{19}, the latter of which is contained within the coding sequence of the open reading frame transcribed from p_5. Each of the RNA transcripts exists in both spliced and unspliced versions, leading to the production of four Rep proteins,

[28] N. R. Blacklow, M. D. Hoggan, A. Z. Kapikian, J. B. Austin, and W. P. Rowe, *Am. J. Epidem.* **94,** 359 (1971).

[29] N. R. Blacklow, *in* "Parvoviruses and Human Disease" (J. Pattison, ed.), pp. 165–174. CRC Press, Boca Raton, Florida, 1985.

[30] K. I. Berns, *in* "Virology" (B. N. Fields, D. M. Knipe, et al., eds.), pp. 1743–1764. Raven Press, New York, 1990.

[31] B. J. Carter, *in* "Handbook of Parvoviruses" (P. Tjissen, ed.), Vol. 1, pp. 155–168. CRC Press, Boca Raton, Florida, 1990.

[32] B. J. Carter, E. Mendelson, and J. P. Trempe, *in* "Handbook of Parvoviruses" (P. Tjissen, ed.), Vol. 1, pp. 169–226. CRC Press, Boca Raton, Florida, 1990.

[33] C. A. Laughlin, C. B. Cardellichio, and H. C. Coon, *J. Virol.* **60,** 515 (1986).

[34] M. D. Hoggan, G. F. Thomas, and F. B. Johnson, *in* "Proc. 4th LePetit Colloq.," Cocoyac, Mexico, pp. 243–253. North-Holland Publishers, Amsterdam.

[35] A. M. K. Cheung, M. D. Hoggan, W. W. Hauswirth, and D. I. Berns, *J. Virol.* **33,** 739 (1980).

[36] R. M. Kotin, M. Siniscalco, R. J. Samulski, X. Zhu, L. Hunter, C. A. Laughlin, S. McLaughlin, N. Muzyczka, M. Rocchi, and K. I. Berns, *Proc. Natl. Acad. Sci. U.S.A.* **87,** 2210 (1990).

[37] R. M. Kotin, R. M. Linden, and K. I. Berns, *EMBO J.* **11,** 5071 (1992).

[38] R. M. Kotin, J. C. Menninger, D. C. Ward, and K. I. Berns, *Genomics* **10,** 831 (1991).

[39] R. J. Samulski, X. Zhu, X. Xiao, J. D. Brook, D. E. Housman, N. Epstein, and L. A. Hunter, *EMBO J.* **10,** 3941 (1991).

[40] R, J. Samulski, *Curr. Opin. Biotech.* **3,** 74 (1993).

[41] C. Giraud, E. Winocour, and K. I. Berns, *Proc. Natl. Acad. Sci. U.S.A.* **91,** 10039 (1994).

[42] B. J. Carter, *Curr. Opin. Biotech.* **3,** 533 (1992).

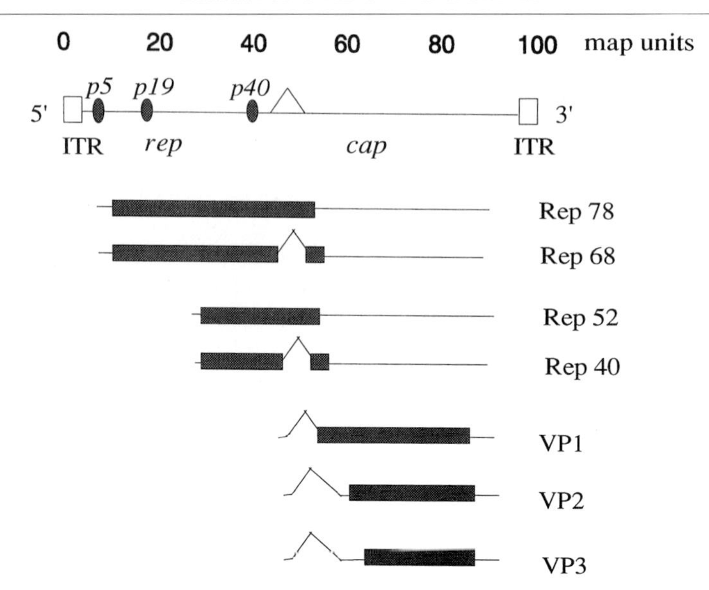

FIG. 2. Structure of the AAV Genome. The AAV2 genome is represented with a 100-map-unit scale (1 map unit equals 1% of genome size, approximately 47 bp). The open boxes represent the inverted terminal repeats (ITRs). The transcription promoters (p_5, p_{19}, p_{40}) are depicted as solid circles. The polyadenylation signal is at map position 96. RNA transcripts from AAV promoters are shown below the DNA map with the introns indicated by carets. Protein coding regions are depicted as solid boxes. The three capsid proteins are VP1, VP2, and VP3; the four Rep proteins are Rep78, Rep68, Rep52, and Rep40.

Rep78, Rep68, Rep52, and Rep40, based on their apparent molecular weights.[43] Rep proteins have multiple functions including transcriptional regulation of AAV promoters and heterologous promoters[44]; site-specific, strand-specific endonuclease (nicking) activity, which is essential for resolution of AAV termini during replication[45]; DNA helicase activity[45]; and an ability to simultaneously bind both the AAV–ITR and the chromosomal AAVS1 site to form a complex that could serve as an intermediate in site-specific integration.[46] The *cap* gene encodes the structural capsid proteins, VP1, VP2, and VP3. The AAV–ITRs serve as the origins for DNA replication in a productive infection, and as packaging signals.

AAV vectors are produced by substituting the foreign gene of interest

[43] E. Mendelson, J. P. Trempe, and B. J. Carter *J. Virol.* **60,** 823 (1986).

[44] A. Beaton, P. Palumbo, and K. I. Berns, *J. Virol.* **63,** 4450 (1989).

[45] D.-S. Im and N. Muzyczka, *Cell* **61,** 3095 (1990).

[46] M. D. Weitmann, S. R. M. Kyöstiö, R. M. Kotin, and R. A. Owens, *Proc. Natl. Acad. Sci. U.S.A.* **91,** 5808 (1994).

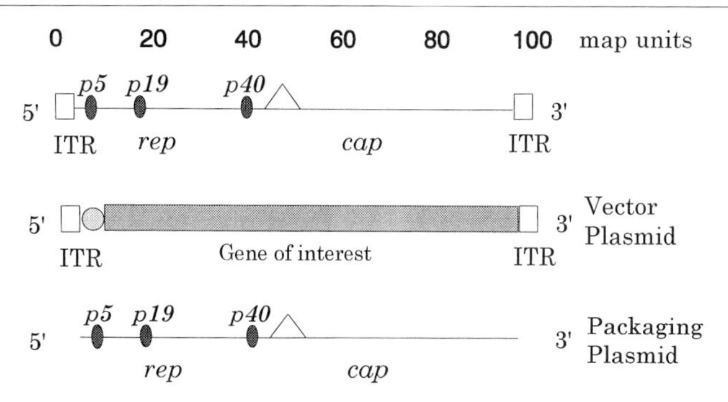

FIG. 3. Organization of AAV2-based vectors. The AAV genome with a map-unit scale is depicted above, with the ITRs as open boxes and the transcription promoters (p_5, p_{19}, p_{40}) as dark shaded circles. Vector plasmids (middle drawing) are constructed by inserting the foreign gene of interest (lightly shaded bar) and a promoter (lightly shaded circle) between the ITRs, which serve as replication origins and packaging signals. Vector DNA is packaged into infectious AAV particles after cotransfection with an ITR-deleted packaging plasmid (bottom drawing) into adenovirus-infected cells.

in place of *rep* and/or *cap,* but still flanked by ITRs (Fig. 3). If *rep* and *cap* are supplied *in trans* within adenovirus-infected cells, any suitably sized ($\leq$4.7 kb, excluding ITRs), ITR-flanked insert will be replicated and packaged into infectious AAV virions. These virions can then be separated from the helper Ad by a combination of CsCl density gradient ultracentrifugation and heat inactivation.[47,48]

AAV vectors have been used to transfer genes into a wide range of mammalian cell types *in vitro,* including K562 erythroleukemia cells,[49] $CD4^+$ lymphocyte cell lines,[50] $CD34^+$ hematopoietic stem cells,[51] and a CF bronchial epithelial cell line.[52,53] Initial studies in which AAV vector stocks contained wild-type AAV as well showed relatively low transduction effi-

[47] R. J. Samulski, L.-S. Chang, and T. Shenk, *J. Virol.* **63,** 3822 (1989).

[48] T. R. Flotte, X. Barraza-Ortiz, R. Solow, S. A. Afione, B. J. Carter, and W. B. Guggino, *Gene Ther.* **2,** 29 (1995).

[49] C. E. Walsh, J. M. Liu, X. Xiao, et al. *Proc. Natl. Acad. Sci. U.S.A.* **89,** 7257 (1992).

[50] S. Chatterjee, P. R. Johnson, and K. K. Wong, *Science* **258,** 1485 (1992).

[51] S. Goodman, X. Xiao, R. E. Donahue, A. Moulten, J. Miller, C. Walsh, N. S. Young, R. I. Samulski, A. W. Nienhuis, *Blood* **84,** 1492 (1994).

[52] T. R. Flotte, R. Solow, R. A. Owens, S. Afione, P. L. Zeitlin, and B. J. Carter, *Am. J. Resp. Cell Mol. Biol.* **7,** 349 (1992).

[53] T. R. Flotte, S. A. Afione, C. K. Conrad, S. A. McGrath, R. Solow, H. Oka, P. L. Zeitlin, W. B. Guggino, and B. J. Carter, *Proc. Natl. Acad. Sci. U.S.A.* **90,** 10613 (1993).

ciencies, but stable gene transfer was achieved.[54–56] Subsequent studies with wild-type-free preparations have indicated gene transfer efficiencies as high as 70%.[47,52]

It was initially assumed that AAV vectors were integrating in a manner similar to wild-type AAV. Evidence indicates, however, that *rep*-deleted AAV vectors may not integrate within the AAVS1 site at high frequency. The integration that does occur[57] appears to be present predominantly at other sites within the genome. Furthermore, there may be a greater proportion of vector DNA that remains in an episomal state in certain cell types.[58,59] The role of host cell division in determining permissiveness for AAV vector transduction is also complex. AAV vectors have been found to express more efficiently in rapidly proliferating cell lines than in nondividing or primary cells.[60–63] Although this is not an absolute obstacle, the relatively limited expression in some cells may reflect the need for AAV to be converted from the ssDNA form present in the virion to the dsDNA form required for expression.[59]

Adeno-Associated Virus Vector-Mediated Complementation of Cystic Fibrosis Defect

AAV vectors have been used to complement the CF defect in the CF bronchial epithelial cell line, IB3-1.[53] AAV vectors expressing the CFTR cDNA from either the intact p_5 promoter or from an *Inr*-like promoter sequence within the ITR were used either by direct lipofection or by packaged virus particle-mediated transduction. CFTR protein expression occurred in transfected and transduced clones, as detected by CFTR immunofluorescence staining. CFTR function was then assessed by determining the ability of cell cultures to increase their rate efflux of the radioisotope tracer $^{36}Cl^-$ in response to stimulation by forskolin, an activator of adenylate cyclase. Since CFTR regulates all cAMP-mediated stimulation of chloride efflux, this provides a direct assessment of CFTR function. As is shown in

[54] P. L. Hermonat and N. Muzyczka, *Proc. Natl. Acad. Sci. U.S.A.* **81,** 6466 (1984).
[55] J.-D. Tratschin, M. H. P. West, R. Sandbank, and B. J. Carter, *Mol. Cell Biol.* **4,** 2072 (1984).
[56] J.-D. Tratschin, I. L. Miller, M. G. Smith, and B. J. Carter, *Mol. Cell Biol.* **5,** 3251 (1985).
[57] S. K. McLaughlin, P. Collis, P. L. Hermonat, and N. Muzyczka, *J. Virol.* **62,** 1963 (1988).
[58] W. G. Kearns, S. Afione, S. B. Fulmer, J. Caruso, P. Pearson, T. Flotte, and G. R. Cutting, *Am. J. Hum. Genet.* **55,** A225 (1994).
[59] K. J. Fisher, G. P. Gao, M. D. Weitzman, R. DeMatteo, J. F. Burda, and J. M. Wilson, *J. Virol.* **70,** 520 (1996).
[60] I. E. Alexander, D. W. Russell, and A. D. Miller, *J. Virol.* **68,** 8282 (1994).
[61] D. W. Russell, A. D. Miller, and I. E. Alexander, *Proc. Natl. Acad. Sci. U.S.A.* **91,** 8915 (1994).
[62] T. R. Flotte, S. A. Afione, and P. L. Zeitlin, *Am. J. Resp. Cell Mol. Biol.* **11,** 517 (1994).
[63] C. L. Halbert, I. E. Alexander, G. M. Wolgamot, and A. D. Miller, *J. Virol.* **69,** 1473 (1995).

Fig. 4, a high-expressing clone, such as A35 was found to have as much as a four-fold stimulation of efflux rate.

The complemented clones were studied in more detail by single-channel patch-clamp recordings[6] and by whole-cell current recordings.[7] In each of these studies, cells complemented with AAV–CFTR were found to have restoration of cAMP-mediated regulation of both the small 10-pS chloride conductance associated with recombinant CFTR expression and the previously described 40-pS outwardly rectifying chloride channel (ORCC).

AAV–CFTR Vectors for *in vivo* Gene Transfer

The ability of AAV–CFTR vectors to transfer the CFTR cDNA to bronchial epithelial cells *in situ* was tested in the New Zealand White rabbit model.[64] Although this animal does not have the CF defect, it is useful for testing the feasibility of CF gene therapy, since it is large enough to undergo fiberoptic bronchoscopy and because the pattern of native CFTR expression is known.[65] For these studies a version of the AAV–CFTR vector was used in which there was a unique epitope of 28 amino acids fused to the amino terminus of the vector-expressed version of CFTR. Vector-specific oligonucleotide primers from this sequence were used to detect AAV–CFTR DNA and RNA within the lungs of vector-treated animals, and a polyclonal chicken antibody raised to recognize this epitope was used to detect vector-derived protein. Doses of 10^{10} total AAV–CFTR particles were instilled by fiber optic bronchoscopy into the right lower lobe (RLL) of a series of animals. Animals were then sacrificed at timed intervals ranging from 3 days to 6 months after vector administration. As Fig. 5 shows, vector RNA expression was detectable from each of these time points, indicating that AAV–CFTR had transduced the lung tissue and that vector RNA expression was stable over time. Furthermore, there was no evidence of inflammation, neoplasia, or any other vector-related complications.

Similar studies have more recently been performed in rhesus macques.[66] In one study doses of AAV–CFTR ranging from 5×10^8 total particles up to 10^{11} total particles were instilled into the RLL as had been done in the rabbit study.[67] Once again, vector RNA expression was detectable for up to 6 months after vector administration, without any evidence of vector-

[64] T. R. Flotte, S. A. Afione, R. Solow, M. L. Drumm, D. Markakis, W. B. Guggino, P. L. Zeitlin, and B. J. Carter, *J. Biol. Chem.* **268,** 3781 (1993).

[65] S. A. McGrath, A. Basu, and P. L. Zeitlin, *Am. J. Resp. Cell Mol. Biol.* **8,** 201 (1993).

[66] T. R. Flotte, C. Conrad, T. Reynolds, S. Afione, R. Adams, S. Allen, W. B. Guggino, and B. J. Carter, *J. Cell Biochem.* **21A,** 364 (1995).

[67] C. Conrad, S. Afione, W. B. Guggino, B. J. Carter, and T. Flotte, *Am. J. Resp. Crit. Care Med.* **149,** A236 (1994).

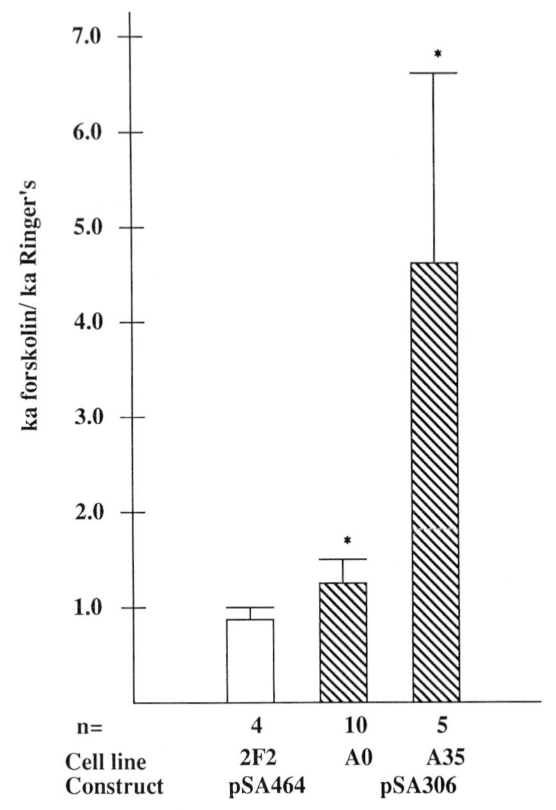

FIG. 4. AAV–CFTR complements defective cAMP-activation of chloride efflux in the CF bronchial epithelial cell line, IB3-1. To determine whether CFTR function was established by AAV–CFTR transduction, cultures of CF-defective cells were infected with packaged virions containing either the AAV–CFTR construct, pSA306, or a frameshift mutant version of AAV–CFTR, pSA464, at a multiplicity of 500 particles per cell. The first-order rate constant, k_a, of efflux of $^{36}Cl^-$ from each cell culture was determined under basal conditions (in a Ringer's isotonic saline solution) and in the presence of the adenylyl cyclase activator, forskolin. The ratio of the forskolin-stimulated rate divided by the basal rate (k_a forskolin/k_a Ringer's) serves as an index of CFTR function and is depicted on the *y* axis. A rate ratio of 1.0 indicates the absence of CFTR function, while cell lines that naturally express CFTR have rate ratios from 2.0 to 4.0. In the case of the AAV–CFTR (pSA306 construct) data are shown both from a pooled cell population (A0) and from a single clone (A35), which was found to express high levels of CFTR. A single clone (2F2) transduced with the mutant vector (pSA464) is shown as a control. (* indicates significant difference from control, $P < 0.05$.) [Reproduced with permission from T. R. Flotte, S. A. Afione, R. Solow, M. L. Drumm, D. Markakis, W. B. Guggino, P. L. Zeitlin, and B. J. Carter. *J. Biol. Chem.* **268,** 3781 (1993).]

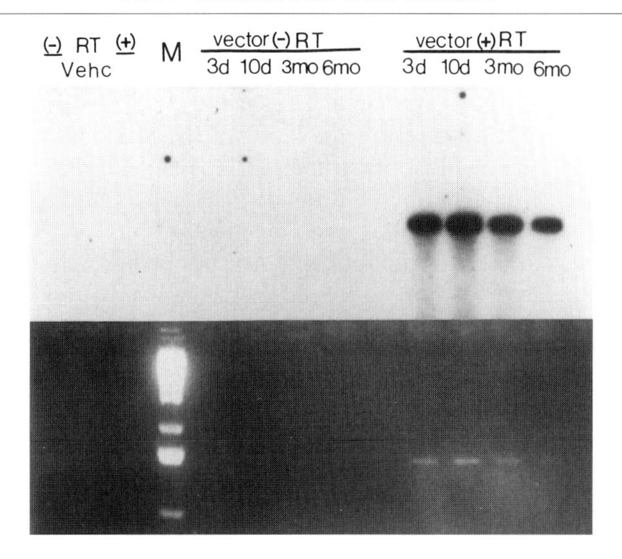

FIG. 5. RT-PCR indicates CFTR mRNA expression in rabbit lungs for 6 months after instillation of AAV–CFTR. RT-PCR was performed on DNase-treated lung RNA extracts harvested at 3 days, 10 days, 3 months, and 6 months from New Zealand White rabbits treated with 10^{10} total particles of AAV–CFTR via fiber optic bronchoscopy. PCR products were analyzed by ethidium bromide-stained agarose gel electrophoresis (*bottom*) and Southern blot analysis (*top*) with an internal CFTR-specific probe. The presence of the expected 0.9-kb fragment (see comparison with 1-kb marker, lane M) indicates that vector RNA expression is present. The absence of signal in samples handled without reverse transcriptase (−RT) indicates that signals are due to RNA expresssion rather than vector DNA contamination. The absence of signal with the vehicle control (Vehc) lung RNA extract indicates the vector specificity of the primers and probes. [Reproduced with permission from T. R. Flotte, S. A. Afione, C. K. Conrad, S. A. McGrath, R. Solow, H. Oka, P. L. Zeitlin, W. B. Guggino, and B. J. Carter. *Proc. Natl. Acad. Sci. U.S.A.* **90,** 10613 (1993).]

related inflammation or other toxicity. In a related study, the persistence and rescue of AAV–CFTR vector DNA *in vivo* was studied in the rhesus model. Vector DNA was readily detectable in double-stranded episomal form and it could be rescued from cells *in vitro* or *in vivo.*[68]

Clinical Trials of AAV–CFTR

Based on the lack of vector-related toxicity with AAV–CFTR, and its ability to accomplish persistent DNA transfer, two clinical trials of AAV–CFTR administration have been initiated. The first of these trials

[68] S. A. Afione, C. K. Conrad, W. G. Kearns, S. Chunduru, R. Adams, T. C. Reynolds, W. B. Guggino, G. R. Cutting, B. J. Carter, and T. R. Flotte, *J. Virol.* in press, 1996.

involves combined nasal and endobronchial administration of AAV–CFTR in a placebo-controlled, dose escalation, phase I trial that is blinded for the nasal administration component. The second study involves administration of AAV–CFTR to the epithelial lining of the maxillary sinus through an in-dwelling catheter. In both studies vector DNA transfer and RNA expression will be detected by PCR, while vector-mediated complementation of CFTR function will be assessed by *in vivo* transepithelial potential difference measurement. Attempts will also be made to characterize the state of vector DNA in cells recovered from the instillation site by cytologic brushing.

It is possible that the decreased efficiency of vector DNA integration in the absence of Rep or the relatively slow rate of conversion from ssDNA to dsDNA in the absence of rapid cell division could be impediments to successful CF gene therapy with AAV vectors. However, a reasonable expectation is that an improved understanding of the mechanisms of AAV vector persistence and expression will allow for modifications in vector design or delivery that could overcome these potential problems. If the safety profile of AAV-CFTR observed in animals holds true in humans, then further studies to enhance vector expression and persistence will clearly be warranted.

[54] Sulfonylurea Receptors and ATP-Sensitive Potassium Ion Channels

By LYDIA AGUILAR-BRYAN, JOHN P. CLEMENT IV, and DANIEL A. NELSON

Introduction

Glucose-stimulated insulin secretion from pancreatic beta (β) cells depends on the closure of ATP-sensitive K^+ channels (K_{ATP}).[1] These channels are regulated by changes in the ratio of [ADP] to [ATP] that result from glucose metabolism. They are pharmacologically regulated by sulfonylurea agents, used in the treatment of non-insulin-dependent diabetes (NIDDM), and diazoxide, used in the treatment of hypoglycemic states such as familial hyperinsulinism. Functional reconstitution of the K_{ATP} channel has been obtained by coexpression of the high affinity sulfonylurea receptor, SUR1,

[1] L. Aguilar-Bryan and J. Bryan, *Diabetes Rev.* **4**(3), 336 (1996) and S. Seino, N. Inagaki, N. Namba, and T. Gonoi, *Diabetes Rev.* **4**(2), 177 (1996).

METHODS IN ENZYMOLOGY, VOL. 292
0076-6879/98 $25.00

a member of the ATP-binding cassette (ABC) traffic ATPase superfamily with multiple transmembrane-spanning domains and two nucleotide-binding folds, with K_{IR} 6.2, a silent member of the inwardly rectifying family of K^+ channels.

The goal of this article is to detail the different biochemical and molecular steps used to identify, purify, and clone the sulfonylurea receptor. Methods are also provided for expression and partial purification of a histidine-tagged SUR1 and for reconstitution of the β-cell ATP-sensitive K^+ channel from SUR1 and K_{IR}6.2.

Materials

Radiochemicals and Chemicals

Unlabeled and ^{125}I-Labeled Iodoglyburide. ^{125}I-labeled iodoglyburide is prepared as follows and used for receptor identification and purification.

Chemicals. Salicylic acid, *p*-(β-aminoethyl)benzenesulfonamide, triethylamine, ethyl chloroformate, cyclohexyl isocyanate, and dry acetone are purchased from Aldrich (Milwaukee, WI).

Step 1. Synthesis of 4-[β-(2-Hydroxybenzenecarboxamido)ethyl]benzenesulfonamide. Salicylic acid (4.1 g or 0.031 mol) is dissolved in 50 ml of dry acetone; 4.3 ml (0.031 mol) of triethylamine is added and the solution cooled to −20°. To this solution 3 ml (0.031 mol) of ethyl chloroformate is added dropwise while stirring. The resulting mixed anhydride is held at −20° for 15 min with occasional stirring.

A suspension is prepared from 6 g (0.031 mol) of *p*-(β-aminoethyl)benzenesulfonamide, 4.3 ml of triethylamine, and 50 ml of dry acetone and cooled to −20°. This suspension is added rapidly, while stirring, to the mixed anhydride already prepared. The mixture is stirred for 2 hr at 4° and then at room temperature (23°) for 4 hr. The acetone is removed by vacuum distillation and the residue acidified with 0.2 *N* HCl. The resulting precipitate is collected by filtration, washed with H_2O, air dried, and recrystallized from ethanol.

Step 2. Synthesis of N-{4-[β-(2-Hydroxybenzenecarboxamido)ethyl]benzenesulfonyl}-N′-cyclohexylurea. Dissolve 12 g of 4[β-(2-hydroxybenzenecarboxamido)ethyl]benzenesulfonamide in 90 ml of acetone at 4°; 1.8 g of NaOH, dissolved in 10 ml of H_2O, is then added, with 5.75 ml of cyclohexyl isocyanate in four equal aliquots over a period of 80 min. After stirring overnight, 150 ml of H_2O is added, stirred for 10 min, then filtered to remove a small amount of precipitate. Acidification by the addition of 75 ml of 2 *N* HCl produces a precipitate that is collected by filtration after 60 min, air dried, and recrystallized from ethanol.

Step 3: Synthesis of N-{4-[β-(2-Hydroxy-5-iodobenzene carboxamido) ethyl]benzenesulfonyl}-N′-cyclohexylurea. Dissolve 0.5 g (1.13 mmol) of *N*-{4-[β-(2-hydroxybenzene carboxamido)ethyl]-benzenesulfonyl}-*N′*-cyclohexylurea and 0.26 g of NaI in 5 ml of dimethylformamide. Chloramine-T (0.39 g) dissolved in 2 ml of dimethylformamide is then added to start the reaction. After stirring for 1 hr at 23°, the reaction is terminated by addition of 25 ml of 0.1 *N* HCl. The iodinated compound (iodoglyburide) is extracted into ethyl acetate and washed sequentially with equal volumes of 50 m*M* sodium bisulfite, H_2O, 0.5 *M* NaCl (2×), and H_2O (2×). The volume of the ethyl acetate phase is reduced to approximately 5–10 ml and the iodinated product collected and recrystallized from ethyl acetate with an approximate yield of 70%.

Step 4: Radioiodination of N-{4-[β-(2-Hydroxybenzenecarboxamido) ethyl]benzenesulfonyl}-N′-cyclohexylurea. Dissolve 1 μl of a 10 m*M* solution of *N′*-{4-[β-(2-hydroxybenzenecarboxamido)ethyl]benzenesulfonyl}-*N′*-cyclohexylurea in dimethylformamide and add to 2–4 μl of Na ^{125}I (1–4 mCi) in a solution of dilute NaOH. Iodination is initiated by the addition of 1 μl of 10 m*M* chloramine-T dissolved in dimethylformamide. The reaction is terminated after 10 min at 23° by addition of 1 μl of 14 *M* 2-mercaptoethanol and 20 μl of 50% methanol in H_2O. This mixture is absorbed to a μBondapak C_{18} column (Waters/Millipore Bedford, MA) equilibrated with 50% methanol, and then separated by high-performance liquid chromatography (HPLC) using a 50–90% methanol gradient with a flow rate of 1 ml/min. A slightly concave gradient was used for increased resolution. The uniodinated and iodinated species are well resolved with retention times of 9.9 and 13.3 min, respectively. The fractions (0.5 ml) containing the ^{125}I-labeled iodoglyburide are pooled and stored at −20°. The concentration of the drug, based on the specific activity of the ^{125}I, is determined immediately after purification. The material gives a single spot on reversed-phase thin-layer chromatography when analyzed on Whatman $KC_{18}F$ plates using 80% methanol/20% 0.5 *M* NaCl as a solvent. The iodinated material cochromatographs with the unlabeled iodinated compound are synthesized as described in step 3. A flow diagram for the synthesis of iodoglyburide is shown in Fig. 1.

Preparation of Digitonin. Digitonin (Sigma, St. Louis, MO) is purchased as a powder and not purified further. For receptor solubilization, 20% digitonin (w/v) is prepared fresh each day by adding the detergent to deionized water, vortexing for a few seconds, and boiling it in a screw-top tube until clear, approximately 2 min. This clear product is suitable for addition to receptor products.

Other Reagents and Chemicals. All other reagents and chemicals are of the highest purity available.

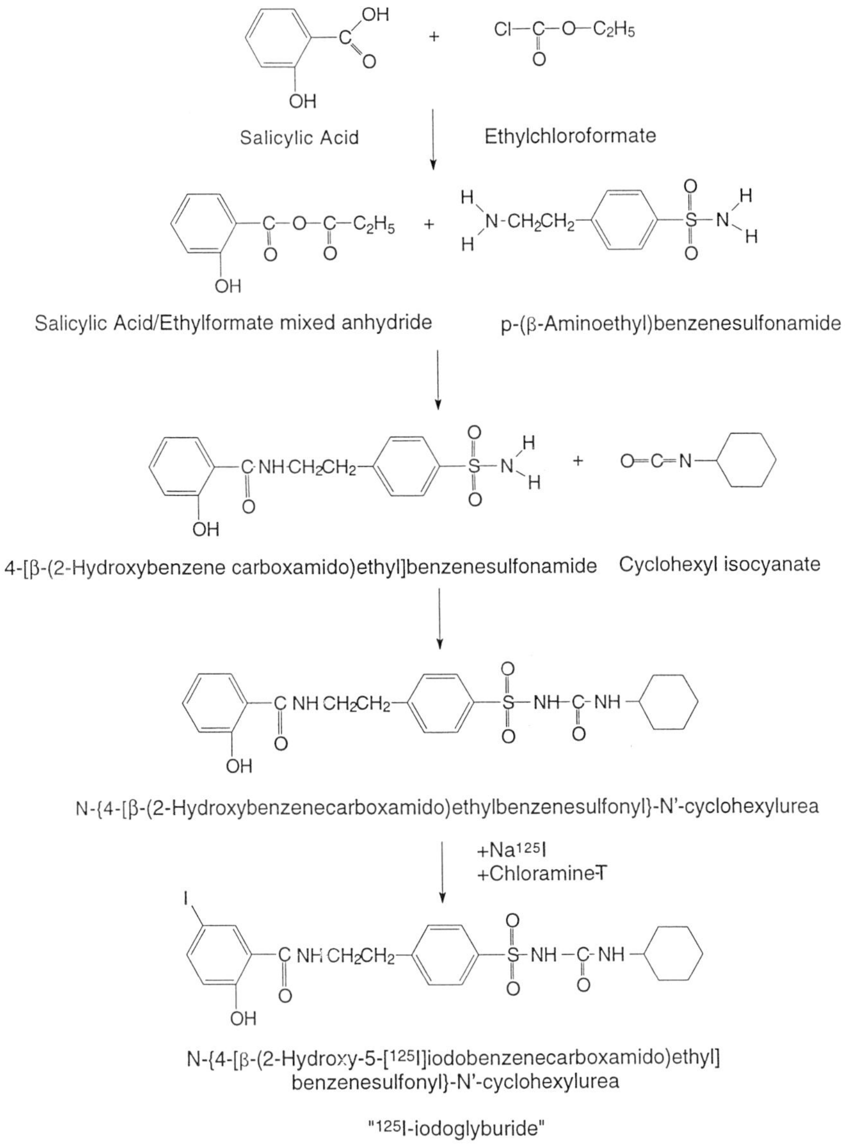

Salicylic Acid + Ethylchloroformate

↓

Salicylic Acid/Ethylformate mixed anhydride + p-(β-Aminoethyl)benzenesulfonamide

↓

4-[β-(2-Hydroxybenzene carboxamido)ethyl]benzenesulfonamide + Cyclohexyl isocyanate

↓

N-{4-[β-(2-Hydroxybenzenecarboxamido)ethylbenzenesulfonyl}-N'-cyclohexylurea

↓ $+Na^{125}I$ +Chloramine-T

N-{4-[β-(2-Hydroxy-5-[^{125}I]iodobenzenecarboxamido)ethyl] benzenesulfonyl}-N'-cyclohexylurea

"^{125}I-iodoglyburide"

FIG. 1. Flow diagram showing an overview of the synthesis of iodoglyburide.

Biologicals

Maintenance of Tissue Culture Cells. Tissue culture cells producing SUR1 include HIT-T15[2] (passage 65–75; CRL1777, American Type Culture Collection, Rockville, MD), RINm5f,[3] and αTC-6[4] cells. COSm6 cells do not produce SUR1 and are used for clonal receptor and channel expression. Cells are maintained in T-175 culture flasks (Falcon, Fisher Scientific, Pittsburgh, PA) as monolayers in Dulbecco's modified Eagle's medium (DMEM)-HG medium supplemented with 10% fetal bovine serum (FBS), 100 U/ml penicillin, and 0.1 mg/ml streptomycin (all purchased from GIBCO Grand Island, NY). Cells are grown in 5% (v/v) CO_2 at 37°, subcultured weekly, and fed three times a week. To subculture, confluent cells are detached with 0.05% trypsin/1 m*M* EDTA (GIBCO), resuspended in supplemented DMEM-HG medium and replated at one-tenth the original density. For expression experiments, COSm6 cells are plated at 50–60% confluence on appropriate size petri dishes 1 day prior to transient transfection (see section on expression in COSm6 cells for details).

Growth in Roller Bottles. For large-scale cell culture, cells are seeded in roller bottles (850 cm^2; Falcon #3007) at 50 × 10^6 cells/bottle in DMEM-HG medium (100 ml) containing 10% FBS. Cells are fed with 200 ml of medium plus serum four to five times over a period of 2 weeks, until the bottles are confluent. After plating and after each feeding, bottles are gassed under the hood with 5% CO_2 prior to capping. Cells are harvested after 2–2.5 weeks.

Isolation of Cells. After discarding the media, cells in confluent roller bottles are washed with 25 ml phosphate-buffered saline (PBS: 0.14 *M* NaCl, 3 m*M* KCl, 2 m*M* KH_2PO_4, 1 m*M* Na_2HPO_4, pH 6.8) and then incubated at room temperature with 25 ml of PBS plus 2 m*M* EDTA until cells detach from the sides of the bottles. Cells are pelleted at 900*g* for 10 min at 4° and washed once with PBS without EDTA prior to preparation of membranes.

Analytical Procedures and Assays

Protein Determinations

Protein concentrations are determined by the method of Bradford[5] using the Bio-Rad (Richmond, CA) protein assay. To reduce interference,

[2] L. Aguilar-Bryan, D. A. Nelson, Q. A. Vu, M. B. Humphrey, and A. E. Boyd III, *J. Biol. Chem.* **265,** 8218 (1990).

[3] D. A. Nelson, J. Bryan, S. Wechsler, J. P. Clement IV, and L. Aguilar-Bryan, *Biochemstry* **35,** 14793 (1996).

[4] A. S. Rajan, L. Aguilar-Bryan, D. A. Nelson, C. G. Nichols, S. W. Wechsler, J. Lechago, and J. Bryan, *J. Biol. Chem.* **268,** 15221 (1993).

[5] M. M. Bradford, *Anal. Biochem.* **22,** 248 (1976).

sample digitonin concentrations must be at or below 1% (w/v). Bovine serum albumin (BSA) is used as a control protein for standard curves, and control samples are measured in the same detergent concentrations as the sample.

Electrophoresis

Sample Preparation. SUR1 aggregates irreversibly if boiled in standard SDS–PAGE sample buffer.[6] Two simple methods are used to circumvent this problem: (1) addition of 2× pH 6.8 sample buffer to an equal volume of receptor sample, followed by incubation at room temperature for 2 min (no boiling) or (2) addition of 2× pH 9.0 sample buffer (no pH adjustment) to an equal volume of sample, followed by boiling for 2 min.

SDS–PAGE. Proteins are electrophoresed on SDS–polyacrylamide gels as described previously.[7] The slab gels consist of a pH 6.8 stacking gel and a separating gel ranging from 5.5 to 10% acrylamide (w/v). Gels are 17- × 17-cm in size. Electrophoresis is at room temperature at constant voltage (130–150 V) until the moving boundary reaches the bottom of the gel.

Staining and Autoradiography. The gels are stained with Coomassie Brilliant Blue R, destained (5/1/4: H_2O/acetic acid/methanol) and vacuum dried at 80°. Dried gels are taped into Kodak cassettes, overlayed with Kodak (Rochester, NY) XAR-5 X-ray film, and placed at −80° for 1–2 days. Film is developed in an X-ray film processor.

Molecular Weight Standards. High molecular weight protein standards (29,000 to 200,000 g/mol) for SDS gel electrophoresis are purchased from Sigma.

Receptor Identification

Photoaffinity Labeling of Proteins in Membrane Preparations. ^{125}I-Labeled iodoglyburide is added to samples (typically to a concentration of 5–10 nM) and the samples incubated at 23° for 30 min. Sample concentration is at 5–10 mg protein/ml. Samples are then transferred onto Parafilm and irradiated at 312 nm in a UV cross-linker (Fisher). Figure 2 illustrates the photolabeling of receptor forms in RINm5f cell membranes with increasing energy (time) of photolabeling. In typical experiments, 1.0–1.5 J/cm^2 is used.

Whole-Cell Photolabeling. The receptor may be photolabeled and identified in whole cells. Attached, confluent cells in petri dishes are incubated at 23° in the dark for 30 min with 1–10 nM ^{125}I-labeled iodoglyburide in PBS supplemented with 0.9 mM $CaCl_2$, 0.5 mM $MgCl_2$, and 10 mM glucose.

[6] D. A. Nelson, L. Aguilar-Bryan, and J. Bryan, *J. Biol. Chem.* **267,** 14928 (1992).
[7] D. E. Garfin, *Methods Enzymol.* **182,** 425 (1990).

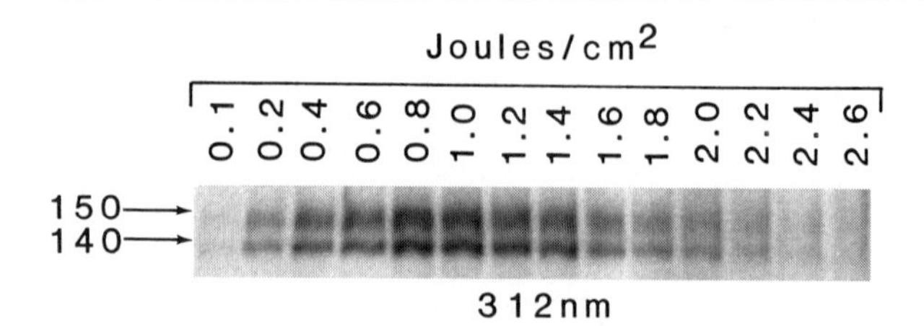

FIG. 2. Photocoupling of RINm5f cell receptors to ^{125}I-labeled iodoglyburide. RINm5f cell membranes at 10 mg/ml were photolabeled at 0.1–2.6 J/cm^2, 2× SDS gel sample buffer (pH 9.0) added, the samples boiled and electrophoresed on a 6% gel. The autoradiogram shows only the 140- to 150-kDa region of the gel.

Just enough of the solution is added to cover the bottom of the petri dish. Cells are irradiated as described for membranes and rinsed three times with PBS prior to further manipulations.

Photolabeling Specificity. Receptor characterization and purification may be confounded by the presence of low-affinity sulfonylurea binding proteins in whole cells and cell membrane preparations. High-affinity receptor proteins may be identified by displacement of ^{125}I-labeled iodoglyburide with glyburide, or unlabeled iodoglyburide prior to photolabeling.[2,6] For example, HIT-T15 cell membrane protein (100 μg/ml) is incubated with 5 n*M* radiolabeled ligand and displaced with 0–500 μM glyburide (Fig. 3). The autoradiogram visually shows that in HIT-T15 cells, only the SUR1 (140 kDa) binds iodoglyburide with high affinity (~7 n*M*).

Purification of Receptor

140- and 150-kDa Proteins. In some cell lines, two forms of SUR1 are observed. The receptors are N-linked glycoproteins that appear to have the same polypeptide chain, but differ in the extent of glycosylation.[3] Lectin affinity chromatography is the key step in purifying each of these proteins.[3,8]

Noted that 140 and 150 kDa refer to the molecular masses of SUR1 forms as deduced by SDS–PAGE using the high molecular weight protein markers as standards. Relative to these standards, the mobility of the receptor is anomalous, since the polypeptide for each receptor form is known to have a calculated mass of 177 kDa.[8]

Step 1: Preparation and Storage of Membranes. All steps are carried out at 0–4°. Cell pellets are resuspended in 5 m*M* Tris, 2 m*M* EDTA, 0.1 m*M* phenylmethylsulfonyl fluoride (PMSF), pH 7.4, using approxi-

[8] L. Aguilar-Bryan, C. G. Nichols, S. W. Wechsler, J. P. Clement IV, A. E. Boyd III, G. Gonzalez, H. Herrera-Sosa, K. Nguy, J. Bryan, and D. A. Nelson, *Science* **268,** 423 (1995).

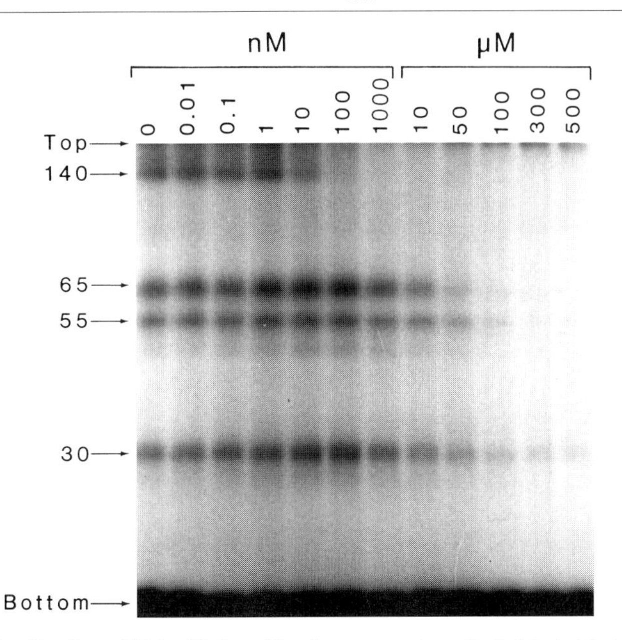

FIG. 3. Identification of high-affinity sulfonylurea receptors. Radiolabeled iodoglyburide was displaced from sulfonylurea binding proteins with increasing concentrations of glyburide. After photolabeling, samples were concentrated by ultracentrifugation for 1 hr at 100,000*g*, and resuspended in SDS sample buffer (pH 9.0). Aliquots were electrophoresed on an SDS–10% acrylamide gel and an autoradiogram prepared.

mately 5 ml of buffer for each roller bottle. Cells are placed on ice for 40 min to allow swelling and then homogenized with 10 strokes of a motorized glass–Teflon homogenizer (500 rpm). The homogenate is centrifuged at 1000*g* for 10 min to remove nuclei and cellular debris, and the supernatant transferred to 30-ml Beckman polycarbonate, screw-cap ultracentrifuge tubes (Beckman, Palo Alto, CA). Supernatants are centrifuged at 100,000*g* for 60 min in a Beckman 60 Ti rotor. The pelleted crude membrane preparation is resuspended in membrane storage buffer [10 m*M* Tris, 100 m*M* NaCl, 2 m*M* EDTA, 20% (v/v) glycerol, 0.1 m*M* PMSF, pH 7.4] and aliquots stored at −80° (5–10 mg protein/ml).

Step 2: Digitonin Solubilization. Membrane isolation is followed by photolabeling approximately 20–25 mg of membrane proteins with ^{125}I-labeled iodoglyburide, as above. The labeled membranes are mixed with approximately 200–300 mg of unlabeled membrane protein and digitonin added to 1%. All subsequent steps are performed at room temperature in the presence of a cocktail of protease inhibitors (0.1 m*M* PMSF, 0.1 m*M* phenanthroline, and 0.1 m*M* iodoacetamide). Membranes are solubilized for 15 min, then ultracentrifuged for 1 hr at 100,000*g*.

Step 3: Lectin Affinity Chromatography. The supernatant is divided into 4-ml aliquots and each cycled twice over 1 ml concanavalin A (Con A)-Sepharose (Sigma) columns equilibrated with 25 m*M* Tris (pH 7.5), 0.1 *M* NaCl, 2 m*M* EDTA, and 1% digitonin (w/v). Columns are washed with 8 ml of the equilibrating buffer and eluted with 4 ml of the equilibrating buffer containing 0.5 *M* methyl-α-D-mannopyranoside. The eluted protein is stored at this stage at $-80°$.

Wheat germ agglutinin (WGA; Sigma)-Sepharose is used instead of concanavalin A-Sepharose for the purification of the 150-kDa receptor. The receptor is eluted with 0.3 *M* *N*-acetylglucosamine. All other manipulations are as described for the purification of the 140-kDa protein.

Step 4: Reactive Green-19 Affinity Chromatography. The eluates from three lectin columns are combined and cycled twice over a 1-ml column of Reactive Green 19-agarose (Sigma) equilibrated with 50 m*M* HEPES (pH 8.5), 2 m*M* EDTA, and 0.2% digitonin. After washing with 8 ml of the equilibrating buffer, and 8 ml of the equilibrating buffer plus 0.4 *M* NaCl, the protein is eluted with 4 ml of 1.5 *M* NaCl in the equilibrating buffer.

Step 5: Phenylboronate-10 Affinity Chromatography. Two eluates from the Reactive Green-19 column are pooled, diluted 1 : 1 with the HEPES equilibrating buffer (equilibrating buffer as in step 4) to reduce the ionic strength, then cycled twice over a 1 ml phenylboronate-10 Sepharose (Amicon, Danvers, MA) column. The phenylboronate column is washed with 8 ml of the HEPES buffer, followed by 2 ml of 0.1 *M* Tris (pH 7.5), 2 m*M* EDTA, and 0.1% digitonin. Protein is eluted with 4 ml of 0.1 *M* Tris (pH 7.5), 2 m*M* EDTA, and 0.1% SDS.

Step 6: Concentration of Receptor. Pooled samples from two phenylboronate-10 columns are concentrated by centrifugation (3000*g*) to 0.5 ml using Amicon 100,000 molecular weight cutoff filters. Filters are pretreated overnight at 4° with 5% Tween 20 (v/v) to prevent loss of protein.

Step 7: Preparative Electrophoresis and Electroelution. After addition of 2× sample buffer, the concentrated protein from Step 6 is loaded onto a single 5-cm-wide lane of a 5.5% polyacrylamide SDS gel, separated by gel electrophoresis, stained with Coomassie blue, and destained. The receptor band is excised with a razor blade, electroeluted into a 14,000 molecular weight cutoff dialysis bag, and concentrated by Amicon filtration. A typical yield is 8–16 pmol (1–2 μg of purified receptor) as indicated in Table I for purification of the 140-kDa HIT-T15 cell protein.

Histidine-Tagged Receptor. Sulfonylurea receptors with a His tag at the N terminus have been engineered, greatly facilitating purification.[3] For details, see the section on expression of COSm6 cells.

TABLE I
PURIFICATION OF 140-kDa SUR1 FROM HIT-T15 CELLS

Step	Total volume (ml)	Total protein (mg)	Receptor[a] (pmol)	Receptor[a] (pmol/mg)	Purification (-fold)	Yield (%)
Crude membrane	90	200	320	1.6[b]	1	100
Supernatant	90	150	240	1.6	1	75
Con A-Sepharose	48	10.2	80	7.8	4.9	25
Green 19-agarose	16	1.80	56	31.1	19.5	18
Phenylboronate-agarose	4	0.56	45	80.4	50.4	14
SDS–PAGE/Electroelute	0.2	~0.002	11	5600	3500	3

[a] The amount of receptor, yields, and purification (fold) after each step are based on the cpm in the 140-kDa band after electrophoresis (the band was excised and counted in a gamma counter), relative to the amount of protein from that sample loaded on a gel lane (as determined using the Bio-Rad protein assay).

[b] The HIT cell membrane starting material contains approximately 1.6 pmol of receptor per mg of membrane protein [L. Aguilar-Bryan, C. G. Nichols, A. S. Rajan, C. Parker, and J. Bryan, *J. Biol. Chem.* **267,** 14934 (1992)].

Receptor Assays

Deglycosylation Assay. Purified, radiolabeled 140-kDa receptor is prepared in 1% *n*-octylglucoside and incubated with endoglycosidase F/N–glycosidase F (Boehringer-Mannheim; 6×10^{-3} units/μl) at 37° for 1 hr for deglycosylation.

Proteolytic Degradation Assay. Native and deglycosylated receptors are incubated with V8 protease (Sigma; 0.1 mg/ml) at 37° for 30 min. Deglycosylated 140- and 150-kDa receptors yield diagnostic, radiolabeled N-terminal degradation products with putative molecular masses of 63 and 43 kDa, as shown in Fig. 4A.

Immunoaffinity Assay. Polyclonal antibodies produced against SUR1 immunoprecipitate both receptor forms, as shown in Fig. 4B. For these experiments, membranes containing photolabeled receptor are solubilized with 1% digitonin and centrifuged at 100,000*g* for 1 hr at 4°. Soluble protein (0.5 ml) is incubated for 1 hr at room temperature with 25 μl of immune serum. Protein A-Sepharose (Sigma) is added and after a further 1-hr incubation the beads washed three times with 25 m*M* Tris (pH 7.5), 0.1 *M* NaCl, 2 m*M* EDTA, and 0.1% digitonin, then heated to 90° for 5 min in the presence of pH 9.0 sample buffer. Samples are electrophoresed as described.

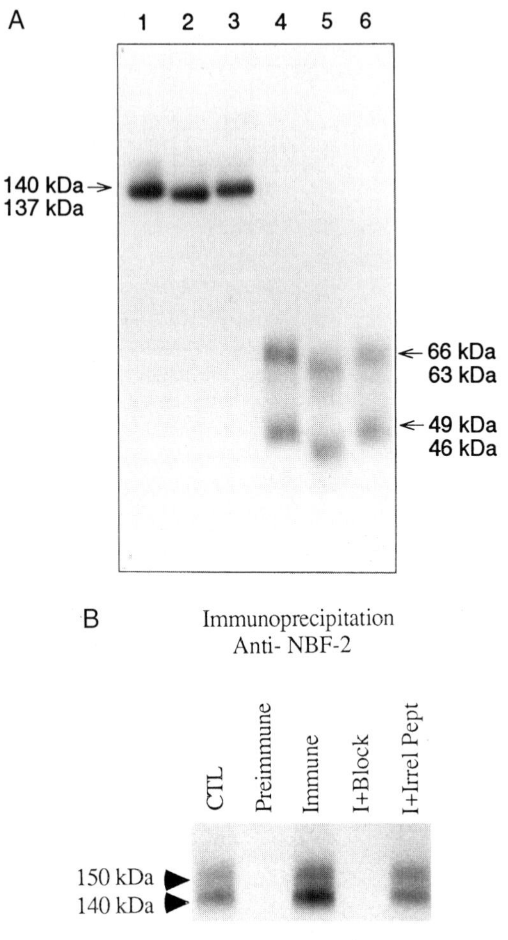

FIG. 4. (A) V8 and endoglycosidase cleavage of the HIT-T15 140-kDa SUR1. The radiolabeled 140-kDa receptor (lanes 1 and 3) was cleaved with endoglycosidase F/N–glycosidase F, increasing the mobility of the protein by approximately 3 kDa (lane 2). Partial V8 protease digestion (lanes 4 and 6) yielded radiolabeled fragments (66 and 49 kDa) that increase mobility with endoglycosidase treatment (lane 5). (B) Immunoprecipitation of the 140- and 150-kDa SUR1 forms from RINm5f cells using antibodies directed against the second nucleotide-binding domain (NBF-2). CTL, radiolabeled 140- and 150-kDa receptors from solubilized RINm5f cell membranes. Immunoprecipitation was with preimmune serum, immune serum, immune serum + NBF-2 fusion protein (I + Block), and immune serum + irrelevant peptide (I + Irrel Pept).

SUR1 and K_{ATP} Expression and Isolation from COSm6 Cells

Generation of Functional Histidine Tagged SUR1. To generate pECE-$SUR1_{N\text{-}6X\text{-}HIS}$, the 3′ *Eco*RI and *Sac*I sites of the hamster SUR1 cDNA (pECE-haSUR1)[9] are deleted by two rounds of partial cleavage, fill-in with Klenow, and religation. PCR primers are designed to amplify a specific fragment to introduce a histidine tag. The forward primer reads 5′-GTC AGA ATT CGC CGC CAT GCA TCA CCA TCA CCA TCA CAT GCC CTT GGC CTT CTG CG-3′ and contains an internal *Eco*RI site, a Kozac[10] consensus sequence, a start codon, and codons encoding six tandem histidine residues followed by 5′ SUR1 cDNA sequence. The reverse primer reads 5′-GCT GTG GTG GAT GTG CAC C-3′. The two primers amplify an approximately 250-bp sequence when pECE-SUR1 is used as a template. The 250-bp product was cleaved with *Eco*RI and *Sac*I and ligated into identically digested pECE-haSUR1.

Cell Culture and Transfections. COSm6 cells are plated at a density of 1×10^6 per dish (150-mm diameter) and cultured in DMEM-HG (high glucose) supplemented with 10% fetal calf serum. Transfections are done as follows. For $^{86}Rb^+$ efflux studies, 3-day-old cultures of COSm6 cells are tripsinized and replated at a density of 2.0×10^5 cells per 35-mm well (6-well dish) and allowed to attach overnight. Typically, 5 μg of a SUR1 plasmid is mixed with 5 μg of a K_{IR} plasmid and brought up to 7.5 μl final volume in TBS (8 g/liter NaCl; 0.38 g/liter KCl; 0.2 g/liter Na_2HPO_4; 3.0 g/liter Tris base; 0.15 g/liter $CaCl_2$; 0.1 g/liter $MgCl_2$, pH 7.5) before addition of DEAE-dextran (30 μl of a 5 mg/ml solution in TBS). The samples are vortexed, collected by briefly spinning in a microfuge, then incubated for 15 min at room temp before addition of 500 μl 10% NuSerum (Collaborative Research Incorporated, Bedford, MA) in TBS. Cells are washed twice with Hanks' balanced salt solution (HBSS), the DNA mix is added, and the cells are maintained in a 37° CO_2 incubator. After 4 hr the DNA mix is decanted and the cells shocked for 2 min in 1 ml HBSS + 10% dimethyl sulfoxide (DMSO), then placed in 1.5 ml (v/v) of DMEM-HG + 2% FBS + 10 μM chloroquine and kept in a 37° CO_2 incubator. After 4 hr, the cells are washed twice with HBSS and incubated in normal growth media until assayed (usually 36–48 hr posttransfection). SUR1 is photolabeled in whole cells and in membrane.[8,9]

Rubidium Efflux Assays. At 24–36 hr posttransfection, cells are placed in fresh media containing approximately 1 μCi/ml ^{86}RbCl for 12–24 hr and assayed as follows: Cells are incubated for 30 min at 25° in Krebs–Ringer

[9] N. Inagaki, T. Gonoi, J. P. Clement IV, N. Namba, J. Inazawa, G. Gonzalez, L. Aguilar-Bryan, S. Seino, and J. Bryan, *Science* **17,** 1166 (1995).
[10] M. Kozak, *Nucl. Acid Res.* **15,** 8125 (1987).

solution under one of three conditions: no additions (basal), with oligomycin (2.5 μg/ml) and 2-dexoy-D-glucose (1 mM) (poisoned) or with oligomycin and deoxyglucose plus 1 μM glyburide (K_{ATP} inhibited). Cells are washed once in $^{86}Rb^{+}$-free Krebs–Ringer solution, with or without the added inhibitors, then time points are taken by removing all the medium from the cells and replacing it with fresh medium at the indicated times. Equal portions of the medium from each time point are counted, and the values are summed to determine flux.[1]

Membrane Preparation. COSm6 cells are transfected for membrane preparations as described earlier with the following modifications: Cells are plated at a density of 1.5×10^6 cells per 100-mm dish. K_{IR} plasmid (100 μg) is added to 100 μg of SUR1 plasmid and the mixture used to transfect ten 100-mm plates. The DNA mix is brought up to 500 μl with TBS and 2 ml of DEAE-dextran (5 mg/ml) is added. The mixture is vortexed and allowed to incubate for 30 min at room temp. Twenty-eight milliliters of 10% NuSerum in TBS is added to the DNA mix, and 3 ml of this mixture is added to each 100 mm plate for 4 hr. The transfections proceed as above except that 5 ml is the operating volume for each 100-mm plate. Membranes are prepared as described 60–72 hr posttransfection from ten to twenty 100-mm dishes.

Histidine-Tagged SUR1/Copurification of Kir6.2

Chromatography. $SUR1_{N\text{-}6X\text{-}HIS}/K_{IR}6.x$ complexes are partially purified by chromatography on a 500-μl column of Ni^{2+}-agarose (Qiagen, Valencia, CA) equilibrated in solubilization buffer (1.0% digitonin, 150 mM NaCl, 25 mM Tris, pH 7.4) plus 4 mM imidazole, pH 7.4. Approximately 150 μg of membranes prepared from COSm6 cells transfected with the appropriate inward rectifier and SUR1 plasmids are photolabeled with 10 nM ^{125}I-labeled azidoglibenclamide, pH 6.5 (gift of Dr. M. Schwanstecher[11] Institute Für Pharmakdogie und Toxikologie, Universität Braunschweig, Germany). The labeled membranes are then solubilized in 200 μl of solubilization buffer for 30 min on ice, then spun for 30 min at 100,000g (Model Tl-100 Beckman Instruments) to remove insoluble material, and passed over the Ni^{2+} column four times. The column is washed with 20 ml (40 times column volume) of 0.2% digitonin wash buffer, then eluted with wash buffer containing 100 mM imidazole, pH 7.4.

Complexes are separated on wheat germ agglutinin and concanavalin A-Sepharose as described by Nelson *et al.*[3]

[11] M. Schwanstecher, S. Loser, F. Chudziak, and U. Panten, *J. Biol. Chem.* **269,** 17768 (1994).

[55] Peptide Transport Assay for TAP Function

By YE WANG, DAVID S. GUTTOH, and MATTHEW J. ANDROLEWICZ

Introduction

Antigenic peptides are transported from the cell cytosol into the endoplasmic reticulum (ER) by the transporters associated with antigen processing (TAP molecules).[1-3] The TAP molecules provide peptides for the major histocompatibility complex (MHC) class I antigen presentation pathway.[4,5] The peptides transported by TAP are ultimately displayed on the cell surface in association with class I molecules, where they are recognized by effector cells of the immune system. A supply of peptides appears to be critical for class I expression, because cells that contain mutant or deleted TAP genes have an impaired ability to express class I molecules.[6,7] TAP proteins are members of the ATP-binding cassette (ABC) transporter superfamily. However, unlike many of its eukaryotic counterparts, TAP is not expressed as a single polypeptide, but as a heterodimer of two homologous subunits termed TAP.1 and TAP.2. The TAP heterodimer maintains the structural features that are hallmarks of ABC transporters, that is, a hydrophobic domain that possesses approximately 12 membrane-spanning regions and two hydrophilic ATP-binding domains.

To facilitate the study of TAP function, two types of peptide transport assays were developed.[8-10] One involves the use of streptolysin O (SLO)-permeabilized cells to gain access to internal cell membranes,[8,9] such as the

[1] J. Trowsdale, I. Hanson, I. Mockridge, S. Beck, A. Townsend, and A. Kelly, *Nature* **348,** 741 (1990).

[2] T. Spies, M. Bresnahan, S. Bahram, D. Arnold, G. Blanck, E. Mellins, D. Pious, and R. DeMars, *Nature* **348,** 744 (1990).

[3] J. J. Monaco, S. Cho, and M. Attaya, *Science* **250,** 1723 (1990).

[4] T. Spies, V. Cerundolo, M. Colonna, P. Cresswell, A. Townsend, and R. DeMars, *Nature* **355,** 644 (1992).

[5] D. Arnold, J. Driscoll, M. Androlewicz, E. Hughes, P. Cresswell, and T. Spies, *Nature* **360,** 171 (1992).

[6] R. D. Salter and P. Cresswell, *EMBO J.* **5,** 943 (1986).

[7] H.-G. Ljunggren, S. Paabo, M. Cochet, G. Kling, P. Kourilsky, and K. Karre, *J. Immunol.* **142,** 2911 (1989).

[8] J. J. Neefjes, F. Momburg, and G. J. Hammerling, *Science* **261,** 769 (1993).

[9] M. J. Androlewicz, K. S. Anderson, and P. Cresswell, *Proc. Natl. Acad. Sci. U.S.A.* **90,** 9130 (1993).

[10] J. C. Shepherd, T. N. M. Schumacher, P. G. Ashton-Rickardt, S. Imaeda, H. L. Ploegh, C. A. Janeway, Jr., and S. Tonegawa, *Cell* **74,** 577 (1993).

0076-6879/98 $25.00

ER, and the other involves the use of microsome preparations.[10] This article focuses on the peptide transport assay that utilizes SLO-permeabilized cells. However, many of the peptides and reagents described here can also be used for the microsomal peptide transport assay as well. The general scheme of the transport assay is to introduce a radiolabeled peptide, which contains an N-linked glycosylation site, into the cytosolic compartment of permeabilized cells, and measure its TAP-dependent transport into the ER by measuring the amount of peptide glycosylation.

Materials

Streptolysin O (SLO) (Murex, Norcross GA) 40 U
Concanavalin A-Sepharose 4B (Sigma, St. Louis, MO)
Dithiothreitol (DTT) (Sigma), minimum 99%
Bovine Serum Albumin (BSA) (Sigma), fraction V, 96–99%

Buffer

Intracellular Transport Buffer (ICT)[11]: 50 m*M* HEPES–KOH, pH 7.0, 78 m*M* KCl, 4 m*M* $MgCl_2$, 8.37 m*M* $CaCl_2$, 10 m*M* EGTA

Methods

Preparation of Streptolysin O Stock Solution

Rehydrate the lyophilized powder from one vial of SLO with 2 ml of distilled water. This results in a stock solution of 20 U/ml. Store aliquots of the SLO solution at −70°.

Preactivation of Streptolysin O

The SLO preparation must be preactivated before use by incubating in the presence of 4 m*M* DTT for 10 min at 37° (water bath). Prepare a 100× stock solution of 0.4 *M* DTT in water, and store aliquots at −20°.

Incubation of Cells with Streptolysin O

Cells are harvested and washed once in serum-free medium [the medium used to culture the cells; in this case Iscove's modified Dulbecco's medium (IMDM, GIBCO, Grand Island, NY)]. Cells that are adherent should be

[11] B. Podbilewicz and I. Mellman, *EMBO J.* **9,** 3477 (1990).

trypsinized by standard procedures[12] prior to permeabilization. The washed cells are resuspended in fresh serum-free medium at a concentration of 5–10 million cells per milliliter, and divided into 1-ml portions (use 1.5-ml microfuge tubes). With the cell suspensions kept on ice, add 50 μl of the preactivated SLO solution to give a final concentration of 1 U/ml SLO. Invert the tubes twice to mix the SLO (do not vortex cells), and incubate the samples on ice for 10 min. Incubation on ice restricts the binding of SLO to the cell surface only, and thus internal membranes are protected from permeabilization. The cells, containing bound SLO, are then pelleted by brief centrifugation (15 sec) at 1300*g* in a microfuge. The supernatants are removed, and the cells washed twice with 1 ml cold serum-free medium to remove the unbound SLO (use the same centrifugation conditions as above). The centrifugation of cells can be performed at room temperature as long as they are kept on ice at all other times during the washing period. (*Note:* Approximately 80% of the cells are permeabilized under the above conditions, as measured by the uptake of trypan blue, for the majority of cell types. However, each cell type used for SLO permeabilization studies should be tested for permeabilization prior to the start of experiments.) The steps involved in SLO permeabilization of cells are shown schematically in Fig. 1.

Cell Permeabilization and Peptide Transport

The washed cell pellets are removed from the ice, and 1 ml of warm (37°) ICT buffer containing 1 m*M* DTT and 4 mg/ml BSA is added to each tube. Immediately after the addition of ICT/DTT/BSA, approximately 10 μl of radiolabeled reporter peptide stock (50 n*M* final concentration) is added to each tube, the tubes inverted once, and incubated at 37° (water bath) for 5 min, to allow for cell permeabilization and peptide transport. (Reporter peptides are iodinated by standard procedures using chloramine-T[13] to a specific activity of approximately 60–90 cpm/fmol.) After incubation, the cells are again pelleted by brief centrifugation at 1300*g*, and the supernant is removed and discarded. The cell pellets are washed once in 1 ml of ICT buffer (room temperature) containing DTT and BSA as described earlier.

Cell Solubilization and Precipitation of Reporter Peptide

Standard procedures for cell solubilization in detergent are used at this point.[14] Typically, 5–10 million permeabilized cells are extracted in 1 ml of

[12] J. A. McAteer and J. Davis, *in* "Basic Cell Culture: A Practical Approach" (J. A. Davis, ed.), p. 93. Oxford University Press, New York, 1994.

[13] P. A. Roche and P. Cresswell, *J. Immunol.* **144,** 1849 (1990).

[14] C. E. Machamer and P. Cresswell, *Proc. Natl. Acad. Sci. U.S.A.* **81,** 1287 (1984).

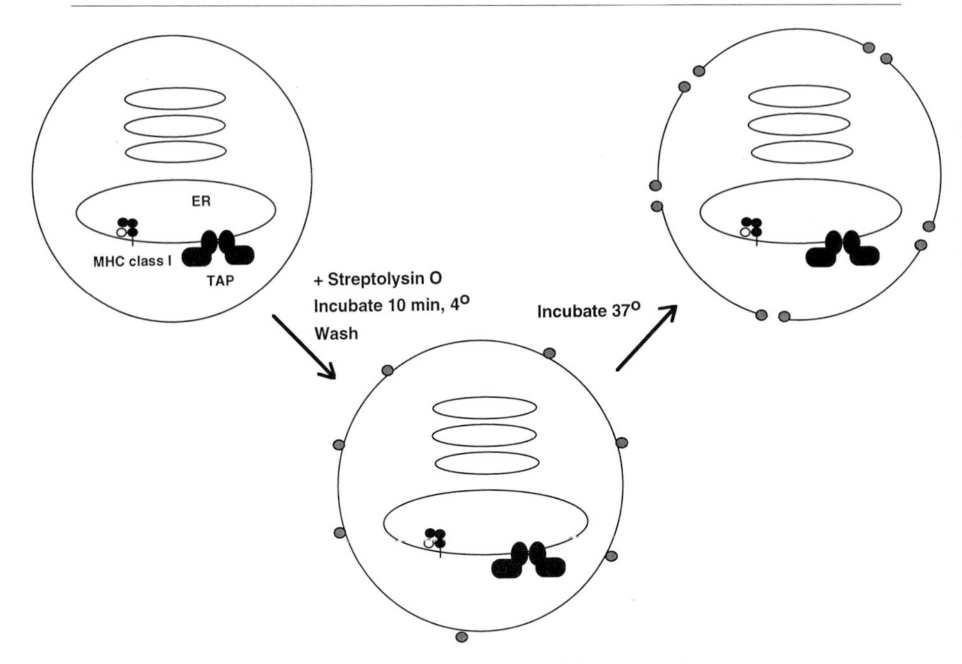

FIG. 1. Permeabilization of cells with streptolysin O.

cold (4°) buffer (10 m*M* Tris-HCl, pH 7.4, 150 m*M* NaCl) containing 1% Triton X-100, 0.5 m*M* phenylmethylsulfonyl fluoride (PMSF), and 5 m*M* iodoacetamide. After extraction and clarification (centrifugation at 15,000*g*), the glycosylated reporter peptide is precipitated from the extracts via the addition of concanavalin A (Con A)-Sepharose (25 μl, 50% suspension). The extracts are incubated with Con A-Sepharose for approximately 1 hr at 4° (with rotation). This is followed by washing the Sepharose beads, which contain the bound glycopeptide, three times with 1 ml of buffer (10 m*M* Tris-HCL, pH 7.4, 150 m*M* NaCl) containing 0.5% Triton X-100. Finally, the washed beads are counted in a gamma counter.

Summary of Steps in Peptide Transport Assay

1. Harvest cells and wash in cold serum-free medium. Resuspend in fresh serum-free medium to give a concentration of 5–10 million cells per milliliter.
2. Divide cells into 1-ml aliquots and keep on ice. Add 50 μl of preactivated SLO (20 U/ml stock) per milliliter of cells, mix by inversion, and incubate on ice for 10 min.
3. Wash SLO-treated cells twice, by brief centrifugation (15 sec) at 1300*g*, with 1 ml of cold (4°) serum-free medium.

4. Resuspend each cell pellet in 1 ml of warm (37°) ICT buffer containing 4 mg/ml BSA and 1 m*M* DTT. Immediately add the radiolabeled reporter peptide, mix by inversion, and incubate the samples in a 37° water bath for 5 min.
5. Pellet cells by brief centrifugation at 1300*g*, and remove the supernatant. Wash cells once with 1 ml of ICT (room temperature) containing DTT and BSA, as described earlier.
6. Extract cell pellets, containing the glycosylated reporter peptide, in detergent, by conventional methods. Precipitate the reporter peptide from the extracts by the addition of Con A-Sepharose. Measure peptide transport by the level of counts associated with the Sepharose beads.

Results

Peptide Transport in Permeabilized Human Melanoma Cell Line

To illustrate the peptide transport assay, we performed a set of experiments utilizing the human melanoma cell line WM98.1[15] (kindly provided by Dr. Meenhard Herlyn, Wistar Institute, Philadelphia, PA). The WM98.1 line was previously shown to possess a relatively high level of TAP-dependent peptide transport (M. Androlewicz, unpublished observations). The results of peptide transport experiments performed on permeabilized and unpermeabilized WM98.1 cells are shown in Fig. 2. The results indicated that the B27#3* reporter peptide (see Table I) is readily taken up in permeabilized cells in an SLO and TAP-dependent manner. In permeabilized cells that were pretreated with apyrase (an ATPase), peptide transport was severely reduced, indicating that the transport is most likely through TAP due to the dependency on ATP. Unpermeabilized cells gave only background levels of counts, indicating that permeabilization with SLO is critical in order to achieve peptide transport.

Peptide Competition Experiments

One utility of the peptide transport assay is that the relative transport capacities of a panel of unlabeled peptides can be determined through competition for reporter peptide transport. In this type of assay different amounts of unlabeled peptide are added to permeabilized cells before the addition of the labeled reporter peptide. Test peptides that are transported

[15] M. Herlyn, W. H. Clark, Jr., M. J. Mastrangelo, D. Guerry IV, D. E. Elder, D. LaRossa, R. Hamilton, E. Bondi, R. Tuthill, Z. Steplewski, and H. Koprowski, *Cancer Res.* **40,** 3602 (1980).

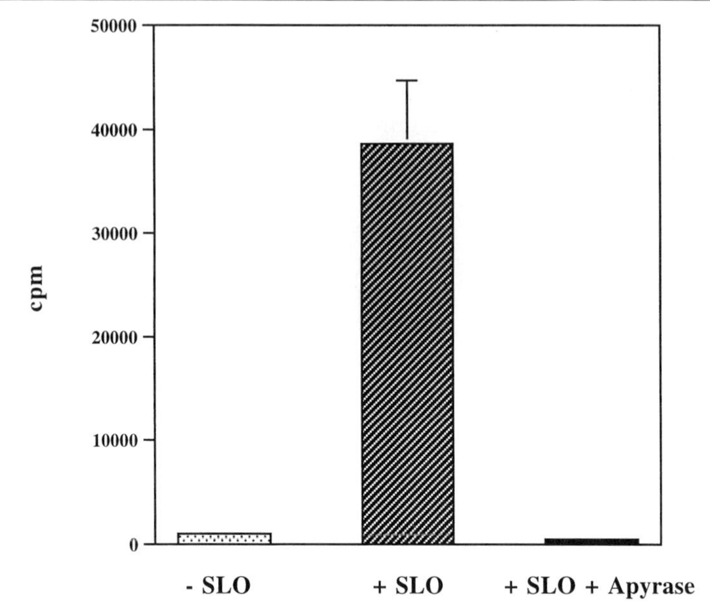

FIG. 2. ^{125}I-Labeled B27#3* transport and glycosylation in permeabilized WM98.1 cells. WM98.1 cells were treated with SLO and incubated with the ^{125}I-labeled B27#3* reporter peptide as described in the text. Cells that were not treated with SLO served as a negative control for peptide transport. Transported (glycosylated) peptide was precipitated from cellular extracts with Con A-Sepharose. The cpm associated with the Con A beads (*y* axis) is indicative of the amount of peptide transport. Cells that were treated with apyrase were preincubated with 25 U/ml apyrase (Sigma) for 3 min at 37° prior to the addition of radiolabeled peptide. Values shown represent the mean of duplicate points.

TABLE I
COMPETITION OF REPORTER PEPTIDE UPTAKE IN PERMEABILIZED WM98.1 CELLS[a]

Peptide	Sequence	Length	IC_{50} (μM)
B27#3*	RRYQNSTEL	9	1.9
Nef7B	QVPLRPMTYK	10	1.0
B27#1	GRIDKPILK	9	1.0
HEL	NTDGSTDYGILQINSRY	17	5.2
I chain	ATKYGNMTEDHVMHLLQNA	19	12
CLIP	LPKPPKPVSKMRMATPLLMQALPM	24	0.6

[a] Competition assays using the ^{125}I-labeled B27#3* reporter peptide were performed as described in the text. The underlined sequence in the B27#3* peptide is the N-linked glycosylation site. HEL, hen egg lysozyme; I chain, invariant chain (131–149); CLIP, class II-associated invariant chain peptide (81–104).

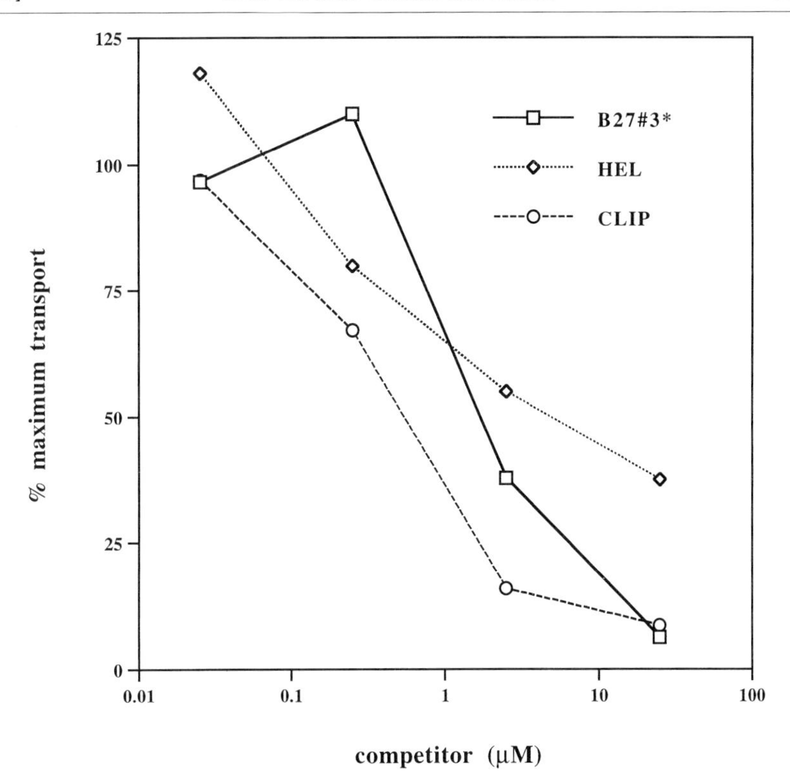

FIG. 3. Competition for ^{125}I-labeled B27#3* transport in permeabilized WM98.1 cells. Cells were treated with SLO and incubated with ^{125}I-labeled B27#3* as described in the text. However, prior to the addition of radiolabeled peptide, unlabeled competitor peptide was added at varying concentrations (0.025, 0.25, 2.5, and 25 μM) (x axis). The labeled reporter peptide was precipitated from cellular extracts with Con A-Sepharose. The level of competition was measured by the degree of inhibition of reporter peptide transport, where the level of reporter peptide transport in the absence of competing peptides was normalized to 100% (% maximum transport (y axis).

well by TAP will significantly reduce the level of reporter peptide transport. Competition curves for such an experiment are shown in Fig. 3. The B27#3* peptide competes with itself as the reporter peptide, and should provide a high degree of competition. The CLIP peptide, which is derived from the MHC class II-associated invariant chain,[16] gave a surprisingly high degree of competition in spite of its length (24 amino acids). The HEL peptide derived from hen egg lysozyme gave a moderate degree of competition.

[16] J. M. Riberdy, J. R. Newcomb, M. J. Surman, J. A. Barbosa, and P. Cresswell, *Nature* **360,** 474 (1992).

To quantitate the competition data, IC_{50} values (concentration of competitor peptide which gives 50% inhibition) were determined. The IC_{50} provides an indication of how well the competitor peptide is transported by TAP. The IC_{50} values for the peptides shown in Fig. 3, as well as additional peptides, are shown in Table I. Good substrates for TAP possess IC_{50} values of 1–2 μM or less, while poor substrates possess IC_{50} values of 100 μM or more. As shown in Table I, the B27#3*, Nef7B, B27#1, and CLIP peptides show good transport capacity (IC_{50} values of 1.9, 1.0, 1.0 and 0.6 μM, respectively), while the HEL and I chain peptides show moderate transport capacity (IC_{50} values of 5.2 and 12.0, respectively).

These results are consistent with previous competition assays performed on permeabilized B-lymphoblastoid cells using the same peptides.[17] In general, peptides of 8–12 amino acids in length are transported well by TAP, while longer peptides show a reduced capacity for transport. This is consistent for the transport of the HEL and I chain peptides, of length 17 and 19 amino acids, respectively. However, CLIP (length of 24 amino acids) transport is not consistent with this generality. In fact, CLIP gave the lowest IC_{50} (0.6 μM), which would indicate that it has a high capacity for transport. Previous data, however, revealed that CLIP did not compete well for peptide binding to TAP,[17] which suggests that the competition shown in the peptide transport assay may be due to an indirect effect.

Comments

The peptide transport assay described by Neefjes *et al.*[8] requires that ATP be added to the permeabilized cells during transport. However, the peptide transport assay described here does not require the addition of ATP to achieve peptide translocation. It is not clear why this difference exists. It is possible that in the procedure described by Neefjes *et al.* the permeabilized cells are washed prior to the addition of radiolabeled peptide. This would account for the need to add ATP back to the system. In our system there seems to be enough ATP after cell permeabilization to support TAP-dependent peptide transport.

A common criticism with permeabilized cell peptide transport systems is that many of the peptides used in these studies may be subject to degradation by cellular proteases that remain active after permeabilization. We propose that for the majority of peptides this is not the case. First, the incubation periods for the transport assay are short (5 min or less), which would tend to minimize proteolysis. Second, the results of peptide competition from past studies using permeabilized cells have been confirmed in

[17] M. J. Androlewicz and P. Cresswell, *Immunity* **1,** 7 (1994).

studies that use a purified microsome system to measure peptide binding to TAP,[18] where protease contamination would be kept to a minimum. Third, many of the peptides that are good substrates for TAP in the permeabilized cell system also compete for the photolabeling of TAP in cell membrane preparations, where protease contamination, again, would be low (M. Androlewicz, unpublished results). Taken together, these observations suggest that the proteolytic breakdown of peptides plays a minimal role in the permeabilized cell peptide transport system.

Peptide transport by TAP can also be measured using nonglycosylatable forms of peptides.[9,10] These peptides, however, must be able to bind to MHC class I molecules present within the ER or microsome to allow for stable accumulation. Peptides that do not bind to class I molecules, or are not glycosylated, do not accumulate within vesicles, and are thought to be specifically exported by a retrograde peptide pump.[19] Peptide transport and association with class I molecules has been shown in SLO-permeabilized cells using the HLA-A3 binding peptide Nef7B.[9] However, the amount of peptide that binds to class I is substantially less than the total amount of peptide transported, therefore the signal achieved is much lower than that of glycosylatable reporter peptides. In general, the measurement of peptide transport by TAP using nonglycosylated peptides is not as direct or reliable as measuring transport using glycosylatable peptides.

[18] P. M. van Endert, D. Riganelli, G. Greco, K. Fleischhauer, J. Sidney, A. Sette, and J.-F. Bach, *J. Exp. Med.* **182,** 1883 (1995).

[19] T. N. M. Schumacher, D. V. Kantesaria, M.-T. Heemels, P. G. Ashton-Rickardt, J. C. Sheperd, K. Fruh, Y. Yang, P. A. Peterson, S. Tonegawa, and H. L. Ploegh, *J. Exp. Med.* **179,** 533 (1994).

[56] Peroxisomal ABC Transporters

By NOAM SHANI and DAVID VALLE

Peroxisomes

History and Functions

Although virtually all eukaryotic cells contain peroxisomes, the existence of these single membrane-limited subcellular organelles was not recognized until the late 1950s by Rhodin[1] and de Duve and co-workers[2] in

[1] J. Rhodin, *Aktiebologet Godvil,* Sweden, (1954).

[2] C. de Duve and P. Baudhuin, *Physiol. Rev.* **46,** 323 (1966).

0076-6879/98 $25.00

their exploration of the "light mitochondrial" fraction. Progress in understanding peroxisomal function was slow initially but accelerated with recognition of a group of genetic diseases resulting from defects in peroxisome biogenesis and function.[3–5] In the last two decades both yeast and mammalian peroxisomal systems have been the subject of intense study. We now know peroxisomes contain more than 50 different matrix proteins, mainly enzymes essential for various metabolic processes including β-oxidation of very long-chain fatty acids (VLCFAs); oxidation of polyamines, D-amino acids, and, in nonprimates, uric acid; and synthesis of plasmalogens, cholesterol, and bile acids.[5,6] Several peroxisomal oxidases generate H_2O_2, which is rapidly decomposed *in situ* by peroxisomal catalase.[6]

Biogenesis

Induction. The number of peroxisomes per cell varies with metabolic state.[7] In rodents, certain hypolidemic drugs and plasticizers increase peroxisome number and induce expression of genes encoding many peroxisomal membrane and matrix proteins.[8] This coordinated induction is mediated by members of the steroid receptor superfamily known as peroxisome proliferator-activated receptors (PPARs).[9] Increase in peroxisome number appears to involve maturation of individual organelles with uptake of both membrane and matrix proteins followed by fission into daughter organelles.[5]

Matrix Proteins. Peroxisomal matrix proteins are synthesized on free cytoplasmic polysomes and targeted posttranslationally to the organelle by specific receptors that recognize *cis*-acting sequences (peroxisomal targeting signals or PTSs) in the primary peptide sequence.[10–15] Most matrix proteins

[3] S. Goldfischer, C. L. Moore, A. B. Johnson, A. J. Spiro, M. P. Valsmis, H. K. Wisniewski, R. H. Ritch, W. T. Norton, I. Rapin, and L. M. Gartner, *Science* **182,** 62 (1973).
[4] P. Borst, *TIBS* **8,** 269 (1983).
[5] P. B. Lazarow and H. W. Moser, *in* "The Metabolic and Molecular Bases of Inherited Disease" (C. Scriver, A. Beaudet, W. Sly, and D. Valle, eds), p. 2287, McGraw-Hill, New York, 1995.
[6] H. van den Bosch, R. B. H. Schutgens, R. J. A. Wanders, and J. M. Tager, *Annu. Rev. Biochem.* **61,** 157 (1992).
[7] S. Subramani, *Physiol. Rev.* **78,** 171 (1998).
[8] J. K. Reddy, S. K. Goel, M. R. Nemali, J. J. Carrino, T. G. Laffer, M. K. Reddy, S. J. Sperbeck, T. Osumi, T. Hashimoto, N. D. Lalwani, and M. S. Rao, *Proc. Natl. Acad. Sci. U.S.A.* **83,** 1747 (1986).
[9] S. Green and W. Wahli, *Mol. Cell Endocrinol.* **100,** 149 (1994).
[10] P. E. Purdue and P. B. Lazarow, *J. Biol. Chem.* **269,** 30065 (1994).
[11] S. J. Gould, G. A. Keller, and S. Subramani, *J. Cell Biol.* **105,** 2923 (1987).
[12] B. W. Swinkels, S. J. Gould, A. G. Bodnar, R. A. Rachubinski, and S. Subramani, *EMBO J.* **10,** 3255 (1991).
[13] R. A. Rachubinski and S. Subramani, *Cell* **83,** 525 (1995).
[14] J. A. McNew and J. M. Goodman, *TIBS* **21,** 54 (1996).

are targeted by PTS1, a C-terminal -SKL or conservative variant thereof. A few matrix proteins are targeted by PTS2, a degenerate sequence (-R/KLX$_5$Q/HL-) located 2–12 residues from the N terminus. Other less well understood PTSs are utilized by a few proteins.[13–15] The yeast and mammalian receptors for PTS1 and PTS2 have been cloned and characterized.[16–24a] The former is a tetratricopeptide repeat protein[25–27]; the latter a WD40 repeat protein.[28,29] Both are primarily cytosolic, an observation that suggests they bind their ligands in the cytosol and transport them to the peroxisome.[30]

Membrane Proteins. The peroxisome membrane contains more than 10 integral peroxisome membrane proteins (PMPs), specific for the organelle, including C_3HC_4 zinc-binding proteins involved in matrix protein import (PEX2 and PEX10),[31–36] an SH3-containing protein (PEX13) required for

---

[15] P. E. Purdue and P. B. Lazarow, *J. Cell Biol.* **134,** 849 (1996).
[16] D. McCollum, E. Monosov, and S. Subramani, *J. Cell Biol.* **121,** 761 (1993).
[17] I. Van der Leij, M. M. Franse, Y. Elgersma, B. Distel, and H. F. Tabak, *Proc. Natl. Acad. Sci. U.S.A.* **90,** 11782 (1993).
[18] G. Dodt, N. Braverman, C. Wong, A. Moser, H. W. Moser, P. Watkins, D. Valle, and S. J. Gould, *Nature Genet.* **9,** 115 (1995).
[19] E. A. C. Wiemer, W. M. Nuttley, B. L. Bertolaet, X. Li, U. Franke, M. J. Wheelock, U. K. Anné, K. R. Johnson, and S. Subramani, *J. Cell Biol.* **130,** 51 (1995).
[20] M. Fransen, C. Brees, E. Baumgart, J. C. T. Vanhooren, M. Baes, G. P. Mannaerts, and P. P. Van Veldhoven, *J. Biol. Chem.* **270,** 7731 (1995).
[21] M. Marzioch, R. Erdmann, M. Veenhuis, and W. H. Kunau, *EMBO J.* **13,** 4908 (1994).
[22] J. W. Zhang and P. B. Lazarow, *J. Cell Biol.* **129,** 65 (1995).
[23] J. W. Zhang and P. B. Lazarow, *J. Cell Biol.* **132,** 325 (1996).
[24] P. Rehling, M. Marzioch, F. Niesen, E. Wittke, M. Veenhuis, and W.-H. Kunau, *EMBO J.* **15,** 2901 (1996).
[24a] N. Braverman, G. Steel, C. Obie, A. Moser, H. Moser, S. J. Gould, and D. Valle. *Nature Genet.* **15,** 369 (1997).
[25] R. S. Sikorski, W. A. Michaud, J. C. Wootton, M. S. Boguski, C. Connelly, and P. Hieter, in *Cold Spring Harbor Symp. Quant. Biol.,* **LVI,** 663 (1991).
[26] J. R. Lamb, S. Tugendreich, and P. Hieter, *TIBS* **20,** 257 (1995).
[27] M. Goebl and M. Yanagida, *TIBS* **16,** 173 (1991).
[28] L. van der Voorn and H. Ploegh, *FEBS* **307,** 131 (1992).
[29] I. Garcia-Higuera, C. Caitatzes, T. F. Smith, and E. J. Neer, *J. Biol. Chem.* **273,** 9041 (1998).
[30] N. Braverman, C. Dodt, S. J. Gould, and D. Valle, *Hum. Mol. Genet.* **4,** 1791 (1995).
[31] N. Schimozawa, T. Tsukamoto, Y. Suzuki, T. Orii, Y. Shirayoshi, T. Mori, and Y. Fujiki, *Science* **255,** 1132 (1992).
[32] V. Berteaux-Lecellier, M. Picard, C. Thompson-Coffe, D. Zickler, A. Panvier-Adoutte, and J.-M. Simonet, *Cell* **81,** 1043 (1995).
[33] H. R. Waterham, Y. de Vries, K. A. Russel, W. Xie, M. Veenhuis, and J. M. Cregg, *Mol. Cell Biol.* **16,** 2527 (1996).
[34] X. Tan, H. R. Waterham, M. Veenhuis, and J. M. Creeg, *J. Cell Biol.* **128,** 307 (1995).
[35] J. E. Kalish, C. Theda, J. C. Morrell, J. M. Berg, and S. J. Gould, *Mol. Cell Biol.* **15,** 6406 (1995).
[36] T. Tsukamoto, S. Miura, and Y. Fujiki, *Nature* **350,** 77 (1991).

docking of the PTS1 receptor,[37–39] a variety of small molecule transporters,[40–42] and multiple members of a family of half ABC transporters (see below).[43–49] Like peroxisomal matrix proteins, PMPs are synthesized on free cytoplasmic polyribosomes and directed to the organelle by *cis*-acting targeting sequences (membrane peroxisome targeting signal or mPTS) about which we know relatively little.[14,50] Recent work identified an internal mPTS between the fourth and fifth membrane-spanning segment of PMP47, the sole peroxisomal member of the mitochondrial solute transporter family.[40] Peroxisomal targeting of PMP70 requires an undefined sequence in the N-terminal third of the protein.[50] Unlike the targeting of peroxisomal matrix proteins, there is no evidence suggesting involvement of cytosolic receptors in the targetting of PMPs.

Genetic Disease and Peroxisomes

Peroxisome Biogenesis Disorders. Not surprisingly, in view of the broad involvement of peroxisomal functions in normal development and physiologic homeostasis, mutations in genes encoding peroxisome components

[37] Y. Elgersma, L. Kwast, K. A. T. Voorn-Brouwer, M. van den Berg, B. Metzig, T. America, H. F. Tabak, and B. Distel, *J. Cell Biol.* **135,** 97 (1996).
[38] R. Erdmann and G. Blobel, *J. Cell Biol.* **135,** 111 (1996).
[39] S. J. Gould, J. E. Kalish, J. C. Morrell, J. Bjorkman, A. J. Urquhart, and D. I. Crane, *J. Cell Biol.* **135,** 85 (1996).
[40] M. T. McCammon, J. A. McNew, P. J. Willy, and J. M. Goodman, *J. Cell Biol.* **124,** 915 (1994).
[41] Y. Elgersma, W. T. van Roermund, R. J. A. Wanders, and H. F. Tabak, *EMBO J.* **14,** 3472 (1995).
[42] C. W. T. van Roermund, Y. Elgersma, N. Singh, R. J. A. Wanders, and H. F. Tabak, *EMBO J.* **14,** (1995).
[43] N. Shani and D. Valle, *Proc. Natl. Acad. Sci. U.S.A.* **93,** 11901 (1996).
[44] N. Shani, P. A. Watkins, and D. Valle, *Proc. Natl. Acad. Sci. U.S.A.* **92,** 6012 (1995).
[45] J. Mosser, A. M. Douar, C. O. Sarde, P. Kioschis, R. Feil, H. Moser, A. M. Poustka, J. L. Mandel, and P. Aubourg, *Nature* **361,** 726 (1993).
[46] J. Mosser, Y. Lutz, M. E. Stoeckel, C. O. Sarde, C. Kretz, A. M. Douar, J. Lopez, P. Aubourg, and J. L. Mandel, *Hum. Mol. Genet.* **3,** 265 (1994).
[47] G. Lombard-Platet, S. Savary, C.-O. Sarde, and J.-L. Mandel, *Proc. Natl. Acad. Sci. U.S.A.* **93,** 1265 (1996).
[48] K. Kamijo, S. Taketani, S. Yokata, T. Osumi, and T. Hashimoto, *J. Biol. Chem.* **265,** 4534 (1990).
[49] J. Gärtner, H. Moser, and D. Valle, *Nature Genet.* **1,** 16 (1992).
[50] T. Imanaka, Y. Shiina, T. Takano, T. Hashimoto, and T. Osumi, *J. Biol. Chem.* **271,** 3706 (1996).

cause genetic diseases with profound phenotypic consequences. The most severe of these are the peroxisome biogenesis disorders (PBDs), a genetically heterogeneous group of autosomal recessive disorders with deficiencies of multiple peroxisome functions. The phenotypic paradigm of the PBD is Zellweger syndrome, a lethal developmental and metabolic disorder with progressive involvement of the central nervous system, liver, kidneys, and skeleton. Cell fusion studies using fibroblasts from PBD patients show at least 11 PBD complementation groups. The genes responsible for five of these have been identified[18,19,24a,30,31,51,52,52a] and a uniform nomenclature for the PBD genes of yeast and mammalian cells has recently been adopted.[53]

A second class of peroxisomal disorders is those in which a single function is deficient. The exemplar is X-linked adrenoleukodystrophy, a progressive neurologic disorder caused by mutations in the ALD gene. The clinical phenotype of adrenoleukodystrophy is variable and includes the childhood cerebral form with onset of neurologic symptoms in midchildhood followed by progressive neurodegeneration and death within a few years. A milder phenotype, known as adrenomyeloneuropathy, is characterized by slowly progressive paraparesis beginning early in adult life.[54] The ALD gene encodes a 75-kDa peroxisomal integral membrane protein (ALDP)-half ABC transporter (see below). Aside from defective oxidation of VLCFAs, peroxisome function and structure are normal in adrenoleukodystrophy.

[51] C. M. Fraser, J. D. Gocayne, O. White, M. D. Adams, R. A. Clayton, R. D. Fleischmann, C. J. Bult, A. R. Kerlavage, G. Sutton, J. M. Kelley, J. L. Fritchman, J. F. Weidman, K. V. Small, M. Sandusky, J. Fuhrmann, D. Nguyen, T. R. Utterback, D. M. Saudek, C. A. Phillips, J. M. Merrick, J.-F. Tomb, B. A. Doughterty, K. F. Bott, P.-C. Hu, T. S. Lucier, S. N. Peterson, H. O. Smith, C. A. Hutchinson III, and J. C. Venter, *Science* **270,** 397 (1995).

[52] S. Fukuda, N. Shimozawa, Y. Suzuki, Z. Zhang, S. Tomatsu, T. Tsukamoto, N. Hashiguchi, T. Osumi, M. Masuno, K. Imaizumi, Y. Kuroki, Y. Fujiki, T. Orii, and N. Kondo, *Am. J. Hum. Genet.* **59,** 1210 (1996).

[52a] S. Subramani, *Nature Genet.* **15,** 331 (1997).

[53] B. Distel, R. Erdmann, S. J. Gould, G. Blobel, D. I. Crane, J. M. Cregg, G. Dodt, Y. Yujiki, J. M. Goodman, W. W. Just, J. A. K. W. Kiel, W.-H. Kunau, P. B. Lazarow, G. P. Mannaerts, H. W. Moser, T. Osumi, R. A. Rachubinski, A. Roscher, S. Subramani, H. F. Tabak, T. Tsukamoto, D. Valle, I. van der Klei, P. P. van Veldhoven, and M. Veenhuis, *J. Cell Biol.* **135,** 1 (1996).

[54] H. Moser, K. Smith, and A. Moser, *in* "The Metabolic and Molecular Bases of Inherited Disease" (C. Scriver, A. Beaudet, W. Sly, and D. Valle, eds), p. 2325, McGraw-Hill, New York, 1995.

Mammalian Peroxisomal ABC Transporters

PMP70

Following induction by administration of hypolipidemic agents, the two most abundant PMPs in rat liver are PMP22 and PMP70.[55,56] In 1990 Kamijo *et al.*[48] cloned the rat PMP70 gene and showed it encoded a 659-amino-acid "half" ABC transporter with one transmembrane domain (TMD) containing six putative transmembrane segments and a cytosolic nucleotide binding domain (NBD) with Walker A and B motifs (Fig. 1). Kinetic studies show that following synthesis as a mature polypeptide on cytosolic ribosomes, PMP70 quickly associates with the peroxisome membrane (half-time of 3 min in rat hepatoma cells) and is completely integrated into the membrane by 30 min. This process is peroxisome specific, ATP independent, and mediated by protein(s) on the surface of the organelle membrane.[50]

On the basis of its peroxisomal location and membership in the ABC transporter family, PMP70 was initially proposed as a candidate PBD gene.[49] To investigate this possibility, Gärtner *et al.* cloned the human PMP70 gene, mapped it to 1p21-22, determined its organization and examined it for mutations in PBD patients.[49,57,58,58a] Two complementation group 1 PBD patients were shown to be heterozygous for PMP70 mutations: a missense mutation (G17D); and a donor splice site change in the last intron, which results in replacement of the C-terminal 25 amino acids by a new sequence of 23 residues. Neither of these mutations was found in more than 200 control PMP70 genes and the splice site change was a *de novo* alteration in the proband. Subsequent studies showed that the mutant PMP70 protein produced by the splice site allele does have altered function in terms of its ability to suppress the cellular phenotype PMP35 mutation in cultured CHO cells[59]; however, the role that PMP70 plays in peroxisome biogenesis remains uncertain.

ALDP

In 1993 Mosser *et al.*[45] reported successful positional cloning of identifying the gene (*ALD*) responsible for adrenoleukodystrophy. Earlier stud-

[55] F. U. Hartl and W. W. Just, *Arch. Biochem. Biophys.* **255,** 109 (1987).

[56] T. Hashimoto, K. T. N. Usuda, and T. Nagata, *J. Biochem.* **100,** 301 (1986).

[57] J. Gärtner and D. Valle, *Semin. Cell Biol.* **4,** 45 (1993).

[58] J. Gärtner, W. Kearns, C. Rosenberg, P. Pearson, N. G. Copeland, D. J. Gilbert, N. Jenkins, and D. Valle, *Genomics* **15,** 412 (1993).

[58a] J. Gärtner, G. Jimenez-Sanchez, P. Roerig, and D. Valle, *Genomics* **48,** 203 (1998).

[59] J. Gärtner, C. Obie, P. Watkins, and D. Valle, *J. Inhert. Met. Dis.* **17,** 327 (1994).

ies localized *ALD* to Xq28.[60] Because VLCFA activation is impaired in adrenoleukodystrophy,[61] it was anticipated that the gene responsible for this disorder would encode a VLCFA CoA ligase. Surprisingly, the protein encoded by the *ALD*, designated ALDP, is a second peroxisomal half ABC transporter more similar to PMP70 (38% overall amino acid identity) than to any other ABC transporter known at the time (Fig. 1). Ultrastructural studies confirmed a peroxisomal location for ALDP and showed an orientation like PMP70 with the NBD on the cytosolic side of the peroxisomal membrane.[46] Several studies have shown that virtually all adrenoleukodystrophy patients have mutations in the *ALD* gene, and more than 50 mutant *ALD* alleles have been identified and characterized providing the most extensive set of mutations in a peroxisomal membrane half ABC transporter.[62–66] The majority of *ALD* missense mutations result in the complete lack of immunologically detectable ALDP, suggesting they either destabilize the protein or interfere with targeting of the protein to the peroxisomal membrane.[62,65] This large series of *ALD* mutations plus the observation that expression of wild-type ALDP in adrenoleukodystrophy patient cells restores β-oxidation of VLCFA[67] make it clear that adrenoleukodystrophy is caused by *ALD* mutations. However, the extreme variability in the clinical phenotype of adrenoleukodystrophy, even in individuals in the same family with the same *ALD* genotype, indicates that other factors strongly influence the phenotype. The distribution of phenotypes within families is consistent with a model in which a highly polymorphic genetic modifier plays a major role in determining phenotypic severity.[68] The recent cloning of the mouse

[60] B. R. Migeon, H. W. Moser, A. B. Moser, J. Axelman, D. Sillence, and R. A. Norum, *Proc. Natl. Acad. Sci. U.S.A.* **78,** 5066 (1981).

[61] O. Lazo, M. Contreras, M. Hashmi, W. Stanley, C. Irazu, and I. Singh, *Proc. Natl. Acad. Sci. U.S.A.* **85,** 7647 (1988).

[62] P. A. Watkins, S. J. Gould, M. A. Smith, L. T. Braiterman, H.-M. Wei, F. Kok, A. B. Moser, H. W. Moser, and K. D. Smith, *Am. J. Hum. Genet.* **57,** 292 (1995).

[63] P. Fok, S. Neumann, C.-O. Sarde, S. Zheng, K.-H. Wu, H.-M. Wei, J. Bergin, P. Watkins, S. Gould, G. Sack, H. Moser, J.-L. Mandel, and K. Smith, *Hum. Mut.* **6,** 104 (1995).

[64] P. Fanen, S. Guidoux, C.-O. Sarde, J.-L. Mandel, M. Goosens, and P. Aubourg, *J. Clin. Invest.* **94,** 516 (1994).

[65] V. Feigenbaum, G. Lombard-Platet, S. Guidoux, C.-O. Sarde, J.-L. Mandel, and P. Aubourg, *Am. J. Hum. Genet.* **58,** 1135 (1996).

[66] M. J. L. Ligtenberg, S. Kemp, C.-O. Sarde, B. M. van Geel, W. J. Kleijer, P. G. Barth, J.-L. Mandel, B. A. van Oost, and P. A. Bolhuis, *Am. J. Hum. Genet.* **56,** 44 (1995).

[67] N. Cartier, J. Lopez, P. Moullier, F. Rocchiccioli, M.-O. Rolland, P. Jorge, J. Mosser, J.-L. Mandel, P.-F. Bougneres, O. Danos, and P. Aubourg, *Proc. Natl. Acad. Sci. U.S.A.* **92,** 1674 (1995).

[68] H. W. Moser, A. B. Moser, K. D. Smith, A. Bergin, J. Borel, J. Shankroff, O. C. Stine, C. Merette, J. Ott, W. Krivit, and E. Shapiro, *J. Inher. Metab. Dis.* **15,** 645 (1992).

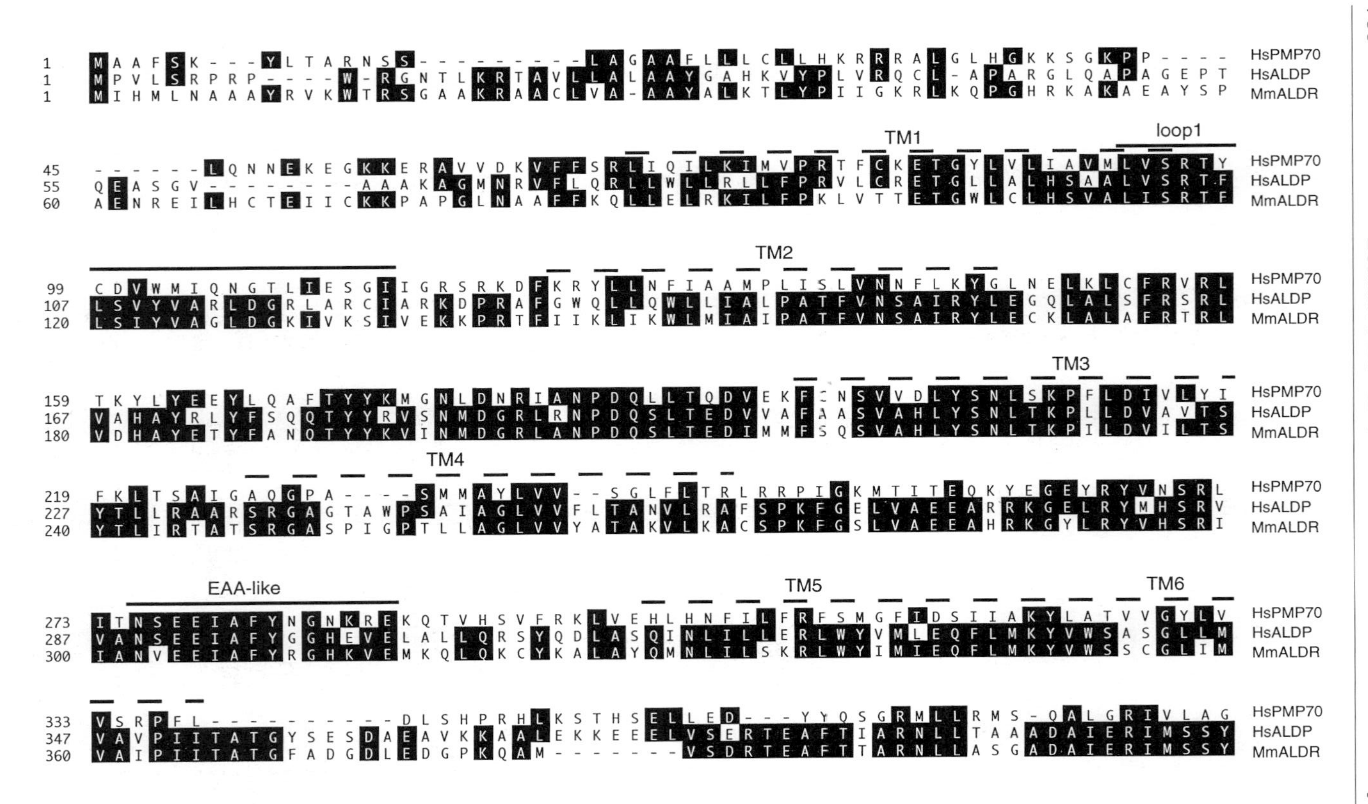
HsPMP70
HsALDP
MmALDR
TM1
loop1
TM2
TM3
TM4
EAA-like
TM5
TM6

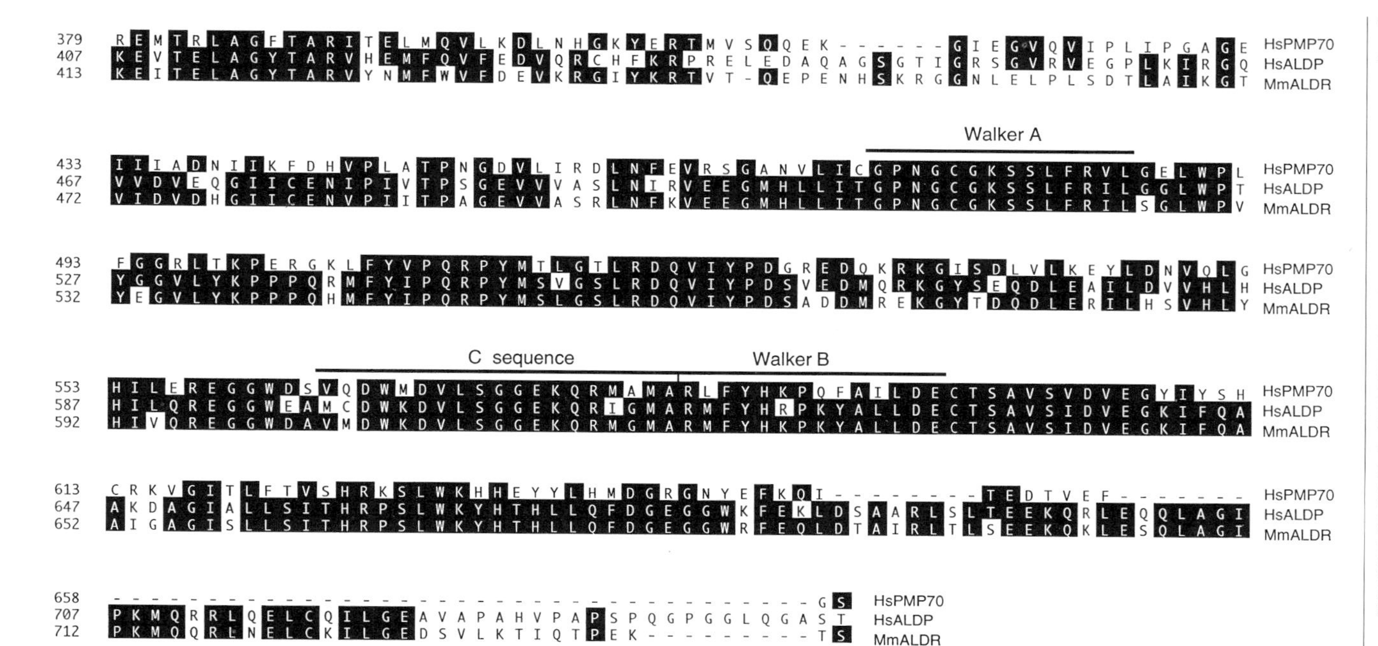

FIG. 1. Alignment of mammalian peroxisomal ABC transporters. The amino acid sequences of human PMP70 and ALDP (HsPMP70 and HsALDP) and murine adrenoleukodystrophy-related (MmALDR) were aligned using the MegAlign program from DNA-STAR. The complete sequence of human ALDR is not available. Identical amino acid are boxed in black. The six putative transmembrane domain segments are indicated by dashed lines. The loop and the EAA-like protein motifs, as well as the Walker A, Walker B and the C sequence of the nucleotide binding fold are indicated by overlines.

ortholog ALD gene,[69] an initial step in production of a mouse model, provides an opportunity to better understand the pathophysiology of adrenoleukodystrophy.

A third mammalian peroxisomal half ABC transporter gene encoding a 741-amino-acid protein highly homologous to ALDP and designated ALD-related or ALDR has recently been described.[47] The chromosomal location of the ALDR structural gene has not been reported. The ALDR protein has 66% identity with ALDP and 38% with PMP70 (Fig. 1). A peroxisomal location was confirmed by immunohistochemical studies. Northern blot analyses in mouse indicate that ALDR gene expression is limited mainly to brain and adrenal, a pattern that is much more restricted than either ALDP or PMP70, which are both widely expressed.[47] To date, no disease associations are known for ALDR. Theoretically, it could interact with and modify ALDP expression or function as a PBD gene or have some function that was similar to but nonoverlapping with ALDP.

Finally, we have very recently identified a fourth human peroxisomal half ABC transporter (designated P70R) that encodes a protein most similar to PMP70, which we refer to as P70R.[70,70a,b] The characterization of this gene is still in progress but immunohistochemical studies show that the protein product localizes to peroxisomes. Thus, there are at least four different half ABC transporters in the mammalian peroxisome membrane. The existence of this collection of half ABC transporters in the mammalian peroxisomal membrane raises the possibility of formation of functional heterodimeric transporters with different ligand specificity by combinatorial assembly as discussed below. In this regard it is interesting that *in situ* studies in rat brain indicate that ALD and PMP70 mRNAs have different spatial and temporal expression in postnatal development.[71]

Saccharomyces cerevisiae PXA Transporter

PXA1 Gene

Many studies on the function and biogenesis of peroxisomes have taken advantage of yeast systems. In contrast to mammalian cells, peroxisomes

[69] C.-O. Sarde, J. Thomas, H. Sadoulet, J.-M. Garnier, and J.-L. Mandel, *Mammal Genome* **5,** 810 (1994).

[70] N. Shani, G. Jimenez-Sanchez, G. Steel, M. Dean, and D. Valle, *Hum. Molec. Genet.* **6,** 1925 (1997).

[70a] A. Holzinger, S. Kammerer, and A. A. Roscher, *Biochem. Biophys. Res. Comm.* **237,** 152 (1997).

[70b] A. Holzinger, A. A. Roscher, P. Landgraf, P. Lichtner, and S. Kammerer, *FEBS Lett.* **426,** 238 (1998).

[71] H. Pollard, J. Moreau, and P. Aubourg, *J. Neurosci. Res.* **42,** 433 (1995).

in yeast are easily induced and isolated (see below). Moreover, fatty acid β-oxidation in yeast is limited to peroxisomes in contrast to mammalian cells, which have overlapping mitochondrial and peroxisomal β-oxidation systems.[72] This characteristic provides a powerful phenotypic selection strategy for mutants of peroxisome biogenesis and function in yeast.[72,73] Several *S. cerevisiae* PMPs are involved in the import of matrix proteins.[14,53] Mutations in the genes encoding these proteins cause peroxisome assembly (pas) phenotypes.[53,72,73] None of the peroxisome assembly mutants characterized to date involves an ABC transporter.

To develop a tractable system to better understand the structure and function of the peroxisomal half ABC transporters, we set out to clone yeast genes encoding orthologs of the mammalian proteins. Utilizing degenerate PCR primers corresponding to the conserved Walker A and B sequences of PMP70 and ALDP and reverse transcribed total RNA isolated from oleic acid-induced *S. cerevisiae,* we amplified a fragment corresponding to the NBD. Sequencing individual components from the pool of amplified products we found that several encoded a sequence homologous to the mammalian PMP70 and ALDP. Using this sequence as a probe to screen an *S. cerevisiae* genomic library, we cloned a newly identified gene, which we designated *PXA1* (for peroxisomal ABC transporter 1).[44] The predicted protein product of *PXA1* (Pxa1p) has 758 amino acids, a predicted molecular mass of 87 kDa, and is slightly more similar to ALDP than to PMP70 (28% and 27% overall amino acid identity, respectively). Like ALDP it is associated with the peroxisomes (Fig. 2). Disruption of the *PXA1* gene results in a growth phenotype on oleic acid medium intermediate between the parental wild-type yeast and a strain in which we disrupted PEX1, a gene essential for peroxisome biogenesis.[53,74] Direct measurement of β-oxidation with radiolabeled oleic acid reveals a similar 50% reduction in the *pxa1* mutant as compared to wild type. Electron microscopy of the *pxa1* shows intact peroxisomes.[44] Thus, like ALDP, Pxa1p is involved in fatty acid β-oxidation and is not required for peroxisomal biogenesis.

More recently, *PXA1* was cloned independently by Swartzman *et al.*[75] who named it *PAL1*. They found that *PXA1* expression is induced by oleic acid and that Pxa1p is an integral peroxisome membrane protein. They also described a potential N-terminal extension of the open reading frame, taking advantage of an upstream in frame methione codon that would add

[72] W. H. Kunau, A. Beyer, T. Franken, K. Gotte, M. Marzioch, J. Saidowsky, A. Skaletz-Rorowski, and F. F. Wiebel, *Biochemie* **75,** 209 (1993).

[73] S. J. Gould, D. McCollum, A. P. Spong, J. A. Heyman, and S. Subramani, *Yeast* **8,** 613 (1992).

[74] R. Erdmann, F. F. Wiebel, A. Flessau, J. Rytka, A. Beyer, K. U. Frohlich, and W. H. Kunau, *Cell* **64,** 499 (1991).

[75] E. E. Swartzman, M. N. Viswanathan, and J. Thorner, *J. Cell Biol.* **132,** 549 (1996).

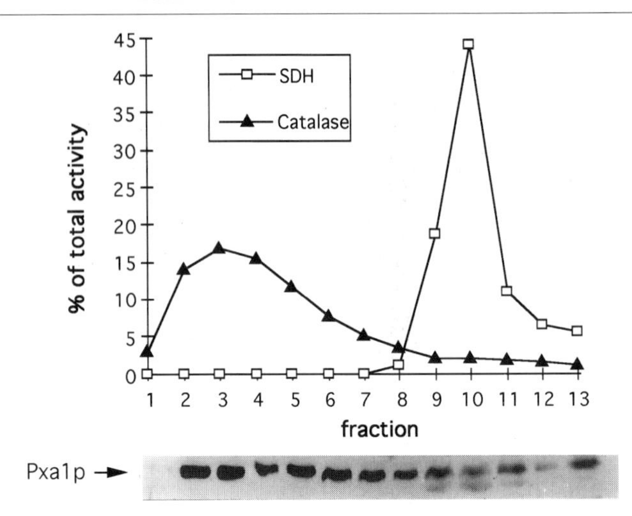

FIG. 2. Localization of Pxa1p to peroxisomes. The organelle pellet of oleic acid-induced wild-type cells was fractionated on a Nycodenz density gradient. *Upper:* Gradient fractions were assayed for activity of catalase and succinate dehydrogenase (SDH). *Lower:* Equal volumes of each fraction were analyzed by immunoblotting with rabbit anti-Pxa1p antiserum.

an additional 112 codons to the 5′ end of the reading frame and increase the predicted molecular mass of the protein to ~100 kDa. This N-terminal sequence has no homology to other ABC transporters and is not required to complement the growth phenotype of a *PXA1* disruption strain on oleic acid medium.

Aligning the sequence of Pxa1p to that of other ABC transporters reveals several regions of high amino acid conservation in addition to the ATP-binding motif. One of these is located between the putative transmembrane segments 4 and 5 and is similar in sequence and location to a motif previously described only in prokaryotic ABC transporters known as the EAA motif.[76–78] For these reasons, we called the eukaryotic counterpart the EAA-like motif[44] (Figs. 1 and 3). Disease-producing mutations in the EAA-like motif of ALDP in four unrelated adrenoleukodystrophy patients (two missense and two single codon deletions) indicate that the region is also important in eukaryotic transporters.[44] We analyzed its function in more detail by site-directed mutagenesis of the PXA1 EAA-like motif (Figs. 3 and 4). As in prokaryotic ABC transporters, the central conserved

[76] W. Saurin, W. Koster, and E. Dassa, *Mol. Microbiol.* **12,** 993 (1994).
[77] W. Koster and B. Bohm, *Mol. Gen. Genet.* **232,** 399 (1992).
[78] R. E. Kerppola and G. F. Ames, *J. Biol. Chem.* **267,** 2329 (1992).

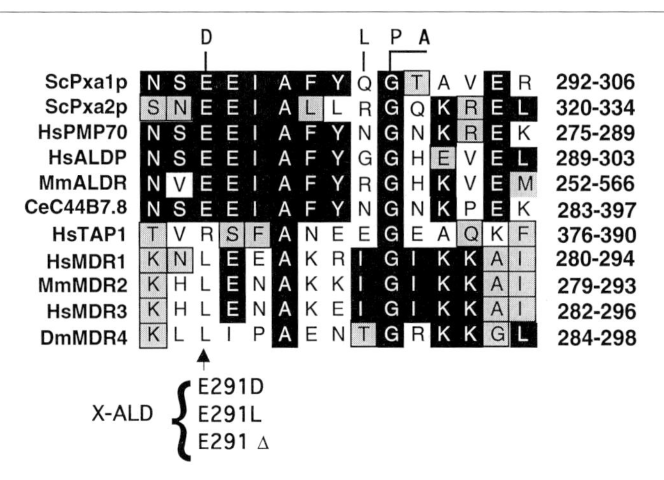

FIG. 3. Alignment of EAA-like motifs in ABC transporters and the location of mutations made in Pxa1p. Amino acid sequences from eukaryotic ABC transporters were aligned using the MegAlign program from DNA-STAR. The proteins aligned are the *S. cerevisiae* Pxa1p and Pxa2p (ScPxa1p and ScPxa2p), human PMP70, ALDP, TAP1, MDR1 and MDR3 (HsPMP70, HsALDP, HsTAP1, HsMDR1 and HsMDR3), murine ALDR and MDR2 (MmALDR and MmMDR2), the *C. elegans* open reading frame C44B7.8 (CeC44B7.8) and the *Drosophila* MDR4 (DmMDR4). Amino acids present in a plurality of these sequences are boxed in black and conservative amino acid substitutions from this plurality are boxed in gray. Mutations in ALDP that cause adrenoleukodystrophy are indicated below. Missense mutations introduced in the *PXA1* gene are indicated above.

glycine residue (Gly301 in Pxa1p) is vital for activity[79] (Figs. 4A and B). Furthermore, a conservative substitution in the first of the two conserved glutamate residues (E294D), completely inactivates Pxa1p[79] (Figs. 4A and B). Missense mutations altering the corresponding glutamic acid residue of ALDP (E291) have been described in two unrelated adrenoleukodystrophy patients (E291D, E291L).[62,66,80,81] Although these results are interesting, the precise function of the EAA motif in prokaryotes or the EAA-like motif in eukaryotes remains to be determined. It does not seem to be involved in targeting of the protein to peroxisomes; the nonfunctional mutant Pxa1p's described above (E294D and G301P) still associate with peroxisomes.[79] On the basis of sequence conservation and the results of mutagenesis experiments, the prokaryotic EAA motif was suggested to

[79] N. Shani, A. Sapag, and D. Valle, *J. Biol. Chem.* **271,** 8725 (1996).

[80] R. Koike, O. Onodera, H. Tabe, K. Keneko, T. Miyatake, J. Mosser, C. Sarde, C. Mandel, and S. Tsuji, *Am. J. Hum. Genet.* **55,** A228 (1994).

[81] N. Cartier, C. O. Sarde, A. M. Douar, J. Mosser, J. L. Mandel, and P. Aubourg, *Hum. Mol. Genet.* **2,** 1949 (1993).

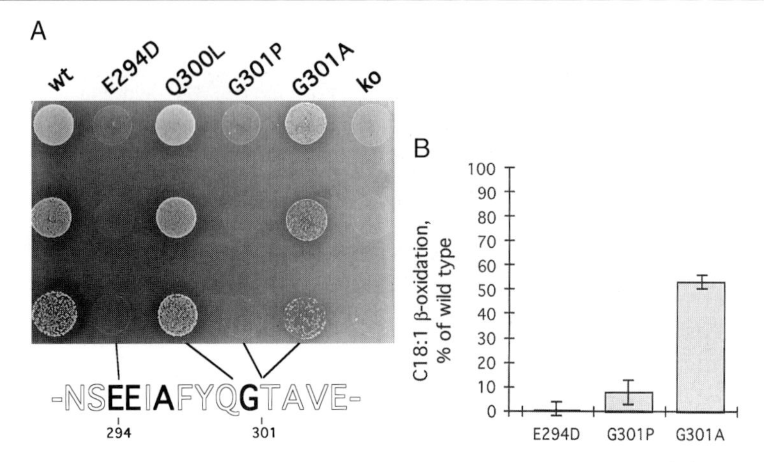

FIG. 4. Phenotype of missense mutations in the EAA-like motif of Pxa1p. (A) Growth of *pxa1::LEU2* knockout yeast transformed with, from left to right, wild type, E294D, Q300L, G301P or G301A alleles of PXA1 or vector alone. Ten-fold dilutions (top to bottom) of the transformants were inoculated on medium with 0.1% oleic acid as a sole carbon source. The amino acid sequence of the EAA-like motif from Pxa1p is indicated on the bottom with conserved residues in bold letters. (B) β-Oxidation of [^{14}C]oleic acid by *pxa1::LEU2* knockout yeast expressing from left to right: E294D, G301P or G301A alleles of PXA1.

play a role in the interaction between the TMD and the NBD.[77,78] We extended this hypothesis suggesting that the EAA-like motif might interact with the C region, a highly conserved amino acid sequence located just N-terminal to the Walker B motif of the NBD.[79]

A second region of conservation between Pxa1p and other peroxisomal ABC transporters is located between the first and second putative transmembrane segments of peroxisomal ABC transporters on a loop extending into the peroxisomal matrix.[79] We designated this sequence the loop 1 motif (Fig. 1) and found that changing a conserved arginine in the motif results in Pxa1p dysfunction. A missense mutation in the corresponding residue in ALDP (Arg-104) causes adrenoleukodystrophy.[66] The function of this domain remains to be determined. Finally, we identified a third region in Pxa1p (N-terminal sequence leading up to and through the first transmembrane segment) with weak sequence homology to the binding motifs of fatty acid binding proteins.[44,82] This is intriguing in view of Pxa1p presumed function in transporting fatty acyl CoAs but no further studies have been done to investigate this possibility. In summary, Pxa1p has similarities in sequence, location, biochemical phenotype, and response to mutation with ALDP.

[82] S. Petrou, R. Ordway, J. Singer, and J. Walsh, *TIBS* **18,** 41 (1993).

PXA2

In 1994 the yeast genome sequencing project identified an open reading frame (YKL741) encoding a putative half ABC transporter with homology to the mammalian PMP70 and ALDP.[83] We disrupted the YKL741 gene and found that the growth phenotype on oleic acid was identical to that of *pxa1* mutant yeast. Moreover, the phenotype of the double mutant, disrupted in both *PXA1* and YKL741, was nonadditive and identical to the phenotype of either single mutant.[43,44] Based on this result and the precedent of the heterodimeric TAP transporter,[84] we suggested that the protein product of YLK741 dimerizes with Pxa1p to form a complete ABC transporter in the yeast peroxisome membrane.[44] Subsequently we showed that the protein product of YKL741 is associated with the peroxisomes[43] and designated this gene as *PXA2*. Similar results involving Pxa2p in the oxidation of long-chain fatty acids were reported recently by Hettema *et al.*[85] (who called the gene *PAT1*). Interestingly, we found that Pxa1p is destabilized in a strain containing a disruption of *PXA2* with a three-fold reduction in half-life, suggesting a physical interaction between Pxa1p and Pxa2p.[43] The converse is not true; Pxa2p stability is not altered in the absence of Pxa1p. This may reflect greater intrinsic stability of Pxa2p or it may interact with other proteins in addition to Pxa1p. Finally, in coimmunoprecipitation experiments, we showed direct physical evidence for heterodimerization of Pxa1p and Pxa2p.[43]

Thus, Pxa1p and Pxa2p are yeast homologs of the mammalian peroxisomal half ABC transporters and are located in the yeast peroxisome membrane. Both proteins are involved in the β-oxidation of fatty acids and their mutant phenotype is identical and nonadditive. Pxa1p is destabilized in the absence of its counterpart. Based on these genetic results and on the physical interaction shown by coimmunoprecipitation of both proteins, we suggested that Pxa1p and Pxa2p assemble to form a complete ABC transporter in the yeast peroxisomal membrane. We named this transporter the PXA transporter.[43] Based on growth and β-oxidation experiments we and others proposed that the PXA transporter functions to transport long-chain (C_{16-18}) fats or fatty acyl CoAs into the peroxisomal matrix.[43,85] Hettema *et al.*[85] found that activation of long-chain fats in yeast takes place in an extraperoxisomal location supporting the hypothesis that the substrates for the PXA transporter are the CoA derivitives of long-chain fats. A recent survey of the complete *S. cerevisiae* genome did not find any

[83] P. Bossier, L. Pernandes, C. Vilela, and C. Rodrigues-Pousada, *Yeast* **10,** 681 (1994).

[84] A. Hill and A. Ploegh, *Proc. Natl. Acad. Sci. U.S.A.* **92,** 341 (1995).

[85] E. H. Hettema, C. W. T. van Roermund, B. Distel, M. van den Berg, C. Vilela, C. Rodrigues-Pousada, R. J. A. Wanders, and H. F. Tabak, *EMBO J.* **15,** 3813 (1996).

other genes coding proteins with high homology to the human or yeast peroxisomal ABC transporters. This suggests that, in contrast to mammalian peroxisomes, the PXA transporter may be suffficient for the metabolic requirements of *S. cerevisiae.*

Other Putative Peroxisomal ABC Transporters

In addition to the mammalian and yeast proteins discussed earlier, a search of GenBank reveals potential orthologs of the peroxisomal ABC transporters in *Caenorhabditis elegans.* Two open reading frames (C44B7.8 and C44B7.9) have 53% and 34% overall amino acid identity to PMP70 and ALDP, respectively. Interestingly, an open reading frame with relatively high homology to the peroxisomal ABC transporters (26% overall amino acid identity to ALDP) is present in the prokaryotic cyanobacterium *Synechocystic* sp. This homology may provide a clue to the evolutionary origin of peroxisomes and merits additional study (Figs. 5A and B).

Structure and Function of Peroxisomal ABC Transporters

Peroxisomal ABC transporters are a distinct subset of ABC transporters characterized by high sequence homology, location in the peroxisomal membrane, and a "half transporter" structure. Several lines of evidence suggest that the two halves of eukaryotic ABC transporters have separate functions. First, there is no well-documented evidence for homodimerization of half ABC transporters. By contrast, there is genetic and/or physical evidence for heterodimerization of eukaryotic half ABC transporters (e.g., the mammalian TAP1 and TAP2 proteins[86]; the *white, brown,* and *scarlet* proteins in *Drosophila*[87] and Pxa1p and Pxa2p[43]). Second, functional analysis of mutant CFTR proteins showed that the two NBDs have different roles in transporter function: ATP hydrolysis by NBD1 starts channel activity while hydrolysis by NBD2 terminates it.[88] Third, for several ABC transporters it has been shown that function is not reconstituted by expression of one half in the absence of the other (shown for Ste6,[89] MDR1,[90] CFTR,[91] and TAP[92]). Taken together, these observations suggest that heterodimerization is the rule for half ABC transporters. Furthermore, genetic studies

[86] A. Kelly, S. H. Powis, L. A. Kerr, L. Mockridge, T. Elliott, J. Bastin, B. U. Ziegler, A. Ziegler, J. Trowsdale, and A. Townsend, *Nature* **355,** 641 (1992).
[87] G. D. Ewart, D. Cannell, G. B. Cox, and A. J. Howell, *J. Biol. Chem.* **269,** 10370 (1994).
[88] M. R. Carson, S. M. Travis, and M. J. Welsh, *J. Biol. Chem.* **270,** 1711 (1995).
[89] C. Berkower and S. Michaelis, *EMBO J.* **10,** 3777 (1991).
[90] T. W. Loo and D. M. Clarke, *J. Biol. Chem.* **269,** 7243 (1994).
[91] L. S. Ostedgaard, D. P. Rich, L. G. DeBerg, and M. J. Welsh, *Mol. Biol. Cell* **5,** 191 (1994).
[92] T. H. Meyer, P. M. van Endert, S. Uebel, B. Ehring, and R. Tampe, *FEBS* **351,** 443 (1994).

with the guanine/tryptophan transporters of *Drosophila* pigment cells suggest that different combinations of the half ABC transporters *white, brown,* and *scarlet* proteins form transporters with different substrate specificity. The *white* protein appears to partner with *brown* to form a guanine transporter and with *scarlet* to form a tryptophan transporter.[87] A similar possibility was predicted for the mammalian peroxisomal half ABC transporters.[47,93]

The four half ABC transporters so far identified in the mammalian peroxisome, PMP70, ALDP, ALDR, and P70R have different yet overlapping expression patterns.[47,70,71] Based on the precedent from *Drosophila* and the number of available partners, we suggest that in mammalian cells various combinations of peroxisomal half ABC transporters assemble into functional transporters with different properties (Fig. 6). This would provide a mechanism for cells and/or tissues to adjust the metabolic capabilities of their peroxisomes. In *S. cerevisiae* there are only two peroxisomal half ABC transporters, Pxa1p and Pxa2p, that heterodimerize to form the functional PXA transporter. We note, however, that in contrast to Pxa1p, Pxa2p is not destabilized by the absence of its counterpart. This may reflect interactions of Pxa2p with other protein(s) (Fig. 6).

The precise function(s) and ligand(s) are not known for any of the peroxisomal ABC transporters. Some indication can be deduced from the biochemical phenotypes caused by their dysfunction. The biochemical hallmark of X-linked adrenoleukodystrophy is elevated tissue and plasma levels of saturated VLCFA due to reduced peroxisomal β-oxidation of these substances.[54] The disease results from ALD mutations.[45,63] Thus, the reduction in β-oxidation is not due to a defect in the peroxisomal β-oxidation machinery but rather to the initial steps of fatty acid activation or transport.[54,61,94] To be oxidized in the peroxisome matrix, fatty acids must be activated by esterification to CoA. In mammalian cells, VLCFA CoA synthetase appears to be located on the cytoplasmic side of the peroxisome membrane.[94,95] Therefore, a reasonable prediction of the function of ALDP is transport of CoA-esterified VLCFA into the peroxisome matrix. The very recent report[96] of cloning of cDNA encoding a rat VLCFA CoA synthetase should stimulate direct studies of the interaction of this protein with ALDP.

[93] J. Gärtner and D. Valle, *Nature* **361,** 682 (1993).

[94] I. Singh, O. Lazo, G. S. Dhaunsi, and M. Contreras, *J. Biol. Chem.* **267,** 13306 (1992).

[95] W. Lageweg, J. M. Tager, and R. J. A. Wanders, *Biochem. J.* **276,** 53 (1991).

[96] A. Uchiyama, T. Aoyama, K. Kamijo, Y. Uchida, N. Kondo, T. Orii, and T. Hashimoto, *J. Biol. Chem.* **271,** 30360 (1996).

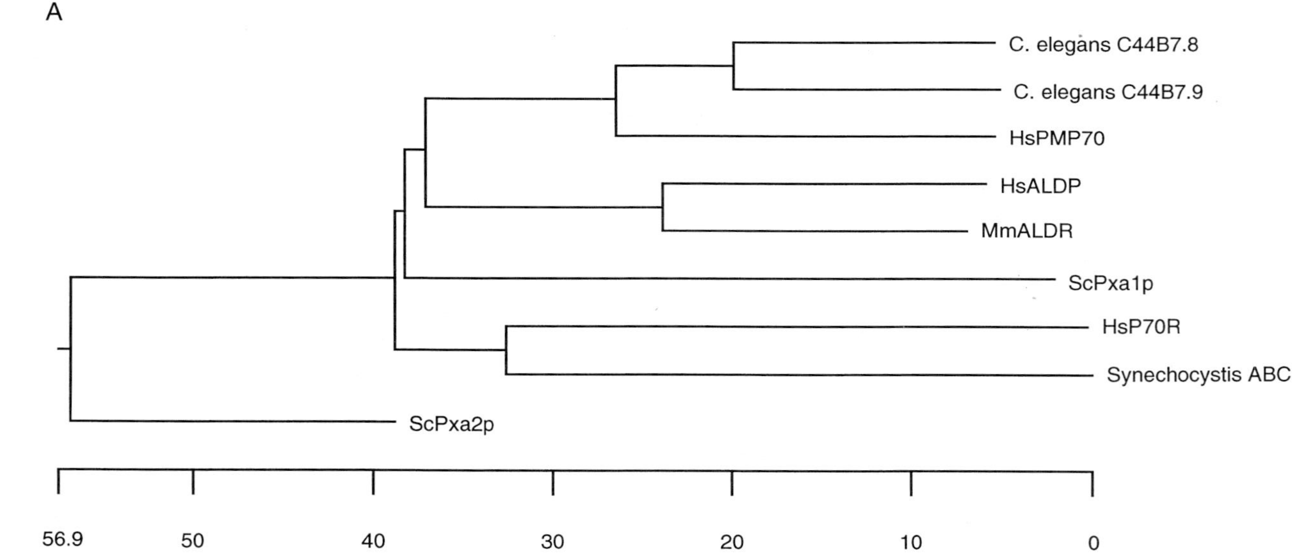

FIG. 5. Alignment of the NBDs and a phylogenetic tree of peroxisomal ABC transporters. Both the alignment and the phylogenetic tree were made using the MegAlign program from DNA-STAR. (A) A phylogenetic tree of the peroxisomal ABC transporters from human (HsPMP70 and HsALDP), murine (MmALDR), *S. cerevisiae* (ScPxa1p and ScPxa2p), and three open reading frames coding for putative peroxisomal ABC transporters from *C. elegans* (CeC44B7.8 and CeC44B7.9) and from the cyanobacterium *Synechocytis* sp (*Synechocystic* ABC). A third human protein (P70R, see text). The length of each pair of branches represents the distance between protein pairs. The scale beneath the tree measures the distance between proteins and the units indicate the number of substitution events. (B) The NBD of the proteins from (A) were aligned. Amino acids present in three or more of these sequences are boxed in black.

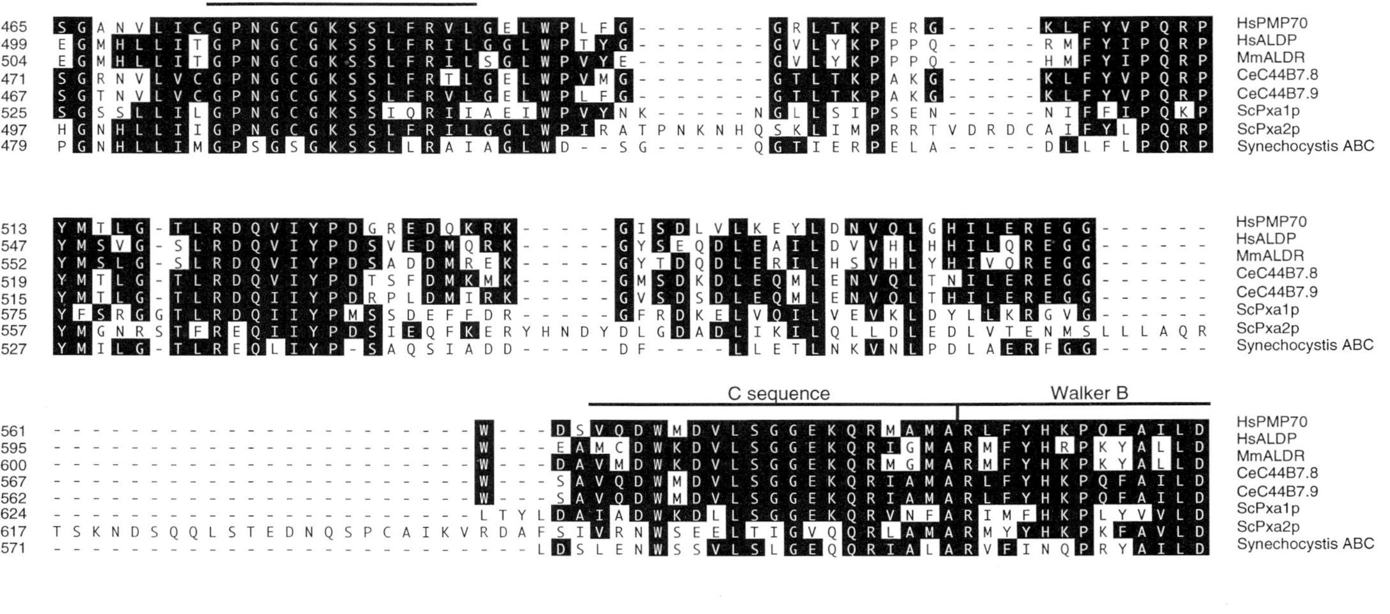

FIG. 5. (*continued*)

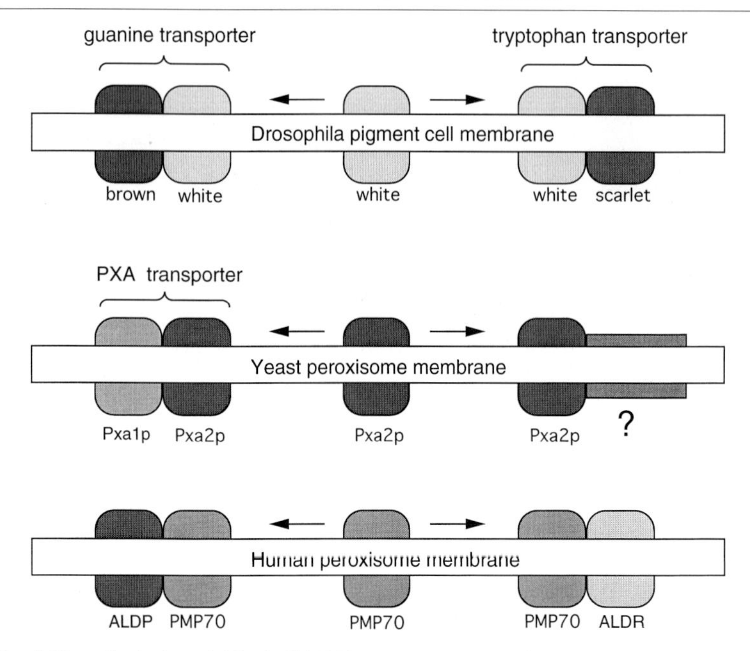

FIG. 6. Hypothetical model for half ABC transporter heterodimerization. A model illustrating formation of different complete ABC transporters by heterodimerization of one half-molecule with two different partners. The possibility that Pxa2p interacts with some other protein is indicated by the "?."

Methods

Yeast Culture Medium Containing Oleic Acid as Sole Carbon Source

In *S. cerevisiae*, β-oxidation of fatty acids is limited to peroxisomes.[97] Therefore, yeast growth on medium containing a fatty acid (e.g., oleic acid) as a sole carbon source requires functional peroxisomes. Medium containing oleic acid as a sole carbon source was used for the isolation and study of several peroxisomal assembly (pas) mutants and in our investigations of the PXA transporter (Fig. 3).[43,44,79] For 20–25 petri plates (100 mm) we prepare oleic acid medium in final volume of 500 ml containing the following:

Yeast nitrogen base (without amino acids)	3.35 g (0.67% final concentration)
Yeast extract	0.5 g (0.1%)

[97] W. H. Kunau, S. Buhne, M. de la Garza, C. Kionka, M. Mateblowski, U. Schultz-Borchard, and R. Thieringer, *Biochem. Soc. Trans.* **16,** 418 (1988).

Agar	10 g (2%)
Oleic acid (Merck, extra pure)	0.5 ml (0.1%)
Tween 40 (Sigma, St. Louis, MO)	2.5 ml (0.5%)

The oleic acid and Tween-40 are mixed in 10 ml of water, vortexed to form an emulsion, and added to the final mixture.

All auxotrophic requirements are added from stock solutions. For strain CH1305 we add the following:

Adenine	40 mg/liter (final concentration)
Uracil	20 mg/liter
Leucine	60 mg/liter
Lysine	30 mg/liter
Histidine	20 mg/liter

The medium is sterilized by autoclaving for 20 min, cooked by shaking at room temperature, and decanted into 20–25 petri dishes, which are left at room temperature for several hours. Oleic acid plates dry very quickly, therefore it is important to carefully seal the plate edges with Parafilm during culture. Compared to rich media, yeast grow slowly on oleic acid. Wild-type yeast (strain CH1305) will form a medium-size colony on rich medium within 1–2 days at 30° and on oleic acid medium in 2–3 weeks. If an aqueous suspension of cells (OD_{600} of 2) is placed in 6-μl drops on the oleic acid medium, differences in growth between wild-type and mutant yeast can be observed within 2–3 days.

Purification of S. cerevisiae Peroxisomes

In *S. cerevisiae,* peroxisomal biogenesis can be induced by growing the yeast in the presence of oleic acid.[7] To isolate the peroxisomes we first separate the high-density organelles (mitochondria and peroxisomes) from the remainder of the cellular components using differential centrifugation. This is followed by centrifugation of the high-density pellet on a Nycodenz gradient, which separates the two organelles based on their density, the more dense peroxisomes sedimenting more quickly.

Induction of Peroxisome Biogenesis

For induction of peroxisome biogenesis we grow yeast in 0.1% oleic acid medium. We transfer a "match head" of yeast from a plate into 50 ml of YPD and incubate overnight on a shaker at 30°. In the morning we dilute the entire overnight solution with 450 ml of YPD and incubate for an additional 8 hr. We then pellet the cells at 1500*g* for 10 min in sterilized tubes, resuspend in 1 liter of induction media, and incubate overnight.

Induction Medium. To 1 liter of water we add 10 g yeast extract (BRL, Gaithersburg, MD), 20 g peptone (BRL), 1 ml oleic acid (Merck, extra pure), and 5 ml Tween 40 (Sigma). The oleic acid and Tween 40 are added as an emulsion in water as above. The medium is sterilized by autoclaving for 20 min and cooled at room temperature for several hours.

Isolation of High-Density Organelles from Induced Yeast

High-density organelles from oleic acid-induced yeast contain mostly mitochondria and peroxisomes.[98] To isolate the organelles, we pellet the cells from 1 liter of induced yeast at 1500*g* for 10 min and wash them once with 500 ml of water. The cells are then resuspended in 50 ml of TEB buffer (100 m*M* Tris, pH 7.5, 50 m*M* EDTA, and 35 μl 2-mercaptoethanol) and incubated for 30 min at room temperature. At this stage the OD_{600} is measured. We then pellet the cells again and wash them with 40 ml of SPP buffer (1.2 *M* sorbitol and 20 m*M* potassium phosphate, pH 7.5). The pellet is resuspended in 15 ml of SPP, and Zymolyase-T20 (ICN, Costa Mesa, CA) is added (0.48 μl/1 OD_{600} of cells from a fresh stock solution of 12 mg Zymolyase/ml SPP). After digestion of the cell wall with Zymolyase for 30 min at 30°, the spheroplasts are pelleted at 1500*g* for 10 min and resuspended by vortexing in 6 ml of Dounce buffer. Formation of spheroplasts can be followed microscopically under phase contrast using a 40× objective. Spheroplast formation is complete when essentially all the cells swell and burst in water but are small, round, bright refractive bodies in 1 *M* sorbitol.

Dounce Buffer. 0.6 *M* Sorbitol, 5 m*M* MES buffer, pH 6.0, 1 m*M* KCl, 0.1% (v/v) ethanol, 0.5 m*M* EDTA, 0.2 mg/ml sodium fluoride. To the cold buffer we add 2.5 mg/ml leupeptin and aprotinin (from a 10 mg/ml stock solution) and 10 μl/ml of freshly prepared stock phenylmethylsulfonyl fluoride (PMSF) (100 m*M* in 2-propanol).

We then homogenize the spheroplasts with 10 strokes on ice, using a Dounce homogenizer, B pestle, and pellet the cellular debris by centrifugation at 1500*g* for 10 min in screw-capped tubes (Oakridge) in a Beckman SS34 rotor. The supernatant is transferred to a new tube and the pellet resuspended in 6 ml of Dounce buffer, rehomogenized as above, and pelleted. The two supernatants are combined and remaining cellular debris removed by pelleting. The supernatant is transferred to a new tube and high-density organelles are collected by centrifugation at 25,000*g* for 30 min in the SS34 rotor. If the organelles are being harvested for subsequent isolation of peroxisomes on a Nycodenz gradient, we pellet them on a 4-ml Nycodenz cushion (see next section). The organelles are collected from the

[98] D. I. Crane, J. E. Kalish, and S. J. Gould, *J. Biol. Chem.* **269,** 21835 (1994).

interphase with a Pasteur pipette and the volume is brought to 4 ml with Dounce buffer. If organelles are to be used directly without subsequent purification, the pellet is resuspended in 1 ml of Dounce buffer and kept at 4°.

Fractionation of Organelles on a Nycodenz Gradient

The Nycodenz gradient is made in a Beckman polyallomer Quick-seal tube 1 day prior to use and stored at 4° overnight. We place 3 ml of Nycodenz (Puredenz from Accurate, Chemical and Scientific, Westbury, NY) at the bottom of the tube and 3 ml of a cushion solution (see tabulation below) over it. A 30-ml continous 15–30% gradient is then poured over the cushion (2 × 15 ml with a 1 ml/min flow rate) (see tabulation below).

Gradient Stock Solutions

50% (w/v) Nycodenz (Accudenz from Accurate)
85% Sucrose (w/v), heat to suspend
10× gradient buffer: 50 m*M* MES buffer; 10 m*M* EDTA; 10 m*M* KCl; 1% (v/v) ethanol

	15% Solution	35% Solution	Cushion
Nycodenz (50% w/v)	12 ml	28 ml	17 ml
Sucrose (85% w/v)	4 ml	2 ml	1 ml
10× buffer	4 ml	4 ml	2 ml
H_2O (double distilled)	20 ml	6 ml	—

The isolated yeast organelles (4 ml) are loaded carefully on top of the gradient; the tube is then sealed and centrifuged in a VTI 50 rotor (Beckman) at 30,000 rpm for 2.5 hr at 4° with low acceleration and deceleration. The gradient is collected from the bottom into 18 fractions of 2 ml and stored at −20°.

Gradient Analysis

In our experience the above gradient efficiently separates peroxisomes and mitochondria. Most of the peroxisomes are found in fractions 2–5, whereas mitochondria are found in fractions 9–13. From each fraction we use 10 μl to assay the activity of the peroxisomal marker catalase and 20 μl to assay the mitochondrial marker succinate dehydrogenase by standard methods.[98] To quantitate proteins, we add 100 μl TCA (100%, w/v) to 1 ml of each fraction. After 1 hr on ice the proteins are sedimented by centrifugation in a microfuge at 13,000g for 15 min at 4°. The pellet is resuspended in 50 μl of SDS–PAGE loading buffer containing 75 mM Tris added as Tris base. Equal volume from the different fractions can be electrophoresed and immunoblotted to compare protein content (Fig. 2).

[57] Mitochondrial ABC Transporters

By JONATHAN LEIGHTON

Introduction

The discovery of an ABC transporter in the mitochondria of the yeast *Saccharomyces cerevisiae*[1] was the result of a specific hunt for one, based on what turned out to be a correct guess that mitochondria would be as likely to contain ABC transporters as other organelles where they had already been found. When the project began, ABC transporters were already known to exist in the mammalian endoplasmic reticulum (ER)[2] and peroxisome[3] and to be encoded by a chloroplast genome,[4] and were soon also found in the fission yeast vacuole.[5] Mitochondria are widely believed to be the descendants of a bacterial endosymbiont,[6] and given the large number of ABC transporters that have been found in bacteria, it seemed rather unlikely that mitochondria would not themselves have retained their own ABC transporters. Mitochondria are metabolically highly active organelles that are involved in continuous substrate exchange with the rest of the cell. We hoped that the discovery of ABC transporters in mitochondria would give us greater insight into how mitochondria transport some substrates across their membranes, and perhaps how they transmit signals to the nucleus, which encodes the vast majority of mitochondrial proteins.

Several approaches were available. We could have attempted to detect an ABC transporter directly in mitochondrial membranes by labeling them with azido-ATP, or by blotting with ABC-specific antibodies. We opted for a somewhat less direct approach, namely, searching via PCR (polymer-

[1] J. Leighton and G. Schatz, *EMBO J.* **14,** 188 (1995).
[2] J. J. Monaco, *Immunol. Today* **13,** 173 (1992).
[3] K. Kamijo, S. Taketani, S. Yokota, T. Osumi, and T. Hashimoto, *J. Biol. Chem.* **265,** 4534 (1990).
[4] K. Ohyama, H. Fukuzawa, T. Kohchi, T. Sano, S. Sano, H. Shirai, K. Umesono, Y. Shiki, M. Takeuchi, Z. Chang, S. Aota, H. Inokuchi, H. Ozeki, *J. Mol. Biol.* **203,** 281 (1988).
[5] D. F. Ortiz, L. Kreppel, D. M. Speiser, G. Scheel, G. McDonald, and D. W. Ow, *EMBO J.* **11,** 3491 (1992).
[6] T. Cavalier-Smith, *Ann. N.Y. Acad. Sci.* **503,** 55 (1987).

0076-6879/98 $25.00

ase chain reaction) for new ABC genes in the yeast genome and then attempting to identify those that code for mitochondrial proteins. The advantage of this approach was that gene fragments could be obtained quickly and immediately used for further experiments. This review elaborates on how certain experimental procedures were used in the context of the project.

Using PCR to Search for New Genes

The presence of typical sequence motifs in classes of proteins is often taken advantage of to identify new members of the same family. This is the approach we took to identify new ABC transporters in yeast, an approach also used successfully by other groups to identify ABC genes in yeast[7,8] and humans.[9] When the project was begun, only two ABC transporters were known in *S. cerevisiae,* Ste6p,[10,11] the exporter of the yeast mating pheromone **a**-factor, and the *ADP1*[12] gene discovered during the sequencing of chromosome III. The size of the yeast genome suggested, however, that many more might be found.

We began by comparing the amino acid sequences of the ATP-binding cassettes of several ABC transporters and determining which regions showed sufficient conservation for use in designing degenerate oligonucleotide primers. Two conserved regions with sequences typified by SGSGKST and DEATSALD were selected. These sequences flank the ATP-binding cassette and thus allow the rapid identification of new ABC genes by sequencing of the cloned PCR fragments. Serine and leucine residues complicate primer design because each can be encoded by six different codons, and it is therefore best to choose sequences with a minimal number of these residues, when possible.

To maximize the number of new genes amplified while still maintaining specificity, we have designed separate primers to encompass the two classes of serine codons (TCT/C/A/G and AGT/C) for one serine in each sequence. Mixes of two nucleotides are used at several positions, as appropriate, and inosine (I) is used at positions where three or four nucleotides could be present, because it is neutral in terms of binding affinity and does

[7] K. Kuchler, H. M. Göransson, M. N. Viswanathan, and J. Thorner, *Cold Spring Harbor Symp. Quant. Biol.* **57,** 579 (1992).

[8] M. Dean, R. Allikmets, B. Gerrard, C. Stewart, A. Kistler, B. Shafer, S. Michaelis, and J. Strathern, *Yeast* **10,** 377 (1994).

[9] M. F. Luciani, F. Denizot, S. Savary, M. G. Mattei, and G. Chimini, *Genomics* **21,** 150 (1994).

[10] J. P. McGrath and A. Varshavsky, *Nature* **340,** 400 (1989).

[11] K. Kuchler, R. E. Sterne, and J. Thorner, *EMBO J.* **8,** 3973 (1989).

[12] B. Purnelle, J. Skala, and A. Goffeau, *Yeast* **7,** 867 (1991).

not increase the degeneracy of the oligonucleotides, which is kept in the range of 8- to 32-fold. To facilitate the cloning of the PCR fragments, we incorporate *Bam*HI and *Eco*RI restriction sites into the 5′ ends of the upstream and downstream primers, respectively. We also include additional nucleotides before the restriction sites to improve the efficiency of digestion. The use of restriction sites allows the cloning of the PCR products both with high efficiency and unidirectionally. A disadvantage is the potential for PCR fragments to be cleaved internally by a restriction enzyme, thereby reducing the size of the cloned fragment. This problem can be reduced by using enzymes that cut rarely (e.g., 8-base cutters) and an appropriate vector. Alternatively, one can fill in the ends of the PCR fragments and then clone by blunt-end ligation, or use one of the commercially available "T-tailed" vectors intended for cloning PCR fragments (PT7Blue, Novagen, Madison, WI; TA Cloning System, Invitrogen, Carlsbad, CA).

PCR reactions are performed in a volume of 100 μl, containing 10 ng of yeast genomic DNA,[13,14] 200 μM of each dNTP, 10 μl of a standard 10× PCR buffer (100 mM Tris–HCl, 15 mM $MgCl_2$, 500 mM KCl, 1 mg/ml gelatin, pH 8.3) and pooled upstream and downstream oligonucleotide primers at concentrations such that 5 pmol of each unique oligonucleotide is present. The more degenerate oligonucleotides are thus added in proportionately higher amounts. The reaction mixtures are overlaid with paraffin oil and heated at 94° for 2 min; this "hot start" is helpful in reducing background. *Taq* polymerase (2.5 units) is then added and a program performed of 30 cycles of 1 min at 94°, 2 min at 45°, and 3 min at 72°, followed by a final 7 min at 72°. The wide range of T_m's among the oligonucleotide pool made it impossible to choose an ideal annealing temperature, but 45° is below the T_m of every oligonucleotide used and proved to be optimal. Among the 10 ABC-homologous PCR fragments that were finally identified, four correspond to genes whose sequences were subsequently published by us or others. A comparison of the oligonucleotide primers with the corresponding sequences of the four genes reveals that single-nucleotide mismatches in one or both primers are tolerated under our amplification conditions.

The PCR products are electrophoresed in 2% agarose gels with wide wells and containing ethidium bromide. The bands are visualized with ultraviolet light, and those of approximately the expected size are excised individually. The DNA is then purified using either electroelution onto

[13] H. Riezman, T. Hase, A. P. G. M. van Loon, L. A. Grivell, K. Suda, and G. Schatz, *EMBO J.* **2,** 2161 (1983).

[14] A. Wach, H. Pick, and P. Philippsen, *in* "Molecular Genetics of Yeast: a Practical Approach" (J. R. Johnston, ed.), p. 1. Oxford University Press, Oxford, 1994.

DEAE paper (Schleicher and Schüll, Keene, NH) or the Gene Clean kit (Bio 101, La Jolla, CA). Kits from Qiagen (Hilden, Germany) for the purification of DNA from gel slices are also practical. Restriction digestion is performed before or after running the PCR products on a gel, although it is preferable to visualize the PCR products before digesting. The DNA is then cloned into the pUC18 vector.[15] Bacterial transformations are spread onto plates containing X-Gal and IPTG (isopropyl-β-D-thiogalactopyranoside), and white colonies are picked. Clones are analyzed by restriction digestion with *Bam*HI and *Eco*RI, to confirm the presence and size of inserts, and by sequencing with the pUC reverse sequencing primer, using the United States Biochemical DNA sequencing kit (Cleveland, OH).

Gene Disruptions Using PCR Fragments

The 10 cloned PCR fragments with homology to the ABC family allow us to probe the role of the corresponding proteins by performing gene disruptions, without having to clone the complete genes. All but one of the cloned fragments are longer than 300 base pairs, and are thus potentially long enough to obtain disruptions in the corresponding genes at reasonable frequencies. Two approaches are used: In one approach, the PCR fragment is subcloned via the *Bam*HI and *Eco*RI sites into the yeast integrating vector YIP5,[16] containing the *URA3* marker. This construct is cut at a unique restriction site as close as possible to the middle of the PCR fragment, to ensure a sufficiently long stretch of genomic DNA on either side. The DNA is then used to transform wild type yeast,[17] and Ura^+ transformants are selected. In the other approach, a selectable marker such as *URA3* or *LEU2* is cloned into a unique restriction site of the PCR fragment while in pUC18, again choosing a site near the middle of the fragment. The PCR fragment bearing the marker is then excised from the vector using *Bam*HI and *Eco*RI and used to transform yeast. With both approaches, transformants are tested for disruption of the intended gene by Southern blotting, using the original PCR fragment as a probe [prepared with the ECL (enhanced chemiluminescence) kit, Amersham (Little Chalfont, UK)].

Hybridization is performed in a solution of 10% (w/v) dextran sulfate, 1% (w/v) SDS, and 1 *M* NaCl, overnight at 68°. Two low-stringency washes were performed of 15 min in 2× SSC, 1% SDS, and 15 min in 2× SSC, 0.1% SDS, both at room temperature, before a high-stringency wash of

[15] C. Yanisch-Perron, J. Vieira, and J. Messing, *Gene* **33,** 103 (1985).

[16] K. Struhl, D. T. Stinchcomb, S. Scherer, and R. W. Davis, *Proc. Natl. Acad. Sci. U.S.A.* **76,** 1035 (1979).

[17] D. Gietz, A. St. Jean, R. A. Woods, and R. H. Schiestl, *Nucl. Acids Res.* **20,** 1425 (1992).

1 hr at 65° in 0.5× SSC, 0.1% SDS (20× SSC is 3 *M* NaCl, 0.3 *M* trisodium citrate.) Despite the sequence homology between the fragments, the hybridization and wash conditions used are sufficiently stringent that, in most cases, only the gene of interest is detected.

The two approaches used here for gene disruptions involve additional subcloning and require the presence of unique restriction sites within the PCR fragments. An alternative approach that has been used successfully involves the synthesis of long oligonucleotides with about 35–50 nucleotides of sequence of the gene to be disrupted at their 5′ ends, and sequence from a marker gene at their 3′ ends. The marker is amplified by PCR and the product can be used directly to transform yeast. Although this approach is costlier, no cloning steps are involved and there is no need for unique restriction sites.[18] The availability of both auxotrophic markers and dominant antibiotic-resistance markers of nonyeast origin can boost the efficiency of integration at the gene of interest to close to 100%.[19]

We have successfully disrupted five of the eight genes we attempted to disrupt. Transformants from the unsuccessful disruption experiments yield ambiguous Southern blot patterns, suggesting that some PCR fragments preferentially integrate elsewhere within the genome.

Although an unambiguous determination of the phenotype of a null mutant requires the disruption of a diploid, followed by sporulation and the dissection of tetrads, we initially looked at the phenotype of haploid disruptants on plates containing either glucose or nonfermentable carbon sources (ethanol and glycerol; lactate), growth on the latter requiring respiring mitochondria. None of the viable haploid disruptants showed a growth defect on either carbon source. However, we were unable to obtain viable haploid disruptants for one gene. The dissection of tetrads from a diploid disrupted in one copy of this gene showed a profound reduction in growth in spores containing the disrupting marker. We therefore decided to clone and sequence the gene, later named *ATM1,* and then localized the protein product within the cell.

Epitope Tagging

To localize the protein within the cell by immunofluorescence and biochemical fractionation, we placed a c-*myc* epitope at its C terminus. The epitope tagging approach allows the use of a commercially available,

[18] M. C. Lorenz, R. S. Muir, E. Lim, J. McElver, S. C. Weber, and J. Heitman, *Gene* **158,** 113 (1995).

[19] A. Wach, A. Brachat, R. Pöhlmann, and P. Philippsen, *Yeast* **10,** 1793 (1994).

highly specific monoclonal antibody of known quality, eliminating problems of cross-reactivity as well as the delay in obtaining a clean polyclonal antiserum against the protein of interest. One can also easily perform double-label immunofluorescence, using a rabbit antiserum against another marker. A disadvantage is the relatively low avidity of a monoclonal antibody. For this reason, it is not always as effective as a polyclonal antiserum for some purposes, for example, immunoprecipitation. Furthermore, there is the risk that the tag will lead to mislocalization of the protein. This likelihood can be reduced by demonstrating that the tagged protein is functional, which is, of course, not always possible, particularly if there is no null-mutant phenotype. Tagging the protein in more than one place and showing the same localization can provide strong evidence against an artefact.

To epitope-tag *ATM1,* we use PCR to amplify a short 3′-terminal region of the gene, using a downstream primer that encodes the six C-terminal amino acids of Atm1p, the 10 amino acids EQKLISEEDL from c-*myc,* which are recognized by monoclonal antibody Myc1-9E10,[20] and a stop codon. The 3′-terminal region of the gene, which had been cloned with its own promoter into both single- and multicopy yeast vectors, is excised and replaced with the tagged version, using appropriate restriction sites that had been incorporated into both primers.

Alternative approaches for epitope tagging include a more general four-primer PCR method used for performing site-directed mutagenesis, traditional oligonucleotide-based site-directed mutagenesis, as well as specialized vectors into which the gene to be tagged can be directly cloned,[21] or from which an epitope can be excised and then inserted in-frame at an internal restriction site within the gene.[22] We placed the epitope tag at the C terminus of Atm1p because others have tagged ABC transporters at the C terminus without abolishing function[23,24]; we subsequently showed that tagged Atm1p is indeed functional in rescuing the slow growth of the null mutant. However, other proteins may need to be tagged elsewhere to avoid mislocalization, particularly if a C-terminal targeting or retention signal is present. The approach used to tag Atm1p involved removing the entire downstream region of the gene, thereby eliminating the authentic transcription terminator. Although terminators are often not essential for expression,

[20] G. I. Evan, G. K. Lewis, G. Ramsay, and J. M. Bishop, *Mol. Cell Biol.* **5,** 3610 (1985).

[21] P. Reisdorf, A. C. Maarse, and B. Daignan-Fornier, *Curr. Genet.* **23,** 181 (1993).

[22] P. Surdej and M. Jacobs-Lorena, *BioTechniques* **17,** 560 (1994).

[23] P. Juranka, F. Zhang, J. Kulpa, J. Endicott, M. Blight, I. B. Holland, and V. Ling, *J. Biol. Chem.* **267,** 3764 (1992).

[24] K. Kuchler, H. G. Dohlman, and J. Thorner, *J. Cell Biol.* **120,** 1203 (1993).

and indeed tagged Atm1p is expressed without one, it is safer to include a terminator. R&D Systems (Abingdon, UK) makes a line of yeast expression vectors with different promoters, all containing a 3′ transcription terminator and, furthermore, a hemagglutinin tag, which can be attached to the 3′ end of the gene by appropriate cloning.

Immunofluorescence

Although mammalian cells are more amenable than yeast to immunofluorescence, satisfactory results with yeast are attainable and an unambiguous determination of the intracellular location of a protein is possible, provided good antibodies against different intracellular marker proteins are available as controls. Immunofluorescence on yeast expressing c-*myc*-tagged Atm1p from a multicopy 2μ plasmid demonstrated a clear colocalization with mitochondrial DNA (stained with DAPI (4′,6-diamidino-2-phenylindole), see below) and with the mitochondrial outer membrane protein porin. Expression could not be detected from a single-copy CEN plasmid. The immunofluorescence protocol used was adapted from that provided by Hall *et al.*[25]

Fixing Cells

1. Inoculate 20 ml of medium with 1 ml of overnight culture. Incubate with aeration at 30° to OD_{600} of 0.2–0.6 (about 4 hr). Alternatively, inoculate the night before, aiming for the same OD_{600}.
2. Add 2.5 ml of 1 *M* KPO_4 buffer, pH 6.5 (46.6 ml 1 *M* K_2HPO_4, 53.4 ml 1 *M* KH_2PO_4), and 2.5 ml of 37% (w/v) formaldehyde. Incubate at room temperature for 2 hr with occasional gentle shaking.
3. Spin 5 min at 3000 rpm. Resuspend in 1 ml SP (1.2 *M* sorbitol, 0.1 *M* KPO_4 buffer, pH 6.5), transfer to an Eppendorf tube, spin 5 sec at about 7000 rpm, repeat the wash 2×. Resuspend in 1 ml SPβ (SP + 100 m*M* 2-mercaptoethanol = 7 μl/ml).
4. Add 10 μl Zymolyase 20T (Seikagaku, Tokyo, Japan), 12.5 mg/ml in SPβ. Incubate at 37° for 1 hr to digest the cell walls.
5. Harvest (5 sec at 7000 rpm). Wash gently with SP, resuspend in 1 ml SP. Store at 4° if necessary.

Treatment of Slides

1. Wash slides (Dynatech, Chantilly, VA; 7-mm wells) with ethanol. Add 10 μl 0.1% poly(L-lysine) (Sigma, St. Louis, MO, molecular weight > 300,000; 1 mg/ml in H_2O) to each well, leave 10 sec or longer. Aspirate. Let dry. Wash 3× with H_2O. Let air dry completely.

[25] M. N. Hall, C. Craik, and Y. Hiraoka, *Proc. Natl. Acad. Sci. U.S.A.* **87,** 6954 (1990).

Antibody Treatment

Note: Keep slides in a humid chamber (plastic box with wet paper towel) during subsequent treatments. Always add subsequent solutions to each well immediately after aspirating, so that wells do not dry out.

1. Apply 10 μl of fixed cells to each well of a treated glass slide. Allow 30 min for cells to settle and adhere.
2. Rinse 3× with PBS (phosphate-buffered saline). Add 10 μl PBT [PBS, 1% bovine serum albumin (BSA), 0.1% Triton X-100, 0.1% azide]. Incubate 5 min.
3. Aspirate. Add 10μl primary antibody diluted in PBT. Incubate for 30 min to 1 hr.
4. Rinse 10× with PBT. Add 10 μl conjugated (rhodamine, fluorescein) secondary antibody diluted in PBT. Incubate 30 min to 1 hr in the dark.
5. Rinse 10× with PBS. Add 5 μl of mounting solution and cover with a cover slide; store at −20° in the dark. [Glycerol works fine as a mounting solution. It can be helpful to include an antibleaching agent, especially with fluorescein: Dissolve 100 mg phenylenediamine in 10 ml PBS, add 90 ml 87% (v/v) glycerol, adjust to pH 8.0 with a few drops 0.5 *M* Na_2CO_3, pH 9.0; store at −20° in the dark. Before use of the mounting solution, it is useful to add 2 μl of 10 mg/ml DAPI per ml of mounting solution for visualization of nuclear and mitochondrial DNA.]
6. View with fluorescent microscope equipped with proper filters, using a 63× or 100× objective.

Submitochondrial Localization

Immunoblotting of Nycodenz-purified mitochondria[26] from cells expressing c-*myc*-tagged Atm1p confirmed the mitochondrial localization of Atm1p. To try to determine the intramitochondrial localization of Atm1p, we determined the protease accessibility of the tagged protein in mitoplasts (mitochondria subjected to osmotic shock to disrupt the outer membrane), in intact mitochondria, and in mitochondria treated with detergent to disrupt both membranes. A procedure for determining the submitochondrial location of a protein is available elsewhere[27]; the following protocol was adapted for the purpose of localizing c-*myc*-tagged Atm1p.

[26] B. S. Glick and L. A. Pon, *Methods Enzymol.* **260,** 213 (1995).
[27] B. S. Glick, *Methods Enzymol.* **260,** 224 (1995).

Crude mitochondria (250 μg; for determining the submitochondrial location of Atm1p, purification of mitochondria on a Nycodenz gradient is unnecessary), stored frozen with 10 mg/ml fatty acid-free BSA (bovine serum albumin), are washed in mitochondrial breaking buffer (0.6 *M* sorbitol, 20 *M* K^+-HEPES, pH 7.4) and resuspended in 450 μl import buffer.[27] Four 100-μl aliquots are transferred to new microfuge tubes, to be treated as follows: Sample 1, untreated; sample 2, protease-treated; sample 3, subjected to detergent and protease; and sample 4, subjected to osmotic shock in the presence of protease. Sample 3 is first washed again with breaking buffer to remove the BSA and resuspended in 100 μl breaking buffer. Sample 1 is left untreated; sample 2 receives 1 μl of 10 mg/ml proteinase K; sample 3 receives 100 μl of 2% (w/v) octylglucoside and 2 μl of proteinase K; and sample 4 receives 700 μl of mitoplasting buffer (20 m*M* K^+-HEPES, pH 7.4) containing 8 μl of 10 mg/ml proteinase K. Samples are incubated for 30 min on ice. To inactivate the proteinase K, samples 2, 3, and 4 receive 1 m*M* PMSF (phenylmethylsulfonyl fluoride) from a freshly prepared 200 m*M* solution in ethanol. Samples 1, 2, and 4 are spun for 5 min, the supernatants aspirated, and the pellets resuspended in 180 μl breaking buffer containing 0.2 m*M* PMSF.

All four samples are transferred to new tubes, and TCA (trichloroacetic acid) is added to 5% (w/v) from a 50% stock solution. The tubes are heated for 5 min at 60° (to inactivate any residual protease activity irreversibly), left 5 min on ice, and then spun for 10 min. The supernatants are completely aspirated to remove the TCA, and the pellets are resuspended in 40 μl of SDS-containing sample loading buffer containing 1 m*M* PMSF and 50 m*M* Na^+-PIPES, pH 7.5. Samples are heated for 5 min at 55° (to avoid possible aggregation of Atm1p at higher temperatures) and then subjected to 10% SDS–PAGE. Gels are blotted and probed with the c-*myc* antibody, as well as with antisera against the following proteins: porin, an outer membrane protein that is inherently protease resistant, even in the presence of detergent, and which served as a control for the amount of protein loaded; cytochrome b_2, an intermembrane space protein that is released by mitoplasting; and α-ketoglutarate dehydrogenase, a matrix protein that is highly protease sensitive and served as a control for the intactness of the inner membrane. In mitoplasts treated with protease, tagged Atm1p remains intact and is still recognized by the c-*myc* antibody, while the tag is fully digested in the presence of the detergent. This result indicates that the C terminus bearing the epitope tag is located in the matrix, and that the protein must therefore be situated in the inner membrane.

This procedure allows an unambiguous localization of Atm1p and a determination of its membrane orientation by taking advantage of the specific detection of the C terminus of the protein. However, had the c-*myc* epitope been protease digested in mitoplasts as well, this result would have indicated only that the C terminus faced the intermembrane space, and would have necessitated the preparation of inner and outer membrane vesicles[28] to determine the membrane localization. An alternative approach to determining the submitochondrial location of a protein involves the selective rupture of the outer membrane at low detergent concentrations.[29]

Determining Function

Several observations made of the *atm1* disruptant provided possible clues to the function of Atm1p:

1. The homology of Atm1p to ABC export proteins, and the orientation of Atm1p in the mitochondrial inner membrane, lead us to predict that Atm1p is involved in the *export* of a substrate from the mitochondrial matrix. This assumption narrows the range of possible functions.
2. The initial slow growth of the disruptant after tetrad dissection is mitigated within days, reflecting adaptation or the appearance of a spontaneous suppressor activity. However, no multicopy suppressor of the slow-growth phenotype (other than *ATM1* itself) could be identified by transforming the disruptant with a 2μ yeast genomic library.
3. The *atm1* disruptant has an unstable mitochondrial genome (rapidly becoming ρ^-) and has a white color, reflecting a loss of all major cytochromes. This loss includes cytochrome *c*, which does not require the mitochondrial genome for its synthesis.
4. Although the lack of cytochromes suggested that Atm1p might function as an exporter of heme from the mitochondrial matrix, a heme assay[30] showed no accumulation of heme in mitochondria from the disruptant, and exogenous heme did not improve the growth of the disruptant. Also, an *atm1* disruptant carrying the *hem1* mutation (which inactivates the first enzyme in heme synthesis) showed no improvement in growth over the *atm1* disruptant. (These experiments were carried out in the presence of exogenous heme that allowed the *hem1* disruptant to grow well.) Thus, the growth defect in *atm1* cells is unlikely to be due either to a toxic accumulation of heme within the matrix or to an inavailability of heme in other parts of the cell.
5. The disruptant is more sensitive to the iron chelator BPS (bathophenanthroline sulfate) than the corresponding ρ^- wild type. This observation

[28] M. Horst, P. Jenö, and N. G. Kronidou, *Methods Enzymol.* **260,** 232 (1995).
[29] K. Hahne, V. Haucke, L. Ramage, and G. Schatz, *Cell* **79,** 829 (1994).
[30] J. S. Rieske, *Methods Enzymol.* **10,** 488 (1967).

is consistent with a (secondary) defect in heme biosynthesis that is exacerbated by reduced iron availability.

6. Preliminary measurements of heavy metals in purified mitochondria (performed at Ciba, Basel) suggested no significant increase in the disruptant compared with the wild type, arguing against a role for Atm1p in heavy metal ion export.

7. There appears to be no significant difference in lipid content in mitochondrial membranes from the disruptant and wild type (performed by T. de Kroon, Utrecht), arguing against a role for Atm1p as a lipid "flippase" (analogous to a proposed role for the Mdr1 protein,[31] whose absence would lead to distortions in the membrane compositions.

8. *atm1* disruptants have a partial leucine requirement. Cultures of *atm1* containing all but one of the amino acids found in casamino acids (Difco, Detroit, MI) grew unless leucine was omitted. Since valine and leucine share the first three mitochondrial steps in their biosynthetic pathways, the cells seem to be blocked in the leucine pathway after its divergence from thc valine pathway.

9. α-Ketoisocaproate, which is imported into mitochondria before being converted to leucine, can substitute for leucine in allowing growth of *atm1,* suggesting a block in the pathway prior to this import step. This effect was only achieved after lowering the pH of the medium to 3.0; even at pH 3.6 α-ketoisocaproate could not substitute for leucine, probably because a charged molecule cannot readily enter yeast cells. Further experiments to examine the ability of α-isopropylmalate, the intermediate immediately preceding α-ketoisocaproate, to substitute for leucine were hampered by the inability of α-isopropylmalate to enter the yeast cells, even at low pH, as determined by the inability of this compound to complement the leucine auxotrophy of a strain lacking isopropylmalate synthase.

Conclusion

Although it was finally not possible during the course of the project to identify the substrate transported by Atm1p, the observations made suggest other experiments which could be carried out and, more generally, reveal how the function of other (yeast) ABC transporters could be addressed, including ABC transporters discovered through the now-completed yeast sequencing project, and other known and yet-to-be-discovered ABC transporters in other organisms. It is very likely that additonal ABC transporters will be found in the mitochondria of yeast and other organisms. Some of

[31] C. F. Higgins and M. M. Gottesman, *Trends Biochem. Sci.* **17,** 18 (1992).

the approaches described here may be useful in approaching the function of these proteins.

Acknowledgments

I thank Jeff Schatz, in whose laboratory this work was performed, for his support, as well as for reviewing the manuscript.

Author Index

Numbers in parentheses are footnote reference numbers and indicate that an author's work is referred to although the name is not cited in the text.

A

Abbott, B. J., 483
Abbracchino, M. P., 664
Able, C., 558
Abraham, E. H., 494, 665, 673
Abrahams, J. P., 17
Abram, T. S., 608, 615(20)
Abramson, T., 279
Ackerley, C. A., 617, 618(7), 625(7), 630, 637(5), 641(5), 650(5)
Ackerman, M. J., 368
Adam, M. A., 548, 723
Adames, N., 198
Adams, J. M., 575
Adams, M. D., 113, 757
Adams, R., 729, 731
Aebersold, P., 558
Afione, S., 369, 540, 653, 659(8), 720, 727, 728, 728(52, 53), 729, 729(6), 730, 731
Afoine, A., 676
Agbayani, R., Jr., 16, 17(76), 662(37), 663
Agresti, M., 178, 226, 293, 294, 294(15), 299, 300(15), 303, 303(17), 304, 304(33), 305(17, 49), 306, 306(17), 316, 401, 406, 478
Aguila, H. L., 550
Aguilar-Bryan, L., 110, 117, 132, 154(19, 20), 665, 673, 732, 736, 737, 738, 738(2, 3, 6), 740(3), 741, 743, 743(8), 744(1, 3)
Ahmad, S., 427(13), 428
Ahmed, K., 655
Ahmed, S., 40
Ahn, K., 120
Aihara, M., 262
Aihara, Y., 262
Aiken, P. M., 573, 589(9)
Air, G. M., 651
Akitama, S., 266
Akiyama, S., 238, 249, 259, 293, 303(19), 560
Aksentijevich, I., 226, 254, 255, 256(10), 257(10), 477, 523, 525, 526, 526(8), 527(7), 528(8), 530(7, 10), 538, 546, 547, 548, 557, 573, 574, 581, 589(11), 591
Alarco, A. M., 143(72), 156
Al-Awqati, Q., 653
Albritton, L. M., 242
Aldwin, L., 372
Aleksandrov, A. A., 616, 622
Aleksandrov, L. A., 622
Alexander, I. E., 728
Alfonzo, M., 21
Aljader, N., 676
Alkhatib, G., 473, 625, 641
Allen, S. S., 118, 624, 642, 650(37), 653, 659(8), 729
Alley, M. C., 483
Allikmets, R., 116, 117(2), 121(6, 8), 126, 130, 135, 139(39), 143(81), 152(39), 158, 160(39), 777
Allison, M. J., 662(37), 663
Alloing, G., 10
Almquist, K. C., 117, 133, 183, 493, 600, 602(19), 604, 607, 608, 609(1), 611(17), 613, 613(17), 615(17)
Alon, N., 117, 132, 154(15), 263, 343, 411, 617, 618, 619(4), 620(14), 621(14), 629, 630(2), 631(2), 633(2), 635(2), 642, 645(34), 646(34), 650(34), 687, 715
Al-Shawi, M. K., 318, 397, 398(11, 13), 401(11), 406(11, 13), 409(11, 13), 410(11, 13), 411, 411(11, 13), 487, 503, 514, 515, 517, 521(14, 16, 17, 19), 620
Altamirano, J., 360, 361(11), 365(11), 367(11), 368(11), 369, 369(11)
Altenberg, G. A., 360
Altenberg, G. S., 368
Alton, E. W. F. W., 707, 722

Altschul, S. F., 122, 138
Alvarez, L. J., 37, 38(41)
Alvarez-Leefmans, F. J., 360, 361(11), 365(11), 367(11), 368(11), 369(11)
Alvarez-Leefmans, F. V., 369
Amador, M. V., 183, 187(7)
Amara, J. F., 653, 656(7), 661(7), 677
Amariglio, N., 475
Amato, J. M., 306
Ambudkar, S. V., 16, 17(76), 21, 40, 45(8), 131, 318, 319(6, 7), 320, 320(7), 321, 321(7), 326(7), 328, 341, 342(18), 353, 355, 368, 372, 397, 398(8), 409(8), 410(8), 411(8), 427, 441, 443(2), 446(2), 450(2), 453(2), 473, 480, 481, 487(2), 492, 493, 493(3), 495, 496, 496(20, 22), 497(19, 20), 500(22, 23), 501, 503(19, 30), 504, 505, 505(3), 506(9), 511(12), 512, 513, 513(9), 515, 523, 574, 584, 656, 655, 659(24, 29), 660(29), 662(37), 663, 663(29), 668
Amemura, M., 43, 44
America, T., 756
American Type Culture Collection, 633, 637, 641, 642, 643, 643(30)
Ames, G. F., 5, 7, 8, 8(13), 9, 10, 10(39), 11, 13(31), 15, 16, 17(76), 19(31), 30, 32, 101, 112, 113, 114, 117, 131, 764, 765(78)
Ames, G. F.-L., 662(37), 663
Amicon, 634
Ämmälä, C., 365, 369(40)
Amman, E., 263
Ammerer, G., 204
Amzel, L. M., 500, 507
Anantharam, V., 501, 503(30), 662(37), 663
Anderie, I., 360
Andersen, J. S., 347, 348(24)
Anderson, K. S., 745, 753(9)
Anderson, M. P., 617, 617(8), 618, 647, 650(60), 652, 653, 656(7), 661(7), 676, 677, 714
Anderson, W. F., 40, 523, 548, 558
Anderson, W. R., 255
Andreeff, M., 546
Andrews, D. W., 266, 268, 268(17), 270(17)
Androlewicz, M. J., 745, 752, 753, 753(9)
Andrulis, I. L., 237
Anklesaria, P., 559
Annable, T., 590
Anné, U. K., 755, 757(19)
Antman, K., 538, 547
Antonarakis, S. E., 676
Aoyama, T., 769
Apasov, S. G., 546, 586
Appel, M., 473
Aran, J. M., 523, 537, 538, 540, 544(13), 545(13), 557
Arceci, R. J., 262, 266, 494, 590
Argast, M., 41, 47(11, 15)
Arias, I. M., 397
Arion, W. J., 655
Arispe, N., 113, 689, 690(7)
Ariyoshi, K., 531
Armentano, D., 714
Arnheim, N., 679
Arnold, D., 117, 118(15), 163, 745
Arnold, E., 474
Arnold, G. F., 474
Arnold, L. D., 421
Aronson, P. S., 654
Arrowsmith, C. H., 53, 61(13), 62(13)
Asami, N., 571
Ashbourne, K. J. D., 410
Ashby, M. N., 198
Ashcroft, F. M., 365, 369(40)
Ashcroft, S. J. H., 365, 369(40)
Asher, C. J., 84
Ashfield, R., 365, 369(40)
Ashton-Rickardt, P. G., 745, 746(10), 753(10)
Aspedon, A., 370
Assaraf, Y. G., 410
Aszalos, A., 365, 367(38)
Attaya, M., 745
Aubourg, P., 117, 132, 153, 756, 758(45), 759, 762, 765, 769(45, 46, 71)
Audigier, Y., 266
Audonnet, J.-C., 473
Auer, K., 47
Auerbach, W., 645
Auffray, C., 122
Augstein, J., 608, 615(18)
August, J. T., 655, 659(24)
Augustinas, O., 620, 630, 633(6), 637(6), 641(6), 642(6), 647(6), 649(6), 650(6)
Ausiello, D. A., 343, 367, 645
Austin, D., 40
Austin, J. B., 725
Ausubel, F. M., 596, 598(15), 599(15), 603(15)
Averboukh, L., 596
Averbuch, S. D., 294, 301(28)
Aviv, H., 598

Axelman, J., 759
Axenovich, S., 225, 227, 227(2), 239(2)
Azzaria, M., 220, 318, 380, 414, 421(3), 422(3)

B

Baas, F., 110, 262, 266, 607, 613
Bach, N. J., 300
Bae, H.-R., 663
Baes, M., 755
Bahram, S., 117, 118(15), 163, 745
Baichwal, V., 8, 10, 32, 113
Bailey, J. L., 321, 463, 506
Bairoch, A., 101
Baker, B. S., 214
Baker, K., 68
Baker, N., 293, 294(22), 301(22), 303(22)
Baker, R. M., 239, 246(26)
Bakker, C. T., 608, 609(13)
Bakker, C. T. M., 128, 132
Bakos, E., 226, 398, 427(16), 428, 430(16), 431(16), 439(16), 440(16)
Baldari, C., 596
Baldwin, T. O., 679
Balm, A. J. M., 262
Baltimore, D., 230, 232(21), 243
Balzi, E., 143(51, 53, 54), 151, 619(40), 626
Banderali, U., 359, 363
Banerjee, D., 546
Baneyx, F., 373
Bank, A., 538, 547, 551, 551(11)
Bankiatis, V., 37
Banks, T. C., 687
Banting, G., 649, 650(67)
Bao, K., 11
Barasch, J., 653, 721
Barbosa, J. A., 751
Barca, S., 262
Bargon, J., 647, 723
Baric, R. S., 109, 177
Barnard, E. A., 664
Barnes, D. A., 176
Barnouin, K., 607, 608(10), 609(10), 611(10), 612(10), 613(10), 614(10), 615(10)
Barone, L. R., 293, 300(26), 302(25), 303(25, 26)
Barras, F., 67
Barraza-Ortiz, X., 727
Barrell, B. G., 131, 134(14), 137(14)
Barrett, J. T., 252, 494
Barth, P. G., 759, 765(66), 766(66)
Barthelemy, M., 72
Bassett, D. E. J., 124, 134, 135, 135(35), 161(35, 38)
Bassett, J., 134, 135(34, 35), 161(33, 34)
Bassford, P. J., 30, 31(6)
Bassford, P. J., Jr., 15
Bastin, J., 768
Basu, A., 729
Baudard, M., 538, 540, 544(13), 545(13)
Baudhuin, P., 753
Bauer, K., 47
Baumann, U., 70
Baumgart, E., 755
Bavoil, P., 31
Beall, E., 10
Bear, C. E., 361, 617, 617(9), 618, 618(3, 7), 619(3), 620, 623(3), 624(3), 625, 625(7), 626, 630, 637(5), 641(5), 645, 646, 650(5), 652, 663(4), 676, 677, 691, 714
Beaton, A., 726
Beaudet, A. L., 687, 688(3), 696, 715
Beaudet, L., 397, 398, 400(23), 401(23), 414, 415(6), 423, 425, 473
Bebagliati, M. R., 270
Bebok, Z., 696
Beck, C., 237
Beck, J., 608
Beck, S., 163, 745
Beck, W. T., 227, 239(16), 259, 293, 294(18), 300, 301(18), 302(18), 303, 303(18), 304(18), 308
Beckman Instruments, 638
Beckwith, J., 6, 31, 35(45, 46), 39
Becq, F., 620, 645, 646(53)
Beddington, R., 576, 577(21)
Bedouelle, H., 25
Bedwell, D. M., 157, 195, 210(12), 211
Begg, M., 538, 547
Behr, J. P., 698, 704(3)
Behringer, R. R., 618, 622(15)
Beijnen, J. H., 110, 133, 151(26), 613
Beja, O., 266, 287, 308, 370, 374, 376, 378
Belas, R., 68
Belfield, G. P., 143(48), 145, 158(48)
Bell, A. W., 8, 116, 158
Bell, D. R., 117, 118(11)
Bell, L. A., 213, 213(5), 214, 214(3), 215(5), 218(5)

Bellamy, W. T., 572
Belley, M., 608, 613(19), 615(19)
Belote, J. M., 214
Beltzer, J. P., 279
Belzai, E., 618, 626(10)
Bendtzen, K., 721
Bennett, H. P. J., 338
Bennink, J. R., 120, 453
Benos, D. J., 689, 690(7)
Benz, R., 47
Beppu, T., 414
Berensci, K., 724
Berenson, R. J., 554
Berg, J. M., 755
Berg, P., 421
Berger, H. A., 212, 676
Berger, M., 720
Bergin, A., 759
Bergin, J., 759, 769(63)
Bergman, P. J., 357
Berkower, C., 130, 135(4), 136(4), 139(4), 143(72, 95), 145(4), 147(4), 156, 157, 768
Berlyn, M. K. B., 43
Bernards, A., 184
Bernelot-Moens, C., 474
Berns, A. J. M., 133, 151(26), 494
Berns, D. I., 725
Berns, K. I., 539, 559, 725, 726
Bernstein, J., 44
Berry, A. K., 691
Berta, G., 691, 722
Berteaux-Lecellier, V., 755
Berthelette, P. B., 714
Bertino, J. R., 259, 546
Bertolaet, B. L., 755, 757(19)
Bertuzzi, G., 343, 676
Beverley, S. M., 183, 184, 185(14), 187(14), 188, 189, 189(24)
Beyer, A., 763
Bhagat, S., 410, 518, 522
Bhardwaj, G., 117, 133, 183, 250, 594, 595, 600, 602, 602(19), 607, 609(1)
Bhardwaj, S. G., 493
Bhattacharjee, H., 82, 91
Bibi, E., 137, 266, 287, 308, 370, 371, 374, 374(1), 375, 376, 378, 379
Bick, T., 475
Biedler, J. L., 259, 290, 291, 298(11), 300(11), 311
Binet, R., 13, 75
Bingham, P. M., 214, 224
Birmingham, J., 14
Birnboim, H. C., 186
Bisbal, C., 152
Bishop, D. F., 525
Bishop, J. M., 170, 171(24), 204, 781
Bishop, L., 16, 17(76), 662(37), 663
Bissinger, P. H., 143(50), 151
Bissonnette, B. M., 289, 300(4)
Biwersi, J., 653
Bjorkman, J., 756
Blacklow, N. R., 725
Blacklow, S. C., 479
Blaese, R. M., 523, 558, 559(7), 566
Blanc, V., 118
Blanck, G., 117, 118(15), 163, 745
Blankenship, N. M., 610
Blattner, F. R., 84
Blight, M., 781
Blobcl, G., 266, 267, 272, 279, 756, 757, 763(53)
Bloch, J., 596
Blume, K. G., 262
Blundell, K., 198
Boat, T. F., 687, 688(3), 715
Bode, W., 68
Bodine, D. M., 547, 548
Bodnar, A. G., 754
Boer, R., 293, 303(24), 317
Bogenschutz, M. P., 175, 177(4)
Boguski, M. S., 122, 124, 134, 135, 135(34, 35), 160, 161(35, 38), 755
Bohen, S. P., 143(90), 160
Bohl, E., 13, 33
Bohm, B., 764, 765(77)
Böhme, M., 610, 611
Bolhuis, P. A., 759, 765(66), 766(66)
Bolhuis, R. L. H., 265
Bond, T., 646
Bond, T. D., 359, 360, 361, 361(11), 365(11), 367(11), 368(11), 369(11)
Bondi, E., 749
Bonetti, B., 157, 211
Bonner, W. M., 298
Boone, C., 198
Boos, W., 3, 4, 4(3), 7(3), 8, 9(28), 13, 14, 18(3), 21, 25, 26(21), 30, 32, 33, 37(21), 39(21), 40, 41, 42, 42(17), 43, 43(17, 20, 21), 44(20), 45, 45(14, 18, 20, 21), 46, 46(26), 47, 47(11, 15), 49, 49(46), 50

Booth, A. G., 654
Booth, I. R., 17
Boothroyd, J. C., 184
Borbolla, M. G., 84
Borchers, C., 293, 303(24), 317
Border, R., 698, 699(4), 704(6), 707(4)
Borel, J., 759
Borgese, F., 368
Borgnia, M. J., 410
Borst, P., 105, 110, 110(13), 128, 132, 133, 151(26), 154, 155(63), 167, 182, 183, 183(2), 184, 185(12, 16), 187(9, 12), 189(2), 191(2), 262, 266, 397, 411, 494, 607, 608, 608(9), 609(13), 611(9), 613, 614(9), 754
Borxmeyer, H. E., 554
Bosch, I., 105
Bosma, P. J., 128, 132, 608, 609(13)
Bossier, P., 143(92), 160, 767
Bott, K. F., 757
Bottega, R., 700
Boucher, R. C., 369, 397, 398(6), 409(6), 410(6), 411(6), 412(6), 427(5, 7), 428, 437(5), 438(5), 439(5), 440(5), 441(5, 7), 467, 487, 507, 618, 626(10), 651, 699, 714, 718
Bougneres, P.-F., 759
Bourne, H., 152
Boutterin, M., 661
Boyd, A. E., 132, 154(19)
Boyd, A. E. III, 736, 738, 738(2), 743(8)
Boyd, D., 6, 30, 31, 31(6), 35(15, 45, 46), 39, 46, 116, 121(6, 8), 126, 130, 135, 139(39), 143(81), 145(89), 152(39), 158, 159(53, 54), 160, 160(39), 193, 194, 195, 196(6, 10), 197(6), 202, 202(6), 203, 204(18), 210, 210(6), 221, 267, 369, 373, 378
Boyd, M. R., 483
Boyer, J. L., 610
Boyle, A. L., 579, 580(24)
Boyle, W. J., 354, 355(30)
Bozon, D., 676
Brachat, A., 133, 780
Bradely, G., 508, 509(18)
Bradford, M. M., 489, 736
Bradley, A., 474, 696
Bradley, G., 183, 260, 262, 263(25, 26), 298, 434, 453, 463, 497, 514, 515, 584, 620, 649
Bradley, J., 625
Bradley, N. A., 722
Brady, R. O., 525, 526(8), 528(8), 538, 557
Braell, W. A., 267, 279
Bragin, A., 288
Braiterman, L. T., 759, 765(62)
Brake, A. J., 206
Brandt, J. F., 548
Brandt, S. J., 547
Brasitus, T. A., 289, 300(4)
Brass, J. M., 14
Brault, M., 157, 398, 505
Braun, V., 6, 49
Braverman, N., 755, 757(18, 30)
Bravo, R., 476
Breathnach, R., 596
Brees, C., 755
Brendel, M., 143(52), 151
Brent, R., 596, 598(15), 599(15), 603(15)
Bresnahan, M., 117, 118(15), 163, 745
Breuer, W., 343, 645
Brickman, E., 30, 31(6)
Briddell, R. A., 548
Bridges, R. J., 625, 645, 652, 663(4), 676, 691, 714
Briedis, D., 473, 625, 641
Bright, P., 109
Brigman, K. K., 618, 626(10)
Brinste, R. L., 618, 622(15)
Brinster, R. L., 576
Brizzio, V., 211
Brodeur, G., 250
Brodeur, G. M., 555
Brody, S. L., 724
Broeks, A., 105
Brom, M., 132, 154, 608, 609(12), 610(12), 611(12), 614(12), 615(12)
Bronduk, W. H., 417
Brook, J. D., 725
Brooker, R. J., 411
Broome, C. V., 445
Brose, K., 558
Brossi, A., 421
Brown, C. S., 698, 709(5), 712(5)
Brown, D., 279(11), 280
Brown, R. E., 24
Browne, B. L., 195, 210(12)
Browne, C. A., 338
Browne, P. C., 344, 345(16), 348(16), 349(16), 356(16)
Broxterman, H. J., 262, 264, 607
Brualt, M., 211, 414, 415(8), 427(8)

Bruggemann, E. P., 304, 305(48), 306, 309, 311, 311(10), 312(10), 313, 313(17), 314, 314(10), 315(10), 319, 322(8), 327, 437, 478, 497, 498(28), 499, 508, 584
Brugger, W., 554
Bruhn, D. F., 83
Brun, R., 188
Brundage, L. A., 37
Brundate, L. A., 27
Brunner, J., 308
Bruno, M. S., 698, 709(5), 712(5)
Bryan, J., 110, 117, 132, 154(19, 20), 369, 732, 736, 737, 738, 738(3, 6), 740(3), 741, 743, 743(8), 744(1, 3)
Bryk, D., 293, 303(17), 305(17), 306(17)
Brzoska, P., 41, 42, 43(20), 44(20), 45(14, 20)
Brzostek, K., 41, 42, 43(20), 44(20), 45(14, 20)
Buchanan, J. A., 135, 629, 630(3), 652, 675(3), 676
Buchholz, U., 607, 607(11), 608, 608(7, 8, 10), 609(7, 8, 10, 16), 610(7, 8, 16), 611(7, 8, 10, 11, 16), 612(8, 10), 613(7, 8;10;11), 614(10), 615(8, 10, 11, 16), 616(8)
Buchler, M., 128, 132, 154, 607, 608, 609(12), 610(12), 611(12), 614(12), 615(12)
Bucholz, U., 154
Buchwald, M., 135, 629, 630(1, 3), 633(1), 645, 646(53), 647, 652, 655, 656(25), 659(24), 661(25), 675, 675(3), 676
Buckel, S. D., 8, 116, 158
Budding, M., 167
Buhne, S., 772
Bujard, H., 628
Bukau, B., 14, 46, 49
Buller, R. M. L., 445
Bult, C. J., 757
Bulte, J. W. M., 262
Bunnell, B. A., 558, 559(7)
Buoncore, L., 545
Burda, J. F., 728
Burland, V., 84
Burnstock, G., 664
Buschman, E., 418
Busconi, L., 627
Bussey, H., 131, 134(14), 137(14)
Bustamente, C., 704, 708(9)
Butcher, G. W., 163
Butt, A. G., 361
Buttimore, C., 560

C

Cabantchik, I. Z., 343
Cabantchik, Z. I., 367, 645
Cahalan, M. D., 359, 360, 361, 361(19), 365(19), 367(19)
Calafat, J., 105, 184, 185(16)
Calamia, J., 268
Calayag, M. C., 266, 267(6), 268(6, 17), 270(17), 271(6), 278(6), 287, 288(18)
Callaghan, R., 342, 355, 368
Callahan, H. L., 189, 191
Candia, O. A., 37, 38(41)
Canessa, C., 369
Canfield, C., 266, 288
Canfield, V., 414
Cannell, D., 118, 136, 150(42), 215, 220(18), 469(87), 768
Cantiello, H. F., 494
Cantrot, Y., 493
Caplen, N. J., 707, 722
Cardarelli, C. O., 131, 226, 227, 248, 254, 259, 264, 304, 305(48), 309, 311(10), 312(10), 314(10), 315(10), 321, 341, 342(18), 355, 368, 372, 397, 398(8), 409(8), 410(8), 411(8), 454, 471, 474, 481, 487(2), 492, 495, 496(22), 500(22), 503, 505, 506, 506(9), 511(12), 513, 513(9), 523, 546, 574, 586, 590(37), 591, 662(37), 663
Cardellichio, C. B., 725
Cardon-Cardo, C., 259
Carlin, A., 84
Carlsen, S. A., 328
Carlson, W., 655
Carrino, J. J., 754
Carroll, M. W., 453
Carroll, T. P., 624, 642, 650(37)
Carson, M. R., 768
Carter, B. J., 369, 539, 540, 544, 653, 659(8), 676, 715, 720, 725, 726, 727, 728, 728(52, 53), 729, 729(6), 730, 731
Cartier, N., 759, 765
Caruso, J., 728
Carvajal, E., 627
Casadaban, M. J., 30, 31(6)
Casals, D., 259
Case, C. C., 46
Casey, P., 427(11), 428, 441(11)
Castanys, S., 183, 187(7)

Catty, P., 143(53), 151, 159(53)
Cavalier-Smith, T., 776
Cenciarelli, C., 262
Center, M., 342, 357(7), 607, 607(11), 608, 608(7), 609(7), 610(7), 611(7, 11), 613(3, 7, 11), 615(11)
Cerundolo, V., 745
Chaidez, J., 270
Chait, B. T., 347, 348(25)
Chakravarti, A., 135, 629, 630(3), 652, 675(3), 676
Chambers, T. C., 328, 329, 333(4), 335(4), 339(4), 341, 342(18), 353, 355, 357, 368, 372, 503, 505, 506(9), 512, 513(9)
Chambon, P., 596
Champion, E., 608, 613(19), 615(19)
Champlin, R., 547
Chan, A., 607
Chan, H. S., 264
Chan, H. W., 697, 704(1)
Chanda, E. R., 606
Chang, B., 43
Chang, C. T., 684
Chang, K.-P., 183
Chang, L.-S., 541, 543(16), 727, 728(47)
Chang, X.-B., 288, 343, 616, 618, 619(13), 620, 620(13, 14), 621(14), 622, 622(24), 626, 642, 645, 645(34, 36), 646, 646(34, 52, 53), 650(34–37), 653, 722
Change, X., 677
Chao, A. C., 360
Chao, N. J., 262
Chapman, A. E., 293, 300(26), 303(26)
Charette, L., 608, 613(19), 615(19)
Charnet, P., 279
Chassagne, J., 264
Chatterjee, A. K., 67
Chatterjee, S., 540, 543(15), 727
Chaudhary, P. M., 242, 259, 546
Chaudhary, V., 313, 327, 497, 498(28), 499, 508, 584
Chavkin, C., 625
Cheadle, J. P., 676
Cheek, D., 68
Chen, C., 484, 514, 561, 619(54), 629
Chen, C.-J., 167, 226, 316, 318, 330, 334(7), 339(7), 431, 493, 504, 577
Chen, C. M., 83
Chen, D., 318, 319(7), 320, 320(7), 321(7), 326(7), 427, 441, 473, 480, 574
Chen, G., 360
Chen, H., 446
Chen, J. L., 21, 28(7)
Chen, M. A., 288
Chen, P., 194, 200(8), 202, 202(8)
Chen, Y., 82
Cheng, L., 654
Cheng, S. H., 136, 281, 343, 617, 620, 622, 628(27), 641, 642(42), 647, 650(59, 60), 652, 655, 676, 697, 712, 713(12, 13), 714
Cheraux, C., 61
Chernick, M. S., 687
Chernov, I., 118
Chervonsky, A. V., 546, 586
Cheung, A. M. K., 725
Cheung, I., 64
Cheung, R., 369
Chevallier-Multon, M.-C., 260
Chiang, Y., 538, 547
Chiba, P., 255
Chiesa, R., 489
Chifflet, S., 489
Childs, S., 102, 116, 117(1), 279, 493
Chimini, G., 116, 160, 777
Chin, C. E., 493, 504
Chin, J. E., 167, 237, 318, 330, 334(7), 339(7), 431, 494, 503(14), 505, 514
Chin, K.-V., 204, 546, 573, 575, 576(4), 580(4), 589(4), 594
Chiquero, M. J., 183, 187(7)
Chiu, M. L., 427(10), 428
Cho, S., 745
Choi, J. C., 194, 200(8), 202, 202(8)
Choi, K., 226, 299, 304(33), 316, 431
Chomczynski, P., 580
Chou, C.-Y., 365
Chou, J., 675, 687
Chou, J.-L., 117, 132, 154(15), 343, 411, 617, 619(4), 629, 630(2), 631(2), 633(2), 635(2), 715
Chow, C. Y., 698, 709(5), 712(5)
Chow, L. M. C., 183
Christensen, O., 361
Chu, C.-S., 687, 724
Chu, F., 109, 177
Chu, J. W. K., 397, 398(14), 409(14), 411(14)
Chu, T. M., 262
Chuck, A. S., 559
Chudziak, F., 744
Chundru, S., 731

Cianfriglia, M., 262
Cirtain, M. C., 300
Civan, M. M., 369
Clapham, D. E., 365, 368
Clapp, D. W., 554
Clark, D. P., 167, 318, 330, 334(7), 339(7), 431, 493, 504, 514, 577
Clark, S. C., 552
Clark, W. H., Jr., 749
Clarke, D. M., 266, 288, 304, 316, 318, 322, 361, 397, 398, 398(9), 409(15), 414, 427(12), 428, 456, 478, 480, 481, 482, 483, 483(11), 484, 484(13), 486, 486(6, 13), 487(8), 490, 491(6), 505, 623, 768
Clarke, M. F., 559
Clarke, S., 202
Claverys, J. P., 10
Clawson, R. E., 294, 301(28)
Clayton, R. A., 757
Cle, S. P. C., 608, 611(17), 613(17), 615(17)
Clement, J. P., 132, 154(19), 369
Clement, J. P. IV, 732, 736, 738, 738(3), 740(3), 743, 743(8), 744(1, 3)
Cleveland, D. W., 312
Cliff, W. H., 361, 617, 721, 723
Cline, A., 547
Coadwell, W. J., 163
Coburn, C., 188, 189(24)
Coca-Prados, M., 369
Cochet, M., 745
Coderre, J. A., 184
Cohen, D., 178, 307, 309(2)
Cohen, M. E., 655
Cohen, P., 342
Cohn, J. A., 343, 369, 647, 649(66), 656, 661(30), 676, 679(62), 718
Cohn, M., 30
Cole, J. L., 629, 630(1), 633(1), 652, 675
Cole, S. T., 40, 42
Cole, S. P. C., 117, 133, 154, 183, 250, 414, 415(8), 427(8), 493, 575(7), 594, 595, 600, 602, 602(19), 604, 606, 606(11, 25), 607, 607(25), 608, 608(8), 609(1, 8), 610(8), 611(8, 11), 612(8), 613, 613(8, 11), 615(8, 11), 616(8)
Coles, B., 365, 369(40)
Colledge, W. H., 369
Collins, F. S., 117, 122, 132, 154(15), 343, 411, 617, 624, 629, 630(1, 2), 631(2), 633(1, 2), 635(2), 652, 675, 676, 677(16), 687, 715, 721, 723
Collins, K. I., 316, 414, 423(2), 478, 590
Collins, M. G., 664
Collis, P., 728
Colognola, R., 265
Colonna, M., 745
Colston, M. J., 475
Connelly, C., 755
Conrad, C. K., 725(53), 727, 729, 731
Conrad, J., 293, 303(24), 317
Conrad, M., 699
Contreras, M., 759, 769, 769(61)
Coon, H. C., 725
Cooper, N., 446, 461
Copeland, N. G., 116
Cormier, M. J., 204
Cornetta, K., 548, 558
Cornish, V. W., 474
Cornwell, M. M., 259, 290, 299, 300(34), 302(34), 330, 436
Cory, S., 547(20), 548, 554, 575
Cossman, J., 250, 555
Costantini, F., 576, 577(21)
Cote, G. J., 110, 117, 132, 154(20)
Cottler-Fox, M., 538, 547
Coulson, A. R., 601, 680
Court, D., 116, 121(8)
Coutelle, C., 707, 722
Couture, L. A., 724
Covitz, K. M., 12, 32(33), 33, 34(33)
Cowan, K. H., 538, 547
Cowherd, R., 558
Cowman, A. F., 109, 157, 167, 175, 176, 176(5, 7), 177, 178(9), 181(9), 211, 415, 427(13)
Cox, B. C., 118
Cox, G. B., 43, 136, 150(42), 215, 220(18), 221, 224, 469(87), 768
Cox, T. K., 135, 629, 630(3), 652, 675(3), 676
Cox, W. I., 473
Craig, E. A., 627
Craik, C., 782
Crane, D. I., 756, 757, 763(53), 774, 775(98)
Crawford, I., 652, 653, 655, 655(10), 656(10), 657(10), 658(10), 661(10), 662(37), 663
Crawford, J. M., 328, 329, 332(5)
Creeg, J. M., 755
Creeger, E. S., 74
Cregg, J. M., 757, 763(53)

Cremer, K., 445
Cresswell, P., 744, 745, 751, 752, 753(9), 756(14)
Crist, W., 250, 555
Croop, J. M., 101, 105, 109, 117, 163, 165, 165(1), 167, 178, 227, 262, 431, 514
Cross, G. A. M., 184
Crowe, W. E., 369
Cruze, J. A., 399
Crystal, R. G., 647, 687, 723, 724
Cui, Z., 155
Cukerman, E., 128
Cullen, M. H., 304, 309
Culvenor, J. G., 175, 176(7)
Culver, K. W., 523, 558
Cunningham, J. M., 242
Cunningham, K., 27, 37
Cuppoletti, J., 413, 618, 626(10)
Curotta de Lafaille, M. A., 189
Currier, S. J., 227, 262, 313, 314, 327, 454, 474, 478
Cussans, G. W., 163
Cuthbert, N. H., 608, 615(20)
Cutting, G. R., 118, 624, 642, 650(37), 653, 655, 659(8), 676, 720, 728, 729(7), 731
Cyr, D. M., 627
Czerwinski, M. J., 483

D

Dahler, G. S., 67
Daignan-Fornier, B., 781
Dalemans, W., 647, 661, 723
Dalton, W. S., 264, 414, 546, 572
Dalziel, H. H., 664
Daniel, C., 724
Daniels, D. L., 84
Daniels, L. B., 204
Dano, K., 259
Danos, O., 759
Darovsky, B., 558
Dassa, E., 6, 7, 31, 35(14, 16), 46, 764
Datzmann, D., 143(68), 155
Davey, M. W., 381
Davidson, A. L., 5, 9, 15, 16, 17, 17(35, 77), 18(35), 19, 20, 21, 22, 22(8), 24(4), 27(4), 28(4, 8), 29, 32, 33(25), 37(25), 38(25), 79, 113, 131
Davidson, N., 625
Davidson, P. N., 279
Davis, J., 744
Davis, N. G., 203
Davis, P. B., 722
Davis, R. W., 131, 134(14), 137(14), 779
Dawson, D. C., 624, 676, 677(16)
Dean, D. A., 10, 12, 15, 20(54), 21, 28(6)
Dean, M., 116, 117(2), 629, 630(1), 633(1), 652, 675, 762, 777
DeAngelis, R., 328, 329, 332(5)
Deb, S., 74
DeBerg, L. G., 768
De Bruijn, E. A., 367
Debs, R. J., 698, 709(5), 712(5)
Deckert, J., 665
Decottignies, A., 143(53, 54), 151, 159(53, 54)
de Duve, C., 753
Deeley, A. M. V., 493
Deeley, R. G., 117, 133, 154, 183, 250, 414, 415(8), 427(8), 493, 594, 595, 596, 597, 600, 602, 602(19), 603, 604, 606(25), 607, 607(11, 25), 608, 608(8), 609(1, 8), 610(8), 611(8, 11, 17), 612(8), 613, 613(8, 11, 17), 615(8, 11, 17), 616(8)
Degand, H., 143(53), 151, 159(53)
de Graaf, D., 225, 229, 574
De Greef, C., 367
De Greef, D., 360
de Haas, M., 607
DeHaven, R. N., 608, 613(19), 615(19)
Deisseroth, A. B., 547
Deitmer, J. W., 360
de Jonge, H. R., 360, 367(15), 369(15)
Dekker, H., 110, 309, 613
de la Flor-Weiss, E., 547, 551(11)
de la Garza, M., 772
Delannoy, M. R., 678, 682, 683, 684, 685
DeLeij, L., 680
Delepelaire, P., 52, 67, 68, 69, 72, 73, 73(14, 18), 74(27), 76, 79(32), 80(26), 81(32)
Delepierre, M., 72, 80(26)
Dellarco, V. L., 474
Delling, U., 157, 211, 398
DeMars, R., 117, 118(15), 163, 745
DeMatteo, R., 728
Demeneix, B., 698, 704(3)
Demmer, A., 227, 423
Demolombe, S., 131

Denicoff, A., 538, 547
Denizot, F., 116, 160, 777
Denning, G. M., 653, 655, 656(7), 661(7), 677
Derbyshire, K. M., 479
Derr, J. E., 38
Desnick, R. J., 525
Detke, S., 183
Deuchars, K., 263
Devault, A., 399, 403(28), 422, 426(22), 431, 481
Devereaux, J., 145(88), 160
Deverson, E. V., 163
Devine, S. E., 226, 304
de Vries, E. G. E., 154, 607, 608(9), 611(9), 614(9)
de Vries, Y., 755
Dey, C. R., 629
Dey, S., 82, 84, 84(3), 86, 86(1, 3), 88, 88(1), 91, 93(1), 94, 95(28), 183, 189, 190, 190(3, 28), 191(28), 318, 319(6)
Dhaunsi, G. S., 769
Dhir, R., 227, 316, 418
Dho, S., 617(9), 618
Diaz, M., 360, 362(7), 363(16), 365(7)
DiCapua, F., 574
Dicke, F. P., 606
DiDiodato, G., 410
Dignard, D., 143(72), 156, 385, 415
Dinchuk, J. E., 579, 580(24)
Dinh, T., 68
Dionet, C., 264
Director, E., 558
Distel, B., 755, 756, 757, 763(53), 767
Dittrich, K. L., 647, 649(66)
Djaballah, H., 120
Dobberstein, B., 266, 267
Dodt, C., 755, 757(30)
Dodt, G., 755, 757, 757(18), 763(53)
Doh, G., 712, 713(13)
Dohlman, H. G., 781
Doige, C. A., 30, 113, 131, 397, 398(10, 14), 409(10, 14), 410(10), 411(10, 14), 487
Dolle, M., 262, 266, 435, 481
Dolnick, B. J., 239, 246(26)
Doly, J., 186
Donahue, R. E., 544, 558, 559(7), 727
Donella, D. A., 347
Dong, J., 259, 264
Dong, Y. J., 360
Donowitz, M., 655
Doolittle, R. F., 138, 279
Doranz, B., 723(22), 724
Dorin, J. R., 722
do Rosario, V. E., 176
Dou, D., 86, 91, 94, 95(28), 183, 190(3)
Douar, A.-M., 117, 132, 153, 756, 758(45), 759(46), 765, 769(45)
Doughterty, B. A., 757
Douglas, M. G., 627
Dowdle, E. B., 71
Downie, J. A., 8, 30, 31(8), 116, 158
Drach, D., 546
Drach, J., 255, 546
Dreesen, T. D., 214, 215(13)
Dreyer, D., 661
Driscoll, J., 745
Droogmans, G., 359, 360, 367, 368(3), 369(3)
Drumm, M. L., 117, 122, 132, 154(15), 343, 411, 540, 617, 624, 629, 630(1, 2), 631(2), 633(1, 2), 635(2), 652, 675, 676, 677(16), 687, 715, 721, 723, 729, 730
D'Souza, M. P., 655, 659(24)
Dudeja, P. K., 289, 300(4)
Dudier, R., 10
Dudler, R., 114, 162, 166, 168(19), 172, 173(25)
Duguay, F., 617, 618(3), 619(3), 623(3), 624(3), 645
Dujon, B., 131, 134(14), 137(14)
Dulhanty, A. M., 620, 636, 642, 645, 646(52)
Dunbar, C. E., 538, 547
Duncan, A. M., 117, 133, 183, 493, 600, 602(19), 607, 609(1)
Dunn, J. J., 373
Duong, F., 68
Duplay, P., 25
Dupuis, M. L., 262
Duran, G. E., 360
Durham, S. R., 722
Durie, B. G. M., 264
Durr, F. E., 590
Duthie, M., 266, 267(9), 278(9), 281, 283(14), 284(14), 285(14), 286(27), 288
DuVall, M. D., 629
Dybing, J., 559

E

Earl, P. E., 446, 461
Earl, P. L., 456
Eaves, A. C., 547

Eaves, C. J., 547
Ebner, K. E., 272
Ebright, R. H., 474
Eckenrode, V. K., 204
Ecker, G., 255
Economou, A., 278
Edelman, A., 649, 650(67)
Edgar, R., 137, 370, 379
Edgell, M. H., 479
Edman, P., 681
Efstratiadis, A., 576
Egan, M. E., 118, 369, 653, 659(8), 664, 673, 674, 676, 720, 729(6)
Eggermont, J., 360, 367
Eglitis, M. A., 255, 523
Egner, R., 151
Ehmann, U., 14
Ehring, B., 768
Ehring, G. R., 360, 361(19), 365(19), 367(19)
Ehrle, R., 13
Ehrlich, P., 82, 93(2)
Ehrmann, M., 6, 13, 49
Eiden, M. V., 560
Eiglmeier, K., 40, 42
Eijdems, E. W. H. M., 607
Eilon, G., 342
Eisenberg, M. T., 220
Eissa, N. T., 724
Elder, D. E., 749
Elferink, O., 132
Elgersma, Y., 755, 756
Elia, L., 211
Elliott, T., 768
Elmes, L., 40
Elroy-Stein, O., 443, 446(10), 457, 459, 460(11), 461(11)
Elvin, C. M., 40, 45(7)
Emma, F., 359, 363(4), 369(4)
Emr, S. D., 206, 628
Emsley, P., 9, 116, 136
Endicott, J. A., 133, 258, 279, 289, 781
Engelhardt, J. F., 647, 679(62), 714, 718, 723(22), 724
Englander, S. W., 651
Englund, P. T., 186
Epand, R. M., 700
Ephrussi, B., 213
Epping, E. A., 143(53), 151, 159(53)
Epstein, N., 725
Erdmann, R., 755, 756, 757, 763, 763(53)
Erickson, A., 620
Erickson, R. P., 474, 475
Erlich, H. A., 679
Ernst, S. A., 718
Escande, D., 131
Evan, G. I., 170, 171(24), 204, 781
Evans, G. L., 320, 427, 441, 473, 480, 572, 573, 574, 576(5), 579
Evans, I. J., 8, 30, 31(8), 116, 158
Evans, M. J., 369
Evernden-Porelle, D., 260, 262, 263(25), 298, 434, 584, 620, 649
Evers, R., 154, 155(63), 613
Ewart, G. D., 118, 136, 150(42), 213, 215, 220, 220(18), 469(87), 768
Eytan, G. D., 410

F

Fairbanks, D. J., 218
Fairchild, C. R., 250
Falk, E., 611
Faloona, F., 679
Fanen, P., 759
Farhood, H., 700
Farmer, J. B., 608, 615(18)
Farmer, J. L., 218
Farr, A. L., 635
Fasbender, A. J., 699, 707(6), 708(6)
Fase-Fowler, F., 183, 184, 185(12, 16), 187(9, 12)
Fath, M. J., 3, 4, 4(5), 51, 67, 130
Fausto, N., 576, 582(19)
Fearon, K., 157, 211
Feigenbaum, V., 759
Feil, R., 117, 132, 756, 758(45), 769(45)
Feillel, V., 264
Feldmann, H., 131, 134(14), 137(14)
Felgner, J. H., 698, 699(4), 704(6), 707(4)
Felgner, P. L., 697, 698, 699(4), 704, 704(1, 6), 707(4), 708(9), 712, 713(13)
Felmlee, T., 164, 165
Felsenstein, J., 105
Felsted, R. L., 291, 294, 298(11), 299, 300(11, 34), 301(28), 302(12, 34), 311
Ferenci, T., 11, 14, 17, 25
Ferkol, T., 722
Fernandes, L., 143(92), 160
Ferrans, V. J., 687

Ferry, D. R., 304, 309
Fetell, M., 538, 547
Fievet, B., 368
Fikes, J. D., 15, 37
Finbarr Tobin, J., 189
Fine, D. L., 483
Fine, R. L., 357, 412, 427(9, 11), 428, 441(9, 11)
Fink, G. R., 133, 159(27), 195
Finkelstein, A., 278, 288
Fischer, S. G., 312
Fisher, K. J., 728
Fitzsimmons, S. C., 715
Flaherty, K. M., 70
Fleishmann, R. D., 757
Fleming, G. F., 306
Fleming, W. F., 550
Flens, M. J., 607
Flessau, A., 763
Fleuren, G. J., 265
Flint, N., 266
Flotte, T. R., 369, 539, 540, 544(13), 545(13), 624, 642, 650(37), 653, 659(8), 676, 715, 720, 727, 728, 728(52, 53), 729, 729(6, 7), 730, 731
Flugge, U. I., 497
Fojo, A., 238, 249, 250, 259, 494, 503(14), 505, 555, 560, 581
Fok, P., 759, 769(63)
Fokt, I., 293, 294(16), 301(16)
Folcher, M., 118
Foley, M., 14
Foley, P. L., 453
Fomsgaard, A., 721
Fonck, Y., 264
Foninson, I., 266
Foote, S. J., 167, 175, 176, 176(5), 178(9), 181(9)
Forbush, B. III, 617
Ford, J. M., 306
Ford-Hutchinson, A. W., 608, 613(19), 615(19)
Fordis, C. M., 476
Forsyth, K., 176, 178(9), 181(9)
Foskett, J. K., 617(9), 618
Fowler, A., 25
Fox, D. T., 474
Fox, R. R., 576
Foxwell, B. M. J., 293, 303(20), 308
Franke, U., 755, 757(19)
Franken, T., 763
Franse, M. M., 755
Fransen, M., 755
Frasch, A. C. C., 184
Fraser, C. M., 757
Fraser, K., 604, 606(25), 607(25)
Fraser, M. J., 128, 439(17)
Freeman, S., 558
Frenette, R., 608, 613(19), 615(19)
Frick, G. P., 475
Friedlander, M., 266
Fritchman, J. L., 757
Fritsch, E. F., 59, 164, 165(17), 209, 245, 385, 431, 446, 563, 571(24), 574, 596, 598(14), 599(14), 603(14), 636, 680
Frizzell, R. A., 361, 617, 624, 629, 676, 677(16), 689, 690(7), 721, 723
Frohlich, K. U., 763
Fromter, E., 646
Froshauer, S., 6, 31, 267
Früh, K., 120
Fu, L., 211
Fuchey, C., 661
Fuchs, R., 658
Fuerst, T. R., 443, 444(8), 456, 457
Fuhrmann, J., 757
Fujii, Y., 414
Fujiki, Y., 117, 755, 757, 757(31)
Fujimoto, E. K., 401, 517
Fujimura-Kamada, K., 196, 197(14)
Fukayama, M., 647, 723
Fukuda, S., 757
Fuller, R. S., 206
Fulmer, S. B., 118, 624, 642, 650(37), 653, 659(8), 728
Fung, Y. W., 128
Furht, E. F., 724
Furkuzawa, H., 776
Futcher, B., 203
Futscher, B. W., 546

G

Gabillat, N., 368
Gadek, T. R., 697, 704(1)
Gadzar, A., 555
Gage, D., 94
Gagel, R. F., 110, 117, 132, 154(20)
Gaillard, D., 661
Galatis, D., 175, 176, 176(7), 177

Galibert, F., 131, 134(14), 137(14)
Galietta, L. J. V., 360, 367(18)
Gallagher, M. P., 6, 9, 17, 116, 136
Galski, H., 250, 546, 547, 555, 573, 575, 576(4), 580(4, 7), 589(4, 7, 11), 590
Gamarro, F., 183, 187(7)
Gambill, B. D., 627
Gammie, E., 211
Ganesan, A. K., 11
Gannon, F., 596
Gao, G. P., 728
Gao, X., 697, 698(2), 704(2), 707, 722
Garcia, J. V., 560, 561(19)
Gardner, P., 360, 608, 615(20), 676
Gardos, G., 439
Garen, A., 42
Garfin, D. E., 737
Gariepy, J., 183, 260, 263(26), 453, 463, 497, 508, 509(18), 515
Garland, P. B., 14
Garnier, J.-M., 762
Garotta, G., 168, 169(22)
Gartner, F. H., 401, 517
Gartner, J., 112, 118, 132, 756, 758, 758(49), 769
Gartner, L. M., 754
Gatmaitan, Z., 397
Gatter, K. C., 653, 655(10), 656(10), 657(10), 658(10), 661(10)
Gattringer, C., 546
Gaughan, K. T., 607
Gauldie, S. D., 604
Gauthier, J. Y., 608, 613(19), 615(19)
Gay, N. J., 3, 7(1), 101, 397, 415, 514, 676
Gazdar, A., 237, 250
Gazit, E., 676
Geddes, D., 707, 722
Gee, A., 15
Gefter, M. L., 637
Geger, R., 367
Gehring, G., 214
Gehring, K., 11, 15, 25
Gekeler, V., 255, 293, 303(24), 317, 608
Gelfand, E. W., 369
Geller, D., 137, 379
Gentschev, I., 52, 64(3)
George, A. M., 381
Georges, E., 183, 260, 263(26), 266, 293, 294(22, 23), 301(22, 23), 303(22, 23), 322, 404, 453, 463, 497, 508, 509(18), 515, 522
Georgiou, G., 373
Gerena, L., 175
Gerlach, J. H., 117, 118(11), 133, 183, 250, 493, 595, 600, 602(19), 606, 607, 609(1)
Gerloczi, A., 439
Germann, U. A., 204, 226, 255, 262, 311, 313, 313(17), 319, 320, 320(10, 10b), 322(8), 327, 328, 341, 342(18), 353, 355, 368, 372, 397, 398, 398(6), 409(6), 410(6), 411(6), 412(6), 427, 427(5, 7, 16), 428, 430(4, 16), 431, 431(4, 16), 434, 435(2, 4, 32), 436(4), 437, 437(4, 5), 438(5), 439(5, 16), 440(5, 16), 441(5–7), 465, 467, 473, 478, 480, 483, 487, 503, 504, 505, 505(4, 5), 506(9), 507, 512, 513(9), 523, 546, 557, 574, 574(2)
Germeraad, W. T., 571
Gerrard, B., 116, 116(10), 117, 121(6, 8), 135, 139(39), 143(81), 152(39), 158, 160(39), 777
Gerweck, L., 494
Ghigo, J. M., 68, 72, 75, 80(26)
Giacomini, J. C., 295
Giacomini, K. M., 295
Giesler, S. C., 474
Gietz, D., 779
Gilardi, P., 723
Gilbert, D. J., 116
Gilbert, W., 576
Gileadi, U., 6, 9, 136
Gill, D. R., 8, 9, 30, 31(8), 116, 136, 158, 360, 361(21), 362(7), 363, 363(8), 365, 365(7), 368(8, 21, 34), 369(21), 494, 665
Gilliam, T. C., 118
Gilson, E., 7, 10, 31
Gindin, E., 303, 309, 311(11), 312(11), 406
Girard, P. R., 330
Giraud, C., 725
Gish, W., 122, 138
Gitschier, J., 165
Glass, D. B., 341
Glauner, B., 74
Glavač, D., 116
Glavy, J. S., 342, 344, 352, 353(18)
Glazer, R. I., 427(13), 428
Glick, B. S., 783, 784(27)
Glickstein, H., 343, 645
Glover, C. J., 291, 294, 298(11), 300(11), 301(28), 311
Gluzman, I. Y., 174, 176
Gocayne, J. D., 757

Godwin, A., 607
Goebel, M., 134
Goebel, W., 52, 64(3)
Goebl, M., 124, 135(35), 161(35), 755
Goehle, S., 558
Goeke, N. M., 401, 517
Goel, S. K., 754
Goffeau, A., 131, 134(14), 137(14), 143(51, 53, 54, 56), 151, 152, 159(53, 54), 618, 619(40), 626, 626(10), 627, 777
Goldberg, A. L., 620, 622(25), 627, 627(25), 645
Goldberg, M., 214
Goldenberg, S., 248, 255, 506
Goldfischer, S., 754
Goldman, G., 279
Goldman, P., 597
Goldsmith, M. E., 250
Goldspiel, B., 538, 547
Goldstein, A., 388
Goldstein, L. J., 250, 547, 555, 573, 589(11)
Golin, J., 143(68), 155, 627
Gollapudi, S. V., 360
Gonoi, T., 369, 732, 743, 744(1)
Gonzalez, G., 132, 154(19), 369, 738, 743, 743(8)
Gönzcöl, E., 724
Goodfellow, H. R., 342, 355, 360, 361(21), 368, 368(21), 369(21), 494
Goodman, H. M., 475
Goodman, J. M., 754, 755(14), 756, 757, 762(14), 763(53)
Goodman, R. H., 576
Goodman, S., 544, 727
Goosens, M., 759
Göransson, H. M., 160, 777
Gorbalenya, A. E., 101
Gordon, A. S., 49
Gordon, E. A., 365, 368
Gordon, J. L., 637, 656, 664, 665(1)
Gorelick, F. S., 647, 656, 661(30)
Gorman, C. M., 542, 561, 710
Gossen, M., 628
Gotte, K., 763
Gottesman, M. M., 131, 133(7), 167, 175, 204, 226, 227, 238, 248, 249, 250, 252, 254, 255, 256(10), 257, 257(10), 258(2), 259, 262, 264, 265, 279, 289, 290, 290(2), 299, 300(5, 34), 302(34), 304, 305(48), 306, 307, 309, 311, 311(10), 312(10), 313, 313(17), 314, 314(10), 315(10), 318, 319, 319(6, 7), 320, 320(7, 10), 321, 321(7), 322(8), 326(7), 327, 328, 330, 334(7), 339(7), 341, 342(18), 353, 355, 368, 372, 397, 398, 398(8), 409(8), 410(8), 411(8), 427, 427(6, 7, 16), 428, 430, 430(4, 16), 431, 431(4, 16), 435(1, 4), 436, 436(4), 437, 437(4), 439(16), 440(16), 441, 441(3, 6, 7), 442, 443, 443(2), 446(2, 11), 450(2), 453(2, 11), 454, 456, 457, 459, 459(10), 461(9, 10), 465, 471, 473, 473(1, 10), 474, 476, 477, 478, 480, 481, 483, 487(2), 492, 493, 493(2, 3), 494, 495, 496(22), 497, 498(28), 499, 500(22), 503, 503(14), 504, 505, 505(3, 4), 506, 506(9), 508, 511(12), 512, 513(9), 514, 523, 525, 526, 526(8), 527(7), 528(8), 530(7, 10), 537, 537(11), 538, 540, 540(1), 543, 544(13), 545(13), 546, 547, 548, 551(11), 555, 557, 558, 560, 561(2), 570, 570(2), 571(26), 572, 573, 574, 574(2), 575, 576(4, 5), 577, 579, 580(4, 7), 581, 583, 584, 585(32), 586, 589(4, 7–9, 11), 590, 590(37), 591, 594, 595, 662(37), 663, 786
Gottesman, S., 169
Götz, F., 84
Gould, S. J., 754, 755, 756, 757, 757(18, 30), 759, 763, 763(53), 765(62), 769(63), 774, 775(98)
Gow, I. R., 163
Graber, P., 411
Gracy, R. W., 338
Graham, S. M., 724
Granadea, A. G. III, 679
Granett, S., 46
Grant, C., 250
Grant, C. E., 117, 133, 183, 250, 414, 415(8), 427(8), 493, 594, 600, 602(19), 604, 606(25), 607(25), 613
Grant, C. R., 607, 609(1)
Gravitt, K. R., 357
Gray, L., 8, 68, 116, 158
Green, A., 250, 555
Green, G. N., 6
Green, M. M., 214
Green, M. R., 270
Green, N., 267
Green, S., 754
Greenberger, L. M., 303, 304, 305(47, 48), 307, 309, 309(2), 311(10, 11), 312(10, 11), 313,

314(10), 315(10), 316, 319, 327, 346, 406, 414, 423(2), 437, 478, 590
Greengard, P., 343, 676
Greenwood, F. C., 639
Gregory, R. J., 136, 281, 343, 617, 617(8), 618, 620, 622, 628(27), 641, 642(42), 647, 650(59, 60), 652, 653, 676, 714, 724
Greig, D. I., 53, 55(12)
Grell, E. H., 219, 220(21), 221(21)
Gress, R., 538, 547
Gribble, F., 365, 369(40)
Griffin, M. S., 167
Griffith, B., 237
Grill, E., 163
Grindley, N. D. F., 479
Grinstein, S., 620, 622(24), 642, 645, 645(36), 646, 650(36), 653
Grivell, L. A., 778
Grizzuti, K., 178, 304, 316, 401
Groarke, J., 16
Grodberg, J., 373
Groen, A. K., 391, 494
Grogan, T. M., 264, 546
Grogl, M., 183
Groisman, E. A., 370
Grondin, K., 94, 95(28), 182, 183, 187(13), 190(3)
Gros, P., 109, 117, 134, 135, 157, 157(36, 37), 159, 163, 165, 165(1), 177, 178, 194, 211, 220, 227, 266, 288, 290, 304, 316, 318, 342, 355, 368, 371, 374, 374(1), 375, 380, 382, 383, 384(10), 385(6), 388(10), 390(10), 395(11), 397, 398, 398(2), 399, 400(23), 401, 401(23), 403(28), 406(20), 413(21), 414, 415, 415(6, 8), 418, 421, 421(3), 422, 422(3), 423, 425, 426, 426(22), 427(8), 431, 473, 481, 494, 505, 514, 515, 574
Gross, R., 52
Grossman, M., 723(22), 724
Grout, M., 721
Grubb, B. R., 714
Gruenert, D. C., 360, 367(18), 653, 655, 674, 707
Grunden, S., 361
Grussenmeyer, T., 43
Grzelczak, Z., 117, 132, 154(15), 343, 411, 617, 619(4), 629, 630(2), 631(2), 633(2), 635(2), 675, 687, 715
Gu, H., 474
Guan, C., 679
Guarente, L., 30, 31(6)
Gudkov, A. V., 225, 227, 227(2), 239(2, 16)
Gueiros-Filho, F. J., 188
Guerry, D. IV, 749
Guggino, B., 730, 731
Guggino, W. B., 118, 369, 540, 624, 642, 647, 650(37), 653, 655, 655(10), 656(10), 657(10), 658(10), 659(8), 661(10), 664, 673, 676, 720, 723, 725(53), 727, 729, 729(6, 7)
Guidotti, G., 494
Guidoux, S., 759
Gunning, K. B., 576
Gupta, K. P., 357
Gupta, S., 360
Guthrie, C., 133, 159(27), 195
Guttoh, D. S., 745
Guzman, R., 15
Guzzo, J., 68
Gwadz, R. W., 176

H

Haas, S., 293, 303(24), 317
Haase, E., 143(52), 151
Habener, J. F., 576
Haber, J. E., 400
Hadam, M. R., 227, 423
Haddad, B., 110, 117, 132, 154(20)
Haddad, G., 264
Hadfield, C., 214
Haeberli, P., 145(88), 160
Haeuptle, M. T., 266
Hagiwara, K., 342
Hahne, K., 785
Haigh, R., 68
Haimeur, A., 182
Hait, W. N., 306
Haizlip, J. E., 699
Halbert, C. L., 620, 628(20), 728
Hall, J. A., 3, 11, 20, 21, 28(7), 493
Hall, M. N., 782
Hall, R. E., 49
Hallstrom, T., 143(68), 155
Halpern, E., 169
Hamada, H., 259, 260, 261, 262, 262(30), 263, 263(30), 264, 264(30), 290, 342, 397, 398(7), 436, 453, 469, 508, 531, 554, 580
Hamilton, C., 44

Hamilton, R., 749
Hammer, R. E., 576
Hämmerling, G. J., 132, 745, 752(8)
Hamosh, A., 655
Hamrick, M., 369
Han, E.-K., 342, 344, 345(16), 348(16), 349(16), 356(16)
Han, E. S., 360, 368
Hanchett, L. A., 239, 246(26)
Handy, J., 109, 177
Hanna, M., 397, 398(2), 505
Hanover, J. A., 238, 249, 560
Hanrahan, J. W., 343, 617, 618, 618(3, 7), 619(3, 13), 620, 620(13, 14), 621(14), 623, 623(3), 624(3), 625(7), 630, 637(5), 641(5), 642, 645, 645(34), 646(34, 52, 53), 650(5, 34, 35)
Hansen, C. T., 576
Hansen, M. B., 721
Hanson, I., 163, 745
Hanten, G., 576
Hara, P., 559
Harder, W., 399
Hardy, C. M., 40, 45(7)
Hardy, S. P., 359, 360, 361(11, 21), 365, 365(11), 367(11), 368(11, 21), 369(11, 21), 494
Harkins, R. N., 576, 582(19)
Harlow, E., 312, 636, 637(18), 639(18), 650(18)
Harold, F. M., 39
Harper, K. L., 360, 367(10)
Harris, A., 653, 655(10), 656(10), 657(10), 658(10), 661(10)
Harris, D. J., 712, 713(12)
Harris, S. L., 400
Harrison, R. G., 399
Harrison, S. D., 357
Hartikka, J., 712, 713(13)
Hartl, F. U., 627, 758
Hartman, E., 268
Hartman, J., 113, 689, 690(7), 695(6)
Hartshorn, M. J., 9, 116, 136
Hartstein, M., 178
Hase, T., 778
Hashiguchi, N., 757
Hashimoto, T., 754, 756, 758, 758(48, 50), 769, 776
Hashmi, M., 759, 769(61)
Hasty, P., 474, 696
Haucke, V., 785
Hauswirth, W. W., 725
Hayashi, S., 40
Hazelrigg, T., 214
Heap, J., 163
Heath, T. D., 698, 709(5), 712(5)
Heddle, J. A., 580
Hedgpeth, J., 270
Hehir, K., 281, 620
Heike, Y., 259, 264(16), 265
Heilmeyer, L. M. G., Jr., 340
Heimfeld, S., 547, 548(17), 552, 554, 558
Heitman, J., 414, 415(8), 427(8), 780
Hekstra, D., 32, 45, 46(38), 49(38)
Helenius, A., 658
Hellen, C. U., 475
Hemenway, C., 414, 415(8), 427(8)
Henderson, D. M., 183
Henderson, P. J. F., 6
Hendrickson, N., 183
Hengge, R., 21
Henikoff, S., 214, 215(13), 358
Henkel, R. D., 38
Henning, D., 625, 641
Henning, U., 373
Heppel, L. A., 25
Herath, J. F., 580
Herlyn, M., 749
Herman, P. K., 628
Hermann, F., 257, 557
Hermanson, G. T., 401, 517
Hermes, J. D., 479
Hermodson, M. A., 116, 158
Hermodsson, A., 8
Hermonat, P. L., 728
Herrera-Sosa, H., 132, 154(19), 738, 743(8)
Herrmann, F., 546, 583, 585(32)
Herskowitz, I., 178, 193, 196(7), 198, 198(7), 225
Hertig, C., 166, 168(19)
Hesdorffer, C., 538, 547
Hess, H. H., 38
Hettema, E., 183, 185(12), 187, 187(12), 767
Heussen, C., 71
Heyman, J. A., 763
Heyneker, H. L., 479
Hi, T. H., 291
Hicke, L., 203, 627

Hidaka, N., 629, 630(1), 633(1), 652, 675
Hieter, P., 124, 134, 135, 135(34, 35), 161(33, 35, 38), 207, 208(32), 755
Higgins, C. F., 3, 4(2), 6, 7, 8, 9, 14, 17, 21, 30, 31(8), 101, 110, 116, 117, 130, 135(2), 136, 136(2), 152(12), 158, 308, 342, 343, 355, 359, 360, 361(11, 21), 362(7), 363, 363(8, 16), 365, 365(7, 11), 367(11), 368, 368(8, 11, 21, 34), 369, 369(11, 21), 493, 494, 514, 653, 655(10), 656(10), 657(10), 658(10), 661(10), 665, 673, 786
Higuchi, R., 460
Hildebrandt, J.-P., 360, 367(15), 369(15)
Hilden, S. A., 654, 657(16)
Hiles, I., 8, 30, 31(8), 116, 158
Hill, A., 120, 767
Hilliard, K. A., 720
Himelstein, A., 547, 551(11)
Hinnebusch, A. G., 143(49), 145, 158(49)
Hinnrasky, J., 661
Hipfner, D. R., 604, 613
Hirai, M., 259
Hirai, S. I., 476
Hirai, Y., 552
Hiraoka, Y., 782
Hirata, D., 143(91), 155, 160
Hiratsuka, T., 681
Hirshberg, C., 267, 274(24)
Hirt, B., 186
Ho, J. P., 602
Hobson, A. C., 9
Hodgson, C. P., 559
Hodson, M. D., 722
Hoestra, D., 9
Hoffman, E. K., 361
Hoffman, R., 548
Hoffmann, E. K., 359, 380(1)
Hoffmann, H. J., 40
Hoffmann, P. E., 347
Hoffman-Posorske, E., 340
Hofmann, E., 13
Hofmann, F., 312
Hofnung, M., 7, 10, 25, 30, 31, 35(14), 47
Hogan, B., 576, 577(21)
Hoggan, M. D., 725
Hoheisel, J. D., 131, 134(14), 137(14)
Hohneker, K. W., 724
Hoiby, N., 721
Holbrook, S. R., 5, 9, 10(39), 101
Holden, K. A., 399
Holland, B., 116
Holland, I. B., 8, 53, 61, 61(11), 68, 158, 781
Hollenberg, C. P., 143(96), 160, 203, 209
Holló, Z., 427(6, 7, 14), 428, 430(14), 440(14), 441(7, 16)
Holm, M., 697, 704(1)
Holmes, C., 649, 650(67)
Holmes, K. C., 677
Holzmayer, T. A., 225, 227(2), 237, 239(2)
Homoloya, L., 427(6, 7), 428, 441(6, 7)
Hong, J.-S., 15
Honore, N., 40
Hood, L. E., 331, 681
Hoof, T., 227, 423
Hoogenboom, H. R., 474
Hope, C. L., 83
Hor, L.-I., 12, 20(54), 21, 28(6), 32(32, 33), 33, 34(33)
Horecka, H., 203
Hori, R., 259
Horie, T., 132, 154, 608, 609(12), 610(12), 611(12), 614(12), 615(12)
Horio, M., 259, 289, 300(5), 495
Horn, G. T., 679
Hörner zu Bentrup, K., 9, 29
Horst, M., 785
Horwitz, S. B., 178, 303, 307, 309, 309(2), 311, 311(11), 312(11), 313, 319, 342, 344, 345(16), 346, 346(17), 348(16), 349(16), 352, 353(18), 356(16), 406, 431
Hoshino, T., 16, 17(80)
Hou, Y.-X., 288, 343, 618, 620, 620(14), 621(14), 622, 642, 645, 645(34), 646(34), 650(34, 37a)
Hourani, S. M. O., 664
Housman, D., 117, 163, 165, 165(1), 290, 431, 514, 725
Howard, B. H., 265, 542, 561, 710
Howard, J. C., 163
Howell, A. J., 768, 769(87)
Howell, K., 282
Howells, A. J., 118, 136, 150(42), 213, 214, 215, 220(18)
Hresko, R. C., 288
Hrycyna, C. A., 202, 255, 318, 319(7), 320, 320(7), 321(7), 326(7), 427, 441, 441(3), 442, 443, 443(2), 446(2, 11), 450(2), 453(2, 11), 456, 457, 459(10), 461(10), 473,

473(1, 10), 480, 504, 505(4), 523, 526, 530(10), 557, 570, 571(26), 574, 574(2)
Hsu, C. M., 89, 90(19)
Hsu, S. I., 178, 344, 346(17), 431
Hu, P.-C., 757
Hu, X. F., 259
Huang, L., 697, 698(2), 700, 704(2), 707, 722
Huang, P., 413, 618, 626, 626(10)
Huang, Z., 689, 691, 695(6), 696, 696(10)
Hubbard, C. E., 399
Hubbard, R. E., 9, 116, 136
Huber, H., 546
Huber, M., 611
Hude, S. C., 653, 655(10), 656(10), 657(10), 658(10), 661(10)
Hudes, G. R., 293, 300(26), 303(26)
Huganir, R., 655
Hughes, C., 52, 53, 55, 61(10), 68
Hughes, E., 745
Hullihen, J. M., 677, 681(20)
Humphrey, M. B., 736, 738(2)
Humphries, R. K., 547
Hunkapiller, M. W., 331, 681
Hunke, S., 9
Hunt, A. G., 15
Hunt, T., 272
Hunter, C. P., 204
Hunter, E., 691, 696, 696(10)
Hunter, L., 559, 725
Hunter, T., 342, 354, 355(30)
Hunter, W. M., 639
Hurley, T. R., 346
Hursey, M. L., 483
Hutchinson, A., 135, 139(39), 152(39), 160(39)
Hutchinson, C. A., 479
Hutchinson, C. A. III, 757
Hwang, D. M., 128
Hwang, T. H., 118, 653, 659(8)
Hwu, P., 558
Hyde, C. F., 360, 363(8), 368(8)
Hyde, S. C., 6, 9, 17, 116, 136, 360, 362(7), 363, 365, 365(7), 368(34)
Hyland, K. J., 24

I

Iannuzzi, M. C., 117, 132, 154(15), 343, 411, 629, 630(1, 2), 633(1, 2), 652, 675, 687, 715
Ichikawa, M., 266
Ihle, J. N., 552
Iida-Saito, H., 259
Ikebuchi, K., 552
Illsley, N. P., 714, 715(20)
Im, D.-S., 726
Imaeda, S., 745, 746(10), 753(10)
Imaizumi, K., 757
Imanaka, T., 756, 758(50)
Imundo, L., 721
Inaba, M., 259
Inagaki, N., 369, 665, 673, 732, 743, 744(1)
Inazawa, J., 369, 743
Innerarity, T. L., 631
Innis, M. A., 204
Inouye, H., 679
Inouye, M., 474
Irazu, C., 759, 769(61)
Ise, W., 293, 303(24), 317, 608
Ishidate, K., 74
Ishii, H., 523
Ishii, S., 259, 263
Ishikawa, T., 608, 609(21), 611, 611(21), 615(21)
Israel, M. A., 250
Itaya, K., 37, 38(39)
Ito, H., 385
Ito, K., 27, 37
Ivy, S. P., 250
Iwahashi, T., 531
Izquierdo, M. A., 607

J

Jaalat, S., 723
Jackson, P. S., 359, 363(4), 369(4)
Jackson, R., 272
Jacob, T. J. C., 360, 365, 369(14)
Jacobs-Lorena, M., 781
Jacq, C., 131, 134(14), 137(14)
Jaffe, H. A., 724
Jagus, R., 281
Jakes, K., 278, 288
Jamieson, J. D., 655
Janeway, C. A., Jr., 745, 746(10), 753(10)
Jang, E., 698, 709(5), 712(5)
Jansen, P. L. M., 154, 607, 608(9), 611(9), 614(9)
Janssen, H. W., 105

Jarvis, K. L., 24
Jeang, K. T., 558, 561(2), 570(2)
Jedlitschky, G., 128, 607, 607(11), 608, 608(7, 8, 10), 609(7, 8, 10, 16), 610, 610(7, 8, 16), 611(7, 8, 10, 11, 16), 612(8, 10), 613(7–11), 614(10), 615(8, 10, 11, 16), 616(8)
Jefferson, D. M., 617, 714
Jeffery, P. K., 722
Jenkins, N. A., 116
Jenkins, R. B., 580
Jennings, M. A., 608, 615(20)
Jenö, P., 785
Jensen, T. J., 288, 343, 617, 618, 618(7), 620, 620(14), 621(14), 622, 622(25), 625, 625(7), 626, 627(25), 630, 633(6), 637(5, 6), 641(5, 6), 642, 642(6), 645, 645(34), 646(34, 53), 647(6), 649(6), 650(5, 6, 34, 37a), 652, 663(4), 676, 677, 691, 714
Jentsch, T. J., 361
Jerabek, L., 550
Jhappan, C., 573, 576, 576(5), 579, 582(19)
Ji, G., 84
Ji, I., 291
Jiang, C., 712, 713(12), 714
Jilling, T., 687, 688(4), 689, 695(6), 696, 722
Joenje, H., 264
Jogn, G. H., 724
Johdo, O., 264
Johns, C. A., 654, 657(16)
Johnson, A. B., 754
Johnson, D., 120
Johnson, D. H., 214, 215(13)
Johnson, F. B., 725
Johnson, K. R., 755, 757(19)
Johnson, K. S., 630, 633(4), 635(4)
Johnson, L. G., 714, 722
Johnson, P. R., 727
Johnson, R. H., 554
Johnson, R. K., 259
Johnson, S., 163
Johnston, M., 131, 134(14), 137(14)
Johnston, T. C., 679
Jones, E. W., 204
Jones, T. R., 608, 613(19), 615(19)
Jorge, P., 759
Judah, J. D., 655
Jugovic, P., 120
Juliano, R. L., 259, 493
Juranka, P., 60, 514, 781
Jurman, M., 279
Jurus, L., 559
Just, W. W., 757, 758, 763(53)
Juvvadi, S. R., 344, 353(18)

K

Kaback, H. R., 371, 374, 374(1), 375
Kabat, D., 558
Kabsch, W., 677
Kadowaki, Y., 599
Kaetzel, C. S., 722
Kaiser, C., 196, 424
Kajiji, S., 178, 294, 304, 316, 401, 574
Kalish, J. E., 755, 756, 774, 775(98)
Kamijo, K., 756, 758(48), 769, 776
Kandror, O., 627
Kane, S. E., 227, 265, 454, 474, 476, 543
Kanebo, A., 361, 362(24)
Kantoff, P. W., 255, 523
Kanz, L., 554
Kao-Shan, C. S., 250
Kapikian, A. Z., 725
Kaplan, J. H., 118
Kaplan-Stern, R., 299, 304(33)
Kapler, G. M., 188, 189(24)
Karazian, H. H., 676
Karcz, S., 109, 175, 176(7), 177
Karre, K., 745
Kartenbeck, J., 128, 132, 154, 607, 608, 609(12), 610(12), 611(12), 614(12), 615(12)
Kartner, N., 117, 118(11), 260, 262, 263, 263(25), 298, 343, 434, 584, 616, 617, 617(9), 618, 618(3, 7), 619(3), 620, 620(14), 621(14), 622(24), 623(3), 624(3), 625(7), 629, 630, 633(6), 637(5, 6), 641(5, 6), 642, 642(6), 645, 645(34, 36), 646, 646(34), 647(6), 649, 649(6), 650(5, 6, 34, 36), 652, 653, 663(4), 691, 714, 722
Kasahara, M., 44
Kasch, L. M., 676
Kasid, A., 558
Kast, C., 266, 288, 414, 415(8), 427(8), 505
Katakura, K., 183
Katsura, Y., 571
Katz, S. L., 445
Katzmann, D., 627
Kaufmann, A., 373
Kaur, P., 90

Kavanagh, J., 547
Kavanaugh, M. P., 558
Kawabata, H., 259, 263
Kawasaki, E. S., 168
Kawinski, E., 262
Kazarov, A. R., 225, 227, 227(2), 239(2, 16), 546, 586
Kearns, W. G., 540, 544(13), 545(13), 728, 731, 758
Keeffe, E. B., 610
Keen, N. T., 67
Keizer, H. G., 264
Keller, B. H., 618, 626(10)
Keller, G. A., 754
Keller, T. A., 47
Kellermann, O., 31
Kelley, J. M., 757
Kelley, K., 579, 580(24)
Kellum, R., 215
Kelly, A., 163, 745, 768
Kemp, D. J., 167, 175, 176, 176(5), 178(9), 181(9)
Kemp, S., 435, 481, 759, 765(66), 766(66)
Keneko, K., 765
Kennedy, D., 617(9), 618, 629, 630(1), 633(1), 652, 675, 676
Kennelly, P. J., 335
Kennet, R. H., 261, 637
Kenny, A. J., 654
Kenny, B., 53, 61, 61(11), 68
Kent, S. B., 347, 348(25)
Keppler, D., 128, 132, 154, 607, 607(11), 608, 608(7, 8, 10), 609(7, 8, 10, 12, 16, 21), 610, 610(7, 8, 12, 16), 611, 611(7, 8, 10–12, 16, 21), 612(8, 10), 613(7, 8, 10, 11), 614(10, 12), 615(8, 10, 12, 16, 21), 616(8)
Kerem, B. S., 117, 132, 135, 154(15), 343, 411, 617, 619(4), 629, 630(1–3), 631(2), 633(1, 2), 635(2), 652, 675, 675(3), 676, 687, 715
Kerlavage, A. R., 757
Kerppola, R. E., 5, 7, 8(13), 114, 764, 765(78)
Kerr, D. J., 309
Kerr, L. A., 768
Kessler, S. W., 548
Kiel, J. A. K. W., 757, 763(53)
Kiem, H.-P., 558, 559
Kikuchi, H., 293, 303(19)
Kim, R., 135, 161(38)
Kimura, A., 385
King, C. R., 596
King, L. A., 128, 430(20)
King, S. A., 686
Kingston, R. E., 596, 598(15), 599(15), 603(15)
Kinrade, E., 707
Kioka, N., 259
Kionka, C., 772
Kioschis, P., 117, 132, 756, 758(45), 769(45)
Kirk, J., 363, 368(34)
Kirk, K., 687, 688(4), 691, 696, 696(10)
Kirschling, D. J., 225, 227(2), 239(2)
Kirschner, L. S., 178
Kirschner, M. W., 312
Kiser, G. L., 616, 618, 626(10)
Kiss, B., 653
Kistler, A., 116, 121(6), 143(81), 158, 777
Klebba, P., 5, 8(13)
Kleijer, W. J., 759, 765(66), 766(66)
Klein, P., 367
Klemm, K., 293, 303(24), 317
Klenk, D. C., 401, 517
Klimecki, W. T., 546
Kling, G., 745
Klinger, K., 281, 617, 620
Klionsky, D. J., 628
Kloser, A. W., 616, 618, 626(10)
Klotz, U., 25
Klünemann, C., 611
Knight, R., 163
Knowles, J. R., 479
Knowles, M. R., 724
Ko, Y. H., 675, 677, 678, 682, 683, 684, 685
Koch, J. P., 40
Koerberl, D. D., 620, 628(20)
Kohchi, T., 776
Kohiyama, M., 16
Kohler, D., 538, 547
Köhler, G., 637
Koike, R., 765
Kok, F., 759, 765(62)
Kolaczkowski, M., 143(54), 151, 159(54)
Kolch, U., 126
Kolling, R., 143(96), 160, 180, 203, 209
Kolodner, R., 576
Kolter, R., 3, 4, 4(5), 51, 67, 130
Komano, T., 259, 414
Kondo, N., 757, 769
König, J., 132, 154, 608, 609(12), 610(12), 611(12), 614(12), 615(12)
Konishi, H., 414
Konstan, M. W., 720

Koonin, E. V., 101
Koppel, H., 360, 369(14)
Koprowski, H., 749
Koronakis, E., 52
Koronakis, V., 53, 61(10), 68
Korte, H., 347
Kose-Terai, K., 16, 17(80)
Koslowsky, T., 367
Kossmann, M., 42, 43(21), 45(21)
Koster, H., 479
Köster, W., 6, 7, 46, 49, 764, 765(77)
Köstiö, S. R. M., 726
Kotin, R. M., 539, 559, 725, 726
Kourilsky, P., 745
Kouyama, K., 360
Koval, D., 699
Koyama, T., 259, 264(16)
Kozak, M., 602, 743
Kozak, S. L., 558
Kralli, A., 143(90), 160
Kramer, R. R., 494
Krapivinsky, G. B., 365, 368
Krapivinsky, L. D., 365, 368
Krebs, E. G., 335
Kreppel, L., 163, 776
Kretz, C., 153, 756, 759(46)
Krieg, P. A., 270
Kriegler, M., 226, 316, 431
Krishnamachary, N., 607, 613(3)
Krivit, W., 759
Krogstad, D. J., 174, 176
Krohn, R. I., 401, 517
Kronauge, J. F., 427(10), 428
Kronborg, G., 721
Kronidou, N. G., 785
Krouse, M. E., 360, 367(10)
Kruh, G. D., 607
Krumlauf, R., 474
Krumm, A., 620, 628(20)
Kruse, M., 288
Krust, A., 596
Ksenzenko, M. Y., 91
Kubo, M., 361
Kuchler, K., 143(50, 71), 151, 156, 160, 163, 178, 193, 200(3), 202(3), 414, 473, 777, 781
Kühnau, S., 50
Kuiper, C. M., 262
Kukuda, Y., 385
Kulpa, J., 781
Kumar, R., 698, 699(4), 704(6), 707(4)
Kunau, W.-H., 755, 757, 763, 763(53), 772
Kunkel, T. A., 482
Kunua, W. H., 755
Kunzelmann, K., 367
Kuo, J. F., 329, 330, 341, 353, 357
Kuroda, M., 82, 88
Kuroki, Y., 757
Kuroko, C., 259, 264(16)
Kurz, E. U., 117, 133, 183, 493, 600, 602, 602(19), 607, 609(1)
Kurz, G., 607, 608(10), 609(10), 611, 611(10), 612(10), 613(10), 614(10), 615(10)
Kustu, S. G., 11
Kuwano, M., 293, 303(19)
Kuwazuru, Y., 266
Kwan, T., 505
Kwast, L., 756
Kwiatkowski, F., 264
Kyle, D. E., 109, 174, 175, 176, 177(11), 178(9), 181(9), 183
Kyostio, S. R. M., 544
Kyte, J., 138, 279

L

Laban, A., 189
Labarca, C. G., 279
Labia, R., 72
Lacy, E., 576, 577(21)
Lacy, L. R., 220
Laemmli, U. K., 89, 298, 312, 330, 346, 354(20), 401, 450, 463, 465(15), 497, 507, 634, 635(11), 656, 680
Laffer, T. G., 754
Lageweg, W., 769
Laghaeian, S. S., 17, 22
Lai, S.-L., 250, 555
Laing, N., 293, 300(26), 302(25), 303(25, 26)
Lalumiere, M., 625, 641
Lalwani, N. D., 754
Lamb, J. R., 755
Lambert, L., 143(53), 151, 159(53)
Lam J. S., 558
Lampen, J. O., 388
Landick, R., 11
Landolt-Marticorena, C., 274
Lane, D. P., 312, 636, 637(18), 639(18), 650(18)
Lane, M., 712, 713(12)

Lange, A. J., 655
Lankelma, J., 110, 262, 264, 309, 613
Lansdorp, P. M., 547
Lanzetta, P. A., 37, 38(41)
LaRossa, D., 749
Larson, T. H., 40, 45(8)
Larson, T. J., 40, 42
Laskey, A., 298
Lasky, L. A., 165
Latour, A. M., 618, 626(10)
Laughlin, C. A., 559, 725
Laurenssen, C., 128
Lautier, D., 493
Laver, W. G., 651
Law, T. C., 360, 367(10)
Law, W., 558
Lawn, R. M., 165
Lazarow, P. B., 754, 754(15), 755, 757, 763(53)
Lazarowski, E., 699
Lazdunski, A., 68
Lazo, O., 759, 769, 769(61)
Lebleu, B., 152
Lecar, H., 117
Lechago, J., 736
Lecocq, J.-P., 647, 723
Ledbetter, D. H., 579
Leder, P., 598
Lee, C., 267
Lee, C.-H., 286(27), 288
Lee, C. Y., 128
Lee, C. G. L., 557, 558, 561(2), 570(2)
Lee, E. R., 712, 713(12)
Lee, K.-S., 11
Lee, R. T., 558
Lee, S. T., 191
Lee, T. B., 608, 615(18)
Leemhuis, T., 548
Légaré, D., 182, 183, 187, 187(7)
Leger, S., 608, 613(19), 615(19)
Leiber, M., 555
Leier, I., 128, 154, 607, 607(11), 608, 608(7, 8, 10), 609(7, 8, 10, 16), 610, 610(7, 8, 16), 611(7, 8, 10, 11, 16), 612(8, 10), 613(7, 8, 10, 11), 614(10), 615(8, 10, 11, 16), 616(8)
Leighton, J., 116(9), 117, 143(80), 157, 776
Leitman, S. F., 538, 547
Lelong, I. H., 131, 321, 397, 398(8), 409(8), 410(8), 411(8), 481, 487(2), 492, 495, 496(22), 500(22), 505, 511(12), 662(37), 663
Lemischka, J. R., 547
Lennon, G. G., 122
Lenschow, D., 558
Leonard, D., 475
Leonard, J. L., 475
Leonard, R. J., 279
Le Pecq, J.-B., 260
Lerman, L. S., 479
Lerner, C. G., 474
Leslie, A. G. W., 17
Lesser, C., 268
Lester, H. A., 279, 625
Leterme, S., 143(51), 151, 619(40), 626
Létoffé, S., 68, 69, 72, 73(14, 18), 75
Leuschner, U., 608
Levenson, R., 266, 288, 414
Lever, J. E., 311
Levin, D., 625, 641
Levinson, A. D., 476
Levinthal, C., 42
Levis, R., 214
Lewis, G., 204
Lewis, G. K., 170, 171(24), 781
Lewis, K., 118
Lewis, R. S., 359, 361
Lewis, S. E., 474, 475
Li, C., 620, 625, 626, 645, 652, 663(4), 677, 691, 714
Li, F., 526, 537(11), 557
Li, J., 83, 91
Li, M.-X., 546
Li, P., 679
Li, X., 755, 757(19)
Li, Y., 368
Li, Z., 143(67), 155
Li, Z.-H., 554
Liang, C. T., 654
Liang, P., 596
Liang, X., 712, 713(13)
Licht, T., 255, 256(10), 257, 257(10), 341, 342(18), 355, 368, 372, 503, 505, 506, 506(9), 513(9), 546, 548, 557, 573, 576(5), 583, 585(32), 589(8)
Lieber, M., 250
Lieberman, M. A., 15
Liew, C. C., 128
Light, T., 248
Lightbody, J., 94, 190
Ligtenberg, M. J. L., 759, 765(66), 766(66)
Lill, R., 27, 37

Lilly, F., 576
Lim, E., 780
Lin, E. C. C., 40
Lin, R., 370
Lin, T. H., 262
Lincke, C. R., 105, 167
Linden, R. M., 725
Lindsay, D. L., 219, 220(21), 221(21)
Ling, V., 30, 51, 52, 53, 53(7), 55, 55(12), 56, 58, 61(13, 14), 62(13), 63(14), 64, 64(14), 102, 116, 117, 117(1), 118(11), 133, 137, 183, 226, 258, 259, 260, 262, 263, 263(25, 26), 264, 266, 267(8, 9), 278(8, 9), 279, 281, 282(17), 283(14), 284(14), 285(14), 286(27, 30), 287, 288, 289, 293, 298, 303(20), 304, 308, 322, 328, 397, 398(12), 401, 404, 410(12), 411(12), 434, 453, 463, 493, 497, 503, 508, 509(18), 514, 515, 522, 584, 620, 649, 781
Lingappa, J. R., 272
Lingappa, V., 266, 267, 267(6, 7), 268, 268(6, 7, 17), 270, 270(17), 271(6, 7, 20), 272, 277, 278, 278(6), 287, 288(18)
Linsdell, P., 623
Linton, G. F., 526, 537(11), 557
Lipinski, M., 260
Lipman, D. J., 122, 138
Lipp, J., 266
Lippincott, J., 32
Lisanti, C. J., 313, 319
Liu, D., 8, 32, 113
Liu, H. Y., 191
Liu, J. M., 544, 727
Liu, Q., 696
Liu, Z., 293, 294(23), 301(23), 303(23)
Ljunggren, H.-G., 745
Loayza, D., 143(95), 157, 160, 203, 204(18), 206, 210, 212
Lodish, H. F., 267, 268, 279
Loe, D. W., 604, 606(25), 607(25), 608, 611(17), 613, 613(17), 615(17)
Loeffler, J. P., 698, 704(3)
Logan, J., 688, 691, 694(5), 695(5), 696, 696(10)
Lok, S., 117, 132, 154(15), 343, 411, 617, 619(4), 629, 630(2), 631(2), 633(2), 635(2), 675, 687, 715
Lombard-Platet, G., 756, 759, 762(47), 769(47)
Lomedico, P. T., 576
Lomovskaya, O., 118
Loo, M. A., 620, 622(25), 627(25), 645
Loo, T. W., 266, 288, 304, 316, 318, 322, 361, 397, 398, 398(9), 409(15), 414, 427(12), 428, 456, 478, 480, 481, 482, 483, 483(11), 484, 484(13), 486, 486(6, 13), 487(8), 490, 491(6), 505, 623, 768
Lopez, C. D., 278
Lopez, J., 153, 756, 759, 759(46)
LoPresti, M. B., 328, 329, 332(5)
Lorenz, M. C., 474, 780
Loser, S., 744
Lothstein, L., 178, 344, 346, 346(17), 431
Lotze, M., 558
Louis, E. J., 131, 134(14), 137(14)
Lovelace, E., 430, 523
Low, B., 43
Low, M. J., 576
Lowan, K. H., 250
Lowe, T. M., 122, 160
Lowrie, D. B., 475
Lowry, O. H., 635
Lu, G. Y., 11, 677
Lu, L., 554, 655
Lu, X., 627
Lu, Y., 143(67), 155
Lucht, J. M., 3, 4, 4(3), 7(3), 18(3), 30, 45
Luciani, M., 116, 160, 777
Lucier, T. S., 757
Luckie, D. B., 360, 367(10)
Luckow, V. A., 128, 429(19, 21), 430, 430(19, 20), 431(21), 433(21), 435(21), 619(33), 625
Ludtke, D., 41, 44
Lugtenberg, B., 41, 47(10)
Lujan, E., 331
Lukacs, G. L., 620, 622(24), 642, 645, 645(36), 646, 650(36), 653, 722
Luo, K. X., 346
Lutsenko, S., 118
Lutter, R., 17
Lutz, Y., 153, 756, 759(46)
Lynch, C. M., 560, 561(19)
Lynch, S. M., 559

M

Ma, L. D., 342, 357(7)
Maarse, A. C., 781

MacAlister, T. J., 74
Macara, I. G., 417
MacFerrin, K. D., 679
Machamer, C. E., 744, 756(14)
Machie, J. E., 117
Mack, E., 369
Mackensen, D. G., 264
Mackett, M., 442
Mackie, A., 293, 303(20), 308
Mackie, J. E., 133, 183, 250, 493, 600, 602(19), 604, 607, 609(1)
MacKinnon, R., 279
Mackman, N., 68
MacLeod, J., 369
Madias, N. E., 654, 657(16)
Magee, B. B., 143(72), 156
Mahadevia, R., 546
Mahanty, S. K., 413, 618, 626(10)
Mahley, R. W., 631
Maina, C. V., 679
Makino, K. H., 43, 44
Malech, H. L., 526, 537(11), 557
Malera, P. W., 304
Malkhandi, P. J., 309
Mallia, A. K., 401, 517
Maloney, P. C., 16, 17(76), 21, 39, 40, 45(8), 495, 496(20), 497(19, 20), 501, 503(19, 30), 652, 653, 655, 655(10), 656, 656(10), 657(10), 658(10), 659(24, 29), 660(29), 661(10), 662(37), 663, 663(29), 668
Malouf, N. N., 618, 626(10)
Manavalan, P., 620, 647, 650(60), 676
Mandel, C., 765
Mandel, J.-L., 117, 132, 153, 756, 758(45), 759, 759(46), 762, 762(47), 765, 765(66), 766(66), 769(45, 47, 63)
Maniatis, T., 59, 164, 165(17), 209, 245, 270, 385, 431, 446, 479, 563, 571(24), 574, 596, 598(14), 599(14), 603(14), 636, 680
Mannaerts, G. P., 402, 520, 755, 757, 763(53)
Mannering, D. E., 17, 22
Mannherz, H. G., 677
Manoil, C., 31, 35(15), 267, 268, 378
Manthorpe, M., 704, 708(9), 712, 713(13)
Marangos, P. J., 665
Marcel, Y. L., 631
Marchase, R. B., 659, 661(33)
Marchuk, D., 122
Margulies, D. H., 637
Mariani, M., 262
Marino, C. R., 647, 656, 661, 661(30), 718
Maris, D., 11
Markakis, D., 540, 729, 730
Markiewicz, D., 135, 629, 630(3), 652, 675(3), 676
Marquardt, D., 342, 357(7)
Marsh, L., 211
Marsh, W., 342
Marshall, J., 136, 281, 343, 620, 622, 628(27), 641, 642(42), 647, 650(59), 652, 653, 656(7), 661(7), 676, 677, 697, 712, 713(12, 13), 714
Marston, F. A. O., 680
Martens, C. L., 676
Martin, M., 558, 561(2), 570(2), 698, 699(4), 704(6), 707(4)
Martin, R. K., 109, 176, 177(11), 178(9), 181(9), 183
Martin, S. K., 174
Martinand, C., 152
Martinoia, E., 154, 155(63), 613
Marton, M. J., 143(49), 145, 158(49)
Marvin, S. A., 617
Marzioch, M., 755, 763
Masiakowski, P., 596
Masson, P., 608, 613(19), 615(19)
Masters, P. S., 15
Mastrangeli, A. A., 724
Mastrangelo, M. J., 749
Masuno, M., 757
Mateblowski, M., 772
Mathew, P. M., 110, 117, 132, 154(20)
Mathews, C., 620
Matile, H., 168, 169(22)
Matovcik, L. M., 647, 656, 661(30)
Mattei, M., 116, 160, 777
Matteucci, M. D., 479
Matthews, C., 645, 646(52)
Matthias, P., 198
Maxey, R. J., 608, 615(20)
Maxwell, I. H., 618, 622(15)
Mayer, R., 128, 607, 608
Mayo, J. G., 483
Mazda, O., 571
Mazo, I. A., 225, 227, 227(2), 239(2)
Mazzaferro, P. K., 399
Mazzei, G. J., 330
McAteer, J. A., 744
McCammon, M. T., 756
McCann, J. D., 617

McCarter, L., 475
McCarty, N. A., 625
McClendon, V., 157, 195, 210(12), 211
McCollum, D., 755, 763
McCombie, W. R., 399
McCormick, F., 152
McCue, K. F., 192
McDonagh, K. T., 538, 547, 548
McDonald, G., 163, 776
McDonald, T. V., 676
McElvaney, N. G., 724
McElver, J., 780
McFarlane, S., 608, 613(19), 615(19)
McGovern, K., 6, 267
McGrath, J. P., 118, 143(72), 156, 163, 167, 178, 193, 777
McGrath, S. A., 725(53), 727, 729, 731
McGuire, T. M., 627
McGurl, B., 163
McIntyre, J. C., 391, 392(17)
McKay, D. B., 70
McKeown, M., 214
McKinney, L., 365, 367(38)
McLachlin, J. R., 255, 523
McLaughlin, S., 559, 725, 728
McNaughton, P. A., 342, 355, 368
McNew, J. A., 754, 755(14), 756, 762(14)
McPherson, J. M., 620
McReynolds, L. A., 679
Mears, G., 538, 547
Mechetner, E. B., 229, 233, 266, 453, 508, 546
Medina-Acosta, E., 184
Mehta, N. D., 226, 293, 294(15), 299, 300(15), 303, 304(33), 316, 406, 478
Meier, P. J., 610
Meijer, C., 154, 607, 608(9), 611(9), 614(9)
Meijer, G., 154
Meijer, H. J., 262
Meissonnier, F., 264
Melamed, M. R., 259
Meldolesi, J., 655
Melera, P. W., 226, 250
Melhus, O., 647, 649(66)
Mellins, E., 117, 118(15), 163, 745
Mellman, I., 658, 746
Melmer, G., 629, 630(1), 633(1), 652, 675
Meltin, D. A., 270
Meltzer, P. S., 264
Mendel, D., 474
Mendelson, E., 725, 726
Menestrina, G., 51
Menninger, J. C., 539, 725
Merchetner, E. B., 262, 264
Meredith, A. L., 676
Merette, C., 759
Merino, G. T., 13, 33
Merlino, G. T., 542, 546, 561, 573, 575, 576, 576(4, 5), 579, 580(4, 7), 582(19), 589(4, 7–9), 590
Merrick, J. M., 757
Merritt, S. C., 177
Mertelsmann, R., 554
Messing, J., 679, 779
Metzig, B., 756
Mewes, H. W., 131, 134(14), 137(14)
Meyer, H. E., 312, 340, 347
Meyer, T. H., 768
Meyers, M. B., 262, 264, 291, 298(11), 300(11), 311
Michaelis, S., 116, 121(6), 130, 135(4), 136(4), 137, 139(4), 143(72, 81, 95), 145(4, 89), 147(4), 156, 157, 158, 160, 178, 193, 194, 196, 196(6, 7, 10), 197(6, 14), 198(7), 200(8), 202, 202(6, 8), 203, 204(18), 210, 210(6), 211, 212, 221, 379, 424, 768, 777
Michaud, W. A., 755
Mickisch, G. H., 546, 547, 572, 573, 580(7), 589(7–9), 590
Micro Ultrasonic Disrupter, 636
Middleton, P. G., 722
Migeon, B. R., 759
Mikami, K., 259
Milhous, W. K., 109, 174, 175, 176, 177(11), 183
Milisav, I., 52
Millan, J., 267
Miller, A. D., 229, 538, 548, 558, 560, 561(19), 620, 628(20), 723, 728
Miller, D. G., 558, 560, 561(19), 723
Miller, I. L., 728
Miller, J., 544
Miller, J. H., 72
Miller, J. L., 544
Miller, L. K., 128, 429(21), 430(21), 431(21), 433(21), 435(21)
Miller, T. P., 264
Miller, W., 122, 138
Mills, J. W., 361
Milne, R. W., 631
Mimmack, M. L., 6, 17

Mimmack, M. M., 9, 116, 136
Mimura, C. S., 5, 9, 10(39), 101
Min, C., 625
Mineishi, S., 546
Mintenig, G. M., 360, 363, 363(8), 365, 368(8, 34)
Mir, A. A., 381
Mirski, S. E. L., 595, 606
Misra, L. M., 206
Misra, T., 83
Mitchell, A., 196, 424
Mittereder, N., 629
Miura, S., 755
Miwa, A., 360, 365(20), 367(20), 368(20)
Miyahara, K., 143(91), 160
Miyakawa, T., 143(91), 155, 160
Miyatake, T., 765
Mobley, H. L. T., 84
Mobraaten, L. E., 576
Mocklin, W., 554
Mockridge, I., 163, 745
Mockridge, L., 768
Modi, W., 116
Moffat, L. F., 710
Mohamed, A., 620, 622(24), 642, 645, 645(36), 650(36, 37a)
Mol, C. A. A., 133, 151(26), 391, 494
Momburg, F., 132, 745, 752(8)
Monaco, J. J., 163, 745, 776
Monk, B. C., 402, 407(35), 412(35)
Monks, A., 483
Monosov, E., 755
Montrose, M. H., 655
Moon, J., 211
Moore, C. L., 754
Moore, D. D., 596, 598(15), 599(15), 603(15)
Moore, M. A. S., 552
Moorhouse, A., 365, 369(40)
Morales, M. M., 624, 642, 650(37)
Moran, L. S., 679
Morbach, S., 31, 32(18)
Moreau, J., 762, 769(71)
Morecki, S., 558
Moreno, F., 46
Morgan, P. F., 665
Morgan, R. A., 558, 559(7), 566
Mori, T., 117, 755, 757(31)
Morimoto, H., 25
Morin, X. K., 361
Morishima, S., 361
Moritani, S., 260
Mornet, C., 154, 155(63), 613
Morrell, J. C., 755, 756
Morris, A. P., 696
Morris, D. I., 293, 303(21), 304, 305(48), 309, 311(10), 312(10), 314(10), 315(10)
Morrone, S., 262
Morse, B. S., 226, 398, 427(16), 428, 430(16), 431(16), 439(16), 440(16)
Moser, A., 755, 757, 757(18), 759, 765(62), 769(54)
Moser, C., 51
Moser, H., 112, 117, 132, 756, 757, 758(49), 759, 769(54, 63)
Moser, H. W., 754, 755, 757, 757(18), 759, 763(53), 765(62)
Moss, B., 442, 443, 444(8), 445, 446, 446(10), 453, 453(6), 455, 456, 457, 457(6), 459, 460(11), 461, 461(11), 473, 626
Moss, S. R., 163
Mosser, H., 132, 756, 758(45), 759(46), 769(45)
Mosser, J., 117, 132, 153, 756, 758(45), 759, 765, 769(45)
Motais, R., 368
Moullier, P., 759
Moulton, A., 544
Mount, S. M., 214
Moye-Rowley, W. S., 143(53, 64, 68, 93), 151, 154, 155, 159(53), 160, 627
Muccio, D. D., 689, 695(6)
Muda, M., 44
Mueckler, M., 288
Mueller, T. J., 259
Muir, M., 11, 17
Muir, R. S., 780
Muir, S., 6, 31, 35(16)
Mukhopadhyay, R., 94
Mulder, N. H., 154, 607, 608(9), 611(9), 614(9)
Mullen, G. P., 682, 683, 684, 685
Müller, M., 154, 226, 398, 427(14, 16), 428, 430(14, 16), 431(16), 439(16), 440(14, 16), 607, 608(9), 610, 611, 611(9), 614(9)
Mulligan, R. C., 558, 653, 676
Mullis, K. B., 679
Murakami, Y., 131, 134(14), 137(14)
Murata, K., 385
Murgier, M., 68

Murphy, C., 214
Murray, G. J., 525, 526(8), 528(8), 538, 557
Murray, J. F., 718, 719(4)
Mutul, J. P., 698, 704(3)
Muul, L. M., 558, 559(7)
Muzyczka, N., 539, 540(12), 559, 725, 726, 728
Myers, E. W., 122, 138
Myers, R. M., 278, 479

N

Nagata, T., 758
Nagayoshi, K., 547(19), 548
Nagy, A., 474
Nairn, A. C., 343, 661, 676
Naismith, A. L., 617, 618(3, 7), 619(3), 623(3), 624(3), 625(7), 630, 633(6), 637(5, 6), 641(5, 6), 642(6), 645, 647(6), 649(6), 650(5, 6)
Naito, M., 258, 259, 261, 263, 263(32), 264, 264(16), 357, 588
Nakajima, T., 342
Nakamoto, R. K., 159, 382(9), 383, 384(9), 386(9), 388(9), 628
Nakata, A., 43, 44
Nakuchi, H., 547(19), 548
Namba, N., 369, 732, 743, 744(1)
Nambudripad, R., 755
Nare, B., 293, 294(23), 301(23), 303(23)
Nash, R., 203
Nastaincyzk, W., 312
Natarajan, A. T., 475
Nechay, B. R., 80
Neefjes, J. J., 132, 745, 752(8)
Neer, E. J., 755
Neidhardt, F. C., 14
Neissel, O. M., 14
Nelles, L., 399
Nelson, D. A., 132, 154(19), 732, 736, 737, 738, 738(2, 3, 6), 740(3), 743(8), 744(3)
Nelson, D. M., 566
Nelson, R. J., 627
Nelson, W. L., 295
Nemali, M. R., 754
Neu, H. C., 25
Neumann, S., 759, 769(63)
Newcomb, J. R., 751
Newman, M. J., 23
Ney, P. A., 544
Neyfakh, A. A., 239, 546, 586
Ng, W. F., 263
Nghiem, P. T., 676
Nguy, K., 132, 154(19), 738, 743(8)
Nguyen, D., 757
Ni, B., 320, 427, 441, 441(3), 442, 456, 473, 473(1), 480, 523, 574
Nichols, C. G., 132, 154(19), 736, 738, 741, 743(8)
Nichols, D. A., 736
Nichols, M. R., 712, 713(12, 13)
Nicklen, S., 601
Nienhuis, A. W., 538, 544, 547, 548
Nies, A. T., 397
Niesen, F., 755
Nietupski, J. B., 712, 713(12, 13)
Nieves, E., 342, 344, 345(16), 348(16), 349(16), 356(16)
Nijbroek, G. L., 157, 193, 198, 206, 210, 211
Nikaido, H., 3, 4, 5, 7, 9, 10, 11, 12, 15, 16, 17(35, 77), 18(35), 19, 20, 20(54), 21, 22, 22(8), 24(4), 27(4), 28(4, 6–8), 31, 32, 33(25), 37(25), 38(25), 79, 113, 131, 493
Nikaido, K., 16
Niklen, S., 680
Niles, E. G., 443, 444(8), 456
Nilius, B., 359, 360, 367, 368(3), 369(3)
Nilsson, I., 274
Nishi, K., 414
Nishikawa, S.-I., 547(19), 548
Nishino, H., 580
Nishiyama, K., 278
Nitecki, D., 372
Noguchi, K., 263
Nolan, G. P., 230, 232(21), 243
Nolte, D. J., 220
Nonnan, K. E., 237
Noone, M., 538, 547
Nordeen, S. K., 479
Norman, J., 704, 708(9), 712, 713(13)
Northrop, J. P., 697, 704(1)
Norton, E., 473
Norton, W. T., 754
Norum, R. A., 759
Norvell, T. M., 720
Notkins, A. L., 445
Nouvet, F. J., 196, 197(14)

Novick, P. J., 382, 383(8), 384(8), 386(8), 388(8), 628
Nunes, F. A., 724
Nuttley, W. M., 755, 757(19)

O

Obie, C., 758
O'Brian, C. A., 357
O'Brien, J. A., 369
O'Brien, J. P., 259, 262, 264
O'Brien, S., 126
Ockner, R. K., 610
O'Connor, B. M., 344, 345(16), 348(16), 349(16), 356(16)
O'Connor, S. P., 714
Odula, A. M. J., 183
Oduola, A. M. J., 174, 175, 176, 178(9), 181(9)
Offerhaus, G. J., 391, 494
Ogawa, M., 552, 554
O'Gorman, S., 474
O'Hare, K., 214
Oh-Hara, T., 259, 263, 588
Ohyama, K., 776
Oka, H., 725(53), 727, 731
Okada, S., 547(19), 548
Okada, Y., 360, 361, 365(20), 367(20), 368(20)
Okamura, N., 259
Okayama, H., 561
Okayama, M., 484, 619(54), 629
Okochi, E., 263, 531
Okumura, K., 265
Oldfield, E. H., 523
Oliver, D., 27, 37
Oliver, S. G., 131, 134(14), 137(14)
Olsen, J. C., 369, 722
Olson, B. J., 401, 517
Omitowoju, G. O., 175
O'Neal, S., 21
O'Neal, W. K., 696
Ono, K., 531
Onodera, M., 566
Onodera, O., 765
Ordway, R., 766
O'Reilly, D. R., 128, 429(21), 430(21), 431(21), 433(21), 435(21)
Orii, T., 117, 755, 757, 757(31), 769
O'Riordan, C. R., 620, 622, 628(27), 647, 650(59), 652, 676
Orozco-Cardenas, M., 163
Orr, G. A., 342, 344, 345(16), 348(16), 349(16), 352, 353(18), 356(16)
Ortiz, D. F., 163, 192, 776
Ortlepp, H., 611
Osada, H., 155
Osborne, W. R., 558
Osgood, C. J., 220
O'Shaughnessy, J. A., 538, 547
Osipchuk, Y. V., 360, 361(19), 365(19), 367(19)
Ostedgaard, L. S., 212, 281, 620, 655, 676, 768
Ostrander, F., 331
Osumi, T., 754, 756, 757, 758(48, 50), 763(53), 776
Ott, J., 759
Ottenhoff, R., 391, 494
Otter, M., 128, 132
Oude Elferink, R. P., 110, 128, 391, 494, 608, 609(13), 613
Ouellette, M., 94, 95(28), 105, 110(13), 182, 183, 183(1, 2), 185(1, 12), 187, 187(7, 9, 12, 13), 189, 189(2), 190, 190(3, 28), 191(2, 28), 411
Overbeek, P. A., 576
Overduin, P., 41, 42, 42(17), 43(17)
Ow, D. W., 163, 192, 776
Owens, R. A., 544, 726, 727, 728(52)
Owolabi, J. B., 83, 86
Oxender, D. L., 11, 44, 113

P

Paabo, S., 745
Paakko, P. K., 687, 723
Pack, S., 579
Paddon, C., 212
Padmanabhan, R., 265, 446
Page, L. J., 647, 649(66)
Pai, E. F., 677
Pai, L. H., 573
Palade, G., 282, 655
Palmiter, R. D., 576, 618, 622(15)
Palsson, B. O., 559
Palumbo, P., 726
Panagiotidis, C. H., 8, 9(28), 12, 30, 32, 32(33), 33, 34(33), 37(21), 39(21)
Pang, M. G., 540, 544(13), 545(13)
Panicali, D., 442

Panten, U., 744
Panton, L. J., 176
Panvier-Adoutte, A., 755
Paoletti, E., 442, 473
Papadopoulou, B., 94, 95(28), 182, 183, 183(1), 185(1), 187, 187(13), 189, 190, 190(3, 28), 191(28)
Paradiso, A. M., 699
Pardee, A. B., 596
Parekh, S. M., 479
Parish, R. W., 114
Park, F. J., 204
Parker, C., 741
Parker, K. A., 399
Parkin, J. D., 259
Paro, R., 214
Parra-Lopez, C., 370
Parsonage, D., 222
Pashinsky, I., 513
Pastan, I., 131, 133(7), 167, 175, 204, 226, 227, 238, 248, 249, 250, 252, 254, 255, 256(10), 257, 257(10), 258(2), 259, 264, 265, 279, 289, 290, 290(2), 299, 300(5, 34), 302(34), 304, 305(48), 306, 307, 309, 311, 311(10), 312(10), 313, 313(17), 314, 314(10), 315(10), 318, 319, 319(6, 7), 320, 320(7, 10), 321, 321(7), 322(8), 326(7), 327, 328, 330, 334(7), 339(7), 341, 342(18), 353, 355, 368, 372, 397, 398(8), 409(8), 410(8), 411(8), 427, 427(6), 428, 430, 430(4), 431, 431(4), 435(1, 4), 436, 436(4), 437, 437(4), 441, 441(3, 6), 442, 443, 443(2)7, 446(2, 11), 450(2), 453(2, 11), 454, 456, 457, 459, 459(10), 461(9, 10), 465, 471, 473, 473(1, 10), 474, 476, 477, 478, 480, 481, 483, 487(2), 492, 493, 493(2, 3), 494, 495, 496(22), 497, 498(28), 499, 500(22), 503, 503(14), 504, 505, 505(3, 4), 506, 506(9), 508, 512, 513(9), 514, 523, 525, 526, 526(8), 527(7), 528(8), 530(7, 10), 537, 538, 540, 540(1), 542, 544(13), 545(13), 546, 547, 548, 551(11), 555, 557, 558, 560, 561, 561(2), 570, 570(2), 571(26), 572, 573, 574, 574(2), 575, 576(4, 5), 577, 580(4, 7), 581, 583, 584, 585(32), 586, 589(4, 7–9, 11), 590, 590(37), 591, 594, 595, 662(37), 663
Paton, D. R., 213, 213(5), 214, 214(3), 215(5), 218(5)
Patterson, Y., 651
Paul, S., 281, 617, 622, 628(27), 647, 650(59, 60), 652, 653
Paulmichl, M., 368
Paulsen, I. T., 68
Paulusma, C. C., 110, 128, 132, 608, 609(13), 613
Pavirani, A., 647, 661, 723
Pear, W. S., 230, 232(21), 243
Pearce, G., 163
Pearce, H. L., 300
Pearce, S. R., 6, 9, 17, 136
Pearson, P., 728, 758
Pedersen, P. L., 500, 507, 675, 677, 678, 681(19, 20), 682, 683, 684, 685
Peel, S. A., 109, 177
Peetz, A., 475
Pelkey Ho, J., 590
Pellet, S., 51
Pellett, S., 164, 165
Pelton, J. G., 11, 25
Penefsky, H. S., 401
Peng, S., 689, 691, 695(6), 696, 696(10)
Pepling, M., 214
Peralta, E., 368
Perara, E., 268, 270
Perez-Soler, R., 293, 294(16), 301(16)
Perkus, M. E., 473
Perlin, D. S., 400, 402, 407(35), 412(35)
Pernandes, L., 767
Perricaudet, M., 647, 723
Peschel, P., 84
Pestov, D. G., 225, 227(2), 239(2)
Petersen, C. C. H., 361
Petersen-Yantorna, K., 369
Peterson, G. L., 635
Peterson, P. A., 120
Peterson, S. N., 757
Petithory, J., 16
Petrou, S., 766
Petrukhin, K., 118
Pharmingen, 647
Phelps, D. A., 399
Philippsen, P., 131, 133, 134(14), 137(14)
Phillippsen, P., 778, 780
Phillips, C. A., 757
Picard, M., 755
Picciotto, M. R., 343, 676
Pick, C., 13
Pick, H., 778
Pickles, R. J., 714

Piechuta, H., 608, 613(19), 615(19)
Pier, G. B., 721
Pind, S., 620, 622(23, 25), 627(25), 642, 645, 646, 650(37a)
Pindeo, H. M., 262, 264, 309
Pines, O., 137, 379
Pinkoski, M. J., 595
Pious, D., 117, 118(15), 163, 745
Pirker, R., 250, 555
Pirrotta, V., 214, 596
Piwnica-Worms, D., 427(10), 428
Plagne, R., 264
Plasterk, R. H., 105
Plavsic, N., 117, 132, 154(15), 343, 411, 617, 619(4), 629, 630(2), 631(2), 633(2), 635(2), 687, 715
Plavsik, J., 675
Plewe, M., 712, 713(13)
Ploegh, A., 767
Ploegh, H. L., 120, 745, 746(10), 753(10)
Ploton, D., 661
Plunkett, G. III, 84
Podbilewicz, B., 746
Podda, S., 547, 551, 551(11)
Podulso, S., 126
Pohl, J., 341, 353
Pöhlmann, R., 133, 780
Pokrovskaya, I. D., 11
Pollard, H., 113, 689, 690(7), 762, 769(71)
Pollard, J. W., 421
Polymeropoulos, M., 122
Pon, L. A., 783
Pong, S. S., 608, 613(19), 615(19)
Poonian, M. S., 375
Pope, H. A., 617, 721, 723
Poplack, D., 581
Porteous, D. J., 722
Possee, R. D., 128, 430(20)
Potenz, R. H., 399
Poustka, A.-M., 117, 132, 756, 758(45), 769(45)
Powis, S. H., 768
Prasad, R., 272
Prasher, D. C., 204
Prat, A. G., 494
Prendergast, F. G., 204
Pretorius, G., 214
Price, E. M., 397, 398(6), 409(6), 410(6), 411(6), 412(6), 427(5, 7), 428, 437(5), 438(5), 439(5), 440(5), 441(7), 467, 487, 507
Price, H. D., 49
Prichard, R. K., 293, 294(22, 23), 301(22, 23), 303(22, 23)
Priebe, W., 293, 294(16), 301(16)
Prince, A., 653, 721
Prince, L. S., 659, 661(33)
Proks, P., 365, 369(40)
Prossnitz, E., 15
Protter, A. A., 631
Provenzano, M. D., 401, 517
Prusiner, S. B., 278
Pryer, N. K., 626, 627(39)
Przewloka, T., 293, 294(16), 301(16)
Przybylski, M., 293, 303(24), 317
Puchelle, E., 661
Pugsley, A. P., 67
Purdue, P. E., 754, 754(15), 755
Purnelle, B., 143(56), 152, 777
Pusch, M., 361
Pyliotis, N. A., 213

Q

Qian, X.-D., 293, 294(18), 301(18), 302(18), 303, 303(18), 304(18), 308
Quaife, C. J., 618, 622(15)
Quak, J. J., 262
Qui, X.-Q., 278, 288
Quick, M., 625
Quinton, P. M., 343, 676
Quiocho, F. A., 10, 11, 677

R

Rabl, W., 110, 117, 132, 154(20)
Rachubinski, R. A., 754, 755(13), 757, 763(53)
Racker, E., 21
Rado, T. A., 689, 690(7), 695(6)
Raeymaekers, L., 360, 367
Raibaud, O., 79
Rajan, A. S., 736, 741
Rajewsky, K., 474
Ram, Z., 523
Ramachandra, M., 318, 319(6, 7), 320(7), 321(7), 326(7), 441, 441(3), 442, 443, 443(2), 446(2, 11), 450(2), 453(2, 11), 456, 457, 459(10), 461(9, 10), 473(1, 10), 570, 571(26)

Ramage, L., 785
Ramirez-Solis, R., 474
Ramjeesingh, M., 625, 626, 645, 652, 663(4), 676, 677, 691, 714
Ramond, M., 290
Ramsay, G., 170, 171(24), 204, 781
Ramsey, P., 698, 699(4), 704(6), 707(4)
Randall, R. J., 635
Randall-Hazelbauer, L. L., 31
Rao, K. R., 382(9), 383, 384(9), 386(9), 388(9)
Rao, M. S., 754
Rao, N. N., 44
Rao, R., 159, 627, 628, 628(42)
Rao, U. S., 398, 412, 427(8, 9), 428, 441(8, 9)
Rao, V. V., 427(10), 428
Rapin, I., 754
Rapoport, T. A., 268
Rasola, A., 360, 367(18)
Ratcliff, R. A., 369
Raus, J. C., 474
Raymond, M., 109, 134, 135, 157(36, 37), 178, 194, 382, 385(6), 398, 406(20), 414, 415, 415(10), 426(10)
Raynor, R. L., 353
Rea, P., 143(67), 155
Reay, P. F., 84
Rechavi, G., 475
Reddy, J. K., 754
Reddy, M. K., 754
Reddy, M. M., 676
Reenstra, W. R., 663
Reeves, M. E., 558
Reeves, R., 124, 134, 135, 135(35), 161(35, 38)
Regulla, S., 312
Rehling, P., 755
Rehmark, S., 578
Reinach, P. S., 37, 38(41)
Reinhard, D. H., 476
Reisdorf, P., 781
Reisin, I. L., 665, 673
Reithmeier, R., 274
Reizer, A., 4, 5(8), 7(8)
Reizer, J., 4, 5(8), 7(8), 10
Resnick, D. A., 474
Reue, K., 578
Reuss, L., 360, 368
Reyes, E., 617, 618(7), 625(7), 626, 630, 637(5), 641(5), 645, 650(5), 677
Reyes, L., 11
Reyes, M., 8, 9(28), 12, 22, 32, 32(33), 33, 34(33), 36, 37(21), 39(21), 49, 50
Reynolds, T., 729, 731
Rhee, S. Y., 662(37), 663
Rhodin, J., 753
Riberdy, J. M., 751
Rich, D. P., 136, 212, 281, 343, 617, 617(8), 618, 620, 641, 642(42), 647, 650(60), 652, 676, 714, 768
Richardson, C., 473, 547, 551, 551(11), 625, 641, 680
Richarme, G., 15, 16
Richert, N. D., 259, 372, 436, 494, 503(14), 505
Richet, E., 79
Rieske, J. S., 655, 785
Riezman, H., 203, 627, 778
Riggs, P. D., 679
Rijkers, T., 475
Rimmele, M., 41, 42, 43(20), 44(20), 45(14, 20)
Rine, J. D., 193
Rinehart, C. A., 699
Ringold, G. M., 697, 704(1)
Riordan, J. R., 117, 132, 154(15), 227, 263, 288, 343, 401, 411, 423, 616, 617, 617(9), 618, 618(3, 7), 619(3, 4, 13), 620, 620(13, 14), 621(14), 622, 622(23–25), 623, 623(3), 624(3), 625, 625(7), 626(10), 627(25), 629, 630, 630(1, 2), 631(2), 633(1, 2, 6), 635(2), 636, 637(5, 6), 641(5, 6), 642, 642(6), 645, 645(36), 646(53), 647(6), 649(6), 650(5, 6, 35, 36, 37a), 652, 653, 663(4), 675, 676, 687, 691, 714, 715
Ritch, R. H., 754
Ritke, M. K., 590
Rittman-Grauer, L., 259, 262, 264
Riviere, I., 558
Roach, P. J., 347
Robanus-Maandag, E. C., 133, 151(26)
Robbins, J. D., 293, 302(25), 303(25)
Roberto, F., 83
Roberts, D. B., 220
Rocchi, M., 559, 725
Rocchiccioli, F., 759
Roche, P. A., 744
Roder, H., 651
Roderick, T. H., 576
Rodgers, M., 183
Rodrigues-Pousada, C., 143(92), 160, 767
Rodseth, L. E., 11
Roelofsen, B., 397
Rohinson, I. B., 398
Rohrbach, M. R., 49
Rojas, E., 113, 689, 690(7)

Rokach, J., 608, 613(19), 615(19)
Rolland, M.-O., 759
Roman, R., 697, 704(1)
Romani, C., 262
Romeo, G., 360, 367(18)
Rommens, J. H., 645, 646(53)
Rommens, J. M., 117, 132, 135, 154(15), 343, 411, 617, 617(9), 618, 618(7), 619(4), 620, 625(7), 626, 629, 630, 630(1–3), 631(2), 633(1, 2), 635(2), 637(5), 641(5), 650(5), 652, 675, 675(3), 676, 677, 715, 721, 723
Rommens, R. M., 687
Roninson, I. B., 167, 225, 226, 227, 227(2), 229, 233, 237, 239(2, 16), 242, 259, 262, 264, 290, 299, 304(33), 316, 318, 330, 334(7), 339(7), 427(16), 428, 430(16), 431, 431(16), 439(16), 440(16), 453, 492, 493, 494, 503(14), 504, 505, 508, 546, 574, 577
Roninson, I. G., 514
Rose, J. K., 545
Rose, M. D., 206, 207, 208(32), 211
Rose, W. P., 725
Rosebrough, N. J., 635
Rosen, B. P., 82, 83, 84, 84(3), 86, 86(3, 3), 88, 88(1), 89, 90, 90(19), 91, 93(1), 94, 95(28), 183, 189, 190, 190(3, 28), 191(28)
Rosen, N., 250
Rosenberg, C., 758
Rosenberg, H., 40, 43, 45(7), 221, 224
Rosenberg, S. A., 558
Rosenbloom, I., 654
Rosenbusch, J. P., 47
Rosenfeld, M. A., 647, 723, 724
Rosenstein, B. J., 676
Rosenstein, R., 84
Rosenthal, E. R., 647, 723
Rosenthal, N., 576
Roskoski, R. J., 353
Rosman, G. J., 229, 560
Ross, B. M., 118
Ross, P. E., 361
Rossant, J., 474
Rossier, B., 369
Rossman, T. G., 94
Roth, A. F., 203
Rotherford, A. V., 523
Rothfield, L. I., 74
Rothman, J. H., 204
Rothman, R. E., 266, 268(17), 270(17)
Rothstein, A., 369
Rotman, B., 15
Rotstein, O. D., 646, 653
Roy, G., 94, 95(28), 183, 187, 189, 190(3, 28), 191(28), 359, 363
Roy, S. N., 311, 342
Rozenblatt, S., 453, 473, 626
Rozmahel, R., 117, 132, 154(15), 343, 411, 617, 619(4), 629, 630(1, 2), 631(2), 633(1, 2), 635(2), 652, 675, 687, 715
Ruan, Z.-S., 662(37), 663
Rubin, G. M., 214
Rucareaunu, C., 360, 363(16)
Rudd, K. E., 43
Rudenko, G., 435, 481
Rudnick, G., 658
Ruetz, S., 135, 157, 157(36, 37), 159, 211, 342, 355, 368, 382, 383, 384(10), 385(6), 388(10), 390(10), 395(11), 397, 398, 406(20), 413(21), 414, 415(8, 10), 426, 426(10), 427(8), 494, 515
Rugman, P. A., 14
Runswick, J., 676
Runswick, M. J., 101, 397, 415, 514
Ruoho, A. E., 293, 303(21)
Ruppel, K. M., 474
Ruscitti, T., 192
Russ, G., 120
Russel, K. A., 755
Russell, D. W., 728
Russell, M. A., 304, 309
Ruth, A., 225
Ruther, U., 475
Rutherford, A. V., 262, 430
Ryan, C. A., 163
Ryan, J., 37
Ryan, U. S., 664
Rybak, S. M., 474
Ryffel, B., 293, 303(20), 308
Ryseck, R. P., 476
Rytka, J., 763

S

Saari, G. C., 204
Sabolic, I., 663
Sacchi, N., 580
Sachs, C. W., 357, 427(11), 428, 441(11)
Sachs, D. H., 547, 573, 589(11)
Sack, G., 759, 769(63)

Sacktor, B., 654
Sadoulet, H., 762
Saeki, T., 259
Safa, A. R., 178, 226, 264, 289, 291, 293, 294, 294(14–16), 298(11), 299, 300, 300(3, 4, 11, 15, 34), 301(16), 302(12, 14, 34), 303, 303(17, 35), 304, 304(33, 35), 305(17, 43, 49), 306, 306(17, 43), 309, 311, 316, 327, 357, 401, 404, 406, 427(13), 428, 437, 478
Saidowsky, J., 763
Saier, M. H., Jr., 4, 5, 5(8), 6(10), 7(8), 10, 10(10), 68
Saiki, R. K., 679
Saiman, L., 653
Sakura, H., 365, 369(40)
Salah-Bey, K., 118
Salako, L. A., 175
Salehzada, T., 152
Salmon, S. E., 264
Salmond, G., 8, 30, 31(8), 116, 158
Salter, R. D., 745
Salvo, J. J., 479
Sambrook, J., 59, 164, 165(17), 209, 245, 385, 431, 446, 563, 571(24), 574, 596, 598(14), 599(14), 603(14), 636, 680
Samulski, R. J., 539, 541, 543(16), 544, 544(10), 559, 725, 727, 728(47)
Sandbank, R., 728
Sanders, D. A., 152
Sanders, K. H., 608
Sanders, V., 264
Sandusky, M., 757
San Francisco, M. J. D., 83
Sanger, F., 601, 680
Sankaran, B., 410, 518, 521, 522, 522(22)
Sano, T., 776
Sansom, S., 617
Santi, D. V., 184
Santoni, A., 262
Santos, H., 42, 43(21), 45(21)
Sapag, A., 134, 143(31), 153(31), 764, 765(79), 766(79)
Sapperstein, S. S., 194, 200(8), 202, 202(8)
Sarangi, F., 263
Saraste, M., 3, 7(1), 101, 397, 415, 514, 676
Sarde, C. O., 117, 132, 153, 756, 758(45), 759, 759(46), 762, 762(47), 765, 765(66), 766(66), 769(45, 47, 63)
Sardini, A., 342, 355, 368
Sarkadi, B., 226, 369, 397, 398, 398(6), 409(6), 410(6), 411(5–7), 412(6), 427(5, 6, 7, 14, 16), 428, 430(14, 16), 431(16), 437(5), 438(5), 439, 439(5, 16), 440(5, 14, 16), 467, 487, 507, 699, 722
Sarkar, H. K., 375
Sarthy, A., 46
Sato, K., 16, 17(80)
Sato, M., 263
Sato, W., 357
Saudek, D. M., 757
Saulino, A., 122
Saurin, W., 7, 25, 46, 764
Savary, S., 116, 160, 756, 762(47), 769(47), 777
Savoia, A., 645, 646(53)
Scadden, D., 242
Scarborough, G. A., 397, 398, 398(6), 409(6), 410(6), 412, 412(6), 413, 427(5, 8, 9, 15), 428, 437(5), 438(5), 439(5), 440(5), 441(5, 6, 8, 9, 15), 467, 487, 507, 618, 626, 626(10)
Schaaper, R. M., 229
Schaffner, W., 497, 655
Schagger, H., 315, 333
Scharf, S., 679
Scharff, M. D., 637
Scharschmidt, B. F., 610
Schatz, G., 116(9), 117, 143(80), 157, 776, 778, 785
Schaub, T., 608, 609(21), 611, 611(21), 615(21)
Schedl, P., 215
Scheel, G., 163, 776
Scheffer, G. L., 128, 132, 607, 608, 609(13)
Schekman, R., 382, 626, 627(39)
Scheper, R. J., 128, 132, 154, 262, 264, 607, 608, 608(9), 609(13), 611(9), 614(9)
Scherer, G., 596
Scherer, S., 779
Scheule, R. K., 620, 697, 712, 713(12)
Schhid, D., 255
Schiavi, S. C., 714
Schiestl, R. H., 779
Schimke, R. T., 184, 229
Schimozawa, N., 755, 757(31)
Schindler, M., 455
Schinkel, A. H., 133, 151(26), 154, 155(63), 262, 266, 391, 397, 435, 481, 494, 514, 613
Schinkel, J., 262
Schirmer, T., 47
Schlamowitz, M., 639
Schlegel, R., 699

Schlesinger, P. H., 174
Schlor, S., 52
Schmid, M., 475
Schmidhauser, C., 114
Schmidt, 209
Schmidt, C. J., 755
Schmidt, T., 114
Schmidt-Wolf, G., 262
Schmidt-Wolf, I., 262
Schneider, E., 9, 29, 31, 32(18)
Schneider, T., 312
Schodl, A., 293, 303(24), 317
Schoenlein, P. V., 252, 255, 494, 504, 505(4), 523, 547, 557, 573, 574(2), 589(11)
Schônenberger, M., 188
Schoonen, W. G. E. J., 264
Schreiber, S. L., 679
Schrer, J., 359, 368(3), 369(3)
Schroder, U. H., 646
Schroeder, A. C., 576
Schroeijers, A. B., 607
Schuening, F. G., 558
Schulein, R., 52
Schultz, P. G., 474
Schultz-Borchard, U., 772
Schulz, I., 360, 367(15), 369(15)
Schumacher, G., 40
Schumacher, T. N. M., 745, 746(10), 753(10)
Schurr, E., 109, 157, 177, 211, 220, 318, 380, 398, 414, 421(3), 422(3)
Schutgens, R. B. H., 754
Schuurhius, G. J., 264
Schwanstecher, M., 744
Schwartz, M., 14, 21, 25, 26(21), 30, 31, 31(6), 46, 47, 49(46), 68, 73(14)
Schwartz, V., 30, 31(6)
Schweitzer, B. I., 546
Schweizer, H. P., 40, 41, 43, 44, 46(26)
Schwiebert, E. M., 118, 361, 653, 659(8), 664, 665, 667, 668, 673, 674
Scott, M. L., 230, 232(21), 243
Scudiero, D. A., 483
Seamon, K. B., 293, 302(25), 303(21, 25), 304, 305(48), 309, 311(10), 312(10), 314(10), 315(10)
Sefton, B. M., 346
Segal, D. M., 265
Seibert, F. S., 616, 620, 623, 645, 646(52)
Seidel, N. E., 548
Seidmon, J. G., 596, 598(15), 599(15), 603(15)
Seimiya, H., 263, 264
Seino, S., 369, 732, 743, 744(1)
Sekhsaria, S., 526, 537(11), 557
Sempé, P., 120
Seneveratne, T., 494
Senior, A. E., 222, 318, 397, 398(11, 13), 401(11), 406(11, 13), 409(11, 13), 410, 410(11, 13), 411, 411(11, 13), 487, 503, 514, 515, 517, 518, 521, 521(14, 16, 17, 19), 522, 522(22), 620
Sepródi, J., 427(7), 428, 441(7)
Sepúlveda, F. V., 359, 360, 361(21), 362(7), 363, 363(8, 16), 365(7), 368(8, 21, 34), 369, 369(21), 494
Serpinskaya, A. S., 546, 586
Serrano, A. E., 175, 177(4)
Serrano, R., 406, 407, 412, 412(42)
Servos, J., 143(52), 151
Seto-Young, D., 400, 402, 407(35), 412(35)
Sewell, J. L., 311
Shafer, B., 116, 121(6), 143(81), 158, 777
Shani, N., 134, 143(30, 31), 153, 153(30, 31), 753, 756, 762, 763(44), 764, 764(44), 765(79), 766(44, 79), 767(43, 44), 768(43), 772(43, 44)
Shankar, A. H., 175, 177(4)
Shankroff, J., 759
Shapiro, A. B., 397, 398(12), 410(12), 411(12), 503, 514
Shapiro, E., 759
Shapiro, T. A., 186
Sharff, A. J., 11
Sharma, R. C., 229
Sharom, F. J., 397, 398(10, 14), 409(10, 14), 410, 410(10), 411(10, 14), 487, 515
Sheard, P., 608, 615(18)
Sheila, S. A., 721, 723
Shelby, M. D., 474
Shelling, A. N., 544
Shen, D.-W., 252, 259, 494, 503(14), 505
Shen, M.-R., 365
Shenbagamurthi, P., 677, 681(19, 20)
Shenk, T., 541, 543(16), 727, 728(47)
Shepherd, J. C., 745, 746(10), 753(10)
Sheps, J. A., 30, 51, 52, 53(7), 55, 56, 58, 61(14), 63(14), 64, 64(14)
Sherman, F., 424
Sherman, M., 627
Shi, L.-B., 268
Shi, W., 84

Shiina, Y., 756, 758(50)
Shimabuku, A. M., 414
Shimagawa, H., 44
Shimozawa, N., 117, 757
Shin, J., 291
Shinagawa, H., 43
Shirai, A., 259
Shirai, H., 776
Shirayoshi, Y., 117, 755, 757(31)
Shlyakhter, D., 431, 435(32)
Shoemaker, R. H., 483
Shoshani, T., 474, 505
Showalter, R., 475
Shulman, M., 637
Shuman, H. A., 8, 9, 9(28), 12, 13, 17(35), 18(35), 19, 20(54), 21, 22, 25, 28(5, 6), 30, 31, 31(6), 32, 32(32, 33), 33, 33(25, 30), 34(30, 33, 34), 36, 37(21, 25), 38(25), 39(21), 49, 50, 113, 131
Shumyatsky, G. B., 696
Shustik, C., 414
Shyamala, V., 5, 8(13), 10, 101
Sibbald, P. R., 101
Sidler, M., 162, 166, 172, 173(25)
Siegel, C. S., 712, 713(12)
Siegfried, W., 723
Siegsmund, M., 581
Sievertsen, A., 8, 9(28), 32, 37(21), 39(21), 50
Sifri, C. D., 183
Sigimoto, Y., 523
Sigma Chemical Co., 635
Sikic, B. I., 262, 360
Sikorski, R. S., 755
Silhavy, T. J., 8, 14, 25, 26(21), 30, 31, 31(6), 41, 47, 49(46)
Silhol, M., 152
Sillence, D., 759
Silva, C. L., 475
Silva, J. T., 313, 319
Silver, S., 83, 84
Silverman, M., 475
Simon, R. H., 723(22), 724
Simon, S. M., 455
Simonet, J.-M., 755
Simoni, R., 279(11), 280
Simonsen, C. C., 476
Simonsen, L. O., 359, 380(1)
Singer, J., 766
Singer, M., 43
Singh, I., 759, 769, 769(61)
Singh, N., 756
Sinicrope, F. A., 289, 300(4)
Siniscalco, M., 559, 725
Skach, W. R., 265, 266, 267, 267(6, 7), 268, 268(6, 7), 271(6, 7, 20), 277, 278(6), 287, 288, 288(18)
Skala, J., 143(56), 152, 777
Skaletz-Rorowski, A., 763
Slamon, D., 581
Slatin, S. L., 278, 288
Slatko, B. E., 679
Slayman, C. W., 159, 382(9), 383, 384(9), 386(9), 388(9), 411, 627, 628, 628(42)
Sleight, R. G., 391, 392(17)
Slotki, I. N., 367
Slovak, M. L., 590, 602
Small, K. V., 757
Smit, J., 266
Smit, J. J., 391
Smit, J. J. M., 133, 151(26), 262, 397, 494, 607
Smit, L. S., 624, 676, 677(16)
Smith, A., 281
Smith, A. D., 474
Smith, A. E., 136, 343, 617, 617(8), 618, 620, 641, 642(42), 647, 650(59, 60), 652, 653, 655, 656(7), 661(7), 676, 677, 697, 712, 713(12, 13), 714, 724
Smith, A. J., 397, 607
Smith, C. D., 293, 300(26), 303(26), 357
Smith, D. G., 630, 633(4), 635(4)
Smith, G. E., 128, 429, 430, 439(17), 692
Smith, G. H., 576, 582(19)
Smith, G. L., 442, 445
Smith, H. O., 757
Smith, J. A., 596, 598(15), 599(15), 603(15)
Smith, J. J., 714
Smith, K., 757, 759, 765(62), 769(54, 63)
Smith, L., 547, 551(11)
Smith, M. A., 759, 765(62)
Smith, M. G., 544, 728
Smith, P. A., 365, 369(40)
Smith, P. K., 401, 517
Smith, T. F., 755
Smith-Gill, S. S., 651
Smithies, O., 145(88), 160, 618, 626(10)
Sneider, M., 267, 274(24)
Snouwaert, J. N., 618, 626(10)
Soares, M. B., 122
Sofia, H. J., 84
Sokolic, R. A., 526, 537(11), 557

Solaiman, F., 559
Solari, R., 212
Solc, C. K., 361
Solodin, I., 698, 709(5), 712(5)
Solomon, S., 338
Solow, R., 369, 540, 653, 659(8), 676, 720, 727, 728(52, 53), 729, 729(6), 730, 731
Sommer, S. S., 580
Sommerfelt, M. A., 691, 696, 696(10)
Sonoda, S., 639
Sood, R., 645
Sorgi, F., 707
Sorrentino, B. P., 538, 547
Sorscher, E. J., 113, 686, 689, 690(7), 691, 695(6), 696, 696(10)
Southern, P. J., 421
Souza, A. E., 676
Souza, D. W., 622, 628(27), 647, 650(59, 60), 652, 653
Sowunmi, A., 175
Spangrude, G. J., 547, 548(17), 552
Sparks, K. E., 604, 606(25), 607(25)
Speicher, L. A., 293, 300(26), 302(25), 303(21, 25, 26)
Speiser, D. M., 9, 13(31), 19(31), 163
Spencer, F., 124, 134, 135, 135(35), 161(35, 38)
Sperbeck, S. J., 754
Spies, M., 266
Spies, T., 117, 118(15), 163, 745
Spieser, D. M., 776
Spiess, M., 279, 279(12), 280
Spilman, C., 631
Spiro, A. J., 754
Spoelstra, E., 309
Spong, A. P., 763
Sprague, G. F., Jr., 203, 424
Spring, H., 132, 154, 608, 609(12), 610(12), 611(12), 614(12), 615(12)
Spudich, E. N., 8
Spudich, J. A., 474
Spurlino, J. C., 11, 677
Sreekrishna, K., 399
Sridhar, C. N., 698, 699(4), 704(6), 707(4)
Srour, E. F., 548
St. Jean, A., 779
Staehle, C., 547
Staehlin, T., 637, 656
Stahle, C., 573, 576, 582(19), 589(11)
Stanley, P., 53, 61(10)
Stanley, W., 759, 769(61)
Stanners, C. P., 421
Stanton, B. A., 361
Starr, M. P., 67
States, D. J., 122
Steed, P. M., 44
Steel, G., 762
Steff, A., 152
Stein, W. D., 226, 471, 513
Steinhart, R., 89
Steplewski, Z., 749
Stern, R. K., 226, 316
Sterne, R. E., 143(71), 156, 193, 200(3), 202(3), 777
Stevens, T. H., 204
Stevenson, B. J., 722
Stewart, A. J., 117, 133, 183, 493, 600, 602(19), 607, 609(1)
Stewart, C., 116, 116(10), 117, 121(6), 143(81), 158, 576, 777
Stieglitz, K., 262
Stier, L. E., 647, 723
Stierhof, Y.-D., 373
Stiles, G. L., 664
Stinchcomb, D. T., 779
Stine, O. C., 759
Stöcker, W., 68
Stoeckel, M. E., 153, 756, 759(46)
Stone, K. L., 328, 329, 332(5)
Storb, R., 558
Strange, K., 359, 363(4), 369(4)
Stratford-Perricaudet, L., 647, 723
Strathern, J., 116, 121(6), 143(81), 158, 777
Strauch, K. L., 373
Striessnig, J., 360, 367(15), 369(15)
Stroffekova, K., 413, 618, 626(10)
Strominger, J. L., 163
Strong, T. V., 624, 676, 677(16)
Strube, M., 288
Struhl, K., 596, 598(15), 599(15), 603(15), 779
Strydom, D., 163
Stüber, D., 168, 169(22)
Studier, F. W., 443, 444(8), 456
Stutts, M. J., 369, 665, 673
Su, T. Z., 44
Subramani, S., 754, 755, 755(13), 757, 757(19), 763, 763(53), 773(7)
Sucharita, P., 620, 676
Suck, D., 677
Suda, K., 778
Suda, T., 547(19), 548

Sugawara, I., 259
Sugimoto, U., 505
Sugimoto, Y., 263, 459, 474, 477, 523, 525, 526, 526(8), 527(7), 528(8), 530(7, 10), 537(11), 538, 546, 557, 588
Sukhu, L., 704, 708(9), 712, 713(13)
Sullivan, D. T., 213, 213(4, 5), 214, 214(3), 215(4, 5), 218(5)
Sullivan, M., 546, 573, 575, 576(4), 580(4), 589(4)
Sullivan, M. C., 213, 213(4), 214, 214(3), 215(4)
Sumizawa, T., 293, 303(19)
Summers, K. M., 213
Summers, M. D., 128, 429(19), 430, 430(19), 439(17), 619(33), 625, 692
Sun, S., 617, 618(7), 625(7), 630, 637(5), 641(5), 650(5)
Surdej, P., 781
Surman, M. J., 751
Sutter, G., 453
Suttle, D. P., 227, 239(16)
Sutton, G., 757
Suzuki, S., 361, 362(24)
Suzuki, T., 278
Suzuki, Y., 117, 755, 757, 757(31)
Svenson, M., 721
Svensson, B., 347, 348(24)
Swang, J., 259
Swartzman, E. E., 134, 143(32), 763
Swiger, R. R., 580
Swinkles, B. W., 754
Szabó, K., 427(14), 428, 430(14), 440(14)
Szasz, I., 439
Szczypka, M., 143(64, 67, 93), 154, 155, 160
Szilvassy, S. J., 547, 547(20), 548, 554
Szmelcman, S., 14, 25, 26(21), 30, 31, 47, 49(46)

T

Tabak, H. F., 755, 756, 757, 763(53), 767
Tabcharani, J. A., 343, 618, 619(13), 620, 620(13, 14), 621(14), 642, 645, 645(34), 646(34, 52), 650(34, 35)
Tabe, H., 765
Tabor, S., 680
Tachibana, M., 361, 362(24)
Tager, J. M., 754, 769
Tagliamonte, J. A., 679
Taglicht, D., 130, 137, 157, 210, 379
Tahara, T., 293, 303(19)
Takabarake, Y., 260
Takano, T., 756, 758(50)
Takebe, S., 414
Takemoto, L., 342, 357(7)
Taketani, S., 756, 758(48), 776
Taketo, M., 576
Talbot, F., 178, 227, 304, 316, 374, 401
Tam, A., 137, 157, 379
Tam, R., 5, 6(10), 10(10)
Tamai, I., 226, 260, 289, 293, 299, 300(3), 303, 303(17), 304, 304(33), 305(17, 43), 306(17, 43), 309, 316, 406
Tampé, R., 120, 768
Tanaka, S., 313, 327
Tang-Wai, D. F., 374, 421, 574
Tanigawara, Y., 259
Tanzer, L. R., 259
Tartaglia, J., 473
Tascon, R. E., 475
Tatsuta, T., 259, 264, 264(16)
Tattersall, M. L., 608, 615(18)
Taylor, C. W., 264
Taylor, J., 360, 361(11), 365(11), 367(11), 368(11), 369(11), 473
Taylor, S., 53, 61(11)
Tearle, R. G., 214
Tebbe, S., 9, 31, 32(18)
Teem, J. L., 212
Teixeira de Mattos, M. J., 14
Telford, J., 21, 596
Tell, B., 255
Tempest, D. W., 14
Tenda, Y., 260
ten Have, J. F. M., 214
Terasaki, T., 260
te Riele, H. P. J., 133, 151(26)
Terranova, M. P., 679
Tettelin, H., 131, 134(14), 137(14)
Tew, K. D., 293, 300(26), 302(25), 303(21, 25, 26)
Tezuka, K., 263
Thaithong, S., 109, 176, 177(11)
The, I., 105, 167
Theda, C., 755
Theodore, L. J., 295
Thévenod, F., 360, 367(15), 369(15)
Thiebaut, F., 259, 262

Thiele, D., 143(64, 67, 93), 154, 155, 160
Thiemann, A., 361
Thieringer, R., 772
Thierry, A. R., 540, 544(13), 545(13)
Thimmapaya, R., 225, 227, 227(2), 239(2, 16)
Thomas, D. Y., 109, 134, 135, 143(72), 156, 157(36, 37), 178, 382, 385, 385(6), 398, 406(20), 414, 415, 415(10), 426(10)
Thomas, G. F., 725
Thomas, J., 762
Thomas, P., 132, 154(20)
Thomas, P. J., 677, 678, 681(19, 20), 682, 683, 684, 685
Thomas, P. M., 110, 117
Thompson, A. M., 608, 615(20)
Thompson, C. J., 118
Thompson, J. K., 175, 176(5), 177
Thompson, L. H., 515
Thompson, R. B., 679
Thompson, S., 653, 676
Thompson-Coffe, C., 755
Thorner, J., 134, 143(32, 71), 156, 160, 163, 178, 193, 200(3), 202(3), 206, 414, 473, 763, 777, 781
Thorner, P. S., 264
Till, J. E., 328
Tilly, B. C., 676
Timmermans-Hereijgers, J. L. P. M., 397
Tisa, L. S., 83, 91
Tizard, R., 576
Tokiwa, G., 203
Tokuda, H., 278
Tolosa, S., 489
Tolstoshev, C. M., 122, 160
Tomatsu, S., 757
Tomb, J.-F., 757
Tombesi, M., 262
Tomida, A., 259, 264(16)
Tominaga, M., 360, 361, 365(20), 367(20), 368(20)
Tominaga, T., 360, 365(20), 367(20), 368(20)
Tommasini, R., 154, 155(63), 613
Tommassen, J., 9, 32, 41, 42, 42(17), 43(17), 45, 46(38), 47, 47(10), 49(38), 50
Tonegawa, S., 745, 746(10), 753(10)
Torok-Storb, B., 558
Torriani, A., 44
Torriglia, A., 489
Towbin, H., 637, 656
Townsend, A., 163, 745, 768
Trapnell, B. C., 629, 647, 687, 723
Tratschin, J.-D., 728
Traut, T. W., 101
Travis, S. M., 768
Traxler, B., 32, 35(46), 39, 378
Trempe, J. P., 725, 726
Trent, J., 263
Treptow, N. A., 12, 21, 28(5), 32, 32(33), 33, 33(30), 34(30, 33), 36
Trezise, A. E. O., 365, 647, 655, 656(25), 659(24)655, 661(25)
Triglia, T., 167
Trisler, P., 169
Trowsdale, J., 163, 745, 768
Trumble, W. R., 375
Tsai, Y. J., 698, 699(4), 704, 704(6), 707(4), 708(9), 712, 713(13)
Tseng, L., 242
Tsuchiya, E., 155
Tsuge, H., 259, 264(16), 588
Tsuhi, A., 260
Tsui, L.-C., 117, 132, 135, 154(15), 212, 343, 411, 617, 617(9), 618, 618(7), 625(7), 629, 630, 630(1–3), 631(2), 633(1, 2), 635(2), 637(5), 641(5), 645, 646(53), 650(5), 652, 675, 675(3), 676, 687, 688(3), 715, 721, 723
Tsui, S., 128
Tsuji, S., 765
Tsukahara, S., 263
Tsukamoto, T., 117, 755, 757, 757(31), 763(53)
Tsuruo, T., 258, 258(3), 259, 260, 261, 262, 262(30), 263, 263(30, 32), 264, 264(16, 30), 265, 266, 290, 302, 397, 398(7), 436, 453, 469, 508, 523, 531, 554, 580, 588
Tucker, J. D., 580
Tudhope, S. R., 608, 615(20)
Tugendreich, S., 755
Tuite, M. F., 143(48), 145, 158(48)
Tümmler, B., 227, 423
Turley, H., 653, 655(10), 656(10), 657(10), 658(10), 661(10)
Tuthill, R., 749
Twentyman, P. R., 264
Tyers, M., 203
Tzagoloff, A., 655

U

Uchida, K., 681
Uchida, N., 550
Uchida, Y., 769

Uchiyama, A., 769
Uda, T., 266
Udell, D. S., 596
Uebel, S., 768
Ueda, K., 167, 255, 259, 313, 318, 327, 330, 334(7), 339(7), 414, 430, 431, 493, 504, 514, 523, 577, 581
Uedi, K., 481
Uezono, Y., 625
Ui, M., 37, 38(39)
Ullman, B., 183
Ulrich, W.-R., 293, 303(24), 317, 608
Umbarger, H. E., 14
Urbatsch, I. L., 318, 397, 398(13), 406(13), 409(13), 410, 410(13), 411, 411(13), 473, 503, 514, 517, 518, 521, 521(16, 17, 19), 522, 522(22)
Urquhart, A. J., 756
Usuda, K. T. N., 758
Utakoji, T., 259
Utterback, T. R., 757

V

Vahabi, S., 303, 406
Valdimarsson, G., 604, 613
Valle, D., 112, 118, 132, 134, 143(30, 31), 153, 153(30, 31), 161(33), 753, 755, 756, 757, 757(18, 30), 758, 758(49), 762, 763(44, 53), 764, 764(44), 765(79), 766(44, 79), 767(43, 44), 768(43), 769, 772(42, 44)
Valls, V. A., 204
Valsmis, M. P., 754
Valtz, N., 198
Valverde, M. A., 359, 360, 361(11, 21), 362(7), 363, 363(8, 16), 365, 365(7, 11), 367(11), 368(8, 11, 21, 34), 369, 369(11, 21), 494
Van Acker, K., 360
Vanacore, L., 661
Van Arsdell, J. N., 204
van Besien, K., 548
van Blitterswijk, W. J., 397
Van de Berg, J. L., 38
van Deemter, L., 133, 151(26), 391, 494
van de Griend, R. J., 265
van den Berg, M., 756, 767
van den Bosch, H., 754
van Dendert, P. M., 120
Van Der Geer, P., 354, 355(30)
van der Groep, P., 607
Van der Heyden, S., 367
van de Rijn, M., 552
van der Klei, I. J., 399, 757, 763(53)
Van der Leij, I., 755
van der Ley, P., 47
van der Lugt, N. M. T., 391, 494
van der Ploeg, L. H. T., 184
Van der Schoot, E., 262
van der Valk, M. A., 133, 151(26), 391, 494
van der Valk, P., 607
Van Dijken, J. P., 399
van Duk, J., 265
Van Dyck, L., 143(51), 151, 619(40), 626
Van Dyke-Phillips, V., 178, 304, 316, 401
van Endert, P. M., 768
van Es, H. H. G., 109, 177
van Geel, B. M., 759, 765(66), 766(66)
Vangelista, L., 212
Van Groenigen, M., 167
Van Heerikhuizen, H., 680
Vanhooren, J. C. T., 755
van Loon, A. P. G. M., 778
van Luenen, H., 184, 185(16)
van Oost, B. A., 759, 765(66), 766(66)
Vanouye, C. G., 360, 368
van Rijn, J., 264
van Roermund, C. W. T., 756, 767
van Roon, M. A., 391, 494
van Tellingen, O., 110, 133, 151(26), 613
van Veldhoven, P. P., 402, 520, 755, 757, 763(53)
Váradi, A., 226, 398, 427(14, 16), 428, 430(14, 16), 431(16), 439(16), 440(14, 16)
Varshavsky, A., 118, 163, 167, 178, 193, 777
Vazquez de Aldana, C. R., 143(49), 145, 158(49)
Veenhuis, M., 399, 755, 757, 763(53)
Veinot-Drebot, L., 263
Velimirov, B., 365
Venet, T., 641
Venter, J. C., 757
Verdine, G. L., 679
Verkman, A. S., 268, 653, 663, 714, 715(20)
Vernet, T., 385, 415, 625
Verrelle, P., 264
Versantvoort, C. H. M., 607
Vialard, J., 625, 641
Viana, F., 360, 367
Vick, R. N., 714
Vida, T. A., 206
Vidal, S., 109, 177

Vieira, J., 679, 779
Vigna, S. A., 647, 649(66)
Viitanen, P. V., 375
Vilela, C., 143(92), 160, 767
Villarejo, M. R., 46
Violand, B., 21
Viswanathan, M. N., 134, 143(32), 160, 763, 777
Vlak, J. M., 429
Voet, M., 143(68), 155
Vogel, B., 172, 173(25)
Vogel, E. W., 475
Vogel, G., 89
Vogel, J. P., 206
Vögeli-Lange, R., 163
Vogt, E., 154, 155(63), 613
Vogt, K., 479
Volckaert, G. X., 143(68), 155
Volkman, S. K., 109, 157, 174, 176, 177(11), 178, 180, 211, 415, 427(13)
von Heijne, G., 274
Von Hoff, D. D., 237
von Jagow, G., 315, 333
von Kalle, C., 558
Vonoye, C. G., 360
von Suhr, N., 560
Voorn-Brouwer, K. A. T., 756
Vu, Q. A., 736, 738(2)

W

Wach, A., 133, 411, 778, 780
Wadsworth, S. C., 714
Wagenaar, E., 133, 151(26), 262, 266, 391, 435, 481, 494
Wagner, G. J., 163
Wagner, M., 627
Wahl, G. M., 474
Wahli, W., 754
Walker, J. E., 3, 7(1), 17, 101, 397, 415, 514, 649, 650(67), 676
Walker-Jonah, A., 176
Walkin, B. K., 655
Wall, D. M., 259
Wallbridge, S., 523
Walsh, C. E., 544, 727
Walsh, J., 766
Walsh, R. A., 38
Walter, C., 9, 29
Walter, P., 266, 272
Walworth, N. C., 382, 383(8), 384(8), 386(8), 388(8), 628
Wan, N. C., 712, 713(12)
Wanders, R. J. A., 754, 756, 767, 769
Wandersman, C., 3, 4(4, 6), 13, 30, 46, 52, 67, 68, 68(3), 69, 72, 73, 73(14, 18), 74(27), 75, 80(26)
Wang, A. M., 168
Wang, K. X., 620, 712, 713(12)
Wang, M., 143(51), 151, 619(40), 626
Wang, R. C., 52
Wang, Y., 745
Wang, Y. F., 47
Wang, Y. H., 347
Wang, Z., 94
Wanner, B. L., 43, 44, 44(24), 46
Ward, D. C., 539, 725
Ward, J. F., 475
Ward, M., 547, 551, 551(11)
Ward, N. E., 357
Ward, W. W., 204
Warnaar, S. O., 265
Wasley, A., 175, 177(4)
Wassif, C., 68
Watanabe, H., 531
Watanabe, M., 263
Watanabe, T., 588
Waterham, H. R., 755
Watkins, P., 134, 143(30), 153(30), 755, 756, 757(18), 758, 759, 763(44), 764(44), 765(62), 766(44), 767(44), 769(63), 772(44)
Watson, J., 649, 650(67)
Weatherwax, R., 9
Weaver, J. L., 365, 367(38)
Weaver, T., 135, 161(38)
Webb, D. C., 43, 221, 224
Webb, T. E., 664
Weber, J., 410, 521, 522(22)
Weber, S. C., 780
Weber-Pendleton, S., 723(22), 724
Webster, P., 661
Webster, R. G., 651
Wechsler, S. W., 132, 154(19), 736, 738, 738(3), 740(3), 743(8), 744(3)
Wei, H.-M., 759, 765(62), 769(63)
Weidman, J. F., 757
Weinberg, R., 473
Weiner, J. H., 40
Weissborn, D. L., 40
Weissman, C., 655

Weissman, I. L., 547, 548(17), 550, 552
Weissmann, C., 497
Weitzmann, M. D., 544, 726, 728
Welch, M. J., 676
Welch, R., 51
Welch, R. A., 164, 165
Welchlin, H., 566
Welker, E., 226, 398, 427(14, 16), 428, 430(14, 16), 431(16), 439(16), 440(14, 16)
Wellems, T. E., 176
Welsh, M. J., 136, 212, 281, 343, 617, 617(8), 618, 620, 641, 642(42), 647, 650(60), 652, 653, 655, 656(7), 661(7), 676, 677, 687, 688(3), 699, 707(6), 708(6), 714, 715, 724, 768
Wemmer, D. E., 11, 25
Wemmie, J. A., 143(64, 93), 154, 160
Wenz, M., 697, 704(1)
Wert, S. E., 629
Wessel, D., 497
Wessels, H., 266, 279, 279(12), 280
West, M. H. P., 728
Westerhoff, H., 309
Westfall, D. P., 664
Wettenhall, R. E., 114
Whang-Peng, J., 250
Wharton, D. C., 655
Wheeler, C. J., 698, 699(4), 704, 704(6), 707(4), 708(9), 712, 713(13)
Wheelock, M. J., 755, 757(19)
White, G. A., 622, 628(27), 647, 650(59), 652
White, O., 757
White, T. C., 184, 185(16)
Whiteway, M., 109, 134, 157(36), 178, 398, 415
Whiting-Theobald, N., 526, 537(11), 557
Whitsett, J. A., 629
Whitt, M. A., 545
Wickman, K., 365, 368
Wickner, W., 27, 37, 278
Wiebel, F. F., 763
Wieder, R., 548
Wieland, B., 84
Wiemer, E. A. C., 755, 757(19)
Wiesmeyer, H., 30
Wilden, C. D., 637
Wilke-Mounts, S., 222
Wilkinson, D. J., 624, 676, 677(16)
Williams, D. B., 620, 622(23, 25), 627(25), 642, 645, 650(37a)
Williams, H., 608, 613(19), 615(19)
Williams, K. R., 328, 329, 332(5)
Williams, L., 17
Williams, P. G., 11, 25
Williams, S. S., 346
Williamson, R., 707, 722
Willingham, M. C., 227, 250, 259, 262, 264, 265, 319, 320(10), 427, 430, 430(4), 431(4), 435(4), 436, 436(4), 437(4), 454, 465, 474, 483, 523, 542, 546, 555, 561, 573, 575, 576(4), 580(4), 589(4)
Willman, C. L., 237
Willy, P. J., 756
Wilmott, R. W., 629
Wilson, C., 109, 175, 176, 177(4, 11), 180, 560
Wilson, G. M., 604, 606(25), 607(25)
Wilson, J. M., 617, 647, 679(62), 714, 718, 721, 723, 723(22), 724, 728
Wilson, K. L., 193
Wilson, T. H., 23
Wilson, W., 538, 547
Wimmer, E., 475
Wine, J. J., 360, 361, 367(10)
Winnacker, E.-L., 163
Winocour, E., 725
Winston, F., 207, 208(32)
Winter, M. A., 300
Winter, M. C., 676
Wirth, D. F., 109, 157, 174, 175, 176, 177(4, 11), 178, 180, 183, 189, 211, 415, 427(13)
Wiskocil, R., 597
Wisniewski, H. K., 754
Withold, B., 680
Wittinghofer, A., 101
Wittke, E., 755
Woel, S., 655
Wohllk, N., 110, 117, 132, 154(20)
Wolf, D. C., 303
Wolff, N., 72, 80(26)
Wolf-Ledbetter, B., 579
Wolfson, M., 342
Wolgamot, G. M., 728
Wong, A. K. C., 183
Wong, C., 755, 757(18)
Wong, G. G., 552
Wong, K. K., 540, 543(15), 727
Wood, P. J., 399
Wood, W. I., 165
Woods, R. A., 779
Woolford, C. A., 204
Wootton, J. C., 755
Workman, R. B., Jr., 659, 661(33)
Worrell, R. T., 361, 624, 676, 677(16)

Wu, C.-S. C., 684
Wu, C.-T., 167
Wu, J., 360, 369(14)
Wu, K.-H., 759, 769(63)
Wu, S., 70
Wu, S.-N., 365
Wuestehube, L. J., 626, 627(39)
Wunder, J. S., 237
Wust, D., 183, 185(12), 187(12)
Wyatt, L. S., 453, 473, 626
Wysock, W., 143(68), 155

X

Xavier, K. B., 42, 43(21), 45(21)
Xiao, M., 554
Xiao, X., 544, 725, 727
Xie, W., 755
Xu, N., 94

Y

Yachie, A., 566
Yahav, J., 676
Yalowich, J. C., 590
Yamaguchi, K., 263
Yamamoto, K. R., 143(90), 160
Yamamoto, T., 599
Yamashima, T., 260
Yamashita, J., 260
Yanagida, M., 755
Yang, C. P., 344, 345(16), 348(16), 349(16), 356(16), 406
Yang, C.-P. H., 303, 309, 311(11), 312(11), 342
Yang, G., 704, 708(9)
Yang, J. T., 684
Yang, P. C., 655
Yang, Y., 120, 718, 723(22), 724
Yanisch-Perron, C., 679, 779
Yaniv, M., 476
Yankaskas, J. R., 647, 679(62), 699, 714, 718
Yano, K., 143(91), 160
Yasuhara, M., 259
Yasusada, M., 547(19), 548
Ye, H., 714
Ye, S., 40
Yei, S., 629
Yellen, G., 279
Yew, N. S., 712, 713(12)
Yewdell, J., 120
Yin, Y., 53, 61(13), 62(13)
Yokata, S., 756, 758(48)
Yokota, S., 776
Yoneyama, K., 647, 723
Yong, M. A., 264
York, I., 120
Yoshida, M., 414
Yoshimura, A., 266, 293, 303(19)
Yoshimura, K., 647, 687, 723
Yost, C. S., 270, 278
Youle, R. J., 474
Young, N. S., 544
Young, R. M., 608, 613(19), 615(19)
Yount, B., 109
Ysern, X., 677, 681(19)
Yu, X., 131, 397, 398(10, 14), 409(10, 14), 410, 410(10), 411(10, 14), 487
Yujiki, Y., 757, 763(53)
Yusa, K., 259, 302, 357

Z

Zabin, I., 25
Zabner, J., 699, 707(6), 708(6)
Zabner, L., 724
Zachar, Z., 224
Zaidi, T. S., 721
Zalcberg, J. R., 259
Zaman, G. J., 110, 128, 132, 154, 155(63), 411, 607, 608, 608(9), 609(13), 611(9), 613, 614(9)
Zamboni, R., 608, 613(19), 615(19)
Zamora, J. M., 300
Zang, F., 30
Zastawny, R. L., 263
Zeitlin, P. L., 369, 540, 653, 655, 655(10), 656(10), 657(10), 658(10), 659(8), 661(10), 667, 720, 727, 728, 728(52, 53), 729, 729(6), 730, 731
Zelnick, C. R., 225, 227, 227(2), 239(2), 239(16)
Zenk, M. H., 163
Zepeda, M., 647, 679(62), 723(22), 724
Zhang, A., 474
Zhang, F., 51, 52, 53, 53(7), 55, 55(12), 56, 58, 61(13, 14), 62(13), 63(14), 64(14), 781

Zhang, J., 131, 267, 321, 397, 398(8), 409(8), 410(8), 411(8), 481, 487(2), 492, 495, 496(22), 500(22), 505, 511(12), 662(37), 663
Zhang, J. J., 360, 365, 369(14)
Zhang, J. T., 137, 266, 267(8, 9), 278(8, 9), 279, 281, 282(17), 283(14), 284(14), 285(14), 286(27, 30), 287, 288, 322, 404, 522
Zhang, J. W., 755
Zhang, L., 427(11), 428, 441(11)
Zhang, M., 288
Zhang, S., 441(3), 442, 456, 473(1), 474, 505
Zhang, W., 558
Zhang, X., 414, 423(2), 474, 478
Zhang, X. P., 316, 590
Zhang, Y., 455
Zhang, Z., 757
Zhao, S., 546
Zheng, B., 329, 357
Zheng, S., 759, 769(63)
Zhou, L., 629
Zhou, T., 88, 91
Zhou, Y., 474
Zhou, Z., 724
Zhu, X. D., 559, 725
Zickler, D., 755
Ziegler, A., 768
Ziegler, B. U., 768
Ziegler, R. J., 712, 713(12)
Zielenski, J., 117, 132, 154(15), 343, 411, 617, 619(4), 629, 630(2), 631(2), 633(2), 635(2), 675, 676, 687, 715
Zietlin, P. L., 676
Zilfou, J. T., 357
Zinn, K., 270, 625
Zobrist, R. H., 295
Zou, Y. R., 474
Zrike, J., 74
Zsiga, M., 629, 630(1), 633(1), 652, 675
Zuhn, D., 225

Subject Index

A

ABC transporter, *see* ATP-binding cassette transporter
Adeno-associated virus vector, gene transfer
  cystic fibrosis transmembrane regulator
    in vitro transfer, 728–729
    in vivo transfer
      clinical trials, 731–732
      macaque, 729, 731
      rabbit, 729
  genomic features of virus, 725–726
  insertion locus, 539, 725, 728
  integration of *rep* negative vectors, 544–545
  inverted terminal repeat sequences, 539, 544
  latency, 725
  MDR1
    recombinant virion production, 540–543
    titration of vector, 543
    transduction and selection of expressing cells, 543, 545
  packaging size and site, 539–540, 545, 727
  proliferating cell transfection efficiency, 728
  transfer efficiencies in various cells, 727–728
ADP1, functions in *Saccharomyces cerevisiae*, 152
Adrenoleukodystrophy
  ALD gene
    homology with PMP70, 759–761
    locus, 759
    mutation in disease, 111–112, 118, 134, 757, 759
  ALD-related protein, 762
  phenotypes, 757, 759
a-factor
  export by Ste6p, 156, 193
  halo assays of export
    patch halo assays, 201
    principle, 198
    sensitivity, 198–199
    spot assays, 199–201
  immunoprecipitation, 201–202
  posttranslational processing, 194
ALD, *see* Adrenoleukodystrophy
Antimony
  pumps, *see* Arsenite pump; PgpA
  therapy in *Leishmania* infection, 182–183, 192–193
Arsenite pump, *see also* PgpA
  ars operon and prokaryotic proteins, 82–84
  ArsA ATPase
    allosteric activation by antimonite or arsenite, 90–91
    ArsB interactions, assay, 91–93
    assay, 89–90
    purification, 88–89
  assays in eukaryotes
    Chinese hamster cells, 94–95
    Leishmania, *see also* PgpA
      plasma membrane-enriched vesicle assays, 95–96
      promastigote assays, 95
  assays in prokaryotes
    arsenite extrusion from intact cells, 84–85
    arsenite uptake in intact cells, 86
    ATP-dependent arsenite uptake in everted membrane vesicles, 86–87
    NADH-dependent arsenite uptake in everted membrane vesicles, 88
  evolution in prokaryotes and eukaryotes, 82, 96–97, 102
Atm1p
  epitope tagging, 780–782
  functions in yeast, 157–158, 785–786
  immunofluorescence
    antibody treatment, 783
    fixing cells, 782
    treatment of slides, 782
  submitochondrial localization, 783–785

ATP
agonist behaviors, 664, 674
intracellular concentration and gradient, 665
photoaffinity labeling, *see* 8-Azidoadenosine-5′-triphosphate
purinergic receptors, 664
release
assays
bioluminescence assay, 668–673
cell culture, 666–667
patch-clamp analysis, 673–674
radioactive trapping assay, 667–668, 672–673
cystic fibrosis transmembrane regulator role, 665
ATPase, ATP-binding cassette proteins
ArsA ATPase
allosteric activation by antimonite or arsenite, 90–91
ArsB interactions, assay, 91–93
assay, 89–90
purification, 88–89
Chinese hamster P-glycoprotein ATPase
assays
lactate dehydrogenase coupled assay, 520
malachite green assay, 519–520
kinetic parameters, 520–521
modification of essential residues, 521
overexpression in CR1R12 cells, 515
plasma membrane preparation, 516–517
purification of P-glycoprotein
dye affinity chromatography, 518–519
reconstitution, 519
solubilization, 517–518
stimulators, 521
suspension culture of CR1R12 cells, 516
vanadate trapping of nucleotide, 521–523
ligand transporters, 17–18
malachite green assay, 38–39, 519–520
maltose transporter
activity and assays, 9, 17–18, 27–29, 37–39
mutants with uncoupled ATPase activity
ATP hydrolysis assays, 37–39
selection for restored transport in *malE* strain, 33–34
transport property characterization, 34, 36–37
MDR1 ATPase
assay of reconstituted protein, 489–490, 492
assays, 438–441, 467–469, 489–490, 499–500, 507
baculovirus–insect cell protein characterization, 438–441
characterization of expressed protein from rodent cultured cells
activity dependence on ATPase expression level, 511–512
contaminating ATPase inhibition, 510–511, 513
crude membrane preparation, 505–506
inhibitor analysis, 509–511
inorganic phosphate colorimetric assay, 507
phosphorylation-deficient mutants, 512–513
kinetic parameters, 397–398
purified or reconstituted proteins, 499–500
site-directed mutant characterization, 490, 492, 505
Mdr3 ATPase
assay, 402–403
contaminating ATPase inhibition, 411–412
kinetic parameters, 410–411
pH optimum, 406–407
valinomycin stimulation, 409
vanadate inhibition, 410–411
verapamil stimulation, 408–409
vinblastine stimulation, 409
PrtD ATPase
activity in various preparations, 79–81
assay, 79
inhibitors, 81
reconstition, 81
ATP-binding cassette transporter, *see also specific proteins*
ATP-binding proteins
ATPase activity, 9
clustering of domains in *Saccharomyces cerevisiae*, 139, 145
membrane topology, 8–9

sequence alignment in *Saccharomyces cerevisiae*, 146–150
sequence homology in eukaryotes, 101–102
sequence homology with other proteins, 3, 7–8
structure, overview, 135–136
three-dimensional structure, 9–10
classification
eukaryotes, 103, 105, 109–110
humans, 129–130
prokaryotes, 3–4
Saccharomyces cerevisiae, 137, 139, 145, 148–159
clinical importance, overview, 132–133, 161–162
domain organization
eukaryotes, 102, 111–112, 116, 130–131, 136
prokaryotes, 5–6, 67–68, 113
functions in bacteria, 30–31, 67, 112
gene cloning strategies
antibody screening, 127
criteria for ATP-binding cassette transporter gene definition, 118
database searching, 122–127, 130, 137–138
degenerate polymerase chain reaction, 121–122
function/positional cloning, 117–118, 120
hybridization, 120–121
yeast two-hybrid system, 127
half-molecules, 111–112, 145, 768–769
heterologous expression in yeast, 159
ligand transport
assay systems
accumulation in intact cells, 13–14
ATP addition to inside-out vesicles, 16
ATP hydrolysis, 17–18
binding protein addition to cells from external medium, 14
binding protein addition to right-side-out vesicles, 15
reconstitution into proteoliposomes, 16–17
ligand-binding protein function and mutation studies, 11–13, 31
mechanism, 18–20
ligand-binding proteins, 10–11
membrane channel proteins, 6–7, 136–137, 145, 148–149
multicomponent exporters, 114–115, 131
phylogenetic analysis, 52–53, 102, 104, 113–114, 144–145
ATPGP1
antibody generation, 167
degenerate polymerase chain reaction in gene cloning
probe design, 164
screening of libraries, 165–166
epitope tagging
c-MYC fusion protein generation, 170–171
expression and purification of fusion protein, 169–170
plasmid construction, 167–169
sequence homology with other P-glycoproteins, 166–167
Western blot analysis
extraction of protein, 171
gel electrophoresis and blotting, 171
immunostaining, 171–172
transgenic plant expression analysis, 173
wild-type plant expression analysis, 172–173
ATPGP2
degenerate polymerase chain reaction in gene cloning
probe design, 164
screening of libraries, 165–166
sequence homology with other P-glycoproteins, 166–167
8-Azidoadenosine-5′-triphosphate, photolabeling
Chinese hamster P-glycoprotein, 521
MDR1 expresed in vaccinia virus–T7 RNA polymerase system, 464–465
Azidopine, photoaffinity labeling of P-glycoprotein, 294, 302, 304, 309–311, 437–438

B

Baculovirus–insect cell expression system
cystic fibrosis transmembrane regulator–glutathione *S*-transferase fusion protein
cell purification, 692–693

fluorescence-activated cell sorting in selection, 692
glutathione affinity chromatography, 693
transfection, 691–692
vector construction, 691–692
verification of expression, 693
MDR1
advantages of expression system, 428–429
ATPase characterization, 438–441
cell infection and culture, 434–435
histidine-tagged protein
expression system, 319–320
insect cell membrane preparation, 320–321
metal affinity chromatography, 321–322
membrane isolation, 436–437
recombinant baculovirus construction
amplification of virus, 433–434
calcium phosphate tranfection, 432
DNA preparation, 431–432
purification of virus, 433
site-directed mutants, 483–484
vector construction for expression, 429–431
BLAST, *see* Expressed sequence tag
Bone marrow, gene therapy, *see* Gene therapy, *MDR1*
Brown protein, *Drosophila melanogaster*
eye pigmentation role, 213–215
gene cloning, 214
hydropathy plot, 215, 218
sequence alignment with other transport proteins, 215–217
site-directed mutagenesis, 215, 218–219
White protein interaction sites, 221–223

C

Cationic lipid: plasmid DNA
advantages in gene delivery, 697–698
cystic fibrosis transmembrane regulator transfection into airway epithelial cells
cell culture, 714
fluorescence assay of channel activity, 714–716
neutral lipid enhancement of transfection, 698
optimization of formulations
in vitro
cell culture, 700
β-galactosidase reporter gene assay, 700–701
liposome preparation, 699–700
overview, 698–699
ratios for optimal transfection efficiency, 701, 704–705
structures of cationic lipids, 701–702
in vivo
chloramphenicol acetyltransferase reporter gene assay, 709–710
correlation to *in vitro* studies, 712
intranasal instillation in mice, 709–710
naked DNA control, 710–711
ratios for optimal transfection efficiency, 711–712
transfection efficiency, factors affecting
in vitro
cell lines, 708
contact time, 708
excipient selection, 707–708
precipitation of complexes, 705, 707
size of complex, 707
vector promoters, 708–709
in vivo
DNA purity, 713
dose, 713
toxicity, 713
CD, *see* Circular dichroism
Cell volume regulation, *see* Chloride channel, swelling-activated
CFTR, *see* Cystic fibrosis transmembrane regulator
Chloride channel, swelling-activated
P-glycoprotein regulation of cell volume
channel activation effects, 365
channel activity measurement, 361
channel sensitivity to hypoosmotic solutions, P-glycoprotein effects, 367
electrophysiologic characterization, 361, 363
evidence of interaction, 360–361
mechanism of channel activity modulation, 367–368
pharmacologic characterization, 363, 365

phosphorylative modulation, 369
physiologic significance, 368–370
regulatory volume decrease, overview, 359–360, 369
Circular dichroism, cystic fibrosis transmembrane regulator, nucleotide binding domain-1, 681, 684, 694
Colchicine, photoaffinity labeling of P-glycoprotein with analogs, 293–294, 300–301
Cystic fibrosis transmembrane regulator
antibody production
applications of antibodies, 630, 652
cross-reactivity characterization, 648, 650–652
fusion protein immunogen preparation, 630–631, 633–635, 648
immunization schedule, 635–636
monoclonal antibody
hybridoma production, 637
hybridoma screening, 637–639, 648–649
purification, 639
nucleotide binding domain-1, 696
polyclonal antibody, 636
ATP release
assays
bioluminescence assay, 668–673
cell culture, 666–667
patch-clamp analysis, 673–674
radioactive trapping assay, 667–668, 672–673
functions, 665, 674
copy number, 616
cystic fibrosis pathogenesis, 652–653, 675–676, 687, 717, 720–722
developmental activity, 720
expression in *Saccharomyces cerevisiae*, 626–628
fluorescence assay of channel activity, 714–716
full-length complementary DNA assembly, 617
gene therapy, *see* Gene therapy, cystic fibrosis transmembrane regulator
immunolocalization, 646–647
immunoprecipitation, 645–646
inhibition by antibodies, 646
nucleotide binding domain-1
ATP analog binding, fluorescence studies, 681, 683, 694
circular dichroism spectroscopy of wild-type and ΔF508 mutant proteins, 681, 684, 694
glutathione *S*-transferase fusion protein expressed in baculovirus–insect cell system
cell purification, 692–693
fluorescence-activated cell sorting in selection, 692
glutathione affinity chromatography, 693
transfection, 691–692
vector construction, 691–692
verification of expression, 693
glutathione *S*-transferase fusion protein expressed in *Escherichia coli*
gel purification, 690, 693–694
glutathione affinity chromatography, 690
inclusion body isolation and solubilization, 690
overexpression system, 689–690
storage, 690
thrombin cleavage, 690
vector construction, 689
maltose-binding protein fusion protein
overexpression in *Escherichia coli*, 680–682
proteolytic cleavage with Factor X_a, 686
purification by amylose affinity chromatography, 680–681
transformation, 680
vector construction, 678–680
mutation in cystic fibrosis, 676–677, 688
phosphorylation, 343
posttranslational processing, 622, 641–642, 650
renal endosome protein
endosomal acidification assay, 655, 658–659
physiological function, 661, 663
renal membrane isolation, 653–654
solubilization and partial purification, 656, 659–661, 663
subcellular localization
immunohistochemistry, 657–658, 661
marker enzyme analysis, 655–657, 661
Western blot analysis, 655–657

sequence homology with other ATP-binding cassette transporters, 109–110, 162
stable expression in mammalian cells
advantages and disadvantages, 622–623
electrophysiologic characterization of Chinese hamster ovary cells, 620, 622
histidine tagging for purification, 622
plasmids and selection, 618, 620, 628
tissue distribution, 653
topology, 687–688
transgenic mice, 628–629
transient expression in mammalian cells, 623
Western blot analysis, 641–643, 645, 650
Xenopus oocyte expression after RNA microinjection, 623–625

D

Daunorubicin, photoaffinity labeling of P-glycoprotein with analogs, 294, 301
Degenerate polymerase chain reaction, *see* Polymerase chain reaction
Dihydropyridine, photoaffinity labeling of P-glycoprotein with azidopine analog, 294, 302, 304, 309–311, 437–438
Doxorubicin, photoaffinity labeling of P-glycoprotein with analogs, 294, 301
Drosopterin
Drosophila melanogaster eye color role, 213
transport, *see* Brown protein; White protein

E

Elongation factor-3
evolutionary relationship to ATP-binding cassette transporter, 115
yeast protein YEF3 functions, 153
EST, *see* Expressed sequence tag
Expressed sequence tag
ATP-binding cassette transporter gene identification
contig assembly, 123–124, 127
full-length sequences, obtaining, 125–126
human genes deduced from yeast sequences, 160–162
pitfalls, 126–127, 130
sequence searching with BLAST, 122–123, 137–138
Xref database searching, 124–125
human libraries with bacterial contamination, 126

F

Fabry disease, gene therapy, *see* Gene therapy, *MDR1*
FACS, *see* Fluorescence-activated cell sorting
FISH, *see* Fluorescence *in situ* hybridization
Fluorescence *in situ* hybridization, homozygous transgenic mouse identification, 578–579
Fluorescence-activated cell sorting
cystic fibrosis transmembrane regulator-expressing cells, 692
hematopoietic stem cells, 547–548, 550
MDR1-expressing cells, 239, 242–243, 453, 469–471, 508–509, 531
gene therapy products, 531, 533, 547–548, 550
transgenic mice, 580–581, 583–584
Forskolin, photoaffinity labeling of P-glycoprotein with analogs, 309

G

Gene therapy, cystic fibrosis transmembrane regulator
adeno-associated virus vectors
genomic features of virus, 725–726
insertion locus, 725, 728
latency, 725
proliferating cell transfection efficiency, 728
size limitations, 727
transfer efficiencies in various cells, 727–728
in vitro transfer, 728–729
in vivo transfer
clinical trials, 731–732
macaque, 729, 731
rabbit, 729

adenovirus vectors, 723–724
airway epithelium factors
electrolyte transport, pumps and channels, 719–720
histology, 718–719
pathogen-binding theory and transfection efficiency requirements, 720–722
cationic lipid: plasmid DNA formulations in gene delivery
advantages, 697–698
neutral lipid enhancement of transfection, 698
optimization of formulations, *in vitro*
cell culture, 700
β-galactosidase reporter gene assay, 700–701
liposome preparation, 699–700
overview, 698–699
ratios for optimal transfection efficiency, 701, 704–705
structures of cationic lipids, 701–702
optimization of formulations, *in vivo*
chloramphenicol acetyltransferase reporter gene assay, 709–710
correlation to *in vitro* studies, 712
intranasal instillation in mice, 709–710
naked DNA control, 710–711
ratios for optimal transfection efficiency, 711–712
transfection, airway epithelial cells
cell culture, 714
fluorescence assay of channel activity, 714–716
transfection efficiency, factors affecting *in vitro*
cell lines, 708
contact time, 708
excipient selection, 707–708
precipitation of complexes, 705, 707
size of complex, 707
vector promoters, 708–709
transfection efficiency, factors affecting *in vivo*
DNA purity, 713
dose, 713
toxicity, 713
Moloney murine leukemia virus vector, 722–723
Gene therapy, *MDR1*
adeno-associated virus vectors
integration locus, 539
integration of *rep* negative vectors, 544–545
inverted terminal repeat sequences, 539, 544
packaging size and site, 539–540, 545
recombinant virion production, 540–543
titration of vector, 543
transduction and selection of expressing cells, 543, 545
application for cancer chemotherapy patients, 524, 526, 538, 547, 557, 572
bicistronic vectors
cell culture, 530
construction
bicistronic vectors, 528–530
subcloning vectors, 526–528
fluorescence activated cell sorting analysis of expression, 531, 533
α-galactosidase bicistronic vector
assay of α-galactosidase, 532
Fabry disease treatment rationale, 525–526
protein expression analysis, 533–535
herpes simplex virus thymidine kinase bicistronic vector
ganciclovir selection, 524–525
protein expression analysis, 535–536
transduction and transplantation of bone marrow cells, 532, 536–537
internal ribosome entry site utilization, 525–526
NADPH oxidase gp91 bicistronic vector for X-linked chronic granulomatous disease therapy, 537
overview of vectors, 525–526, 537
retrovirus transduction, 531, 533
transfection with calcium phosphate, 531, 533
vincristine sensitivity assay, 530, 534–535
bone marrow targeting, 524–525
dilution effect, 523
retroviral vectors
limitations, 538
transfer of human P-glycoprotein to human T lymphocytes
applications, 557–558

cell culture, 560
characterization of transduced cells, 570–571
packaging cells, 558, 560–561
safety, 571–572
screening of producer clones, 565–567
transduction efficiency, 558–559, 571
transduction of packaging cells, 564–565
transduction of T cells, 567, 569–571
transfection of producer cells, 561, 563
transfer of human P-glycoprotein to murine hematopoietic stem cells
cell culture, 255–256, 550–551
colony assays of multidrug-resistant stem cells, 257–258
fluorescence-activated cell sorting, 547–548, 550
gene transfer, 256, 550–552
immature cell phenotype, preservation in culture, 552, 554
immunomagnetic cell fractionation, 256, 548, 550
protein expression assays in transduced bone marrow cells, 257, 554–555
protein expression assays in transplanted transduced bone marrow cells, 555–556
Glycerol 3-phosphate, *see also* Uptake glycerol phosphate system
metabolism in bacteria, 40
uptake systems in bacteria, 40–41
Glycosylation
peptide antigens, 752–753
P-glycoprotein
assay for defective mutants, 486–487
N-linked glycosylation detection, 274–275, 277, 283–284
sulfonylurea receptor assay, 741

H

Hemolysin B
hemolysin A transport
assays
gel electrophoresis of secreted hemolysin A, 57–59
hemolytic zone assay, 55–57
secreted hemolysin A measurement in liquid culture by absorption spectroscopy, 57
energetics, 68
overview, 51–52, 55
signal sequence
genetic complementation analysis of hemolysin B interaction, 53–55, 60–61, 63–66
modeling of hemolysin B interaction, 65–66
mutations resulting in transport deficiency, 60–61, 64–65
structure, 53
phylogenetic analysis, 52–53
Histidine transporter, *Salmonella typhimurium*
ATP-binding protein
membrane topology, 8–9
sequence homology with other proteins, 3, 7–8
three-dimensional structure, 9–10
domain organization, 5–6
ligand transport
assay systems
ATP addition to inside-out vesicles, 16
reconstitution into proteoliposomes, 16–17
ligand-binding protein mutation studies, 13
membrane channel protein, 6–7
subunits, 5, 32

I

Immunomagnetic cell fractionation, hematopoietic stem cells, 256, 548, 550
Immunoprecipitation
a-factor, 201–202
cystic fibrosis transmembrane regulator, 645–646
MDR1, 297–298
Ste6p, 205–206
sulfonylurea receptor, 741

Iodoglyburide
photoaffinity labeling of sulfonylurea receptor
membrane preparations, 737, 739
specificity, 738
whole cells, 737–738
synthesis of radiolabeled iodine-125 compound, 733–734

L

Liposome gene delivery, *see* Cationic lipid: plasmid DNA

M

Major histocompatibility peptide transport system, *see* Transporters associated with antigen processing
Malachite green, dye-binding assay for ATP hydrolysis, 38–39, 519–520
Maltose transporter, *Escherichia coli*
ATP-binding protein
ATPase activity and assays, 9, 17–18, 27–29, 37–39
membrane topology, 8–9
sequence homology with other proteins, 3, 7–8
domain organization, 5–6
inner membrane transport complex purification
bacteria growth, 22
high-performance liquid chromatography, 24
solubilization, 23–24
vesicle preparation, 23
ligand transport
assay systems
accumulation in intact cells, 13–14
ATP addition to inside-out vesicles, 16
binding protein addition to cells from external medium, 14
binding protein addition to right-side-out vesicles, 15
substrate uptake in proteoliposomes, 16–17, 27–29
mechanism, 18–20
maltose-binding protein
binding modes, 10–11, 21
functional exchange with UgpC, 49–50
mutational analysis, 11–13, 21
purification, 24–26
membrane channel proteins, 6
mutants with uncoupled ATPase activity
ATP hydrolysis assays, 37–39
selection for restored transport in *malE* strain, 33–34
transport property characterization, 34, 36–37
reconstitution into proteoliposomes, 16–17, 21, 26
subunits, 5, 21, 31–32
Mass spectrometry, phosphopeptides
data analysis, 350–353
electrospray ionization, 348, 355
sensitivity, 348
sequencing, 347
MDR1, *see also* P-glycoprotein
assays of expression and drug transport
colony assay for rhodamine-123 efflux, 246–247
fluorescence-activated cell sorting, 239, 242–243, 453, 469–471, 508–509, 531
indirect immunofluorescence and flow cytometry, 233–234
radiolabeled drug accumulation, 471–472, 500–503
ATPase
assays, 438–441, 467–469, 489–490, 499–500, 507
baculovirus–insect cell protein characterization, 438–441
characterization of expressed protein from rodent cultured cells
activity dependence on ATPase expression level, 511–512
contaminating ATPase inhibition, 510–511, 513
crude membrane preparation, 505–506
inhibitor analysis, 509–511
inorganic phosphate colorimetric assay, 507
phosphorylation-deficient mutants, 512–513
kinetic parameters, 397–398

purified or reconstituted proteins, 499–500
site-directed mutant characterization, 490, 492, 505
dihydrofolate reductase fusion protein
mammalian cell transfection library of mutants, 479–480
methotrexate resistance in selection, 475, 477
polymerase chain reaction mutagenesis, 478–479
vector construction, 476–477
drug specificity, 226–227, 492–493
expression in baculovirus–insect cell system
advantages, 428–429
ATPase characterization, 438–441
cell infection and culture, 434–435
histidine-tagged protein
expression system, 319–320
insect cell membrane preparation, 320–321
metal affinity chromatography, 321–322
membrane isolation, 436–437
recombinant baculovirus construction
amplification of virus, 433–434
calcium phosphate tranfection, 432
DNA preparation, 431–432
purification of virus, 433
site-directed mutants, 483–484
vector construction for expression, 429–431
expression in *Escherichia coli*, 380–381
expression in vaccinia virus–T7 RNA polymerase system
advantages, 442–443, 455, 457
characterization of expressed protein, 453, 463–473
disadvantages, 454–455, 473
efficiency, 443–444, 455–457
infection of cells, 452–453, 460–462
membrane preparation, 462–463
plasmid construction, 444, 446, 459–460
principle, 443
recombinant virus screening, 449–450
safety issues, 445–446, 453
transfection with lipofectin, 447–448, 461
virus amplification and purification, 450–451
virus generation, 444, 446–447
virus isolation by plaque purification, 444, 448–450
virus titer determination, 451–452
gene, *see MDR1*
heterologous expression systems, overview, 473
monoclonal antibodies
applications
detection of drug-resistant cells, 264
fluorescence-activated cell sorting of expressing cells, 469–470
gene cloning, 263
modulation of transport, 263–264
protein purification, 263
selective killing of cells, 264–265
generation, 260–263
mutation and drug transport specificity, 226–227
phosphorylation
ATPase activity of mutants, 512–513
confirmation of site phosphorylation by site-directed mutagenesis, 341–342
peptide mapping
amino acid sequence of Lys-C peptide, 334–335
proteolysis following reduction and carboxamidomethylation, 332–333
reversed-phase high-performance liquid chromatography, 333–335
starting material requirements, 328–330
tryptic phosphopeptides, purification and sequences, 336, 338–340
protein kinase C phosphorylation reaction, 329–331, 341
purification of phosphorylated protein by gel electrophoresis, 331–332
photoaffinity labeling in drug-binding site identification
amino acid sequence analysis, 312
assumptions, 290, 308–309
chimeric protein studies, 316–317
denaturing polyacrylamide gel electrophoresis, 297–298, 310, 324
immunological mapping
high-resolution mapping, 314–315
low-resolution mapping, 313–314
principle, 312–313

immunoprecipitiation, 297–298
inhibition with unlabeled drugs, 298–300
intact cell labeling, 466
membrane labeling, 467
membrane preparations, 311
peptide mapping, 305–306, 312
probe selection, 308–309
site-directed mutagenesis, 304–305, 315–316
synthesis of analogs and labeling
ATP, 464–465
colchicine, 293–294, 300–301
daunorubicin, 294, 301
dihydropyridine, 294, 302, 304, 309–311, 437–438, 465–467
doxorubicin, 294, 301
forskolin, 309
labeling reactions, 296–297, 310, 322, 324, 465–467
phenothiazines, 302–303
prazosin, 305–306, 309, 313–314, 322, 324–325, 327, 437–438, 465–467
verapamil, 294–296, 302
vinblastine, 291, 293, 299–300, 304
vanadate trapping, effects on labeling, 325–328
purification from multidrug-resistant KB-V1 cells
anion-exchange chromatography, 495–496, 498, 503
characterization of purified protein, 499–504
orientation determination in proteoliposomes, 499
plasma membrane vesicle isolation, 494–495
reconstitution, 496–497, 499
solubilization, 495
storage, 503
Western blot analysis, 497–499
wheat germ agglutinin affinity chromatography, 495–496, 498, 503
site-directed mutagenesis
applicability of MDR1 to mutagenesis studies, 480–481
assay for glycosylation-defective mutants, 486–487
baculovirus–insect cell system expression, 483–484
overview of mutagenesis, 481–482
stable expression in mammalian cells, 482–483, 505
transient expression in human embryonic kidney cells
advantages, 484
ATPase assay of reconstituted protein, 489–490, 492
purification of histidine-tagged mutants, 487–489
transfection, 484–485
Western blot analysis, 463–464, 507–508
MDR1
gene therapy, *see* Gene therapy, *MDR1*
genetic suppressor elements
definition, 225
isolation from retroviral expression libraries
functional testing in selection, 239, 242–243, 246–247
library construction, 239–242
overview, 227–228, 238–239
recovery of complementary DNA, 243–246
mutant isolation
drug selection and recovery of complementary DNA inserts, 234–237
infection of recipient cells, 233
mutagenized plasmid preparation, 230–231
overview, 229–230
sequencing of clones, 237–238
transfection of packaging cells, 232
protein, *see* MDR1
transduction into bone marrow cells, 255–257
transgenic mice
actin promoter and stability of expression, 573–575
Eμ transgenic mice production, overview, 572, 575–576
expression assays
immunofluorescence and fluorescence-activated cell sorting, 580–581, 583–584
Northern blot analysis of RNA, 580
ribonuclease protection assay, 581–582
Western blot analysis, 584
functional assays
drug resistance measurements in white blood cells, 589–593

drug sources, preparation, and doses, 586, 588–589
rhodamine-123 efflux, 584–586, 593
homozygous transgenic mouse identification, 578–580
protection against toxic substrates, 546, 572, 589–593
reversal agent testing, 588, 593
screening of founder mice, 577–578
strain selection, 576
tissue expression and targeting, 574, 582–583
mdr1, murine
expression in *Escherichia coli*
advantages, 370–371
bioenergetics analysis, 379–382
drug susceptibility testing, 376–377
functional assays
leaky mutant generation for outer membrane permeability, 375–376
photolabeling, 373–374
transport assay with tetraphenylarsonium, 374–375
limitations, 371–372
plasmids, 372
reversal of drug resistance, 377–378
strains and growth conditions, 372–373
topology studies, 378–379
expression in yeast, *see* Secretory vesicle, *Saccharomyces cerevisiae*
genes, 290
knockout mice, 573
linker region of protein
expression and purification of recombinant Mdr1a, 353
mass spectrometry of phosphopeptides
data analysis, 350–353
electrospray ionization, 348, 355
sensitivity, 348
sequence analysis, 347
peptide mapping, 353–355
phosphorylation *in vitro* with kinases, 353
sequences, 343–344
site-directed mutagenesis of phosphorylation sites, 355, 357–358
phosphorylation of protein
cyanogen bromide digestion, 344–347
labeling *in vitro* with kinases, 344–346, 350
peptide mapping, 348–349
phosphorous-32, metabolic labeling, 345–346
mdr2, murine
expression in yeast, *see* Secretory vesicle, *Saccharomyces cerevisiae*
phosphatidylcholine translocation assay in yeast secretory vesicles
fluorescent phosphatidylcholine, 391–392
principle, 391–393
reaction conditions, 395–396
secretory vesicle preparation with outer leaflet label, 393, 395
mdr3, murine
ATPase
assay, 402–403
contaminating ATPase inhibition, 411–412
kinetic parameters, 410–411
pH optimum, 406–407
valinomycin stimulation, 409
vanadate inhibition, 410–411
verapamil stimulation, 408–409
vinblastine stimulation, 409
expression in yeast
Pichia pastoris
advantages, 413
alcohol oxidase promoter in overexpression, 399, 403, 412–413
contaminating ATPase inhibition, 411–412
efficiency of expression system, 412–413
induction, 404
membrane preparation, 400–401, 406–407
plasmid construction, 399, 403
screening of clones, 403–404
transformation, 399–400
Saccharomyces cerevisiae, see also Secretory vesicle, *Saccharomyces cerevisiae*
chimeric amino-terminal nucleotide-binding domain protein construction, 415–421
membrane isolation, 425–426
site-directed mutagenesis, 417–418

functional assay by Chinese hamster ovary cell transfection, 421–423
functional replacement of *STE6*, 134–135, 157, 159, 194, 211, 382, 398, 421, 423–425
photoaffinity labeling of protein with iodoarylazidoprazosin, 401–402, 404–406
solubilization and reconstitution of protein in *Escherichia coli* lipids, 402, 407–408, 412
Metalloprotease transporter, *Erwinia chrysanthemi*
assays by protease detection
cup–plate assay, 71
gelatin hydrolysis following gel electrophoresis, 71
skim milk agar plate assay, 70
cell compartment markers in secretion specificity determination, 71–72
comparison to HasA secretion system, 75
domain organization, 67–68
function, 67
gene organization, 68–69
metalloprotease signal sequence, 72
metalloprotease specificity, 68, 70–72
proteins
antibody studies, 73–74
expression in *Escherichia coli*, 73–74, 76
topology, 75
PrtD
ATPase
activity in various preparations, 79–81
assay, 79
inhibitors, 81
reconstition, 81
overexpression system, 76
purification
cytoplasmic membrane vesicles, 76–77
phosphocellulose chromatography, 78–79
solubilization, 77–78
site-directed mutants of Walker A domain conserved lysine, 76
Microsomal membranes, cell-free translation of P-glycoprotein, 272–273, 281
Mitochondria, ATP-binding cassette proteins
Atm1p
epitope tagging, 780–782
functions in yeast, 785–786
immunofluorescence
antibody treatment, 783
fixing cells, 782
treatment of slides, 782
submitochondrial localization, 783–785
gene disruption using polymerase chain reaction fragments, 779–780, 785–786
polymerase chain reaction search for genes, 776–779
MRP, *see* Multidrug-resistance associated protein
Multidrug resistance, subfamily in *Saccharomyces cerevisiae*, 155–158
Multidrug-resistance associated protein
discovery, 595
functional assays of transfected HeLa cells
ATP depletion assay, 606–607
chemosensitivity testing, 604–605
drug accumulation and efflux assays, 605–606
functional homology with P-glycoprotein, 594–595
gene cloning
differential display, 596–597
differential hybridization, 596–600
full-length complementary DNA isolation, 601–602
messenger RNA probes and screening, 598–600
Northern blot, 603–604
positive clone analysis, 600
subtractive hybridization, 596
Western blot analysis, 603–604
gene transfection
calcium phosphate transfection, 603
vector construction, 602–603
inhibitors, 614–616
isoforms, 607–608
kinetic properties of MRP1 transport, 612–614
leukotriene C_4, assay of ATP-dependent transport, 608, 611–612
physiological function, 607

plasma membrane vesicle preparation, 609–611
sequence homology with other ATP-binding cassette transporters, 109–110, 162, 595
substrate specificity, 608–609, 612–614
tissue distribution, 607
Multidrug-resistance associated protein/ Cystic fibrosis transmembrane regulator subfamily, ATP-binding cassette transporters in *Saccharomyces cerevisiae*, 154–155
Multidrug-resistant cells
colony assays of stem cells, 257–258
MDR1 transduction into bone marrow cells, 255–257
selection
clones
cross-resistance screening, 253–254
drug concentration and handling, 251–252
expansion, 252
freezing, 253
maintenance, 254
media, 251
overview, 248–249
plating, 250–251
mass populations and multifactorial multidrug-resistant cells, 249–250
starting cell lines, 250
transfection and transduction of multidrug resistance genes, 254–255

N

Northern blot analysis
MDR1 RNA expression in transgenic mice, 580
multidrug-resistance associated protein messenger RNA expression, 603–604

P

PCR, *see* Polymerase chain reaction
PDR subfamily, *see* Pleiotropic drug resistance subfamily
Peptide mapping
MDR1
drug-binding site, 305–306, 312
phosphorylation site
amino acid sequence of Lys-C peptide, 334–335
proteolysis following reduction and carboxamidomethylation, 332–333
reversed-phase high-performance liquid chromatography, 333–335
starting material requirements, 328–330
tryptic phosphopeptides, purification and sequences, 336, 338–340
Mdr1 phosphorylation sites, 348–349, 353–355
P-glycoprotein, drug-binding site identification, 305–306
Peroxisome
ATP-binding cassette proteins
functions, 769
half-ATP-binding cassette proteins, 758–759, 762, 768
phylogenetic analysis, 768, 770
sequence alignment, 760–761, 769, 771
biogenesis
disorders, *see also* Adrenoleukodystrophy
classification, 756–757
Zellweger syndrome, 757
induction, 754
matrix proteins, 754–755
membrane proteins, 755–756
discovery, 753–754
Saccharomyces cerevisiae, *see also* PXA1; PXA2
culture using oleic acid as sole carbon source, 772–773
induction of peroxisome biogenesis, 773–774
purification of peroxisomes
fractionation on Nycodenz gradient, 775
gradient analysis, 775–776
high-density organelle isolation, 774–775
overview, 773
suitability for peroxisome studies, 762–763

Peroxisome membrane protein-70
  biological function, 758
  homology with other ATP-binding cassette proteins, 759–762
pfmdr1
  copy number and drug sensitivity, 176–177
  functional analysis by heterologous expression
    Chinese hamster ovary cells, 177–178
    functional complementation of Ste6p, 174, 178–180, 427
  mutation analysis, 178
  role in *Plasmodium falciparum* drug resistance, 174–177, 180–181
  sequence homology with other P-glycoproteins, 105, 109
  subcellular localization of protein, 175, 179–181
  substrate specificity, 175–176
  transcripts, 180–181
Pgh1 gene, *see pfmdr1*
P-glycoprotein, *see also specific proteins*
  Arabidopsis thaliana, *see* ATPGP1; ATPGP2
  Caenorhabditis elegans, 105, 768
  cell volume regulation via swelling-activated chloride channels
    channel activation effects, 365
    channel activity measurement, 361
    channel sensitivity to hypoosmotic solutions, P-glycoprotein effects, 367
    electrophysiologic characterization, 361, 363
    evidence of interaction, 360–361
    mechanism of channel activity modulation, 367–368
    pharmacologic characterization, 363, 365
    phosphorylative modulation, 369
    physiologic significance, 368–370
    regulatory volume decrease, overview, 359–360, 369
  Chinese hamster ATPase
    assays
      lactate dehydrogenase coupled assay, 520
      malachite green assay, 519–520
    kinetic parameters, 520–521
    modification of essential residues, 521
    overexpression in CR1R12 cells, 515
    plasma membrane preparation, 516–517
    purification of P-glycoprotein
      dye affinity chromatography, 518–519
      reconstitution, 519
      solubilization, 517–518
    stimulators, 521
    suspension culture of CR1R12 cells, 516
    vanadate trapping of nucleotide, 521–523
  deletion in adrenoleukodystrophy, 111–112
  domain organization, 307–308, 318, 397, 415, 493, 504, 514
  Drosophila, 105
  Entoamoeba histolytica, 109
  gene therapy, *see* Gene therapy, *MDR1*
  genes
    classification, 103, 397
    overview, 290
  half-molecules, 111–112
  heterologous expression systems, overview, 398
  human, *see* MDR1; *MDR1*
  monoclonal antibodies
    applications
      detection of drug-resistant cells, 264
      gene cloning, 263
      modulation of transport, 263–264
      protein purification, 263
      selective killing of cells, 264–265
    Chinese hamster, 260, 263
    human, 260–263
  mouse, *see mdr1*, murine; *mdr2*, murine; *mdr3*, murine
  photoaffinity labeling in drug-binding site identification
    amino acid sequence analysis, 312
    assumptions, 290, 308–309
    chimeric protein studies, 316–317
    denaturing polyacrylamide gel electrophoresis, 297–298, 310, 324
    immunological mapping
      high-resolution mapping, 314–315

low-resolution mapping, 313–314
principle, 312–313
immunoprecipitiation, 297–298
inhibition with unlabeled drugs, 298–300
labeling reaction, 296–297, 310, 322, 324
membrane preparations, 311
peptide mapping, 305–306, 312
probe selection, 308–309
site-directed mutagenesis, 304–305, 315–316
synthesis of analogs and labeling
colchicine, 293–294, 300–301
daunorubicin, 294, 301
dihydropyridine, 294, 302, 304, 309–311
doxorubicin, 294, 301
forskolin, 309
phenothiazines, 302–303
prazosin, 305–306, 309, 313–314, 322, 324–325, 327
verapamil, 294–296, 302
vinblastine, 291, 293, 299–300, 304
vanadate trapping, effects on labeling, 325–328
Plasmodium falciparum, *see pfmdr1*
reconstitution, 326–327
reversing agents of multidrug resistance, mechanism and structure, 300, 304
sequence homology between classes and species, 103, 105, 109, 111–112, 162–163, 290, 514
steps in drug transport, 318
substrate specificity, 259
tissue distribution and functions, 259–260
topology
alternative topologies, 286–289
C terminus translocation reporters
criteria for reporter domains, 268, 283
fusion protein generation, 268–270
overview, 267–268, 282–283
definition, 265
determinants
characterization, 270–271
types of sequences, 266
expression systems for study
microsomal membranes, 272–273, 281
rabbit reticulocyte lysate, 272, 274, 281
RNA digestion in cell-free systems, 273
transcription *in vitro* using SP6 RNA polymerase, 280
wheat germ extract, 281, 286–287
Xenopus laevis oocyte microinjection, 273–274
folding steps, 266–267, 285–288
mapping, 277–278
membrane stripping with alkaline buffer, 281–282
N-linked glycosylation detection, 274–275, 277, 283–284
protease digestion assay, 275–277, 285
yeast homolog, *see* Ste6p
Pgp, *see* P-glycoprotein
PgpA
gene amplification
response to drugs, 183
detection of amplification
agarose gel electrophoresis, 184–186
circular amplicon isolation, 186–187
DNA extraction, 184
Leishmania growth, 184
linear amplicon isolation, 187–188
gene disruption studies of resistance, 190
gene locus, 183
gene transfection studies of resistance, 188–189
metal specificity, 182–183
resistance mechanisms in *Leishmania*, 190–192
thiol conjugation and pump activity, 190–191
Phosphatidylcholine, translocation assay with Mdr2 in yeast secretory vesicles
fluorescent phosphatidylcholine, 391–392
principle, 391–393
reaction conditions, 395–396
secretory vesicle preparation with outer leaflet label, 393, 395
Photoaffinity labeling
P-glycoprotein, drug-binding site identification
assumptions, 290
denaturing polyacrylamide gel electrophoresis, 297–298

immunoprecipitiation, 297–298
inhibition with unlabeled drugs, 298–300
labeling reactions, 296–297
peptide mapping, 305–306
site-directed mutagenesis, 304–305
synthesis of analogs
colchicine, 293–294, 300–301
daunorubicin, 294, 301
dihydropyridine, 294, 302, 304
doxorubicin, 294, 301
phenothiazines, 302–303
verapamil, 294–296, 302
vinblastine, 291, 293, 299–300, 304
sulfonylurea receptor, labeling with iodoglyburide
membrane preparations, 737, 739
specificity, 738
whole cells, 737–738
Pleiotropic drug resistance subfamily, ATP-binding cassette transporters in *Saccharomyces cerevisiae*, 150–152
PMP70, *see* Peroxisome membrane protein-70
Polymerase chain reaction
degenerate polymerase chain reaction in gene cloning
overview, 121–122
probe design, 164
screening of libraries, 165–166
MDR1, random mutagenesis, 478–479
mitochondria ATP-binding cassette genes
disruption using polymerase chain reaction fragments, 779–780, 785–786
search for genes, 776–779
Prazosin, photoaffinity labeling of P-glycoprotein with analogs, 305–306, 309, 313–314, 322, 324–325, 327, 401–402, 404–406, 437–438
PrtD, *see* Metalloprotease transporter
PXA1
functions in yeast, 153, 763, 766
gene cloning, 763
sequence homology to other ATP-binding cassette proteins, 133–134, 153, 764–766
site-directed mutagenesis, 764–766
PXA2
functions in yeast, 153, 767–768
heterodimerization with PXA1, 767
homology to human P-glycoproteins, 133–134, 153

R

Rabbit reticulocyte lysate, cell-free translation of P-glycoprotein, 272, 274, 281
Regulatory volume decrease, *see* Chloride channel, swelling-activated
Rhodamine-123, efflux assays
colony assay, 246–247
MDR1 transgenic mice, 584–586, 593
Ribonuclease L inhibitor subfamily, ATP-binding cassette transporters in *Saccharomyces cerevisiae*, 152–153
RLI subfamily, *see* Ribonuclease L inhibitor subfamily
RRL, *see* Rabbit reticulocyte lysate
RVD, *see* Regulatory volume decrease

S

Scarlet protein, *Drosophila melanogaster*
eye pigmentation role, 213–215
gene cloning, 214
hydropathy plot, 215, 218
sequence alignment with other transport proteins, 215–217
site-directed mutagenesis, 215, 218–219
White protein interactions, 223
Secretory vesicle, *Saccharomyces cerevisiae*
advantages in membrane protein expression, 383–384, 396
cystic fibrosis transmembrane regulator expression system, 628
electrochemical potential, 384, 388
invertase as marker, 388
morphological characterization, 387
murine P-glycoprotein expression
applications, 426
drug transport studies
buffers, 389
calculations, 390
cold stop solution, 389–391
filter assay, 388–391
incubation conditions, 390–391
isolation of secretory vesicles, 386–387

phosphatidylcholine translocation assay with Mdr2
fluorescent phosphatidylcholine, 391–392
principle, 391–393
reaction conditions, 395–396
secretory vesicle preparation with outer leaflet label, 393, 395
transformation, 385–386
yeast strains and growth, 384–385
Site-directed mutagenesis
Brown protein, 215, 218–219
MDR1
applicability of MDR1 to mutagenesis studies, 480–481
assay for glycosylation-defective mutants, 486–487
ATPase, 490, 492, 505
baculovirus–insect cell system expression, 483–484
overview of mutagenesis, 481–482
phosphorylation site, 341–342
stable expression in mammalian cells, 482–483, 505
transient expression in human embryonic kidney cells
advantages, 484
ATPase assay of reconstituted protein, 489–490, 492
purification of histidine-tagged mutants, 487–489
transfection, 484–485
Mdr1 phosphorylation site, 355, 357–358
Mdr3, 417–418
photoaffinity labeling sites, 304–305, 315–316
PrtD, 76
PXA1, 764–766
Scarlet protein, 215, 218–219
Ste6p, 210–211
White protein, 215, 218–219, 224
SNQ2, functions in *Saccharomyces cerevisiae*, 151
Southern blot analysis, screening of *MDR1* transgenic mice, 577–578
Ste6p
a-factor export role, 156, 193
assays
halo assays of **a**-factor export
patch halo assays, 201
principle, 198
sensitivity, 198–199
spot assays, 199–201
immunoprecipitation of **a**-factor, 201–202
mating assay of **a**-factor export
filter assay, 196–197
overview, 156, 195–196
patch mating assay, 197–198
plate assay, 197
epitope tagging, 203–204
functional replacement
chimeric molecules, 212
mouse Mdr3, 134–135, 157, 159, 194, 211, 382, 398, 421, 423–425
pfmdr1, 174, 178–180, 211
half-life, 203
half-molecule reconstitution studies, 209–210
heterologous expression in *Escherichia coli* and topology studies, 379, 381
immunofluorescence assay in subcellular localization, 206–208
immunoprecipitation, 205–206
pulse-chase metabolic labeling, 204–205
sequence homology with P-glycoproteins, 109
site-directed and random mutagenesis analysis, 210–211
Western blot analysis, 208–209
Sulfonylurea receptor
deglycosylation assay, 741
immunoprecipitation, 741
insulin secretion role, 132, 154, 732
iodoglyburide
photoaffinity labeling
membrane preparations, 737, 739
specificity, 738
whole cells, 737–738
synthesis of radiolabeled iodine-125 compound, 733–734
proteolytic degradation assay, 741
purification from cultured cells
cell culture and isolation, 736
digitonin solubilization, 739–740
dye affinity chromatography, 740
lectin affinity chromatography, 740
membrane preparation and storage, 738–739

polyacrylamide gel electrophoresis
analytical, 737–738
preparative, 740
purification of histidine-tagged protein from COSm6 cells
cell culture and transfection, 743
membrane preparation, 744
nickel affinity chromatography, 744
rubidium efflux assays, 743–744
vector construction, 743
sequence homology with other ATP-binding cassette transporters, 109–110
SUR, *see* Sulfonylurea receptor

T

TAP, *see* Transporters associated with antigen processing
Topology, *see* Transmembrane topology
Traffic ATPase, *see* ATP-binding cassette transporter
Transgenic mice, *MDR1*
actin promoter and stability of expression, 573–575
Eμ transgenic mice production, overview, 572, 575–576
expression assays
immunofluorescence and fluorescence-activated cell sorting, 580–581, 583–584
Northern blot analysis of RNA, 580
ribonuclease protection assay, 581–582
Western blot analysis, 584
functional assays
drug resistance measurements in white blood cells, 589–593
drug sources, preparation, and doses, 586, 588–589
rhodamine-123 efflux, 584–586, 593
homozygous transgenic mouse identification, 578–580
protection against toxic substrates, 546, 572, 589–593
reversal agent testing, 588, 593
screening of founder mice, 577–578
strain selection, 576
tissue expression and targeting, 574, 582–583
Transmembrane topology
ATP-binding proteins, 8–9
definition, 265
heterologous expression of P-glycoproteins in *Escherichia coli*
Mdr1, 378–379
Ste6p, 379
metalloprotease transporter, *Erwinia chrysanthemi*, 75
P-glycoprotein
alternative topologies, 286–289
C terminus translocation reporters
criteria for reporter domains, 268, 283
fusion protein generation, 268–270
overview, 267–268, 282–283
determinants
characterization, 270–271
types of sequences, 266
expression systems for study
microsomal membranes, 272–273, 281
rabbit reticulocyte lysate, 272, 274, 281
RNA digestion in cell-free systems, 273
transcription *in vitro* using SP6 RNA polymerase, 280
wheat germ extract, 281, 286–287
Xenopus laevis oocyte microinjection, 273–274
folding steps, 266–267, 285–288
mapping, 277–278
membrane stripping with alkaline buffer, 281–282
N-linked glycosylation detection, 274–275, 277, 283–284
protease digestion assay, 275–277, 285
Transporters associated with antigen processing
components of major histocompatibility peptide transport system, 111
gene inactivation by herpes simplex virus, 118, 120
sequence homology of TAP proteins with P-glycoprotein, 111
subunit structure, 745
transport assay
competition experiments, 749, 751–753
incubation conditions, 747, 748, 752

melanoma cell line WM98.1, 749
peptide proteolysis and glycosylation, 752–753
precipitation of reporter peptide, 748–749
solubilization of cells, 747–748
streptolysin O permeabilization of cells, 746–748

U

Ugp system, *see* Uptake glycerol phosphate system
Uptake glycerol phosphate system
appearance in suppressor strains, 40–41
assay of transport, 42–43
ATP-binding protein
functional exchange with MalK, 49–50
regulation of *pho* regulon, 50–51
genes and regulation, 43–45
ligand-binding protein
binding assays, 47–49
purification, 47
phosphate scavenger role, 45
phosphoglycerol diester phosphodiesterase activity, 42
proteins
UgpA, 46
UgpB, 45–49
UgpC, 46, 49–51
UgpE, 46
substrates, 41–42
UvrA, evolutionary relationship to ATP-binding cassette transporter, 115

V

Valinomycin, Mdr3 ATPase stimulation, 409
Vanadate
MDR1 ATPase inhibition, 438–441
Mdr3 ATPase inhibition, 410–411
trapping of ATPase
Chinese hamster P-glycoprotein, 521–523
effects on P-glycoprotein photoaffinity labeling, 325–328
Verapamil
MDR1 ATPase stimulation, 439–440
Mdr3 ATPase stimulation, 408–409
photoaffinity labeling of P-glycoprotein with analogs, 294–296, 302
Vinblastine
Mdr3 ATPase stimulation, 409
photoaffinity labeling of P-glycoprotein with analogs, 291, 293, 299–300, 304
radiolabeled drug accumulation assays of MDR1
intact cells, 471–472
reconstituted proteoliposomes, 500–503

W

Western blot analysis
ATPGP1
extraction of protein, 171
gel electrophoresis and blotting, 171
immunostaining, 171–172
transgenic plant expression analysis, 173
wild-type plant expression analysis, 172–173
cystic fibrosis transmembrane regulator, 641–643, 645, 650, 655–657
MDR1
multidrug-resistant KB-V1 cell protein, 497–499
rodent cell expression, 507–508
transgenic mouse expression, 584
vaccinia virus–T7 RNA polymerase system protein, 463–464
multidrug-resistance associated protein expression, 603–604
Ste6p, 208–209
Wheat germ extract, cell-free translation of P-glycoprotein, 281, 286–287
White protein, *Drosophila melanogaster*
ATP-binding domain, 219–220
Brown protein interaction sites, 221–223
eye pigmentation role, 213–215
gene cloning, 214
hydropathy plot, 215, 218
Scarlet protein interactions, 223
sequence alignment with other transport proteins, 215–217

site-directed mutagenesis, 215, 218–219, 224

X

Xanthommatin
- *Drosophila melanogaster* eye color role, 213
- transport, *see* Scarlet protein; White protein

Xenopus laevis oocyte
- cystic fibrosis transmembrane regulator expression after RNA microinjection, 623–625
- microinjection of P-glycoprotein messenger RNA, 273–274

X-linked chronic granulomatous disease, gene therapy, *see* Gene therapy, *MDR1*

Y

YCF1
- functions in yeast, 154
- multidrug-resistance protein, functional homology, 154–155
- subcellular localization, 155

YEF3, *see* Elongation factor-3

YOR1, function in yeast, 155

ISBN 0-12-182193-5

90038

9 780121 821937